PENGUIN BOOKS

ON THE SHOULDERS OF GIANTS

Stephen Hawking is considered the most brilliant theoretical physicist since Einstein. He has also done much to popularize science. His book *A Brief History of Time* sold more than 10 million copies in 40 languages, achieving the kind of success almost unheard of in the history of science writing. His subsequent books, *The Universe in A Nutshell* and *The Future of Spacetime*, with Kip S. Thorne and others, have also been well received.

He was born in Oxford on January 8, 1942 (300 years after the death of Galileo). He studied physics at University College, Oxford, received his Ph.D. in Cosmology at Cambridge, and since 1979 has held the post of Lucasian Professor of Mathematics. The chair was founded in 1663 with money left in the will of the Reverend Henry Lucas, who had been the Member of Parliament for the university. It was first held by Isaac Barrow, and then in 1663 by Isaac Newton. It is reserved for those individuals considered the most brilliant thinkers of their time.

Professor Hawking has worked on the basic laws that govern the universe. With Roger Penrose, he showed that Einstein's General Theory of Relativity implied space and time would have a beginning in the Big Bang and an end in black holes. The results indicated it was necessary to unify General Relativity with Quantum Theory, the other great scientific development of the first half of the twentieth century. One consequence of such a unification was that black holes should not be completely black but should emit radiation and eventually disappear. Another conjecture is that the universe has no edge or boundary in imaginary time.

Stephen Hawking has twelve honorary degrees, and is the recipient of many awards, medals and prizes. He is a Fellow of the Royal Society and a Member of the US National Academy of Sciences. He continues to combine family life (he has three children and one grandchild) and his research into theoretical physics together with an extensive programme of travel and public lectures.

KA

D1434204

ON THE
SHOULDERS
of
GIANTS

THE GREAT WORKS OF PHYSICS AND ASTRONOMY

EDITED, WITH COMMENTARY, BY

STEPHEN
HAWKING

PENGUIN BOOKS

PENGUIN BOOKS

Published by the Penguin Group
Penguin Books Ltd, 80 Strand, London WC2R 0RL, England
Penguin Putnam Inc., 375 Hudson Street, New York, New York 10014, USA
Penguin Books Australia Ltd, 250 Camberwell Road, Camberwell, Victoria 3124, Australia
Penguin Books Canada Ltd, 10 Alcorn Avenue, Toronto, Ontario, Canada M4V 3B2
Penguin Books India (P) Ltd, 11 Community Centre, Panchsheel Park, New Delhi – 110 017, India
Penguin Books (NZ) Ltd, Cnr Rosedale and Airborne Roads, Albany, Auckland, New Zealand
Penguin Books (South Africa) (Pty) Ltd, 24 Sturdee Avenue, Rosebank 2196, South Africa

Penguin Books Ltd, Registered Offices: 80 Strand, London WC2R 0RL, England

www.penguin.com

First published in the USA by Running Press 2002
First published in Great Britain by Penguin Books 2003
9

Text of *On the Revolutions of Heavenly Spheres* courtesy of Annapolis:
St. John's Bookstore, copyright 1939.
Text of *Harmonies of the World* courtesy of Annapolis:
St. John's Bookstore, copyright 1939.
Text of *Dialogues Concerning Two New Sciences* courtesy
of Dover Publications.
Text of *Principia* and *System of the World* courtesy of
New York: Daniel Adee, copyright 1848.
Selections from *The Principle of Relativity: A Collection
of Papers on the Special and General Theory of Relativity*,
courtesy of Dover Publications.

Printed in England by Clays Ltd, St Ives plc

www.greenpenguin.co.uk

Penguin Books is committed to a sustainable future
for our business, our readers and our planet.
The book in your hands is made from paper
certified by the Forest Stewardship Council.

CONTENTS

A NOTE ON THE TEXTS

The texts in this book are based on translations of the original, printed editions. We have made no attempt to modernize the authors' own distinct usage, spelling or punctuation, or to make the texts consistent with each other in this regard. Here are other relevant details:

On the Revolutions of Heavenly Spheres, by Nicolaus Copernicus, was first published in 1543 under the title *De revolutionibus orbium coelestium*. This translation is by Charles Glen Wallis.

Dialogues Concerning Two New Sciences, by Galileo Galilei, was originally published in 1638 under the title *Discorsi e Dimostrazioni Matematiche, intorno à due nuoue scienze*, by the Dutch publish Louis Elzevir. Our text is based on the translation by Henry Crew and Alfonso deSalvio.

We have selected Book Five of the five-book *Harmonies of the World* by Johannes Kepler. Kepler completed the work on May 27, 1618, publishing it under the title, *Harmonices Mundi*. This translation is by Charles Glen Wallis.

The Principia, by Isaac Newton, was originally published in 1687 under the title of *Philosophiae naturalis principia mathematica* (*The Mathematical Principles of Natural Philosophy*). This translation is by Andrew Motte.

We haven chosen seven works by Albert Einstein from *The Principles of Relativity: A Collection of Original Papers on the Special Theory of Relativity*, by H.A. Lorentz, A. Einstein, H. Minkowski and H. Weyl. The entire collection was originally published in German, under the title "Des Relativitatsprinzip" in 1922. Our text comes from the translation by W. Perrett and G.B. Jeffery.

The Editors

INTRODUCTION

IF I HAVE SEEN FARTHER, IT IS BY STANDING ON THE SHOULDERS OF GIANTS, WROTE ISAAC NEWTON IN A LETTER TO ROBERT HOOKE IN 1676. ALTHOUGH NEWTON WAS REFERRING TO HIS DISCOVERIES IN OPTICS RATHER THAN HIS MORE IMPORTANT WORK ON GRAVITY AND THE LAWS OF MOTION, IT IS AN APT COMMENT ON HOW SCIENCE, AND INDEED THE WHOLE OF CIVILIZATION, IS A SERIES OF INCREMENTAL ADVANCES, EACH BUILDING ON WHAT WENT BEFORE. THIS IS THE THEME OF THIS FASCINATING VOLUME, WHICH USES THE ORIGINAL TEXTS TO TRACE THE EVOLUTION OF OUR PICTURE OF THE HEAVENS FROM THE REVOLUTIONARY CLAIM OF NICOLAUS COPERNICUS THAT THE EARTH ORBITS THE SUN TO THE EQUALLY REVOLUTIONARY PROPOSAL OF ALBERT EINSTEIN THAT SPACE AND TIME ARE CURVED AND WARPED BY MASS AND ENERGY. IT IS A COMPELLING STORY BECAUSE BOTH COPERNICUS AND EINSTEIN HAVE BROUGHT ABOUT PROFOUND CHANGES IN WHAT WE SEE AS OUR POSITION IN THE ORDER OF THINGS. GONE IS OUR PRIVILEGED PLACE AT THE CENTER OF THE UNIVERSE, GONE ARE ETERNITY AND CERTAINTY, AND GONE ARE ABSOLUTE SPACE AND TIME TO BE REPLACED BY RUBBER SHEETS.

IT IS NO WONDER BOTH THEORIES ENCOUNTERED VIOLENT OPPOSITION: THE INQUISITION IN THE CASE OF THE COPERNICAN THEORY AND THE NAZIS IN THE CASE OF RELATIVITY. WE NOW HAVE A TENDENCY

TO DISMISS AS PRIMITIVE THE EARLIER WORLD PICTURE OF ARISTOTLE AND PTOLEMY IN WHICH THE EARTH WAS AT THE CENTER AND THE SUN WENT ROUND IT. HOWEVER WE SHOULD NOT BE TOO SCORNFUL OF THEIR MODEL, WHICH WAS ANYTHING BUT SIMPLE-MINDED. IT INCORPORATED ARISTOTLE'S DEDUCTION THAT THE EARTH IS A ROUND BALL RATHER THAN A FLAT PLATE AND IT WAS REASONABLY ACCURATE IN ITS MAIN FUNCTION, THAT OF PREDICTING THE APPARENT POSITIONS OF THE HEAVENLY BODIES IN THE SKY FOR ASTROLOGICAL PURPOSES. IN FACT, IT WAS ABOUT AS ACCURATE AS THE HERETICAL SUGGESTION PUT FOR-WARD IN 1543 BY COPERNICUS THAT THE EARTH AND THE PLANETS MOVED IN CIRCULAR ORBITS AROUND THE SUN.

GALILEO FOUND COPERNICUS' PROPOSAL CON-VINCING NOT BECAUSE IT BETTER FIT THE OBSERVA-TIONS OF PLANETARY POSITIONS BUT BECAUSE OF ITS SIMPLICITY AND ELEGANCE, IN CONTRAST TO THE COMPLICATED EPICYCLES OF THE PTOLEMAIC MODEL. IN *DIALOGUES CONCERNING TWO SCIENCES*, GALILEO'S CHARACTERS, SALVIATI AND SAGREDO, PUT FORWARD PERSUASIVE ARGUMENTS IN SUPPORT OF COPERNICUS. YET, IT WAS STILL POSSIBLE FOR HIS THIRD CHARACTER, SIMPLICIO, TO DEFEND ARISTOTLE AND PTOLEMY AND TO MAINTAIN THAT IN REALITY THE EARTH WAS AT REST AND THE SUN WENT ROUND THE EARTH.

IT WAS NOT UNTIL KEPLER'S WORK MADE THE SUN-CENTERED MODEL MORE ACCURATE AND NEWTON GAVE IT LAWS OF MOTION THAT THE EARTH-

CENTERED PICTURE FINALLY LOST ALL CREDIBILITY. IT WAS QUITE A SHIFT IN OUR VIEW OF THE UNIVERSE: IF WE ARE NOT AT THE CENTER, IS OUR EXISTENCE OF ANY IMPORTANCE? WHY SHOULD GOD OR THE LAWS OF NATURE CARE ABOUT WHAT HAPPENS ON THE THIRD ROCK FROM THE SUN, WHICH IS WHERE COPERNICUS HAS LEFT US? MODERN SCIENTISTS HAVE OUT-COPERNICUSED COPERNICUS BY SEEKING AN ACCOUNT OF THE UNIVERSE IN WHICH MAN (IN THE OLD PRE-POLITICALLY CORRECT SENSE) PLAYED NO ROLE. ALTHOUGH THIS APPROACH HAS SUCCEEDED IN FINDING OBJECTIVE IMPERSONAL LAWS THAT GOVERN THE UNIVERSE, IT HAS NOT (SO FAR AT LEAST) EXPLAINED WHY THE UNIVERSE IS THE WAY IT IS RATHER THAN BEING ONE OF THE MANY OTHER POSSIBLE UNIVERSES THAT WOULD ALSO BE CONSISTENT WITH THE LAWS.

SOME SCIENTISTS WOULD CLAIM THAT THIS FAILURE IS ONLY PROVISIONAL, THAT WHEN WE FIND THE ULTIMATE UNIFIED THEORY, IT WILL UNIQUELY PRE-SCRIBE THE STATE OF THE UNIVERSE, THE STRENGTH OF GRAVITY, THE MASS AND CHARGE OF THE ELECTRON AND SO ON. HOWEVER, MANY FEATURES OF THE UNI-VERSE (LIKE THE FACT THAT WE ARE ON THE THIRD ROCK, RATHER THAN THE SECOND OR FOURTH) SEEM ARBITRARY AND ACCIDENTAL AND NOT THE PREDIC-TIONS OF A MASTER EQUATION. MANY PEOPLE (MYSELF INCLUDED) FEEL THAT THE APPEARANCE OF SUCH A COMPLEX AND STRUCTURED UNIVERSE FROM SIMPLE LAWS REQUIRES THE INVOCATION OF SOMETHING CALLED THE ANTHROPIC PRINCIPLE, WHICH RESTORES

US TO THE CENTRAL POSITION WE HAVE BEEN TOO MODEST TO CLAIM SINCE THE TIME OF COPERNICUS. THE ANTHROPIC PRINCIPLE IS BASED ON THE SELF-EVIDENT FACT THAT WE WOULDN'T BE ASKING QUESTIONS ABOUT THE NATURE OF THE UNIVERSE IF THE UNIVERSE HADN'T CONTAINED STARS, PLANETS AND STABLE CHEMICAL COMPOUNDS, AMONG OTHER PREREQUISITES OF (INTELLIGENT?) LIFE AS WE KNOW IT. IF THE ULTIMATE THEORY MADE A UNIQUE PREDICTION FOR THE STATE OF THE UNIVERSE AND ITS CONTENTS, IT WOULD BE A REMARKABLE COINCIDENCE THAT THIS STATE WAS IN THE SMALL SUBSET THAT ALLOW LIFE.

HOWEVER THE WORK OF THE LAST THINKER IN THIS VOLUME, ALBERT EINSTEIN, RAISES A NEW POSSIBILITY. EINSTEIN PLAYED AN IMPORTANT ROLE IN THE DEVELOPMENT OF QUANTUM THEORY WHICH SAYS THAT A SYSTEM DOESN'T JUST HAVE A SINGLE HISTORY AS ONE MIGHT HAVE THOUGHT. RATHER IT HAS EVERY POSSIBLE HISTORY WITH SOME PROBABILITY. EINSTEIN WAS ALSO ALMOST SOLELY RESPONSIBLE FOR THE GENERAL THEORY OF RELATIVITY IN WHICH SPACE AND TIME ARE CURVED AND BECOME DYNAMIC. THIS MEANS THAT THEY ARE SUBJECT TO QUANTUM THEORY AND THAT THE UNIVERSE ITSELF HAS EVERY POSSIBLE SHAPE AND HISTORY. MOST OF THESE HISTORIES WILL BE QUITE UNSUITABLE FOR THE DEVELOPMENT OF LIFE BUT A VERY FEW HAVE ALL THE CONDITIONS NEEDED. IT DOESN'T MATTER IF THESE FEW HAVE A VERY LOW PROBABILITY RELATIVE TO THE OTHERS: THE LIFELESS UNIVERSES WILL HAVE NO ONE TO OBSERVE THEM. IT IS

SUFFICIENT THAT THERE IS AT LEAST ONE HISTORY IN WHICH LIFE DEVELOPS, AND WE OURSELVES ARE EVIDENCE FOR THAT, THOUGH MAYBE NOT FOR INTELLIGENCE. NEWTON SAID HE WAS "*STANDING ON THE SHOULDERS OF GIANTS.*" BUT AS THIS VOLUME ILLUSTRATES SO WELL, OUR UNDERSTANDING DOESN'T ADVANCE JUST BY SLOW AND STEADY BUILDING ON PREVIOUS WORK. SOMETIMES AS WITH COPERNICUS AND EINSTEIN, WE HAVE TO MAKE THE INTELLECTUAL LEAP TO A NEW WORLD PICTURE. MAYBE NEWTON SHOULD HAVE SAID, "*I USED THE SHOULDERS OF GIANTS AS A SPRINGBOARD.*"

Nicolaus Copernicus

(1473-1543)

HIS LIFE AND WORK

Nicolaus Copernicus, a sixteenth-century Polish priest and mathematician, is often referred to as the founder of modern astronomy. That credit goes to him because he was the first to conclude that the planets and Sun did not revolve around the earth. Certainly speculation that a heliocentric—or sun-centered—universe had existed as far back as Aristarchus (d. 230 B.C.), but the idea was not seriously considered before Copernicus. Yet to understand the contributions of Copernicus, it is important to consider the religious and cultural implications of scientific discovery in his time.

As far back as the fourth century B.C., the Greek thinker and philosopher Aristotle (384-322 B.C.) devised a planetary system in his book, *On the Heavens* (*De Caelo*), and concluded that because the Earth's shadow on the Moon during eclipses was always round, the world was spherical in shape rather than flat. He also surmised the Earth was round because when one watched a ship sail out to sea one noticed that the hull disappeared over the horizon before the sails did.

In Aristotle's geocentric vision, the earth was stationary and the planets Mercury, Venus, Mars, Jupiter, and Saturn, as well as the sun and the moon, performed circular orbits around the earth. Aristotle also believed the stars were fixed to the celestial sphere, and his scale of the universe purported these fixed stars to be not much further beyond the orbit of Saturn. He believed in perfect circular motions and had good evidence to

believe the earth to be at rest. A stone dropped from a tower fell straight down. It did not fall to the west, as we would expect it to do if the earth rotated from west to east. (Aristotle did not consider that the stone might partake in the Earth's rotation.) In an attempt to combine physics with the metaphysical, Aristotle devised his theory of a "prime mover," which held that a mystical force behind the fixed stars caused the circular motions he observed. This model of the universe was accepted and embraced by theologians, who often interpreted prime movers as angels, and Aristotle's vision endured for centuries. Many modern scholars believe universal acceptance of this theory by religious authorities hindered the progress of science, as to challenge Aristotle's theories was to call into question the authority of the church itself.

Five centuries after Aristotle's death, an Egyptian named Claudius Ptolemaeus (Ptolemy, 87-150 A.D.), created a model for the universe that more accurately predicted the movements and actions of spheres in the heavens. Like Aristotle, Ptolemy believed the earth was stationary. Objects fell to the center of the earth, he reasoned, because the earth must be fixed at the center of the universe. Ptolemy ultimately elaborated a system in which the celestial bodies moved around the circumference of their own epicycles (a circle in which a planet moves and which has a center that is itself carried around at the same time on the circumference of a larger circle). To accomplish this, he put the Earth slightly off center of the universe and called this new center the "equant"—an imaginary point that helped him account for observable planetary movements. By custom designing the sizes of circles, Ptolemy was better able to predict the motions of celestial bodies. Western Christendom had little quarrel with Ptolemy's geocentric system, which left room in the universe behind the fixed stars to accommodate a heaven and a hell, and so the church adopted the Ptolemaic model of the universe as truth.

Aristotle and Ptolemy's picture of the cosmos reigned, with few significant modifications, for well over a thousand years. It wasn't until 1514 that the Polish priest Nicolaus Copernicus revived the heliocentric model of the universe. Copernicus proposed it merely as a model for calculating planetary positions because he was concerned that the church might label him a heretic if he proposed it as a description of reality. Copernicus became convinced, through his own study of planetary motions, that the earth was merely another planet and the sun was the center of the universe. This hypothesis became known as a heliocentric model. Copernicus' breakthrough marked one of the greatest paradigm shifts in world history, opening the way to modern astronomy and broadly affecting science, philosophy and religion. The elderly priest was hesitant to divulge his theory, lest it provoke church authorities to any angry response, and so he withheld his work from all but a few astronomers. Copernicus' landmark *De Revolutionibus* was published while he was on his deathbed, in 1543. He

did not live long enough to witness the chaos his heliocentric theory would cause.

Copernicus was born on February 19, 1473 in Torun, Poland, into a family of merchants and municipal officials who placed a high priority on education. His uncle, Lukasz Watzenrode, prince-bishop of Ermland, ensured that his nephew received the best academic training available in Poland. In 1491, Copernicus enrolled at Cracow University, where he pursued a course of general studies for four years before traveling to Italy to study law and medicine, as was common practice among Polish elites at the time. While studying at the University of Bologna (where he would eventually become a professor of astronomy), Copernicus boarded at the home of Domenico Maria de Novara, the renowned mathematician of whom Copernicus would ultimately become a disciple. Novara was a critic of Ptolemy, whose second-century astronomy he regarded with skepticism. In November 1500, Copernicus observed a lunar eclipse in Rome. Although he spent the next few years in Italy studying medicine, he never lost his passion for astronomy.

After receiving the degree of Doctor of Canon Law, Copernicus practiced medicine at the episcopal court of Heilsberg, where his uncle lived. Royalty and high clergy requested his medical services, but Copernicus spent most of his time in service of the poor. In 1503, he returned to Poland and moved into his uncle's bishopric palace in Lidzbark Warminski. There he tended to the administrative matters of the diocese, as well as serving as an advisor to his uncle. After his uncle's death in 1512, Copernicus moved permanently to Frauenburg and would spend the rest of his life in priestly service. But the man who was a scholar in mathematics, medicine and theology was only beginning the work for which he would become best known.

In March of 1513, Copernicus purchased 800 building stones, and a barrel of lime from his chapter so that he could build an observation tower. There, he made use of astronomical instruments such as quadrants, parallactics and astrolabes to observe the sun, moon and stars. The following year, he wrote a brief *Commentary on the Theories of the Motions of Heavenly Objects from Their Arrangements* (*De hypothesibus motuum coelestium a se constitutis commentariolus*), but he refused to publish the manuscript and only discreetly circulated it among his most trusted friends. The *Commentary* was a first attempt to propound an astronomical theory that the earth moves and the sun remains at rest. Copernicus had become dissatisfied with the Aristotelian-Ptolemaic astronomical system that had dominated Western thought for centuries. The center of the earth, he thought, was not the center of the universe, but merely the center of the Moon's orbit. Copernicus had come to believe that apparent perturbations in the observable motion of the planets was a result of the earth's own rotation around its axis and of its travel in orbit. "We revolve around the Sun," he concluded in *Commentary*, "like any other planet."

Despite speculation about a sun-centered universe as far back as the third century B.C. by Aristarchus, theologians and intellectuals felt more comfortable with a geocentric theory, and the premise was barely challenged in earnest. Copernicus prudently abstained from disclosing any of his views in public, preferring to develop his ideas quietly by exploring mathematical calculations and drawing elaborate diagrams, and to keep his theories from circulating outside of a select group of friends. When, in 1514, Pope Leo X summoned Bishop Paul of Fossombrone to recruit Copernicus to offer an opinion on reforming the ecclesiastical calendar, the Polish astronomer replied that knowledge of the motions of the sun and moon in relation to the length of the year was insufficient to have any bearing on reform. The challenge must have preoccupied Copernicus, however, for he later wrote to Pope Paul III, the same Pope who commissioned Michaelangelo to paint the Sistine Chapel, with some relevant observations, which later served to form the foundation of the Gregorian calendar seventy years later.

Still, Copernicus feared exposing himself to the contempt of the populace and the church, and he spent years working privately to amend and expand the *Commentary*. The result was *On the Revolutions of Heavenly Spheres (De Revolutionibus Orbium Coelestium)* which he completed in 1530, but withheld from publication for thirteen years. The risk of the church's condemnation was not, however, the only reason for Copernicus' hesitancy to publish. Copernicus was a perfectionist and considered his observations in constant need of verification and revision. He continued to lecture on these principles of his planetary theory, even appearing before Pope Clement VII, who approved of his work. In 1536, Clement formally requested that Copernicus publish his theories. But it took a former pupil, 25-year-old Georg Joachim Rheticus of Germany, who relinquished his chair in mathematics in Wittenberg so that he could study under Copernicus, to persuade his master to publish *On the Revolutions*. In 1540, Rheticus assisted in the editing of the work and presented the manuscript to a Lutheran printer in Nuremberg, ultimately giving birth to the Copernican Revolution.

When *On the Revolutions* appeared in 1543, it was attacked by Protestant theologians who held the premise of a heliocentric universe to be unbiblical. Copernicus' theories, they reasoned, might lead people to believe that they are simply part of a natural order, and not the masters of nature, the center around which nature was ordered. Because of this clerical opposition, and perhaps also general incredulity at the prospect of a non-geocentric universe, between 1543 and 1600, fewer than a dozen scientists embraced Copernican theory. Still, Copernicus had done nothing to resolve the major problem facing any system in which the earth rotated on its axis (and revolved around the sun), namely, how it is that terrestrial bodies stay with the rotating Earth. The answer was proposed by Giordano Bruno, an Italian scientist and avowed Copernican, who suggested that space might have no boundaries and that the solar system might be

4

one of many such systems in the universe. Bruno also expanded on some purely speculative areas of astronomy that Copernicus did not explore in *On the Revolutions*. In his writings and lectures, the Italian scientist held that there were infinite worlds in the universe with intelligent life, some perhaps with beings superior to humans. Such audacity brought Bruno to the attention of the Inquisition, which tried and condemned him for his heretical beliefs. He was burned at the stake in 1600.

On the whole, however, the book did not have an immediate impact on modern astronomic study. In *On the Revolutions,* Copernicus did not actually put forth a heliocentric system, but rather a heliostatic one. He considered the Sun to be not precisely at the center of the universe, but only close to it, so as to account for variations in observable retrogression and brightness. The earth, he asserted, made one full rotation on its axis daily, and orbited around the sun once yearly. In the first section of the book's six sections, he took issue with the Ptolemaic system, which placed all heavenly bodies in orbit around the earth, and established the correct heliocentric order: Mercury, Venus, Mars, Jupiter and Saturn (the six planets known at the time). In the second section, Copernicus used mathematics (namely epicycles and equants) to explain the motions of the stars and planets, and reasoned that the sun's motion coincided with that of the earth. The third section gives a mathematical explanation of the precession of the equinoxes, which Copernicus attributes to the Earth's gyration around its axis. The remaining sections of *On the Revolutions* focus on the motions of the planets and the moon.

Copernicus was the first to position Venus and Mercury correctly, establishing with remarkable accuracy the order and distance of the known planets. He saw these two planets (Venus and Mercury) as being closer to the sun, and noticed that they revolved at a faster rate inside the Earth's orbit.

Before Copernicus, the sun was thought to be another planet. Placing the sun at the virtual center of the planetary system was the beginning of the Copernican revolution. By moving the Earth away from the center of the universe, where it was presumed to anchor all heavenly bodies, Copernicus was forced to address theories of gravity. Pre-Copernican gravitational explanations had posited a single center of gravity (the earth), but Copernicus theorized that each heavenly body might have its own gravitational qualities and asserted that heavy objects everywhere tended toward their own center. This insight would eventually lead to the theory of universal gravitation, but its impact was not immediate.

By 1543, Copernicus had become paralyzed on his right side and weakened both physically and mentally. The man who was clearly a perfectionist had no choice but to surrender control of his manuscript, *On the Revolutions,* in the last stages of printing. He entrusted his student, George Rheticus, with the manuscript, but when Rheticus

was forced to leave Nuremberg, the manuscript fell into the hands of Lutheran theologian Andreas Osiander. Osiander, hoping to appease advocates of the geocentric theory, made several alterations without Copernicus's knowledge and consent. Osiander placed the word "hypothesis" on the title page, deleted important passages, and added his own sentences which diluted the impact and certainty of the work. Copernicus was said to have received a copy of the printed book in Frauenburg on his deathbed, unaware of Osiander's revisions. His ideas lingered in relative obscurity for nearly one hundred years, but the seventeenth century would see men like Galileo Galilei, Johannes Kepler and Isaac Newton build on his theories of a heliocentric universe, effectively obliterating Aristotelian ideas. Many have written about the unassuming Polish priest who would change the way people saw the universe, but the German writer and scientist Johann Wolfgang von Goethe may have been the most eloquent when he wrote of the contributions of Copernicus:

> *Of all discoveries and opinions, none may have exerted a greater effect on the human spirit than the doctrine of Copernicus. The world had scarcely become known as round and complete in itself when it was asked to waive the tremendous privilege of being the center of the universe. Never, perhaps, was a greater demand made on mankind — for by this admission so many things vanished in mist and smoke! What became of Eden, our world of innocence, piety and poetry; the testimony of the senses; the conviction of a poetic-religious faith? No wonder his contemporaries did not wish to let all this go and offered every possible resistance to a doctrine which in its converts authorized and demanded a freedom of view and greatness of thought so far unknown, indeed not even dreamed of.*

> — *Johann Wolfgang von Goethe*

INTRODUCTION

To the Reader Concerning the Hypotheses of this Work[1]

[*b*][2] Since the newness of the hypotheses of this work—which sets the earth in motion and puts an immovable sun at the centre of the universe—has already received a great deal of publicity, I have no doubt that certain of the savants have taken grave offense and think it wrong to raise any disturbance among liberal disciplines which have had the right set-up for a long time now. If, however, they are willing to weigh the matter scrupulously, they will find that the author of this work has done nothing which merits blame. For it is the job of the astronomer to use painstaking and skilled observation in gathering together the history of the celestial movements, and then since he cannot by any line of reasoning reach the true causes of these movements—to think up or construct whatever causes or hypotheses he pleases such that, by the assumption of these causes, those same movements can be calculated from the principles of geometry for the past and for the future too. This artist is markedly outstanding in both of these respects: for it is not necessary that these hypotheses should be true, or even probably; but it is enough if they provide a calculus which fits the observations—unless by some chance there is anyone so ignorant of geometry and optics as to hold the epicycle of Venus as probable and to believe this to be a cause why Venus alternately precedes and follows the sun at an angular distance of up to 40° or more. For who does not see that it necessarily follows from this assumption that the diameter of the planet in its perigee should appear more than four times greater, and the body of the planet more than sixteen times greater, than in its apogee? Nevertheless the experience of all the ages is opposed to that.[3] There are also other things in this discipline which are just as absurd, but it is not necessary to examine them right now. For it is sufficiently clear that this art is absolutely and profoundly ignorant of the causes of the apparent irregular movements. And if it constructs and thinks up causes—and it has certainly thought up a good many—nevertheless it does not think them up in order to persuade anyone of their truth but only in order that they may provide a correct basis for calculation. But since for one and the same movement varying hypotheses are proposed

---

[1] This foreword, at first ascribed to Copernicus, is held to have been written by Andrew Osiander, a Lutheran theologian and friend of Copernicus, who saw the *De Revolutionibus* through the press.

[2] The numbers within the brackets refer to the pages of the first edition, published in 1543 at Nuremberg.

[3] Ptolemy makes Venus move on an epicycle the ratio of whose radius to the radius of the eccentric circle carrying the epicycle itself is nearly three to four. Hence the apparent magnitude of the planet would be expected to vary with the varying distance of the planet from the Earth, in the ratios stated by Osiander.

Moreover, it was found that, whenever the planet happened to be on the epicycle, the mean position of the sun appeared in line with *EPA*. And so, granted the ratios of epicycle and eccentric, Venus would never appear from the Earth to be at an angular distance of much more than 40° from the centre of her epicycle, that is to say, from the mean position of the sun, as it turned out by observation.

from time to time, as eccentricity or epicycle for the movement of the sun, the astronomer much prefers to take the one which is easiest to [*ii*ᵃ] grasp. Maybe the philosopher demands probability instead; but neither of them will grasp anything certain or hand it on, unless it has been divinely revealed to him. Therefore let us permit these new hypotheses to make a public appearance among old ones which are themselves no more probable, especially since they are wonderful and easy and bring with them a vast storehouse of learned observations. And as far as hypotheses go, let no one expect anything in the way of certainty from astronomy, since astronomy can offer us nothing certain, lest, if anyone take as true that which has been constructed for another use, he go away from this discipline a bigger fool than when he came to it. Farewell.

PREFACE AND DEDICATION TO POPE PAUL III

[*ii*ᵇ] I can reckon easily enough, Most Holy Father, that as soon as certain people learn that in these books of mine which I have written about the revolutions of the spheres of the world I attribute certain motions to the terrestrial globe, they will immediately shout to have me and my opinion hooted off the stage. For my own works do not please me so much that I do not weigh what judgments others will pronounce concerning them. And although I realize that the conceptions of a philosopher are placed beyond the judgment of the crowd, because it is his loving duty to seek the truth in all things, in so far as God has granted that to human reason; nevertheless I think we should avoid opinions utterly foreign to rightness. And when I considered how absurd this "lecture" would be held by those who know that the opinion that the Earth rests immovable in the middle of the heavens as if their centre had been confirmed by the judgments of many ages—if I were to assert to the contrary that the Earth moves; for a long time I was in great difficulty as to whether I should bring to light my commentaries written to demonstrate the Earth's movement, or whether it would not be better to follow the example of the Pythagoreans and certain others who used to hand down the mysteries of their philosophy not in writing but by word of mouth and only to their relatives and friends—witness the letter of Lysis to Hipparchus. They however seem to me to have done that not, as some judge, out of a jealous unwillingness to communicate their doctrines but in order that things of very great beauty which have been investigated by the loving care of great men should not be scorned by those who find it a bother to expend any great energy on letters—except on the money-making variety— or who are provoked by the exhortations and examples of others to the liberal study of philosophy but on account of their natural [*iii*ᵃ] stupidity hold the position among philosophers that drones hold among bees. Therefore, when I weighed these things in my mind, the scorn which I had to fear on account of the newness and absurdity of my opinion almost drove me to abandon a work already undertaken.

But my friends made me change my course in spite of my long-continued hesitation and even resistance. First among them was Nicholas Schonberg, Cardinal of Capua, a man distinguished in all branches of learning; next to him was my devoted friend Tiedeman Giese, Bishop of Culm, a man filled with the greatest zeal for the divine and liberal arts: for he in particular urged me frequently and even spurred me on by added reproaches into publishing this book and letting come to light a work which I had kept hidden among my things for not merely nine years, but for almost four times nine years. Not a few other learned and distinguished men demanded the same thing of me, urging me to refuse no longer—on account of the fear which I felt— to contribute my work to the common utility of those who are really interested in mathematics: they said that the absurder my teaching about the movement of the Earth now seems to very many persons, the more wonder and thanksgiving will it be the object of, when after the publication of my commentaries those same persons see the fog of absurdity dissipated by my luminous demonstrations. Accordingly I was led by such persuasion and by that hope finally to permit my friends to undertake the publication of a work which they had long sought from me.

But perhaps Your Holiness will not be so much surprised at my giving the results of my nocturnal study to the light—after having taken such care in working them out that I did not hesitate to put in writing my conceptions as to the movement of the Earth—as you will be eager to hear from me what came into my mind that in opposition to the general opinion of mathematicians and almost in opposition to common sense I should dare to imagine some movement of the Earth. And so I am unwilling to hide from Your Holiness that nothing except my knowledge that mathematicians have not agreed with one another in their researches moved me to think out a different scheme of drawing up the movements of the spheres of the world. For in the first place mathematicians are so uncertain about the movements of the sun and moon that they can neither demonstrate nor observe the unchanging magnitude of the [iii^b] revolving year. Then in setting up the solar and lunar movements and those of the other five wandering stars, they do not employ the same principles, assumptions, or demonstrations for the revolutions and apparent movements. For some make use of homocentric circles only, others of eccentric circles and epicycles, by means of which however they do not fully attain what they seek. For although those who have put their trust in homocentric circles have shown that various different movements can be composed of such circles, nevertheless they have not been able to establish anything for certain that would fully correspond to the phenomena. But even if those who have thought up eccentric circles seem to have been able for the most part to compute the apparent movements numerically by those means, they have in the meanwhile admitted a great deal which seems to contradict the first principles of regularity of movement. Moreover, they have

9

not been able to discover or to infer the chief point of all, *i.e.*, the form of the world and the certain commensurability of its parts. But they are in exactly the same fix as someone taking from different places hands, feet, head, and the other limbs—shaped very beautifully but not with reference to one body and without correspondence to one another—so that such parts made up a monster rather than a man. And so, in the process of demonstration which they call "method," they are found either to have omitted something necessary or to have admitted something foreign which by no means pertains to the matter; and they would by no means have been in this fix, if they had followed sure principles. For if the hypotheses they assumed were not false, everything which followed from the hypotheses would have been verified without fail; and though what I am saying may be obscure right now, nevertheless it will become clearer in the proper place.

Accordingly, when I had meditated upon this lack of certitude in the traditional mathematics concerning the composition of movements of the spheres of the world, I began to be annoyed that the philosophers, who in other respects had made a very careful scrutiny of the least details of the world, had discovered no sure scheme for the movements of the machinery of the world, which has been built for us by the Best and Most Orderly Workman of all. Wherefore I took the trouble to reread all the books by philosophers which I could get hold of, to see if any of them even supposed that the movements of the spheres of the world [*iv*ᵃ] were different from those laid down by those who taught mathematics in the schools. And as a matter of fact, I found first in Cicero that Nicetas thought that the Earth moved. And afterwards I found in Plutarch that there were some others of the same opinion: I shall copy out his words here, so that they may be known to all:

> *Some think that the Earth is at rest; but Philolaus the Pythagorean says that it moves around the fire with an obliquely circular motion, like the sun and moon. Herakleides of Pontus and Ekphantus the Pythagorean do not give the Earth any movement of locomotion, but rather a limited movement of rising and setting around its centre, like a wheel.*[1]

Therefore I also, having found occasion, began to meditate upon the mobility of the Earth. And although the opinion seemed absurd, nevertheless because I knew that others before me had been granted the liberty of constructing whatever circles they pleased in order to demonstrate astral phenomena, I thought that I too would be readily permitted to test whether or not, by the laying down that the Earth had some movement, demonstrations less shaky than those of my predecessors could be found for the revolutions of the celestial spheres.

[1] *De placitis philosophorum*, III, 13.

And so, having laid down the movements which I attribute to the Earth farther on in the work, I finally discovered by the help of long and numerous observations that if the movements of the other wandering stars are correlated with the circular movement of the Earth, and if the movements are computed in accordance with the revolution of each planet, not only do all their phenomena follow from that but also this correlation binds together so closely the order and magnitudes of all the planets and of their spheres or orbital circles and the heavens themselves that nothing can be shifted around in any part of them without disrupting the remaining parts and the universe as a whole.

Accordingly, in composing my work I adopted the following order: in the first book I describe all the locations of the spheres or orbital circles together with the movements which I attribute to the earth, so that this book contains as it were the general set-up of the universe. But afterwards in the remaining books I correlate all the movements of the other planets and their spheres or orbital circles with the mobility of the Earth, so that it can be gathered from that how far the apparent movements of the remaining planets and their orbital circles can be saved by being correlated with the movements of the Earth. And I have no doubt that talented and learned mathematicians will agree with me, if—as philosophy [iv^b] demands in the first place—they are willing to give not superficial but profound thought and effort to what I bring forward in this work in demonstrating these things. And in order that the unlearned as well as the learned might see that I was not seeking to flee from the judgment of any man, I preferred to dedicate these results of my nocturnal study to Your Holiness rather than to anyone else; because, even in this remote corner of the earth where I live, you are held to be most eminent both in the dignity of your order and in your love of letters and even of mathematics; hence, by the authority of your judgment you can easily provide a guard against the bites of slanderers, despite the proverb that there is no medicine for the bite of a sycophant.

But if perchance there are certain "idle talkers" who take it upon themselves to pronounce judgment, although wholly ignorant of mathematics, and if by shamelessly distorting the sense of some passage in Holy Writ to suit their purpose, they dare to reprehend and to attack my work; they worry me so little that I shall even scorn their judgments as foolhardy. For it is not unknown that Lactantius, otherwise a distinguished writer but hardly a mathematician, speaks in an utterly childish fashion concerning the shape of the Earth, when he laughs at those who have affirmed that the Earth has the form of a globe. And so the studious need not be surprised if people like that laugh at us. Mathematics is written for mathematicians; and among them, if I am not mistaken, my labours will be seen to contribute something to the ecclesiastical commonwealth, the principate of which Your Holiness now holds. For not many years ago under Leo X when the Lateran Council was considering the question of reforming

the Ecclesiastical Calendar, no decision was reached, for the sole reason that the magnitude of the year and the months and the movements of the sun and moon had not yet been measured with sufficient accuracy. From that time on I gave attention to making more exact observations of these things and was encouraged to do so by that most distinguished man, Paul, Bishop of Fossombrone, who had been present at those deliberations. But what have I accomplished in this matter I leave to the judgment of Your Holiness in particular and to that of all other learned mathematicians. And so as not to appear to Your Holiness to make more promises concerning the utility of this book than I can fulfill, I now pass on to the body of the work.

BOOK ONE[1]

Among the many and varied literary and artistic studies upon which the natural talents of man are nourished, I think that those above all should be embraced and pursued with the most loving care which have to do with things that are very beautiful and very worthy of knowledge. Such studies are those which deal with the godlike circular movements of the world and the course of the stars, their magnitudes, distances, risings and settings, and the causes of the other appearances in the heavens; and which finally explicate the whole form. For what could be more beautiful than the heavens which contain all beautiful things? Their very names make this clear: *Caelum* (heavens) by naming that which is beautifully carved; and *Mundus* (world), purity and elegance. Many philosophers have called the world a visible god on account of its extraordinary excellence. So if the worth of the arts were measured by the matter with which they deal, this art—which some call astronomy, others astrology, and many of the ancients the consummation of mathematics—would be by far the most outstanding. This art which is as it were the head of all the liberal arts and the one most worthy of a free man leans upon nearly all the other branches of mathematics. Arithmetic, geometry, optics, geodesy, mechanics, and whatever others, all offer themselves in its service. And since a property of all good arts is to draw the mind of man away from the vices and direct it to better things, these arts can do that more plentifully, over and above the unbelievable pleasure of mind (which they furnish). For who, after applying himself to things which he sees established in the best order and directed by divine ruling, would not through diligent contemplation of them and through a certain habituation be awakened to that which is best and would not wonder at the Artificer of all things, in Whom is all happiness and every good? For the divine Psalmist surely did not say gratuitously that he took pleasure in the workings of God and rejoiced in the works of His hands, unless by means of these things as by some sort of vehicle we are transported to the contemplation of the highest Good.

Now as regards the utility and ornament which they confer upon a commonwealth—to pass over the innumerable advantages they give to private citizens—Plato makes an extremely good point, for in the seventh book of the *Laws* he says that this study should be pursued in especial, that through it the orderly arrangement of days into months and years and the determination of the times for solemnities and sacrifices should keep the state alive and watchful; and he says that if anyone denies that this study is necessary for a man who is going to take up any of the highest branches of learning, then such a person is thinking foolishly; and he thinks that it is impossible for anyone to become godlike or be called so who has no necessary knowledge of the sun, moon, and the other stars.

[1] The three introductory paragraphs are found in the Thorn centenary and Warsaw editions

However, this more divine than human science, which inquires into the highest things, is not lacking in difficulties. And in particular we see that as regards its principles and assumptions, which the Greeks call "hypotheses," many of those who undertook to deal with them were not in accord and hence did not employ the same methods of calculation. In addition, the courses of the planets and the revolution of the stars cannot be determined by exact calculations and reduced to perfect knowledge unless, through the passage of time and with the help of many prior observations, they can, so to speak, be handed down to posterity. For even if Claud Ptolemy of Alexandria, who stands far in front of all the others on account of his wonderful care and industry, with the help of more than forty years of observations brought this art to such a high point that there seemed to be nothing left which he had not touched upon; nevertheless we see that very many things are not in accord with the movements which should follow from his doctrine but rather with movements which were discovered later and were unknown to him. Whence even Plutarch in speaking of the revolving solar year says, "So far the movement of the stars has overcome the ingenuity of the mathematicians." Now to take the year itself as my example, I believe it is well known how many different opinions there are about it, so that many people have given up hope of risking an exact determination of it. Similarly, in the case of the other planets I shall try—with the help of God, without Whom we can do nothing—to make a more detailed inquiry concerning them, since the greater the interval of time between us and the founders of this art—whose discoveries we can compare with the new ones made by us—the more means we have of supporting our own theory. Furthermore, I confess that I shall expound many things differently from my predecessors—although with their aid, for it was they who first opened the road of inquiry into these things.

1. THE WORLD IS SPHERICAL

[1a] In the beginning we should remark that the world is globe-shaped; whether because this figure is the most perfect of all, as it is an integral whole and needs no joints; or because this figure is the one having the greatest volume and thus is especially suitable for that which is going to comprehend and conserve all things; or even because the separate parts of the world i.e., the sun, moon, and stars are viewed under such a form; or because everything in the world tends to be delimited by this form, as is apparent in the case of drops of water and other liquid bodies, when they become delimited of themselves. And so no one would hesitate to say that this form belongs to the heavenly bodies.

2. THE EARTH IS SPHERICAL TOO

The Earth is globe-shaped too, since on every side it rests upon its centre. But it is not perceived straightway to be a perfect sphere, on account of the great height of its mountains and the lowness of its valleys, though they modify its universal roundness to only a very small extent.

That is made clear in this way. For when people journey northward from anywhere, the northern vertex of the axis of daily revolution gradually moves overhead, and the other moves downward to the same extent; and many stars situated to the north are seen not to set, and many to the south are seen not to rise any more. So Italy does not see Canopus, which is visible to Egypt. And Italy sees the last star of Fluvius, which is not visible to this region situated in a more frigid zone. Conversely, for people who travel southward, the second group of stars becomes higher in the sky; while those become lower which for us are high up.

Moreover, the inclinations of the poles have everywhere the same ratio with places at equal distances from the poles of the Earth and that [1b] happens in no other figure except the spherical. Whence it is manifest that the Earth itself is contained between the vertices and is therefore a globe.

Add to this the fact that the inhabitants of the East do not perceive the evening eclipses of the sun and moon; nor the inhabitants of the West, the morning eclipses; while of those who live in the middle region—some see them earlier and some later.

Furthermore, voyagers perceive that the waters too are fixed within this figure; for example, when land is not visible from the deck of a ship, it may be seen from the top of the mast, and conversely, if something shining is attached to the top of the mast, it appears to those remaining on the shore to come down gradually, as the ship moves from the land, until finally it becomes hidden, as if setting.

Moreover, it is admitted that water, which by its nature flows, always seeks lower places—the same way as earth—and does not climb up the shore any farther than the convexity of the shore allows. That is why the land is so much higher where it rises up from the ocean.

3. HOW LAND AND WATER MAKE UP A SINGLE GLOBE

And so the ocean encircling the land pours forth its waters everywhere and fills up the deeper hollows with them. Accordingly it was necessary for there to be less water than land, so as not to have the whole earth soaked with water—since both of them tend toward the same centre on account of their weight—and so as to leave some portions of land—such as the islands discernible here and there—for the preservation of living creatures. For what is the continent itself and the *orbis terrarum* except an island which is larger than the rest? We should not listen to certain Peripatetics who maintain that there is ten times more water than land and who arrive at that conclusion because in the transmutation of the elements the liquefaction of one part of earth results in ten

parts of water. And they say that land has emerged for a certain distance because, having hollow spaces inside, it does not balance everywhere with respect to weight and so the centre of gravity is different from the centre of magnitude. But they fall into error through ignorance of geometry; for they do not know that there cannot be seven times more water than land and some part of the land still remain dry, unless the land abandon its centre of gravity and give place to the waters as being heavier. For spheres are to one another as the cubes of their diameters. If therefore there were seven parts of water and one part of land, [2a] the diameter of the land could not be greater than the radius of the globe of the waters. So it is even less possible that the water should be ten times greater. It can be gathered that there is no difference between the centres of magnitude and of gravity of the Earth from the fact that the convexity of the land spreading out from the ocean does not swell continuously, for in that case it would repulse the sea-waters as much as possible and would not in any way allow interior seas and huge gulfs to break through. Moreover, from the seashore outward the depth of the abyss would not stop increasing, and so no island or reef or any spot of land would be met with by people voyaging out very far. Now it is well known that there is not quite the distance of two miles—at practically the centre of the *orbis terrarum*—between the Egyptian and the Red Sea. And on the contrary, Ptolemy in his *Cosmography* extends inhabitable lands as far as the median circle, and he leaves that part of the Earth as unknown, where the moderns have added Cathay and other vast regions as far as 60° longitude, so that inhabited land extends in longitude farther than the rest of the ocean does. And if you add to these the islands discovered in our time under the princes of Spain and Portugal and especially America—named after the ship's captain who discovered her—which they consider a second *orbis terrarum* on account of her so far unmeasured magnitude—besides many other islands heretofore unknown, we would not be greatly surprised if there were antiphodes or antichthones. For reasons of geometry compel us to believe that America is situated diametrically opposite to the India of the Ganges.

And from all that I think it is manifest that the land and the water rest upon one centre of gravity; that this is the same as the centre of magnitude of the land, since land is the heavier; that parts of land which are as it were yawning are filled with water; and that accordingly there is little water in comparison with the land, even if more of the surface appears to be covered by water.

Now it is necessary that the land and the surrounding waters have the figure which the shadow of the Earth casts, for it eclipses the moon by projecting a perfect circle upon it. Therefore the Earth is not a plane, as Empedocles and Anaximenes opined; or a tympanoid, as Leucippus; or a scaphoid, as Heracleitus; or hollowed out in any other way, as Democritus; or again a cylinder, as Anaximander; and it is not infinite in its

lower part, with the density increasing rootwards, as Xenophanes thought; but it is perfectly round, as the philosophers perceived.

4. THE MOVEMENT OF THE CELESTIAL BODIES IS REGULAR, CIRCULAR, AND EVERLASTING—OR ELSE COMPOUNDED OF CIRCULAR MOVEMENTS

[2^b] After this we will recall that the movement of the celestial bodies is circular. For the motion of a sphere is to turn in a circle; by this very act expressing its form, in the most simple body, where beginning and end cannot be discovered or distinguished from one another, while it moves through the same parts in itself.

But there are many movements on account of the multitude of spheres or orbital circles.[1] The most obvious of all is the daily revolution—which the Greeks call νυχθήμερν; i.e., having the temporal span of a day and a night. By means of this movement the whole world—with the exception of the Earth—is supposed to be borne from east to west. This movement is taken as the common measure of all movements, since we measure even time itself principally by the number of days.

Next, we see other as it were antagonistic revolutions; i.e., from west to east, on the part of the sun, moon, and the wandering stars. In this way the sun gives us the year, the moon the months—the most common periods of time; and each of the other five planets follows its own cycle. Nevertheless these movements are manifoldly different from the first movement. First, in that they do not revolve around the same poles as the first movement but follow the oblique ecliptic; next, in that they do not seem to move in their circuit regularly. For the sun and moon are caught moving at times more slowly and at times more quickly. And we perceive the five wandering stars sometimes even to retrograde and to come to a stop between these two movements. And though the sun always proceeds straight ahead along its route, they wander in various ways, straying sometimes towards the south, and at other times towards the north—whence they are called "planets." Add to this the fact that sometimes they are nearer the Earth—and are then said to be at their perigee—and at other times are farther away—and are said to be at their apogee.

We must however confess that these movements are circular or are composed of many circular movements, in that they maintain these irregularities in accordance with a constant law and with fixed periodic returns: and that could not take place, if they were not circular. For it is only the circle which can bring back what is past and over with; and in this way, for example, the sun by a movement composed of circular movements brings back to us the inequality of days and nights and the four seasons of the

---

[1] The "orbital circle" (orbis) is the great circle whereon the planet moves in its sphere (sphaera). Copernicus; uses the word orbis which designates a circle primarily rather than a sphere because, while the sphere may be necessary for the mechanical explanation of the movement, only the circle is necessary for the mathematical.

year. [3ª] Many movements are recognized in that movement, since it is impossible that a simple heavenly body should be moved irregularly by a single sphere. For that would have to take place either on account of the inconstancy of the motor virtue—whether by reason of an extrinsic cause or its intrinsic nature—or on account of the inequality between it and the moved body. But since the mind shudders at either of these suppositions, and since it is quite unfitting to suppose that such a state of affairs exists among things which are established in the best system, it is agreed that their regular movements appear to us as irregular, whether on account of their circles having different poles or even because the earth is not at the centre of the circles in which they revolve. And so for us watching from the Earth, it happens that the transits of the planets, on account of being at unequal distances from the Earth, appear greater when they are nearer than when they are farther away, as has been shown in optics: thus in the case of equal arcs of an orbital circle which are seen at different distances there will appear to be unequal movements in equal times. For this reason I think it necessary above all that we should note carefully what the relation of the Earth to the heavens is, so as not—when we wish to scrutinize the highest things—to be ignorant of those which are nearest to us, and so as not—by the same error—to attribute to the celestial bodies what belongs to the Earth.

5. DOES THE EARTH HAVE A CIRCULAR MOVEMENT? AND OF ITS PLACE

Now that it has been shown that the Earth too has the form of a globe, I think we must see whether or not a movement follows upon its form and what the place of the Earth is in the universe. For without doing that it will not be possible to find a sure reason for the movements appearing in the heavens. Although there are so many authorities for saying that the Earth rests in the centre of the world that people think the contrary supposition inopinable and even ridiculous; if however we consider the thing attentively, we will see that the question has not yet been decided and accordingly is by no means to be scorned. For every apparent change in place occurs on account of the movement either of the thing seen or of the spectator, or on account of the necessarily unequal movement of both. For no movement is perceptible relatively to things moved equally in the same directions—I mean relatively to the thing seen and the spectator. Now it is from the Earth that the celestial circuit is beheld and presented to our sight. Therefore, if some movement should belong to the Earth [3ᵇ] it will appear, in the parts of the universe which are outside, as the same movement but in the opposite direction, as though the things outside were passing over. And the daily revolution in especial is such a movement. For the daily revolution appears to carry the whole universe along, with the exception of the Earth and the things around it. And if

you admit that the heavens possess none of this movement but that the Earth turns from west to east, you will find—if you make a serious examination—that as regards the apparent rising and setting of the sun, moon, and stars the case is so. And since it is the heavens which contain and embrace all things as the place common to the universe, it will not be clear at once why movement should not be assigned to the contained rather than to the container, to the thing placed rather than to the thing providing the place.

As a matter of fact, the Pythagoreans Herakleides and Ekphantus were of this opinion and so was Hicetas the Syracusan in Cicero; they made the Earth to revolve at the centre of the world. For they believed that the stars set by reason of the interposition of the Earth and that with cessation of that they rose again. Now upon this assumption there follow other things, and a no smaller problem concerning the place of the Earth, though it is taken for granted and believed by nearly all that the Earth is the centre of the world. For if anyone denies that the Earth occupies the midpoint or centre of the world yet does not admit that the distance (between the two) is great enough to be compared with (the distance to) the sphere of the fixed stars but is considerable and quite apparent in relation to the orbital circles of the sun and the planets; and if for that reason he thought that their movements appeared irregular because they are organized around a different centre from the centre of the Earth, he might perhaps be able to bring forward a perfectly sound reason for movement which appears irregular. For the fact that the wandering stars are seen to be sometimes nearer the Earth and at other times farther away necessarily argues that the centre of the Earth is not the centre of their circles. It is not yet clear whether the Earth draws near to them and moves away or they draw near to the Earth and move away.

And so it would not be very surprising if someone attributed some other movement to the earth in addition to the daily revolution. As a matter of fact, Philolaus the Pythagorean—no ordinary mathematician, whom Plato's biographers say Plato went to Italy for the sake of seeing—is supposed to have held that the Earth moved in a circle and wandered in some other movements and was one of the planets.

Many however have believed that they could show by geometrical reasoning that the Earth is in the middle of the world; that it has the proportionality of a point in relation to the immensity of the heavens, occupies the central position, and for this reason is immovable, because, when the universe moves, the centre [4a] remains unmoved and the things which are closest to the centre are moved the most slowly.

6. ON THE IMMENSITY OF THE HEAVENS IN RELATION TO THE MAGNITUDE OF THE EARTH

It can be understood that this great mass which is the Earth is not comparable with the magnitude of the heavens, from the fact that the boundary circles—for that is the translation of the Greek ὁρίζοντες—cut the whole celestial sphere into two halves; for that could not take place if the magnitude of the Earth in comparison with the heavens, or its distance from the centre of the world, were considerable. For the circle bisecting a sphere goes through the centre of the sphere, and is the greatest circle which it is possible to circumscribe.

Now let the horizon be the circle *ABCD*, and let the Earth, where our point of view is, be *E*, the centre of the horizon by which the visible stars are separated from those which are not visible. Now with a dioptra or horoscope or level placed at *E*, the beginning of Cancer is seen to rise at point *C*; and at the same moment the beginning of Capricorn appears to set at *A*. Therefore, since *AEC* is in a straight line with the dioptra, it is clear that this line is a diameter of the ecliptic, because the six signs bound a semicircle, whose centre *E* is the same as that of the horizon. But when a revolution has taken place and the beginning of Capricorn arises at *B*, then the setting of Cancer 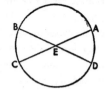 will be visible at *D*, and *BED* will be a straight line and a diameter of the ecliptic. But it has already been seen that the line *AEC* is a diameter of the same circle; therefore, at their common section, point *E* will be their centre. So in this way the horizon always bisects the ecliptic, which is a great circle of the sphere. But on a sphere, if a circle bisects one of the great circles, then the circle bisecting is a great circle. Therefore the horizon is a great circle; and its centre is the same as that of the ecliptic, as far as appearance goes; although nevertheless the line passing through the centre of the Earth and the line touching to the surface are necessarily different; but on account of their immensity in comparison with the Earth they are like parallel lines, which on account of the great distance between the termini appear to be one line, when the space contained between them [4b] is in no perceptible ratio to their length, as has been shown in optics.

From this argument it is certainly clear enough that the heavens are immense in comparison with the Earth and present the aspect of an infinite magnitude, and that in the judgment of sense-perception the Earth is to the heavens as a point to a body and as a finite to an infinite magnitude. But we see that nothing more than that has been shown, and it does not follow that the Earth must rest at the centre of the world. And we should be even more surprised if such a vast world should wheel completely around during the space of twenty-four hours rather than that its least part, the Earth, should. For saying that the centre is immovable and that those things which are closest to the centre are moved least does not argue that the Earth rests at the centre of the

world. That is no different from saying that the heavens revolve but the poles are at rest and those things which are closest to the poles are moved least. In this way Cynosura (the pole star) is seen to move much more slowly than Aquila or Canicula because, being very near to the pole, it describes a smaller circle, since they are all on a single sphere, the movement of which stops at its axis and which does not allow any of its parts to have movements which are equal to one another. And nevertheless the revolution of the whole brings them round in equal times but not over equal spaces.

The argument which maintains that the Earth, as a part of the celestial sphere and as sharing in the same form and movement, moves very little because very near to its centre advances to the following position: therefore the Earth will move, as being a body and not a centre, and will describe in the same time arcs similar to, but smaller than, the arcs of the celestical circle. It is clearer than daylight how false that is; for there would necessarily always be noon at one place and midnight at another, and so the daily risings and settings could not take place, since the movement of the whole and the part would be one and inseparable.

But the ratio between things separated by diversity of nature is so entirely different that those which describe a smaller circle turn more quickly than those which describe a greater circle. In this way Saturn, the highest of the wandering stars, completes its revolution in thirty years, and the moon which is without doubt the closest to the Earth completes its circuit in a month, and finally the Earth itself will be considered to complete a circular movement in the space of a day and a night. So this same problem concerning the daily revolution comes up again. And also the question about the place of the Earth becomes even less certain on account of what was just said. For that demonstration proves nothing except that the heavens are of an indefinite magnitude with respect to the Earth. But it is not at all clear how far this immensity stretches out. On the contrary, since the minimal and indivisible corpuscles, which are called atoms, are not perceptible to sense, they do not, when taken in twos or in some small number, constitute a visible body; but they can be taken in such a large quantity that there will at last be enough to form a visible magnitude. So it is as regards the place of the earth; for although it is not at the centre of the world, nevertheless the distance is as nothing, particularly in comparison with the sphere of the fixed stars.

7. WHY THE ANCIENTS THOUGHT THE EARTH WAS AT REST AT THE MIDDLE OF THE WORLD AS ITS CENTRE

[5a] Wherefore for other reasons the ancient philosophers have tried to affirm that the Earth is at rest at the middle of the world, and as principal cause they put forward heaviness and lightness. For Earth is the heaviest element; and all things of any weight are borne towards it and strive to move towards the very centre of it.

For since the Earth is a globe towards which from every direction heavy things by their own nature are borne at right angles to its surface, the heavy things would fall on one another at the centre if they were not held back at the surface; since a straight line making right angles with a plane surface where it touches a sphere leads to the centre. And those things which are borne toward the centre seem to follow along in order to be at rest at the centre. All the more then will the Earth be at rest at the centre; and, as being the receptacle for falling bodies, it will remain immovable because of its weight.

They strive similarly to prove this by reason of movement and its nature. For Aristotle says that the movement of a body which is one and simple is simple, and the simple movements are the rectilinear and the circular. And of rectilinear movements, one is upward, and the other is downward. As a consequence, every simple movement is either toward the centre, *i.e.*, downward, or away from the centre, *i.e.*, upward, or around the centre, *i.e.*, circular. Now it belongs to earth and water, which are considered heavy, to be borne downward, *i.e.*, to seek the centre: for air and fire, which are endowed with lightness, move upward, *i.e.*, away from the centre. It seems fitting to grant rectilinear movement to these four elements and to give the heavenly bodies a circular movement around the centre. So Aristotle. Therefore, said Ptolemy of Alexandria, if the Earth moved, even if only by its daily rotation, the contrary of what was said above would necessarily take place. For this movement which would traverse the total circuit of the Earth in twenty-four hours would necessarily be very headlong and of an unsurpassable velocity. Now things which are suddenly and violently whirled around are seen to be utterly unfitted for reuniting, and the more unified are seen to become dispersed, unless some constant force constrains them to stick together. And a long time ago, he says, the scattered Earth would have passed beyond the heavens, as is certainly ridiculous; [5b] and *a fortiori* so would all the living creatures and all the other separate masses which could by no means remain unshaken. Moreover, freely falling bodies would not arrive at the places appointed them, and certainly not along the perpendicular line which they assume so quickly. And we would see clouds and other things floating in the air always borne toward the west.

8. ANSWER TO THE AFORESAID REASONS AND THEIR INADEQUACY

For these and similar reasons they say that the Earth remains at rest at the middle of the world and that there is no doubt about this. But if someone opines that the Earth revolves, he will also say that the movement is natural and not violent. Now things which are according to nature produce effects contrary to those which are violent. For things to which force or violence is applied get broken up and are unable to subsist for a long time. But things which are caused by nature are in a right condition and are kept in their best organization. Therefore Ptolemy had no reason to fear that

the Earth and all things on the Earth would be scattered in a revolution caused by the efficacy of nature, which is greatly different from that of art or from that which can result from the genius of man. But why didn't he feel anxiety about the world instead, whose movement must necessarily be of greater velocity, the greater the heavens are than the Earth? Or have the heavens become so immense, because an unspeakably vehement motion has pulled them away from the centre, and because the heavens would fall if they came to rest anywhere else?

Surely if this reasoning were tenable, the magnitude of the heavens would extend infinitely. For the farther the movement is borne upward by the vehement force, the faster will the movement be, on account of the ever-increasing circumference which must be traversed every twenty-four hours: and conversely, the immensity of the sky would increase with the increase in movement. In this way, the velocity would make the magnitude increase infinitely, and the magnitude the velocity. And in accordance with the axiom of physics that *that which is infinite cannot be traversed or moved in any way*, then the heavens will necessarily come to rest.

But they say that beyond the heavens there isn't any body or place or void or anything at all; and accordingly it is not possible for the heavens to move outward; in that case it is rather surprising that something can be held together by nothing. But if the heavens were infinite and were finite only with respect to a hollow space inside, then it will be said with more truth that there is nothing outside the heavens, since anything [6ᵃ] which occupied any space would be in them; but the heavens will remain immobile. For movement is the most powerful reason wherewith they try to conclude that the universe is finite.

But let us leave to the philosophers of nature the dispute as to whether the world is finite or infinite, and let us hold as certain that the Earth is held together between its two poles and terminates in a spherical surface. Why therefore should we hesitate any longer to grant to it the movement which accords naturally with its form, rather than put the whole world in a commotion—the world whose limits we do not and cannot know? And why not admit that the appearance of daily revolution belongs to the heavens but the reality belongs to the Earth? And things are as when Aeneas said in Virgil: "We sail out of the harbor, and the land and the cities move away." As a matter of fact, when a ship floats on over a tranquil sea, all the things outside seem to the voyagers to be moving in a movement which is the image of their own, and they think on the contrary that they themselves and all the things with them are at rest. So it can easily happen in the case of the movement of the Earth that the whole world should be believed to be moving in a circle. Then what would we say about the clouds and the other things floating in the air or falling or rising up, except that not only the Earth and the watery element with which it is conjoined are moved in this way but also no

small part of the air and whatever other things have a similar kinship with the Earth? whether because the neighbouring air, which is mixed with earthly and watery matter, obeys the same nature as the Earth or because the movement of the air is an acquired one, in which it participates without resistance on account of the contiguity and perpetual rotation of the Earth. Conversely, it is no less astonishing for them to say that the highest region of the air follows the celestial movement, as is shown by those stars which appear suddenly—I mean those called "comets" or "bearded stars" by the Greeks. For that place is assigned for their generation; and like all the other stars they rise and set. We can say that that part of the air is deprived of terrestrial motion on account of its great distance from the Earth. Hence the air which is nearest to the Earth and the things floating in it will appear tranquil, unless they are driven to and fro by the wind or some other force, as happens. For how is the wind in the air different from a current in the sea?

But we must confess that in comparison with the world the movement of falling and of rising bodies is twofold and is in general compounded of the rectilinear and the circular. As regards things which move downward on account of their weight [6b] because they have very much earth in them, doubtless their parts possess the same nature as the whole, and it is for the same reason that fiery bodies are drawn upward with force. For even this earthly fire feeds principally on earthly matter; and they define flame as glowing smoke. Now it is a property of fire to make that which it invades to expand; and it does this with such force that it can be stopped by no means or contrivance from breaking prison and completing its job. Now expanding movement moves away from the centre to the circumference; and so if some part of the Earth caught on fire, it would be borne away from the centre and upward. Accordingly, as they say, a simple body possesses a simple movement—this is first verified in the case of circular movement—as long as the simple body remain in its unity in its natural place. In this place, in fact, its movement is none other than the circular, which remains entirely in itself, as though at rest. Rectilinear movement, however, is added to those bodies which journey away from their natural place or are shoved out of it or are outside it somehow. But nothing is more repugnant to the order of the whole and to the form of the world than for anything to be outside of its place. Therefore rectilinear movement belongs only to bodies which are not in the right condition and are not perfectly conformed to their nature—when they are separated from their whole and abandon its unity. Furthermore, bodies which are moved upward or downward do not possess a simple, uniform, and regular movement—even without taking into account circular movement. For they cannot be in equilibrium with their lightness or their force of weight. And those which fall downward possess a slow movement at the beginning but increase their velocity as they fall. And conversely we note that this earthly fire—

and we have experience of no other—when carried high up immediately dies down, as if through the acknowledged agency of the violence of earthly matter.

Now circular movement always goes on regularly, for it has an unfailing cause; but (in rectilinear movement) the acceleration stops, because, when the bodies have reached their own place, they are no longer heavy or light, and so the movement ends. Therefore, since circular movement belongs to wholes and rectilinear to parts, we can say that the circular movement stands with the rectilinear, as does animal with sick. And the fact that Aristotle divided simple movement into three genera: away from the centre, toward the centre, and around the centre, will be considered merely as an act of reason, just as we distinguish between line, point, and surface, though none of them can subsist without the others or [7ª] without body.

In addition, there is the fact that the state of immobility is regarded as more noble and godlike than that of change and instability, which for that reason should belong to the Earth rather than to the world. I add that it seems rather absurd to ascribe movement to the container or to that which provides the place and not rather to that which is contained and has a place, *i.e.*, the Earth. And lastly, since it is clear that the wandering stars are sometimes nearer and sometimes farther away from the Earth, then the movement of one and the same body around the centre—and they mean the centre of the Earth—will be both away from the centre and toward the centre. Therefore it is necessary that movement around the centre should be taken more generally; and it should be enough if each movement is in accord with its own centre. You see therefore that for all these reasons it is more probably that the Earth moves than that it is at rest—especially in the case of the daily revolution, as it is the Earth's very own. And I think that is enough as regards the first part of the question.

9. WHETHER MANY MOVEMENTS CAN BE ATTRIBUTED TO THE EARTH, AND CONCERNING THE CENTRE OF THE WORLD

Therefore, since nothing hinders the mobility of the Earth, I think we should now see whether more than one movement belongs to it, so that it can be regarded as one of the wandering stars. For the apparent irregular movement of the planets and their variable distances from the Earth—which cannot be understood as occurring in circles homocentric with the Earth—make it clear that the Earth is not the centre of their circular movements. Therefore, since there are many centres, it is not foolhardy to doubt whether the centre of gravity of the Earth rather than some other is the centre of the world. I myself think that gravity or heaviness is nothing except a certain natural appetency implanted in the parts by the divine providence of the universal Artisan, in order that they should unite with one another in their oneness and wholeness and come

together in the form of a globe. It is believable that this affect is present in the sun, moon, and the other bright planets and that through its efficacy they remain in the spherical figure in which they are visible, though they nevertheless accomplish their circular movements in many different ways. Therefore if the Earth too possesses movements different from the one around its centre, then they will necessarily be movements which similarly appear on the outside in the many bodies; and we find the yearly revolution among these movements. For if the annual revolution were changed from being solar to being terrestrial, and immobility were granted to the sun, [7b] the risings and settings of the signs and of the fixed stars—whereby they become morning or evening stars—will appear in the same way; and it will be seen that the stoppings, retrogressions, and progressions of the wandering stars are not their own, but are a movement of the Earth and that they borrow the appearances of this movement. Lastly, the sun will be regarded as occupying the centre of the world. And the ratio of order in which these bodies succeed one another and the harmony of the whole world teaches us their truth, if only—as they say—we would look at the thing with both eyes.

10. ON THE ORDER OF THE CELESTIAL ORBITAL CIRCLES

I know of no one who doubts that the heavens of the fixed stars is the highest up of all visible things. We see that the ancient philosophers wished to take the order of the planets according to the magnitude of their revolutions, for the reason that among things which are moved with equal speed those which are the more distant seem to be borne along more slowly, as Euclid proves in his *Optics*. And so they think that the moon traverses its circle in the shortest period of time, because being next to the Earth, it revolves in the smallest circle. But they think that Saturn, which completes the longest circuit in the longest period of time, is the highest. Beneath Saturn, Jupiter. After Jupiter, Mars.

There are different opinions about Venus and Mercury, in that they do not have the full range of angular elongations from the sun that the others do.[1] Wherefore some place them above the sun, as Timaeus does in Plato; some, beneath the sun, as Ptolemy and a good many moderns. Alpetragius makes Venus higher than the sun and Mercury lower. Accordingly, as the followers of Plato suppose that all the planets—which are otherwise dark bodies—shine with light received from the sun, they think that if the planets were below the sun, they would on account of their slight distance from the sun be viewed as only half—or at any rate as only partly—spherical. For the light which they receive is reflected by them upward for the most part, *i.e.*, towards the sun, as we see in the case of the new moon or the old. Moreover, they say that necessarily the sun would sometimes be obscured through their interposition and that its light would be

[1]The greatest angular elongation of Venus from the sun is approximately 45°; that of Mercury, approximately 24°; while Saturn, Jupiter, and Mars have the full range of possible angular elongation, *i.e.*, up to 180°.

eclipsed in proportion to their magnitude; and as that has never appeared to take place, they think that these planets cannot by any means be below the sun.[1]

On the contrary, those who place Venus and Mercury below the sun claim as a reason the amplitude of the space which they find between the sun and the moon. [8a] For they find that the greatest distance between the Earth and the moon, *i.e.*, $64\frac{1}{6}$ units, whereof the radius of the Earth is one, is contained almost 18 times in the least distance between the sun and the Earth. This distance is 1160 such units, and therefore the distance between the sun and the moon is 1096 such units. And then, in order for such a vast space not to remain empty, they find that the intervals between the perigees and apogees—according to which they reason out the thickness of the spheres[2]—add up to approximately the same sum: in such fashion that the apogee of the moon may be succeeded by the perigee of Mercury, that the apogee of Mercury may be followed by the perigee of Venus, and that finally the apogee of Venus may nearly touch the perigee of the sun. In fact they calculate that the interval between the perigee and the apogee of Mercury contains approximately $177\frac{1}{2}$ of the aforesaid units and that the remaining space is nearly filled by the 910 units of the interval between the perigee and apogee of Venus.[3] Therefore they do not admit that these planets have a certain opacity, like that of the moon; but that they shine either by their own proper light or because their entire bodies are impregnated with sunlight, and that accordingly they do not obscure the sun, because it is an extremely rare occurrence for them to be interposed between our sight and the sun, as they usually withdraw (from the sun) latitudinally. In addition, there is the fact that they are small bodies in comparison with the sun, since Venus even though larger than Mercury can cover scarcely one one-hundredth part of the sun, as al-Battani the Harranite maintains, who holds that the diameter of the sun is ten times greater, and therefore it would not be easy to see such a little speck in the midst of such beaming light. Averroes, however, in his paraphrase of Ptolemy records having seen something blackish, when he observed the

---

[1] The transit of Venus across the face of the sun was first observed—by means of a telescope—in 1639.
[2] That is to say, the thickness of the sphere would measured by the ratio of the diameter of the epicycle to the diameter of the sphere, or, in the accompanying diagram, by the distance between the inmost and the outmost of the three homocentric circles.
[3] The succession of the orbital circles according to their perigees and apogees may be represented in the following diagram, which has been drawn to scale.

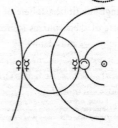

conjunction of the sun and Mercury which he had computed. And so they judge that these two planets move below the solar circle.

But how uncertain and shaky this reasoning is, is clear from the fact that though the shortest distance of the moon is 38 units whereof the radius of the Earth is one unit—according to Ptolemy, but more than 49 such units by a truer evaluation, as will be shown below—nevertheless we do not know that this great space contains anything except air, or if you prefer, what they call the fiery element.

Moreover, there is the fact that the diameter of the epicycle of Venus—by reason of which Venus has an angular digression of approximately 45° on either side of the sun—would have to be six times greater than the distance from the centre of the Earth to its perigee, as will be shown in the proper place.[1] Then what will they say is contained in all this space, which [8ᵇ] is so great as to take in the Earth, air, ether, moon and Mercury, and which moreover the vast epicycle of Venus would occupy if it revolved around an immobile Earth?

Furthermore, how unconvincing is Ptolemy's argument that the sun must occupy the middle position between those planets which have the full range of angular elongation from the sun and those which do not is clear from the fact that the moon's full range of angular elongation proves its falsity.

But what cause will those who place Venus below the sun, and Mercury next, or separate them in some other order—what cause will they allege why these planets do not also make longitudinal circuits separate and independent of the sun, like the other planets[2]—if indeed the ratio of speed or slowness does not falsify their order? Therefore it will be necessary either for the Earth not to be the centre to which the order of the planets and their orbital circles is referred, or for there to be no sure reason for their order and for it not to be apparent why the highest place is due to Saturn rather than to Jupiter or some other planet. Wherefore I judge that what Martianus Capella—who wrote the *Encyclopedia*—and some other Latins took to be the case is by no means to be despised. For they hold that Venus and Mercury circle around the sun as a centre; and they hold that for this reason Venus and Mercury do not have any farther elongation from the sun than the convexity of their orbital circles permits; for they do not make a circle around the earth as do the others, but have perigee and apogee interchangeable (in the sphere of the fixed stars). Now what do they mean except that the centre of their spheres is around the sun? Thus the orbital circle of Mercury will be

[1]According to Ptolemy, the ratio of the radius of Venus' epicycle to the radius of its eccentric is between 2 to 3 and 3 to 4, or approximately 43¹⁄₆ to 60. Now since at perigee the epicycle subtracts from the mean distance, or radius of the eccentric circle, that which at apogee it adds to the mean distance, the ratio of Venus' distance at perigee to its distance at apogee is approximately 1 to 6. That is to say, in the passage from apogee to perigee, the ratio of increase, in the apparent magnitude of the planet should be approximately 36 to 1, as the apparent magnitude varies inversely in the ratio of the square of the distance. But no such increase in the magnitude of the planet is apparent. This opposition between an appearance and the consequences of an hypothesis made to save another appearance is still present within Copernicus' own scheme.

[2]Ptolemy makes the centres of the epicycles of Venus and Mercury travel around the Earth longitudinally at the same rate as the mean sun, and in such fashion that the mean sun is always on the straight line extending from the centre of the Earth through the centres of their epicycles, while the centres of the epicycles of the upper planets may be at any angular distance from the mean sun.

enclosed within the orbital circle of Venus—which would have to be more than twice as large—and will find adequate room for itself within that amplitude.[1] Therefore if anyone should take this as an occasion to refer Saturn, Jupiter, and Mars also to this same centre, provided he understands the magnitude of those orbital circles to be such as to comprehend and encircle the Earth remaining within them, he would not be in error, as the table of ratios of their movements makes clear.[2] For it is manifest that the planets are always nearer the Earth at the time of their evening rising, *i.e.*, when they are opposite to the sun and the Earth is in the middle between them and the sun. But they are farthest away from the Earth at the time of their evening setting, *i.e.*, when they are occulted in the neighbourhood of the sun, namely, when we have the sun between them and the Earth. All that shows clearly enough that their centre is more directly related to the sun and is the same as that to which Venus and Mercury refer

[1]As in the following diagram which has been drawn to scale.

[2]Take the case of Mars. In Ptolemy, the ratio of its epicycle to its eccentric is $39^1/_2$ to 60, or approximately 2 to 3. Mars has 37 cycles of anomaly, or movement on the epicycle, and 42 cycles of longitude, or movement of the epicycle on the eccentric, in 79 solar years; or for the sake of easiness let us say that the ratio of the sun's movement to either of the planets' two movements is 2 to 1. Copernicus is here suggesting that if the centre of the planet's movement is placed around the moving sun, then the Ptolemaic cycles of anomaly will represent the number of times the sun has overtaken the planet in longitude: thus the 37 cycles of anomaly plus the 42 cycles of longitude add up to the 79 solar revolutions. That is to say, the sun will now be traveling around the Earth on a circle which has the same relative magnitude as the Martian epicycle in Ptolemy and bears an epicycle having the same relative magnitude as Ptolemy's Martian eccentric circle, on which epicycle Mars travels in the opposite direction at half the speed of the sun. Under both hypotheses the appearances from the Earth will be the same, as can be seen in the following diagrams.

For according to the Ptolemaic hypothesis, let the Earth be at the center of the approximately homocentric circles of the sun, Maps, and the ecliptic. Let the radius of the planet's epicycle be to the radius of the planet's eccentric as 2 to 3. Now, first, let the sun be viewed at the beginning of Leo, and let the planet at the perigee of its epicycle be viewed at the beginning of Aquarius, in opposition to the sun. Next, let the sun move 240° eastwards, to the beginning of Aries; and during the same interval let the epicycle move 120° eastwards, to the beginning of Gemini, and the planet 120° eastwards on the epicycle. Now the planet will be found to appear in Taurus, about 36° west of the sun.

But if according to the semi-Copernican hypothesis, the sun is made to revolve around the Earth on a circle having the same relative magnitude as Mars' Ptolemaic epicycle, while Mars is placed on an epicycle which has the same relative magnitude as its Ptolemaic eccentric and has its centre at the sun; and if the apparent positions of Mars and the sun are first the same as before, and the sun moves 240° eastwards, bearing along the deferent of Mars, while Mars moves 120° westwards on its epicycle; then Mars will once more be found to appear in Taurus, approximately 36° west of the sun.

PTOLEMAIC HYPOTHESIS

Movement of Sun=240°
Movement of Eccentric=120°
Movement of Epicycle=120°

SEMI-COPERNICAN HYPOTHESIS

Movement of Sun=240°
Movement of Mars=120°

their revolutions.[1] But as they all have one common centre, it is necessary that the space left between the convex orbital circle of Venus and the concave orbital circle of Mars should be viewed as an orbital circle [9ᵃ] or sphere homocentric with them in respect to both surfaces, and that it should receive the Earth and its satellite the moon and whatever is contained beneath the lunar globe. For we can by no means separate the moon from the Earth, as the moon is incontestably very near to the Earth—especially since we find in this expanse a place for the moon which is proper enough and sufficiently large. Therefore we are not ashamed to maintain that this totality—which the moon embraces—and the centre of the Earth too traverse that great orbital circle among the other wandering stars in an annual revolution around the sun; and that the centre of the world is around the sun. I also say that the sun remains forever immobile and that whatever apparent movement belongs to it can be verified of the mobility of the Earth; that the magnitude of the world is such that, although the distance from the sun to the Earth in relation to whatsoever planetary sphere you please possesses magnitude which is sufficiently manifest in proportion to these dimensions, this distance, as compared with the sphere of the fixed stars, is imperceptible. I find it much more easy to grant that than to unhinge the understanding by an almost infinite multitude of spheres—as those who keep the earth at the centre of the world are forced to do. But we should rather follow the wisdom of nature, which, as it takes very great care not to have produced anything superfluous or useless, often prefers to endow one thing with many effects. And though all these things are difficult, almost inconceivable, and quite contrary to the opinion of the multitude, nevertheless in what follows we will with God's help make them clearer than day—at least for those who are not ignorant of the art of mathematics.

Therefore if the first law is still safe—for no one will bring forward a better one than that the magnitude of the orbital circles should be measured by the magnitude of time—then the order of the spheres will follow in this way—beginning with the highest: the first and highest of all is the sphere of the fixed stars, which comprehends itself and all things, and is accordingly immovable. In fact it is the place of the universe, *i.e.*, it is that to which the movement and position of all the other stars are referred. For in the deduction of terrestrial movement, we will however give the cause why there are appearances such as to make people believe that even the sphere of the fixed stars somehow moves. Saturn, the first of the wandering stars follows; it completes its circuit in

Conjunction Opposition
ACCORDING TO PTOLEMY

[1]Copernicus is asking what reason there is why the planets are always found to be at their apogees at the time of conjunction with the sun, and at their perigees at the time of opposition, since according to the Ptolemaic scheme the reverse is also possible—as is evident from the accompanying diagram.

But if the sun and not the Earth is the centre of the planet's movements, the reason is obvious.

Conjunction Opposition
ACCORDING TO COPERNICUS

30 years. After it comes Jupiter moving in a 12-year period of revolution. Then Mars, which completes a revolution every 2 years. The place fourth in order is occupied by the annual revolution [9b] in which we said the Earth together with the orbital circle of the moon as an epicycle is comprehended. In the fifth place, Venus, which completes its revolution in 7½ months. The sixth and final place is occupied by Mercury, which completes its revolution in a period of 88 days.[1] In the center of all rests the sun. For who would place this lamp of a very beautiful temple in another or better place than this wherefrom it can illuminate everything at the same time? As a matter of fact, not unhappily do some call it the lantern; others, the mind and still others, the pilot of the world. Trismegistus calls

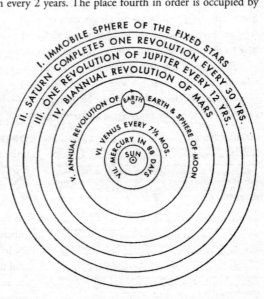

	Cycles of anomaly	Cycles of longitude	Solar years
Mercury	145	46 +	46 +
Venus	5	8 —	8 —

It is noteworthy that the number of cycles of longitude in one year is equal to the number of solar cycles. Moreover, the two planets have a limited angular elongation from the sun. In order to explain these two peculiar appearances Copernicus sets the Earth in motion on the circumference of a circle which encloses the orbits of Venus and Mercury, with the sun at the centre of all three orbits. Thus the planet's cycles of anomaly in so many years become the number of times the planet has overtaken the Earth, as they revolve around the sun. That is to say, in so many solar years the planet will have traveled around the sun a number of times which is equal to the sum of its cycles of anomaly and its cycles in longitude. Thus, for example, Venus travels around the sun approximately 13 times in 8 solar years; hence its period of revolution is approximately 7½ months; and similarly, that of Mercury is approximately 88 days—although for some obscure reason Copernicus actually writes down 9 months for Venus (*nono mense reducino*) and 80 days for Mercury (*octaginta dierum spatio circumcurrens*).

The reader may intuit from the following diagrams the equipollence, with respect to the appearances, of the Ptolemaic and the Copernican explanations of the movement of Venus.

Now, on Ptolemy's hypothesis, let the Earth be placed at the centre of the ecliptic, the solar circle, and the orbital circle of Venus, which carries the planetary epicycle. The radius of the epicycle is to that of the orbital circle approximately as 3 is to 4. First let the sun be situated at the middle of Scorpio, and let Venus be in conjunction with the sun and at the perigee of its epicycle. Next let the sun move 180° eastwards to the middle of Taurus, and similarly the centre of the epicycle; during this same interval the planet will move 112½° eastwards on its epicycle and will be found to appear in the middle of Aries approximately, or 30° west of the sun.

But according to the Copernican hypothesis, let us place the sun at the centre of the orbital circles of Venus and the Earth, which preserve the relative magnitudes of the Ptolemaic epicycle and orbital circle of Venus, but let us keep the Earth at the centre of the ecliptic, as far as appearances go, since the distance between the Earth and the sun is imperceptible in comparison with the magnitude of the sphere of the fixed stars. Now if the Earth is placed in the middle of Taurus, as viewed from the sun, and the planet at its perigee between the Earth and the sun, in such fashion that Venus and the sun would appear in the middle of Scorpio, while Venus moves eastwards 292½°, then the sun will be found to appear in the middle of Taurus, and the planet itself in middle of Aries or 30° west of the sun.

But let us turn to the three upper planets.

	Cycles of anomaly	Cycles of longitude	Solar years
Mars	37	42 +	79
Jupiter	65	6 +	71 —
Saturn	57	2 +	59 —

It is here noteworthy that according to the Ptolemaic hypothesis the sum of the revolutions of the eccentric circle and the revolutions in anomaly is equal to the number of solar cycles; and also that, the conjunctions with the sun take place at the planet's apogee, and the oppositions at its perigee.

(continued p. 26)

it a "visible god"; Sophocles' Electra, "that which gazes upon all things." And so the sun, as if resting on a kingly throne, governs the family of stars which wheel around. Moreover, the Earth is by no means cheated of the services of the moon; but, as Aristotle says in the *De Animalibus*, the earth has the closest kinship with the moon. The Earth moreover is fertilized by the sun and conceives offspring every year.

Therefore in this ordering we find [10^a] that the world has a wonderful commensurability and that there is a sure bond of harmony for the movement and magnitude of the orbital circles such as cannot be found in any other way.[1] For now the careful observer can note why progression and retrogradation appear greater in Jupiter than in

But according to Copernicus the Ptolemaic cycles of anomaly will now represent the number of times the Earth has overtaken the planet; and the period of revolution in longitude will stay the alone. Thus, for example, Saturn will have two revolutions in longitude in 59 years, or one revolution around the sun in about 30 years. The planet will be revolving directly on its eccentric circle instead of on its Ptolemaic epicycle, and the Earth will now be revolving on an inner circle which has the same relative magnitude as the former epicycle. The two hypotheses, of course, are equipollent here too, with respect to appearances.

In other words, in constructing a theory to account for four coincidences which were left unexplained by Ptolemy, namely, (1) the equality between the number of cycles in longitude and the solar cycles, in the two lower planets; (2) the equality between the solar cycles and the sum of the cycles of anomaly and longitude, in the upper planets; (3) the limited angular digressions of Mercury and Venus away from the sun; and (4) the apogeal conjunctions and perigeal oppositions of Saturn, Jupiter, and Mars; Copernicus has telescoped the eccentric circle of Venus and that of Mercury into one circle carrying the Earth; and he has furthermore collapsed the three epicycles of Saturn, Jupiter, and Mars into this same one circle. That is to say, one circle is now doing the work of five.

COPERNICAN HYPOTHESIS

Movement of Earth=180°
Movement of Venus=292½°

[1] Let us recall the Ptolemaic ratios between the radius of the epicycle and that of the eccentric circle, and also the eccentricity.

	Epicycle	Eccentric	Eccentricity
Mercury	$22^1/_2$	60	3
Venus	$43^1/_6$	60	$1^1/_4$
Mars	$39^1/_2$	60	6
Jupiter	$11^1/_2$	60	$2^2/_5$
Saturn	$6^1/_2$	60	$3^1/_4$

By the Ptolemaic scheme it is impossible to compute the magnitudes of the eccentric circles themselves relative to one another, as there is no common measure. But now that the eccentric circles of Mercury and Venus and the epicycles of Mars, Jupiter, and Saturn have all been reduced to the orbital circle of the Earth, it is easy to calculate the relative magnitudes of the orbital circles—heretofore the epicycles of the lower planets and the eccentric circles of the upper—since, by reason of the necessary commensurability between epicycle and eccentric, they are all commensurable with the orbital circle of the Earth. Thus, for example, if we take the distance from the Earth to the sun as 1, the planets will observe the following approximate distances from the sun.

Mercury	$1/_3$	Earth	1	Jupiter	5
Venus	$3/_4$	Mars	$1^1/_2$	Saturn	9

Saturn and smaller than in Mars; and in turn greater in Venus than in Mercury.[1] And why these reciprocal events appear more often in Saturn than in Jupiter, and even less often in Mars and Venus than in Mercury.[2] In addition, why when Saturn, Jupiter, and Mars are in opposition (to the mean position of the sun) they are nearer to the Earth than at the time of their occultation and their reappearance. And especially why at the times when Mars is in opposition to the sun, it seems to equal Jupiter in magnitude and to be distinguished from Jupiter only by a reddish color, but when discovered through careful observation by means of a sextant is found with difficulty among the stars of second magnitude?[3] All these things proceed from the same cause, which resides in the movement of the Earth.

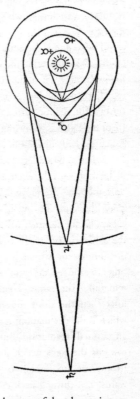

But that there are no such appearances among the fixed stars argues that they are at an immense height away, which makes the circle of annual movement or its image disappear from before our eyes since every visible thing has a certain distance beyond which it is no longer seen, as is shown in optics. For the brilliance of their lights shows that there is a very great distance between Saturn the highest of the planets and the sphere of the fixed stars. It is by this mark in particular that they are distinguished from the planets, as it is proper to have the greatest difference between the moved and the unmoved. How exceedingly fine is the godlike work of the Best and Greatest Artist!

11. A Demonstration of the Threefold Movement of the Earth

Therefore since so much and such great testimony on the part of the planets is consonant with the mobility of the Earth, we shall now give a summary of its movement,

[1] In the three upper planets, the angles which measure the apparent progression and retrogradation have as their vertex the centre of the planet and as their sides the tangents drawn to the orbital circle of the Earth. In the two lower planets, however, the vertex of the angle is at the centre of the Earth and the sides are the tangents drawn to the orbital circle of the planet. It is easy to see that, on account of the relative magnitudes of the orbital circles, the arcs of progression and retrogradation will appear smaller in Saturn than in Jupiter, and smaller in Jupiter than in Mars, and greater in Venus than in Mercury.

[2] The interchanges of progression and retrogradation are proportional to the number of times the Earth overtakes the outer planets and the inner planets overtake the Earth. Now the Earth overtakes Saturn more often than Jupiter, Jupiter more often than Mars, Mars more often than Venus, and overtaken less often by Venus than by Mercury. Hence the frequency of progression and retrogradation is in that order.

[3] According to the Ptolemaic scheme, it can be inferred only from the changes in magnitude of the planet Mars what its relative distances from the Earth are at perigee and apogee. But according to the Copernican scheme, it follows from the relative distances of the planet at perigee and at apogee—which are as 1 to 5—that the apparent diameter of the planet should vary inversely in that ratio—assuming that the planet could be seen when in conjunction with the sun.

insofar as the appearances can be shown forth by its movement as by an hypothesis. We must allow a threefold movement altogether.

The first—which we said the Greeks called νυχθημέρινος—is the proper circuit of day and night, which goes around the axis of the earth from west to east—as the world is held to move in the opposite direction—and describes the equator or the equinoctial circle—which some, imitating the Greek expression [10ᵇ] ἰσηέρινος call the equidial.

The second is the annual movement of the centre, which describes the circle of the (zodiacal) signs around the sun similarly from west to east, *i.e.*, towards the signs which follow (from Aries to Taurus) and moves along between Venus and Mars, as we said, together with the bodies accompanying it. So it happens that the sun itself seems to traverse the ecliptic with a similar movement. In this way, for example, when the centre of the Earth is traversing Capricorn, the sun seems to be crossing Cancer; and when Aquarius, Leo, and so on, as we were saying.

It has to be understood that the equator and the axis of the Earth have a variable inclination with the circle and the plane of the ecliptic. For if they remained fixed and only followed the movement of the centre simply, no inequality of days and nights would be apparent, but it would always be the summer solstice or the winter solstice or the equinox, or summer or winter, or some other season of the year always remaining the same. There follows then the third movement, which is the declination: it is also an annual revolution but one towards the signs which precede (from Aries to Pisces), or westwards, *i.e.*, turning back counter to the movement of the centre; and as a consequence of these two movements which are nearly equal to one another but in oppo-

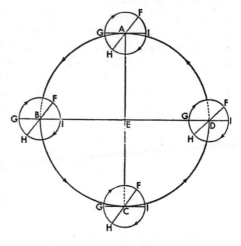

site directions, it follows that the axis of the Earth and the greatest of the parallel circles on it, the equator, always look towards approximately the same quarter of the world, just as if they remained immobile. The sun in the meanwhile is seen to move along the oblique ecliptic with that movement with which the centre of the earth moves, just as if the centre of the earth were the centre of the world—provided you

remember that the distance between the sun and the earth in comparison with the sphere of the fixed stars is imperceptible to us.

Since these things are such that they need to be presented to sight rather than merely to be talked about, let us draw the circle *ABCD*, which will represent the annual circuit of the centre of the earth in the plane of the ecliptic, and let *E* be the sun around its centre. I will cut this circle into four equal parts by means of the diameters *AEC* and *BED*. Let the point *A* be the beginning of Cancer; *B* of Libra; *E* of Capricorn; and *D* of Aries. Now let us put the centre of the earth first at *A*, around which we shall describe the terrestrial equator *FGHI*, but not in the same plane (as the ecliptic) except that the diameter *GAI* is the common section of the circles, *i.e.*, of the equator and the ecliptic. Also let the diameter *FAH* be drawn at right angles to *GAI*; and let *F* be the limit of the greatest southward declination (of the equator), and *H* of the northward declination. With this set-up, the Earth-dweller will see the sun—which is at the centre *E*—at the point of the winter solstice in Capricorn—[11ᵃ] which is caused by the greatest northward declination at *H* being turned toward the sun; since the inclination of the equator with respect to line *AE* describes by means of the daily revolution the winter tropic, which is parallel to the equator at the distance comprehended by the angle of inclination *EAH*. Now let the centre of the Earth proceed from west to east; and let *F*, the limit of greatest declination, have just as great a movement from east to west, until at *B* both of them have traversed quadrants of circles. Meanwhile, on account of the equality of the revolutions, angle *EAI* will always remain equal to angle *AEB*; the diameters will always stay parallel to one another—*FAH* to *FBH* and *GAI* to *GBI*; and the equator will remain parallel to the equator. And by reason of the cause spoken of many times already, these lines will appear in the immensity of the sky as the same. Therefore from the point *B* the beginning of Libra, *E* will appear to be in Aries, and the common section of the two circles (of the ecliptic and the equator) will fall upon line *GBIE*, in respect to which the daily revolution has no declination; but every declination will be on one side or the other of this line. And so the sun will be soon in the spring equinox. Let the centre of the Earth advance under the same conditions; and when it has completed [11ᵇ] a semicircle at *C*, the sun will appear to be entering Cancer. But since *F* the southward declination of the equator is now turned toward the sun, the result is that the sun is seen in the north, traversing the summer tropic in accordance with angle of inclination *ECF*. Again, when *F* moves on through the third quadrant of the circle, the common section *GI* will fall on line *ED*, whence the sun, seen in Libra, will appear to have reached the autumn equinox. But then as, in the same progressive movement, *HF* gradually turns in the direction of the sun, it will make the situation at the beginning return, which was our point of departure.

In another way: Again in the underlying plane let *AEC* be both the diameter (of the ecliptic) and its common section with the circle perpendicular to its plane. In this circle let *DGFI*, the meridian passing through the poles of the Earth be described around *A* and *C*, in turn, *i.e.*, in Cancer and in Capricorn. And let the axis of the Earth be *DF*, the north

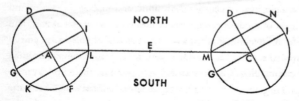

pole *D*, the south pole *F*, and *GI* the diameter of the equator. Therefore when *F* is turned in the direction of the sun, which is at E, and the incli-

nation of the equator is northward in proportion to angle *IAE*, then the movement around the axis will describe—with the diameter *KL* and at the distance *LI*—parallel to the equator the southern circle, which appears with respect to the sun as the tropic of Capricorn. Or—to speak more correctly—this movement around the axis describes, in the direction of *AE*, a conic surface, which has the centre of the earth as its vertex and a circle parallel to the equator as its base.[1] Moreover in the opposite sign, *C*, the same things take place but conversely. Therefore it is clear how the two mutually opposing movements, *i.e.*, that of the centre and that of the inclination, force the axis of the Earth to remain balanced in the same way and to keep a similar position, and how they make all things appear as if they were movements of the sun.

Now we said that the yearly revolutions of the centre and of the declination were approximately equal, because if they were exactly so, then the points of equinox and solstice and the obliquity of the ecliptic in relation to the sphere of the fixed stars could not change at all. But as the difference is very slight, [12ª] it is not revealed except as it increases with time: as a matter of fact, from the time of Ptolemy to ours there has been a precession of the equinoxes and solstices of about 21°. For that reason some have believed that the sphere of the fixed stars was moving, and so they choose a ninth higher sphere. And when that was not enough, the moderns added a tenth, but without attaining the end which we hope we shall attain by means of the movement of the Earth. We shall use this movement as a principle and a hypothesis in demonstrating other things.

12. ON THE STRAIGHT LINES IN A CIRCLE

Because the proofs which we shall use in almost the entire work deal with straight lines and arcs, with plane and spherical triangles, and because Euclid's *Elements*, although they clear up much of this, do not have what is here most required, namely, how to find the sides from the angles and the angles from the sides, since the angle

[1] Or, in other words, the axis of the terrestrial equator describes around the axis of the terrestrial ecliptic a double conic surface having its vertices at the centre of the Earth, in a period of revolution equal approximately to that of the Earth's centre.

does not measure the subtending straight line—just as the line does not measure the angle—but the arc does, there has accordingly been found a method whereby the lines subtending any arc may become known. By means of these lines, or chords, it is possible to determine the arc corresponding to the angle: and conversely by means of the arc to determine the straight line, or chord, which subtends the angle. So it does not seem irrelevant, if we treat of these lines, and also of the sides and angles of plane and spherical triangles—which Ptolemy discussed a few at a time here and there—in order that these questions may be answered here once and for all and that what we are going to teach may become clearer. Now, by the common agreement of mathematicians, we divide the circle into 360 degrees. Now the ancients employed a diameter of 120 parts. But in order to avoid the complication of minutes and seconds in the multiplication and division of the numbers attached to the lines, as the lines are usually incommensurable in length, and often in square too; some of their successors established a rational diameter of 1,200,000 parts or of 2,000,000 parts, or of some other rational quantity—from the time when Arabic numerals came into general use. This mathematical notation surpasses any other—Greek or Latin—[12^b] in a certain singular ease of employment and readily accommodates itself to every class of computation. For that reason we too have taken a division of the diameter into 200,000 parts as sufficient to exclude any very noticeable error. For as regards things which are not related as number to number, it is enough to attain a close approximation. But we will unfold this in six theorems and a problem—following Ptolemy fairly closely.

FIRST THEOREM

The diameter of a circle being given, the sides of the triangle, tetragon, hexagon, and decagon, which the same circle circumscribes, are also given.

Half the diameter, or the radius, is equal to the side of the hexagon, (Euclid, IV, 15); the square on the side of the triangle is three times the square on the side of the hexagon, (Euclid, XIII, 12); and the square on the side of the tetragon is twice the square on the side of the hexagon, Euclid as is shown in Euclid's *Elements* (IV, 9 and I, 47). Therefore the side of a hexagon is given in length as 100,000 parts, that of the tetragon as 141,422 parts, and that of the triangle as 173,205 parts.

Now, let *AB* be the side of the hexagon; and by Euclid, II, 11, or VI, 30, let it be cut in mean and extreme ratio at point *C*; and let *CB* be the greater segment to which its equal *BD* is added. Therefore the whole *ABD* will have been cut in extreme and mean ratio, and the lesser segment *BD* will be the side of the decagon inscribed in the

circle, and *AB* will be the side of the inscribed hexagon, as is made clear by Euclid, XIII, 5 and 9.

But *BD* will be given in this way: let *AB* be bisected at *E*, and it will be clear from Euclid, XIII, 3 that

$$\text{sq. } EBD = 5 \text{ sq. } EB.$$

But

$$EB = 50,000.$$

Whence

$$5 \text{ sq. } EB \text{ is given.}$$

Hence

$$EBD = 111,803.$$

And

$$BD = EBD - EB = 111,803 - 50,000 = 61,803,$$

which is the side of the decagon sought.

Moreover the side of the pentagon, the square on which is equal to the sum of the squares on the side of the hexagon and on the side of the decagon (*Elements*, XIII, 10), is given as 117,557 parts.

Therefore the diameter of the circle being given, the sides of the triangle, tetragon, pentagon, hexagon and decagon, which may be inscribed in the same circle, have been given—as was to be shown.

PORISM

Furthermore, it is clear that when the chord subtending an arc has been given, that chord too can be found which subtends the rest [13ª] of the semicircle..

Since the angle in a semicircle is right, and in right triangles the square on the chord subtending the right angle, *i.e.*, the square on the diameter, is equal to the sum of the squares on the sides comprehending the right angle; therefore—since the side of the decagon, which subtends 36° of the circumference, has been shown to have 61,803 parts whereof the diameter has 200,000 parts—the chord which subtends the remaining 144° of the semicircle has 190,211 parts.

And in the case of the side of the pentagon, which is equal to 117,557 parts of the diameter and subtends an arc of 72°, a straight line of 161,803 parts is given, and it subtends remaining 108° of the circle.

SECOND THEOREM

If a quadrilateral is inscribed in a circle, the rectangle comprehended by the diagonals is equal to the two rectangles which are comprehended by the two pairs of opposite sides.

For let the quadrilateral *ABCD* be inscribed in a circle; I say that the rectangle

comprehended by the diagonals *AC* and *DB* is equal to those comprehended by *AB*, *CD* and by *AD*, *BC*.

For let us make

$$\text{angle } ABE = \text{angle } CBD.$$

Therefore by addition

$$\text{angle } ABD = \text{angle } EBC,$$

taking angle *EBD* as common to both. Moreover

$$\text{angle } ACB = \text{angle } BDA$$

because they stand on the same segment of the circle; and accordingly the two similar triangles *BCE* and *BDA* will have their sides proportional. Hence

$$BC : BD = EC : AD.$$

And

$$\text{rect. } EC, BD = \text{rect. } BC, AD.$$

But also the triangles *ABE* and *CBD* are similar, because

$$\text{angle } ABE = \text{angle } CBD.$$

And

$$\text{angle } BAC = \text{angle } BDC,$$

because they intercept the same arc of the circle.
So again,

$$AB : BD = AE : CD$$

And

$$\text{rect. } AB, DC = \text{rect. } AE, BD.$$

But it has already been made clear that

$$\text{rect. } AD, BC = \text{rect. } BD, EC.$$

Accordingly, taken as a whole,

$$\text{rect. } BD, AC = \text{rect. } AD, BC + \text{rect. } AB, CD,$$

as it was opportune to have shown.

THIRD THEOREM

Hence if straight lines subtending unequal arcs in a semicircle are given, the chord subtending the arc whereby the greater arc exceeds the smaller is also given.

[13b] In the semicircle *ABCD* with diameter *AD*, let the straight lines *AB* and *AC*

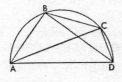

subtending unequal arcs be given. To us, who wish to discover the chord subtending *BC*, there are given by means of the aforesaid the chords *BD* and *CD* subtending the remaining arcs of the semicircle, and these chords bound the quadrilateral *ABCD* in the semicircle. The diagonals *AC*

and *BD* have been given together with the three sides *AB*, *AD*, and *CD*. And, as has already been shown,

$$\text{rect. } AC, BD = \text{rect. } AB, CD + \text{rect. } AD, BC.$$

Therefore,

$$\text{rect. } AD, BC = \text{rect. } AC, BD - \text{rect. } AB, CD.$$

Accordingly, in so far as the division may be carried out,

$$(AC \cdot BD - AB \cdot CD) \div AD = BC,$$

which was sought.

Further when, for example, the sides of the pentagon and hexagon are given from the above, by this computation a line is given subtending 12°—which is the difference between the arcs—and it is equal to 20,905 parts of the diameter.

FOURTH THEOREM

Given a chord subtending any arc, the chord subtending half of the arc is also given.

Let us describe the circle *ABC*, whose diameter is *AC*, and let the arc *BC* be given together with the chord subtending it, and let the line *EF* from the centre *E* cut *BC* at right angles. Accordingly by Euclid, III, 3, it will bisect chord *BC* at *F*, and the arc at *D*. Let the chords subtending arcs *AB* and *BD* be drawn. Since the triangles *ABC* and *EFC* are right and also similar—for they have angle *ECF* in common; therefore, as

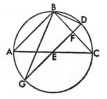

$$CF = {}^1/_2 \; BFC,$$

so

$$EF = {}^1/_2 \; AB.$$

But chord *AB* is given, for it subtends the remaining arc of the semicircle. Therefore *EF* is given; and so is line *DF* the remainder of the radius. Let the diameter *DEG* be completed, and let *BG* be joined. Therefore in triangle *BDG* line *BF* falls from the right angle at *B* perpendicular to the base. Accordingly,

$$\text{rect. } GD, DF = \text{sq. } BD.$$

Therefore *BD* is given in length, and it subtends half of the arc *BDC*.

And since a chord subtending 12° has already been given, the chord subtending 6° is given as 10,467 parts; that subtending 3°, as 5235 parts; that subtending $1^1/_2$°, as 2618 parts; and that subtending 45', as 1309 parts.

[14ª] FIFTH THEOREM

Again, when chords are given subtending two arcs, the chord subtending the whole arc made up of them is also given.

Let there be given in the circle the two chords subtending the arcs *AB* and *BC*; I say that the chord subtending the whole arc *ABC* is also given.

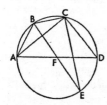

For let the diameters *AFD* and *BFE* be drawn, and also the chords *BD* and *CE*, which are given by means of the foregoing, on account of chords *AB* and *BC* being given; and

chord *DE* = chord *AB*.

The joining of *CD* completes the quadrilateral *BCDE*, whose diagonals *BD* and *CE* are given together with the three sides *BC*, *DE*, and *BE*; and the remaining side *CD* will be given by the second theorem; accordingly chord *CA* which subtends the remaining part of the semicircle will be given, and it subtends the whole arc *ABC* and is what was sought.

Furthermore, since so far there have been discovered chords which subtend 3°, $1^1/_2$°, and $^3/_4$°; by means of these intervals a table can be constructed with the most exact ratios. Nevertheless if we ascend through the degrees and add one arc to another arc either by halves or by some other mode, there is not unjustified doubt concerning the chords subtending those arcs, as the graphical ratios by which they can be shown are lacking to us. Nothing, however, prevents us from going on with that by some mode which is this side of error perceptible to sense and which is least unconsonant with the assumed number. This was what Ptolemy too sought as regards the chords subtending arcs of 1° or of $^1/_2$°; and he admonished us in the first place.

SIXTH THEOREM

The ratio of the arcs is greater than the ratio of the greater to the smaller of the chords.

Let there be in a circle two unequal successive arcs *AB* and *BC*, and let *BC* be the greater. I say that

arc *BC* : arc *AB* > chord *BC* : chord *AB*.

These chords comprehend angle *B*, and let that be bisected by line *BD*. And let *AC* be joined, which cuts *BD* at point *E*. Similarly let *AD* and *CD* be joined; then

AD = CD,

because they subtend equal arcs.

Accordingly, since in triangle *ABC*, the line which bisects the angle also cuts *AC* [14b] at *E*, then

EC, segment of base : *AE* = *BC* : *AB* (Euclid, VI, 3)

and since

BC > AB,

then

EC > EA.

41

Let DF be erected perpendicular to AC; it will bisect AC at point F. And F must necessarily be found in the greater segment EC. And since in every triangle the greater angle is subtended by the greater side, in the triangle DEF

$$\text{side } DE > \text{side } DF,$$

and further,

$$AD > DE,$$

wheretofore the circumference described with D as center and DE as radius will cut AD and pass beyond DF. Therefore let it cut AD at H, and let it be extended in the straight line DFI. Since

$$\text{sect. } EDI > \text{trgl. } EDF,$$

while

$$\text{trgl. } DEA > \text{sect. } DEH,$$

therefore

$$\text{trgl. } DEF : \text{trgl. } DEA < \text{sect. } DEI : \text{sect. } DEH.$$

But sectors are proportional to their arcs or to the angles at the centre; while triangles under the same vertex are proportional to their bases. Accordingly

$$\text{angle } EDF : \text{angle } ADE > \text{base } EF : \text{base } AE.$$

Therefore, *componendo*,

$$\text{angle } FDA : \text{angle } ADE > \text{base } AF : \text{base } AE.$$

And, in the same way,

$$\text{angle } CDA : \text{angle } ADE > \text{base } AC : \text{base } AE.$$

But, *separando*,

$$\text{angle } CDE : \text{angle } EDA > \text{base } CE : \text{base } EA.$$

But

$$\text{angle } CDE : \text{angle } EDA = \text{arc } CB : \text{arc } AB.$$

And

$$\text{base } CE : \text{base } AE = \text{chord } CB : \text{chord } AB.$$

Therefore

$$\text{arc } CB : \text{arc } AB > \text{chord } BC : \text{chord } AB,$$

as was to be shown.

PROBLEM

But since the arc is always greater than the straight line subtending it—as the straight line is the shortest of those lines which have the same termini—nevertheless in going from greater to lesser sections of the circle, the inequality approaches equality, so that finally the circular line and the straight line go out of existence simultaneously at the point of tangency on the circle. Therefore it is necessary that just before that moment they differ from one another by no discernible difference.

For example, let arc AB be 3° and arc AC 1½°. It has been shown that

$$\text{ch. } AB = 5235,$$

where diameter = 200,000,

and that

$$\text{ch. } AC = 2618.$$

And though

$$\text{arc } AB = 2 \text{ [15}^a\text{] arc } AC,$$

yet

$$\text{ch. } AB < 2 \text{ ch. } AC$$

and

$$\text{ch. } AC - 2617 = 1.$$

But if we make

$$\text{arc } AB = 1\frac{1}{2}°$$

and

$$\text{arc } AC = \frac{3}{4}°,$$

then

$$\text{ch. } AB = 2618$$

and

$$\text{ch. } AC = 1309,$$

and even though chord AC ought to be greater than half of chord AD, it is seen to be no different from the half. And the ratios of the arcs and the straight lines are now apparently the same. Therefore, since we see that we have come so far that the difference between the straight and the circular line evades sense-perception as completely as if there were only one line, we do not hesitate to take 1309 as subtending ¾° and in the same ratio to fit the chord to the degree and to the remaining parts (of the degree); and so with the addition of ¼° to the ¾° we establish 1° as subtended by 1745, ½° by 872½, and ⅓° by approximately 582.

Nevertheless I think it will be enough if in the table we give only the halves of the chords subtending twice the arc, whereby we may concisely comprehend in the quadrant what it used to be necessary to spread out over the semicircle; and especially because the halves come more frequently into use in demonstration and calculation than the whole chords do. Now we have set forth a table increasing by ⅙'s and having three columns. In the first column are the degrees and sixth parts of a degree. The second contains the numerical length of half the chord subtending twice the arc. The third contains the difference between the numerical lengths of each half chord, and by means of these differences we can make proportional additions in taking half-chords of a particular number of minutes. The table follows:

TABLE OF THE CHORDS IN A CIRCLE

Arcs		Halves of the chords subtending twice the arcs	Differences between each half-chord	Arcs		Halves of the chords subtending twice the arcs	Differences between each half-chord	Arcs		Halves of the chords subtending twice the arcs	Differences between each half-chord
Deg.	Min.			Deg.	Min.			Deg.	Min.		
0	10	291	291	7	40	13341	288	15	10	26163	280
0	20	582	291	7	50	13629	288	15	20	26443	281
0	30	873	290	8	0	13917	288	15	30	26724	280
0	40	1163	291	8	10	14205	288	15	40	27004	280
0	50	1454	291	8	20	14493	288	15	50	27284	280
1	0	1745	291	8	30	14781	288	16	0	27564	279
1	10	2036	291	8	40	15069	287	16	10	27843	279
1	20	2327	290	8	50	15356	287	16	20	28122	279
1	30	2617	291	9	0	15643	288	16	30	28401	279
1	40	2908	291	9	10	15931	287	16	40	28680	279
1	50	3199	291	9	20	16218	287	16	50	28959	278
2	0	3490	291	9	30	16505	287	17	0	29237	278
2	10	3781	290	9	40	16792	286	17	10	29515	278
2	20	4071	291	9	50	17078	287	17	20	29793	278
2	30	4362	291	10	0	17365	286	17	30	30071	277
2	40	4653	290	10	10	17651	286	17	40	30348	277
2	50	4943	291	10	20	17937	286	17	50	30625	277
3	0	5234	290	10	30	18223	286	18	0	30902	276
3	10	5524	290	10	40	18509	286	18	10	31178	276
3	20	5814	291	10	50	18795	286	18	20	31454	276
3	30	6105	290	11	0	19081	285	18	30	31730	276
3	40	6395	290	11	10	19366	286	18	40	32006	276
3	50	6685	290	11	20	19652	285	18	50	32282	275
4	0	6975	290	11	30	19937	285	19	0	32557	275
4	10	7265	290	11	40	20222	285	19	10	32832	274
4	20	7555	290	11	50	20507	284	19	20	33106	275
4	30	7845	290	12	0	20791	285	19	30	33381	274
4	40	8135	290	12	10	21076	284	19	40	33655	274
4	50	8425	290	12	20	21360	284	19	50	33929	273
5	0	8715	290	12	30	21644	284	20	0	34202	273
5	10	9005	290	12	40	21928	284	20	10	34475	273
5	20	9295	290	12	50	22212	283	20	20	34748	273
5	30	9585	289	13	0	22495	283	20	30	35021	272
5	40	9874	290	13	10	22778	284	20	40	35293	272
5	50	10164	289	13	20	23062	282	20	50	35565	272
6	0	10453	289	13	30	23344	283	21	0	35837	271
6	10	10742	289	13	40	23627	283	21	10	36108	271
6	20	11031	289	13	50	23910	282	21	20	36379	271
6	30	11320	289	14	0	24192	282	21	30	36650	270
6	40	11609	289	14	10	24474	282	21	40	36920	270
6	50	11898	289	14	20	24756	282	21	50	37190	270
7	0	12187	289	14	30	25038	281	22	0	37460	270
7	10	12476	288	14	40	25319	282	22	10	37730	269
7	20	12764	289	14	50	25601	281	22	20	37999	269
7	30	13053	288	15	0	25882	281	22	30	38268	269

TABLE OF THE CHORDS IN A CIRCLE

Arcs		Halves of the chords subtending twice the arcs	Differences between each half-chord	Arcs		Halves of the chords subtending twice the arcs	Differences between each half-chord	Arcs		Halves of the chords subtending twice the arcs	Differences between each half-chord
Deg.	Min.			Deg.	Min.			Deg.	Min.		
22	40	38587	268	30	10	50252	251	37	40	61107	230
22	50	38805	268	30	20	50503	251	37	50	61337	229
23	0	39073	268	30	30	50754	250	38	0	61566	229
23	10	39341	267	30	40	51004	250	38	10	61795	229
23	20	39608	267	30	50	51254	250	38	20	62024	227
23	30	39875	266	31	0	51504	249	38	30	62251	228
23	40	40141	267	31	10	51753	249	38	40	62479	227
23	50	40408	266	31	20	52002	248	38	50	62706	226
24	0	40674	265	31	30	52250	248	39	0	62932	226
24	10	40939	265	31	40	52498	247	39	10	63158	225
24	20	41204	265	31	50	52745	247	39	20	63383	225
24	30	41469	265	32	0	52992	246	39	30	63608	224
24	40	41734	264	32	10	53238	246	39	40	63832	224
24	50	41998	264	32	20	53484	246	39	50	64056	223
25	0	42262	263	32	30	53730	245	40	0	64279	222
25	10	42525	263	32	40	53975	245	40	10	64501	222
25	20	42788	263	32	50	54220	244	40	20	64723	222
25	30	43051	262	33	0	54464	244	40	30	64945	221
25	40	43313	262	33	10	54708	243	40	40	65166	220
25	50	43575	262	33	20	54951	243	40	50	65386	220
26	0	43837	261	33	30	55194	242	41	0	65606	219
26	10	44098	261	33	40	55436	242	41	10	65825	219
26	20	44359	261	33	50	55678	241	41	20	66044	218
26	30	44620	260	34	0	55919	241	41	30	66262	218
26	40	44880	260	34	10	56160	240	41	40	66480	217
26	50	45140	259	34	20	56400	241	41	50	66697	216
27	0	45399	259	34	30	56641	239	42	0	66913	216
27	10	45658	259	34	40	56880	239	42	10	67129	215
27	20	45917	258	34	50	57119	239	42	20	67344	215
27	30	46175	258	35	0	57358	238	42	30	67559	214
27	40	46433	257	35	10	57596	237	42	40	67773	214
27	50	46690	257	35	20	57833	237	42	50	67987	213
28	0	46947	257	35	30	58070	237	43	0	68200	212
28	10	47204	256	35	40	58307	236	43	10	68412	212
28	20	47460	256	35	50	58543	236	43	20	68624	211
28	30	47716	255	36	0	58779	235	43	30	68835	211
28	40	47971	255	36	10	59014	234	43	40	69046	210
28	50	48226	255	36	20	59248	234	43	50	69256	210
29	0	48481	254	36	30	59482	234	44	0	69466	209
29	10	48735	254	36	40	59716	233	44	10	69675	208
29	20	48989	253	36	50	59949	232	44	20	69883	208
29	30	49242	253	37	0	60181	232	44	30	70091	207
29	40	49495	253	37	10	60413	232	44	40	70298	207
29	50	49748	252	37	20	60645	231	44	50	70505	206
30	0	50000	252	37	30	60876	231	45	0	70711	205

Arcs		Halves of the chords subtending twice the arcs	Differences between each half-chord	Arcs		Halves of the chords subtending twice the arcs	Differences between each half-chord	Arcs		Halves of the chords subtending twice the arcs	Differences between each half-chord
Deg.	Min.			Deg.	Min.			Deg.	Min.		
45	10	70916	205	52	40	79512	176	60	10	86747	145
45	20	71121	204	52	50	79688	176	60	20	86892	144
45	30	71325	204	53	0	79864	174	60	30	87036	142
45	40	71529	203	53	10	80038	174	60	40	87178	142
45	50	71732	202	53	20	80212	174	60	50	87320	142
46	0	71934	202	53	30	80386	172	61	0	87462	141
46	10	72136	201	53	40	80558	172	61	10	87603	140
46	20	72337	200	53	50	80730	172	61	20	87743	139
46	30	72537	200	54	0	80902	170	61	30	87882	138
46	40	72737	199	54	10	81072	170	61	40	88020	138
46	50	72936	199	54	20	81242	169	61	50	88158	137
47	0	73135	198	54	30	81411	169	62	0	88295	136
47	10	73333	198	54	40	81580	168	62	10	88431	135
47	20	73531	197	54	50	81748	167	62	20	88566	135
47	30	73728	196	55	0	81915	167	62	30	88701	134
47	40	73924	195	55	10	82082	166	62	40	88835	133
47	50	74119	195	55	20	82248	165	62	50	88968	133
48	0	74314	194	55	30	82413	164	63	0	89101	131
48	10	74508	194	55	40	82577	164	63	10	89232	131
48	20	74702	194	55	50	82741	163	63	20	89363	130
48	30	74896	194	56	0	82904	162	63	30	89493	129
48	40	75088	192	56	10	83066	162	63	40	89622	129
48	50	75280	191	56	20	83228	161	63	50	89751	128
49	0	75471	190	56	30	83389	160	64	0	89879	127
49	10	75661	190	56	40	83549	159	64	10	90006	127
49	20	75851	189	56	50	83708	159	64	20	90133	125
49	30	76040	189	57	0	83867	158	64	30	90258	125
49	40	76299	188	57	10	84025	157	64	40	90383	124
49	50	76417	187	57	20	84182	157	64	50	90507	124
50	0	76604	187	57	30	84339	156	65	0	90631	122
50	10	76791	186	57	40	84495	155	65	10	90753	122
50	20	76977	185	57	50	84650	155	65	20	90875	121
50	30	77162	185	58	0	84805	154	65	30	90996	120
50	40	77347	184	58	10	84959	153	65	40	91116	119
50	50	77531	184	58	20	85112	152	65	50	91235	119
51	0	77715	182	58	30	85264	151	66	0	91354	118
51	10	77897	182	58	40	85415	151	66	10	91472	118
51	20	78079	182	58	50	85566	151	66	20	91590	116
51	30	78261	181	59	0	85717	149	66	30	91706	116
51	40	78442	180	59	10	85866	149	66	40	91822	114
51	50	78622	179	59	20	86015	148	66	50	91936	114
52	0	78801	179	59	30	86163	147	67	0	92050	114
52	10	78980	178	59	40	86310	147	67	10	92164	112
52	20	79158	177	59	50	86457	145	67	20	92276	112
52	30	79335	177	60	0	86602	145	67	30	92388	111

TABLE OF THE CHORDS IN A CIRCLE

Arcs		Halves of the chords subtending twice the arcs	Differences between each half-chord	Arcs		Halves of the chords subtending twice the arcs	Differences between each half-chord	Arcs		Halves of the chords subtending twice the arcs	Differences between each half-chord
Deg.	Min.			Deg.	Min.			Deg.	Min.		
67	40	92499	110	75	10	96667	75	82	40	99182	37
67	50	92609	109	75	20	96742	73	82	50	99219	36
68	0	92718	109	75	30	96815	72	83	0	99255	35
68	10	92827	108	75	40	96887	72	83	10	99290	34
68	20	92935	107	75	50	96959	71	83	20	99324	33
68	30	93042	106	76	0	97030	69	83	30	99357	32
68	40	93148	105	76	10	97099	70	83	40	99389	32
68	50	93253	105	76	20	97169	68	83	50	99421	31
69	0	93358	104	76	30	97237	67	84	0	99452	30
69	10	93462	103	76	40	97304	67	84	10	99482	29
69	20	93565	102	76	50	97371	66	84	20	99511	28
69	30	93667	102	77	0	97437	65	84	30	99539	28
69	40	93769	101	77	10	97502	64	84	40	99567	27
69	50	93870	99	77	20	97566	64	84	50	99594	26
70	0	93969	99	77	30	97630	62	85	0	99620	24
70	10	94068	99	77	40	97692	62	85	10	99644	24
70	20	94167	97	77	50	97754	61	85	20	99668	24
70	30	94264	97	78	0	97815	60	85	30	99692	22
70	40	94361	96	78	10	97875	59	85	40	99714	22
70	50	94457	95	78	20	97934	58	85	50	99736	20
71	0	94552	94	78	30	97992	58	86	0	99756	20
71	10	94646	93	78	40	98050	57	86	10	99776	19
71	20	94739	93	78	50	98107	56	86	20	99795	18
71	30	94832	92	79	0	98163	55	86	30	99813	17
71	40	94924	91	79	10	98218	54	86	40	99830	17
71	50	95015	90	79	20	98272	53	86	50	99847	16
72	0	95105	90	79	30	98325	53	87	0	99863	15
72	10	95195	89	79	40	98378	52	87	10	99878	14
72	20	95284	88	79	50	98430	51	87	20	99892	13
72	30	95372	87	80	0	98481	50	87	30	99905	12
72	40	95459	86	80	10	98531	49	87	40	99917	11
72	50	95545	85	80	20	98580	49	87	50	99928	11
73	0	95630	85	80	30	98629	47	88	0	99939	10
73	10	95715	84	80	40	98676	47	88	10	99949	9
73	20	95799	83	80	50	98723	46	88	20	99958	8
73	30	95882	82	81	0	98769	45	88	30	99966	7
73	40	95964	81	81	10	98814	44	88	40	99973	6
73	50	96045	81	81	20	98858	44	88	50	99979	6
74	0	96126	80	81	30	98902	42	89	0	99985	4
74	10	96206	79	81	40	98944	42	89	10	99989	4
74	20	96285	78	81	50	98986	41	89	20	99993	3
74	30	96363	77	82	0	99027	40	89	30	99996	2
74	40	96440	77	82	10	99067	39	89	40	99998	1
74	50	96517	75	82	20	99106	38	89	50	99999	1
75	0	96592	75	82	30	99144	38	90	0	100000	0

13. ON THE SIDES AND ANGLES OF PLANE RECTILINEAR TRIANGLES

I

[19ᵇ] The sides of a triangle whose angles are given are given.

I say let there be the triangle *ABC*, around which a circle is circumscribed, by Euclid, IV, 5. Therefore arcs *AB*, *BC*, and *CA* will be given in degrees whereof 360° are equal to two right angles. Now given the arcs, the subtending sides of the triangle inscribed in the circle are also given by the table drawn up, where the diameter is assumed to have 200,000 parts.

II

But if two sides of the triangle are given together with one of the angles, the remaining side and the remaining angles may become known.

For the given sides are either equal or unequal. But the given angle is either right or acute or obtuse. Again, the given sides either comprehend the angle or they do not comprehend it.

Therefore in triangle *ABC* first let the two given sides *AB* and *AC* be equal, and let them comprehend the given angle *A*.

Therefore the remaining angles at base *BC* are also given—since they are equal—as half of the remainder, when *A* is subtracted from two right angles. And if the angle given first was at the base, then its equal is soon given, and from the two of them the remaining angle that goes to make up two right angles. But given the angles of a triangle, the sides are given; and moreover the base *BC* is given from the table in the parts whereof *AB* or *AC* as radius has 100,000 parts or whereof the diameter has 200,000 parts.

III

But if the angle BAC comprehended by the given sides is right, the same thing will result.

Since it is obvious that

[20ᵃ] sq. *AB* + sq. *AC* = sq. *BC*;

therefore *BC* is given in length and the sides in their ratio to one another. But the segment of a circle which comprehends a right triangle is a semicircle, and base *BC* is the diameter. Therefore *AB* and *AC* as subtending the remaining angles *C* and *B* will be

48

given in the parts whereof *BC* has 200,000 parts. And the ratio of the table will reveal the angles in the degrees whereof 180° are equal to two right angles.

The same thing will result if *BC* is given together with one of the sides comprehending the right angle, as I judge has been clearly established.

IV

But now let the given angle ABC be acute, and also let it be comprehended by the given sides AB and BC.

And from point *A* drop a perpendicular to *BC* extended, if necessary, according as it falls inside or outside the triangle, and let it be *AD*. By this perpendicular the two right triangles *ABD* and *ADC* are distinguished, and since the angles in *ABD* are given—for *D* is a right angle, and *B* is given by hypothesis; therefore *AD* and *BD* are given by the table as subtending angles *A* and *B* in the parts whereof *AB*, the diameter of the circle, has 200,000 parts. And in the same ratio wherein *AB* was given in length, *AD* and *BD* are given similarly; and *CD*, which is the difference between *BC* and *BD*, is given also.

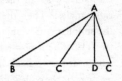

Therefore in the right triangle *ADC*, the sides *AD* and *CD* being given, *AC* the side sought and angle *ACD* are given according to what has been shown above.

V

And it will not turn out differently, if angle *B* is obtuse.

For the perpendicular *AD* dropped from point *A* to straight line *BC* extended makes the triangle *ABD* have its angles given. For angle *ABD*, which is exterior to angle *ABC*, is given; and

angle *D* = 90°.

Therefore sides *BD* and *AD* are given in the parts whereof *AB* has 200,000. And since *BA* and *BC* have a given ratio to one another, therefore *AB* too is given in the same parts, wherein *BD* and the whole *CBD* are given.

Accordingly in the right triangle *ADC*, since the two sides *AD* and *CD* are given, side *AC* and angles *BAC* and *ACB*, which were sought for, are also given.

VI

But let either of the given sides, *AC* or *AB*, be the one subtending the given angle *B*. [20[b]] Therefore *AC* is given by the table in parts whereof the diameter of the circle circumscribing the triangle *ABC* has 200,000 and according to the given ratio of *AC*

to *AB. AB* is given in similar parts, and by the table the angle *ACB* is given together with the remaining angle *BAC,* by which chord *CB* is also given. And by this ratio they are given in any magnitude.

VII

Given all the sides of the triangle, the angles are given.

It is too well known to be worth mentioning that each angle of an equilateral triangle is one third of two right angles.

It is also clear in the case of an isosceles triangle. For each of the equal sides is to the third side as half of the diameter is to the side subtending the arc by which the angle comprehended by the equal sides is given according to the table, wherein the 360° around the centre are equal to four right angles.[1] Then the two angles at the base are given as half of the supplementary angle.

Therefore it now remains to show this in the case of scalene triangles, which we divide in the same way into right triangles. Therefore let there be the scalene triangle *ABC* of which the sides are given, and upon the side which is the longest, namely *BC*, drop the perpendicular *AD*. Now Euclid, II, 13 tells us that if *AB* subtends the acute angle, then

(sq. *AC* + sq. *BC*)–sq. *AB* = 2 rect. *BC, CD.*

Now it is necessary for angle *C* to be acute; for otherwise *AB* would be the longest side contrary to the hypotheses, according to Euclid, I, 17 - 19. Therefore *BD* and *DC* are given, and there will be the right triangles *ABC* and *ADC* with their sides and angles given—as has so often happened before—and so the angles of triangle *ABC* which were sought become established.

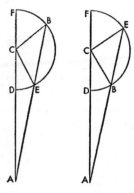

Another way. Similarly Euclid, III, 36 will perhaps give us an easy method, if with *BC* the shorter side as radius and with point *C* as centre, we describe a circle which will cut either one or both of the remaining sides.

First, let it cut both: *AB* at point *E* and *AC* at *D;* and let line *ADC* extended to point *F* in order to complete the diameter *DCF.* And with this construction it is clear from that proposition of Euclid that

[21a] rect. *FA, AD* = rect. *BA, AE,*

since each is equal to the square on the tangent to the circle from *A.* But the whole *AF* is given, as all its segments are given, since

[1]As in the subjoined figure:

radius CF = radius CD = BC,

and

$$AD = CA - CD.$$

Wherefore, as the rectangle BA, AE is given, AE also is given in length; and so is the remainder BE subtending arc BE. By joining EC we shall have the isosceles triangle BCE with all its sides given. Therefore the angle EBC is given. Hence in the triangle ABC the remaining angles at C and at A may become known by means of what has been shown above.

However, let the circle not cut AB as in the other figure, where AB falls upon the concave circumference; nevertheless BE will be given, and in the isosceles triangle BCE angle CBE will be given and also the exterior angle ABC. And by the same method as before the remaining angles are given.

And we have said enough concerning rectilinear triangles, in which a great part of geodesy consists. Now let us turn to spherical triangles.

14. ON SPHERICAL TRIANGLES

In this place we take that triangle as spherical which is comprehended by three arcs of great circles on a spherical surface. But we take the difference and magnitude of the angles from the arc of a great circle, *i.e.*, a great circle described with the point of section as a pole; and this arc is the arc intercepted by the quadrants of the circles comprehending the angle. For as the arc thus intercepted is to the whole circumference, so is the angle of section to four right angles—which we have said contain 360 equal degrees.

I

[21ᵇ] If there are three arcs of the great circles of a sphere, and if any two of them joined together are longer than the third; it is clear that a spherical triangle can be constructed from them.

For Euclid, XI, 23 shows in the case of angles what is here proposed in the case of arcs. Since there is the same ratio between angles as between arcs, and since the great circles are those circles which pass through the centre of the sphere; it is manifest that those three sectors of circles, *i.e.*, the sectors to which the three arcs belong, form a solid angle at the centre of the sphere. Therefore what was proposed has been established.

II

Any arc of a (spherical) triangle must be less than a semicircle.

For the semicircle makes no angle at the centre but falls upon it in a straight line.

But the remaining two angles which intercept the arcs cannot complete a solid angle at the centre, and so they cannot complete a spherical triangle.

And I think this is the reason why Ptolemy in his exposition of triangles of this genus, especially as regards the figure of the spherical sector, argues that none of the arcs taken together must be greater than a semicircle.

III

In spherical triangles having a right angle, the chord subtending twice the side opposite the right angle is to a chord subtending twice one of the sides comprehending the right angle as the diameter of the sphere is to the chord which subtends the angle comprehended in the great circle of the sphere by the first side and by the remaining side.

For let there be the spherical triangle *ABC*, of which the angle at *C* is right. I say that

ch. 2 *AB* : ch. 2 *BC* = dmt. sph. : ch. 2 *BAC* gr. circ. sph.

With *A* as a pole draw *DE* the arc of a great circle, and let *ABD* and *ACE* the quadrants of the circles be completed. And from the centre *F* of the sphere draw the common sections of the circles: *FA* the common section of circles *ABD* and *ACE*, [22ª] *FE* of circles *ACE* and *DE*, and *FD* of circles *ABD* and *DE*; and moreover, *FC* of the circles *AC* and *BC*. Then draw *BG* at right angles to *FA*, *BI* at right angles to *FC*, and *DK* at right angles to *FE*; and let *GI* be joined.

Since if a circle cuts a circle described through its poles, it cuts it at right angles; therefore the angle *AED* will be right; and angle *ACB* is right by hypothesis; and each of the planes *EDF* and *BCF* is perpendicular to plane *AEF*. Wherefore if a line be erected in the underlying plane of *AFE* at right angles to point *K* in the common section, this line and *KD* will comprehend a right angle, by the definition of planes which are perpendicular to one another. Wherefore, by Euclid, XI, 4, line *KD* is perpendicular to circle *AEF*. But *BI* was erected in the same relation to the same plane; and so by Euclid, XI, 6, *DK* is parallel to *BI* and *FD* is parallel to *GB*, because

angle *FGB* = angle *GFD* = 90°.

And by Euclid, XI, 10,

angle *FDK* = angle *GBI*.

But

angle *FKD* = 90°,

and by definition

GI is perpendicular to *IB*.

Accordingly the sides of similar triangles are proportional; and

$$DF : BG = DK : BI.$$

But

$$BI = {}^1/_2 \text{ ch. } 2 \text{ } CB,$$

since it is at right angles to the radius from center F; and for the same reason,

$$BG = {}^1/_2 \text{ ch. } 2 \text{ } BA,$$
$$DK = {}^1/_2 \text{ ch. } 2 \text{ } DE, \text{ or } {}^1/_2 \text{ ch. } 2 \text{ } DAE,$$

and

$$DF = {}^1/_2 \text{ dmt. sph.,}$$

Therefore it is clear that

$$\text{ch. } 2 \text{ } AB : \text{ch. } 2 \text{ } BC = \text{dmt.} : \text{ch. } 2 \text{ } DAE \text{ (or ch. } 2 \text{ } DE),$$

as it was time to show.

IV

In any triangle having a right angle, if another angle and any side are given, the remaining angle and the remaining sides will be given.

For let there be the triangle ABC having the angle A right and having one of the other two angles, namely B, given.

Let us take three cases of the given side. For it is either adjacent to both the given angles, as AB, or only to the right angle, as AC, or is opposite the right angle, as BC.

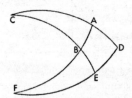

Therefore let AB be the side given first; and with C as a pole let arc [22b] DE of the great circle be described. Let the quadrants CAD and CBE be completed; and let AB and DE be extended, until they cut one another at point F. Therefore conversely the pole of CAD will be at F, because

$$\text{angle } A = \text{angle } D = 90°.$$

And since, if in a sphere the great circles cut one another at right angles, they will bisect one another and pass through the poles of one another; therefore ABF and DEF are quadrants of circles. And since AB is given, BF the remainder of the quadrant is also given; and the vertical angle EBF is equal to the given angle ABC. But by what has been shown above

$$\text{ch. } 2 \text{ } BF : \text{ch. } 2 \text{ } EF = \text{dmt. sph.} : \text{ch. } 2 \text{ } EBF.$$

But three of the chords have been given:

$$\text{dmt. sph.,}$$
$$\text{ch. } 2 \text{ } BF,$$
$$\text{ch. } 2 \text{ } EBF,$$

or the half-chords; and therefore by Euclid, VI, 15, there is also given

$^1/_2$ ch. 2 *EF*;

and by the table the arc *EF* itself and *DE* the remainder of the quadrant, or the angle at *C*, which was sought. Similarly and alternately,

ch. 2 *DE* : ch. 2 *AB* = ch. 2 *EBC* : ch. 2 *CB*.

But *DE*, *AB*, and *CE* on the quadrants of the circle have already been given; and therefore the fourth chord, subtending twice arc *CB*, will be given, and also the side *CB*, which was sought.

And since

ch. 2 *CB* : ch. 2 *CA* = ch. 2 *BF* : ch. 2 *EF*,

because they both have the ratio of

dmt. sph. : ch. 2 *CBA*,

and because things which have the same ratio to one and the same thing have the same ratio to one another; therefore with the three chords *BF*, *EF*, and *CB* given, the fourth chord *CA* is also given; and arc *CA* is the third side of the triangle *ABC*.

But now let *AC* be the side assumed as given, and let our problem be to find the sides *AB* and *BC* together with the remaining angle *C*. Again similarly and by inversion,

ch. 2 *CA* : ch. 2 *CB* = ch. 2 *ABC* : dmt.

Hence the side *CB* is given, and also *AD* and *BE* the remainders of the quadrants of the circles. And so again,

ch. 2 *AD* : ch. 2 *BE* = ch. 2 *ABF*, *i.e.*, dmt., : ch. 2 *BF*.

Therefore arc *BF* is given, and the side *AB*, which is the remainder.
And similarly,

ch. 2 *BC* : ch. 2 *AB* = 2 ch. *CBE* : ch. 2 *DE*.

Hence arc *DE*, or twice the remaining angle at *C*, will be given.

Furthermore, if it was *BC* which was assumed, again as before, *AC* and the remainders *AD* and *BE* will be given. Hence arc *BF* and the remaining side *AB* are given by means of the diameter and the chords [23ª] subtending them, as has often been said. And as in the preceding theorem, by means of arcs *BC*, *AB*, and *CBE* being given, the arc *ED*, *i.e.*, the remaining angle at *C*, which we were seeking, is discovered.

And so again in the triangle *ABC* with two angles *A* and *B* given, of which *A* is right, and with one of the three sides given, the third angle and the remaining sides are given, as was to be shown.

V

The sides of a right triangle, of which the angles are given, are also given.

Let the preceding diagram be kept. On account of the angle *C* being given, the arc *DE* and *EF* the remainder of the quadrant are given. And since *BEF* is a right angle, because *BE* was let fall from the pole of arc *DEF*; and since angle *EBF* is equal to its vertical angle, which was given; therefore the triangle *BEF*, having the right angle *E* and the angle at *B* given together with the side *EF*, has its sides and angles given by the preceding theorem. Therefore *BF* is given, and so is *AB* the remainder of the quadrant. And similarly in the triangle *ABC* the remaining sides *AC* and *BC* are shown as above.

VI

If in the same sphere two triangles have right angles and another angle equal to another angle and one side equal to one side—whether the sides be adjacent to the equal angles or lie opposite one of the equal angles—they will have the remaining sides equal to the remaining sides and the remaining angle equal to the remaining angle.

Let there be the hemisphere *ABC*, in which the two triangles *ABD* and *CEF* are taken. Let the angles at *A* and *C* be right; and furthermore let the angle *ADB* be equal to *CEF*, and one side to one side. And first let the equal sides be adjacent to the equal angles, *i.e.*, let *AD* be equal to *CE*. I say moreover that side *AB* is equal to side *CF*, side *BD* to *EF*, and the remaining angle *ABD* to the remaining angle *CFE*.

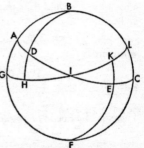

For with *B* and *F* as poles, draw *GHI* and *IKL* the quadrants of the great circles. And let quadrants *ADI* and *CEI* be completed. They necessarily cut one another at the pole of the hemisphere, point *I*, [23b] because

$$\text{angle } A = \text{angle } C = 90°$$

and quadrants *GHI* and *CEI* have been drawn through the poles of the circle *ABC*.

Therefore, since it has been assumed that

$$\text{side } AD = \text{side } CE$$

then by subtraction

$$\text{arc } DI = \text{arc } EI.$$

And

$$\text{angle } IDH = \text{angle } IEK;$$

for they are placed at the vertices of the angles assumed as equal; and

$$\text{angle } H = \text{angle } K = 90°.$$

As things which have the same ratio to the same are in the same ratio; and since by Theorem III in this chapter,

$$\text{ch. 2 } ID : \text{ch. 2 } HI = \text{dmt. sph.} : \text{ch. 2 } IDH,$$

and

$$\text{ch. } EI : \text{ch. 2 } KI = \text{dmt. sph.} : \text{ch. 2 } IEK;$$

therefore

$$\text{ch. 2 } ID : \text{ch. 2 } HI = \text{ch. 2 } EI : \text{ch. 2 } IK.$$

And by Euclid's *Elements*, V, 14, since

$$\text{ch. 2 } DI = \text{ch. 2 } IE$$

therefore

$$\text{ch. 2 } HI = \text{ch. 2 } IK.$$

And as in equal circles equal chords cut off equal arcs, and as the parts of multiples are in the same ratio (as the multiples); therefore the plain arcs *IH* and *IK* will be equal; and so will *GH* and *KL* the remainders of the quadrants. Whence it is clear that

$$\text{angle } B = \text{angle } F,$$

and since, by the inverse of the third theorem,

$$\text{ch. 2 } AD : \text{ch. 2 } BD = \text{ch. 2 } HG : \text{ch 2 } BDH, \text{ or dmt.,}$$

and

$$\text{ch. 2 } EC : \text{ch. 2 } EF = \text{ch. 2 } KL : \text{ch. 2 } FEK, \text{ or dmt.,}$$

wherefore

$$\text{ch. 2 } AD : \text{ch. 2 } BD = \text{ch. 2 } EC : \text{ch. 2 } EF$$

and

$$AD = CE.$$

Therefore, by Euclid's *Elements*, V, 14,

$$\text{arc } BD = \text{arc } EF,$$

on account of the chords subtending twice the area being equal.

In the same way with *BD* and *EF* equal, we will show that the remaining sides and angles are equal.

And in turn, if sides *AB* and *CF* are assumed to be equal, the results will follow the same identity of ratio.

VII

Now also even if there is no right angle, but provided that the sides which are adjacent to the equal angles are equal to one another, the same thing will be shown.

In this way if in the two triangles *ABD* and *CEF*

$$\text{angle } B = \text{angle } F$$

and

$$\text{angle } D = \text{angle } E,$$

and if side *BD* is adjacent to the equal [24ª], angles and

side *BD* = side *EF*,

I say that again the triangles are equilateral and equiangular.

For once more with *B* and *F* as poles, describe *GH* and *KL*, the arcs of the great circles. And let *AD* and *GH* extended intersect at *N*; and let *EC* and *LK* similarly extended intersect at *M*.

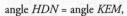

Therefore since in the two triangles *HDN* and *EKM*

angle *HDN* = angle *KEM*,

because they are placed at the vertex of the angles assumed equal; and since

angle *H* = angle *K* = 90°

on account of the intersection of circles described through the poles of one another; and

side *DH* = side *EK*;

therefore the triangles are equiangular and equilateral by the preceding proof. And again because

arc *GH* = arc *KL*

on account of its being assumed that

angle *B* = angle *F*;

therefore by addition

arc *GHN* = arc *MKL*,

by the axiom concerning the addition of equals. And therefore there are these two triangles *AGN* and *MCL* where

side *GN* = side *ML*,
angle *ANG* = angle *CML*,

and

angle *G* = angle *L* = 90°.

So the triangles will have their sides and angles equal. Therefore when equals have been subtracted from equals, the remainders will be equal:

arc *AD* = arc *CE*,
arc *AB* = arc *CF*,

and

angle *BAD* = angle *ECF*,

as was to be shown.

VIII

Now further, if two triangles have two sides equal to two sides and an angle equal to an angle, whether the angle which the equal sides comprehend, or an angle at the base, they will also have base equal to base and the remaining angles equal to the remaining angles.

As in the preceding diagram, let

$$\text{side } AB = \text{side } CF$$

and

$$\text{side } AD = \text{side } CE.$$

And first let

$$\text{angle } A = \text{angle } C,$$

which is comprehended by the equal sides. I say also that

$$\text{base } BD = \text{base } EF,$$
$$\text{angle } B = \text{angle } F,$$

and

$$\text{angle } BDA = \text{angle } CEF.$$

For we shall have the two triangles AGN and CLM, where

$$\text{angle } G = \text{angle } L = 90°.$$

And since

$$\text{angle } GAN = 180° - \text{angle } BAD,$$

and

$$\text{angle } MCL = 180° - \text{angle } ECF,$$

then

$$\text{angle } GAN = \text{angle } MCL.$$

Therefore the triangles are equiangular and equilateral.
Wherefore since

$$\text{arc } AN = \text{arc } CM$$

and

$$\text{arc } AD = \text{arc } CE,$$

then by subtraction

$$\text{arc } DN = \text{arc } ME.$$

But it has already been made clear that

$$\text{angle } DNH = \text{angle } EMK,$$

and

$$\text{angle } H = \text{angle } K = 90°.$$

Therefore the two triangles DHN and EMK will also be equiangular and equilateral. [24b] Hence (by the subtraction of equals)

$$\text{arc } BD = \text{arc } EF$$

and

$$\text{arc } GH = \text{arc } KL.$$

Hence

$$\text{angle } B = \text{angle } F,$$

and

$$\text{angle } ADB = \text{angle } FEC.$$

But if instead of sides *AD* and *CE* it be assumed that

$$\text{base } BD = \text{base } EF,$$

which are opposite the equal angles; and if the rest stays the same; then the proof will be similar. For since

$$\text{exterior angle } GAN = \text{exterior angle } MCL,$$
$$\text{angle } G = \text{angle } L = 90°,$$

and

$$\text{side } AG = \text{side } CL;$$

in the same way as before we shall have the two triangles *AGN* and *MCL* as equiangular and equilateral. And moreover, as parts of them,

$$\text{trgl. } DNH = \text{trgl. } MEK,$$

because

$$\text{angle } H = \text{angle } K = 90°,$$
$$\text{angle } DNH = \text{angle } KME,$$

and by subtraction from the quadrant

$$\text{side } DH = \text{side } EK.$$

Whence the same things follow as before.

IX

Moreover, in isosceles spherical triangles the angles at the base are equal to one another.

Let there be triangle *ABC*, where

$$\text{side } AB = \text{side } AC.$$

I say that on the base angle *ABC* = angle *ACB*.

From the vertex *A* drop a great circle which will cut the base at right angles, *i.e.*, a circle through the poles of the base; and let this circle be *AD*. Therefore, since in the two triangles *ABD* and *ADC*

$$\text{side } BA = \text{side } AC,$$

and

$$\text{side } AD = \text{side } AD,$$

and

$$\text{angle } BDA = \text{angle } CDA = 90°,$$

it is clear from what was shown above that

$$\text{angle } ABC = \text{angle } ACB,$$

as was to be shown.

PORISM

Hence it follows that the arc from the vertex of an isosceles triangle which falls at right angles upon the base will at the same time bisect the base and the angle comprehended by the equal sides, and vice versa. And that is clear from what has been shown above.

X

If two triangles in the same sphere have the sides of the one severally equal to the sides of the other, they will have the angles of the one severally equal to the angles of the other.

For in each triangle the three segments of great circles form pyramids which have as their apexes the centre of the sphere and as their bases the plane triangles which are comprehended by the straight lines subtending the arcs of the convex triangles. And those pyramids are similar and [25a] equal by the definition of similar and equal solid figures (Euclid, XI, Def. 10); now the ratio of similarity is that the angles taken in any order will be severally equal to one another. Therefore the triangles will have their angles equal to one another.

In particular, those who define similarity of figures more generally say that similar figures are those which have similar declinations, and have corresponding angles equal to one another. Whence I think it is manifest that in a sphere the triangles which are equilateral are similar, just as in the case of plane triangles.

XI

Every triangle which has two sides and an angle given will have the remaining sides and angles given.

For if the two sides are given as equal, the angles at the base will be equal, and by drawing an arc from the vertex at right angles to the base, what is sought will easily be found by means of the Porism to the ninth theorem.

But if however the sides given are unequal, as in triangle *ABC*, where angle *A* is

given together with two sides, the sides either comprehend the given angle or do not comprehend it: First, let the given sides *AB* and *AC* comprehend it. And with *C* as a pole draw arc *DEF* of a great circle; and let the quadrants *CAD* and *CBE* be completed; and let *AB* extended cut *DE* at point *F*. So also in the triangle *ADF*,

$$\text{side } AD = 90° - \text{arc } AC;$$

and

$$\text{angle } BAD = 180° - \text{angle } CAB.$$

For the ratios and dimensions of these angles are the same as those of angles occurring at the intersection of straight lines and planes. And

$$\text{angle } D = 90°.$$

Therefore by the fourth theorem of this chapter, triangle *ADF* will have its sides and angles given. And again in triangle *BEF* angle *F* has been found, and

$$\text{angle } E = 90°$$

on account of the intersection of circles through the poles of one another; and

$$\text{side } BF = \text{arc } ABF - \text{arc } AB.$$

Therefore by the same theorem triangle *BEF* also will have its angles and sides given. Whence *BC* the side sought is given, as

$$BC = 90° - BE,$$

and *BC* is the side sought. And

$$\text{arc } DE = \text{arc } DEF - \text{arc } EF.$$

And so angle *C* is given. Any by means of angle *EBF*, the vertical angle *ABC*, which was sought, is given.

But if in place of side *AB*, side *CB* which is opposite to the given angle is assumed, the same thing will result. For *AD* and *BE* the remainders of quadrants are given; and by the same argument the two triangles *ADF* and *BEF* will have their sides and angles given, as before.

Whence, as was intended, *ABC* the triangle set before us will have its sides and angles given.

[25ᵇ] XII

Furthermore, if any two angles are given together with one side, there will be the same result.

For let the construction in the previous figure stay; and in triangle *ABC* let the two angles *ACB* and *BAC* be given together with side *AC*, which is adjacent to both angles. Now if one of the angles given were right, then everything else would follow from the ratios by the preceding fourth theorem. But we wish to keep the theorems different and to have neither of the angles right. Therefore

$$AD = 90° - AC.$$

And

$$\text{angle } BAD = 180° - \text{angle } BAC.$$

And

$$\text{angle } D = 90°.$$

Therefore by the fourth theorem in this chapter, triangle *AFD* will have its angles and sides given. But through angle *C* being given, the arc *DE* is given, and so is the remainder

$$\text{arc } EF = 90° - \text{arc } DE.$$

And

$$\text{angle } BEF = 90°;$$

and

$$\text{angle } F = \text{angle } F.$$

In the same way by the fourth theorem BE and BF are given; and through them we can discover sides AB and BC, which were sought.

Moreover, if one of the given angles is opposite the given side, namely if angle ABC is given in place of angle ACB, and if the rest stayed the same, then it can be shown in similar fashion that the whole triangle ADF will be established as having its sides and angles given; and similarly the part of it which is triangle BEF; since on account of angle F being common to both, angle EBF being at the vertex of the given angle, and angle E being right, it is shown as above that all the sides are given. And from that there follows what I said. For all these things are always tied together by a mutual and perpetual bond, as befits the form of a globe.

XIII

Finally, all the sides of a triangle being given, the angles are given.

Let all the sides of triangle ABC be given: I say that all the angles too are found.

For the triangle either will have equal sides or it will not. First therefore let AB and AC be equal. It is clear that the halves of chords subtending twice those sides will be equal. And let these halves be BE and CE, which on account of being at an equal distance from the centre of the sphere will cut one another at point E in DE the common section of the circles, as is clear from Euclid, III, Def. 4, [26ᵃ] and its converse.

But by Euclid, III, 3, in plane ABD

$$\text{angle } DEB = 90°;$$

and in plane ACD similarly

$$\text{angle } DEC = 90°.$$

Therefore by Euclid, XI, Def. 3, BEC is the angle of inclination of the planes; and we shall find it as follows; for since there is a straight line subtending BC, we shall have a rectilinear triangle BEC with its sides given on account of their arcs being given; and then since the angles may be found, we shall have the angle BEC, which was sought, *i.e.*, we shall have the spherical angle BAC; and we shall have the others as above.

But if the triangle is scalene, as in the second figure, it is clear that the halves of the chords subtending twice the sides will by no means touch one another. For if

$$\text{arc } AC > \text{arc } AB,$$

then, as

$$CF = \frac{1}{2} \text{ ch. } 2\,AC,$$

CF will fall lower down. But if

$$\text{arc } AC < \text{arc } AB,$$

then *CF* will fall higher up, according as such lines become nearer and farther away from the centre, by Euclid, III, 15. Now however let *FG* be drawn parallel to *BE*; and at point *G* let it cut *BD* the common section of the two circles (*AB* and *BC*). And let *GC* be joined. Therefore it is clear that

$$\text{angle } EFG = \text{angle } AEB = 90°$$

And too

$$\text{angle } EFC = 90°;$$

for

$$CF = \frac{1}{2} \text{ ch. } 2\,AC.$$

Therefore angle *CFG* will be the angle of section of circles *AB* and *AC*; and we shall find this angle too. For

$$DF : FG = DE : EB,$$

since triangles *DFG* and *DEB* are similar. Therefore *FG* is given in the parts wherein *FC* is also given; and

$$DG : DB = DE : EB.$$

Hence *DG* will be given in the same parts whereof *DC* has 100,000. But as the angle *GDC* is given through the arc *BC*, therefore by the second theorem on plane triangles the side *GC* is given in the same parts wherein the remaining sides of the plane triangle *GFC* are given. Therefore by the last theorem on plane triangles we shall have the angle *GFC*, *i.e.*, the spherical angle *BAC*, which was sought; and then we shall find the remaining angles by the eleventh theorem on spherical triangles.

XIV

If a given arc of a circle is cut anywhere so that both of the segments together are less than a semicircle, and if the ratio of half of the chord subtending twice one segment to the half

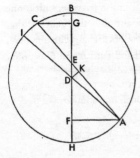

of the chord subtending twice the other segment is given, [26b] *the arcs of those segments will also be given.*

For let arc *ABC* be given, around centre *D*; and let *ABC* be cut at point *B* anywhere, but in such a way that the segments are less than a semicircle; and let

$$\frac{1}{2} \text{ ch. } 2\,AB : \frac{1}{2} \text{ ch. } 2\,BC$$

be somehow given in length: I say that the axes *AB* and *BC* are also given.

For let the straight line *AC* be drawn, which the diameter cuts at point *E*; and from the termini *A* and *C* let the perpendiculars *AF* and *CG* fall upon the diameter. And of necessity

$$AF = \tfrac{1}{2} \text{ ch. } 2 \ AB$$

and

$$CG = \tfrac{1}{2} \text{ ch. } 2 \ BC.$$

Therefore in the right triangles *AEF* and *CEG*

$$\text{angle } AEF = \text{angle } CEG,$$

because they are vertical angles. And the triangles which are therefore equiangular and similar have the sides opposite the equal angles proportional:

$$AF : CG = AE : EC.$$

Therefore we shall have *AE* and *EC* in the parts wherein *AF* or *GC* has been given. But the chord subtending arc *ABC* is given in the parts wherein the radius *DEB*, *AK* the half of chord *AC*, and the remainder *EK* are given. Let *DA* and *DK* be joined, and they will be given in the parts wherein *BD* is given: *DK* will be given as half of the chord subtending the remaining segment which is supplementary to arc *ABC* and is comprehended by angle *DAK*. And therefore angle *ADK* is given, which comprehends half of arc *ABC*. But in the triangle *EDK* having two sides given and angle *EKD* right, angle *EDK* will also be given. Hence the whole angle *EDA* comprehending the arc *AB* will be given. Thereby also the remainder *CB* will be manifest. And it was this that we were trying to show.

XV

If all the angles of a triangle are given, even though now is a right angle, all the sides are given.

Let there be the triangle *ABC*, all the angles of which are given but none of which is right. I say that all the sides are given too.

For from some one of the angles, say *A*, drop the arc *AD* through the poles of *CB*. *AD* will cut *BC* at right angles, and it will fall within the triangle, unless one of the angles at the base—angle *B* or angle *C*—is obtuse and the other acute. If that were the case, the arc would have to be drawn from the obtuse angle to the base. So with the quadrants *BAF*, *CAG*, and *DAE* completed and with *B* and *C* as poles, let the arcs *EF* and *EG* [27a] be drawn.

Therefore

$$\text{angle } F = \text{angle } G = 90°.$$

Therefore in the right triangle *EAF*

$$\tfrac{1}{2} \text{ ch. } 2 \ AE : \tfrac{1}{2} \text{ ch. } 2 \ EF = \tfrac{1}{2} \text{ dmt. sph. } : \tfrac{1}{2} \text{ ch. } 2 \ EAF$$

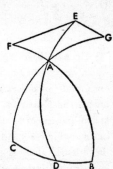

Similarly in right triangle *AEG*

$^1/_2$ ch. 2 *AE* : $^1/_2$ ch. 2 *EG* = $^1/_2$ dmt. sph. : $^1/_2$ ch. 2 *EAG*.

Therefore, *ex aequali*,

$^1/_2$ ch. 2 *EF* : $^1/_2$ ch. 2 *EG* = $^1/_2$ ch. 2 *EAF* : $^1/_2$ ch. 2 *EAG*.

And because arcs *FE* and *EG* arc given, since

arc *FE* = 90° − angle *B*

and

arc *EG* = 90° − angle *C*;

thence we shall have the ratio between angles *EAF* and *EAG* given, *i.e.*, the ratio between *BAD* and *CAD*, which are their vertical angles. Now the whole angle *BAC* has been given; therefore by the foregoing theorem, angles *BAD* and *CAD* will also be given. Then by the fifth theorem we shall determine sides *AB*, *BD*, *AC*, *CD*, and the whole of arc *BC*.

This much said enroute concerning triangles, according as they are necessary for our undertaking, will be sufficient. For if they had to be treated in greater detail, the work would be of unusual size.

[27ᵇ] BOOK TWO

Since we have expounded briefly the three terrestrial movements, by means of which we promised to demonstrate all the planetary appearances, now we shall fulfill our promise by proceeding from the whole to the parts and examining and investigating particular questions to the extent of our powers. Now we shall begin with the best-known movement of all, the revolution of day and night—which we said the Greeks called νυχθὴμερος and which we have taken as belonging wholly and immediately to the terrestrial globe, since from this movement arise the months, years, and other variously named periods of time, as number from unity. Therefore we shall say only a few words about the inequality of days and nights, the rising and setting of the sun and of the parts of the ecliptic and the signs, and the consequences of this type of revolution; for many people have written about these subjects copiously enough and what they say is in harmony and agreement with our conceptions. It is of no importance if we take up in an opposite fashion what others have demonstrated by means of a motionless earth and a giddy world and race with them toward the same goal, since things related reciprocally happen to be inversely in harmony with one another. Nevertheless we shall omit nothing necessary. But no one should be surprised if we still speak of the rising and setting of the sun and stars, *et cetera*; but he should realize that we are speaking in the usual manner of speech which can be recognized by all and that we are nevertheless always keeping in mind that: "To us who are being carried by the Earth, the sun and the moon seem to pass over; and the stars return to their former positions and again move away."

1. ON THE CIRCLES AND THEIR NAMES

We have said that the equator is the greatest of the parallel circles on the terrestrial globe described around the axis of its daily revolution and that the ecliptic is the circle through the middle [28ᵃ] of the signs under which the centre of the Earth moves in a circle in its annual revolution.

But since the ecliptic crosses the equator obliquely; in proportion to the inclination of the axis of the Earth to it, it describes in the course of the daily revolution two circles which touch it on either side of the equator, as if the farthest limits to its obliquity. These circles are called the tropics. For on them the sun appears to make its "tropes," *i.e.*, its winter and summer changes of direction. Whence the northern circle used to be called the tropic of the summer solstice and the other the tropic of the shortest day, as was set forth in our summary exposition of the circular movements of the Earth.

Next follows the so-called horizon, which the Latins call the boundary circle; for it is the boundary between that part of the world which is visible to us and that part which lies hidden. All stars which set are seen to have their rising on it; and it has its centre on the surface of the Earth and its pole at the point directly overhead. But since it is impossible to compare the Earth with the immensity of the heavens—for according to our hypothesis even the total distance between the sun and the moon is indiscernible beside the magnitude of the heavens—the circle of the horizon appears to bisect the heavens, as if it went through the centre of the world, as we demonstrated in the beginning.

But when the horizon is oblique to the equator, it too touches on either side of the equator twin parallel circles, *i.e.*, the northern circle of the always visible stars and the southern circle of the always hidden stars. The first circle was called the arctic, and the second the antarctic by Proclus and the Greeks; and they become greater or smaller in proportion to the obliquity of the horizon or the elevation of the pole of the equator.[1]

There remains the meridian circle which passes through the poles of the horizon and through the poles of the equator too and hence is perpendicular to both circles. The sun's reaching it gives us midday and midnight.

But these two circles which have their centres on the surface of the Earth, *i.e.*, the horizon and the meridian, are wholly consequent upon the movement of the Earth and upon our sight at some particular place. For the eye everywhere becomes as it were the centre of the sphere of all things which are visible to it on all sides.

Furthermore all the circles assumed on the Earth produce circles in the heavens as their likenesses and images, as will be shown more clearly in cosmography and in connection with the dimensions of the Earth. And these circles at any rate are the ones having proper names, though there are infinite ways of designating and naming others.

2. ON THE OBLIQUITY OF THE ECLIPTIC AND THE DISTANCE OF THE TROPICS AND HOW THEY ARE DETERMINED

[28b] Since the ecliptic lies between the tropics and crosses the equator obliquely, I therefore think that we should now try to observe what the distance between the tropics is and hence what the angle of section between the equator and the ecliptic is. For in order to perceive this by sense with the help of artificial instruments, by means of which the job can be done best, it is necessary to have a wooden square prepared, or preferably a square made from some other more solid material, from stone or metal; for the wood might not stay in the same condition on account of some alteration in the atmosphere and might mislead the observer. Now one surface of it should be very

[1]That is to say, the magnitude of the circle of the always visible stars varies inversely with the obliquity of the horizon and directly with the elevation of the poles of the equator.

carefully planed, and it should be of sufficient area to admit being divided into sections, that is, a side should be about 5 or 6 feet long. Now with one of the corners (of the square) as centre and with a side as radius, let a quadrant of a circle be drawn and divided into 90 equal degrees; and let each of the degrees be subdivided into 60 minutes, or whatever number can be taken. Next let a cylindrical pointer which has been well turned on a lathe be set up at the centre (of the quadrant) and fixed in such a way as to be perpendicular to the surface and to extend out from it a little, say perhaps a finger's width or less.

When the instrument has been prepared in this way, the next thing to do is to exhibit the line of the meridian on a piece of flooring which lies in the plane of the horizon and which has been made even as carefully as is possible by means of a hydroscope or ground-level, so as not to have a slope in any part of it. The piece of flooring should have a circle drawn on it and a cylinder erected at the center of the circle: we shall take observations and mark the point where at some time before midday the extremity of the shadow of the cylinder touches the circumference of the circle, We shall do the same thing in the afternoon, and then shall bisect the arc of the circle lying between the two points we have already marked. In this way a straight line drawn from the centre through the point of section will indicate infallibly for us the south and the north.

Accordingly the plane surface of the instrument should be set up on this piece of flooring as a base and fixed perpendicular to it with the centre (of the quadrant) to the south, so that a plumb-line from the centre would fall exactly at right angles to the meridian line. For it comes about in this way that the surface of the instrument exhibits the meridian circle. Hence on the days of summer and winter solstice the shadows of the sun at noon [29a] are to be observed according as they are cast by the pointer, or cylinder, from the centre (of the quadrant); and some mark is to be made on the arc of the quadrant, so that the place of the shadow may be kept more surely. And we shall note down the centre of the shadow in degrees and minutes as accurately as is possible. For if we do this, the arc between the marked shadows—the summer—and winter—solstitial shadows—will be found and will show us the distance between the tropics and also the total obliquity of the ecliptic.[1] By taking half of the arc, we shall have the distance of the tropics from the equator, and it will be clear what the angle of inclination is between the equator and the ecliptic.

Now Ptolemy took the interval between the aforesaid limits—the northern and the southern—as 47°42'40", whereof the circle has 360°, as he found had been observed by Hipparchus and Eratosthenes before his time; and there are 11ᵖ whereof the whole circle has 83ᵖ. Hence half the arc—and half the arc has 23°51'20", whereof

---

[1] Since the distance between the sun and the Earth is imperceptible in relation to the radius of the sphere of the fixed stars, the centre of the quadrant may be taken as the centre of the sphere of the fixed stars.

the circle has 360°—showed the distance of the tropics from the equator and what the angle of section with the ecliptic was. Accordingly Ptolemy believed that these things were invariably such and would always remain so. But these distances have been found to have decreased continually from that time down to ours. For it has already been discovered by us and some of our contemporaries that the distance between the tropics is not more than 46°58' approximately and that the angle of section is 23°29'. Hence it is clear enough that the obliquity of the ecliptic is not fixed. More on this below, where we shall show by a probable enough conclusion that it was never greater than 23°52' and will not ever be less than 23°28'.

3. On the Arcs and Angles of the Intersections of the Equator, Ecliptic, and Meridian, by Means of Which Declinations and Right Ascensions are Determined, and on the Computation of These Arcs and Angles

Accordingly as we were saying in the case of the horizon that the parts of the world have their risings and settings on it, we say that the meridian circle [29ᵇ] halves the heavens. During the space of twenty-four hours this circle is crossed by both the ecliptic and the equator and divides both of their circumferences by cutting them at the spring and at the autumnal intersection and in turn has its circumference divided by the arc intercepted by those two circles. Since they are all great circles, they form a spherical right triangle; for the angle is right where the meridian circle by definition cuts the equator described through its poles. Now the arc of the meridian circle, or any arc of a circle passing through the poles (of the equator) and intercepted in this way is called the declination of a segment of the ecliptic; and the corresponding arc on the equator is called the right ascension occurring at the same time as the similar arc on the ecliptic.

All this is easily demonstrated in a convex triangle. For let the circle *ABCD* be a circle passing simultaneously through the poles of the ecliptic and of the equator— most people call this circle the "colure"—let the semi- circle of the ecliptic be *AEC*, the semicircle of the equator *BED*; let the spring equinox be at point *E*, the summer solstice at *A*, and the winter solstice at *C*. Now let *F* be taken as the pole of daily revolution, and on the ecliptic let

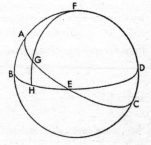

arc *EG* = 30°,

for example, and let it be cut off by *FGH* the quad- rant of a circle.

Then it is clear that in triangle EGH

$$\text{side } EG = 30°,$$
$$\text{angle } GEH \text{ is given,}$$

since at its least, in conformity with the greatest declination AB,

$$\text{angle } GFH = 23°28'$$

where 4 rt. angles = 360°;

and

$$\text{angle } GHE = 90°.$$

Therefore by the fourth theorem on sphericals, triangle EHG will have its sides and angles given. For it was shown that

$$\text{ch. } 2\ EG : \text{ch. } 2\ GH = \text{ch. } 2\ AGE, \text{ or dmt. sph. : ch. } 2\ AB$$

and their halves are in the same ratio. And since

$$^1/_2 \text{ ch. } 2\ AGE = \text{radius} = 100,000,$$
$$^1/_2 \text{ ch. } 2\ AB = 39,822,$$

and

$$^1/_2 \text{ ch. } 2\ EG = 50,000;$$

and since, if four numbers are proportional, the product of the means is equal to the product of the extremes; therefore

$$^1/_2 \text{ ch. } 2\ GH = 19,911,$$

and hence, by the table,

$$\text{arc } GH = 11°29',$$

which is the declination of segment EG,

$$\text{side } FG = 78°31',$$
$$\text{side } AG = 60°,$$

since they are the remainders of the quadrants, and

$$\text{angle } FAG = 90°.$$

In the same way

[30ª] $^1/_2$ ch. $2\ FG : {}^1/_2$ ch. $2\ AG = {}^1/_2$ ch. $2\ FGH : {}^1/_2$ ch. $2\ BH$.

Now since three of these chords are given, the fourth will also be given, that is to say,

$$\text{arc } BH = 62°6',$$

which is the right ascension from the summer solstice, and

$$HE = 27°54'$$

from the spring equinox. Similarly, since

$$\text{side } FG = 78°31',$$
$$\text{side } AF = 64°30',$$

and

$$AGE = 90°;$$

then, since angles AGF and HGE are vertical angles,

angle *AGF* = angle *HGE* = 63°29$^1/_2$'.

In the rest we shall do as in this example. But we should not be ignorant of the fact that the meridian circle cuts the ecliptic at right angles in the signs where the ecliptic touches the tropics, for then the meridian circle cuts it through its poles, as we said. But at the equinoctial points the meridian makes an angle less than a right angle by the angle of inclination of the ecliptic, so that in conformity with the least inclination of the ecliptic it makes an angle of 66°32'.

Moreover we should note that equal sides and equal angles of the triangles follow upon equal arcs of the ecliptic being taken from the points of solstice or equinox. In this way if we

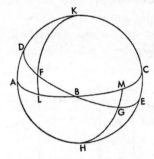

draw the equatorial arc *ABC* and the ecliptic *BDE* as intersecting at point *B*, where the equinox is, and if we take as equal the arcs *FB* and *BG* and also arcs *KFL* and *HGM* two quadrants of circle described through the poles of daily revolution; there will be the two triangles *FLB* and *BMG*, wherein

side *BF* = side *BG*

angle *FLB* = angle *GBM*

and

angle *FLB* = angle *GMB* = 90°.

Therefore by the sixth theorem on spherical triangles the sides and angles are equal. Hence

declination *FL* = declination *GM*

rt. ascension *LB* = rt. ascension *BM*

and

angle *LFB* = angle *MGB*.

This same fact will be manifest upon the assumption of equal arcs described from a point of solstice, for example, when *AB* and *BC* on different sides of their point of contact *B*, the solstice, are equally distant from it. For when the arcs *DA* and *DB* (and *DC*) have been drawn from the pole of the equator, there will similarly be the two triangles *ABD* and *DBC*.

Then

base *AB* = base *BC*

side *BD* is common

and

angle *ABD* = angle *CBD* = 90°.

Accordingly by the eighth theorem on sphericals the triangles will be shown to have equal sides and angles. It is clear from this that such angles and arcs laid out in one quadrant of the ecliptic [30^b] are in accord with the remaining quadrants of the full circle.

We shall subjoin an example of these things in the tables. In the first column are placed the degrees of the ecliptic; in the following column, the declinations answering to those degrees; and in the third column the minutes, which are the differences between the particular declinations and the declinations which occur at the time of greatest obliquity of the ecliptic: the greatest of these differences is 24'.

We shall do the same thing in the table of ascensions and the table of meridian angles. For it is necessary for all things which are consequences of the obliquity of the ecliptic to be changed with a change in it. Furthermore, in the right ascensions an extremely slight difference is found, one which does not exceed $1/10$ "time" and which in the space of an hour makes only $1/150$ "time."—The ancients give the name of "time" to the parts of the equator which arise together with the parts of the ecliptic. Each of these circles, as we have often said, has 360 parts; but in order to distinguish between them, most of the ancients called the parts of ecliptic "degrees" and those of

TABLE OF DECLINATIONS OF THE DEGREES OF THE ECLIPTIC

Ecliptic Deg.	Declination Deg.	Min.	Difference Min.	Ecliptic Deg.	Declination Deg.	Min.	Difference Min.	Ecliptic Deg.	Declination Deg.	Min.	Difference Min.
1	0	24	0	31	11	50	11	61	20	23	20
2	0	48	1	32	12	11	12	62	20	35	21
3	1	12	1	33	12	32	12	63	20	47	21
4	1	36	2	34	12	52	13	64	20	58	21
5	2	0	2	35	13	12	13	65	21	9	21
6	2	23	2	36	13	32	14	66	21	20	22
7	2	47	3	37	13	52	14	67	21	30	22
8	3	11	3	38	14	12	14	68	21	40	22
9	3	35	4	39	14	31	14	69	21	49	22
10	3	58	4	40	14	50	14	70	21	58	22
11	4	22	4	41	15	9	15	71	22	7	22
12	4	45	4	42	15	27	15	72	22	15	23
13	5	9	4	43	15	46	16	73	22	23	23
14	5	32	5	44	16	4	16	74	22	30	23
15	5	55	5	45	16	22	16	75	22	37	23
16	6	19	6	46	16	39	17	76	22	44	23
17	6	41	6	47	16	56	17	77	22	50	23
18	7	4	7	48	17	13	17	78	22	55	23
19	7	27	7	49	17	30	18	79	23	1	24
20	7	49	8	50	17	46	18	80	23	5	24
21	8	12	8	51	18	1	18	81	23	10	24
22	8	34	8	52	18	17	18	82	23	13	24
23	8	57	9	53	18	32	19	83	23	17	24
24	9	19	9	54	18	47	19	84	23	20	24
25	9	41	9	55	19	2	19	85	23	22	24
26	10	3	10	56	19	16	19	86	23	24	24
27	10	25	10	57	19	30	20	87	23	26	24
28	10	46	10	58	19	44	20	88	23	27	24
29	11	8	10	59	19	57	20	89	23	28	24
30	11	29	11	60	20	10	20	90	23	28	24

the equator "times"; and we will copy them for the remainder of the work.—Therefore since the difference is so small that it can be properly neglected, we are not peeved at having to place it in a separate column.

Hence these tables can be made to apply to any other obliquity of the ecliptic, if in conformity with the ratio of difference between the least and greatest obliquity of the ecliptic we make the proper corrections. For example, if with an obliquity of 23°34' we wish to know how great a declination follows from taking it distance of 30° from the equator along the ecliptic, we find that in the table there are 11°29' in the column of declinations and 11' in the column of differences. These 11' would be all added in the case of the greatest obliquity of the ecliptic, which is, as we said, an obliquity of 23°52'. But it has already been laid down that the obliquity is 23°34' and is accordingly greater than the least obliquity by 6', which are one quarter of 24', which is the excess of the greatest obliquity over the least. Now

TABLE OF RIGHT ASCENSIONS

Ecliptic Deg.	Equator Deg.	Equator Min.	Difference Min.	Ecliptic Deg.	Equator Deg.	Equator Min.	Difference Min.	Ecliptic Deg.	Equator Deg.	Equator Min.	Difference Min.
1	0	55	0	31	28	54	4	61	58	51	4
2	1	50	0	32	29	51	4	62	59	54	4
3	2	45	0	33	30	50	4	63	60	57	4
4	3	40	0	34	31	46	4	64	62	0	4
5	4	35	0	35	32	45	4	65	63	3	4
6	5	30	0	36	33	43	5	66	64	6	3
7	6	25	1	37	34	41	5	67	65	9	3
8	7	20	1	38	35	40	5	68	66	13	3
9	8	15	1	39	36	38	5	69	67	17	3
10	9	11	1	40	37	37	5	70	68	21	3
11	10	6	1	41	38	36	5	71	69	25	3
12	11	0	2	42	39	35	5	72	70	29	3
13	11	57	2	43	40	34	5	73	71	33	3
14	12	52	2	44	41	33	6	74	72	38	2
15	13	48	2	45	42	32	6	75	73	43	2
16	14	43	2	46	43	31	6	76	74	47	2
17	15	39	2	47	44	32	5	77	75	52	2
18	16	34	3	48	45	32	5	78	76	57	2
19	17	31	3	49	46	32	5	79	78	2	2
20	18	27	3	50	47	33	5	80	79	7	2
21	19	23	3	51	48	34	5	81	80	12	1
22	20	19	3	52	49	35	5	82	81	17	1
23	21	15	3	53	50	36	5	83	82	22	1
24	22	10	4	54	51	37	5	84	83	27	1
25	23	9	4	55	52	38	4	85	84	33	1
26	24	6	4	56	53	41	4	86	85	38	0
27	25	3	4	57	54	43	4	87	86	43	0
28	26	0	4	58	55	45	4	88	87	48	0
29	26	57	4	59	56	46	4	89	88	54	0
30	27	54	4	60	57	48	4	90	90	0	0

$$3' : 11 \fallingdotseq 6' : 24'.$$

When I add 3' to the 11°29', I shall have 11°32', which will then measure the declination of the arc of the ecliptic 30° from the equator.

The same thing can be done in the table of meridian angles and right ascensions, except that we must always add the differences in the case of right ascensions but subtract them in the case of the meridian angles, so that everything may proceed correctly in conformity with the time.

TABLE OF THE MERIDIAN ANGLES

Ecliptic	Angle		Difference	Ecliptic	Angle		Difference	Ecliptic	Angle		Difference
Deg.	Deg.	Min.	Min.	Deg.	Deg.	Min.	Min.	Deg.	Deg.	Min.	Min.
1	66	32	24	31	69	35	21	61	78	7	12
2	66	33	24	32	69	48	21	62	78	29	12
3	66	34	24	33	70	0	20	63	78	51	11
4	66	35	24	34	70	13	20	64	79	14	11
5	66	37	24	35	70	26	20	65	79	36	11
6	66	39	24	36	70	39	20	66	79	59	10
7	66	42	24	37	70	53	20	67	80	22	10
8	66	44	24	38	71	7	19	68	80	45	10
9	66	47	24	39	71	22	19	69	81	9	9
10	66	51	24	40	71	36	19	70	81	33	9
11	66	55	24	41	71	52	19	71	81	58	8
12	66	59	24	42	72	8	18	72	82	22	8
13	67	4	23	43	72	24	18	73	82	46	7
14	67	10	23	44	72	39	18	74	83	11	7
15	67	15	23	45	72	55	17	75	83	35	6
16	67	21	23	46	73	11	17	76	84	0	6
17	67	27	23	47	73	28	17	77	84	25	6
18	67	34	23	48	73	47	17	78	84	50	5
19	67	41	23	49	74	6	16	79	85	15	5
20	67	49	23	50	74	24	16	80	85	40	4
21	67	56	23	51	74	42	16	81	86	5	4
22	68	4	22	52	75	1	15	82	86	30	3
23	68	13	22	53	75	21	15	83	86	55	3
24	68	22	22	54	75	40	15	84	87	19	3
25	68	32	22	55	76	1	14	85	87	53	2
26	68	41	22	56	76	21	14	86	88	17	2
27	68	51	22	57	76	42	14	87	88	41	1
28	69	2	21	58	77	3	13	88	89	6	1
29	69	13	21	59	77	24	13	89	89	33	0
30	69	24	21	60	77	45	13	90	90	0	0

4. HOW TO DETERMINE THE DECLINATION AND RIGHT ASCENSION OF ANY STAR WHICH IS PLACED OUTSIDE THE ECLIPTIC BUT WHOSE LONGITUDE AND LATITUDE HAVE BEEN ESTABLISHED; AND WITH WHAT DEGREE OF THE ECLIPTIC IT HALVES THE HEAVENS

[32b] These things have been set down concerning the ecliptic and the equator and their intersections. But as regards the daily revolution, it is of interest not only to know what parts of the ecliptic appear, by means of which the causes of the sun's appearing where it does are discovered, but also to know that there is a similar demonstration of the declination from the equator and of the right ascension in the case of those fixed or wandering stars which are outside the ecliptic but whose longitude and latitude have been given.

Therefore let the circle *ABCD* be described through the poles of the equator and of the ecliptic; let *AEC* be the semicircle of the equa-
tor above pole *F*; let *BED* be the semicircle of the ecliptic about pole *G*; and let its intersection with the equator be at point *E*. Now from the pole *G* let the arc *GHKL* be drawn through a star, and let the position of the star be given as point *H*, and let *FHMN* a quadrant of a circle fall through *H* from the pole of daily movement.

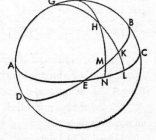

Then it is clear that the star which is at *H* falls upon the meridian at the same time as points *M* and
N do, and that arc *HMN* is the declination of the star from the equator, and *EN* is its ascension in the right sphere, and those are what we are seeking. Accordingly since in triangle *KEL*

$$\text{side } KE \text{ is given,}$$
$$\text{angle } KEL \text{ is given,}$$

and

$$\text{angle } EKL = 90°,$$

therefore by the fourth theorem on spherical triangles,

$$\text{side } KL \text{ is given,}$$
$$\text{side } EL \text{ is given,}$$

and

$$\text{angle } KLE \text{ is given.}$$

Therefore by addition

$$\text{arc } HKL \text{ is given.}$$

And on that account, in triangle *HLN*,

angle *HLN* is given,

angle *LNH* = 90°,

and

side *HL* is given.

Therefore by the same fourth theorem on sphericals there are also given the remaining sides: *HN* the declination of the star, and *LN*, and the remaining distance *NE*, the right ascension, which measures the distance the sphere turns from the equinox to the star.

—Or in another way. If in the foregoing you take *KE* the arc of the ecliptic as the right ascension of *LE*, conversely *LE* will be given by the table of right ascensions; and so will *LK*, as the declination corresponding to *LE*; [33ª] and the angle *KLE* will be given by the table of meridian angles; and hence the remaining sides and angles, as we have showed, may be learned.—

Then by means of the right ascension *EN*, the number of degrees of *EM* the arc of the ecliptic are given. And in conformity with these things the star together with point *M* halves the heavens.

5. On the Sections of the Horizon

Now the horizon of a right sphere is different from the horizon of an oblique sphere. For the horizon to which the equator is perpendicular, or which passes through the poles of the equator, is called a right horizon.

We call the horizon which has some inclination with the equator the horizon of an oblique sphere.

Therefore on a right horizon all the stars rise and set, and the days are always equal to the nights. For this horizon bisects all the parallel circles described by the diurnal movement, and passes through their poles; and there occurs there what we have already explained in the case of the meridian circle. Here, however, we are taking the day as extending from sunrise to sunset, and not from light to darkness, as the crowds understand it, *i.e.*, from early morning twilight to the first street lights; but we shall say more on this subject in connection with the rising and setting of the signs.

On the contrary, where the axis of the Earth is perpendicular to the horizon there are no risings or settings, but all the stars turn in a gyre and are always visible or hidden, unless they are affected by some other motion, such as the annual movement around the sun. Consequently, there day lasts perpetually for the space of half a year and night for the rest of the time; and there is nothing else to differentiate summer and winter, since there the horizon coincides with the equator.

Furthermore, in an oblique sphere certain stars rise and set; and certain others are always visible or always hidden; and meanwhile the days and nights are unequal there

where an oblique horizon touches two parallel circles in proportion to its inclination. And of these circles, the one nearer the visible pole is the boundary of the always visible stars, and conversely the circle nearer the hidden pole is the boundary for the always hidden stars. Therefore the horizon, as falling completely between these boundaries, cuts all the parallel circles in the middle into unequal arcs, except the equator, which is the greatest of the parallels; and great circles bisect one another. Therefore an oblique horizon in the upper hemisphere cuts off axes of parallels in the direction of the visible pole which are greater than the arcs which are toward the southern and hidden [33b] pole; and the converse is the case in the hidden hemisphere. The sun becomes visible in these horizons by reason of the diurnal movement and causes the inequality of days and nights.

6. WHAT THE DIFFERENCES BETWEEN THE MIDDAY SHADOWS ARE

There are differences between the midday shadows on account of which some people are called periscian, others amphiscian, and still others heteroscian. The periscian are those whom we might call "circumumbratile," that is to say, "throwing the shadow of the sun on every side." And they live where the distance between the vertex, or pole, of the horizon and the pole of the Earth is less or no greater than that between the tropic and the equator. For there the parallels which the horizon touches as the boundaries of the always apparent or always hidden stars are greater than, or equal to, the tropics. And so the summer sun high up among the always apparent stars at that time throws the shadow of a pointer in every direction. But where the horizon touches the tropics, the tropics become the boundaries of the always apparent and the always hidden stars. Wherefore instead of there being midnight the sun at its (winter) solstice seems to graze the Earth, at which time the whole circle of the ecliptic coincides with the horizon; and straightway six signs rise at the same time, and on the opposite side six signs set at the same time, and the pole of the ecliptic coincides with the pole of the horizon.

The amphiscian, who cast midday shadows on both sides, are those who live between the tropics. This is the space which the ancients called the middle zone. And since throughout that whole tract the circle of the ecliptic passes directly over head twice, as is shown in the second theorem of the *Phaenomena* of Euclid, the shadows of pointers are cast in two directions there: for as the sun moves back and forth, the pointers throw their shadows sometimes to the south and sometimes to the north.

The rest of us who inhabit the region between the two others are heteroscian, because we cast our midday shadows in only one direction, *i.e.*, towards the north.

Now the ancient mathematicians were accustomed to divide the world into seven climates, through Meröe, Siona, Alexandria, Rhodes, the Hellespont, the middle of the Pontus, Boristhenes, Byzantium, and so on with the single parallel circles taken according to the differences between the longest days and according to the lengths of the shadows, which they observed by means of pointers at noon on the days of equinoxes and solstices, and [34ᵃ] according to the elevation of the pole or the latitude of some segment. Since these things have partly changed through time, they are not exactly the same as they once were, on account of the variable obliquity of the ecliptic, as we said, of which the ancients were ignorant; or, to speak more correctly, on account of the variable inclination of the equator to the plane of the ecliptic, upon which these things depend. But the elevations of the pole or the latitudes of the places and the equinoctial shadows agree with those which antiquity discovered and made note of. That would necessarily take place, since the equator depends upon the pole of the terrestrial globe. Wherefore those segments are not accurately enough designated and defined by shadows falling on special days, but more correctly by their distances from the equator, which remain perpetually. But although this variability of the tropics, because very slight, admits but slight diversity of days and of shadows in southern places, it becomes more apparent to those who are moving northward. Therefore as regards the shadows of pointers, it is clear that for any given altitude of the sun the length of the shadow is derivable and vice versa.

In this way if there is the pointer AB which casts a shadow BC; since the pointer is perpendicular to the plane of the horizon, angle ABC must always be right, by the definition of lines perpendicular to a plane. Wherefore if AC be joined, we shall have a right triangle ABC; and for a given altitude of the sun we shall have angle ACB given. And by the first theorem on plane triangles the ratio of the pointer AB to its shadow BC will be given, and BC will be given in length. Conversely, moreover, when AB and BC are given, it will be clear from the third theorem on plane triangles what angle ACB is and what the elevation of the sun making that shadow at that time is. In this way the ancients in describing the regions of the terrestrial globe gave the lengths of the midday shadows sometimes at the equinoxes and sometimes at the solstices.

7. HOW THE LONGEST DAY, THE DISTANCE OF RISING, AND THE INCLINATION OF THE SPHERE ARE DERIVED FROM ONE ANOTHER, AND ON THE DIFFERENCES BETWEEN DAYS

[34ᵇ] In this way too for any obliquity of the sphere or inclination of the horizon we will demonstrate simultaneously the longest and the shortest day together with the

distance of rising (of the sun) and the difference of the remaining days. Now the distance of rising is the arc of the horizon intercepted between the summer solstitial and the winter solstitial sunrises, or the sum of the distances of the solstitial from the equinoctial sunrise.

Therefore let $ABCD$ be the meridian circle, and let BED be the semicircle of the horizon in the eastern hemisphere, and let AEC be the similar semicircle of the equa-

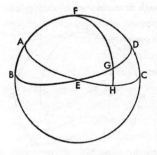

tor with F as its northern pole. Let point G be taken as the rising of the sun at the summer solstice, and let FGH the arc of a great circle be drawn. Therefore since the motion of the terrestrial sphere takes place around pole F of the equator, then necessarily points G an H will fit onto the meridian $ABCD$ at the same time, since parallel circles are around the same poles through which pass the great circles which intercept similar arcs on the parallel circles. Wherefore the selfsame time from the rising at G to midday measures also the arc AEH; and the time from midnight to sunrise measures CH the remaining and subterranean arc of the semicircle. Now AEC is a semicircle; and AE and EC are quadrants of circles, since they are drawn through the pole of $ABCD$. On that account, EH will be half the difference between the longest day and the equinox; and EG will be the distance between the equinoctial and the solstitial sunrise. Therefore since in triangle EHG angle GEH, the obliquity of the sphere, is established by means of arc AB, and angle GHE is right, and side GH is given as the distance from the summer tropic to the equator; the remaining sides are also given by the fourth theorem on sphericals: side EH as half the difference between the longest day and the equinox, and side GE as the distance of sunrise. Moreover, if together with side GH side EH, (half) the difference between the longest day and the equinox, or else EG, is given; angle E of the inclination of the sphere is given, and hence FD the elevation of the pole above the horizon.

But even if it is not the tropic but some other point G in the ecliptic which is taken, nevertheless arcs EG and EH will become manifest: since by the table of declinations set out above GH the arc of declination for that degree of the ecliptic becomes known, and the rest can be demonstrated in the same way.

Hence it also follows that the degrees of the ecliptic which are equally distant from the tropic cut off equal arcs of the horizon [35ª] between the equinoctial sunrise and the same degrees and make the lengths of days and nights inversely equal. And that is so because the parallels which pass through each of those degrees of the ecliptic are equal, since each of the degrees has the same declination.

79

But when equal arcs are taken between the equinoctial intersection and the two degrees (on the ecliptic), again the distances of rising are equal but in different directions; and the duration of days and nights are inversely equal, because on each side of the equinox the durations describe equal arcs of parallels, according as the signs themselves which are equally distant from the equinox have equal declinations from the equator.

For in the same figure let GM and KN the arcs of parallels be described cutting the horizon BED at points G and K, and let LKO a quadrant of a great circle be drawn from the south pole L. Therefore since

declination HG = declination KO,

there will be two triangles DFG and BLK, wherein two sides of the one are equal to two sides of the other:

$$FG = LK$$

and the elevations of the poles are equal,

$$FD = LB,$$

and

$$\text{angle } D = \text{angle } B = 90°.$$

Therefore

$$\text{base } DG = \text{base } BK;$$

and hence, as the distances of sunrise are the remainders of the quadrants

$$GE = EK.$$

Wherefore since here too,

$$\text{side } EG = \text{side } EK,$$
$$\text{side } GH = \text{side } KO,$$

and

$$\text{vertical angle } KEO = \text{vertical angle } GEH;$$
$$\text{side } EH = \text{side } EO.$$

And

$$EH + 90° = OE + 90°.$$

Hence

$$\text{arc } AEH = \text{arc } OEC.$$

But since great circles described through the poles of parallel circles cut off similar arcs, GM and KN will be similar and equal, as was to be shown.

But all this can be shown differently. In the same way let the meridian circle $ABCD$ be described with centre E. Let the diameter of the equator and the common section of the two circles be AEC; let BED be the diameter of the horizon and the meridian line, let LEM be the axis of the sphere; and let L be the apparent pole and M the hidden. Let AF be taken as the distance of the summer solstice or as some other declination; and to AF let GF be drawn as the diameter of a parallel and its common section with the meridian; FG will cut the axis at K and the meridian line at N.

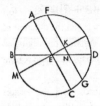

Therefore, [35ᵇ] since by the definition of Posidonius those lines are parallel which neither move toward nor move away from one another but which everywhere make the perpendicular lines between them equal,

$$KE = \frac{1}{2} \text{ ch. } 2 AF.$$

Similarly *KN* will be half of the chord subtending twice the arc of the parallel circle whose radius is *FK*. And twice this arc is the difference between the equinoctial day and the other day. And this is true because all the semicircles of which these lines are the common sections and diameters—namely, *BED* of the oblique horizon, *LEM* of the right horizon, *AEC* of the equator, and *FKG* of the parallel—are perpendicular to the plane of circle *ABCD*. And by Euclid's *Elements*, XI, 19, the common sections which they make with one another are perpendicular to the same plane at points *E*, *K*, and *N*: and by XI, 6, these common sections are parallel to one another.

And *K* is the centre of the parallel circle; and *E* is the centre of the sphere. Wherefore *EN* is half the chord subtending twice the arc of the horizon which is the difference between sunrise on the parallel and the equinoctial sunrise. Therefore, since the declination *AF* was given together with *FL* the remainder of the quadrant, *KE* half of the chord subtending twice arc *AF* and *FK* half the chord subtending twice arc *FL* will be established in terms of the parts whereof *AE* has 100,000. But in the right triangle *EKN* angle *KEN* is given by *DL* the elevation of the pole, and the remaining angle *KNE* is equal to *AEB*, because in the oblique sphere the parallels are equally inclined to the horizon; and the sides are given in the same parts whereof the radius has 100,000. Therefore *KN* will be given in the parts whereof *FK* the radius of the parallel has 100,000; for *KN* is equal to half the chord subtending the arc which measures the distance between the equinoctial day and a day on the parallel; and this arc is similarly given in the degrees whereof the parallel circle has 360°.

From this it is clear that

$$FK : KN = \frac{1}{2} \text{ ch. } 2 FL : \frac{1}{2} \text{ ch. } 2 AF \text{ comp. } \frac{1}{2} \text{ ch. } 2 AB : \frac{1}{2} \text{ ch. } 2 DL$$

and

$$\frac{1}{2} \text{ ch. } 2 FL : \frac{1}{2} \text{ ch. } 2 AF \text{ comp. } \frac{1}{2} \text{ ch. } 2 AB : \frac{1}{2} \text{ ch. } 2 DL = FK : KE \text{ comp. } EK : KN.$$

That is to say, *EK* is taken as a mean between *FK* and *KN*. Similarly too

$$BE : EN = BE : EK \text{ comp. } KE : EN,$$

as Ptolemy shows in greater detail by means of spherical segments. So I think that not only the inequality of days and nights can be determined; but also that in the case of the moon and the stars whose declination on the parallels described through them by the daily movement has been given, the segments (of the parallels) which are above the horizon can be distinguished from those which are below; and hence the risings and settings (of the moon or stars) can be easily understood.

TABLE OF DIFFERENCE OF THE ASCENSIONS IN AN OBLIQUE SPHERE

Declination	Elevation of the Pole										
	31°		32°		33°		34°		35°		36°	
	Times	Min.	Times	Min.	Times	Min.	Times	Min.	Times	Min.	Times	Min.
1	0	36	0	37	0	39	0	40	0	42	0	44
2	1	12	1	15	1	18	1	21	1	24	1	27
3	1	48	1	53	1	57	2	2	2	6	2	11
4	2	24	2	30	2	36	2	42	2	48	2	55
5	3	1	3	8	3	15	3	23	3	31	3	39
6	3	37	3	46	3	55	4	4	4	13	4	23
7	4	14	4	24	4	34	4	45	4	56	5	7
8	4	51	5	2	5	14	5	26	5	39	5	52
9	5	28	5	41	5	54	6	8	6	22	6	36
10	6	5	6	20	6	35	6	50	7	6	7	22
11	6	42	6	59	7	15	7	32	7	49	8	7
12	7	20	7	38	7	56	8	15	8	34	8	53
13	7	58	8	18	8	37	8	58	9	18	9	39
14	8	37	8	58	9	19	9	41	10	3	10	26
15	9	16	9	38	10	1	10	25	10	49	11	14
16	9	55	10	19	10	44	11	9	11	25	12	2
17	10	35	11	1	11	27	11	54	12	22	12	50
18	11	16	11	43	12	11	12	40	13	9	13	39
19	11	56	12	25	12	55	13	26	13	57	14	29
20	12	38	13	9	13	40	14	13	14	46	15	20
21	13	20	13	53	14	26	15	0	15	36	16	12
22	14	3	14	37	15	13	15	49	16	27	17	5
23	14	47	15	23	16	0	16	38	17	17	17	58
24	15	31	16	9	16	48	17	29	18	10	18	52
25	16	16	16	56	17	38	18	20	19	3	19	48
26	17	2	17	45	18	28	19	12	19	58	20	45
27	17	50	18	34	19	19	20	6	20	54	21	44
28	18	38	19	24	20	12	21	1	21	51	22	43
29	19	27	20	16	21	6	21	57	22	50	23	45
30	20	18	21	9	22	1	22	55	23	51	24	48
31	21	10	22	3	22	58	23	55	24	53	25	53
32	22	3	22	59	23	56	24	56	25	57	27	0
33	22	57	23	54	24	19	25	59	27	3	28	9
34	23	55	24	56	25	59	27	4	28	10	29	21
35	24	53	25	57	27	3	28	10	29	21	30	35
36	25	53	27	0	28	9	29	21	30	35	31	52

TABLE OF DIFFERENCE OF THE ASCENSIONS IN AN OBLIQUE SPHERE

Declination	Elevation of the Pole										
	37°		38°		39°		40°		41°		42°	
	Times	Min.	Times	Min.	Times	Min.	Times	Min.	Times	Min.	Times	Min.
1	0	45	0	47	0	49	0	50	0	52	0	54
2	1	31	1	34	1	37	1	41	1	44	1	48
3	2	16	2	21	2	26	2	31	2	37	2	42
4	3	1	3	8	3	15	3	22	3	29	3	37
5	3	47	3	55	4	4	4	13	4	22	4	31
6	4	33	4	43	4	53	5	4	5	15	5	26
7	5	19	5	30	5	42	5	55	6	8	6	21
8	6	5	6	18	6	32	6	46	7	1	7	16
9	6	51	7	6	7	22	7	38	7	55	8	12
10	7	38	7	55	8	13	8	30	8	49	9	8
11	8	25	8	44	9	3	9	23	9	44	10	5
12	9	13	9	34	9	55	10	16	10	39	11	2
13	10	1	10	24	10	46	11	10	11	35	12	0
14	10	50	11	14	11	39	12	5	12	31	12	58
15	11	39	12	5	12	32	13	0	13	28	13	58
16	12	29	12	57	13	26	13	55	14	26	14	58
17	13	19	13	49	14	20	14	52	15	25	15	59
18	14	10	14	42	15	15	15	49	16	24	17	1
19	15	2	15	36	16	11	16	48	17	25	18	4
20	15	55	16	31	17	8	17	47	18	27	19	8
21	16	49	17	27	18	7	18	47	19	30	20	13
22	17	44	18	24	19	6	19	49	20	34	21	20
23	18	39	19	22	20	6	20	52	21	39	22	28
24	19	36	20	21	21	8	21	56	22	46	23	38
25	20	34	21	21	22	11	23	2	23	55	24	50
26	21	34	22	24	23	16	24	10	25	5	26	3
27	22	35	23	28	24	22	25	19	26	17	27	18
28	23	37	24	33	25	30	26	30	27	31	28	36
29	24	41	25	40	26	40	27	43	28	48	29	57
30	25	47	26	49	27	52	28	59	30	7	31	19
31	26	55	28	0	29	7	30	17	31	29	32	45
32	28	5	29	13	30	54	31	31	32	54	34	14
33	29	18	30	29	31	44	33	1	34	22	35	47
34	30	32	31	48	33	6	34	27	35	54	37	24
35	31	51	33	10	34	33	35	59	37	30	39	5
36	33	12	34	35	36	2	37	34	39	10	40	51

TABLE OF DIFFERENCE OF THE ASCENSIONS IN AN OBLIQUE SPHERE

Declin ation	Elevation of the Pole											
	43°		44°		45°		46°		47°		48°	
	Times	Min.	Times	Min.	Times	Min.	Times	Min.	Times	Min.	Times	Min.
1	0	56	0	58	1	0	1	2	1	4	1	7
2	1	52	1	56	2	0	2	4	2	9	2	13
3	2	48	2	54	3	0	3	7	3	13	3	20
4	3	44	3	52	4	1	4	9	4	18	4	27
5	4	41	4	51	5	1	5	12	5	23	5	35
6	5	37	5	50	6	2	6	15	6	28	6	42
7	6	34	6	49	7	3	7	18	7	34	7	50
8	7	32	7	48	8	5	8	22	8	40	8	59
9	8	30	8	48	9	7	9	26	9	47	10	8
10	9	28	9	48	10	9	10	31	10	54	11	18
11	10	27	10	49	11	13	11	37	12	2	12	28
12	11	26	11	51	12	16	12	43	13	11	13	39
13	12	26	12	53	13	21	13	50	14	20	14	51
14	13	27	13	56	14	26	14	58	15	30	16	5
15	14	28	15	0	15	32	16	7	16	42	17	19
16	15	31	16	5	16	40	17	16	17	54	18	34
17	16	34	17	10	17	48	18	27	19	8	19	51
18	17	38	18	17	18	58	19	40	20	23	21	9
19	18	44	19	25	20	9	20	53	21	40	22	29
20	19	50	20	35	21	21	22	8	22	58	23	51
21	20	59	21	46	22	34	23	25	24	18	25	14
22	22	8	22	58	23	50	24	44	25	40	26	40
23	23	19	24	12	25	7	26	5	27	5	28	8
24	24	32	25	28	26	26	27	27	28	31	29	38
25	25	47	26	46	27	48	28	52	30	0	31	12
26	27	3	28	6	29	11	30	20	31	32	32	48
27	28	22	29	29	30	38	31	51	33	7	34	28
28	29	44	30	54	32	7	33	25	34	46	36	12
29	31	8	32	22	33	40	35	2	36	28	38	0
30	32	35	33	53	35	16	36	43	38	15	39	53
31	34	5	35	28	36	56	38	29	40	7	41	52
32	35	38	37	7	38	40	40	19	42	4	43	57
33	37	16	38	50	40	30	42	15	44	8	46	9
34	38	58	40	39	42	25	44	18	46	20	48	31
35	40	46	42	33	44	27	46	23	48	36	51	3
36	42	39	44	33	46	36	48	47	51	11	53	47

TABLE OF DIFFERENCE OF THE ASCENSIONS IN AN OBLIQUE SPHERE

Declination	Elevation of the Pole										
	49°		50°		51°		52°		53°		54°	
	Times	Min.	Times	Min.	Times	Min.	Times	Min.	Times	Min.	Times	Min.
1	1	9	1	12	1	14	1	17	1	20	1	23
2	2	18	2	23	2	28	2	34	2	39	2	45
3	3	27	3	35	3	43	3	51	3	59	4	8
4	4	37	4	47	4	57	5	8	5	19	5	31
5	5	47	5	50	6	12	6	26	6	40	6	55
6	6	57	7	12	7	27	7	44	8	1	8	19
7	8	7	8	25	8	43	9	2	9	23	9	44
8	9	18	9	38	10	0	10	22	10	45	11	9
9	10	30	10	53	11	17	11	42	12	8	12	35
10	11	42	12	8	12	35	13	3	13	32	14	3
11	12	55	13	24	13	53	14	24	14	57	15	31
12	14	9	14	40	15	13	15	47	16	23	17	0
13	15	24	15	58	16	34	17	11	17	50	18	32
14	16	40	17	17	17	56	18	37	19	19	20	4
15	17	57	18	39	19	19	20	4	20	50	21	38
16	19	16	19	59	20	44	21	32	22	22	23	15
17	20	36	21	22	22	11	23	2	23	56	24	53
18	21	57	22	47	23	39	24	34	25	33	26	34
19	23	20	24	14	25	10	26	9	27	11	28	17
20	24	45	25	42	26	43	27	46	28	53	30	4
21	26	12	27	14	28	18	29	26	30	37	31	54
22	27	42	28	47	29	56	31	8	32	25	33	47
23	29	14	30	23	31	37	32	54	34	17	35	45
24	31	4	32	3	33	21	34	44	36	13	37	48
25	32	26	33	46	35	10	36	39	38	14	39	59
26	34	8	35	32	37	2	38	38	40	20	42	10
27	35	53	37	23	39	0	40	42	42	33	44	32
28	37	43	39	19	41	2	42	53	44	53	47	2
29	39	37	41	21	43	12	45	12	47	21	49	44
30	41	37	43	29	45	29	47	39	50	1	52	37
31	43	44	45	44	47	54	50	16	52	53	55	48
32	45	57	48	8	50	30	53	7	56	1	59	19
33	48	19	50	44	53	20	56	13	59	28	63	21
34	50	54	53	30	56	20	59	42	63	31	68	11
35	53	40	56	34	59	58	63	40	68	18	74	32
36	56	42	59	59	63	47	68	26	74	36	90	0

TABLE OF DIFFERENCE OF THE ASCENSIONS IN AN OBLIQUE SPHERE

Declin ation	Elevation of the Pole											
	55°		56°		57°		58°		59°		60°	
	Times	Min.	Times	Min.	Times	Min.	Times	Min.	Times	Min.	Times	Min.
1	1	26	1	29	1	32	1	36	1	40	1	44
2	2	52	2	58	3	5	3	12	3	20	3	28
3	4	17	4	27	4	38	4	49	5	0	5	12
4	5	44	5	57	6	11	6	25	6	41	6	57
5	7	11	7	27	7	44	8	3	8	22	8	43
6	8	38	8	58	9	19	9	41	10	4	10	29
7	10	6	10	29	10	54	11	20	11	47	12	17
8	11	35	12	1	12	30	13	0	13	32	14	5
9	13	4	13	35	14	7	14	41	15	17	15	55
10	14	35	15	9	15	45	16	23	17	4	17	47
11	16	7	16	45	17	25	18	8	18	53	19	41
12	17	40	18	22	19	6	19	53	20	43	21	36
13	19	15	20	1	20	50	21	41	22	36	23	34
14	20	52	21	42	22	35	23	31	24	31	25	35
15	22	30	23	24	24	22	25	23	26	29	27	39
16	24	10	25	9	26	12	27	19	28	30	29	47
17	25	53	26	57	28	5	29	18	30	35	31	59
18	27	39	28	48	30	1	31	20	32	44	34	19
19	29	27	30	41	32	1	33	26	34	58	36	37
20	31	19	32	39	34	5	35	37	37	17	39	5
21	33	15	34	41	36	14	37	54	39	42	41	40
22	35	14	36	48	38	28	40	17	42	15	44	25
23	37	19	39	0	40	49	42	47	44	57	47	20
24	39	29	41	18	43	17	45	26	47	49	50	27
25	41	45	43	44	45	54	48	16	50	54	53	52
26	44	9	46	18	48	41	51	19	54	16	57	39
27	46	41	49	4	51	41	54	38	58	0	61	57
28	49	24	52	1	54	58	58	19	62	14	67	4
29	52	20	55	16	58	36	62	31	67	18	73	46
30	55	32	58	52	62	45	67	31	73	55	90	0
31	59	6	62	58	67	42	74	4	90	0		
32	63	10	67	53	74	12	90	0				
33	68	1	74	19	90	0			*The vacant spaces go to*			
34	74	33	90	0					*stars which neither rise*			
35	90	0							*nor set*			
36												

8. On the Hours and Parts of the Day and Night

[38^b] Accordingly it is clear from this that if from the table we take the difference of days which correspond to the declination of the sun and is found under the given elevation of the pole and add it to a quadrant of a circle in the case of a northern declination and subtract it in the case of a southern declination, and then double the result, we shall have the length of that day and the span of night, which is the remainder of the circle.

Any of these segments divided by 15 "times" will show how many equal hours there are (in that day). But by taking a twelfth part of the segment we shall have the duration of one seasonal hour. Now the hours get their name from their day, whereof each hour is always the twelfth part. Hence the hours are found to have been called summer-solstitial, equinoctial, and winter-solstitial by the ancients.

But there were not any others in use at first except the twelve hours from sunrise to sunset; and they divided the night into four vigils or watches. This set-up of the hours lasted a long time by the tacit consent of mankind. And for its sake were water-clocks invented: by the addition and subtraction of dripping water people adjusted the hours to the different lengths of days, so as not to have distinctions in time obscured by a cloud. But afterwards when equal hours common to day and night came into general use, as making it easier to tell the time, then the seasonal hours became obsolete, so that if you asked any ordinary person whether it was the first, third or sixth, ninth, or eleventh hour of the day, he would not have any answer to make or would make one which had nothing to do with the matter. Furthermore, at present some measure the number of equal hours from noon, some from sunset, some from midnight, and others from sunrise, according as it is instituted by the state.

9. On the Oblique Ascension of the Parts of the Ecliptic and How the Degree Which Is in the Middle of the Heavens Is Determined with Respect to the Degree Which Is Rising

[39^a] Now that the lengths and differences of days and nights have been expounded, there follows in proper order an exposition of oblique ascensions, that is to say, together with what "times" (of the equator) the dodekatemoria, *i.e.*, the twelve parts of the ecliptic, or some other arcs of it, cross the horizon. For the differences between right and oblique ascensions are the same as the differences between the equinox and a different day, as we set forth. Furthermore, the ancients borrowed the names of animals for twelve constellations of unmoving stars, and, beginning at the spring equinox, called them Aries, Taurus, Gemini, Cancer, and so on in order.

Therefore for the sake of greater clearness let the meridian circle *ABCD* be repeated; and the equatorial semicircle *AEC* and the horizon *BED*, which cut one another at point *E*. Now let point *H* be taken as the equinox. Let the ecliptic *FHI* pass through this point and cut the horizon at *L*; and through this intersection let *KLM* the quadrant of a great circle fall from *K* the pole of the equator. Thus it is perfectly clear that arc *HL* of the ecliptic and arc *HE* of the equator cross the horizon together, but that in the right sphere (arc *HL*) was rising together with arc *HEM*. Arc *EM* is the difference between these ascensions; and we have already

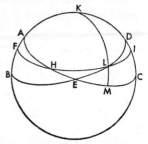

shown that it is half the difference between the equinox and the different day. But in a northern declination what was there added (to the quadrant of a circle) is here subtracted (from the right ascension); but in a southern declination it is added to the right ascension, so that the ascension may become oblique. And hence the extent that the whole sign or some other arc of the ecliptic has emerged may become manifest by means of the numbered ascensions from beginning to end.

From this it follows that when some degree of the ecliptic is given, the rising of which has been measured from the equinox, the degree which is in the middle of the heavens is also given. For when the declination of a degree rising at *L* has been given as corresponding to arc *HL* the distance from the equinox, and arc *HEM* is the right ascension, and the whole *AHEM* is the arc of half a day: then the remainder *AH* is given. And *AH* is the right ascension of arc *FH*, which is also given by the table, or because angle *AHF* the angle of section is given together with side *AH*, and angle *FAH* is right. And so *FHL* the whole arc between the degree rising and the degree in the middle of the heavens is given.

Conversely, if the degree which is in the middle of the heavens, namely arc *FH*, is given first, we shall also know the sign which [39b] is rising. For arc *AF* the declination will be known; and by means of the angle of obliquity of the sphere, arc *AFB* and arc *FB* the remainder will become known. Now in triangle *BFL* angle *BFL* and side *FB* are given by the above, and angle *FBL* is right. Therefore side *FHL*, which was sought, is given; or by a different method, as below.

10. ON THE ANGLE OF SECTION OF THE ECLIPTIC WITH THE HORIZON

Moreover, as the ecliptic is oblique to the axis of the sphere, it makes various angles with the horizon. For we have already said in the case of the differences of the shadows that two opposite degrees of the ecliptic pass through the axis of the horizon of those

who live between the tropics. But I think it will be sufficient for our purpose if we demonstrate the angles which we the heteroscian inhabitants find. By means of these angles the universal ratio of them may easily be understood. Accordingly I think it is clear enough that in the oblique sphere when the equinox or the beginning of Aries is rising, the more the greatest southward declination increases—and this declination is measured from the beginning of Capricorn which is in the middle of the heavens at this time—the more the ecliptic is inclined and verges towards the horizon; and conversely, when the ecliptic has a greater elevation (above the horizon), it makes a greater eastern angle, when the beginning of Libra is emerging and the beginning of Cancer is in the middle of the heavens; since these three circles, the equator, the ecliptic, and the horizon, coincide in one common section at the poles of the meridian circle, and the arcs of the meridian circle intercepted by them show how great the angle of rising should be judged to be.

But in order that the way of taking the measurements of the other parts of the ecliptic may be clear, again let *ABC* be the meridian circle, let *BED* be the semicircle of the horizon, let *AEC* be the semi-circle of the ecliptic, and let any degree of the ecliptic be rising at *E*.

Our problem is to find how great angle *AEB* is, according as four right angles are equal to 360°. Therefore since *E* is given as the rising degree, there are given by the foregoing the degree which is in the middle of the heavens and the arc *AE*.

And since

angle ABE = 90°;

ch. 2 AE : ch. 2 AB = dmt. sph. : ch. 2 AEB.

[40^a] Therefore too

angle AEB is given.

But if the degree given is not rising but the degree in the middle of the heavens—and let it be *A*—nevertheless angle *AEB* will be the measure of the eastern angle or angle of rising. For with *E* as pole, let *FGH* the quadrant of a great circle be described, and let the quadrants *EAG* and *EBH* be completed.

Therefore since

meridian altitude AB is given,

and

$$AF = 90° — AB,$$

and by the foregoing

angle FAG is given,

and

89

$$\text{angle } FGA = 90°;$$

therefore

$$\text{arc } FG \text{ is given,}$$

and

$$90°—FG = GH,$$

which measures the sought angle of rising. Similarly, it is also made evident here how for the degree which is in the middle of the heavens the degree which is rising is given, because

$$\text{ch. } 2 \text{ } GH : 2 \text{ } AB = \text{dmt. sph.} : \text{ch. } 2 \text{ } AE,$$

as in spherical triangles.

We are subjoining three sets of tables of these things. The first will be the table of ascensions in the right sphere, beginning with Aries and increasing by sixtieth parts of the ecliptic. The second will be that of ascensions in the oblique sphere, proceeding by steps of 6° in the ecliptic, from the parallel for which there is a polar elevation of 39° to the parallel which has a polar elevation of 57°—increasing the elevation by 3°s each time. The remaining table contains the angles made with the horizon and proceeds through the ecliptic by steps of 6° beneath the same seven segments. These tables have been set up in accordance with the least obliquity of the ecliptic, namely 23°28', which is approximately right for our age.

TABLE OF THE ASCENSIONS OF THE SIGNS IN THE REVOLUTION
OF THE RIGHT SPHERE

Ecliptic Signs	Deg.	Ascensions Times	Min.	One Degree Times	Min.	Ecliptic Signs	Deg.	Ascensions Times	Min.	One Degree Times	Min.
Aries ♈	6	5	30	0	55	Libra ♎	6	185	30	0	55
	12	11	0	0	55		12	191	0	0	55
	18	16	34	0	56		18	196	34	0	56
	24	22	10	0	56		24	202	10	0	56
	30	27	54	0	57		30	207	54	0	57
Taurus ♉	6	33	43	0	58	Scorpio ♏	6	213	43	0	58
	12	39	35	0	59		12	219	35	0	59
	18	45	32	1	0		18	225	32	1	0
	24	51	37	1	1		24	231	37	1	1
	30	57	48	1	2		30	237	48	1	2
Gemini ♊	6	64	6	1	3	Sagittarius ♐	6	244	6	1	3
	12	70	29	1	4		12	250	29	1	4
	18	76	57	1	5		18	256	57	1	5
	24	83	27	1	5		24	263	27	1	5
	30	90	0	1	5		30	270	0	1	5
Cancer ♋	6	96	33	1	5	Capricornus ♑	6	276	33	1	5
	12	103	3	1	5		12	283	3	1	5
	18	109	31	1	5		18	289	31	1	5
	24	115	54	1	4		24	295	54	1	4
	30	122	12	1	3		30	302	12	1	3
Leo ♌	6	128	23	1	2	Aquarius ♒	6	308	23	1	2
	12	134	28	1	1		12	314	28	1	1
	18	140	25	1	0		18	320	25	1	0
	24	146	17	0	59		24	326	17	0	59
	30	152	6	0	58		30	332	6	0	58
Virgo ♍	6	157	50	0	57	Pisces ♓	6	337	50	0	57
	12	163	26	0	56		12	343	26	0	56
	18	169	0	0	56		18	349	0	0	56
	24	174	30	0	55		24	354	30	0	55
	30	180	0	0	55		30	360	0	0	55

TABLE OF THE ASCENSIONS IN THE OBLIQUE SPHERE

Ecliptic Signs	39° Ascension Times	Min.	42° Ascension Times	Min.	45° Ascension Times	Min.	48° Ascension Times	Min.	51° Ascension Times	Min.	54° Ascension Times	Min.	57° Ascension Times	Min.
♈ 6	3	34	3	20	3	6	2	50	2	32	2	12	1	49
12	7	10	6	44	6	15	5	44	5	8	4	27	3	40
18	10	50	10	10	9	27	8	39	7	47	6	44	5	34
24	14	32	13	39	12	43	11	40	10	28	9	7	7	32
30	18	26	17	21	16	11	14	51	13	26	11	40	9	40
♉ 6	22	30	21	12	19	46	18	14	16	25	14	22	11	57
12	26	39	25	10	23	32	21	42	19	38	17	13	14	23
18	31	0	29	20	27	29	25	24	23	2	20	17	17	2
24	35	38	33	47	31	43	29	25	26	47	23	42	20	2
30	40	30	38	30	36	15	33	41	30	49	27	26	23	22
♊ 6	45	39	43	31	41	7	38	23	35	15	31	34	27	7
12	51	8	48	52	46	20	43	27	40	8	36	13	31	26
18	56	56	54	35	51	56	48	56	45	28	41	22	36	20
24	63	0	60	36	57	54	54	49	51	15	47	1	41	49
30	69	25	66	59	64	16	61	10	57	34	53	28	48	2
♋ 6	76	6	73	42	71	0	67	55	64	21	60	7	54	55
12	83	2	80	41	78	2	75	2	71	34	67	28	62	26
18	90	10	87	54	85	22	82	29	79	10	75	15	70	28
24	97	27	95	19	92	55	90	11	87	3	83	22	78	55
30	104	54	102	54	100	39	98	5	95	13	91	50	87	46
♌ 6	112	24	110	33	108	30	106	11	103	33	100	28	96	48
12	119	56	118	16	116	25	114	20	111	58	109	13	105	58
18	127	29	126	0	124	23	122	32	120	28	118	3	115	13
24	135	4	133	46	132	21	130	48	128	59	126	56	124	31
30	142	38	141	33	140	23	139	3	137	38	135	52	133	52
♍ 6	150	11	149	19	148	23	147	20	146	8	144	47	143	12
12	157	41	157	1	156	19	155	29	154	38	153	36	153	24
18	165	7	164	40	164	12	163	41	163	5	162	24	162	47
24	172	34	172	21	172	6	171	51	171	33	171	12	170	49
30	180	0	180	0	180	0	180	0	180	0	180	0	180	0

TABLE OF THE ASCENSIONS IN THE OBLIQUE SPHERE, II

Ecliptic Signs	39° Ascension Times	Min.	42° Ascension Times	Min.	45° Ascension Times	Min.	48° Ascension Times	Min.	51° Ascension Times	Min.	54° Ascension Times	Min.	57° Ascension Times	Min.
♎ 6	187	26	187	39	187	54	188	9	188	27	188	48	189	11
12	194	53	195	19	195	48	196	19	196	55	197	36	198	23
18	202	21	203	0	203	41	204	30	205	24	206	25	207	36
24	209	49	210	41	211	37	212	40	213	52	215	13	216	48
30	217	22	218	27	219	37	220	57	222	22	224	8	226	8
♏ 6	224	56	226	14	227	38	229	12	231	1	233	4	235	29
12	232	56	234	0	235	37	237	28	239	32	241	57	244	47
18	240	31	241	44	243	35	245	40	248	2	250	47	254	2
24	247	36	249	27	251	30	253	49	256	27	259	32	263	12
30	255	36	257	6	259	21	261	52	264	47	268	10	272	14
♐ 6	262	8	264	41	267	5	269	49	272	57	276	38	281	5
12	269	50	272	6	274	38	277	31	280	50	284	45	289	32
18	276	58	279	19	281	58	248	58	288	26	292	32	297	34
24	283	54	286	18	289	0	292	5	295	39	299	53	305	5
30	290	75	293	1	295	45	298	50	302	26	306	42	311	58
♑ 6	297	0	299	24	302	6	305	11	308	45	312	59	318	11
12	303	4	305	25	308	4	311	4	314	32	318	38	323	40
18	308	52	311	8	313	40	316	33	319	52	323	47	328	34
24	314	21	316	29	318	53	321	37	324	45	328	26	332	53
30	319	30	321	30	323	45	326	19	329	11	332	34	336	38
♒ 6	324	21	326	13	328	16	330	35	333	13	336	18	339	58
12	330	0	330	40	332	31	334	36	336	58	339	43	342	58
18	333	21	334	50	336	27	338	18	340	22	342	47	345	37
24	337	30	338	48	340	3	341	46	343	35	345	38	348	3
30	341	34	342	39	343	49	345	9	346	34	348	20	350	20
♓ 6	345	29	346	21	347	17	348	20	349	32	350	53	352	28
12	349	11	349	51	350	33	351	21	352	14	353	16	354	26
18	352	50	353	16	353	45	354	16	354	52	355	33	356	20
24	356	26	356	40	356	23	357	10	357	53	357	48	358	11
30	360	0	360	0	360	0	360	0	360	0	360	0	360	0

TABLE OF THE ANGLES MADE BY THE ECLIPTIC WITH THE HORIZON

Ecliptic Sign		39° Angles Deg. Min.		42° Angles Deg. Min.		45° Angles Deg. Min.		48° Angles Deg. Min.		51° Angles Deg. Min.		54° Angles Deg. Min.		57° Angles Deg. Min.		Ecliptic Sign	
♈	0	27	32	24	32	21	32	18	32	15	32	12	32	9	32	30	
	6	27	37	24	36	21	36	18	36	15	35	12	35	9	35	24	
	12	27	49	24	39	21	48	18	47	15	45	12	43	9	41	18	
	18	28	13	25	9	22	6	19	3	15	59	12	56	9	53	12	
	24	28	45	25	40	22	34	19	29	16	23	13	18	10	13	6	♓
	30	29	27	26	15	23	11	20	5	16	56	13	45	10	31	30	
♉	6	30	19	27	9	23	59	20	48	17	34	14	20	11	2	24	
	12	31	21	28	9	24	56	20	41	18	23	15	3	11	40	18	
	18	32	35	29	20	26	3	22	43	19	21	15	56	12	26	12	
	24	34	5	30	43	27	23	24	2	20	41	16	59	13	20	6	♒
	30	35	40	32	17	28	52	25	26	21	52	18	14	14	26	30	
♊	6	37	29	34	1	30	37	27	5	23	11	19	42	15	48	24	
	12	39	32	36	4	32	32	28	56	25	15	21	25	17	23	18	
	18	41	44	38	14	34	41	31	3	27	18	23	25	19	16	12	
	24	44	8	40	32	37	2	33	22	29	35	25	37	21	26	6	♑
	30	46	41	43	11	39	33	35	53	32	5	28	6	23	52	30	
♋	6	49	18	45	51	42	15	38	35	34	44	30	50	26	36	24	
	12	52	3	48	34	45	0	41	8	37	55	33	43	29	34	18	
	18	54	44	51	20	47	48	44	13	40	31	36	40	32	39	12	
	24	57	30	54	5	50	38	47	6	43	33	39	43	35	50	6	♐
	30	60	4	56	42	53	22	49	54	46	21	42	43	38	56	30	
♌	6	62	40	59	27	56	0	52	34	49	9	45	37	41	57	24	
	12	64	59	61	44	58	26	55	7	51	46	48	19	44	48	18	
	18	67	7	63	56	60	20	57	26	54	6	50	47	47	24	12	
	24	68	59	65	52	62	42	59	30	56	17	53	7	49	47	6	♏
	30	70	38	67	27	64	18	61	17	58	9	54	50	52	38	30	
♍	6	72	0	68	53	65	51	62	46	59	37	56	27	53	16	24	
	12	73	4	70	2	66	59	63	56	60	53	57	50	54	46	18	
	18	73	51	70	50	67	49	64	48	61	46	58	45	55	44	12	
	24	74	19	71	20	68	20	65	19	62	18	59	17	56	16	6	♎
	30	74	28	71	28	68	28	65	28	62	28	59	28	56	28	0	

11. ON THE USE OF THESE TABLES

[42^b] Now the use of these tables is clear from the demonstrations, since if we take the right ascension corresponding to the known degree of the sun and if for every equal hour measured from noon we add 15 "times" to it—not counting the 360° of the whole circle, if there is more than that—the sum of the right ascensions will show the degree of the ecliptic in the middle of the heavens at the proposed hour.

Similarly if you do the same thing in the case of the oblique ascension of your region, you will have the rising degree of the ecliptic for the hour measured from sunrise.

Moreover, in the case of certain stars which are outside the ecliptic but of which the right ascension has been established—as we taught above—by their right ascension from the beginning of Aries the degrees of the ecliptic which are in the middle of the heavens together with them are given according to the table; and by their oblique ascension the degree of the ecliptic which arises with them, according as the ascensions and parts of the ecliptic are placed in corresponding regions of the tables. It is possible to operate with the setting similarly but by means of the position opposite.

Moreover, if to the right ascension in the middle of the heavens a quadrant of a circle is added, the sum is the oblique ascension of the rising degree. Wherefore the rising degree is given by means of the degree in the middle of the heavens, and vice versa.

There follows the table of the angles of the ecliptic and the horizon, which are measured at the rising degree of the ecliptic. Hence it is understood how great the elevation of the 90th degree of the ecliptic is above the horizon; and it is particularly necessary to know that in the case of solar eclipses.

12. ON THE ANGLES AND ARCS OF THE CIRCLES WHICH PASS THROUGH THE POLES OF THE HORIZON AND INTERSECT THE SAME CIRCLE OF THE ECLIPTIC

In what follows we shall expound the ratio of the angles and arcs made by the intersection of the ecliptic with the circles through the vertex of the horizon, in the cases wherein the intersections have some altitude above the horizon. But we spoke above concerning the meridian altitude of the sun or of any degree of the ecliptic which is in the middle of the heavens, and concerning the angle of section with the meridian, since [43^a] the meridian circle is also one of those circles which pass through the vertex of the horizon. Moreover we have already talked about the angle of the rising sign, the complementary angle to which is the angle which is comprehended by a great circle passing through the vertex of the horizon and by the rising ecliptic. Therefore there remain to be considered the mean sections, that is, the mean sections of the meridian circle with the semicircles of the ecliptic and the horizon.

Let the above figure be repeated. Let *G* be taken as any point on the ecliptic between midday and the point of rising or setting. Through *G* from *F* the pole of the horizon let fall *FGH* a quadrant of a circle.

"Hour" *AGE* is given

as the whole arc of the ecliptic between the meridian and the horizon, and by hypothesis

AG is given.

Similarly, because

meridian altitude *AB* is given,

and

meridian angle *FAG* is given;

therefore

AF is given.

And by what has been shown concerning spherical triangles,

arc *FG* is given.

And hence

altitude of *G* is given,

because

$$90° - FG = GH.$$

And

meridian angle *FAG* is given.

And those are what we were looking for.

En route, we have taken from Ptolemy these truths about the angles and intersections of the ecliptic, and have referred ourselves to the geometry of spherical triangles. If anyone wishes to pursue this study at length, he can find by himself more utilities than we have given examples of.

13. ON THE RISING AND SETTING OF THE STARS

The rising and setting of the stars also seems to depend upon the daily revolution, not only the simple risings and settings of which we have just spoken but also those which occur in the morning or evening; because although their occurrence is affected by the course of the annual revolution, it will be better to speak of them here.

The ancient mathematicians distinguish the true risings and settings from the apparent. The morning rising of the star is true when the star rises at the same time as the sun; and the morning setting is true, when the star sets at sunrise; for morning is said to occur at the midpoint of this time. But the evening rising is true when the star rises at sunset; and the evening setting is true when the star sets at the same time as the

sun; for evening is said to occur at the midpoint of this time, namely the time [43ᵇ] between the time which is beginning and the time which ceases with night.

But the morning rising of a star is apparent when it rises first in the twilight before sunrise and begins to be apparent; and the morning setting is apparent when the star is seen to set very early before the sun rises. The evening rising is apparent when the star is seen to rise first in the evening; and the evening setting is apparent, when the star ceases to be apparent some time after sunset, and the star is occulted by the approach of the sun, until they come forth in their previous order at the morning rising.

This is true of the fixed stars and of the planets Saturn, Jupiter, and Mars. But Venus and Mercury rise and set in a different fashion. For they are not occulted by the approach of the sun, as the higher planets are; and they are not uncovered again by its departure. But, coming in front, they mingle with the radiance of the sun and free themselves. But when the higher planets have an evening rising and a morning setting, they are not obscured at any time, so as not to traverse the night with their illumination. But the lower planets remain hidden indifferently from sunset to sunrise, and cannot be seen anywhere. There is still another difference, namely that in higher planets the true morning risings and settings are prior to the apparent ones; and the evening risings and settings are posterior to the apparent, according as in the morning they precede the rising of the sun and in the evening follow its setting. But in the lower planets the apparent morning and evening risings are posterior to the true, while the apparent settings are prior to the true.

Now it can be understood from the above, where we expounded the oblique ascension of any star having a known position, how (the risings and settings) may be discerned, and together with what degree of the ecliptic the star rises or sets and at what position, or degree opposite—if the sun has become apparent by that time—the star has its true morning or evening rising or setting. The apparent risings and settings differ from the true according to the clarity and magnitude of the star, so that the stars which give a more powerful light are less dimmed by the rays of the sun than those which are less luminous. And the boundaries of occultation and apparition are determined in the lower hemisphere, between the horizon and the sun, on the arcs of circles which pass through the poles of the horizon. And the limits are 12° for the primary stars, 11° for Saturn, 10° for Jupiter, $11^1/_2$° for Mars, 5° for Venus, and 10° for Mercury. But in this whole period during which what is left of daylight yields to night—this period embraces twilight; or dusk—there are 18° of the aforesaid circle. When the sun has traversed these degrees, the smaller stars too begin to be apparent. By this distance the mathematicians determine [44ᵃ] a parallel below the horizon in the lower hemisphere, and they say that when the sun has reached this parallel, day has ended and night has begun. Therefore when we have learned with what degree of the

ecliptic the star rises or sets and what the angle of section of the ecliptic with the horizon at that point is, and if then too we find as many degrees of the ecliptic between the rising degree and the sun as are sufficient to give the sun an altitude below the horizon in accord with the prescribed limits of the star in question; we shall pronounce that the first emergence or occultation of the star has taken place.

But what we have expounded, in the foregoing explanation, in the case of the altitude of the sun above the Earth agrees in all respects with its descent below the Earth. For there is no difference in the corresponding positions; and consequently those stars which are setting in the visible hemisphere are rising in the hidden hemisphere; and everything is the converse, and is easy to understand. What has been said concerning the rising and setting of the stars and the daily revolution of the terrestrial globe shall be sufficient.

14. ON INVESTIGATING THE POSITIONS OF THE STARS AND THE CATALOGUE OF THE FIXED STARS

After the daily revolution of the terrestrial globe and its consequences have been expounded by us, the demonstrations relating to the annual circuit ought to follow now. But since some of the ancient mathematicians thought the phenomena of the fixed stars ought to come first as being the first beginnings of this art, accordingly we decided to act in accordance with this opinion, as among our principles and hypotheses we had assumed that the sphere of the fixed stars, to which the wanderings of all the planets are equally referred, is wholly immobile. But no one should be surprised at our following this order, although Ptolemy in his *Almagest* held that an explanation of the fixed stars could not be given, unless knowledge of the positions of the sun and moon had preceded it, and accordingly he judged that whatever had to do with the fixed stars should be put off till then. We think that this opinion must be opposed. But if you understand it of the numbers with which the apparent motion of the sun and moon is computed perhaps the opinion will stand. For Menelaus the geometer discovered the positions of many stars by means of the numbers relating to their conjunctions with the moon. [44b] But we shall do a much better job if we determine a star by the aid of instruments after examining carefully the positions of the sun and moon, as we will show how in a little while. We are even admonished by the wasted attempt of those who thought that the magnitude of the solar year could be defined simply by the equinoxes or solstices without the fixed stars. We shall never agree with them on that, so much so that there will nowhere be greater discord. Ptolemy called our attention to this: when he had evaluated the solar year in his time not without suspicion of an error which might emerge with the passage of time, he admonished posterity to examine the further certainty of the thing later on. Therefore it seemed to us to be

worth the trouble to show how by means of artificial instruments the positions of the sun and moon may be determined, that is, how far distant they are from the spring equinox or some other cardinal points of the world. The knowledge of these positions will afford us some facilities for investigating the other stars, and thus we shall be able to set forth before your eyes the sphere of the fixed stars and an image of it embroidered with constellations.

Now we have set forth above with what instruments the distance of the tropics, the obliquity of the ecliptic, and the inclination of the sphere, or the altitude of the pole of the equator, may be determined. In the same way we can determine any other altitude of the sun at midday. This altitude will exhibit to us through its difference from the inclination of the sphere how great the declination of the sun from the equator is. Then by means of this declination the position of the sun at midday will become clear as measured from the solstice or the equinox. Now the sun seems to traverse approximately 1° during the space of 24 hours; $2^{1}/_{2}'$ come as the hourly allotment. Hence its position at any other definite hour will easily be determined.

But for observing the positions of the moon and stars another instrument is constructed, which Ptolemy calls the astrolabe. For let two circles, or rather four-sided rims of circles, be constructed in such a way that they may have their concave and convex surfaces at right angles to the plane sides. These rims are to be equal and similar in every respect and of a suitable size, in order not to become hard to handle through being too large, though they must be of sufficient amplitude to be divided into degrees and minutes. Their width and thickness [45^a] should be at least one thirtieth of the diameter. Therefore they are to be fitted together and joined at right angles to one another, having their convex sides as it were on the surface of the same sphere, and their concave sides on the surface of another single sphere. Now one of the circles should have the relative position of the ecliptic; and the other, that of the circle which passes through both poles, i.e., the poles of the equator and of the ecliptic. Therefore the circle of the ecliptic is to be divided along its sides into the conventional number of 360°, which are again to be subdivided according to the capacity of the instrument. Moreover, when quadrants on the other circle have been measured from the ecliptic, the poles of the ecliptic should be marked on it; and when a distance proportionate to the obliquity has been measured from those points, the poles of the equator are also to be marked down. When these circles are finished, two other circles should be prepared and constructed around the same poles of the ecliptic: they will move about these poles, one circle inside and one circle outside. They should be of equal thicknesses between their plane surfaces, and the width of their plane surfaces should be equal to that of the others; and they should be so constructed that at all points the concave surface of the larger will touch the convex surface of the ecliptic; and the convex surface

of the smaller, the concave surface of the ecliptic. Nevertheless do not let their revolutions be impeded, but have them able to traverse freely and easily both the ecliptic together with its meridian circle and one another. Therefore we shall make holes in these circles diametrically opposite the poles of the ecliptic, and pass axles through these holes, so that by means of these axles the circles will be bound together and carried along. Moreover, the inside circle should be divided into 360° in such fashion that the single quadrant of 90° will be at the poles. Furthermore, within the concavity of the inside circle a fifth circle should be placed which can be turned in the same plane and which has an apparatus fixed to its plane surfaces which has openings diametrically opposite and reflectors or eyepieces, through which the light of the sun, as in a dioptra, can break through and go out along the diameter of the circle. And certain appliances or pointers for numbers are fitted on to this fifth circle at opposite points for the sake of observing the latitudes on the container circle. Finally, a sixth circle is to be applied which will embrace and support the whole astrolabe, which is hung on to it by means of fastenings at the poles of the equator; this last circle is to be placed upon some sort of column, or stand, and made to rest upon it perpendicular to the plane of the horizon. Moreover, the poles (of the equator) should be adjusted to the inclination of the sphere, so that the outmost circle will have a position similar to that of a natural meridian and will by no means waver from it.

Therefore after the instrument has been prepared in this way, when we wish to determine the position of some star, then in the evening or at the approach of sunset and at a time when the moon too is visible, we shall adjust the outer circle to the degree of the ecliptic, in which [45^b] we have determined—by the methods spoken of—that the sun is at that time. And we shall turn the intersection of the (ecliptic and the outer) circle towards the sun itself, until both of them—I mean the ecliptic and the outer circle which passes through its poles—cast shadows on themselves evenly.[1] Then we shall turn the inner circle towards the moon; and with the eye placed in its plane we shall mark its position on the ecliptic part of the instrument there where we shall view the moon as opposite, or as it were bisected by the same plane. That will be the position of the moon as seen in longitude. For without the moon there is no way of discovering the positions of the stars, as the moon alone among all is a partaker of both day and night. Then after nightfall, when the star whose position we are seeking is visible, we shall adjust the outer circle to the position of the moon; and thus, as we did in the case of the sun, we shall bring the position of the astrolabe into relation with the moon. Then also we shall turn the inner circle towards the star, until the star seems to be in contact with the plane surfaces of the circle and is viewed through the eyepieces which are on the little circle contained (by the inner circle). For in this way we shall

---

[1] *i.e.,* until the shadows intersect as two straight lines at right angles to one another.

have discovered the longitude and latitude of the star. When this is being done, the degree of the ecliptic which is in the middle of the heavens will be before the eyes; and accordingly it will be obvious at what hour the thing itself was done.[1]

For example, in the 2nd year of the Emperor Antoninus Pius, on the 9th day of Pharmuthi, the 8th month by the Egyptian calendar, Ptolemy, who was then at Alexandria and wished to observe at the time of sunset the position of the star which is in the breast of Leo and is called Basiliscus or Regulus, adjusted his astrolabe to the setting sun at 5 equatorial hours after midday. At this time the sun was at $3^1/_{24}°$ of Pisces, and by moving the inner circle he found that the moon was $92^1/_8°$ east of the sun: hence it was seen that the position of the moon was then at $5^1/_6°$ of Gemini. After half an hour—which made six hours since noon—when the star had already begun to be apparent and 4° of Gemini was in the middle of the heavens, he turned the outer circle of the instrument to the already determined position of the moon. Proceeding with the inner circle, he took the distance of the star from the moon as $57^1/_{10}°$ to the east. Accordingly the moon had been found at $92^1/_8°$ from the setting sun, as was said—which placed the moon at $5^1/_6°$ of Gemini; but it was correct for the moon to have moved $^1/_4°$ in the space of half an hour; since the hourly allotment in the movement of the moon is more or less $^1/_2°$; but on account of the then subtractive parallax of the moon it must have been slightly less than $^1/_4°$, [46^a] that is to say, about $^1/_6°$: hence the moon was at $5^1/_3°$ of Gemini. But when we have discussed the parallaxes of

[1]Legend:
1. Circle through poles of ecliptic
2. Ecliptic
3. Outer circle
4. Inner circle
5. Little circle
6. Meridian circle
A and A_1. Poles of equator
BCC_1B_1. Axis of ecliptic
D. Zenith, or pole of horizon

The astrolabe is constructed as an image of the Ptolemaic heavens, or as a "smaller world." Accordingly the astrolabe in operation is an imitation of the revolving heavens on a reduced scale.

The astrolabe is set up with the meridian circle (6) fixed in the meridian line and with the northern and southern poles of the equator (A and A_1) pointing towards the celestial poles above and below the horizon, as the meridian does not change during the course of the daily revolution. The degrees of the celestial ecliptic are marked off on the ecliptic circle (2), with the solstices or equinoxes at the intersections of the ecliptic (2) and the circle through the poles of the ecliptic (1). The outer circle (3) is turned to that point on the ecliptic where the position of the sun is computed to be, and then this intersection of the outer circle and the ecliptic is turned towards the sun itself, until each circle casts its shadow in the form of a straight line intersecting the other shadow at right angles. Now as the revolution of the outer circle around the axis of the equator makes the axis of the ecliptic, the circle through the poles of the ecliptic, the inner circle, and the little circle swing round the axis of the equator, and as the pole of the ecliptic revolves around the pole of the equator during the daily revolution; the turning towards the sun of the intersection of the two circles serves to bring the yearly and daily movement of the sun into proper ratio with one another; and the cruciform shadow is a sign that the wooden ecliptic occupies the relative position in the astrolabe that the celestial ecliptic occupies at this moment of the daily revolution. The inner circle (4) can now be turned towards the moon in order to mark on the ecliptic the lunar longitude, and the little circle (5) can be wheeled around in the plane of the inner circle, in order to mark the lunar latitude on the graduated inner circle.

COPERNICUS' ASTROLABE

MERIDIAN ———————————— LINE

the moon, the difference will not appear to have been so great; and hence it will be evident enough that the position of the moon viewed was more than $5^1/_3°$ but a little less than $5^2/_5°$. The addition of $57^1/_{10}°$ to this locates the position of the star at 2°30' of Leo at a distance of about $32^1/_2$ from the summer solstice of the sun and with a northern latitude of $^1/_6°$. This was the position of Basiliscus; and consequently the way was laid open to the other fixed stars. This observation of Ptolemy's was made in the year of Our Lord 139 by the Roman calendar, on the 24th day of February, in the 1st year of the 229th Olympiad.

That most outstanding of mathematicians took note of what position at that time each of the stars had in relation to the spring equinox, and catalogued the constellations of the celestial animals. Thus he helps us not a little in this our enterprise and relieves us of some difficult enough labour, so that we, who think that the positions of the stars should not be referred to the equinoxes which change with time but that the equinoxes should be referred to the sphere of the fixed stars, can easily draw up a description of the stars from any other unchanging starting-point. We decided to begin this description with the Ram as being the first sign, and with its first star, which is in its head—so that in this way a configuration which is absolute and always the same will be possessed by those stars which shine together as if fixed and clinging perpetually and at the same time to the throne which they have seized. But by the marvellous care and industry of the ancients the stars were distributed into forty-eight constellations with the exception of those which the circle of the always hidden stars removed from the fourth climate, which passes approximately through Rhodes; and in this way the unconstellated stars remained unknown to them. According to the opinion of Theo the Younger in the *Aratean Treatise* the stars were not arranged in the form of images for any other reason except that their great multitude might be divided into parts and that they might be designated separately by certain names in accordance with an ancient enough custom, since even in Hesiod and Homer we read the names of the Pleiades, Hyas, Arcturus, and Orion. Accordingly in the description of the stars according to longitude we shall not employ the "twelve divisions," or dodekatemoria, which are measured from the equinoxes or solstices, but the simple and conventional number of degrees. We shall follow Ptolemy as to the rest with the exception of a few cases, where we have either found some corruption or a different state of affairs. We shall however teach you in the following book how to find out what their distances are from those cardinal points (*i.e.*, the equinoxes).

CATALOGUE OF THE SIGNS AND OF THE STARS
AND FIRST THOSE OF THE NORTHERN REGION

Constellations	Longitude			Latitude		Magnitude
	Deg.	Min.		Deg.	Min.	
URSA MINOR, OR THE LITTLE BEAR, OR CYNOSURA						
The (star) at the tip of the tail	53	30	N	66	0	3
The (star) to the east in the tail	55	50	N	70	0	4
The (star) at the base of the tail	69	20	N	74	0	4
The more southern (star) on the western side of the quadrilateral	83	0	N	75	20	4
The northern (star) on the same side	87	0	N	77	40	4
The more southern of the stars on the eastern side	100	30	N	72	40	2
The more northern on the same side	109	30	N	74	50	2
7 stars: 2 of second magnitude, 1 of third, 4 of fourth						
The most southern unconstellated star near the Cynosure, in a straight line with the eastern side	103	20	N	71	10	4
URSA MAJOR, OR THE GREAT BEAR						
The star in the muzzle	78	40	N	39	50	4
The western star in the two eyes	79	10	N	43	0	5
The star to the east of that	79	40	N	43	0	5
The more western star of the two in the forehead	79	30	N	47	10	5
The star to the east in the forehead	81	0	N	47	0	5
The western star in the right ear	81	30	N	50	30	5
The more western of the two in the neck	85	50	N	43	50	4
The eastern	92	50	N	44	20	4
NORTHERN SIGNS						
The more northern of the two in the breast	94	20	N	44	0	4
The more southern	93	20	N	42	0	4
The star at the knee of the left foreleg	89	0	N	35	0	3
The more northern of the two in the left forefoot	89	50	N	29	0	3
The more southern	88	40	N	28	30	3
At the knee of the right foreleg	89	0	N	36	0	4
The star below the knee	101	10	N	33	30	4
The star on the shoulder	104	0	N	49	0	2
The star on the flanks	105	30	N	44	30	2
The star at the base of the tail	116	30	N	51	0	3
The star in the left hind leg	117	20	N	46	30	2
The more western of the two in the left hind foot	106	0	N	29	38	3
The star to the east of that	107	30	N	28	15	3
[47ª] The star in the hollow of the left leg	115	0	N	35	15	4
The more northern of the two which are in the right hind foot	123	10	N	25	50	3

NORTHERN SIGNS

Constellations	Longitude			Latitude		Magnitude
	Deg.	Min.		Deg.	Min.	
The more southern	123	40	N	25	0	3
The first of the three in the tail after the base	125	30	N	53	30	2
The middle star	131	20	N	55	40	2
The star which is last and at the tip of the tail	143	10	N	54	0	2
27 stars: 6 of second magnitude, 8 of third,						
8 of fourth, and 5 of fifth						
UNCONSTELLATED STARS NEAR THE GREAT BEAR						
The star to the south of the tail	141	10	N	39	45	3
The more obscure star to the west	133	30	N	41	20	5
The star between the forefeet of the Bear						
and the head of the Lion	98	20	N	17	15	4
The star more to the north than that one	96	40	N	19	10	4
The last of the three obscure stars	99	30	N	20	0	0 obscure
The one to the west of that	95	30	N	22	4	5 obscure
The one more to the west	94	30	N	23	1	5 obscure
The star between the forefeet and the Twins	100	20	N	22	1	5 obscure
8 unconstellated stars: 1 of third magnitude,						
2 of fourth, 1 of fifth, 4 obscure						
DRACO, OR THE DRAGON						
The star in the tongue	200	0	N	76	30	4
On the jaws	215	10	N	78	30	4 greater
Above the eye	216	30	N	75	40	3
In the cheek	229	40	N	75	20	4
Above the head	233	30	N	75	30	3
The most northern star in the first curve of						
the neck	258	40	N	82	20	4
The most southern	295	50	N	78	15	4
The star in between	262	10	N	80	20	4
The star to the east of them at the second curve	282	50	N	81	10	4
The more southern star on the western side						
of the quadrilateral	331	20	N	81	40	4
The more northern star on the same side	343	50	N	83	0	4
The more northern star on the eastern side	1	0	N	78	50	4
The more southern on the same side	346	10	N	77	50	4
The more southern star in the triangle at the						
third curve	4	0	N	80	30	4
The more western of the other two in the triangle	15	0	N	81	40	5
The star to the east	19	30	N	80	15	5
<The star to the east> in the triangle to the west	66	20	N	83	30	4
The more southern of the remaining two in the						
same triangle	43	40	N	83	30	4

NORTHERN SIGNS

Constellations	Longitude			Latitude		Magnitude
	Deg.	Min.		Deg.	Min.	
[47b] The star which is more northern than the two above	35	10	N	84	50	4
Of the small stars west of the triangle, the more eastern	200	0	N	87	30	6
The more western	195	0	N	86	50	6
The most southern of the three which are in a straight line towards the east	152	30	N	81	15	5
The one in the middle	152	50	N	83	0	5
The most northern	151	0	N	84	50	3
The more northern of the two which follow towards the west	153	20	N	78	0	3
The more southern	156	30	N	74	40	4 greater
The star to the west of them, in the coil of the tail	156	0	N	70	0	3
The more western of the two rather distant from that one	120	40	N	64	40	4
The star to the east of it	124	30	N	65	30	3
The star to the east in the tail	192	30	N	61	15	3
At the tip of the tail	186	30	N	56	15	3
Therefore 31 stars: 8 of third magnitude, 16 of fourth, 5 of fifth, 2 of sixth						
CEPHEUS						
In the right foot	28	40	N	75	40	4
In the left foot	26	20	N	64	15	4
On the right side beneath the belt	0	40	N	71	10	4
The star which touches the top of the right shoulder	340	0	N	69	0	3
The star which touches the right joint of the elbow	332	40	N	72	0	4
The star to the east which touches the same elbow	333	20	N	74	0	4
The star on the chest	352	0	N	65	30	5
On the right arm	1	0	N	62	30	4 greater
The most southern of the three on the tiara	339	40	N	60	15	5
The one in the middle	340	40	N	61	15	4
The most northern	342	20	N	61	30	5
11 stars: 1 of third magnitude, 7 of fourth, 3 of fifth						
Of the two unconstellated stars, the one to the west of the tiara	337	0	N	64	0	5
The one to the east of the tiara	344	40	N	59	30	4
BOÖTES, OR ARCTURUS						
The more western of the three in the left hand	145	40	N	58	40	5
The middle one of the three, the more southern	147	30	N	58	20	5
The more eastern of the three	149	0	N	60	10	5
The star in the left joint of the elbow	143	0	N	54	40	5

NORTHERN SIGNS

Constellations	Longitude			Latitude		Magnitude
	Deg.	Min.		Deg.	Min.	
On the left shoulder	163	0	N	49	0	3
On the head	170	0	N	53	50	4 greater
On the right shoulder	179	0	N	48	40	4
[48ª] The more southern of the two on the crook	179	0	N	53	15	4
The star more to the north, at the tip of the crook	178	20	N	57	30	4
The more northern of the two under the shoulder and on the spear	181	0	N	46	10	4 greater
The more southern	181	50	N	45	30	5
At the extremity of the right hand	181	35	N	41	20	5
The more western of the two in the palm	180	0	N	41	40	5
The one to the east	180	20	N	42	30	5
At the extremity of the handle of the crook	181	0	N	40	20	5
On the right leg	183	20	N	40	15	3
The more eastern of the two in the belt	169	0	N	41	40	4
The more western	168	20	N	42	10	4 greater
At the right heel	178	40	N	28	0	3
The more northern of the three on the left ham	164	40	N	28	0	3
The middle one of the three	163	50	N	26	30	4
The more southern of them	164	50	N	25	0	4
22 stars: 4 of third magnitude, 9 of fourth, 9 of fifth						
The unconstellated star between the thighs, which they call Arcturus	170	20	N	31	30	1
CORONA BOREALIS, OR THE NORTHERN CROWN						
The brilliant star in the crown	188	0	N	44	30	2 greater
The most western of all	185	0	N	46	10	4 greater
The eastern star towards the north	185	10	N	48	0	5
The eastern star more to the north	193	0	N	50	30	6
The star to the south-east of the brilliant one	191	30	N	44	45	4
The next star to the east	190	30	N	44	50	4
The star farther to the east	194	40	N	46	10	4
The most eastern of all in the crown	195	0	N	49	20	4
8 stars: 1 of second magnitude, 5 of fourth, 1 of fifth, 1 of sixth						
ENGONASI, OR THE KNEELING MAN						
On the head	221	0	N	37	30	3
At the right arm-pit	207	0	N	43	0	3
On the right arm	205	0	N	40	10	3
In the right flank	201	20	N	37	10	4
On the left shoulder	220	20	N	49	30	4 greater
[48ᵇ] In the left flank	231	0	N	42	0	4

NORTHERN SIGNS

Constellations	Longitude			Latitude		Magnitude
	Deg.	Min.		Deg.	Min.	
<The more eastern> of the three in the left palm	238	50	N	52	50	4 greater
The more northern of the remaining two	235	0	N	54	0	4 greater
The more southern	234	50	N	53	0	4
On the right side	207	10	N	56	10	3
On the left side	213	30	N	53	30	4
On the <lower part of the> left buttock	213	20	N	56	10	5
At the beginning of the left leg	214	30	N	58	30	5
The most western of the three in the left ham	217	20	N	59	50	3
The more eastern	218	40	N	60	20	4
The most eastern	219	40	N	61	15	4
At the left knee	237	10	N	61	0	4
On the <upper part of the> left buttock	225	30	N	69	20	4
The most western of the three in the left foot	188	40	N	70	15	6
The middle star	220	10	N	71	15	6
The most eastern of the three	223	0	N	72	0	6
At the beginning of the right leg	207	0	N	60	15	4 greater
The more northern on the right ham	198	50	N	63	0	4
At the right knee	189	0	N	65	30	4 greater
The more southern of the two under the right knee	186	40	N	63	40	4
The more northern	183	30	N	64	15	4
On the right shin	184	30	N	60	0	4
At the extremity of the right foot, the same as the tip of Boötes' crook	178	20	N	57	30	4
Besides that last one, 28 stars: 6 of third magnitude, 17 of fourth, 2 of fifth, 3 of sixth						
The unconstellated star to the south of the right arm	26	0	N	38	10	5
LYRA, OR THE LYRE						
The brilliant star which is called Lyra or Fidicula	250	40	N	62	0	1
The more northern of the two adjacent stars	253	40	N	62	40	4 greater
The more southern	253	40	N	61	0	4 greater
The star which is at the centre of the beginning of the horns	262	0	N	60	0	4
The more northern of the two which are next and to the east	265	20	N	61	20	4
The more southern	265	0	N	60	20	4
The more northern of the two westerly stars on the cross-piece	254	20	N	56	10	3
The more southern	254	10	N	55	0	4 smaller
The more northern of the two easterly stars on the cross-piece	257	30	N	55	20	3
The more southern	258	20	N	54	45	4 smaller

NORTHERN SIGNS

Constellations	Longitude			Latitude		Magnitude
	Deg.	Min.		Deg.	Min.	
10 stars: 1 of first magnitude, 2 of third magnitude, 7 of fourth						
[49ª] CYGNUS, OR THE SWAN						
At the mouth	267	50	N	49	20	3
On the head	272	20	N	50	30	5
In the middle of the neck	279	20	N	54	30	4 greater
In the breast	291	50	N	56	20	3
The brilliant star in the tail	302	30	N	60	0	2
In the elbow of the right wing	282	40	N	64	40	3
The most southern of the three in the flat of the wing	285	50	N	69	40	4
The middle star	284	30	N	71	30	4 greater
The last of the three, and at the tip of the wing	310	0	N	74	0	4 greater
At the elbow of the left wing	294	10	N	49	30	3
In the middle of the left wing	298	10	N	52	10	4 greater
At the tip of the same	300	0	N	74	0	3
In the left foot	303	20	N	55	10	4 greater
At the left knee	307	50	N	57	0	4
The more western of the two in the right foot	294	30	N	64	0	4
The more eastern	296	0	N	64	30	4
The nebulous star at the right knee	305	30	N	63	45	5
17 stars: 1 of second magnitude, 5 of third, 9 of fourth, 2 of fifth						
AND TWO UNCONSTELLATED STARS NEAR THE SWAN						
The more southern of the two under the left wing	306	0	N	49	40	4
The more northern	307	10	N	51	40	4
CASSIOPEIA						
On the head	1	10	N	45	20	4
On the breast	4	10	N	46	45	3 greater
On the girdle	6	20	N	47	50	4
Above the seat, at the hips	10	0	N	49	0	3 greater
At the knees	13	40	N	45	30	3
On the leg	20	20	N	47	45	4
At the extremity of the foot	355	0	N	48	20	4
On the left arm	8	0	N	44	20	4
On the left forearm	7	40	N	45	0	5
On the right forearm	357	40	N	50	0	6
At the foot of the chair	8	20	N	52	40	4
At the middle of the settle	1	10	N	51	40	3 smaller
At the extremity	27	10	N	51	40	6
13 stars: 4 of third magnitude, 6 of fourth, 1 of fifth, 2 of sixth						

NORTHERN SIGNS

Constellations	Longitude			Latitude		Magnitude
	Deg.	Min.		Deg.	Min.	
[49^b] PERSEUS						
The nebulous star at the extremity of the right hand	21	0	N	40	30	nebulous
On the right forearm	24	30	N	37	30	4
On the right shoulder	26	0	N	34	30	4 smaller
On the left shoulder	20	50	N	32	20	4
On the head, or a nebula	24	0	N	34	30	4
On the shoulder-blades	24	50	N	31	10	4
The brilliant star on the right side	28	10	N	30	0	2
The most western of the three on the same side	28	40	N	27	30	4
The middle one	30	20	N	27	40	4
The remaining one of the three	31	0	N	27	30	3
On the left forearm	24	0	N	27	0	4
The brilliant star in the left hand and in the head of Medusa	23	0	N	23	0	2
The easterly star on the head of the same	22	30	N	21	0	4
The more western on the head of the same	21	0	N	21	0	4
The most western	20	10	N	22	15	4
On the right knee	38	10	N	28	15	4
The one to the west of this one at the knee	37	10	N	28	10	4
The more western of the two on the belly	35	40	N	25	10	4
The more eastern	37	20	N	26	15	4
On the right hip	37	30	N	24	30	5
On the right calf	39	40	N	28	45	5
On the left hip	30	10	N	21	40	4 greater
On the left knee	32	0	N	19	50	3
On the left calf	31	40	N	14	45	3 greater
On the left heel	24	30	N	12	0	3 smaller
On the top part of the left foot	29	40	N	11	0	3 greater
26 stars: 2 of second magnitude, 5 of third, 16 of fourth, 2 of fifth, 1 nebulous						
UNCONSTELLATED STARS AROUND PERSEUS						
To the east of the left hand	34	10	N	31	0	5
To the north of the right hand	38	20	N	31	0	5
To the west of Medusa's head	18	0	N	20	40	obscure
3 stars: 2 of fifth magnitude, 1 obscure						
[50^a] AURIGA, OR THE CHARIOTEER						
The more southern of the two on the head	55	50	N	30	0	4
The more northern	55	40	N	30	50	4
The brilliant star on the left shoulder, which is called Capella	48	20	N	22	30	1
On the right shoulder	56	10	N	20	0	2
On the right forearm	54	30	N	15	15	4
On the palm of the right hand	56	10	N	13	30	4 greater
On the left forearm	45	20	N	20	40	4 greater
The star to the west of the Haedi	45	30	N	18	0	4 smaller

NORTHERN SIGNS

Constellations	Longitude			Latitude		Magnitude
	Deg.	Min.		Deg.	Min.	
The star on the palm of the left hand which						
is to the east of the Haedi	46	0	N	18	0	4 greater
On the left calf	53	10	N	10	10	3 small
On the right calf and at the tip of the						
northern horn of Taurus	49	0	N	5	0	3 greater
At the ankle	49	20	N	8	30	5
On the buttocks	49	40	N	12	20	5
The small star on the left foot	24	0	N	10	20	6
14 stars: 1 of first magnitude, 1 of second,						
2 of third, 7 of fourth, 2 of fifth, 1 of sixth						
OPHIUCHUS, OR THE SERPENT-HOLDER						
On the head	228	10	N	36	0	3
The more western of the two on the right						
shoulder	231	20	N	27	15	4 greater
The more eastern	232	20	N	26	45	4
The more western of the two on the left						
shoulder	216	40	N	33	0	4
The more eastern	218	0	N	31	50	4
At the left elbow	211	40	N	34	30	4
The more western of the two in the left hand	208	20	N	17	0	4
The more eastern	209	20	N	12	30	3
At the right elbow	220	0	N	15	0	4
The more western in the right hand	205	40	N	18	40	4 smaller
The more eastern	207	40	N	14	20	4
At the right knee	224	30	N	4	30	3
On the right shin	227	0	N	2	15	3 greater
The most western of the four on the right foot	226	20	S	2	15	4 greater
The more easterly	227	40	S	1	30	4 greater
The next to the east	228	20	S	0	20	4 greater
The most easterly	229	10	S	1	45	5 greater
The star which touches the heel	229	30	S	1	0	5
[50b] At the left knee	215	30	N	11	50	3
The most northern of the three in a straight						
line on the lower part of the left leg	215	0	N	5	20	5 greater
The middle one	214	0	N	3	10	5
The most southern of the three	213	10	N	1	40	5 greater
The star on the left heel	215	40	N	0	40	5
The star touching the hollow of the left foot	214	0	S	0	45	4
24 stars: 5 of third magnitude, 13 of fourth,						
6 of fifth						
UNCONSTELLATED STARS AROUND OPHIUCHUS						
The most northern of the three to the east						
of the right shoulder	235	20	N	28	10	4
The middle one	236	0	N	26	20	4
The most southern of the three	233	40	N	25	0	4
Another one, farther to the east of the three	237	0	N	27	0	4

NORTHERN SIGNS

Constellations	Longitude			Latitude		Magnitude
	Deg.	Min.		Deg.	Min.	
A star separate from the four, to the north	238	0	N	33	0	4
Therefore 5 unconstellated stars: all of fourth magnitude						
SERPENS OPHIUCHI, OR THE SERPENT						
On the quadrilateral, the star in the cheeks	192	10	N	38	0	4
The star touching the nostrils	201	0	N	40	0	4
On the temples	197	40	N	35	0	3
At the beginning of the neck	195	20	N	34	15	3
At the middle of the quadrilateral, and on the jaws	194	40	N	37	15	4
To the north of the head	201	30	N	42	30	4
At the first curve of the neck	195	0	N	29	15	3
The most northern of the three to the east	198	10	N	26	30	4
The middle one	197	40	N	25	20	3
The most southern of the three	199	40	N	24	0	3
The star to the west of the left hand of Ophiuchus	202	0	N	16	30	4
The star to the east of the same hand	211	30	N	16	15	5
The star to the east of the right hip	227	0	N	10	30	4
The more southern of the two to the east of that	230	20	N	8	30	4 greater
The more northern	231	10	N	10	30	4
To the east of the right hand, in the coil of the tail	237	0	N	20	0	4
Farther east in the tail	242	0	N	21	10	4
At the tip of the tail	251	40	N	27	0	4 greater
18 stars: 5 of third magnitude, 12 of fourth, 1 of fifth						
[51ª] SAGITTA, OR THE ARROW						
At the head	273	30	N	39	20	4
The most eastern of the three on the shaft	270	0	N	39	10	6
The middle one	269	10	N	39	50	5
The most western of the three	268	0	N	39	0	5
At the notch	266	40	N	38	45	5
5 stars: 1 of fourth magnitude, 3 of fifth, 1 of sixth						
AQUILA, OR THE EAGLE						
In the middle of the head	270	30	N	26	50	4
On the neck	268	10	N	27	10	3
The brilliant star on the shoulder-blades, which is called Aquila	267	10	N	29	10	2 greater
The star to the north which is very near	268	0	N	30	0	3 smaller
The more western on the left shoulder	266	30	N	31	30	3
The more eastern	269	20	N	31	30	5
The star to the west in the right shoulder	263	0	N	28	40	5
The star to the east	264	30	N	26	40	5 greater

NORTHERN SIGNS

Constellations	Longitude			Latitude		Magnitude
	Deg.	Min.		Deg.	Min.	
The star in the tail, which touches the milky circle	265	30	N	26	30	3
9 stars: 1 of second magnitude, 4 of third, 1 of fourth, 3 of fifth						
UNCONSTELLATED STARS AROUND AQUILA						
The more western star south of the head	272	0	N	21	40	3
The more eastern	272	20	N	29	10	3
Away from the right shoulder and to the south-west	259	20	N	25	0	4 greater
To the south	261	30	N	20	0	3
Farther south	263	0	N	15	30	5
West of all	254	30	N	18	10	3
6 unconstellated stars: 4 of third magnitude, 1 of fourth, and 1 of fifth						
DELPHINUS, OR THE DOLPHIN						
The most western of the three in the tail	281	0	N	29	10	3 smaller
The more northern of the two remaining	282	0	N	29	0	4 smaller
The more southern	282	0	N	26	40	4
The more southern on the western side of the rhomboid	281	50	N	32	0	3 smaller
The more northern on the same side	283	30	N	33	50	3 smaller
The more southern on the eastern side	284	40	N	32	0	3 smaller
The more northern on the same side	286	50	N	33	10	3 smaller
The most southern of the three between the tail and the rhombus	280	50	N	34	15	6
The more western of the other two to the north	280	50	N	31	50	6
The more eastern	282	20	N	31	30	6
10 stars: 5 of third magnitude, 2 of fourth, 3 of sixth						
[51b] EQUI SECTIO, OR THE SECTION OF THE HORSE						
The more western of the two on the head	289	40	N	20	30	obscure
The more eastern	292	20	N	20	40	obscure
The more western of the two at the mouth	289	40	N	25	30	obscure
The more eastern	291	21	N	25	0	obscure
4 stars: all obscure						
PEGASUS, OR THE WINGED HORSE						
Within the open mouth	298	40	N	21	30	3 greater
The more northern of the two close together on the head	302	40	N	16	50	3
The more southern	301	20	N	16	0	4
The more southern of the two on the mane	314	40	N	15	0	5
The more northern	313	50	N	16	0	5
The more western of the two on the neck	312	10	N	18	0	3
The more eastern	313	50	N	19	0	4

NORTHERN SIGNS

Constellations	Longitude			Latitude		Magnitude
	Deg.	Min.		Deg.	Min.	
On the left pastern	305	40	N	36	30	4 greater
On the left knee	311	0	N	34	15	4 greater
On the right pastern	317	0	N	41	10	4 greater
The more western of the two close together on the breast	319	30	N	29	0	4
The more eastern	320	20	N	29	30	4
The more northern of the two on the right knee	322	20	N	35	0	3
The more southern	321	50	N	24	30	5
The more northern of the two beneath the wing, on the body	327	50	N	25	40	4
The more southern	328	20	N	25	0	4
At the shoulder-blades and juncture of the wing	350	0	N	19	40	2 smaller
On the right shoulder and at the beginning of the leg	325	30	N	31	0	2 smaller
At the tip of the wing	335	30	N	12	30	2 smaller
At the navel, and on the head of Andromeda too	341	10	N	26	0	2 smaller
20 stars: 4 of second magnitude, 4 of third, 9 of fourth, 3 of fifth						
ANDROMEDA						
On the shoulder-blades	348	40	N	24	30	3
On the right shoulder	349	40	N	27	0	4
On the left shoulder	347	40	N	23	0	4
The most southern of the three on the right arm	347	0	N	32	0	4
The most northern	348	0	N	33	30	4
The middle one of the three	348	20	N	32	20	5
The most southern of the three on the top of the right hand	343	0	N	41	0	4
The middle star	344	0	N	42	0	4
[52ᵃ] The most northern of the three	345	30	N	44	0	4
On the left arm	347	30	N	17	30	4
At the left elbow	349	0	N	15	50	3
The most southern of the three on the girdle	357	10	N	25	20	3
The middle one	355	10	N	30	0	3
The most northern	355	20	N	32	30	3
On the left foot	10	10	N	23	0	3
On the right foot	10	30	N	37	10	4 greater
To the south of those two	8	30	N	35	20	4 greater
The more northern of the two under the hamstrings	5	40	N	29	0	4
The more southern	5	20	N	28	0	4
At the right knee	5	30	N	35	30	5
The more northern of the two on the flowing robe	6	0	N	34	30	5
The more southern	7	30	N	32	30	5

NORTHERN SIGNS

Constellations	Longitude Deg.	Min.		Latitude Deg.	Min.	Magnitude
The unconstellated star west of the right hand 23 stars: 7 of third magnitude, 12 of fourth, 4 of fifth	5	0	N	44	0	3
TRIANGULUM, OR THE TRIANGLE						
At the vertex of the triangle	4	20	N	16	30	3
The most western of the three on the base	9	20	N	20	40	3
The middle one	9	30	N	20	20	4
The most eastern of the three	10	10	N	19	0	3
4 stars: 3 of third magnitude, 1 of fourth						

Therefore in the northern region there are 360 stars, all in all: 3 of first magnitude, 18 of second, 81 of third, 177 of fourth, 58 of fifth, 13 of sixth, 1 nebulous, and 9 obscure.

THE SIGNS AND STARS WHICH ARE IN THE MIDDLE AND AROUND THE ECLIPTIC

Constellations	Longitude Deg.	Min.		Latitude Deg.	Min.	Magnitude
AIRES, OR THE RAM						
The star which is first of all and the more western of the two on the horn	0	0	N	7	20	3 smaller
The more eastern on the horn	1	0	N	8	20	3
The more northern of the two in the opening of the jaws	4	20	N	7	40	5
The more southern	4	50	N	6	0	5
On the neck	9	50	N	5	30	5
On the kidneys	10	50	N	6	0	6
At the beginning of the tail	14	40	N	4	50	4
The most western of the three on the tail	17	10	N	1	40	4
The middle one	18	40	N	2	30	4
[52[b]] The most eastern	20	20	N	1	50	4
On the hips	13	0	N	1	10	5
On the ham	11	20	S	1	30	5
At the tip of the hind foot	8	10	S	5	15	4 greater
13 stars: 2 of third magnitude, 4 of fourth, 6 of fifth, 1 of sixth						
UNCONSTELLATED STARS AROUND ARIES						
The brilliant star over the head	3	50	N	10	0	3 greater
The very northerly star above the back	15	0	N	10	10	4
The most northern of the remaining three small stars	14	40	N	12	40	5
The middle one	13	0	N	10	40	5
The most southern	12	30	N	10	40	5
5 stars: 1 of third magnitude, 1 of fourth, 3 of fifth						
TAURUS, OR THE BULL						
The most northern of the four in the section	19	40	S	6	0	4
The next after that	19	20	S	7	15	4
The third	18	0	S	8	30	4

IN THE MIDDLE, AND AROUND THE ECLIPTIC

Constellations	Longitude Deg.	Min.		Latitude Deg.	Min.	Magnitude
The fourth and most southern	17	50	S	9	15	4
On the right shoulder	23	0	S	9	30	5
In the breast	27	0	S	8	0	3
At the right knee	30	0	S	12	40	4
On the right pastern	26	20	S	14	50	4
At the left knee	35	30	S	10	0	4
On the left pastern	36	20	S	13	30	4
Of the five called Hyades and on the face, the one at the nostrils	32	0	S	5	45	3 smaller
Between that star and the northern eye	33	40	S	4	15	3 smaller
Between that same star and the southern eye	34	10	S	8	50	3 smaller
The brilliant star, in the very eye, called Palilicius by the Romans	36	0	S	5	10	1
On the northern eye	35	10	S	3	0	3 smaller
The star south of the horn between the base and the ear	40	30	S	4	0	4
The more southern of the two on the same horn	43	40	S	5	0	4
The more northern	43	20	S	3	30	5
At the extremity of the same	50	30	S	2	30	3
To the north of the base of the horn	49	0	S	4	0	4
At the extremity of the horn and on the right foot of Auriga	49	0	N	5	0	3
The more northern of the two in the north ear	35	20	N	4	30	5
The more southern	35	0	N	4	30	5 — Apogee of Venus: 48°20'
[53ª] The more western of the two small stars on the neck	30	20	N	0	40	5
The more eastern	32	20	N	1	0	6
The more southern on the western side of the quadrilateral on the neck	31	20	N	5	0	5
The more northern on the same	32	10	N	7	10	5
The more southern on the eastern side	35	20	N	3	0	5
The more northern on the same side	35	0	N	5	0	5
The northern limit of the western side of the Pleiades	25	30	N	4	30	5
The southern limit of the same side	25	50	N	4	40	5
The very narrow limit of the eastern side of the Pleiades	27	0	N	5	20	5
A small star of the Pleiades, separated from the limits	26	0	N	3	0	5
32 stars, apart from that which is at the tip of the northern horn: 1 of first magnitude, 6 of third, 11 of fourth, 13 of fifth, 1 of sixth						
UNCONSTELLATED STARS AROUND TAURUS Between the foot and below the shoulder	18	20	S	17	30	4

IN THE MIDDLE, AND AROUND THE ECLIPTIC

Constellations	Longitude			Latitude		Magnitude
	Deg.	Min.		Deg.	Min.	
The most western of the three to the south of the horn	43	20	S	2	0	5
The middle one	47	20	S	1	45	5
The most eastern of the three	49	20	S	2	0	5
The more northern of the two under the tip of the same horn	52	20	S	6	20	5
The more southern	52	20	S	7	40	5
The most western of the five under the northern horn	50	20	N	2	40	5
The next to the east	52	20	N	1	0	5
The third and to the east	54	20	N	1	20	5
The more northern of the remaining two	55	40	N	3	20	5
The more southern	56	40	N	1	15	5
11 unconstellated stars: 1 of fourth magnitude, 10 of fifth						
GEMINI, OR THE TWINS						
On the head of the western Twin, Castor	76	40	N	9	30	2
The reddish star on the head of the eastern Twin, Pollux	79	50	N	6	15	2
At the left elbow of the western Twin	70	0	N	10	0	4
On the left arm	72	0	N	7	20	4
At the shoulder-blades of the same Twin	75	20	N	5	30	4
On the right shoulder of the same	77	20	N	4	50	4
On the left shoulder of the eastern Twin	80	0	N	2	40	4
On the right side of the western Twin	75	0	N	2	40	5
On the left side of the eastern Twin	76	30	N	3	0	5
[53b] At the left knee of the western Twin	66	30	N	1	30	3
At the left knee of the eastern	71	35	S	2	30	3
On the left groin of the same	75	0	S	0	30	3
At the hollow of the right knee of the same	74	40	S	0	40	3
The more western star in the foot of the western Twin	60	0	S	1	30	4 greater
The more eastern star in the same foot	61	30	S	1	15	4
At the extremity of the foot of the western Twin	63	30	S	3	30	4
On the top of the foot of the eastern Twin	65	20	S	7	30	3
On the bottom of the foot of the same	68	0	S	10	30	4
18 stars: 2 of second magnitude, 5 of third, 9 of fourth, 2 of fifth						
UNCONSTELLATED STARS AROUND GEMINI						
The star west of the top of the foot of the western Twin	57	30	8	0	40	4
The brilliant star to the west of the knee of the same	59	50	N	5	50	4 greater
To the west of the left knee of the eastern Twin	68	30	S	2	15	5
The most northern of the three east of the right hand of the eastern Twin	81	40	S	1	20	5

IN THE MIDDLE, AND AROUND THE ECLIPTIC

Constellations	Longitude			Latitude		Magnitude
	Deg.	Min.		Deg.	Min.	
The middle one	79	40	S	3	20	5
The most southern of the three, and in the neighbourhood of the right arm	79	20	S	4	30	5
The brilliant star to the east of the three	84	0	S	2	40	4
7 unconstelled stars; 3 of fourth magnitude, 4 of fifth						
CANCER, OR THE CRAB						
The nebulous star in the breast, which is called Praeses	93	40	N	0	40	nebulous
The more northern of the two west of the quadrilateral	91	0	N	1	15	4 smaller
The more southern	91	20	S	1	10	4 smaller
The more northern of the two to the east, which are called the Asses	93	40	N	2	40	4 greater
The southern Ass	94	40	S	0	10	4 greater
On the claws or the southern arm	99	50	S	5	30	4
On the northern arm	91	40	N	11	50	4
At the extremity of the northern foot	86	0	N	1	0	5
At the extremity of the southern foot	90	30	S	7	30	4 greater
9 stars: 7 of fourth magnitude, 1 of fifth, 1 nebulous						
UNCONSTELLATED STARS AROUND CANCER						
Above the elbow of the southern claw	103	0	S	2	40	4 smaller
East of the extremity of the same claw	105	0	S	5	40	4 smaller
[54ª] The more western of the two above the little cloud	97	20	N	4	50	5
The more eastern	100	20	N	7	15	5
4 unconstellated stars: 2 of fourth magnitude, 2 of fifth						
LEO, OR THE LION						
At the nostrils	101	40	N	10	0	4
At the opening of the jaws	104	30	N	7	30	4
The more northern of the two on the head	107	40	N	12	0	3
The more southern	107	30	N	9	30	3 greater
The most northern of the three on the neck	113	30	N	11	0	3 Apogee of Mars: 109°50'
The middle one	115	30	N	8	30	2
The most southern of the three	114	0	N	4	30	3
At the heart, the star called Basiliscus or Regulus	115	50	N	0	10	1
The more southern of the two on the breast	116	50	S	1	50	4
A little to the west of the star at the heart	113	20	S	0	15	5
At the knee of the right foreleg	110	40		0	0	5
On the right pad	117	30	S	3	40	6
At the knee of the left foreleg	122	30	S	4	10	4
On the left pad	115	50	S	4	15	4

IN THE MIDDLE, AND AROUND THE ECLIPTIC

| Constellations | Longitude | | | Latitude | | Magnitude |
	Deg.	Min.		Deg.	Min.	
At the left arm-pit	122	30	S	0	10	4
The most western of the three on the belly	120	20	N	4	0	6
The more northern of the two to the east	126	20	N	5	20	6
The more southern	125	40	N	2	20	6
The more western of the two on the loins	124	40	N	12	15	5
The more eastern	127	30	N	13	40	2
The more northern of the two on the rump	127	40	N	11	30	5
The more southern	129	40	N	9	40	3
At the hips	133	40	N	5	50	3
At the hollow of the knee	135	0	N	1	15	4
On the lower part of the leg	135	0	S	0	50	4
On the hind foot	134	0	S	3	0	5
At the tip of the tail	137	50	N	11	50	1 smaller
27 stars: 2 of first magnitude, 2 of second, 6 of third, 8 of fourth, 5 of fifth, 4 of sixth						
UNCONSTELLATED STARS AROUND LEO						
The more western of the two above the back	119	20	N	13	20	5
The more eastern	121	30	N	15	30	5
The most northern of the three below the belly	129	50	N	1	10	4 smaller
[54b] The middle one	130	30	S	0	30	5
The most southern of the three	132	20	S	2	40	5
The star farthest north between the extremities of Leo and the nebulous complex called Coma Berenices	138	10	N	30	0	luminous
The more western of the two to the south	133	50	N	25	0	obscure
The star to the east, in the shape of an ivy leaf	141	50	N	25	30	obscure
8 unconstellated stars: 1 of fourth magnitude, 4 of fifth, 1 luminous, 2 obscure						
VIRGO, OR THE VIRGIN						
The more southwestern of the two on the top of the head	139	40	N	4	15	5
The more northeastern	140	20	N	5	40	5
The more northern of the two on the face	144	0	N	8	0	5
The more southern	143	30	N	5	30	5
At the tip of the left and southern wing	142	20	N	6	0	3
The most western of the four on the left wing	151	35	N	1	10	3
The next to the east	156	30	N	2	50	3
The third	160	30	N	2	50	5
The last and most eastward of the four	164	20	N	1	40	4
On the right side beneath the girdle	157	40	N	8	30	3
The most western of the three on the right and northern wing	151	30	N	13	50	5
The more southern of the two remaining	153	30	N	11	40	6 Apogee of Jupiter: 154°20'

IN THE MIDDLE, AND AROUND THE ECLIPTIC

Constellations	Longitude			Latitude		Magnitude
	Deg.	Min.		Deg.	Min.	
The more northern of them, called Vindemiator	155	30	N	15	10	3 greater
On the left hand, called Spica	170	0	S	2	0	1
Beneath the girdle and on the right buttock	168	10	N	8	40	3
The more northern of the two on the western side of the quadrilateral on the left hip	169	40	N	2	20	5
The more southern	170	20	N	0	10	6
The more northern of the two on the eastern side	173	20	N	1	30	4
The more southern	171	20	N	0	20	5
At the left knee	175	0	N	1	30	5
On the posterior side of the right hip	171	20	N	8	30	5
On the flowing robe, in the middle	180	0	N	7	30	4
More to the south	180	40	N	2	40	4
More to the north	181	40	N	11	40	4 Apogee of Mercury: 183°20'
On the left and southern foot	183	20	N	0	30	4
On the right and southern foot	186	0	N	9	50	3
26 Stars: 1 of first magnitude, 7 of third, 6 of fourth, 10 of fifth, 2 of sixth						
UNCONSTELLATED STARS AROUND VIRGO						
[55ª] The most western of the three in a straight line under the left arm	158	0	S	3	30	5
The middle one	162	20	S	3	30	5
The most eastern	165	35	S	3	20	5
The most western of the three in a straight line under Spica	170	30	S	7	20	6
The middle one, which is also a double star	171	30	S	8	20	5
The most eastern of the three	173	20	S	7	50	6
6 unconstellated stars: 4 of fifth magnitude, 2 of sixth						
CHELAE, OR THE CLAWS						
The bright one of the two at the extremity of the southern claw	191	20	N	0	40	2 greater
The more obscure star to the north	190	20	N	2	30	5
The bright one of the two at the extremity of the northern claw	195	30	N	8	30	2
The more obscure star to the west of that	191	0	N	8	30	5
In the middle of the southern claw	197	20	N	1	40	4
In the same claw, but to the west	194	40	N	1	15	4
At the middle of the northern claw	200	50	N	3	45	4
In the same claw, but to the east	206	20	N	4	30	4
8 stars: 2 of second magnitude, 4 of fourth, 2 of fifth						

IN THE MIDDLE, AND AROUND THE ECLIPTIC

Constellations	Longitude			Latitude		Magnitude
	Deg.	Min.		Deg.	Min.	
UNCONSTELLATED STARS AROUND THE CHELAE						
The most western of the three north of the northern claw	199	30	N	9	0	5
The more southern of the two to the east	207	0	N	6	40	4
The more northern	207	40	N	9	15	4
The most eastern of the three between the claws	205	50	N	5	30	6
The more northern of the remaining two to the west	203	40	N	2	0	4
The more southern	204	30	N	1	30	5
The most western of the three beneath the southern claw	196	20	S	7	30	3
The more northern of the remaining two to the east	204	30	S	8	10	4
The more southern	205	20	S	9	40	4
9 unconstellated stars: 1 of third magnitude, 5 of fourth, 2 of fifth, 1 of sixth						
SCORPIO, OR THE SCORPION						
The most northern of the three bright stars on the forehead	209	40	N	1	20	3 greater
The middle one	209	0	S	1	40	3
The most southern of the three	209	0	S	5	0	3
More to the south and in the foot	209	20	S	7	50	3
The more northern of the two adjacent bright stars	210	20	N	1	40	4
The more southern	210	40	N	0	30	4
The most western of the three bright stars on the body	214	0	S	3	45	3
The reddish star in the middle, called Antares	216	0	S	4	0	2 greater
The most eastern of the three	217	50	S	5	30	3
[55b] The more western of the two at the extremity of the foot	212	40	S	6	10	5
The more eastern	213	50	S	6	40	5
At the first vertebra of the body	221	50	S	11	0	3
At the second vertebra	222	10	S	15	0	4
The more northern of the double at the third	223	20	8	18	40	4
The more southern of the double	223	30	S	18	0	3
At the fourth vertebra	226	30	8	19	30	3 Apogee of Saturn: 226°30'
At the fifth	231	30	S	18	50	3
At the sixth vertebra	233	50	S	16	40	3
At the seventh, and next to the sting	232	20	S	15	10	3
The more eastern of the two on the sting	230	50	S	13	20	3
The more western	230	20	S	13	30	4
21 stars: 1 of second magnitude, 13 of third, 5 of fourth, 2 of fifth						

IN THE MIDDLE, AND AROUND THE ECLIPTIC

Constellations	Longitude			Latitude		Magnitude
	Deg.	Min.		Deg.	Min.	
UNCONSTELLATED STARS AROUND SCORPIO						
The nebulous star to the east of the sting	234	30	S	12	15	nebulous
The more western of the two north of the sting	228	50	S	6	10	5
The more eastern	232	50	S	4	10	5
3 unconstellated stars: 2 of fifth magnitude, 1 nebulous						
SAGITTARIUS, OR THE ARCHER						
At the head of the arrow	237	50	S	6	30	3
In the palm of the left hand	241	0	S	6	30	3
On the southern part of the bow	241	20	S	10	50	3
The more southern of the two to the north	242	20	S	1	30	3
More northward, at the extremity of the bow	240	0	N	2	50	4
On the left shoulder	248	40	S	3	10	3
To the west and on the dart	246	20	S	3	50	4
The nebulous double star in the eye	248	30	N	0	45	nebulous
The most western of the three on the head	249	0	N	2	10	4
The middle one	251	0	N	1	30	4 greater
The most eastward	252	30	N	2	0	4
The most southern of the three on the northern garment	254	40	N	2	50	4
The middle one	255	40	N	4	30	4
The most northern	256	10	N	6	30	4
The obscure star east of the three	259	0	N	5	30	6
The most northern of the two on the southern. garment	262	50	N	5	0	5
The more southern	261	0	N	2	0	6
On the right shoulder	255	40	S	1	50	5
[56a] At the right elbow	258	10	S	2	50	5
At the shoulder-blades	253	20	S	2	30	5
At the foreshoulder	251	0	S	4	30	4 greater
Beneath the arm-pit	249	40	S	6	45	3
On the pastern of the left foreleg	251	0	S	23	0	2
At the knee of the same leg	250	20	S	18	0	2
On the pastern of the right foreleg	240	0	S	13	0	3
At the left shoulder blade	260	40	S	13	30	3
At the knee of the right foreleg	260	0	S	20	10	3
The more western on the northern side of the quadrilateral at the beginning of the tail	261	0	S	4	50	5
The more eastern on the same side	261	10	S	4	50	5
The more western on the southern side	261	50	S	5	50	5
The more eastern on the same side	263	0	S	6	50	5
31 stars: 2 of second magnitude, 9 of third, 9 of fourth, 8 of fifth, 2 of sixth, 1 nebulous.						

IN THE MIDDLE, AND AROUND THE ECLIPTIC

Constellations	Longitude Deg.	Min.		Latitude Deg.	Min.	Magnitude
CAPRICORNUS, OR THE GOAT						
The most northern of the three on the western horn	270	40	N	7	30	3
The middle one	271	0	N	6	40	6
The most southern of the three	270	40	N	5	0	3
At the extremity of the eastern horn	272	20	N	8	0	6
The most southern of the three at the opening of the jaws	272	20	N	0	45	6
The more western of the two remaining	272	0	N	1	45	6
The more eastern	272	10	N	1	30	6
Under the right eye	270	30	N	0	40	5
The more northern of the two on the neck	275	0	N	4	50	6
The more southern	275	10	S	0	50	5
At the right knee	274	10	S	6	30	4
At the left knee, which is bent	275	0	S	8	40	4
On the left shoulder	280	0	S	7	40	4
The more western of the two contiguous stars below the belly	283	30	S	6	50	4
The more eastern	283	40	S	6	0	5
The most eastern of the three in the middle of the body	282	0	S	4	15	5
The more southern of the two remaining to the west	280	0	S	7	0	5
The more northern	280	0	S	2	50	5
The more western of the two on the back	280	0		0	0	4
The more eastern	284	20	S	0	50	4
The more western of the two on the southern part of the spine	286	40	S	4	45	4
[56ᵇ] The more eastern	288	20	S	4	30	4
The more western of the two at the base of the tail	288	40	S	2	10	3
The more eastern	289	40	S	2	0	3
The more western of the four in the northern part of the tail	290	10	S	2	20	4
The most northern of the remaining three	292	0	S	5	0	5
The middle one	291	0	S	2	50	5
The most northern, at the extremity of the tail	292	0	N	4	20	5
28 stars: 4 of third magnitude, 9 of fourth, 9 of fifth, 6 of sixth						
AQUARIUS, OR THE WATER-BOY						
On the head	293	40	N	15	45	5
The brighter of the two on the right shoulder	299	40	N	11	0	3
The more obscure	298	30	N	9	40	5
On the left shoulder	290	0	N	8	50	3
Under the arm-pit	290	40	N	6	15	5
The most eastern of the three under the left hand and on the coat	280	0	N	5	30	3
The middle one	279	30	N	8	0	4

IN THE MIDDLE, AND AROUND THE ECLIPTIC

Constellations	Longitude			Latitude		Magnitude
	Deg.	Min.		Deg.	Min.	
The most western of the three	278	0	N	8	30	3
At the right elbow	302	50	N	8	45	3
The farthest north on the right hand	303	0	N	10	45	3
The more western of the two remaining to the south	305	20	N	9	0	3
The more eastern	306	40	N	8	30	3
The more western of the two adjacent stars on the right hip	299	30	N	3	0	4
The more eastern	300	20	N	2	10	5
On the right buttock	302	0	S	0	50	4
The more southern of the two on the left buttock	295	0	S	1	40	4
The more northern	295	30	N	4	0	6
The more southern of the two on the right shin	305	0	S	6	30	3
The more northern	304	40	S	5	0	4
On the left hip	301	0	S	5	40	5
The more southern of the two on the left shin	300	40	S	10	0	5
The northern star beneath the knee	302	10	S	9	0	5
The first star in the fall of water from the hand	303	20	N	2	0	4
More to the south-east	308	10	N	0	10	4
To the east at the first bend in the water	311	0	S	1	10	4
To the east of that	313	20	S	0	30	4
In the second and southern bend	313	50	S	1	40	4
The more northern of the two to the east	312	30	S	3	30	4
The more southern	312	50	S	4	10	4
Farther off to the south	314	10	S	8	15	5
[57a] Eastward, the more western of the two adjacent	316	0	S	11	0	5
The more eastern	316	30	S	10	50	5
The most northern of the three at the third bend in the water	315	0	S	14	0	5
The middle one	316	0	S	14	45	5
The most eastern of the three	316	30	S	15	40	5
The most northern of three in a similar figure to the east	310	20	S	14	10	4
The middle one	310	50	S	15	0	4
The most southern of the three	311	40	S	15	45	4
The most western of the three at the last bend in the water	305	10	S	14	50	4
The more southern of the two to the east	306	0	S	15	20	4
The more northern	306	30	S	14	0	4
The last in the water, and in the mouth of the southern Fish	300	20	S	23	0	1

42 stars: 1 of first magnitude, 9 of third,
18 of fourth, 13 of fifth, 1 of sixth

IN THE MIDDLE, AND AROUND THE ECLIPTIC

Constellations	Longitude			Latitude		Magnitude
	Deg.	Min.		Deg.	Min.	
UNCONSTELLATED STARS AROUND AQUARIUS						
The most western of the three east of the bend in the water	320	0	S	15	30	4
The more northern of the two remaining	323	0	S	14	20	4
The more southern	322	20	S	18	15	4
3 stars: greater than fourth magnitude						
PISCES, OR THE FISH						
In the mouth of the western Fish	315	0	N	9	15	4
The more western of the two on the occiput	317	30	N	7	30	4 greater
The more northern	321	30	N	9	30	4
The more western of the two on the back	319	20	N	9	20	4
The more eastern	324	0	N	7	30	4
The more western one on the belly	319	20	N	4	30	4
The more eastern	323	0	N	2	30	4
On the tail of the same Fish	329	20	N	6	20	4
On the fishing-line, the first star from the tail	334	20	N	5	45	6
To the east of that	336	20	N	2	45	6
The most western of the three bright stars to the east	340	30	N	2	15	4
The middle one	343	50	N	1	10	4
The most eastern	346	20	S	1	20	4
The more northern of the two small stars on the curvature	345	40	S	2	0	6
The more southern	346	20	S	5	0	6
The most western of the three after the curvature	350	20	S	2	20	4
The middle one	352	0	S	4	40	4
The most eastern one	354	0	S	7	45	4
[57b] At the knot of the two fishing-lines	356	0	S	8	30	3
In the northern line, west of the knot	354	0	S	4	20	4
The most southern of the three to the east	353	30	N	1	30	5
The middle one	353	40	N	5	20	3
The most northern of the three and the last in the line	353	50	N	9	0	4
THE EASTERN FISH						
The more northern of the two in the mouth	355	20	N	21	45	5
The more southern	355	0	N	21	30	5
The most eastern of the three small stars on the head	352	0	N	20	0	6
The middle one	351	0	N	19	50	6
The most western of the three	350	20	N	23	0	6
The most western of the three on the southern fin, near the left elbow of Andromeda	349	0	N	14	20	4
The middle one	349	40	N	13	0	4
The most eastern of the three	351	0	N	12	0	4
The more northern of the two on the belly	355	30	N	17	0	4
The more southern	352	40	N	15	20	4

IN THE MIDDLE, AND AROUND THE ECLIPTIC

Constellations	Longitude			Latitude		Magnitude
	Deg.	Min.		Deg.	Min.	
On the eastern fin, near the tail	353	20	N	11	45	4
34 stars: 2 of third magnitude, 22 of fourth, 3 of fifth, 7 of sixth						
UNCONSTELLATED STARS AROUND PISCES						
The more western on the northern side of the quadrilateral under the western Fish	324	30	S	2	40	4
The more eastern	325	35	S	2	30	4
The more western on the southern side	324	0	S	5	50	4
The more eastern	325	40	S	5	30	4
4 unconstellated stars: of fourth magnitude						

Therefore, all in all, there are 348 stars in the zodiac: 5 of first magnitude, 9 of second, 65 of third, 132 of fourth, 105 of fifth, 27 of sixth, 3 nebulous, 2 obscure; and, over and above the count, the Coma, which we said above was called Coma Berenices by Conon the mathematician.

THE STARS OF THE SOUTHERN REGION

Constellations	Longitude			Latitude		Magnitude
	Deg.	Min.		Deg.	Min.	
CETUS, OR THE WHALE						
At the extremity of the nose	11	0	S	7	45	4
The most eastern of the three in the jaws	11	0	S	11	20	3
The middle one, in the middle of the mouth	6	0	S	11	30	3
The most western of the three, on the cheek	3	50	S	14	0	3
In the eye	4	0	S	8	10	4
Northward, in the hair	5	30	S	6	20	4
[58ª] Westward, in the mane	1	0	S	4	10	4
The more northern on the western side of the quadrilateral in the breast	355	20	S	24	30	4
The more southern	356	40	S	28	0	4
The more northern of the two to the east	0	0	S	25	10	4
The more southern	0	0	S	27	30	3
The middle one of the three on the body	345	20	S	25	20	3
The most southern	346	20	S	30	30	4
The most northern of the three	348	20	S	20	0	3
The more eastern of the two at the tail	343	0	S	15	20	3
The more western	338	20	S	15	40	3
The more northern on the eastern side of the quadrilateral in the tail	335	0	S	11	40	5
The more southern	334	0	S	13	40	5
The more northern of the two remaining to the west	332	40	S	13	0	5
The more southern	332	20	S	14	0	5
At the northern extremity of the tail	327	40	8	9	30	3
At the southern extremity of the tail	329	0	S	20	20	3
22 stars: 10 of third magnitude, 8 of fourth, 4 of fifth						

SOUTHERN SIGNS

Constellations	Longitude			Latitude		Magnitude
	Deg.	Min.		Deg.	Min.	
ORION						
The nebulous star on the head	50	20	S	16	30	nebulous
The bright, reddish star on the right shoulder	55	20	S	17	0	1
On the left shoulder	43	40	S	17	30	2 greater
East of that star	48	20	S	18	0	4 smaller
At the right elbow	57	40	S	14	30	4
On the right forearm	59	40	S	11	50	6
The more eastern on the southern side of the quadrilateral in the right hand	59	50	S	10	40	4
The more western	59	20	S	9	45	4
The more eastern on the northern side	60	40	S	8	15	6
The more western on the same side	59	0	S	8	15	6
The more western of the two on the club	55	0	S	3	45	5
The more eastern	57	40	S	3	15	5
The most eastern of the four in a straight line on the back	50	50	S	19	40	4
More western	49	40	S	20	0	6
Still more western	48	40	S	20	20	6
Most western	47	30	S	20	30	5
The most northern of the nine on the shield	43	50	S	8	0	4
The second	42	40	S	8	10	4
The third	41	20	S	10	15	4
The fourth	39	40	S	12	50	4
The fifth	38	30	S	14	15	4
The sixth	37	50	S	15	50	3
[58b] The seventh	38	10	S	17	10	3
The eighth	38	40	S	20	20	3
The last and most southern	39	40	S	21	30	3
The most western of the three bright stars on the sword-belt	48	40	S	24	10	2
The middle one	50	40	S	24	50	2
The most eastern of the three in a straight line	52	40	S	25	30	2
On the hilt of the sword	47	10	S	25	50	3
The most northern of the three on the sword	50	10	S	28	40	4
The middle one	50	0	S	29	30	3
The most southern one	50	20	S	29	50	3 smaller
The more eastern of the two at the tip of the sword	51	0	S	30	30	4
The more western	49	30	S	30	50	4
On the left foot, the bright star which belongs to Fluvius too	42	30	S	31	30	1
On the left shin	44	20	S	30	15	4 greater
At the right heel	46	40	S	31	10	4
At the right knee	53	30	S	33	30	3

38 stars: 2 of first magnitude, 4 of second,
8 of third, 15 of fourth, 3 of fifth, 5 of sixth,
and 1 nebulous

SOUTHERN SIGNS

Constellations	Longitude Deg.	Longitude Min.		Latitude Deg.	Latitude Min.	Magnitude
FLUVIUS, OR THE RIVER						
After the left foot of Orion, and at the beginning of Fluvius	41	40	S	31	50	4
The most northern star within the bend of Orion's leg	42	10	S	28	15	4
The more eastern of the two after that	41	20	S	29	50	4
The more western	38	0	S	28	15	4
The more eastern of the next two	36	30	S	25	15	4
The more western	33	30	S	25	20	4
The most eastern of the three after them	29	40	S	26	0	4
The middle one	29	0	S	27	0	4
The most western of the three	26	18	S	27	50	4
The most eastern of the four after the interval	20	20	S	82	50	3
More western	18	0	S	31	0	4
Still more western	17	30	S	28	50	3
The most western of all four	15	30	S	28	0	3
Again similarly, the most eastward of the four	10	30	S	25	30	3
More westward	8	10	S	23	50	4
Still more westward	5	30	S	23	10	3
The most westward of the four	3	50	S	23	15	4
The star in the bend of Fluvius which touches the breast of Cetus	358	30	S	32	10	4
East of that	359	10	S	34	50	4
The most westward of the three to the seat	2	10	S	38	30	4
[59ª] The middle one	7	10	S	38	10	4
The most eastward of the three	10	50	S	39	0	5
The more northern of the two on the western side of the quadrilateral	14	40	S	41	30	4
The more southern	14	50	S	42	30	4
The more western on the eastern side	15	30	S	43	20	4
The most eastward of those four	18	0	S	43	20	4
The more northern of the two contiguous stars towards the east	27	30	S	50	20	4
The more southern	28	20	S	51	45	4
The more eastern of the two at the bend	21	30	S	53	50	4
The more western	19	10	S	53	10	4
The most eastern of the three in the remaining space	11	10	S	53	0	4
The middle one	8	10	S	53	30	4
The most western of the three	5	10	S	52	0	4
The bright star at the extremity of the river	353	30	S	53	30	1
34 stars; 1 of first magnitude, 5 of third, 27 of fourth, 1 of fifth						
LEPUS, OR THE RABBIT						
The more northern one on the western side of the quadrilateral at the ears	43	0	S	35	0	5
The more southern	43	10	S	36	30	5
The more northern one on the eastern side	44	40	S	35	30	5

SOUTHERN SIGNS

Constellations	Longitude			Latitude		Magnitude
	Deg.	Min.		Deg.	Min.	
The more southern	44	40	S	36	40	5
At the chin	42	30	S	39	40	4 greater
At the extremity of the left forefoot	39	30	S	45	15	4 greater
In the middle of the body	48	50	S	41	30	3
Beneath the belly	48	10	S	44	20	3
The more northern of the two on the hind feet	54	20	S	44	0	4
The more southern	52	20	S	45	50	4
On the loins	53	20	S	38	20	4
At the tip of the tail	56	0	S	38	10	4
12 stars: 2 of third magnitude, 6 of fourth, 4 of fifth						
CANIS, OR THE DOG						
The very bright star called Canis, in the mouth	71	0	S	39	10	1 very great
On the ears	73	0	S	35	0	4
On the head	74	40	S	36	30	5
The more northern of the two on the neck	76	40	S	37	45	4
The more southern	78	40	S	40	0	4
On the breast	73	50	S	42	30	5
The more northern of the two at the right knee	69	30	S	41	15	5
The more southern	69	20	S	42	30	5
At the extremity of the forefoot	64	20	S	41	20	3
[59b] The more western of the two on the left knee	68	0	S	46	30	5
The more eastern	69	30	S	45	50	5
The more eastern of the two on the left shoulder	78	0	S	46	0	4
The more western	75	0	S	47	0	5
On the left hip	80	0	S	48	45	3 smaller
Beneath the belly between the thighs	77	0	S	51	30	3
In the hollow of the right foot	76	20	S	55	10	4
At the extremity of the same foot	77	0	S	55	40	3
At the tip of the tail	85	30	S	50	30	3 smaller
18 stars: 1 of first magnitude, 5 of third, 5 of fourth, 7 of fifth						
UNCONSTELLATED STARS AROUND CANIS						
North of the head of the Dog	72	50	S	25	15	4
The most southern in a straight line under the hind feet	63	20	S	60	30	4
The more northern	64	40	S	58	45	4
Still more northern	66	20	S	57	0	4
The last and farthest north of the four	67	30	S	56	0	4
The most western of the three westward as it were in a straight line	50	20	S	55	30	4
The middle one	53	40	S	57	40	4
The most eastern of the three	55	40	S	59	30	4

SOUTHERN SIGNS

Constellations	Longitude			Latitude		Magnitude
	Deg.	Min.		Deg.	Min.	
The more western of the two bright stars beneath them	52	20	S	59	40	2
The more western	49	20	S	57	40	2
The remaining star, more southern	45	30	S	59	30	4
11 stars: 2 of second magnitude, 9 of fourth						
CANICULA, OR PROCYON, OR THE LITTLE BITCH						
On the neck	78	20	S	14	0	4
The bright star on the thigh, that is, Προκύων or Canicula, the Dog-star	82	30	S	16	10	1
2 stars: 1 of first magnitude, 1 of fourth						
ARGO, OR THE SHIP						
The more western of the two at the extremity of the Ship	93	40	S	42	40	5
The more eastern	97	40	S	43	20	3
The more northern of the two on the stern	92	10	S	45	0	4
The more southern	92	10	S	46	0	4
West of the two	88	40	S	45	30	4
The bright star in the middle of the shield	89	40	S	47	15	4
The most western of the three beneath the shield	88	40	S	49	45	4
The most eastern	92	40	S	49	50	4
The middle one of the three	91	40	S	49	15	4
At the extremity of the rudder	97	20	S	49	50	4
The more northern of the two on the stern keel	87	20	S	53	0	4
The more southern	87	20	S	58	30	3
[60ª] The most northern on the cross-bank of the stem	93	30	S	55	30	5
The most western of the three on the same cross-bank	95	30	S	58	30	5
The middle one	96	40	S	57	15	4
The most eastern	99	50	S	57	45	4
The bright star to the east on the cross bank	104	30	S	58	20	2
The more western of the two obscure stars beneath that	101	30	S	60	0	5
The more eastern	104	20	S	59	20	5
The more western of the two east of the aforesaid bright star	106	30	S	56	40	5
The more eastern	107	40	S	57	0	5
The most northern of the three on the small shields and at the foot of the mast	119	0	S	51	30	4 greater
The middle one	119	30	S	55	30	4 greater
The most southern of the three	117	20	S	57	10	4
The more northern of the two contiguous stars beneath them	122	30	S	60	0	4
The more southern	122	20	S	61	15	4

SOUTHERN SIGNS

Constellations	Longitude Deg.	Min.		Latitude Deg.	Min.	Magnitude
The more southern of the two in the middle of the mast	113	30	S	51	30	4
The more northern	112	40	S	49	0	4
The more western of the two at the top part of the sail	111	40	S	43	20	4
The more eastern	112	20	S	43	30	4
Below the third star east of the shield	98	30	S	54	30	2 smaller
In the section of the bridge	100	50	S	51	15	2
Between the oars in the keel	95	0	S	63	0	4
The obscure star east of that	102	20	S	64	30	6
The bright star, east of that and below the cross-bank	113	20	S	63	50	2
The bright star to the south, more within the keel	121	50	S	69	40	2
The most western of the three to the east of that	128	30	S	65	40	3
The middle one	134	40	S	65	50	3
The most eastern	139	20	S	65	50	2
The more western of the two in the section	144	20	S	62	50	3
The more eastern	151	20	S	62	15	3
The more western in the northwestern oar	57	20	S	65	50	4 greater
The more eastern	73	30	S	65	40	3 greater
The more western one in the remaining oar; Canopus	70	30	S	75	0	1
The remaining star east of that	82	20	S	71	50	3 greater
45 stars: 1 of first magnitude, 6 of second, 8 of third, 22 of fourth, 7 of fifth, 1 of sixth						
HYDRA						
Of the two more western of the five on the head, the more southern, at the nostrils	97	20	S	15	0	4
The more northern of the two, and in the eye	98	40	S	13	40	4
On the occiput, the more northern of the two to the east	99	0	S	11	30	4
[60b] The more southern, and at the jaws	98	50	S	14	45	4
East of all those and on the cheeks	100	50	S	12	15	4
The more western of the two at the beginning of the neck	103	40	S	11	50	5
The more eastern	106	40	S	13	30	4
The middle one of the three at the curve of the neck	111	40	S	15	20	4
East of that	114	0	S	14	50	4
The most southern	111	40	S	17	10	4
The obscure and northern star of the two contiguous to the south	112	30	S	19	45	6
The bright one and to the south-east	113	20	S	20	30	2
The most western of the three after the curve in the neck	119	20	S	26	30	4

SOUTHERN SIGNS

Constellations	Longitude			Latitude		Magnitude
	Deg.	Min.		Deg.	Min.	
The most eastern						
124	30		S	23	15	4
The middle one	122	0	S	24	0	4
The most western of the three in a straight line	131	20	S	24	30	3
The middle one	133	20	S	23	0	4
The most eastern one	136	20	S	23	10	3
The more northern of the two beneath the base of the Cup	144	50	S	25	45	4
The more southern	145	40	S	30	10	4
East of them, the most western of the three on the triangle	155	30	S	31	20	4
The most southern	157	50	S	34	10	4
The most eastern of the same three	159	30	S	31	40	3
East of the Crow, near the tail	173	20	S	13	30	4
At the extremity of the tail	186	50	S	17	30	4
25 stars: 1 of second magnitude, 3 of third, 19 of fourth, 1 of fifth, 1 of sixth						
UNCONSTELLATED STARS AROUND HYDRA						
South of the head	96	0	S	23	15	3
East of those on the neck	124	20	S	26	0	3
2 unconstellated stars: of third magnitude						
CRATER, OR THE CUP						
On the base of the Cup and in Hydra too	139	40	S	23	0	4
The more southern of the two in the middle of the Cup	146	0	S	19	30	4
The more northern of them	143	30	S	18	0	4
On the southern rim of the Cup	150	20	S	18	30	4 greater
On the northern part of the rim	142	40	S	13	40	4
On the southern part of the stem	152	30	S	16	30	4 smaller
On the northern part	145	0	S	11	50	4
7 stars: of fourth magnitude						
[61a] CORVUS, OR THE CROW						
On the beak, and in Hydra too	158	40	S	21	30	5
On the neck	157	40	S	19	40	5
In the breast	160	0	S	18	10	5
On the right wing, the western wing	160	50	S	14	50	3
The more western of the two on the eastern wing	160	0	S	12	30	3
The more eastern	161	20	S	11	45	4
At the extremity of the foot, and in Hydra too	163	50	S	18	10	3
7 Stars: 5 of third magnitude, 1 of fourth, 1 of fifth						
CENTAURUS, OR THE CENTAUR						
The most southern of the four on the head	183	50	S	21	20	5
The more northern	183	20	S	13	50	5

SOUTHERN SIGNS

Constellations	Longitude			Latitude		Magnitude
	Deg.	Min.		Deg.	Min.	
The more western of the two in the middle	182	30	S	20	30	5
The more eastern and last of the four	182	20	S	20	0	5
On the left and western shoulder	179	30	S	25	30	3
On the right shoulder	189	0	S	22	30	3
On the left forearm	182	30	S	17	30	4
The more northern of the two on the western side of the quadrilateral on the shield	191	30	S	22	30	4
The more southern	192	30	S	23	45	4
Of the remaining two, the one at the top of the shield	195	20	S	18	15	4
The more southern	196	50	S	20	50	4
The most western of the three on the right side	186	40	S	28	20	4
The middle one	187	20	S	29	20	4
The most eastern	188	30	S	28	0	4
On the right arm	189	40	S	26	30	4
On the right elbow	196	10	S	25	15	3
At the extremity of the right hand	200	50	S	24	0	4
The bright star at the junction of the human body	191	20	S	33	30	3
The more eastern of the two obscure stars	191	0	S	31	0	5
The more western	189	50	S	30	20	5
At the beginning of the back	185	30	S	33	50	5
West of that, on the horse's back	182	20	S	37	30	5
The most eastern of the three on the loins	179	10	S	40	0	3
The middle one	178	20	S	40	20	4
The most western of the three	176	0	S	41	0	5
The more western of the two contiguous stars on the right hip.	176	0	S	46	10	2
The more eastern	176	40	S	46	45	4
On the breast, beneath the horse's wing	191	40	S	40	45	4
[61b] The more western of the two under the belly	179	50	S	43	0	2
The more eastern	181	0	S	43	45	3
In the hollow of the right hind foot	183	20	S	51	10	2
On the pastern of the same	188	40	S	51	40	2
In the hollow of the left <hind> foot	188	40	S	55	10	4
Under the muscle of the same foot	184	30	S	55	40	4
On top of the right forefoot	181	40	S	41	10	1
At the left knee	197	30	S	45	20	2
The unconstellated star below the right thigh	188	0	S	49	10	3
37 stars: 1 of first magnitude, 5 of second, 7 of third, 15 of fourth, 9 of fifth						
BESTIA QUAM TENET CENTAURUS, OR THE BEAST HELD BY THE CENTAUR—THE WOLF						
At the top of the hind foot and in the hand of the Centaur	201	20	S	24	50	3

SOUTHERN SIGNS

Constellations	Longitude			Latitude		Magnitude
	Deg.	Min.		Deg.	Min.	
On the hollow of the same foot	199	10	S	20	10	3
The more western of the two on the foreshoulder	204	20	S	21	15	4
The more eastern	207	30	S	21	0	4
In the middle of the body	206	20	S	25	10	4
On the belly	203	30	S	27	0	5
On the hip	204	10	S	29	0	5
The more northern of the two at the beginning of the hip	208	0	S	28	30	5
The more southern	207	0	S	30	0	5
The upmost part of the loins	208	40	S	33	10	5
The most southern of the three at the extremity of the tail	195	20	S	31	20	5
The middle one	195	10	S	30	0	4
The most northern of the three	196	20	S	29	20	4
The more southern of the two at the throat	212	10	S	17	0	4
The more northern	212	40	S	15	20	4
The more western of the two at the opening of the jaws	209	0	S	13	30	4
The more eastern	210	0	S	12	50	4
The more southern of the two on the forefoot	240	40	S	11	80	4
The more northern	239	50	S	10	0	4
19 stars: 2 of third magnitude, 11 of fourth, 6 of fifth						
ARA OR THURIBULUM, THE ALTAR OR THE CENSER						
The more northern of the two at the base	231	0	S	22	40	5
The more southern	233	40	S	25	45	4
At the center of the altar	229	30	S	26	30	4
[62ª] The most northern of the three on the hearth	224	0	S	30	20	5
The more southern of the remaining two contiguous stars	228	30	S	34	10	4
The more northern	228	20	S	33	20	4
In the midst of the flames	224	10	S	34	10	4
7 Stars: 5 of fourth magnitude, 2 of fifth						
CORONA AUSTRINA, OR SOUTHERN CROWN						
The more western star on the outer periphery	242	30	S	21	30	4
East of that on the crown	245	0	S	21	0	5
East of that too	246	30	S	20	20	5
Farther east of that also	248	10	S	20	0	4
East of that and west of the knee of Sagittarius	249	30	S	18	30	5
The bright star to the north on the knee	250	40	S	17	10	4
The more northern	250	10	S	16	0	4
Still more northern	249	50	S	15	20	4

SOUTHERN SIGNS

Constellations	Longitude			Latitude		Magnitude
	Deg.	Min.		Deg.	Min.	
The more eastern of the two on the northern part of the periphery	248	30	S	15	50	6
The more western	248	0	S	14	50	6
Some distance west of those	245	10	S	14	40	5
Still west of that	243	0	S	15	50	5
The last star, more towards the south	242	30	S	18	30	5
13 stars: 5 of fourth magnitude, 6 of fifth, 2 of sixth						
PISCIS AUSTRINUS, OR THE SOUTHERN FISH						
In the mouth, and the same as at the extremity of Aqua	300	20	S	23	0	1
The most western of the three on the head	294	0	S	21	20	4
The middle one	297	30	S	22	15	4
The most eastern	299	0	S	22	30	4
At the gills	297	40	S	16	15	4
On the southern and dorsal fin	288	30	S	19	30	5
The more eastern of the two in the belly	294	30	S	15	10	5
The more western	292	10	S	14	30	4
The most eastern of the three on the northern fin	288	30	S	15	15	4
The middle one	285	10	S	16	30	4
The most western of the three	284	20	S	18	10	4
At the extremity of the tail	289	20	S	22	15	4
11 stars beside the first: 9 of fourth magnitude, 2 of fifth						
[62ᵇ] UNCONSTELLATED STARS AROUND PISCIS AUSTRINUS						
The most western of the bright stars west of Piscis	271	20	S	22	20	3
The middle one	274	30	S	22	10	3
The most eastern of the three	277	20	S	21	0	3
The obscure star west of that	275	20	S	20	50	5
The more southern of the two remaining to the north	277	10	S	16	0	4
The more northern	277	10	S	14	50	4
6 stars: 3 of third magnitude, 2 of fourth, 1 of fifth						

In the southern region 316 stars: 7 of first magnitude, 18 of second, 60 of third, 167 of fourth, 54 of fifth, 9 of sixth, and 1 nebulous. And so there are altogether 1024 stars: 15 of first magnitude, 45 of second, 206 of third, 476 of fourth, 217 of fifth, 49 of sixth, 11 obscure, and 5 nebulous.

BOOK THREE

1. ON THE PRECESSIONS OF THE SOLSTICES AND EQUINOXES

[63ᵃ] Having depicted the appearance of the fixed stars in relation to the annual revolution, we must pass on; and we shall treat first of the change of the equinoxes, by reason of which even the fixed stars are believed to move. Now we find that the ancient mathematicians made no distinction between the "turning" or natural year, which begins at an equinox or solstice, and the year which is determined by means of some one of the fixed stars. That is why they thought the Olympic years, which they measured from the rising of Canicula, were the same as the years measured from the summer solstice, since they did not yet know the distinction between the two.

But Hipparchus of Rhodes, a man of wonderful acumen, was the first to call attention to the fact that there was a difference in the length of these two kinds of year. While making careful observations of the magnitude of the year, he found that it was longer as measured from the fixed stars than as measured from the equinoxes or solstices. Hence he believed that the fixed stars too possessed a movement eastward, but one so slow as not to be immediately perceptible. But now through the passage of time, the movement has become very evident. By it we discern a rising and setting of the signs and stars which are already far different from those risings and settings described by the ancients; and we see that the twelve parts of the ecliptic have receded from the signs of the fixed stars by a rather great interval, although in the beginning they agreed in name and in position.

Moreover, an irregular movement has been found; and wishing to assign the cause for its irregularity, astronomers have brought forward different theories. Some maintained that there was a sort of swinging movement of the suspended world—like the movement in latitude which we find in the case of the planets—and that back and forth within fixed limits as far out as the world has gone forward in one direction it will come back again in the other at some time,[1] and that the extent of its digression from the middle on either side was not more than 8°. But this already outdated theory can no longer hold, especially because [63ᵇ] it is already clear enough that the head of the constellation of Aries has become more than three times 8° distant from the spring equinox—and similarly for other stars—and no trace of a regression has been perceived during so many ages. Others indeed have opined that the sphere of the fixed stars moves forward but does so by irregular steps; and nevertheless they have failed to define any fixed mode of movement.

[1]*i.e.*, the sphere of the world has rotated westward and will at some time rotate eastward the same distance.

Moreover, there is an additional surprise of nature, in that the obliquity of the ecliptic does not appear so great to us as before Ptolemy—as we said above.

For the sake of a cause for these facts some have thought up a ninth sphere and others a tenth: they thought these facts could be explained through those spheres; but they were unable to produce what they had promised. Already an eleventh sphere has begun to see the light of day; and in talking of the movement of the Earth we shall easily prove that this number of circles is superfluous.

For, as we have already set out separately in Book I, the two revolutions, that is, of the annual declination and of the centre of the Earth, are not altogether equal, namely because the restoration of the declination slightly anticipates the period of the centre, whence it necessarily follows that the equinoxes seem to arrive before their time—not that the sphere of the fixed stars is moved eastward, but rather that the equator is moved westward, as it is inclined obliquely to the plane of the ecliptic in proportion to the amount of deflexion of the axis of the terrestrial globe. For it seems more accurate to say that the equator is inclined obliquely to the ecliptic than that the ecliptic, a greater circle, is inclined to the equator, a smaller. For the ecliptic, which is described by the distance between the sun and the Earth during the annual circuit, is much greater than the equator, which is described by the daily movement of the Earth around its axis. And in this way the common sections of the equator and the oblique ecliptic are perceived, with the passage of time, to get ahead, while the stars are perceived to lag behind. But the measure of this movement and the ratio of its irregularity were hidden from our predecessors, because the period of revolution was not yet known on account of its surprising slowness—I mean that during the many ages after it was first noticed by men, it has advanced through hardly a fifteenth part of a circle, or 24°. Nevertheless, we shall state things with as much certitude as possible, with the aid of what we have learned concerning these facts from the history of observations down to our own time.

2. History of the Observations Confirming the Irregular Precession of the Equinoxes and Solstices

[64^a] Accordingly in the 36th year of the first of the seventy-six-year periods of Callippus, which was the 30th year after the death of Alexander the Great, Timochares the Alexandrian, who was the first to investigate the positions of the fixed stars, recorded that Spica, which is in the constellation of Virgo, had an angular elongation of $82^1/_3$° from the point of summer solstice with a southern latitude of 2°; and that the star in the forehead of Scorpio which is the most northward of the three and is first in the order of formation of the sign had a latitude of $1^1/_3$° and a longitude of 32° from the autumn equinox.

And again in the 48th year of the same period he found that Spica in Virgo had a longitude of $82^1/_2°$ from the summer solstice but had kept the same latitude.

Now Hipparchus in the 50th year of the third period of Callippus, in the 196th year since the death of Alexander, found that the star called Regulus, which is in the breast of Leo, was $29^5/_6°$ to the east of the summer solstice.

Next Menelaus, the Roman geometer, in the first year of Trajan's reign, i.e., in the 99th year since the birth of Christ, and in the 422nd year since the death of Alexander, recorded that Spica in Virgo had a longitude of $86^1/_4°$ from the (summer) solstice and that the star in the forehead of Scorpio had a longitude of $35^{11}/_{12}°$ from the autumn equinox.

Following them, Ptolemy, in the second year of the reign of Antoninus Pius, in the 462nd year since the death of Alexander, discovered that Regulus in Leo had a longitude of $32^1/_2°$ from the (summer) solstice; Spica, $86^1/_2°$; and that the star in the forehead of Scorpio had a longitude of $36^1/_3°$ from the autumn equinox, with no change in latitude—as was set forth above in drawing up the tables. And we have passed these things in review, just as they were recorded by our predecessors.

After a great lapse of time, however, in the 1202nd year after the death of Alexander, came the observations of al-Battani the Harranite; and we may place the utmost confidence in them. In that year Regulus, or Basiliscus, was seen to have attained a longitude of 44°5' from the (summer) solstice; and the star in the forehead of Scorpio, one of 47°50' [64^b] from the autumn equinox. The latitude of these stars stayed completely the same, so that there is no longer any doubt on that score.

Wherefore in the year of Our Lord 1525, in the year after leap-year by the Roman calendar and 1849 Egyptian years after the death of Alexander, we were taking observations of the often mentioned Spica, at Frauenburg, in Prussia. And the greatest altitude of the star on the meridian circle was seen to be approximately 27°. We found that the latitude of Frauenburg was $54°19^1/_2°$. Wherefore its declination from the equator stood to be 8°40'. Hence its position became known as follows:

For we have described the meridian circle $ABCD$ through the poles of the ecliptic and the equator. Let AEC be the diameter and common section with the equator; and BED is the diameter and common section with the ecliptic. Let F be the north pole of the ecliptic and FEG its axis; and let B be the beginning of Capricorn and D of Cancer. Now let

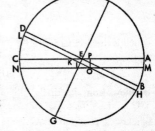

$$\text{arc } BH = 2°,$$

which is the southern latitude of the star. And from point H let HL be drawn parallel to BD; and let HL cut the axis of the ecliptic at I and the equator at K. Moreover, let

$$\text{arc } MA = 8°40',$$

in proportion to the southern declination of the star; and from point M let MN be drawn parallel to AC.

MN will cut HIL the parallel to the ecliptic; therefore let MN cut HIL at point O; and if the straight line OP is drawn at right angles to MN and AC, then

$$OP = {}^1/_2 \text{ ch. } 2\ AM.$$

But the circles having the diameters FG, HL, and MN are perpendicular to plane $ABCD$; and by Euclid's *Elements*, XI, 19, their common sections are at right angles to the same plane in points O and I. Hence by XI, 6, they (the common sections) are parallel to one another. And since I is the centre of the circle whose diameter is HL, therefore line OI will be equal to half the chord subtending twice an arc in a circle of diameter HL— an arc similar to the arc which measures the longitude of the star from the beginning of Libra, and this arc is what we are looking for. It is found in this way:

Since the exterior angle is equal to its interior and opposite,

$$\text{angle } AEB = \text{angle } OKP$$

and

$$\text{angle } OPK = 90°.$$

Accordingly

$$[65^a]\ OP : OK = {}^1/_2 \text{ ch. } 2\ AB : BE = {}^1/_2 \text{ ch. } 2\ AH : HIK.$$

For the lines comprehend triangles similar to OPK.

But

$$\text{arc } AB = 23°28^1/_2{}',$$

and

$$^1/_2 \text{ ch. } 2\ AB = 39,832,$$
$$\text{where } BE = 100,000.$$

And

$$\text{arc } ABH = 25°28^1/_2{}',$$
$$^1/_2 \text{ ch. } 2\ ABH = 43,010,$$
$$\text{arc } MA = 8°40',$$

which is the declination, and

$$^1/_2 \text{ ch. } 2\ MA = 15,069.$$

It follows from this that

$$HIK = 107,978,$$
$$OK = 37,831,$$

and by subtraction

$$HO = 70,147.$$

But

$$HOI = \frac{1}{2} \text{ ch. } HGL$$

and

$$\text{arc } HGL = 176°.$$

Then

$$HOI = 99,939,$$
$$\text{where } BE = 100,000.$$

And therefore by subtraction,

$$OI = HOI - HO = 29,792.$$

But in so far as HOI = radius = 100,000,

$$OI = 29,810$$
$$\frac{1}{2} \text{ ch. 2 arc } 17°21'.$$

This was the distance of Spica in (the constellation) Virgo from the beginning of Libra; and the position of the star was here. Moreover, ten years before, in 1515, we found that it had a declination of 8°36'; and its position was 17°14' distant from the beginning of the Balances.

Now Ptolemy recorded that it had a declination of only $\frac{1}{2}°$. Therefore its position was at 26°40' of the (zodiacal sign) Virgo, which seems to be more or less true in comparison with the previous observations.

Hence it appears clearly enough that during nearly the whole period of 432 years from Timochares to Ptolemy the equinoxes and solstices were moved according to a precession of 1° per 100 years—if a constant ratio is set up between the time and the amount of precession, which added up to $4\frac{1}{3}°$. For in the 266 years between Hipparchus and Ptolemy the longitude of Basiliscus in Leo from the summer solstice moved $2\frac{2}{3}°$, so that here too, by taking the time into comparison, there is found a precession of 1° per 100 years.

Moreover, because during the 782 mean years between the observation of Menelaus and that of al-Battani the first star in the forehead of Scorpio had a change in longitude of 11°55', it will certainly seem that 1° should be assigned not to 100 years but rather to 66 years; but for the 741 years after Ptolemy, 1° to only 65 years.

If finally the remaining space of 645 years is compared with the difference of 9°11' given by our observation, there will be 71 years allotted to 1°.

From this it is clear that the precession of the equinoxes was slower [65^{b}] during the 400 years before Ptolemy than during the time between Ptolemy and al-Battani, and that the precession in this middle period was speedier than in the time from al-Battani to us.

Moreover, there is found a difference in the movement of obliquity, since Aristarchus of Samos found that the obliquity of the ecliptic and the equator was 23°51'20", just as Ptolemy did; al-Battani, 23°35'; 190 years later Arzachel the

Spaniard, 23°34'. And similarly after 230 years Prophatius the Jew found that the obliquity was approximately 2' smaller. And in our time it has not been found greater than 23°28$^1/_2$'. Hence it is also clear that the movement was least from the time of Aristarchus to that of Ptolemy and greatest from that of Ptolemy to that of al-Battani.

3. The Hypotheses by Means of Which the Mutation of the Equinoxes and of the Obliquity of the Ecliptic and the Equator are Shown

Accordingly it seems clear from this that the solstices and equinoxes change around in an irregular movement. No one perhaps will bring forward a better reason for this than that there is a certain deflexion of the axis of the Earth and the poles of the equator. For that seems to follow upon the hypothesis of the movement of the Earth, since it is clear that the ecliptic remains perpetually unchangeable—the constant latitudes of the fixed stars bear witness to that—while the equator moves. For if the movement of the axis of the Earth were simply and exactly in proportion to the movement of the centre, there would not appear at all any precession of the equinoxes and solstices, as we said; but as these movements differ from one another by a variable difference, it was necessary for the solstices and equinoxes to precede the positions of the stars in an irregular movement.

The same thing happens in the case of the movement of declination, which changes the obliquity of the ecliptic irregularly—although this obliquity should be assigned more rightly to the equator.

For this reason you should understand two reciprocal movements belonging wholly to the poles, like hanging balances, since the poles and circles in a sphere imply one another mutually and are in agreement. Therefore there will be one movement which changes the inclination of those circles [66^a] by moving the poles up and down in proportion to the angle of section. There is another which alternately increases and decreases the solstitial and equinoctial precessions by a movement taking place crosswise. Now we call these movements "librations," or "swinging movements," because like hanging bodies swinging over the same course between two limits, they become faster in the middle and very slow at the extremes. And such movements occur very often in connection with the latitudes of the planets, as we shall see in the proper place.

They differ moreover in their periods, because the irregular movement of the equinoxes is restored twice during one restoration of obliquity. But as in every apparent irregular movement, it is necessary to understand a certain mean, through which the ratio of irregularity can be determined; so in this case too it was quite necessary to consider the mean poles and the mean equator and also the mean equinoxes and points of solstice. The poles and the terrestrial equator, by being deflected in opposite directions

away from these mean poles, though within fixed limits, make those regular movements appear to be irregular. And so these two librations competing with one another make the poles of the earth in the passage of time describe certain lines similar to a twisted garland.

But since it is not easy to explain these things adequately with words, and still

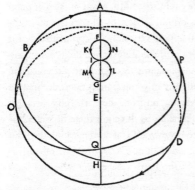

less—I fear—to have them grasped by the hearing, unless they are also viewed by the eyes, therefore let us describe on a sphere circle *ABCD* which is the ecliptic. Let the north pole (of the ecliptic) be *E*, the beginning of Capricornus *A*, that of Cancer *C*, that of Aries *B*, and that of Libra *D*. And through points *A* and *C* and pole *E* let circle *AEC* be drawn. And let the greatest distance between the north poles of the ecliptic and of the equator be *EF*, and the least *EG*. Similarly let *I* be the pole in the middle position, and around *I* let the equator *BHD* be described, and let that be called the mean equator; and let *B* and *D* be called the mean equinoxes.

Let the poles of the equator, the equinoxes, and the equator be all carried around *E* by an always regular movement westward, *i.e.*, counter to the order of the signs in the sphere of the fixed stars, and with a slow movement, as I said. Now let there be understood two reciprocal movements of the terrestrial poles like hanging bodies—one of them between the limits *F* and *G*, which will be called the movement of anomaly,[1]— *i.e.*, irregularity—of declination; the other from westward to eastward, and from eastward to westward. This second movement, which has twice the velocity of the first, we shall call the anomaly of the equinoxes. As both of these movements belong to the poles of the earth, they deflect the poles in a surprising way.

For first with *F* as the north pole of the earth, [66ᵇ] the equator described around the pole will pass through the same sections *B* and *D*, *i.e.*, through the poles of circle *AFEC*. But it will make greater angles of obliquity in proportion to arc *FI*. Now the second movement supervening does not allow the terrestrial pole, which was about to cross from the assumed starting point *F* to the mean obliquity at *I*, to proceed in a straight line along *FI*, but draws it aside in a circular movement towards its farthest eastward latitude, which is at *K*. The intersection of the apparent equator *OQP* described around this position will not be in *B* but to the east of it in *O*, and the precession of the equinoxes will be decreased in proportion to arc *BO*. Changing its direction and

---

[1]The term *anomaly* will be used to designate a regular movement the compounding of which with the principal regular movement being considered makes that principal movement appear irregular.

moving westwards, the pole is carried by the two simultaneously competing movements to the mean position *I*. And the apparent equator is in all respects identical with the regular or mean equator. Crossing there, the pole of the earth moves westward and separates the apparent equator from the mean equator and increases the precession of the equinoxes up to the other limit *L*. There changing its direction again, it subtracts what it had just added to the precession of the equinoxes, until, when situated at point *G*, it causes the least obliquity at the same common section *B*, where once more the movement of the equinoxes and solstices will appear very slow, in approximately the same way as at *F*. At this time the irregularity of the equinoxes stands to have completed its revolution, since it has passed from the mean through both extremes and back to the mean; while the movement of obliquity in going from greatest declination to least has completed only half its circuit. Moving on from there, the pole advances eastward to the farthest limit *M*; and, after reversing its direction there becomes one with the mean pole *I*; and once more it proceeds westward and after reaching the limit *N* finally [67ª] completes what we called the twisted line *FKILGMINF*. And so it is clear that in one cycle of obliquity the pole of the Earth reaches the westward limit twice and the eastward limit twice.

4. How the Reciprocal Movement or Movement of Libration Is Composed of Circular Movements

Accordingly we shall make clear exactly how this movement agrees with the appearances. In the meantime someone will ask how the regularity of these librations is to be understood, since it was said in the beginning that the celestial movement was regular, or composed of regular and circular movements. But here in either case of libration two movements are apparent as one movement between two limits, and the two limits necessarily make a cessation of movement intervene. For we acknowledge that there are twin movements, which are demonstrated from regular movements in this way.

Let there be the straight line *AB*, and let it be cut into four equal parts at points *C*, *D*, and *E*. Let the homocentric circles *ADB* and *CDE* be described around *D* in the same plane. And in the selfsame plane *ADB* and *CDE*, let any point *F* be taken on the circumference of the inner circle; and with *F* as centre and radius equal to *FD* let circle *GHD* be described. And let it cut the straight line *AB* at point *H*; and let the diameter *DFG* be drawn. We have to show that when the twin movements of circles *GHD* and *CFE* compete

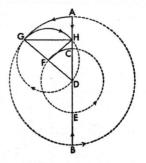

with one another, the movable point H proceeds back and forth along the same straight line AB by a reciprocal motion.

This will take place if we understand that H is moved in a different direction from F and through twice the distance, since the angle same CDF which is situated at the centre of circle CFE and on the circumference of circle GHD comprehends both arcs of the equal circles: arc FC and arc GH which is twice arc FC.

It is laid down that at some time upon the coincidence of the straight lines ACD and DFG the moving point H will be at G, which will then coincide with A; and F will be at C. Now, however, the centre F has moved towards the right along CF, and H has moved along the circumference to the left twice the distance CF [67^b] or vice versa; accordingly H will be deflected along line AB; otherwise the part would be greater than the whole, as it is easy to see. But H has moved away from its first position along the length AH made by the bent line DFH, which is equal to AD; and H has moved for a distance by which the diameter DFG exceeds chord DH. And in this way H will be made to arrive at centre D, which will be the point of tangency of circle DHG with straight line AB, namely when GD is at right angles to AB; and then H will reach the other limit at B, and from that position it will move back again according to the same ratio.

Therefore it is clear that movement along a straight line is compounded of two circular movements which compete with one another in this way; and that a reciprocal and irregular movement is composed of regular movements; as was to be shown. Moreover it follows from this that the straight line GH will always be at right angles to AB; for lines DH and HG, being in a semicircle, will always comprehend a right angle. And accordingly

$$GH = {}^1/_2 \text{ ch. } 2\, AG;$$

and

$$DH = {}^1/_2 \text{ ch. } 2\, (90° - AG),$$

because circle AGB has twice the diameter of circle HGD.

5. A DEMONSTRATION OF THE IRREGULARITY OF THE EQUINOCTIAL PRECESSION AND THE OBLIQUITY

For this reason some call this movement of the circle a movement in width, *i.e.*, along the diameter. But they determine its periodicity and its regularity by means of the circumference, and its magnitude by means of the chords subtending. On that account it is easily shown that the movement appears irregular and faster at the centre and slower [68^a] at the circumference.

For let there be the semicircle ABC with centre D and diameter ADC, and let it be bisected at point B. Now let equal arcs AE and BF be taken, and from points F and E

let *EG* and *FK* be drawn perpendicular to *ADC*. Therefore, since

$$2\ DK = 2\ \text{ch.}\ BF,$$

and

$$2\ EG = 2\ \text{ch.}\ AE,$$

then

$$DK = EG.$$

But by Euclid's *Elements*, III, 7,

$$AG < GE;$$

hence

$$AG < DK.$$

But *GA* and *KD* will take up equal time because

$$\text{arc}\ AE = \text{arc}\ BF;$$

therefore the movement in the neighbourhood of an *A* will (appear to) be slower than in the neighbourhood of the centre *D*.

Having shown this, let us take *L* as the centre of the Earth, so that the straight line *DL* is perpendicular to plane *ABC* of the semicircle; and with *L* as centre and through points *A* and *C* let arc *AMC* of a circle be described; and let *LDM* be drawn in a straight line. Accordingly the pole of the semicircle *ABC* will be at *M*, and *ADC* will be the common section of the circles. Let *LA* and *LC* be joined; and similarly *LK* and *LG* too. And let *LK* and *LG* extended in straight lines cut arc *AMC* at *N* and *O*. Therefore, since angle *LDK* is right, angle *LKD* is acute. Wherefore too the line *LK* is longer than *LD*, and all the more is side *LG* greater than side *LK*, and *LA* than *LG* in the obtuse triangles. Therefore the circle described with *L* as centre and *LK* as radius will fall beyond *LD*, but will cut *LG* and *LA*; let it be described, and let it be *PKRS*. And since

$$\text{trgl.}\ LDK < \text{sect.}\ LPK,$$

while

$$\text{trgl.}\ LGA > \text{sect.}\ LRS,$$

on that account

$$\text{trgl.}\ LDK : \text{sect.}\ LPK < \text{trgl.}\ LGA : \text{sect.}\ LRS.$$

Hence, alternately also,

$$\text{trgl.}\ LDK : \text{trgl.}\ LGA < \text{sect.}\ LPK : \text{sect.}\ LRS.$$

And by Euclid's *Elements*, VI, 1,

$$\text{trgl.}\ LDK : \text{trgl.}\ LGA = \text{base}\ DK : \text{base}\ AG.$$

But

$$\text{sect.}\ LPK : \text{sect.}\ LRS = \text{angle}\ DLK : \text{angle}\ RLS = \text{arc}\ MN : \text{arc}\ OA.$$

Therefore

$$\text{base}\ DK : \text{base}\ GA < \text{arc}\ MN : \text{arc}\ OA.$$

But we have already shown that

$$DK > GA.$$

All the more then

$$[68^b]\ MN > OA.$$

And arcs *MN* and *OA* are understood as having been described during equal intervals of time by the poles of the earth in accordance with the equal arcs of anomaly *AE* and *BF*—as was to be shown. But since the difference between greatest and least obliquity is so slight as not to exceed $^2/_5°$, there will be no sensible difference between the curved line *AMC* and the straight line *ADC*; and so no error will arise if we work simply with line *ADC* and semicircle *ABC*.

Practically the same thing happens in the case of the other movement of the poles, which has to do with the equinoxes, since this movement does not ascend to the mean degree, as will be apparent below. Once more let there be the circle *ABCD* through the poles of the ecliptic and the mean equator. We may call it the mean colure of Cancer. Let the semicircle of the ecliptic be *DEB* and the mean equator *AEC*; and let them cut one another at point *E*, where the mean equinox will be. Now let the pole of the equator be *F*, and let the great circle *FEI* be described through it. On that account it will be the colure of the mean or regular equinoxes.

Therefore for the sake of an easier demonstration let us separate the libration of the equinoxes from the obliquity of the ecliptic. On the colure *EF* let arc *FG* be taken,

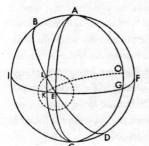

and through that distance let *G* the apparent pole of the equator be understood as removed from *F* the mean pole. And with *G* as a pole let *ALKC* the semicircle of the apparent equator be described. It will cut the ecliptic at *L*. Therefore point *L* will be the apparent equinox; and its distance from the mean equinox will be measured by arc *LE*, which is produced by arc *EK* the equal of *FG*.

But with *K* as a pole we shall describe circle *AGC*; and let it be understood that the equatorial pole during the time in which the libration *FG* takes place does not remain the "true" pole at point *G*, but, driven by another libration or swinging movement, moves away in the direction of the oblique ecliptic through arc *GO*. Therefore while ecliptic *BED* abides, the "true" equator will be changed to the "apparent" in accordance with the transposition of the pole to *O*. And similarly the movement of intersection *L* the apparent equinox will be faster in the neighbourhood of the mean (equinox) *E* and very slow in the neighbourhood of the extreme equinoxes, more or less in proportion to the swinging movement of the poles which we have already demonstrated—as was worth the trouble of our attention.

6. ON THE REGULAR MOVEMENTS OF THE
PRECESSION OF THE EQUINOXES AND OF THE
INCLINATION OF THE ECLIPTIC

[69ª] Now every apparent irregular circular movement passes through four termini: there is the terminus where it appears slow and the terminus where it appears fast,
as if at the extremes, and the terminus where it appears to have a mean velocity, as if at
the means, since from the point which is the end of decrease in velocity and the
beginning of increase it passes on to a mean velocity; and from the mean velocity it
increases till it becomes fast; again after being fast it approaches a mean velocity,
whence for the remainder of the cycle it changes to its former slowness.

By means of that it is possible to know in what part of the circle the position of
the non-uniform movement, or irregularity, is at a given time; and too by means of
these indications the restitution of the irregularity is perceptible.[1] Accordingly in a
quadrisected circle let *A* be the position of greatest slowness, *B* the mean of increasing
velocity, *C* the end of the increase and the beginning of the decrease, and *D* the mean
of decreasing velocity. Therefore, since, as was reported above, the apparent movement
of the precession of the equinoxes was found to be rather slow in the time between
Timochares and Ptolemy in comparison with the other times, and because for a while
it appeared regular and uniform, as is shown by the observations of Aristyllus,
Hipparchus, Agrippa, and Menelaus which were made at the middle of that time; it
argues that the apparent movement of the equinoxes had been simply at its slowest and
at the middle of this time was at the beginning of increase in velocity, when the cessation of the decrease conjoined to the beginning of the increase by reason of mutual
compensation made the movement seem uniform for the time being. Accordingly
Timochares' observation must be placed in the fourth quadrant of the circle along *DA*;
but Ptolemy's falls in the first quadrant along *AB*. Again, because in the second interval, the one between Ptolemy and al-Battani the Harranite the movement is found to
have been faster than in the third, it is clear that the point of
highest velocity was passed during the second interval of time,
and the irregularity had already reached the third quadrant of
the circle along *CD*, and that from the third interval down to us
the restoration of the irregularity was nearly completed and has
nearly returned to its starting-point with Timochares. For if we
divide the cycle of 1819 years between Timochares and us into

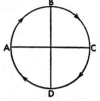

the customary 360 parts, we shall have proportionately an arc of 85¹/₂° for 432 years;
146°51' for 742 years; and the remaining arc of 127°39' for the remaining 645 years.

[1]This circle is not the circle of libration, of course, but a circle which typifies the cycle in velocity which results from compounding the libration with the regular movement of precession of the equinoxes.

We have made these determinations by an obvious and simple inference; [69b] but upon working them over with stricter calculations, to see how exactly they agree with observations, we find that the movement of irregularity during the 1819 Egyptian years has already exceeded its complete revolution by 21°24', and that the time of the period comprehends only 1717 Egyptian years. In accordance with this ratio it is discovered that the first segment of the circle has 90°35'; the second, 155°34'; while the third period of 543 years will comprehend the remaining 113°51' of the circle. Now that these things have been set up in this way, the mean movement of the precession of the equinoxes is also disclosed; and it is 23°57' for these same 1717 years, at the end of which the whole movement of irregularity was restored to its pristine status, since for the 1819 years we have an apparent movement of approximately 25°1'.

But in 102 years after Timochares—the difference between the 1717 years and the 1819 years—the apparent movement must have been about 1°4', because it is probable that the apparent movement was then a little greater than to need only 1° per 100 years, since it was decreasing without yet having reached the end of the decrease. Hence, if we subtract 1°4' from 25°1' there will remain, as we said, for the 1717 Egyptian years a mean and regular movement of 23°57', now corrected according to the apparent and irregular movement: hence the complete and regular revolution of the precession of the equinoxes arises in 25,816 years, during which time 15 cycles of irregularity are traversed and approximately $1/_{28}$ cycle over and above.

Moreover, the movement of obliquity, whose restoration we said was twice as slow as the irregular precession of the equinoxes, accords with this ratio. For the fact that Ptolemy reports that during the 400 years between the time of Aristarchus of Samos and his own the obliquity of 23°51'20" had hardly changed at all indicates that the obliquity was then close to the limit of greatest obliquity, namely when the precession of the equinoxes was in its slowest motion. But now also when the same restoration of slowness is approaching, the inclination of the axis is not at its greatest but is near its least. In the middle of the time in between, al-Battani the Harranite, as was said, found that the inclination was 23°35'; 190 years after him Arzachel the Spaniard, 23°34'; and similarly 230 years later Prophatius the Jew found it approximately 2' less. Finally as regards our own times, we in 30 years of frequent observation found it approximately 23°28²/₅'—from which George Peurbach and John of Monteregium, [70a] our nearest predecessors, differ slightly. Here again it is perfectly obvious that for the 900 years after Ptolemy the change in obliquity was greater than for any other interval of time.

Therefore since we already have the cycle of irregularity of precession in 1717 years, we shall also have half the period of obliquity in that time, and in 3434 years its complete restoration. Wherefore if we divide the 360° by the number of the 3434 years or 180° by 1717, the quotient will be the annual movement of simple anomaly of

6'17"24"'9"". These in turn distributed through the 365 days give a daily movement of 1"2"'2"".

Similarly, when the mean precession of equinoxes has been distributed through the 1717 years—and there was 23°57'—an annual movement of 50"12"'5"" will be the result; and this distributed through the 365 days will give a daily movement of 8"'15"".

But in order that the movements may be more in the open and may be found right at hand when there is need of them, we shall draw up their tables or canons by the continuous and equal addition of annual movement—60 parts always being carried over into the minutes or degrees, if the sum exceeds that—and for the sake of convenience we shall keep on adding until we reach the 60th year, since the same configuration of numbers returns every sixty years, only with the denominations of degrees and minutes moved up, so that what were formerly seconds become minutes and so on.[1] By this abridgement in the form of brief tables, it will be possible merely by a double entry to determine and infer the regular movements for the years in question among the 3600 years. This is also the case with the number of days.

Moreover, in our computations of the celestial movements we shall employ the Egyptian years, which alone among the legal years are found equal. For it is necessary for the measure to agree with the measured; but that is not the case with the years of the Romans, Greeks, and Persians, for intercalations are made not in any single way, but according to the will of the people. But the Egyptian year contains no ambiguity as regards the fixed number of 365 days, in which throughout twelve equal months— which they name in order by these names: Thoth, Phaophi, Athyr, Chiach, Tybi, Mechyr, Phamenoth, Pharmuthi, Pachon, Pauni, Epiphi, and Mesori—in which, I say, six periods of 60 days are comprehended evenly together with the five remaining days, which they call the intercalary days. For that reason Egyptian years are most convenient for calculating regular movements. Any other years are easily reducible to them by resolving the days.

[1] That is to say, the same configurations of numbers return in multiples of sixty years, because the cycle of movement is divided according to the sexagesimal system—jut as it would return in multiples of ten years if the circle were divided according to the decimal system.

REGULAR MOVEMENT OF THE PRECESSION OF THE EQUINOXES IN YEARS AND PERIODS OF SIXTY YEARS

Egyptian Years	60°	°	'	"	'''	Egyptian Years	60°	°	'	"	'''
1	0	0	0	50	12	31	0	0	25	56	14
2	0	0	1	40	24	32	0	0	26	46	26
3	0	0	2	30	36	33	0	0	27	36	38
4	0	0	3	20	48	34	0	0	28	26	50
5	0	0	4	11	0	35	0	0	29	17	2
6	0	0	5	1	12	36	0	0	30	7	15
7	0	0	5	51	24	37	0	0	30	57	27
8	0	0	6	41	36	38	0	0	31	47	38
9	0	0	7	31	48	39	0	0	32	37	51
10	0	0	8	22	0	40	0	0	33	28	3
11	0	0	9	12	12	41	0	0	34	18	15
12	0	0	10	2	25	42	0	0	35	8	27
13	0	0	10	52	37	43	0	0	35	58	39
14	0	0	11	42	49	44	0	0	36	48	51
15	0	0	12	33	1	45	0	0	37	39	3
16	0	0	13	23	13	46	0	0	38	29	15
17	0	0	14	13	25	47	0	0	39	19	27
18	0	0	15	3	37	48	0	0	40	9	40
19	0	0	15	53	49	49	0	0	40	59	52
20	0	0	16	44	1	50	0	0	41	50	4
21	0	0	17	34	13	51	0	0	42	40	16
22	0	0	18	24	25	52	0	0	43	30	28
23	0	0	19	14	37	53	0	0	44	20	40
24	0	0	20	4	50	54	0	0	45	10	52
25	0	0	20	55	2	55	0	0	46	1	4
26	0	0	21	45	14	56	0	0	46	51	16
27	0	0	22	35	26	57	0	0	47	41	28
28	0	0	23	25	38	58	0	0	48	31	40
29	0	0	24	15	50	59	0	0	49	21	52
30	0	0	25	6	2	60	0	0	50	12	5

Position of the Birth of Christ—5'32'

REGULAR MOVEMENT OF THE PRECESSION OF THE EQUINOXES IN YEARS AND PERIODS OF SIXTY YEARS

Days	60°	°	'	"	'''	Days	60°	°	'	"	'''
1	0	0	0	0	8	31	0	0	0	4	15
2	0	0	0	0	16	32	0	0	0	4	24
3	0	0	0	0	24	33	0	0	0	4	32
4	0	0	0	0	33	34	0	0	0	4	40
5	0	0	0	0	41	35	0	0	0	4	48
6	0	0	0	0	49	36	0	0	0	4	57
7	0	0	0	0	57	37	0	0	0	5	5
8	0	0	0	1	6	38	0	0	0	5	13
9	0	0	0	1	14	39	0	0	0	5	21
10	0	0	0	1	22	40	0	0	0	5	30
11	0	0	0	1	30	41	0	0	0	5	38
12	0	0	0	1	39	42	0	0	0	5	46
13	0	0	0	1	47	43	0	0	0	5	54
14	0	0	0	1	55	44	0	0	0	6	3
15	0	0	0	2	3	45	0	0	0	6	11
16	0	0	0	2	12	46	0	0	0	6	19
17	0	0	0	2	20	47	0	0	0	6	27
18	0	0	0	2	28	48	0	0	0	6	36
19	0	0	0	2	36	49	0	0	0	6	44
20	0	0	0	2	45	50	0	0	0	6	52
21	0	0	0	2	53	51	0	0	0	7	0
22	0	0	0	3	1	52	0	0	0	7	9
23	0	0	0	3	9	53	0	0	0	7	17
24	0	0	0	3	18	54	0	0	0	7	25
25	0	0	0	3	26	55	0	0	0	7	33
26	0	0	0	3	34	56	0	0	0	7	42
27	0	0	0	3	42	57	0	0	0	7	50
28	0	0	0	3	51	58	0	0	0	7	58
29	0	0	0	3	59	59	0	0	0	8	6
30	0	0	0	4	7	60	0	0	0	8	15

Position of the Birth of Christ—5'32'

MOVEMENT OF THE SIMPLE ANOMALY OF EQUINOXES IN YEARS AND PERIODS OF SIXTY YEARS

Position of the Birth of Christ −6°45′

Egyptian Years	Longitude					Egyptian Years	Longitude				
	60°	°	′	″	‴		60°	°	′	″	‴
1	0	0	6	17	24	31	0	3	14	59	28
2	0	0	12	34	48	32	0	3	21	16	52
3	0	0	18	52	12	33	0	3	27	34	16
4	0	0	25	9	36	34	0	3	33	51	41
5	0	0	31	27	0	35	0	3	40	9	5
6	0	0	37	44	24	36	0	3	46	26	29
7	0	0	44	1	49	37	0	3	52	43	53
8	0	0	50	19	13	38	0	3	59	1	17
9	0	0	56	36	36	39	0	4	5	18	42
10	0	1	2	54	1	40	0	4	11	36	6
11	0	1	9	11	25	41	0	4	17	53	30
12	0	1	15	28	49	42	0	4	24	10	54
13	0	1	21	46	13	43	0	4	30	28	18
14	0	1	28	3	38	44	0	4	36	45	42
15	0	1	34	21	2	45	0	4	43	3	6
16	0	1	40	38	26	46	0	4	49	20	31
17	0	1	46	55	50	47	0	4	55	37	55
18	0	1	53	13	14	48	0	5	1	55	19
19	0	1	59	30	38	49	0	5	8	12	43
20	0	2	5	48	2	50	0	5	14	30	7
21	0	2	12	5	27	51	0	5	20	47	31
22	0	2	18	22	51	52	0	5	27	4	55
23	0	2	24	40	15	53	0	5	33	22	20
24	0	2	30	57	39	54	0	5	39	39	44
25	0	2	37	15	3	55	0	5	45	57	8
26	0	2	43	32	27	56	0	5	52	14	32
27	0	2	49	49	52	57	0	5	58	31	56
28	0	2	56	7	16	58	0	6	4	49	20
29	0	3	2	24	40	59	0	6	11	6	45
30	0	3	8	42	4	60	0	6	17	24	9

MOVEMENT OF THE SIMPLE ANOMALY OF EQUINOXES IN YEARS AND PERIODS OF SIXTY YEARS

Position of the Birth of Christ −6°45′

Days	Longitude					Days	Longitude				
	60°	°	′	″	‴		60°	°	′	″	‴
1	0	0	0	1	2	31	0	0	0	32	3
2	0	0	0	2	4	32	0	0	0	33	5
3	0	0	0	3	6	33	0	0	0	34	7
4	0	0	0	4	8	34	0	0	0	35	9
5	0	0	0	5	10	35	0	0	0	36	11
6	0	0	0	6	12	36	0	0	0	37	13
7	0	0	0	7	14	37	0	0	0	38	15
8	0	0	0	8	16	38	0	0	0	39	17
9	0	0	0	9	18	39	0	0	0	40	19
10	0	0	0	10	20	40	0	0	0	41	21
11	0	0	0	11	22	41	0	0	0	42	23
12	0	0	0	12	24	42	0	0	0	43	25
13	0	0	0	13	26	43	0	0	0	44	27
14	0	0	0	14	28	44	0	0	0	45	29
15	0	0	0	15	30	45	0	0	0	46	31
16	0	0	0	16	32	46	0	0	0	47	33
17	0	0	0	17	34	47	0	0	0	48	35
18	0	0	0	18	36	48	0	0	0	49	37
19	0	0	0	19	38	49	0	0	0	50	39
20	0	0	0	20	40	50	0	0	0	51	41
21	0	0	0	21	42	51	0	0	0	52	43
22	0	0	0	22	44	52	0	0	0	53	45
23	0	0	0	23	46	53	0	0	0	54	47
24	0	0	0	24	48	54	0	0	0	55	49
25	0	0	0	25	50	55	0	0	0	56	51
26	0	0	0	26	52	56	0	0	0	57	53
27	0	0	0	27	54	57	0	0	0	58	55
28	0	0	0	28	56	58	0	0	0	59	57
29	0	0	0	29	58	59	0	0	1	0	59
30	0	0	0	31	1	60	0	0	1	2	2

7. WHAT THE GREATEST DIFFERENCE IS BETWEEN THE REGULAR AND THE APPARENT PRECESSION OF THE EQUINOXES

[72[b]] Now that the mean movements have been set out in this way, we must inquire what the greatest difference is between the regular and the apparent movement of the equinoxes, or what the diameter of the small circle is, through which the movement of anomaly turns.[1] For when this is known, it will be easy to discern various other differences in the movements. As was written above, between the observation of Timochares, which came first, and that of Ptolemy in the second year of the reign of Antoninus Pius, there were 432 years; and during that time the mean movement was 6° and the apparent 4°20'. So the difference between them is 1°40'. And the movement of double[2] anomaly was 90°35'. Moreover, it seems that at the middle of this period of time or around there the apparent movement reached its peak of greatest slowness. At that time the (position of the) apparent movement necessarily agreed with the mean movement, and the true equinox and the mean equinox occurred at the same section of the circles.[3] Wherefore if we make a distribution of the movement and the time into two equal parts, there will be in each part as differences between the irregular and the regular movement $10/_{12}°$, which the circle of anomaly comprehends on either side beneath an arc of $45°17^1/_2'$. But since all these differences are very small and do not amount to $1^1/_2°$ on the ecliptic, and the straight lines are almost equal to the arcs subtended by them, and there is scarcely any diversity found in the third-minutes: we who are staying within the minutes will make no error if we employ straight lines instead of arcs.

[73[a]] Let ABC be a part of the ecliptic and on it let the mean equinox be B. And with B as pole let there be described the semicircle ADC, and let it cut the ecliptic at points A and C. Moreover let DB be drawn from the pole of the ecliptic, it will bisect the semicircle at D; and let D be understood to be limit of greatest slowness and beginning of the increase.[4] In the quadrant AD let

arc $DE = 45°17^1/_2'$;

and through point E from the pole of the ecliptic, let fall EF; and let

$$BF = 50'.$$

Our problem is to find out from this what the whole BFA is.

---

[1] *i.e.*, what the diameter of the small circle is, along which the libration takes place back and forth.

[2] The anomaly of precession is called the "double" anomaly because it completes two cycles for one cycle of the anomaly of obliquity.

[3] As Copernicus showed in Chapter 4, the movement of the libration, considered above, appears fastest around the centre of the circle. Hence the apparent movement itself will appear slowest when the fastest movement of libration is in opposition to the mean movement with which it is compounded. And the fastest libration is in opposition to the mean movement when the apparent equinox is swinging eastward and is in the neighbourhood of the centre of the circle or the mean equinox.

[4] Thus, circle ADC is the circle of libration transferred from the pole of the ecliptic to around the equinox—as in the last diagram in Chapter 5.

Accordingly it is clear that

$$2\ BF = \text{ch. } 2\ DE.$$

But

$$FB : AFB = 7107 : 10,000 = 50' : 70'.$$

Hence

$$AB = 1°10',$$

and that is the greatest difference between the mean and the apparent movement of the equinoxes, which we were seeking; and the greatest polar deflexion of 28' follows upon it.

[72b] For with this set-up let *ABC* be the arc of the ecliptic, *BDE* the mean equatorial arc, and *B* the mean section of the apparent equinoxes, either Aries or Libra, and

through the poles of *DBE* let fall *BF*. Now along arc *ABC* on both sides let

$$\text{arc } BI = \text{arc } BK = 1°10';$$

hence, by addition,

$$\text{arc } IBK = 2°20'.$$

Moreover, let there be drawn at right angles to *FB* extended to *FBH* the two arcs *IG* and *HK* of the apparent equators. Now I say "at right angles," [73a] though the poles of *IG* and *IK* are usually outside of circle *BF*, since the movement of obliquity gets mixed in, as was seen in the hypothesis, but on account of the distance being very slight—for at its greatest it does not exceed 90°/350—we employ these angles as angles which are right to sense-perception. For no great error will appear on that account. Therefore in triangle *IBG*

$$\text{angle } IBG = 66°20',$$

since its complement, as being the angle of mean obliquity of the ecliptic,

$$\text{angle } DBA = 23°40'.$$

And

$$\text{angle } BGI = 90°.$$

Moreover,

$$\text{angle BIG} \doteqdot \text{angle IBD},$$

because they are alternate angles. And

$$\text{side } IB = 70'.$$

Therefore too

$$\text{arc } BG = 28',$$

and that is the distance between the poles of the mean and the apparent equator.

Similarly in triangle *BHK*,

$$\text{angle } BHK = \text{angle } IGB$$

and

$$\text{angle } HBK = \text{angle } IBG,$$

and

$$\text{side } BK = \text{side } BI.$$

Moreover,

$$BH = BG = 28'.$$

For

$$GB : IB = BH : BK;$$

and the movements will be of the same ratio in the poles as in the intersections.

8. On the Particular Differences in the Movements and the Table of Them

[73ᵇ] Therefore since

$$\text{arc } AB = 70',$$

and since arc AB does not appear to differ from the chord subtending it lengthwise, it will not be difficult to exhibit certain other differences between the mean and the apparent movements. The Greeks call the differences προσθαφαιρέἰς, or "additosubtractions," and later writers "aequationes," by the subtraction or addition of which the apparent movements are made to harmonize (with the mean movements). We shall employ the Greek word as being more fitting. Therefore if

$$\text{arc } ED = 3°,$$

then in accordance with the ratio of AB to the chord BF,

$$\text{arc } BF = 4',$$

which is the additosubtraction. And if

$$ED = 6°,$$

then

$$\text{arc } BF = 7';$$

and if

$$ED = 9°,$$

then

$$BF = 11',$$

and so on.

We think we should use a similar ratio in the case of the change of obliquity also, where, as we said, a difference of 24' has been found between the greatest and the least obliquity. These 24' subtend a semicircle of simple anomaly every 1717 years, and the mean differences subtending a quadrant of a circle will be 12', where the pole of the small circle of this anomaly will be at an obliquity of 23°40'.

And in this way, as we said, we shall extract the remaining parts of difference approximately in proportion to the aforesaid as in the subjoined table. And if through

these demonstrations the apparent movements can be compounded by various modes, nevertheless that mode is better whereby all the particular additosubtractions may be taken separately, that the calculus of their movements may be easier to understand and may agree better with the explanations of what has been demonstrated.

Accordingly we have drawn up a table of sixty rows, increasing by 3°'s. For in this way it will not be spread over too much space, and it will not seem to be compressed into too little—as we shall do in the case of the similar remaining tables. The table will have only four main columns, the first two of which will contain the degrees of both semicircles; and we call them the common numbers, because the obliquity of the circle of signs is taken from the simple number, and twice the number applies to the additosubtractions of the movement of the equinoxes; and the numbers have their commencement at the beginning of the increase [74ᵃ] (in velocity).[1] In the third column will be placed the additosubtractions of the equinoxes corresponding to the single 3°'s; and they are to be added to, or subtracted from, the mean movement—which we measure from the head of Aries at the spring equinox. The subtractive additosubtractions correspond to the numbers in the first semicircle of the anomaly or the first column; and the additive, to those in the second column and the second semicircle. Finally, in the last column are the minutes, which are called the differences in the proportions of obliquity and which go up to 60', since in place of the difference of 24' between greatest and least obliquity we are putting 60', and we adjust the proportional minutes to them in the same ratio in proportion to the other differences of obliquity. On that account we place 60' as corresponding to the beginning and end of the anomaly; but where the difference of obliquity is 22', as in the anomaly of 33°, we put 55' instead. In this way we put 50' in place of 20', as in the anomaly of 48°; and so on for the rest, as in the subjoined table.[2]

[1] i.e., in the foregoing diagram, the first quadrant comprises the arc *DA*; and the fourth quadrant the arc *CD*.

[2] Thus, let line *FIG*, as in Chapter 3, represent the colors of the solstices. Point *F* is the limit of greatest obliquity of the ecliptic, point *G* that of the least; and thus the distance *FG* is 28'. The distance *KN* of the libration of the equinoxes is 2°20'. In the foregoing table the anomalies of precession and of obliquity we taken as starting at point *I* and proceeding along the route *INFKILGM*.

ADDITIONS-AND-SUBTRACTIONS OF EQUINOXES, OBLIQUITY OF THE ECLIPTIC

Common Numbers		Additions-and-Subtractions of Movement of Equinoxes		Proportional Minutes of Obliquity	Common Numbers		Additions-and-Subtractions of Movement of Equinoxes		Proportional Minutes of Obliquity
Deg.	Deg.	Deg.	Min.		Deg.	Deg.	Deg.	Min.	
3	357	0	4	60	93	267	1	10	28
6	354	0	7	60	96	264	1	10	27
9	351	0	11	60	99	261	1	9	25
12	348	0	14	59	102	258	1	9	24
15	345	0	18	59	105	255	1	8	22
18	342	0	21	59	108	252	1	7	21
21	339	0	25	58	111	249	1	5	19
24	336	0	28	57	114	246	1	4	18
27	333	0	32	56	117	243	1	2	16
30	330	0	35	56	120	240	1	1	15
33	327	0	38	55	123	237	0	59	14
36	324	0	41	54	126	234	0	56	12
39	321	0	44	53	129	231	0	54	11
42	318	0	47	52	132	228	0	52	10
45	315	0	49	51	135	225	0	49	9
48	312	0	52	50	138	222	0	47	8
51	309	0	54	49	141	219	0	44	7
54	306	0	56	48	144	216	0	41	6
57	303	0	9	46	147	213	0	38	5
60	300	1	1	45	150	210	0	35	4
63	297	1	2	44	153	207	0	32	3
66	294	1	4	42	156	204	0	28	3
69	291	1	5	41	159	201	0	25	2
72	288	1	7	39	162	198	0	21	1
75	285	1	8	38	165	195	0	18	1
78	282	1	9	36	168	192	0	14	1
81	279	1	9	35	171	189	0	11	0
84	276	1	10	33	174	186	0	7	0
87	273	1	10	32	177	183	0	4	0
90	270	1	10	30	180	180	0	0	0

9. ON THE EXAMINATION AND CORRECTION OF THAT WHICH WAS SET FORTH CONCERNING THE PRECESSION OF THE EQUINOXES

[75ᵃ] But since by an inference we took the beginning of increase in the movement of anomaly as occurring in the middle of the time from the 36th year of the first period of Callippus to the 2nd year of Antoninus, and we take the order of the movement of anomaly from that beginning; it is still necessary for us to test whether we did that correctly and whether it agrees with the observations.

Let us consider again the three observations of the stars made by Timochares, Ptolemy, and al-Battani the Harranite: And it is clear that there were 432 Egyptian years in the first interval and 742 years in the second. The regular movement in the first span of time was 6°; the irregular movement 4°20'; and the movement of double anomaly 90°35', subtracting 1°40' from the regular movement. During the second interval the regular movement was 10°21', the irregular $11^1/_2$°; and the movement of double anomaly was 155°34', adding 1°9' to the regular movement.

Now as before let the arc of the ecliptic be *ABC*, and let *B*—which is to be the mean spring equinox—be taken as a pole; let

$$\text{arc } AB = 1°10',$$

and let the small circle *ADCE* be described. But let the regular movement of *B* be understood as in the direction of *A*, *i.e.*, westward; and let *A* be the westward limit, where the irregular equinox is westernmost; and *C* the eastern limit, where the irregular equinox is easternmost. Furthermore, from the pole of the ecliptic drop *DBE* through point *B*. *DBE* together with the ecliptic will cut the small circle *ADCE* into four equal parts, since circles described through the poles of one another cut one another at right angles. However since the movement in the semicircle *ADC* is eastward, and the move-

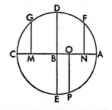

ment remaining in *CEA* is westward, the extreme slowness of the apparent equinox will be at *D* on account of its resistance to the forward movement of *B*; but there will be at *E* the greatest velocity for the movements moving forwards in the same direction.

Moreover on either side of *D* let

$$\text{arc } FD = \text{arc } DG = 45°17^1/_2'$$

Let *F* be the first terminus of the anomaly—the one observed by Timochares; *G* the second—the one observed by Ptolemy; and *P* the third—the one observed by al-Battani. And through these points let fall great circles *FN*, *GM*, and *OP* through the poles of the ecliptic; and they all [75ᵇ] appear in this very small circle rather much like straight lines. Therefore

$$\text{arc } FDG = 99°35',$$

where circle $ADCE = 360°$,

wherefrom

$$-add.\ MN = 1°40',$$
$$\text{where } ABC = 2°20'.$$

And

$$\text{arc } GCEP = 155°34',$$

wherefrom

$$+\ add.\ MBO = 109'.$$

Accordingly, by subtraction,

$$\text{arc } PAF = 113°51'$$

wherefrom

$$+\ add.\ ON = 31'$$
$$\text{where } AB = 70'.$$

But since by addition

$$\text{arc } DGCEP = 200°51'$$

and

$$EP = DGCEP - 180° = 200°51';$$

therefore by the table of chords in a circle, as if a straight line,

$$BO = 356,$$
$$\text{where } AB = 1,000.$$

But

$$BO = 24',$$
$$\text{where } AB = 70';$$

and

$$MB = 50'.$$

Hence, by addition,

$$MBO = 74',$$

and,

$$NO = MN - MBO = 26'.$$

But, in the foregoing

$$MBO = 69',$$

and

$$NO = 31'.$$

Hence NO has a deficiency of 5'; and MO has an excess of 5'. Accordingly the circle $ADCE$ must be revolved, until there is compensation on both sides.

But this will take place if

$$\text{arc } DG = 42^1/_2°,$$

so that by subtraction,

arc $DF = 48°5'$.

For by this both errors will seem to be corrected and everything else will be all right, since—with the beginning at D the limit of greatest slowness—

arc $DGCEPAF = 311°55'$,

which is the movement of anomaly at the first terminus; at the second terminus

arc $DG = 42\frac{1}{2}°$;

and at the third terminus

arc $DGCEP = 198°4'$.

Now since

$AB = 70'$;

at the first terminus

+ add. $BN = 52'$,

by what has been shown; at the second terminus

−add. $MB = 47\frac{1}{2}'$;

and at the third terminus again

+ add. BO 21'.

Therefore during the first interval

arc $MN = 1°40'$,

and during the second interval

arc $MBO = 1°9'$;

and they agree exactly with the observations. Moreover, by those means a simple anomaly of $155°57\frac{1}{2}°$ is made evident at the first terminus; at the second terminus, one of $21°15'$; and at the third terminus, a simple anomaly of $99°2'$—as was to be shown.

10. What the Greatest Difference Is Between the Intersections of the Equator and the Ecliptic

[76ᵃ] In the same way we shall confirm what we expounded concerning the change in obliquity of the ecliptic and the equator and shall find it to be correct. For we have in Ptolemy for the second year of Antoninus Pius a corrected simple anomaly of $21\frac{1}{4}°$; and a greatest obliquity of $23°51'20"$ was found to go with it. From this position down to the observation made by us there have been 1387 years, during which the movement of simple anomaly is reckoned to be $144°04'$; and at this time an obliquity of approximately $23°28\frac{2}{5}'$ is found.

In connection with this let there be drawn again arc ABC of the ecliptic, or instead of it a straight line on account of the shortness of the arc; and above it the semicircle of simple anomaly around pole B, as before. And let A be the limit of greatest declination and C the limit of least declination; and it is the difference between them which we are examining. Therefore in the small circle let

arc AE = 21°15',

and,

arc ED = $AD - AE$ = 68°45';

while, by calculation,

arc EDF = 144°4'

and,

arc DF = $EDF - ED$ = 75°19'.

Drop perpendiculars EG and EK upon the diameter ABC.

Now on the great circle

arc GK = 22'56",

on account of the difference in obliquities from Ptolemy to us. But on account of being like a straight line,

GB = $^1/_2$ ch. 2 ED = 932,

where AC the diameter's image = 2,000.

And also

KB = $^1/_2$ ch. 2 DF = 967.

And

GK = 1899,

where AC = 2,000.

But according as

GK = 22'56",

AC 24',

the difference between greatest and least obliquity which we have been examining. So it is established that the greatest obliquity which occurred during the time between Timochares and Ptolemy was 23°52' and a least obliquity of 23°28' is now approaching. [76b] Hence also whatever mean inclinations of these circles there happen to be are discovered by the same mathematical reasoning we expounded in connection with the precession.

11. ON DETERMINING THE POSITIONS OF THE REGULAR MOVEMENTS OF THE EQUINOXES AND OF THE ANOMALY

With all that unfolded, it remains for us to determine the positions of the movements of the spring equinox. Some people call these positions "roots," because computations may be drawn from them for any given time. Ptolemy considered that the farthest point in history to which our knowledge of this question extends was the beginning of the reign of Nabonassar of the Chaldees, whom many people—taken in by the similarity of the names—have thought to be Nabuchodonoso, and whom the ratio of time and the computation of Ptolemy—which according to the historians falls in the

reign of Shalmaneser of the Chaldees—declare to have been much later. But we, seeking better known times, have judged it sufficient if we start with the first Olympiad, which—measured from the summer solstice—is found to have preceded Nabonassar by 28 years. At this time Canicula was beginning to rise for the Greeks, and the Olympic games were being held, as Censorinus and other trustworthy authors report. Whence, according to the more exact reckoning of the times which is necessary in calculating the heavenly movements, there are 27 years and 247 days from the first Olympiad at noon on the first day of the month Hekatombaion by the Greek calendar to Nabonassar and noon of the first day of the month of Thoth by the Egyptian calendar.

From this to the death of Alexander there are 424 Egyptian years.

But from the decease of Alexander to the beginning of the years of Julius Caesar, there are 278 Egyptian years $118^1/_2$ days up to the midnight before the Kalends of January, which Julius Caesar took as the beginning of the year instituted by him; it was in his third year as Pontifex Maximus and during the consulship of Marcus Aemilius Lepidus that he instituted this year. And so the later years have been called Julian from the year as established by Julius Caesar.

And from the fourth consulship of Caesar to Octavius Augustus there are by the Roman calendar 18 years up to the Kalends of January, although it was on the 16th day before the Kalends of February that Augustus was proclaimed Emperor and son of the deified Julius Caesar by the Senate and the other citizens according to the decree of Numatius Plancus, in the seventh year of the consulship of Marcus Vipsanus and himself. But inasmuch as two years before this the Egyptians came into the power of the Romans after the fall of Antony [77ª] and Cleopatra, the Egyptians reckon 15 years $246^1/_2$ days up to noon of the first day of the month Thoth, which by the Roman calendar was the 3rd day before the Kalends of September.

Accordingly from Augustus to the years of Christ, which begin similarly in January, there are 27 years by the Roman calendar but 29 years $130^1/_2$ days by the Egyptian.

From this to the 2nd year of Antoninus, when, as Claud Ptolemy says, the positions of the stars were observed by him, there are 138 Roman years 55 days. And these years add 34 days to the Egyptian reckoning.

Between the first Olympiad and that moment of time there have been altogether 913 years 101 days, for which time the regular precession of the equinoxes was 12°44', and the simple anomaly was 95°44'.

But in the second year of Antoninus, as has been narrated, the spring equinox was 6°40' to the west of the first of the stars which are in the head of Aries; and since there was a double anomaly of $42^1/_2$°, there was a subtractive difference of 48' between the regular and the apparent movement. And when this difference was restored to the 6°40' of the apparent movement, it made the mean position of the spring equinox to

be at 7°28'. If to this we add the 360° of a circle, and from the sum subtract 12°44', we shall have the mean position of the spring equinox at 354°44'—that is to say, the one which was then 5°16' east of the first star of Aries—for the first Olympiad which began on noon of the first day of the month Hekatombaion among the Athenians.

In the same way if from the 21°15' of simple anomaly 95°45' are subtracted, there will remain a position of simple anomaly of 285°30' for the same beginning of the Olympiads.

And again by a series of additions of movement made in accordance with the lengths of time—the 360° are not counted where there is an excess above that—we shall have the position or root of the regular movement at the death of Alexander as 1°2', and the position of the movement of simple anomaly as 332°52'; at the beginning of the years of Caesar a mean movement of 4°55' and an anomaly of 2°2'; and at the beginning of the years of Christ a position of the mean movement at 5°32' and an anomaly of 6°45'; and in this way we shall determine the roots of movements for whatever beginnings of time are chosen.

12. ON THE COMPUTATION OF THE PRECESSION OF THE SPRING EQUINOX AND THE OBLIQUITY

[77b] Therefore, whenever we wish to determine the position of the spring equinox, if the years from the assumed beginning to the given time are unequal, such as those of the Roman calendar, which we use commonly, we shall reduce them to equal or Egyptian years. For we do not use any other years than the Egyptian in calculating the regular movements, on account of the reason which we mentioned. In so far as the number of years is greater than a period of 60 years, we shall divide it into periods of 60 years; and when we enter the tables of movement through these 60-year periods, we shall pass over as supernumerary the first column appearing in the movements; and beginning with the second column, we shall determine the 60°'s, if there are any, together with the other degrees and minutes, which follow.[1] Next as the second entry and from the first column, as they are found, we shall take the 60°'s, degrees and minutes corresponding to the remaining years. We shall do the same thing in the case of days and the periods of 60 days, since we wish to connect the days with their regular movements according to the table of days and minutes, although in this case the minutes of days or even the days themselves are not wrongly neglected on account of the slowness of their movements, as within the daily movement there is a question only of seconds or third minutes. Therefore when we have made a sum of all these together with their root, by adding single numbers to single numbers within the same species—not counting six 60°'s, if they occur—we shall have the mean position of the spring equinox, its distance to the west of the first star of the Ram, or the distance of that star east of the equinox.

---

[1] That is to say, reading the column of degrees as 60°'s, the minutes as degrees, and so on.

In the same way we shall determine the anomaly too.

But we shall find placed in the last column of the table of additosubtractions and corresponding to the simple anomaly the proportional minutes: we shall set them aside and save them. Then, in the third column of the same table and corresponding to the double anomaly we shall find the additosubtraction, *i.e.*, the degrees and minutes by which the true movement differs from the mean. And if the double anomaly is less than a semicircle, we shall subtract the additosubtraction from the mean movement. But if, by having more than 180°, the double anomaly exceeds a semicircle, we shall add [78ª] the additosubtraction to the mean movement. And that which is thus the sum or remainder will comprehend the true and apparent precession of the spring equinox, or in turn the then angular elongation of the first star of Aries from the spring equinox. But if you seek the position of any other star, add its number as assigned in the catalogue of the stars.

But since things which have to do with the laboratory usually become clearer by means of some examples, let our problem be to find the true position of the spring equinox together with the obliquity of the ecliptic for the 16th day before the Kalends of May in the year of Our Lord 1525, and how great the angular distance of Spica in Virgo from the same equinox is. Therefore, it is clear that in the 1524 Roman years 106 days from the beginning of the years of Our Lord up to this time, there has been an intercalation of 381 days, *i.e.*, 1 year 16 days, which in terms of equal years make 1525 years 122 days: there are twenty-five periods of 60 years and 25 years over, and two periods of 60 days and 2 days over. But to the twenty-five periods of 60 years there correspond in the table of mean movement 20°55'2"; to the 25 years, 20'55"; to the two periods of 60 days, 16"; the remaining 2 days are in third minutes. All these together with their root—which was 5°32'—add up to 26°48', the mean precession of the spring equinox. Similarly, the movement of simple anomaly in the twenty-five periods of 60 years has been two 60°'s and 37°15'3"; in the 25 years, 2°37'15"; in the two periods of 60 days, 2'4"; and in the 2 days, 2". There also, together with the root—which is 6°45'—add up to 166°40', the simple anomaly. I shall save as corresponding to this anomaly the proportional minutes found in the last column of the table of the additosubtractions; for they will come into use in investigating the obliquity; and only 1' is found in this case. Next, as corresponding to the double anomaly of 333°20', I find 32' as the additosubtraction, which is additive because the double anomaly is greater than a semicircle. And when it is added to the mean movement, there comes about a true and apparent precession of the spring equinox of 27°21'. And lastly if to that we add the 170° which is the angular distance of Spica in Virgo from the first star in Aries, I shall have its position [78ᵇ] to the east of the spring equinox at 17°21' of Libra, it was found at approximately the time of our observation.

Now the obliquity of the ecliptic and its declination have the ratio that when there are 60 proportional minutes, the differences located in the table of declinations—I mean the differences at greatest and least obliquity—are added in their entirety to the degrees of the declinations. But in this case, 1' adds only 24" to the obliquity. Wherefore the declinations of the degrees of the ecliptic placed in the table remain as they are throughout this time on account of the least obliquity already approaching as, though at some other time they would be more obviously changeable. In this way, for example, if the simple anomaly were 99°, as it was in the 1380th Egyptian year of Our Lord, there are given by it 25 proportional minutes. But 24' is the difference between greatest and least obliquity and

$$60' : 24' = 25' : 10'.$$

And the addition of 10' to 28' gives an obliquity of 23°38' for that time. If then I should wish to know the declination of any degree on the ecliptic, for example, 3° of Taurus, which is 33° distant from the equinox, I find in the table 12°32', with a difference of 12'. But

$$60' : 25' = 12' : 5';$$

and the addition of 5' to 32' gives 12°37' for 33° of the ecliptic. We can do the same thing in the case of the angles of section of the ecliptic and the equator and the right ascensions—if it is not better to make use of the ratios of spherical triangles—except that it is always necessary to add in the case of the angles of section and to subtract in the case of the right ascensions, so that all things may be corrected to accord with their time.

13. ON THE MAGNITUDE AND DIFFERENCE OF THE SOLAR YEAR

But that this is the way it is with the precession of the equinoxes and solstices—the precession being due to the inclination of the Earth's axis, as we said—will also be confirmed by the annual movement of the centre of the Earth, as it affects the appearance of the sun, which we must now discuss. It follows of absolute necessity that the magnitude of the year, when referred to one of the equinoxes or solstices, is found unequal on account of the irregular change of the termini. For these things imply one another mutually.

Wherefore we must separate and distinguish [79ª] the seasonal year from the sidereal year. For we call that the natural year which times the four seasonal changes of the year for us; and that the sidereal, the revolutions of which are referred to some one of the fixed stars. Now the observations of the ancients make clear in many ways that the natural year, which is also called the revolving year, is unequal. For Callippus, Aristarchus of Samos, and Archimedes of Syracuse determined the year as containing a quarter of a day in addition to the 365 whole days—taking the beginning of the year at the summer solstice, after the Athenian manner.

But Claud Ptolemy, realizing that the apprehension of the solstices was detailed and difficult, did not rely upon their observations very much and went over rather to Hipparchus, who left after him records not so much of the solar solstices as of the equinoxes in Rhodes and reported that there was some small deficiency in the quarter-day; and afterwards Ptolemy decided that the deficiency was $1/_{300}$th part of a day—as follows. For he took the autumn equinox observed as accurately as possible by Hipparchus at Alexandria in the 177th year after the death of Alexander the Great, at midnight of the third intercalary day by the Egyptian calendar—which the fourth intercalary day follows. Then Ptolemy compared it with the equinox as observed by himself at Alexandria in the third year of Antoninus, which was the 463rd year since the death of Alexander, on the 9th day of Athyr, the third month of the Egyptians, at approximately one hour after the rising of the sun. Accordingly between this observation and that of Hipparchus there were 285 Egyptian years 70 days $7^1/_5$ hours, though there should have been 71 days 6 hours, if the revolving year had a full quarter-day in addition to the whole days. Accordingly the 285 years were deficient by $19/_{20}$th of a day, whence it follows that a whole day fell out in 300 years. Moreover, he made a similar inference from the spring equinox. For what he recorded as reported by Hipparchus in the 178th year of Alexander on the 27th day of Mechir, the 6th month by the Egyptian calendar, at sunrise, he himself found in the 463rd year of Alexander on the 7th day of Pachon the 9th month by the Egyptian calendar at a little more than one hour after midday; and in the same way the 285 years were deficient by $19/_{20}$th of a day. By the aid of these indications Ptolemy determined the revolving year as having 365 days 14 min. (of a day) 48 sec. (or 5 hours 55 min. 12 sec.).[1]

Afterwards al-Battani in Arata, Syria, [79^b] in the 1206th year after the death of Alexander observed the autumn equinox with no less diligence and found that it occurred after the 7th day of the month Pachon, approximately $7^2/_5$ hours later in the night, *i.e.*, $4^3/_5$ hours before the light of the 8th day. Accordingly, comparing his own observation with that of Ptolemy made in the third year of Antoninus one hour after sunrise at Alexandria—which is 10° to the west of Arata—he corrected Ptolemy's observation for the meridian at Arata and found the equinox must have occurred at $1^2/_3$ hours after sunrise. Accordingly in the period of 743 equal years the sum of the quarter-days amounted to 178 extra days and $17^3/_5$ hours instead of $185^1/_4$ days. Accordingly since there was deficiency of 7 days $2/_5$ hours, it was seen that the quarter-day was deficient by $1/_{106}$th of a day. Therefore in accordance with the number of years he subtracted one 743rd part of the 7 days $2/_5$ hours (which is 13 min. of an hour 36 sec.) from the quarter-day and recorded the natural year as containing 365 days 5 hours 46 min. 24 sec.

---

[1] *i.e.*, Ptolemy found $1/_{300}$th part of a day lacking to a full quarter-day.

We too made observations of the autumn equinox at Frauenburg in the year of Our Lord 1515 on the 18th day before the Kalends of October: but according to the Egyptian calendar it was the 1840th year after the death of Alexander on the 6th day of the month Phaophi, half an hour after sunrise. But since Arata is about 25° to the east of this spot—which makes $1^2/_3$ hours—therefore during the time between our equinox and that of al-Battani there were 633 Egyptian years and 153 days $6^3/_4$ hours in place of 158 days 6 hours. But between the observation made by Ptolemy at Alexandria and the place and date of our observation, there were 1376 Egyptian years 332 days $^1/_2$ hour. For there is about an hour's difference between us and Alexandria. Therefore during the 633 years between al-Battani and us there have fallen out 4 days $23^3/_4$ hours, or 1 day per 128 years; but during the 1376 years after Ptolemy approximately 12 days, i.e., 1 day per 115 years, and again the year has become unequal on both sides.

[80ª] Moreover, we determined the spring equinox, which occurred in the year of Our Lord 1516, $4^1/_3$ hours after midnight on the 5th day before the Ides of March; and since the spring equinox of Ptolemy—the meridian of Alexandria being corrected for ours—there have been 1376 Egyptian years 332 days $16^1/_3$ hours, in which it is apparent that the distances between the spring and autumn equinoxes are unequal. And so it is of much importance that the solar year as determined in this way should be equal. For the fact that at the autumnal equinoxes between Ptolemy and us, as was shown, in accordance with the equal distribution of years, the quarter-day should be deficient in the 115th part of a day makes the equinox come half a day later than al-Battani's. And the period from al-Battani to us, where the quarter-day must have been deficient in the 128th part of a day, is not consonant with Ptolemy, but the date precedes by a full day the equinox observed by him, and the equinox of Hipparchus by two days. Similarly the time of al-Battani's equinox as measured from Ptolemy's precedes the equinox of Hipparchus by 2 days.

Therefore the equality of the solar year is more correctly measured from the sphere of the fixed stars, as Thebites ben Chora was the first to find; and its magnitude is 365 days 15 minutes (of a day) 23 seconds (which are approximately 6 hours 9 min. 12 sec.) according to a probable argument taken from the fact that the year appears longer in the slower passage of the equinoxes and solstices than in the faster and in accordance with a fixed proportion; and that could not be the case, if there were no equality with reference to the sphere of the fixed stars. Wherefore Ptolemy is not to be listened to in that part where he thinks that it is absurd and irrelevant to measure the annual regularity of the sun through its restitutions with reference to some one of the fixed stars and that this is no more fitting than if someone were to take Jupiter or Saturn as the measure of that regularity. And so there is a ready reason why the seasonal year was longer before Ptolemy and after him became shorter, by a variable difference.

But also in the case of the astral or sidereal year an error can come about, but nevertheless a very slight one and far less than the one which we have already described; and it occurs because this same movement of the centre of the Earth around the sun appears irregular by reason of a twofold irregularity. [80^b] The first and simple irregularity relates to the annual restoration; the other, which varies the first by changing it around, is perceptible not immediately but after a long stretch of time; and accordingly it is not simple or easy to know the ratio of the equality of the year. For if anyone wishes to determine it simply in relation to the fixed distance of some star having a known position—which can be done by using an astrolabe and with the help of the moon, in the way we described in the case of Basiliscus in Leo—he will not avoid error completely, unless at that time the sun on account of the movement of the Earth either has no additosubtraction or else obtains similar and equal additosubtractions at both termini. But unless this happens and unless there is some difference made manifest in accordance with the irregularity, an equal circuit will certainly not seem to have taken place in equal times. But if in both termini the total difference is subtracted or added proportionally, the job will be perfect. Furthermore, the apprehension of the difference requires a prior knowledge of the mean movement, which we are seeking for that reason; and we are versed in this business as in the Archimedean quadrature of the circle.

Nevertheless in order to arrive at the resolution of this knotty problem some time—we find four causes altogether for the appearance of irregularity. The *first* is the irregular precession of the equinoxes, which we have expounded; the *second* is that whereby the sun seems to traverse unequal arcs on the ecliptic, which occurs approximately annually; the *third* is the one which varies this irregularity which we call the second. There remains the *fourth*, which changes the highest and lowest apsides[1] of the centre of the Earth, as will appear below. Of all these only the second was marked by Ptolemy; and it by itself could not produce the inequality of the year but contributes to it through being involved in the others.

But for demonstrating the difference between the regular and the apparent movement of the sun the most accurate ratio of the year does not seem necessary; and it seems to be enough if in the demonstration we take as the magnitude of the year the $365^1/_4$ days, in which the movement of the first irregularity is completed, since that which stands out so little, when taken on the total circle, vanishes utterly when taken on a lesser magnitude. But on account of the excellence of the order and the facility in teaching we are here expounding first the regular movements of the annual revolution of the centre of the Earth by means of necessary demonstrations. And then we shall build up the regular movements together with the difference between the regular and the apparent movement.

---

[1] The apsides are the positions of greatest and least altitudinal distance of a planet from the sun.

14. ON THE REGULAR AND MEAN MOVEMENTS OF THE REVOLUTIONS OF THE CENTRE OF THE EARTH

[81ª] We find that the magnitude of the year and its equality is only 1 second 10 thirds greater than Thebith ben Chora recorded it to be, so that it contains 365 days 15 minutes 24 seconds 10 third-minutes—which amounts to 6 hours 9 minutes 40 seconds, and its fixed equality with reference to the sphere of the fixed stars is disclosed.

Therefore, when we have multiplied the 360° of a circle by 365 days and have divided the sum by 365 days 15 minutes 24 seconds 10 third-minutes, we shall have the movement of an Egyptian year as 359°44'49"7"'4"" and the movement during 60 similar years—not counting the total circles—will be 344°49'7"4"'. Again, if we divide the annual movement by 365 days, we shall have a daily movement of 59'8"11"'22"".

But if we add to these the mean and regular precession of the equinoxes, we shall compose another regular annual movement in seasonal years of 359°45'39"19"'9"" and a daily movement 59'8"19"'37"". And for this reason we can call the former movement of the sun—to use the common expression—the regular and simple movement; and the latter, the regular and composite movement. And we shall set them out in tables, as we did with the precession of the equinoxes. The regular movement of the anomaly of the sun is added to them; but we shall speak of that later.

TABLE OF THE MEAN AND SIMPLE MOVEMENT OF THE SUN IN YEARS AND PERIODS OF SIXTY YEARS

Position of the Birth of Christ—27°31'

Egyptian Years	Movement				
	60°	°	'	"	'''
1	5	59	44	49	7
2	5	59	29	38	14
3	5	59	14	27	21
4	5	58	59	16	28
5	5	58	44	5	35
6	5	58	28	54	42
7	5	58	13	43	49
8	5	57	58	32	56
9	5	57	43	22	3
10	5	57	28	11	10
11	5	57	13	0	17
12	5	56	57	49	24
13	5	56	42	38	31
14	5	56	27	27	38
15	5	56	12	16	46
16	5	55	57	5	53
17	5	55	41	55	0
18	5	55	26	44	7
19	5	55	11	33	14
20	5	54	56	22	21
21	5	54	41	11	28
22	5	54	26	0	35
23	5	54	10	49	42
24	5	53	55	38	49
25	5	53	40	27	56
26	5	53	25	17	3
27	5	53	10	6	10
28	5	52	54	55	17
29	5	52	39	44	24
30	5	52	24	33	32
31	5	52	9	22	39
32	5	51	54	11	46
33	5	51	39	0	53
34	5	51	23	50	0
35	5	51	8	39	7
36	5	50	53	28	14
37	5	50	38	17	21
38	5	50	23	6	28
39	5	50	7	55	35
40	5	49	52	44	42
41	5	49	37	33	49
42	5	49	22	22	56
43	5	49	7	12	3
44	5	48	52	1	10
45	5	48	36	50	18
46	5	48	21	39	25
47	5	48	6	28	32
48	5	47	51	17	39
49	5	47	36	6	46
50	5	47	20	55	53
51	5	47	5	45	0
52	5	46	50	34	7
53	5	46	35	23	14
54	5	46	20	12	21
55	5	46	5	1	28
56	5	45	49	50	35
57	5	45	34	39	42
58	5	45	19	28	49
59	5	45	4	17	56
60	5	44	49	7	4

TABLE OF THE REGULAR AND SIMPLE MOVEMENT OF THE SUN IN DAYS AND PERIODS OF SIXTY DAYS

Position of the Birth of Christ—27°31'

Days	Movement				
	60°	°	'	"	'''
1	0	0	59	8	11
2	0	1	58	16	22
3	0	2	57	24	34
4	0	3	56	32	45
5	0	4	55	40	56
6	0	5	54	49	8
7	0	6	53	57	19
8	0	7	53	5	30
9	0	8	52	13	42
10	0	9	51	21	53
11	0	10	50	30	5
12	0	11	49	38	16
13	0	12	48	46	27
14	0	13	47	54	39
15	0	14	47	2	50
16	0	15	46	11	1
17	0	16	45	19	13
18	0	17	44	27	24
19	0	18	43	35	35
20	0	19	42	43	47
21	0	20	41	51	58
22	0	21	41	0	9
23	0	22	40	8	21
24	0	23	39	16	32
25	0	24	38	24	44
26	0	25	37	32	55
27	0	26	36	41	6
28	0	27	35	49	18
29	0	28	34	57	29
30	0	29	34	5	41
31	0	30	33	13	52
32	0	31	32	22	3
33	0	32	31	30	15
34	0	33	30	38	26
35	0	34	29	46	37
36	0	35	28	54	49
37	0	36	28	3	0
38	0	37	27	11	11
39	0	38	26	19	23
40	0	39	25	27	34
41	0	40	24	35	45
42	0	41	23	43	57
43	0	42	22	52	8
44	0	43	22	0	20
45	0	44	21	8	31
46	0	45	20	16	42
47	0	46	19	24	54
48	0	47	18	33	5
49	0	48	17	41	16
50	0	49	16	49	28
51	0	50	15	57	39
52	0	51	15	5	50
53	0	52	14	14	2
54	0	53	13	22	13
55	0	54	12	30	25
56	0	55	11	38	36
57	0	56	10	46	47
58	0	57	9	54	59
59	0	58	9	3	10
60	0	59	8	11	22

TABLE OF THE REGULAR COMPOSITE MOVEMENT OF THE SUN IN YEARS AND PERIODS OF SIXTY YEARS

Position of the Birth of Christ—278°2'

Egyptian Years	Movement					Egyptian Years	Movement				
	60°	°	'	"	'''		60°	°	'	"	'''
1	5	59	45	39	19	31	5	52	35	18	53
2	5	59	31	18	38	32	5	52	20	58	12
3	5	59	16	57	57	33	5	52	6	37	31
4	5	59	2	37	16	34	5	51	52	16	51
5	5	58	48	16	35	35	5	51	37	56	10
6	5	58	33	55	54	36	5	51	23	35	29
7	5	58	19	35	14	37	5	51	9	14	48
8	5	58	5	14	33	38	5	50	54	54	7
9	5	57	50	53	52	39	5	50	40	33	26
10	5	57	36	33	11	40	5	50	26	12	46
11	5	57	22	12	30	41	5	50	11	51	5
12	5	57	7	51	49	42	5	49	57	31	24
13	5	56	53	31	8	43	5	49	43	10	43
14	5	56	39	10	28	44	5	49	28	49	2
15	5	56	24	49	47	45	5	49	14	29	21
16	5	56	10	29	6	46	5	49	0	8	40
17	5	55	56	8	25	47	5	48	45	47	0
18	5	55	41	47	44	48	5	48	31	27	19
19	5	55	27	27	3	49	5	48	17	6	38
20	5	55	13	6	23	50	5	48	2	45	57
21	5	54	58	45	42	51	5	47	48	25	16
22	5	54	44	25	1	52	5	47	34	4	35
23	5	54	30	4	20	53	5	47	19	43	54
24	5	54	15	43	39	54	5	47	5	23	14
25	5	54	1	22	58	55	5	46	51	2	33
26	5	53	47	2	17	56	5	46	36	41	52
27	5	53	32	41	37	57	5	46	22	21	11
28	5	53	18	20	56	58	5	46	8	0	30
29	5	53	4	0	15	59	5	45	53	39	49
30	5	52	48	39	34	60	5	45	39	19	9

TABLE OF THE REGULAR COMPOSITE MOVEMENT OF THE SUN

Position of the Birth of Christ—278°2'

Egyptian Years	Movement					Egyptian Years	Movement				
	60°	°	'	"	'''		60°	°	'	"	'''
1	0	0	59	8	19	31	0	30	33	18	8
2	0	1	58	16	39	32	0	31	32	26	27
3	0	2	57	24	58	33	0	32	31	34	47
4	0	3	56	33	18	34	0	33	30	43	6
5	0	4	55	41	38	35	0	34	29	51	26
6	0	5	54	49	57	36	0	35	28	59	46
7	0	6	53	58	17	37	0	36	28	8	5
8	0	7	53	6	36	38	0	37	27	16	25
9	0	8	52	14	56	39	0	38	26	24	45
10	0	9	51	23	16	40	0	39	25	33	4
11	0	10	50	31	35	41	0	40	24	41	24
12	0	11	49	39	55	42	0	41	23	49	43
13	0	12	48	48	15	43	0	42	22	58	3
14	0	13	47	56	34	44	0	43	22	6	22
15	0	14	47	4	54	45	0	44	21	14	42
16	0	15	46	13	13	46	0	45	20	23	2
17	0	16	45	21	33	47	0	46	19	31	21
18	0	17	44	29	53	48	0	47	18	39	41
19	0	18	43	38	12	49	0	48	17	48	1
20	0	19	42	46	32	50	0	49	16	56	20
21	0	20	41	54	51	51	0	50	16	4	40
22	0	21	41	3	11	52	0	51	15	12	59
23	0	22	40	11	31	53	0	52	14	21	19
24	0	23	39	19	50	54	0	53	13	29	39
25	0	24	38	28	10	55	0	54	12	37	58
26	0	25	37	36	30	56	0	55	11	46	18
27	0	26	36	44	49	57	0	56	10	54	38
28	0	27	35	53	9	58	0	57	10	2	57
29	0	28	35	1	28	59	0	58	9	11	17
30	0	29	34	9	48	60	0	59	8	19	37

TABLE OF THE REGULAR MOVEMENT OF ANOMALY[1] OF THE SUN IN YEARS AND PERIODS OF SIXTY YEARS

Position of the Birth of Christ—211°19'

Egyptian Years	Movement					Egyptian Years	Movement				
	60°	°	'	"	'''		60°	°	'	"	'''
1	5	59	44	24	46	31	5	51	56	48	11
2	5	59	28	49	33	32	5	51	41	12	58
3	5	59	13	14	20	33	5	51	25	37	45
4	5	58	57	39	7	34	5	51	10	2	32
5	5	58	42	3	54	35	5	50	54	27	19
6	5	58	26	28	41	36	5	50	38	52	6
7	5	58	10	53	27	37	5	50	23	16	52
8	5	57	55	18	14	38	5	50	7	41	39
9	5	57	39	43	1	39	5	49	52	6	26
10	5	57	24	7	48	40	5	49	36	31	13
11	5	57	8	32	35	41	5	49	20	56	0
12	5	56	52	57	22	42	5	49	5	20	47
13	5	56	37	22	8	43	5	48	49	45	33
14	5	56	21	46	55	44	5	48	34	10	20
15	5	56	6	11	42	45	5	48	18	35	7
16	5	55	50	36	29	46	5	48	2	59	54
17	5	55	35	1	16	47	5	47	47	24	41
18	5	55	19	26	3	48	5	47	31	49	28
19	5	55	3	50	49	49	5	47	16	14	14
20	5	54	48	15	36	50	5	47	0	39	1
21	5	54	32	40	23	51	5	46	45	3	48
22	5	54	17	5	10	52	5	46	29	28	35
23	5	54	1	29	57	53	5	46	13	53	22
24	5	53	45	54	44	54	5	45	58	18	9
25	5	53	30	19	30	55	5	45	42	42	55
26	5	53	14	44	17	56	5	45	26	7	42
27	5	52	59	9	4	57	5	45	11	32	29
28	5	52	43	33	51	58	5	44	55	57	16
29	5	52	27	58	38	59	5	44	40	22	3
30	5	52	12	23	25	60	5	44	24	46	50

MOVEMENT OF ANOMALY OF THE SUN IN DAYS AND PERIODS OF SIXTY DAYS

Position of the Birth of Christ—211°19'

Egyptian Years	Movement					Egyptian Years	Movement				
	60°	°	'	"	'''		60°	°	'	"	'''
1	0	0	59	8	7	31	0	30	33	11	48
2	0	1	58	16	14	32	0	31	32	19	55
3	0	2	57	24	22	33	0	32	31	28	3
4	0	3	56	32	29	34	0	33	30	36	10
5	0	4	55	40	36	35	0	34	29	44	17
6	0	5	54	48	44	36	0	35	28	52	25
7	0	6	53	56	51	37	0	36	28	0	32
8	0	7	53	4	58	38	0	37	27	8	39
9	0	8	52	13	6	39	0	38	26	16	47
10	0	9	51	21	13	40	0	39	25	24	54
11	0	10	50	29	21	41	0	40	24	33	2
12	0	11	49	37	28	42	0	41	23	41	9
13	0	12	48	45	35	43	0	42	22	49	16
14	0	13	47	53	43	44	0	43	21	57	24
15	0	14	47	1	50	45	0	44	21	5	31
16	0	15	46	9	57	46	0	45	20	13	38
17	0	16	45	18	5	47	0	46	19	21	46
18	0	17	44	26	12	48	0	47	18	29	53
19	0	18	43	34	19	49	0	48	17	38	0
20	0	19	42	42	27	50	0	49	16	46	8
21	0	20	41	50	34	51	0	50	15	54	15
22	0	21	40	58	42	52	0	51	15	2	23
23	0	22	40	6	49	53	0	52	14	10	30
24	0	23	39	14	56	54	0	53	13	18	37
25	0	24	38	23	4	55	0	54	12	26	45
26	0	25	37	31	11	56	0	55	11	34	52
27	0	26	36	39	18	57	0	56	10	42	59
28	0	27	35	47	26	58	0	57	9	51	7
29	0	28	34	55	33	59	0	58	8	59	14
30	0	29	34	3	41	60	0	59	8	7	22

[1] Any regular movement which, when compounded with a mean movement, causes an appearance of irregularity is called a movement of anomaly. In this case, the regular movement of anomaly is the movement of the eccentric circle, or the first epicycle.

15. THEOREMS PREREQUISITE FOR DEMONSTRATING THE APPARENT IRREGULARITY OF THE MOVEMENT OF THE SUN

[84ᵇ] But for the sake of making a better determination of the apparent irregular movement of the sun we shall now demonstrate more clearly that—with the sun occupying the central position in the world and with the Earth revolving around it as around a centre—if, as we said, there is a distance between the Earth and the sun which cannot be perceived in relation to the immensity of the sphere of the fixed stars; then the sun will be seen to have a regular motion with reference to any point or star in the same sphere (of the fixed stars).

For let *AB* be the greatest circle in the world in the plane of the ecliptic. Let *C* be

its centre, and let the sun be situated there. And in accordance with the distance *CD* between the sun and the Earth—in comparison with which the depth of the world is immense—let the circle *CDE*, in which the annual revolution of the centre of the Earth is located, be described in the same plane of the ecliptic: I say that the sun will seem to have a regular motion with reference to any point or star taken on circle *AB*.

Let some point be taken; and let it be *A*. And to *A* let the view of the sun from the Earth—which is at *D*—be extended as *DCA*. Now let the Earth be moved anywhere through arc *DE*; and let *AE* and *DE* be drawn from *E* the

position of the Earth. Therefore the sun will now be seen from *E* at point *B*. And since *AC* is immense in comparison with *CD* or *CE* its equal, *AE* too will be immense in comparison with *CE*. For let any point *F* be taken on *AC*, and let *EF* be joined. Therefore since two straight lines from the termini *C* and *E* of the base fall outside triangle *EFC* on point *A*; by the converse of Euclid's *Elements*, I, 21,

angle *FAE* < angle *EFC*.

Wherefore the straight lines extended to immensity comprehend at last an angle *CAE* so acute that it is no longer perceptible; and

angle *CAE* = angle *BCA* – angle *AEC*.

Moreover, on account of the slightness of the difference between them angles *BCA* and *AEC* seem to be equal; and lines *AC* and *AE* seem to be parallel; and the sun seems to have [85ᵃ] a regular motion with reference to any point on the sphere of the fixed stars, just as if it were revolving around the centre *E*, as was to be shown.

But its irregular movement is demonstrated, because the movement of the centre of the Earth in its annual revolution is not absolutely around the centre of the sun.

That can be understood in two ways, either through an eccentric circle, *i.e.*, one whose centre is not the centre of the sun, or through an epicycle on a homocentric circle.

Now it is made clear through an eccentric in this way. For let *ABCD* be an eccentric circle in the plane of the ecliptic; and let its centre *E* be no very slight distance away from the centre of the sun or world. Let the centre of the world be *F*; and let *AEFD* be the diameter (of circle *ABCD*) passing through both centres. And let its apogee be at *A*—which is called the highest apsis by the Romans—the place farthest removed from the centre of the world, and *D* the perigee, which is nearest (to the centre of the world) and is the lowest apsis.

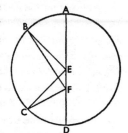

Therefore while the Earth is moved regularly in its orbital circle *ABCD* around its centre *E*, as has been already said, there will appear to be an irregular movement around *F*.

For let

$$\text{arc } AB = \text{arc } CD;$$

and let the straight lines *BE*, *CB*, *BF*, and *CF* be drawn.

$$\text{Angle } AEB = \text{angle } CED,$$

because angles *AEB* and *CED* are intercepting equal arcs around the centre *E*. But angle *CFD* is the angle of sight, and

$$\text{ext. angle } CFD > \text{int. angle } CED.$$

But

$$\text{angle } AEB = \text{angle } CED.$$

Hence

$$\text{angle } CFD > \text{angle } AEB.$$

But also

$$\text{ext. angle } AEB > \text{int. angle } AFB;$$

and hence by so much more

$$\text{angle } CFD > \text{angle } AFB.$$

But an equal time produces both angle *CFD* and angle *AFB* because

$$\text{arc } AB = \text{arc } CD.$$

Therefore the movement will appear regular from around *E* and irregular from around *F*.

Moreover, it is possible to see the same thing more simply, because arc *AB* is farther away from *F* than arc *CD* is. For by Euclid, III, 7, lines *AF* and *BF* by which arc *AB* is intercepted are longer than *CF* and *DF* by which arc *CD* is intercepted, and, as is shown in optics, equal magnitudes which are nearer appear greater than the ones farther away. And so what was proposed in the case of the eccentric circle is manifest.

The same thing will also be made clear by means of an epicycle on a homocentric circle. For let the centre of the homocentric circle *ABCD* and the centre of the world where the sun is be at *E*; and in the same plane let *A* be the centre of epicycle *FG*. And through both centres let the straight line *CEAF* be drawn. Let *F* be the apogee of the epicycle; and *I*, the perigee. Therefore it is clear that there is regularity [85b] in *A*, but apparent irregularity in epicycle *FG*. For if the movement of *A* takes place in the direction of *B*, *i.e.*, eastward, while the movement of the centre of the Earth is from its apogee *F* westward; then in the perigee—which is *I*—*E* will appear to be moving faster, because the two movements of *A* and *I* are in the same direction. But in the apogee, which is *F*, point *E* will seem to be moved more slowly, namely because it is moved only by the excelling movement out of two contraries; and the Earth situated at *G* is to the west of the regular movement but at *K* is to the east of it, and the distance of the

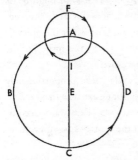

Earth from the regular movement is measured by arcs *AK* and *AG*, in accordance with which the sun will seem to move irregularly.

But whatever things take place by means of the epicycle can happen in the same way by means of the eccentric circle, which the transit of the planet in the epicycle describes equal to the homocentric circle and in the same plane; and the centre of the eccentric circle is at a distance from the centre of the homocentric circle equal to the radius of the epicycle. And all this occurs in three ways, since, if the epicycle on the homo-centric circle and the planet on the epicycle made similar revolutions but with movements opposite to one another, the movement of the planet will trace a fixed eccentric circle, *i.e.*, one whose apogee and perigee possess unchanging locations.

In this way let *ABC* be the homocentric circle, and the centre of the world *D*, the diameter *ADC*. And let us put down that, when the epicycle is at *A*, the planet is in the apogee of the epicycle, which is at *G*, and its radius is in the straight line *DAG*. Now let arc *AB* of the homocentric circle be taken; and with centre *B* and radius equal to *AG*, let the epicycle *EF* be described, and let *BD* and *BE* be extended in a straight line; and let the arc *EF* be similar to arc *AB*, but let arc *EF* be taken in the opposite direction. And let the planet or Earth be in *F*. Let *BF* be joined. And on line *AD* let

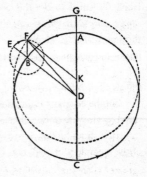

$$DK = BF.$$

Therefore, since

$$\text{angle } EBF = \text{angle } BDA,$$

and for those reasons

$$BF = DK$$

and BF is parallel to DK; and since, if straight lines are joined to equal and parallel straight lines, they are also equal and parallel by Euclid, I, 33; and since

$$DK = AG$$

[86ª] and AK is their common annex;

$$GAK = AKD$$

and therefore

$$GAK = KF.$$

Therefore the circle described with centre K and radius KAG will pass through F. By means of a movement compounded of AB and EF point F describes this circle as eccentric and equal to the homocentric and accordingly fixed too. For when the epicycle makes proportionally equal revolutions with the homocentric circle, the apsides of the eccentric circle so described necessarily remain in the same place.

But if the centre and the circumference of the epicycle make proportionately unequal revolutions, then the movement of the planet will not designate a fixed eccentric circle but one whose centre and apsides are carried westward or eastward, according as the movement of the planet is faster or slower than the centre of its epicycle. In this way if

$$\text{angle } EBF > \text{angle } BDA,$$

let

$$\text{angle } BDM = \text{angle } EBF.$$

It will similarly be shown that if on line DM there be taken DL equal to BF, the circle described with L as centre and with radius LMN equal to AD will pass through planet F. Hence it is clear that by the composite movement of the planet there is described arc NF of the eccentric circle, whose apogee meanwhile travels from point G westward along arc GN.

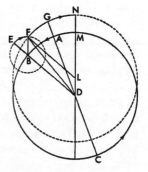

On the contrary, if the movement of the planet in the epicycle were slower, then the centre of the eccentric circle should follow it eastward, whither the centre of the epicycle is carried; that is if

$$\text{angle } EBF = \text{angle } BDM > \text{angle } BDA,$$

it is clear that what we have spoken of will take place.

From all that it is clear that the same irregularity of appearance is always produced

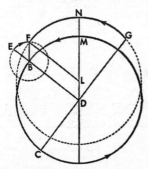

whether by means of an epicycle on a homocentric circle or by means of an eccentric circle equal to the homocentric; and they by no means differ from one another, provided the distance between the centres (of the homocentric and the eccentric) is equal to the radius of the epicycle. Accordingly it is not easy to determine which of them exists in the heavens. Indeed Ptolemy, where he understood simple irregularity and certain and immutable locations for the apsides—as he thought was the case in the sun—judged that the scheme of eccentricity was sufficient. But to the moon and to the five planets which wander in two or more different ways [86^b] he applied eccentric circles carrying epicycles.

From this moreover it is easily demonstrated that the greatest difference between regularity and appearance is seen at the time when the planet appears in the mean position between the highest and the lowest apsis in the case of the eccentric circle, but in the case of the epicycle at its point of contact (with the circle carrying the epicycle), as in Ptolemy.

In the case of the eccentric circle thus: For let there be the circle *ABCD* around the centre *E*; and *AEC* the diameter through *F* the sun, which is off centre. Now let line *BFD* be drawn through *F* at right angles to the diameter; and let *BE* and *ED* be joined. Let *A* be the apogee and *C* the perigee; and let *B* and *D* be the means appearing between them.

I say that no angle greater than angle *B* or *D* can be constructed with its vertex on the circumference and line *EF* as its base.

For let points *G* and *H* be taken on either side of *B*; and let *GD*, *GE*, and *GF* be joined, and also *HE*, *HF*, and *HD*. Since line *FG* is nearer the centre than line *DF*,

$$\text{line } FG > \text{line } DF.$$

And therefore

$$\text{angle } GDF > \text{angle } DGF.$$

But

$$\text{angle } EDG = \text{angle } EGD,$$

because sides *EG* and *ED* falling upon the base are equal. And therefore

$$\text{angle } EDF > \text{angle } EGF.$$

But

$$\text{angle } EDF = \text{angle } EBF.$$

Similarly too

$$\text{line } DF > \text{line } FH;$$

and

$$\text{angle } FHD > \text{angle } FDH.$$

But

$$\text{angle } EHD = \text{angle } EDH,$$

because

$$\text{line } EH = \text{line } ED.$$

Therefore, by subtraction,

$$\text{angle } EDF > \text{angle } EHF.$$

But

$$\text{angle } EDF = \text{angle } EBF.$$

Therefore no angle greater than the angles at points B and D will ever be constructed with line EF as base. And so the greatest difference between regularity and appearance is found in the mean position between apogee and perigee.

16. ON THE APPARENT IRREGULARITY OF THE SUN

These things have been demonstrated generally; and they are applicable not only to the apparent movements of the sun but also to the irregularity of the other planets. Now we shall investigate what relates to the sun and the Earth, first in respect to what has been handed down to us by Ptolemy and the other ancients, and then in respect to what modern times and experience have taught us. Ptolemy found [87a] that $94^{1}/_{2}$ days were comprehended between the spring equinox and the summer solstice, and $92^{1}/_{2}$ between the solstice and the autumn equinox. Therefore in accordance with the ratio of time during the first interval there was a mean and regular movement of 93°9'; during the second interval, one of 91°11'.

Let $ABCD$ be the circle of the year as divided in this way; and let E be the centre. Let

$$\text{arc } AB = 93°9'$$

for the first period of time; and let

$$\text{arc } BC = 91°11'$$

for the second. Let the spring equinox be viewed from A; the summer solstice from B; the autumn equinox from C; and the remaining winter solstice from D. Let AC and BD be joined.

AC and BD cut one another at right angles at F, where we set up the sun.

Therefore since

$$\text{arc } ABC > 180°$$

and too

$$\text{arc } AB > \text{arc } BC;$$

Ptolemy understood from this that the centre of the circle was located between lines *BF* and *FA*, and the apogee between the spring equinox and the summer tropic of the sun. Now let *IEG*, which will cut *BFD* in *L*, be drawn through centre *E* parallel to *AFC*, and let *HEK*, which will cut *AF* in *M*, be drawn parallel to *BFD*. In this way there will be constructed the right parallelogram whose diameter *FE* extended in the straight line *FEN* will indicate the Earth's greatest distance in length from the sun and the position of the apogee in *N*.

Therefore since

$$\text{arc } ABC = 184°19',$$

and

$$\text{arc } AH = {}^1/_2 \text{ arc } ABC = 92°9^1/_2';$$
$$\text{arc } HB = \text{arc } AGB - \text{arc } AH = 59°.$$

Again

$$\text{arc } AG = \text{arc } AH - 90° = 2°10'.$$

Now

$$LF = {}^1/_2 \text{ ch. } 2 AG = 377,$$
$$\text{where radius} = 10,000.$$

But

$$EL = {}^1/_2 \text{ ch. } 2 BH = 172.$$

And, as two sides of triangle *ELF* are given,

$$\text{side } EF = 414 \ {}^1/_{24} \text{ radius } NE$$
$$\text{where radius } NE = 10,000.$$

But

$$EF : EL = NE : {}^1/_2 \text{ ch. } 2 NH.$$

Therefore

$$\text{arc } NH = 24^1/_2°.$$

And thus

$$\text{angle } NEH \text{ is given,}$$

and

$$\text{angle } NEH = \text{angle } LFE,$$

which is the angle of apparent movement. By such an interval therefore did the highest apsis before Ptolemy precede the summer solstice of the sun. But

$$\text{arc } IK - 90°,$$

and

$$[87^b] \text{ arc } IC = \text{arc } AG$$

and

$$\text{arc } DK = \text{arc } HB.$$

Hence

$$\text{arc } CD = \text{arc } IK - (\text{arcs } IC + DK) = 86°51'$$

and

$$\text{arc } DA = \text{arc } CDA - \text{arc } CD = 88°49'.$$

But to the 86°51' there correspond $88^1/_8$ days; and to the 88°49', 90 days and 3 hours—the eighth part of a day. During these periods the sun on account of the regular movement of the Earth seemed to cross from the autumn equinox to the winter solstice and for the remainder of the year to return from the winter solstice to the spring equinox. Indeed Ptolemy testifies that he found these things no different from what were reported by Hipparchus before him. Wherefore he judged that for the remainder of time the highest apsis would be $24^1/_2°$ before the summer tropic and that the eccentricity of—as I said—a 24th part of the radius would remain perpetually.

But now it is found that both of them have changed by a manifest difference. Al-Battani noted it as being 93 days 35 minutes (of a day) from the spring equinox to the summer solstice, and 186 days 37 minutes to the autumn, from which by Ptolemy's rule he elicited an eccentricity of not more than 346 parts whereof the radius has 10,000. Arzachel the Spaniard agrees with him in the ratio of eccentricity but reported an apogee 12°10' west of the solstice, and al-Battani viewed it as 7°43' west of the same solstice. By these tokens it has been grasped that there still remains another irregularity in the movement of the centre of the Earth, as has been attested by the observations of our time also. For during the ten and more years in which we applied our intelligence to investigating these things and especially in the year of Our Lord 1515, we found that there were 186 days $5^1/_2$ minutes from the spring equinox to the autumnal. And so as not to deceive ourselves in determining the solstices—which some suspected had happened in the case of our predecessors—we took certain other positions of the sun into consideration in this business which were not difficult to observe even in comparison with the equinoxes, such as the mean positions in the signs of Taurus, Leo, Scorpio, and Aquarius. Therefore we found that there were 45 days 16 minutes from the autumn equinox to the middle point of Scorpio, and 178 days $53^1/_2$ minutes to the spring equinox. Now the regular movement during the first interval was 44°37'; and during the second interval 176°19'.

[88ª] Now that these preparations have been made, let circle *ABCD* be repeated; and let *A* be the point from which the sun was seen at the spring equinox; *B* the point at which the autumn equinox was viewed; and *C* the midpoint of Scorpio. Let *AB* and *CD*, which cut one another at *F* the centre of the sun, be joined; and let arc *AC* be subtended.

Therefore, since

$$\text{arc } CB = 44°37',$$
$$\text{angle } BAC = 44°37',$$
$$\text{where 2 rt. angles} = 360°.$$

And

$$\text{angle } BFC = 45°,$$
$$\text{where 4 rt. angles} = 360°;$$

and is the angle of apparent movement; but

$$\text{angle } BFC = 90°,$$
$$\text{where 2 rt. angles} = 60°.$$

Hence,

$$\text{angle } ACD = 45°23',$$

because

$$\text{arc } AD = 45°23'.$$

But

$$\text{arc } ACB = 176°19',$$

and

$$\text{arc } AC = \text{arc } ACB - \text{arc } BC = 131°42',$$

and

$$\text{arc } CAD = \text{arc } AC + \text{arc } AD = 177°5'.$$

Therefore, since

$$\text{arc } ACB < 180°,$$

and

$$\text{arc } CAD < 180°,$$

it is clear that the centre of the circle is located in the remainder *BD*. And let the centre be *E*. And through *E* let the diameter *LEFG* be drawn. Let *L* be the apogee and *G* the perigee. Let *EK* be erected perpendicular to *CFD*. But the chords subtending the given arcs are also given by the table:

$$AC = 182,494$$

and

$$CFD = 199,934,$$
$$\text{where diameter} = 200,000.$$

Accordingly, as triangle *ACF* has its angles given, the ratio of the sides will be given by the first rule for plane triangles.

$$CF = 97,697,$$

according as

$$AC = 182,494;$$

and for that reason

$$FK = \frac{1}{2}\ CD - CF = 2,000.$$

And since

$$180° - \text{arc } CAD = 2°55';$$

and since

$$EK = \frac{1}{2}\ \text{ch. } 2°55' = 2,534;$$

then, in triangle *EFK*, as the two sides *FK* and *KE* comprehending the right angle have been given, the triangle will have its sides and angles given:

$$EF = 323,$$
$$\text{where } EL = 10,000;$$

and

$$\text{angle } EFK = 51\frac{2}{3}°$$
$$\text{where 4 rt. angles} = 360°.$$

Therefore, by addition,

$$\text{angle } AFL = 96\frac{2}{3}°$$

and, by subtraction,

$$\text{angle } BFL = 83\frac{1}{3}°.$$

But

$$EF1^{P}56',$$
$$\text{where } EL = 60^{P}.$$

This is the distance of the sun from the centre of the orbital circle: and it has now become approximately $\frac{1}{31}$st (of the radius of the orbital circle), [88ᵇ] though to Ptolemy it seemed to he $\frac{1}{24}$th. And the apogee, which was at that time $24\frac{1}{2}°$ to the west of the summer solstice, is now $6\frac{2}{3}°$ to the east of it.

17. DEMONSTRATION OF THE FIRST AND ANNUAL IRREGULARITY OF THE SUN TOGETHER WITH ITS PARTICULAR DIFFERENCES

Therefore since many differences of the irregular movement of the sun are found, we judge that the difference which occurs annually and is more known than the rest should be deduced first.

Accordingly let circle *ABC* be constructed again, around centre *E* with diameter

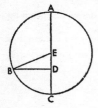

AEC; apogee at *A*, perigee at *C*; and the sun at *D*. Now it has been shown that the greatest difference between regular and apparent movement occurs at the position which with respect to the apparent movement is midway between the apsides. For that reason let *BD* be erected perpendicular to *AEC*, and let it cut the circumference in point *B*, and let *BE* be joined. Therefore, since

in right triangle *BDE* two sides have been given, namely *BE* which is the radius of the circle and *DE* the distance of the sun from the centre; the triangle will have its angles given. And angle *DBE* will be given, which is the difference between the angle *BEA* of regular movement and right angle *EDB* the angle of apparent movement.

But as *DE* is made greater or less, the whole species of the triangle changes. Thus, before Ptolemy

$$\text{angle } B = 2°23';$$

in the time of al-Battani and Arzachel

$$\text{angle } B = 1°59';$$

but at present

$$\text{angle } B = 1°51'.$$

And for Ptolemy

$$\text{arc } AB = 92°23',$$

which is intercepted by angle *AEB*, and

$$\text{arc } BC = 87°37'.$$

For al-Battani

$$\text{arc } AB = 91°59'$$

and

$$\text{arc } BC = 88°1'.$$

And at present

$$\text{arc } AB = 91°51'$$

and

$$\text{arc } BC = 88°9'.$$

Whence too the remaining differences are manifest. For let any other arc *AB* be taken, as in the following figure: and let angle *AEB* be given, and the interior angle *BED*, and the two sides *BE* and *ED*. By the calculus of plane triangles there will be given [89ᵃ] angle *EBD*, the additosubtraction, the difference between the regular and the apparent movement. And it is necessary for these differences to change on account of the change of side *ED*, as has already been said.

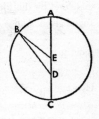

18. ON THE EXAMINATION OF THE REGULAR MOVEMENT IN LONGITUDE

These things have been set forth concerning the annual irregularity of the sun, but not by means of the simple difference so appearing but by means of the difference still mingled with that which the length of time has disclosed.

We shall distinguish them from one another later on. Meanwhile, the mean and regular movement of the centre of the Earth will be given in numbers which will be the more certain the more that movement is separated from any differences of irregularity and the more it extends in time. Now that will be established in this way.

We have taken that autumn equinox which was observed by Hipparchus at Alexandria in the 32nd year of the third period of Callippus—which, as was said above, was the 177th year after the death of Alexander—at midnight after the third intercalary day, which the fourth day followed. But according as Alexandria is approximately 1 hour to the east of Cracow in longitude, it was approximately 1 hour before midnight. Therefore according to the calculation handed on above the position of the autumn equinox in the sphere of the fixed stars was 176°10' from the head of Aries and that was the apparent position of the sun. It was $114^1/_2°$ distant from the highest apsis.

In accordance with this model let there be traced around centre D the circle ABC which the centre of the Earth describes. Let ADC be the diameter; and let the sun be situated on the diameter at point E; the apogee in A; and the perigee in C. But let B be the point where the sun appears in the autumn equinox, and let the straight lines BD and BE be joined.

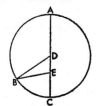

Since

$$\text{angle } DEB = 114^1/_2°,$$

and is seen to measure the distance of the sun from the apogee; and

$$\text{side } DE = 414,$$

$$\text{where } BD = 10,000;$$

therefore, by the fourth theorem on plane triangles, triangle BDE has its sides and angles given. And

$$\text{angle } BDE = \text{angle } BDA = \text{angle } BED = 2°10'.$$

[89^b] But

$$\text{angle } BED = 114°30'.$$

Hence

$$\text{angle } BDA = 116°40';$$

and the mean or regular position of the sun is 178°20' from the head of the Ram in the sphere of the fixed stars.

With this we have compared the autumn equinox observed by us in Frauenburg under the same meridian of Cracow in the year of Our Lord 1515, on the 18th day before the Kalends of October, in the 1840th year since the death of Alexander, on the 6th day of Phaophi the second month by the Egyptian calendar, half an hour after sunrise. At this time the position of the autumn equinox by calculation and observation was 152°45' in the sphere of the fixed stars and was 83°29' distant from the highest apsis in accordance with the preceding demonstration.

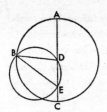

Now let

$$\text{angle } BEA = 83°20',$$

where 2 rt. angles = 180°;

and two sides of the triangle are given:

$$BD = 10,000$$

and

$$DE = 323.$$

By the fourth theorem on plane triangles

$$\text{angle } DBE \ 1°50'.$$

For, if a circle circumscribes triangle BDE, then, as on the circumference,

$$\text{angle } BED = 166°40',$$

where 2 rt. angles = 360°.

And

$$\text{ch. } BD = 19,864$$

where dmt. = 20,000.

And as

$$BD : DE \text{ is given,}$$

$$\text{ch. } DE \ 640,$$

and

$$DE = \text{ch. } DBE$$

and, as on circumference,

$$\text{angle } DBE = 1°50';$$

but, as at centre,

$$\text{angle } DBE = 3°40'.$$

And this was the additosubtraction and difference between the regular and the apparent movement. And

$$\text{angle } BDA = \text{angle } DBE + \text{angle } BED = 1°50' + 83°20' = 85°10',$$

the distance of the regular movement from the apogee, and hence the mean position of the sun is 154°35' in the sphere of the fixed stars.

Therefore in the time between both observations there are 1662 Egyptian years 37 days 18 minutes (of a day) 45 seconds. And the mean and regular movement over and above the whole revolutions—of which there were 1660—is approximately 336°15', which is consonant with the number which we set out in the table of regular movements.

19. ON DETERMINING THE POSITIONS OF THE REGULAR MOVEMENT OF THE SUN AT THE BEGINNINGS (OF YEARS)

[90ª] Accordingly in the flow of time between the death of Alexander the Great and the observation made by Hipparchus there were 176 years 362 days 27½ minutes,

in which the mean movement was 312°43', according to calculation. When these degrees are subtracted from the sum of the 178°20' of Hipparchus' observation and from the 360° of the circle, there will remain, for noon of the first day of Thoth the first month of the Egyptians at the beginning of the years named after the death of Alexander, a position of 225°37' beneath the meridian of Cracow and of Frauenburg, the place of our observation.

From this to the beginning of the Roman years of Julius Caesar in 278 years 118$^1/_2$ days the mean movement is 46°27' over and above the complete revolutions. The addition of these degrees to the degrees of the position of Alexander gives 272°4' as Caesar's position at midnight before the Kalends of January, from which the Romans are accustomed to take the beginning of their years and days.

Then in 45 years 12 days, or in 323 years 130$^1/_2$ days from the death of Alexander the Great, there arises the position of Christ at 272°31'.

And since Christ was born in the third year of the 194th Olympiad, the calculations which give 775 years and 12$^1/_2$ days from the beginning of the year of the first Olympiad to midnight before the Kalends of January similarly give 96°16' as the position of the first Olympiad at noon of the first day of the month Hekatombaion, the anniversary of which day is now the Kalends of July according to the Roman calendar.

In this way the beginnings of the simple movement of the sun are determined with respect to the sphere of the fixed stars. Moreover, the positions of the composite movement are given by the addition of the precession of the equinoxes and similarly to the others: the Olympic position at 90°59'; the position of Alexander at 226°38'; that of Caesar at 276°59'; and that of Christ at 278°2'. All these things, as we said, are taken with respect to the Cracow meridian.

20. ON THE SECOND AND TWOFOLD IRREGULARITY WHICH OCCURS IN THE CASE OF THE SUN ON ACCOUNT OF THE CHANGE OF THE APSIDES

[90^b] But there is now a greater difficulty in connexion with the inconstancy of the apsis of the sun, since, although Ptolemy thought it to be fixed, others have thought it to follow the movement of the starry sphere, according as they judged that the fixed stars moved too. Arzachel opined that this movement also was irregular, that is to say, as happening to retrograde—from the token that, although, as was said, al-Battani had found the apogee 7°44' to the west of the solstice (for previously during the 740 years after Ptolemy it had progressed approximately 17°), it seemed to Arzachel 193 years later to have retrograded approximately 4$^1/_2$°. And accordingly he thought there was some other movement made by the centre of the annual orbital circle in a small circle, in accordance with which (movement) the apogee was deflected back and

forth and the centre of the circle (of the year) was at unequal distances from the centre of the world. That was a good enough device, but it was not accordingly accepted, because upon a universal comparison it is not consonant with the rest; that is to say, if the succession in the order of movement is considered: namely, that at some time before Ptolemy the movement came to a standstill, that during 740 years or thereabouts it traversed 17°, that in the 200 years thereafter it retrograded 4° or 5°, that in the time remaining down to us it progressed, and that no other retrogradation was perceived during the total time, and no more standstills, though they necessarily intervene in the case of contrary movements back and forth. And this can by no means be understood as occurring in uniform and circular movement. Wherefore it is believed by many that some error had crept into their observations. But each mathematician is alike in his care and industry, so that it is doubtful which one we should follow in preference to the other. At all events, I confess that nowhere is there greater difficulty than in determining the apogee of the sun, where we ratiocinate with very small and hardly perceptible magnitudes, since in the neighbourhood of the perigee and apogee (a movement of) 1° effects only a variation of approximately 2' in the additosubtraction, but in the neighbourhood of the mean apsides (a movement of) 1' effects 5° or 6° (in the additosubtraction); and so a slight [91ᵃ] error can propagate itself greatly. Hence in placing the apogee at $6^2/_3$° of Cancer, we were not content to rely upon the instruments of the horoscope, unless the eclipses of the sun and moon gave us more certainty, since if any error lay concealed in our observations, the eclipses would uncover it without fail. Therefore, in accordance with most likelihood, we can apply our intelligence to conceiving the movement as a whole: it is eastward, but irregular, since after that standstill between Hipparchus and Ptolemy the apogee has appeared to be in continuous, orderly, and increased progression down to our time, with the exception of the movement which occurred erroneously—it is believed—between al-Battani and Arzachel, as all the rest seems to be in harmony. For it seems to follow from the same ratio of circular movement that the additosubtraction (of the movement) of the sun similarly does not stop decreasing and that corrections are made for these two irregularities in conjunction with the first and simple anomaly of the obliquity of the ecliptic or something similar.

But in order for this to become more clear, let the circle *AB* around centre *C* be in the plane of the ecliptic. And let the diameter be *ACB*, and on *ACB* let *D* be the globe of the sun as it were at the centre of the world; and let another quite small circle *EF* be described around centre *C* in such a way as not to comprehend the sun. And let it be understood that the centre of the annual revolution of the Earth moves around this small circle with a rather slow progress. And since the small circle *EF* together with line *AD* has a rather slow movement eastward and the centre of the annual revolution has

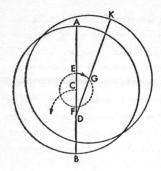

a rather slow movement westward along circle *EF*, sometimes the centre of the annual orbital circle will be found at its greatest distance which is *DE*, and sometimes at its least which is *DF*, and with a slower movement at the greatest distance and a faster movement at the least. And along the middle curves the small circle makes the distance between the centres increase and decrease with time, and it makes the highest apsis alternately precede and follow the apsis or apogee which is on line *ACD* as if in the middle position. In this way, if arc *EG* is taken and with *G* as centre a circle equal to *AB* is described, the then highest apsis will be on line *DGK* and *DG* will be a shorter distance than *DE*, by Euclid, III, 8.

And these things are demonstrated by means of a circle eccentric to an eccentric circle as above; and by means of the epicycle [91^b] on the epicycle as follows: Let circle *AB* be homocentric with the world and with the sun, and let *ABC* be the diameter, whereon the highest apsis is. And with *A* as centre let the epicycle *DE* be described; and again with *D* as centre, the epicycle *FG*, whereon the Earth revolves. And all in the same plane of the ecliptic. Let the movement of the first epicycle be eastward and approximately annual; and that of the second too, *i.e.*, *D*, be similar but westward. And let both have proportionately equal revolutions with respect to line *AC*. Moreover, let the centre of the Earth moving westward from *F* add a little movement to *D*.

From this it is clear that when the Earth is at *F*, it will make the apogee of the sun to be farthest away; and when at *G* it will make the apogee to be nearest; but in the mean arcs of epicycle *FG*, it will make the apogee precede or follow, increased or decreased, greater or less; and hence it will make the movement appear irregular, as has been demonstrated before of the epicycle and the eccentric circle.

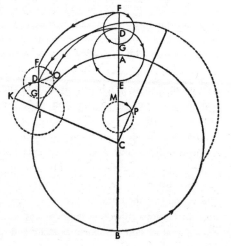

Now let arc *AI* be taken. And with point *I* as centre let the epicyclical epicycle be taken again. And let *CI* be joined and extended in the straight line *CIK*.

angle *KID* = angle *ACI*

on account of the revolutions being proportionately equal. Therefore, as we demonstrated above, point D will describe around centre L an eccentric circle equal to homocentric circle AB, and with an eccentricity CL equal to DI; and F will describe an eccentric circle having an eccentricity CLM equal to IDF; and G similarly an eccentric circle having an eccentricity CN equal to IG. Meanwhile, if the centre of the Earth has by now measured [92ª] any arc FO on its second epicycle, point O will not describe an eccentric circle whose centre is on line AC but one whose centre is on a line parallel (to DO), such as LP. But if OI and CP are joined,

$$OI = CP,$$

but

$$OI < IF$$

and

$$CP < CM.$$

And

$$\text{angle } DIO = \text{angle } LCP,$$

by Euclid, I, 8. And that is the interval whereby the apogee of the sun on line CP will be seen to precede A.

From this moreover it is clear that the same thing occurs through the eccentric circle having an epicycle, since with the eccentric circle alone pre-existing which epicycle D describes around centre L, the centre of the Earth revolves through arc FO in accordance with the aforesaid conditions, *i.e.*, (in a movement) less than the annual revolution. For it will describe, as before, another circle eccentric to the first, around centre P; and the same things will occur again. And since so many ways lead to the same number, I could not really say which one is right, except that the perpetual harmony of numbers and appearances compels us to believe that it is some one of them.

21. HOW GREAT THE SECOND DIFFERENCE IN THE IRREGULARITY OF THE SUN IS

Therefore, since it has already been seen that the second irregularity follows after that first and simple anomaly of the obliquity of the ecliptic or its similitude, we shall have its fixed differences, if some error on the part of past observers does not stand in the way. For according to calculation we have a simple anomaly of approximately 165°39' for the year of Our Lord 1515, and also its beginning by a calculation backwards to approximately 64 years before the birth of Christ, from which time to us there has been a passage of 1580 years. Now the greatest eccentricity of that beginning has been found by us to be 414, whereof the radius is 10,000. But the eccentricity of our time, as was shown, is 323.

Now let *AB* be a straight line, and on it let *B* be the sun and centre of the world. Let *AB* be the greatest eccentricity and *BD* the least. Let a small circle be described whose diameter is *AD*, and let

arc *AC* - 165°39',

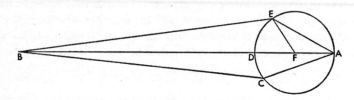

in proportion to the first simple anomaly. Since line *AB* has [92ᵇ] been found at the beginning of the simple anomaly, *i.e.*, at *A*, and

AB = 414,

and now

line *BC* = 323,

therefore we shall now have triangle *ABC* with sides *AB* and *BC* given; and also one angle *CAD* given, because *CD* the remaining arc of the semicircle is given, *i.e.*,

arc *CD* = 14°21'.

Therefore, by what we have shown concerning plane triangles, there are given the remaining side *AC* and angle *ABC*, the difference between the mean and the irregular movement of the apogee; and inasmuch as line *AC* subtends the given arc, diameter *AD* of circle *ACD* will also be given. For since

angle *CAD* = 14°21',

CB = 2496

where diameter of circle circumscribing triangle = 20,000,

and since

BC : *AB* is given,

AB = 3225 = ch. *ACB* = ch. 341°26'.

Hence by subtraction

angle *CBD* = 4°13',

where 2 rt. angles = 360°.

And

ch. *CBD* = *AC* = 735.

Therefore

AC 95,

where *AB* = 414.

And according as AC subtends the given arc, it will have a ratio to AD as to a diameter. Therefore
$$AD = 96,$$
$$\text{where } ADB = 414;$$
and, by subtraction,
$$DB = 321,$$
and that is the distance of the least eccentricity. But, as on the circumference
$$\text{angle } CBD = 4°13',$$
and as at the centre
$$\text{angle } CBD = 2°6\frac{1}{2}',$$
which is the additosubtraction to be subtracted from the regular movement of AB around centre B. Now let there be drawn the straight line BE touching the circle at point E; and with centre F taken, let EF be joined. Therefore, since in right triangle BEF,
$$\text{side } EF = 48,$$
and
$$\text{side } BDF = 369,$$
$$EF = 1300,$$
$$\text{where radius } FB = 10,000.$$
And
$$EF = \frac{1}{2} \text{ ch. 2 } EBF;$$
and
$$\text{angle } EBF = 7°28',$$
$$\text{where 4 rt. angles} = 360°;$$
and that is the greatest additosubtraction between the regular movement at F and the apparent at E.

Hence the remaining and particular differences can be discovered: for instance, if
$$\text{angle } AFE = 6°.$$
For we shall have the triangle with sides EF and FB and angle EFB given. Hence
$$\text{angle } EBF = 41',$$
which is the additosubtraction. [93a] But if
$$\text{angle } AFE = 12°,$$
$$\text{add.} = 1°23'.$$
And if
$$\text{angle } AFE = 18°,$$
$$\text{add.} = 2°3';$$
and so for the rest in this way, as was said above in the case of the additosubtractions for the annual revolution.

22. HOW THE REGULAR MOVEMENT OF THE APOGEE OF
THE SUN AND THE IRREGULAR MOVEMENT ARE UNFOLDED

Therefore, since the time in which the greatest eccentricity coincided with the beginning of the first and simple anomaly was the third year of the 178th Olympiad but the 259th year of Alexander the Great by the Egyptian calendar, and on that account the simultaneously true and mean position of the apogee was at $5^1/_2$° of Gemini, *i.e.*, $65^1/_2$° from the spring equinox; and since the precession of the equinoxes—the true at that time coinciding with the mean—was 4°38': the subtraction of 4°38' from $65^1/_2$° leaves 60°52' from the head of Aries in the sphere of the fixed stars as the position of the apogee.

Again in the second year of the 573rd Olympiad and in the 1515th year of Our Lord, the position of the apogee was found at $6^2/_3$° of Cancer. But by calculation the precession of the spring equinox was $27^1/_4$°; and the subtraction of $27^1/_4$° from 96°40' leaves 69°25'. Now it was shown that with a first anomaly of 165°39' existing at that time there was an additosubtraction of 2°7', by which the true locus preceded the mean. Wherefore it was clear that the mean locus of the apogee of the sun was 71°32'.

Therefore during the middle 1580 Egyptian years the mean and regular movement of the apogee was 10°41'. And when we have divided that by the number of the years, we shall have an annual rate of 24"20'"14"".

23. ON THE CORRECTION OF THE ANOMALY OF THE SUN
AND THE DETERMINATION OF ITS PRIOR POSITIONS

[93^b] If we subtract these 24"20'"14"" from the simple annual movement, which was 359°44'49"7'"4"" there will remain an annual regular movement of anomaly of 359°44'24"46'"50"". Again, the distribution of 359°44'24"46'"50"" through the 365 days will give a daily rate of 59'8"7'"22"" in accord with what was set out above in the tables. Hence we shall have the positions at the established beginnings of years—starting at the 1st Olympiad. For it was shown that on the 18th day before the Kalends of October in the second year of the 573rd Olympiad at half an hour after sunrise the mean apogee of the sun was at 71°32', from which the sun had a distance of 82°58'. And from the first Olympiad there have been 2290 Egyptian years 281 days 46 minutes, during which the movement of anomaly—the whole cycles not being counted—was 42°33'. The subtraction of 42°33' from 82°58' leaves 40°25' as the position of anomaly for the first Olympiad.

And similarly, as above, the position for the Alexander years is 166°38'; for the Caesar years, 211°11'; and for the years of Our Lord, 211°19'.

24. TABLE OF THE DIFFERENCES BETWEEN REGULAR AND APPARENT MOVEMENT

But in order that those things which we have shown concerning the (additive and subtractive) differences between the regular and apparent movements of the sun may be better fitted up for use, we shall also set out a table of them, having sixty rows and six orders of columns.

For the two first columns of both semicircles—that is to say, of the ascending and the descending semicircles—will contain numbers increasing by 3°'s, as above in the case of the movements of the equinoxes.

In the third column will be inscribed the degrees of additosubtraction arising from the movement [94ª] or anomaly of the solar apogee; and this additosubtraction ascends to the height of approximately $7^{1}/_{2}°$, according as it fits each row of degrees.

The fourth place is given over to the proportional minutes, which go up to 60'; and they are reckoned according to the differences between the greater and the lesser additosubtractions arising from the simple anomaly. For since the greatest of these differences is 32', the sixtieth part will be 32". Therefore in accordance with the magnitude of the difference, which we derive from the eccentricity by the mode described above, we put down a number up to 60 to correspond to the single items in the column of the 3°'s.

In the fifth column the single additosubtractions arising from the annual and first anomaly are set up in accordance with the least distance of the sun from the centre.

In the sixth and final column, the differences between these additosubtractions and the additosubtractions which occur at greatest eccentricity.[1] The table is as follows:

---

[1] The movements on the homocentric circle, on the first epicycle, and on the second epicycle we proportionately equal. Hence, from the first two columns of the table are to be taken the movement on the second epicycle, or arc KJ; and from the third column, the additosubtraction to be applied to the annual anomaly or movement of the first epicycle; this additosubtraction, angle KFJ, corrects the mean anomaly from H to I. (Proportional minutes corresponding to arc KJ are to be saved.) Then from the fifth column is to be taken the additosubtraction GEI, corresponding to angle GFI. But since the true position of the sun, is not at I but at J, the additosubtraction most be corrected for the difference between angle FEI and angle FIJ. The proportional minutes which have been saved enable one to adjust the final difference, angle IEJ, according as chord FJ varies in length between FL and FK: that is to say, the change in eccentricity according to the movement around circle CO may be considered as it were a variation in the length of the radius of the corrected epicycle HK.

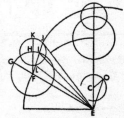

TABLE OF THE ADDITIONS-AND-SUBTRACTIONS OF THE MOVEMENT OF THE SUN

Common Numbers	Additions-and-subtractions arising from movement of the centre	Proportional minutes	Additions-and-subtractions arising from eccentric orbital circle or first epicycle	Differences	Common Numbers	Additions-and-subtractions arising from movement of the centre	Proportional minutes	Additions-and-subtractions arising from eccentric orbital circle or first epicycle	Differences
Deg. Deg.	Deg. Min.		Deg. Min.	Min.	Deg. Deg.	Deg. Min.		Deg. Min.	Min.
3 357	0 21	60	0 6	1	93 267	7 24	30	1 50	32
6 354	0 41	60	0 11	3	96 264	7 24	29	1 50	33
9 351	1 2	60	0 17	4	99 261	7 24	27	1 50	32
12 348	1 23	60	0 22	6	102 258	7 23	26	1 49	32
15 345	1 44	60	0 27	7	105 255	7 21	24	1 48	31
18 342	2 3	59	0 33	9	108 252	7 18	23	1 47	31
21 339	2 24	59	0 38	11	111 249	7 13	21	1 45	31
24 336	2 44	59	0 43	13	114 246	7 6	20	1 43	30
27 333	3 4	58	0 48	14	117 243	6 58	18	1 40	30
30 330	3 23	57	0 53	16	120 240	6 49	16	1 38	29
33 327	3 41	57	0 58	17	123 237	6 37	15	1 35	28
36 324	4 0	56	1 3	18	126 234	6 25	14	1 32	27
39 321	4 18	55	1 7	20	129 231	6 14	12	1 29	25
42 318	4 35	54	1 12	21	132 228	5 59	11	1 25	24
45 315	4 51	53	1 16	22	135 225	5 44	10	1 21	23
48 312	5 6	51	1 20	23	138 222	5 28	9	1 17	22
51 309	5 20	50	1 24	24	141 219	5 19	7	1 12	21
54 306	5 34	49	1 28	25	144 216	4 51	6	1 7	20
57 303	5 47	47	1 31	27	147 213	4 30	5	1 3	18
60 300	6 3	46	1 34	28	150 210	4 9	4	0 58	17
63 297	6 12	44	1 37	29	153 207	3 46	3	0 53	14
66 294	6 27	42	1 39	29	156 204	3 23	3	0 47	13
69 291	6 33	41	1 42	30	159 201	3 1	2	0 42	12
72 288	6 42	40	1 44	30	162 198	2 37	1	0 36	10
75 285	6 51	39	1 46	30	165 195	2 12	1	0 30	9
78 282	6 58	38	1 48	31	168 192	1 47	1	0 24	7
81 279	7 5	36	1 49	31	171 189	1 21	0	0 18	5
84 276	7 11	35	1 49	31	174 186	0 54	0	0 12	4
87 273	7 16	33	1 50	31	177 183	0 27	0	0 6	2
90 270	7 21	32	1 51	32	180 180	0 0	0	0 0	0

25. ON THE CALCULATION OF THE APPARENT MOVEMENT
OF THE SUN

[95^b] From that, I think, it is now sufficiently clear how the apparent position of the sun is calculated for any given time. For we must seek the true position of the spring equinox for that time or its precession together with its first and simple anomaly, as we have set forth above, and then the mean simple movement of the centre of the Earth—or you may call it the movement of the sun—and the annual anomaly, by means of the tables of regular movements; and they are added to their established beginnings. Accordingly you will take the number of the first simple anomaly found in the first or second column of the preceding table; and in the third column[1] you will find the corresponding additosubtraction for correcting the annual anomaly, and in the following column the proportional minutes; save the proportional minutes. Now add the additosubtraction to the annual anomaly, if the first (and simple anomaly) or its number contained in the first column—is less than a semicircle; otherwise subtract. For the remainder or aggregate will be the corrected anomaly of the sun; now by means of this take the additosubtraction arising from the annual (eccentric) orbital circle (or first epicycle)—which is found in the fifth column—and the difference in the following column. If this difference, when adjusted to the proportional minutes you have saved, amounts to something, it is always added to this additosubtraction, and the additosubtraction thus becomes corrected and is subtracted from the mean position of the sun, if the number of the annual anomaly is found in the first column or is less than a semicircle, but it is added, if the annual anomaly is greater or is found in one of the other columns of numbers. For that which in this way becomes the remainder or aggregate will determine the true position of the sun as measured from the head of the constellation of Aries, and if finally the true precession of the spring equinox is added to this (position of the sun), it will straightway show the distance of the sun from the equinox in degrees of the ecliptic among the twelve signs.

But if you wish to do that in another way, take the regular composite movement instead of the simple, and do the other things we spoke of, except that instead of the precession of the equinox, you add or subtract merely its additosubtraction, as the case demands. And so the rational explanation of the appearance of the sun by means of the mobility of the Earth is consonant with ancient and modern findings; and it is all the more [96^a] presumed to hold for the future.

But furthermore, we are not ignorant of the fact that, if anyone thought that the centre of the annual revolutions were fixed as the centre of the world but that the sun moved in accordance with two movements similar and equal to those which we

---

[1] *i.e.,* since the movements on the first epicycle and on the second epicycle are proportionately equal to one another.

demonstrated in the case of the centre of the eccentric circle, everything will be manifest which was manifest before—the same numbers and the same demonstrations—since nothing else is changed in them except their situation, especially those which have to do with the sun. For then the movement of the centre of the Earth round the centre of the world would be absolute and simple, as the other two movements would be attributed to the sun itself. And on that account there will still remain some doubt as to which of these centres is the centre of the world, as we said ambiguously in the beginning that the centre of the world was at the sun or around the sun. But we shall say more about this question in our explanation of the five wandering stars; and we shall decide the issue to the extent that we are able, holding it enough, if we apply to the apparent movement of the sun calculations which have certitude and are not misleading.

26. On the Nyxohmepon, *i.e.*, the Difference of the Natural Day

In connection with the sun there still remains something to be said about the inequality of the natural day. This time is comprehended by the space of twenty-four hours, which up to now we have used as the common and certain measure of the celestial movements. But some, like the Chaldees and the ancient Jews, define such a day as the time between two sunrises; others, like the Athenians, as that between two sunsets; or like the Romans, from midnight to midnight; or like the Egyptians, from noon to noon. Now it is clear that during this time the revolution proper to the terrestrial globe is completed together with that which is added by the annual revolution in accordance with the apparent movement of the sun.[1] The apparent irregular course of the sun in especial shows that this addition is unequal, as does the fact that the natural day takes place with respect to the poles of the equator, but the year with respect to the ecliptic. Wherefore that apparent time cannot be the common and certain measure of movement, since day does not accord with day in every respect; and so it was necessary to choose among them some mean and equal day, by which it would be possible [96b]

---

[1] In Ptolemy the daily revolution and the annual movement were in opposite directions, and thus the solar day was slightly longer than the sidereal day. Here the daily revolution of the Earth and its annual movement are both of them in the same direction, *i.e.*, eastward, and the solar day remains longer than the sidereal day on account of the third movement of the Earth, *i.e.*, the declination of the pole of the Earth, which is approximately equal to the annual revolution but in the opposite direction.

Let *A* be the sun, *CF* and *DEF* the Earth with centre *B* and *G*. And let *FBC* and *FGD* be the same meridian line. Let the centre of the Earth move from *B* to *G* during the space of 24 equatorial hours. As the movement of declination keeps the axis of the Earth parallel to itself, so too the meridian line *FBC* or *FGD* will be parallel to itself at the end of one daily revolution, but it will not be one with *GEA*, the line from the centre of the Earth to the centre of the sun, until the Earth has further revolved through arc *DE*. That is to sky, the solar day is equal to the 360° of sidereal day *DEFD* plus arc *DE*.

to measure regularity of movement without trouble. Therefore since in the circle of the total year there are 365 revolutions around the poles of the Earth, to which there accretes approximately one whole supernumerary revolution on account of the daily addition made by the apparent progress of the sun: consequently one 365th part of that would fill out the natural day upon an equal basis.

Wherefore we must define and separate the equal day from the apparent and irregular. Accordingly we call that the equal day which comprehends the whole revolution of the equator and over and above that the portion which the sun is seen to traverse with regular movement during that time; but the unequal and apparent day that which comprehends the 360 "times"[1] of one revolution of the equator and in addition that which ascends in the horizon or meridian together with the apparent progress of the sun. Although the difference between these days is very slight and not immediately perceptible, nevertheless it becomes evident after the passage of a certain number of days.

There are two causes for this: the irregularity of the apparent movement of the sun and the unequal ascension of the oblique ecliptic. The first cause, which exists by reason of the irregular apparent movement of the sun, has already been explained, since in the case of the semicircle in which the highest apsis holds the midpoint there is a deficiency of $4^3/_4$ "times" with respect to the ecliptic, according to Ptolemy, and in the case of the other semicircle, in which the lowest apsis is, there is a similar excess of the same amount. Accordingly the total excess of one semicircle over the other was $9^1/_2$ "times."

But in the case of the other cause—which has to do with the rising and setting—the greatest difference occurs between the semicircles comprehending each solstice. This is the difference which exists between the shortest and the longest day and which is most variable, as being particular to each region. The difference which is measured from noon or midnight is comprehended by four termini everywhere, since from 16° of Taurus to 14° of Leo, 88° (of the ecliptic) cross the meridian together with approximately 93 "times"; and from 14° of Leo to 16° of Scorpio, 92° (of the ecliptic) and 87 "times" pass over the meridian, so that in the latter case there is a deficiency of 5 "times" and in the former case an excess of 5 "times." And so the sum of the days in the first segment exceeds those in the second by ten "times"—which make two thirds of one hour; and the same thing takes place conversely in the other semicircle within the remaining termini set diametrically opposite to these. Now the mathematicians chose [97ª] to take the natural day from noon or midnight rather than from sunrise or sunset. For the difference which is taken from the horizon is more manifold; for it extends to a certain number of hours, and moreover it is not everywhere the same but varies manifoldly according to the obliquity of the sphere. But the one which pertains to the meridian is everywhere the same and is more simple. Therefore the total difference, which is constituted by the aforesaid causes: the apparent

---

[1] The unit parts of the equator we called "times" instead of degrees.

irregular progress of the sun and the irregular passage over the meridian, in the time before Ptolemy, took its beginning of decrease at the midpoint of Aquarius and, increasing from the beginning of Scorpio, added up to $8^1/_3$ "times"; and now decreasing from 20° of Aquarius or thereabouts to 10° of Scorpio and increasing from 10° of Scorpio to 20° of Aquarius, it has contracted to 7 "times" 48'. For these things too are changed on account of the inconstancy of the perigee and the eccentricity with the passage of time. Finally, moreover, if the greatest difference in the precession of the equinoxes is taken into account, the total difference of the natural days can extend itself to above 10 "times" for a period of years. In this the third cause of the inequality of days was hidden up to now, because the revolution of the equator was found regular in respect to the mean and regular equinox but not in respect to the apparent equinoxes, which—as is clear enough—are not wholly regular. Therefore the doubling of the 10 "times" makes $1^1/_3$ hours, by which sometimes the longer days can exceed the shorter.

These things can perhaps be neglected this side of manifest error in connexion with the annual progress of the sun and rather slow movement of the fixed stars; but on account of the speed of the moon—by reason of which an inexactitude of $5/_6$° in the movement of the sun can cause error—they are by no means to be neglected. Accordingly, the method of reducing the irregular and apparent time—wherein all differences agree—to the equal time, is as follows.

For any period of time proposed there must be sought in each limit of the time—I mean in the beginning and the end—the mean position of the sun with respect to the mean equinox according to its regular movement which we called composite, and also the true apparent position with respect to the true equinox; and we must consider how many "times" the right ascensions [97b] at midday or midnight have amounted to, or how many "times" intervened between the first true position and the second true position. For if they are equal to the degrees between the two mean positions, then the apparent time assumed will be equal to the mean time. But if the "times" exceed, the excess should be added to the given time; and if they are deficient, the deficiency should be subtracted from the apparent time. For if we take the sums and remainders, we shall have the time reduced to equality by taking for one "time" four minutes of an hour or ten seconds of a minute of a day. But if the equal time is given, and you want to know how much apparent time corresponds to it, you will do the contrary.

Now for the first Olympiad we have the mean position of the sun at 90°59' in relation to the mean spring equinox, on noon of the first day of Hekatombaion, the first month by the Athenian calendar, and at 0°36' of Cancer in relation to the apparent equinox. But for the years of Our Lord we have the mean movement of the sun at 8°2' of Capricorn and the true movement at 8°48' of the same. Therefore 178 "times" 54' ascend in the right sphere from 0°36' of Cancer to 8°48' of Capricorn, and they exceed the distance of the mean positions by 1 "times" 51', which make 7 minutes of an hour. And so for the rest, by means of which the course of the moon can be examined most accurately: we shall speak of that in the following book.

BOOK FOUR

[98ª] Since in the preceding book, to the extent that our mediocrity was able, we explained the appearances due to the movement of the Earth around the sun, and we proposed by that same means to determine the movements of all the planets; the circular movement of the moon interrupts us now and does so of necessity because through her in particular, who shares in both night and day, the positions of the stars are apprehended and examined; then, because she alone of all the planets refers her revolutions however irregular directly to the centre of the earth and is most closely akin to the earth. And on that account, in so far as she is considered in herself, she does not indicate anything about the mobility of the Earth—except perhaps in the case of the daily movement; and for that reason the ancients believed that the Earth was the centre of the world and the centre common to all revolutions. In our explanation of the circular movement of the moon we do not differ from the ancients as regards the opinion that it takes place around the Earth. But we shall bring forward certain things which are different from what we received from our elders and are more consonant; by means of them we shall try to set up the movement of the moon with more certitude, in so far as that is possible.

1. The Hypotheses of the Circles of the Moon According to the Opinion of the Ancients

Accordingly the movement of the moon has the following property: it does not follow the ecliptic but follows an incline proper to itself, which bisects the ecliptic and is in turn bisected by it, and from this line of intersection the moon crosses over into both latitudes. These facts are as firmly established as the solstices in the annual movement of the sun. As the year belongs to the sun, so the month belongs to the moon. Now the middle positions at the sections are called (by some) ecliptic; by others, nodes—and the conjunctions and oppositions of the sun and moon occurring at those positions are called ecliptic. [98ᵇ] For there are not any other points common to both circles except these in which the eclipses of the sun and moon can take place. For in other places the divagation of the moon keeps the sun and moon from opposing one another with their lights; but, as they pass by, they do not block one another. Moreover, the orbital circle of the moon with its four "hinges" or cardinal points revolves obliquely around the centre of the Earth in a regular movement of approximately 3' per day, and it completes its revolution in 19 years. Accordingly the moon is perceived always to move eastward in this orbital circle and in its plane, but sometimes with least velocity and at other times with greatest velocity. For it is slower, the higher up it is; and faster, the nearer to Earth; and this fact can be apprehended more easily in the case of the moon than in that of any other planet on account of the nearness of the moon.

Accordingly the ancients understood that change in velocity to occur on account of an epicycle; in running around this epicycle the moon, when in the upper semicircle, subtracts from the regular movement, but when in the lower semicircle, it adds the same amount to it. Besides, it has been demonstrated that those things which take place through an epicycle can take place through an eccentric circle. But the ancients chose the epicycle because the moon seemed to admit to a twofold irregularity. For when it was at the highest or the lowest apsis of the epicycle, there was no apparent difference from regular movement. But around the point of contact of the epicycle and the greater circle there was a variable difference, for the difference was far greater when the half moon was waxing or waning than when there was a full moon; and this in a fixed and orderly succession. Wherefore they thought that the circle in which the epicycle moved was not homocentric with the Earth; but that there was an eccentric circle carrying an epicycle in which the moon was moved in accordance with the law that in all mean oppositions and conjunctions of the sun and moon the epicycle should be at the apogee of the eccentric circle but in the mean quadrants of the (synodic) circle[1] at the perigee of the eccentric circle. Therefore they imagined two equal and mutually opposing movements around the centre of the Earth—namely, that of the epicycle eastward and that of the centre of the eccentric circle and its apsides westward, with the line of the mean position of the sun always half-way between them. And in this way the epicycle traverses the eccentric circle twice a month.

And in order that these things may be brought before our eyes, let *ABCD* be the oblique lunar circle homocentric with the Earth. Let it be quadrisected by the diameters *AEC* and *BED*; and let *E* be the centre of the Earth. Now on line *AC* there will be the mean conjunction of the sun and moon and at the same position and time the apogee of the eccentric circle—whose centre is *F*—and the centre [99a] of the epicycle *MN*. Now let the apogee of the eccentric circle be moved as far westward as the epicycle eastward, and let them both move regularly around *E* in regular and monthly revolutions as measured by the mean conjunctions or oppositions of the sun. And let line *AEC* of the mean position of the sun be always half way between them; and furthermore let the moon move westward from the apogee of the

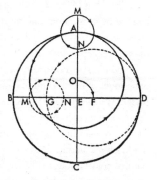

epicycle. For astronomers think that the appearances agree with this set-up. For since the epicycle in the time of half a month moves the distance of a semicircle away from the sun but completes a full revolution from the apogee of the eccentric circle; as a consequence,

---

[1] The synodic circle is the lunar cycle of revolution with respect to the sun.

at the midpoint of this time—when the moon is half full—the moon and the apogee are in opposition to one another along diameter *BD*; and the epicycle is at the perigee of the eccentric circle, as at point *G* where—having become nearer to the Earth—it makes greater differences of irregularity. For when equal magnitudes are set out at unequal intervals, the one which is nearer to the eye appears the greater. Accordingly the differences will be least when the epicycle is at *A*, but greatest when it is at *G*, since the diameter *MN* of the epicycle will be to line *AE* in its least ratio but will be to *GE* in a greater ratio than to all the other lines which are found in the rest of the positions, since *GE* is the shortest line and *AE* or its equal *DE* the longest of all those lines which can be extended from the centre of the Earth to the eccentric circle.

2. ON THE INADEQUACY OF THOSE ASSUMPTIONS

Our predecessors assumed that such a composition of circles was consonant with lunar appearances. But if we consider the thing itself rather carefully, we shall not find this hypothesis very fitting or adequate, as we can prove by reason and sense. For while they admit that the movement of the centre of the epicycle is regular around the centre of the Earth, they must also admit that the movement is irregular on its own eccentric circle which it describes. If—for example—it is assumed that

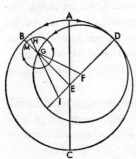

angle *AEB* = angle *AED* = 45°,

so that by addition

angle *BED* = 90°;

and if the centre of the epicycle is taken in *G* [99ᵇ] and *GF* is joined; it is clear that

angle *GFD* > angle *GEF*,

the exterior than the interior and opposite angle. Wherefore the dissimilar arcs *DAB* and *DG* are both described during one period of time, so that when

arc *DAB* = 90°;

arc *DG* > 9W,

and arc *DG* has been described by the centre of the epicycle during this same time. But it is clear at half moon

arc *DAB* = arc *DG* = 180° :

therefore the movement of the epicycle on the eccentric circle which it describes is not regular.

But if this is so, what shall we reply to the axiom: *The movement of the heavenly bodies is regular except for seeming irregular with respect to appearances;* if the apparent regular movement of the epicycle is really irregular and takes place utterly contrary to the principle set up and assumed? But if you say that the epicycle moves regularly

around the centre of the Earth and that that takes care sufficiently of the regularity, then what sort of regularity will that be which occurs in a circle foreign to the epicycle, in which its movement does not exist, and not in its own eccentric circle?

We also are amazed at the fact that they mean the regularity of the moon in its epicycle to be understood not in relation to the centre of the Earth, namely, in respect to line *EGM*, to which the regularity having to do with the centre of the epicycle should rightly be referred, but in relation to some other different point, which has the Earth midway between it and the centre of the eccentric circle, and that line *IGH* is, as it were, the index of the regularity of the moon in the epicycle. And that shows well enough that this movement is really irregular. For the appearances which in part follow upon this hypothesis force this admission. And now that the moon traverses its own epicycle irregularly, we may mark what the line of reasoning would be like if we should try to confirm the irregularity of apparent movement by means of real irregularities. For what else shall we be doing except giving a hold to those who detract from this art?

Furthermore, experience and sense-perception teach us that the parallaxes of the moon are not consonant with those which the ratio of the circles promises. For the parallaxes, which are called commutations, take place on account of the magnitude of the Earth being evident in the neighbourhood of the moon. For since the straight lines which are extended from the centre of the Earth and its surface do not appear parallel but [100ᵃ] in accord with a manifest inclination cut one another in the body of the moon, they are necessarily able to make for irregularity in the apparent movement of the moon, so that the moon is seen in a different position by those viewing it obliquely along the convexity of the Earth and by those who behold the moon from the centre or vertex (of the Earth). Accordingly such parallaxes vary in proportion to the distance of the moon from the Earth. For by the consensus of all the mathematicians the greatest distance is $64^1/_6$ units whereof the radius of the Earth is one unit; but in accordance with the commensurability of these things the least distance should be 33ᴾ33', so that the moon would move towards us through approximately half the total distance—and by the ensuing proportion it was necessary for the parallaxes at greatest and least distance to differ from one another in the ratio of the squares.[1] But we see that those parallaxes, which occur at the time of the half moon waxing or waning, even in the perigee of the epicycle differ slightly or not at all from those which occur at the eclipses of the sun and moon, as we shall show satisfactorily in the proper place.

[1]Literally, in duplicate ratio.

But the body itself of the moon makes perfectly clear that error, because for the same reason it would appear twice as large and twice as small in its diameter. But just as circles are in the ratio of the squares[1] of their diameters, the moon should seem almost four times greater in its quadratures when nearest the earth than when opposite the sun, if it were a full moon shining; but since a half moon is shining, nevertheless it should shine with twice the area of light as a full moon there—although the contrary of this is self-evident. If someone who is not content with simple sight wishes to make an experiment with the dioptra of Hipparchus or some other instruments by which the diameter of the moon may be determined, he will find that the diameter does not vary except in so far as the epicycle without the eccentric circle demands. For that reason Menelaus and Timochares in investigating the fixed stars by means of the positions of the moon did not hesitate to use the same lunar diameter always as $1/2°$, which the moon was seen to occupy most of the time.

3. ANOTHER THEORY OF THE MOVEMENT OF THE MOON

In this way it is perfectly clear that it is not an eccentricity which makes the epicycle appear greater and smaller, but some other relation of circles. [100^b] For let AB be the epicycle which

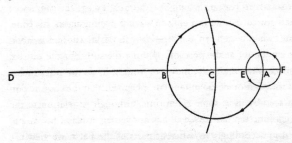

we shall call first and greater; and let C be its centre. Let D be the centre of the Earth, and from D let the straight line DC be extended to the highest apsis of the epicycle; and with A as centre let another small

epicycle EF also be described—and all this in the same plane of the oblique circle of the moon. Now let C be moved eastward but A westward; and again let the moon be moved eastward from F the upper part of EF. And let such an order be kept that when line DE is one with line of the mean position of the sun, the moon is always nearest to centre C, i.e., is in point E, but in the quadratures is farthest away at F. I say that the lunar appearances agree with this set-up.

For it follows that twice a month the moon runs around epicycle EF, during which time C makes one revolution with respect to the mean position of the sun. And the new and the full moon will be seen to cause the least circle, namely, that whereof the radius is CE; but the moon in its quadratures will cause the greatest circle with radius CF; and thus again in the conjunctions and oppositions it will make lesser differences between the regular and the apparent movement, but in the quadratures greater differences by

means of similar but unequal arcs around centre *E*. And since the centre of the epicycle is always in a circle homocentric with the Earth, it will not exhibit such diverse parallaxes but parallaxes in conformity with the epicycle. And the reason will be evident why the body of the moon is seen somehow similar to itself; and all the other things which are perceived in the movement of the moon will come about in this way.

We shall demonstrate them successively by means of our hypothesis, although the same things can take place through eccentric circles, as in the case of the sun—the due proportion being kept. Now we shall take our start from the regular movements, as we did above, without which the irregular movement cannot be separated out. But here there is no small difficulty on account of the parallaxes, which we mentioned; and for that reason the position (of the moon) is not observable by means of astrolabes and other such instruments. But the kindness of nature makes provision for human longing even in this respect, so that the position (of the moon) is more surely determinable through its eclipses than by means of instruments and without any suspicion of error. [101ᵃ] For since the other parts of the world are pure and are filled with the light of day, it stands to reason that night is nothing except the shadow of the Earth, which has the figure of a cone and ends in a point. Falling upon this cone, the moon is dimmed; and, when placed at the midpoint of its darkness, is understood to have arrived without any doubt at the position opposite the sun. But the eclipses of the sun, which take place when the moon moves in front of it, do not offer such a certain determination of the position of the moon. For it happens that the conjunction of the sun and moon is seen by us at some given time, although in relation to the centre of the Earth the conjunction has passed or has not yet taken place, because of the aforesaid parallax. And accordingly in different parts of the Earth we view the same eclipse of the sun as unequal in magnitude and duration and not similar in all respects. But in the case of the eclipses of the moon no such hindrance occurs, since the Earth transmits the axis of that blotting shadow through its centre from the sun; and for that reason the eclipses of the moon are best fitted for determining the course of the moon with the utmost certainty.

4. ON THE REVOLUTIONS OF THE MOON AND ITS PARTICULAR MOVEMENTS

Among the ancients who cared to hand these things down to posterity by means of numbers was found Meton the Anthenian, who flourished around the 37th Olympiad. He reported that in 19 solar years there were 235 full months, whence that great ἐννεαδεκάτερις year, *i.e.*, year of nineteen years, was called the Metontic year. That number was so suitable that it was set up publicly in the market place at Athens

and other famous cities, and even down to the present it has remained in common use, because they think that through it the beginnings and ends of the months are established in a sure order and that through it also the solar year of 365¹/₄ days is commensurable with the months. Hence the Callippic period of 76 years, in which there is an intercalation of 19 days and which they call the Callippic year.

But Hipparchus discovered through careful study that in 304 years there was an excess of a total day, and that (the Callippic year) was verifiable only when the solar year was ¹/₃₀₀ of a day smaller. And so by some men that year was called the great year of Hipparchus, [101ᵇ] in which there were 3760 full months. These years are called more simply and crassly, so to speak, Minerva's, when the recurrences of anomaly and of latitude are also sought, for the sake of which that same Hipparchus was making further investigations. For by comparing the readings which he took in making careful observation of the lunar eclipses with those which he had got from the Chaldees, he determined the time in which the revolutions of the months and of the anomaly recurred simultaneously to be 345 Egyptian years 82 days and 1 hour; and during that time there were 4267 full months and 4573 cycles of anomaly. Therefore when the number of months has been reduced to days and there are 126,007 days 1 hour, one month is found equal to 29 days 31 minutes 50 seconds 8 thirds 9 fourths 20 fifths. According to that ratio the movement during any time is manifest. For the division of the 360° of revolution of one month by the number of days in a month produces a daily movement of the moon in relation to the sum of 12°11'26"41'''20''''18'''''. The multiplication of that by 365 makes—in addition to the 12 revolutions—an annual movement of 129°37'21"28'''29''''.

Furthermore, since the 4267 months and 4573 cycles of anomaly are given in numbers which are composite with respect to one another, that is, as being numbered by the common measure of 17, the ratio of 4267 months to 4573 cycles of anomaly will in least terms be the ratio of 251 to 269; and by Euclid, x, 15, we shall have the proportion of the revolution of the moon to the movement of anomaly in that ratio. Accordingly, when we have multiplied the (annual) movement of the moon by 269 and divided the product by 251, the quotient will be annual movement of anomaly of 13 full revolutions and 88°43'8"40'''20'''' and hence a daily movement of 13°3'53"56'''29''''.

But the revolution in latitude has another ratio. For it does not agree with the prescribed time in which the anomaly has recurred; but we understand that the latitude of the moon has returned only at that time when a later eclipse of the moon is in every respect similar and equal to an earlier, so that both obscurations are in the same part of the moon and are equal, i.e., in magnitude and duration. And this happens when the distances of the moon from the highest or the lowest apsis are

equal. For then the moon is understood to have traversed equal shadows in equal time. [102ª] Now according to Hipparchus such a returning occurs once in 5458 months, to which there correspond 5923 revolutions of latitude. And by that ratio the particular movements of latitude in years and days will be established, as in the case of the others. For when we have multiplied the movement of the moon away from the sun by 5923 months and divided the product by 5458, we shall have an annual movement of the moon in latitude of 13 revolutions 148°42'46"49"'3"" and a daily movement of 13°13'45"39"'40"".

Hipparchus gave this as the rate of the regular movements of the moon, and no one before him had made a closer approximation. Nevertheless the succeeding ages did not show these movements absolved by all the same numbers. For Ptolemy found the same mean movement away from the sun as did Hipparchus, but an annual movement of anomaly deficient with respect to the former in 1"11"'39"" and an annual movement of latitude with an excess of 53"'41"". But now after the passage of many ages since Hipparchus we also found a mean annual movement deficient in 1" 2"'49"" and a movement of anomaly deficient in only 24"'49"". Moreover, there is an excess of 1"1"'44"" in the movement in latitude. And so the regular movement of the moon, whereby it differs from the terrestrial movement, will be an annual movement of 129°37'22"32"'40"", a movement of anomaly of 88°43'9"5"'9"" and a movement in latitude of 148°42'45"17"'21"".

MOVEMENT OF THE MOON IN YEARS AND PERIODS OF SIXTY YEARS

Egyptian Years	Movement					Egyptian Years	Movement				
	60°	°	'	"	'''		60°	°	'	"	'''
1	2	9	37	22	36	31	0	58	18	40	48
2	4	19	14	45	12	32	3	7	56	3	25
3	0	28	52	7	49	33	5	17	33	26	1
4	2	38	29	30	25	34	1	27	10	48	38
5	4	48	6	53	2	35	3	36	48	11	14
6	0	57	44	15	38	36	5	46	25	33	51
7	3	7	21	38	14	37	1	56	2	56	27
8	5	16	59	0	51	38	4	5	40	19	3
9	1	26	36	23	27	39	0	15	17	41	40
10	3	36	13	46	4	40	2	24	55	4	16
11	5	45	51	8	40	41	4	34	32	26	53
12	1	55	28	31	17	42	0	44	9	49	29
13	4	5	5	53	53	43	2	53	47	12	5
14	0	14	43	16	29	44	5	3	24	34	42
15	2	24	20	39	6	45	1	13	1	57	18
16	4	33	58	1	42	46	3	22	39	19	55
17	0	43	35	24	19	47	5	32	16	42	31
18	2	53	12	46	55	48	1	41	54	5	8
19	5	2	50	9	31	49	3	51	31	27	44
20	1	12	27	32	8	50	0	1	8	50	20
21	3	22	4	54	44	51	2	10	46	12	57
22	5	31	42	17	21	52	4	20	23	35	33
23	1	41	19	39	57	53	0	30	0	58	10
24	3	50	57	2	34	54	2	39	38	20	46
25	0	0	34	25	10	55	4	49	15	43	22
26	2	10	11	47	46	56	0	58	53	5	59
27	4	19	49	10	23	57	3	8	30	28	35
28	0	29	26	32	59	58	5	18	7	51	12
29	2	39	3	55	36	59	1	27	45	13	48
30	4	49	41	18	12	60	3	37	22	36	25

Position of the Birth of Christ—209°58'

MOVEMENT OF THE MOON IN DAYS AND PERIODS OF SIXTY DAYS

Days	Movement					Days	Movement				
	60°	°	'	"	'''		60°	°	'	"	'''
1	0	12	11	26	41	31	6	17	54	47	26
2	0	24	22	53	23	32	6	30	6	14	8
3	0	36	34	20	4	33	6	42	17	40	49
4	0	48	45	46	46	34	6	54	29	7	31
5	1	0	57	13	27	35	7	6	40	34	12
6	1	13	8	40	9	36	7	18	52	0	54
7	1	25	20	6	50	37	7	31	3	27	35
8	1	37	31	33	32	38	7	43	14	54	17
9	1	49	43	0	13	39	7	55	26	20	58
10	2	1	54	26	55	40	8	7	37	47	40
11	2	14	5	53	36	41	8	19	49	14	21
12	2	26	17	20	18	42	8	32	0	41	3
13	2	38	28	46	59	43	8	44	12	7	44
14	2	50	40	13	41	44	8	56	23	34	26
15	3	2	51	40	22	45	9	8	35	1	7
16	3	15	3	7	4	46	9	20	46	27	49
17	3	27	14	33	45	47	9	32	57	54	30
18	3	39	26	0	27	48	9	45	9	21	12
19	3	51	37	27	8	49	9	57	20	47	53
20	4	3	48	53	50	50	10	9	32	14	35
21	4	16	0	20	31	51	10	21	43	41	16
22	4	28	11	47	13	52	10	33	55	7	58
23	4	40	23	13	54	53	10	46	6	34	39
24	4	52	34	40	36	54	10	58	18	1	21
25	5	4	46	7	17	55	11	10	29	28	2
26	5	16	57	33	59	56	11	22	40	54	44
27	5	29	9	0	40	57	11	34	52	21	25
28	5	41	20	27	22	58	11	47	3	48	7
29	5	53	31	54	3	59	11	59	15	14	48
30	6	5	43	20	45	60	12	11	26	41	31

Position of the Birth of Christ—209°58'

Movement of Anomaly of the Moon in Years and Periods of Sixty Years

Egyptian Years	*Movement*						*Egyptian Years*	*Movement*				
	60°	°	′	″	‴			*60°*	°	′	″	‴
1	1	28	43	9	7	Position of the Birth of Christ–207°7′	31	3	50	17	42	44
2	2	57	26	18	14		32	5	19	0	51	52
3	4	26	9	27	21		33	0	47	43	0	59
4	5	54	52	36	29		34	2	16	27	10	6
5	1	23	35	45	36		35	3	45	10	19	13
6	2	52	18	54	43		36	5	13	53	28	21
7	4	21	2	3	59		37	0	42	36	37	28
8	5	49	45	12	58		38	2	11	19	46	35
9	1	18	28	22	5		39	3	40	2	55	42
10	2	47	11	31	12		40	5	8	46	4	50
11	4	15	54	40	19		41	0	37	29	13	57
12	5	44	37	49	27		42	2	6	12	23	4
13	1	13	20	58	34		43	3	34	55	32	11
14	2	42	4	7	41		44	5	3	38	41	19
15	4	10	47	16	48		45	0	32	21	50	26
16	5	39	30	25	56		46	2	1	4	59	33
17	1	8	13	35	3		47	3	29	48	8	40
18	2	36	56	44	10		48	4	58	31	17	48
19	4	5	39	53	17		49	0	27	14	26	55
20	5	34	23	2	25		50	1	55	57	36	2
21	1	3	6	11	32		51	3	24	40	45	9
22	2	31	49	20	39		52	4	53	23	54	17
23	4	0	32	29	46		53	0	22	7	3	24
24	5	29	15	38	54		54	1	50	50	12	31
25	0	57	58	48	1		55	3	19	33	21	38
26	2	26	41	57	8		56	4	48	16	30	46
27	3	55	25	6	15		57	0	16	59	39	53
28	5	24	8	15	23		58	1	45	42	49	0
29	0	52	51	24	30		59	3	14	25	58	7
30	2	21	34	33	37		60	4	43	9	7	15

Movement of Lunar Anomaly in Periods of Sixty Days

Days	*Movement*						*Days*	*Movement*				
	60°	°	′	″	‴			*60°*	°	′	″	‴
1	0	13	3	53	56	Position of the Birth of Christ–207°7′	31	6	45	0	52	11
2	0	26	7	47	53		32	5	58	4	46	8
3	0	39	11	41	49		33	7	11	8	40	4
4	0	52	15	35	46		34	7	24	12	34	1
5	1	5	19	29	42		35	7	37	16	27	57
6	1	18	23	23	39		36	7	50	20	21	54
7	1	31	27	17	35		37	8	3	24	15	50
8	1	44	31	11	32		38	8	16	28	9	47
9	1	57	35	5	28		39	8	29	32	3	43
10	2	10	38	59	25		40	8	42	35	57	40
11	2	23	42	53	21		41	8	55	39	51	36
12	2	36	46	47	18		42	9	8	43	45	33
13	2	49	50	41	14		43	9	21	47	39	29
14	3	2	54	35	11		44	9	34	51	33	26
15	3	15	58	29	7		45	9	47	55	27	22
16	3	29	2	23	4		46	10	0	59	21	19
17	3	42	6	17	0		47	10	14	3	15	15
18	3	55	10	10	57		48	10	27	7	9	12
19	4	8	14	4	53		49	10	40	11	3	8
20	4	21	17	58	50		50	10	53	14	57	5
21	4	34	21	52	46		51	11	6	18	51	1
22	4	47	25	46	43		52	11	19	22	44	58
23	5	0	29	40	39		53	11	32	26	38	54
24	5	13	33	34	36		54	11	45	30	32	51
25	5	26	37	28	32		55	11	58	34	26	47
26	5	39	41	22	29		56	12	11	38	20	44
27	5	52	45	16	25		57	12	24	42	14	40
28	6	5	49	10	22		58	12	37	46	8	37
29	6	18	53	4	18		59	12	50	50	2	33
30	6	31	56	58	15		60	13	3	53	56	30

LUNAR MOVEMENT IN LATITUDE IN YEARS AND PERIODS OF SIXTY YEARS

Center label (vertical, between the two halves): *Position of the Birth of Christ—129°45'*

Egyptian Years	60°	°	'	"	'''		Egyptian Years	60°	°	'	"	'''
1	2	28	42	45	17		31	4	50	5	23	57
2	4	57	25	30	34		32	1	18	48	9	14
3	1	26	8	15	52		33	3	47	30	54	32
4	3	54	51	1	9		34	0	16	13	39	48
5	0	23	33	46	26		35	2	44	56	25	6
6	2	52	16	31	44		36	5	13	39	10	24
7	5	20	59	17	1		37	1	42	21	55	41
8	1	49	42	2	18		38	4	11	4	40	58
9	4	18	24	47	36		39	0	39	47	26	16
10	0	47	7	32	53		40	3	8	30	11	33
11	3	15	50	18	10		41	5	37	12	56	50
12	2	44	33	3	28		42	2	5	55	42	8
13	2	13	15	48	45		43	4	34	38	27	25
14	4	41	58	34	2		44	1	3	21	12	42
15	1	10	41	19	20		45	3	32	3	58	0
16	3	39	24	4	37		46	0	0	46	43	17
17	0	8	6	49	54		47	2	29	29	28	34
18	2	36	49	35	12		48	5	58	12	13	52
19	5	5	32	20	29		49	1	26	54	59	8
20	1	34	15	5	46		50	4	55	37	44	26
21	4	2	57	51	4		51	0	24	20	29	44
22	0	31	40	36	21		52	2	53	3	15	1
23	3	0	23	21	38		53	5	21	46	0	18
24	5	29	6	6	56		54	1	50	28	45	36
25	1	57	48	52	13		55	4	19	11	30	53
26	4	26	31	37	30		56	0	47	54	16	10
27	0	55	14	22	48		57	3	16	37	1	28
28	3	23	57	8	5		58	5	45	19	46	45
29	5	52	39	53	22		59	2	14	2	32	2
30	2	21	12	38	40		60	4	42	45	17	21

MOVEMENT IN LATITUDE OF THE MOON IN DAYS AND PERIODS OF SIXTY DAYS

Center label (vertical, between the two halves): *Position of the Birth of Christ—129°45'*

Days	60°	°	'	"	'''		Days	60°	°	'	"	'''
1	0	13	13	45	39		31	6	50	6	35	20
2	0	26	27	31	18		32	7	3	20	20	59
3	0	39	41	16	58		33	7	16	34	6	39
4	0	52	55	2	37		34	7	29	47	52	18
5	1	6	8	48	16		35	7	43	1	37	58
6	1	19	22	33	56		36	7	56	15	23	37
7	1	32	36	19	35		37	8	9	29	9	16
8	1	45	50	5	14		38	8	22	42	54	56
9	1	59	3	50	54		39	8	35	56	40	35
10	2	12	17	36	33		40	8	49	10	26	14
11	2	25	31	22	13		41	9	2	24	11	54
12	2	38	45	7	52		42	9	15	37	57	33
13	2	51	58	53	31		43	9	28	51	43	13
14	3	5	12	39	11		44	9	42	5	28	52
15	3	18	26	24	50		45	9	55	19	14	31
16	3	31	40	10	29		46	10	8	33	0	11
17	3	44	53	56	9		47	10	21	46	45	50
18	3	58	7	41	48		48	10	35	0	31	29
19	4	11	21	27	28		49	10	48	14	17	9
20	4	24	35	13	7		50	11	1	28	2	48
21	4	37	48	58	46		51	11	14	41	48	28
22	4	51	2	44	26		52	11	27	55	34	7
23	5	4	16	30	5		53	11	41	9	19	46
24	5	17	30	15	44		54	11	54	23	5	26
25	5	30	44	1	24		55	12	7	36	51	5
26	5	43	57	47	3		56	12	20	50	36	44
27	5	57	11	32	43		57	12	34	3	22	24
28	6	10	25	18	22		58	12	47	18	8	3
29	6	23	39	4	1		59	13	0	31	53	43
30	6	36	52	49	41		60	13	13	45	39	22

5. DEMONSTRATION OF THE FIRST IRREGULARITY OF THE MOON WHICH OCCURS AT THE NEW AND AT THE FULL MOON

[105b] We have set out the regular movements of the moon, according as they can be known by us at present. Now we must approach the ratio of irregularity which we shall demonstrate by way of the epicycle, and first the irregularity which occurs in the conjunction and oppositions with the sun, in connexion with which the ancient mathematicians exercised their amazing genius in triads of lunar eclipses. We shall also follow the road thus prepared for us by them, and we shall take three eclipses carefully observed by Ptolemy and compare them with three others noted with no less care, in order to examine the regular movements already set out, to see if they have been set out correctly. In explaining them we shall in imitation of the ancients employ as regular the mean movement of the sun and moon away from the position of the spring equinox, since the variation which occurs on account of the irregular precession of equinoxes is not perceptible in such a short time or even in ten years.

Accordingly, Ptolemy took as first the eclipse occurring in the 17th year of Hadrian's reign, after the close of the 20th day of the month Pauni by the Egyptian calendar; and it was the year of Our Lord 133 on the 6th day of May or the day before the Nones. There was a total eclipse, the midtime of which was a quarter of an equal hour before midnight at Alexandria; but at Frauenburg or Cracow it was an hour and a quarter before the midnight which the seventh day followed; and the sun was at $12^{1}/_{4}°$ of Taurus, but according to the mean movement at 12°21' of Taurus.

He says that the second occurred in the 19th year of Hadrian, when two days of Chiach—the fourth Egyptian month had passed: that was in the year of Our Lord 134, 13 days before the Kalends of November. There was an eclipse from the north covering ten twelfths of its diameter. The midtime was one equatorial hour before midnight at Alexandria, but two hours before midnight at Cracow; and the sun was at $25^{1}/_{6}°$ of Libra but by its mean movement at 26°43' of the same.

The third eclipse occurred in the 20th year of Hadrian, when 19 days of Pharmuthi—the eighth Egyptian month—had passed; in the year of Our Lord [106a] 135, when the 6th day of March had passed. The moon was again eclipsed in the north to the extent of half its diameter. The midtime was four equatorial hours past midnight at Alexandria, but at Cracow it was three hours after midnight, that morning being the Nones of March. At that time the sun was at $14^{1}/_{2}°$ of Pisces, but by its mean movement at 11°44' of Pisces.

Now it is clear that in the middle space of time between the first and the second eclipse the moon traversed as much space as the sun in its apparent movement—not

counting the full circles—*i.e.*, 161°55'; and between the second and the third eclipse, 138°55'. Now in the first interval there were 1 year 166 days 23³/₄ equal hours according to apparent time, but by corrected time 23⁵/₈ hours; but in the second interval 1 year 137 days 5 hours simply, but 5¹/₂ hours correctly.

And during the first interval the regular movement of the sun and the moon measured as one—not counting the circles—was 169°37', and there was a movement of anomaly of 110°21'; in the second interval the similarly regular movement of the sun and the moon was 137°34' and there was a movement of anomaly of 81°36'. Therefore it is clear that during the first interval the 110°21' of the epicycle subtract 7°42' from the mean movement of the moon; and during the second interval the 81°36' of the epicycle add 1°21'.

With these things thus before us, let there be described the lunar epicycle *ABC*, in which the first eclipse of the moon is at *A*, the second at *B*, and the remaining one at

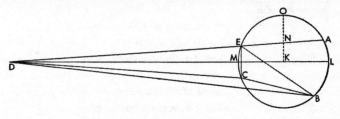

C, and in the order as above let the transit of the moon be understood as occurring westward. And let

$$\text{arc } AB = 110°21',$$

hence

$$- \text{add. } AB = 7°42',$$

as we said; and let

$$\text{arc } BC = 81°36',$$

hence

$$+ \text{add. } BDC = 1°21'.$$

And, as the remainder of the circle,

$$\text{arc } CA = 168°3'$$

and it adds the remainder of the additosubtraction, *i.e.*,

$$+ \text{add. } CDA = 6°21'.$$

Since on the ecliptic

$$\text{arc } AB = 7°42',$$

therefore

$$\text{angle } ADB = 7°42',$$

where 2 rt. angles = 180°.

But

angle $ADB = 15°24'$,

[106^b] where 2 rt. angles = 360°.

And, as on the circumference and as an exterior angle of triangle BDE,

angle $AEB = 110°21'$:

therefore

angle $EBD = 94°57'$.

But the sides of triangles whose angles are given are themselves given:

$DE = 147,396$

and

$BE = 26,798$,

where diameter of circle circumscribing triangle = 200,000.

Again, since on the ecliptic

arc $AEC = 6°21'$,

angle $EDC = 6°21'$,

where 2 rt. angles = 180°.

But

angle $EDC = 12°42'$,

where 2 rt. angles = 360°.

And

angle $AEC = 191°57'$.

And

angle ECD = angle AEC − angle $CDE = 179°15'$.

Therefore the sides are given:

$DE = 199,996$

and

$CE = 22,120$,

where the diameter of the circle circumscribing triangle = 200,000.

But

$CE = 16,302$

and

$BE = 26,798$,

where $DE = 147,396$.

Again, since in triangle BEC

side BE is given,

side EC is given,

and

angle CEB = 81°36'

and hence

arc BC = 81°36' :

therefore, by the proofs concerning plane triangles,

side BC = 17,960.

But since the diameter of the epicycle = 200,000,
and

arc BC = 81°36',
chord BC = 130,694.

And in accordance with the ratio given

ED = 1,072,684

and

CE = 118,637,

and

arc CB = 72°46'10".

But, by construction,

arc CEA = 168°3'.

Therefore, by subtraction,

arc EA = 95°16'50"

and

chord EA = 147,786.

Hence by addition

line AED = 1,220,470.

But since segment EA is less than a semicircle, the centre of the epicycle will not be in [107ª] it but in the remainder $ABCE$. Therefore let K be the centre, and let $DMKL$ be drawn through both apsides, and let L be the highest apsis and M the lowest, Now, by Euclid, III, 36, it is clear that

rect. AD, DE = rect. LD, DM.

Now since LM, the diameter of the circle—to which DM is added in a straight line—is bisected at K, then

rect. LD, DM + sq. KM = sq. DK.

Therefore

DK = 1,148,556
where KL = 100,000;

and on that account,

LK = 8,706
where DKL = 100,000

211

and *LK* is the radius of the epicycle. Having done that, draw *KNO* perpendicular to *AD*. Since *KD*, *DE*, and *EA* have their ratios to one another given in the parts whereof *LK* = 100,000, and since

$$NE = {}^1/_2 \, AE = 73,893 :$$

therefore, by addition,

$$DEN = 1,146,577.$$

But in triangle *DKN*

side *DK* is given,
side *ND* is given,

and

angle *N* = 90°;

on that account, at the centre,

angle *NKD* = 86°38$^1/_2$'

and

arc *MEO* = 86°38$^1/_2$'.

Hence,

arc *LAO* = 180° – arc *NEO* = 93°21$^1/_2$'.

Now

arc *OA* = $^1/_2$ arc *AOE* = 47°38$^1/_2$';

and

arc *LA* = arc *LAO* – arc *OA* = 45°43',

which is the distance—or position of anomaly—of the moon from the highest apsis of the epicycle at the first eclipse. But

arc *AB* = 110°21'.

Accordingly, by subtraction,

arc *LB* = 64°38',

which is the anomaly at the second eclipse. And by addition

arc *LBC* = 146°14',

where the third eclipse falls. Now it will also be clear that since

angle *DKN* = 86°38$^1/_2$',
where 4 rt. angles = 360°,
angle *KDN* = 90° - angle *DKN* = 3°21$^1/_2$';

and that is the additosubtraction which the anomaly adds at the first eclipse. Now

angle *ADB* = 7°42';

therefore, by subtraction,

angle *LDB* = 4°20$^1/_2$',

which arc *LB* subtracts from the regular movement of the moon at the second eclipse. And since

[107^b] angle BDC = 1°21',

and therefore, by subtraction,

angle CDM = 2°49',

the subtractive additosubtraction caused by arc LBC at the third eclipse; therefore the mean position of the moon, *i.e.*, of centre K, at the first eclipse was 9°53' of Scorpio, because its apparent position was at 13°15' of Scorpio; and that was the number of degrees of the sun diametrically opposite in Taurus. And thus the mean movement of the moon at the second eclipse was at 29^1/$_2$° of Aries; and in the third eclipse, at 17°4' of Virgo. Moreover, the regular distances of the moon from the sun were 177°33' for the first eclipse, 182°47' for the second, 185°20' for the last. So Ptolemy.

Following his example, let us now proceed to a third trinity of eclipses of the moon, which were painstakingly observed by us. The first was in the year of Our Lord 1511, after October 6th had passed. The moon began to be eclipsed 1^1/$_8$ equal hours before midnight, and was completely restored 2^1/$_3$ hours after midnight, and in this way the middle of the eclipse was at ^{7}/$_{12}$ hours after midnight—the morning following being the Nones of October, the 7th. There was a total eclipse, while the sun was in 22°25' of Libra but by regular movement at 24°13' of Libra.

We observed the second eclipse in the year of Our Lord 1522, in the month of September, after the lapse of five days. The eclipse was total, and began at ^{2}/$_5$ equal hours before midnight, but its midpoint occurred 1^1/$_3$ hours after midnight, which the 6th day followed—the 8th day before the Ides of September. The sun was in 22^1/$_5$° of Virgo but, according to its regular movement, in 23°59' of Virgo.

We observed the third in the year of Our Lord 1523, at the close of August 25th. It began 2^4/$_5$ hours after midnight, was a total eclipse, and the midtime was 4^5/$_{12}$ hours after the midnight prior to the 7th day before the Kalends of September. The sun was in 11°21' of Virgo but according to its mean movement at 13°2' of Virgo.

And here it is also manifest that the distance between the true positions of the sun and the moon from the first eclipse to the second was 329°47', [108^a] but from the second to the third it was 349°9'. Now the time from the first eclipse to the second was 10 equal years 337 days ^{3}/$_4$ hours according to apparent time, but by corrected equal time ^{4}/$_5$ hours. From the second to the third there were 354 days 3 hours 5 minutes; but according to equal time 3 hours 9 minutes.

During the first interval the mean movement of the sun and the moon measured as one—not counting the complete circles—amounted to 334°47', and the movement of anomaly to 250°36', subtracting approximately 5° from the regular movement; in the second interval the mean movement of the sun and moon was 346°10'; and the movement of anomaly was 306°43', adding 2°59' to the mean movement.

Now let *ABC* be the epicycle, and let 21 be the position of the moon at the middle of the first eclipse, *B* at the second, *C* at the third. And let the movement of the

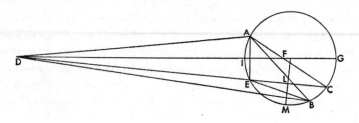

epicycle be understood as proceeding from *C* to *B* and from *B* to *A*, *i.e.*, from above, westward, and from below, eastward. And

$$\text{arc } ACB = 250°36',$$

and, as we said, it subtracts 5° from the mean movement during the first interval of time. But

$$\text{arc } BAC = 306°43',$$

which adds 2°59' to the mean movement of the moon; and accordingly by subtraction the remainder

$$\text{arc } AC = 197°19',$$

which subtracts the remaining 2°1'. But since arc *AC* is greater than a semicircle and is subtractive, then it must contain the highest apsis. For the highest apsis cannot be in area *BA* or *CBA*, which are additive and each less than a semicircle; but the lesser movement is placed by the apogee. Therefore let *D* be taken opposite as the centre of the Earth; and let *AD*, *DB*, *DEC*, *AB*, *AE*, and *EB* be joined.

Now since in triangle *DBE*

$$\text{exterior angle } CEB = 53°17',$$

because angle *CEB* intercepts arc *CB*, and

$$\text{arc } CB = 360° - \text{arc } BAC;$$

and since, as at the centre,

$$\text{angle } BDE = 2°59',$$

but, as at the circumference,

$$\text{angle } BDE = 5°58';$$

and since, therefore, by subtraction,

$$\text{angle } EBD = 47°19';$$

wherefore

$$\text{side } BE = 1042$$

and

$$\text{side } DE = 8024$$

= [108ᵇ] where radius of circle circumscribing the triangle = 10,000. Similarly, as standing on arc *AC* of the circumference,

angle *AEC* = 197°19',

and, as at the centre,

angle *ADC* = 2°1',

but, as on the circumference,

angle *ADC* = 4°2';

therefore, by subtraction,

angle *DAE* = 193°17',

where 2 rt. angles = 360°.

Therefore the sides are also given in the parts whereof the radius of the circle circumscribing triangle *ADE* = 10,000:

AE = 702

and

DE = 19,865 :

but whereas

DE = 8,024,

AE = 283

and

BE = 1042.

Therefore once more we shall have triangle *ABE*, wherein

side *AE* is given,

side *EB* is given,

and

angle *AEB* = 250°36',

where 2 rt. angles = 360°.

Accordingly by what we have shown concerning plane triangles

AB = 1,227

where *EB* = 1,042.

Accordingly in this way we have got hold of the ratios of the three lines *AB*, *EB*, and *ED*; and hence they will become manifest in terms of the parts whereof the radius of the epicycle = 10,000:

ch. *AB* = 16,323,

ED = 106,751,

and

ch. *EB* = 13,853.

Whence also

arc *EB* = 87°41';

215

and

$$\text{arc } EBC = \text{arc } EB + \text{arc } BC = 140°58';$$

and

$$\text{ch. } CE = 18,851,$$

and, by addition,

$$CED = 125,602.$$

Now let the centre of the epicycle be set forth: it necessarily falls in segment *EAC* as being greater than a semicircle. And let *F* be the centre; and let *DIFG* be extended in a straight line through both apsides, *I* the lowest and *G* the highest. Again it is clear that

$$\text{rect. } CD, DE = \text{rect. } GD, DI.$$

But

$$\text{rect. } GD, DI + \text{sq. } FI = \text{sq. } DF.$$

Therefore

$$DIF = 116,226,$$
$$\text{where } FG = 10,000.$$

Accordingly

$$FG = 8,604,$$
$$\text{where } DF = 100,000,$$

—which agrees with what we find reported by most of our predecessors after Ptolemy's time.

[109ª] Now from centre *F* let *FL* be drawn at right angles to *EC* and extended in the straight line *FLM*. It will bisect *CE* at point *I*. Now since

$$\text{line } ED = 106,751$$

and

$$^1/_2CE = LF = 9,426;$$

therefore, by addition,

$$DEL = 116,177$$
$$\text{where } FG = 10,000$$
$$\text{and where } DF = 116,226.$$

Therefore, in triangle *DFL*

$$\text{side } DF \text{ is given,}$$
$$\text{side } DL \text{ is given,}$$
$$\text{angle } DFL = 88°21',$$

and, by subtraction,

$$\text{angle } FDL = 1°39';$$

and similarly

$$\text{arc } IEM = 88°21'$$

216

and
$$\text{arc } MC = {}^1/_2 \text{ arc } EBC = 70°29';$$
hence, by addition,
$$\text{arc } IMC = 158°50',$$
and
$$\text{arc } GC = 180° - \text{arc } IMC = 21°10'.$$
And this was the distance of the moon from the apogee of the epicycle, or the position of anomaly at the third eclipse. And at the second eclipse
$$\text{arc } GCB = 74°27';$$
and at the first eclipse
$$\text{arc } GBA = 183°51'.$$
Again at the third eclipse, and as at the centre,
$$\text{angle } IDE = 1°39',$$
which is the subtractive additosubtraction. And at the second eclipse
$$\text{angle } IDB = 4°38',$$
which is still a subtractive addition-and-subtraction, because
$$\text{angle } IDB = \text{angle } GDC + \text{angle } CDB = 1°39' + 2°59'.$$
And accordingly
$$\text{angle } ADI = \text{angle } ADB - \text{angle } IDB = 5° - 4°38' = 22'$$
which are added to the regular movement at the first eclipse.

For that reason the position of regular movement of the moon in the first eclipse was 22°3' of Aries, but the position of the apparent movement was at 22°25'; and the sun was opposite, at the same number of degrees of Libra. In this way too the mean position of the moon in the second eclipse was at 26°50' of Pisces, but in the third eclipse, at 13° of Pisces, and the mean lunar movement by which it is separated from the annual movement of the Earth, was 177°50' at the first eclipse; at the second eclipse, 182°51'; and at the third eclipse, 179°58'.

6. CONFIRMATION OF WHAT HAS BEEN SET OUT CONCERNING THE MOON'S MOVEMENTS OF ANOMALY IN LONGITUDE

Moreover, by means of these things which are set out concerning the eclipses of the moon, it will be possible to test whether we have set out the regular movements of the moon correctly. For it was shown that in the second of the two eclipses the distance of the moon from the sun was 182°47', and (the movement) of anomaly was 64°38'; [109[b]] but in the second of those eclipses occurring in our time the movement of the moon away from the sun was 182°51' but (the movement) of anomaly was 74°27'. It is clear that in the intervening time there were 17,166 full months and as it were a

movement of 4' and a movement of anomaly—not counting the whole cycles—of 9°49'. Now the time which intervenes between the 19th year of Hadrian on the 2nd day of the Egyptian month Chiach 2 hours before midnight, followed by the 3rd day of the month, and the 1522nd year of Our Lord on September 5th, $1^1/_3$ hours after midnight amounts to 1388 Egyptian years 302 days $3^1/_3$ hours by apparent time; and when corrected, 3 hours 34 minutes after midnight.

And in that time after the 17,165 complete revolutions of equal months there was according to Ptolemy and Hipparchus a movement away from the sun of 359°38'. And according to Hipparchus the movement of anomaly was 9°39', but according to Ptolemy 9°11'. Accordingly the lunar movement away from the sun calculated by Hipparchus and Ptolemy is deficient in 26', and the movement of anomaly of Ptolemy and of Hipparchus is deficient in 38'. These minutes swell our movements, and are consonant with the numbers which we have set out.

7. ON THE POSITIONS OF THE MOON IN LONGITUDE AND ANOMALY

Now we shall speak of these things, as above; and here we are to determine positions for the established beginnings of calendar years of the Olympiads, of the years of Alexander, Caesar, and Our Lord, and any additional one desired. Therefore if we consider the second of the three ancient eclipses—the one which occurred in the 19th year of Hadrian, on the 2nd day of the Egyptian month Chiach, one equatorial hour before midnight at Alexandria but for us under the Cracow meridian at 2 hours before midnight—we shall find from the beginning of the years of Our Lord to this movement 133 Egyptian years 325 days 22 hours simply, but 21 hours 37 minutes correctly. During this time the movement of the moon according to our calculation was 332°49' and (the movement) of anomaly was 217°32'. [110ª] And when they have been subtracted from the findings for the eclipse, each from its own kind, there remain 209°58' as the mean position of the moon away from the sun, and a position of anomaly of 207°7' at the beginning of the years of Our Lord at midnight before the Kalends of January.

Again (from the 1st Olympiad) to the beginning of the years of Our Lord, there are 193 Olympiads 2 years $194^1/_2$ days, which make 775 Egyptian years $12^1/_2$ days, but by corrected time 12 hours 11 minutes. Similarly from the death of Alexander to the birth of Christ, they compute 323 Egyptian years $130^1/_2$ days by apparent time, but by corrected time 12 hours 16 minutes. And from Caesar to Christ there are 45 Egyptian years 12 days, in which the ratios of equal and apparent time agree.

Accordingly when we have deducted the movements corresponding to the intervals of time from the positions at the birth of Christ, by subtracting single items from single items, we shall have for noon of the 1st day of the month Hekatombaion of the

1st Olympiad a regular lunar distance from the sun of 39°43' and a distance of anomaly of 46°20'.

At the beginning of the years of Alexander at noon on the first day of the month Thoth the moon was 310°44' distant from the sun, and the movement of anomaly was 85°41'.

And at the beginning of the years of Julius Caesar at midnight before the Kalends of January the moon was 350°39' distant from the sun, and the movement of anomaly was 17°58'. All this with reference to the Cracow meridian, since Gynopolis—commonly called Frauenburg—where we took our observations at the mouth of the Vistula, lies under this meridian, as the eclipses of the sun and moon observed in both places at the same time teach us; and Dyrrhachium in Macedonia—which was called Epidamnum in antiquity—is also under this meridian.

8. On the Second Irregularity of the Moon and What Ratio the First Epicycle Has to the Second

Accordingly, in this way the regular movement of the moon together with its first irregularity has been demonstrated. Now we must inquire into what ratio the first epicycle has to the second and both of them to the distance of the centre of the Earth. But, as we said, the greatest difference (between regular and apparent movement) is found in the mean quadratures when the half moon is waxing or waning, and that difference is $7^2/_3$°, [110^b] as even the observations of the ancients record. For they were making observations of the time in which the half moon had nearly reached the mean distance of the epicycle and was in the neighbourhood of the tangent from the centre of the Earth—and that is easily perceptible by means of the calculus set forth above. And as the moon was then at about 90° of the ecliptic measured from its rising or setting, they were aware of the error which the parallax could bring into the movement of longitude. For at that time the circle through the vertex of the horizon divides the ecliptic at right angles and does not admit any parallax in longitude but the parallax falls wholly in latitude. Then by means of the astrolabe they determined the position of the moon in relation to the sun. When they made their comparison, the moon was found to differ from its regular movement by $7^2/_3$°, as we said, instead of by 5°.

Now let epicycle AB be described; and let its centre be C. Let the centre of the Earth be D, and from D let the straight line $DBCA$ be extended. Let A be the apogee of the epicycle, B the perigee; and let DE be drawn tangent to the epicycle, and let CE be joined. Accordingly since the greatest additosubtraction is at the tangent and in this case is 7°40', and hence

$$\text{angle } BDE = 7°40',$$

and

$$\text{angle } CED = 90°,$$

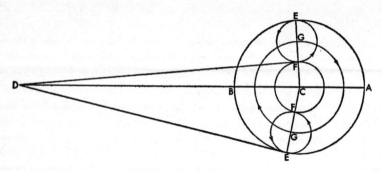

as being at the point of tangency of circle *AB*:
wherefore

$$CE = 1334,$$
where radius $CD = 10,000$.

But at the full moon it was much shorter, that is,

$$CE \doteq 860.$$

Let *CE* be such again, and let

$$CF = 860.$$

Point *F*, which the new moon and the full moon occupy, will be circumcurrent around the same centre; and accordingly, by subtraction,

$$FE = 474$$

and is the diameter of the second epicycle. Let *FE* be bisected at centre *G*, and by addition

$$CFG = 1097$$

and will be the radius of the circle which the centre of the second epicycle describes. And so it is established that

$$CG : GE = 1097 : 237,$$
where $CD = 10,000$.

9. ON THE REMAINING DIFFERENCE, BY REASON OF WHICH THE MOON SEEMS TO MOVE IRREGULARLY AWAY FROM THE HIGHEST APSIS OF ITS EPICYCLE

[111ª] By this induction it is given to understand how the moon is moved irregularly in its first epicycle, and that its greatest difference occurs when the half moon is horn-shaped or gibbous. Once more, let *AB* be that first epicycle, which the centre of the second epicycle describes through its mean movement; let *C* be the centre, *A* the highest apsis, and *B* the lowest. Let point *E* be taken anywhere in the circumference, and let *CE* be joined. Now let

$$CE : EF = 1097 : 237.$$

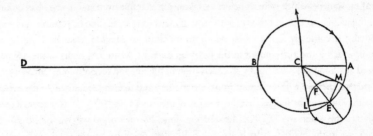

And with radius *EF* let the second epicycle be described around centre *E*. And let the straight lines *CL* and *CM* be drawn tangent to it on both sides. Let the movement of the small epicycle be from *A* to *E*, *i.e.*, from above, westward; and let the movement of the moon be from *F* to *L*, still westward. Accordingly it is clear that, since movement *AE* is regular, the second epicycle by virtue of its motion *FL* adds arc *EL* to the regular movement and by virtue of *MF* subtracts from the regular movement. But since in triangle *CEL*

$$\text{angle } L = 90°,$$

and

$$EL = 237,$$
$$\text{where } CE = 1,097.$$

Therefore

$$EL = 2,160,$$
$$\text{where } CE = 10,000.$$

And by the table

$$EL = \frac{1}{2} \text{ ch. } 2 \text{ } ECL.$$

And

$$\text{angle } ECL = \text{angle } MCF,$$

since the triangles are similar and equal. And that is the greatest difference by which the moon varies in its movement from the highest apsis of the first epicycle. It occurs when the moon by its mean movement is 38°46' distant on either side of the line of mean movement of the Earth. And so it is perfectly clear that these greatest addito-subtractions occur at a mean distance of 38°46' between the sun and moon, and at the same distance on either side of the mean opposition.

10. HOW THE APPARENT MOVEMENT OF THE MOON IS DEMONSTRATED FROM THE REGULAR MOVEMENTS

[111^b] Having seen all that first, we now wish to show by means of diagrams how the regular and apparent movements of the moon are separated out from those regular

lunar movements which were set before us, taking our example from among the observations of Hipparchus, so that in this way our teaching may at the same time be confirmed experimentally. Accordingly in the 197th year from the death of Alexander, on the 17th day of Pauni, which is the 10th month in the Egyptian calendar, when $9^1/_3$ hours of the day had passed at Rhodes, Hipparchus by an observation of the sun and moon through an astrolabe found that they were 48°6' distant from one another, and that the moon was to the east of the sun. And since he judged that the position of the sun was in $10^9/_{10}$° of Cancer, as a consequence the moon was at 29° of Leo. At that time 29° of Scorpio was rising, and 10° of Virgo was in the middle of the heavens over Rhodes, which has an elevation of the north pole of 36°. By this argument it is clear that the moon, which was situated at 90° of the ecliptic from the meridian had at that time admitted no parallax in vision or else one imperceptible in longitude. But since this observation was made $3^1/_3$ hours after midday of the 17th—which corresponds at Rhodes to four equatorial hours—at Cracow it was $3^1/_6$ equatorial hours after midday in accordance with the distance which makes Rhodes a sixth of an hour nearer to us than Alexandria. Accordingly from the death of Alexander there were 196 years, 286 days, $3^1/_6$ hours simply, but $3^1/_3$ hours by equal time. At that time the sun by its mean movement had arrived at 12°3' of Cancer, but by its apparent movement at 10°40' of Cancer, whence the moon appeared in truth to be at 28°37' of Leo. But the regular movement of the moon according to the monthly revolution was at 45°9', and the movement of anomaly was 333° away from the highest apsis by our calculations.

With this example before us let us describe the first epicycle *AB*. Let *C* be its centre. [112ᵃ] Let *ACB* be its diameter, and let *ACB* be extended as *ABD* in a straight line to the centre of the Earth. And in the epicycle, let

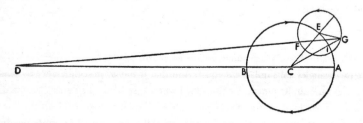

arc *ABE* = 333°.

Let *CE* be joined and again cut in *F*, so that

EF = 237,

where EC = 1,097.

And with *E* as centre and *EF* as radius, let *FG* the epicycle of the epicycle be described. Let the moon be at point *G*; and let

arc *FG* = 90°18',

in the ratio of double to the regular movement away from the sun, which was 45°9'.
And let *CG*, *EG*, and *DG* be joined. Accordingly, since in triangle *CEG* two sides
are given:

$$CE = 1,097$$

and

$$EG = EF = 237;$$

and

$$\text{angle } GEC = 90°18';$$

therefore, by what we have shown concerning plane triangles

$$\text{side } CG = 1,123$$

and

$$\text{angle } ECG = 12°11'.$$

By this means there are determined arc *EI* and the additive additosubtraction caused
by the anomaly; and, by addition,

$$\text{arc } ABEI = 345°11'.$$

And, by subtraction,

$$\text{angle } GCA = 14°49',$$

and is the true distance of the moon from the highest apsis of epicycle *AB*; and

$$\text{angle } BCG = 165°11'.$$

Wherefore in triangle *GDC* two sides are given also.

$$GC = 1,123,$$
$$\text{where } CD = 10,000;$$

and

$$\text{angle } GCD = 165°11'.$$

Hence

$$\text{angle } CDG = 1°29',$$

the additosubtraction which was added to the mean movement of the moon. Hence
the true distance of the moon from the mean movement of the sun is 46°34', and its
apparent position is at 28°37' of Leo, and is 47°57' distant from the true position of
the sun. And there is a deficiency of 9' according to Hipparchus' observation.

But in order that no one on that account should suspect that either his investigation or ours is wrong—though the deficiency is very slight—nevertheless I shall show
that neither he nor we committed any error but that this is the way things rightly are.
For if we recollect that the lunar circle which the moon itself follows is oblique, we will
admit that it produces some sort of longitudinal irregularity in the ecliptic, especially
around the mean positions, which lie between the northern and the southern limits of
latitude and the ecliptic sections, in approximately the same way as between the
oblique [112b] ecliptic and the equator, as we expounded in connection with the

inequality of the natural day. And so if we transfer the ratios to the orbital circle of the moon, which Ptolemy recorded as being inclined to the ecliptic, we find that at those positions the ratios cause a 7' difference in longitude in relation to the ecliptic—and twice that difference is 14'; and the difference increases and decreases proportionally, since when the sun and moon are a quadrant of a circle distant from one another, and if the limit of northern or southern latitude is at the midpoint between them, then the arc intercepted on the ecliptic will be 14' greater than a quadrant of the lunar circle; and conversely in the other quadrants, which the ecliptic sections halve, the circles through the poles of the ecliptic intercept an arc that much less than a quadrant. So in the present case. Since the moon was in the neighbourhood of the mean position between the southern limit of latitude and the ascending ecliptic section—which the moderns call the head of the Dragon—and the sun had already passed by the other descending section—which they call the tail; it is not surprising if when the moon's distance of 47°57' in its own orbital circle was referred to the ecliptic, it increased by at least 7', without the fact of the sun declining in the west causing any subtractive parallax of vision. We shall speak more clearly of all that in our explanation of the parallaxes.

And so the distance of the luminaries, which Hipparchus determined by his instrument as being 48°6', agrees with our calculation perfectly and as it were unanimously.

11. ON THE TABLE OF THE LUNAR ADDITIONS-AND-SUBTRACTIONS OR *AEQUATIONES*

Accordingly, I judge that the mode of determining the motions of the moon is understood generally from this example, since in triangle CEG, the two sides GE and CE always remain the same. But we determine the remaining side GC together with angle ECG—which is the additosubtraction to be used in correcting the anomaly—according to angle GEC which changes continually but which is given. Then in triangle CDG, since the two sides DC and GC together with angle DCG have been computed, in the same way angle D at the centre of the Earth, the angular difference between the true and the regular movement, becomes established.

So that these things may be at hand, [113a] we shall set out a table of the additosubtractions, which will contain six columns. For after the two columns of common numbers of the circle, in the third column will come the additosubtractions which are caused by the small epicycle and vary the regular movement of the first epicycle in accordance with the bi-monthly revolution. Then, we shall leave the fourth column vacant for the time being, and fill up the fifth column first, in which we shall inscribe the additosubtractions caused by the first and greater epicycle which occur at the mean conjunctions and oppositions of the sun and moon, and the greatest is 4°56'. In the next to the last column will be placed the numbers whereby the additosubtractions

which occur at half moon exceed the former additosubtractions, and the greatest of these excesses is 2°44'. But in order that the other excesses may be evaluated, the proportional minutes have been worked out, and this is the ratio of them. For we have taken 2°44' as 60 minutes in relation to any other excesses occurring at the point of tangency of the (small) epicycle (with the line from the centre of the Earth).

In this way, in the same example,

$$CG = 1123,$$
$$\text{where } CD = 10,000.$$

And that makes the greatest additosubtraction at the point of tangency of the (small) epicycle (with the line from the centre of the Earth) to be 6°29', which exceeds the first additosubtraction by 1°33'. But

$$2°44' : 1°33' = 60° : 34°;$$

and so we have the ratio of the excess which occurs at the semicircle of the small epicycle to the excess corresponding to the given arc of 90°18'. Therefore we shall write down 34 minutes in that part of the table corresponding to 90°. In this way we shall find the minutes which are proportional to the arcs inscribed in the table; and we shall set them out in the fourth column.

Finally we have added the degrees of northern and southern latitude in the last column, and we shall speak of them below. For convenience and ease of operation advise us to put them in this order.

TABLE OF ADDITIONS-AND-SUBTRACTIONS OF THE MOON

Common Numbers		Additions-and-subtractions caused by small epicycle		Proportional Minutes	Additions-and-subtractions caused by great epicycle		Excesses		Degrees of Northern Latitudude	
Deg.	Deg.	Deg.	Min.		Deg.	Min.	Deg.	Min.	Deg.	Min.
3	357	0	51	0	0	14	0	7	4	59
6	354	1	40	0	0	28	0	14	4	58
9	351	2	28	1	0	43	0	21	4	56
12	348	3	15	1	0	57	0	28	4	53
15	345	4	1	2	1	11	0	35	4	50
18	342	4	47	3	1	24	0	43	4	45
21	339	5	31	3	1	38	0	50	4	40
24	336	6	13	4	1	51	0	56	4	34
27	333	6	54	5	2	5	1	4	4	27
30	330	7	34	5	2	17	1	12	4	20
33	327	8	10	6	2	30	1	18	4	12
36	324	8	44	7	2	42	1	25	4	3
39	321	9	16	8	2	54	1	30	3	53
42	318	9	47	10	3	6	1	37	3	43
45	315	10	14	11	3	17	1	42	3	32
48	312	10	30	12	3	27	1	48	3	20
51	309	11	0	13	3	38	1	52	3	8
54	306	11	21	15	3	47	1	57	2	56
57	303	11	38	16	3	56	2	2	2	44
60	300	11	50	18	4	5	2	6	2	30
63	297	12	2	19	4	13	2	10	2	16
66	294	12	12	21	4	20	2	15	2	2
69	291	12	18	22	4	27	2	18	1	47
72	288	12	23	24	4	33	2	21	1	33
75	285	12	27	25	4	39	2	25	1	18
78	282	12	28	27	4	43	2	28	1	2
81	279	12	26	28	4	47	2	30	0	47
84	276	12	23	30	4	51	2	34	0	31
87	273	12	17	32	4	53	2	37	0	16
90	270	12	12	34	4	55	2	40	0	0
93	267	12	3	35	4	56	2	42	0	16
96	264	11	53	37	4	56	2	42	0	31
99	261	11	41	38	4	55	2	43	0	47
102	258	11	27	39	4	54	2	43	1	2
105	255	11	10	41	4	51	2	44	1	18
108	252	10	52	42	4	48	2	44	1	33
111	249	10	35	43	4	44	2	43	1	47
114	246	10	17	45	4	39	2	41	2	2
117	243	9	57	46	4	34	2	38	2	16
120	240	9	35	47	4	27	2	35	2	30
123	237	9	13	48	4	20	2	31	2	44
126	234	8	50	49	4	11	2	27	2	56
129	231	8	25	50	4	2	2	22	3	9
132	228	7	59	51	3	53	2	18	3	21
135	225	7	33	52	3	42	2	13	3	32
138	222	7	7	53	3	31	2	8	3	43
141	219	6	38	54	3	19	2	1	3	53
144	216	6	9	55	3	7	1	53	4	3
147	213	5	40	56	2	53	1	46	4	12
150	210	5	11	57	2	40	1	37	4	20
153	207	4	42	57	2	25	1	28	4	27
156	204	4	11	58	2	10	1	20	4	34
159	201	3	41	58	1	55	1	12	4	40
162	198	3	10	59	1	39	1	4	4	45
165	195	2	39	59	1	23	1	0	4	50
168	192	2	7	59	1	7	0	53	4	53
171	189	1	36	60	0	51	0	33	4	56
174	186	1	4	60	0	34	0	22	4	58
177	183	0	32	60	0	17	0	11	4	59
180	180	0	0	60	0	0	0	0	5	0

12. ON THE COMPUTATION OF THE COURSE OF THE MOON

[114b] Accordingly the method of computing the apparent movement of the moon is clear from what has been shown and is as follows. We shall reduce to equal time the time for which we are seeking the position of the moon proposed to us. By means of the time we shall deduce the mean movements of longitude, anomaly, and latitude—which last we shall also define soon—as we did in the case of the sun, from the given beginning of the years of Our Lord, or from some other beginning, and we shall declare the positions of the single movements at the time set before us. Then we shall seek in the table twice the regular longitude of the moon or twice its angular distance from the sun and[1] the corresponding additosubtraction found in the third column; and we shall note the proportional minutes which are in the next column. Accordingly if the number with which we entered upon the table was found in the first column or is less than 180°, we shall add the additosubtraction to the lunar anomaly; but if it is greater than 180° or is in the second column, the additosubtraction will be subtracted from the anomaly; and we shall have the corrected anomaly of the moon and its true angular distance from the highest apsis.

And entering the table again with this (distance) we shall determine the corresponding additosubtraction in the fifth column and the excess which follows in the sixth column, which the second (the small) epicycle adds (to the additosubtraction), over and above the first epicycle. The proportional part of this excess taken in accordance with the ratio of the 60 minutes is always added to this additosubtraction. The sum is subtracted from the mean movement of longitude or latitude, if the corrected anomaly is less than 180° or a semicircle; and it is added, if the anomaly is greater. And in this way we shall have the true distance of the moon from the mean position of the sun and the corrected movement of latitude. Wherefore the true position of the moon will not be unknown, either its distance from the first star of Aries in the case of the simple movement of the sun or its distance from the spring equinox in the case of the composite movement or the addition of the precession. Finally by means of the corrected movement in latitude we shall have in the seventh and last place of the table the degrees of latitude which measure the distance of the moon from the ecliptic. That latitude will be northern at the time when the movement of latitude is found in the first part of the table, [115a] i.e., if it is less than 90° or greater than 270°; otherwise it will be following a southern latitude. And so the moon will be coming down from the north to 180°, and afterwards it will be going up from the southern limit, until it has completed the remaining parts of the circle. Thus the apparent course of the moon has somehow as many affairs around the centre of the Earth as the Earth has around the sun.

[1] Because the moon traverses the small epicycle twice during one synodic month, the time of one revolution with respect to the sun.

13. HOW THE MOVEMENT OF LUNAR LATITUDE IS EXAMINED AND DEMONSTRATED

Now too we must give the ratio of the lunar movement in latitude, and it seems more difficult to discover, as it is complicated by more attendant circumstances. For, as we said before, if two eclipses of the moon were similar and equal in all respects, *i.e.*, with the parts eclipsed having the same position to the north or to the south and at the same ascending or descending ecliptic section: its distance from the Earth or from the highest apsis would be equal, since in this harmony the moon is understood to have completed its whole circles of latitude by true movement. For since the shadow of the Earth is conoid, and if a right cone is cut in a plane parallel with the base, the section is a circle which is smaller the greater the distance from the base and greater the shorter the distance from the base, and similarly equal at an equal distance. And so the moon at equal distances from the Earth traverses equal circles of shadow and presents to our vision equal disks of itself. Hence the moon, standing out with equal parts in the same direction according to an equal distance from the centre of the shadow makes us certain of equal latitudes, from which it necessarily follows that the moon has returned to its former position in latitude and is now distant from the same ecliptic node by an equal interval. But that is especially true if the position fulfils two of those conditions. For its approach to the Earth or withdrawal from it changes the total magnitude of the shadow, [115b] but so slightly that it can hardly be grasped. Accordingly the greater the interval of time between both eclipses, the more definite can we have the movement in latitude of the moon, as was said in the case of the sun.

But since you rarely find two eclipses agreeing in these conditions—and up to now none have come our way—nevertheless we note there is another method which will give us the same result, since—the other conditions remaining—if the moon is eclipsed in different directions and at opposite sections, then it will signify that at the second eclipse the moon has arrived at a position diametrically opposite to the former and in addition to the whole circles has described a semicircle; and that will seem to be satisfactory for investigating the thing.

Accordingly we have found two eclipses fairly close in these respects: the first in the 7th year of Ptolemy Philometor, which was the 150th year of Alexander when—as Claud says—27 days of Phamenoth the 7th month of the Egyptians had passed, in the night which the 28th day followed. And the moon was eclipsed from the beginning of the 8th hour till the end of the 10th hour in Alexandrian nocturnal seasonal hours, to the extent of seven-twelfths of the lunar diameter, and it was eclipsed from the north around a descending section. Therefore the midtime of the eclipse was, he says, 2 seasonal hours after midnight, which make $2^1/_3$ equatorial hours, since the sun was at 6° of Taurus, but $1^1/_3$ hours after midnight at Cracow.

We have taken the second eclipse beneath the same Cracow meridian in the year of Our Lord 1519, after the 4th day before the Nones of June, when the sun was at 21° of Gemini. The midtime of the eclipse was $11^3/_5$ equatorial hours after midday; and the moon was eclipsed for approximately eight-twelfths of its diameter, from the south, at an ascending section.

Accordingly, from the beginning of the years of Alexander (to the first eclipse) there are 149 Egyptian years 206 days $14^1/_3$ hours at Alexandria, but at Cracow $13^1/_3$ hours according to apparent time, but $13^1/_2$ upon correction. At that time the position of anomaly by our calculation, which agreed approximately with Ptolemy's, was at 163°33' of regular movement; and there was a subtractive additosubtraction of 1°23', by which the true position of the moon was exceeded by the regular. But from the established beginning of the years of Alexander to the second eclipse [116ª] there are 1832 Egyptian years 295 days 11 hours 45 minutes by apparent time, but by equal time 11 hours 55 minutes: whence the regular movement of the moon was 182°18'. The position of anomaly was 159°55', but as corrected it was 161°13'; and the additive additosubtraction, by which the regular movement was exceeded by the apparent, was 1°44'.

Accordingly it is clear that in both eclipses the distance of the moon from the Earth was equal, and the sun was approximately at the apogee in both cases, but there was a difference of one-twelfth in the eclipses. But since the diameter of the moon usually occupies approximately $1/_2$°, as we will show afterwards, its twelfth part will be $2^1/_2$', which corresponds to approximately $1/_2$' in the oblique circle of the moon at the ecliptic sections. And so the moon was $1/_2$° farther away from the ascending section at the second eclipse than from the descending section at the first eclipse. Hence it is clear that the true movement in latitude of the moon was $179^1/_2$° after the complete revolutions. But the lunar anomaly between the first and second eclipse adds 21'—which is the difference between the additosubtractions—to the regular (movement). Accordingly we shall have a regular lunar movement in latitude of 179°51' after the full circles. Now the time between the two eclipses was 1683 years 88 days 22 hours 35 minutes by apparent time, which agreed with the equal (time). During that time there were 40,577 complete equal revolutions and 179°51', which agree with the numbers which we have already set down.

14. ON THE POSITIONS OF LUNAR ANOMALY IN LATITUDE

However, in order to determine the positions of the moon's movement in relation to the established beginnings of calendar years, we have here also assumed two lunar eclipses, not at the same section and not at diametrically opposite parts, as in the foregoing, but at equal distances north or south, and fulfilling all the other requirements,

[116^b] as we said, in accordance with Ptolemy's rule, and in this way we shall solve our problem without any error.

Accordingly, the first eclipse, which we have already used in investigating other movements of the moon, is the one which we said was observed by Claud Ptolemy in the 19th year of Hadrian when two days of the month Chiach had passed, one equatorial hour before midnight at Alexandria, but at Cracow two hours before midnight, which the third day followed. The moon was eclipsed at the midpoint of the eclipse to the extent of ten-twelfths of the diameter, *i.e.*, ten-twelfths from the north, while the sun was at 25°10' of Libra, and the position of lunar anomaly was 64°38' and its subtractive additosubtraction was 4°20' around the descending section.

We made careful observations of the other eclipse at Rome, in the year of Our Lord 1500, after the Nones of November, 2 hours after midnight, and it was the 8th daybreak before the Ides of November. But at Cracow which is 5° to the east, it was 2²/₅ hours after midnight, while the sun was at 23°16' of Scorpio; and once more there was a ten-twelfths eclipse from the north.

Therefore, since the death of Alexander there have passed 1824 Egyptian years 84 days 14 hours 20 minutes by apparent time, but by equal time 14 hours 16 minutes. Accordingly the mean movement of the moon was 174°14', and the lunar anomaly was 294°44', but as corrected it was 291°35'; and there was an additive additosubtraction of 4°27'.

Accordingly it is clear that at both these eclipses the distances of the moon from the highest apsis was approximately equal, and at both times the sun was at its mean apsis, and the magnitude of the shadows was equal. All that makes clear that the latitude of the moon was southern and equal; and hence that the moon was at an equal distance from the sections but was ascending at the second eclipse and descending at the first. Accordingly between both eclipses there are 1366 Egyptian years 358 days 4 hours 20 minutes by apparent time, but by equal time 4 hours 24 minutes, wherein the movement in latitude was 159°55'.

Now let *ABCD* be the oblique circle of the moon; and let *AB* be its diameter and common section with the ecliptic. Let *C* be the northern limit, and *D* the southern; [117^a] *A* the ecliptic section descending, and *B* the ecliptic section ascending. Now let there be taken *AF* and *BE* two equal arcs in the south, according as the first eclipse was at point *F* and the second at *E*. And again let *FK* be the subtractive additosubtraction at the first eclipse, and *EL* the additive additosubtraction at the second.

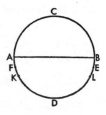

Accordingly, since

$$\text{arc } KL = 159°55'$$

and

$$\text{arc } FK = 4°20'$$

and

$$\text{arc } EL = 4°27';$$
$$\text{arc } FKLE = \text{arcs } FK + KL + LE = 168°42'.$$

And

$$180° - 168°42' = 11°18°.$$

Now

$$\text{arc } AF = \text{arc } BE = \frac{1}{2}(11°18') = 5°39',$$

which is the true distance of the moon from section AB, and on that account

$$\text{arc } AFK = 9°59'.$$

Hence it is clear that K the mean position in latitude is 99°59' away from the northern limit.

From the death of Alexander to this position and time of Ptolemy's observation there are 457 Egyptian years 91 days 10 hours by apparent time, but by equal time 9 hours 54 minutes, during which the mean movement in latitude is 50°59'. And when 50°59' is subtracted from 99°59', there remain 49° for noon on the first day of Thoth the first month by the Egyptian calendar at the beginning of the years of Alexander but on the Cracow meridian. Hence for each of the other beginnings there are given in accordance with the differences of time the positions of the course of the moon in latitude as taken in relation to the northern limit, from which we measure the movement.

Now from the first Olympiad to the death of Alexander there are 451 Egyptian years 247 days—from which in accordance with equality of time 7 minutes of an hour are subtracted—and during that time the progress in latitude was 136°57'. Again from the first Olympiad to Caesar there are 780 Egyptian years 12 hours, but 10 minutes of an hour are added to the equal time; and during that time the movement is 206°53'. Then come 45 years 12 days up to Christ. Accordingly if 136°57' are subtracted from 49° plus the 360° of the circle, there remain 272°3' for noon of the first day of the month Hekatombaion of the first Olympiad.

Now if 206°53' are added to 272°3', the sum will be 118°56' for midnight before the Kalends of January [117[b]] at the beginning of the Julian years.

Finally, with the addition of 10°49' the sum becomes 129°45' as the position (at the beginning of the years) of Our Lord similarly at midnight before the Kalends of January.

15. CONSTRUCTION OF THE INSTRUMENT FOR OBSERVING PARALLAXES

But chance and the hindrance of the lunar parallaxes did not grant to us, as it had to Ptolemy, the occasion of discovering experimentally that the greatest latitude of the moon—in accordance with the angle of section of its orbital circle and the ecliptic—is 5°, whereof the circle is 360°. For he was watching at Alexandria—which has 30°58' as the elevation of the north pole—until the moon should come most near to the vertex of the horizon, namely when it was at the beginning of Cancer and at the northern limit, which can be foreknown by means of calculations. Therefore at that time by means of an instrument which he called the parallacticon and which was constructed for measuring the parallaxes of the moon, he found that the least distance was only $2^1/_8$° from the vertex, and if any parallax had occurred at this distance it would necessarily have been very slight in such a small spatial interval. Accordingly by the subtraction of $2^1/_8$° from 30°58'[1] the remainder is 28°50$^1/_2$', which exceeds the greatest obliquity of the ecliptic—which at that time was 23°51'20"—by approximately 5°; and this latitude for the moon is found to agree in every respect with the other particulars.

But the instrument for observing parallaxes consists of three straight-edges. Two of them are of equal length and are at least eight or nine feet long; the third is somewhat longer. The latter and one of the former two are joined to both extremities of the remaining straight-edge, by carefully making holes and fitting cylinders or pivots into them in such a way that while the straight-edges are movable in a plane surface they do not wobble at all at the joints. Now in the longer straight-edge a straight line should be drawn from the centre of its place of joining through its total length, and the line is made equal to the distance between the places of joining (on the other straight-edge) measured as accurately as possible. This line is divided into 1,000 equal parts—or into more, if that can be done; and the division of the remainder should be carried on [118ª] in the same unit parts, until it reaches 1414 parts, which subtend the side of a square that may be inscribed in a circle whose radius has 1,000 parts. It will be all right to cut off as superfluous the remainder of the straight-edge over and above this. In the other ruler too, there should be drawn from the centre of the joining-place a line equal to those 1,000 parts or to the distance between the centres of the two joining-places; the ruler should have eyepieces fastened to it on one side, as in a dioptra, which sight may have passage through. The eyepieces should be so adjusted that the sight-passages do not at all swerve away from the line already drawn the length of the straight-edge, but keep at an equal distance; and provided also that the line as extended from its terminus to the longer ruler can touch the divided line. And in this way by means of the rulers an isosceles triangle is made, the base of which will be along the parts of the divided line.

[1]The elevation of the north pole above the horizon is equal to the declination of the vertex of the horizon from the equator.

Then a pole, which has been divided crosswise in the best manner and well smoothed, should be erected on a firm base. The ruler which has the two joining-places should be affixed to this pole by means of pivots, around which the instrument may swing, like a swinging door, but in such a way that the straight line through the centres of the joining-places will always correspond to the plumbline of the ruler and point towards the vertex of the horizon, as if its axis. Accordingly when a person who wishes to find the distance of some star from the vertex of the horizon has the star itself in full view along a straight line through the eyepieces, then by the application underneath of the ruler with the divided line, he should learn how many unit parts—whereof the diameter of the circle has 20,000—subtend the angle between the line of vision and the axis of the horizon; and by means of the table he will get the sought arc of the great circle passing through the star and the vertex of the horizon.

16. HOW THE PARALLAXES OF THE MOON ARE DETERMINED

By means of this instrument, as we said, Ptolemy found the greatest latitude of the moon to be 5°. Next he turned to observing the parallax and said he discovered that at Alexandria it was 1°7', while the sun was at 5°28' of Libra and the mean movement of the moon away from the sun was 78°13', the regular anomaly was 262°20'; the movement in latitude was 354°40'; the additive additosubtraction was 7°26'; [118b] and accordingly the position of the moon was at 3°9' of Capricorn. The corrected movement in latitude was 2°6'; the northern latitude of the moon was 4°59'; its declination from the equator was 23°49'; and the latitude of Alexandria was 30°58'. The moon, he says, as seen through the instrument, was approximately in the meridian circle at 50°45' from the vertex of the horizon, i.e., 1°7' more than the computation demanded. Hence by the rule of the ancients concerning the eccentric circle and the epicycle, he shows that the distance of the moon from the centre of the Earth was then 39P45' whereof the radius of the Earth is 1P; and what next follows from the ratio of the circles, namely that the greatest distance of the moon from the Earth—which they say occurs at a new and at a full moon in the apogee of the epicycle—is 64P10', but the least distance—at the quadratures and at the half moon in the perigee of the epicycle—is only 33P33'. Hence he even evaluated the parallaxes, which occur at about 90° from the vertex (of the horizon): the least at 53'34", and the greatest at 1°43'—as it is possible to see in a broad outline what he built up concerning them. But now it is perfectly obvious to those wishing to consider the question that these things are far otherwise, as we have found out experimentally very often.

However, we shall review two observations, by which it is again made clear that our hypotheses as to the moon have more certitude than his, because they are found to

agree better with the appearances and to leave nothing in doubt. In the 1522nd year since the birth of Christ, on the 5th day before the Kalends of October, after the passage of $5^2/_3$ equal hours since midday, at about sunset at Frauenburg we found by means of the parallactic instrument that the centre of the moon, which was in the meridian circle, was 82°50' distant from the vertex of the horizon. Accordingly from the beginning of the years of Our Lord to this hour there were 1522 Egyptian years 284 days $17^2/_3$ hours by apparent time but by equal time 17 hours 24 minutes. Wherefore the apparent position of the sun was by calculation at 13°29' of Libra and the regular movement of the moon away from the sun [119ª] was 87°6'; the regular anomaly was 357°39'; but the true (the corrected) anomaly was 358°40', and it added 7'; and thus the true position of the moon was at 12°32' of Aries. The mean movement in latitude was 197°1' from the northern limit, the true was 197°8'; the southern latitude of the moon was 4°47'; the moon had a declination of 27°41' from the equator; the latitude of the place of our observation was 54°19', and the addition of 54°19' to the lunar declination makes the true distance of the moon from the pole of the horizon to be 82°. Accordingly the 50' not accounted for belong to the parallax, which by Ptolemy's teaching should be 1°17'.

Once more we made another observation at the same place in the 1524th year of Our Lord on the 7th day before the Ides of August 6 hours after midday; and we saw through the same instrument the moon at 82° from the vertex of the horizon. Accordingly from the beginning of the years of Our Lord to this hour there were 1524 Egyptian years 234 days 18 hours (by apparent time) and also 18 hours by exact time. The position of the sun was by calculation at 24°14' of Leo; the mean movement of the moon away from the sun was 97°6'; the regular anomaly was 242°10'; the corrected anomaly was 239°43', adding approximately 7° to the mean movement; wherefore the true position of the moon was at 9°39' of Sagittarius; the mean movement of latitude was 193°19'; the true, 200°17'; the southern latitude of the moon was 4°41'; the southern declination was 26°36', and the addition of 26°36' to 54°19' of the latitude of the place of observation makes the distance of the moon from the pole of the horizon to be 80°55'. But there appeared to be 82°. Accordingly the difference of 1°5' came from the lunar parallax, which according to Ptolemy should have been 1°38' and also according to the theory of the ancients, as the harmonic ratio, which follows from their hypotheses, forces you to admit.

17. DISTANCE OF THE MOON FROM THE EARTH AND DEMONSTRATION OF THEIR RATIO IN PARTS WHEREOF THE RADIUS OF THE EARTH IS THE UNIT

[119ᵇ] From this it will now be made apparent how great the distance of the moon from the earth is. And without this distance a sure ratio cannot be given for the parallaxes, for they are mutually related. And it will be established in this way. Let *AB* be a

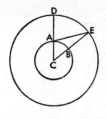

great circle of the Earth, and let C be its centre. Around C let another circle be described in comparison with which the Earth has considerable magnitude, and let this circle be DE. Let D be the pole of the horizon, and let the centre of the moon be at E, so that DE its distance from the vertex is known. Accordingly, since at the first observation,

$$\text{angle } DAE = 82°50'$$

and by calculation

$$\text{angle } ACE = 82°$$

and hence

$$\text{angle } DAE - \text{angle } ACE = 50',$$

which belonged to the parallax; we have triangle ACE with its angles given and therefore with its sides given. For since

$$\text{angle } CAE \text{ is given,}$$
$$\text{side } CE = 99{,}219$$

where diameter of circle circumscribing triangle AEC = 100,000 and

$$AC = 1{,}454;$$

and

$$CE 68^P,$$

where AC, radius of Earth, = 1^P.

And this was the distance of the moon from the centre of the Earth at the first observation.

But at the second observation

$$\text{angle } DAE = 82°,$$

as the apparent movement; and, by calculation,

$$\text{angle } ACE = 80°55;$$

and, by subtraction,

$$\text{angle } AEC = 1°5'.$$

Accordingly,

$$\text{side } EC = 99{,}027$$

and

$$\text{side } AC = 1894$$

where diameter of circle circumscribing triangle = 100,000.

And so

$$CB = 56^P42',$$

where the radius of Earth = 1^P.

And that was the distance of the moon.

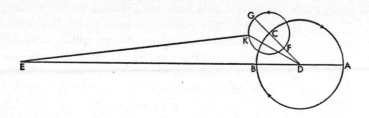

But now let *ABC* be the greater epicycle of the moon; and let its centre be *D*. Let *E* be taken as the centre of the Earth, and from *E* let the straight line *EBDA* be drawn, so that *A* is the apogee and *B* the perigee. Now let

$$\text{arc } ABC = 242°10',$$

in accordance with the computed regularity of the lunar anomaly.
And with *C* as centre let epicycle *FGK* be described, whereon

$$\text{arc } FGK = 194°10',$$

twice the distance of the moon from the sun. And let *DK* be joined.

Thus,

$$\text{angle } GDK = - \text{ add. } [120^a] \; 2°27';$$

and, by subtraction,

$$\text{corr. anomaly} = 59°43',$$

since

$$\text{arc } CDB = \text{arc } ABC - 180° = 62°10',$$

and

$$\text{angle } BEK = 7°.$$

Therefore, in triangle *KDE* the angles are given in the degrees whereof 2 rt. angles = 180°; and the ratio of the sides is also given:

$$DE = 91,856$$

and

$$EK = 86,354,$$

where diameter of circle circumscribing triangle *KDE* = 100,000.

But

$$KE = 94,010$$
$$\text{where } DE = 100,000.$$

Now it was shown above that

$$DF = 8,600$$

and

$$DFG = 13,340.$$

236

Accordingly it follows from the given ratio that when, as was shown,

$$EK = 56\text{P}42',$$

where radius of the Earth = 1P;

then

$$DE = 60\text{P}18',$$
$$DF = 5\text{P}11',$$
$$DFG = 8\text{P}2';$$

and hence, as extended in a straight line.

$$EDG = 68^{1}/_{3}\text{P};$$

and that is the greatest altitude of the half moon. Furthermore,

$$ED - DG = 52°17',$$

which is its least distance. And thus, at its greatest,

$$EDF = 65^{1}/_{2}\text{P},$$

which is the altitude occurring at the bright, full moon; and at its least,

$$EDF - DF = 55\text{P}8'.$$

And we should not be moved by the fact that others—and especially those to whom the parallaxes of the moon could not become known except partially, on account of the location of their places—estimate the greatest distance of the new moon and the full moon to be 64P10'. But the greater nearness of the moon to the horizon—for it is clear that the parallaxes are filled out in relation to the horizon—has allowed us to perceive them more perfectly, and we have not found the parallaxes to differ by more than 1' on account of the difference caused by the nearness of the moon to the horizon.

18. ON THE DIAMETER OF THE MOON AND ON THE DIAMETER OF THE TERRESTRIAL SHADOW IN THE PLACE OF PASSAGE OF THE MOON

[120^{b}] Moreover, the apparent diameters of the moon and the shadow vary with the distance of the moon from the Earth. Wherefore it is pertinent to speak of them. And although the diameters of the sun and the moon are rightly determined through the dioptra of Hipparchus, nevertheless in the case of the moon astronomers judge that this is done with more certainty through some particular eclipses of the moon, in which the moon is at an equal distance from its highest or lowest apsis, especially if at that time the sun too is in the same relative situation, so that the circle of shadow which the moon passes through is found equal—unless the eclipses themselves are unequal in extent. For it is clear that the comparison of the difference in extent of the eclipses with the latitude of the moon shows how much of the circle around the centre of the Earth the diameter of the moon subtends. When that has been perceived, the semidiameter of the shadow is also known.

All this will be made clearer by an example. In this way at the midpoint of the first eclipse $3/_{12}$ of the diameter of the moon was eclipsed; and the moon had a latitude of 47'54"; but at the other eclipse $10/_{12}$ of the diameter was eclipsed, and the latitude was 29'37". The difference between the extent of the eclipses is $7/_{12}$ of the diameter; the difference in latitude is 18'17"; and the $12/_{12}$ are proportional to the 31'20" which the diameter of the moon subtends. Accordingly it is clear that the centre of the moon at the midpoint of the first eclipse was about a quarter of the moon's diameter—or 7'50" of latitude—beyond the shadow if these 7'50" are subtracted from the 47'54" of the total latitude, 40'4" remain as the semidiameter of the shadow; just as at the other eclipse the shadow occupied—in proportion to $1/_3$ of the lunar diameter—10'27" more than the latitude of the centre of the moon. The addition of 29'37" to 10'27" similarly makes the semidiameter of the shadow to be 40'4". And so in accordance with Ptolemy's conclusion, when the sun and moon are in conjunction or opposition at their greatest distance from the Earth, the diameter of the moon is 31'20"—[121ᵃ] as he admits he found the sun's diameter to be through the dioptra of Hipparchus—but the diameter of the shadow is 1°21'20"; and he believed that the diameters were in the ratio of 13 to 5, *i.e.*, the ratio of double plus three-fifths.

19. How the Distances of the Sun and Moon from the Earth, Their Diameters and That of the Shadow at the Place of Crossing of the Moon, and the Axis of the Shadow Are Demonstrated Simultaneously

But even the sun has some parallax; and since it is very slight, it is not perceived so easily, except that the following things are related reciprocally: namely the distance of the sun and moon from the Earth, their diameters and that of the shadow at the crossing of the moon, and the axis of the shadow; and for that reason they are mutually productive of one another in analytical demonstrations. First we shall review Ptolemy's conclusions on these things, and how he demonstrated them, and we shall draw out from them what seems the most true. He assumed $31^1/_3$' as the apparent diameter of the sun, which he employed without any qualification. He assumed as equal to that the diameter of the full and new moon when at its apogee, which he says was at a distance of 64ᴾ10', whereof the radius of the Earth is 1ᴾ.

From that he demonstrated the rest in this way: Let *ABC* be the circle of the solar globe around centre *D*; and let *EFG* be the circle of the terrestrial globe around its centre *K* at its greatest distance from the sun. Let *AG* and *CE* be straight lines touching both circles, and let them as extended meet at the apex of the shadow, as at point *S*. And let *DKS* be a line through the centres of the sun and the Earth. Moreover, let *AK* and *KC* be drawn, and let *AC* and *GE* be joined, which should hardly differ at all from

the diameters on account of the great distance between them. Now on *DKS* let equal segments *LK* and *KM* be taken in proportion to the distance of the moon in the apogee when new and when full: in his opinion 64ᴾ10', where *EK* is 1ᴾ. Let *QMR* be the diameter of the shadow at this crossing of the moon; and let *NLO* be the diameter of the moon at right angles to *DK*, and let it be extended as *LOP*.

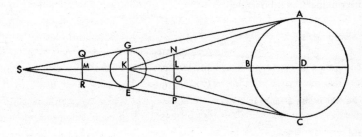

The first problem is to find

$$DK : KE.$$

Accordingly, since

$$\text{angle } NKD = 31^{1}/_{3}',$$
$$\text{where 4 rt. angles} = 360°$$

and [121ᵇ]

$$\text{angle } LKO = {}^{1}/_{2} \text{ angle } NKO,$$
$$\text{angle } LKO = 15^{2}/_{3}'.$$

And

$$\text{angle } L = 90°.$$

Accordingly, in triangle *LKO*, which has its angles given,

$$KL : LO \text{ is given,}$$

and

$$LO = 17\text{ᴾ}33',$$

and

$$LK = 64\text{ᴾ}10',$$
$$\text{where } KE = 1\text{ᴾ}$$

And because

$$LO : MR = 5 : 13,$$
$$MR = 45'38''.$$

But since *LOP* and *MR* are parallel to *KE* at equal intervals, on that account

$$LOP + MR = 2 KE.$$

And

$$OP = LOP - (MR + LO) - 56'49''.$$

Now by Euclid, VI, 2,

$$EC : PC = KC : OC = KD : LD = KE : OP = 60' : 56'49".$$

Hence

$$LD = 56'49",$$
$$\text{where } DLK = 1\text{P}.$$

And accordingly, by subtraction,

$$KL = 3'11".$$

But according as

$$KL = 64\text{P}10',$$
$$\text{where } FK = 1\text{P},$$
$$KD = 1210\text{P}.$$

Now it is also clear that

$$MR = 45'38".$$

Hence

$$KE : MR \text{ is given}$$

and

$$KMS : MS \text{ is given.}$$

And of the whole KMS

$$KM = 14'22".$$

And, *separando*,

$$KMS = 268\text{P}$$
$$\text{where } KM = 64\text{P}10'.$$

So in truth Ptolemy.

But others after Ptolemy, since they found that these things did not agree sufficiently with the appearances, published other things concerning all this. None the less they admit that the greatest distance of the full moon and the new moon from the Earth is 64P10'; and that the apparent diameter of the sun at its apogee is $31^1/_3$'. They even grant that the diameter of the shadow at the place of crossing of the moon is as 13 to 5, even as Ptolemy himself. Nevertheless they deny that the apparent diameter of the moon at that time is greater than $29^1/_2$'; and for that reason they put the diameter of the shadow at approximately $1°16^3/_4$'. They hold that it follows from this that at its apogee the distance of the sun from the Earth is 1146P and that the axis of the shadow is 254P, whereof the radius of the Earth is 1P. [122ª] And astronomers attribute these things to the Harranite philosopher (al-Battani) as the discoverer, although they cannot be joined together at all reasonably. We considered that these things must be adjusted and corrected as follows, since we put the apparent diameter of the sun in its apogee at 31'40"—for it should be somewhat greater now than before Ptolemy—and that of the full or the new moon in its highest apsis at 30' and the diameter of the shadow in its

crossing at $80^3/_5'$. For astronomers should have a slightly greater ratio than that of 5 to 13, that is to say 150 to 403. And the whole sun is not covered by the moon, unless the moon is at a lesser distance from the Earth than 62P, whereof the radius of the Earth is 1P. For when these things are put down in this way they seem to be connected with one another and the rest in a sure fashion and to be consonant with the apparent eclipses of the sun and moon. And in accordance with the foregoing demonstration:

$$LO = 17'85"$$

where KE radius of Earth = 1P.

And for that reason

$$MR = 46'1",$$

and

$$OP = 56'51".$$

And

$$DLK = 1179 \text{ P},$$

the distance from the Earth of the sun at its apogee; and

$$KMS = 265,$$

which is the axis of the shadow.

20. ON THE MAGNITUDE OF THESE THREE CELESTIAL BODIES: THE SUN, MOON, AND EARTH, AND THEIR COMPARISON WITH ONE ANOTHER

Hence it is also manifest that

$$LK : KD = 1 : 18$$

and

$$LO : DC = 1 : 18.$$

Now

$$1 : 18 = 17'8" : 5\text{P}27'$$

where $KE = 1\text{P}$.

And

$$SK : KE = 265\text{P} : 1\text{P} = SKD : DC = 1444\text{P} : 5\text{P}27'.$$

For they are all proportional; and that will be the ratio of the diameters of the sun and Earth. But, as globes are in the ratio of the cubes[1] of their diameters, accordingly

$$(5\text{P}27')^3 - 161^7/_8\text{P};$$

and the sun is $161^7/_8$ greater than the terrestrial globe.

Again, since

$$\text{moon's radius} = 17'9"$$

[1] *literally*, in the triplicate ratio.

where KE = 1P,

[122ᵇ] Earth's diameter: moon's diameter = 7 : 2, *i.e.*, in the triple sesquialter ratio. When the cube[1] of that ratio is taken, it shows that the Earth is $42^7/_8$ greater than the moon.

And hence the sun will be $6,999^{62}/_{63}$ greater than the moon.

21. ON THE APPARENT DIAMETER OF THE SUN AND ITS PARALLAXES

But since the same magnitude when farther away appears smaller than when nearer; for that reason it happens that the sun, moon, and the shadow of the Earth vary with their unequal distances from the Earth no less than do their parallaxes. By means of the aforesaid, all these things are easily determinable for any elongation whatsoever. That is first made manifest in the case of the sun. For since we have shown that the Earth at its farthest is 10,323 parts distant from the sun, whereof the radius of the orbital circle of annual revolution = 10,000; and at its nearest the Earth has a distance of 9,678 parts of the remainder of the diameter: accordingly the highest apsis is 1179P whereof the radius of the Earth is 1P, the lowest apsis will be 1105P, and so the mean apsis will be 1142P. Accordingly in the right triangle[2]

$$1,000,000 \div 1179 = 848^3 = {}^1/_2 \text{ ch. 2 (2'55")},$$

which is the small angle of greatest parallax, and that is found around the horizon. Similarly, as the least distance is 1105P,

$$1,000,000 - 1105 = 905^4 = {}^1/_2 \text{ ch. 2 (3'7")};$$

and 3'7" measures the angle of greatest parallax of the lowest apsis. Now it was shown that the diameter of the sun is 5P27', whereof the diameter of the Earth is 1P, and that it appears at the highest apsis as 31'48". For

$$1179P : 5P27' = 2,000,000 : 9,245;$$

$$\text{where diameter of circle} = 2,000,000,$$

and

$${}^1/_2 \text{ ch. 2(31'48")} = 9245.$$

It follows that at the least distance of 1105P there is an apparent diameter of 33'54". Therefore the difference between them is 2'6"; but there is a difference of only 12" [123ᵃ] between the parallaxes. Ptolemy considered that both of these differences should be ignored on account of their smallness; for 1' or 2' is not easily perceptible to the senses, much less than are a few seconds perceptible. Wherefore if we keep the greatest parallax of the sun at 3' everywhere, we shall be seen to have made no error.

[1] *literally*, the triplicate.
[2] *i.e.*, the right triangle formed by the line joining the centres of the sun and the Earth, the tangent from the centre of the sun to the Earth's surface, and the radius of the Earth to that point of tangency.
[3] *i.e.*, when the highest apsis = 1179P, 1P = 848
whereof radius of circle = 1,000,000.
[4] Similarily, where the lowest apsis = 1105P, 1P = 905 whereof radius of circle = 1,000,000.

Now we shall determine the mean apparent diameters of the sun through its mean distances; or, as do others, through the apparent hourly movement of the sun, which they believe to be to its diameter as 5 to 66 or as 1 to $14^1/_5$. For its hourly movement is approximately proportional to its distance.

22. ON THE UNEQUAL APPARENT DIAMETER OF THE MOON AND ITS PARALLAXES

A greater diversity in the apparent diameter and parallaxes appears in the case of the moon as being the nearest planet. For since its greatest distance from the Earth is $65^1/_2$P at new moon and full moon, its least distance—will by the above demonstrations be 55P8'; and the greatest (altitudinal) elongation of the half moon will be 68P21', and the least 52P17'. Accordingly we shall have the parallaxes of the setting or rising moon at these four termini, when we have divided the radius of the circle by the distances of the moon from the Earth: the parallax of the farthest half moon will be 50'18" and that of the farthest new or full moon will be 52'24"; the parallax of the nearest full or new moon will be 62'21" and that of the nearest half moon 65'45".

Furthermore by this the apparent diameters of the moon are established. For it was shown that the diameter of the Earth is to the diameter of the moon as 7 to 2, and the radius of the Earth will be to the diameter of the moon as 7 to 4. Moreover, the parallaxes are in that ratio to the apparent diameters of the moon, since the straight lines, which comprehend the angles of the greater parallaxes, do not differ at all from the apparent diameters at the same crossing of the moon; and the angles, (or arcs of parallax) are approximately proportional to the chords subtending them; and their difference is not perceptible to sense. By this summary it is clear that at the first limit of the parallaxes which have been already set forth the apparent [123b] diameter of the moon will be $28^3/_4$'; at the second, approximately 30'; at the third, 35'38"; and at the last limit, 37'34". By the hypothesis of Ptolemy and others the diameter would have been approximately 1°, and so it ought to have been, as the half moon at that time was shedding as much light on the Earth as the full moon would.

23. WHAT THE RATIO OF DIFFERENCE BETWEEN THE SHADOWS OF THE EARTH IS

We have already made clear that

shadow's diameter; moon's diameter = 403 : 150.

For that reason at a full or a new moon, when the sun is at its apogee, the shadow's is found to be 80'36" at its least and 95'44" at its greatest; and the greatest difference is 15'8". Moreover, the shadow of the Earth varies, even in the same place of crossing of the moon, on account of the unequal distance of the Earth from the sun, as follows:

For, as in the foregoing diagram, let *DKS* the straight line through the centres of the sun and the Earth be drawn again, and also *CES* the line of tangency. As was shown, when

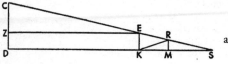

distance $DK = 1179^P$,

where $KE = 1^P$,

and

$KM = 62^P$

then the semidiameter of the shadow

$MR = 46'1''$

where $KE = 1^P$;

and (*KR* being joined)

angle $MKR = 42'32''$,

which is the angle of sight, and the axis of the shadow

$KMS = 265^P.$

Now when the Earth is nearest to the sun, so that

$DK = 1105^P,$

we shall evaluate the shadow of the Earth at the same crossing of the moon, as follows: For let *EZ* be drawn parallel to *DK*. Then

$CZ : ZE = EK : KS.$

But

$CZ = 4^P27'$

and

$ZE = 1105^P.$

For

$ZE = DK$

and

$DZ = KE,$

as *KZ* is a parallelogram. Accordingly

$KS = 248^P19'$

where $KE = 1^P.$

Now

$KM = 62;$

and accordingly, by subtraction,

$MS = 186^P19'.$

But since

$SM : MR = SK : KE,$

therefore

$$MR = 45'1'',$$
where [124a] $KE = 1^P.$

And hence

$$\text{angle } MKR = 41'35'',$$

which is the angle of sight. Whence it happens that on account of the approach and withdrawal of the sun and the Earth, the greatest difference in the diameters of the shadow at the same place of crossing of the moon is 1', whereof $EK = 1^P$, in proportion to an angle of sight of 57", whereof 4 rt. angles = 360°.

Furthermore, in the first case

$$\text{shadow's diameter : moon's diameter} > 13 : 5;$$

but here

$$\text{shadow's diameter : moon's diameter} < 13 : 5,$$

as 13 : 5 is a sort of mean ratio. Wherefore we shall make but slight error if we employ it as everywhere the same, thus saving labour and following the judgment of the ancients.

24. ON THE TABLE OF THE PARTICULAR PARALLAXES IN THE CIRCLE PASSING THROUGH THE POLES OF THE HORIZON

Moreover, it will not be difficult now to determine all the single parallaxes of the sun and the moon. For let there be drawn again AB the terrestrial circle through the vertex of the horizon, with C as its centre. And in the same plane let DE be the orbital circle of the moon, FG that of the sun, CDF the line through the vertex of the horizon; and let line CEG be drawn, in which the true positions of the sun and the moon are understood to be, and let the lines of sight AG and AE be joined to those points. Therefore the parallaxes of the sun are measured by angle AGC, those of the moon 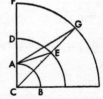 by angle AEC. Moreover, there is a parallax between the sun and moon which is measured by angle GAE, which is determined according to the difference between angles AGC and AEC. Now let us take angle ACG, with which we wish to compare those angles; and, for example, let

$$\text{angle } ACG = 30°.$$

Now it is clear from what we have shown concerning plane triangles that when

$$\text{line } CG = 1142^P,$$
$$\text{where } AC = 1^P,$$
$$\text{angle } AGC = 1\frac{1}{2}',$$

which is the difference between the true and the seeming altitude of the sun.

But when

$$\text{angle } ACG = 60°,$$
$$\text{angle } AGC = 2'36''.$$

Everything will be similarly clear as regards the remaining angles.

But in the case of the moon at its four limits: If at the greatest lunar distance from the Earth, wherein, as we said,

$$CE = 68^P21',$$
[124b] where $CA = 1^P,$
angle $DCE = 30°,$
where 4 rt. angles = 360°,

we shall have triangle ACE in which the two sides AC and CE together with angle ACE have been given. From that we find that

$$\text{parallax } AEC = 25'28".$$

And when

$$CE = 65^1/_2{}^P$$
angle $AEC = 26'36".$

Similarly in the third case when

$$CE = 55^P8',$$
parallax $AEC = 31'42".$

Finally, at the least distance when

$$CE = 52^P17',$$
angle $AEC = 33'27".$

Again, when

$$\text{arc } DE = 60°,$$

the parallaxes in the same order will be as follows:

$$\text{First parallax} = 43'55",$$
$$\text{second parallax} = 45'51",$$
$$\text{third parallax} = 54^1/_2{}',$$

and

$$\text{fourth parallax} = 57^1/_2{}'.$$

We shall inscribe all these things after the order of the subjoined table, which for the sake of convenience we shall extend like all the other tables into a series of thirty rows but proceeding by 6°'s by which twice the arcs from the vertex of the horizon— of which the greatest is 90°—are given to be understood. But we have divided the table into nine columns. For in the first and second will be found the common numbers of the circle. We shall put the parallaxes of the sun in the third, and in the next the lunar parallaxes, and in the fifth column the differences, by which the least parallaxes, which occur at the half moon and at the apogee, are deficient as measured by the parallaxes occurring at the apogee of the full moon or the new moon. The sixth column will

246

contain the parallaxes which the full or bright moon produces at its perigee; and in the next column are the minutes of difference, by which the parallaxes which occur at half

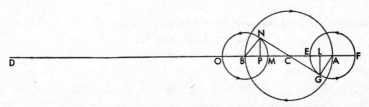

moon when the moon is nearest to us exceed those occurring at half moon in the apogee. Then, the two spaces which are left are reserved for the proportional minutes, by which the parallaxes between these four limits can be computed. We shall set forth these parallaxes, and first in connection with the apogee and the parallaxes which are between the first two limits—as follows.

Let circle [125ª] AB be the first epicycle of the moon, and let C be its centre. With D taken as the centre of the Earth, let the straight line $DBCA$ be drawn; and with apogee A as centre let the second epicycle EFG be described. Now let

$$\text{arc } EG = 60°,$$

and let AG and CG be joined. Accordingly, since in the foregoing it was shown that

$$\text{straight line } CE = 5^{\text{P}}11',$$

where radius of Earth = 1^{P},

$$\text{straight line } DC = 60^{\text{P}} \, 18',$$

and

$$\text{straight line } EF = 2^{\text{P}}51',$$

then in triangle ACG

$$\text{side } GA = 1^{\text{P}}25'$$

and

$$\text{side } AC = 6^{\text{P}}36';$$

and

$$\text{angle } CAG \text{ is given,}$$

which is the angle comprehended by GA and AC. Accordingly, by what has been shown concerning plane triangles,

$$\text{side } CG = 6^{\text{P}}7'.$$

Accordingly, as extended in a straight line,

$$DCG = DCL = 66^{\text{P}}25'.$$

But

$$DCE = 65\tfrac{1}{2}\text{P}.$$

Therefore, by subtraction,

$$EL\,55\tfrac{1}{2}',$$

and that is the excess. Moreover, by this given ratio, when

$$DCE = 60\text{P};$$
$$EF = 2\text{P}37',$$

and

$$EL = 46'.$$

Therefore, according as,

$$EF = 60',$$
$$\text{excess } EL \fallingdotseq 18'.$$

We shall mark these down in the eighth column of the table as corresponding to 60° (in the first column).

We shall show something similar in the case of perigee B. Let the second epicycle MNO be drawn again around centre B, and let

$$\text{angle } MBN = 60°.$$

For, as before, triangle BCN will have its sides and angles given, and similarly

$$\text{excess } MP = 55\tfrac{1}{2}'$$
$$\text{where Earth's radius} = 1\text{P}.$$

But that is because

$$DBM = 55\text{P}8'.$$

If

$$DBM = 60\text{P};$$
$$MBO = 3\text{P}7',$$

and

$$\text{excess } MP = 55'.$$

Now

$$3\text{P}7' : 55' = 60' : 18';$$

and so on, the same as before. Nevertheless there is a difference of a few seconds. We shall do this for the rest; and thus we shall fill out the eighth column of the table. But if we were to employ instead of them, those (proportional minutes) which were set

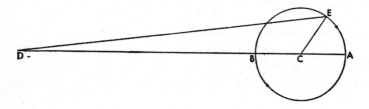

out in the table of additosubtractions, we shall make no error. For they are approximately the same—and it is a question of very small numbers. [125b] There remain the proportional minutes which occur at the mean termini, namely between the second and the third termini.

Now let circle AB be the first epicycle at the new or full moon. Let C be its centre; and let D be taken as the centre of the Earth. And let the straight line DBCA be extended. Now from apogee A let some arc be taken: for instance, let

$$\text{arc } AE = 60°;$$

and let DE and CE be joined. For we shall have the triangle DCE, in which two sides are given:

$$CD = 60\text{P}19'$$

and

$$CE = 5\text{P}11'.$$

Now angle DCE is an interior angle, and

$$\text{angle } DCE = 180° - \text{angle } ACE.$$

Accordingly, by what we have shown concerning triangles,

$$DE = 63\text{P}4'.$$

But

$$DBA = 65^1/_2\text{P},$$

and

$$DBA - ED = 2\text{P}26'.$$

Now

$$AB = 10\text{P}22';$$

and

$$10\text{P}22' : 2\text{P}26' = 60' : 14'.$$

And they are inscribed in the table in the ninth column opposite 60°. Following this example, we have completed the rest and filled out the table, which follows. And we have added another table of the semidiameters of the sun and the moon, and the shadow of the Earth, so that as far as possible they may be at hand.

TABLE OF THE PARALLAXES OF THE SUN AND MOON

Common Numbers		Parallaxes of the sun		Parallax of the moon at the second limit		Differences between the first and second limit of the moon: to be subtracted		Parallax of the moon at the third limit		Differences between the third and forth lunar limit: to be added		Proportional minutes of the smaller epicycle	Proportional minutes of the greater epicycle
Deg.	Deg.	Min.	Sec.	Min.	Sec.	Min.	Sec.	Min.	Sec.	Min.	Sec.		
6	354	0	10	2	46	0	7	3	18	0	12	0	0
12	348	0	19	5	33	0	14	6	36	0	23	1	0
18	342	0	29	8	19	0	21	9	53	0	34	3	1
24	336	0	38	11	4	0	28	13	10	0	45	4	2
30	330	0	47	13	49	0	35	16	26	0	56	5	3
36	324	0	56	16	32	0	42	19	40	1	6	7	5
42	318	1	5	19	5	0	48	22	47	1	16	10	7
48	312	1	13	21	39	0	55	25	47	1	26	12	9
54	306	1	22	24	9	1	1	28	49	1	35	15	12
60	300	1	31	26	36	1	8	31	42	1	45	18	14
66	294	1	39	28	57	1	14	34	31	1	54	21	17
72	288	1	46	31	14	1	19	37	14	2	3	24	20
78	282	1	53	33	25	1	24	39	50	2	11	27	23
84	276	2	0	35	31	1	29	42	19	2	19	30	26
90	270	2	7	37	31	1	34	44	40	2	26	34	29
96	264	2	13	39	24	1	39	46	54	2	33	37	32
102	258	2	20	41	10	1	44	49	0	2	40	39	35
108	252	2	26	42	50	1	48	50	59	2	46	42	38
114	246	2	31	44	24	1	52	52	49	2	53	45	41
120	240	2	36	45	51	1	56	54	30	3	0	47	44
126	234	2	40	47	8	2	0	56	2	3	6	49	47
132	228	2	44	48	15	2	2	57	23	3	11	51	49
138	222	2	49	49	15	2	3	58	36	3	14	53	52
134	216	2	52	50	10	2	4	59	39	3	17	55	54
150	210	2	54	50	55	2	4	60	31	3	20	57	56
156	204	2	56	51	29	2	5	61	12	3	22	58	57
162	198	2	58	51	56	2	5	61	47	3	23	59	58
168	192	2	59	52	13	2	6	62	9	3	23	59	59
174	186	3	0	52	22	2	6	62	19	3	24	60	60
180	180	3	0	52	24	2	6	62	21	3	24	60	60

TABLE OF THE SEMIDIAMETERS OF THE SUN, MOON, AND SHADOW

Common Numbers		Semidiameter of the Sun		Semidiameter of the Moon		Semidiameter of the Shadow		Variation of the Shadow
Deg.	Deg.	Min.	Sec.	Min.	Sec.	Min.	Sec.	Min.
6	354	15	50	15	0	40	18	0
12	358	15	50	15	1	40	21	0
18	342	15	51	15	3	40	26	1
24	336	15	52	15	6	40	34	2
30	330	15	53	15	9	40	42	3
36	324	15	55	15	14	40	56	4
42	318	15	57	15	19	41	10	6
48	312	16	0	15	25	41	26	9
54	306	16	3	15	32	41	44	11
60	300	16	6	15	39	42	2	14
66	294	16	9	15	47	42	24	16
72	288	16	12	15	56	42	40	19
78	282	16	15	16	5	43	13	22
84	276	16	19	16	13	43	34	25
90	270	16	22	16	22	43	58	27
96	264	16	26	16	30	44	20	31
102	258	16	29	16	39	44	44	33
108	252	16	32	16	47	45	6	36
114	246	16	36	16	55	45	20	39
120	240	16	39	17	4	45	52	42
126	234	16	42	17	12	46	13	45
132	228	16	45	17	19	46	32	47
138	222	16	48	17	26	46	51	49
144	216	16	50	17	32	47	7	51
150	210	16	53	17	38	47	23	53
156	204	16	54	17	41	47	31	54
162	198	16	55	17	44	47	39	55
168	192	16	56	17	46	47	44	56
174	186	16	57	17	48	47	49	56
180	180	16	57	17	49	47	52	57

25. ON COMPUTING THE PARALLAX OF THE SUN AND MOON

[127ᵃ] We shall also set out briefly the mode of computing the parallaxes of the sun and moon by the table. If for the distance of the sun or twice the distance of the moon from the vertex of the horizon we take the corresponding parallaxes in the table—the solar parallaxes simply but the lunar parallaxes at the four limits—and if we take the first proportional minutes corresponding to twice the movement of the moon or twice its distance from the sun; by means of these minutes we shall determine the parts of the difference between the first and the last terminus which are proportional to sixty minutes; we shall always subtract these parts from the parallaxes following next, and we shall always add the later parts to the parallax at the next to the last limit. And we shall have two corrected parallaxes of the moon at the apogee and perigee; the lesser epicycle increases or decreases these parallaxes. Then we shall take the last proportional minutes corresponding to the lunar anomaly; and by means of them we shall determine the proportional part of the difference between the two parallaxes found nearest; we shall always add this proportional part to the first corrected parallax, the parallax at the apogee; and the result will be the parallax of the moon sought for that place and time, as in the following example.

Let the distance of the moon from the vertex (of the horizon) be 54°, the mean movement of the moon 15°, and the corrected anomaly 100°: I wish to find from them by means of the table the lunar parallax. I double the degrees of distance, and the result is 108°, to which in the table there correspond a difference of 1'48" between the first and second limit, a parallax of 42'50" at the second limit, a parallax of 50'69" at the third limit, and a difference of 2'46" between the third and the fourth limit—which I shall mark down separately. Doubling the movement of the moon makes 30°; I find five of the first proportional minutes corresponding to it, and with them I determine 9" to be the part of the first difference which is proportional to sixty minutes: I subtract these 9" from the 42'50" of the parallax, and the remainder is 42'41". Similarly the proportional part of the second difference—which was 2'46"—is 14"; and I add it to the 50'59" of the parallax at the third limit; the sum is 51'13". The difference between these parallaxes is 8'32". After this I take the last proportional minutes corresponding to the corrected anomaly, and there are 39'. By means of them I take 4'50" as the proportional part of the difference of 8'32"; [127ᵇ] I add this 4'50" to the first corrected parallax, and the sum is 47'31", which will be the sought parallax of the moon in the circle of altitude.

But since any other parallaxes of the moon differ very little from the parallaxes at full moon and new moon, it would seem to be sufficient if we kept within the mean

limits everywhere, for we have great need of them for the sake of predicting eclipses. The rest do not require such great examination, which will be held to offer perhaps less in the way of utility than in the satisfaction of curiosity.

26. HOW THE PARALLAXES OF LONGITUDE AND LATITUDE ARE DISTINGUISHED

Now the parallax is divided simply into the parallax of longitude and that of latitude, or the parallax between the sun and moon is distinguished according to the arcs and angles of the intersection of the ecliptic and the circle through the poles of the horizon; since it is clear that when this circle falls at right angles upon the ecliptic, it makes no parallax in longitude, but the parallax is transferred wholly to latitude, as the circle is wholly a circle of latitude and altitude. But where conversely the ecliptic falls at right angles upon the horizon and becomes wholly the same as the circle of altitude; then, if the moon has no latitude, it does not admit anything except a parallax in longitude, but if it has a digression in latitude, it does not escape some parallax in latitude.

In this way let circle *ABC* be the ecliptic, and let it be at right angles to the horizon, and let *A* be the pole of the horizon. Accordingly circle *ABC* will be the same as the circle of altitude of a moon without latitude. Let *B* be the position of the moon, and *BC* its total parallax in longitude.

But when it also has latitude, let *DBE* be the circle described through the poles of the ecliptic and with *DB* or *BE* as the latitude of the moon, it is clear that side *AD* or *AE* will not be equal to *AB*; and the angle at *D* or *E* will not be right, since *DA* and *EA* are not circles through the poles of *DBE*; and the parallax will participate in latitude, and it will do so all the more the nearer the moon is to the vertex. For let triangle *ADE* keep the same base, but let sides *AD* and *AE* be shorter and comprehend acuter angles at the base; the greater the distance of the moon from the vertex is, the more like right angles will the angles be.

Now let *ABC* be the ecliptic, and *DBE* the oblique circle of altitude of a moon not having latitude, as being at an ecliptic section. [128ª] Let *B* be the ecliptic section, and *BE* the parallax in the circle of altitude. Let there be drawn *EF* the arc of a circle through the poles of *ABC*. Accordingly since in triangle *BEF* angle *EBF* is given, as was shown above, and

angle *F* = 90°,

and side *BE* is also given: by what has been shown concerning spherical triangles, the remaining sides are given: *BF* the parallax in longitude and *FE* the parallax in latitude,

which agree with parallax *BE*. But since *BE*, *EF*, and *FB* on account of their shortness differ but slightly and imperceptibly from straight lines, we shall not make an error if we use the right triangle as rectilinear; and on that account the ratio will become easy.

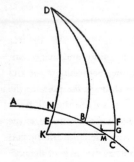

For let circle *ABC* be drawn as the ecliptic, and let *DB* the oblique circle through the poles of the horizon fall upon it. Let *B* be the position in longitude of the moon and *FB* the northern latitude or *BE* the southern. Let the vertex of the horizon be *D*, and from *D* let fall on the moon the circles of altitude *DEK* or *DFC*, whereon are the parallaxes *EK* and *FG*. For the true positions of the moon in longitude and latitude will be at points *E* or *F*, but the seeming positions will be at *K* or *G*. And from *K* and *G* let arcs *KM* and *LG* be drawn at right angles to the ecliptic *ABC*. Accordingly, since the longitude and latitude of the moon have been established together with the latitude of the region; in triangle *DEB* two sides *DB* and *BE* will be known and also *ABD* the angle of section and

angle *DBE* = angle *ABD* + 90°.

Accordingly the remaining side *DE* will be given together with angle *DEB*.

Similarly in triangle *DBF* since two sides *DB* and *BF* are given together with angle *DBF*, which with angle *ABD* makes up a right angle, *DF* will also be given together with angle *DFB*. Accordingly parallaxes *EK* and *FG* on arcs *DE* and *DF* are given by the table, and so is *DE* or *DF*, the true distance of the moon from the vertex, and similarly the seeming distance *DEK* or *DFG*. But in triangle *EBN*, which has the intersection of *DE* with the ecliptic at point *N*, angle *NEB* is given, and base *BE* is given, and angle *NBE* is right: the remaining angle *BNE* will become known too together with the remaining sides *BN* and *NE*. And similarly in the whole triangle *NKM*, as angles *M* and *N* and the whole side *KEN* are given, the base *KM* will be established. *KM* is the seeming southern latitude of the moon, and its excess over *EB* is the parallax of latitude: and the remaining side *NBM* is given: by the subtraction of *NB* from *NBM* the remainder *BM* will be the parallax in longitude.

Moreover, thus in the northern triangle *BFC*, since side *BF* is given together with angle *BFC* [128^b] and angle *B* is right, the remaining sides *BLC* and *FGC* together with the remaining angle *C* are given; and after the subtraction of *FG* from *FGC* there remains *GC*, which is a side given in triangle *GLC* together with the angle *LCG* and angle *CLG*, which is right. Accordingly the remaining sides are given: *GL* and *LC*, and hence *BL* which is the remainder of *BC* and is the parallax in longitude; and *GL* is the seeming latitude, and its parallax is the excess of *BF* the true latitude over *GL*.

As you see however, this computation, which deals with very small magnitudes, contains more labour than fruitfulness. For it will be sufficient if we use angle ABD instead of DCB and angle DBF instead of DEB, and as before, simply the mean DB instead of arcs DE and DF—neglecting the latitude of the moon. For no error will appear because of that, especially in regions of the northern part of the Earth; but in very southern regions, where B touches the vertex of the horizon at the greatest latitude of 5° and when the moon is at its perigee, there is a difference of approximately 6′. But in the ecliptic conjunctions with the sun, where the latitude of the moon cannot exceed $1/_2$°, there can be a difference of merely $1^3/_4$′. Accordingly it is clear from this that the parallax is always added to the true position of the moon in the eastern quadrant of the ecliptic and is always subtracted in the other quadrant, so that we may have the seeming longitude of the moon; and may have the seeming latitude through the parallax in latitude, since if the true latitude and the parallax are in the same direction, they are added together; if in different directions, the lesser is subtracted from the greater, and the remainder is the seeming latitude of that same part, to which the greater falls away.

27. CONFIRMATION OF WHAT HAS BEEN EXPOUNDED CONCERNING THE PARALLAXES OF THE MOON

Accordingly we can confirm by many other observations (such as the following one) that the parallaxes of the moon as set forth above are in conformity with the appearances. We made this observation at Bologna, after sunset on the seventh day before the Ides of March, in the year of Christ 1497. For we observed how long [129ª] the moon would occult the bright star of the Hyades (which the Romans call Paliticium), and with this in mind, we saw the star brought into contact with the shadowy part of the lunar body and already lying hidden between the horns of the moon at the end of the fifth hour of the night, though the star was nearer the southern horn by three quarters as it were of the width or diameter of the moon. And since according to the tables the star was at 2°52′ of Gemini with a southerly latitude of $5^1/_6$°, it was clear that to eyesight the centre of the moon was half a diameter to the west of the star, and accordingly its seen position was 2°36′ in longitude and approximately 5°6′ in latitude.

Accordingly from the beginning of the years of Christ there have been 1497 Egyptian years 76 days 23 hours at Bologna; but at Cracow, which is approximately 8° farther east, 23 hours 36 minutes, to which equal time adds 4 minutes. For the sun was at $28^1/_2$° of Pisces and therefore the regular movement of the moon away from the sun was 74°, the regular anomaly was 111°10′, the true position of the moon was at 3°24′ of Gemini, the southerly latitude 4°35′, for the true movement of latitude was 203°41′. Moreover, at that time at Bologna, 26° of Scorpio was rising, with an angle of $57^1/_2$°;

and the moon was 83° from the vertex of the horizon, and the angle of section between the circle of altitude and the ecliptic was approximately 29°, the parallax of the moon was 1°51' in longitude and 30' in latitude. Those things agree perfectly with the observations; and all the less will anyone doubt that our hypotheses and what results from them are correct.

28. ON THE MEAN OPPOSITIONS AND CONJUNCTIONS OF THE SUN AND MOON

The method of investigating the conjunctions and oppositions of the sun and the moon is clear from what has been said so far concerning their movement. For in relation to that approaching time at which we think this or that conjunction or opposition will take place, we shall seek the regular movement of the moon; and if we find that the regular movement has already completed a circle, we understand a full conjunction at the semicircle. [129b] But since that rarely presents itself, we shall have to observe the distance between the sun and the moon; and when we have divided it by the daily movement of the moon, we shall know by how much time the one of them is in advance of the other, or how far off in the future the conjunction or the opposition is. Therefore we shall seek out the movements and positions for this time, and with them we shall set up ratios for the true new moons and full moons; and we shall distinguish the ecliptic conjunctions from the others, as we shall indicate below. When we have got these things set up, it will be possible to go on into any number of months and continue through some number of years by means of the table of twelve months, which contains the time and the regular movements of the anomaly of the sun and the moon and the regular movement of the moon in latitude—joining single movements to the single movements already found. But, we shall put down the anomaly of the sun as true, so as to have it as corrected immediately. For its difference is not perceptible to sense in one or more years on account of its slowness at its beginning, *i.e.*, at its highest apsis.

TABLE OF THE CONJUNCTION AND OPPOSITION OF SUN AND MOON

Months	Divisions of Time				Movements of Lunar Anomaly				Movement in Latitude of the Moon			
	Days	Min. of Day	Sec.	Thirds	60°	Deg.	Min.	Sec.	60°	Deg.	Min.	Sec.
1	29	31	50	8	0	25	49	0	0	30	40	13
2	59	3	40	16	0	51	38	0	1	1	20	27
3	88	35	30	24	1	17	27	0	1	32	0	41
4	118	7	20	32	1	43	16	0	2	2	40	55
5	147	39	10	40	2	9	5	0	2	33	21	9
6	177	11	0	48	2	34	34	0	3	4	1	23
7	206	42	50	57	3	0	43	0	3	34	41	36
8	236	14	41	25	3	26	32	0	4	5	21	50
9	265	46	31	13	3	52	21	0	4	36	2	4
10	295	18	21	21	4	18	10	0	5	6	42	18
11	324	50	11	29	4	4	59	0	5	37	22	32
12	354	22	1	37	5	9	48	0	0	8	2	46

THE HALF MONTH BETWEEN THE FULL AND NEW MOON

1/2	14	45	55	4	3	12	54	30	3	15	20	6

MOVEMENT OF SOLAR ANOMALY

Months	60°	°	'	"	Months	60°	°	'	"
1	0	29	6	18	7	3	23	44	6
2	0	58	12	36	8	3	52	50	24
3	1	27	18	54	9	4	21	56	42
4	1	56	25	12	10	4	51	3	0
5	2	25	31	30	11	5	20	9	19
6	2	54	57	48	12	5	49	15	37

THE HALF MONTH

	1/2	0	14	33	9

29. ON THE CLOSE EXAMINATION OF THE TRUE CONJUNCTIONS AND OPPOSITIONS OF THE SUN AND MOON

[130^b] Since we possess, as was said, the time of mean conjunction or opposition of these heavenly bodies together with their movements, then the true distance between them, whereby they precede or follow one another, will be necessary in order to find their true (conjunctions and oppositions). For if the (true) moon is prior to the sun in (mean) conjunction or opposition, it is clear that the true one will be in the future; but if the sun, then it is already past the true one which we are seeking. This is made clear by the additosubtractions in the case of both of them, since if there were no additosubtractions, or if they were equal and of the same quality, *viz.*, both additive or both subtractive, it is clear that at the same moment the true conjunctions or oppositions and the mean ones coincide. But if they are unequal, the difference indicates what their distance is and that the star to which the additive or subtractive difference belongs precedes or follows. But when they are in different parts (of their circles) that star all the more precedes whose additosubtraction is subtractive; and the adding together of the additosubtractions shows what the distance between them is. In connection with this we shall decide how many whole hours can be is traversed by the moon—taking two hours for every degree of distance.

In this way, if there were about 6° of distance, we should take 12 hours as corresponding to them. Therefore we shall seek the true movement of the moon away from the sun for the interval of time thus set up; and we shall do that easily, when we know that the mean movement of the moon is 1°1' per 2 hours, but that the true hourly movement of anomaly around the full moon or the new moon is approximately 50'. In 6 hours that makes the regular movement to be 3°3', and the true movement of anomaly 5°; and in the table of lunar additosubtractions we shall note the difference between the additosubtractions and add it to the mean movement—if the anomaly is in the lower part of the circle—and subtract it if the anomaly is in the upper. For the sum or the remainder is the true movement of the moon for the hours taken. Therefore that movement, if equal to the distance first existing, is sufficient. Otherwise the distance multiplied by the number of estimated hours should be divided by this movement; or else we shall divide the true simple distance by the hourly movement taken. [131^a] For the quotient will be the true difference in time in hours and minutes between the mean and the true conjunction or opposition. We shall add this difference to the mean time of conjunction or opposition, if the moon is west of the sun, or to the position of the sun diametrically opposite: or we shall subtract, if the moon is eastward; and we shall have the time of true conjunction or opposition, although we must

confess that the anomaly of the sun too adds or subtracts something, but it is rightly neglected, as in the whole tract and at greatest elongation—which extends beyond 7°—the anomaly cannot fill 1'; and the method of evaluating the lunar movements is more certain.

For those who rely only upon the hourly movement of the moon, which they call the hourly excelling movement, make mistakes sometimes and are forced rather often to repeat their calculations. For the moon is changeable even from hour to hour and does not stay like itself. Accordingly, for the time of true conjunction or opposition, we shall work out the true movement in latitude, so as to learn the latitude of the moon and work out the true position of the sun in relation to the spring equinox, *i.e.*, in the signs, whereby the true position of the moon is known to lie the same or opposite to it. And since time is here understood as mean and equal with respect to the Cracow meridian, we shall reduce it to apparent time by the method described above. But if we should wish to set this up for any other place than Cracow, we shall note its longitude and take four minutes of an hour for each degree of longitude and four seconds of an hour for each minute of longitude; and we shall add them to the Cracow time, if the other place is to the east, and subtract them, if it is to the west. And the sum or the remainder will be the time of conjunction or opposition of the sun and moon.

30. HOW THE ECLIPTIC CONJUNCTIONS AND OPPOSITIONS OF THE SUN AND MOON ARE DISTINGUISHED FROM THE OTHERS

In the case of the moon it is easily discernible whether or not they are ecliptic; since, if the latitude of the moon is less than half the diameters of the moon and the shadow, it will undergo an eclipse, but if greater, it will not. But there is more than enough bother in the case of the sun, as the parallax of each of them, by which for the most part the visible conjunction differs from the true, is mixed up in it. Accordingly when we have examined [131^b] what the parallax in longitude between the sun and moon is at the time of true conjunction, similarly we shall look for the apparent (angular) elongation of the moon from the sun at the interval of an hour before the true conjunction in the eastern quarter of the ecliptic or after the true conjunction in the western quarter, in order to understand how far the moon seems to move away from the sun in one hour. Therefore when we have divided the parallax by this hourly movement, we shall have the difference in time between the true and the seen conjunction, When that is subtracted from the time of the true conjunction in the eastern part of the ecliptic or added in the western—for in the eastern part the seen conjunction precedes the true, and in the western it follows it—the result will be the time of seen conjunction which we were looking for. Therefore we shall reckon the seen latitude of the

moon in relation to the sun for this time, or the distance between the centres of the sun and the moon at the seen conjunction, after deducting the parallax of the sun. If this latitude is greater than half the diameters of the sun and moon, the sun will not undergo an eclipse; but if smaller, it will. From this it is clear that if the moon at the time of true conjunction does not have any parallax in longitude, the seen and the true conjunction will be the same, and the conjunction will take place at 90° of the ecliptic as measured from the east or the west.

31. HOW GREAT AN ECLIPSE OF THE SUN OR MOON WILL BE

Therefore, after we have learned that the sun or moon will undergo an eclipse, we shall easily come to know how great the eclipse will be—in the case of the sun by means of the seen latitude between the sun and moon at the time of seen conjunction. For if we subtract the latitude from half the diameters of the sun and the moon, the remainder is the eclipse of the sun as measured along its diameter; and when we have multiplied that by twelve and divided the product by the diameter of the sun, we shall have the number of twelfths of the eclipse of the sun. But if there is no latitude between the sun and the moon, there will be a total eclipse of the sun or as much of it as the moon can cover.

Approximately the same method (is used) in the case of a lunar eclipse, except that instead of the seen latitude we employ the simple latitude. When the latitude is subtracted from half the diameters of the moon and shadow, the remainder is the part of [132ª] the moon eclipsed, provided the latitude of the moon is not less than half the diameters of the moon and shadow, as taken along the diameter of the moon. For then there will be a total eclipse. And furthermore the lesser latitude even adds some delay in the darkness; and the delay will be greatest when there is no latitude—as I think is perfectly clear to those who consider it. Accordingly, in the case of a particular eclipse of the moon, when we have multiplied the eclipsed part by twelve and divided the product by the diameter of the moon, we shall have the number of twelfths of the eclipse—just as in the case of the sun.

32. HOW TO KNOW BEFOREHAND HOW LONG AN ECLIPSE WILL LAST

It remains to see how long an eclipse will last. It should be noted that we use the arcs which occur in the case of the sun, moon, and shadow as straight lines; for they are so small that they do not seem to be different from straight lines.

Accordingly let us take point A as the centre of the sun or of the shadow, and line BC as the passage of the orb of the moon. And let B be the centre of the moon touching

the sun or shadow at the beginning of incidence and C at the end of its transit. Let AB and BC be joined, and let fall AD perpendicular to BC.

It is clear that when the centre of the moon is at D, it will be the middle of the eclipse. For AD is the shortest of the lines falling from A, and

$$BD = DC,$$

since

$$AB = AC,$$

and AB or AC is equal to half the sum of the diameters of the sun and the moon in a solar eclipse and to that of the diameters of the moon and shadow in a lunar eclipse; and AD is the true latitude of the moon or the seen latitude at the middle of the eclipse. Accordingly, when we have subtracted the square on AD from the square on AB, the remainder is the square on BD. Therefore BD will be given in length. When we have divided it by the true hourly movement of the moon during the eclipse of the moon, or by the visible movement in the case of a solar eclipse, we shall have the time of half the duration. But the moon very often delays in the middle of the darkness—that happens when half the sum of the diameters of the moon and the shadow exceeds the latitude of the moon by more than the moon's diameter, as we said. Accordingly when we have placed E the centre of the moon at the starting-point of the total [132b] obscuration, when the moon touches the concave circumference of the shadow, and F at the other point of contact, when the moon first emerges, and have joined AE and AF, it will be made clear in the same way as before that ED and DF are the halves of the delay in darkness, because AD is the known latitude of the moon, and AE or AF is that whereby the half of the diameter of the shadow is greater than half the diameter of the moon. Therefore DE or DF will be established; and once more when we have divided it by the true hourly movement of the moon, we shall have the time of half the delay, which we were looking for.

Nevertheless we must notice here that since the moon moves in its own orbital circle, it does not, by the mediation of the circles passing through the poles of the ecliptic, cut arcs of longitude on the ecliptic wholly equal to the arcs in its own orbital circle. But the difference is very slight, so that at the total distance of 12° from the ecliptic section, which is approximately the farthest limit of the eclipses of the sun and moon, the arcs of the circles do not differ from one another by 2', which makes $^1/_{15}$ hour; on that account we often use one instead of the other as if they were the same. So too we use the same latitude of the moon at the limits of the eclipses as at the middle of the eclipse, although the latitude of the moon is always increasing or decreasing, and on that account the intervals of incidence and withdrawal are not wholly equal, but the difference is so slight that it seems a waste of time to examine them more closely.

In this way the times, durations, and magnitudes of eclipses have been unfolded with respect to the diameters.

But since it is the opinion of many persons that the parts eclipsed should be distinguished not with respect to the diameters but with respect to the surfaces, for it is

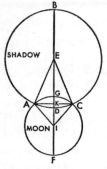

not lines but surfaces which are eclipsed: accordingly let *ABCD* be the circle of the sun or of the shadow, and let *E* be its centre. Let *AFCG* be the lunar circle, and let *I* be its centre. Let the circles cut one another in points *A* and *C*, let the straight line *BEIF* be drawn through the centres of both, and let *AE, EC, IA, IC* be joined and line *AKC* at right angles to *BF*. By means of this we wish to examine how great *ADCG* the surface obscured is, or how many twelfths

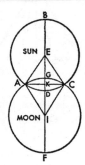

of the whole surface of the orb of the moon or sun belong to the part eclipsed. Accordingly since the semidiameters *AE* and *AI* of each circle are given by the above, and also *EI* the distance between their centres or the lunar latitude, we shall have [133ᵃ] triangle *AEI* with its sides given; and for that reason with its angles given by the demonstrations above; and angle *AEI* is similar and equal to angle *EIC*. Accordingly

arcs *ADC* and *AGC* will be given,

where the circumference = 360°.

Furthermore, in the measurement of the circle Archimedes of Syracuse records that

circumference : diameter < $3^{1}/_{7}$: 1

but

circumference : diameter > $3^{10}/_{71}$: 1.

Ptolemy assumed as a mean between these

3P8'30" : 1P.

By means of this ratio

arcs *AGC* and *ADG* will be given,

in terms of the same parts as the semidiameters *AE* and *AI*.
And

quad. *EA, AD* = sector *AEC*,

and

quad. *IA, AG* = sector *AIC*.

But in the isosceles triangles *AEC* and *AIC* the common base *AKC* and the perpendiculars *EK* and *KI* are given. And accordingly the quadrilateral *AK, KE* is given, which is the area of triangle *AEC*—and similarly the quadrilateral *AK, KI* is the area of triangle *ACI*. Accordingly

$$\text{sect. } AFCK - \text{trgl. } AIC = \text{seg. circ. } AFC$$

and

$$\text{sect. } ABCK - \text{trgl. } AEC = \text{seg. circ. } ABC$$

and hence,

figure $ADCG$ is given,

which was sought.

And moreover, the total area of the circle—which is comprehended by BE and BAD in a solar eclipse or by FI and FAG in a lunar eclipse—was given. Accordingly it will be manifest how many twelfths of the total circle of the sun or moon was eclipsed in $ADCG$. Let all this—which has been treated in more detail by others—be enough now concerning the moon: for we are in a hurry to get to the revolutions of the remaining five planets, which will be spoken of in the books following.

BOOK FIVE

[133^b] Up to now we have been explaining to the best of our ability the revolutions of the Earth around the sun and of the moon around the Earth. Now we are turning to the movements of the five wandering stars: the mobility of the Earth binds together the order and magnitude of their orbital circles in a wonderful harmony and sure commensurability, as we said in our brief survey in the first book, when we showed that the orbital circles do not have their centres around the Earth but rather around the sun. Accordingly it remains for us to demonstrate all these things singly and with greater clarity; and let us fulfil our promises adequately, in so far as we can, particularly by measuring the appearances by the experiments which we have got from the ancients or from our own times, in order that the ratio of the movements may be held with greater certainty. Now in Plato's *Timaeus* each of these five stars is named in accordance with its visible aspect: Saturn, Phaenon—as if to say "shining" or "appearing," for Saturn is hidden less than the others, and emerges more quickly after undergoing occultation by the sun; Jupiter, Phaeton from his radiance; Mars, Pyrois from his fiery glow; Venus sometimes φωσφόρος and sometimes ἕσπερος, *i.e.*, Lucifer and Vesperugo, according as she shines at morning or evening; and finally Mercury, Stilbon from his sparkling and twinkling light. Moreover the planets have greater irregularities in longitude and in latitude than the moon.

1. ON THEIR REVOLUTIONS AND MEAN MOVEMENTS

Two longitudinal movements which are quite different appear in the planets. One of them is on account of the movement of the Earth, as we said; and the other is proper to each planet. We may rightly call the first the movement of parallax, since it is the one which makes the planets appear to have stoppings, progressions, and retrogradations— [134^a] not that the planet which always progresses by its own movement, is pulled in different directions, but that it appears to do so by reason of the parallax caused by the movement of the Earth taken in relation to the differing magnitudes of their orbital circles.

Accordingly it is clear that the true position of Saturn, Jupiter, and Mars become visible to us only at the time when they are in opposition to the sun; and that occurs approximately in the middle of their retrogradations. For at that time they fall on a straight line with the mean position of the sun, and lay aside their parallax.

Furthermore there is a different ratio in the case of Venus and Mercury: for they are hidden at the time they are in conjunction with the sun, and they show only the digressions which they make on either side away from the sun: hence they are never found without parallax.

Therefore the revolution of parallax—I mean the movement of the Earth with respect to the planet—is private to each planet; and the planet and the Earth are mutually explanatory of it. For we say that the movement of parallax is nothing except that wherein the regular movement of the Earth exceeds their movement, as in the case of Saturn, Jupiter, and Mars, or is exceeded by it, as in the case of Venus and Mercury. But since such periods of parallax are found unequal by a manifest difference, the ancients recognized that the movements of these planets too were irregular and had apsides; of circles to which their irregularity returned, and the ancients supposed that these apsides had perpetual seats in the sphere of the fixed stars. By that argument the road is opened for learning their mean movements and equal periods. For when the ancients had recorded in memory the position of some planet with respect to its exact distance from the sun and a fixed star, and after an interval of time found that it had arrived at the same position with an equal distance from the sun; the planet was seen to have completed its whole movement of irregularity and to have returned through all to its former relationship with the Earth. And so by means of the time which intervened they calculated the number of whole and equal revolutions and from them the particular movements of the planet. Ptolemy surveyed the circuits through a number of years, according as, he acknowledged, he got them from Hipparchus. Now he means solar years to be understood as the years measured from an equinox or solstice. But it has already been made clear that such years are not quite equal; on that account we shall use years measured from the fixed stars, and by means of them the movements of these five planets have been reconstituted more correctly by us, according as in our time [134^b] we found there was some deficiency in them or excess, as follows.

For the Earth has 57 revolutions in respect to Saturn—we call this the movement of parallax—in 59 of our solar years 1 day 6 minutes of a day 48 seconds approximately: during this time the planet has by its own movement completed two circuits plus 1°6'6".

Jupiter is outrun by the Earth 65 times in 71 solar years minus 5 days 45 minutes 27 seconds: during this time the planet by its own movement has 6 revolutions minus 5°41'2$^1/_2$".

Mars has 37 revolutions of parallax in 79 solar years 2 days 27 minutes 3 seconds: during this time the planet by its own movement completes 42 periods plus 2°24'56".

Venus outruns the movement of the Earth 5 times in 8 solar years minus 2 days 26 minutes 46 seconds. And during this time it has 13 revolutions minus 2°24'40" around the sun.

Finally, Mercury completes 145 periods of parallax, by which it outruns the movement of the Earth, in 46 solar years plus 34 minutes of a day 23 seconds. And it has 191 revolutions around the sun in that time plus 34 minutes of a day 23 seconds approximately.

Accordingly the single circuits of parallax are as follows: for the single planets;

Saturn:	378 days	5 min.	32 sec.	11 thirds
Jupiter:	398 days	23 min.	25 sec.	56 thirds
Mars:	779 days	56 min.	19 sec.	7 thirds
Venus:	583 days	45 min.	17 sec.	24 thirds
Mercury:	115 days	52 min.	42 sec.	12 thirds

When we have reduced these circuits to the degrees of a circle and multiplied by the ratio of 365 to the number of days and minutes, we shall have as the annual movements (of parallax):

Saturn:	347°32'2"54"'12""
Jupiter:	329°25'8"15"'6""
Mars:	168°28'29"13"'12""
Venus:	225°1'48"54"'30""
Mercury:	3 (360°) + 53°56'46"54"'40""

[135ᵃ] The three-hundred-sixty-fifth part of these is the daily movement:

Saturn:	57'7"44"'
Jupiter:	54'9"3"'49""
Mars:	27'41"40"'8""
Venus:	36'49"28"'35""
Mercury:	3°6'24"7"'43""

according as they are set out in the following tables, like the mean movement of the sun and moon.

But we thought it unnecessary to set down their proper movements in this way. For the proper movements are determined by the subtraction of the movements of parallax from the mean movement of the sun, as the proper movement of the planet and the mean movement of parallax compose the mean movement of the sun. For the proper annual movements in relation to the sphere of the fixed stars are as follows for the upper planets:

Saturn:	12°12'46"12"'52""
Jupiter:	30°19'40"51"'58""
Mars:	191°16'19"53"'52""

But in the case of Venus and Mercury, since their proper movements are not apparent to us,[1] the movement of the sun itself is used by us instead; and it furnishes a way of investigating and demonstrating their apparent movements, in the following tables.

---

[1] The proper movements of Venus and Mercury are not apparent to us in that their positions are never viewed without parallax.

SATURN'S MOVEMENT OF PARALLAX IN YEARS AND PERIODS OF SIXTY YEARS

Egyptian Years	Movement					Egyptian Years	Movement				
	60°	°	'	''	'''		60°	°	'	''	'''
1	5	47	32	3	9	31	5	33	33	37	59
2	5	35	4	6	19	32	5	11	5	41	9
3	5	22	36	9	29	33	5	8	37	44	19
4	5	10	8	12	38	34	4	56	9	47	28
5	4	57	40	15	48	35	4	43	41	50	38
6	4	45	12	18	58	36	4	31	13	53	48
7	4	32	44	22	7	37	4	18	45	56	57
8	4	20	16	25	17	38	4	6	18	0	7
9	4	7	48	28	27	39	3	53	50	3	17
10	3	55	20	31	36	40	3	41	22	6	26
11	3	42	52	34	46	41	3	18	54	9	36
12	3	30	24	37	56	42	3	16	26	12	46
13	3	17	56	41	5	43	3	3	58	15	55
14	3	5	28	44	15	44	2	51	30	19	5
15	2	53	0	47	25	45	2	39	2	22	15
16	2	40	32	50	34	46	2	26	34	25	24
17	2	28	4	53	44	47	2	14	6	28	34
18	2	15	36	56	54	48	1	1	38	31	44
19	2	3	9	0	3	49	1	49	10	34	53
20	1	50	41	3	13	50	1	36	42	38	3
21	1	38	13	6	23	51	1	24	14	41	13
22	1	25	45	9	32	52	1	11	46	44	22
23	1	13	17	12	42	53	0	59	18	47	32
24	1	0	49	15	52	54	0	46	50	50	42
25	0	48	21	19	1	55	0	34	22	43	51
26	0	35	53	22	11	56	0	21	54	57	1
27	0	23	25	25	21	57	0	9	27	0	11
28	0	10	57	28	30	58	5	56	59	3	20
29	5	58	29	31	40	59	5	44	31	6	30
30	5	46	1	34	50	60	5	32	3	9	40

SATURN'S MOVEMENT OF PARALLAX IN PERIODS OF SIXTY DAYS

Days	Movement					Days	Movement				
	60°	°	'	''	'''		60°	°	'	''	'''
1	0	0	57	7	44	31	0	29	30	59	46
2	0	1	54	15	28	32	0	30	28	7	30
3	0	2	51	23	12	33	0	31	25	15	14
4	0	3	48	30	56	34	0	32	22	22	58
5	0	4	45	38	40	35	0	33	19	30	42
6	0	5	42	46	24	36	0	34	16	38	26
7	0	6	39	54	8	37	0	35	13	46	10
8	0	7	37	1	52	38	0	36	10	53	55
9	0	8	34	9	36	39	0	37	8	1	39
10	0	9	31	17	20	40	0	38	5	9	23
11	0	10	28	25	4	41	0	39	2	17	7
12	0	11	25	32	49	42	0	39	59	24	51
13	0	12	22	40	33	43	0	40	56	32	35
14	0	13	19	48	17	44	0	41	53	40	19
15	0	14	16	56	1	45	0	42	50	48	3
16	0	15	14	3	45	46	0	43	47	55	47
17	0	16	11	11	29	47	0	44	45	3	31
18	0	17	8	19	13	48	0	45	42	11	16
19	0	18	5	26	57	49	0	46	39	19	0
20	0	19	2	34	41	50	0	47	36	26	44
21	0	19	59	42	25	51	0	48	33	34	28
22	0	20	56	50	9	52	0	49	30	42	12
23	0	21	53	57	53	53	0	50	27	49	56
24	0	22	51	5	38	54	0	51	24	57	40
25	0	23	48	13	22	55	0	52	22	5	24
26	0	24	45	21	6	56	0	53	19	13	8
27	0	25	42	28	50	57	0	54	16	20	52
28	0	26	39	36	34	58	0	55	13	28	36
29	0	27	36	44	18	59	0	56	10	36	20
30	0	28	33	52	2	60	0	57	7	44	5

JUPITER'S MOVEMENT OF PARALLAX IN YEARS AND PERIODS OF SIXTY YEARS

Egyptian Years	Movement					Egyptian Years	Movement				
	60°	°	′	″	‴		60°	°	′	″	‴
1	5	29	25	8	15	31	2	11	59	15	48
2	4	58	50	16	30	32	1	41	24	24	3
3	4	28	15	24	45	33	1	10	49	32	18
4	3	57	40	33	0	34	0	40	14	40	33
5	3	27	5	41	15	35	0	9	39	48	48
6	2	56	30	49	30	36	5	39	4	57	8
7	2	25	55	57	45	37	4	8	30	5	18
8	1	55	21	6	0	38	4	37	55	13	33
9	1	24	46	14	15	39	3	7	20	21	48
10	0	54	11	22	31	40	3	36	45	30	4
11	0	23	36	30	46	41	2	6	10	38	19
12	5	53	1	39	1	42	2	35	35	46	34
13	5	22	25	47	16	43	1	5	0	54	49
14	4	51	51	55	31	44	1	34	26	3	4
15	4	21	17	3	46	45	0	3	51	11	19
16	3	50	42	12	1	46	0	33	16	19	34
17	3	20	7	20	16	47	5	2	41	27	49
18	2	49	32	28	31	48	5	32	6	36	4
19	2	18	57	35	46	49	4	1	31	44	19
20	1	48	22	45	2	50	4	30	56	52	34
21	1	17	47	58	17	51	3	0	22	0	50
22	0	47	13	1	32	52	3	29	47	9	5
23	0	16	38	9	47	53	2	59	12	17	20
24	5	45	3	18	2	54	2	28	37	25	35
25	5	15	28	26	17	55	1	58	2	33	50
26	4	44	53	34	32	56	1	27	27	42	5
27	4	14	18	42	47	57	0	56	52	50	20
28	3	43	43	51	2	58	0	26	17	58	35
29	3	13	8	59	17	59	5	55	43	6	50
30	2	42	34	7	33	60	5	25	8	15	6

JUPITER'S MOVEMENT OF PARALLAX IN PERIODS OR SIXTY DAYS

Days	Movement					Days	Movement				
	60°	°	′	″	‴		60°	°	′	″	‴
1	0	0	54	9	3	31	0	27	58	40	58
2	0	1	49	18	7	32	0	28	52	50	2
3	0	2	42	27	11	33	0	29	46	59	5
4	0	3	36	36	15	34	0	30	41	8	9
5	0	4	30	45	19	35	0	31	35	17	13
6	0	5	24	54	22	36	0	32	29	26	17
7	0	6	19	3	26	37	0	33	23	35	21
8	0	7	13	12	30	38	0	34	17	44	25
9	0	8	7	21	34	39	0	35	11	53	29
10	0	9	1	30	38	40	0	36	6	2	32
11	0	9	55	39	41	41	0	37	0	11	36
12	0	10	49	48	45	42	0	37	54	20	40
13	0	11	43	57	49	43	0	38	48	29	44
14	0	12	38	6	53	44	0	39	42	38	47
15	0	13	32	15	57	45	0	40	36	47	51
16	0	14	26	25	1	46	0	41	30	56	55
17	0	15	20	34	4	47	0	42	25	5	59
18	0	16	14	43	8	48	0	43	19	15	3
19	0	17	8	52	12	49	0	44	13	24	6
20	0	18	3	1	16	60	0	45	7	33	10
21	0	18	57	10	20	51	0	46	1	42	14
22	0	19	51	19	23	52	0	46	55	51	18
23	0	20	45	28	27	53	0	47	50	0	22
24	0	21	39	37	31	64	0	48	44	9	26
25	0	22	33	46	35	55	0	49	38	18	29
26	0	23	27	55	39	56	0	50	32	27	33
27	0	24	22	4	43	57	0	51	26	36	37
28	0	25	16	13	46	58	0	52	20	45	41
29	0	26	10	22	50	59	0	53	14	54	45
30	0	27	4	31	54	60	0	54	9	3	49

MARS' MOVEMENT OF PARALLAX IN YEARS AND PERIODS OF SIXTY YEARS

Egyptian Years	Movement					Egyptian Years	Movement				
	60°	°	′	″	‴		60°	°	′	″	‴
1	2	48	28	30	36	31	3	2	43	48	38
2	5	36	57	1	12	32	5	51	12	19	14
3	2	25	25	31	48	33	2	39	40	49	50
4	5	13	54	2	24	34	5	28	9	20	26
5	2	2	22	33	0	35	2	16	37	51	2
6	4	50	51	3	36	36	5	5	6	21	38
7	1	39	19	34	12	37	1	53	34	52	14
8	4	27	48	4	48	38	4	42	3	22	50
9	1	16	16	35	24	39	1	30	31	53	26
10	4	4	45	6	0	40	4	19	0	24	2
11	0	53	13	36	36	41	1	7	28	54	38
12	3	41	42	7	12	42	3	55	57	25	14
13	0	30	10	37	46	43	0	44	25	55	50
14	3	18	39	8	24	44	3	32	54	26	26
15	0	7	7	39	1	45	0	21	22	57	3
16	2	55	36	9	37	46	3	9	51	27	39
17	5	44	4	40	13	47	5	58	19	58	15
18	2	32	33	10	49	48	2	46	48	28	51
19	5	21	1	41	25	49	5	35	16	59	27
20	2	9	30	12	1	50	2	23	45	30	3
21	4	57	58	42	37	51	5	12	14	0	39
22	1	46	27	13	13	52	2	0	42	31	15
23	4	34	55	43	49	53	4	49	11	1	51
24	1	23	24	14	25	54	1	37	39	32	27
25	4	11	52	45	1	55	4	26	8	3	3
26	1	0	21	15	37	56	1	14	36	33	39
27	3	48	49	46	13	57	4	3	5	4	15
28	0	37	18	16	49	58	0	51	33	34	51
29	3	25	46	47	25	59	3	40	2	5	27
30	0	14	15	18	2	60	0	28	30	36	4

MARS' MOVEMENT OF PARALLAX IN PERIODS OF SIXTY DAYS

Days	Movement					Days	Movement				
	60°	°	′	″	‴		60°	°	′	″	‴
1	0	0	27	41	40	31	0	14	18	31	51
2	0	0	55	23	20	32	0	14	46	13	31
3	0	1	23	5	1	33	0	15	14	55	12
4	0	1	50	46	41	34	0	15	41	36	52
5	0	2	18	28	21	35	0	16	9	18	32
6	0	2	46	10	2	36	0	16	37	0	13
7	0	3	13	51	42	37	0	17	4	41	53
8	0	3	41	33	22	38	0	17	32	23	33
9	0	4	9	15	3	39	0	18	0	5	14
10	0	4	36	35	43	40	0	18	27	46	54
11	0	5	4	38	24	41	0	18	55	28	35
12	0	5	32	20	4	42	0	19	23	10	15
13	0	6	0	1	44	43	0	19	50	51	55
14	0	6	27	43	25	44	0	20	18	33	36
15	0	6	55	25	5	45	0	20	46	15	16
16	0	7	23	6	45	46	0	21	13	56	56
17	0	7	50	48	26	47	0	21	41	38	37
18	0	8	18	30	6	48	0	22	9	20	17
19	0	8	46	11	47	49	0	22	37	1	57
20	0	9	13	53	27	50	0	23	4	43	38
21	0	9	41	35	7	51	0	23	32	25	18
22	0	10	9	16	48	52	0	24	0	6	59
23	0	10	36	58	28	53	0	24	27	48	39
24	0	11	4	40	8	54	0	24	55	30	19
25	0	11	32	21	48	55	0	25	23	12	0
26	0	12	0	3	29	56	0	25	50	53	40
27	0	12	27	45	9	57	0	26	18	35	20
28	0	12	59	25	50	58	0	26	46	17	1
29	0	13	23	8	30	59	0	27	13	58	41
30	0	13	50	50	11	60	0	27	41	40	22

VENUS' MOVEMENT OF PARALLAX IN YEARS AND PERIODS OF SIXTY YEARS

Egyptian Years	60°	°	'	"	'''
1	3	45	1	45	3
2	1	30	3	30	7
3	5	15	5	15	11
4	3	0	7	0	14
5	0	45	8	45	18
6	4	30	10	30	22
7	2	15	12	15	25
8	0	0	14	0	29
9	3	45	15	45	33
10	1	30	17	30	36
11	5	15	19	15	40
12	3	0	21	0	44
13	0	45	22	45	47
14	4	30	24	30	51
15	2	15	26	15	55
16	0	0	28	0	58
17	3	45	29	46	2
18	1	30	31	31	6
19	5	15	33	16	9
20	3	0	35	1	13
21	0	45	36	46	17
22	4	30	38	31	20
23	2	15	40	16	24
24	0	0	42	1	28
25	3	45	43	46	31
26	1	30	45	31	35
27	5	15	47	16	39
28	3	0	49	1	42
29	0	45	50	46	46
30	4	20	52	31	50
31	2	15	54	16	53
32	0	0	56	1	57
33	3	45	57	47	1
34	1	30	59	32	4
35	5	16	1	17	8
36	3	1	3	2	12
37	0	46	4	47	15
38	4	31	6	32	19
39	2	16	8	17	23
40	0	0	10	2	26
41	3	46	11	47	30
42	1	31	13	32	34
43	5	16	15	17	37
44	3	1	17	2	41
45	0	46	18	47	45
46	4	31	20	32	48
47	2	16	22	17	52
48	0	1	24	2	56
49	3	46	25	47	59
50	1	31	27	33	3
51	5	16	29	18	7
52	3	1	31	3	10
53	0	46	32	48	14
54	4	31	34	33	18
55	2	16	36	18	21
56	0	1	38	3	25
57	3	46	39	48	29
58	1	31	41	33	32
59	5	16	43	18	36
60	3	1	45	3	40

VENUS' MOVEMENT OF PARALLAX IN PERIODS OF SIXTY DAYS

Days	60°	°	'	"	'''
1	0	0	36	59	28
2	0	1	13	58	57
3	0	1	50	58	25
4	0	2	27	57	54
5	0	3	4	57	22
6	0	3	41	56	51
7	0	4	18	56	20
8	0	4	55	55	48
9	0	5	32	55	17
10	0	6	9	54	45
11	0	6	46	54	14
12	0	7	23	53	43
13	0	8	0	53	11
14	0	8	37	52	40
15	0	9	14	52	8
16	0	9	51	51	37
17	0	10	28	51	5
18	0	11	5	50	34
19	0	11	42	50	2
20	0	12	19	49	31
21	0	12	56	48	59
22	0	13	33	48	28
23	0	14	0	47	57
24	0	14	47	47	26
25	0	15	24	46	54
26	0	16	1	46	23
27	0	16	38	45	51
28	0	17	15	45	20
29	0	17	52	44	48
30	0	18	29	44	17
31	0	19	6	43	46
32	0	19	43	43	14
33	0	20	20	42	43
34	0	20	57	42	11
35	0	21	34	41	40
36	0	22	11	41	9
37	0	22	48	40	37
38	0	23	25	40	6
39	0	24	2	39	34
40	0	24	39	39	3
41	0	25	16	38	31
42	0	25	53	38	0
43	0	26	30	37	29
44	0	27	7	36	57
45	0	27	44	36	26
46	0	28	21	35	54
47	0	28	58	35	23
48	0	29	35	34	52
49	0	30	12	34	20
50	0	30	49	33	49
51	0	31	26	33	17
52	0	32	3	32	46
53	0	32	40	32	14
54	0	33	17	31	43
55	0	33	54	31	12
56	0	34	31	30	40
57	0	35	8	30	9
58	0	35	45	29	37
59	0	36	22	29	6
60	0	36	59	28	35

MERCURY'S MOVEMENT OF PARALLAX IN YEARS AND PERIODS OF SIXTY YEARS

Egyptian Years	Movement					Egyptian Years	Movement				
	60°	°	′	″	‴		60°	°	′	″	‴
1	0	53	57	23	6	31	3	52	38	56	21
2	1	47	54	46	13	32	4	46	36	19	28
3	2	41	52	9	19	33	5	40	33	42	34
4	3	35	49	32	26	34	0	34	31	5	41
5	4	29	46	55	32	35	1	28	28	28	47
6	5	23	44	18	39	36	2	22	25	51	54
7	0	17	41	41	45	37	3	16	23	15	0
8	1	11	39	4	52	38	4	10	20	38	7
9	2	5	36	27	58	39	5	4	18	1	13
10	2	59	33	51	5	40	5	58	15	24	20
11	3	53	31	14	11	41	0	52	12	47	26
12	4	47	28	37	18	42	1	46	10	10	33
13	5	41	26	0	24	43	2	40	7	33	39
14	0	35	23	23	31	44	3	34	4	56	46
15	1	29	20	46	37	45	4	28	2	19	52
16	2	23	18	9	44	46	5	21	59	42	59
17	3	17	15	32	50	47	0	15	57	6	5
18	4	11	12	55	57	48	1	9	54	29	12
19	5	5	10	19	3	49	2	3	51	52	18
20	5	59	7	42	10	50	2	57	49	15	25
21	0	53	5	5	16	51	3	51	46	38	31
22	1	47	2	28	23	52	4	45	44	1	38
23	2	40	59	51	29	53	5	39	41	24	44
24	3	34	57	14	36	54	0	33	38	47	51
25	4	28	54	37	42	55	1	27	36	10	57
26	5	22	52	0	49	56	2	21	33	34	4
27	0	16	49	23	55	57	3	15	30	57	10
28	1	10	46	47	2	58	4	9	28	20	17
29	2	4	44	10	8	59	5	3	25	43	23
30	2	58	41	33	15	60	5	57	23	6	30

MERCURY'S MOVEMENT OF PARALLAX IN PERIODS OF SIXTY DAYS

Days	Movement					Days	Movement				
	60°	°	′	″	‴		60°	°	′	″	‴
1	0	3	6	24	13	31	1	36	18	31	3
2	0	6	12	48	27	32	1	39	24	55	17
3	0	9	19	12	41	33	1	42	31	19	31
4	0	12	25	36	54	34	1	45	37	43	44
5	0	15	32	1	8	35	1	48	44	7	58
6	0	18	38	25	22	36	1	51	50	32	12
7	0	21	44	49	35	37	1	54	56	56	25
8	0	24	51	13	49	38	1	58	3	20	39
9	0	27	57	38	3	39	2	1	9	44	53
10	0	31	4	2	16	40	2	4	16	9	6
11	0	34	10	26	30	41	2	7	22	33	20
12	0	37	16	50	44	42	2	10	28	57	34
13	0	40	23	14	57	43	2	13	35	21	47
14	0	43	29	39	11	44	2	16	41	46	1
15	0	46	36	3	25	45	2	19	48	10	15
16	0	49	42	27	38	46	2	22	54	34	28
17	0	52	48	51	52	47	2	26	0	58	42
18	0	55	55	16	6	48	2	29	7	22	56
19	0	59	1	40	19	49	2	32	13	47	9
20	1	2	8	4	33	50	2	35	20	11	23
21	1	5	14	28	47	51	2	38	26	35	37
22	1	8	20	53	0	52	2	41	32	59	50
23	1	11	27	17	14	53	2	44	39	24	4
24	1	14	33	41	28	54	2	47	45	48	18
25	1	17	40	5	41	55	2	50	52	12	31
26	1	20	46	29	55	56	2	53	58	36	45
27	1	23	52	54	9	57	2	57	5	0	59
28	1	26	59	18	22	58	3	0	11	25	12
29	1	30	5	42	36	59	3	3	17	49	26
30	1	33	12	6	50	60	3	6	24	13	40

2. DEMONSTRATION OF THE REGULAR AND APPARENT MOVEMENTS OF THESE PLANETS ACCORDING TO THE THEORY OF THE ANCIENTS

[140[b]] Accordingly this is the way the mean movements are. Now let us turn to the apparent irregularity. The ancient mathematicians who kept the earth immobile imagined in the case of Saturn, Jupiter, Mars, and Venus eccentric circles bearing epicycles, and another further eccentric circle, with respect to which the epicycle and the planet in the epicycle should move regularly.

In this way let *AB* be an eccentric circle, and let its centre be at *C*. Let *ABC* be its diameter, whereon *D* is the centre of the Earth, so that *A* is the apogee and *B* the perigee. Let *DC* be bisected at *E*, and with *E* as centre let *FG* another eccentric circle to the first be described.

Let *H* be anywhere on this eccentric circle, and with *H* as centre let the epicycle *IK* be described. Through its centre let there be drawn the straight line *IHKC* and similarly *LHME*. Now let it be understood that on account of the latitudes of the planet the eccentric circles are inclined to the plane of the ecliptic and similarly the epicycle to the plane of the eccentric circle; but here they are represented as if in one plane for the sake of ease of demonstration. Accordingly the ancients say that this whole plane together with points *E* and *C* moves around *D*—the centre of the ecliptic—in the movement of the sphere of the fixed stars: by this they mean that these

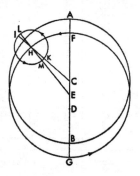

points have unchanging positions in the sphere of the fixed stars. And they say that the epicycle moves eastward in circle *FHG* but in accordance with line *IHC*, and in relation to this line the planet revolves regularly in epicycle *IK*. But it is clear that the regularity of the epicycle should occur in relation to *E* the centre of its deferent,[1] and the revolution of the planet in relation to line *LME*. Accordingly they concede that in this case the regularity of the circular movement can occur with respect to a foreign and not the proper centre; similarly and more so in the case of Mercury. But I think I have already made a sufficient refutation of that in the case of the moon. These and similar things furnished us with an occasion for working out the mobility of the Earth and some other ways by which regularity and the principles of this art might be preserved, and the ratio of apparent irregularity rendered more constant.

3. GENERAL DEMONSTRATION OF APPARENT IRREGULARITY ON ACCOUNT OF THE MOVEMENT OF THE EARTH

[141[a]] Accordingly there are two reasons why the regular movement of a planet should appear irregular: on account of the movement of the Earth and on account of

---

[1] The deferent of an epicycle is the circle on the circumference of which the centre of the epicycle moves.

its proper movement. We shall make both of them clear generally and separately by ocular demonstration, whereby they can be better distinguished from one another; and we shall begin with the movement which mixes itself with all of them on account of the movement of the Earth: and first in the case of Venus and Mercury, which are comprehended by the (orbital) circle of the Earth.

Therefore let *AB* be the circle eccentric to the sun, which the centre of the Earth describes during its annual circuit in the way we explained above; and let *C* be its centre. But now let us put down that the planet has no other irregularity except this one;

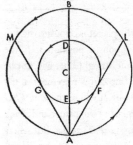

and that will be the case if we make *DE*, the orbital circle of Venus or Mercury, homocentric with *AB*; and *DE* should be inclined to *AB* on account of its latitude. But for the sake of ease of demonstration they can be thought of as if in the same plane. Let the Earth be assumed at point *A*; and from *A* let there be drawn the lines of sight *AFL* and *AGM* touching the circle of the planet at points *F* and *G*; and let *ACB* the diameter common to both circles be drawn.

Now let the movement of both the Earth and the planet be in the same direction, *i.e.*, eastward, but with greater velocity in the case of the planet than in that of the Earth. Therefore *C* and line *ACB* will appear to the eye borne along at *A* to move in accordance with the mean movement of the sun; but the planet in circle *DFG* as in an epicycle will traverse arc *FDG* eastward in greater time than it will the remaining arc *GEF* westward; and in the upper arc it will add the total angle *FAG* to the mean movement of the sun, and in the lower arc will subtract the same. Accordingly where the subtractive movement of the planet, especially around *E* the perigee, is greater than the additive (movement) of *C*, it will seem to *A* to retrograde in proportion to the excelling (movement)—as happens in these planets, when line *CE* has a greater ratio to line *AE* than the movement at *A* has to the movement of the planet, according to the demonstrations of Apollonius of Perga, as will be said later. But where the additive movement is equal to the subtractive, [141b] the planet will seem to come to a stop on account of the mutual equilibrium; all this agrees with the appearances.

Accordingly if there were no other irregularity in the movement of the planet, as Apollonius opined, this would be sufficient. But the greatest angular elongations from the mean movement of the sun, which these planets have in the morning and evening and which are understood by angles *FAE* and *GAE*, are not everywhere equal, neither the one to the other, nor are the sums of the two equal; for the apparent reason that the route of these planets is not along circles homocentric with the terrestrial circle but along certain others, by which they effect the second irregularity.

The same thing is also demonstrated in the case of the three upper planets, Saturn, Jupiter, and Mars, which circle around the Earth. For let the former circle of the Earth be drawn again, and let *DE* be as an exterior homocentric circle in the same plane: let the position of the planet be taken anywhere, at point *D*; and from *D* let there be drawn *DACBE* the common diameter and *DF* and *DG* straight lines touching the orbital circle of the Earth at points *F* and *G*. It is manifest that from point *A* only will the true position of the planet in *DE*

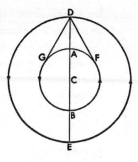

the line of mean movement of the sun be apparent, when the planet is opposite the sun and is nearest to the Earth. For when the Earth is in the opposite position at *B*, the opposition (of the planet and the sun), although in the same straight line, will not be at all apparent on account of the closeness of the sun to *C*. But as the movement of the Earth is speedier, so that it outruns the movement of the planet, it will seem along *FBG* the arc of apogee to add the total angle *GDF* to the movement of the planet and along the remaining arc *GAF* to subtract the same, according as arc *GAF* is smaller. But where the subtractive movement of the Earth excels the additive movement of the planet, especially in the neighbourhood of *A*, the planet will seem to be left behind by the Earth, to move westward and to come to a stop at the place where there is least difference between the movements which are contrary according to sight.

And so it is once more manifest that all these apparent movements—which the ancients were looking into by means of the epicycles of the individual planets—occur on account of the movement of the Earth. But since in spite of the opinion of Apollonius and the ancients the movement of the planet is not found regular, as the irregular revolution of the Earth with respect to the planet produces that; accordingly the planets are not carried in a homocentric circle but in some other which we shall demonstrate straightway.

4. WHY THE PROPER MOVEMENTS OF THE PLANETS APPEAR IRREGULAR

[142ª] But since their proper movements in longitude follow approximately the same mode except for Mercury, which is seen to differ from them, we shall treat of those four planets together, but another place has been given over to Mercury. Accordingly as the ancients placed one movement in two eccentric circles, as was shown, we have decreed two regular movements out of which the apparent irregularity is compounded either by a circle eccentric to an eccentric circle, or by the epicycle of an epicycle or by a combination of an eccentric circle carrying an epicycle. For they can

all effect the same irregularity, as we demonstrated above in the case of the sun and the moon.

Accordingly let *AB* be an eccentric circle around centre *C*. Let *ACB* be the diameter drawn through the highest and lowest apsis of the planet and containing the mean position of the sun. On *ACB* let *D* be the centre of the orbital circle of the Earth; and with the highest apsis *A* as centre and the third part of *CD* as radius, let epicycle *EF* be described. Let *F* be its perigee, and let the planet be placed there. Now let the movement of the

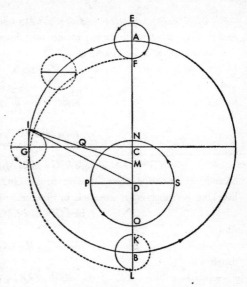

epicycle along eccentric circle *AB* take place eastward; and let the movement of the planet in the upper arc (of the epicycle) take place similarly eastward [142ᵇ] but in the remaining arc westward; and let the revolutions of the epicycle and the planet be proportionately equal to one another.

On that account when the epicycle is at the highest apsis of the eccentric circle and the planet on the contrary is at the perigee of the epicycle, the relation between their movements is reversed[1] with respect to one another, since both the planet and the epicycle have traversed their semicircle. But in both mean quadrants each will have its mean apsis, and then only will the diameter[2] of the epicycle be parallel to line *AB*; and at the midpoints (between the mean quadrants and the perigee or apogee) the diameter will be perpendicular to *AB*: the rest of the time always moving towards *AB* or moving away. All that is easily understood as following from the movements.

Hence it will also be demonstrated that by this composite movement the planet does not describe a perfect circle in accordance with the theory of the ancient mathematicians but a curve differing imperceptibly from one.

For let the same epicycle *KL* be drawn again, and let *B* be its centre. Let *AG* the quadrant of a circle be assumed, and let *HI* be an epicycle around *G*. Let *CD* be cut into three equal parts, and let

---

[1]That is to say, during the hemicycle of movement wherein the epicycle is passing from the lowest to the highest apsis of the eccentric circle and the planet is passing from the apogee to the perigee of the epicycle, the movement on the epicycle adds to the movement on the eccentric circle; but during the hemicycle wherein the epicycle is passing from the highest to the lowest apsis, the movement on the epicycle subtracts from the movement on the eccentric circle.

[2]In this passage Copernicus is speaking as if the planet were borne around the epicycle by the revolving diameter, although he usually speaks as if the diameter of the epicycle pointed perpetually at the centre of the homocentric circle.

$$CM = {}^1/_3CD = GI.$$

And let *GC* and *IM*, which cut one another at *Q*, be joined.

Accordingly since, by hypothesis

$$\text{arc } AG = \text{arc } HI$$

and

$$\text{angle } ACG = 90°;$$

then

$$\text{angle } HGI = 90°.$$

And

$$\text{angle } IQG = \text{angle } MQC,$$

because they are vertical angles. Therefore triangles *GIQ* and *QCM* are equiangular; and they have correspondingly equal sides, since by hypothesis

$$\text{base } GI = \text{base } CM.$$

And

$$QI > QC = QI > QG;$$

therefore

$$IQM > GQC,$$

but

$$FM = ML = AC = CG.$$

Therefore the circle which is described around centre *M* through points *F* and *L* and is hence equal to circle *AB* will cut line *IM*. The same demonstration will hold in the opposite quadrant. Accordingly by the regular movements of the epicycle in the eccentric circle the planet in the epicycle will not describe a perfect circle but a quasi-circle—as was to be demonstrated.[1]

Now around centre *D* let *NO* the annual orbital circle of the Earth be described; let *IDR* be extended; and let *PDS* be drawn parallel to *CG*. Accordingly *IDR* will be the straight line of the true movement of the planet; *GC*, the straight line of the mean and regular movement. And *R* will be the true apogee of the Earth with respect to the planet; and *S*, the mean apogee. Accordingly angle *RDS* or *IDP* is the difference between the regular and the apparent movement of both, namely between angle *ACG* and angle *CDI*.

But in place of eccentric circle *AB* we may take an equal homocentric circle around *D* as the deferent of the epicycle, whose radius is equal to *DC* and which is the deferent of the other epicycle, whose semi-diameter is half *MD*.[2] New let the first epicycle be moved [143ª] eastward, but the second in the opposite direction; and lastly let the planet on it (*i.e.*,

---

[1] As has been pointed out, if in the foregoing diagram we consider a point *X* so situated on semi-diameter *CA* that *CX* is equal to *GI* (and consequently *DM* is equal to *MX*), then since the planet on reaching point *I* has expended one quarter of its periodic time and has traversed one quarter of a full revolution about point *X*, evidently point *X* is analogous to the centre of a Ptolemaic equant, point *M* (the centre of the quasi-circle) to the centre of the Ptolemaic deferent, and point *D* (the centre of the sun for Copernicus) to the centre of the Earth.

[2] As in the accompanying diagram:

on the second epicycle) be deflected by the twofold movement. The same things will happen as before and no differently from in the moon, or by some other of the aforesaid modes.

But here we have chosen the eccentric circle bearing the epicycle, because by remaining always between the sun and C centre D is meantime found to have changed, as was shown in the case of solar appearances. But as the remaining appearances do not accord proportionately with this change there must be some other irregularity in those planetary movements: this irregularity, although very slight, is perceptible in the case of Mars and Venus, as will be seen in the right place.

Accordingly we shall soon demonstrate from observations that these hypotheses are sufficient for the appearances; and we shall do that first in the case of Saturn, Jupiter, and Mars: in them the position of the apogee and the distance CD are very difficult to find and of the greatest importance, since the rest is easily demonstrable by means of the apogee and the distance CD. Now in this case we shall use the method we used concerning the moon, namely a comparison of three ancient solar oppositions with the same number of modern ones, which the Greeks call their "acronychial gleams" and we the "deeps of the night," namely when the planet opposite the sun falls upon the straight line of the mean movement of the sun, where it throws off all that irregularity which the movement of the Earth brings to it. Such positions are determined by observations with an astrolabe and by computation of the oppositions of the sun, until it is clear that the planet has arrived at a point opposite the Sun.

5. DEMONSTRATIONS OF THE MOVEMENT OF SATURN

Accordingly we shall begin with Saturn by taking three oppositions once observed by Ptolemy. The first of them occurred in the 11th year of Hadrian on the 7th day of the month Pachom at the first hour of night; in the year of Our Lord 127 on the 7th day before the Kalends of April, 17 equal hours after midnight in relation to the Cracow meridian, which we find an hour distant from Alexandria. Now the position of the planet in relation to the sphere of the fixed stars, to which as to the starting-point of the regular movement we are referring all these things, was found to be at approximately 174°40', since [143b] the sun by its simple movement was then opposite at 354°40' from the horn of Aries, the starting-point assumed.

The second opposition was in the 17th year of Hadrian on the 18th day of the month Epiphi by the Egyptian calendar; but by the Roman, in the year of Our Lord 133 on the 3rd day before the Nones of June, 11 equatorial hours after midnight: he found the planet at 243°3', while by its mean movement the sun was at 63°3', 15 hours after midnight.

He recorded the third as occurring in the 20th year of Hadrian on the 24th day of the month Mesori by the Egyptian calendar; which was in the year of Our Lord 136 on the 8th day before the Ides of July, 11 hours after midnight (similarly according to the Cracow meridian) at 277°37', while by its mean movement the sun was at 97°37'.

Accordingly in the first interval there are 6 years 70 days 55 minutes (of a day), during which the planet is moved 62°23' in relation to sight, and the mean movement of the Earth with respect to the planet, *i.e.*, the movement of parallax, is 352°44'. Accordingly the 7°16' in which the circle is deficient belong to the mean movement of the planet, so that it is 75°39'.

In the second interval there are 3 Egyptian years 35 days 50 minutes; the apparent movement of the planet is 34°34', (the movement) of parallax 356°43'; and the remaining 3°17' of a circle are added to the apparent movement of the planet, so that the mean movement is 37°51'.

After this survey let *ABC* the eccentric circle of the planet be described. Let *D* be its centre, and *FDG* its diameter, whereon *E* is the centre of the great orbital circle of the Earth. Now let *A* be the centre of the epicycle at the first opposition to the sun; *B*, at the second; and *C*, at the third; and around them let the same epicycle be described with a radius equal to one-third of *DE*. Let the centres *A*, *B*, and *C* be joined to *D* and *E* by straight lines, which will cut the

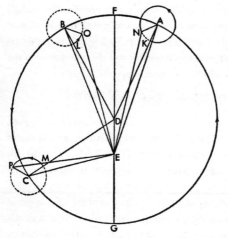

circumference of the epicycle at points *K*, *L*, and *M*. And let there be taken arc *KN* similar to *AF*, arc *LO* similar to *BF*, and *MP* similar to *FBC*; and let *EN*, *EO*, and *EP* be joined. Therefore by computation

arc *AB* = 75°39'

and

arc *BC* = 37°51';

and of the angles of apparent movement,

angle *NEO* = 68°23'

and

angle *OEP* = 34°34'.

Our problem is to examine the positions of highest and lowest apsis, *i.e.*, of *F* and *G*, together with the distance *DE* between the centres, without which there is no way of discerning the regular and the apparent [144ᵃ] movement.

But here too we run into as great a difficulty as in this part of Ptolemy, since, if the given angle *NEO* comprehended the given arc *AB*, and (angle) *OEP* (arc) *BC*, the entrance to demonstrating what we are looking for would be already opened. But the known arc *AB* subtends the unknown angle *AEB*, and similarly the unknown angle

BEC is subtended by the known arc *BC*; for it was necessary for both of them to be known. But *AEN*, *BEO*, and *CEP*, the differences between the angles, cannot be perceived, unless arcs *AF*, *FB*, and *FBC* are first set up as similar to those on the epicycle; accordingly these things are mutually dependent so as to be simultaneously known or unknown. Therefore those who were destitute of the means of demonstration relied upon detours and the *a posteriori* method, as the straightforward and *a priori* approach was not open. So Ptolemy in this investigation expended his energies in a prolix argument and a great multitude of calculations, which I judge boring and supererogatory to review, especially as in our calculations, which follow, we shall copy the same method approximately.

Finally in going over his calculations again he found that

$$\text{arc } AF = 57°1',$$
$$\text{arc } BF = 18°37',$$

and

$$\text{arc } FBC = 56\frac{1}{2}°.$$

But

$$\text{ecc.} = 6^\text{P}50',$$
$$\text{where } DF = 60^\text{P}.$$

But

$$\text{ecc.} = 1139,$$
$$\text{where } DF = 10,000.$$

Now

$$\frac{3}{4}(1139) \doteqdot 854,$$

and

$$\frac{1}{4}(1139) \doteqdot 1285.$$

Hence

$$DE = 854$$

and

$$\text{rad. ep.} = 285.$$

Making these assumptions and borrowings for our hypothesis, [144^b] we shall show that these things agree with the appearances observed.

Now at the first solar opposition, in triangle *ADE*,

$$\text{side } AD = 10,000,$$

and

$$\text{side } DE = 854;$$

and

$$\text{angle } ADE = 180° - \text{angle } ADF.$$

Hence, by means of what we have shown concerning plane triangles,

side *AE* = 10,489,

and

angle *DEA* = 53°6',

and

angle *DAE* = 3°55',
where 4 rt. angles = 360°.

But

angle *KAN* = angle *ADF* = 57°1'.

Therefore by addition

angle *NAE* = 60°56'.

Accordingly in triangle *NAE* two sides are given:

side *AE* = 10,489,
side *NA* = 285
where *AD* = 10,000,

and

angle *NAE* is given.

Hence

angle *AEN* = 1°22';

and, by subtraction,

angle *NED* = 51°44',
where 4 rt. angles = 360°.

Similarly at the second solar opposition. For in triangle *BDE*

side *DE* = 854,
where *BD* = 10,000;

and

angle *BDE* = 180° − *BDF* = 161°22'.

So triangle *BDE* too has its sides and angles given:

side *BE* = 10,812,
where *BD* = 10,000,

and

angle *DBE* = 1°27',

and

angle *BED* = 17°11'.

But

angle *OBL* = angle *BDF* = 18°36'.

Therefore, by addition

angle *EBO* = 20°3'.

Accordingly in triangle *EBO* two sides are given together with angle *EBO*:

$$BE = 10,812$$

and

$$BO = 285.$$

By what we have shown concerning plane triangles

$$\text{angle } BEO = 32'.$$

Hence

$$\text{angle } BED = 16°39'.$$

Moreover, in the third solar opposition, in triangle *CDE*, as before,

side *CD* is given

and

side *DE* is given;

and

$$\text{angle } CDE = 180° - 56°29'.$$

By the fourth rule for plane triangles

$$\text{base } CE = 10,512,$$
$$\text{where } CD = 10,000;$$

and

$$\text{angle } DCE = 3°53'$$

and, by subtraction,

$$\text{angle } CED = 52°36'.$$

Therefore, by addition,

$$\text{angle } ECP = 60°22',$$
$$\text{where 4 rt. angles} = 360°.$$

So also in triangle *ECP* two sides are given together with angle *ECP*; furthermore,

$$\text{angle } CEP = 1°22',$$

whence, by subtraction,

$$\text{angle } PED = 51°14'.$$

Hence, of the total angles of apparent movement,

$$\text{angle } OEN = 68°23',$$

and

$$\text{angle } OEP = 34°35',$$

which agree with the observations. And the position of the highest apsis of the eccentric circle

$$F = 226°20'$$

from the head of Aries. And as the then existing precession of [145a] the spring equinox was 6°40',

$$226°20' + 6°40' = 23° \text{ of Scorpio,}$$

in accordance with Ptolemy's conclusion. For the apparent position of the planet at this third solar opposition, as was reported above, was 227°37'. And as the angle of apparent movement,

$$\text{angle } PED = 51°14'.$$

Hence

$$227°37' - 51°14' = 226°23',$$

which is the position of the highest apsis of the eccentric circle.

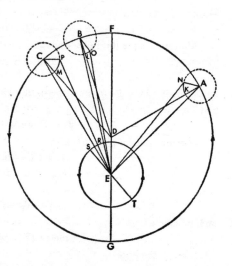

Now let there be described *RST* the annual orbital circle of the Earth, which will cut line *PE* at point *R*; and let the diameter *SET* be drawn parallel to the line of mean movement of the planet. Accordingly, as

$$\text{angle } SED = \text{angle } CDF,$$

angle *SER* will be the difference and the additosubtraction between the apparent and mean movement, *i.e.*, between angles *CDF* and *PED*, and

$$\text{angle } SER = 5°16'.$$

And there is the same difference between the mean and the true movements of parallax. Now

$$\text{arc } RT = 180° - \text{arc } SER = 174°44',$$

which is the regular movement of parallax from starting-point *T*, *i.e.*, from the mean conjunction of the sun and the planet, to this third solar opposition or true opposition of the Earth and the planet.

Accordingly at the time of this observation, namely in the 20th year of the reign of Hadrian, but in the 136th year of Our Lord on the 8th day before the Ides of July, 11 hours after midnight, we have the movement of anomaly of Saturn from the highest apsis of its eccentric circle as $56\frac{1}{2}°$, and the mean movement of parallax as 174°44', as was timely to demonstrate on account of what follows.

6. ON THREE OTHER SOLAR OPPOSITIONS OF SATURN RECENTLY OBSERVED

[145b] Now since the computation of the movement of Saturn handed down by Ptolemy has no small discrepancy with our times, and since it cannot be understood right away in what quarter the error lies, we are forced to make new observations, out of which we have again taken three solar oppositions. The first opposition was in the

year of Our Lord 1514, on the 3rd day before the Nones of May $1^1/_5$ hours before midnight, at which time Saturn was discovered at 205°24'.

The second was in the year of Our Lord 1520 on the third day before the Ides of July at midday, and the planet was at 273°25'.

The third was in the year of Our Lord 1527 on the 6th day before the Ides of October $6^2/_5$ hours after midnight; and Saturn appeared at 7' from the horn of Aries.

Accordingly between the first and second solar oppositions there are 6 Egyptian years 70 days 33 minutes (of a day), during which time the apparent movement of Saturn is 68°1'.

From the second to the third there are 7 Egyptian years 89 days 46 minutes, and the apparent movement of the planet is 86°42'; and the mean movement during the first interval is 75°39'; and during the second, 88°29'. Accordingly in investigating the

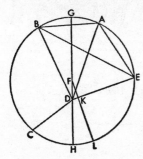

highest apsis and the eccentricity, we must at first abide by the rule of Ptolemy, just as if the planet moved in a simple eccentric circle; and although that is not sufficient, nevertheless we shall be led fairly near and shall arrive at the truth more easily.

Accordingly, let *ABC* be the circle in which the planet is moved regularly: and let the first opposition be at *A*, the second at *B*, and the third at *C*. Let the centre of the orbital circle of the Earth be taken within it as *D*. Let *AD*, *BD*, and *CD* be joined, and let any one of them be extended in a straight line to the opposite part of the circumference—say *CDE*—and let *AE* and *BE* be joined.

Accordingly, since

$$\text{angle } BDC = 86°42',$$
$$\text{where 2 rt. angles} = 180°;$$
$$\text{angle } BDE = [146^a] \ 93°18';$$

but

$$\text{angle } BDE = 186°36',$$
$$\text{where 2 rt. angles} = 360°.$$

And, as intercepting arc *BC*,

$$\text{angle } BED = 88°29',$$

and

$$\text{angle } DBE = 84°55'.$$

Accordingly, as the angles of triangle *BDE* are given, the sides are given by the table:
$$BE = 19,953$$

and

$$DE = 13,501$$

where diameter of circle circumscribing triangle = 20,000.

Similarly in triangle *ADE*, since

$$\text{angle } ADC = 154°43',$$

where 2 rt. angles = 180°;

$$\text{angle } ADE = 180° - \text{angle } ADC = 25°17';$$

but

$$\text{angle } ADE = 50°34',$$

where 2 rt. angles = 360°.

And, as intercepting arc *ABC*,

$$\text{angle } AED = 164°8',$$

and

$$\text{angle } DAE = 145°18' :$$

hence the sides are established:

$$DE = 19,090$$

and

$$AE = 8,542,$$

where diameter of circle circumscribing triangle *ADE* = 20,000.

But

$$AE = 6,043$$

where *DE* = 13,501 and

$$BE = 19,953.$$

Hence too, in triangle *ABE*, these two sides *BE* and *EA* have been given; and, as intercepting arc *AB*,

$$\text{angle } AEB = 75°39'.$$

Accordingly by what we have shown concerning plane triangles,

$$AB = 15,647$$

where *BE* = 19,968.

But according as

$$\text{ch. } AB = 12,266,$$

where diameter of eccentric circle = 20,000;

$$EB = 15,664$$

and

$$DE = 10,599.$$

Accordingly, in proportion to chord *BE*,

$$\text{arc } BAE = 103°7'.$$

Hence, by addition,

$$\text{arc } EABC = 191°36';$$

and

$$\text{arc } CE = 360° - \text{arc } EABC = 168°24';$$

and hence

$$\text{ch. } CDE = 19,898.$$

And

$$CD = CDE - DE = 9,299.$$

And now it is manifest that, if CDE were the diameter of the eccentric circle, the positions of highest and lowest apsis would fall upon it, and the distance between the centres would be evident; but because segment $EABC$ is greater, the centre will be in it. Let F be the centre, and let the diameter $GFDG$ be extended through F and D, and let FKL be drawn at right angles to CDE.

Now it is manifest that

$$\text{rect. } CD, DE = \text{rect. } GD, DH.$$

But

$$\text{rect. } GD, DH + \text{sq. } FD = \text{sq. } (^1/_2GDH) = \text{sq. } FDH.$$

Accordingly

$$\text{sq. } FDH - \text{rect. } CD, DE = \text{sq. } FD.$$

Therefore

$$FD = 1,200$$
$$\text{where radius } GF = 10,000;$$

but

$$FD = 7^{\text{P}}12'$$
$$\text{where radius } = 60^{\text{P}},$$

[146^{b}] which differs little from Ptolemy.

But since

$$CDK = {}^1/_2CDE = 9,949$$

and

$$CD = 9,299,$$

therefore

$$DK = CDK - CD = 650,$$
$$\text{where } GF = 10,000$$
$$\text{and } FD = 1,200.$$

But

$$DK = 5,411$$
$$\text{where } FD = 10,000.$$

And since

$$DK = {}^1/_2 \text{ ch. } 2 \text{ } DFK,$$
$$\text{angle } DFK = 32°45',$$

where 4 rt. angles = 360°;

and as standing at the centre of the circle it intercepts a similar chord and arc *HL* on the circumference.

But

$$\text{arc } CHL = {}^{1}/_{2}CLE = 84°13';$$

therefore

$$\text{arc } CH = CHL - HL = 51°28',$$

which is the distance from the third opposition to the perigee. Now

$$180° - 51°28' = CBG = 128°32'$$

from the highest apsis to the third opposition. And since

$$\text{arc } CB = 88°29',$$

$$\text{arc } BG = CBG - CB = 40°3',$$

from the highest apsis to the second solar opposition. Then, as

$$\text{arc } BGA = 75°39',$$

$$\text{arc } GA = BGA - BG = 35°36'$$

from the first opposition to the apogee *G*.

Now let *ABC* be a circle with diameter *FDEG*, centre *D*, apogee *F*, perigee *G*. Let

arc *AF* = 35°36',

arc *FB* = 40°3',

and

arc *FBC* = 128°32'.

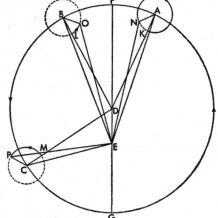

Now let *DE* be taken as three quarters of what has already been shown to be the distance between the centres, *i.e.*, let

DE = 900;

and

quarter distance = 300

where radius = 10,000.

And with that quarter distance as radius, let the epicycle be described around centres *A*, *B*, and *C*—and let the figure be completed according to the hypothesis set before us. But if with this layout we wish to elicit the observed positions of Saturn [147ª] by the method handed down above and soon to be repeated, we shall find some discrepancy.

And—to speak briefly, so as not to burden the reader with many words or seem to have laboured more in indicating by-ways than in pointing out the high road—these

things will, by means of what we have shown concerning triangles, necessarily lead us to the conclusion that

$$\text{angle } NEO = 67°35'$$

and

$$\text{angle } OEP = 87°12'.$$

But angle OEP is $1/2°$ greater than the apparent angle, and the angle NEO is 26' smaller. And we find that they square with one another only if we move the apogee forward a little, and set up

$$\text{arc } AF = 38°50',$$
$$\text{arc } FB = 36°49',$$
$$\text{arc } FBC = 125°18',$$
$$DE = 854,$$

which is the distance between the centres, and

$$\text{rad. ep.} = 285,$$

where $FD = 10,000$;

and that agrees approximately with Ptolemy, as set out above. For it is clear from this that these magnitudes agree with the three apparent solar oppositions observed.

Since at the first opposition, in triangle ADE,

$$\text{side } DE = 854,$$
$$\text{where } AD = 10,000,$$

and

$$\text{angle } ADE = 141°10',$$
$$\text{where angle } ADE + \text{angle } ADF = 2 \text{ rt. angles;}$$

hence it is shown that

$$\text{side } AE = 10,679,$$
$$\text{where radius } FD = 10,000,$$
$$\text{angle } DAE = 2°52',$$

and

$$\text{angle } DEA = 35°58'.$$

Similarly in triangle AEN, since

$$\text{angle } KAN = \text{angle } ADF;$$
$$\text{angle } EAN = 41°42',$$

and

$$\text{side } AN = 285,$$
$$\text{where } AE = 10,679.$$

Hence

$$\text{angle } AEN = 1°3'.$$

But

$$\text{angle } DEA = 35°58';$$

accordingly, by subtraction,

$$\text{angle } DEN = 34°55'.$$

In the second solar opposition triangle DEB has two sides given:

$$DE = 854,$$
$$\text{where } DB = 10,000$$

and

$$\text{angle } BDE = 153°11'.$$

Accordingly

$$BE = 10,697,$$
$$\text{angle } DBE = 2°45',$$

and

$$\text{angle } BED = 34°4'.$$

But

$$\text{angle } LBO = \text{angle } BDF;$$

therefore, as at the centre,

$$\text{angle } EBO = 39°34'.$$

Now this angle is comprehended by the given sides

$$BO = 285$$

and

$$BE = 10,697;$$

hence

$$\text{angle } BEO = 59'.$$

And

$$\text{angle } OED = \text{angle } BED - \text{angle } BEO \; 33°5'.$$

But in the first solar opposition it has already been shown that

$$\text{angle } DEN = 34°55'.$$

Therefore by addition

$$\text{angle OEN} = 68°$$

by which the distance of the first solar opposition from the second becomes apparent; and it harmonizes with the observations.

The same thing will be shown at the third opposition.
In triangle CDE

$$\text{angle } CDE = 54°42',$$
$$\text{side } CD = 10,000,$$

and

$$\text{side } DE = 854;$$

[147b] hence

$$\text{side } EC = 9{,}532,$$
$$\text{angle } CED = 121°5',$$

and

$$\text{angle } DCE = 4°13';$$

therefore by addition

$$\text{angle } PCE = 129°31'.$$

So again in triangle EPC

$$\text{side } CE = 9{,}532$$

and

$$\text{side } PC = 285,$$

and

$$\text{angle } PCE = 129°31' :$$

hence

$$\text{angle } PEC = 1°18'.$$

And

$$\text{angle } PED = \text{angle } CED - \text{angle } PEO = 119°47'$$

from the highest apsis of the eccentric circle to the position of the planet at the third opposition.

Now it was shown that there were 33°5' to the second solar opposition: accordingly between the second and third solar oppositions of Saturn there remain 86°42', which agree with the observations. Now the position of Saturn was found by observation at that time to be at 7' from the assumed starting-point of the first star of Aries, and it was shown that there were 60°13' from it to the lowest apsis of the eccentric circle: accordingly the lowest apsis is approximately $60^{1}/_{3}°$, and the position of the highest apsis is diametrically opposite at $240^{1}/_{3}°$.

Now let RST the great orbital circle of the Earth be set around its centre E, and let its diameter SET be parallel to CD the line of mean movement; and let

$$\text{angle } FDC = \text{angle } DES.$$

Therefore the Earth and our point of sight will be on line PE, namely at point R. Now

$$\text{angle } PES = 5°31',$$

and angle PES or arc RS is the difference between FDC the angle of regular movement and DEP the angle of apparent movement.

Now

$$\text{arc } RT = 180° - 5°31' = 174°29'$$

which is the distance of the planet from the apogee of the orbital circle, *i.e.*, from T, as if from the mean position of the sun.

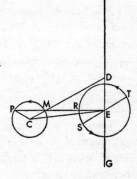

And so we have demonstrated that in the year of Our Lord 1527 on the sixth day before the Ides of October at $6^2/_5$ hours after midnight, the movement of anomaly of Saturn from the highest apsis of the eccentric circle was 125°18', the movement of parallax was 174°29', and the position of the highest apsis was at 240°21' from the first star of Aries in the sphere of the fixed stars.

7. ON THE EXAMINATION OF THE MOVEMENT OF SATURN

[148ª] Now it was shown that Saturn at the time of the last of the three observations of Ptolemy was by its movement of parallax at 174°44', and the position of the highest apsis of the eccentric circle was at 226°23', from the head of the constellation of Aries. Accordingly it is clear that in the midtime between the two observations Saturn has completed 1344 revolutions minus $^1/_4$° of regular parallaxes.

Now from the 20th year of Hadrian on the 24th day of the Egyptian month Mesori one hour before midday to the year of Our Lord 1527 on the 6th day before the Ides of October at 6 hours (after midnight, the time) of this observation, there are 1392 Egyptian years 75 days 48 minutes (of a day).

Hence if we wish to get the movement itself from the table, we shall similarly find 359°45', the movement beyond the 1343 revolutions of parallax. Accordingly what was set down concerning the mean movements of Saturn is correct. Moreover during that time the simple movement of the sun is 82°30'. If 359°45' are subtracted from 82°30', the remainder is the 82°45' of the mean movement of Saturn, which are already being added up in its 47th revolution, in harmony with the computation. Meanwhile too the position of the highest apsis of the eccentric circle has been moved forward to 13°58' in the sphere of the fixed stars. Ptolemy believed it to be fixed in the same way, but now it appears to move approximately 1° per 100 years.

8. ON DETERMINING THE POSITIONS OF SATURN

Now from the beginning of the years of Our Lord to the 20th of Hadrian on the 24th day of the month Mesori at 1 hour before midday, the time of Ptolemy's observation, there are 135 Egyptian years 222 days 27 minutes (of a day), during which time Saturn's movement of parallax was 328°55'. The subtraction of 328°55' from 174°44' leaves 205°49' [148ᵇ] as the locus of distance of the mean position of the sun from the mean (position) of Saturn, and as its movement of parallax at midnight before the Kalends of January.

From the first Olympiad to this locus 775 Egyptian years $12^1/_2$ days comprehend a movement of 70°55' besides the whole revolutions. The subtraction of 70°55' from 205°49' leaves 134°54' for the beginning of the Olympiads at noon on the 1st day of the month Hekatombaion.

Then after 451 years 247 days there are 13°7' besides the whole revolutions: the addition of 13°7' to 134°54' puts the locus (of the years) of Alexander the Great at 148°1' on noon of the 1st day of the month Thoth by the Egyptian calendar; and there are 278 years 118^1/$_2$ days to (years of) Caesar; the movement is 247°20', and it sets up the locus at 35°21' on midnight before the Kalends of January.

9. ON THE PARALLAXES OF SATURN, WHICH ARISE FROM THE ANNUAL ORBITAL CIRCLE OF THE EARTH, AND HOW GREAT THE DISTANCE OF SATURN IS (FROM THE EARTH)

In this way it has been demonstrated that the regular movements of Saturn in longitude are at one with the apparent. For the other apparent movements which occur in the case of Saturn are, as we said, parallaxes arising from the annual orbital circle of the Earth, since, as the magnitude of the Earth in relation to the distance of the moon causes parallaxes, so too its orbital circle, in which it revolves annually, should in the case of the five wandering stars cause (parallaxes) which are far more evident in proportion to the magnitude of the orbital circle. Now such parallaxes cannot be determined, unless the altitude of the planet—which, however, it is possible to apprehend through any one observation of a parallax—becomes known first.

We have such (an observation) in the case of Saturn in the year of Our Lord 1514 on the sixth day before the Kalends of May 5 equatorial hours after the preceding midnight. For Saturn was seen to be in a straight line with the stars in the forehead of Scorpio, namely with the second and third stars, which have the same longitude and are at 209° of the sphere of the fixed stars. Accordingly the position of Saturn is made evident through them. Now there are 1514 Egyptian years 61 days 13 minutes (of a day) from the beginning of the years of Our Lord to this time; and according to [149a] calculation the mean position of the sun was at 315°41', the anomaly of parallax of

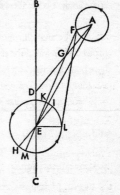

Saturn was at 116°31', and for that reason the mean position of Saturn was 199°10' and that of the highest apsis of the eccentric circle was at approximately 240^1/$_3$°.

Now in accordance with our problem, let ABC be the eccentric circle: let D be its centre, and on the diameter BDC let B be the apogee, C the perigee, and E the centre of the orbital circle of the Earth. Let AD and AE be joined, and with A as centre and ^{1}/$_3$ DE as radius let the epicycle be drawn. On the epicycle let F be the position of the planet; and let

angle DAF = angle ADB.

And through E the centre of the orbital circle of the Earth let HI be drawn, as if in the same plane with circle ABC, and as

a diameter, parallel to *AD*, so as to have it understood that with respect to the planet the apogee of the orbital circle is at *H* and the perigee at *I*.

Now on the orbital circle let

$$\text{arc } HL = 116°31'$$

in accordance with the computation of the anomaly of parallax; let *FL* and *EL* be joined, and let *FKEM* produced cut both arcs of the orbital circle.

Accordingly since by hypothesis

$$\text{angle } ADB = \text{angle } DAF = 41°10',$$

and

$$\text{angle } ADE = 180° - ADB = 138°50';$$

and

$$DE = 854$$
$$\text{where } AD = 10,000:$$

whence in triangle *ADE*

$$\text{side } AE = 10,667,$$
$$\text{angle } DEA = 38°9',$$

and

$$\text{angle } EAD = 3°1':$$

therefore by addition

$$\text{angle } EAF = 44°12'.$$

So again in triangle *FAE*

$$\text{side } FA = 285$$
$$\text{where } AE = 10,667,$$
$$\text{side } FKE = 10,465,$$

and

$$\text{angle } AEF = 1°5':$$

accordingly it is manifest that

$$\text{angle } AEF + \text{angle } DAE = 4°6',$$

which is the total difference or additosubtraction between the mean and the true position of the planet. Wherefore if the position of the Earth had been at *K* or *M*, the position of Saturn would have been apparent as if from centre *E* and would have been seen to be at 203°16' from the constellation of Aries. But with the Earth at *L*, Saturn is seen to be at 209°. The difference [149b] of 5°44' goes to the parallax in accord with angle *KFL*. But by calculation of the regular movement

$$\text{arc } HL = 116°31',$$

and

$$\text{arc } ML = \text{arc } HL - \text{add. } HM = 112°25'.$$

And by subtraction[1]

[1]Arc *MLIK* = 180°.

292

$$\text{arc } LIK = 67°35' :$$

hence

$$\text{angle } KEL = 67°35'.$$

Wherefore in triangle *FEL* the angles are given, and the ratio of the sides is given too:
Hence

$$EL = 1,090$$
$$\text{where } EF = 10,465,$$
$$\text{and } AD = BD = 10,000;$$

but

$$EL = 6\text{P}32',$$
$$\text{where } BD = 60\text{P},$$

by usage of the ancients;

and there is very little difference between that and what Ptolemy gave.

Accordingly

$$BDE = 10,854,$$

and, as the remainder of the diameter

$$CE = 9,146.$$

But since the epicycle when at *B* always subtracts 285 from the altitude of the planet, but adds the same amount, *i.e.*, its radius, when at *C*; on that account the greatest distance of Saturn from centre *E* will be 10,569, and the least 9,431, where *BD* = 10,000. By this ratio the altitude of the apogee of Saturn is 9P42', where the radius of the orbital circle of the Earth = 1P; and the altitude of the perigee is 8P39': hence it is quite evident by the mode set forth above in the case of the small parallaxes of the moon that the parallaxes of Saturn can be greater. And when Saturn is at the apogee,

$$\text{greatest parallax} = 5°45';$$

and when at the perigee,

$$\text{greatest parallax} = 6°39';$$

and they differ from one another by 44'—measuring the angles by the lines coming from the planet and tangent to the orbital circle of the Earth. In this way the particular differences in the movement of Saturn have been found, and we shall afterwards set them out simultaneously and in conjunction with those of the five planets.

10. DEMONSTRATIONS OF THE MOVEMENT OF JUPITER

Having solved the problems concerning Saturn, we shall use the same method and order of demonstration in the case of the movement of Jupiter too, and first we shall repeat three positions reported and demonstrated by Ptolemy, and by the foreshown transformation of circles we shall reconstitute them as the same or as very little different.

The first of the solar oppositions was in the 17th year of Hadrian on the 1st day of the month Epiphi by the Egyptian calendar 1 hour before the following midnight [150ᵃ] at 23°11' of Scorpio, as he says, but after deducting the precession of the equinoxes, at 226°33'.

He recorded the second as occurring on the 21st year of Hadrian on the 13th day of the month Phaophi by the Egyptian calendar 2 hours before the following midnight, at 7°54' of Pisces; but with respect to the sphere of the fixed stars it was 331°16'.

The third was during the 1st year of Antoninus in the month Athyr during the night following the 20th day of the month 5 hours after midnight, at 7°45' in the sphere of the fixed stars.

Accordingly from the first opposition to the second there were 3 Egyptian years 106 days 23 hours, and the apparent movement of the planet was 104°43'. From the second to the third opposition there was 1 year 37 days 7 hours, and the apparent movement of the planet was 36°29'. During the first interval of time the mean movement was 99°55'; during the second it was 33°26'.

Now he found that the arc of the eccentric circle from the highest apsis to the first opposition was 77°15'; and next, 2°50' from the second opposition to the lowest apsis; and from that to the third opposition, 30°36'. Now the eccentricity of the whole circle was $5^{1}/_{2}$ᵖ whereof the radius is 60ᵖ; but it is 917, whereof the radius would be 10,000; and all that corresponds approximately to the observations.

Now let *ABC* be the circle; and from the first opposition to the second let

　　　arc *AB* = 99°55';

and let

　　　arc *BC* = 33°26'.

Through the centre *D* let diameter *FDG* be drawn, so that from the highest apsis *F*

　　　FA = *77°15'*,
　　　FAB = *177°10'*,

and

　　　GC = 30°36'.

Now let *E* be taken as the centre of the orbital circle of the Earth. Let the distance between the centres be equal to three-quarters 917, *i.e.*, let

　　　DE = 687;

let

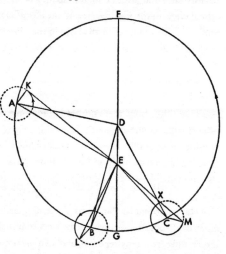

rad. ep. = 229,

which is one-quarter distance, and let the epicycle be described at points *A*, *B*, and *C*.
Let *AD*, *BD*, *CD*, *AE*, *BE*, and *CE* be joined; and in the epicycles let *AK*, *BL*, and *BM*
be joined in such a way that

angle *DAK* = angle *ADF*,

angle *DBL* = angle *FDB*,

and

angle *DCM* = angle *FDC*.

Finally let *K*, *L*, and *M* be joined to *E* by straight lines.

Accordingly, since in triangle *ADE*

angle *ADE* = 102°45',

because angle *ADF* is given; and

side *DE* = 687,

where *AD* = 10,000;

side *AE* = 10,174,

angle *EAD* = 3°48',

and

angle *DEA* = 73°27';

and by addition

angle *EAK* = 81°3'.

Accordingly in [150ᵇ] triangle *AEK* two sides have been given:

EA = 10,174

and

AK = 229,

and

angle *EAK* = 81°3';

it will be clear that

angle *AEK* = 1°17'.

Hence, by subtraction,

angle *KEO* = 72°10'.

Something similar will be shown in triangle *BED*. For the sides *BD* and *DE* always
remain equal to the corresponding sides in the first triangle; but

angle *BDE* = 2°50'.

For that reason

base *BE* = 9,314,

where *DB* = 10,000;

and

angle *DBE* = 12'.

So once more, in triangle *ELB* two sides are given; and
$$\text{angle } EBL = 177°22';$$
moreover
$$\text{angle } LEB = 4'.$$
But
$$\text{angle } FEL = \text{angle } FDB - 16' = 176°54'.$$
And as
$$\text{angle } KED = 72°10';$$
$$\text{angle } KEL = \text{angle } FEL - \text{angle } KED = 104°44',$$
which is the angle of apparent movement between the first and the second termini observed; and there is approximate agreement.

Similarly at the third opposition, in triangle *CDE* two sides *CD* and *DE* have been given, and
$$\text{angle } CDE = 30°36';$$
$$\text{base } EC = 9,410$$
and
$$\text{angle } DCE = 2°8'.$$
Whence in triangle *ECM*
$$\text{angle } ECM = 147°49';$$
hence
$$\text{angle } CEM = 39';$$
and because the exterior angle is equal to the sum of the interior and opposite angles
$$\text{angle } DXE = \text{angle } ECX + \text{angle } CEX = 2°47'$$
and
$$\text{angle } FDC - \text{angle } DEM = 2°47'.$$
Hence
$$\text{angle } GEM = 180° - \text{angle } DEM = 33°23';$$
and, by addition,
$$\text{angle } LEM = [151^a]\ 36°29',$$
which is the distance from the second opposition to the third; and that agrees with the observations. But since this third solar opposition was found to be at 7°45' (in the sphere of the fixed stars) and 33°23' to the east of the lowest apsis; the remainder of the semicircle gives us the position of the highest apsis as 154°22' in the sphere of the fixed stars.

Now around *E* let there be drawn *RST* the annual orbital circle of the Earth with diameter *SET* parallel to line *DC*. Now it has been made clear that
$$\text{angle } GDC = \text{angle } GER = 30°36';$$
and

angle *DXE* = angle *RES* = arc *RS* = 2°47',
the distance of the planet from the mean perigee of the
orbital circle. Hence by addition

arc *TSR* = 182°47',

which is the distance from the highest apsis of the orbital circle.

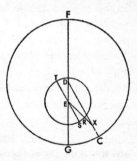

And by this we have confirmation of the fact that at
the time of the third opposition of Jupiter during the first
year of Antoninus on the 20th day of the month Athyr
by the Egyptian calendar 5 hours after the following midnight the planet Jupiter by its anomaly of parallax was at
182°47'. Its regular position in longitude was at 4°58', and the position of the highest
apsis of the eccentric circle was at 154°22'. All these things are in perfect agreement
with our hypothesis of the mobility of the Earth and absolute regularity (of movement).

11. ON THREE OTHER OPPOSITIONS OF JUPITER
RECENTLY OBSERVED

Having recorded three positions of the planet Jupiter and evaluated them in this
way, we shall set up three others in their place, which we observed with greatest care at
the solar oppositions of Jupiter.

The first was in the year of Our Lord 1520 on the day before the Kalends of May
11 hours after the preceding midnight, at 220°18' of the sphere of the fixed stars.

The second was in the year of Our Lord 1526 on the fourth day before the
Kalends of December 3 hours after midnight, at 48°34'.

But the third opposition was in the year of Our Lord 1529 on the Kalends of
February 18 hours after midnight, at 113°44'.

From the first [151b] to the second there are 6 years 212 days 40 minutes (of a
day), during which time the apparent movement of Jupiter was 208°6'. From the second to the third opposition there are 2 Egyptian years 66 days 39 minutes (of a day),
and the apparent movement of the planet is 65°10'. But the regular movement of the
planet during the first interval is 199°40', and during
the second 66°10'.

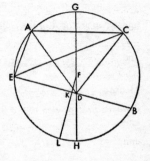

With this as a paradigm let eccentric circle *ABC* be
described, in which the planet is assumed to move simply and regularly. And let the three positions observed
be designated in the order of the letters *A*, *B*, and *C* in
such a way that

arc *AB* = 199°40'

and

$$\text{arc } BC = 66°10',$$

on that account

$$\text{arc } AC = 360° - (AB + BC) = 94°10'.$$

Moreover let D be taken as the centre of the annual orbit of the Earth. Let AD, BD and CD be joined; and let any of them, say DB, be extended in a straight line BDE to both arcs of the circle; and let AC, AE and CE be joined.

Accordingly, since

$$\text{angle } BDC = 65°10',$$

$$\text{where 4 rt. angles at centre} = 360°;$$

and that is the angle of apparent movement, and since

$$\text{angle } CDE = 180° - 65°10' = 114°50',$$

but

$$\text{angle } CDE = 229°40',$$

$$\text{where 2 rt. angles at circumference } 360°;$$

and since, as standing on arc BC of circumference,

$$\text{angle } CED = 66°10',$$

and accordingly

$$\text{angle } DCE = 64°10';$$

therefore, as triangle CDE has its angles given, it has its sides given too:

$$CE = 18,150$$

and

$$ED = 10,918$$

$$\text{where diameter of circle circumscribing triangle} = 20,000.$$

Similarly, in triangle ADE, since

$$\text{angle } ADB = 151°54',$$

which is the remainder of the circle after the subtraction of the given distance between the first opposition and the second; accordingly

$$\text{angle } ADE = 180° - 151°54' = 28°6',$$

as at the centre, but as on the circumference

$$\text{angle } ADE = 56°12';$$

and, as on arc BCA of the circumference

$$\text{angle } AED = 160°20';$$

and

$$\text{angle } EAD = 143°28'.$$

Hence

$$\text{side } AE = 9,420$$

and

$$\text{side } ED = 18,992$$

where diameter of circle circumscribing triangle ADE = 20,000.

But

$$AE = 5,415$$
$$\text{where } ED = 10,918$$
$$\text{and } CE = 18,150$$

Again therefore we shall have triangle EAC, of which the two sides EA and EC are given; and, as standing on arc AC of the circumference

$$\text{angle } AEC = 94°10'.$$

[152ᵃ] Hence it will be shown that, as standing on arc AE,

$$\text{angle } ACE = 30°40',$$
$$\text{angle } ACE + \text{arc } AC = 124°50',$$

and

$$CE = \text{ch. } EAC = 17,727$$
$$\text{where diameter of eccentric circle} = 20,000.$$

And by the ratio given before,

$$DE = 10,665,$$

and

$$\text{arc } BCAE = 191°.$$

It follows that

$$\text{arc } EB = 360° - 191° = 169°$$

and

$$BDE = \text{ch. } EB = 19,908$$

and by subtraction

$$BD = 9,243.$$

Accordingly, since $BCAE$ is the greater segment, it will contain F the centre of the circle. Now let the diameter $GFDH$ be drawn. It is manifest that

$$\text{rect. } ED, DB = \text{rect. } GD, DH,$$

which is therefore also given. But

$$\text{rect. } GD, DH + \text{sq. } FD = \text{sq. } FDH.$$

Now

$$\text{sq. } FDH - \text{rect. } GD, DH = \text{sq. } FD.$$

Therefore

$$FD = 1,193,$$
$$\text{where } FG = 10,000,$$

but

$$FD = 7^{\text{P}}9',$$
$$\text{where } FG = 60^{\text{P}}.$$

Now let *BE* be bisected at *K*, and let *FKL* be extended; accordingly *FKL* will be at rt. angles to *BE*. And since

$$BDK = \frac{1}{2} BE = 9,954$$

and

$$DB = 9,243,$$

then, by subtraction,

$$DK = 711.$$

Accordingly in triangle *DFK*, which has its sides given,

$$\text{angle } DFK = 36°35',$$

and similarly

$$\text{arc } HL = 36°35'.$$

But

$$\text{arc } LHB = 84\frac{1}{2}°;$$

and, by subtraction,

$$\text{arc } BH = 47°55',$$

which is the distance of the second position from the perigee.
And

$$\text{arc } BCG = 180° - 47°55' = 132°5',$$

which is the distance of the apogee from the second position.
And

$$\text{arc } BCG - \text{arc } BC = 132°5' - 66°10' = 65°55',$$

which is the distance from the third position to the apogee *G*.
Now

$$99°10' - 65°55' = 28°15',$$

which is the distance from the apogee to the first position of the epicycle. That harmonizes too little with the appearances, as the planet does not run through the proposed eccentric circle: hence this method of demonstration which is based upon an uncertain principle cannot give us any certainty. One sign of this among others is that Ptolemy in the case of Saturn recorded a too great distance between the centres and in the case of Jupiter a too small distance; but the same thing seemed a great enough distance to us, so that evidently upon the assumption of different arcs of circles for the same planet [152b] that which is sought does not come about in the same way. Not otherwise was it possible to compound the apparent and the regular movements at the three proposed termini and then at all the termini, unless we kept the total egression of eccentricity of the centres which was recorded by Ptolemy as 5ᴾ30', whereof the radius of the eccentric circle is 60ᴾ, but which is 917 parts, whereof the radius is 10,000. And let the arc from the highest apsis to the first opposition be 45°2'; from the lowest apsis to the second opposition 64°42'; and from the third opposition to the highest apsis 49°8'.

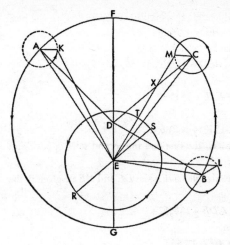

For let the above figure of the eccentric circle carrying an epicycle be repeated, inasmuch as it fits this example. So by our hypothesis

$$DE = 687,$$

which is three-quarters of the total distance between the centres. And

radius of epicycle = 229,

where $FD = 10,000$,

which is the remaining quarter of the distance. Accordingly, since

angle $ADF = 45°2'$,

triangle ADE will have the two sides AD and DE given, together with angle ADE; hence it is shown that

side $AE = 10,496$,

where $AD = 10,000$;

and

angle $DAE = 2°39'$.

And since

angle DAK = angle ADF,

by addition

angle $EAK = 47°41'$.

Also in triangle AEK the two sides AK and AE are given. Hence

angle $AEK = 57'$.

Now

angle KED = angle ADF − (angle AEK + angle DAE) = 41°26',

as the angle of apparent movement at the first solar opposition.

[153ª] In triangle BDE a similar thing will be shown. Since the two sides BD and DE are given, and

angle $BDE = 64°42'$;

side $BE = 9,725$

where $BD = 10,000$

and

angle $BDE = 3°40'$.

Furthermore, in triangle BEL the two sides BE and BL are also given, and

angle $EBL = 118°58'$:

angle $BEL = 1°10'$;

301

and hence

$$\text{angle } DEL = 110°28'.$$

But it has already been made clear that

$$\text{angle } AED = 41°26',$$

therefore, by addition,

$$\text{angle } KEL = 151°54'.$$

Hence

$$360° - 151°54' = 208°6',$$

the angle of apparent movement between the first and the second solar oppositions; and that agrees with the observations.

Finally, at the third opposition, in triangle CDE, sides DC and DE are given in the same way; and

$$\text{angle } CDE = 130°52'.$$

On account of angle FDC being given,

$$\text{side } CE = 10,463,$$
$$\text{where } CD = 10,000,$$

and

$$\text{angle } DCE = 2°51'.$$

Therefore, by addition,

$$\text{angle } ECM = 51°59'.$$

Now in triangle ECM the two sides CM and CE are given together with angle MCE:

$$\text{angle } MEC = 1°,$$

and

$$\text{angle } MEC + \text{angle } DCE = \text{angle } FDC - \text{angle } DEM,$$

and angles FDC and DEM are the angles of regular and apparent movement. And hence, at the third solar opposition,

$$\text{angle } DEM = 45°17'.$$

But it has already been shown that

$$\text{angle } DEL = 90°28';$$

accordingly

$$\text{angle } LEM = 65°10',$$

which is the distance between the second and the third solar oppositions observed; and that agrees with the observations. But since the third position of Jupiter was viewed at 113°44' of the sphere of the fixed stars, it shows that the position of the highest Jovial apsis is at approximately 159°.

But if around centre E we now describe RST the orbital circle of the Earth, of which the diameter RES is parallel to DC, then it will be manifest that at the third opposition of Jupiter

angle FDX = angle DES = 49°8',

and that the apogee of the regular movement in parallax is at R.

But now that the Earth has passed through 180° plus arc ST, it is in conjunction with Jupiter at its solar opposition; and

arc ST = 3°51',

according as angle SET has been shown to be of the same magnitude.

And so it is clear from this that in the year of Our Lord, 1529, on the Kalends of February 19 hours after midnight, [153[b]] the regular movement of anomaly of parallax of Jupiter was at 183°51', but by its proper movement Jupiter was at 109°52'; and the apogee of the eccentric circle is approximately 159° from the horn of the constellation of the Ram, as was to be investigated.

12. CONFIRMATION OF THE REGULAR MOVEMENT OF JUPITER

But it has already been seen above that at the last of the three solar oppositions observed by Ptolemy the planet Jupiter by its proper movement was at 4°58' with an anomaly of parallax of 182°47'. Hence it is clear that during the time between the two observations the movement of parallax of Jupiter was 1°5' besides the full revolutions and its proper movement was approximately 104°54'. The time, however, which flowed between the 1st year of Antoninus on the 20th day of the month Athyr by the Egyptian calendar at 5 hours after the following midnight and the year of Our Lord 1529 on the Kalends of February 18 hours after the preceding midnight was 1392 Egyptian years 99 days 37 minutes of a day, to which time there similarly corresponds according to the above calculation 1°5' besides the whole revolutions, by which the regular revolutions of the Earth has anticipated Jupiter 1267 times; and so the number is seen to harmonize with the observations and is held as certain and exact.

And it is also manifest that during this time the highest and lowest apsides of the eccentric circle moved $4^1/_2$° to the east. The equal distribution (of the movement) yields approximately 1° per 300 years.

13. POSITIONS TO BE ASSIGNED TO THE MOVEMENT OF JUPITER

But the time from the last of the three observations, in the 1st year of Antoninus on the 20th day of the month Athyr at 4 hours after the following midnight, going back to the beginning of the years of Our Lord, amounts to 136 Egyptian years 314 days 10 minutes (of a day), during which time the mean movement of [154[a]] parallax was 84°31'. The subtraction of 84°31' from 182°47' leaves 98°16' for the movement up to midnight on the Kalends of January at the beginning of the years of Our Lord.

Backward to the first Olympiad there were 775 Egyptian years $12^1/_2$ days, during which time a movement of 70°58' was reckoned besides the whole revolutions. The subtraction of 70°58' from 98°16' leaves 27°18' as the position for the Olympiad.

Coming down from there for 451 years 247 days, there are 110°52', which together with the movement for the first Olympiad amount to 138°10' for the position of the years of Alexander at noon of the 1st day of the month Thoth by the Egyptian calendar. And so for any others.

14. ON INVESTIGATING THE PARALLAXES OF JUPITER AND ITS ALTITUDE IN RELATION TO THE ORBITAL CIRCLE OF TERRESTRIAL REVOLUTION

In order to investigate the remaining apparent movements of parallax in the case of Jupiter, we carefully observed its position in the year of Our Lord 1520 on the 12th day before the Kalends of March 6 hours before noon, and we perceived through the instrument that Jupiter was 4°31' to the west of the first bright star in the forehead of Scorpio; and since the position of the fixed star was at 209°40', it is clear that the position of Jupiter was at 205°9' in the sphere of the fixed stars.

Accordingly from the beginning of the years of Christ to the time of this observation there were 1520 equal years 62 days 15 minutes (of a day), during which time the mean movement of the sun is calculated to have been 309°16', and the anomaly of parallax 111°15', whereby the mean position of the planet Jupiter is put at 198°1'. And since at this our time the position of the highest apsis of the eccentric circle was found to be at 159°, the anomaly of the eccentric circle of Jupiter was 39°1'.

Following this example, let eccentric circle *ABC* be described with centre *D* and diameter *ADC*. Let *A* be the apogee, and *C* the perigee; and for that reason let *E* the centre of the annual orbital circle of the Earth be on *DC*. Now let

arc *AB* = 39°1';

and with *B* as centre let the epicycle be described with *BF* as radius equal to one third of distance *DE*. Let

angle *DBF* = angle [154b] *ADB*;

and let the straight lines *BD*, *BE*, and *FE* be joined. Accordingly since in triangle *BDE* two sides are given:

DE = 687,

where *BD* = 10,000;

and since these two sides comprehend the given angle *BDE*, and

angle *BDE* = 140°59' :

it will be shown that

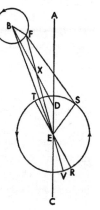

$$\text{base } BE = 10,543$$

and

$$\text{angle } DBE = \text{angle } ADB - \text{angle } BED = 2°21',$$

which is the difference between angle BED and angle ADB. Therefore, by addition,

$$\text{angle } EBF = 41°22'.$$

Accordingly in triangle EBF angle EBF is given together with the two sides comprehending it:

$$EB = 10,543$$

and

$$BF = {}^1/_3 \, DE = 229,$$
$$\text{where } BD = 10,000.$$

It follows from this that

$$\text{side } FE = 10,373$$

and

$$\text{angle } BEF = 50'.$$

Now as lines BD and FE out one another in point X,

$$\text{angle } DXE = \text{angle } BDA - \text{angle } FED$$

and angles FED and BDA are the angles of mean and true movement; and

$$\text{angle } DXE = \text{angle } DBE + \text{angle } BEF = 3°11'.$$

Now

$$\text{angle } FED = 39°1' - 3°11' = 35°50'$$

from the highest apsis of the eccentric circle to the planet.
But the position of the highest apsis was at 159°; and

$$159° + 35°50' = 194°50',$$

which was the true position of Jupiter with respect to centre E, but the apparent position was at 205°9'. Accordingly, the difference of 10°19' is due to the parallax.

Now let RST the orbital circle of the Earth be described around centre E; and let its diameter RET be parallel to DB, so that R is the apogee of parallax. Moreover, in accordance with the measure of the mean anomaly of parallax, let

$$\text{arc } RS = 111°15';$$

and let FEV be extended in a straight line to both arcs of the orbital circle of the Earth. The true apogee of the planet will be at V; angle REV is equal to the difference between regular and apparent movement; and

$$\text{angle } REV = \text{angle } DXE.$$

Hence, by addition,

$$\text{arc } VRS = 114°26',$$

and by subtraction

angle $FES = 65°34'$.

[155ª] But since

angle $EFS = 10°19'$;

angle $FSE = 104°7'$.

Hence, as in triangle EFS the angles are given, the ratio of the sides will be given too:

$$FE : ES = 9,698 : 1,791.$$

Accordingly

$$FE = 10,373$$

and

$$ES = 1,916,$$

where $BD = 10,000$.

For Ptolemy however

$$ES = 11^p30',$$

where radius of eccentric circle = 60^p;

and that is approximately the same ratio as

$$1,916 : 10,000;$$

and therein we do not seem to differ from Ptolemy at all.

Accordingly

$$\text{dmtr. } ADC : \text{dmtr. } RET = 5^p13' : 1^p.$$

Similarly

$$AD : ES = AD : RE = 5^p13'9" : 1^p$$

thus

$$DE = 21'9"$$

and

$$BF = 7'10".$$

Accordingly, when Jupiter is at apogee,

$$(ADF - BF) : \text{radius of orbital circle of Earth} = 5^p27'29" : 1^p.$$

And when Jupiter is at perigee;

$$(EC + BF) : \text{radius of orbital circle of Earth} = 4^p58'49" : 1^p;$$

and when in the mean positions, as is proportional. Hence it is gathered that Jupiter at apogee has a greatest parallax of $10°35'$, and at perigee $11°35'$: there is a difference of $1°$ between them. So the regular movements of Jupiter have been demonstrated to be at one with the apparent.

15. ON THE PLANET MARS

We must now inspect the revolutions of Mars by taking three ancient solar oppositions, with which we shall connect the mobility of the Earth in antiquity. Accordingly of those oppositions which Ptolemy recorded, the first was in the 15th year of Hadrian

on the 26th day of Tybi the 5th month by the Egyptian calendar 1 equatorial hour after the midnight following. And he says that it was at 21° of Gemini, but in relation to the sphere of the fixed stars was at 84°20'.

He noted the second opposition [155ᵇ] as occurring in the 19th year of Hadrian on the 6th day of Pharmuthi the 8th month by the Egyptian calendar 3 hours before the following midnight at 28°50' of Leo but at 142°10' in the sphere of the fixed stars.

The third was in the 2nd year of Antoninus on the 12th day of Epiphi the 11th month by the Egyptian calendar 2 equatorial hours before the following midnight at 2°34' of Sagittarius but at 235°54' of the sphere of the fixed stars.

Accordingly, between the first and second oppositions there are 4 Egyptian years 69 days 20 hours or 50 minutes of a day, and the apparent movement of the planet was 67°50' besides the whole revolutions. From the second opposition to the third there were 4 years 96 days and 1 hour, and the apparent movement of the star was 93°44'. Now during the first interval the mean movement was 81°44' besides the complete revolutions; during the second interval it was 95°28'. Then he found that the total distance between the centres was 12ᴾ, whereof the radius of the eccentric circle was 60ᴾ; but it was 2,000 whereof the radius was 10,000. And the mean movement from the first opposition to the highest apsis was 41°33'; and then it was 40°11' from the highest apsis to the second opposition; and from the third opposition to the lowest apsis it was 44°21': But by our hypothesis of regular movements there will be three-quarters of that distance, *i.e.*, 1,500 between the centres of the eccentric circle and the orbital circle of the Earth, and the remaining quarter of 500 will be the radius of the epicycle.

Now thus let eccentric circle *ABC* be described with centre *D*, and with *FDG* as the diameter through both apsides; and let *E* the centre of the orbital circle of annual

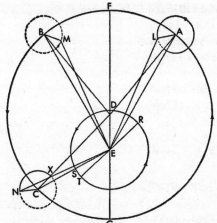

revolution be on the diameter. Let *A*, *B*, and *C* be the points of solar opposition, in that order; and let

arc *AF* = 41°34',
arc *FB* = 40°11',

and

arc *CG* = 44°21'.

At the separate points *A*, *B*, and *C* let the epicycle be described with one-third of distance *DE* as radius. And let *AD*, *BD*, *CD*, *AE*, *BE*, and *CE* be joined. On the epicycle let *AL*, *BM*, and *CN* be joined, but in such a way that

angle DAL = angle ADF,

angle DBM = angle BDF,

and

angle DCN = angle CDF.

Accordingly, since in triangle ADE

angle ADE = 138°26',

because angle FDA was given and also the two sides, *viz.*,

DE = 1,500

where AD = 10,000;

it follows from this that

side AE = 11,172,

and

angle DAE = 5°7'.

[156ª] Hence, by addition,

angle EAL = 46°41'.

So also in triangle EAL angle EAL is given, together with the two sides:

AE = 11,172

and

AL = 500,

where AD = 10,000.

Moreover,

angle AEL = 1°56';

and

angle AEL + angle DAE = 7°3',

which is the total difference between angles ADF and LED; and hence

angle DEL = 34$^1/_2$°.

Similarly at the second opposition: in triangle BDE

angle BDE = 139°49'

and

side DE = 1,500

where BD = 10,000.

Hence

side BE = 11,188,

angle BED = 35°13',

and

angle DBE = 4°58'.

Therefore

angle EBM = 45°13'

and is comprehended by the given sides *BE* and *BM*, from which it follows that

$$\text{angle } BEM = 1°53',$$

and by subtraction

$$\text{angle } DEM = 33°20'.$$

Accordingly

$$\text{angle } LEM = 47°50',$$

whereby the movement of the planet from the first solar opposition to the second is apparent; and the number is consonant with experience.

Again at the third solar opposition: triangle *CDE* has the two sides *CD* and *DE* given, which comprehend angle *CDE*. And

$$\text{angle } CDE = 44°21';$$

hence

$$\text{base } CE = 8,988,$$
$$\text{where } CD = 10,000$$
$$\text{and } DE = 1,500;$$

and

$$\text{angle } CED = 135°39',$$

and

$$\text{angle } DCE = 6°42'.$$

This again in triangle *CEN*

$$\text{angle } ECN = 142°21'$$

and is comprehended by the known sides *EC* and *CN* : hence too

$$\text{angle } CEN = 1°52'.$$

[156^b] Therefore by subtraction

$$\text{angle } NED = 127°5'$$

at the third solar opposition. But it has already been shown that

$$\text{angle } DEM = 33°20'.$$

Hence by subtraction

$$\text{angle } MEN = 93°45',$$

and is the angle of apparent movement between the second and the third solar oppositions, wherein the calculation agrees sufficiently with the observations. But at this last observed opposition of Mars the planet was seen at 235°54', being 127°5' distant from the apogee of the eccentric circle, as was shown: therefore the position of the apogee of the eccentric circle of Mars was at 108°50' in the sphere of the fixed stars.

Now let *RST* the annual orbital circle of the Earth be described around centre *E* with diameter *RET* parallel to *DC*, so that *R* is the apogee of parallax and *T* the perigee. Accordingly the planet was seen on *EX* at 235°54', in longitude, and it was shown that

$$\text{angle } DXE = 8°34',$$

the difference between the regular and the apparent movement; and on that account

$$\text{mean movement} = 244^1/_2°;$$

but, at the centre,

$$\text{angle } SET = \text{angle } DXE = 8°34'.$$

Accordingly

$$\text{arc } RS = \text{arc } RT - \text{arc } ST = 180° - 8°34' = 171°26',$$

the mean movement of parallax of the planet. Furthermore among other things we have demonstrated by this hypothesis of the mobility of the Earth that in the 2nd year of Antoninus on the 12th day of the month Epiphi by the Egyptian calendar 10 equal hours after midday the planet Mars by its mean movement in longitude was at $244^1/_2°$, and the anomaly of parallax was at $171°26'$.

16. ON THREE OTHER SOLAR OPPOSITIONS OF MARS WHICH HAVE BEEN OBSERVED RECENTLY

We have compared these three of Ptolemy's observations of Mars with three other observations, which we did not take carelessly. The first was in the year of Our Lord 1512 on the Nones of June, 1 hour after midnight, and the position of Mars was found to be at $235°33'$, according as the sun was opposite [157a] at $55°33'$ from the first star of Aries in the sphere of the fixed stars as a starting point.

The second was in the year of Our Lord 1518 on the day before the Ides of December 8 hours after midday; and the planet was apparent at $63°2'$.

The third was in the year of Our Lord 1523 on the 8th day before the Kalends of March 8 hours before Boon, at $183°20'$.

Accordingly from the first to the second opposition there were 6 Egyptian years 191 days 45 minutes (of a day); from the second to the third 4 years 72 days 23 minutes.

During the first interval of time the apparent movement was $187°29'$, and the regular movement was $168°7'$. During the second interval of time the apparent movement was $80°18'$, and the regular was $83°$.

Now let the eccentric circle of Mars be repeated again, except that here

$$\text{arc } AB = 168°7'$$

and

$$\text{arc } BC = 83°.$$

Accordingly, by the same method which we employed in the case of Saturn and Jupiter—let us pass over in silence the multitude, complication, and boredom of the calculations—we finally find that the apogee of Mars is on arc BC. For it is manifest that the apogee cannot be in arc AB because there the apparent movement is $19°22'$ greater than the mean. Again the apogee cannot be in arc CA, because even if BC the arc preceding CA is the lesser, nevertheless arc BC exceeds the apparent movement by

a greater difference than arc *CA* does. But it was shown above that in the eccentric circle the lesser and decreased movement takes place around the apogee. Accordingly the apogee will be held correctly to be in arc *BC*.

Let the apogee be *F*; and let *FDG* be the diameter of the circle. And let the centre of the orbital circle of the Earth be on the diameter. Accordingly, we find that

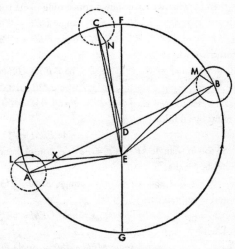

arc *FCA* = 125°29',

arc *BF* = 66°18',

and

arc *FC* = 16°36';

but

DE = 1,460,

where radius *DE* = 10,000 which is the distance between the centres; and semi-diameter of epicycle = 500 : whence the apparent and regular movements are shown to be consonant with one another and to agree with experiments.

Therefore let the figure be filled out, as before. For it will be shown that, since in triangle *ADE* two sides *AD* and *DE* are known and

angle *ADE* = 54°31'

from the first opposition of Mars to the perigee;

angle *DAE* = 7°24',

and by subtraction

angle *AED* = 118°5';

and

side *AE* = 9,229.

Now by hypothesis

angle *DAL* = angle *FDA*.

Accordingly by addition

angle *EAL* = 132°53'.

So too in triangle *EAL* the two sides *EA* and *AL* comprehending the given angle at *A* are themselves given; [157ᵇ] accordingly

angle *AEL* = 2°12',

and

angle *LED* = 115°53'.

311

Similarly it will be shown in the case of the second opposition that, since in triangle BDE the two given sides DB and BE comprehend angle BDE, and

$$\text{angle } BDE = 113°35',$$

then by what we have shown concerning plane triangles

$$\text{angle } DBE = 7°11',$$
$$\text{angle } DEB = 59°13',$$

and

$$\text{base } BE = 10,668,$$
$$\text{where } DB = 10,000$$
$$\text{and } BM = 500.$$

And by addition

$$\text{angle } EBM = 73°36'.$$

So too in triangle EBM, since the sides comprehending the given angle are given, it will be shown that

$$\text{angle } BEM = 2°36';$$

and by subtraction

$$\text{angle } DEM = 56°38'.$$

Then

$$\text{angle } MEG = 180° - \text{angle } DEM = 123°22'.$$

But it has already been shown that

$$\text{angle } LED = 115°53';$$

hence

$$\text{angle } LEG = 64°7';$$

and

$$\text{angle } LEG + \text{angle } GEM = 187°29',$$
$$\text{where 4 rt. angles} = 360°.$$

And that agrees with the apparent distance from the first opposition to the second.

A similar thing can be seen at the third opposition. For it has been shown that

$$\text{angle } DCE = 2°6',$$

and

$$\text{side } EC = 11,407,$$
$$\text{where } CD = 10,000.$$

Accordingly, as

$$\text{angle } ECN = 18°42',$$

and as sides CE and CN of triangle ECN have already been given, it will be clear [158ᵃ] that

$$\text{angle } CEN = 50'.$$

And

angle CEN + angle DCE = 2°56',

which is the difference by which angle DEN of apparent movement is exceeded by angle FDC of regular movement. Therefore

angle DEN = 13°40',

which agrees approximately with the apparent movement observed between the second and the third oppositions.

Accordingly, since the planet Mars, as we told you, was apparent in this position at 133°20' from the head of the constellation of Aries; and it has been shown that

angle $FEN \fallingdotseq$ 13°40';

it is manifest upon calculation backward that the position of the apogee of the eccentric circle at this last observation was at 119°40' in the sphere of the fixed stars.

At the time of Antoninus, Ptolemy found it at 108°50', and so during the time between then and now it has moved $10^{10}/_{12}$° eastward. Moreover we have found a lesser distance between the centres, *i.e.*, 40, whereof the radius of the eccentric circle is given as 10,000—not because either Ptolemy or ourselves made a slip, but manifestly because the centre of this orbital circle of the Earth has approached the centre of the orbital circle of Mars, while the sun has remained immobile. For these things correspond approximately to one another, as will be shown below clearer than day.

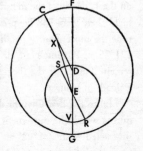

Now let the annual orbital circle of the Earth be described around the centre E; and let its diameter SER be parallel to CD on account of the equality of revolutions. Let R be the regular apogee with respect to the planet, S the perigee, and T the Earth. Now let ET be extended; the line of sight of the planet will thus cut CD at point X. Now the line of sight along ETX, as was said at the last opposition, is at 133°20' of longitude.

Moreover, it has been shown that

angle DXE = 2°56',

for angle DXE is the difference by which angle XDF of mean movement exceeds angle XED of apparent movement. But angle SET is equal to its alternate angle DXE, and is the additosubtraction arising from the parallax. Now

$$180° - 2°56' = 177°4',$$

which is the regular movement of the anomaly of parallax from R the apogee of the regular movement—and hence we have shown here that in the year of Our Lord 1523 on the 8th day before the Kalends of March, 7 equatorial hours before noon, the planet Mars by its mean movement in longitude was at 136°16'; and its regular anomaly of parallax was at 177°4', and the highest apsis of the eccentric circle was at 119°40', as was to be shown.

17. Confirmation of the Movement of Mars

[158b] Now it was made clear above that in the last of Ptolemy's three observations Mars by its mean movement was at 244^1/$_2$°, and its anomaly of parallax was at 171°26'. Accordingly during the year between there was a movement of 5°38' besides the complete revolutions. Now for the 2nd year of Antoninus on the 12th day of Epiphi the 11 month by the Egyptian calendar 9 hours after midday, *i.e.*, 3 equatorial hours before the following midnight, with respect to the Cracow meridian, to the year of Our Lord 1523 on the 8th day before the Kalends of March 7 hours before noon, there were 1384 Egyptian years 251 days 19 minutes (of a day). During that time there were by the above calculation 5°38' and 648 complete revolutions of anomaly of parallax. Now the regular movement of the sun was held to be 257^1/$_2$°. The subtraction from 257^1/$_2$° of the 5°38' of the movement of parallax leaves 251°52' as the mean movement of Mars in longitude. And all that agrees approximately with what was set down just now.

18. Determination of the Position of Mars

Now from the beginning of the years of Our Lord to the 2nd year of Antoninus on the 12th day of the month Epiphi by the Egyptian calendar at 3 hours before midnight there were 138 Egyptian years 180 days 52 minutes (of a day), and during that time the movement of parallax was 293°4'. And when 293°4' is subtracted from the 171°26' of Ptolemy's last observation—a complete revolution being borrowed—there remain 238°22' at the (beginning of) the first year of Our Lord on midnight of the Kalends of January.

From the first Olympiad to this time there were 775 Egyptian years 12^1/$_2$ days, during which the movement of parallax was 254°1'. When 254°1' has similarly been subtracted from 238°22' and a revolution borrowed, its [159a] position at the first Olympiad remains as 344°21'.

Similarly by calculating the movements according to the other intervals of time we shall have its position at the beginning of the years of Alexander as 120°39' and at the beginning of the years of Caesar as 211°25'.

19. How Great the Orbital Circle of Mars Is in Terms of the Parts of Which the Annual Orbital Circle of the Earth Is the Unit

Moreover we took observations of the conjunction of Mars with the first bright star of the Chelae—called the southern Claw—which occurred in 1512 on the Kalends of January, For on the morning of that day 6 equatorial hours before noon we saw Mars

$1/_4°$ distant from the fixed star but deflected towards the solstitial rising, by which it was signified that Mars was already in longitude $1/_8°$ to the east of the star but $1/_5°$ distant in northern latitude. Now it was established that the position of the star was 191°20' with a northern latitude of 40'. So it was clear that the position of Mars was at 191°28' with a northern latitude of 51'. But at this time the anomaly of parallax by calculation was 98°28'; the mean position of the sun was at 262°, and the mean position of Mars at 163°32', and the anomaly of the eccentric circle was 43°52'.

With that before us let the eccentric circle *ABC* be described. Let *D* be its centre, *ADC* its diameter, *A* the apogee, *C* the perigee, and let

$$ecc. \ DE = 1,460,$$
$$where \ AD = 10,000.$$

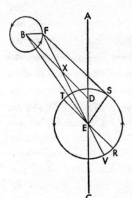

Now let

$$arc \ AB = 43°52'.$$

Now with *B* as centre and radius *BF* of 500, whereof *AD* is 10,000, let the epicycle be described. Let

$$angle \ DBF = angle \ ADB;$$

and let *BD*, *BE*, *BF*, and *FE* be joined. Moreover, let *RST* the great orbital circle of the Earth be described around centre *E* with its diameter *RET* parallel to *BD*; and on the diameter let *R* be the apogee of the planet's regular movement of parallax, and *T* the perigee. Now let the Earth be at *S*; and in accordance with the regular anomaly of parallax as computed, let

$$arc \ RS = 98°28'.$$

Let *FE* be extended in the straight line *FEV*, which will cut *BD* at point *X* and the convex arc of the orbital circle of the Earth at *V*, where the true apogee of parallax is.

Accordingly, in triangle *BDE* [159^b] two sides are given:

$$DE = 1,460,$$
$$where, \ BD = 10,000;$$

and they comprehend angle *BDE*. And

$$angle \ ADB = 43°52';$$

now

$$angle \ BDE = 180° − 43°52' = 136°8'.$$

Hence it will be shown that

$$base \ BE = 11,097$$

and

$$angle \ DBE = 5°13'.$$

But by hypothesis

$$\text{angle } DBF = \text{angle } ABD;$$

by addition

$$\text{angle } EBF = 49°5',$$

and is comprehended by the given sides EB and BF. On that account

$$\text{angle } BEF = 2°$$

and

$$\text{side } FE = 10{,}776,$$
$$\text{where } DB = 10{,}000.$$

Accordingly

$$\text{angle } DXE = 7°13',$$

because

$$\text{angle } DXE = \text{angle } XBE + \text{angle } XEB,$$

the interior and opposite angles. Angle DXE is the subtractive additosubtraction, the difference by which angle ADB exceeds angle XED, and by which the mean position of Mars exceeds the true. Now the mean position is reckoned as 163°32'; therefore the true position is to the west at 156°19'. But its position appears to be at 191°28' to those viewing it from S. Therefore its parallax, or commutation, is 35°9' eastward. Therefore it is clear that

$$\text{angle } EFS = 35°9'.$$

Now as RT is parallel to BD,

$$\text{angle } DXE = \text{angle } REV,$$

and similarly

$$\text{arc } RV = 7°13'.$$

Thus, by addition,

$$\text{arc } VRS = 105°41',$$

which is the corrected anomaly of parallax, and hence angle VES exterior to triangle FES is given. Hence, as being interior and opposite

$$\text{angle } FSE = 70°32',$$
$$\text{where 2 rt. angles} = 180°.$$

But as the angles of the triangle are given, the ratio of the sides is given too: therefore

$$FE = 9{,}428$$

and

$$ES = 5{,}727$$

where diameter of circle circumscribing triangle = 10,000.

Accordingly

$$[160^a]\ ES \fallingdotseq 6{,}580,$$
$$\text{where } EF = 10{,}776$$
$$\text{and } BD = 10{,}000;$$

and that is approximately the same as Ptolemy's findings. But by addition

$$ADE = 11,460,$$

and by subtraction

$$EC = 8,540.$$

And at the lowest apsis of the eccentric circle the epicycle adds the 500 which it subtracts at A the highest apsis, so that the remainder at the highest apsis is 10,960, and the sum at the lowest apsis is 9,040. Accordingly, in so far as the radius of the orbital circle of the Earth is 1P, Mars will have a greatest distance of 1P39'57" at its apogee, a least distance of 1P22'26", and a mean distance of 1P39'11". So too in the case of Mars the movements, magnitudes, and distances have been explicated in a fixed ratio by means of the movement of the Earth.

20. ON THE PLANET VENUS

Now that we have set out the movements of the three higher planets Saturn, Jupiter, and Mars which circle around the Earth, it is time to speak of the planets which the Earth circles around. And first of Venus, which admits an easier and clearer demonstration of its movement than does Mercury, if only the necessary observations of some positions are not wanting; since if its greatest distances, *i.e.*, at morning and at evening, in either direction from the mean position of the sun are found equal to one another, then we have as certain that the highest or lowest apsis of the eccentric circle of Venus is at the midpoint between these two positions of the sun. The apsides are distinguished from one another by the fact that such equal (angular) elongations are smaller when they take place around the apogee and greater when they take place around the perigee. Finally at its other positions we perceive through the differences by which the angular elongations exceed one another how far distant the orb of Venus is from the highest or lowest apsis and also what its eccentricity is, according as these things have been passed on to us by Ptolemy with great clarity, so that there is no need to repeat them separately, except in so far as things from Ptolemy's observations are applicable to our hypothesis of terrestrial mobility.

He took as his first observation one made by the mathematician Theo of Alexandria in the 16th year of Hadrian, he tells us, on the 21st day of the month Pharmuthi, at the first hour of the following night; and that was in the year of Our Lord 132 on the evening of the 8th day before the Ides of March. And Venus was seen at its greatest evening distance of $47^1/_4°$ from the mean position of the sun, [160b] while the mean position of the sun was by calculation at 337°41' in the sphere of the fixed stars. With this observation he compared one of his own which he said he made in the 4th year of Antoninus on the 12th day of the month Thoth at daybreak, *i.e.*, in the year of Our Lord 142 on the early morning of the third day before the Kalends of August.

317

He says that at this time the greatest morning elongation of Venus was equal to the previous elongation and was 47°15' from the mean position of the sun, which was at 119° in the sphere of the fixed stars and which on the previous date had been at 337°41'. Now it is manifest that midway between these mean positions are the apsides diametrically opposite one another at 48$^1/_3$° and 228$^1/_3$°. When the 6$^2/_3$° of the precession of the equinoxes has been added to both of them, they will fall upon 25° of Taurus and of Scorpio according to Ptolemy, and the diametrically opposite highest and lowest apsides of Venus must be at those positions.

Once more for the further confirmation of the thing, he assumed another observation made by Theo in the 4th year of Hadrian at morning twilight on the 20th day of the month Athyr, which was in the year of Our Lord 119 on the morning of the fourth day before the Ides of October, at which time Venus was again found at a great distance of 47°32' from the mean position of the sun at 181°13'. With that he connected his own observation made in the 21st year of Hadrian, which was the year of Our Lord 136, on the 9th day of the month Mechyr by the Egyptian calendar but by the Roman calendar the 8th day before the Kalends of January, at the first hour of the following night, and the evening distance was found to be 47°32' from the mean position of the sun at 265°25'. But in the preceding observation made by Theo the mean position of the sun was at 191°13'. Again the apsides fall midway between these positions, at 48°20' and at 228°20' approximately, where the apogee and the perigee must be. And they are distant from the equinoxes at 25° of Taurus and at 25° of Scorpio, and Ptolemy separated them by two other observations as follows.

The first observation was made by Theo in the 13th year of Hadrian on the 3rd day of the month Epiphi, but in the year of Our Lord 129 at early morning on the 12th day before the Kalends of January, and he found the farthest morning elongation of Venus to be 44°48', while the sun by its mean movement was at 48$^{10}/_{12}$°, and Venus was apparent at 4° of the sphere of the fixed stars. Ptolemy himself made the other observation in the 21st year of Hadrian on the 2nd day of the month [161^a] Tybi by the Egyptian calendar, which was by the Roman calendar the year of Our Lord 136 on the 5th day before the Kalends of January at the 1st hour of the following night, while the sun by its mean movement was at 228°54', from which Venus had a greatest evening elongation of 47°16' and was itself apparent at 276$^1/_6$°. Hence the apsides are distinguished from one another, that is to say, the highest is put at 48$^1/_3$°, where the shorter wanderings of Venus are, and the lowest at 228$^1/_3$°, where the greater wanderings are—as was to be demonstrated.

21. WHAT THE RATIO OF THE DIAMETERS OF THE ORBITAL CIRCLE OF THE EARTH AND OF VENUS IS

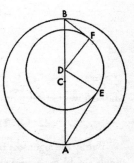

Furthermore from these last two observations the ratio of the diameters of the orbital circles of the Earth and Venus will be apparent. For let *AB* the orbital circle of the Earth be described around centre *C*. Let *ACB* be its diameter through both apsides; and on *ACB* let *D* be taken as the centre of the orbital circle of Venus which is eccentric to circle *AB*. Now let *A* be the position of the apogee; and when the Earth is there, the centre of the orbital circle of Venus is at its greatest distance, while *AB* is the line of mean movement of the sun—$48^1/_3°$ at *A* and $228^1/_3°$ at *B*. Now let the straight lines *AE* and *BF* be drawn touching the orbital circle of Venus at points *E* and *F*, and let *DE* and *DF* be joined. Accordingly, since as at the centre

$$\text{angle } DAE = 44^4/_5°,$$

and

$$\text{angle } AED = 90°,$$

then triangle *DAE* will have its angles given and hence its sides:

$$DE = {}^1/_2 \text{ ch. } 2\,DAE = 7{,}046,$$
$$\text{where } AD = 10{,}000.$$

In the same way in the right triangle *BDF*

$$\text{angle } DBF = 47^1/_3°,$$

and

$$\text{ch. } DF = 7{,}346,$$
$$\text{where } BD = 10{,}000.$$

Accordingly

$$BD = 9{,}582,$$
$$\text{where } DF = DE = 7{,}046.$$

Hence by addition

$$ACB = 19{,}582$$

and

$$AC = {}^1/_2 ACB = 9{,}791;$$

and by subtraction

$$CD = 209.$$

Accordingly, in so far as

$$[161^b]\ AC = 1^P,$$

$$DE = 43\frac{1}{6}',$$

and

$$CD \doteqdot 1\frac{1}{4}';$$

and

$$DE = DF \doteqdot 7,193$$

and

$$CD \doteqdot 213,$$
$$\text{where } AC = 10,000.$$

And that was to be demonstrated.

22. ON THE TWOFOLD MOVEMENT OF VENUS

But by the argument from two of Ptolemy's observations, Venus does not have a simple regular movement around D. He made the first observation in the 18th year of Hadrian on the 2nd day of the month Pharmuthi by the Egyptian calendar, but by the Roman calendar it was the year of Our Lord 134 at early morning on the 12th day before the Kalends of March. For at that time the sun by its mean movement was at $318\frac{10}{12}°$; and Venus, which was apparent in the morning at $275\frac{1}{4}°$ of the ecliptic, had reached a farthest limit of elongation of $43°35'$.

He made the second in the 3rd year of Antoninus on the 4th day of the month Pharmuthi by the Egyptian calendar, which by the Roman calendar was in the year of Our Lord 140 on the evening of the 12th day before the Kalends of March. And at that time the mean position of the sun was at $318\frac{10}{12}°$; and Venus was at a greatest evening elongation of $48\frac{1}{3}°$ and was visible at $7\frac{10}{12}°$ in longitude.

With that set out, let point G be taken as the position of the Earth in the same terrestrial orbital circle, so that arc AG is a quadrant of a circle—the quadrant which measures how far the sun diametrically opposite at both observations according to its mean

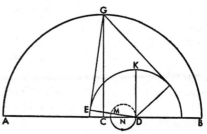

movement was seen to be west of the apogee of the eccentric circle of Venus. Let GC be joined, and let DK be drawn parallel to GC. Let GE and GF be drawn touching the orbital circle of Venus; and let DE, DF, and DG be joined. Accordingly, since,

$$\text{angle } EGC = 43°35',$$

which was the morning elongation at the time of the first observation, and since

$$\text{angle } CGF = 48\frac{1}{3}°,$$

which was the evening elongation at the time of the second observation;

angle EGF = angle EGC + angle CGF = $91^{11}/_{12}°$;

and accordingly

angle DGF = $^1/_2 EGF$ = $45°47^1/_2'$;

and by subtraction

angle CGD = $2°23'$.

But

angle DCG = $90°$.

Accordingly, as the angles of triangle CGD are given, the ratios of the sides are given too; and

$$CD = 416,$$
$$\text{where } CG = 10,000.$$

Now it has already been shown that the distance between the centres was 208; and now the distance has become approximately twice as great. Accordingly, if CD is bisected at point M, similarly

[162ª] DM = 208,

the total variation in this approach and withdrawal. Again if DM is bisected at N, it will be seen to be the mean and regular point in this movement.

Hence, as in the case of the three higher planets, the movement of Venus happens to be compounded of two regular movements, either by reason of the epicycle of an eccentric circle, as above, or by any other of the aforesaid modes. This planet however is somewhat different from the others in the order and commensurability of its movements; and, as I opine, there will be an easier and more convenient demonstration by means of the eccentric circle of an eccentric circle. In this way let us take N as centre and DN as radius and describe a small circle, on which (the centre of) the orbital circle of Venus is borne and moved around according to the law that whenever the Earth falls upon diameter ACB, on which the highest and lowest apsides of the eccentric circle are, the centre of orbital circle of the planet will always be at least distance, *i.e.*, at point M; and when the Earth is at its mean apsis, *i.e.*, at G, the centre of the orbital circle of the planet will reach point D and the greatest distance CD. Hence you are given to understand that at the time when the Earth has made one orbital circuit, the centre of the orbital circle of the planet has made two revolutions around centre N in the same direction as the Earth, *i.e.*, eastward. For according to such an hypothesis in the case of Venus, all the regular and apparent movements agree with the observations, as will be shown later. Now all this which has so far been demonstrated concerning Venus is found to be consonant with our times, except that the eccentricity has decreased approximately one sixth, so that what before was 416 is now 350, as many observations teach us.

23. ON THE EXAMINATION OF THE MOVEMENT OF VENUS

In this connection I have taken two positions observed very accurately, the first by Timochares in the 13th year of Ptolemy Philadelphus the 52nd year after the death of Alexander in the early morning [162^b] of the 18th day of Mesori the eight month by the Egyptian calendar; and it was recorded that Venus was seen to have occupied the position of the fixed star which is westernmost of the four stars in the left wing of Virgo and is sixth in the description of the sign; its longitude 151^1/$_2$°, its northern latitude 1^1/$_6$°; and it is of third magnitude. Accordingly the position of Venus was made manifest in this way; and the mean position of the sun was by calculation at 194°23'.

With this as an example, let the figure be drawn with point A still at 48°20':

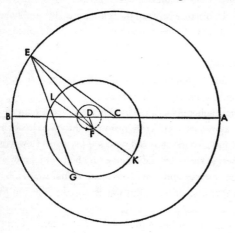

arc AE = 146°3',

and by subtraction

arc BE = 33°57'.

Angle CEG = 42°53',

which is the angular distance of the planet from the mean position of the sun. Accordingly, since

line CD = 312,

where CE = 10,000,

and

angle BCE = 33°57';

in triangle CDE

angle CED = 1°1',

and

base DE = 9,743.

But

angle CDF = 2BCE = 67°54';

and

angle BDF = 180° − 67°54' = 112°6'.

And, as the exterior angle of triangle CDE,

angle BDE = 33°57'.

Hence it is clear that

angle EDF = 144°4',

DF = 104,

where DE = 9,743.

So in triangle DEF

angle DEF = 20',

and by addition

angle CEF = 1°21',

and

side EF = 9,831.

But it has already been shown that

angle CEG = 42°53';

accordingly, by subtraction,

angle FEG = 41°32';

and, as radius of the orbital circle,

FG = 7,193,

where EF = 9,831

Accordingly, since in triangle EFG angle FEG and the ratios of the sides are given, the remaining angles are given too. And

[163ª] angle EFG = 72°5'.

Arc KLG = 180° + angle EFG = 252°5',

measured from the highest apsis of the orbital circle. And so we have shown that in the 13th year of Ptolemy Philadelphus on the 18th day of the month Mesori the anomaly of parallax of Venus was 252°5'.

We ourselves made observations of a second position of Venus in the year of Our Lord 1529 on the 4th day before the Ides of March, 1 hour after sunset and at the beginning of the 8th hour after midday. We saw the moon begin to occult Venus at the midpoint of the dark part between the horns, and the occultation lasted till the end of the hour or a little later, until the planet was seen to emerge towards the west on the other side at the midpoint of the gibbosity of the horns. Accordingly it is clear that at the middle of the hour or thereabouts, the centres of the moon and Venus were in conjunction, and we had a full view at Frauenburg. Venus was still in her evening increase (of elongation) and this side of the point of tangency of the orbital circle with a line from the Earth. Accordingly from the birth of Christ there have been 1529 Egyptian years 87 days $7^{1}/_{2}$ hours by apparent time but by equal time 7 hours 34 minutes, and the mean position of the sun considered simply had reached 332°11'; and the precession of the equinoxes was 27°24'. The regular movement of the moon was 33°57' away from the sun, the regular movement of anomaly was 205°1', and the movement in latitude was 71°59'. Hence it is reckoned that the true position of the moon was at 10°, but measured from the equinox it was at 7°24' of Taurus with a northern latitude of 1°13'. But since the 15° of Libra were rising, on that account the lunar parallax in longitude was 48°32', and so the apparent position (of the moon) was at 6°26' of Taurus; but its longitude in the sphere of the fixed stars was 9°11' with a northern latitude of 41'; and the apparent position of Venus was at an evening distance of 37°1' from the

mean position of the sun, and the distance of the Earth from the highest apsis of Venus was 76°9' to the west.

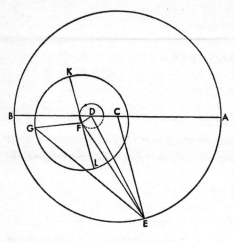

Now let the figure be drawn again according to the previous mode of construction, except that

angle *ECA* = 76°9',

and

angle *CDF* = 2*ECA* = 152°18';

and

ecc. *CD* = 246,

as it is found today; and

DF = 104,

where *CE* = 10,000.

Therefore in triangle *CDE*, by subtraction

angle *DCE* = 103°51'

and is comprehended by the given sides. From that it will be shown that

angle *CED* = 1°15',

and

base *DE* = 10,056,

and

angle [163ᵇ] *CDE* = 74°54'.

But

angle *CDF* = 2*ACE* = 152°18'.

And

angle *EDF* = angle *CDF* - angle *CDE* = 77°24'.

So again in triangle *DEF* two sides are given;

DF = 104,

where *DE* = 10,056;

and they comprehend the given angle *EDF*. Moreover,

angle *DEF* = 35',

and

base *EF* = 10,034.

Hence by addition

angle *CEF* = 1°50'.

Furthermore,

angle *CEG* = 37°1',

which measures the apparent distance of the planet from the mean position of the sun.

Now

$$\text{angle } FEG = \text{angle } CEG - \text{angle } CEF = 35°11'.$$

Similarly in triangle *EFG* two sides are given:

$$EF = 10,034,$$
$$\text{where } FG = 7,193$$

and the angle at *E* is given: hence too the remaining angles are calculable;

$$\text{angle } EGF = 53^1/_2°$$

and

$$\text{angle } EFG = 91°19',$$

which is the distance of the planet from the true perigee of its orbital circle.

But since diameter *KFL* has been drawn parallel to *CE*, so that *K* is the apogee of the regular movement and *L* is the perigee, and since

$$\text{angle } EFL = \text{angle } CEF;$$

then

$$\text{angle } LFG = \text{angle } EFG - \text{angle } EFL = 89°29';$$

and

$$\text{arc } KG = 180° - 89°29' = 90°31',$$

which is the planet's anomaly of parallax as measured from the highest apsis of regular movement of the orbital circle—and that is what we were investigating at the time of this observation of ours.

But at the time of the observation made by Timochares the anomaly was 252°5'; accordingly during the years between there was a movement of 198°26' besides the 1115 complete revolutions. Now the time from the 13th year of Ptolemy [164ᵃ] Philadelphus at early morning on the 18th day of the month Mesori to the year of Our Lord 1529 on the 4th day before the Ides of March 7¹/₂ hours after midday was 1800 Egyptian years 236 days 40 minutes (of a day) approximately.

Accordingly when we have multiplied the movement of 1115 revolutions and 198°26' by 365 days, and divided the product by 1800 years 226 days 40 minutes, we shall have an annual movement of 225°1'45"3'"40"".

Once more the distribution of this through 365 days leaves a daily movement of 36'59"28'", which were added to the table which we set out above.

24. On the Positions of the Anomaly of Venus

Now from the first Olympiad to the 13th year of Ptolemy Philadelphus at early morning of the 18th day of the month Mesori there are 503 Egyptian years 228 days 40 minutes (of a day), during which time the movement was reckoned to be 290°39'. But if 290°39' is subtracted from 252°5' and 360° is borrowed, the remainder will be 321°26', the position of the movement at the beginning of the first Olympiad.

The remaining positions are in proportion to the movement and time so often spoken of: 81°52' at the beginning of the years of Alexander; 70°26' at the beginning of the years of Caesar; and 126°45' at the beginning of the years of Our Lord.

25. ON MERCURY

It has been shown how Venus is bound up with the movement of the Earth and in what ratio of circles the regularity of its movement is concealed. Mercury remains, and without fail will also submit to the principle assumed, although it has more complicated wanderings than Venus or any of the aforesaid planets. It has been established experimentally by ancient observations that in the sign of Libra, Mercury has its least angular elongations from the sun and has *greater* elongations in the opposite sign, as is right. But Mercury does not have its *greatest* elongations in this position but in some other positions higher and beyond, as in Gemini and Aquarius, particularly at the time of Antoninus according to Ptolemy; and that occurs in the case of no other planet. When the ancient mathematicians, who supposed the reason for this [164b] to be the immobility of the Earth and the movement of Mercury in its great epicycle along an eccentric circle, had noticed that one simple eccentric circle could not account satisfactorily for these appearances, not only did they grant that the movement on the eccentric circle was not around its own centre but around a foreign centre, but they were also compelled to admit that this same eccentric circle carrying the epicycle moved along another small circle, as they admitted the moon's eccentric circle did. And so there were three centres, namely that of the eccentric circle carrying the epicycle, that of the small circle, and that of the circle which the moderns call the equant. They passed over the first two circles and acknowledged that the epicycle did not move regularly except around the centre of the equant, which was the most foreign to the true centre, to its ratio, and to both the centres already extant. But they judged that the appearances of this planet could be saved by no other scheme, as Ptolemy declares at great length in his *Composition*.

But in order that this last planet may be freed from the liability to injury and disparagement and that the regularity of its movement in relation to the mobility of the Earth may be no less clear than in the case of the other preceding planets; we shall assign to it too a circle eccentric to an eccentric circle instead of the epicycle which the ancients assumed, but in a way different from that of Venus. And nevertheless an epicycle does move on the eccentric circle, but the planet is not borne on its circumference but up and down along its diameter: that can take place through regular circular movements, as was set forth above in connection with the precession of the equinoxes. And it is not surprising, since Proclus in his commentary on the *Elements* of Euclid admits that a straight line can be described by many movements, by all of which movements its appearance will be demonstrable.

But in order that the hypothesis may be grasped more perfectly, let *AB* be the great orbital circle of the Earth with centre *C* and diameter *ACB*. And on *ACB* between points *B* and *C* let *D* be taken as a centre and with one-third *CD* as radius let the small circle *EF* be described, so that *F* is its greatest distance from *C*, and *E* its least. Let *HI*

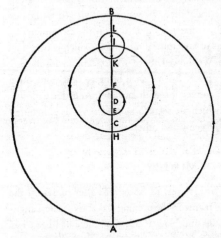

the orbital circle of Mercury be described around centre *F*; and then with *I* the highest apsis as centre let the epicycle which the planet traverses be added. Let *HI* be the orbital circle which is eccentric to an eccentric circle and carries the epicycle. When the figure has been drawn in this way, all those things will fall in order on the straight line *AHCEDFKILB*.

Meanwhile let the planet be set up at *K*, *i.e.*, at the least distance *KF* from centre *F*. [165ª] Now with that point established as the starting-point of the revolutions of Mercury, let it be understood that the centre *F* makes two revolutions for every one of the Earth, and in the same direction too, *i.e.*, eastward; similarly too the planet on *LK*, but along the diameter, up and down with respect to the centre of circle *HI*.

For it follows from this that whenever the Earth is at *A* or *B*, the centre of the orbital circle of Mercury is at *F*, which is the position farthest away from *C*; but whenever the Earth is at the middle quadrants, the centre is at *E* which is the position nearest (to *C*), and this in a manner contrary to that of Venus. Moreover, according to this law Mercury traversing the diameter of epicycle *KL* is nearest to the centre of the orbital circle carrying the epicycle, *i.e.*, is at *K*, when the Earth falls upon the diameter *AB*; and when the Earth is at its mean positions, the planet will be at *L* the most distant position. In this way there take place the two twin revolutions of the centre of the orbital circle on the circumference of the small circle *EF* and of the planet along the diameter *LK* which are equal to one another and commensurable with the annual movement of the Earth.

But meanwhile let the epicycle or line *FI* be moved by its own proper movement along orbital circle *HI*, and let its centre move regularly, completing one revolution simply and with respect to the sphere of the fixed stars in approximately 88 days. But by that movement, whereby it outruns the movement of the Earth and which we call the movement of parallax, it has with respect to the Earth one revolution in 116 days,

as can be derived more exactly [165^b] from the table of mean movements. Hence it follows that Mercury by its own proper movement does not always describe the same circumference of a circle, but in proportion to its distance from the centre of its orbital circle it describes a circumference of greatly varying magnitude, least at point *K*, greatest at *L*, and middling around *I*, in practically the same way as was viewed in the case of the lunar epicycle on an epicycle. But what the moon did on the circumference, Mercury does on the diameter by means of a reciprocal movement which is nevertheless compounded of regular movements. We showed above in connection with the precession of the equinoxes how that takes place. But we shall bring to bear some other things concerning this, farther down in connection with latitudes. And this hypothesis is sufficient for all the appearances which are seen to be Mercury's, as is manifest from the history of the observations of Ptolemy and others.

26. On the Positions of the Highest and Lowest Apsides of Mercury

For Ptolemy observed Mercury in the 1st year of Antoninus after sunset on the 20th day of the month Epiphi, while the planet in the evening was at its greatest angular distance from the mean position of the sun. Now at this time there had been 137 Christian years 188 days 42$^1/_2$ minutes (of a day) at Cracow, and accordingly the mean position of the sun according to our calculation was at 63°50', and the planet observed through the instrument was at 7° of Cancer, he says. But after the subtraction of the precession of the equinoxes, which at that time was 6°40', it was shown that the position of Mercury was at 90°20' from the beginning of Aries in the sphere of the fixed stars, and its greatest angular elongation from the mean position of the sun was 26$^1/_2$°.

He made a second observation in the 4th year of Antoninus on the early morning of the 19th day of the month Phamenoth when 140 years 67 days 12 minutes (of a day) approximately had passed since the beginning of the years of Christ, and the mean position of the sun was at 303°19'. Now Mercury was apparent through the instrument at 13$^1/_2$° of Capricorn, but it was at approximately 276°49' from the fixed beginning of Aries, and accordingly the greatest morning distance was similarly 26$^1/_2$°. Accordingly since the limits of elongation on either side of the mean position of the sun are equal, it is necessary that the apsides of Mercury be either way at the midpoint between these positions, *i.e.*, between 226°49' and 90°20'. And they are 3°34' and 183°34' diametrically opposite, where the highest and the lowest apsides [166^a] of Mercury must be.

As in the case of Venus, the apsides are distinguished through two observations. He made the first in the 19th year of Hadrian on the early morning of the 15th day of the month Athyr, while the mean position of the sun was at 182°38'. The greatest

morning distance of Mercury from the sun was 19°3', since the apparent position of Mercury was at 163°35'. And in the same 19th year of Hadrian, which was the year of Our Lord 135, at dusk of the 19th day of the month Pachon by the Egyptian calendar, Mercury was found by the aid of the instrument at 27°43' in the sphere of the fixed stars, while the sun by its mean movement was at 4°28'. Again it was shown that the greatest evening distance of the planet was 23°15'—which is greater than the previous distance—whence it was clear enough that the apogee of Mercury at that time could be only at approximately 183$^1/_2$°—as was to be taken note of.

27. How Great the Eccentricity of Mercury Is and What the Commensurability of Its Circles Is

Moreover through this the distance between the centres and the magnitudes of the orbital circles are demonstrated simultaneously, For let AB be the straight line passing through A the highest and B the lowest apsis of Mercury, and also the diameter of the great circle, whose centre is D. And with D taken as centre let the orbital circle of the planet be described. Therefore let lines AE and BF be drawn touching the orbital circle, and let DE and DF be joined.

Accordingly, since at the first of the two preceding observations the greatest morning distance of the planet was seen to be 19°3',

$$\text{angle } CAE = 19°3'.$$

But at the second observation the greatest evening distance was seen to be 23$^1/_4$°. Accordingly, as both of the right triangles AED and BFD have their angles given, [166^b] the ratios of their sides will be given too, so that, as radius of the orbital circle,

$$ED = 32,639,$$
$$\text{where } AD = 100,000.$$

But

$$FD = 39,474,$$
$$\text{where } BD = 100,000.$$

But according as

$$FD = ED$$

as radius of the orbital circle,

$$FD = 32,639,$$
$$\text{where } AD = 100,000;$$

and by subtraction

$$DB = 82,685$$

Hence

$$AC = {}^1/_2 AB = 91,342;$$

and by subtraction

$$CD = 8,658,$$

which is the distance between the centres. And the radius of the orbital circle of Mercury will be 21'26", where $AC = 1^P = 60'$;
and

$$CD = 5'4'';$$

and

$$DF = 35,733$$

and

$$CD = 9,479,$$

where $AC = 100,000$

as was to be demonstrated.

But these magnitudes also do not stay everywhere the same; but are quite different from those found in connection with the mean apsides, as the apparent morning and evening longitudes observed at those positions and recorded by Theo and Ptolemy teach us. For Theo observed the evening limit of Mercury in the 14th year of Hadrian on the 18th day of the month Mesori after sunset; and at 129 years 216 days 45 minutes (of a day) after the birth of Christ, while the mean position of the sun was $93^1/_2°$, *i.e.*, approximately at the mean apsis of Mercury. Now the planet was seen through the instrument to be $3^{10}/_{12}°$ to the west of Basiliscus in Leo; and on that account its position was $119^3/_4°$ and its greatest evening distance was $26^3/_4°$.

Ptolemy reported that the second limit was observed by him in the 2nd year of Antoninus on the 21st day of the month Mesori at early morning, at which time there had been 138 Christian years 219 days 12 minutes, and the mean position of the sun was similarly 93°39'; [167ᵃ] and the greatest morning distance of Mercury from that was found to be $20^1/_4°$. For Mercury was visible at $73^2/_5°$ in the sphere of the fixed stars.

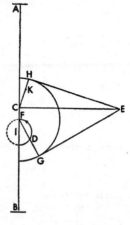

Therefore let *ACDB* which is the diameter of the great orbital circle (of the Earth) and which passes through the apsides of Mercury, be again drawn, as before; and at point *C* let *CE* the line of mean movement of the sun be erected at right angles. Let point *F* be taken between *C* and *D*; and let the orbital circle of Mercury be described around *F*. Let the straight lines *EH* and *EG* touch this small circle; and let *FG*, *FH*, and *EF* be joined.

Now once more our problem is to find point F, and what ratio radius FG has to AC. For since

$$\text{angle } CEG = 26^1/_4°,$$

and

$$\text{angle } CEH = 20^1/_4°;$$

accordingly by addition

$$\text{angle } HEG = 46^1/_2°.$$

And

$$\text{angle } HEF = {}^1/_2 HEG = 23^1/_4°.$$

And so by subtraction

$$\text{angle } CEF = 3°.$$

For that reason the sides of the right triangle CEF are given:

$$CF = 524$$

and

$$FE = 10,014,$$
$$\text{where } CE = AC = 10,000.$$

But it has been shown already that

$$CD = 948,$$

while the Earth is at the highest or lowest apsis of the planet. Hence DF will be the excess (of CD over CF) and the diameter of the small circle which the centre of the orbital circle of Mercury describes.

$$DF = 424,$$

and

$$\text{radius } IF = 212.$$

Hence by addition

$$CFI = 736.$$

Similarly, as in triangle HEF

$$\text{angle } H = 90°,$$

and

$$\text{angle } HEF = 23^1/_4° :$$

hence

$$FH = 3,947,$$
$$\text{where } EF = 10,000.$$

But

$$FH = 3,953,$$
$$\text{where } EF = 10,014$$
$$\text{and } CE = 10,000.$$

Now it has been shown above that

$$FK = 3,573.$$

Therefore by subtraction

$$HK = 380,$$

which is the greatest difference in the planet's distance from F the centre of its orbital circle; and this greatest difference is found when the planet is between its highest or lowest apsis and its mean apsis. On account of this varying distance from F the centre of its orbital circle, the planet describes unequal circles in proportion to the varying distances—the least distance being 3,573, the greatest 3,953, and the mean 3,763—as was to be demonstrated.

28. Why the Angular Digressions of Mercury at Around 60° from the Perigee Appear Greater Than Those at the Perigee

Hence too it will seem less surprising that Mercury has greater angular digressions at a distance of 60° from the perigee than when at the perigee, since they are also greater than the ones which we have already demonstrated; consequently it was held by the ancients that in one revolution [167b] of the Earth, Mercury's orb was twice very near to the Earth.

For let the construction be made such that

$$\text{angle } BCE = 60°.$$

On that account

$$\text{angle } BIF = 120°$$

For F is put down as making two revolutions for one of E the Earth. Therefore let EF and EI be joined. Accordingly, since it has been shown that

$$CI = 736,$$
$$\text{where } EC = 10,000,$$

and

$$\text{angle } ECI = 60°;$$

hence in triangle ECI

$$\text{base } EI = 9,655;$$

and

$$\text{angle } CEI \, 3°47',$$

which is the difference between angle ACE and angle CIE. But

$$\text{angle } ACE = 120°.$$

Accordingly

$$\text{angle } CIE = 116°13'.$$

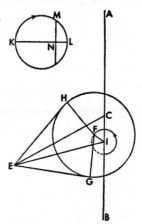

But also

$$\text{angle } FIB = 120°,$$

since by construction

$$\text{angle } FIB = 2 \ ECI;$$

and

$$\text{angle } CIF = 180° - 120° = 60°;$$

[and

$$\text{angle } BIE = 63°47']$$

hence by subtraction

$$\text{angle } EIF = 56°13'.$$

But it was shown that

$$IF = 212,$$
$$\text{where } EI = 9,655;$$

and EI and IF comprehend the given angle EIF. Hence it is inferred that

$$\text{angle } FEI = 1°4';$$

and by subtraction

$$\text{angle } CEF = 2°44',$$

which is the difference between the centre of the orbital circle of the planet and the mean position of the sun; and

$$\text{side } EF = 9,540.$$

Now let GH the orbital circle of Mercury be described around centre F; and from E let EG be drawn touching the orbital circle, and let FG and FH be joined.

We must first examine how great the radius FG or FH is under these circumstances; and we shall do that as follows:

For let a small circle be taken, whose

$$\text{diameter} = 380,$$
$$\text{where } AC = 10,000.$$

And let it be understood that the planet on straight line FG or FH approaches and recedes from centre F along that diameter or along a line equal to it, as we set forth above in connection with the precession of the equinoxes. And in accordance with our hypothesis, wherein angle BCE intercepts 60° of the circumference, let

$$\text{arc } KM = 120°;$$

and let MN be drawn at right angles to KL. And since

$$MN = {}^1/_2 \text{ ch. } 2ML = {}^1/_2 \text{ ch. } 2 \ KM,$$

then

$$LN = 95,$$

which is one quarter of the diameter—as is shown by [168ª] Euclid's *Elements*, XIII, 12 and V, 15. Accordingly,

$$KN = {}^3/_4 \, KL = 285.$$

Line *KN* and the least distance of the planet added together make the distance sought for this position, *i.e.*,

$$FG = FH = 3,858,$$
$$\text{where } AC = 10,000$$
$$\text{and } EF = 9,540.$$

Wherefore two sides of right triangle *FEG* or *FEH* have been given: so angle *FEG* or *FEH* will also be given. For

$$FG = FH = 4,044,$$
$$\text{where } EF = 10,000;$$

and

$$FG = FH = \text{ch. } 23°52',$$

so that by addition

$$\text{angle } GEH = 47°44'.$$

But at the lowest apsis only $46^1/_2°$ is seen, and at the mean apsis similarly $46^1/_2°$. Accordingly the elongation here becomes 1°14' greater, not because the orbital circle of the planet is nearer to the Earth than it was at the perigee, but because the planet is here describing a greater circle than there. All these things are consonant with both present and past observations, and follow from the regular movements.

29. EXAMINATION OF THE MEAN MOVEMENTS OF MERCURY

For it is found by the ancient observations that in the 21st year of Ptolemy Philadelphus in the morning twilight of the 19th day of the month Thoth by the Egyptian calendar, Mercury was apparent on the straight line passing through the first and second of the stars in the forehead of Scorpio and was two lunar diameters distant to the east but was separated from the first star by one lunar diameter to the north. Now it is known that the position of the first star is $209^2/_3°$ in longitude and $1^1/_3°$ in northern latitude; and the position of the second is 209° in longitude and $1^5/_6°$ in southern latitude. From that it was concluded that the position of Mercury was $110^2/_3°$ in longitude and approximately $1^5/_6°$ in northern latitude. Now there were 59 years 17 days 45 minutes (of a day) since the death of Alexander; and the mean position of the sun according to our calculation was 228°8'; and the morning distance of the star was 17°28' and was still increasing, as was noted during the four following days. Hence it was certain that the planet had not yet arrived at the farthest morning limit or at the point of tangency of its orbital circle, but was still moving in the lower part of the circumference nearer to the Earth. But since the highest apsis was at 183°20', there were 44°48' to the mean position of the sun.

[168ᵇ] Therefore again let *ACB* be the diameter of the great orbital circle, as above; and from centre *C* let *CE* the line of mean movement of the sun be drawn, in such fashion that

angle *ACE* = 44°48'.

And let there be described around centre *I* the small circle on which the centre *F* of the eccentric circle is home. And since by hypothesis

angle *BIF* = 2 angle *ACE*,

let

angle *BIF* = 89°36'.

And let *EF* and *EI* be joined.

Accordingly, in triangle *ECI* two sides have been given:

CI = 736¹/₂,

where *CE* = 10,000.

And sides *CI* and *CE* comprehend the given angle *ECI*. And

angle *ECI* = 180° – angle *ACE* = 135°12'; side

EI = 10,534;

and

angle *CEI* = 2°49',

which is the excess of angle *ACE* over angle *EIC*.

Therefore too

angle *CIE* = 41°59'.

But

angle *CIF* = 180° – angle *BIF* = 90°24'.

Therefore by addition

angle *EIF* = 132°23';

and angle *EIF* is comprehended by the given sides *EI* and *IF* of triangle *EFI*, and

side *EI* = 10,534

and

side *IF* = 211¹/₂,

where *AC* = 10,000,

Hence

angle *FEI* = 50';

and

side *EF* = 10,678

And by subtraction

335

angle $CEF = 1°59'$.

Now let the small circle LM be taken; and let

diameter $LM = 380$,

where $AC = 10,000$.

And in accordance with the hypothesis let

arc $LN = 89°36'$.

Let chord LN also be drawn; and let NR be drawn perpendicular to LM. Accordingly, since

sq. LN = rect. LM, LR;

that ratio being given,

side $LR \doteqdot 189$,

where diameter $LM = 380$.

That straight line, *i.e.*, LR, measures the distance of the planet from F the centre of its orbital circle at the time when line EC has completed angle ACE. Accordingly, by the addition of this [169a] line to the least distance

$$189 + 3,573 = 3,672,$$

which is the distance at this position.

Accordingly, with the centre F and radius 3,762, let a circle be described; and let EG be drawn cutting the convex circumference at point G, in such a way that

angle $CEG = 17°28'$,

which is the apparent angular elongation of the planet from the mean position of the sun. Let FG be joined; and let FK be drawn parallel to CE. Now

angle FEG = angle CEG – angle $CEF = 15°29'$.

Hence in triangle EFG two sides have been given;

$$EF = 10,678,$$

and

$$FG = 3,762,$$

and

angle $FEG = 15°29'$:

whence it will be clear that

angle $EFG = 3346$.

Now, since

angle EFK = angle CEF,

angle KFG = angle EFG – angle $RFK = 31°48'$;

and

arc $KG = 31°48'$,

which is the distance of the planet from K the mean perigee of its orbital circle.

arc $KG + 180° = 211°48'$,

which was the mean movement of the anomaly of parallax at the time of this observation—as was to be shown.

30. ON THREE MODERN OBSERVATIONS OF THE MOVEMENTS OF MERCURY

The ancients have directed us to this method of examining the movement of this planet, but they were favoured by a clearer atmosphere at a place, where the Nile—so they say—does not give out vapours as the Vistula does among us. For nature has denied that convenience to us who inhabit a colder region, where fair weather is rarer; and furthermore on account of the great obliquity of the sphere it is less frequently possible to see Mercury, as its rising does not fall within our vision at its greatest distance from the sun when it is in Aries or Pisces, and its setting in Virgo and Libra is not visible; and it is not apparent in Cancer or Gemini at evening or early morning, and never at night, except when the sun has receded through the greater part of Leo. On this account the planet has made us take many detours and undergo much labour in order to examine its wanderings. On this account we have borrowed three positions from those which have been carefully observed at Nuremburg.

The first observation was taken by Bernhard Walther, a pupil of Regiomontanus, in the year of Our Lord 1491 on the 9th of September, the fifth day before the Ides, 5 equal hours after midnight, by means of an astrolabe brought into relation with the Hyades. And he saw Mercury at $13^1/_2°$ [169b] of Virgo with a northern latitude of $1^5/_6°$; and at that time the planet was at the beginning of its morning occultation, while during the preceding days its morning (elongation) had decreased continuously. Accordingly there were 1491 Egyptian years 258 days $12^1/_2$ minutes (of a day) since the beginning of the years of Our Lord; the simple mean position of the sun was at from the spring equinox but in 26°47' of Virgo, wherein the position of Mercury was approximately $13^1/_2°$.

The second was taken by Johann Schöner in the year of Our Lord 1504 on the 5th day before the Ides of January $6^1/_2$ hours after midnight, when 10° of Scorpio was in the middle of the heavens over Nuremburg; and the planet was apparent at $3^1/_3°$ of Capricorn with a northern latitude of 45'. Now by our calculation the mean position of the sun away from the spring equinox was at 27°7' of Capricorn and a morning Mercury was 23°42' to the west of that.

The third observation was taken by this same Johann Schöner in the same year 1504 on the 15th day before the Kalends of April, at which time he found Mercury at $26^1/_{10}°$ of Aries with a northern latitude of approximately 3°, while 25° of Cancer was in the middle of the heavens over Nuremburg—as seen through an astrolabe brought into relation with the Hyades, at $12^1/_2$ hours after midday, at which time the mean

position of the sun away from the spring equinox was at 5°39' of Aries, and an evening Mercury was 21°17' away from the sun.

Accordingly from the first position to the second, there are 12 Egyptian years 125 days 3 minutes (of a day) 45 seconds, during which time the simple movement of the sun was 120°14', and Mercury's movement of anomaly of parallax was 316°1'. During the second interval there were 69 days 31 minutes 45 seconds the simple mean position of the sun was 68°32', and Mercury's mean anomaly of parallax was 216°.

Accordingly we wish to examine the movements of Mercury during our time by means of these three observations, and I think we must grant that the commensurability of the circles has remained from Ptolemy's time to now, since in the case of the other planets the good authorities who preceded us are not found to have been mistaken here. If we have the position of the apsis of the eccentric circle together with these observations, nothing further should be desired in the case of the apparent movement of this planet. Now we have taken the position of the highest apsis as $211\frac{1}{2}°$, i.e., at $28\frac{1}{2}°$ of Scorpio; for it was not possible to take it as less without prejudice to the observations. And so we shall have the anomaly of the eccentric circle—I mean [170ª] the distance of the mean movement of the sun from the apogee—as 298°15' at the first terminus, as 58°29' at the second, and as 127°1' at the third.

Therefore let the figure be constructed as before, except that

angle $ACE = 61°45'$,

which measures the westward distance of the line of mean movement of the sun from the apogee at the time of the first observation; and then the rest according to the hypothesis. And since

$$IC = 736\frac{1}{2},$$

where $AC = 10,000$;

and angle ECI in triangle ECI is also given; then

angle $CEI = 3°35'$,

and

side $IE = 10,369$,

where $EC = 10,000$;

and

$$IF = 211\frac{1}{2}.$$

So in triangle EFI also there are two sides having a given ratio; and since by construction

angle $BIF = 2$ angle ACE;

angle $BIF = 123\frac{1}{2}°$,

and

$$\text{angle } CIF = 180° - 123^1/_2° = 56^1/_2°.$$

Therefore by addition

$$\text{angle } EIF = 114°40'.$$

Accordingly

$$\text{angle } IEF = 1°5',$$

and

$$\text{side } EF = 10,371.$$

Hence

$$\text{angle } CEF = 2^1/_2°.$$

But in order that we may know how greatly the orbital circle, whose centre is F, is increased by the movement of approach and withdrawal from the apogee or perigee, let a small circle be drawn and quadrisected by the diameters LM, NR at centre O. And let

$$\text{angle } POM = 2 \text{ angle } ACE = 123^1/_2°;$$

and from point P let PS be drawn perpendicular to LM. Accordingly by the ratio given,

$$OP : OS = LO : OS = 10,000 : 8,349 = 190 : 105.$$

Whence

$$LS = 295,$$
$$\text{where } [170^\text{b}] \; AC = 10,000;$$

and LS measures the farther removal of the planet from centre F. As the least distance is 3,573,

$$LS + 3,573 = 3,868,$$

which is the present distance.

And with 3,868 as radius and F as centre, let circle HG be drawn. Let EG be joined; and let EF be extended in the straight line EFH. Accordingly, it has been shown that

$$\text{angle } CEF = 2^1/_2°,$$

and by observation

$$\text{angle } GEC = 13^1/_4°,$$

which is the morning distance of the planet from the mean sun. Therefore by addition

$$\text{angle } FEG = 15^3/_4°.$$

But in triangle EFG

$$EF : EG = 10,371 : 3,868;$$

and angle EFG is also given; that shows us that

$$\text{angle } EGF = 49°8'.$$

Hence

$$\text{angle } GFH = 64°53',$$

as it is the exterior angle; and

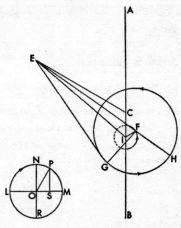

339

$$360° - \text{angle } GFH = 295°7',$$

which is the true anomaly of parallax. And

$$295°7' + \text{angle } CEF = 297°37',$$

the mean and regular anomaly of parallax—which is what we were looking for. And

$$297°37' + 316°1' = 253°38',$$

which is the regular anomaly of parallax at the second observation—and we shall show that this number is certain and is consonant with the observations.

For let us make

$$\text{angle } ACE = 58°29'$$

in accordance with the second movement of anomaly of the eccentric circle. Then also in triangle CEI two sides are given:

$$IC = 736,$$
$$\text{where } EC = 10,000;$$

and IC and EC comprehend angle ECI, and

$$\text{angle } ECI = 121°31';$$

accordingly

$$\text{side } EI = 10,404$$

and

$$\text{angle } CEI = 3°28'.$$

Similarly, since in triangle EIF
angle $EIF = 118°3'$,
and

$$\text{side } IF = 211^{1}/_{2}°,$$
$$\text{where } IE = 10,404 :$$
$$\text{side } EF = 10,505,$$

and

$$\text{angle } IEF = 61'.$$

And so by subtraction

$$\text{angle } FEC = 2°27',$$

which is the additive additosubtraction of the eccentric circle; and the addition of angle FEC to the mean movement of parallax makes the true movement to be 256°5'.

Now also in the epicycle of approach [171a] and withdrawal let us take

$$\text{angle } LOP = 2 \text{ angle } ACE = 116°58'.$$

Then too, as in right triangle OPS

$$OP : OS = 1,000 : 455;$$
$$OS = 85,$$
$$\text{where } OP = OL = 190.$$

And by addition

$$LOS = 276.$$

The addition of *LOS* to the least distance of 3,573 makes 3,849.

With 3,849 as radius let circle *HG* be described around centre *F*, so that the apogee of parallax is at point *H* from which the planet has the westward distance of 103°55' of arc *HG*, which measures the difference between a full revolution and the 256°5' of the movement of corrected parallax. And on that account

$$\text{angle } EFG = 180° - 103°55' = 76°5'.$$

So again in triangle *EFG* two sides are given:

$$FG = 3,849,$$
$$\text{where } EF = 10,505.$$

On that account

$$\text{angle } FEG = 21°19';$$

and

$$\text{angle } CEG = \text{angle } FEG + \text{angle } CEF = 23°46'.$$

That is the apparent distance between *C* the centre of the great orbital circle and *G* the planet; and it differs very little from the observation.

All this will be further confirmed by the third example, wherein we have set down that

$$\text{angle } ACE = 127°1',$$

or

$$\text{angle } BCE = 180° - 127°1' = 52°59'.$$

Hence it is shown that

$$\text{angle } CEI = 3°31',$$

and

$$\text{side } IE = 9,575,$$
$$\text{where } EC = 10,000.$$

And since by construction

$$\text{angle } EIF = 49°28',$$

and the sides comprehending angle *EIF* are given:

$$FI = 211^{1}/_{2},$$
$$\text{where } EI = 9,575;$$
$$\text{side } EF = 9,440,$$

and

$$\text{angle } IEF - 59'.$$

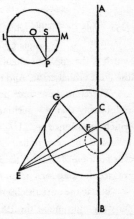

$$\text{Angle } FEC = \text{angle } IEC - 59' = 2°32',$$

which is the subtractive additosubtraction of the anomaly of the eccentric circle. When the 2°32' has been added to the mean anomaly of parallax, which we reckoned as 109°38' after adding the 216° of the second (movement of anomaly), the sum will be the 112°10' of the true anomaly of parallax.

Now in the epicycle let
$$\text{angle } LOP = 2 \text{ angle } ECI = 105°58';$$
here too by the ratio $PO : OS$,
$$OS = 52,$$
so that by addition
$$LOS = 242.$$
Now the least distance is 3,573; and
$$3,573 + 242 = 3,815,$$
which is the corrected distance.

With 3,815 as radius and F as centre, let the circle be described, in which the highest apsis of parallax is H, which is on the straight line made by extending line EFH. And in proportion to the true anomaly of parallax [171b] let
$$\text{arc } HG = 112°10';$$
and let GF be joined. Therefore
$$\text{angle } GFE = 180° - 112°10' = 67°50',$$
which is comprehended by the given sides:
$$GF = 3,815$$
and
$$EF = 9,440.$$
Hence
$$\text{angle } FEG = 23°50'.$$
Now angle CEF is the additosubtraction; and
$$\text{angle } CEG = \text{angle } FEG - \text{angle } CEF = 21°18',$$
which is the apparent angular distance between the evening planet and the centre of the great orbital circle. And that is approximately the distance found by observation.

Therefore these three positions which are in agreement with observations testify indubitably that the position of the highest apsis, of the eccentric circle is the one which we assumed at $211\frac{1}{2}°$ in the sphere of the fixed stars for our time; and that what follows is also certain, namely, that the regular anomaly of parallax was 297°37' at the first position, 253°38' at the second, and 109°38' at the third position; and that is what we were looking for.

But at the ancient observation made in the 21st year of Ptolemy Philadelphus in the early morning of the 19th day of Thoth, the 1st month of the Egyptian calendar the position of the highest apsis of the eccentric circle was 182°20' according to Ptolemy, and the position of regular anomaly of parallax was 211°47'. Now the time between this latest and that ancient observation amounts to 1768 Egyptian years 200 days 33 minutes (of a day), during which time the highest apsis of the eccentric circle moved 28°10' in the sphere of the fixed stars, and the movement of parallax was

257°51' besides the 5,570 complete [172^a] revolutions—as approximately 63 periods are completed in 20 years and that amounts to 5,544 periods in 1,760 years and 26 revolutions in the remaining 8 years 200 days. Similarly in the 1,768 years 200 days 33 minutes there are in addition to the 5,570 revolutions 257°51', which is the distance between the position observed in ancient times and the one observed by us. That agrees with the numbers which we set out in the tables. Now when we have compared the 28°10' with the time during which the apogee of the eccentric circle has moved, it will be seen to have moved 1° per 63 years, if only the movement were regular.

31. ON DETERMINING THE FORMER POSITIONS OF MERCURY

Accordingly there have been 1504 Egyptian years 87 days 48 minutes (of a day) from the beginning of the years of Our Lord to the hour of the last observation, during which time Mercury's movement of anomaly of parallax was 63°14'—not counting the complete revolutions. When 63°14' has been subtracted from 109°38', it will leave 46°24' as the position of the movement of anomaly at the beginning of the years of Our Lord.

Again between that time and the beginning of the first Olympiad there are 775 Egyptian years 12$^{1}/_{2}$ days, during which the movement was calculated to be 95°3' besides the whole revolutions.

If 95°3' is subtracted from the position at the beginning of the years of our Lord and one revolution is borrowed, 311°21' will be left as the position at the time of the first Olympiad.

Moreover between this and the death of Alexander there are 451 years 247 days, and by computation the position is 213°3'.

32. ON ANOTHER EXPLANATION OF APPROACH AND WITHDRAWAL

But before we leave Mercury, let us survey another method no less credible than the former, by which that approach and withdrawal can take place and can be understood.

For let *GHKP* be a circle quadrisected at centre *F*, and around centre *F* let *LM* a small homocentric circle be inscribed. And again, with *L* as centre and radius *LFO* equal to *FG* or *FH*, let another circle *OR* be described.

Now let it be postulated that this whole configuration of circles [172^b] together with its sections *GFR*

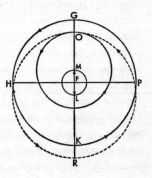

and *HFP* moves eastward around centre *F* in a daily movement of approximately 2°7'—namely, a movement as great as that whereby the movement of parallax of the planet exceeds the movement of the Earth in the ecliptic—away from the apogee of the eccentric circle of the planet; and the planet meanwhile furnishes the remaining movement away from *G* along *OR* the proper circle of parallax; and this movement is approximately equal to the terrestrial movement. Let it also be assumed that in this same annual revolution the centre of *OR*, the orbital circle which carries the planet, is borne by a movement of libration along a diameter *LFM* which is twice as great as the one which we laid down at first; and let it move back and forth, as was said above.

With this as the set-up, we have placed the Earth by its mean movement in a position corresponding to the apogee of the eccentric circle of the planet; and at that time the centre of the orbital circle carrying the planet, at *L*; and the planet itself, at point *O*. The planet then being at its least distance from *F* will describe by its whole movement its least circle, whose radius is *FO*. Consequently, when the Earth is in the neighbourhood of the mean apsis, the planet falling upon point *H* at its greatest distance from *F* will describe its greatest arcs, namely in proportion to the circle having *F* as centre. For at that time *OR* the deferent coincides with circle *GH* on account of the unity of centre at *F*. Hence as the Earth advances in the direction of the perigee and the centre of the orbital circle *OR* towards *M* the other extreme, the orbital circle itself is placed beyond *GK*, and the planet at *R* again is at its least distance from *F*, and the same things occur as in the beginning. For here the three revolutions—namely, that of the Earth through the apogee of the eccentric orbital circle of Mercury, the libration of the centre along diameter *LM*, and the movement of the planet in the same direction away from line *FG*—are equal to one another, and only the movement of the sections *GH* and *KP* away from the apsis of the eccentric circle is different from those revolutions, as we said.

And so in the case of this planet nature has sported a wonderful variety, but one which she has confirmed by a perpetual, certain, and unchanging order. And we should note here that the planet does not traverse the middle spaces of quadrants *GH* and *KP* without any irregularity in longitude, provided that the diversity of the centres, which comes into play, necessarily makes some additosubtraction; but the instability of that centre prevents that. For if, for example, the centre abided at *L* and the planet proceeded from *O*, then it would admit greatest irregularity at *H* in proportion [173ª] to the eccentricity *FL*. But it follows from the assumptions that the planet proceeding from *O* begins and promises to cause the irregularity which is due to *FL* the distance between the centres; but as the mobile centre approaches the midpoint *F*, more and more is taken away from the promised irregularity, and it is made void to such an extent that it wholly vanishes at the mean sections *H* and *P*, where you would expect

it to be greatest. And nevertheless, as we acknowledge, when the planet is made small beneath the rays of the sun,[1] it undergoes occultation; and when the planet is rising or setting in the morning or evening, it is not discernible on the curves of the circle. And we are unwilling to pass over this method which is no less rationale than the former, and which will come into open use in connection with the movements in latitude.

33. ON THE TABLES OF THE ADDITOSUBTRACTIONS OF THE FIVE WANDERING STARS

These things have been demonstrated concerning the regular and apparent movements of Mercury and the other planets and set out in numbers. And by their example the way to calculating differences of movement for certain other positions will be clear. And for this use we have made ready separate tables for each planet: six column, thirty rows, ascending by triads of 3°, as is usual. The first two columns will contain the common numbers—both of the anomaly of the eccentric circle and of the parallax. The third, the sums of the additosubtractions of the eccentric circle—I mean the total differences occurring between the regular and irregular movements of the orbital circles. The fourth, the proportional minutes—and they go up to 60'—by which the parallaxes are increased or diminished on account of the greater or lesser distance of the Earth. The fifth, the additosubtractions which are the parallaxes occurring at the highest apsis of the eccentric circle and which arise from the great orbital circle. The sixth and last, their excesses over the parallaxes which take place at the lowest apsis of the eccentric circle. And the tables are as follows:

---

[1] *i.e.,* when the planet is in conjunction with the sm.

ADDITIONS-AND-SUBTRACTIONS OF SATURN

Common Numbers		Addito-subtractions of the eccentric circle		Proportional Minutes	Parallaxes of the orbital circle of the Earth		Excesses over the parallax of the lowest apsis	
Deg.	Deg.	Deg.	Min.	Min.	Deg.	Min.	Deg.	Min.
3	357	0	20	0	0	17	0	2
6	354	0	40	0	0	34	0	4
9	351	0	58	0	0	51	0	6
12	348	1	17	0	1	3	0	8
15	345	1	36	1	1	23	0	10
18	342	1	55	1	1	40	0	12
21	339	2	13	1	1	56	0	14
24	336	2	31	2	2	11	0	16
27	333	2	49	3	2	26	0	18
30	330	3	6	3	2	42	0	19
33	327	3	33	4	2	56	0	21
36	324	3	39	4	3	10	0	23
39	321	3	55	4	3	25	0	24
42	318	4	10	5	3	38	0	26
45	315	4	25	6	3	52	0	27
48	312	4	39	7	4	5	0	29
51	309	4	52	8	4	17	0	31
54	306	5	5	9	4	28	0	33
57	303	5	17	10	4	38	0	34
60	300	5	29	11	4	49	0	35
63	297	5	41	12	4	59	0	36
66	294	5	50	13	5	8	0	37
69	291	5	59	14	5	17	0	38
72	288	6	7	16	5	24	0	38
75	285	6	14	17	5	31	0	39
78	282	6	19	18	5	37	0	39
81	279	6	23	19	5	42	0	40
84	276	6	27	21	5	46	0	41
87	273	6	29	22	5	50	0	42
90	270	6	31	23	5	52	0	42
93	267	6	31	25	5	52	0	43
96	264	6	30	27	5	53	0	44
99	261	6	28	29	5	53	0	45
102	258	6	26	31	5	51	0	46
105	255	6	22	32	5	48	0	46
108	252	6	17	34	5	45	0	45
111	249	6	12	35	5	40	0	45
114	246	6	6	36	5	36	0	44
117	243	5	58	38	5	29	0	43
120	240	5	49	39	5	22	0	42
123	237	5	40	41	5	13	0	41
126	234	5	28	42	5	3	0	40
129	231	5	16	44	4	52	0	39
132	228	5	3	46	4	41	0	37
135	225	4	48	47	4	29	0	35
138	222	4	33	48	4	15	0	34
141	219	4	17	50	4	1	0	32
144	216	4	0	51	3	46	0	30
147	213	3	42	52	3	30	0	28
150	210	3	24	53	3	13	0	26
153	207	3	6	54	2	56	0	24
156	204	2	46	55	2	38	0	22
159	201	2	27	56	2	21	0	19
162	198	2	7	57	2	2	0	17
165	195	1	46	58	1	42	0	14
168	192	1	25	59	1	22	0	12
171	189	1	4	59	1	2	0	9
174	186	0	43	60	0	42	0	7
177	183	0	22	60	0	21	0	4
180	180	0	0	60	0	0	0	0

ADDITIONS-AND-SUBTRACTIONS OF JUPITER

Common Numbers		Addito-subtractions of the eccentric circle		Proportional Minutes		Parallaxes of the orbital circle of the Earth		Excesses over the parallax of the lowest apsis	
Deg.	Deg.	Deg.	Min.	Min.	Sec.	Deg.	Min.	Deg.	Min.
3	357	0	16	0	3	0	28	0	2
6	354	0	31	0	12	0	56	0	4
9	351	0	47	0	18	1	25	0	6
12	348	1	2	0	30	1	53	0	8
15	345	1	18	0	45	2	19	0	10
18	342	1	33	1	3	2	46	0	13
21	339	1	48	1	23	3	13	0	15
24	336	2	2	1	48	3	40	0	17
27	333	2	17	2	18	4	6	0	19
30	330	2	31	2	50	4	32	0	21
33	327	2	44	3	26	4	57	0	23
36	324	2	58	4	10	5	22	0	25
39	321	3	11	5	40	5	47	0	27
42	318	3	23	6	43	6	11	0	29
45	315	3	35	7	48	6	34	0	31
48	312	3	47	8	50	6	56	0	34
51	309	3	58	9	53	7	18	0	36
54	306	4	8	10	57	7	39	0	38
57	303	4	17	12	0	7	58	0	40
60	300	4	26	13	10	8	17	0	42
63	297	4	35	14	20	8	35	0	44
66	294	4	42	15	30	8	52	0	46
69	291	4	50	16	50	9	8	0	48
72	288	4	56	18	10	9	22	0	50
75	285	5	1	19	17	9	35	0	52
78	282	5	5	20	40	9	47	0	54
81	279	5	9	22	20	9	59	0	55
84	276	5	12	23	50	10	8	0	56
87	273	5	14	25	23	10	17	0	57
90	270	5	15	26	57	10	24	0	58
93	267	5	15	28	33	10	25	0	59
96	264	5	15	30	12	10	33	1	0
99	261	5	14	31	43	10	34	1	1
102	258	5	12	33	17	10	34	1	1
105	255	5	10	34	50	10	33	1	2
108	252	5	6	36	21	10	29	1	3
111	249	5	1	37	47	10	23	1	3
114	246	4	55	39	0	10	15	1	3
117	243	4	49	40	25	10	5	1	3
120	240	4	41	41	50	9	54	1	2
123	237	4	32	43	18	9	41	1	1
126	234	4	23	44	46	9	25	1	0
129	231	4	13	46	11	9	8	0	59
132	228	4	2	47	37	8	56	0	58
135	225	3	50	49	2	8	27	0	57
138	222	3	38	50	22	8	5	0	55
141	219	3	25	51	46	7	39	0	53
144	216	3	13	53	6	7	12	0	50
147	213	2	59	54	10	6	43	0	47
150	210	2	45	55	15	6	13	0	43
153	207	2	30	56	12	5	41	0	39
156	204	2	15	57	0	5	7	0	35
159	201	1	59	57	37	4	32	0	31
162	198	1	43	58	6	3	56	0	27
165	195	1	27	58	34	3	18	0	23
168	192	1	11	59	3	2	40	0	19
171	189	0	53	59	36	2	0	0	15
174	186	0	35	59	58	1	20	0	11
177	183	0	17	60	0	0	40	0	6
180	180	0	0	60	0	0	0	0	0

ADDITIONS-AND-SUBTRACTIONS OF MARS

Common Numbers		Addito-subtractions of the eccentric circle		Proportional Minutes		Parallaxes of the orbital circle of the Earth		Excesses over the parallax of the lowest apsis	
Deg.	Deg.	Deg.	Min.	Min.	Sec.	Deg.	Min.	Deg.	Min.
3	357	0	32	0	0	1	8	0	8
6	354	1	5	0	2	2	16	0	17
9	351	1	37	0	7	3	24	0	25
12	348	2	8	0	15	4	31	0	33
15	345	2	39	0	28	5	38	0	41
18	342	3	10	0	42	6	45	0	50
21	339	3	41	0	57	7	52	0	59
24	336	4	11	1	13	8	58	1	8
27	333	4	41	1	34	10	5	1	16
30	330	5	10	2	1	11	11	1	25
33	327	5	38	2	31	12	16	1	34
36	324	6	6	3	2	13	22	1	43
39	321	6	32	3	32	14	26	1	52
42	318	6	58	4	3	15	31	2	2
45	315	7	22	4	37	16	35	2	11
48	312	7	47	5	16	17	39	2	20
51	309	8	10	6	2	18	42	2	30
54	306	8	32	6	50	19	45	2	40
57	303	8	53	7	39	20	47	2	50
60	300	9	12	8	30	21	49	3	0
63	297	9	30	9	27	22	50	3	11
66	294	9	47	10	25	23	48	3	22
69	291	10	3	11	28	24	47	3	34
72	288	10	19	12	33	25	44	3	46
75	285	10	32	13	38	26	40	3	59
78	282	10	42	14	46	27	35	4	11
81	279	10	50	16	4	28	29	4	24
84	276	10	56	17	24	29	21	4	36
87	273	11	1	18	45	30	12	4	50
90	270	11	2	20	8	31	0	5	20
93	267	11	8	21	38	31	45	5	35
96	264	11	5	22	58	32	30	5	51
99	261	11	1	24	32	33	13	6	7
102	258	10	56	26	7	33	53	6	25
105	255	10	45	27	43	34	30	6	45
108	252	10	33	29	21	35	3	7	4
111	249	10	11	31	2	35	34	7	25
114	246	10	7	32	46	35	59	7	46
117	243	9	51	34	41	36	21	8	11
120	240	9	33	36	16	36	37	8	34
123	237	9	13	38	1	36	49	8	59
126	234	8	50	39	46	36	54	9	24
129	231	8	27	41	30	36	53	9	49
132	228	8	2	43	12	36	45	10	17
135	225	7	36	44	50	36	30	10	47
138	222	7	7	46	26	36	5	11	15
141	219	6	37	48	1	35	34	11	45
144	216	6	7	49	35	34	59	12	12
147	213	5	34	51	2	34	3	12	35
150	210	5	0	52	22	33	26	12	54
153	207	4	25	53	38	32	28	13	28
156	204	3	49	54	50	30	51	13	7
159	201	3	12	56	0	28	32	12	47
162	198	2	35	57	6	26	26	12	12
165	195	1	57	58	54	23	5	10	59
168	192	1	18	58	22	20	8	9	1
171	189	1	0	59	50	16	28	6	40
174	186	0	39	59	11	13	51	3	28
177	183	0	0	59	44	8	32	3	28
180	180	0	0	60	0	4	0	0	0

ADDITIONS-AND-SUBTRACTIONS OF VENUS

Common Numbers		Additosubtractions of the eccentric circle		Proportional Minutes		Parallaxes of the orbital circle of the Earth		Excesses over the parallax of the lowest apsis	
Deg.	Deg.	Deg.	Min.	Min.	Sec.	Deg.	Min.	Deg.	Min.
3	357	0	6	0	0	1	15	0	1
6	354	0	13	0	0	2	30	0	2
9	351	0	19	0	10	3	45	0	3
12	348	0	25	0	39	4	59	0	5
15	345	0	31	0	58	6	13	0	6
18	342	0	36	1	20	7	28	0	7
21	339	0	42	1	39	8	42	0	9
24	336	0	48	2	23	9	56	0	11
27	333	0	53	2	59	11	10	0	12
30	330	0	59	3	38	12	24	0	13
33	327	1	4	4	18	13	37	0	14
36	324	1	10	5	3	14	50	0	16
39	321	1	15	5	45	16	3	0	17
42	318	1	20	6	32	17	16	0	18
45	315	1	25	7	22	18	28	0	20
48	312	1	29	8	18	19	40	0	21
51	309	1	33	9	31	20	52	0	22
54	306	1	36	10	48	22	3	0	24
57	303	1	40	12	8	23	14	0	26
60	300	1	43	13	32	24	24	0	27
63	297	1	46	15	8	25	34	0	28
66	294	1	49	16	35	26	43	0	30
69	291	1	52	18	0	27	52	0	32
72	288	1	54	19	33	28	57	0	34
75	285	1	56	21	8	30	4	0	36
78	282	1	58	22	32	31	9	0	38
81	279	1	59	24	7	32	13	0	41
84	276	2	0	25	30	33	17	0	43
87	273	2	0	27	5	34	20	0	45
90	270	2	0	28	28	35	21	0	47
93	267	2	0	29	58	36	20	0	50
96	264	2	0	31	28	37	17	0	53
99	261	1	59	32	57	38	13	0	55
102	258	1	58	34	26	39	7	0	58
105	255	1	57	35	55	40	0	1	0
108	252	1	55	37	23	40	49	1	4
111	249	1	53	38	52	41	36	1	8
114	246	1	51	40	19	42	18	1	11
117	243	1	48	41	45	42	59	1	14
120	240	1	45	43	10	43	35	1	18
123	237	1	42	44	37	44	7	1	22
126	234	1	39	46	6	44	32	1	26
129	231	1	35	47	36	44	49	1	50
132	228	1	31	49	6	45	4	1	36
135	225	1	27	50	12	45	10	1	41
138	222	1	22	51	17	45	5	1	47
141	219	1	17	52	33	44	51	1	53
144	216	1	12	53	48	44	22	2	0
147	213	1	7	54	28	43	36	2	6
150	210	1	1	55	0	42	34	2	13
153	207	0	55	55	57	41	12	2	19
156	204	0	49	56	47	39	20	2	34
159	201	0	43	57	33	36	58	2	27
162	198	0	37	58	16	33	58	2	27
165	195	0	31	58	59	30	14	2	27
168	192	0	25	59	39	25	42	1	16
171	189	0	19	59	48	20	20	1	56
174	186	0	13	59	54	14	7	1	26
177	183	0	7	59	58	7	16	0	46
180	180	0	0	60	0	0	16	0	0

ADDITIONS-AND-SUBTRACTIONS OF MERCURY

Common Numbers		Additosubtractions of the eccentric circle		Proportional Minutes		Parallaxes of the orbital circle of the Earth		Excesses over the parallax of the lowest apsis	
Deg.	Deg.	Deg.	Min.	Min.	Sec.	Deg.	Min.	Deg.	Min.
3	357	0	8	0	3	0	44	0	8
6	354	0	17	0	12	1	28	0	15
9	351	0	26	0	24	2	12	0	23
12	348	0	34	0	50	2	56	0	31
15	345	0	43	1	43	3	41	0	38
18	342	0	51	2	42	4	25	0	45
21	339	0	59	3	51	5	8	0	53
24	336	1	8	5	10	5	51	1	1
27	333	1	16	6	41	6	34	1	8
30	330	1	24	8	29	7	15	1	16
33	327	1	32	10	35	7	57	1	24
36	324	1	39	12	50	8	38	1	32
39	321	1	46	15	7	9	18	1	40
42	318	1	53	17	26	9	59	1	47
45	315	2	0	19	47	10	38	1	55
48	312	2	6	22	8	11	17	2	2
51	309	2	12	24	31	11	54	2	10
54	306	2	18	26	17	12	31	2	18
57	303	2	24	29	17	13	7	2	26
60	300	2	29	31	39	13	41	2	34
63	297	2	34	33	59	14	14	2	42
66	294	2	38	36	12	14	46	2	51
69	291	2	43	38	29	15	17	2	59
72	288	2	47	40	45	15	46	3	8
75	285	2	50	42	58	16	14	3	16
78	282	2	53	45	6	16	40	3	24
81	279	2	56	46	59	17	4	3	32
84	276	2	58	48	50	17	27	3	40
87	273	2	59	50	36	17	48	3	48
90	270	3	0	52	2	18	6	3	56
93	267	3	0	53	43	18	23	4	3
96	264	3	1	55	4	18	37	4	11
99	261	3	0	56	14	18	48	4	19
102	258	2	59	57	14	18	56	4	27
105	255	2	58	58	1	19	2	4	34
108	252	2	56	58	40	19	3	4	42
111	249	2	55	59	14	19	3	4	49
114	246	2	53	59	40	18	59	4	54
117	243	2	49	59	57	18	53	4	58
120	240	2	44	60	0	18	42	5	2
123	237	2	39	59	49	18	27	5	4
126	234	2	34	59	35	18	8	5	6
129	231	2	28	59	19	17	44	5	9
132	228	2	22	58	59	17	17	5	9
135	225	2	16	58	32	16	44	5	6
138	222	2	10	57	56	16	7	4	59
141	219	2	3	56	41	15	25	4	52
144	216	1	55	55	27	14	38	4	41
147	213	1	47	54	55	13	47	4	26
150	210	1	38	54	25	12	52	4	10
153	207	1	29	53	54	11	51	3	53
156	204	1	19	53	23	10	44	3	33
159	201	1	10	52	54	9	34	3	10
162	198	1	0	52	33	8	20	2	43
165	195	0	51	52	18	7	4	2	14
168	192	0	41	52	8	5	43	1	43
171	189	0	31	52	3	4	19	1	9
174	186	0	21	52	2	2	54	0	43
177	183	0	10	52	2	1	27	0	35
180	180	0	0	52	2	0	0	0	0

34. HOW THE POSITIONS IN LONGITUDE OF THE FIVE PLANETS ARE CALCULATED

[178ᵇ] Therefore by means of the tables drawn up in this way by us we shall calculate without any difficulty the positions in longitude of the five wandering stars. There is approximately the same method of computation in all of them, though the three outer planets differ slightly from Venus and Mercury in this respect.

Therefore let us speak of Saturn, Jupiter, and Mars first. In their case the calculation is such that the mean movements—that is, the simple movement of the sun and the movement of parallax of the planet—are sought for any given time by the method described above. Next, the position of the highest apsis of the eccentric circle is subtracted from the simple position of the sun, and the movement of parallax is subtracted from the remainder; the first remainder is the anomaly of the eccentric circle of the planet. We shall look it up among the common numbers in one of the first two columns of the table, and correspondingly in the third column we shall take the additosubtraction of the eccentric circle, and the proportional minutes in the following column. We shall add this additosubtraction to the movement of anomaly of parallax and subtract it from the anomaly of the eccentric circle, if the number whereby we entered (the table) was found in the first column; and conversely we shall subtract it from the anomaly of the eccentric circle—if the number was found in the second column. The sum or remainder will be the corrected anomaly of parallax or the corrected anomaly of the eccentric circle—the proportional minutes being reserved for a use we shall speak of soon. Then we shall look up this corrected anomaly (of parallax) in the first two columns of common numbers; and from the corresponding place in the fifth column we shall take the additosubtraction arising from the movement of parallax, together with its excess found in the last column; and of that excess we shall take the proportional part in accordance with the number of proportional minutes; and we shall always add this proportional part to the additosubtraction. The sum will be the true parallax of the planet; and is to be subtracted from the corrected anomaly of parallax, if the (corrected anomaly) is less than a semicircle, or added, if greater than a semicircle. For in this way we shall have the true and apparent distance of the planet westward from the mean position of the sun; and when we have subtracted that distance from the mean position of the sun, the remainder will be the sought position of the planet [179ᵃ] in the sphere of the fixed stars, and the addition of the precession of the equinoxes will determine the position of the planet in relation to the spring equinox.

In the case of Venus and Mercury we shall use the distance from the highest apsis to the mean position of the sun as the anomaly of the eccentric circle; and by means of this anomaly we shall correct the movement of parallax and the anomaly of the

eccentric circle, as was said already. But if the additosubtraction of the eccentric circle and the corrected parallax are of the same quality or species (*i.e.*, are both additive or both subtractive), they are simultaneously added to or subtracted from the mean position of the sun. But if they are of different species, the lesser is subtracted from the greater; and by means of the remainder there will take place that which we have just mentioned, according to the additive or subtractive property of the greater number; and the final result will be the position which we are looking for.

35. ON THE STATIONS AND RETROGRADATIONS OF THE FIVE WANDERING STARS

Moreover, the knowledge of where and when the stations, retrogradations, and returns take place and how great they are seems also to pertain to the account of movement in longitude. The mathematicians, especially Apollonius of Perga, have dealt a good deal with them; but they have done so under the assumption of only one irregular movement, namely, that whereby the planets are moved with respect to the sun and which we have called the parallax due to the great orbital circle of the Earth.

For if the circles of the planets—whereon all the planets are borne with unequal periods of revolution but in the same direction, *i.e.*, towards the east—are homocentric with the great orbital circle of the Earth, and some planet on its own orbital circle and within the great orbital circle, such as Venus or Mercury, has greater velocity than the movement of the Earth has; *and if a straight line drawn from the Earth cuts the orbital circle of the planet in each a way that half the segment comprised within the orbital circle has the same ratio to the line which extends from our point of vision the Earth to the lower and convex are of the intersected orbital circle, as does the movement of the Earth to the velocity of the planet then, if a point is made at the extremity of this line drawn to the arc which is at the perigee of the circle of the planet, the point will separate the retrogradation from the progression, so that when the planet is at that position, it will have the appearance of stopping.*

Similarly in the case of the three outer planets which have a movement slower than the velocity [179^{b}] of the Earth, *if a straight line drawn through our point of vision cuts the great orbital circle, in such a way that half the segment comprised within the orbital circle has the same ratio to the line which extends from the planet to our point of vision located on the nearer and convex surface of the orbital circle, as does the movement of the planet to the velocity of the Earth; then the planet when in that position will present to our vision the appearance of stopping.*

But if half the segment comprised within the circle, as was said, *has a greater ratio to the remaining external segment than the velocity of the Earth has to the velocity of Venus or Mercury, or than the movement of any of the three upper planets has to the velocity of the*

Earth; then the planet will progress eastward; but if the ratio is less, then it will retrograde westward.

In order to demonstrate all this, Apollonius took a certain lemma, which was in accord with the hypothesis of the immobility of the Earth but which none the less squares with our principle of terrestrial mobility and which for that reason we too shall employ. And we can enunciate it in this form: *if the greater side of a triangle is so cut that one of the segments is not less than the adjoining side, then this segment will have a greater ratio to the remaining segment than the angles on the side cut, taken in reverse order will have to one another.*

For let *BC* be the greater side of triangle *ABC*; and if on side *BC*

$$CD < AC,$$

then I say that

$$CD : BD > \text{angle } ABC : \text{angle } BCA.$$

Now it is demonstrated as follows. Let the parallelogram *ADCE* be completed; and *BA* and *CE* extended will meet at point *E*. Accordingly since

$$AE < AC,$$

the circle described with centre *A* and radius *AE* will pass through *C* or beyond it. Now let *GEC* be the circle, and let it pass through *C*. Since

$$\text{trgl. } AEF > \text{sect. } AEG,$$

while

$$\text{trgl. } AEC < \text{sect. } AEC;$$

then

$$\text{trgl. } AEF : \text{trgl. } AEC > \text{sect. } AEG : \text{sect. } AEC.$$

But

$$\text{trgl. } AEF : \text{trgl. } AEC = \text{base } FE : \text{base } EC.$$

Therefore

$$FE : EC > \text{angle } FAE : \text{angle } EAG.$$

But

$$FE : EC = CD : DB.$$

And

$$\text{angle } FAE = \text{angle } ABC;$$

and

$$\text{angle } EAC = \text{angle } BCA.$$

Accordingly

$$[180^a] \quad CD : DB > \text{angle } ABC : \text{angle } ACB.$$

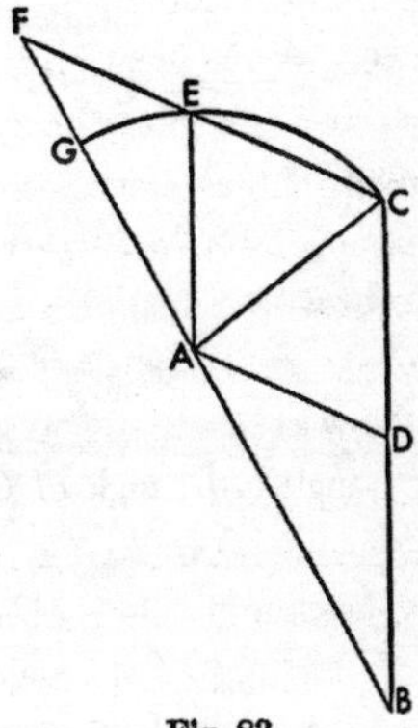

Now it is manifest that the ratio will be much greater if it is not assumed that

$$CD = AC = AE$$

but that

$$CD > AE.$$

Fig. 83

Now let *ABC* be the circle of Venus or Mercury around centre *D*; and let the Earth *E* outside the circle be movable around the same centre *D*. From *E* our point of vision let the straight line *ECDA* be drawn through the centre of the circle; and let *A* be the position farthest from the Earth, and *C* the nearest. And let *DC* be put down as having a greater ratio to *CE* than the movement of the point of vision has to the velocity of the planet. Accordingly it is possible to find a line *EFB* such that half *BF* has the same ratio to *FE* that the movement of the point of vision has to the movement of the planet. For let line *EFB* be moved away from centre *D* and be decreased along *FB* and increased along *EF*, until we meet with what is demanded.

I say that *when the planet is set up at point F, it will present to us the appearance of stopping; and that whatever size of the arc we take on either side of F, we shall find the planet progressing, if the arc is taken in the direction of the apogee, and retrograding, if in as direction of the perigee.*

For first let the arc *FG* be taken in the direction of the apogee: let *EGK* be extended, and let *BG*, *DG*, and *DF* be joined. Accordingly since in triangle *BGE* segment *BF* of the greater side *BE* is greater than *BG*, then

$$BF : EF > \text{angle } FEG : \text{angle } GBF.$$

Furthermore,

$$\tfrac{1}{2}BF : FE > \text{angle } FEG : 2 \text{ angle } GBF,$$

i.e.,

$$\tfrac{1}{2}BF : FE > \text{angle } FEG : \text{angle } GDF.$$

But

$$\tfrac{1}{2}BF : FE = \text{movement of Earth : movement of planet.}$$

Therefore

$$\text{angle } FEG : \text{angle } GDF < \text{velocity of Earth : velocity of planet.}$$

Now let

$$\text{angle } FEL : \text{angle } FDG = \text{movement of Earth : movement of planet.}$$

Therefore

$$\text{angle } FEL > \text{angle } FEG.$$

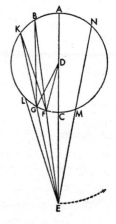

Accordingly, during the time in which the planet traverses arc *GF* of the orbital circle, our line of sight [180ᵇ] will be thought to have traversed during that time the contrary space between line *EF* and line *EL*. It is manifest that in the same time in which to our sight arc *GF* transports the planet westward in accordance with the smaller angle *FEG*, the passage of the Earth drags it back eastward in accordance with the greater angle *FEL*, so that the planet will go on increasing its angular distance eastward by angle *GEL* and will not seem to have come to a stop yet.

354

Now it is manifest that the opposite of this can be shown by the same means. If in the same diagram we put down that

$$1/_2GK : GE = \text{movement of Earth : velocity of planet;}$$

and if we take arc *GF* in the direction of the perigee and away from straight line *EK*, and join *KF* and make triangle *KEF*, where

$$GE > EF;$$

then

$$KG : GE < \text{angle } FEG : \text{angle } FKG.$$

Thus too

$$1/_2KG : GE < \text{angle } FEG : 2 \text{ angle } FKG,$$

i.e.,

$$1/_2KG : GE < \text{angle } FEG : \text{angle } GDF,$$

conversely to what was shown before. And it is inferred by the same means that

angle *GDF* : angle *FEG* < velocity of planet : velocity of line of sight.

Accordingly, when angle *GDF* has been made greater, so that the angles have the same ratio, then the planet will complete a greater movement westwards than progression demands.

Hence it is also manifest that if we make

arc *FC* = arc *CM*

the second station will be at point *M*; and if line *EMN* is drawn,

$$1/_2MN : ME = 1/_2BF : FE = \text{velocity of Earth : velocity of planet;}$$

and accordingly points *M* and *F* will designate the two stations and will determine the whole arc *FCM* as retrogressive and the remainder of the circle as progressive.

Moreover, it follows that at certain distances

$$DC : CE > \text{velocity of Earth : velocity of planet;}$$

and it will not be possible to draw another straight line in the ratio (which the velocity of the Earth has to the velocity of the planet); and the planet will not seem to stop or to retrograde. For since it was assumed that in triangle *DEG*

$$DC < EG;$$
$$\text{angle } CEG : \text{angle } CDG < DC : CE$$

but

$$DC : CE > \text{velocity of Earth : velocity of planet.}$$

Therefore also

angle *CEG* : angle *CDG* < velocity of Earth : velocity of planet.

Where that occurs, the planet will progress; [181ᵃ] and we shall not find anywhere in the orbital circle of the planet an arc through which it seems to retrograde. All this concerning Venus and Mercury, which are inside the great orbital circle (of the Earth).

We can demonstrate this concerning the three outer planets by the same method and with the same diagrams—merely by reversing the names, so that we put down

ABC as the great orbital circle of the Earth and as the circuit of our point of vision and the planet at *E*, whose movement in its own orbital circle is less than the speed of our point of vision in the great orbital circle. The rest of the demonstration will proceed as before.

36. How the Times, Positions, and Arcs of the Retrogradations are Determined

Now if the orbital circles which bear the wandering stars were homocentric with the great orbital circle, it would be easy to establish that which the demonstrations promise, as the ratio of the velocity of the planet to the velocity of the point of vision would always be the same. But the orbital circles are eccentric, and hence their movements appear as irregular. For that reason it will be necessary for us to assume irregular and corrected movements everywhere as the differences of velocity and to employ them in the demonstrations, and not the simple and regular movements, except when the planet happens to be at its mean longitudes, the only place where it seems to be carried in its orbital circle with a mean movement.

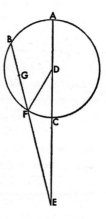

Now we shall show this in the case of Mars, so that the retrogradations of the other planets may become clearer by means of this example. For let *ABC* be the great orbital circle, on which our point of vision revolves; and let the planet be at point *E*. From the planet let the straight line *ECDA* be drawn through the centre of the orbital circle; and let *EFB* also be drawn. Half of chord *BF—i.e.*, chord *GF*—will have the ratio to line *EF* which the varying velocity of the planet has to the velocity of the line of sight, whereby it exceeds the planet. Our problem is to find arc *FC* of half the retrogradation, or *ABF*, so as to know at what distance from its farthest position from *A* the planet becomes stationary and what the angle comprehended by *FEC* is. For by means of this we shall foretell the time and position of such an affection of the planet.

Now let the planet be placed at the mean apsis of the eccentric circle, where the movements of longitude and of anomaly differ very little from the regular movements. Therefore in the case of the planet Mars, since its mean movement (that is half of the line *BF*) [181b] is 1P8°7", the motion of parallax, which is the relation of our vision to the mean movement of the star, consists of one part and is the straight line *EF*. Hence

$$EB = 3\text{P}16'14"$$

and likewise

$$\text{rect. } BE, EF = 3\text{P}16'14".$$

Now we have shown that

$$DA = 6,580,$$
$$\text{where } DE = 10,000,$$

and *DA* is the radius of the orbital circle.
But

$$DA = 39\text{P}29',$$
$$= \text{where } DE = 60\text{P};$$

and

$$AE : EC = 99\text{P}29' : 20\text{P}31'.$$

And

$$\text{rect. } AE, EC = \text{rect. } BE, EF = 2,041\text{P}4'.$$

Accordingly by reduction

$$2,041\text{P}4' \div 3\text{P}16'14'' = 624\text{P}4',$$

and similarly

$$\text{side } EF = 24\text{P}58'52'',$$
$$\text{where } DE = 60\text{P}.$$

But

$$EF = 4,163$$
$$\text{where } DE = 10,000$$
$$\text{and } DF = 6,580.$$

Accordingly, as the sides of triangle *DEF* are given,

$$\text{angle } DEF = 27°15',$$

which is the angular retrogradation of the planet, and

$$\text{angle } CDF = 16°50',$$

which is the angular anomaly of parallax. Accordingly, since the planet, when first stationary, appeared on line *EF*; and the planet, when opposite the sun, on line *EC*; if the planet is not moved eastward, the 16°15' of arc *CF* will comprehend the 27°25' of angle *AEF* found to be the retrogradation; but according to the ratio set forth of the velocity of the planet to the velocity of our line of sight, 16°5' corresponds to the section of the anomaly of parallax and approximately 19°6'39" corresponds (to the section of the anomaly) of longitude of the planet. Now

$$27°15' - 19°6'39'' = 8°8',$$

which is the distance from the other station to the solar opposition—and there are approximately $36\frac{1}{2}$ days during which the anomaly in longitude is 19°6'39"—and hence the total retrogradation is 16°16' in 73 days. These things which have been demonstrated for the mean longitudes of the eccentric circle can be similarly demonstrated for other positions—the planet being credited with an always varying velocity, according as its position demands, as we said.

Hence in Saturn, Jupiter, and Mars the same way of demonstration is open, provided we take the point of sight instead of the planet and the planet instead of the point of sight. Now the reverse of what occurs in the orbital circles which the Earth encloses occurs in the orbital circles which enclose the Earth; and let that be enough, so that we won't have to repeat the same old song. Nevertheless, since the variable movement of the planet with respect to the point of sight and to the ambiguity of the stationary points—of which the theorem of Apollonius does not relieve us—give no little difficulty; I do not know whether it would not be better to investigate the stations simply and in connection with the nearest position, by the method whereby we investigate by means of the known numbers of their movements the conjunction of the planet, when opposite the sun, with the line of mean movement of the sun, or the conjunction of any of the planets. And we shall leave that to your pleasure.

BOOK SIX

[182ª] We have indicated to the best of our ability what power and effect the assumption of the revolution of the Earth has in the case of the apparent movement in longitude of the wandering stars and in what a sure and necessary order it places all the appearances. It remains for us to occupy ourselves with the movements of the planets by which they digress in latitude and to show how in this case too the selfsame mobility of the Earth exercises its command and prescribes laws for them here also. Moreover this is a necessary part of the science, as the digressions of these planets cause no little variation in the rising and setting, apparitions and occultations, and the other appearances of which there has been a general exposition above. And their true positions are said to be known only when their longitude together with their latitude in relation to the ecliptic has been established. Accordingly by means of the assumption of the mobility of the Earth we shall do with perhaps greater compactness and more becomingly what the ancient mathematicians thought to have demonstrated by means of the immobility of the Earth.

1. GENERAL EXPOSITION OF THE DIGRESSION IN LATITUDE OF THE FIVE WANDERING STARS

The ancients found in all the planets two digressions in latitude answering to their twofold irregularity in longitude—one digression taking place by reason of the eccentricity of the orbital circles, and the other in accordance with the epicycles. In place of the epicycles, as has been often repeated, we have taken the single great orbital circle of the Earth—not that the orbital circle has some inclination with respect to the plane of the ecliptic fixed once and forever, since they are the same, but that the orbital circles of the planets are inclined to this plane [182ᵇ] with a variable obliquity, and this variability is regulated according to the movement and revolutions of the great orbital circle of the Earth.

But since the three higher planets, Saturn, Jupiter, and Mars, move longitudinally under different laws from those under which the remaining two do, so also they differ not a little in their latitudinal movement. Accordingly, the ancients first examined where and how great their farthest northern limits in latitude were. Ptolemy found the limits in the case of Saturn and Jupiter around the beginning of Libra, but in the case of Mars around the end of Cancer near the apogee of the eccentric circle. But in our time we found this northern limit in the case of Saturn at 7° of Scorpio, in the case of Jupiter at 27° of Libra, in the case of Mars at 27° of Leo, according as the apogees have been changing around down to our time; for the inclinations and the cardinal points of latitude follow upon the movement of those orbital circles. At corrected or apparent

distances of 90° between these limits, they seem to be making no digression in latitude, wherever the Earth happens to be at that time. Therefore, when they are at these mean longitudes, they are understood to be at the common section of their orbital circles with the ecliptic, just as the moon was at the ecliptic sections. Ptolemy calls these points the nodes: the ascending node, after which the planet enters upon northern latitudes; and the descending node, after which the planet crosses over into southern latitudes— not that the great orbital circle of the Earth, which always remains the same in the plane of the ecliptic, gives them any latitude; but every digression in latitude is measured from the nodes and varies greatly in positions different from the nodes. And according as the Earth approaches other positions, where the planets are seen to be opposite the sun and *acronycti*, the planets always move with a greater digression than in any other position of the Earth: in the northern semicircle to the north, and in the southern to the south, and with greater variation than the approach or withdrawal of the Earth demands. By that happening, it is known that the inclination of their orbital circles is not fixed, but that it changes in a certain movement of libration commensurable with the revolutions of the great circle of the Earth, as will be said a little farther on.

Now Venus and Mercury seem to digress somewhat differently but under a fixed law which has been observed to hold at the mean, highest, and lowest apsides. For at the mean longitudes, namely when the line of the mean movement of the sun is at a quadrant's distance from their highest or lowest apsis, and the planets as evening or morning stars are themselves at a distance of a quadrant of their orbital circle from the same line of mean movement of the sun; [183ª] the ancients found that the planets had not digressed from the ecliptic, and hence the ancients understood them to be at that time the common section of their separate orbital circles and the ecliptic. This section passes through their apogees and perigees; and accordingly when they are higher or lower than the Earth, they then make manifest digressions—the greatest digressions at their greatest distances from the Earth, *i.e.*, at the evening apparition or at the morning occultation, when Venus is farthest north, and Mercury farthest south. And conversely at a position nearer to the Earth, when they undergo occultation in the evening or emerge in the morning, Venus is to the south, and Mercury to the north. Vice versa, when the Earth is at the position opposite to this and at the other mean apsis namely, when the anomaly of the eccentric circle is 270°—Venus is apparent at its greater southern distance from the Earth, and Mercury is to the north, and at a nearer position of the Earth Venus is to the north, and Mercury to the south. At the solstice of the Earth at the apogee of these planets, Ptolemy found that the latitude of Venus the morning star was northern, and of Venus the evening star, southern; and inversely in the case of Mercury: southern when the morning star, and northern when the evening star. These relations are similarly reversed at the opposite position of the perigee, so that

Venus Lucifer is seen in the south, and Venus Vesperugo in the north; but Mercury as morning star in the north, and Mercury as evening star in the south. And the ancients found that at both these positions the northern digression of Venus was always greater than the southern, and that the southern digression of Mercury was greater than the northern.

Taking this as an occasion, the ancients reasoned out a twofold latitude for this position, and a threefold latitude universally. They called the first latitude, which occurs at the mean longitudes, the inclination; the second, which occurs at the highest and lowest apsides, the obliquation; and the third one, which occurs in conjunction with the second, the deviation: it is always northern in the case of Venus and southern in the case of Mercury. Between these four limits the latitudes are mixed with one another, and alternately increase and decrease and yield mutually; and we shall give the right causes for all that.

2. HYPOTHESES OF THE CIRCLES ON WHICH THE PLANETS ARE MOVED IN LATITUDE

Accordingly in the case of these five planets we must assume that their orbital circles are inclined to the plane of the ecliptic—the common section being through the diameter of the ecliptic—by a variable but regular inclination, [183^b] since in Saturn, Jupiter, and Mars the angle of section receives a certain libration around that section as around an axis, like the libration which we demonstrated in the case of the precession of the equinoxes, but simple and commensurable with the movement of parallax. The angle of section is increased and decreased by this libration within a fixed period, so that, whenever the Earth is nearest to the planet, *i.e.*, to the planet in opposition to the sun, the greatest inclination of the orbital circle of the planet occurs; at the contrary position the least inclination; at the mean position, the mean inclination: consequently, when the planet is at its farthest limit of northern or southern latitude, its latitude appears much greater at the nearness of the Earth than at its greatest distance from the Earth. And although this irregularity can be caused only by the unequal distances of the Earth, in accordance with which things nearer seem greater than things farther away; nevertheless there is a rather great difference between the excess and the deficiency of these planetary latitudes: and that cannot take place unless the orbital circles too have a movement of libration with respect to their obliquity. But, as we said before, in the case of things which are undergoing a libration, we must take a certain mean between the extremes.

In order that this may be clearer, let *ABCD* be the Earth's great orbital circle in the plane of the ecliptic with centre *E*; and let *FGKL* the orbital circle of the planet be inclined to *ABCD* in a mean and permanent declination whereof *F* is the northern limit in latitude, *K* the southern, *G* the descending node of section, and *BED* the

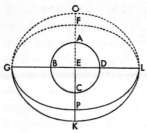

common section, which is extended in the straight lines *GB* and *DL*. And those four termini do not change, except along with the movement of the apsides. Let it be understood however that the movement of the planet in longitude takes place not on the plane of circle *FG* but on *OP*, another circle which is inclined to and homocentric with *FG*. These two circles cut one another in that same [184ª] straight line *GBDL*. Therefore, while the planet is being borne on orbital circle *OP*, it meanwhile falls upon plane *FK* by the movement of libration, goes beyond plane *FK* in either direction, and on that account makes the latitude appear variable.

For first let the planet be at point *O* at its greatest northern latitude and at its position nearest to the Earth in *A*; then the latitude of the planet will increase in proportion to *OGF* the angle of greatest inclination of orbital circle *OGP*. This movement (of libration) is a movement of approach and withdrawal, because by hypothesis it is commensurable with the movement of parallax: if then the Earth is at *B*, point *O* will coincide with *F*, and the latitude of the planet will appear less in the same position than before; and it will be much less if the Earth is at point *C*. For *O* will cross over to the farthest and most diverse part of its libration, and will leave only as much latitude as is in excess over the subtractive libration of the northern latitude, namely over the angle equal to *OGF*. Hence the latitude of the planet around *F* in the north will increase throughout the remaining semicircle *CDA*, until the Earth returns to the first point *A*, from which it set out. There will be the same way of progress for the meridian planet set up around point *K*—the movement of the Earth starting from *C*. But if the planet, in opposition to the sun or hidden by it, is at one of the nodes *G* or *L*, even though at that time the orbital circles *FK* and *OP* have their greatest inclination to one another, on that account no planetary latitude is perceptible, namely because the planet is at the common section of the orbital circles. From that, I judge, it is easily understood how the northern latitude of the planet decreases from *F* to *G*; and the southern latitude increases from *G* to *K*, but vanishes totally at *L* and becomes northern. And this is the way with those three higher planets.

Venus and Mercury differ from them no little in their latitudes, as in longitude, because they have the common sections of the orbital circles located through the apogee and perigee. Now their greatest inclinations at the mean apsides become changeable by a movement of libration, as in the case of the higher planets; but they undergo furthermore a libration dissimilar to the first. Nevertheless both librations are commensurable with the revolutions of the Earth, but not in the same way. For the first libration has the following property; when there has been one revolution of the Earth

with respect to the apsides of the planets, there have been two revolutions of the movement of libration having as an immobile axis the section through the apogee and the perigee, which we spoke of; so that whenever the line of mean movement of the sun is at the perigee or apogee of the planets, the greatest angle of section occurs; while the least angle occurs at the mean longitudes. [184^b] But the second libration supervening upon this one differs from it in that, by possessing a movable axis, it has the following effect: namely, that when the Earth is located at a mean longitude, the planet of Venus or Mercury is always on the axis, *i.e.*, at the common section of this libration, but shows its greatest deviation when the Earth is in line with its apogee or perigee—Venus always being to the north, as was said, and Mercury to the south; although on account of the former simple inclination they should at this time be lacking latitude.

For example, when the mean movement of the sun is at the apogee of Venus and Venus is in the same position, it is manifest that, in accordance with the simple inclination and the first libration, Venus, being at the common section of its orbital circle with the plane of the ecliptic, would at that time have had no latitude; but the second libration, which has its section or axis along the transverse diameter of the eccentric orbital circle and cuts at right angles the diameter passing through the highest and lowest apsis, adds its greatest deviation to the planet. But if at this time Venus is in one of the other quadrants and around the mean apsides of its orbital circle, then the axis of this libration will coincide with the line of mean movement of the sun, and Venus itself will add to the northern obliquity the greatest deviation, which it subtracts from the southern obliquity and leaves smaller. In this way the libration of deviation is made commensurate with the movement of the Earth.

In order that these things may be grasped more easily, let *ABCD* be drawn again as the great orbital circle. Let *FGK* be the orbital circle of Venus or Mercury: it is eccentric to circle *ABC* and inclined to it in accordance with the equal inclination *FGK*. Let

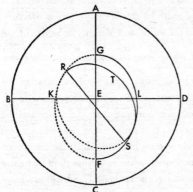

FG be the common section of these two circles through *F* the apogee of the orbital circle and *G* the perigee. First for the sake of an easier demonstration let us put down *GKF* the inclination of the eccentric orbital circle as simple and fixed, or, if you prefer, as midway between the greatest and the least inclination, except that the common section *FG* [185^a] changes according to the movement of the perigee and apogee. When the Earth is on this common section, *i.e.*, at *A* or *C*, and the planet is on the

same line, it is manifest that at that time the planet would have no latitude, since all latitude is sideways in the semicircles *GKF* and *FLG*, whereon the planet effects its northern or southern approaches, as has been said, in proportion to the inclination of circle *FKG* to the plane of the ecliptic. Now some call this digression of the planet the obliquation; others, the reflexion. But when the Earth is at *B* or *D*, *i.e.*, at the mean apsides of the planet, there will be the same latitudes *FKG* and *GFL* above and below; and they call them the declinations. And so these latitudes differ nominally rather than really from the former latitudes, and even the names are interchanged at the middle positions. But since the angle of inclination of these circles is found to be greater in the obliquation than in the declination, the ancients understood this as taking place through a certain libration, curving itself around section *FG* as an axis, as was said above. Accordingly since the angle of section is known in both cases, it will be easy to understand from the difference between them how great the libration from least to greatest inclination is.

Now let there be understood another circle, the circle of deviation, which is inclined to circle *GKFL* and homocentric in the case of Venus but eccentric to the eccentric circle in the case of Mercury, as will be said later: And let *RS* be their common section as axis of libration, an axis movable in a circle, in such fashion that when the Earth is at *A* or *B*, the planet is at the farthest limit of deviation, wherever that is, as at point *T*; and as far as the Earth has advanced from *A*, so far away from *T* let the planet be understood to have moved, while the inclination of the circle of deviation decreases, so that when the Earth bas measured the quadrant *AB*, the planet should be understood as having arrived at the node of this latitude, *i.e.*, at *R*. But as at this time the planes coincide at the mean movement of libration and are tending in different directions, the remaining semicircle of the deviation, which before was southerly, becomes northern; and as Venus passes into this semicircle, Venus avoids the south and seeks the north again, never to seek the south by this libration, just as Mercury, by crossing in the opposite direction, stays in the south; and Mercury also differs from Venus in that its libration takes place not in a circle homocentric with an eccentric circle but in a circle eccentric to an eccentric circle.

We employed an epicycle instead of this eccentric circle in demonstrating the irregularity in the movement in longitude. But since there we were considering longitude without latitude, and here latitude [185b] without longitude, and as one and the same revolution comprehends and brings them to pass equally; it is clear enough that it is one and the same movement and one and the same libration which can cause both irregularities and be eccentric and have an inclination at the same time; and that there is no other hypothesis besides this which we have just spoken of and will say more about below.

3. HOW GREAT THE INCLINATIONS OF THE ORBITAL CIRCLES OF SATURN, JUPITER, AND MARS ARE

After setting out our hypothesis for the digressions of the five planets, we must descend to the things themselves and discern singulars; and first how great the inclinations of the single circles are. We measure these inclinations against the great circle which passes through the poles of the circle having the inclination and is at right angles to the ecliptic; the transits in latitude are observed in relation to this great circle. For when we have apprehended these (inclinations), the way of learning the latitudes of each planet will be disclosed. Beginning once more with the three higher planets, we find that according to Ptolemy the digression of Saturn in opposition to the sun at the farthest limits of southern latitude was 3°5', the digression of Jupiter 2°7', that of Mars 7°; but in opposite positions, namely when they were in conjunction with the sun, the digression of Saturn was 2°2', that of Jupiter 1°5', and that of Mars only 5', so that it almost touched the ecliptic—according as it is possible to mark the latitudes from the observations which he took in the neighbourhood of their occultations and apparitions.

Let that be kept before us; and in the plane which is at right angles to the ecliptic and through its centre, let *AB* be the common section (of the plane) with the ecliptic, and *CD* the common section (of the plane) with any of the three eccentric circles

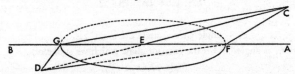

through the greatest northern and southern limits. Moreover, let *E* be the centre of the ecliptic, and *FEG* the diameter of the great orbital circle of the Earth. Now let *D* be the southern latitude and *C* the northern; and let *CF*, *CG*, *DF*, and *DG* be joined.

But the ratios of *EG* the great orbital circle of the Earth to *ED* the eccentric circle of the planet at any of their given positions have already been demonstrated above in the cases of the single (planets). But the positions of greatest latitudes have been given by the observations. Therefore, since angle *BGD*, the angle of greatest southern latitude and an exterior angle of triangle *EGD*, has been given, the interior and opposite angle *GED*, the angle of greatest southern inclination of the eccentric circle to the plane of the ecliptic, will also be given by what has been shown concerning plane triangles.

Similarly we shall demonstrate the least inclination by mean of the least southern latitude, namely by means of angle [186ª] *EFD*. Since in triangle *EFD* the ratio of side *EF* to side *ED* is given together with angle *EFD*, we shall have *GED* given, the exterior angle and angle of least southern inclination: hence from the difference between

both declinations we shall have the total libration of the eccentric circle in relation to the ecliptic. Moreover against these angles of inclination we shall measure the opposite northern latitudes, that is to say, angles *AFC* and *EGC*; and if they agree with the observations, it will be a sign that we have not erred at all.

Now as our example we shall take Mars, which has a greater digression in latitude than any of the others. Ptolemy marked the greatest southern latitude as being approximately 7° in the case of the perigee of Mars, and the greatest northern latitude as 4°20' at the apogee. But as we have assumed that

$$\text{angle } BGD = 6°50',$$

we shall find that correspondingly

$$\text{angle } AFC \fallingdotseq 4°30'.$$

For since

$$EG : ED = 1^P : 1^P22'26''$$

and since

$$\text{angle } BCD = 6°50';$$
$$\text{angle } DEG \fallingdotseq 1°51',$$

which is the angle of greatest southern inclination.
And since

$$EF : CE = 1^P : 1^P39'57'',$$

and

$$\text{angle } CEF = \text{angle } DEG = 1°51',$$

it follows that, as angle *CFA* is the exterior angle which we spoke of,

$$\text{angle } CFA = 4^1/_2°,$$

when the planet is in opposition to the sun.

Similarly, in the opposite position where it is in conjunction with the sun, if we assume that

$$\text{angle } DFE = 5',$$

then, since sides *DE* and *EF* and angle *EFD* are given,

$$\text{angle } EDF = 4',$$

and, as exterior angle,

$$\text{angle } DEG \fallingdotseq 9',$$

which is the angle of least inclination. And that will show us that

$$\text{angle } CGE = 6',$$

which is the angle of northern latitude. Therefore, by the subtraction of the least inclination from the greatest,

$$1°5' - 9' = 1°42'$$

which is the libration of this inclination, and

$$^1/_2(1°42') \fallingdotseq 50^1/_2'.$$

In the case of the other two, Jupiter and Saturn, there is a similar method for discovering the angles of the inclinations together with the latitudes; for the greatest inclination of Jupiter is 1°42', and the least 1°18'; [186ᵇ] so that its total libration does not comprehend more than 24'. Now the greatest inclination of Saturn is 2°44', and the least 2°16'; and the libration between them is 19'. Hence by means of the least angles of inclination, which occur at the opposite position, when the planets are hidden beneath the sun, their digressions in latitude away from the ecliptic will be exhibited: that of Saturn as 2°3' and that of Jupiter as 1°6'—as were to be shown and reserved for the tables to be drawn up below.[1]

4. On the Exposition of the Other Latitudes in Particular and in General

Now that these things have been shown, the latitudes of these three planets will be made clear in general and in particular. For as before, let *AB* the line through the farthest limits of digression be the common section of the plane perpendicular to the ecliptic. And let the northern limit be at *A*; and let *CD*, which cuts *AB* in point *D*, be

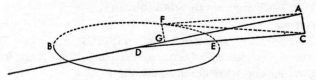

the perpendicular common section of the orbital circle of the planet. And with *D* as centre let *EF* the great orbital circle of the Earth be described. From the opposition, which is at *E*, let any known arc, such as *EF*, be measured, and from *F* and from *C*, the position of the planet, let the perpendiculars *CA* and *FG* be drawn to *AB*; and let *FA* and *FC* be joined.

We are first looking to see how great *ADC* the angle of inclination of the eccentric circle is, with this set-up. Now it has been shown that the inclination was greatest when the Earth was at point *E*. Moreover it has been made clear that the total libration is commensurate with the revolution of the Earth on circle *EF* in relation to the diameter *BE*, as the nature of libration demands. Therefore on account of arc *EF* being given, the ratio of *ED* to *EG* will be given; and that is the ratio of the total libration to which angle *ADC* has just now decreased. For that reason angle *ADC* is given in this case. Accordingly triangle *ADC* has all its angles given together with its sides. But since by the foregoing, *CD* has a given ratio to *ED*, the ratio of *CD* to the remainder *DG* is given. Accordingly the ratios of *CD* and *AD* to *GD* are given. And hence the remainder *AG* is given. Hence too *FG* is given; for

[1]p. 379.

$$FG = \frac{1}{2} \text{ ch. 2 } EF.$$

Therefore as two sides of the right triangle AGF have been given, side AF is given, and the ratio of AF to AC. Finally as two sides of right triangle ACF [187ª] have been given, angle AFC will be given; and that is the angle of apparent latitude, which we were looking for.

Once more we shall take Mars as our example of this. Let its limit of greatest southern latitude be around A, which is approximately at its lowest apsis. Now let the position of the planet be at C, where—as has been demonstrated—the angle of inclination was greatest, *i.e.*, 1°50', when the Earth was at point E. Now let us put the Earth at point F and the movement of parallax at 45° in accordance with arc EF: therefore

$$\text{line } FG = 7{,}071,$$
$$\text{where } ED = 10{,}000,$$

and

$$GE = 10{,}000 - 7{,}071 = 2{,}929,$$

which is the remainder of the radius. Now it has been shown that

$$\frac{1}{2} \text{ libration of angle } ADC = 50\frac{1}{2}°;$$

and half of the libration has the following ratio of increase and decrease in this case,

$$DE : GE = 50\frac{1}{2}' : 15'.$$

Now at present

$$\text{angle } ADC = 1°50' - 15' = 1°35',$$

which is the angle of inclination. On that account triangle ADC will have its sides and angles given; and since it has been shown above that

$$CD = 9{,}040,$$
$$\text{where } ED = 6{,}580;$$
$$FG = 4{,}653,$$
$$AD = 9{,}036,$$

and by subtraction

$$AEG = 4{,}383,$$

and

$$AC = 249\frac{1}{2}.$$

Accordingly, in right triangle AFG, since

$$\text{perpendicular } AG = 4{,}383$$

and

$$\text{base } FG = 4{,}653$$
$$\text{side } AF = 6{,}392.$$

Thus finally in triangle ACF, whereof

$$\text{angle } CAF = 90°$$

and sides AC and AF are given,

$$\text{angle } ACF = 2°15',$$

which is the angle of apparent latitude in relation to the Earth placed at *F*. We shall apply similar reasoning in the case of Saturn and Jupiter.

5. On the Latitudes of Venus and Mercury

Venus and Mercury remain, and their transits in latitude will be demonstrated, as I said, by means of three simultaneous and complicated latitudinal divagations. [187^b] In order that they may be discerned separately, we shall begin with the one which the ancients call declination, as if from a simpler handling of it. And it happens to the declination alone to be sometimes separate from the others; and that occurs around the mean longitudes and around the nodes in accordance with the exact movements in longitude when the Earth has moved through a quadrant of a circle from the apogee or perigee of the planet. For when the Earth is very near, a northern or southern latitude of 6°22' is found in the case of Venus, and 4°5' in the case of Mercury; but at the greatest distance from the Earth, 1°2' in the case of Venus; and in the case of Mercury, 1°45'. Thereby the angles of inclination at this position are made manifest by means of the tables of additosubtractions which have been drawn up; and for Venus in that position at its greatest distance from the Earth the latitude is 1°2', and at its least distance 6°22', and on either side (of the mean latitude) the arc of the circle (through the poles of the orbital circle and perpendicular to the plane of the ecliptic) is approximately 2¹/₂°; but in the case of Mercury the 1°45' at its greatest distance and the 4°5' at its least demand 6¹/₄° as the (total) arc of its circle: consequently the angle of inclination of the circles of Venus is 2°30', and that of Mercury is 6¹/₄°, whereof four right angles are equal to 360°. By means of these (angles) the particular latitudes of declination can be unfolded, as we shall demonstrate, and first in the case of Venus.

For in the plane of the ecliptic and through the centre of the perpendicular plane, let *ABC* be the common section (of the two planes) and *DBE* the common section (of the perpendicular plane) with the plane of the orbital circle of Venus. And let *A* be the

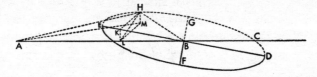

centre of the Earth, *B* the centre of the orbital circle of the planet, and *ABE* the angle of inclination of the orbital circle to the ecliptic. Let circle *DFEG* be described around *B*, and let diameter *FBG* be drawn perpendicular to diameter *DE*. Now let it be understood that the plane of the circle is so related to the assumed perpendicular plane that lines in the plane of the circle which are drawn at right angles to *DE* are parallel to one

another and to the plane of the ecliptic; and in the plane of the circle line *FBG* alone has been drawn.

Now our problem is to find out, by means of the given straight lines *AB* and *BC* together with angle *ABE* the given angle of inclinations, how far distant in latitude the planet is, when, for example, [188ª] it is 45° distant from *E* the point nearest to the Earth; and, following Ptolemy, we have chosen this position so that it may become apparent whether the inclination of the orbital circle adds any difference in longitude to Venus or Mercury. For such differences should be most visible around the positions midway between the limits *D*, *F*, *E*, and *G*, because the planet when situated at these four limits has the same longitude as it would have without declination, as is manifest of itself.

Therefore, as was said, let us assume that

$$\text{arc } EH = 45°;$$

and let *HK* be drawn perpendicular to *BE*, and *KL* and *HM* perpendicular to the plane of the ecliptic; and let *HB*, *LM*, *AM*, and *AH* be joined. We shall have the right parallelogram *LKHM*, as *HK* is parallel to the plane of the ecliptic. For angle *LAM* comprehends the additosubtraction in longitude; and angle *HAM* comprehends the transit in latitude, since *HM* also falls perpendicular upon the same plane of the ecliptic. Accordingly, since

$$\text{angle } HBE = 45°;$$
$$HK = \frac{1}{2} \text{ ch. 2 } HE = 7{,}071,$$
$$\text{where } EB = 10{,}000.$$

Similarly in triangle *KBL*

$$\text{angle } BKL = 2\frac{1}{2}°$$

and

$$\text{angle } BLK = 90°,$$

and

$$\text{side } BK = 7{,}071,$$
$$\text{where } BE = 10{,}000;$$

hence

$$\text{side } KL = 308$$

and

$$\text{side } BL = 7{,}064.$$

But since, by what was shown above,

$$AB : BE \doteqdot 10{,}000 : 7{,}193;$$

then

$$HK = 5{,}086,$$

and

$$HM = KL = 221,$$

and

$$BL = 5,081;$$

hence, by subtraction,

$$LA = 4,919.$$

Moreover, as in triangle ALM side AL is given,

$$LM = HK,$$

and

$$\text{angle } ALM = 90°;$$

then

$$\text{side } AM = 7,075$$

and

$$\text{angle } MAL = 45°57',$$

which is the additosubtraction or great parallax of Venus according to calculation. Similarly, as in triangle MAH

$$\text{side } AM = 7,075$$

and

$$\text{side } MH = KL;$$
$$\text{angle } MAH = 1°47',$$

which is the angular declination in latitude.

And if it is not boring to examine what difference in the longitude of Venus is caused by this inclination, let us take triangle ALH, as we understand side LH to be the diagonal of parallelogram $LKHM$. For

$$LH = 5,091,$$
$$\text{where } AL = 4,919$$

and

$$\text{angle } ALK = 90° :$$

hence

$$\text{side } AH = 7,079.$$

Accordingly, as the ratio of the sides is given,

$$\text{angle } HAL = 45°59'.$$

But it has been shown that

$$\text{angle } MAL = 45°57';$$

therefore there is a difference of only 2', as was to be shown.

Again, in the case of Mercury, [188^{b}] with a similar scheme of declination we shall demonstrate the latitudes with the help of a diagram similar to the foregoing: wherein

$$\text{arc } EH = 45°,$$

so that again

$$HK = KB = 7,071,$$
where side AB = 10,000.

Accordingly, as can be gathered from the differences in longitude which have already been demonstrated, in this case

$$BK = KH = 2,975,$$
where radius BH = 3,953
and AB = 9,964.

And since it has been shown that

angle of inclination ABE = 6°15',
where 4 rt. angles = 360°;

accordingly, as the angles of right triangle BKL are given,

base KL = 304

and

perpendicular BL = 2,778.

And so by subtraction

$$AL = 7,186.$$

But also

$$LM = HK = 2,795;$$

accordingly, as in triangle ALM

angle L = 90°

and sides AL and LM have been given;

side AM = 7,710

and

angle LAM = 21°16',

which is the additosubtraction calculated.

Similarly, since in triangle AMH side AM has been given,

side MH = KL

and

angle M = 90°,

which is comprehended by sides AM and MH;

angle MAH = 2°16',

which is the latitude sought for. But if we wish to inquire how much is due to the true and the apparent additosubtraction, let us take LH the diagonal of the parallelogram: we deduce from the sides (of the parallelogram) that

$$LH = 2,811.$$

And

$$AL = 7,186.$$

Hence

angle *LAH* = 21°23',

which is the additosubtraction of apparent movement and has an excess of approximately 7' over the previously reckoned difference, (angle *LAM*), as was to be shown.

6. ON THE SECOND TRANSIT IN LATITUDE OF VENUS AND MERCURY ACCORDING TO THE OBLIQUATION OF THEIR ORBITAL CIRCLES IN THE APOGEE AND THE PERIGEE

That is enough on the transit in latitude of these planets, which occurs around the mean longitudes of their orbital circles: we have said that these latitudes are called the declinations. Now we must speak of those latitudes which occur at the perigee and apogee and to which the third digression, the deviation, is conjoined—not as the latitudes occur in the three higher planets, but as follows, in order that the third digression may be more easily separated and discerned by reason. For Ptolemy observed that these latitudes appeared greatest at the time when the planets were on the straight lines from the centre of the Earth which touch the orbital circles; and that occurs, [189ª] as we said, at their greatest morning and evening distances from the sun. He found that the northern latitudes of Venus were $1/3$° greater than the southern, but that the southern latitudes of Mercury were approximately $1/2$° greater than the northern. But, wishing to reduce the difficulty and labour of calculations, he took in accordance with a certain mean ratio $2^1/_2$° in different directions of latitude; the latitude themselves subtend these degrees in the circle perpendicular to the ecliptic and around the Earth, against which circle the latitudes are measured—especially as he did not think the error would on that account be very great, as we shall soon show. But if we take only $2^1/_2$° as the equal digression on each side of the ecliptic and exclude the deviation for the time being, until we have determined the latitudes of the obliquations, our demonstrations will be simpler and easier. Accordingly we must first show that this latitudinal digression is greatest around the point of tangency of the eccentric circle, where the additosubtractions in longitude are also greatest.

For let there be drawn the common section of the plane of the ecliptic and the plane of the eccentric circle of Venus or Mercury—the common section through the apogee and the perigee; and on it let *A* be taken as the position of the Earth and *B* as

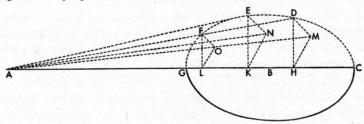

the centre of eccentric circle *CDEFG* which is inclined to the ecliptic, so that straight lines drawn anywhere at right angles to *CG* comprehend angles equal to the obliquation; and let *AE* be drawn tangent to the circle, and *AD* as cutting it somewhere. Moreover, from points *D*, *E*, and *F* let *DH*, *EK*, and *FL* be drawn perpendicular to line *CG*, and *DM*, *EN*, and *FO* perpendicular to the underlying plane of the ecliptic; and let *MH*, *NK*, and *OL* be joined, and also *AN* and *AOM*; for *AOM* is a straight line, since its three points are each in two planes—namely, in the plane of the ecliptic and in the plane *ADM* perpendicular to the plane of the ecliptic.

Accordingly since in the present obliquation the angles *HAM* and *KAN* comprehend the additosubtractions of these planets; and the angles *DAM* and *EAN* are the digressions in latitude: [189^b] I say, first, that angle *EAN*, the angle situated at the point of tangency, where the additosubtraction in longitude is also approximately greatest, is the greatest of all the angles of latitude.

For since angle *EAK* is greater than any of the others,

$$KE : EA > HD : DA$$

and

$$KE : EA > LF : FA.$$

But

$$EK : EN = HD : DM = LF : FO.$$

For, as we said,

$$\text{angle } EKN = \text{angle } HDM = \text{angle } LFO;$$

and

$$\text{angle } M = \text{angle } N = \text{angle } O = 90°.$$

Therefore

$$NE : EA > MD : DA$$

and

$$NE : EA > DF : FA;$$

and again

$$\text{angle } DMA = \text{angle } ENA = \text{angle } OFA = 90°.$$

Accordingly

$$\text{angle } EAN > \text{angle } DAM,$$

and angle *EAN* is greater than each of the other angles constructed in this way. Whence it is manifest that among the differences occurring between the additosubtractions and arising from the obliquation in longitude, the difference which is determined at point *E* in the greatest transit is the greatest. For

$$HD : HM = KE : KN = LF : FO,$$

on account of their subtending equal angles (in similar triangles). And since these lines are in the same ratio as the differences between them,

$$EK - KN : EA > HD - HM : AD$$

and

$$EK - KN : EA > LF - FO : AF.$$

Hence it is also clear that the additosubtractions in longitude of the segments of the eccentric circle will have the same ratio to the transits in latitude as the greatest additosubtraction in longitude has to the greatest transit in latitude, since

$$KE : EN = LF : FO = HD : DM,$$

—as was set before us to be demonstrated.

7. HOW GREAT THE ANGLES OF OBLIQUATION OF VENUS AND MERCURY ARE

Having first noted all that, let us see how great an angle is comprehended by the obliquation of the planes of either planet; and let us repeat what was said before: each planet has 5° between its greatest and least distance (in latitude), so that for the most part they become more northern or southern at contrary times and in accordance with their position on the orbital circle, for when the transit or manifest difference of Venus makes a digression greater or less than 5° through the apogee or perigee of the eccentric circle, the transit of Mercury however is more or less at $^1/_2°$.

[190^a] Accordingly as before, let ABC be the common section of the ecliptic and the eccentric circle; and let the orbital circle of the planet be described around centre

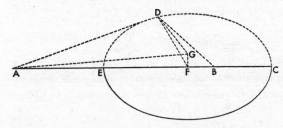

B oblique to the plane of the ecliptic in the way set forth. Now from the centre of the Earth let straight line AD be drawn touching the orbital circle at point D; and from D let DF be

drawn perpendicular to CBE and DG perpendicular to the underlying plane of the ecliptic; and let BD, FG, and AG be joined. Moreover, let it be assumed that

$$\text{angle } DAG = 2^1/_2°,$$

where 4 rt. angles = 360°,

which is half the difference in latitude set forth for each planet.

Our problem is to find how great the angle of obliquation between the planes is, *i.e.*, to find angle DFG.

Accordingly, since in the case of the planet Venus it has been shown that

the greater distance, at the apogee = 10,208,

where radius = 7,193,

and

the lesser, at the perigee = 9,792,

and

the mean distance = 10,000,

which Ptolemy decided to assume in this demonstration, as he wished to avoid labour and difficulty and to make an epitome; for where the extremes do not cause any great difference, it is better to use the mean. Accordingly,

$$AB : BD = 10,000 : 7,193,$$

and

angle ADB = 90°.

Therefore

side AD = 6,947.

Again, since

angle $DAG = 2^1/_2°$

and

angle AGD = 90°;

accordingly, the angles of triangle AGD are given, and

side DG = 303,

where AD = 6,947.

Thus also, two sides DF and DG have been given, and

angle DGF = 90°;

hence

angle DFG = 3°29',

which is the angle of inclination or obliquation. But since the excess of angle DAF over angle FAG comprehends the difference made by the parallax in longitude, hence that difference is to be determined by the measurement of those magnitudes. For it has been shown that

$$AD = 6,947$$

and

$$DF = 4,997,$$

where DG = 303.

Now

sq. AD – sq. DG = sq. AG,

and

sq. FD – sq. DG = sq. GF;

therefore

$$AG = 6,940$$

and

$$FG = 4,988.$$

But

$$FG = 7,187,$$
$$\text{where } AG = 10,000,$$

and

$$\text{angle } FAG = 45°57';$$

and

$$DF = 7,193,$$
$$\text{where } AD = 10,000,$$

and

$$\text{angle } DA \doteqdot F46°.$$

Therefore at the greatest obliquation the additosubtraction of the parallax is deficient by approximately 3'. [190^b] Now it was made clear that at the mean apsis the angle of inclination of the orbital circles was $2^1/_2°$; but here it has increased by approximately 1°, which the first movement of libration—of which we have spoken—has added to it.

There is a similar demonstration in the case of Mercury. For the greatest distance of the orbital circle from the Earth is 10,948, where the radius of the orbital circle is 3,573; the least is 9,052; and the mean between these is 10,000. Moreover

$$AB : BD = 10,000 : 3,573.$$

Therefore

$$\text{side } AD = 9,340;$$

and since

$$BD : BF = AB : AD;$$

therefore

$$DF = 3,337.$$

And since

$$\text{angle } DAG = 2^1/_2°,$$

which is the angle of latitude;

$$DG = 407,$$
$$\text{where } DF = 3,337.$$

And so in triangle DFG the ratio of these two sides is given,
and

$$\text{angle } G = 90°;$$

hence

$$\text{angle } DF \doteqdot G7°.$$

And that is the angle of inclination or obliquation between the orbital circle of Mercury and the plane of the ecliptic. But it has been shown that around the mean longitudes or quadrants the angle of inclination was 6°15'. Therefore 45' have now been added to it by the movement of libration.

There is a similar argument in picking out the additosubtractions and their differences, after it has been shown that

$$DG = 407,$$
$$\text{where } AD = 9,340$$
$$\text{and } DF = 3,337$$

Accordingly,

$$\text{sq. } AD - \text{sq. } DG = \text{sq. } AG,$$

and

$$\text{sq. } DF - \text{sq. } DG = \text{sq. } FG.$$

Therefore

$$AG = 9,331$$

and

$$FG = 3,314;$$

hence it is inferred that

$$\text{angle } GAF = 20°48',$$

which is the additosubtraction, and

$$\text{angle } DAF = 20°56',$$

which is approximately 8' greater than the angle proportionate to the obliquation. It still remains for us to see if such angles of obliquation and the latitudes in accordance with the greatest and least distance of the orbital circle are found to be in conformity with those gathered from observation.

Wherefore once more with the same diagram; and first at the greatest distance of the orbital circle of Venus, let

$$AB : BD = 10,208 : 7,193.$$

And since

$$\text{angle } ADB = 90°;$$
$$DF = 5,102.$$

[191ª] But it has been found that

$$\text{angle } DFG = 3°29',$$

which is the angle of obliquation; hence

$$\text{side } DG = 309,$$
$$\text{where } AD = 7,238.$$

Accordingly,

$$DG = 427,$$
$$\text{where } AD = 10,000;$$

whence it is concluded that at the greatest distance from the Earth

$$\text{angle } DAG = 2°27'.$$

But at the least distance

$$AB = 9,792,$$

where radius of orbital circle = 7,193.

And

$$AD = 6,644,$$

which is perpendicular to the radius; and similarly, since

$$BD : DF = AB : AD,$$
$$DF = 4,883.$$

But

$$\text{angle } DFG = 3°28';$$

therefore

$$DG = 297,$$

where $AD = 6,644$.

And as the sides of the triangle have been given,

$$\text{angle } DAG = 2°34'.$$

But neither 3' nor 4' is large enough to be measured by means of an astrolabe; therefore that which was considered to be the greatest latitude of obliquation of the planet Venus is correct.

Again, let the greatest distance of the orbital circle of Mercury be taken, *i.e.*, let

$$AB : AD = 10,948 : 3,573;$$

consequently, by demonstrations similar to the foregoing, we still infer that

$$AD = 9,452$$

and

$$DF = 3,085.$$

But here too we have it recorded that

$$\text{angle } DFG = 7°,$$

which is the angle of obliquation; hence

$$DG = 376,$$

where $DF = 3,085$

and $DA = 9,452$.

Accordingly, as the sides of right triangle DAG are given,

$$\text{angle } DAG \doteqdot 2°17',$$

which is the greatest digression in latitude. But at the least distance

$$AB : BD = 9,052 : 3,573;$$

therefore

$$AD = 8,317$$

and

$$DF = 3,283.$$

Now since by reason of this same (obliquation)

$$DF : DG = 3,283 : 400,$$
$$\text{where } AD = 8,317;$$

whence

$$\text{angle } DAG = 2°45'.$$

Accordingly there is a difference of at least 13' between the $2^1/_2$° of the digression in latitude according to the mean ratio and the digression at the apogee; and at the most a difference of 15' between the mean digression and that at the perigee. And in making our calculations according to the mean ratio we shall use $^1/_4$° as the difference; for it is not sensibly diverse from the observed differences.

Having demonstrated these things and also that the greatest additosubtractions in longitude have the same ratio to the greatest transit in latitude as the additosubtractions in the remaining sections of the orbital circle have to the particular transits in latitude, we shall have at hand the numbers of all the latitudes, which occur on account of the obliquation of the orbital circle of Venus and Mercury. But we have calculated only those latitudes which occur midway between the apogee and the perigee, as we said; and it was shown that the greatest of these latitudes is $2^1/_2$°, and the greatest [191^b] additosubtraction in the case of Venus is 46° and that in the case of Mercury about 22°. And in the tables of irregular movements we have already placed the additosubtractions opposite the particular sections of the orbital circles. Accordingly in the case of each of the two planets we shall take from the $2^1/_2$° a part proportionate to the excess of the greatest additosubtraction over each of the lesser additosubtractions; we shall inscribe it in the table to be drawn up below with all its numbers; and in this way we shall have unfolded all the particular latitudes of the obliquations which occur when the Earth is at their highest apsis and at their lowest—just as we set forth the latitudes of the declinations in the case of the mean quadrants and mean longitudes. The latitudes which occur between these four limits can be unfolded by the subtle art of mathematics with the help of the proposed hypothesis of circles but not without labour. Now Ptolemy— who is compendious wherever he can be so—seeing that each of these aspects of latitude as a whole and in all its parts increased and decreased proportionally, like the latitude of the moon, accordingly took twelve parts of it, since their greatest latitude is 5° and that number is a twelfth part of 60, and made proportional minutes out of them, to be used not only in the case of these two planets but also in that of the three higher planets, as will be made clear below.

8. ON THE THIRD ASPECT OF THE LATITUDE OF VENUS AND MERCURY, WHICH THEY CALL THE DEVIATION

Now that these things have been set forth, it still remains to say something about the third movement in latitude, which is the deviation. The ancients, who held the

Earth down at the centre of the world, believed that the deviation took place by reason of the inclination of an eccentric circle which has an epicycle and which revolves around the centre of the Earth—the deviation occurring most greatly when the epicycle is at the apogee or perigee and being always $1/6°$ to the north in the case of Venus and $3/4°$ to the south in the case of Mercury, as we said before. It is not however sufficiently clear whether they meant the inclination of the orbital circles to be equal and always the same: for their numbers indicate that, when they order a sixth part of the proportional minutes to be taken as the deviation of Venus, and three parts out of four as that of Mercury. That does not hold, unless the angle of inclination always remains [192ª] the same, as is demanded by the ratio of the minutes, which they take as their base. But if the angle remains the same, it is impossible to understand how the latitude of the planets suddenly springs back from the common section into the same latitude which it had just left, unless you say that takes place in the manner of refraction of light, as in optics. But here we are dealing with movement, which is not instantaneous but is by its own nature measured by time. Accordingly we must acknowledge that a libration such as we have expounded is present in those (circles) and makes the parts of the circle move over in different directions: And that necessarily follows, as the numbers differ $1/5°$ in the case of Mercury. That should seem less surprising, if in accordance with our hypothesis this latitude is variable and not wholly simple but does not produce any apparent error, as is to be seen in the case of all differences, as follows:

For in the plane perpendicular to the ecliptic let (ABC) be the common section (of the two planes), and in the common section let A be the centre of the Earth and B the centre of the circle CDF at greatest or least distance from the Earth and as it were through the poles of the inclined orbital circle. And when the centre of the orbital circle is at the apogee or

the perigee, *i.e.*, on line AB, the planet, wherever it is, is at its greatest deviation, in accordance with the circle parallel to the orbital circle; and DF is the diameter parallel to CBE, the diameter of the orbital circle. And DF and CBE are put down as the common sections of the planes perpendicular to plane CDF. Now let DF be bisected at G, which will be the centre of the parallel circle; and let BG, AG, AD, and AF be joined. Let us put down that

angle BAG = 10',

as in the greatest deviation of Venus. Accordingly, in triangle ABG

angle B = 90°;

and we have the following ratio for the sides:

$$AB : BG = 10,000 : 29.$$

But

$$\text{line } ABC = 17,193;$$

and by subtraction

$$AE = 2,807;$$

and

$$^{1}/_{2} \text{ ch. } 2 \ CD = {}^{1}/_{2} \text{ ch. } 2 \ EF = BG.$$

Accordingly

$$\text{angle } CAD = 6'$$

and

$$\text{angle } EAF \fallingdotseq 15'.$$

Now

$$\text{angle } BAG - \text{angle } CAD = 4',$$

while

$$\text{angle } EAF - \text{angle } BAG = 5';$$

and those differences can be neglected on account of their smallness. Accordingly, when the Earth is situated at its apogee or perigee, the apparent deviation of Venus will be slightly more or less than 10', [192ᵇ] in whatever part of its orbital circle the planet is.

But in the case of Mercury when

$$\text{angle } BAG = 45',$$

and

$$AB : BG = 10,000 : 131,$$

and

$$ABC = 13,573,$$

and by subtraction

$$AE = 6,827;$$

then

$$\text{angle } CAD = 33°$$

and

$$\text{angle } EA \fallingdotseq F70'.$$

Accordingly, angle CAD has a deficiency of 12', and angle EAF has an excess of 25'. But these differences are practically obliterated beneath the rays of the sun before Mercury emerges into our sight, wherefore the ancients considered only its apparent and as it were simple deviation. But if anyone wishes to examine with least labour the precise ratio of their passages when hidden beneath the sun, we shall show how that takes place, as follows. We shall take Mercury as our example, because it makes a more considerable deviation than Venus.

For let *AB* be the straight line in the common section of the orbital circle of the planet and the ecliptic, while the Earth—which is at *A*—is at the apogee or the perigee

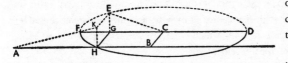

of the planet's orbital circle. Now let us put down that

$$AB = 10,000$$

indifferently, as if the mean between greatest and least distance, as we did in the case of the obliquation. Now around centre *C* let there be described circle *DEF*, which is parallel to the eccentric orbital circle at a distance *CB*; and the planet on this parallel circle be understood as being at this time at its greatest deviation. Let *DCF* be the diameter of this circle, which is also necessarily parallel to *AB*; and *DCF* and *AB* are in the same plane perpendicular to the orbital circle of the planet. Therefore, for example, let

$$\text{arc } EF = 45°,$$

in relation to which we shall examine the deviation of the planet. And let *EG* be drawn perpendicular to *CF*, and *EK* and *GH* perpendicular to the underlying plane of the orbital circle. Let the right parallelogram be completed by joining *HK*; and let *AE*, *AK*, and *EC* also be joined.

Now in the greatest deviation of Mercury

$$BC = 131,$$
$$\text{where } AB = 10,000.$$

And

$$CE = 3,573;$$

and the angles of the right triangle *EGC* are given; hence

$$\text{side } EG = KH = 2526.$$

And since

$$BH = EG = CG,$$
$$AH = BA - BH = 7,474.$$

Accordingly, since in triangle *AHK*

$$\text{angle } H = 90°$$

and the sides of comprehending angle *H* are given;

$$\text{side } AK = 7,889.$$

But

$$\text{side } KE = CB = GH = 131.$$

Accordingly, since in triangle [193ᵃ] *AKE* the two sides *AK* and *KE* comprehending the right angle *K* have been given, angle *KAE* is given, which answers to the deviation we were seeking for the postulated arc *EF* and differs very little from the angle observed. We shall do similarly in the case of Venus and the other planets; and we shall inscribe our findings in the subjoined table.

Having made this exposition, we shall work out proportional minutes for the deviations between these limits. For let *ABC* be the eccentric orbital circle of Venus or Mercury; and let *A* and *C* be the nodes of this movement in latitude, and *B* the limit of greatest deviation. And with *B* as centre let there be described the small circle *DFG*,

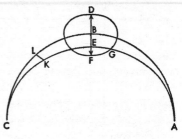

with the diameter *DBF* across it, along which diameter the libration of the movement of deviation takes place. And since it has been laid down that when the Earth is at the apogee or the perigee of the eccentric orbital circle of the planet, the planet itself is at its greatest deviation, namely in point *F*, where at this time the circle carrying the planet touches the small circle. Now let the Earth be somewhere removed from the apogee or perigee of the eccentric circle of the planet; and in accordance with this movement let a similar arc *FG* be taken on the small circle. Let circle *AGC* be described, which bears the planet and will cut the small circle, and the diameter *DF* at point *E*; and let the planet be taken as being on this circle at point *K* in accordance with arc *EK* which is by hypothesis similar to arc *FG*; and let *KL* be drawn perpendicular to circle *ABC*.

Our problem is to find by means of *FG*, *EK*, and *BE* the magnitude of *KL*, i.e., the distance of the planet from circle *ABC*. For since by means of arc *FG* arc *EG* will be given as a straight line hardly different from a circular or convex line, and *EF* will similarly be given in terms of the parts, whereof *BF* and the remainder *BE* will be given; for

$$BF : BE = \text{ch. } 2\ CE : \text{ch. } 2\ CK = BE : KL.$$

Accordingly if we put down *BF* and the radius of circle *CE* in terms of the same number, sixty, we shall have from them the number which picks out *BE*. When that number has been multiplied by itself and the product divided by sixty, we shall have *KL*, the minutes proportional to arc *EK*; and we shall inscribe them similarly in the fifth and last column of the table which follows:

LATITUDES OF SATURN, JUPITER, AND MARS

Common Numbers		SATURN				JUPITER				MARS				Proportional Minutes	
		Northern		Southern		Northern		Southern		Northern		Southern			
Deg.	Deg.	Deg.	Min.	Deg.	Min.	Deg.	Min.	Deg.	Min.	Deg.	Min.	Deg.	Min.	Deg.	Min.
3	357	2	3	2	2	1	6	1	5	0	6	0	5	59	48
6	354	2	4	2	2	1	7	1	5	0	7	0	5	59	36
9	351	2	4	2	3	1	7	1	5	0	9	0	6	59	6
12	348	2	5	2	3	1	8	1	6	0	9	0	6	58	36
15	345	2	5	2	3	1	8	1	6	0	10	0	8	57	48
18	342	2	6	2	3	1	8	1	6	0	11	0	8	57	0
21	339	2	6	2	4	1	9	1	7	0	12	0	9	56	48
24	336	2	7	2	4	1	9	1	7	0	13	0	9	54	36
27	333	2	8	2	5	1	10	1	8	0	14	0	10	53	18
30	330	2	8	2	5	1	10	1	8	0	14	0	11	52	0
33	327	2	9	2	6	1	11	1	9	0	15	0	11	50	12
36	324	2	10	2	7	1	11	1	9	0	16	0	12	48	24
39	321	2	10	2	7	1	12	1	10	0	17	0	12	46	24
42	318	2	11	2	8	1	12	1	10	0	18	0	13	44	24
45	315	2	11	2	9	1	13	1	11	0	19	0	15	42	12
48	312	2	12	2	10	1	13	1	11	0	20	0	16	40	0
51	309	2	13	2	11	1	14	1	12	0	22	0	18	37	36
54	306	2	14	2	12	1	14	1	13	0	23	0	20	35	12
57	303	2	15	2	13	1	15	1	14	0	25	0	22	32	36
60	300	2	16	2	15	1	16	1	15	0	27	0	24	30	0
63	297	2	17	2	16	1	17	1	17	0	29	0	25	27	12
66	294	2	18	2	18	1	18	1	18	0	31	0	27	24	24
69	291	2	20	2	19	1	19	1	19	0	33	0	29	21	24
72	288	2	21	2	21	1	21	1	21	0	35	0	31	18	24
75	285	2	22	2	22	1	22	1	22	0	37	0	34	15	24
78	282	2	24	2	24	1	24	1	24	0	40	0	37	12	24
81	279	2	25	2	26	1	25	1	25	0	42	0	39	9	24
84	276	2	27	2	27	1	27	1	27	0	45	0	42	6	24
87	273	2	28	2	28	1	28	1	28	0	48	0	45	3	12
90	270	2	30	2	30	1	30	1	30	0	51	0	49	0	0
93	267	2	31	2	31	1	31	1	31	0	55	0	52	3	12
96	264	2	33	2	33	1	33	1	33	0	59	0	56	6	24
99	261	2	34	2	34	1	34	1	34	1	2	1	0	9	9
102	258	2	36	2	36	1	36	1	36	1	6	1	4	12	12
105	255	2	37	2	37	1	37	1	37	1	11	1	8	15	15
108	252	2	39	2	39	1	39	1	39	1	15	1	12	18	18
111	249	2	40	2	40	1	40	1	40	1	19	1	17	21	21
114	246	2	42	2	42	1	42	1	42	1	25	1	22	24	24
117	243	2	43	2	43	1	43	1	43	1	31	1	28	27	12
120	240	2	45	2	45	1	44	1	44	1	36	1	34	30	0
123	237	2	46	2	46	1	46	1	46	1	41	1	40	32	37
126	234	2	47	2	48	1	47	1	47	1	47	1	47	35	12
129	231	2	49	2	49	1	49	1	49	1	54	1	55	37	36
132	228	2	50	2	51	1	50	1	51	2	2	2	5	40	6
135	225	2	52	2	53	1	53	1	53	2	10	2	15	42	12
138	222	2	53	2	54	1	52	1	54	2	19	2	26	44	24
141	219	2	54	2	55	1	53	1	55	2	29	2	38	47	24
144	216	2	55	2	56	1	55	1	57	2	37	2	48	48	24
147	213	2	56	2	57	1	56	1	58	2	47	3	4	50	12
150	210	2	57	2	58	1	58	1	59	2	51	3	20	52	0
153	207	2	58	2	59	1	59	2	1	3	12	3	32	53	18
156	204	2	59	3	0	2	0	2	2	3	23	3	52	54	36
159	201	2	59	3	1	2	1	2	3	3	34	4	13	55	48
162	198	3	0	3	2	2	2	2	4	3	46	4	36	57	0
165	195	3	0	3	2	2	2	2	5	3	57	5	0	57	48
168	192	3	1	3	3	2	3	2	5	4	9	5	23	58	36
171	189	3	1	3	3	2	3	2	6	4	17	5	48	59	6
174	186	3	2	3	4	2	4	2	6	4	23	6	15	59	36
177	183	3	2	3	4	2	4	2	7	4	27	6	35	59	48
180	180	3	2	3	5	2	4	2	7	4	30	6	50	60	0

Latitudes of Venus and Mercury

Common Numbers		VENUS				MERCURY				Deviation of Venus		Deviation of Mercury		Proportional Minutes of the Deviation	
		Declination		Obliquity		Declination		Obliquity							
Deg.	Deg.	Deg.	Min.	Deg.	Min.	Deg.	Min.	Deg.	Min.	Deg.	Min.	Deg.	Min.	Deg.	Min.
3	357	1	2	0	4	0	7	1	45	0	5	0	33	59	36
6	354	1	2	0	8	0	7	1	45	0	11	0	33	59	12
9	351	1	1	0	12	0	7	1	45	0	16	0	33	58	25
12	348	1	1	0	16	0	7	1	44	0	22	0	33	57	14
15	345	1	0	0	21	0	7	1	44	0	27	0	33	55	41
18	342	1	0	0	25	0	7	1	43	0	33	0	33	54	9
21	339	0	59	0	29	0	7	1	42	0	38	0	33	52	12
24	336	0	59	0	33	0	7	1	40	0	44	0	34	49	43
27	333	0	58	0	37	0	7	1	38	0	49	0	34	47	21
30	330	0	57	0	41	0	8	1	36	0	55	0	34	45	4
33	327	0	56	0	45	0	8	1	34	1	0	0	34	42	0
36	324	0	55	0	49	0	8	1	30	1	6	0	34	39	15
39	321	0	53	0	53	0	8	1	27	1	11	0	35	35	53
42	318	0	51	0	57	0	8	1	23	1	16	0	35	32	51
45	315	0	49	1	1	0	8	1	19	1	21	0	35	29	41
48	312	0	46	1	5	0	8	1	15	1	26	0	36	23	40
51	309	0	44	1	9	0	8	1	11	1	31	0	36	26	34
54	306	0	41	1	13	0	8	1	8	1	35	0	36	30	39
57	303	0	38	1	17	0	8	1	4	1	40	0	37	17	40
60	300	0	35	1	20	0	8	0	59	1	44	0	38	15	0
63	297	0	32	1	24	0	8	0	54	1	48	0	38	12	20
66	294	0	29	1	28	0	9	0	49	1	52	0	39	9	55
69	291	0	26	1	32	0	9	0	44	1	56	0	39	7	38
72	288	0	23	1	35	0	9	0	38	2	0	0	40	5	39
75	285	0	20	1	38	0	9	0	32	2	3	0	41	3	57
78	282	0	16	1	42	0	9	0	26	2	7	0	42	2	34
81	279	0	12	1	46	0	9	0	21	2	10	0	42	1	28
84	276	0	8	1	50	0	10	0	16	2	14	0	43	0	40
87	273	0	4	1	54	0	10	0	8	2	17	0	44	0	10
90	270	0	0	1	57	0	10	0	0	2	20	0	45	0	0
93	267	0	5	2	0	0	10	0	8	2	23	0	45	0	10
96	264	0	10	2	3	0	10	0	15	2	25	0	46	0	40
99	261	0	15	2	6	0	10	0	23	2	27	0	47	1	28
102	258	0	20	2	9	0	11	0	31	2	28	0	48	2	34
105	255	0	26	2	12	0	11	0	40	2	29	0	48	3	57
108	252	0	32	2	15	0	11	0	48	2	29	0	49	5	39
111	249	0	38	2	17	0	11	0	57	2	30	0	50	7	38
114	246	0	44	2	20	0	11	1	6	2	30	0	51	9	55
117	243	0	50	2	22	0	11	1	16	2	30	0	51	12	20
120	240	0	59	2	24	0	12	1	25	2	29	0	52	15	0
123	237	1	8	2	26	0	12	1	35	2	28	0	53	17	40
126	234	1	18	2	27	0	12	1	45	2	26	0	54	20	39
129	231	1	28	2	29	0	12	1	55	2	23	0	55	23	34
132	228	1	38	2	30	0	12	2	6	2	20	0	56	26	40
135	225	1	48	2	30	0	13	2	16	2	16	0	57	29	41
138	222	1	59	2	30	0	13	2	27	2	11	0	57	32	51
141	219	2	11	2	29	0	13	2	37	2	6	0	58	35	53
144	216	2	25	2	28	0	13	2	47	2	0	0	59	39	25
147	213	2	43	2	26	0	13	2	57	1	53	1	0	42	0
150	210	3	3	2	22	0	13	3	7	1	46	1	1	45	4
153	207	3	23	2	18	0	13	3	17	1	38	1	2	47	21
156	204	3	44	2	12	0	14	3	26	1	29	1	3	49	43
159	201	4	5	2	4	0	14	3	34	1	20	1	4	52	12
162	198	4	26	1	55	0	14	3	42	1	10	1	5	54	9
165	195	4	49	1	42	0	14	3	48	0	59	1	6	55	41
168	192	5	13	1	27	0	14	3	54	0	48	1	7	57	14
171	189	5	36	1	9	0	14	3	58	0	36	1	7	58	25
174	186	5	52	0	48	0	14	4	2	0	24	1	8	59	12
177	183	6	7	0	25	0	14	4	4	0	12	1	9	59	36
180	180	6	22	0	0	0	14	4	5	0	0	1	10	60	0

9. ON THE CALCULATION OF THE LATITUDES OF THE FIVE WANDERING STARS

[195ᵇ] Now this is the method of calculating the latitudes of the five wandering stars by means of these tables. For in the case of Saturn, Jupiter, and Mars we shall take the discrete, or corrected, anomaly of the eccentric circle among the common numbers: in the case of Mars, the anomaly as is; in that of Jupiter, after the subtraction of 20°; and in that of Saturn, after the addition of 50°. Accordingly we shall note the numbers which occur in the region of the 60's, in the proportional minutes placed in the last column. Similarly by means of the corrected anomaly of parallax we shall determine the proper number of each planet, corresponding to the latitude: the first and northern latitude, if the proportional minutes are in the first half of the column—which happens when the anomaly of the eccentric circle is less than 90° or more than 270°; the second and southern latitude, if the proportional minutes are in the second half of the column, *i.e.*, if the anomaly of the eccentric circle, whereby the table was entered upon, was more than 90° or less than 270°. Accordingly if we adjust one of these latitudes to its 60's, the result will be the distance north or south of the ecliptic in accordance with the denomination of the circles assumed.

But in the case of Venus and Mercury the three latitudes of declination, obliquation, and deviation, which are marked down separately, are to be taken first by means of the corrected anomaly of parallax, except that in the case of Mercury one tenth of the obliquation is to be subtracted, if the anomaly of the eccentric circle and its number are found in the first column of the table, or merely added, if in the second column of the table; and the remainder or sum is to be kept.

And we must discern whether their denominations are northern or southern, since if the corrected anomaly of parallax is in the apogeal semicircle, *i.e.*, is less than 90° or more than 270° and the anomaly of the eccentric circle is also less than a semicircle; or again, if the anomaly of parallax is in the perigeal arc, *i.e.*, is more than 90° and less than 270° and the anomaly of the eccentric circle is greater than a semicircle; the declination of Venus will be northern and that of Mercury southern. But if the anomaly of parallax is in the perigeal arc and the anomaly of the eccentric circle is less than a semicircle; [196ᵃ] or if the anomaly of parallax is in the apogeal arc and the anomaly of the eccentric circle is more than a semicircle; conversely the declination of Venus will be southern and that of Mercury northern. But in the case of the obliquation, if the anomaly of parallax is less than a semicircle and the anomaly of the eccentric circle is apogeal; or if the anomaly of parallax is greater than a semicircle and the anomaly of the eccentric circle is perigeal; the obliquation of Venus will be to the north and that of Mercury to the south; and vice versa. But the deviations of Venus always remain northern and those of Mercury southern.

Then, corresponding to the corrected anomaly of the eccentric circle, the proportional minutes should be taken which are common to all the five planets, although they are ascribed to the three higher planets. These are assigned to the obliquation and lastly to the deviation. After this, when we have added 90° to the same anomaly of the eccentric circle, we shall once more take the sum and find the common proportional minutes which correspond to it and assign them to the latitude of declination. Having placed these things in this order, we shall adjust each of the three particular latitudes set forth to their proportional minutes; and the result will be the corrected latitude for the position and time, so that at last we may have the sum of the three latitudes of the two planets. If all the latitudes are of one denomination, they are added together; but if not, only the two are added which have the same denomination; and according as the sum is greater or less than the third latitude, which is different from there, there will be a subtraction; and the remainder will be the predominant latitude sought for.

Galileo Galilei

(1564-1642)

HIS LIFE AND WORK

In 1633, ninety years after the death of Copernicus, the Italian astronomer and mathematician Galileo Galilei was taken to Rome to stand trial before the Inquisition for heresy. The charge stemmed from the publication of Galileo's *Dialogue Concerning the Two Chief World Systems: Ptolemaic and Copernican* (*Dialogo sopra i due massimi sistemi del mondo: tolemaico, e copernicono*). In this book, Galileo forcefully asserted, in defiance of a 1616 edict against the propagation of Copernican doctrine, that the heliocentric system was not just a hypothesis but was the truth. The outcome of the trial was never in doubt. Galileo admitted that he might have gone too far in his arguments for the Copernican system, despite previous warnings by the Roman Catholic church. A majority of the cardinals in the tribunal found him "vehemently suspected of heresy" for supporting and teaching the idea that the earth moves and is not the center of the universe, and they sentenced him to life imprisonment.

Galileo was also forced to sign a handwritten confession and to renounce his beliefs publicly. On his knees, and with his hands on the Bible, he pronounced this abjuration in Latin:

> *I, Galileo Galilei, son of the late Vincenzio Galilei of Florence, aged 70 years, tried personally by this court, and kneeling before You, the most Eminent and Reverend Lord Cardinals, Inquisitors-General*

throughout the Christian Republic against heretical depravity, having before my eyes the Most Holy Gospels, and laying on them my own hands; I swear that I have always believed, I believe now, and with God's help I will in future believe all which the Holy Catholic and Apostolic Church doth hold, preach, and teach.

But since I, after having been admonished by this Holy Office entirely to abandon the false opinion that the sun was the centre of the universe and immoveable, and that the earth was not the centre of the same and that it moved, and that I was neither to hold, defend, nor teach in any manner whatever, either orally or in writing, the said false doctrine; and after having received a notification that the said doctrine is contrary to Holy Writ, I did write and cause to be printed a book in which I treat of the said already condemned doctrine, and bring forward arguments of much efficacy in its favour, without arriving at any solution: I have been judged vehemently suspected of heresy, that is, of having held and believed that the sun is the centre of the universe and immoveable, and that the earth is not the centre of the same, and that it does move.

Nevertheless, wishing to remove from the minds of your Eminences and all faithful Christians this vehement suspicion reasonably conceived against me, I abjure with sincere heart and unfeigned faith, I curse and detest the said errors and heresies, and generally all and every error and sect contrary to the Holy Catholic Church. And I swear that for the future I will neither say nor assert in speaking or writing such things as may bring upon me similar suspicion; and if I know any heretic, or one suspected of heresy, I will denounce him to this Holy Office, or to the Inquisitor and Ordinary of the place in which I may be.

I also swear and promise to adopt and observe entirely all the penances which have been or may be by this Holy Office imposed on me. And if I contravene any of these said promises, protests, or oaths (which God forbid!) I submit myself to all the pains and penalties which by the Sacred Canons and other Decrees general and particular are against such offenders imposed and promulgated. So help me God and the

Holy Gospels, which I touch with my own hands.

I Galileo Galilei aforesaid have abjured, sworn, and promised, and hold myself bound as above; and in token of the truth, with my own hand have subscribed the present schedule of my abjuration, and have recited it word by word. In Rome, at the Convent della Minerva, this 22nd day of June, 1633. I, Galileo Galilei, have abjured as above, with my own hand.

Legend has it that as Galileo rose to his feet, he uttered under his breath, "Eppur si muove"—"And yet, it moves." The remark captivated scientists and scholars for centuries, as it represented defiance of obscurantism and nobility of purpose in the search for truth under the most adverse circumstances. Although an oil portrait of Galileo dating from 1640 has been discovered bearing the inscription "Eppur si muove," most historians regard the story as myth. Still, it is entirely within Galileo's character to have only paid lip service to the church's demands in his abjuration and then to have returned to his scientific studies, whether they adhered to non-Copernican principles or not. After all, what had brought Galileo before the Inquisition was his publication of *Two Chief World Systems*, a direct challenge to the church's 1616 edict forbidding him from teaching the Copernican theory of the earth in motion around the sun as anything but a hypothesis. "Eppur si mouve" may not have concluded his trial and abjuration, but the phrase certainly punctuated Galileo's life and accomplishments.

Born in Pisa on February 18, 1564, Galileo Galilei was the son of Vincenzo Galilei, a musician and mathematician. The family moved to Florence when Galileo was young, and there he began his education in a monastery. Although from an early age Galileo demonstrated a penchant for mathematics and mechanical pursuits, his father was adamant that he enter a more useful field, and so in 1581 Galileo enrolled in the University of Pisa to study medicine and the philosophy of Aristotle. It was in Pisa that Galileo's rebelliousness emerged. He had little or no interest in medicine and began to study mathematics with a passion. It is believed that while observing the oscillations of a hanging lamp in the cathedral of Pisa, Galileo discovered the isochronism of the pendulum—the period of swing is independent of its amplitude—which he would apply a half-century later in building an astronomical clock.

Galileo persuaded his father to allow him to leave the university without a degree, and he returned to Florence to study and teach mathematics. By 1586, he had begun to question the science and philosophy of Aristotle, preferring to reexamine the work of the great mathematician Archimedes, who was also known for discovering and perfecting methods of integration for calculating areas and volumes. Archimedes also gained a

reputation for his invention of many machines ultimately used as engines of war, such as giant catapults to hurl boulders at an advancing army and large cranes to topple ships. Galileo was inspired mainly by Archimedes' mathematical genius, but he too was swept up in the spirit of invention, designing a hydrostatic balance to determine an object's density when weighed in water.

In 1589, Galileo became a professor of mathematics at the University of Pisa, where he was required to teach Ptolemaic astronomy—the theory that the sun and the planets revolve around the earth. It was in Pisa, at the age of twenty-five, that Galileo obtained a deeper understanding of astronomy and began to break with Aristotle and Ptolemy. Lecture notes recovered from this period show that Galileo had adopted the Archimedean approach to motion; specifically, he was teaching that the density of a falling object, not its weight, as Aristotle had maintained, was proportional to the speed at which it fell. Galileo is said to have demonstrated his theory by dropping objects of the different weights but the same density from atop the leaning tower of Pisa. In Pisa, too, he wrote *On motion* (*De motu*), a book that contradicted the Aristotelian theories of motion and established Galileo as a leader in scientific reformation.

After his father's death in 1592, Galileo did not see much of a future for himself in Pisa. The pay was dismal, and with the help of a family friend, Guidobaldo del Monte, Galileo was appointed to the chair in mathematics at the University of Padua, in the Venetian Republic. There, Galileo's reputation blossomed. He remained at Padua for eighteen years, lecturing on geometry and astronomy as well as giving private lessons on cosmography, optics, arithmetic and the use of the sector in military engineering. In 1593, he assembled treatises on fortifications and mechanics for his private students and invented a pump that could raise water under power of a single horse.

In 1597, Galileo invented a calculating compass that proved useful to mechanical engineers and military men. He also began a correspondence with Johannes Kepler, whose book *Mystery of the Cosmos* (*Mysterium cosmographicum*) Galileo had read. Galileo sympathized with Kepler's Copernican views, and Kepler hoped that Galileo would openly support the theory of a heliocentric earth. But Galileo's scientific interests were still focused on mechanical theories, and he did not follow Kepler's wishes. Also at that time Galileo had developed a personal interest in Marina Gamba, a Venetian woman by whom he had a son and two daughters. The eldest daughter, Virginia, born in 1600, maintained a very close relationship with her father, mainly through an exchange of correspondence, for she spent most of her short adult life in a convent, taking the name Maria Celeste in tribute to her father's interest in celestial matters.

In the first years of the seventeenth century, Galileo experimented with the pendulum and explored its association with the phenomenon of natural acceleration. He

also began work on a mathematical model describing the motion of falling bodies, which he studied by measuring the time it took balls to roll various distances down inclined planes. In 1604, a supernova observed in the night sky above Padua renewed questions about Aristotle's model of the unchanging heavens. Galileo thrust himself into the forefront of the debate, delivering several provocative lectures, but he was hesitant to publish his theories. In October 1608, a Dutchman by the name of Hans Lipperhey applied for a patent on a spyglass that could make faraway objects appear closer. Upon hearing of the invention, Galileo set about attempting to improve it. Soon he had designed a nine-power telescope, three times more powerful than Lipperhey's device, and within a year, he had produced a thirty-power telescope. When he pointed the scope toward the skies in January 1610, the heavens literally opened up to humankind. The moon no longer appeared to be a perfectly smooth disc but was seen to be a mountainous and full of craters. Through his telescope, Galileo determined that the Milky Way was actually a vast gathering of separate stars. But most important, he sighted four moons around Jupiter, a discovery that had tremendous implications for many of the geocentrically inclined, who held that all heavenly bodies revolved exclusively around the earth. That same year, he published *The Starry Messenger* (*Sidereus Nuncius*), in which he announced his discoveries and which put him in the forefront of contemporary astronomy. He felt unable to continue teaching Aristotelian theories, and his renown enabled him to take a position in Florence as mathematician and philosopher to the grand duke of Tuscany.

Once free from the responsibilities of teaching, Galileo was able to devote himself to telescopy. He soon observed the phases of Venus, which confirmed Copernicus' theory that the planet revolved around the sun. He also noted Saturn 's oblong shape, which he attributed to numerous moons revolving around the planet, for his telescope was unable to detect Saturn's rings.

The Roman Catholic Church affirmed and praised Galileo's discoveries but did not agree with his interpretations of them. In 1613, Galileo published *Letters on Sunspots*, marking the first time in print that he had defended the Copernican system of a heliocentric universe. The work was immediately attacked and its author denounced, and the Holy Inquisition soon took notice. When in 1616 Galileo published a theory of tides, which he believed was proof that the earth moved, he was summoned to Rome to answer for his views. A council of theologians issued an edict that Galileo was practicing bad science when he taught the Copernican system as fact. But Galileo was never officially condemned. A meeting with Pope Paul V led him to believe that the pontiff held him in esteem and that he could continue to lecture under the pontiff's protection. He was, however, strongly warned that Copernican theories ran contrary to the Scriptures and that they may only be presented as hypotheses.

When upon Paul's death in 1623 one of Galileo's friends and supporters, Cardinal Barberini, was elected pope, taking the name Urban VIII, Galileo presumed that the 1616 edict would be reversed. Urban told Galileo that he himself was responsible for omitting the word "heresy" from the edict and that as long as Galileo treated Copernican doctrine as hypothesis and not truth, he would be free to publish. With this assurance, over the next six years Galileo worked on *Dialogue Concerning the Two Chief World Systems*, the book that would lead to his imprisonment.

Two Chief World Systems takes the form of a polemic between an advocate of Aristotle and Ptolemy and a supporter of Copernicus, who seek to win an educated everyman over to the respective philosophies. Galileo prefaced the book with a statement in support of the 1616 edict against him, and by presenting the theories through the book's characters, he is able to avoid openly declaring his allegiance to either side. The public clearly perceived, nonetheless, that in *Two Chief World Systems* Galileo was disparaging Aristotelianism. In the polemic, Aristotle's cosmology is only weakly defended by its simpleminded supporter and is viciously attacked by the forceful and persuasive Copernican. The book achieved a great success, despite being the subject of massive protest upon publication. By writing it in vernacular Italian rather than Latin, Galileo made it accessible to a broad range of literate Italians, not just to churchmen and scholars. Galileo's Ptolemaic rivals were furious at the dismissive treatment that their scientific views had been given. In Simplicio, the defender of the Ptolemaic system, many readers recognized a caricature of Simplicius, a sixth-century Aristotelian commentator. Pope Urban VIII, meanwhile, thought that Simplicio was meant as a caricature of himself. He felt misled by Galileo, who apparently had neglected to inform him of any injunction in the 1616 edict when he sought permission to write the book. Galileo, on the other hand, never received a written injunction, and seemed to be unaware of any violations on his part.

By March 1632, the church had ordered the book's printer to discontinue publication, and Galileo was summoned to Rome to defend himself. Pleading serious illness, Galileo refused to travel, but the pope insisted, threatening to have Galileo removed in chains. Eleven months later, Galileo appeared in Rome for trial. He was made to abjure the heresy of the Copernican theory and was sentenced to life imprisonment. Galileo's life sentence was soon commuted to gentle house arrest in Siena under the guard of Archbishop Ascanio Piccolomini, a former student of Galileo's. Piccolomini permitted and even encouraged Galileo to resume writing. There, Galileo began his final work, *Dialogues Concerning Two New Sciences*, an examination of his accomplishments in physics. But the following year, when Rome got word of the preferential treatment Galileo was receiving from Piccolomini, it had him removed to another home, in the

hills above Florence. Some historians believe that it was upon his transfer that Galileo actually said "Eppur si muove," rather than at his public abjuration following the trial.

The transfer brought Galileo closer to his daughter Virginia, but soon she died, after a brief illness, in 1634. The loss devastated Galileo, but he eventually he was able to resume working on *Two New Sciences*, and he finished the book within a year. However, the Congregation of the Index, the church censor, would not allow Galileo to publish it. The manuscript had to be smuggled out of Italy to Leiden, in Protestant northern Europe, by Louis Elsevier, a Dutch publisher, before it could appear in print in 1638. *Dialogues Concerning Two New Sciences*, which set out the laws of accelerated motion governing falling bodies, is widely held to be the cornerstone of modern physics. In this book, Galileo reviewed and refined his previous studies of motion, as well as the principles of mechanics. The two new sciences Galileo focuses on are the study of the strength of materials (a branch of engineering), and the study of motion (kinematics, a branch of mathematics). In the first half of the book, Galileo described his inclined-plane experiments in accelerated motion. In the second half, Galileo took on the intractable problem of calculating the path of a projectile fired from a cannon. At first it had been thought that, in keeping with Aristotelian principles, a projectile followed a straight line until it lost its "impetus" and fell straight to the ground. Later, observers noticed that it actually returned to earth on a curved path, but the reason this happened and an exact description of the curve no one could say—until Galileo. He concluded that the projectile's path is determined by two motions—one vertical, caused by gravity, which forces the projectile down, and one horizontal, governed by the principle of inertia.

Galileo demonstrated that the combination of these two independent motions determined the projectile's course along a mathematically describable curve. He showed this by rolling a bronze ball coated in ink down an inclined plane and onto a table, whence it fell freely off the edge and onto the floor. The inked ball left a mark on the floor where it hit, always some distance out from the table's edge. Thus Galileo proved that the ball continued to move horizontally, at a constant speed, while gravity pulled it down vertically. He found that the distance increased in proportion to the square of the time elapsed. The curve achieved a precise mathematical shape, which the ancient Greeks had termed a parabola.

So great a contribution to physics was *Two New Sciences* that scholars have long maintained that the book anticipated Isaac Newton's laws of motion. By the time of its publication, however, Galileo had gone blind. He lived out the remaining years of his life in Arcetri, where he died on January 8, 1642. Galileo's contributions to humanity were never understated. Albert Einstein recognized this when he wrote: "Propositions arrived at purely by logical means are completely empty as regards reality. Because

Galileo saw this, and particularly because he drummed it into the scientific world, he is the father of modern physics—indeed of modern science."

In 1979, Pope John Paul II stated that the Roman Catholic church may have mistakenly condemned Galileo, and he called for a commission specifically to reopen the case. Four years later, the commission reported that Galileo should not have been condemned, and the church published all the documents relevant to his trial. In 1992, the pope endorsed the commission's conclusion.

DIALOGUES CONCERNING TWO SCIENCES

FIRST DAY

Interlocutors: Salviati, Sagredo And Simplicio

Salv. The constant activity which you Venetians display in your famous arsenal suggests to the studious mind a large field for investigation, especially that part of the work which involves mechanics; for in this department all types of instruments and machines are constantly being constructed by many artisans, among whom there must be some who, partly by inherited experience and partly by their own observations, have become highly expert and clever in explanation.

Sagr. You are quite right. Indeed, I myself, being curious by nature, frequently visit this place for the mere pleasure of observing the work of those who, on account of their superiority over other artisans, we call "first rank men." Conference with them has often helped me in the investigation of certain effects including not only those which are striking, but also those which are recondite and almost incredible. At times also I have been put to confusion and driven to despair of ever explaining something for which I could not account, but which my senses told me to be true. And notwithstanding the fact that what the old man told us a little while ago is proverbial and commonly accepted, yet it seemed to me altogether false, like many another saying which is current among the ignorant; for I think they introduce these expressions in order to give the appearance of knowing something about matters which they do not understand.

Salv. You refer, perhaps, to that last remark of his when we asked the reason why they employed stocks, scaffolding and bracing of larger dimensions for launching a big

vessel than they do for a small one; and he answered that they did this in order to avoid the danger of the ship parting under its own heavy weight [*vasta mole*], a danger to which small boats are not subject?

Sagr. Yes, that is what I mean; and I refer especially to his last assertion which I have always regarded as a false, though current, opinion; namely, that in speaking of these and other similar machines one cannot argue from the small to the large, because many devices which succeed on a small scale do not work on a large scale. Now, since mechanics has its foundation in geometry, where mere size cuts no figure, I do not see that the properties of circles, triangles, cylinders, cones and other solid figures will change with their size. If, therefore, a large machine be constructed in such a way that its parts bear to one another the same ratio as in a smaller one, and if the smaller is sufficiently strong for the purpose for which it was designed, I do not see why the larger also should not be able to withstand any severe and destructive tests to which it may be subjected.

Salv. The common opinion is here absolutely wrong. Indeed, it is so far wrong that precisely the opposite is true, namely, that many machines can be constructed even more perfectly on a large scale than on a small; thus, for instance, a clock which indicates and strikes the hour can be made more accurate on a large scale than on a small. There are some intelligent people who maintain this same opinion, but on more reasonable grounds, when they cut loose from geometry and argue that the better performance of the large machine is owing to the imperfections and variations of the material. Here I trust you will not charge

[51]

me with arrogance if I say that imperfections in the material, even those which are great enough to invalidate the clearest mathematical proof, are not sufficient to explain the deviations observed between machines in the concrete and in the abstract. Yet I shall say it and will affirm that, even if the imperfections did not exist and matter were absolutely perfect, unalterable and free from all accidental variations, still the mere fact that it is matter makes the larger machine, built of the same material and in the same proportion as the smaller, correspond with exactness to the smaller in every respect except that it will not be so strong or so resistant against violent treatment; the larger the machine, the greater its weakness. Since I assume matter to be unchangeable and always the same, it is clear that we are no less able to treat this constant and invariable property in a rigid manner than if it belonged to simple and pure mathematics. Therefore, Sagredo, you would do well to change the opinion which you, and perhaps also many other students of mechanics, have entertained concerning the ability of machines and structures to resist external disturbances, thinking that when they are

built of the same material and maintain the same ratio between parts, they are able equally, or rather proportionally, to resist or yield to such external disturbances and blows. For we can demonstrate by geometry that the large machine is not proportionately stronger than the small. Finally, we may say that, for every machine and structure, whether artificial or natural, there is set a necessary limit beyond which neither art nor nature can pass; it is here understood, of course, that the material is the same and the proportion preserved.

Sagr. My brain already reels. My mind, like a cloud momentarily illuminated by a lightning-flash, is for an instant filled with an unusual light, which now beckons to me and which now suddenly mingles and obscures strange, crude ideas. From what you have said it appears to me impossible to build two similar structures of the same material, but of different sizes and have them proportionately strong; and if this were so, it would

[52]

not be possible to find two single poles made of the same wood which shall be alike in strength and resistance but unlike in size.

Salv. So it is, Sagredo. And to make sure that we understand each other, I say that if we take a wooden rod of a certain length and size, fitted, say, into a wall at right angles, i. e., parallel to the horizon, it may be reduced to such a length that it will just support itself; so that if a hair's breadth be added to its length it will break under its own weight and will be the only rod of the kind in the world.[1] Thus if, for instance, its length be a hundred times its breadth, you will not be able to find another rod whose length is also a hundred times its breadth and which, like the former, is just able to sustain its own weight and no more: all the larger ones will break while all the shorter ones will be strong enough to support something more than their own weight. And this which I have said about the ability to support itself must be understood to apply also to other tests; so that if a piece of scantling [corrente] will carry the weight of ten similar to itself, a beam [trave] having the same proportions will not be able to support ten similar beams.

Please observe, gentlemen, how facts which at first seem improbable will, even on scant explanation, drop the cloak which has hidden them and stand forth in naked and simple beauty. Who does not know that a horse falling from a height of three or four cubits will break his bones, while a dog falling from the same height or a cat from a height of eight or ten cubits will suffer no injury? Equally harmless would be the fall of a grasshopper from a tower or the fall of an ant from the distance of the moon. Do not children fall with impunity from heights which would cost their elders a broken

1. The author here apparently means that the solution is unique. [Trans.]

leg or perhaps a fractured skull? And just as smaller animals are proportionately stronger and more robust than the larger, so also smaller plants are able to stand up better than larger. I am certain you both know that an oak two hundred cubits [*braccia*] high would not be able to sustain its own branches if they were distributed as in a tree of ordinary size; and that nature cannot produce a horse as large as twenty ordinary horses or a giant ten times taller than an

[53]

ordinary man unless by miracle or by greatly altering the proportions of his limbs and especially of his bones, which would have to be considerably enlarged over the ordinary. Likewise the current belief that, in the case of artificial machines the very large and the small are equally feasible and lasting is a manifest error. Thus, for example, a small obelisk or column or other solid figure can certainly be laid down or set up without danger of breaking, while the very large ones will go to pieces under the slightest provocation, and that purely on account of their own weight. And here I must relate a circumstance which is worthy of your attention as indeed are all events which happen contrary to expectation, especially when a precautionary measure turns out to be a cause of disaster. A large marble column was laid out so that its two ends rested each upon a piece of beam; a little later it occurred to a mechanic that, in order to be doubly sure of its not breaking in the middle by its own weight, it would be wise to lay a third support midway; this seemed to all an excellent idea; but the sequel showed that it was quite the opposite, for not many months passed before the column was found cracked and broken exactly above the new middle support.

Simp. A very remarkable and thoroughly unexpected accident, especially if caused by placing that new support in the middle.

Salv. Surely this is the explanation, and the moment the cause is known our surprise vanishes; for when the two pieces of the column were placed on level ground it was observed that one of the end beams had, after a long while, become decayed and sunken, but that the middle one remained hard and strong, thus causing one half of the column to project in the air without any support. Under these circumstances the body therefore behaved differently from what it would have done if supported only upon the first beams; because no matter how much they might have sunken the column would have gone with them. This is an accident which could not possibly have happened to a small column, even though made of the same stone and having a length corresponding to its thickness, i. e., preserving the ratio between thickness and length found in the large pillar.

[54]

Sagr. I am quite convinced of the facts of the case, but I do not understand why

the strength and resistance are not multiplied in the same proportion as the material; and I am the more puzzled because, on the contrary, I have noticed in other cases that the strength and resistance against breaking increase in a larger ratio than the amount of material. Thus, for instance, if two nails be driven into a wall, the one which is twice as big as the other will support not only twice as much weight as the other, but three or four times as much.

Salv. Indeed you will not be far wrong if you say eight times as much; nor does this phenomenon contradict the other even though in appearance they seem so different.

Sagr. Will you not then, Salviati, remove these difficulties and clear away these obscurities if possible: for I imagine that this problem of resistance opens up a field of beautiful and useful ideas; and if you are pleased to make this the subject of to-day's discourse you will place Simplicio and me under many obligations.

Salv. I am at your service if only I can call to mind what I learned from our Academician[2] who had thought much upon this subject and according to his custom had demonstrated everything by geometrical methods so that one might fairly call this a new science. For, although some of his conclusions had been reached by others, first of all by Aristotle, these are not the most beautiful and, what is more important, they had not been proven in a rigid manner from fundamental principles. Now, since I wish to convince you by demonstrative reasoning rather than to persuade you by mere probabilities, I shall suppose that you are familiar with present-day mechanics so far as it is needed in our discussion. First of all it is necessary to consider what happens when a piece of wood or any other solid which coheres firmly is broken; for this is the fundamental fact, involving the first and simple principle which we must take for granted as well known.

Fig. 1

To grasp this more clearly, imagine a cylinder or prism, AB, made of wood or other solid coherent material. Fasten the upper end, A, so that the cylinder hangs vertically. To the lower end, B, attach the weight C. It is clear that however great they may be, the tenacity and coherence [*tenacità e*

[55]

coerenza] between the parts of this solid, so long as they are not infinite, can be overcome by the pull of the weight C, a weight which can be increased indefinitely until finally the solid breaks like a rope. And as in the case of the rope whose strength we know to be derived from a multitude of hemp threads which compose it, so in the case

2. I. e. Galileo: The author frequently refers to himself under this name. [*Trans.*]

of the wood, we observe its fibres and filaments run lengthwise and render it much stronger than a hemp rope of the same thickness. But in the case of a stone or metallic cylinder where the coherence seems to be still greater the cement which holds the parts together must be something other than filaments and fibres; and yet even this can be broken by a strong pull.

Simp. If this matter be as you say I can well understand that the fibres of the wood, being as long as the piece of wood itself, render it strong and resistant against large forces tending to break it, But how can one make a rope one hundred cubits long out of hempen fibres which are not more than two or three cubits long, and still give it so much strength? Besides, I should be glad to hear your opinion as to the manner in which the parts of metal, stone, and other materials not showing a filamentous structure are put together; for, if I mistake not, they exhibit even greater tenacity.

Salv. To solve the problems which you raise it will be necessary to make a digression into subjects which have little bearing upon our present purpose.

Sagr. But if, by digressions, we can reach new truth, what harm is there in making one now, so that we may not lose this knowledge, remembering that such an opportunity, once omitted, may not return; remembering also that we are not tied down to a fixed and brief method but that we meet solely for our own entertainment? Indeed, who knows but that we may thus

[56]

frequently discover something more interesting and beautiful than the solution originally sought? I beg of you, therefore, to grant the request of Simplicio, which is also mine; for I am no less curious and desirous than he to learn what is the binding material which holds together the parts of solids so that they can scarcely be separated. This information is also needed to understand the coherence of the parts of fibres themselves of which some solids are built up.

Salv. I am at your service, since you desire it. The first question is, How are fibres, each not more than two or three cubits in length, so tightly bound together in the case of a rope one hundred cubits long that great force [*violenza*] is required to break it?

Now tell me, Simplicio, can you not hold a hempen fibre so tightly between your fingers that I, pulling by the other end, would break it before drawing it away from you? Certainly you can. And now when the fibres of hemp are held not only at the ends, but are grasped by tile surrounding medium throughout their entire length is it not manifestly more difficult to tear them loose from what holds them than to break them? But in the case of the rope the very act of twisting causes the threads to bind one another in such a way that when the rope is stretched with a great force the fibres break rather than separate from each other.

At the point where a rope parts the fibres are, as everyone knows, very short, nothing like a cubit long, as they would be if the parting of the rope occurred, not by the breaking of the filaments, but by their slipping one over the other.

Sagr. In confirmation of this it may be remarked that ropes sometimes break not by a lengthwise pull but by excessive twisting. This, it seems to me, is a conclusive argument because tile threads bind one another so tightly that the compressing fibres do not permit those which are compressed to lengthen the spirals even that little bit by which it is necessary for them to lengthen in order to surround the rope which, on twisting, grows shorter and thicker.

[57]

Salv. You are quite right. Now see how one fact suggests another. The thread held between the fingers does not yield to one who wishes to draw it away even when pulled with considerable force, but resists because it is held back by a double compression, seeing that the upper finger presses against the lower as hard as the lower against the upper. Now, if we could retain only one of these pressures there is no doubt that only half the original resistance would remain; but since we are not able, by lifting, say, the upper finger, to remove one of these pressures without also removing the other, it becomes necessary to preserve one of them by means of a new device which causes the thread to press itself against the finger or against some other solid body upon which it rests; and thus it is brought about that the very force which pulls it in order to snatch it away compresses it more and more as the pull increases. This is accomplished by wrapping the thread around the solid in the manner of a spiral; and will be better understood by means of a figure. Let AB and CD be two cylinders between which is stretched the thread EF: and for the sake of greater clearness we will imagine it to be a small cord. If these two cylinders be pressed strongly together, the cord EF, when drawn by the end F, will undoubt-

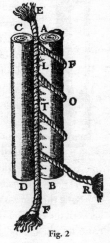

Fig. 2

edly stand a considerable pull before it slips between the two compressing solids. But if we remove one of these cylinders the cord, though remaining in contact with the other, will not thereby be prevented from slipping freely. On the other hand, if one holds the cord loosely against the top of the cylinder A, winds it in the spiral form AFLOTR, and then pulls it by the end R, it is evident that the cord will begin to bind the cylinder; the greater the number of spirals the more tightly will the cord be pressed against the cylinder by any given pull. Thus as the number of turns increases, the line

of contact becomes longer and in consequence more resistant; so that the cord slips and yields to the tractive force with increasing difficulty.

[58]

Is it not clear that this is precisely the kind of resistance which one meets in the case of a thick hemp rope where the fibres form thousands and thousands of similar spirals? And, indeed, the binding effect of these turns is so great that a few short rushes woven together into a few interlacing spirals form one of the strongest of ropes which I believe they call pack rope [*susta*].

Sagr. What you say has cleared up two points which I did not previously understand. One fact is how two, or at most three, turns of a rope around the axle of a windlass cannot only hold it fast, but can also prevent it from slipping when pulled by the immense force of the weight [*forza del peso*] which it sustains; and moreover how, by turning the windlass, this same axle, by mere friction of the rope around it, can wind up and lift huge stones while a mere boy is able to handle the slack of the rope. The other fact has to do with a simple but clever device, invented by a young kinsman of mine, for the purpose of descending from a window by means of a rope without lacerating the palms of his hands, as had happened to him shortly before and greatly to his discomfort. A small sketch will make this clear. He took a wooden cylinder, AB, about as thick as a walking stick and about one span long: on this he cut a spiral channel of about one turn and a half, and large enough to just receive the rope which he wished to use. Having introduced the rope at the end A and led it out again at the end B, he enclosed both the cylinder and the rope in a case of wood or tin, hinged along the side so that it could be easily opened and closed. After he had fastened the rope to a firm support above, he could, on grasping and squeezing the case with both hands, hang by his arms. The pressure on the rope, lying between the case and the cylinder, was such that he could, at will, either grasp the case more tightly and hold himself from slipping, or slacken his hold and descend as slowly as he wished.

Fig. 3

[59]

Salv. A truly ingenious device! I feel, however, that for a complete explanation other considerations might well enter; yet I must not now digress upon this particular topic since you are waiting to hear what I think about the breaking strength of other materials which, unlike ropes and most woods, do not show a filamentous structure.

The coherence of these bodies is, in my estimation, produced by other causes which may be grouped under two heads. One is that much-talked-of repugnance which nature exhibits towards a vacuum; but this horror of a vacuum not being sufficient, it is necessary to introduce another cause in the form of a gluey or viscous substance which binds firmly together the component parts of the body.

First I shall speak of the vacuum, demonstrating by definite experiment the quality and quantity of its force [*virtù*]. If you take two highly polished and smooth plates of marble, metal, or glass and place them face to face, one will slide over the other with the greatest ease, showing conclusively that there is nothing of a viscous nature between them. But when you attempt to separate them and keep them at a constant distance apart, you find the plates exhibit such a repugnance to separation that the upper one will carry the lower one with it and keep it lifted indefinitely, even when the latter is big and heavy.

This experiment shows the aversion of nature for empty space, even during the brief moment required for the outside air to rush in and fill up the region between the two plates. It is also observed that if two plates are not thoroughly polished, their contact is imperfect so that when you attempt to separate them slowly the only resistance offered is that of weight; if, however, the pull be sudden, then the lower plate rises, but quickly falls back, having followed the upper plate only for that very short interval of time required for the expansion of the small amount of air remaining between the plates, in consequence of their not fitting, and for the entrance of the surrounding air. This resistance which is exhibited between the two plates is doubtless likewise present between the parts of a solid, and enters, at least in part, as a concomitant cause of their coherence.

[60]

Sagr. Allow me to interrupt you for a moment, please; for I want to speak of something which just occurs to me, namely, when I see how the lower plate follows the upper one and how rapidly it is lifted, I feel sure that, contrary to the opinion of many philosophers, including perhaps even Aristotle himself, motion in a vacuum is not instantaneous. If this were so the two plates mentioned above would separate without any resistance whatever, seeing that the same instant of time would suffice for their separation and for the surrounding medium to rush in and fill the vacuum between them. The fact that the lower plate follows the upper one allows us to infer, not only that motion in a vacuum is not instantaneous, but also that, between the two plates, a vacuum really exists, at least for a very short time, sufficient to allow the surrounding medium to rush in and fill the vacuum; for if there were no vacuum there would be no need of any motion in the medium. One must admit then that a vacuum is sometimes

produced by violent motion [*violenza*] or contrary to the laws of nature, (although in my opinion nothing occurs contrary to nature except the impossible, and that never occurs).

But here another difficulty arises. While experiment convinces me of the correctness of this conclusion, my mind is not entirely satisfied as to the cause to which this effect is to be attributed. For the separation of the plates precedes the formation of the vacuum which is produced as a consequence of this separation; and since it appears to me that, in the order of nature, the cause must precede the effect, even though it appears to follow in point of time, and since every positive effect must have a positive cause, I do not see how the adhesion of two plates and their resistance to separation—actual facts—can be referred to a vacuum as cause when this vacuum is yet to follow. According to the infallible maxim of the Philosopher, the non-existent can produce no effect.

Simp. Seeing that you accept this axiom of Aristotle, I hardly think you will reject another excellent and reliable maxim of his, namely, Nature undertakes only that which happens without resistance; and in this saying, it appears to me, you will find the solution of your difficulty. Since nature abhors a vacuum, she prevents that from which a vacuum would follow as a necessary consequence. Thus it happens that nature prevents the separation of the two plates.

[61]

Sagr. Now admitting that what Simplicio says is an adequate solution of my difficulty, it seems to me, if I may be allowed to resume my former argument, that this very resistance to a vacuum ought to be sufficient to hold together the parts either of stone or of metal or the parts of any other solid which is knit together more strongly and which is more resistant to separation. If for one effect there be only one cause, or if, more being assigned, they can be reduced to one, then why is not this vacuum which really exists a sufficient cause for all kinds of resistance?

Salv. I do not wish just now to enter this discussion as to whether the vacuum alone is sufficient to hold together the separate parts of a solid body; but I assure you that the vacuum which acts as a sufficient cause in the case of the two plates is not alone sufficient to bind together the parts of a solid cylinder of marble or metal which, when pulled violently, separates and divides. And now if I find a method of distinguishing this well known resistance, depending upon the vacuum, from every other kind which might increase the coherence, and if I show you that the aforesaid resistance alone is not nearly sufficient for such an effect, will you not grant that we are bound to introduce another cause? Help him, Simplicio, since he does not know what reply to make.

Simp. Surely, Sagredo's hesitation must be owing to another reason, for there can be no doubt concerning a conclusion which is at once so clear and logical.

Sagr. You have guessed rightly, Simplicio. I was wondering whether, if a million of gold each year from Spain were not sufficient to pay the army, it might not be necessary to make provision other than small coin for the pay of the soldiers.[3]

But go ahead, Salviati; assume that I admit your conclusion and show us your method of separating the action of the vacuum from other causes; and by measuring it show us how it is not sufficient to produce the effect in question.

Salv. Your good angel assist you. I will tell you how to separate the force of the vacuum from the others, and afterwards how to measure it. For this purpose let us consider a continuous substance whose parts lack all resistance to separation except that derived from a vacuum, such as is the case with water, a fact fully demonstrated by our Academician in one of his treatises. Whenever a cylinder of water is subjected to a pull and

[62]

offers a resistance to the separation of its parts this can be attributed to no other cause than the resistance of the vacuum. In order to try such an experiment I have invented a device which I can better explain by means of a sketch than by mere

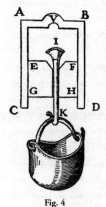

Fig. 4

words. Let CABD represent the cross section of a cylinder either of metal or, preferably, of glass, hollow inside and accurately turned. Into this is introduced a perfect fitting cylinder of wood, represented in cross section by EGHF, and capable of up-and-down motion. Through the middle of this cylinder is bored a hole to receive an iron wire, carrying a hook at the end K, while the upper end of the wire, I, is provided with a conical head. The wooden cylinder is countersunk at the top so as to receive, with a perfect fit, the conical head I of the wire, IK, when pulled down by the end K.

Now insert the wooden cylinder EH in the hollow cylinder AD, so as not to touch the upper end of the latter but to leave free a space of two or three finger-breadths; this space is to be filled with water by holding the vessel with the mouth CD upwards, pushing down on the stopper EH, and at the same time keeping the conical head of the wire, I, away from the hollow portion of the wooden cylinder. The air is thus allowed to escape alongside the iron wire (which does not make a close fit) as soon as one presses down on the wooden stopper. The air having been allowed to escape and the iron wire having been drawn back so that it fits snugly against the conical depression in the wood,

3. The bearing of this remark becomes clear on reading what Salviati says below. [Trans.]

invert the vessel, bringing it mouth downwards, and hang on the hook K a vessel which can be filled with sand or any heavy material in quantity sufficient to finally separate the upper surface of the stopper, EF, from the lower surface of the water to which it was attached only by the resistance of the vacuum. Next weigh the stopper and wire together with the attached vessel and its contents; we shall then have the force of the vacuum [*forza del vacuo*].

[63]

If one attaches to a cylinder of marble or glass a weight which, together with the weight of the marble or glass itself, is just equal to the sum of the weights before mentioned, and if breaking occurs we shall then be justified in saying that the vacuum alone holds the parts of the marble and glass together; but if this weight does not suffice and if breaking occurs only after adding, say, four times this weight, we shall then be compelled to say that the vacuum furnishes only one fifth of the total resistance [*resistenza*].

Simp. No one can doubt the cleverness of the device; yet it presents many difficulties which make me doubt its reliability. For who will assure us that the air does not creep in between the glass and stopper even if it is well packed with tow or other yielding material? I question also whether oiling with wax or turpentine will suffice to make the cone, I, fit snugly on its seat. Besides, may not the parts of the water expand and dilate? Why may not the air or exhalations or some other more subtile substances penetrate the pores of the wood, or even of the glass itself?

Salv. With great skill indeed has Simplicio laid before us the difficulties; and he has even partly suggested how to prevent the air from penetrating the wood or passing between the wood and the glass. But now let me point out that, as our experience increases, we shall learn whether or not these alleged difficulties really exist. For if, as is the case with air, water is by nature expansible, although only under severe treatment, we shall see the stopper descend; and if we put a small excavation in the upper part of the glass vessel, such as indicated by V, then the air or any other tenuous and gaseous substance, which might penetrate the pores of glass or wood, would pass through the water and collect in this receptacle V. But if these things do not happen we may rest assured that our experiment has been performed with proper caution; and we shall discover that water does not dilate and that glass does not allow any material, however tenuous, to penetrate it.

Sagr. Thanks to this discussion, I have learned the cause of a certain effect which I have long wondered at and despaired of understanding. I once saw a cistern which had been provided with a pump under the mistaken impression that the water might thus be drawn with less effort or in greater quantity than by means of the ordinary bucket.

[64]

The stock of the pump carried its sucker and valve in the upper part so that the water was lifted by attraction and not by a push as is the case with pumps in which the sucker is placed lower down. This pump worked perfectly so long as the water in the cistern stood above a certain level; but below this level the pump failed to work. When I first noticed this phenomenon I thought the machine was out of order; but the workman whom I called in to repair it told me the defect was not in the pump but in the water which had fallen too low to be raised through such a height; and he added that it was not possible, either by a pump or by any other machine working on the principle of attraction, to lift water a hair's breadth above eighteen cubits; whether the pump be large or small this is the extreme limit of the lift. Up to this time I had been so thoughtless that, although I knew a rope, or rod of wood, or of iron, if sufficiently long, would break by its own weight when held by the upper end, it never occurred to me that the same thing would happen, only much more easily, to a column of water. And really is not that thing which is attracted in the pump a column of water attached at the upper end and stretched more and more until finally a point is reached where it breaks, like a rope, on account of its excessive weight?

Salv. That is precisely the way it works; this fixed elevation of eighteen cubits is true for any quantity of water whatever, be the pump large or small or even as fine as a straw. We may therefore say that, on weighing the water contained in a tube eighteen cubits long, no matter what the diameter, we shall obtain the value of the resistance of the vacuum in a cylinder of any solid material having a bore of this same diameter. And having gone so far, let us see how easy it is to find to what length cylinders of metal, stone, wood, glass, etc., of any diameter can be elongated without breaking by their own weight.

[65]

Take for instance a copper wire of any length and thickness; fix the upper end and to the other end attach a greater and greater load until finally the wire breaks; let the maximum load be, say, fifty pounds. Then it is clear that if fifty pounds of copper, in addition to the weight of the wire itself which maybe, say, 1/8 ounce, is drawn out into wire of this same size we shall have the greatest length of this kind of wire which can sustain its own weight. Suppose the wire which breaks to be one cubit in length and 1/8 ounce in weight; then since it supports 50 lbs. in addition to its own weight, i. e., 4800 eighths-of-an-ounce, it follows that all copper wires, independent of size, can sustain themselves up to a length of 4801 cubits and no more. Since then a copper rod can sustain its own weight up to a length of 4801 cubits it follows that that part of the

breaking strength [*resistenza*] which depends upon the vacuum, comparing it with the remaining factors of resistance, is equal to the weight of a rod of water, eighteen cubits long and as thick as the copper rod. If, for example, copper is nine times as heavy as water, the breaking strength [*resistenza allo strapparsi*] of any copper rod, in so far as it depends upon the vacuum, is equal to the weight of two cubits of this same rod. By a similar method one can find the maximum length of wire or rod of any material which will just sustain its own weight, and can at the same time discover the part which the vacuum plays in its breaking strength.

Sagr. It still remains for you to tell us upon what depends the resistance to breaking, other than that of the vacuum; what is the gluey or viscous substance which cements together the parts of the solid? For I cannot imagine a glue that will not burn up in a highly heated furnace in two or three months, or certainly within ten or a hundred. For if gold, silver and glass are kept for a long while in the molten state and are removed from the furnace, their parts, on cooling, immediately reunite and bind themselves together as before. Not only so, but whatever difficulty arises with respect to the cementation of the parts of the glass arises also with regard to the parts of the glue; in other words, what is that which holds these parts together so firmly?

[66]

Salv. A little while ago, I expressed the hope that your good angel might assist you. I now find myself in the same straits. Experiment leaves no doubt that the reason why two plates cannot be separated, except with violent effort, is that they are held together by the resistance of the vacuum; and the same can be said of two large pieces of a marble or bronze column. This being so, I do not see why this same cause may not explain the coherence of smaller parts and indeed of the very smallest particles of these materials. Now, since each effect must have one true and sufficient cause and since I find no other cement, am I not justified in trying to discover whether the vacuum is not a sufficient cause?

Simp. But seeing that you have already proved that the resistance which the large vacuum offers to the separation of two large parts of a solid is really very small in comparison with that cohesive force which binds together the most minute parts, why do you hesitate to regard this latter as something very different from the former?

Salv. Sagredo has already answered this question when he remarked that each individual soldier was being paid from coin collected by a general tax of pennies and farthings, while even a million of gold would not suffice to pay the entire army. And who knows but that there may be other extremely minute vacua which affect the smallest particles so that that which binds together the contiguous parts is throughout of the same mintage? Let me tell you something which has just occurred to me and which I do not

offer as an absolute fact, but rather as a passing thought, still immature and calling for more careful consideration. You may take of it what you like; and judge the rest as you see fit. Sometimes when I have observed how fire winds its way in between the most minute particles of this or that metal and, even though these are solidly cemented together, tears them apart and separates them, and when I have observed that, on removing the fire, these particles reunite with the same tenacity as at first, without any loss of quantity in the case of gold and with little loss in the case of other metals, even though these parts have been separated for a long while, I have thought that the explanation might lie in the fact that the extremely fine particles of fire, penetrating the slender pores of the metal (too small to admit even the finest particles of air or of many other fluids), would fill the small intervening vacua and would set free these small particles from the attraction which these same vacua exert upon them and which prevents their separation.

[67]

Thus the particles are able to move freely so that the mass [*massa*] becomes fluid and remains so as long as the particles of fire remain inside; but if they depart and leave the former vacua then the original attraction [*attrazzione*] returns and the parts are again cemented together.

In reply to the question raised by Simplicio, one may say that although each particular vacuum is exceedingly minute and therefore easily overcome, yet their number is so extraordinarily great that their combined resistance is, so to speak, multiplied almost without limit. The nature and the amount of force [*forza*] which results [*risulta*] from adding together an immense number of small forces [*debolissimi momenti*] is clearly illustrated by the fact that a weight of millions of pounds, suspended by great cables, is overcome and lifted, when the south wind carries innumerable atoms of water, suspended in thin mist, which moving through the air penetrate between the fibres of the tense ropes in spite of the tremendous force of the hanging weight. When these particles enter the narrow pores they swell the ropes, thereby shorten them, and perforce lift the heavy mass [*mole*].

Sagr. There can be no doubt that any resistance, so long as it is not infinite, may be overcome by a multitude of minute forces. Thus a vast number of ants might carry ashore a ship laden with grain. And since experience shows us daily that one ant can easily carry one grain, it is clear that the number of grains in the ship is not infinite, but falls below a certain limit. If you take another number four or six times as great, and if you set to work a corresponding number of ants they will carry the grain ashore and the boat also. It is true that this will call for a prodigious number of ants, but in my opinion this is precisely the case with the vacua which bind together the least particles of a metal.

Salv. But even if this demanded an infinite number would you still think it impossible?
Sagr. Not if the mass [*mole*] of metal were infinite; otherwise. . .

[68]

Salv. Otherwise what? Now since we have arrived at paradoxes let us see if we cannot prove that within a finite extent it is possible to discover an infinite number of vacua. At the same time we shall at least reach a solution of the most remarkable of all that list of problems which Aristotle himself calls wonderful; I refer to his *Questions in*

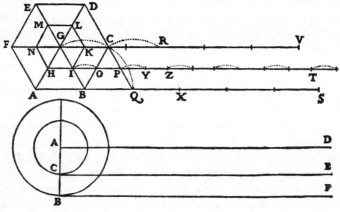

Fig. 5

Mechanics. This solution may be no less clear and conclusive than that which he himself gives and quite different also from that so cleverly expounded by the most learned Monsignor di Guevara.[4]

First it is necessary to consider a proposition, not treated by others, but upon which depends the solution of the problem and from which, if I mistake not, we shall derive other new and remarkable facts. For the sake of clearness let us draw an accurate figure. About G as a center describe an equiangular and equilateral polygon of any number of sides, say the hexagon ABCDEF. Similar to this and concentric with it, describe another smaller one which we shall call HIKLMN. Prolong the side AB, of the larger hexagon, indefinitely toward S; in like manner prolong the corresponding side HI of the smaller hexagon, in the same direction, so that the line HT is parallel to AS; and through the center draw the line GV parallel to the other two.

4. Bishop of Teano; b. 1561, d. 1641. *[Trans.]*

[69]

This done, imagine the larger polygon to roll upon the line AS, carrying with it the smaller polygon. It is evident that, if the point B, the end of the side AB remains fixed at the beginning of the rotation, the point A will rise and the point C will fall describing the arc CQ until the side BC coincides with the line BQ, equal to BC. But during this rotation the point I, on the smaller polygon, will rise above the line IT because IB is oblique to AS; and it will not again return to the line IT until the point C shall have reached the position Q. The point I, having described the arc IO above the line HT, will reach the position O at the same time the side IK assumes the position OP; but in the meantime the center G has traversed a path above GV and does not return to it until it has completed the arc GC. This step having been taken, the larger polygon has been brought to rest with its side BC coinciding with the line BQ while the side IK of the smaller polygon has been made to coincide with the line OP, having passed over the portion IO without touching it; also the center G will have reached the position C after having traversed all its course above the parallel line GV. And finally the entire figure will assume a position similar to the first, so that if we continue the rotation and come to the next step, the side DC of the larger polygon will coincide with the portion QX and the side KL of the smaller polygon, having first skipped the arc PY, will fall on YZ, while the center still keeping above the line GV will return to it at R after having jumped the interval CR. At the end of one complete rotation the larger polygon will have traced upon the line AS, without break, six lines together equal to its perimeter; the lesser polygon will likewise have imprinted six lines equal to its perimeter, but separated by the interposition of five arcs, whose chords represent the parts of HT not touched by the polygon: the center G never reaches the line GV except at six points. From this it is clear that the space traversed by the smaller polygon is almost equal to that traversed by the larger, that is, the line HT approximates the line AS, differing from it only by the length of one chord of one of these arcs, provided we understand the line HT to include the five skipped arcs.

[70]

Now this exposition which I have given in the case of these hexagons must be understood to be applicable to all other polygons, whatever the number of sides, provided only they are similar, concentric, and rigidly connected, so that when the greater one rotates the lesser will also turn however small it may be. You must also understand that the lines described by these two are nearly equal provided we include in the space traversed by the smaller one the intervals which are not touched by any part of the perimeter of this smaller polygon.

Let a large polygon of, say, one thousand sides make one complete rotation and thus lay off a line equal to its perimeter; at the same time the small one will pass over an approximately equal distance, made up of a thousand small portions each equal to one of its sides, but interrupted by a thousand spaces which, in contrast with the portions that coincide with the sides of the polygon, we may call empty. So far the matter is free from difficulty or doubt.

But now suppose that about any center, say A, we describe two concentric and rigidly connected circles; and suppose that from the points C and B, on their radii, there are drawn the tangents CE and BF and that through the center A the line AD is drawn parallel to them, then if the large circle makes one complete rotation along the line BF, equal not only to its circumference but also to the other two lines CE and AD, tell me what the smaller circle will do and also what the center will do. As to the center it will certainly traverse and touch the entire line AD while the circumference of the smaller circle will have measured off by its points of contact the entire line CE, just as was done by the above mentioned polygons. The only difference is that the line HT was not at every point in contact with the perimeter of the smaller polygon, but there were left untouched as many vacant spaces as there were spaces coinciding with the sides. But herein the case of the circles the circumference of the smaller one never leaves the line CE, so that no part of the latter is left untouched, nor is there ever a time when some point on the circle is not in contact with the straight line. How now can the smaller circle traverse a length greater than its circumference unless it go by jumps?

Sagr. It seems to me that one may say that just as the center of the circle, by itself, carried along the line AD is constantly in contact with it, although it is only a single point, so the points on the circumference of the smaller circle, carried along by the motion of the larger circle, would slide over some small parts of the line CE.

[71]

Salv. There are two reasons why this cannot happen. First because there is no ground for thinking that one point of contact, such as that at C, rather than another, should slip over certain portions of the line CE. But if such slidings along CE did occur they would be infinite in number since the points of contact (being mere points) are infinite in number: an infinite number of finite slips will however make an infinitely long line, while as a matter of fact the line CE is finite. The other reason is that as the greater circle, in its rotation, changes its point of contact continuously the lesser circle must do the same because B is the only point from which a straight line can be drawn to A and pass through C. Accordingly the small circle must change its point of contact whenever the large one changes: no point of the small circle touches the straight line CE in more than one point. Not only so, but even in the rotation of the polygons there

was no point on the perimeter of the smaller which coincided with more than one point on the line traversed by that perimeter; this is at once clear when you remember that the line IK is parallel to BC and that therefore IK will remain above IP until BC coincides with BQ, and that IK will not lie upon IP except at the very instant when BC occupies the position BQ; at this instant the entire line IK coincides with OP and immediately afterwards rises above it.

Sagr. This is a very intricate matter. I see no solution. Pray explain it to us.

Salv. Let us return to the consideration of the above mentioned polygons whose behavior we already understand. Now in the case of polygons with 100000 sides, the line traversed by the perimeter of the greater, i. e., the line laid down by its 100000 sides one after another, is equal to the line traced out by the 100000 sides of the smaller, provided we include the 100000 vacant spaces interspersed. So in the case of the circles, polygons having an infinitude of sides, the line traversed by the continuously distributed [*continuamente disposti*] infinitude of sides is in the greater circle equal to the line laid down by the infinitude of sides in the smaller circle but with the exception that these latter alternate with empty spaces; and since the sides are not finite in number, but infinite, so also are the intervening empty spaces not finite but infinite.

[72]

The line traversed by the larger circle consists then of an infinite number of points which completely fill it; while that which is traced by the smaller circle consists of an infinite number of points which leave empty spaces and only partly fill the line. And here I wish you to observe that after dividing and resolving a line into a finite number of parts, that is, into a number which can be counted, it is not possible to arrange them again into a greater length than that which they occupied when they formed a *continuum* [*continuate*] and were connected without the interposition of as many empty spaces. But if we consider the line resolved into an infinite number of infinitely small and indivisible parts, we shall be able to conceive the line extended indefinitely by the interposition, not of a finite, but of an infinite number of infinitely small indivisible empty spaces.

Now this which has been said concerning simple lines must be understood to hold also in the case of surfaces and solid bodies, it being assumed that they are made up of an infinite, not a finite, number of atoms. Such a body once divided into a finite number of parts cannot be reassembled so as to occupy more space than before unless we interpose a finite number of empty spaces, that is to say, spaces free from the substance of which the solid is made. But if we imagine the body, by some extreme and final analysis, resolved into its primary elements, infinite in number, then we shall be able to think of them as indefinitely extended in space, not by the interposition of a finite,

but of an infinite number of empty spaces. Thus one can easily imagine a small ball of gold expanded into a very large space without the introduction of a finite number of empty spaces, always provided the gold is made up of an infinite number of indivisible parts.

Simp. It seems to me that you are travelling along toward those vacua advocated by a certain ancient philosopher.

Salv. But you have failed to add, "who denied Divine Providence," an inapt remark made on a similar occasion by a certain antagonist of our Academician.

Simp. I noticed, and not without indignation, the rancor of this ill-natured opponent; further references to these affairs I omit, not only as a matter of good form, but also because I know how unpleasant they are to the good tempered and well ordered mind of one so religious and pious, so orthodox and God-fearing as you.

But to return to our subject, your previous discourse leaves with me many difficulties which I am unable to solve. First among these is that, if the circumferences of the two circles are equal to the two straight lines, CE and BF, the latter considered as a *continuum*, the former as interrupted with an infinity of empty points, I do not see how it is possible to say that the line AD described by the center, and made up of an infinity of points, is equal to this center which is a single point. Besides, this building up of lines out of points, divisibles out of indivisibles, and finites out of infinites, offers me an obstacle difficult to avoid; and the necessity of introducing a vacuum, so conclusively refuted by Aristotle, presents the same difficulty.

[73]

Salv. These difficulties are real; and they are not the only ones. But let us remember that we are dealing with infinities and indivisibles, both of which transcend our finite understanding, the former on account of their magnitude, the latter because of their smallness. In spite of this, men cannot refrain from discussing them, even though it must be done in a roundabout way.

Therefore I also should like to take the liberty to present some of my ideas which, though not necessarily convincing, would, on account of their novelty, at least, prove somewhat startling. But such a diversion might perhaps carry us too far away from the subject under discussion and might therefore appear to you inopportune and not very pleasing.

Sagr. Pray let us enjoy the advantages and privileges which come from conversation between friends, especially upon subjects freely chosen and not forced upon us, a matter vastly different from dealing with dead books which give rise to many doubts but remove none. Share with us, therefore, the thoughts which our discussion has suggested to you; for since we are free from urgent business there will be abundant time

to pursue the topics already mentioned; and in particular the objections raised by Simplicio ought not in anywise to be neglected.

Salv. Granted, since you so desire. The first question was, How can a single point be equal to a line? Since I cannot do more at present I shall attempt to remove, or at least diminish, one improbability by introducing a similar or a greater one, just as sometimes a wonder is diminished by a miracle.[5]

And this I shall do by showing you two equal surfaces, together with two equal solids located upon these same surfaces as bases, all four of which diminish continuously and uniformly in such a way that their remainders always preserve equality among themselves, and finally both the surfaces and the solids terminate their previous constant equality by degenerating, the one solid and the one surface into a very long line, the other solid and the other surface into a single point; that is, the latter to one point, the former to an infinite number of points.

[74]

Sagr. This proposition appears to me wonderful, indeed; but let us hear the explanation and demonstration.

Salv. Since the proof is purely geometrical we shall need a figure. Let AFB be a semicircle with center at C; about it describe the rectangle ADEB and from the center draw the straight lines CD and CE to the points D and E. Imagine the radius CF to be drawn perpendicular to either of the lines AB or DE, and the entire figure to rotate

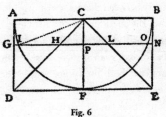

Fig. 6

about this radius as an axis. It is clear that the rectangle ADEB will thus describe a cylinder, the semicircle AFB a hemisphere, and the triangle CDE, a cone. Next let us remove the hemisphere but leave the cone and the rest of the cylinder, which, on account of its shape, we will call a "bowl." First we shall prove that the bowl and the cone are equal; then we shall show that a plane drawn parallel to the circle which forms the base of the bowl and which has the line DE for diameter and F for a center—a plane whose trace is GN—cuts the bowl in the points G, I, O, N, and the cone in the points H, L, so that the part of the cone indicated by CHL is always equal to the part of the bowl whose profile is represented by the triangles GAI and BON. Besides this we shall prove that the base of the cone, i. e., the circle whose diameter is HL, is equal to the circular surface which forms the base of this portion of the bowl, or as one might say, equal to a ribbon whose width is GI. (Note by the way the nature of mathematical

5. Cf. page 421 below. [Trans.]

definitions which consist merely in the imposition of names or, if you prefer, abbreviations of speech established and introduced in order to avoid the tedious drudgery which you and I now experience simply because we have not agreed to call this surface a "circular band" and that sharp solid portion of the bowl a "round razor.")

[75]

Now call them by what name you please, it suffices to understand that the plane, drawn at any height whatever, so long as it is parallel to the base, i. e., to the circle whose diameter is DE, always cuts the two solids so that the portion CHL of the cone is equal to the upper portion of the bowl; likewise the two areas which are the bases of these solids, namely the band and the circle HL, are also equal. Here we have the miracle mentioned above; as the cutting plane approaches the line AB the portions of the solids cut off are always equal, so also the areas of their bases. And as the cutting plane comes near the top, the two solids (always equal) as well as their bases (areas which are also equal) finally vanish, one pair of them degenerating into the circumference of a circle, the other into a single point, namely, the upper edge of the bowl and the apex of the cone. Now, since as these solids diminish equality is maintained between them up to the very last, we are justified in saying that, at the extreme and final end of this diminution, they are still equal and that one is not infinitely greater than the other. It appears therefore that we may equate the circumference of a large circle to a single point. And this which is true of the solids is true also of the surfaces which form their bases; for these also preserve equality between themselves throughout their diminution and in the end vanish, the one into the circumference of a circle, the other into a single point. Shall we not then call them equal seeing that they are the last traces and remnants of equal magnitudes? Note also that, even if these vessels were large enough to contain immense celestial hemispheres, both their upper edges and the apexes of the cones therein contained would always remain equal and would vanish, the former into circles having the dimensions of the largest celestial orbits, the latter into single points. Hence in conformity with the preceding we may say that all circumferences of circles, however different, are equal to each other, and are each equal to a single point.

Sagr. This presentation strikes me as so clever and novel that, even if I were able, I would not be willing to oppose it; for to deface so beautiful a structure by a blunt pedantic attack would be nothing short of sinful. But for our complete satisfaction pray give us this geometrical proof that there is always equality between these solids and between their bases; for it cannot, I think, fail to be very ingenious, seeing how subtle is the philosophical argument based upon this result.

[76]

Salv. The demonstration is both short and easy. Referring to the preceding figure, since IPC is a right angle the square of the radius IC is equal to the sum of the squares on the two sides IP, PC; but the radius IC is equal to AC and also to GP, while CP is equal to PH. Hence the square of the line GP is equal to the sum of the squares of IP and PH, or multiplying through by 4, we have the square of the diameter GN equal to the sum of the squares on IO and HL. And, since the areas of circles are to each other as the squares of their diameters, it follows that the area of the circle whose diameter is GN is equal to the sum of the areas of circles having diameters IO and HL, so that if we remove the common area of the circle having IO for diameter the remaining area of the circle GN will be equal to the area of the circle whose diameter is HL. So much for the first part. As for the other part, we leave its demonstration for the present, partly because those who wish to follow it will find it in the twelfth proposition of the second book of *De centro gravitatis solidorum* by the Archimedes of our age, Luca Valerio,[6] who made use of it for a different object, and partly because, for our purpose, it suffices to have seen that the above-mentioned surfaces are always equal and that, as they keep on diminishing uniformly, they degenerate, the one into a single point, the other into the circumference of a circle larger than any assignable; in this fact lies our miracle.[7]

Sagr. The demonstration is ingenious and the inferences drawn from it are remarkable. And now let us hear something concerning the other difficulty raised by Simplicio, if you have anything special to say, which, however, seems to me hardly possible, since the matter has already been so thoroughly discussed.

[77]

Salv. But I do have something special to say, and will first of all repeat what I said a little while ago, namely, that infinity and indivisibility are in their very nature incomprehensible to us; imagine then what they are when combined. Yet if we wish to build up a line out of indivisible points, we must take an infinite number of them, and are, therefore, bound to understand both the infinite and the indivisible at the same time. Many ideas have passed through my mind concerning this subject, some of which, possibly the more important, I may not be able to recall on the spur of the moment; but in the course of our discussion it may happen that I shall awaken in you, and especially in Simplicio, objections and difficulties which in turn will bring to memory that which, without such stimulus, would have lain dormant in my mind. Allow me therefore the customary liberty of introducing some of our human fancies, for indeed we

6. Distinguished Italian mathematician; born at Ferrara about 1552; admitted to the Accademia dei Lincei 1612; died 1618. *[Trans.]*
7. *Cf.* above. *[Trans.]*

may so call them in comparison with supernatural truth which furnishes the one true and safe recourse for decision in our discussions and which is an infallible guide in the dark and dubious paths of thought.

One of the main objections urged against this building up of continuous quantities out of indivisible quantities [*continuo d' indivisibili*] is that the addition of one indivisible to another cannot produce a divisible, for if this were so it would render the indivisible divisible. Thus if two indivisibles, say two points, can be united to form a quantity, say a divisible line, then an even more divisible line might be formed by the union of three, five, seven, or any other odd number of points. Since however these lines can be cut into two equal parts, it becomes possible to cut the indivisible which lies exactly in the middle of the line. In answer to this and other objections of the same type we reply that a divisible magnitude cannot be constructed out of two or ten or a hundred or a thousand indivisibles, but requires an infinite number of them.

Simp. Here a difficulty presents itself which appears to me insoluble. Since it is clear that we may have one line greater than another, each containing an infinite number of points, we are forced to admit that, within one and the same class, we may have something greater than infinity, because the infinity of points in the long line is greater than the infinity of points in the short line. This assigning to an infinite quantity a value greater than infinity is quite beyond my comprehension.

[78]

Salv. This is one of the difficulties which arise when we attempt, with our finite minds, to discuss the infinite, assigning to it those properties which we give to the finite and limited; but this I think is wrong, for we cannot speak of infinite quantities as being the one greater or less than or equal to another. To prove this I have in mind an argument which, for the sake of clearness, I shall put in the form of questions to Simplicio who raised this difficulty.

I take it for granted that you know which of the numbers are squares and which are not.

Simp. I am quite aware that a squared number is one which results from the multiplication of another number by itself; thus 4, 9, etc., are squared numbers which come from multiplying 2, 3, etc., by themselves.

Salv. Very well; and you also know that just as the products are called squares so the factors are called sides or roots; while on the other hand those numbers which do not consist of two equal factors are not squares. Therefore if I assert that all numbers, including both squares and non-squares, are more than the squares alone, I shall speak the truth, shall I not?

Simp. Most certainly.

Salv. If I should ask further how many squares there are one might reply truly that there are as many as the corresponding number of roots, since every square has its own root and every root its own square while no square has more than one root and no root more than one square.

Simp. Precisely so.

Salv. But if I inquire how many roots there are, it cannot be denied that there are as many as there are numbers because every number is a root of some square. This being granted we must say that there are as many squares as there are numbers because they are just as numerous as their roots, and all the numbers are roots. Yet at the outset we said there are many more numbers than squares, since the larger portion of them are not squares. Not only so, but the proportionate number of squares diminishes as we pass to larger numbers. Thus up to 100 we have 10 squares, that is, the squares constitute 1/10 part of all the numbers; up to 10000, we find only 1/100 part to be squares; and up to a million only 1/1000 part; on the other hand in an infinite number, if one could conceive of such a thing, he would be forced to admit that there are as many squares as there are numbers all taken together.

[79]

Sagr. What then must one conclude under these circumstances?

Salv. So far as I see we can only infer that the totality of all numbers is infinite, that the number of squares is infinite, and that the number of their roots is infinite; neither is the number of squares less than the totality of all numbers, nor the latter greater than the former; and finally the attributes "equal," "greater," and "less," are not applicable to infinite, but only to finite, quantities. When therefore Simplicio introduces several lines of different lengths and asks me how it is possible that the longer ones do not contain more points than the shorter, I answer him that one line does not contain more or less or just as many points as another, but that each line contains an infinite number. Or if I had replied to him that the points in one line were equal in number to the squares; in another, greater than the totality of numbers; and in the little one, as many as the number of cubes, might I not, indeed, have satisfied him by thus placing more points in one line than in another and yet maintaining an infinite number in each? So much for the first difficulty.

Sagr. Pray stop a moment and let me add to what has already been said an idea which just occurs to me. If the preceding be true, it seems to me impossible to say either that one infinite number is greater than another or even that it is greater than a finite number, because if the infinite number were greater than, say, a million it would follow that on passing from the million to higher and higher numbers we would be approaching the infinite; but this is not so; on the contrary, the larger the number to

which we pass, the more we recede from [this property of] infinity, because the greater the numbers the fewer [relatively] are the squares contained in them; but the squares in infinity cannot be less than the totality of all the numbers, as we have just agreed; hence the approach to greater and greater numbers means a departure from infinity.[8]

[80]

Salv. And thus from your ingenious argument we are led to conclude that the attributes "larger," "smaller," and "equal" have no place either in comparing infinite quantities with each other or in comparing infinite with finite quantities.

I pass now to another consideration. Since lines and all continuous quantities are divisible into parts which are themselves divisible without end, I do not see how it is possible to avoid the conclusion that these lines are built up of an infinite number of indivisible quantities because a division and a subdivision which can be carried on indefinitely presupposes that the parts are infinite in number, otherwise the subdivision would reach an end; and if the parts are infinite in number, we must conclude that they are not finite in size, because an infinite number of finite quantities would give an infinite magnitude. And thus we have a continuous quantity built up of an infinite number of indivisibles.

Simp. But if we ran carry on indefinitely the division into finite parts what necessity is there then for the introduction of non-finite parts?

Salv. The very fact that one is able to continue, without end, the division into finite parts [in parti quante] makes it necessary to regard the quantity as composed of an infinite number of immeasurably small elements [di infiniti non quanti]. Now in order to settle this matter I shall ask you to tell me whether, in your opinion, a continuum is made up of a finite or of an infinite number of finite parts [parti quante].

Simp. My answer is that their number is both infinite and finite; potentially infinite but actually finite [infinite, in potenza; e finite, in atto]; that is to say, potentially infinite before division and actually finite after division; because parts cannot be said to exist in a body which is not yet divided or at least marked out; if this is not done we say that they exist potentially.

Salv. So that a line which is, for instance, twenty spans long is not said to contain actually twenty lines each one span in length except after division into twenty equal parts; before division it is said to contain them only potentially. Suppose the facts are as you say; tell me then whether, when the division is once made, the size of the original quantity is thereby increased, diminished, or unaffected.

Simp. It neither increases nor diminishes.

---

8. A certain confusion of thought appears to be introduced here through a failure to distinguish between the *number* n and the *class* of the first n numbers; and likewise from a failure to distinguish infinity as a number from infinity as the class of all numbers. *[Trans.]*

Salv. That is my opinion also. Therefore the finite parts [*parti quante*] in a *continuum*, whether actually or potentially present, do not make the quantity either larger or smaller; but it is perfectly clear that, if the number of finite parts actually contained in the whole is infinite in number, they will make the magnitude infinite. Hence the number of finite parts, although existing only potentially, cannot be infinite unless the magnitude containing them be infinite; and conversely if the magnitude is finite it cannot contain an infinite number of finite parts either actually or potentially.

[81]

Sagr. How then is it possible to divide a *continuum* without limit into parts which are themselves always capable of subdivision?

Salv. This distinction of yours between actual and potential appears to render easy by one method what would be impossible by another. But I shall endeavor to reconcile these matters in another way; and as to the query whether the finite parts of a limited *continuum* [*continuo terminato*] are finite or infinite in number I will, contrary to the opinion of Simplicio, answer that they are neither finite nor infinite.

Simp. This answer would never have occurred to me since I did not think that there existed any intermediate step between the finite and the infinite, so that the classification or distinction which assumes that a thing must be either finite or infinite is faulty and defective.

Salv. So it seems to me. And if we consider discrete quantities I think there is, between finite and infinite quantities, a third intermediate term which corresponds to every assigned number; so that if asked, as in the present case, whether the finite parts of a *continuum* are finite or infinite in number the best reply is that they are neither finite nor infinite but correspond to every assigned number. In order that this may be possible, it is necessary that those parts should not be included within a limited number, for in that case they would not correspond to a number which is greater; nor can they be infinite in number since no assigned number is infinite; and thus at the pleasure of the questioner we may, to any given line, assign a hundred finite parts, a thousand, a hundred thousand, or indeed any number we may please so long as it be not infinite. I grant, therefore, to the philosophers, that the *continuum* contains as many finite parts as they please and I concede also that it contains them, either actually or potentially, as they may like; but I must add that just as a line ten fathoms [*canne*] in length contains ten lines each of one fathom and forty lines each of one cubit [*braccia*] and eighty lines each of half a cubit, etc., so it contains an infinite number of points; call them actual or potential, as you like, for as to this detail, Simplicio, I defer to your opinion and to your judgment.

[82]

Simp. I cannot help admiring your discussion; but I fear that this parallelism between the points and the finite parts contained in a line will not prove satisfactory, and that you will not find it so easy to divide a given line into an infinite number of points as the philosophers do to cut it into ten fathoms or forty cubits; not only so, but such a division is quite impossible to realize in practice, so that this will be one of those potentialities which cannot be reduced to actuality.

Salv. The fact that something can be done only with effort or diligence or with great expenditure of time does not render it impossible; for I think that you yourself could not easily divide a line into a thousand parts, and much less if the number of parts were 937 or any other large prime number. But if I were to accomplish this division which you deem impossible as readily as another person would divide the line into forty parts would you then be more willing, in our discussion, to concede the possibility of such a division?

Simp. In general I enjoy greatly your method; and replying to your query, I answer that it would be more than sufficient if it prove not more difficult to resolve a line into points than to divide it into a thousand parts.

Salv. I will now say something which may perhaps astonish you; it refers to the possibility of dividing a line into its infinitely small elements by following the same order which one employs in dividing the same line into forty, sixty, or a hundred parts, that is, by dividing it into two, four, etc. He who thinks that, by following this method, he can reach an infinite number of points is greatly mistaken; for if this process were followed to eternity there would still remain finite parts which were undivided.

Indeed by such a method one is very far from reaching the goal of indivisibility; on the contrary he recedes from it and while he thinks that, by continuing this division and by multiplying the multitude of parts, he will approach infinity, he is, in my opinion, getting farther and farther away from it. My reason is this. In the preceding discussion we concluded that, in an infinite number, it is necessary that the squares and cubes should be as numerous as the totality of the natural numbers [*tutti i numeri*], because both of these are as numerous as their roots which constitute the totality of the natural numbers. Next we saw that the larger the numbers taken the more sparsely distributed were the squares, and still more sparsely the cubes; therefore it is clear that the larger the numbers to which we pass the farther we recede from the infinite number; hence it follows that, since this process carries us farther and farther from the end sought, if on turning back we shall find that any number can be said to be infinite, it must be unity. Here indeed are satisfied all those conditions which are requisite for an infinite number; I mean that unity contains in itself as many squares as there are cubes and natural numbers [*tutti i numeri*].

GALILEO GALILEI

[83]

Simp. I do not quite grasp the meaning of this.

Salv. There is no difficulty in the matter because unity is at once a square, a cube, a square of a square and all the other powers [*dignità*]; nor is there any essential peculiarity in squares or cubes which does not belong to unity; as, for example, the property of two square numbers that they have between them a mean proportional; take any square number you please as the first term and unity for the other, then you will always find a number which is a mean proportional. Consider the two square numbers, 9 and

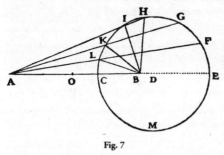

4; then 3 is the mean proportional between 9 and 1; while 2 is a mean proportional between 4 and 1; between 9 and 4 we have 6 as a mean proportional. A property of cubes is that they must have between them two mean proportional numbers; take 8 and 27; between them lie 12 and 18; while between 1 and 8 we have 2 and 4 intervening; and

Fig. 7

between 1 and 27 there lie 3 and 9. Therefore we conclude that unity is the only infinite number. These are some of the marvels which our imagination cannot grasp and which should warn us against the serious error of those who attempt to discuss the infinite by assigning to it the same properties which we employ for the finite, the natures of the two having nothing in common.

With regard to this subject I must tell you of a remarkable property which just now occurs to me and which will explain the vast alteration and change of character which a finite quantity would undergo in passing to infinity. Let us draw the straight line AB of arbitrary length and let the point C divide it into two unequal parts; then I say that, if pairs of lines be drawn, one from each of the terminal points A and B, and if the ratio between the lengths of these lines is the same as that between AC and CB, their points of intersection will all lie upon the circumference of one and the same circle.

[84]

Thus, for example, AL and BL drawn from A and B, meeting at the point L, bearing to one another the same ratio as AC to BC, and the pair AK and BK meeting at K also bearing to one another the same ratio, and likewise the pairs AI, BI, AH, BH, AG, BG, AF, BF, AE, BE, have their points of intersection L, K, I, H, G, F, E, all lying upon the circumference of one and the same circle. Accordingly if we imagine the point C to

move continuously in such a manner that the lines drawn from it to the fixed terminal points, A and B, always maintain the same ratio between their lengths as exists between the original parts, AC and CB, then the point C will, as I shall presently prove, describe a circle. And the circle thus described will increase in size without limit as the point C approaches the middle point which we may call O; but it will diminish in size as C approaches the end B. So that the infinite number of points located in the line OB will, if the motion be as explained above, describe circles of every size, some smaller than the pupil of the eye of a flea, others larger than the celestial equator. Now if we move any of the points lying between the two ends O and B they will all describe circles, those nearest O, immense circles; but if we move the point O itself, and continue to move it according to the aforesaid law, namely, that the lines drawn from O to the terminal points, A and B, maintain the same ratio as the original lines AO and OB, what kind of a line will be produced? A circle will be drawn larger than the largest of the others, a circle which is therefore infinite, But from the point O a straight line will also be drawn perpendicular to BA and extending to infinity without ever turning, as did the others, to join its last end with its first; for the point C, with its limited motion, having described the upper semicircle, CHE, proceeds to describe the lower semicircle EMC, thus returning to the starting point.

[85]

But the point O having started to describe its circle, as did all the other points in the line AB, (for the points in the other portion OA describe their circles also, the largest being those nearest the point O) is unable to return to its starting point because the circle it describes, being the largest of all, is infinite; in fact, it describes an infinite straight line as circumference of its infinite circle. Think now what a difference there is between a finite and an infinite circle since the latter changes character in such a manner that it loses not only its existence but also its possibility of existence; indeed, we already clearly understand that there can be no such thing as an infinite circle; similarly there can be no infinite sphere, no infinite body, and no infinite surface of any shape. Now what shall we say concerning this metamorphosis in the transition from finite to infinite? And why should we feel greater repugnance, seeing that, in our search after the infinite among numbers we found it in unity? Having broken up a solid into many parts, having reduced it to the finest of powder and having resolved it into its infinitely small indivisible atoms why may we not say that this solid has been reduced to a single *continuum* [*un solo continuo*] perhaps a fluid like water or mercury or even a liquified metal? And do we not see stones melt into glass and the glass itself under strong heat become more fluid than water?

Sagr. Are we then to believe that substances become fluid in virtue of being resolved into their infinitely small indivisible components?

Salv. I am not able to find any better means of accounting for certain phenomena of which the following is one. When I take a hard substance such as stone or metal and when I reduce it by means of a hammer or fine file to the most minute and impalpable powder, it is clear that its finest particles, although when taken one by one are, on account of their smallness, imperceptible to our sight and touch, are nevertheless finite in size, possess shape, and capability of being counted. It is also true that when once heaped up they remain in a heap; and if an excavation be made within limits the cavity will remain and the surrounding particles will not rush in to fill it; if shaken the particles come to rest immediately after the external disturbing agent is removed; the same effects are observed in all piles of larger and larger particles, of any shape, even if spherical, as is the case with piles of millet, wheat, lead shot, and every other material.

[86]

But if we attempt to discover such properties in water we do not find them; for when once heaped up it immediately flattens out unless held up by some vessel or other external retaining body; when hollowed out it quickly rushes in to fill the cavity; and when disturbed it fluctuates for a long time and sends out its waves through great distances.

Seeing that water has less firmness [*consistenza*] than the finest of powder, in fact has no consistence whatever, we may, it seems to me, very reasonably conclude that the smallest particles into which it can be resolved are quite different from finite and divisible particles; indeed the only difference I am able to discover is that the former are indivisible. The exquisite transparency of water also favors this view; for the most transparent crystal when broken and ground and reduced to powder loses its transparency; the finer the grinding the greater the loss; but in the case of water where the attrition is of the highest degree we have extreme transparency. Gold and silver when pulverized with acids [*acque forti*] more finely than is possible with any file still remain powders,[9] and do not become fluids until the finest particles [*gl' indivisibili*] of fire or of the rays of the sun dissolve them, as I think, into their ultimate, indivisible, and infinitely small components.

Sagr. This phenomenon of light which you mention is one which I have many times remarked with astonishment. I have, for instance, seen lead melted instantly by means of a concave mirror only three hands [*palmi*] in diameter. Hence I think that if the mirror were very large, well-polished and of a parabolic figure, it would just as readily and quickly melt any other metal, seeing that the small mirror, which was not well

9. It is not clear what Galileo here means by saying that gold and silver when treated with acids still remain powders. [*Trans.*]

polished and had only a spherical shape, was able so energetically to melt lead and burn every combustible substance. Such effects as these render credible to me the marvels accomplished by the mirrors of Archimedes.

[87]

Salv. Speaking of the effects produced by the mirrors of Archimedes, it was his own books (which I had already read and studied with infinite astonishment) that rendered credible to me all the miracles described by various writers. And if any doubt had remained the book which Father Buonaventura Cavalieri[10] has recently published on the subject of the burning glass [*specchio ustòrio*] and which I have read with admiration would have removed the last difficulty.

Sagr. I also have seen this treatise and have read it with pleasure and astonishment; and knowing the author I was confirmed in the opinion which I had already formed of him that he was destined to become one of the leading mathematicians of our age. But now, with regard to the surprising effect of solar rays in melting metals, must we believe that such a furious action is devoid of motion or that it is accompanied by the most rapid of motions?

Salv. We observe that other combustions and resolutions are accompanied by motion, and that, the most rapid; note the action of lightning and of powder as used in mines and petards; note also how the charcoal flame, mixed as it is with heavy and impure vapors, increases its power to liquify metals whenever quickened by a pair of bellows. Hence I do not understand how the action of light, although very pure, can be devoid of motion and that of the swiftest type.

Sagr. But of what kind and how great must we consider this speed of light to be? Is it instantaneous or momentary or does it like other motions require time? Can we not decide this by experiment.

Simp. Everyday experience shows that the propagation of light is instantaneous; for when we see a piece of artillery fired, at great distance, the flash reaches our eyes without lapse of time; but the sound reaches the ear only after a noticeable interval.

Sagr. Well, Simplicio, the only thing I am able to infer from this familiar bit of experience is that sound, in reaching our ear, travels more slowly than light; it does not inform me whether the coming of the light is instantaneous or whether, although extremely rapid, it still occupies time. An observation of this kind tells us nothing more than one in which it is claimed that "As soon as the sun reaches the horizon its light reaches our eyes"; but who will assure me that these rays had not reached this limit earlier than they reached our vision?

10. One of the most active investigators among Galileo's contemporaries; born at Milan 1598; died at Bologna 1647; a Jesuit father, first to introduce the use of logarithms into Italy and first to derive the expression for the focal length of a lens having unequal radii of curvature. His "method of indivisibles" is to be reckoned as a precursor of the infinitesimal calculus. [*Trans.*]

[88]

Salv. The small conclusiveness of these and other similar observations once led me to devise a method by which one might accurately ascertain whether illumination, i. e., the propagation of light, is really instantaneous. The fact that the speed of sound is as high as it is, assures us that the motion of light cannot fail to be extraordinarily swift. The experiment which I devised was as follows:

Let each of two persons take a light contained in a lantern, or other receptacle, such that by the interposition of the hand, the one can shut off or admit the light to the vision of the other. Next let them stand opposite each other at a distance of a few cubits and practice until they acquire such skill in uncovering and occulting their lights that the instant one sees the light of his companion he will uncover his own. After a few trials the response will be so prompt that without sensible error [*svario*] the uncovering of one light is immediately followed by the uncovering of the other, so that as soon as one exposes his light he will instantly see that of the other. Having acquired skill at this short distance let the two experimenters, equipped as before, take up positions separated by a distance of two or three miles and let them perform the same experiment at night, noting carefully whether the exposures and occultations occur in the same manner as at short distances; if they do, we may safely conclude that the propagation of light is instantaneous; but if time is required at a distance of three miles which, considering the going of one light and the coming of the other, really amounts to six, then the delay ought to be easily observable. If the experiment is to be made at still greater distances, say eight or ten miles, telescopes may be employed, each observer adjusting one for himself at the place where he is to make the experiment at night; then although the lights are not large and are therefore invisible to the naked eye at so great a distance, they can readily be covered and uncovered since by aid of the telescopes, once adjusted and fixed, they will become easily visible.

Sagr. This experiment strikes me as a clever and reliable invention. But tell us what you conclude from the results.

Salv. In fact I have tried the experiment only at a short distance, less than a mile, from which I have not been able to ascertain with certainty whether the appearance of the opposite light was instantaneous or not; but if not instantaneous it is extraordinarily rapid—I should call it momentary; and for the present I should compare it to motion which we see in the lightning flash between clouds eight or ten miles distant from us. We see the beginning of this light—I might say its head and source—located at a particular place among the clouds; but it immediately spreads to the surrounding ones, which seems to be an argument that at least some time is required for propagation; for if the illumination were instantaneous and not gradual, we should not be able to distinguish its origin—its center, so to speak—from its outlying portions. What a sea

431

we are gradually slipping into without knowing it! With vacua and infinities and indivisibles and instantaneous motions, shall we ever be able, even by means of a thousand discussions, to reach dry land?

[89]

Sagr. Really these matters lie far beyond our grasp. Just think; when we seek the infinite among numbers we find it in unity; that which is ever divisible is derived from indivisibles; the vacuum is found inseparably connected with the plenum; indeed the views commonly held concerning the nature of these matters are so reversed that even the circumference of a circle turns out to be an infinite straight line, a fact which, if my memory serves me correctly, you, Salviati, were intending to demonstrate geometrically. Please therefore proceed without further digression.

Salv. I am at your service; but for the sake of greater clearness let me first demonstrate the following problem:

> *Given a straight line divided into unequal parts which bear to each*
> *other any ratio whatever, to describe a circle such that two straight*
> *lines drawn from the ends of the given line to any point on the*
> *circumference will bear to each other the same ratio as the two parts*
> *of the given line, thus making those lines which are drawn from the*
> *same terminal points homologous.*

[90]

Let AB represent the given straight line divided into any two unequal parts by the point C; the problem is to describe a circle such that two straight lines drawn from the terminal points, A and B, to any point on the circumference will bear to each other the same ratio as the part AC bears to BC, so that lines drawn from the same terminal points are homologous. About C as center describe a circle having the shorter part CB of the given line, as radius. Through A draw a straight line AD which shall be tangent to the circle at D and indefinitely prolonged toward E. Draw the radius CD which will be perpendicular to AE. At B erect a perpendicular to AB; this perpendicular will intersect AE at some point since the angle at A is acute; call this point of intersection E, and from it draw a perpendicular to AE which will intersect AB prolonged in F. Now I say the two straight lines FE and FC are equal. For if we join E and C, we shall have two triangles, DEC and BEC, in which the two sides of the one, DE and EC, are equal to the two sides of the other, BE and EC, both DE and EB being tangents to the circle DB while the bases DC and CB are likewise equal; hence the two angles, DEC and

432

BEC, will be equal. Now since the angle BCE differs from a right angle by the angle CEB, and the angle CEF also differs from a right angle by the angle CED, and since

these differences are equal, it follows that the angle FCE is equal to CEF; consequently the sides FE and FC are equal. If we describe a circle with F as center and FE as radius it will pass through the point C; let CEG be such a circle. This is the circle sought, for if we draw lines from the terminal points A and B to any point on its circumference they will bear to each other the same ratio as the two portions AC and BC which meet at

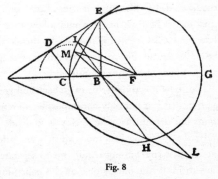

Fig. 8

the point C. This is manifest in the case of the two lines AE and BE, meeting at the point E, because the angle E of the triangle AEB is bisected by the line CE, and therefore AC:CB=AE:BE. The same may be proved of the two lines AG and BG terminating in the point G. For since the triangles AFE and EFB are similar, we have AF:FE=EF:FB, or AF:FC=CF:FB, and *dividendo* AC:CF=CB:BF, or AC:FG=CB:BF; also *componendo* we have both AB:BG=CB:BF and AG:GB=CF:FB=AE:EB=AC:BC. Q. E. D.

[91]

Take now any other point in the circumference, say H, where the two lines AH and BH intersect; in like manner we shall have AC:CB=AH:HB. Prolong HB until it meets the circumference at I and join IF; and since we have already found that AB:BG=CB:BF it follows that the rectangle AB.BF is equal to the rectangle CB.BG or IB.BH. Hence AB:BH=IB:BF. But the angles at B are equal and therefore AH:HB=IF:FB=EF:FB=AE:EB.

Besides, I may add, that it is impossible for lines which maintain this same ratio and which are drawn from the terminal points, A and B, to meet at any point either inside or outside the circle, CEG. For suppose this were possible; let AL and BL be two such lines intersecting at the point L outside the circle: prolong LB till it meets the circumference at M and join MF. If AL:BL=AC:BC=MF:FB, then we shall have two triangles ALB and MFB which have the sides about the two angles proportional, the angles at the vertex, B, equal, and the two remaining angles, FMB and LAB, less than right angles (because the right angle at M has for its base the entire diameter CG and not merely a part BF: and the other angle at the point A is acute because the line AL

the homologue of AC, is greater than BL, the homologue of BC). From this it follows that the triangles ABL and MBF are similar and therefore AB:BL=MB:BF, making the rectangle AB.BF=MB.BL; but it has been demonstrated that the rectangle AB.BF is equal to CB.BG; whence it would follow that the rectangle MB.BL is equal to the rectangle CB.BG which is impossible; therefore the intersection cannot fall outside the circle. And in like manner we can show that it cannot fall inside; hence all these intersections fall on the circumference.

But now it is time for us to go back and grant the request of Simplicio by showing him that it is not only not impossible to resolve a line into an infinite number of points but that this is quite as easy as to divide it into its finite parts. This I will do under the following condition which I am sure, Simplicio, you will not deny me, namely, that you will not require me to separate the points, one from the other, and show them to you, one by one, on this paper; for I should be content that you, without separating the four or six parts of a line from one another, should show me the marked divisions or at most that you should fold them at angles forming a square or a hexagon: for, then, I am certain you would consider the division distinctly and actually accomplished.

Simp. I certainly should.

Salv. If now the change which takes place when you bend a line at angles so as to form now a square, now an octagon, now a polygon of forty, a hundred or a thousand angles, is sufficient to bring into actuality the four, eight, forty, hundred, and thousand parts which, according to you, existed at first only potentially in the straight line, may I not say, with equal right, that, when I have bent the straight line into a polygon having an infinite number of sides, i. e., into a circle, I have reduced to actuality that infinite number of parts which you claimed, while it was straight, were contained in it only potentially? Nor can one deny that the division into an infinite number of points is just as truly accomplished as the one into four parts when the square is formed or into a thousand parts when the millagon is formed; for in such a division the same conditions are satisfied as in the case of a polygon of a thousand or a hundred thousand sides. Such a polygon laid upon a straight line touches it with one of its sides, i. e., with one of its hundred thousand parts; while the circle which is a polygon of an infinite number of sides touches the same straight line with one of its sides which is a single point different from all its neighbors and therefore separate and distinct in no less degree than is one side of a polygon from the other sides. And just as a polygon, when rolled along a plane, marks out upon this plane, by the successive

contacts of its sides, a straight line equal to its perimeter, so the circle rolled upon such a plane also traces by its infinite succession of contacts a straight line equal in length to its own circumference. I am willing, Simplicio, at the outset, to grant to the Peripatetics the truth of their opinion that a continuous quantity [*il continuo*] is divisible only into parts which are still further divisible so that however far the division and subdivision be continued no end will be reached; but I am not so certain that they will concede to me that none of these divisions of theirs can be a final one, as is surely the fact, because there always remains "another"; the final and ultimate division is rather one which resolves a continuous quantity into an infinite number of indivisible quantities, a result which I grant can never be reached by successive division into an ever-increasing number of parts.

[93]

But if they employ the method which I propose for separating and resolving the whole of infinity [*tutta la infinità*], at a single stroke (an artifice which surely ought not to be denied me), I think that they would be contented to admit that a continuous quantity is built up out of absolutely indivisible atoms, especially since this method, perhaps better than any other, enables us to avoid many intricate labyrinths, such as cohesion in solids, already mentioned, and the question of expansion and contraction, without forcing upon us the objectionable admission of empty spaces [in solids] which carries with it the penetrability of bodies. Both of these objections, it appears to me, are avoided if we accept the above-mentioned view of indivisible constituents.

Simp. I hardly know what the Peripatetics would say since the views advanced by you would strike them as mostly new, and as such we must consider them. It is however not unlikely that they would find answers and solutions for these problems which I, for want of time and critical ability, am at present unable to solve. Leaving this to one side for the moment, I should like to hear how the introduction of these indivisible quantities helps us to understand contraction and expansion avoiding at the same time the vacuum and the penetrability of bodies.

Sagr. I also shall listen with keen interest to this same matter which is far from clear in my mind; provided I am allowed to hear what, a moment ago, Simplicio suggested we omit, namely, the reasons which Aristotle offers against the existence of the vacuum and the arguments which you must advance in rebuttal.

Salv. I will do both. And first, just as, for the production of expansion, we employ the line described by the small circle during one rotation of the large one—a line greater than the circumference of the small circle—so, in order to explain contraction, we point out that, during each rotation of the smaller circle, the larger one describes a straight line which is shorter than its circumference.

[94]

For the better understanding of this we proceed to the consideration of what happens in the case of polygons. Employing a figure similar to the earlier one, construct the two hexagons, ABC and HIK, about the common center L and let them roll along the parallel lines HOM and ABc. Now holding the vertex I fixed, allow the smaller

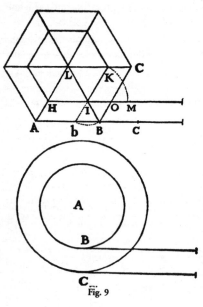

Fig. 9

polygon to rotate until the side IK lies upon the parallel, during which motion the point K will describe the arc KM, and the side KI will coincide with IM. Let us see what, in the meantime, the side CB of the larger polygon has been doing. Since the rotation is about the point I, the terminal point B, of the line IB, moving backwards, will describe the arc B*b* underneath the parallel *c*A so that when the side KI coincides with the line MI, the side BC will coincide with *bc*, having advanced only through the distance B*c*, but having retreated through a portion of the line BA which subtends the arc B*b*. If we allow the rotation of the smaller polygon to go on it will traverse and describe along its parallel a line equal to its perimeter; while the larger one will traverse and describe a line less than its perimeter by as many times the length *b*B as there are sides less one; this line is approximately equal to that described by the smaller polygon exceeding it only by the distance *b*B. Here now we see, without any difficulty, why the larger polygon, when carried by the smaller, does not measure off with its sides a line longer than that traversed by the smaller one; this is because a portion of each side is superposed upon its immediately preceding neighbor.

[95]

Let us next consider two circles, having a common center at A, and lying upon their respective parallels, the smaller being tangent to its parallel at the point B; the larger, at the point C. Here when the small circle commences to roll the point B does not remain at rest for a while so as to allow BC to move backward and carry with it the

point C, as happened in the case of the polygons, where the point I remained fixed until the side KI coincided with MI and the line IB carried the terminal point B backward as far as *b*, so that the side BC fell upon *bc*, thus superposing upon the line BA, the portion B*b*, and advancing by an amount B*c*, equal to MI, that is, to one side of the smaller polygon. On account of these superpositions, which are the excesses of the sides of the larger over the smaller polygon, each net advance is equal to one side of the smaller polygon and, during one complete rotation, these amount to a straight line equal in length to the perimeter of the smaller polygon.

But now reasoning in the same way concerning the circles, we must observe that whereas the number of sides in any polygon is comprised within a certain limit, the number of sides in a circle is infinite; the former are finite and divisible; the latter infinite and indivisible. In the case of the polygon, the vertices remain at rest during an interval of time which bears to the period of one complete rotation the same ratio which one side bears to the perimeter; likewise, in the case of the circles, the delay of each of the infinite number of vertices is merely instantaneous, because an instant is such a fraction of a finite interval as a point is of a line which contains an infinite number of points. The retrogression of the sides of the larger polygon is not equal to the length of one of its sides but merely to the excess of such a side over one side of the smaller polygon, the net advance being equal to this smaller side; but in the circle, the point or side C, during the instantaneous rest of B, recedes by an amount equal to its excess over the side B, making a net progress equal to B itself. In short the infinite number of indivisible sides of the greater circle with their infinite number of indivisible retrogressions, made during the infinite number of instantaneous delays of the infinite number of vertices of the smaller circle, together with the infinite number of progressions, equal to the infinite number of sides in the smaller circle—all these, I say, add up to a line equal to that described by the smaller circle, a line which contains an infinite number of infinitely small superpositions, thus bringing about a thickening or contraction without any overlapping or interpenetration of finite parts.

[96]

This result could not be obtained in the case of a line divided into finite parts such as is the perimeter of any polygon, which when laid out in a straight line cannot be shortened except by the overlapping and interpenetration of its sides. This contraction of an infinite number of infinitely small parts without the interpenetration or overlapping of finite parts and the previously mentioned [see above] expansion of an infinite number of indivisible parts by the interposition of indivisible vacua is, in my opinion, the most that can be said concerning the contraction and rarefaction of bodies, unless we give up the impenetrability of matter and introduce empty spaces of finite size.

If you find anything here that you consider worth while, pray use it; if not regard it, together with my remarks, as idle talk; but this remember, we are dealing with the infinite and the indivisible.

Sagr. I frankly confess that your idea is subtle and that it impresses me as new and strange; but whether, as a matter of fact, nature actually behaves according to such a law I am unable to determine; however, until I find a more satisfactory explanation I shall hold fast to this one. Perhaps Simplicio can tell us something which I have not yet heard, namely, how to explain the explanation which the philosophers have given of this abstruse matter; for, indeed, all that I have hitherto read concerning contraction is so dense and that concerning expansion so thin that my poor brain can neither penetrate the former nor grasp the latter.

Simp. I am all at sea and find difficulties in following either path, especially this new one; because according to this theory an ounce of gold might be rarefied and expanded until its size would exceed that of the earth, while the earth, in turn, might be condensed and reduced until it would become smaller than a walnut, something which I do not believe; nor do I believe that you believe it. The arguments and demonstrations which you have advanced are mathematical, abstract, and far removed from concrete matter; and I do not believe that when applied to the physical and natural world these laws will hold.

[97]

Salv. I am not able to render the invisible visible, nor do I think that you will ask this. But now that you mention gold, do not our senses tell us that that metal can be immensely expanded? I do not know whether you have observed the method employed by those who are skilled in drawing gold wire, of which really only the surface is gold, the inside material being silver. The way they draw it is as follows: they take a cylinder or, if you please, a rod of silver, about half a cubit long and three or four times as wide as one's thumb; this rod they cover with gold-leaf which is so thin that it almost floats in air, putting on not more than eight or ten thicknesses. Once gilded they begin to pull it, with great force, through the holes of a draw-plate; again and again it is made to pass through smaller and smaller holes, until, after very many passages, it is reduced to the fineness of a lady's hair, or perhaps even finer; yet the surface remains gilded. Imagine now how the substance of this gold has been expanded and to what fineness it has been reduced.

Simp. I do not see that this process would produce, as a consequence, that marvellous thinning of the substance of the gold which you suggest: first, because the original gilding consisting of ten layers of gold-leaf has a sensible thickness; secondly, because in drawing out the silver it grows in length but at the same time diminishes proportionally

in thickness; and, since one dimension thus compensates the other, the area will not be so increased as to make it necessary during the process of gilding to reduce the thinness of the gold beyond that of the original leaves.

Salv. You are greatly mistaken, Simplicio, because the surface increases directly as the square root of the length, a fact which I can demonstrate geometrically.

Sagr. Please give us the demonstration not only for my own sake but also for Simplicio, provided you think we can understand it.

[98]

Salv. I'll see if I can recall it on the spur of the moment. At the outset, it is clear that the original thick rod of silver and the wire drawn out to an enormous length are two cylinders of the same volume, since they are the same body of silver. So that, if I

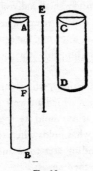

determine the ratio between the surfaces of cylinders of the same volume, the problem will be solved. I say then,

The areas of cylinders of equal volumes, neglecting the bases, bear to each other a ratio which is the square root of the ratio of their lengths.

Take two cylinders of equal volume having the altitudes AB and CD, between which the line E is a mean proportional. Then I claim that, omitting the bases of each cylinder, the surface of the cylinder AB is to that of the cylinder CD as the length AB is to the line E, that is, as the square root of AB is to the square root of CD. Now cut off the cylinder AB at F so that the altitude AF is equal to CD. Then since the bases of cylin-

Fig. 10

ders of equal volume bear to one another the inverse ratio of their heights, it follows that the area of the circular base of the cylinder CD will be to the area of the circular base of AB as the altitude BA is to DC: moreover, since circles are to one another as the squares of their diameters, the said squares will be to each other as BA is to CD. But BA is to CD as the square of BA is to the square of E: and, therefore, these four squares will form a proportion; and likewise their sides; so the line AB is to E as the diameter of circle C is to the diameter of the circle A. But the diameters are proportional to the circumferences and the circumferences are proportional to the areas of cylinders of equal height; hence the line AB is to E as the surface of the cylinder CD is to the surface of the cylinder AF. Now since the height AF is to AB as the surface of AF is to the surface of AB; and since the height AB is to the line E as the surface CD is to AF, it follows, *ex æquali in proportione perturbata*,[11] that the height AF is to E as the surface CD is to the surface AB, and *convertendo*, the surface of the cylinder AB is to the surface of the cylinder CD as the line E is to AF, i. e., to CD, or as AB is to E which is the square root of the ratio of AB to CD. Q. E. D.

[99]

If now we apply these results to the case in hand, and assume that the silver cylinder at the time of gilding had a length of only half a cubit and a thickness three or four times that of one's thumb, we shall find that, when the wire has been reduced to the fineness of a hair and has been drawn out to a length of twenty thousand cubits (and perhaps more), the area of its surface will have been increased not less than two hundred times. Consequently the ten leaves of gold which were laid on have been extended over a surface two hundred times greater, assuring us that the thickness of the gold which now covers the surface of so many cubits of wire cannot be greater than one twentieth that of an ordinary leaf of beaten gold. Consider now what degree of fineness it must have and whether one could conceive it to happen in any other way than by enormous expansion of parts; consider also whether this experiment does not suggest that physical bodies [*materie fisiche*] are composed of infinitely small indivisible particles, a view which is supported by other more striking and conclusive examples.

Sagr. This demonstration is so beautiful that, even if it does not have the cogency originally intended;—although to my mind, it is very forceful—the short time devoted to it has nevertheless been most happily spent.

Salv. Since you are so fond of these geometrical demonstrations, which carry with them distinct gain, I will give you a companion theorem which answers an extremely interesting query. We have seen above what relations hold between equal cylinders of different height or length; let us now see what holds when the cylinders are equal in area but unequal in height, understanding area to include the curved surface, but not the upper and lower bases. The theorem is:

> *The volumes of right cylinders having equal curved surfaces are inversely proportional to their altitudes.*

[100]

Let the surfaces of the two cylinders, AE and CF, be equal but let the height of the latter, CD, be greater than that of the former, AB: then I say that the volume of the cylinder AE is to that of the cylinder CF as the height CD is to AB. Now since the surface of CF is equal to the surface of AE, it follows that the volume of CF is less than that of AE; for, if they were equal the surface of CF would, by the preceding proposition, exceed that of AE, and the excess would be so much the greater if the volume of the cylinder CF were greater than that of AE. Let us now take a cylinder ID having a volume equal

11. See *Euclid*, Book V, Def. 20., Tadhunter's Ed., p. 137 (London, 1877.) *[Trans.]*

to that of AE; then, according to the preceding theorem, the surface of the cylinder ID is to the surface of AE as the altitude IF is to the mean proportional between IF and AB. But since one datum of the problem is that the surface of AE is equal to that of CF, and since the surface ID is to the surface CF as the altitude IF is to the altitude CD, it follows that CD is a mean proportional between IF and AB. Not only so, but since the volume of the cylinder ID is equal to that of AE, each will bear the same ratio to the volume of the cylinder CF; but the volume ID is to the volume CF as the altitude IF is to the altitude CD; hence the volume of AE is to the volume of CF as the length IF is to the length CD, that is, as the length CD is to the length AB.   Q. E. D.

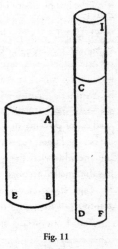

Fig. 11

This explains a phenomenon upon which the common people always look with wonder, namely, if we have a piece of stuff which has one side longer than the other, we can make from it a cornsack, using the customary wooden base, which will hold more when the short side of the cloth is used for the height of the sack and the long side is wrapped around the wooden base, than with the alternative arrangement. So that, for instance, from a piece of cloth which is six cubits on one side and twelve on the other, a sack can be made which will hold more when the side of twelve cubits is wrapped around the wooden base, leaving the sack six cubits high than when the six cubit side is put around the base making the sack twelve cubits high. From what has been proven above we learn not only the general fact that one sack holds more than the other, but we also get specific and particular information as to how much more, namely, just in proportion as the altitude of the sack diminishes the contents increase and *vice versa*.

[101]

Thus if we use the figures given which make the cloth twice as long as wide and if we use the long side for the seam, the volume of the sack will be just one-half as great as with the opposite arrangement. Likewise if we have a piece of matting which measures 7 x 25 cubits and make from it a basket, the contents of the basket will, when the seam is lengthwise, be seven as compared with twenty-five when the seam runs endwise.

Sagr. It is with great pleasure that we continue thus to acquire new and useful information. But as regards the subject just discussed, I really believe that, among those who are not already familiar with geometry, you would scarcely find four persons in a hundred who would not, at first sight, make the mistake of believing that bodies

having equal surfaces would be equal in other respects. Speaking of areas, the same error is made when one attempts, as often happens, to determine the sizes of various cities by measuring their boundary lines, forgetting that the circuit of one may equal to the circuit of another while the area of the one is much greater than that of the other. And this is true not only in the case of irregular, but also of regular surfaces, where the polygon having the greater number of sides always contains a larger area than the one with the less number of sides, so that finally the circle which is a polygon of an infinite number of sides contains the largest area of all polygons of equal perimeter. I remember with particular pleasure having seen this demonstration when I was studying the sphere of Sacrobosco[12] with the aid of a learned commentary.

[102]

Salv. Very true! I too came across the same passage which suggested to me a method of showing how, by a single short demonstration, one can prove that the circle has the largest content of all regular isoperimetric figures; and that, of other figures, the one which has the larger number of sides contains a greater area than that which has the smaller number.

Sagr. Being exceedingly fond of choice and uncommon propositions, I beseech you to let us have your demonstration.

Salv. I can do this in a few words by proving the following theorem:

> The area of a circle is a mean proportional between any two regular and similar polygons of which one circumscribes it and the other is isoperimetric with it. In addition, the area of the circle is less than that of any circumscribed polygon and greater than that of any isoperimetric polygon. And further, of these circumscribed polygons, the one which has the greater number of sides is smaller than the one which has a less number; but, on the other hand, that isoperimetric polygon which has the greater number of sides is the larger.

Let A and B be two similar polygons of which A circumscribes the given circle and B is isoperimetric with it. The area of the circle will then be a mean proportional between the areas of the polygons. For if we indicate

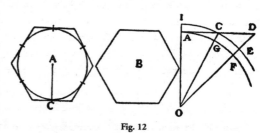

Fig. 12

12. See interesting biographical note on Sacrobosco [John Holywood] in *Ency. Brit.*, 11th Ed. *[Trans.]*

the radius of the circle by AC and if we remember that the area of the circle is equal to that of a right-angled triangle in which one of the sides about the right angle is equal to the radius, AC, and the other to the circumference; and if likewise we remember that the area of the polygon A is equal to the area of a right-angled triangle one of

[103]

whose sides about the right angle has the same length as AC and the other is equal to the perimeter of the polygon itself; it is then manifest that the circumscribed polygon bears to the circle the same ratio which its perimeter bears to the circumference of the circle, or to the perimeter of the polygon B which is, by hypothesis, equal to the circumference of the circle. But since the polygons A and B are similar their areas are to each other as the squares of their perimeters; hence the area of the circle A is a mean proportional between the areas of the two polygons A and B. And since the area of the polygon A is greater than that of the circle A, it is clear that the area of the circle A is greater than that of the isoperimetric polygon B, and is therefore the greatest of all regular polygons having the same perimeter as the circle.

We now demonstrate the remaining portion of the theorem, which is to prove that, in the case of polygons circumscribing a given circle, the one having the smaller number of sides has a larger area than one having a greater number of sides; but that on the other hand, in the case of isoperimetric polygons, the one having the more sides has a larger area than the one with less sides. To the circle which has O for center and OA for radius draw the tangent AD; and on this tangent lay off, say, AD which shall represent one-half of the side of a circumscribed pentagon and AC which shall represent one-half of the side of a heptagon; draw the straight lines OGC and OFD; then with O as a center and OC as radius draw the arc ECI. Now since the triangle DOC is greater than the sector EOC and since the sector COI is greater than the triangle COA, it follows that the triangle DOC bears to the triangle COA a greater ratio than the sector EOC bears to the sector COI, that is, than the sector FOG bears to the sector GOA. Hence, *componendo et permutando*, the triangle DOA bears to the sector FOA a greater ratio than that which the triangle COA bears to the sector GOA, and also 10 such triangles DOA bear to 10 such sectors FOA a greater ratio than 14 such triangles COA bear to 14 such sectors GOA, that is to say, the circumscribed pentagon bears to the circle a greater ratio than does the heptagon. Hence the pentagon exceeds the heptagon in area.

[104]

But now let us assume that both the heptagon and the pentagon have the same perimeter as that of a given circle. Then I say the heptagon will contain a larger area

than the pentagon. For since the area of the circle is a mean proportional between areas of the circumscribed and of the isoperimetric pentagons, and since likewise it is a mean proportional between the circumscribed and isoperimetric heptagons, and since also we have proved that the circumscribed pentagon is larger than the circumscribed hexagon, it follows that this circumscribed pentagon bears to the circle a larger ratio than does the heptagon, that is, the circle will bear to its isoperimetric pentagon a greater ratio than to its isoperimetric heptagon. Hence the pentagon is smaller than its isoperimetric heptagon.   Q. E. D.

Sagr. A very clever and elegant demonstration! But how did we come to plunge into geometry while discussing the objections urged by Simplicio, objections of great moment, especially that one referring to density which strikes me as particularly difficult?

Salv. If contraction and expansion [*condensazione e rarefazzione*] consist in contrary motions, one ought to find for each great expansion a correspondingly large contraction. But our surprise is increased when, every day, we see enormous expansions taking place almost instantaneously. Think what a tremendous expansion occurs when a small quantity of gunpowder flares up into a vast volume of fire! Think too of the almost limitless expansion of the light which it produces! Imagine the contraction which would take place if this fire and this light were to reunite, which, indeed, is not impossible since only a little while ago they were located together in this small space. You will find, upon observation, a thousand such expansions for they are more obvious than contractions since dense matter is more palpable and accessible to our senses.

[105]

We can take wood and see it go up in fire and light, but we do not see them recombine to form wood; we see fruits and flowers and a thousand other solid bodies dissolve largely into odors, but we do not observe these fragrant atoms coming together to form fragrant solids. But where the senses fail us reason must step in; for it will enable us to understand the motion involved in the condensation of extremely rarefied and tenuous substances just as clearly as that involved in the expansion and dissolution of solids. Moreover we are trying to find out how it is possible to produce expansion and contraction in bodies which are capable of such changes without introducing vacua and without giving up the impenetrability of matter; but this does not exclude the possibility of there being materials which possess no such properties and do not, therefore, carry with them consequences which you call inconvenient and impossible. And finally, Simplicio, I have, for the sake of you philosophers, taken pains to find an explanation of how expansion and contraction can take place without our admitting the penetrability of matter and introducing vacua, properties which you deny and

dislike; if you were to admit them, I should not oppose you so vigorously. Now either admit these difficulties or accept my views or suggest something better.

Sagr. I quite agree with the peripatetic philosophers in denying the penetrability of matter. As to the vacua I should like to hear a thorough discussion of Aristotle's demonstration in which he opposes them, and what you, Salviati, have to say in reply. I beg of you, Simplicio, that you give us the precise proof of the Philosopher and that you, Salviati, give us the reply.

Simp. So far as I remember, Aristotle inveighs against the ancient view that a vacuum is a necessary prerequisite for motion and that the latter could not occur without the former. In opposition to this view Aristotle shows that it is precisely the phenomenon of motion, as we shall see, which renders untenable the idea of a vacuum. His method is to divide the argument into two parts. He first supposes bodies of different weights to move in the same medium; then supposes, one and the same body to move in different media.

[106]

In the first case, he supposes bodies of different weight to move in one and the same medium with different speeds which stand to one another in the same ratio as the weights; so that, for example, a body which is ten times as heavy as another will move ten times as rapidly as the other. In the second case he assumes that the speeds of one and the same body moving in different media are in inverse ratio to the densities of these media; thus, for instance, if the density of water were ten times that of air, the speed in air would be ten times greater than in water. From this second supposition, he shows that, since the tenuity of a vacuum differs infinitely from that of any medium filled with matter however rare, any body which moves in a plenum through a certain space in a certain time ought to move through a vacuum instantaneously; but instantaneous motion is an impossibility; it is therefore impossible that a vacuum should be produced by motion.

Salv. The argument is, as you see, *ad hominem*, that is, it is directed against those who thought the vacuum a prerequisite for motion. Now if I admit the argument to be conclusive and concede also that motion cannot take place in a vacuum, the assumption of a vacuum considered absolutely and not with reference to motion, is not thereby invalidated. But to tell you what the ancients might possibly have replied and in order to better understand just how conclusive Aristotle's demonstration is, we may, in my opinion, deny both of his assumptions. And as to the first, I greatly doubt that Aristotle ever tested by experiment whether it be true that two stones, one weighing ten times as much as the other, if allowed to fall, at the same instant, from a height of, say, 100 cubits, would so differ in speed that when the heavier had reached the ground, the other would not have fallen more than 10 cubits.

Simp. His language would seem to indicate that he had tried the experiment, because he says: *We see the heavier*; now the word *see* shows that he had made the experiment.

[107]

Sagr. But I, Simplicio, who have made the test can assure you that a cannon ball weighing one or two hundred pounds, or even more, will not reach the ground by as much as a span ahead of a musket ball weighing only half a pound, provided both are dropped from a height of 200 cubits.

Salv. But, even without further experiment, it is possible to prove clearly, by means of a short and conclusive argument, that a heavier body does not move more rapidly than a lighter one provided both bodies are of the same material and in short such as those mentioned by Aristotle. But tell me, Simplicio, whether you admit that each falling body acquires a definite speed fixed by nature, a velocity which cannot be increased or diminished except by the use of force [*violenza*] or resistance.

Simp. There can be no doubt but that one and the same body moving in a single medium has a fixed velocity which is determined by nature and which cannot be increased except by the addition of momentum [*impeto*] or diminished except by some resistance which retards it.

Salv. If then we take two bodies whose natural speeds are different, it is clear that on uniting the two, the more rapid one will be partly retarded by the slower, and the slower will be somewhat hastened by the swifter. Do you not agree with me in this opinion?

Simp. You are unquestionably right.

Salv. But if this is true, and if a large stone moves with a speed of, say, eight while a smaller moves with a speed of four, then when they are united, the system will move with a speed less than eight; but the two stones when tied together make a stone larger than that which before moved with a speed of eight. Hence the heavier body moves with less speed than the lighter; an effect which is contrary to your supposition. Thus you see how, from your assumption that the heavier body moves more rapidly than the lighter one, I infer that the heavier body moves more slowly.

[108]

Simp. I am all at sea because it appears to me that the smaller stone when added to the larger increases its weight and by adding weight I do not see how it can fail to increase its speed or, at least, not to diminish it.

Salv. Here again you are in error, Simplicio, because it is not true that the smaller stone adds weight to the larger.

Simp. This is, indeed, quite beyond my comprehension.

Salv. It will not be beyond you when I have once shown you the mistake under which you are laboring. Note that it is necessary to distinguish between heavy bodies in motion and the same bodies at rest. A large stone placed in a balance not only acquires additional weight by having another stone placed upon it, but even by the addition of a handful of hemp its weight is augmented six to ten ounces according to the quantity of hemp. But if you tie the hemp to the stone and allow them to fall freely from some height, do you believe that the hemp will press down upon the stone and thus accelerate its motion or do you think the motion will be retarded by a partial upward pressure? One always feels the pressure upon his shoulders when he prevents the motion of a load resting upon him; but if one descends just as rapidly as the load would fall how can it gravitate or press upon him? Do you not see that this would be the same as trying to strike a man with a lance when he is running away from you with a speed which is equal to, or even greater than, that with which you are following him? You must therefore conclude that, during free and natural fall, the small stone does not press upon the larger and consequently does not increase its weight as it does when at rest.

[109]

Simp. But what if we should place the larger stone upon the smaller?

Salv. Its weight would be increased if the larger stone moved more rapidly; but we have already concluded that when the small stone moves more slowly it retards to some extent the speed of the larger, so that the combination of the two, which is a heavier body than the larger of the two stones, would move less rapidly, a conclusion which is contrary to your hypothesis. We infer therefore that large and small bodies move with the same speed provided they are of the same specific gravity.

Simp. Your discussion is really admirable; yet I do not find it easy to believe that a bird-shot falls as swiftly as a cannon ball.

Salv. Why not say a grain of sand as rapidly as a grindstone? But, Simplicio, I trust you will not follow the example of many others who divert the discussion from its main intent and fasten upon some statement of mine which lacks a hair's-breadth of the truth and, under this hair, hide the fault of another which is as big as a ship's cable. Aristotle says that "an iron ball of one hundred pounds falling from a height of one hundred cubits reaches the ground before a one-pound ball has fallen a single cubit." I say that they arrive at the same time. You find, on making the experiment, that the larger outstrips the smaller by two finger-breadths, that is, when the larger has reached

the ground, the other is short of it by two finger-breadths; now you would not hide behind these two fingers the ninety-nine cubits of Aristotle, nor would you mention my small error and at the same time pass over in silence his very large one. Aristotle declares that bodies of different weights, in the same medium, travel (in so far as their motion depends upon gravity) with speeds which are proportional to their weights; this he illustrates by use of bodies in which it is possible to perceive the pure and unadulterated effect of gravity, eliminating other considerations, for example, figure as being of small importance [*minimi momenti*], influences which are greatly dependent upon the medium which modifies the single effect of gravity alone. Thus we observe that gold, the densest of all substances, when beaten out into a very thin leaf, goes floating through the air; the same thing happens with stone when ground into a very fine powder. But if you wish to maintain the general proposition you will have to show that the same ratio of speeds is preserved in the case of all heavy bodies, and that a stone of twenty pounds moves ten times as rapidly as one of two; but I claim that this is false and that, if they fall from a height of fifty or a hundred cubits, they will reach the earth at the same moment.

[110]

Simp. Perhaps the result would be different if the fall took place not from a few cubits but from some thousands of cubits.

Salv. If this were what Aristotle meant you would burden him with another error which would amount to a falsehood; because, since there is no such sheer height available on earth, it is clear that Aristotle could not have made the experiment; yet he wishes to give us the impression of his having performed it when he speaks of such an effect as one which we see.

Simp. In fact, Aristotle does not employ this principle, but uses the other one which is not, I believe, subject to these same difficulties.

Salv. But the one is as false as the other; and I am surprised that you yourself do not see the fallacy and that you do not perceive that if it were true that, in media of different densities and different resistances, such as water and air, one and the same body moved in air more rapidly than in water, in proportion as the density of water is greater than that of air, then it would follow that any body which falls through air ought also to fall through water. But this conclusion is false inasmuch as many bodies which descend in air not only do not descend in water, but actually rise.

Simp. I do not understand the necessity of your inference; and in addition I will say that Aristotle discusses only those bodies which fall in both media, not those which fall in air but rise in water.

[111]

Salv. The arguments which you advance for the Philosopher are such as he himself would have certainly avoided so as not to aggravate his first mistake. But tell me now whether the density [*corpulenza*] of the water, or whatever it may be that retards the motion, bears a definite ratio to the density of air which is less retardative; and if so fix a value for it at your pleasure.

Simp. Such a ratio does exist; let us assume it to be ten; then, for a body which falls in both these media, the speed in water will be ten times slower than in air.

Salv. I shall now take one of those bodies which fall in air but not in water, say a wooden ball, and I shall ask you to assign to it any speed you please for its descent through air.

Simp. Let us suppose it moves with a speed of twenty.

Salv. Very well. Then it is clear that this speed bears to some smaller speed the same ratio as the density of water bears to that of air; and the value of this smaller speed is two. So that really if we follow exactly the assumption of Aristotle we ought to infer that the wooden ball which falls in air, a substance ten times less-resisting than water, with a speed of twenty would fall in water with a speed of two, instead of coming to the surface from the bottom as it does; unless perhaps you wish to reply, which I do not believe you will, that the rising of the wood through the water is the same as its falling with a speed of two. But since the wooden ball does not go to the bottom, I think you will agree with me that we can find a ball of another material, not wood, which does fall in water with a speed of two.

Simp. Undoubtedly we can; but it must be of a substance considerably heavier than wood.

Salv. That is it exactly. But if this second ball falls in water with a speed of two, what will be its speed of descent in air? If you hold to the rule of Aristotle you must reply that it will move at the rate of twenty; but twenty is the speed which you yourself have already assigned to the wooden ball; hence this and the other heavier ball will each move through air with the same speed. But now how does the Philosopher harmonize this result with his other, namely, that bodies of different weight move through the same medium with different speeds—speeds which are proportional to their weights? But without going into the matter more deeply, how have these common and obvious properties escaped your notice?

[112]

Have you not observed that two bodies which fall in water, one with a speed a hundred times as great as that of the other, will fall in air with speeds so nearly equal that

one will not surpass the other by as much as one hundredth part? Thus, for example, an egg made of marble will descend in water one hundred times more rapidly than a hen's egg, while in air falling from a height of twenty cubits the one will fall short of the other by less than four finger breadths. In short, a heavy body which sinks through ten cubits of water in three hours will traverse ten cubits of air in one or two pulse-beats; and if the heavy body be a ball of lead it will easily traverse the ten cubits of water in less than double the time required for ten cubits of air. And here, I am sure, Simplicio, you find no ground for difference or objection. We conclude, therefore, that the argument does not bear against the existence of a vacuum; but if it did, it would only do away with vacua of considerable size which neither I nor, in my opinion, the ancients ever believed to exist in nature, although they might possibly be produced by force [*violenza*] as may be gathered from various experiments whose description would here occupy too much time.

Sagr. Seeing that Simplicio is silent, I will take the opportunity of saying something. Since you have clearly demonstrated that bodies of different weights do not move in one and the same medium with velocities proportional to their weights, but that they all move with the same speed, understanding of course that they are of the same substance or at least of the same specific gravity; certainly not of different specific gravities, for I hardly think you would have us believe a ball of cork moves with the same speed as one of lead; and again since you have clearly demonstrated that one and the same body moving through differently resisting media does not acquire speeds which are inversely proportional to the resistances, I am curious to learn what are the ratios actually observed in these cases.

[113]

Salv. These are interesting questions and I have thought much concerning them. I will give you the method of approach and the result which I finally reached. Having once established the falsity of the proposition that one and the same body moving through differently resisting media acquires speeds which are inversely proportional to the resistances of these media, and having also disproved the statement that in the same medium bodies of different weight acquire velocities proportional to their weights (understanding that this applies also to bodies which differ merely in specific gravity), I then began to combine these two facts and to consider what would happen if bodies of different weight were placed in media of different resistances; and I found that the differences in speed were greater in those media which were more resistant, that is, less yielding. This difference was such that two bodies which differed scarcely at all in their speed through air would, in water, fall the one with a speed ten times as great as that of the other. Further, there are bodies which will fall rapidly in air, whereas if placed in

water not only will not sink but will remain at rest or will even rise to the top: for it is possible to find some kinds of wood, such as knots and roots, which remain at rest in water but fall rapidly in air.

Sagr. I have often tried with the utmost patience to add grains of sand to a ball of wax until it should acquire the same specific gravity as water and would therefore remain at rest in this medium. But with all my care I was never able to accomplish this. Indeed, I do not know whether there is any solid substance whose specific gravity is, by nature, so nearly equal to that of water that if placed anywhere in water it will remain at rest.

[114]

Salv. In this, as in a thousand other operations, men are surpassed by animals. In this problem of yours one may learn much from the fish which are very skillful in maintaining their equilibrium nor only in one kind of water, but also in waters which are notably different either by their own nature or by some accidental muddiness or through salinity, each of which produces a marked change. So perfectly indeed can fish keep their equilibrium that they are able to remain motionless in any position. This they accomplish, I believe, by means of an apparatus especially provided by nature, namely, a bladder located in the body and communicating with the mouth by means of a narrow tube through which they are able, at will, to expel a portion of the air contained in the bladder: by rising to the surface they can take in more air; thus they make themselves heavier or lighter than water at will and maintain equilibrium.

Sagr. By means of another device I was able to deceive some friends to whom I had boasted that I could make up a ball of wax that would be in equilibrium in water. In the bottom of a vessel I placed some salt water and upon this some fresh water; then I showed them that the ball stopped in the middle of the water, and that, when pushed to the bottom or lifted to the top, would not remain in either of these places but would return to the middle.

Salv. This experiment is not without usefulness. For when physicians are testing the various qualities of waters, especially their specific gravities, they employ a ball of this kind so adjusted that, in certain water, it will neither rise nor fall. Then in testing another water, differing ever so slightly in specific gravity [peso], the ball will sink if this water be lighter and rise if it be heavier. And so exact is this experiment that the addition of two grains of salt to six pounds of water is sufficient to make the ball rise to the surface from the bottom to which it had fallen. To illustrate the precision of this experiment and also to clearly demonstrate the non-resistance of water to division, I wish to add that this notable difference in specific gravity can be produced not only by solution of some heavier substance, but also by merely heating or cooling; and so sensitive

is water to this process that by simply adding four drops of another water which is slightly warmer or cooler than the six pounds one can cause the ball to sink or rise; it will sink when the warm water is poured in and will rise upon the addition of cold water. Now you can see how mistaken are those philosophers who ascribe to water viscosity or some other coherence of parts which offers resistance to separation of parts and to penetration.

[115]

Sagr. With regard to this question I have found many convincing arguments in a treatise by our Academician; but there is one great difficulty of which I have not been able to rid myself, namely, if there be no tenacity or coherence between the particles of water how is it possible for those large drops of water to stand out in relief upon cabbage leaves without scattering or spreading out?

Salv. Although those who are in possession of the truth are able to solve all objections raised, I would not arrogate to myself such power; nevertheless my inability should not be allowed to becloud the truth. To begin with let me confess that I do not understand how these large globules of water stand out and hold themselves up, although I know for a certainty, that it is not owing to any internal tenacity acting between the particles of water; whence it must follow that the cause of this effect is external. Beside the experiments already shown to prove that the cause is not internal, I can offer another which is very convincing. If the particles of water which sustain themselves in a heap, while surrounded by air, did so in virtue of an internal cause then they would sustain themselves much more easily when surrounded by a medium in which they exhibit less tendency to fall than they do in air; such a medium would be any fluid heavier than air, as, for instance, wine: and therefore if some wine be poured about such a drop of water, the wine might rise until the drop was entirely covered, without the particles of water, held together by this internal coherence, ever parting company.

[116]

But this is not the fact; for as soon as the wine touches the water, the latter without waiting to be covered scatters and spreads out underneath the wine if it be red. The cause of this effect is therefore external and is possibly to be found in the surrounding air. Indeed there appears to be a considerable antagonism between air and water as I have observed in the following experiment. Having taken a glass globe which had a mouth of about the same diameter as a straw, I filled it with water and turned it mouth downwards; nevertheless, the water, although quite heavy and prone to descend, and the air, which is very light and disposed to rise through the water, refused, the one to

descend and the other to ascend through the opening, but both remained stubborn and defiant. On the other hand, as soon as I apply to this opening a glass of red wine, which is almost inappreciably lighter than water, red streaks are immediately observed to ascend slowly through the water while the water with equal slowness descends through the wine without mixing, until finally the globe is completely filled with wine and the water has all gone down into the vessel below. What then can we say except that there exists, between water and air, a certain incompatibility which I do not understand, but perhaps. . .

Simp. I feel almost like laughing at the great antipathy which Salviati exhibits against the use of the word antipathy; and yet it is excellently adapted to explain the difficulty.

Salv. Alright, if it please Simplicio, let this word antipathy be the solution of our difficulty. Returning from this digression, let us again take up our problem. We have already seen that the difference of speed between bodies of different specific gravities is most marked in those media which are the most resistant: thus, in a medium of quicksilver, gold not merely sinks to the bottom more rapidly than lead but it is the only substance that will descend at all; all other metals and stones rise to the surface and float. On the other hand the variation of speed in air between balls of gold, lead, copper, porphyry, and other heavy materials is so slight that in a fall of 100 cubits a ball of gold would surely not outstrip one of copper by as much as four fingers. Having observed this I came to the conclusion that in a medium totally devoid of resistance all bodies would fall with the same speed.

Simp. This is a remarkable statement, Salviati. But I shall never believe that even in a vacuum, if motion in such a place were possible, a lock of wool and a bit of lead can fall with the same velocity.

[117]

Salv. A little more slowly, Simplicio. Your difficulty is not so recondite nor am I so imprudent as to warrant you in believing that I have not already considered this matter and found the proper solution. Hence for my justification and for your enlightenment hear what I have to say. Our problem is to find out what happens to bodies of different weight moving in a medium devoid of resistance, so that the only difference in speed is that which arises from inequality of weight. Since no medium except one entirely free from air and other bodies, be it ever so tenuous and yielding, can furnish our senses with the evidence we are looking for, and since such a medium is not available, we shall observe what happens in the rarest and least resistant media as compared with what happens in denser and more resistant media. Because if we find as a fact that the variation of speed among bodies of different specific gravities is less and less according

as the medium becomes more and more yielding, and if finally in a medium of extreme tenuity, though not a perfect vacuum, we find that, in spite of great diversity of specific gravity [*peso*], the difference in speed is very small and almost inappreciable, then we are justified in believing it highly probable that in a vacuum all bodies would fall with the same speed. Let us, in view of this, consider what takes place in air, where for the sake of a definite figure and light material imagine an inflated bladder. The air in this bladder when surrounded by air will weigh little or nothing, since it can be only slightly compressed; its weight then is small being merely that of the skin which does not amount to the thousandth part of a mass of lead having the same size as the inflated bladder. Now, Simplicio, if we allow these two bodies to fall from a height of four or six cubits, by what distance do you imagine the lead will anticipate the bladder? You may be sure that the lead will not travel three times, or even twice, as swiftly as the bladder, although you would have made it move a thousand times as rapidly.

Simp. It may be as you say during the first four or six cubits of the fall; but after the motion has continued a long while, I believe that the lead will have left the bladder behind not only six out of twelve parts of the distance but even eight or ten.

[118]

Salv. I quite agree with you and doubt not that, in very long distances, the lead might cover one hundred miles while the bladder was traversing one; but, my dear Simplicio, this phenomenon which you adduce against my proposition is precisely the one which confirms it. Let me once more explain that the variation of speed observed in bodies of different specific gravities is not caused by the difference of specific gravity but depends upon external circumstances and, in particular, upon the resistance of the medium, so that if this is removed all bodies would fall with the same velocity; and this result I deduce mainly from the fact which you have just admitted and which is very true, namely, that, in the case of bodies which differ widely in weight, their velocities differ more and more as the spaces traversed increase, something which would not occur if the effect depended upon differences of specific gravity. For since these specific gravities remain constant, the ratio between the distances traversed ought to remain constant whereas the fact is that this ratio keeps on increasing as the motion continues. Thus a very heavy body in a fall of one cubit will not anticipate a very light one by so much as the tenth part of this space; but in a fall of twelve cubits the heavy body would outstrip the other by one-third, and in a fall of one hundred cubits by 90/100, etc.

Simp. Very well: but, following your own line of argument, if differences of weight in bodies of different specific gravities cannot produce a change in the ratio of their speeds, on the ground that their specific gravities do not change, how is it possible for the medium, which also we suppose to remain constant, to bring about any change in the ratio of these velocities?

Salv. This objection with which you oppose my statement is clever; and I must meet it. I begin by saying that a heavy body has an inherent tendency to move with a constantly and uniformly accelerated motion toward the common center of gravity, that is, toward the center of our earth, so that during equal intervals of time it receives equal increments of momentum and velocity. This, you must understand, holds whenever all external and accidental hindrances have been removed; but of these there is one which we can never remove, namely, the medium which must be penetrated and thrust aside by the falling body. This quiet, yielding, fluid medium opposes motion through it with a resistance which is proportional to the rapidity with which the medium must give way to the passage of the body; which body, as I have said, is by nature continuously accelerated so that it meets with more and more resistance in the medium and hence a diminution in its rate of gain of speed until finally the speed reaches such a point and the resistance of the medium becomes so great that, balancing each other, they prevent any further acceleration and reduce the motion of the body to one which is uniform and which will thereafter maintain a constant value. There is, therefore, an increase in the resistance of the medium, not on account of any change in its essential properties, but on account of the change in rapidity with which it must yield and give way laterally to the passage of the falling body which is being constantly accelerated.

[119]

Now seeing how great is the resistance which the air offers to the slight momentum [*momento*] of the bladder and how small that which it offers to the large weight [*peso*] of the lead, I am convinced that, if the medium were entirely removed, the advantage received by the bladder would be so great and that coming to the lead so small that their speeds would be equalized. Assuming this principle, that all falling bodies acquire equal speeds in a medium which, on account of a vacuum or something else, offers no resistance to the speed of the motion, we shall be able accordingly to determine the ratios of the speeds of both similar and dissimilar bodies moving either through one and the same medium or through different space-filling, and therefore resistant, media. This result we may obtain by observing how much the weight of the medium detracts from the weight of the moving body, which weight is the means employed by the falling body to open a path for itself and to push aside the parts of the medium, something which does not happen in a vacuum where, therefore, no difference [of speed] is to be expected from a difference of specific gravity. And since it is known that the effect of the medium is to diminish the weight of the body by the weight of the medium displaced, we may accomplish our purpose by diminishing in just this proportion the speeds of the falling bodies, which in a non-resisting medium we have assumed to be equal.

[120]

Thus, for example, imagine lead to be ten thousand times as heavy as air while ebony is only one thousand times as heavy. Here we have two substances whose speeds of fall in a medium devoid of resistance are equal: but, when air is the medium, it will subtract from the speed of the lead one part in ten thousand, and from the speed of the ebony one part in one thousand, i. e. ten parts in ten thousand. While therefore lead and ebony would fall from any given height in the same interval of time, provided the retarding effect of the air were removed, the lead will, in air, lose in speed one part in ten thousand; and the ebony, ten parts in ten thousand. In other words, if the elevation from which the bodies start be divided into ten thousand parts, the lead will reach the ground leaving the ebony behind by as much as ten, or at least nine, of these parts. Is it not clear then that a leaden ball allowed to fall from a tower two hundred cubits high will outstrip an ebony ball by less than four inches? Now ebony weighs a thousand times as much as air but this inflated bladder only four times as much; therefore air diminishes the inherent and natural speed of ebony by one part in a thousand; while that of the bladder which, if free from hindrance, would be the same, experiences a diminution in air amounting to one part in four. So that when the ebony ball, falling from the tower, has reached the earth, the bladder will have traversed only three-quarters of this distance. Lead is twelve times as heavy as water; but ivory is only twice as heavy. The speeds of these two substances which, when entirely unhindered, are equal will be diminished in water, that of lead by one part in twelve, that of ivory by half. Accordingly when the lead has fallen through eleven cubits of water the ivory will have fallen through only six. Employing this principle we shall, I believe, find a much closer agreement of experiment with our computation than with that of Aristotle.

In a similar manner we may find the ratio of the speeds of one and the same body in different fluid media, not by comparing the different resistances of the media, but by considering the excess of the specific gravity of the body above those of the media. Thus, for example, tin is one thousand times heavier than air and ten times heavier than water; hence, if we divide its unhindered speed into 1000 parts, air will rob it of one of these parts so that it will fall with a speed of 999, while in water its speed will be 900, seeing that water diminishes its weight by one part in ten while air by only one part in a thousand.

[121]

Again take a solid a little heavier than water, such as oak, a ball of which will weigh let us say 1000 drachms; suppose an equal volume of water to weigh 950, and an equal volume of air, 2; then it is clear that if the unhindered speed of the ball is 1000, its

speed in air will be 998, but in water only 50, seeing that the water removes 950 of the 1000 parts which the body weighs, leaving only 50.

Such a solid would therefore move almost twenty times as fast in air as in water, since its specific gravity exceeds that of water by one part in twenty. And here we must consider the fact that only those substances which have a specific gravity greater than water can fall through it—substances which must, therefore, be hundreds of times heavier than air; hence when we try to obtain the ratio of the speed in air to that in water, we may, without appreciable error, assume that air does not, to any considerable extent, diminish the free weight [*assoluta gravità*], and consequently the unhindered speed [*assoluta velocità*] of such substances. Having thus easily found the excess of the weight of these substances over that of water, we can say that their speed in air is to their speed in water as their free weight [*totale gravità*] is to the excess of this weight over that of water. For example, a ball of ivory weighs 20 ounces; an equal volume of water weighs 17 ounces; hence the speed of ivory in air bears to its speed in water the approximate ratio of 20:3.

Sagr. I have made a great step forward in this truly interesting subject upon which I have long labored in vain. In order to put these theories into practice we need only discover a method of determining the specific gravity of air with reference to water and hence with reference to other heavy substances.

Simp. But if we find that air has levity instead of gravity what then shall we say of the foregoing discussion which, in other respects, is very clever?

Salv. I should say that it was empty, vain, and trifling. But can you doubt that air has weight when you have the clear testimony of Aristotle affirming that all the elements have weight including air, and excepting only fire? As evidence of this he cites the fact that a leather bottle weighs more when inflated than when collapsed.

Simp. I am inclined to believe that the increase of weight observed in the inflated leather bottle or bladder arises, not from the gravity of the air, but from the many thick vapors mingled with it in these lower regions. To this I would attribute the. increase of weight in the leather bottle.

[122]

Salv. I would not have you say this, and much less attribute it to Aristotle; because, if speaking of the elements, he wished to persuade me by experiment that air has weight and were to say to me: "Take a leather bottle, fill it with heavy vapors and observe how its weight increases," I would reply that the bottle would weigh still more if filled with bran; and would then add that this merely proves that bran and thick vapors are heavy, but in regard to air I should still remain in the same doubt as before. However, the experiment of Aristotle is good and the proposition is true. But I cannot say as much

of a certain other consideration, taken at face value; this consideration was offered by a philosopher whose name slips me; but I know I have read his argument which is that air exhibits greater gravity than levity, because it carries heavy bodies downward more easily than it does light ones upward.

Sagr. Fine indeed! So according to this theory air is much heavier than water, since all heavy bodies are carried downward more easily through air than through water, and all light bodies buoyed up more easily through water than through air; further there is an infinite number of heavy bodies which fall through air but ascend in water and there is an infinite number of substances which rise in water and fall in air. But, Simplicio, the question as to whether the weight of the leather bottle is owing to thick vapors or to pure air does not affect our problem which is to discover how bodies move through this vapor-laden atmosphere of ours. Returning now to the question which interests me more, I should like, for the sake of more complete and thorough knowledge of this matter, not only to be strengthened in my belief that air has weight but also to learn, if possible, how great its specific gravity is. Therefore, Salviati, if you can satisfy my curiosity on this point pray do so.

[123]

Salv. The experiment with the inflated leather bottle of Aristotle proves conclusively that air possesses positive gravity and not, as some have believed, levity, a property possessed possibly by no substance whatever; for if air did possess this quality of absolute and positive levity, it should on compression exhibit greater levity and, hence, a greater tendency to rise; but experiment shows precisely the opposite.

As to the other question, namely, how to determine the specific gravity of air, I have employed the following method. I took a rather large glass bottle with a narrow neck and attached to it a leather cover, binding it tightly about the neck of the bottle: in the top of this cover I inserted and firmly fastened the valve of a leather bottle, through which I forced into the glass bottle, by means of a syringe, a large quantity of air. And since air is easily condensed one can pump into the bottle two or three times its own volume of air. After this I took an accurate balance and weighed this bottle of compressed air with the utmost precision, adjusting the weight with fine sand. I next opened the valve and allowed the compressed air to escape; then replaced the flask upon the balance and found it perceptibly lighter: from the sand which had been used as a counterweight I now removed and laid aside as much as was necessary to again secure balance. Under these conditions there can be no doubt but that the weight of the sand thus laid aside represents the weight of the air which had been forced into the flask and had afterwards escaped. But after all this experiment tells me merely that the weight of the compressed air is the same as that of the sand removed from the balance;

when however it comes to knowing certainly and definitely the weight of air as compared with that of water or any other heavy substance this I cannot hope to do without first measuring the volume [*quantità*] of compressed air; for this measurement I have devised the two following methods.

According to the first method one takes a bottle with a narrow neck similar to the previous one; over the mouth of this bottle is slipped a leather tube which is bound tightly about the neck of the flask; the other end of this tube embraces the valve attached to the first flask and is tightly bound about it. This second flask is provided with a hole in the bottom through which an iron rod can be placed so as to open, at will, the valve above mentioned and thus permit the surplus air of the first to escape after it has once been weighed: but his second bottle must be filled with water. Having prepared everything in the manner above described, open the valve with the rod; the air will rush into the flask containing the water and will drive it through the hole at the bottom, it being clear that the volume [*quantità*] of water thus displaced is equal to the volume [*mole e quantità*] of air escaped from the other vessel. Having set aside this displaced water, weigh the vessel from which the air has escaped (which is supposed to have been weighed previously while containing the compressed air), and remove the surplus of sand as described above; it is then manifest that the weight of this sand is precisely the weight of a volume [*mole*] of air equal to the volume of water displaced and set aside; this water we can weigh and find how many times its weight contains the weight of the removed sand, thus determining definitely how many times heavier water is than air; and we shall find, contrary to the opinion of Aristotle, that this is not 10 times, but, as our experiment shows, more nearly 400 times.

[124]

The second method is more expeditious and can be carried out with a single vessel fitted up as the first was. Here no air is added to that which the vessel naturally contains but water is forced into it without allowing any air to escape; the water thus introduced necessarily compresses the air. Having forced into the vessel as much water as possible, filling it, say, three-fourths full, which does not require any extraordinary effort, place it upon the balance and weigh it accurately; next hold the vessel mouth up, open the valve, and allow the air to escape; the volume of the air thus escaping is precisely equal to the volume of water contained in the flask. Again weigh the vessel which will have diminished in weight on account of the escaped air; this loss in weight represents the weight of a volume of air equal to the volume of water contained in the vessel.

Simp. No one can deny the cleverness and ingenuity of your devices; but while they appear to give complete intellectual satisfaction they confuse me in another direction. For since it is undoubtedly true that the elements when in their proper places

have neither weight nor levity, I cannot understand how it is possible for that portion of air, which appeared to weigh, say, 4 drachms of sand, should really have such a weight in air as the sand which counterbalances it. It seems to me, therefore, that the experiment should be carried out, not in air, but in a medium in which the air could exhibit its property of weight if such it really has.

[125]

Salv. The objection of Simplicio is certainly to the point and must therefore either be unanswerable or demand an equally clear solution. It is perfectly evident that that air which, under compression, weighed as much as the sand, loses this weight when once allowed to escape into its own element, while, indeed, the sand retains its weight. Hence for this experiment it becomes necessary to select a place where air as well as sand can gravitate; because, as has been often remarked, the medium diminishes the weight of any substance immersed in it by an amount equal to the weight of the displaced medium; so that air in air loses all its weight. If therefore this experiment is to be made with accuracy it should be performed in a vacuum where every heavy body exhibits its momentum without the slightest diminution. If then, Simplicio, we were to weigh a portion of air in a vacuum would you then be satisfied and assured of the fact?

Simp. Yes truly: but this is to wish or ask the impossible.

Salv. Your obligation will then be very great if, for your sake, I accomplish the impossible. But I do not want to sell you something which I have already given you; for in the previous experiment we weighed the air in vacuum and not in air or other medium. The fact that any fluid medium diminishes the weight of a mass immersed in it, is due, Simplicio, to the resistance which this medium offers to its being opened up, driven aside, and finally lifted up. The evidence for this is seen in the readiness with which the fluid rushes to fill up any space formerly occupied by the mass; if the medium were not affected by such an immersion then it would not react against the immersed body. Tell me now, when you have a flask, in air, filled with its natural amount of air and then proceed to pump into the vessel more air, does this extra charge in any way separate or divide or change the circumambient air? Does the vessel perhaps expand so that the surrounding medium is displaced in order to give more room? Certainly not.

[126]

Therefore one is able to say that this extra charge of air is not immersed in the surrounding medium for it occupies no space in it, but is, as it were, in a vacuum. Indeed, it is really in a vacuum; for it diffuses into the vacuities which are not completely filled

by the original and uncondensed air. In fact I do not see any difference between the enclosed and the surrounding media: for the surrounding medium does not press upon the enclosed medium and, *vice versa*, the enclosed medium exerts no pressure against the surrounding one; this same relationship exists in the case of any matter in a vacuum, as well as in the case of the extra charge of air compressed into the flask. The weight of this condensed air is therefore the same as that which it would have if set free in a vacuum. It is true of course that the weight of the sand used as a counterpoise would be a little greater *in vacuo* than in free air. We must, then, say that the air is slightly lighter than the sand required to counterbalance it, that is to say, by an amount equal to the weight *in vacuo* of a volume of air equal to the volume of the sand.*

[127]

Simp. The previous experiments, in my opinion, left something to be desired: but now I am fully satisfied.

Salv. The facts set forth by me up to this point and, in particular, the one which shows that difference of weight, even when very great, is without effect in changing the speed of falling bodies, so that as far as weight is concerned they all fall with equal speed: this idea is, I say, so new, and at first glance so remote from fact, that if we do not have the means of making it just as clear as sunlight, it had better not be mentioned; but having once allowed it to pass my lips I must neglect no experiment or argument to establish it.

Sagr. Not only this but also many other of your views are so far removed from the commonly accepted opinions and doctrines that if you were to publish them you would stir up a large number of antagonists; for human nature is such that men do not look with favor upon discoveries—either of truth or fallacy—in their own field, when made by others than themselves. They call him an innovator of doctrine, an unpleasant title, by which they hope to cut those knots which they cannot untie, and by subterranean mines they seek to destroy structures which patient artisans have built with customary tools. But as for ourselves who have no such thoughts, the experiments and

---

* At this point in an annotated copy of the original edition the following note by Galileo is found.

[Sagr. A very clever discussion, solving a wonderful problem, because it demonstrates briefly and concisely the manner in which one may find the weight of a body *in vacuo* by simply weighing it in air. The explanation is as follows: when a heavy body is immersed in air it loses in weight an amount equal to the weight of a volume [*mole*] of air equivalent to the volume [*mole*] of the body itself. Hence if one adds to a body, without expanding it, a quantity of air equal to that which it displaces and weighs it, he will obtain its absolute weight *in vacuo*, since, without increasing it in size, he has increased its weight by just the amount which it lost through immersion in air.

When therefore we force a quantity of water into a vessel which already contains its normal amount of air, without allowing any of this air to escape it is clear that this normal quantity of air will be compressed and condensed into a smaller space in order to make room for the water which is forced in; it is also clear that the volume of air thus compressed is equal to the volume of water added. If now the vessel be weighed in air in this condition, it is manifest that the weight of the water will be increased by that of an equal volume of air; the total weight of water and air thus obtained is equal to the weight of the water alone *in vacuo*.

Now record the weight of the entire vessel and then allow the compressed air to escape; weigh the remainder; the difference of these two weights will be the weight of the compressed air which, in volume, is equal to that of the water. Next find the weight of the water alone and add to it that of the compressed air; we shall then have the water alone *in vacuo*. To find the weight of the water we shall have to remove it from the vessel and weigh the vessel alone; subtract this weight from that of the vessel and water together. It is clear that the remainder will be the weight of the water alone in air.]

[128]

arguments which you have thus far adduced are fully satisfactory; however if you have any experiments which are more direct or any arguments which are more convincing we will hear them with pleasure.

Salv. The experiment made to ascertain whether two bodies, differing greatly in weight will fall from a given height with the same speed offers some difficulty; because, if the height is considerable, the retarding effect of the medium, which must be penetrated and thrust aside by the falling body, will be greater in the case of the small momentum of the very light body than in the case of the great force [*violenza*] of the heavy body; so that, in a long distance, the light body will be left behind; if the height be small, one may well doubt whether there is any difference; and if there be a difference it will be inappreciable.

It occurred to me therefore to repeat many times the fall through a small height in such a way that I might accumulate all those small intervals of time that elapse between the arrival of the heavy and light bodies respectively at their common terminus, so that this sum makes an interval of time which is not only observable, but easily observable. In order to employ the slowest speeds possible and thus reduce the change which the resisting medium produces upon the simple effect of gravity it occurred to me to allow the bodies to fall along a plane slightly inclined to the horizontal. For in such a plane, just as well as in a vertical plane, one may discover how bodies of different weight behave: and besides this, I also wished to rid myself of the resistance which might arise from contact of the moving body with the aforesaid inclined plane. Accordingly I took two balls, one of lead and one of cork, the former more than a hundred times heavier than the latter, and suspended them by means of two equal fine threads, each four or five cubits long.

[129]

Pulling each ball aside from the perpendicular, I let them go at the same instant, and they, falling along the circumferences of circles having these equal strings for semi-diameters, passed beyond the perpendicular and returned along the same path. This free vibration [*per lor medesime le andate e le tornate*] repeated a hundred times showed clearly that the heavy body maintains so nearly the period of the light body that neither in a hundred swings nor even in a thousand will the former anticipate the latter by as much as a single moment [*minimo momento*], so perfectly do they keep step. We can also observe the effect of the medium which, by the resistance which it offers to motion, diminishes the vibration of the cork more than that of the lead, but without altering the frequency of either; even when the arc traversed by the cork did not exceed five or six degrees while that of the lead was fifty or sixty, the swings were performed in equal times.

Simp. If this be so, why is not the speed of the lead greater than that of the cork, seeing that the former traverses sixty degrees in the same interval in which the latter covers scarcely six?

Salv. But what would you say, Simplicio, if both covered their paths in the same time when the cork, drawn aside through thirty degrees, traverses an arc of sixty, while the lead pulled aside only two degrees traverses an arc of four? Would not then the cork be proportionately swifter? And yet such is the experimental fact. But observe this: having pulled aside the pendulum of lead, say through an arc of fifty degrees, and set it free, it swings beyond the perpendicular almost fifty degrees, thus describing an arc of nearly one hundred degrees; on the return swing it describes a little smaller arc; and after a large number of such vibrations it finally comes to rest. Each vibration, whether of ninety, fifty, twenty, ten, or four degrees occupies the same time: accordingly the speed of the moving body keeps on diminishing since in equal intervals of time, it traverses arcs which grow smaller and smaller.

Precisely the same things happen with the pendulum of cork, suspended by a string of equal length, except that a smaller number of vibrations is required to bring it to rest, since on account of its lightness it is less able to overcome the resistance of the air; nevertheless the vibrations, whether large or small, are all performed in time-intervals which are not only equal among themselves, but also equal to the period of the lead pendulum. Hence it is true that, if while the lead is traversing an arc of fifty degrees the cork covers one of only ten, the cork moves more slowly than the lead; but on the other hand it is also true that the cork may cover an arc of fifty while the lead passes over one of only ten or six; thus, at different times, we have now the cork, now the lead, moving more rapidly. But if these same bodies traverse equal arcs in equal times we may rest assured that their speeds are equal.

[130]

Simp. I hesitate to admit the conclusiveness of this argument because of the confusion which arises from your making both bodies move now rapidly, now slowly and now very slowly, which leaves me in doubt as to whether their velocities are always equal.

Sagr. Allow me, if you please, Salviati, to say just a few words. Now tell me, Simplicio, whether you admit that one can say with certainty that the speeds of the cork and the lead are equal whenever both, starting from rest at same moment and descending the same slopes, always traverse equal spaces in equal times?

Simp. This can neither be doubted nor gainsaid.

Sagr. Now it happens, in the case of the pendulums, that each of them traverses now an arc of sixty degrees, now one of fifty, or thirty or ten or eight or four or two, etc.; and when they both swing through an are of sixty degrees they do so in equal

intervals of time; the same thing happens when the are is fifty degrees or thirty or ten or any other number; and therefore we conclude that the speed of the lead in an arc of sixty degrees is equal to the speed of the cork when the latter also swings through an arc of sixty degrees; in the case of a fifty-degree arc these speeds are also equal to each other; so also in the case of other arcs. But this is not saying that the speed which occurs in an arc of sixty is the same as that which occurs in an arc of fifty; nor is the speed in an arc of fifty equal to that in one of thirty, etc.; but the smaller the arcs, the smaller the speeds; the fact observed is that one and the same moving body requires the same time for traversing a large arc of sixty degrees as for a small arc of fifty or even a very small arc of ten; all these arcs, indeed, are covered in the same interval of time. It is true therefore that the lead and the cork each diminish their speed [*moto*] in proportion as their arcs diminish; but this does not contradict the fact that they maintain equal speeds in equal arcs.

[131]

My reason for saying these things has been rather because I wanted to learn whether I had correctly understood Salviati, than because I thought Simplicio had any need of a clearer explanation than that given by Salviati which like everything else of his is extremely lucid, so lucid, indeed, that when he solves questions which are difficult not merely in appearance, but in reality and in fact, he does so with reasons, observations and experiments which are common and familiar to everyone.

In this manner he has, as I have learned from various sources, given occasion to a highly esteemed professor for undervaluing his discoveries on the ground that they are commonplace, and established upon a mean and vulgar basis; as if it were not a most admirable and praiseworthy feature of demonstrative science that it springs from and grows out of principles well-known, understood and conceded by all.

But let us continue with this light diet; and if Simplicio is satisfied to understand and admit that the gravity inherent [*interna gravità*] in various falling bodies has nothing to do with the difference of speed observed among them, and that all bodies, in so far as their speeds depend upon it, would move with the same velocity, pray tell us, Salviati, how you explain the appreciable and evident inequality of motion; please reply also to the objection urged by Simplicio—an objection in which I concur—namely, that a cannon ball falls more rapidly than a bird-shot. From my point of view, one might expect the difference of speed to be small in the case of bodies of the same substance moving through any single medium, whereas the larger ones will descend, during a single pulse-beat, a distance which the smaller ones will not traverse in an hour, or in four, or even in twenty hours; as for instance in the case of stones and fine sand and especially that very fine sand which produces muddy water and which in many hours will

not fall through as much as two cubits, a distance which stones not much larger will traverse in a single pulse-beat.

[132]

Salv. The action of the medium in producing a greater retardation upon those bodies which have a less specific gravity has already been explained by showing that they experience a diminution of weight. But to explain how one and the same medium produces such different retardations in bodies which are made of the same material and have the same shape, but differ only in size, requires a discussion more clever than that by which one explains how a more expanded shape or an opposing motion of the medium retards the speed of the moving body. The solution of the present problem lies, I think, in the roughness and porosity which are generally and almost necessarily found in the surfaces of solid bodies. When the body is in motion these rough places strike the air or other ambient medium. The evidence for this is found in the humming which accompanies the rapid motion of a body through air, even when that body is as round as possible. One hears not only humming, but also hissing and whistling, whenever there is any appreciable cavity or elevation upon the body. We observe also that a round solid body rotating in a lathe produces a current of air. But what more do we need? When a top spins on the ground at its greatest speed do we not hear a distinct buzzing of high pitch? This sibilant note diminishes in pitch as the speed of rotation slackens, which is evidence that these small rugosities on the surface meet resistance in the air. There can be no doubt, therefore, that in the motion of falling bodies these rugosities strike the surrounding fluid and retard the speed; and this they do so much the more in proportion as the surface is larger, which is the case of small bodies as compared with greater.

Simp. Stop a moment please, I am getting confused. For although I understand and admit that friction of the medium upon the surface of the body retards its motion and that, if other things are the same, the larger surface suffers greater retardation, I do not see on what ground you say that the surface of the smaller body is larger. Besides if, as you say, the larger surface suffers greater retardation the larger solid should move more slowly, which is not the fact. But this objection can be easily met by saying that, although the larger body has a larger surface, it has also a greater weight, in comparison with which the resistance of the larger surface is no more than the resistance of the small surface in comparison with its smaller weight; so that the speed of the larger solid does not become less. I therefore see no reason for expecting any difference of speed so long as the driving weight [*gravità movente*] diminishes in the same proportion as the retarding power [*facolta ritardante*] of the surface.

[133]

Salv. I shall answer all your objections at once. You will admit, of course, Simplicio, that if one takes two equal bodies, of the same material and same figure, bodies which would therefore fall with equal speeds, and if he diminishes the weight of one of them in the same proportion as its surface (maintaining the similarity of shape) he would not thereby diminish the speed of this body.

Simp. This inference seems to be in harmony with your theory which states that the weight of a body has no effect in either accelerating or retarding its motion.

Salv. I quite agree with you in this opinion from which it appears to follow that, if the weight of a body is diminished in greater proportion than its surface, the motion is retarded to a certain extent; and this retardation is greater and greater in proportion as the diminution of weight exceeds that of the surface.

Simp. This I admit without hesitation.

Salv. Now you must know, Simplicio, that it is not possible to diminish the surface of a solid body in the same ratio as the weight, and at the same time maintain similarity of figure. For since it is clear that in the case of a diminishing solid the weight grows less in proportion to the volume, and since the volume always diminishes more rapidly than the surface, when the same shape is maintained, the weight must therefore diminish more rapidly than the surface. But geometry teaches us that, in the case of similar solids, the ratio of two volumes is greater than the ratio of their surfaces; which, for the sake of better understanding, I shall illustrate by a particular case.

Take, for example, a cube two inches on a side so that each face has an area of four square inches and the total area, i. e., the sum of the six faces, amounts to twenty-four square inches; now imagine this cube to be sawed through three times so as to divide it into eight smaller cubes, each one inch on the side, each face one inch square, and the total surface of each cube six square inches instead of twenty-four as in the case of the larger cube. It is evident therefore that the surface of the little cube is only one-fourth that of the larger, namely, the ratio of six to twenty-four; but the volume of the solid cube itself is only one-eighth; the volume, and hence also the weight, diminishes therefore much more rapidly than the surface. If we again divide the little cube into eight others we shall have, for the total surface of one of these, one and one-half square inches, which is one-sixteenth of the surface of the original cube; but its volume is only one-sixty-fourth part.

[134]

Thus, by two divisions, you see that the volume is diminished four times as much as the surface. And, if the subdivision be continued until the original solid be reduced to

a fine powder, we shall find that the weight of one of these smallest particles has diminished hundreds and hundreds of times as much as its surface. And this which I have illustrated in the case of cubes holds also in the case of all similar solids, where the volumes stand in sesquialteral ratio to their surfaces. Observe then how much greater the resistance, arising from contact of the surface of the moving body with the medium, in the case of small bodies than in the case of large; and when one considers that the rugosities on the very small surfaces of fine dust particles are perhaps no smaller than those on the surfaces of larger solids which have been carefully polished, he will see how important it is that the medium should be very fluid and offer no resistance to being thrust aside, easily yielding to a small force. You see, therefore, Simplicio, that I was not mistaken when, not long ago, I said that the surface of a small solid is comparatively greater than that of a large one.

Simp. I am quite convinced; and, believe me, if I were again beginning my studies, I should follow the advice of Plato and start with mathematics, a science which proceeds very cautiously and admits nothing as established until it has been rigidly demonstrated.

Sagr. This discussion has afforded me great pleasure; but before proceeding further I should like to hear the explanation of a phrase of yours which is new to me, namely, that similar solids are to each other in the sesquialteral ratio of their surfaces; for although I have seen and understood the proposition in which it is demonstrated that the surfaces of similar solids are in the duplicate ratio of their sides and also the proposition which proves that the volumes are in the triplicate ratio of their sides, yet I have not so much as heard mentioned the ratio of the volume of a solid to its surface.

[135]

Salv. You yourself have suggested the answer to your question and have removed every doubt. For if one quantity is the cube of something of which another quantity is the square does it not follow that the cube is the sesquialteral of the square? Surely. Now if the surface varies as the square of its linear dimensions while the volume varies as the cube of these dimensions may we not say that the volume stands in sesquialteral ratio to the surface?

Sagr. Quite so. And now although there are still some details, in connection with the subject under discussion, concerning which I might ask questions yet, if we keep making one digression after another, it will be long before we reach the main topic which has to do with the variety of properties found in the resistance which solid bodies offer to fracture; and, therefore, if you please, let us return to the subject which we originally proposed to discuss.

Salv. Very well; but the questions which we have already considered are so numerous and so varied, and have taken up so much time that there is not much of this day left to spend upon our main topic which abounds in geometrical demonstrations calling for careful consideration. May I, therefore, suggest that we postpone the meeting until to-morrow, not only for the reason just mentioned but also in order that I may bring with me some papers in which I have set down in an orderly way the theorems and propositions dealing with the various phases of this subject, matters which, from memory alone, I could not present in the proper order.

Sagr. I fully concur in your opinion and all the more willingly because this will leave time to-day to take up some of my difficulties with the subject which we have just been discussing. One question is whether we are to consider the resistance of the medium as sufficient to destroy the acceleration of a body of very heavy material, very large volume, and spherical figure. I say *spherical* in order to select a volume which is contained within a minimum surface and therefore less subject to retardation.

[136]

Another question deals with the vibrations of pendulums which may be regarded from several viewpoints; the first is whether all vibrations, large, medium, and small, are performed in exactly and precisely equal times: another is to find the ratio of the times of vibration of pendulums supported by threads of unequal length.

Salv. These are interesting questions: but I fear that here, as in the case of all other facts, if we take up for discussion any one of them, it will carry in its wake so many other facts and curious consequences that time will not remain to-day for the discussion of all.

Sagr. If these are as full of interest as the foregoing, I would gladly spend as many days as there remain hours between now and nightfall; and I dare say that Simplicio would not be wearied by these discussions.

Simp. Certainly not; especially when the questions pertain to natural science and have not been treated by other philosophers.

Salv. Now taking up the first question, I can assert without hesitation that there is no sphere so large, or composed of material so dense but that the resistance of the medium, although very slight, would check its acceleration and would, in time reduce its motion to uniformity; a statement which is strongly supported by experiment. For if a falling body, as time goes on, were to acquire a speed as great as you please, no such speed, impressed by external forces [*motore esterno*], can be so great but that the body will first acquire it and then, owing to the resisting medium, lose it. Thus, for instance, if a cannon ball, having fallen a distance of four cubits through the air and having acquired a speed of, say, ten units [*gradi*] were to strike the surface of the water, and if

the resistance of the water were not able to chock the momentum [*impeto*] of the shot, it would either increase in speed or maintain a uniform motion until the bottom were reached: but such is not the observed fact; on the contrary, the water when only a few cubits deep hinders and diminishes the motion in such a way that the shot delivers to the bed of the river or lake a very slight impulse.

[137]

Clearly then if a short fall through the water is sufficient to deprive a cannon ball of its speed, this speed cannot be regained by a fall of even a thousand cubits. How could a body acquire, in a fall of a thousand cubits, that which it loses in a fall of four? But what more is needed? Do we not observe that the enormous momentum, delivered to a shot by a cannon, is so deadened by passing through a few cubits of water that the ball, so far from injuring the ship, barely strikes it? Even the air, although a very yielding medium, can also diminish the speed of a falling body, as may be easily understood from similar experiments. For if a gun be fired downwards from the top of a very high tower the shot will make a smaller impression upon the ground than if the gun had been fired from an elevation of only four or six cubits; this is clear evidence that the momentum of the ball, fired from the top of the tower, diminishes continually from the instant it leaves the barrel until it reaches the ground. Therefore a fall from ever so great an altitude will not suffice to give to a body that momentum which it has once lost through resistance of the air, no matter how it was originally acquired. In like manner, the destructive effect produced upon a wall by a shot fired from a gun at a distance of twenty cubits cannot be duplicated by the fall of the same shot from any altitude however great. My opinion is, therefore, that under the circumstances which occur in nature, the acceleration of any body falling from rest reaches an end and that the resistance of the medium finally reduces its speed to a constant value which is thereafter maintained.

Sagr. These experiments are in my opinion much to the purpose; the only question is whether an opponent might not make bold to deny the fact in the case of bodies [moli] which are very large and heavy or to assert that a cannon ball, falling from the distance of the moon or from the upper regions of the atmosphere, would deliver a heavier blow than if just leaving the muzzle of the gun.

[138]

Salv. No doubt many objections may be raised not all of which can be refuted by experiment: however in this particular case the following consideration must be taken into account, namely, that it is very likely that a heavy body falling from a height will,

·

on reaching the ground, have acquired just as much momentum as was necessary to carry it to that height; as may be clearly seen in the case of a rather heavy pendulum which, when pulled aside fifty or sixty degrees from the vertical, will acquire precisely that speed and force which are sufficient to carry it to an equal elevation save only that small portion which it loses through friction on the air. In order to place a cannon ball at such a height as might suffice to give it just that momentum which the powder imparted to it on leaving the gun we need only fire it vertically upwards from the same gun; and we can then observe whether on falling back it delivers a blow equal to that of the gun fired at close range; in my opinion it would be much weaker. The resistance of the air would, therefore, I think, prevent the muzzle velocity from being equalled by a natural fall from rest at any height whatsoever.

We come now to the other questions, relating to pendulums, a subject which may appear to many exceedingly arid, especially to those philosophers who are continually occupied with the more profound questions of nature. Nevertheless, the problem is one which I do not scorn. I am encouraged by the example of Aristotle whom I admire especially because he did not fail to discuss every subject which he thought in any degree worthy of consideration.

Impelled by your queries I may give you some of my ideas concerning certain problems in music, a splendid subject, upon which so many eminent men have written: among these is Aristotle himself who has discussed numerous interesting acoustical questions. Accordingly, if on the basis of some easy and tangible experiments, I shall explain some striking phenomena in the domain of sound, I trust my explanations will meet your approval.

[139]

Sagr. I shall receive them not only gratefully but eagerly. For, although I take pleasure in every kind of musical instrument and have paid considerable attention to harmony, I have never been able to fully understand why some combinations of tones are more pleasing than others, or why certain combinations not only fail to please but are even highly offensive. Then there is the old problem of two stretched strings in unison; when one of them is sounded, the other begins to vibrate and to emit its note; nor do I understand the different ratios of harmony [*forme delle consonanze*] and some other details.

Salv. Let us see whether we cannot derive from the pendulum a satisfactory solution of all these difficulties. And first, as to the question whether one and the same pendulum really performs its vibrations, large, medium, and small, all in exactly the same time, I shall rely upon what I have already heard from our Academician. He has clearly shown that the time of descent is the same along all chords, whatever the arcs which

subtend them, as well along an arc of 180° (i. e., the whole diameter) as along one of 100°, 60°, 10°, 2°, 1/2°, or 4'. It is understood, of course, that these arcs all terminate at the lowest point of the circle, where it touches the horizontal plane.

If now we consider descent along arcs instead of their chords then, provided these do. not exceed 90°, experiment shows that they are all traversed in equal times; but these times are greater for the chord than for the arc, an effect which is all the more remarkable because at first glance one would think just the opposite to be true. For since the terminal points of the two motions are the same and since the straight line included between these two points is the shortest distance between them, it would seem reasonable that motion along this line should be executed in the shortest time; but this is not the case, for the shortest time—and therefore the most rapid motion— is that employed along the arc of which this straight line is the chord.

As to the times of vibration of bodies suspended by threads of different lengths, they bear to each other the same proportion as the square roots of the lengths of the thread; or one might say the lengths are to each other as the squares of the times; so that if one wishes to make the vibration-time of one pendulum twice that of another, he must make its suspension four times as long. In like manner, if one pendulum has a suspension nine times as long as another, this second pendulum will execute three vibrations during each one of the first; from which it follows that the lengths of the suspending cords bear to each other the [inverse] ratio of the squares of the number of vibrations performed in the same time.

[140]

Sagr. Then, if I understand you correctly, I can easily measure the length of a string whose upper end is attached at any height whatever even if this end were invisible and I could see only the lower extremity. For if I attach to the lower end of this string a rather heavy weight and give it a to-and-fro motion, and if I ask a friend to count a number of its vibrations, while I, during the same time interval, count the number of vibrations of a pendulum which is exactly one cubit in length, then knowing the number of vibrations which each pendulum makes in the given interval of time one can determine the length of the string. Suppose, for example, that my friend counts 20 vibrations of the long cord during the same time in which I count 240 of my string which is one cubit in length; taking the squares of the two numbers, 20 and 240, namely 400 and 57600, then, I say, the long string contains 57600 units of such length that my pendulum will contain 400 of them; and since the length of my string is one cubit, I shall divide 57600 by 400 and thus obtain 144. Accordingly I shall call the length of the string 144 cubits.

Salv. Nor will you miss it by as much as a hand's breadth, especially if you observe a large number of vibrations.

Sagr. You give me frequent occasion to admire the wealth and profusion of nature when, from such common and even trivial phenomena, you derive facts which are not only striking and new but which are often far removed from what we would have imagined. Thousands of times I have observed vibrations especially in churches where lamps, suspended by long cords, had been inadvertently set into motion; but the most which I could infer from these observations was that the view of those who think that such vibrations are maintained by the medium is highly improbable: for, in that case, the air must needs have considerable judgment and little else to do but kill time by pushing to and fro a pendent weight with perfect regularity. But I never dreamed of learning that one and the same body, when suspended from a string a hundred cubits long and pulled aside through an arc of 90° or even 1° or 1/2°, would employ the same time in passing through the least as through the largest of these arcs; and, indeed, it still strikes me as somewhat unlikely. Now I am waiting to hear how these same simple phenomena can furnish solutions for those acoustical problems-solutions which will be at least partly satisfactory.

[141]

Salv. First of all one must observe that each pendulum has its own time of vibration so definite and determinate that it is not possible to make it move with any other period [*altro periodo*] than that which nature has given it. For let any one take in his hand the cord to which the weight is attached and try, as much as he pleases, to increase or diminish the frequency [*frequenza*] of its vibrations; it will be time wasted. On the other hand, one can confer motion upon even a heavy pendulum which is at rest by simply blowing against it; by repeating these blasts with a frequency which is the same as that of the pendulum one can impart considerable motion. Suppose that by the first puff we have displaced the pendulum from the vertical by, say, half an inch; then if, after the pendulum has returned and is about to begin the second vibration, we add a second puff, we shall impart additional motion; and so on with other blasts provided they are applied at the right instant, and not when the pendulum is coming toward us since in this case the blast would impede rather than aid the motion. Continuing thus with many impulses [*impulsi*] we impart to the pendulum such momentum [*impeto*] that a greater impulse [*forza*] than that of a single blast will be needed to stop it.

Sagr. Even as a boy, I observed that one man alone by giving these impulses at the right instant was able to ring a bell so large that when four, or even six, men seized the rope and tried to stop it they were lifted from the ground, all of them together being unable to counterbalance the momentum which a single man, by properly-timed pulls, had given it.

[142]

Salv. Your illustration makes my meaning clear and is quite as well fitted, as what I have just said, to explain the wonderful phenomenon of the strings of the cittern [*cetera*] or of the spinet [*cimbalo*], namely, the fact that a vibrating string will set another string in motion and cause it to sound not only when the latter is in unison but even when it differs from the former by an octave or a fifth. A string which has been struck begins to vibrate and continues the motion as long as one hears the sound [*risonanza*]; these vibrations cause the immediately surrounding air to vibrate and quiver; then these ripples in the air expand far into space and strike not only all the strings of the same instrument but even those of neighboring instruments. Since that string which is tuned to unison with the one plucked is capable of vibrating with the same frequency, it acquires, at the first impulse, a slight oscillation; after receiving two, three, twenty, or more impulses, delivered at proper intervals, it finally accumulates a vibratory motion equal to that of the plucked string, as is clearly shown by equality of amplitude in their vibrations. This undulation expands through the air and sets into vibration not only strings, but also any other body which happens to have the same period as that of the plucked string. Accordingly if we attach to the side of an instrument small pieces of bristle or other flexible bodies, we shall observe that, when a spinet is sounded, only those pieces respond that have the same period as the string which has been struck; the remaining pieces do not vibrate in response to this string, nor do the former pieces respond to any other tone.

If one bows the base string on a viola rather smartly and brings near it a goblet of fine, thin glass having the same tone [*tuono*] as that of the string, this goblet will vibrate and audibly resound. That the undulations of the medium are widely dispersed about the sounding body is evinced by the fact that a glass of water may be made to emit a tone merely by the friction of the finger-tip upon the rim of the glass; for in this water is produced a series of regular waves. The same phenomenon is observed to better advantage by fixing the base of the goblet upon the bottom of a rather large vessel of water filled nearly to the edge of the goblet; for if, as before, we sound the glass by friction of the finger, me shall see ripples spreading with the utmost regularity and with high speed to large distances about the glass. I have often remarked, in thus sounding a rather large glass nearly full of water, that at first the waves are spaced with great uniformity, and when, as sometimes happens, the tone of the glass jumps an octave higher I have noted that at this moment each of the aforesaid waves divides into two; a phenomenon which shows clearly that the ratio involved in the octave [*forma dell' ottava*] is two.

[143]

Sagr. More than once have I observed this same thing, much to my delight and also to my profit. For a long time I have been perplexed about these different harmonies since the explanations hitherto given by those learned in music impress me as not sufficiently conclusive. They tell us that the diapason, i. e. the octave, involves the ratio of two, that the diapente which we call the fifth involves a ratio of 3:2, etc.; because if the open string of a monochord be sounded and afterwards a bridge be placed in the middle and the half length be sounded one hears the octave; and if the bridge be placed at 1/3 the length of the string, then on plucking first the open string and afterwards 2/3 of its length the fifth is given; for this reason they say that the octave depends upon the ratio of two to one [contenuta tra'l due e l'uno] and the fifth upon the ratio of three to two. This explanation does not impress me as sufficient to establish 2 and 3/2 as the natural ratios of the octave and the fifth; and my reason for thinking so is as follows. There are three different ways in which the tone of a string may be sharpened, namely, by shortening it, by stretching it and by making it thinner. If the tension and size of the string remain constant one obtains the octave by shortening it to one-half, i. e., by sounding first the open string and then one-half of it; but if length and size remain constant and one attempts to produce the octave by stretching he will find that it does not suffice to double the stretching weight; it must be quadrupled; so that, if the fundamental note is produced by a weight of one pound, four will be required to bring out the octave.

And finally if the length and tension remain constant, while one changes the size[13] of the string he will find that in order to produce the octave the size must be reduced to 1/4 that which gave the fundamental. And what I have said concerning the octave, namely, that its ratio as derived from the tension and size of the string is the square of that derived from the length, applies equally well to all other musical intervals [intervalli musici].

[144]

Thus if one wishes to produce a fifth by changing the length he finds that the ratio of the lengths must be sesquialteral, in other words he sounds first the open string, then two-thirds of it; but if he wishes to produce this same result by stretching or thinning the string then it becomes necessary to square the ratio 3/2 that is by taking 9/4 [dupla sesquiquarta]; accordingly, if the fundamental requires a weight of 4 pounds, the higher note will be produced not by 6, but by 9 pounds; the same is true in regard to size, the string which gives the fundamental is larger than that which yields the fifth in the ratio of 9 to 4.

13. For the exact meaning of "size" see below. [Trans.]

In view of these facts, I see no reason why those wise philosophers should adopt 2 rather than 4 as the ratio of the octave, or why in the case of the fifth they should employ the sesquialteral ratio, 3/2, rather than that of 9/4 Since it is impossible to count the vibrations of a sounding string on account of its high frequency, I should still have been in doubt as to whether a string, emitting the upper octave, made twice as many vibrations in the same time as one giving the fundamental, had it not been for the following fact, namely, that at the instant when the tone jumps to the octave, the waves which constantly accompany the vibrating glass divide up into smaller ones which are precisely half as long as the former.

Salv. This is a beautiful experiment enabling us to distinguish individually the waves which are produced by the vibrations of a sonorous body, which spread through the air, bringing to the tympanum of the ear a stimulus which the mind translates into sound. But since these waves in the water last only so long as the friction of the finger continues and are, even then, not constant but are always forming and disappearing, would it not be a fine thing if one had the ability to produce waves which would persist for a long while, even months and years, so as to easily measure and count them?

Sagr. Such an invention would, I assure you, command my admiration.

[145]

Salv. The device is one which I hit upon by accident; my part consists merely in the observation of it and in the appreciation of its value as a confirmation of something to which I had given profound consideration; and yet the device is, in itself, rather common. As I was scraping a brass plate with a sharp iron chisel in order to remove some spots from it and was running the chisel rather rapidly over it, I once or twice, during many strokes, heard the plate emit a rather strong and clear whistling sound; on looking at the plate more carefully, I noticed a long row of fine streaks parallel and equidistant from one another. Scraping with the chisel over and over again, I noticed that it was only when the plate emitted this hissing noise that any marks were left upon it; when the scraping was not accompanied by this sibilant note there was not the least trace of such marks. Repeating the trick several times and making the stroke, now with greater now with less speed, the whistling followed with a pitch which was correspondingly higher and lower. I noted also that the marks made when the tones were higher were closer together; but when the tones were deeper, they were farther apart. I also observed that when, during a single stroke, the speed increased toward the end the sound became sharper and the streaks grew closer together, but always in such a way as to remain sharply defined and equidistant. Besides whenever the stroke was accompanied by hissing I felt the chisel tremble in my grasp and a sort of shiver run through my hand. In short we see and hear in the case of the chisel precisely that which

is, seen and heard in the case of a whisper followed by a loud voice; for, when the breath is emitted without the production of a tone, one does not feel either in the throat or mouth any motion to speak of in comparison with that which is felt in the larynx and upper part of the throat when the voice is used, especially, when the tones employed are low and strong.

At times I have also observed among the strings of the spinet two which were in unison with two of the tones produced by the aforesaid scraping; and among those which differed most in pitch I found two which were separated by an interval of a perfect fifth. Upon measuring the distance between the markings produced by the two scrapings it was found that the space which contained 45 of one contained 30 of the other, which is precisely the ratio assigned to the fifth.

[146]

But now before proceeding any farther I want to call your attention to the fact that, of the three methods for sharpening a tone, the one which you refer to as the fineness of the string should be attributed to its weight. So long as the material of the string is unchanged, the size and weight vary in the same ratio. Thus in the case of gut-strings, we obtain the octave by making one string 4 times as large as the other; so also in the case of brass one wire must have 4 times the size of the other; but if now we wish to obtain the octave of a gut-string, by use of brass wire, we must make it, not four times as large, but four times as heavy as the gutstring: as regards size therefore the metal string is not four times as big but four times as heavy. The wire may therefore be even thinner than the gut notwithstanding the fact that the latter gives the higher note. Hence if two spinets are strung, one with gold wire the other with brass, and if the corresponding strings each have the same length, diameter, and tension it follows that the instrument strung with gold will have a pitch about one-fifth lower than the other because gold has a density almost twice that of brass. And here it is to be noted that it is the weight rather than the size of a moving body which offers resistance to change of motion [velocità del moto] contrary to what one might at first glance think. For it seems reasonable to believe that a body which is large and light should suffer greater retardation of motion in thrusting aside the medium than would one which is thin and heavy; yet here exactly the opposite is true.

Returning now to the original subject of discussion, I assert that the ratio of a musical interval is not immediately determined either by the length, size, or tension of the strings but rather by the ratio of their frequencies, that is, by the number of pulses of air waves which strike the tympanum of the ear, causing it also to vibrate with the same frequency. This fact established, we may possibly explain why certain pairs of notes, differing in pitch produce a pleasing sensation, others a less pleasant effect and

still others a disagreeable sensation. Such an explanation would be tantamount to an explanation of the more or less perfect consonances and of dissonances. The unpleasant sensation produced by the latter arises, I think, from the discordant vibrations of two different tones which strike the ear out of time [*sproporzionatamente*]. Especially harsh is the dissonance between notes whose frequencies are incommensurable; such a case occurs when one has two strings in unison and sounds one of them open, together with a part of the other which bears the same ratio to its whole length as the side of a square bears to the diagonal; this yields a dissonance similar to the augmented fourth or diminished fifth [*tritono o semidiapente*].

[147]

Agreeable consonances are pairs of tones which strike the car with a certain regularity; this regularity consists in the fact that the pulses delivered by the two tones, in the same interval of time, shall be commensurable in number, so as not to keep the ear drum in perpetual torment, bending in two different directions in order to yield to the ever-discordant impulses.

The first and most pleasing consonance is, therefore, the octave since, for every pulse given to the tympanum by the lower string, the sharp string delivers two; accordingly at every other vibration of the upper string both pulses are delivered simultaneously so that one-half the entire number of pulses are delivered in unison. But when two strings are in unison their vibrations always coincide and the effect is that of a single string; hence we do not refer to it as consonance. The fifth is also a pleasing interval since for every two vibrations of the lower string the upper one gives three, so that considering the entire number of pulses from the upper string one-third of them will strike in unison, i. e., between each pair of concordant vibrations there intervene two single vibrations; and when the interval is a fourth, three single vibrations intervene. In case the interval is a second where the ratio is 9/8 it is only every ninth vibration of the upper string which reaches the ear simultaneously with one of the lower; all the others are discordant and produce a harsh effect upon the recipient ear which interprets them as dissonances.

Simp. Won't you be good enough to explain this argument a little more clearly?

[148]

Salv. Let AB denote the length of a wave [*lo spazio e la dilatazione d'una vibrazione*] emitted by the lower string and CD that of a higher string which is emitting the octave of AB; divide AB in the middle at E. If the two strings begin their motions at A and C, it is clear that when the sharp vibration has reached the end D,

the other vibration will have travelled only as far as E, which, not being a terminal point, will emit no pulse; but there is a blow delivered at D. Accordingly when the one wave comes back from D to C, the other passes on from E to B; hence the two pulses from B and C strike the drum of the ear simultaneously. Seeing that these vibrations are repeated again and again in the same manner, we conclude that each alternate pulse from CD falls in unison with one from AB. But each of the pulsations at the terminal points, A and B, is constantly accompanied by one which leaves always from C or always from D. This is clear because if we suppose the waves to reach A and

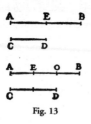

Fig. 13

C at the same instant, then, while one wave travels from A to B, the other will proceed from C to D and back to C, so that waves strike at C and B simultaneously; during the passage of the wave from B back to A the disturbance at C goes to D and again returns to C, so that once more the pulses at A and C are simultaneous.

Next let the vibrations AB and CD be separated by an interval of a fifth, that is, by a ratio of 3/2; choose the points E and O such that they will divide the wave length of the lower string into three equal parts and imagine the vibrations to start at the same instant from each of the terminals A and C. It is evident that when the pulse has been delivered at the terminal D, the wave in AB has travelled only as far as O; the drum of the ear receives, therefore, only the pulse from D. Then during the return of the one vibration from D to C, the other will pass from O to B and then back to O, producing an isolated pulse at B—a pulse which is out of time but one which must be taken into consideration.

Now since we have assumed that the first pulsations started from the terminals A and C at the same instant, it follows that the second pulsation, isolated at D, occurred after an interval of time equal to that required for passage from C to D or, what is the same thing, from A to O; but the next pulsation, the one at B, is separated from the preceding by only half this interval, namely, the time required for passage from O to B. Next while the one vibration travels from O to A, the other travels from C to D, the result of which is that two pulsations occur simultaneously at A and D. Cycles of this kind follow one after another, i. e., one solitary pulse of the lower string interposed between two solitary pulses of the upper string. Let us now imagine time to be divided into very small equal intervals; then if we assume that, during the first two of these intervals, the disturbances which occurred simultaneously at A and C have travelled as far as O and D and have produced a pulse at D; and if we assume that during the third and fourth intervals one disturbance returns from D to C, producing a pulse at C, while the other, passing on from O to B and back to O, produces a pulse at B; and if finally, during the fifth and sixth intervals, the disturbances travel from O and C to A

and D, producing a pulse at each of the latter two, then the sequence in which the pulses strike the ear will be such that, if we begin to count time from any instant where two pulses are simultaneous, the ear drum will, after the lapse of two of the said intervals, receive a solitary pulse; at the end of the third interval, another solitary pulse; so also at the end of the fourth interval; and two intervals later, i. e., at the end of the sixth interval, will be heard two pulses in unison. Here ends the cycle—the anomaly, so to speak—which repeats itself over and over again.

[149]

Sagr. I can no longer remain silent; for I must express to you the great pleasure I have in hearing such a complete explanation of phenomena with regard to which I have so long been in darkness. Now I understand why unison does not differ from a single tone; I understand why the octave is the principal harmony, but so like unison as often to be mistaken for it and also why it occurs with the other harmonies. It resembles unison because the pulsations of strings in unison always occur simultaneously, and those of the lower string of the octave are always accompanied by those of the upper string; and among the latter is interposed a solitary pulse at equal intervals and in such a manner as to produce no disturbance; the result is that such a harmony is rather too much softened and lacks fire. But the fifth is characterized by its displaced beats and by the interposition of two solitary beats of the upper string and one solitary beat of the lower string between each pair of simultaneous pulses; these three solitary pulses are separated by intervals of time equal to half the interval which separates each pair of simultaneous beats from the solitary beats of the upper string. Thus the effect of the fifth is to produce a tickling of the ear drum such that its softness is modified with sprightliness, giving at the same moment the impression of a gentle kiss and of a bite.

Salv. Seeing that you have derived so much pleasure from these novelties, I must show you a method by which the eye may, enjoy the same game as the ear. Suspend three balls of lead, or other heavy material, by means of strings of different length such that while the longest makes two vibrations the shortest will make four and the medium three; this will take place when the longest string measures 16, either in hand breadths or in any other unit, the medium 9 and the shortest 4, all measured in the same unit.

Now pull all these pendulums aside from the perpendicular and release them at the same instant; you will see a curious interplay of the threads passing each other in various manners but such that at the completion of every fourth vibration of the longest pendulum, all three will arrive simultaneously at the same terminus, whence they start over again to repeat the same cycle. This combination of vibrations, when produced on strings is precisely that which yields the interval of the octave and the intermediate fifth.

If we employ the same disposition of apparatus but change the lengths of the threads, always however in such a way that their vibrations correspond to those of agreeable musical intervals, we shall see a different crossing of these threads but always such that, after a definite interval of time and after a definite number of vibrations, all the threads, whether three or four, will reach the same terminus at the same instant, and then begin a repetition of the cycle.

[150]

If however the vibrations of two or more strings are incommensurable so that they never complete a definite number of vibrations at the same instant, or if commensurable they return only after a long interval of time and after a large number of vibrations, then the eye is confused by the disorderly succession of crossed threads. In like manner the ear is pained by an irregular sequence of air waves which strike the tympanum without any fixed order.

But, gentlemen, whither have we drifted during these many hours lured on by various problems and unexpected digressions? The day is already ended and we have scarcely touched the subject proposed for discussion. Indeed we have deviated so far that I remember only with difficulty our early introduction and the little progress made in the way of hypotheses and principles for use in later demonstrations.

Sagr. Let us then adjourn for to-day in order that our minds may find refreshment in sleep and that we may return to-morrow, if so please you, and resume the discussion of the main question.

Salv. I shall not fail to be here to-morrow at the same hour, hoping not only to reader you service but also to enjoy your company.

END OF THE FIRST DAY

[151]

SECOND DAY

Sagr. While Simplicio and I were awaiting your arrival we were trying to recall that last consideration which you advanced as a principle and basis for the results you intended to obtain; this consideration dealt with the resistance which all solids offer to fracture and depended upon a certain cement which held the parts glued together so that they would yield and separate only under considerable pull [*potente attrazzione*]. Later we tried to find the explanation of this coherence, seeking it mainly in the vacuum; this was the occasion of our many digressions which occupied the entire day and led us far afield from the original question which, as I have already stated, was the consideration of the resistance [*resistenza*] that solids offer to fracture.

Salv. I remember it all very well. Resuming the thread of our discourse, whatever the nature of this resistance which solids offer to large tractive forces [*violenta attrazione*] there can at least be no doubt of its existence; and though this resistance is very great in the case of a direct pull, it is found, as a rule, to be less in the case of bending forces [*nel violentargli per traverso*]. Thus, for example, a rod of steel or of glass will sustain a longitudinal pull of a thousand pounds while a weight of fifty pounds would be quite sufficient to break it if the rod were fastened at right angles into a vertical wall.

[152]

It is this second type of resistance which we must consider, seeking to discover in what proportion it is found in prisms and cylinders of the same material whether alike or unlike in shape, length, and thickness. In this discussion I shall take for granted the well-known mechanical principle which has been shown to govern the behavior of a bar, which we call a lever, namely, that the force bears to the resistance the inverse ratio of the distances which separate the fulcrum from the force and resistance respectively.

Simp. This was demonstrated first of all by Aristotle, in his *Mechanics*.

Salv. Yes, I am willing to concede him priority in point of time; but as regards rigor of demonstration the first place must be given to Archimedes, since upon a single proposition proved in his book on Equilibrium[14] depends not only the law of the lever but also those of most other mechanical devices.

Sagr. Since now this principle is fundamental to all the demonstrations which you propose to set forth would it not be advisable to give us a complete and thorough proof of this proposition unless possibly it would take too much time?

Salv. Yes, that would be quite proper, but it is better I think to approach our subject in a manner somewhat different from that employed by Archimedes, namely, by first assuming merely that equal weights placed in a balance of equal arms will produce equilibrium—a principle also assumed by Archimedes—and then proving that it is no less true that unequal weights produce equilibrium when the arms of the steelyard have lengths inversely proportional to the weights suspended from them; in other words, it amounts to the same thing whether one places equal weights at equal distances or unequal weights at distances which bear to each other the inverse ratio of the weights.

In order to make this matter clear imagine a prism or solid cylinder, AB, suspended at each end to the rod [*linea*] HI, and supported by two threads HA and IB; it is evident that if I attach a thread, C, at the middle point of the balance beam HI, the entire prism AB will, according to the principle assumed, hang in equilibrium since one-half its weight lies on one side, and the other half on the other side, of the point of suspension C. Now suppose the prism to be divided into unequal parts by a plane

14. *Works of Archimedes.* Trans. By T. L. Heath. pp. 189-220. *[Trans.]*

through the line D, and let the part DA be the larger and DB the smaller: this division having been made, imagine a thread ED, attached at the point E and supporting the parts AD and DB, in order that these parts may remain in the same position relative to line HI: and since the relative position of the prism and the beam HI remains unchanged, there can be no doubt but that the prism will maintain its former state of equilibrium.

But circumstances would remain the same if that part of the prism which is now held up, at the ends, by the threads AH and DE were supported at the middle by a single thread GL; and likewise the other part DB would not change position if held by a

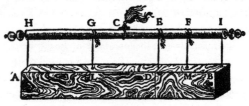

Fig. 14

thread FM placed at its middle point. Suppose now the threads HA, ED, and IB to be removed, leaving only the two GL and FM, then the same equilibrium will be maintained so long as the suspension is at C. Now let us consider that we have here two heavy bodies AD and DB hung at the ends G and F, of a balance beam GF in equilibrium about the point C, so that the line CG is the distance from C to the point of suspension of the heavy body AD, while CF is the distance at which the other heavy body DB is supported. It remains now only to show that these distances bear to each other the inverse ratio of the weights themselves, that is, the distance GC is to the distance CF as the prism DB is to the prism DA—a proposition which we shall prove as follows: Since the line GE is the half of EH, and since EF is the half of EI, the whole length GF will be half of the entire line HI, and therefore equal to CI: if now we subtract the common part CF the remainder GC will be equal to the remainder FI, that is, to FE, and if to each of these we add CE we shall have GE equal to CF: hence GE:EF=FC:CG. But GE and EF bear the same ratio to each other as do their doubles HE and EI, that is, the same ratio as the prism AD to DB. Therefore, by equating ratios we have, *convertendo*, the distance GC is to the distance CF as the weight BD is to the weight DA, which is what I desired to prove.

[154]

If what precedes is clear, you will not hesitate, I think, to admit that the two prisms AD and DB are in equilibrium about the point C since one-half of the whole body AB lies on the right of the suspension C and the other half on the left; in other words, this arrangement is equivalent to two equal weights disposed at equal distances. I do not see how any one can doubt, if the two prisms AD and DB were transformed into cubes, spheres, or into any other figure whatever and if G and F were retained as points of suspension, that they would remain in equilibrium about the point C, for it is only too evident that change of figure does not produce change of weight so long as the mass [*quantità di materià*] does not vary. From this we may derive the general conclusion that any two heavy bodies are in equilibrium at distances which are inversely proportional to their weights.

This principle established, I desire, before passing to any other subject, to call your attention to the fact that these forces, resistances, moments, figures, etc., may be considered either in the abstract, dissociated from matter, or in the concrete, associated with matter. Hence the properties which belong to figures that are merely geometrical and non-material must be modified when we fill these figures with matter and therefore give them weight. Take, for example, the lever BA which, resting upon the support E, is used to lift a heavy stone D. The principle just demonstrated makes it clear that a force applied at the extremity B will just suffice to equilibrate the resistance offered by the heavy body D provided this force [*momento*] bears to the force [*momento*] at D the same ratio as the distance AC bears to the distance CB; and this is true so long as we consider only the moments of the single force at B and of the resistance at D, treating the lever as an immaterial body devoid of weight. But if we take into account the weight of the lever itself—an instrument which may be made either of wood or of iron—it is manifest that, when this weight has been added to the

[155]

force at B, the ratio will be changed and must therefore be expressed in different terms. Hence before going further let us agree to distinguish between these two points of view; when we consider an instrument in the abstract, i. e., apart from the weight of its own material, we shall speak of "taking it in an absolute sense" [*prendere assolutamente*]; but if we fill one of these simple and absolute figures with matter and thus give it weight, we shall refer to such a material figure as a "moment" or "compound force" [*momento o forza composta*].

Sagr. I must break my resolution about not leading you off into a digression; for I cannot concentrate my attention upon what is to follow until a certain doubt is

removed from my mind, namely, you seem to compare the force at B with the total weight of the stone D, a part of which—possibly the greater part—rests upon the horizontal plane: so that . . .

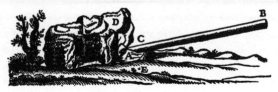

Fig. 15

Salv. I understand perfectly. you need go no further. However please observe that I have not mentioned the total weight of the stone; I spoke only of its force [*momento*] at the point A, the extremity of the lever BA, which force is always less than the total weight of the stone, and varies with its shape and elevation.

Sagr. Good: but there occurs to me another question about which I am curious. For a complete understanding of this matter, I should like you to show me, if possible, how one can determine what part of the total weight is supported by the underlying plane and what part by the end A of the lever.

Salv. The explanation will not delay us long and I shall therefore have pleasure in granting your request. In the accompanying figure, let us understand that the weight having its center of gravity at A rests with the end B upon the horizontal plane and with the other end upon the lever CG. Let N be the fulcrum of a lever to which the force [*potenza*] is applied at G. Let fall the perpendiculars, AO and CF, from the center A and the end C. Then I say, the magnitude [*momento*] of the entire weight bears to the magnitude of the force [*momento della potenza*] at G a ratio compounded of the ratio between the two distances GN and NC and the ratio between FB and BO. Lay off a distance X such that its ratio to NC is the same as that of BO to FB; then, since the total weight A is counterbalanced by the two forces at B and at C, it follows that the force at B is to that at C as the distance FO is to the distance OB.

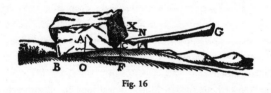

Fig. 16

[156]

Hence, *componendo*, the sum of the forces at B and C, that is, the total weight A [*momento di tutto 'l peso A*], is to the force at C as the line FB is to the line BO, that is, as NC is to X: but the force [*momento della potenza*] applied at C is to the force applied at G as the distance GN is to the distance NC; hence it follows, *ex æquali in proportione perturbata*,[15] that the entire weight A is to the force applied at G as the distance GN is to X. But the ratio of GN to X is compounded of the ratio of GN to NC and of NC to X, that is, of FB to BO; hence the weight A bears to the equilibrating force at G a ratio compounded of that of GN to NC and of FB to BO: which was to be proved.

Let us now return to our original subject; then, if what has hitherto been said is clear, it will be easily understood that,

PROPOSITION I

A prism or solid cylinder of glass, steel, wood or other breakable material which is capable of sustaining a very heavy weight when applied longitudinally is, as previously remarked, easily broken by the transverse application of a weight which may be much smaller in proportion as the length of the cylinder exceeds its thickness.

Let us imagine a solid prism ABCD fastened into a wall at the end AB, and supporting a weight E at the other end; understand also that the wall is vertical and that the prism or cylinder is fastened at right angles to the wall. It is clear that, if the cylinder breaks, fracture will occur at the point B where the edge of the mortise act as a fulcrum for the lever BC, to which the force is applied; the thickness of the solid BA is the other arm of the lever along which is located the resistance. This resistance opposes the separation of the part BD, lying outside the wall, from that portion lying inside. From the preceding, it follows that the magnitude [*momento*] of the force applied at C bears to the magnitude [*momento*] of the resistance, found in the thickness of the prism, i. e., in the attachment of the base BA to its contiguous parts, the same ratio which the length CB bears to half the length BA; if now we define absolute resistance to fracture as that offered to a longitudinal pull (in which case the stretching force acts in the same direction as that through which the body is moved), then it follows that the absolute resistance of the prism BD is to the breaking load placed at the end of the lever BC in the same ratio as the length BC is to the half of AB in the case of a prism, or the semidiameter in the case of a cylinder.

15. For definition of *perturbata* see Todhunter's *Euclid*. Book V, Def. 20. *[Trans.]*

[157]

This is our first proposition.[16] Observe that in what has here been said the weight of the solid BD itself has been left out of consideration, or rather, the prism has been assumed to be devoid of weight. But if the weight of the prism is to be taken account of in conjunction with the weight E, we must add to the weight E one half that of the prism BD: so that if, for example, the latter weighs two pounds and the weight E is ten pounds we must treat the weight E as if it were eleven pounds.

Fig. 17

Simp. Why not twelve?

Salv. The weight E, my dear Simplicio, hanging at the extreme end C acts upon the lever BC with its full moment of ten pounds: so also would the solid BD if suspended at the same point exert its full moment of two pounds; but, as you know, this solid is uniformly distributed throughout its entire length, BC, so that the parts which lie near the end B are less effective than those more remote.

Accordingly if we strike a balance between the two, the weight of the entire prism may be considered as concentrated at its center of gravity which lies midway of the lever BC. But a weight hung at the extremity C exerts a moment twice as great as it would if suspended from the middle: therefore if we consider the moments of both as located at the end C we must add to the weight E one-half that of the prism.

[158]

Simp. I understand perfectly; and moreover, if I mistake not, the force of the two weights BD and E, thus disposed, would exert the same moment as would the entire weight BD together with twice the weight E suspended at the middle of the lever BC.

Salv. Precisely so, and a fact worth remembering. Now we can readily understand

16. The one fundamental error which is implicitly introduced into this proposition and which is carried through the entire discussion of the Second Day consists in a failure to see that, in such a beam, there must be equilibrium between the forces of tension and compression over any cross-section. The correct point of view seems first to have been found by E. Mariotte in 1680 and by A. Parent in 1713. Fortunately this error does not vitiate the conclusions of the subsequent propositions which deal only with proportions—not actual strength—of beams. Following K. Pearson (Todhunter's *History of Elasticity*) one might say that Galileo's mistake lay in supposing the fibres of the strained beam to be inextensible. Or, confessing the anachronism, one might say that the error consisted in taking the lowest fibre of the beam as the neutral axis. *[Trans.]*

PROPOSITION II

How and in what proportion a rod, or rather a prism, whose width is greater than its thickness offers more resistance to fracture when the force is applied in the direction of its breadth than in the direction of its thickness.

For the sake of clearness, take a ruler *ad* whose width is *ac* and whose thickness, *cb*, is much less than its width. The question now is why will the ruler, if stood on edge, as in the first figure, withstand a great weight T, while, when laid flat, as in the second figure, it will not support the weight X which is less than T. The answer is evident when we remember that in the one case the fulcrum is at the line *bc*, and in the other case at *ca*, while the distance at which the force is applied is the same in both cases, namely, the length *bd*: but in the first case the distance of the

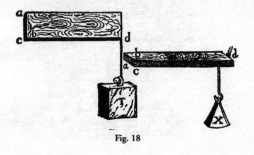

Fig. 18

resistance from the fulcrum-half the line *ca*-is greater than in the other case where it is only half of *bc*. Therefore the weight T is greater than X in the same ratio as half the width *ca* is greater than half the thickness *bc*, since the former acts as a lever arm for *ca*, and the latter for *cb*, against the same resistance, namely, the strength of all the fibres in the cross-section *ab*. We conclude, therefore, that any given ruler, or prism, whose width exceeds its thickness, will offer greater resistance to fracture when standing on edge than when lying flat, and this in the ratio of the width to the thickness.

PROPOSITION III

Considering now the case of a prism or cylinder growing longer in a horizontal direction, we must find out in what ratio the moment of its own weight increases in comparison with its resistance to fracture. This moment I find increases in proportion to the square of the length.

[159]

In order to prove this let AD be a prism or cylinder lying horizontal with its end A firmly fixed in a wall. Let the length of the prism be increased by the addition of the portion BE. It is clear that merely changing the length of the lever from AB to AC will, if we disregard its weight, increase the moment of the force [at the end] tending to pro-

Fig. 19

duce fracture at A in the ratio of CA to BA. But, besides this, the weight of the solid portion BE, added to the weight of the solid AB increases the moment of the total weight in the ratio of the weight of the prism AE to that of the prism AB, which is the same as the ratio of the length AC to AB.

It follows, therefore, that, when the length and weight are simultaneously increased in any given proportion, the moment, which is the product of these two, is increased in a ratio which is the square of the preceding proportion. The conclusion is then that the bending moments due to the weight of prisms and cylinders which have the same thickness but different lengths, bear to each other a ratio which is the square of the ratio of their lengths, or, what is the same thing, the ratio of the squares of their lengths.

[160]

We shall next show in what ratio the resistance to fracture [bending strength], in prisms and cylinders, increases with increase of thickness while the length remains unchanged. Here I say that

488

PROPOSITION IV

*In prisms and cylinders of equal length, but of unequal thicknesses,
the resistance to fracture increases in the same ratio as the cube of the
diameter of the thickness, i. e., of the base.*

Let A and B be two cylinders of equal lengths DG, FH; let their bases be
circular but unequal, having the diameters CD and EF. Then I say that the

resistance to fracture offered by the cylinder
B is to that offered by A as the cube of the
diameter FE is to the cube of the diameter
DC. For, if we consider the resistance to
fracture by longitudinal pull as dependent
upon the bases, i. e., upon the circles EF
and DC, no one can doubt that the
strength [*resistenza*] of the cylinder B is
greater than that of A in the same propor-

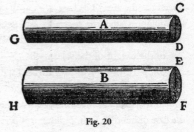

Fig. 20

tion in which the area of the circle EF exceeds that of CD; because it is precisely in
this ratio that the number of fibres binding the parts of the solid together in the one
cylinder exceeds that in the other cylinder.

But in the case of a force acting transversely it must be remembered that we are
employing two levers in which the forces are applied at distances DG, FH, and the
fulcrums are located at the points D and F; but the resistances are applied at
distances which are equal to the radii of the circles DC and EF, since the fibres
distributed over these entire cross-sections act as if concentrated at the centers.
Remembering this and remembering also that the arms, DG and FH, through
which the forces G and H act are, equal, we can understand that the resistance,
located at the center of the base EF, acting against the force at H, is more effective
[*maggiore*] than the resistance at the center of the base CD opposing the force G,
in the ratio of the radius FE to the radius DC. Accordingly the resistance to fracture offered by the cylinder B is greater than that of the cylinder A in a ratio which
is compounded of that of the area of the circles EF and DC and that of their radii,
i. e., of their diameters; but the areas of circles are as the squares of their diameters.
Therefore the ratio of the resistances, being the product of the two preceding ratios,
is the same as that of the cubes of the diameters. This is what I set out to prove. Also
since the volume of a cube varies as the third power of its edge we may say that the
resistance [strength] of a cylinder whose length remains constant varies as the third
power of its diameter.

[161]

From the preceding we are able to conclude that

COROLLARY

The resistance [strength] of a prism or cylinder of constant length varies in the sesquialteral ratio of its volume.

This is evident because the volume of a prism or cylinder of constant altitude varies directly as the area of its base, i. e., as the square of a side or diameter of this base; but, as just demonstrated, the resistance [strength] varies as the cube of this same side or diameter. Hence the resistance varies in the sesquialteral ratio of the volume—consequently also of the weight-of the solid itself.

Simp. Before proceeding further I should like to have one of my difficulties removed. Up to this point you have not taken into consideration a certain other kind of resistance which, it appears to me, diminishes as the solid grows longer, and this is quite as true in the case of bending as in pulling; it is precisely thus that in the case of a rope we observe that a very long one is less able to support a large weight than a short one. Whence, I believe, a short rod of wood or iron will support a greater weight than if it were long, provided the force be always applied longitudinally and not transversely, and provided also that we take into account the weight of the rope itself which increases with its length.

Salv. I fear, Simplicio, if I correctly catch your meaning, that in this particular you are making the same mistake as many others; that is if you mean to say that a long rope, one of perhaps 40 cubits, cannot bold up so great a weight as a shorter length, say one or two cubits, of the same rope.

Simp. That is what I meant, and as far as I see the proposition is highly probable.

Salv. On the contrary, I consider it not merely improbable but false; and I think I can easily convince you of your error. Let AB represent the rope, fastened at the upper end A: at the lower end attach a weight C whose force is just sufficient to break the rope. Now, Simplicio, point out the exact place where you think the break ought to occur.

[162]

Simp. Let us say D.

Salv. And why at D?

Simp. Because at this point the rope is not strong enough to support, say, 100 pounds, made up of the portion of the rope DB and the stone C.

Salv. Accordingly whenever the rope is stretched [*violentata*] with the weight of 100 pounds at D it will break there.

Simp. I think so.

Salv. But tell me, if instead of attaching the weight at the end of the rope, B, one fastens it at a point nearer D, say, at E: or if, instead of fixing the upper end of the rope at A, one fastens it at some point F, just above D, will not the rope, at the point D, be subject to the same pull of 100 pounds?

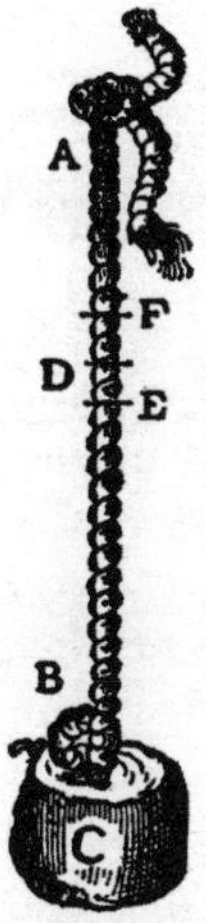

Simp. It would, provided you include with the stone C the portion of rope EB.

Salv. Let us therefore suppose that the rope is stretched at the point D with a weight of 100 pounds, then according to your own admission it will break; but FE is only a small portion of AB; how can you therefore maintain that the long rope is weaker than the short one? Give up then this erroneous view which you share with many very intelligent people, and let us proceed.

Now having demonstrated that, in the case of [uniformly loaded] prisms and cylinders of constant thickness, the moment of force tending to produce fracture [*momento sopra le proprie resistenze*] varies as the square of the length; and having likewise shown that, when the length is constant and the thickness varies, the resistance to fracture varies as the cube of the side, or diameter, of the base, let us pass to the investigation of the case of solids which simultaneously vary in both length and thickness. Here I observe that,

Fig. 21

PROPOSITION V

Prisms and cylinders which differ in both length and thickness offer resistances to fracture [i. e., can support at their ends loads] which are directly proportional to the cubes of the diameters of their bases and inversely proportional to their lengths.

[163]

Let ABC and DEF be two such cylinders; then the resistance [bending strength] of the cylinder AC bears to the resistance of the cylinder DF a ratio which is the product of the cube of the diameter AB divided by the cube of the diameter DE, and of the length EF divided by the length BC. Make EG equal to BC: let H be a third proportional to the lines AB and DE; let I be a fourth proportional, [AB/DE=H/I]: and let I:S=EF:BC.

Now since the resistance of the cylinder AC is to that of the cylinder DG as the cube of AB is to the cube of DE, that is, as the length AB is to the length I; and since the resistance of the cylinder DG is to that of the cylinder DF as the length FE is to EG, that is, as I is to S, it follows that the length AB is to S as the resistance of the

cylinder AC is to that of the cylinder DF. But the line AB bears to S a ratio which is the product of AB/I and I/S. Hence the resistance [bending strength] of the cylinder AC bears to the resistance of the cylinder DF a ratio which is the product of AB/I (that is, AB^3/DE^3) and of I/S (that is, EF/BC): which is what I meant to prove.

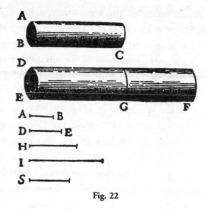

Fig. 22

This proposition having been demonstrated, let us next consider the case of prisms and cylinders which are similar. Concerning these we shall show that,

PROPOSITION VI

In the case of similar cylinders and prisms, the moments [stretching forces] which result from multiplying together their weight and length [i. e., from the moments produced by their own weight and length], which latter acts as a lever-arm, bear to each other a ratio which is the sesquialteral of the ratio between the resistances of their bases.

In order to prove this let us indicate the two similar cylinders by AB and CD: then the magnitude of the force [*momento*] in the cylinder AB, opposing the resistance of its base B, bears to the magnitude [*momento*] of the force at CD, opposing the resistance of its base D, a ratio which is the sesquialteral of the ratio

[164]

between the resistance of the base B and the resistance of the base D. And since the solids AB and CD, are effective in opposing the resistances of their bases B and D, in proportion to their weights and to the mechanical advantages [*forze*] of their lever arms respectively, and since the advantage [*forza*] of the lever arm AB is equal to the advantage [*forza*] of the lever arm CD (this is true because in virtue of the similarity of the cylinders the length AB is to the radius of the base B as the length CD is

Fig. 23

to the radius of the base D), it follows that the total force [*momento*] of the cylinder AB is to the total force [*momento*] of the cylinder CD as the weight alone of the cylinder AB is to the weight alone of the cylinder CD, that is, as the volume of the cylinder AB [*l'istesso cilindro AB*] is to the volume CD [*all'istesso CD*]: but these are as the cubes of the diameters of their bases B and D; and the resistances of the bases, being to each other as their areas, are to each other consequently as the squares of their diameters. Therefore the forces [*momenti*] of the cylinders are to each other in the sesquialteral ratio of the resistance of their bases.[17]

Simp. This proposition strikes me as both new and surprising: at first glance it is very different from anything which I myself should have guessed: for since these figures are similar in all other respects, I should have certainly thought that the forces [*momenti*] and the resistances of these cylinders would have borne to each other the same ratio.

Sagr. This is the proof of the proposition to which I referred, at the very beginning of our discussion, as one imperfectly understood by me.

Salv. For a while, Simplicio, I used to think, as you do, that the resistances of similar solids were similar; but a certain casual observation showed me that similar solids do not exhibit a strength which is proportional to their size, the larger ones being less fitted to undergo rough usage just as tall men are more apt than small children to be injured by a fall. And, as we remarked at the outset, a large beam or column falling

[165]

from a given height will go to pieces when under the same circumstances a small scantling or small marble cylinder will not break. It was this observation which led me to the investigation of the fact which I am about to demonstrate to you: it is a very remarkable thing that, among the infinite variety of solids which are similar one to another, there are no two of which the forces [*momenti*], and the resistances of these solids are related in the same ratio.

Simp. You remind me now of a passage in Aristotle's *Questions in Mechanics* in which he tries to explain why it is that a wooden beam becomes weaker and can be more easily bent as it grows longer, notwithstanding the fact that the shorter beam is thinner and the longer one thicker: and, if I remember correctly, he explains it in terms of the simple lever.

17. The preceding paragraph beginning with Prop. VI is of more than usual interest as illustrating the confusion of terminology current in the time of Galileo. The translation given is literal except in the case of those words for which the Italian is supplied. The facts which Galileo has in mind are so evident that it is difficult to see how one can here interpret "*moment,*" to mean the force "*opposing the resistance of its base,*" unless "*the force of the lever arm .AB*" be taken to mean "*the mechanical advantage of the lever made up of AB and the radius of the base B*"; and similarly for "*the force of the lever arm CD.*" [*Trans.*]

Salv. Very true: but, since this solution seemed to leave room for doubt, Bishop di Guevara,[18] whose truly learned commentaries have greatly enriched and illuminated this work, indulges in additional clever speculations with the hope of thus overcoming all difficulties; nevertheless even he is confused as regards this particular point, namely, whether, when the length and thickness of these solid figures increase in the same ratio, their strength and resistance to fracture, as well as to bending, remain constant. After much thought upon this subject I have reached the following result. First I shall show that,

PROPOSITION VII

Among heavy prisms and cylinders of similar figure, there is one and only one which under the stress of its own weight lies just on the limit between breaking and not breaking: so that every larger one is unable to carry the bad of its own weight and breaks; while every smaller one is able to withstand some additional force tending to break it.

Let AB be a heavy prism, the longest possible that will just sustain its own weight, so that if it be lengthened the least bit it will break. Then, I say, this prism is unique among all similar prisms—infinite in number—in occupying that boundary line between breaking and not breaking; so that every larger one will break under its own weight,

[166]

and every smaller one win not break, but will be able to withstand some force in addition to its own weight.

Let the prism CE be similar to, but larger than, AB: then, I say, it will not remain intact but will break under its own weight. Lay off the portion CD, equal in length to AB. And, since, the resistance [bending strength] of CD is to that of AB as the cube of the thickness of CD is to the cube of the thickness of AB, that is, as the prism CE is to the similar prism AB, it follows that the weight of CE is the utmost load which a prism of the length CD can sustain; but the length of CE is greater; therefore the prism CE will break. Now take another prism FG which is smaller than AB. Let FH equal AB, then it can be shown in a similar manner that the resistance

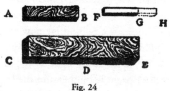

Fig. 24

[bending strength] of FG is to that of AB as the prism FG is to the prism AB provided the distance AB that is FH, is equal to the distance FG; but AB is greater than FG,

and therefore the moment of the prism FG applied at G is not sufficient to break the prism FG.

Sagr. The demonstration is short and clear; while the proposition which, at first glance, appeared improbable is now seen to be both true and inevitable. In order therefore to bring this prism into that limiting condition which separates breaking from not breaking, it would be necessary to change the ratio between thickness and length either by increasing the thickness or by diminishing the length. An investigation of this limiting state will, I believe, demand equal ingenuity.

Salv. Nay, even more; for the question is more difficult; this I know because I spent no small amount of time in its discovery which I now wish to share with you.

PROPOSITION VIII

Given a cylinder or prism of the greatest length consistent with its not breaking under its own weight; and having given a greater length, to find the diameter of another cylinder or prism of this greater length which shall be the only and largest one capable of withstanding its own weight.

Let BC be the largest cylinder capable of sustaining its own weight; and let DE be a length greater than AC: the problem is to find the diameter of the cylinder which, having the length DE, shall be the largest one just able to withstand its own weight.

[167]

Let I be a third proportional to the lengths DE and AC; let the diameter FD be to the diameter BA as DE is to I; draw the cylinder FE; then, among all cylinders having the same proportions, this is the largest and only one just capable of sustaining its own weight.

Let M a third proportional to DE and I: also let O be a fourth proportional to DE, I, and M; lay off FG equal to AC. Now since the diameter FD is to the diameter AB as the length DE is to I, and since O is a fourth proportional to DE, I and M, it follows that $FD^3:BA^3=DE:O$. But the resistance [bending strength] of the cylinder DG

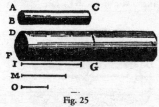

Fig. 25

is to the resistance of the cylinder BC as the cube of FD is to the cube of BA: hence the resistance of the cylinder DG is to that of cylinder BC as the length DE is to O. And since the moment of the cylinder BC is held in equilibrium by [*e equale alla*] its resistance., we shall accomplish our end (which is to prove that the moment of the

cylinder FE is equal to the resistance located at FD), if we show that the moment of the cylinder FE is to the moment of the cylinder BC as the resistance DF is to the resistance BA, that is, as the cube of FD is to the cube of BA, or as the length DE is to O. The moment of the cylinder FE is to the moment of the cylinder DG as the square of DE is to the square of AC, that is, as the length DE is to I; but the moment of the cylinder DG is to the moment of the cylinder BC, as the square of DF is to the square of BA, that is, as the square of DE is to the square of I, or as the square of I is to the square of M, or, as I is to O. Therefore by equating ratios, it results that the moment of the cylinder FE is to the moment of the cylinder BC as the length DE is to O, that is, as the cube of DF is to the cube of BA, or as the resistance of the base DF is to the resistance of the base BA; which was to be proven.

Sagr. This demonstration, Salviati, is rather long and difficult to keep in mind from a single hearing. Will you not, therefore, be good enough to repeat it?

Salv. As you like; but I would suggest instead a more direct and a shorter proof: this will, however, necessitate a different figure.

[168]

Sagr. The favor will be that much greater: nevertheless I hope you will oblige me by putting into written form the argument just given so that I may study it at my leisure.

Salv. I shall gladly do so. Let A denote a cylinder of diameter DC and the largest capable of sustaining its own weight: the problem is to determine a larger cylinder which shall be at once the maximum and the unique one capable of sustaining its own weight.

Let E be such a cylinder, similar to A, having the assigned length, and having a diameter KL. Let MN be a third proportional to the two lengths DC and

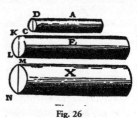

Fig. 26

KL: let MN also be the diameter of another cylinder, X, having the same length as E: then, I say, X is the cylinder sought. Now since the resistance of the base DC is to the resistance of the base KL as the square of DC is to the square of KL, that is, as the square of KL is to the square of MN, or, as the cylinder E is to the cylinder X, that is, as the moment E is to the moment X; and since also the resistance [bending strength] of the base KL is to the resistance of the base MN as the cube of KL is to the cube of MN, that is, as the cube of DC is to the cube of KL, or, as the

cylinder A is to the cylinder E, that is, as the moment of A is to the moment of E; hence it follows, *ex æquali in proportione perturbata*,[19] that the moment of A is to the moment of X as the resistance of the base DC is to the resistance of the base MN; therefore moment and resistance; are related to each other in prism X precisely as they are in prism A.

Let us now generalize the problem; then it will read as follows:

> *Given a cylinder AC in which moment and resistance [bending strength] are related in any manner whatsoever; let DE be the length of another cylinder; then determine what its thickness must be in order that the relation between its moment and resistance shall be identical with that of the cylinder AC.*

Using Fig. 25 in the same manner as above, we may say that, since the moment of the cylinder FE is to the moment of the portion DG as the square of ED is to the square of FG, that is, as the length DE is to I; and since the moment of the cylinder FG is to the moment of the cylinder AC as the square of FD is to the square of AB, or, as the square of ED is to the square of I, or, as the square of I is to the square of M, that is, as the length I is to O; it follows, *ex æquali*, that the moment of the cylinder FE is to

[169]

the moment of the cylinder AC as the length DE is to O, that is, as the cube of DE is to the cube of I, or, as the cube of FD is to the cube of AB, that is, as the resistance of the base FD is to the resistance of the base AB; which was to be proven.

From what has already been demonstrated, you can plainly see the impossibility of increasing the size of structures to vast dimensions either in art or in nature; likewise the impossibility of building ships, palaces, or temples of enormous size in such a way that their oars, yards, beams, iron-bolts, and, in short, all their other parts will hold together; nor can nature produce trees of extraordinary size because the branches would break down under their own weight; so also it would be impossible to build up the bony structures of men, horses, or other animals so as to hold together and perform their normal functions if these animals were to be increased enormously in height; for this increase in height can be accomplished only by employing a material which is harder and stronger than usual, or by enlarging the size of the bones, thus changing their shape until

19. For definition of *perturbata* see Todhunter's *Euclid*, Book V, Def. 20. *[Trans.]*

the form and appearance of the animals suggest a monstrosity. This is perhaps what our wise Poet had in mind, when he says, in describing a huge giant:

> "Impossible it is to reckon his height
> "So beyond measure is his size." [20]

To illustrate briefly, I have sketched a bone whose natural length has been increased three times and whose thickness has been multiplied until, for a correspondingly large animal, it would perform the same function which the small bone performs for its small animal. From the figures here shown you can see how out of proportion the enlarged bone appears. Clearly then if one wishes to maintain in a great giant the same proportion of limb as that found in an ordinary man he must either find a harder and stronger material for making the bones, or he must admit a diminution of

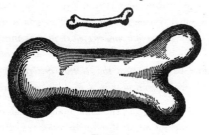

Fig. 27

[170]

strength in comparison with men of medium stature; for if his height be increased inordinately he will fall and be crushed under his own weight. Whereas, if the size of a body be diminished, the strength of that body is not diminished in the same proportion; indeed the smaller the body the greater its relative strength. Thus a small dog could probably carry on his back two or three dogs of his own size; but I believe that a horse could not carry even one of his own size.

Simp. This may be so; but I am led to doubt it on account of the enormous size reached by certain fish, such as the whale which, I understand, is ten times as large as an elephant; yet they all support themselves.

Salv. Your question, Simplicio, suggests another principle, one which had hitherto escaped my attention and which enables giants and other animals of vast size to support themselves and to move about as well as smaller animals do. This result may be secured either by increasing the strength of the bones and other parts intended to carry not only their weight but also the superincumbent load; or, keeping the proportions of the bony structure constant, the skeleton will hold together in the same manner or even more easily, provided one diminishes, in the proper proportion, the weight of the bony material, of the flesh, and of anything else which the skeleton has to carry. It is this

20. *Non si può compartir quanto sia lungo,*
Si smisuratamente è tutto grosso.
Ariosto's *Orlando Furioso*, XVII, 30 *[Trans.]*

second principle which is employed by nature in the structure of fish, making their bones and muscles not merely light but entirely devoid of weight.

Simp. The trend of your argument, Salviati, is evident. Since fish live in water which on account of its density [*corpulenza*] or, as others would say, heaviness [*gravità*] diminishes the weight [*peso*] of bodies immersed in it, you mean to say that, for this reason, the bodies of fish will be devoid of weight and will be supported without injury to their bones. But this is not all; for although the remainder of the body of the fish may be without weight, there can be no question but that their bones have weight. Take the case of a whale's rib, having the dimensions of a beam; who can deny its great weight or its tendency to go to the bottom when placed in water? One would, therefore, hardly expect these great masses to sustain themselves.

[171]

Salv. A very shrewd objection! And now, in reply, tell me whether you have ever seen fish stand motionless at will under water, neither descending to the bottom nor rising to the top, without the exertion of force by swimming?

Simp. This is a well-known phenomenon.

Salv. The fact then that fish are able to remain motionless under water is a conclusive reason for thinking that the material of their bodies has the same specific gravity as that of water; accordingly, if in their make-up there are certain parts which are heavier than water there must be others which are lighter, for otherwise they would not produce equilibrium.

Hence, if the bones are heavier, it is necessary that the muscles or other constituents of the body should be lighter in order that their buoyancy may counterbalance the weight of the bones. In aquatic animals therefore circumstances are just reversed from what they are with land animals inasmuch as, in the latter, the bones sustain not only their own weight but also that of the flesh, while in the former it is the flesh which supports not only its own weight but also that of the bones. We must therefore cease to wonder why these enormously large animals inhabit the water rather than the land, that is to say, the air.

Simp. I am convinced and I only wish to add that what we call land animals ought really to be called air animals, seeing that they live in the air, are surrounded by air, and breathe air.

Sagr. I have enjoyed Simplicio's discussion including both the question raised and its answer. Moreover I can easily understand that one of these giant fish, if pulled ashore, would not perhaps sustain itself for any great length of time, but would be crushed under its own mass as soon as the connections between the bones gave way.

Salv. I am inclined to your opinion; and, indeed, I almost think that the same thing would happen in the case of a very big ship which floats on the sea without going to pieces under its load of merchandise and armament, but which on dry land and in air would probably fall apart. But let us proceed and show how:

[172]

Given a prism or cylinder, also its own weight and the maximum load which it can carry, it is then possible to find a maximum length beyond which the cylinder cannot be prolonged without breaking under its own weight.

Let AC indicate both the prism and its own weight; also let D represent the maximum load which the prism can carry at the end C without fracture; it is required to find the maximum to which the length of the said prism can be increased without breaking. Draw AH of such a length that the weight of the prism AC is to the sum of AC and twice the weight D as the length CA is to AH; and let AG be a mean proportional between CA and AH; then, I say, AG is the length sought. Since the moment of

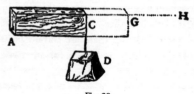

Fig. 28

the weight [*momento gravante*] D attached at the point C is equal to the moment of a weight twice as large as D placed at the middle point AC, through which the weight of the prism AC acts, it follows that the moment of the resistance of the prism AC located at A is equivalent to twice the weight D plus the weight of AC, both acting through the middle point of AC. And since we have agreed that the moment of the weights thus located, namely, twice D plus AC, bears to the moment of AC the same ratio which the length HA bears to CA and since AG is a mean proportional between these two lengths, it follows that the moment of twice D plus AC is to the moment of AC as the square of GA is to the square of CA. But the moment arising from the weight [*momento premente*] of the prism GA is to the moment of AC as the square of GA is to the square of CA; thence AG is the maximum length sought, that is, the length up to which the prism AC may be prolonged and still support itself, but beyond which it will break.

Hitherto we have considered the moments and resistances of prisms and solid cylinders fixed at one end with a weight applied at the other end; three cases were discussed, namely, that in which the applied force was the only one acting, that in which the weight of the prism itself is also taken into consideration, and that in which the weight of the prism alone is taken into consideration. Let us now consider these same

[173]

prisms and cylinders when supported at both ends or at a single point placed some-where between the ends.

In the first place, I remark that a cylinder carrying only its own weight and hav-ing the maximum length, beyond which it will break, will, when supported either in the middle or at both ends, have twice the length of one which is mortised into a wall and supported only at one end. This is very evident because, if we denote the cylinder by ABC and if we assume that one-half of it, AB, is the greatest possible length capable of supporting its own weight with one end fixed at B, then, for the same

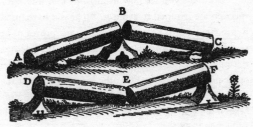

Fig. 29

reason, if the cylinder is carried on the point G, the first half will be counterbalanced by the other half BC. So also in the case of the cylinder DEF, if its length be such that it will support only one-half this length when the end D is held fixed, or the other half when the end F is fixed, then it is evident that when supports, such as H and I, are placed under the ends D and F respectively the moment of any additional force or weight placed at E will produce fracture at this point.

A more intricate and difficult problem is the following: neglect the weight of a solid such as the preceding and find whether the same force or weight which produces fracture when applied at the middle of a cylinder, supported at both ends, win also break the cylinder when applied at some other point nearer one end than the other.

Thus, for example, if one wished to break a stick by holding it with one hand at each end and applying his knee at the middle, would the same force be required to break it in the same manner if the knee were applied, not at the middle, but at some point nearer to one end?

Sagr. This problem, I believe, has been touched upon by Aristotle in his *Questions in Mechanics*.

[174]

Salv. His inquiry however is not quite the same; for he seeks merely to discover why it is that a stick may be more easily broken by taking hold, one hand at each end of the stick, that is, far removed from the knee, than if the hands were closer together. He gives a general explanation, referring it to the lengthened lever arms which are

secured by placing the hands at the ends of the stick. Our inquiry calls for something more: what we want to know is whether, when the hands are retained at the ends of the stick, the same force is required to break it wherever the knee be placed.

Sagr. At first glance this would appear to be so, because the two lever arms exert, in a certain way, the same moment, seeing that as one grows, shorter the other grows correspondingly longer.

Salv. Now you see how readily one falls into error and what caution and circumspection are required to avoid it. What you have just said appears at first glance highly probable, but on closer examination it proves to be quite far from true; as will be seen from the fact that whether the knee—the fulcrum of the two levers—be placed in the middle or not makes such a difference that, if fracture is to be produced at any other point than the middle, the breaking force at the middle, even when multiplied four, ten, a hundred, or a thousand times would not suffice. To begin with we shall offer some general considerations and then pass to the determination of the ratio in which the breaking force must change in order to produce fracture at one point rather than another.

Let AB denote a wooden cylinder which is to be broken in the middle, over the supporting point C, and let DE represent an identical cylinder which is to be broken

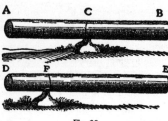

Fig. 30

just over the supporting point F which is not in the middle. First of all it is clear that, since the distances AC and CB are equal, the forces applied at the extremities B and A must also be equal. Secondly since the distance DF is less than the distance AC the moment of any force acting at D is less than the moment of the same force at A, that is, applied at the distance CA; and the moments are less in the ratio of the length DF to AC; consequently it is necessary to increase the force [*momento*] at D in order to overcome, or even to balance, the resistance at F; but in comparison with the length AC the distance DF can be diminished indefinitely: in order therefore to counterbalance the resistance at F it will be necessary to increase indefinitely the force [*forza*] applied at D.

[175]

On the other hand, in proportion as we increase the distance FE over that of CB, we must diminish the force at E in order to counterbalance the resistance at F; but the distance FE, measured in terms of CB, cannot be increased indefinitely by sliding the fulcrum F toward the end D; indeed, it cannot even be made double the length CB.

Therefore the force required at E to balance the resistance at F will always be more than half that required at B. It is clear then that, as the fulcrum F approaches the end D, we must of necessity indefinitely increase the sum of the forces applied at E and D in order to balance, or overcome, the resistance at F.

Sagr. What shall we say, Simplicio? Must we not confess that geometry is the most powerful of all instruments for sharpening the wit and training the mind to think correctly? Was not Plato perfectly right when he wished that his pupils should be first of all well grounded in mathematics? As for myself, I quite understood the property of the lever and how, by increasing or diminishing its length, one can increase or diminish the moment of force and of resistance; and yet, in the solution of the present problem I was not slightly but greatly, deceived.

Simp. Indeed I begin to understand that while logic is an excellent guide in discourse, it does not, as regards stimulation to discovery, compare with the power of sharp distinction which belongs to geometry.

Sagr. Logic, it appears to me, teaches us how to test the conclusiveness of any argument or demonstration already discovered and completed; but I do not believe that it teaches us to discover correct arguments and demonstrations. But it would be better if Salviati were to show us in just what proportion the forces must be increased in order to produce fracture as the fulcrum is moved from one point to another along one and the same wooden rod.

[176]

Salv. The ratio which you desire is determined as follows:

> *If upon a cylinder one marks two points at which fracture is to be produced, then the resistances at these two points will bear to each other the inverse ratio of the rectangles formed by the distances from the respective points to the ends of the cylinder.*

Let A and B denote the least forces which will bring about fracture of the cylinder at C; likewise E and F the smallest forces which will break it at D. Then, I say, that the sum of the forces A and B is to the sum of the forces E and F as the area of the rectangle AD.DB is to the area of the rectangle AC.CB. Because the sum of the forces A and B bears to the sum of the forces E and F a ratio which is the product of the three following ratios, namely, (A+B)/B, B/F, and

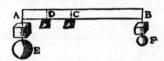

Fig. 31

F/(F+E); but the length BA is to the length CA as the sum of the forces A and B is to

the force B; and, as the length DB is to the length CB, so is the force B to the force F; also as the length AD is to AB, so is the force F to the sum of the forces F and E.

Hence it follows that the sum of the forces A and B bears to the sum of the forces E and F a ratio which is the product of the three following ratios, namely, BA/CA, BD/BC, and AD/AB. But DA/CA is the product of DA/BA and BA/CA. Therefore the sum of the forces A and B bears to the sum of the forces E and F a ratio which is the product of DA:CA and DB:CB. But the rectangle AD.DB bears to the rectangle AC.CB a ratio which is the product of DA/CA and DB/CB. Accordingly the sum of the forces A and B is to the sum of the forces E and F as the rectangle AD.DB is to the rectangle AC.CB, that is, the resistance to fracture at C is to the resistance to fracture at D as the rectangle AD.DB is to the rectangle AC.CB. Q. E. D.

[177]

Another rather interesting problem may be solved as a consequence of this theorem, namely,

> *Given the maximum weight which a cylinder or prism can support at its middle-point where the resistance is a minimum, and given also a larger weight, find that point in the cylinder for which this Larger weight is the maximum load that can be supported.*

Let that one of the given weights which is larger than the maximum weight supported at the middle of the cylinder AB bear to this maximum weight the same ratio which the length E bears to the length F. The problem is to find that point in the cylinder at which this larger weight becomes the maximum that can be supported. Let G be

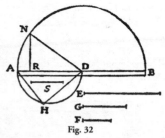

Fig. 32

a mean proportional between the lengths E and F. Draw AD and S so that they bear to each other the same ratio as E to G; accordingly S will be less than AD.

Let AD be the diameter of a semicircle AHD, in which take AH equal to S; join the points H and D and lay off DR equal to HD. Then, I say, R is the point sought, namely, the point at which the given weight, greater than the maximum supported at the middle of the cylinder D, would become the maximum load.

On AB as diameter draw the semicircle ANB: erect the perpendicular RN and join the points N and D. Now since the sum of the squares on NR and RD is equal to the

square of ND, that is, to the square of AD, or to the sum of the squares of AH and HD; and, since the square of HD is equal to the square of DR, it follows that the square of NR, that is, the rectangle AR.RB, is equal to the square of AH, also therefore to the square of S; but the square of S is to the square of AD as the length F is to the length E, that is, as the maximum weight supported at D is to the larger of the two given weights. Hence the latter will be the maximum load which can be carried at the point R; which is the solution sought.

Sagr. Now I understand thoroughly; and I am thinking that, since the prism AB grows constantly stronger and more resistant to the pressure of its load at points which are more and more removed from the middle, we could in the case of large heavy beams cut away a considerable portion near the ends which would notably lessen the weight, and which, in the beam work of large rooms, would prove to be of great utility and convenience.

[178]

It would be a fine thing if one could discover the proper shape to give a solid in order to make it equally resistant at every point, in which case a load placed at the middle would not produce fracture more easily than if placed at any other point.[21]

Salv. I was just on the point of mentioning an interesting and remarkable fact connected with this very question. My meaning will be clearer if I draw a figure. Let DB represent a prism; then, as we have already shown, its resistance to fracture [bending strength] at the end AD, owing to a load placed at the end B, will be less than the resistance at CI in the ratio of the length CB to AB. Now imagine this same prism to be cut through diagonally along the line FB so that the opposite faces will be triangular; the side facing us will be FAB. Such a solid will have properties different from those of the prism; for, if the load remain at B, the resistance against fracture [bending strength] at C will be less than that at A in the ratio of the length CB to the length AB. This is easily proved: for if CNO represents a cross-section parallel to AFD, then the length FA bears to the length CN, in the triangle FAB, the same ratio which the length AB bears to the length CB. Therefore, if we imagine A and C to be the points at which the fulcrum is placed, the lever arms in the two cases BA, AF and BC, CN will be proportional [*simili*]. Hence the moment of any force applied at B and

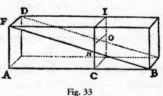

Fig. 33

21. The reader will notice that two different problems are here involved. That which is suggested in the last remark of Sagredo is the following:
To find a beam whose maximum stress has the same value when a constant load moves from one end of the beam to the other.
The second problem—the one which Salviati proceeds to solve—is the following:
To find a beam in all cross-sections of which the maximum stress is the same for a constant load in a fixed position. [*Trans.*]

acting through the arm BA, against a resistance placed at a distance AF will be equal to that of the same force at B acting through the arm BC against the same resistance located at a distance CN. But now, if the force still be applied at B, the resistance to be overcome when the fulcrum is at C, acting through the arm CN, is less than the resistance with the fulcrum at A in the same proportion as the rectangular cross-section CO is less than the rectangular cross-section AD, that is, as the length CN is less than AF, or CB than BA.

Consequently the resistance to fracture at C, offered by the portion OBC, is less than the resistance to fracture at A, offered by the entire block DAB, in the same proportion as the length CB is smaller than the length AB.

By this diagonal saw-cut we have now removed from the beam, or prism DB, a portion, i. e., a half, and have left the wedge, or triangular prism, FBA. We thus have two solids possessing opposite properties; one body grows stronger as it is shortened

[179]

while the other grows weaker. This being so it would seem not merely reasonable, but inevitable, that there exists a line of section such that, when the superfluous material his been removed, there will remain a solid of such figure that it will offer the same resistance [strength] at all points.

Simp. Evidently one must, in passing from greater to less, encounter equality.

Sagr. But now the question is what path the saw should follow in making the cut.

Simp. It seems to me that this ought not to be a difficult task: for if by sawing the prism along the diagonal line and removing half of the material, the remainder acquires a property just the opposite to that of the entire prism, so that at every point where the latter gains strength the former becomes weaker, then it seems to me that by taking a middle path, i. e., by removing half the former half, or one-quarter of the whole, the strength of the remaining figure will be constant at a those points where, in the two previous figures, the gain in one was equal to the loss in the other.

Salv. You have missed the mark, Simplicio. For, as I shall presently show you, the amount which you can remove from the prism without weakening it is not a quarter but a third. It now remains, as suggested by Sagredo, to discover the path along which the saw must travel: this, as I shall prove, must be a parabola. But it is first necessary to demonstrate the following lemma:

> *If the fulcrums are so, placed under two levers or balances that the arms through which the forces act are to each other in the same ratio as the squares of the arms through which the resistances act, and if these resistances are to each other in the same ratio as the arms through which they act, then the forces will be equal.*

[180]

Let AB and CD represent two levers whose lengths are divided by their fulcrums in such a way as to make the distance EB bear to the distance FD a ratio which is equal to the square of the ratio between the distances EA and FC. Let the resistances located at A and C be to each other as EA is to FC. Then, I say, the forces which must be applied at B and D in order to hold in equilibrium the

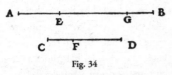

Fig. 34

resistances at A and C are equal. Let EG be a mean proportional between EB and FD. Then we shall have BE:EG=EG:FD=AE:CF. But this last ratio is precisely that which we have assumed to exist between the resistances at A and C. And since EG:FD=AE:CF, it follows, *permutando*, that EG:AE=FD:CF. Seeing that the distances DC and GA are divided in the same ratio by the points F and E, it follows that the same force which, when applied at D, will equilibrate the resistance at C, would if applied at G equilibrate at A a resistance equal to that found at C.

But one datum of the problem is that the resistance at A is to the resistance at C as the distance AE is to the distance CF, or as BE is to EG. Therefore the force applied at G, or rather at D, will, when applied at B, just balance the resistance located at A. Q. E. D.

This being clear draw the parabola FNB in the face FB of the prism DB. Let the prism be sawed along this parabola whose vertex is at B. The portion of the solid which

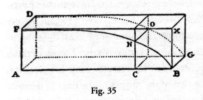

Fig. 35

remains will be included between the base AD, the rectangular plane AG, the straight line BG and the surface DGBF, whose curvature is identical with that of the parabola FNB. This solid will have, I say, the same strength at every point. Let the solid be cut by a plane CO parallel to the plane AD. Imagine the points A and C to be the fulcrums of two levers of which one will have the arms BA and AF; the other BC and CN. Then since in the parabola FBA, we have BA:BC=AF^2:CN^2, it is clear that the arm BA of one lever is to the arm BC of the other lever as the square of the arm AF is to the square of the other arm CN. Since the resistance to be balanced by the lever BA is to the resistance to be balanced by the lever BC in the same ratio as the rectangle DA is to the rectangle OC, that is as the length AF is to the length CN, which two lengths are the other arms of the levers, it follows, by the lemma just demonstrated, that the same force which, when applied at BG will equilibrate the resistance at DA, will also balance the resistance at CO. The same is true for any other section. Therefore this parabolic solid is equally strong throughout.

[181]

It can now be shown that, if the prism be sawed along the line of the parabola FNB, one-third part of it will be removed; because the rectangle FB and the surface FNBA bounded by the parabola are the bases of two solids included between two parallel planes, i. e., between the rectangles FB and DG; consequently the volumes of these two solids bear to each other the same ratio as their bases. But the area of the rectangle is one and a half times as large as the area FNBA under the parabola; hence by cutting the prism along the parabola we remove one-third of the volume. It is thus seen how one can diminish the weight of a beam by as much as thirty-three per cent without diminishing its strength; a fact of no small utility in the construction of large vessels, and especially in supporting the decks, since in such structures lightness is of prime importance.

Sagr. The advantages derived from this fact are so numerous that it would be both wearisome and impossible to mention them all; but leaving this matter to one side, I should like to learn just how it happens that diminution of weight is possible in the ratio above stated. I can readily understand that, when a section is made along the diagonal, one-half the weight is removed; but, as for the parabolic section removing one-third of the prism, this I can only accept on the word of Salviati who is always reliable; however I prefer first-hand knowledge to the word of another.

Salv. You would like then a demonstration of the fact that the excess of the volume of a prism over the volume of what we have called the parabolic solid is one-third of the entire prism. This I have already given you on a previous occasion; however I shall now try to recall the demonstration in which I remember having used a certain lemma from Archimedes' book *On Spirals*,[22] namely, given any number of lines, differing in length one from another by a common difference which is equal to the shortest of these lines; and given also an equal number of lines each of which has the same length as the longest of the first mentioned series; then the sum of the squares of the lines of this second group will be less than three times the sum of the squares of the lines in the first group. But the sum of the squares of the second group will be greater than three times the sum of the squares of all excepting the longest of the first group.

[182]

Assuming this, inscribe in the rectangle ACBP the parabola AB. We have now to prove that the mixed triangle BAP whose sides are BP and PA, and whose base is the parabola BA, is a third part of the entire rectangle CP. If this is not true it will be either greater or less than a third. Suppose it to be less by an area which is represented by X.

22. For demonstration of the theorem here cited, see "*Works of Archimedes*" translated by T. L. Heath (Camb. Univ. Press 1897) p. 107 and p. 162. *[Trans.]*

By drawing lines parallel to the sides BP and CA, we can divide the rectangle CP into equal parts; and if the process be continued we shall finally reach a division into parts so small that each of them will be smaller than the area X; let the rectangle OB represent one of these parts and, through the points where the other parallels cut the parabola, draw lines parallel to AP. Let us now describe about our "mixed triangle" a figure made up of rectangles such as BO, IN, HM, FL, EK, and GA; this figure will also be less than a third part of the rectangle CP because the excess of this figure above the area of

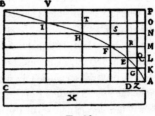

Fig. 36

the "mixed triangle" is much smaller than the rectangle BO which we have already made smaller than X.

Sagr. More slowly, please; for I do not see how the excess of this figure described about the "mixed triangle" is much smaller than the rectangle BO.

Salv. Does not the rectangle BO have an area which is equal to the sum of the areas of all the little rectangles through which the parabola passes? I mean the rectangles BI, IH, HF, FE, EG, and GA of which only a part lies outside the "mixed triangle." Have we not taken the rectangle BO smaller than the area X? Therefore if, as our opponent might say, the triangle plus X is equal to a third part of this rectangle CP, the circumscribed figure, which adds to the triangle an area less than X, will still remain smaller than a third part of the rectangle, CP. But this cannot be, because this circumscribed figure is larger than a third of the area. Hence it is not true that our "mixed triangle" is less than a third of the rectangle.

[183]

Sagr. You have cleared up my difficulty; but it still remains to be shown that the circumscribed figure is larger than a third part of the rectangle CP, a task which will not, I believe, prove so easy.

Salv. There is nothing very difficult about it. Since in the parabola $DE^2:ZG^2=DA:AZ=$ rectangle KE: rectangle AG, seeing that the altitudes of these two rectangles, AK and KL, are equal, it follows that $ED^2:ZG^2=LA^2:AK^2=$ rectangle KE: rectangle KZ. In precisely the same manner it may be shown that the other rectangles LF, MH, NI, OB, stand to one another in the same ratio as the squares of the lines MA, NA, OA, PA.

Let us now consider the circumscribed figure, composed of areas which bear to each other the same ratio as the squares of a series of lines whose common difference in length is equal to the shortest one in the series; note also that the rectangle CP is

made up of an equal number of areas each equal to the largest and each equal to the rectangle OB. Consequently, according to the lemma of Archimedes, the circumscribed figure is larger than a third part of the rectangle CP; but it was also smaller, which is impossible. Hence the "mixed triangle" is not less than a third part of the rectangle CP.

Likewise, I say, it cannot be greater. For, let us suppose that it is greater than a third part of the rectangle CP and let the area X represent the excess of the triangle over the third part of the rectangle CP; subdivide the rectangle into equal rectangles and continue the process until one of these subdivisions is smaller than the area X. Let BO represent such a rectangle smaller than X. Using the above figure, we have in the "mixed triangle" an inscribed figure, made up of the rectangles VO, TN, SM, RL, and QK, which will not be less than a third part of the large rectangle CP.

For the "mixed triangle" exceeds the inscribed figure by a quantity less than that by which it exceeds the third part of the rectangle CP; to see that this is true we have only to remember that the excess of the triangle over the third part of the rectangle CP is equal to the area X, which is less than the rectangle BO, which in turn is much less than the excess of the triangle over the inscribed figure. For the rectangle BO is made

[184]

up of the small rectangles AG, GF, EF, FH, HI, and IB; and the excess of the triangle over the inscribed figure is less than half the sum of these little rectangles. Thus since the triangle exceeds the third part of the rectangle CP by an amount X, which is more than that by which it exceeds the inscribed figure, the latter will also exceed the third part of the rectangle, CP. But, by the lemma which we have assumed, it is smaller. For the rectangle CP, being the sum of the largest rectangles, bears to the component rectangles of the inscribed figure the same ratio which the sum of all the squares of the lines equal to the longest bears to the squares of the lines which have a common difference, after the square of the longest has been subtracted.

Therefore, as in the case of squares, the sum total of the largest rectangles, i. e., the rectangle CP, is greater than three times the sum total of those having a common difference minus the largest; but these last make up the inscribed figure. Hence the "mixed triangle" is neither greater nor less than the third part of rectangle CP; it is therefore equal to it.

Sagr. A fine, clever demonstration; and all the more so because it gives us the quadrature of the parabola, proving it to be four-thirds of the inscribed[23] triangle, a fact which Archimedes demonstrates by means of two different, but admirable, series of many propositions. This same theorem has also been recently established

23. Distinguish carefully between this triangle and the "mixed triangle" above mentioned. *[Trans.]*

510

by Luca Valerio,[24] the Archimedes of our age; his demonstration is to be found in his book dealing with the centers of gravity of solids.

Salv. A book which, indeed, is not to be placed second to any produced by the most eminent geometers either of the present or of the past; a book which, as soon as it fell into the hands of our Academician, led him to abandon his own researches along these lines; for he saw how happily everything had been treated and demonstrated by Valerio.

[185]

Sagr. When I was informed of this event by the Academician himself, I begged of him to show the demonstrations which he had discovered before seeing Valerio's book; but in this I did not succeed.

Salv. I have a copy of them and will show them to you; for you will enjoy the diversity of method employed by these two authors in reaching and proving the same conclusions; you will also find that some of these conclusions are explained in different ways, although both are in fact equally correct.

Sagr. I shall be much pleased to see them and will consider it a great favor if you will bring them to our regular meeting. But in the meantime, considering the strength of a solid formed from a prism by means of a parabolic section, would it not, in view of the fact that this result promises to be both interesting and useful in many mechanical operations, be a fine thing if you were to give some quick and easy rule by which a mechanician might draw a parabola upon a plane surface?

Salv. There are many ways of tracing these curves; I will mention merely the two which are the quickest of all. One of these is really remarkable; because by it I can trace thirty or forty parabolic curves with no less neatness and precision, and in a shorter time than another man can, by the aid of a compass, neatly draw four or six circles of different sizes upon paper. I take a perfectly round brass ball about the size of a walnut and project it along the surface of a metallic mirror held in a nearly upright position, so that the ball in its motion will press slightly upon the mirror and trace out a fine sharp parabolic line; this parabola will grow longer and narrower as the angle of elevation increases. The above experiment furnishes clear and tangible evidence that the path of a projectile is a parabola; a fact first observed by our friend and demonstrated by him in his book on motion which we shall take up at our next meeting. In the execution of this method, it is advisable to slightly heat and moisten the ball by rolling in the hand in order that its trace upon the minor may be more distinct.

24. An eminent Italian mathematician, contemporary with Galileo. *[Trans.]*

[186]

The other method of drawing the desired curve upon the face of the prism is the following: Drive two nails into a wall at a convenient height and at the same level; make the distance between these nails twice the width of the rectangle upon which it is desired to trace the semiparabola. Over these two nails hang a light chain of such a length that the depth of its sag is equal to the length of the prism. This chain will assume the form of a parabola,[25] so that if this form be marked by points on the wall we shall have described a complete parabola which can be divided into two equal parts by drawing a vertical line through a point midway between the two nails. The transfer of this curve to the two opposing faces of the prism is a matter of no difficulty; any ordinary mechanic will know how to do it.

By use of the geometrical lines drawn upon our friend's compass,[26] one may easily lay off those points which will locate this same curve upon the same face of the prism.

Hitherto we have demonstrated numerous conclusions pertaining to the resistance which solids offer to fracture. As a starting point for this science, we assumed that the resistance offered by the solid to a straight-away pull was known; from this base one might proceed to the discovery of many other results and their demonstrations; of these results the number to be found in nature is infinite. But, in order to bring our daily conference to an end, I wish to discuss the strength of hollow solids, which are employed in art—and still oftener in nature—in a thousand operations for the purpose of greatly increasing strength without adding to weight; examples of these are seen in the bones of birds and in many kinds of reeds which are light and highly resistant both to bending and breaking. For if a stem of straw which carries a head of wheat heavier than the entire stalk were made up of the same amount of material in solid form, it would

[187]

offer less resistance to bending and breaking. This is an experience which has been verified and confirmed in practice where it is found that a hollow lance or a tube of wood or metal is much stronger than would be a solid one of the same length and weight, one which would necessarily be thinner; men have discovered, therefore, that in order to make lances strong as well as light they must make them hollow. We shall now show that:

> In the case of two cylinders, one hollow the other solid but having
> equal volumes and equal lengths, their resistances [bending strengths]
> are to each other in the ratio of their diameters.

25. It is now well known that this curve is not a parabola but a catenary the equation of which was first given, 49 years after Galileo's death, by James Bernoulli. *[Trans.]*
26. The geometrical and military compass of Galileo, described in Nat. Ed. Vol. 2. *[Trans.]*

Let AE denote a hollow cylinder and IN a solid one of the same weight and length; then, I say, that the resistance against fracture exhibited by the tube AE bears to that of the solid cylinder IN the same ratio as the diameter AB to the diameter IL. This is very evident; for since the tube and the solid cylinder IN have the same volume and length, the area of the circular base IL will

Fig. 37

be equal to that of the annulus AB which is the base of the tube AE. (By annulus is here meant the area which lies between two concentric circles of different radii.) Hence their resistances to a straight-away pull are equal; but in producing fracture by a transverse pull we employ, in the case of the cylinder IN, the length LN as one lever arm, the point L as a fulcrum, and the diameter LI, or its half, as the opposing lever arm: while in the case of the tube, the length BE which plays the part of the first lever arm is equal to LN, the opposing lever arm beyond the fulcrum, B, is the diameter AB, or its half. Manifestly then the resistence [bending strength] of the tube exceeds that of the solid cylinder in the proportion in which the diameter AB exceeds the diameter IL which is the desired result.

[188]

Thus the strength of a hollow tube exceeds that of a solid cylinder in the ratio of their diameters whenever the two are made of the same material and have the same weight and length.

It may be well next to investiage the general case of tubes and solid cylinders of constant length, but with the weight and the hollow portion variable. First we shall show that:

> *Given a hollow tube, a solid cylinder may be determined which will be equal [eguale] to it.*

The method is very simple. Let AB denote the external and CD the internal diameter of the tube. In the larger circle lay off the line AE equal in length to the diameter CD; join the points E and B. Now since the angle at E inscribed in a semicircle, AEB, is a right-angle, the area of the circle whose diameter is AB is equal to the sum of the areas of the two circles whose respective diameters are AE and EB. But AE is the diameter of the hollow portion of the tube. Therefore the area of the circle whose diameter is EB is the same as the area of the annulus ACBD. Hence a solid cylinder of circular base having a diamter EB will have the same volume as the walls of the tube of equal length.

By use of this theorem, it is easy:

To find the ratio between the resistance [bending strength] of any tube and that of any cylinder of equal length.

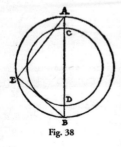

Fig. 38

Let ABE denote a tube and RSM a cylinder of equal length: it is required to find the ratio between their resistances. Using the preceding proposition, determine a cylinder ILN which shall have the same volume and length as the tube. Draw a line V of such a length that it will be related to IL and RS (diameters of the bases of the cylinders IN and RM), as follows: V:RS=RS:IL. Then, I say, the resistance of the tube AE is to that of the cylinder RM as the length of the line AB is to the length

[189]

V. For, since the tube AE is equal both in volume and length, to the cylinder IN, the resistance of the tube will bear to the resistance of the cylinder the same ratio as the line AB to IL; but the resistance of the cylinder IN is to that of the cylinder RM as the cube of IL is to the cube of RS, that is, as the length IL is to length V: therefore, *ex æquali*, the resistance [bending strength] of the tube AE bears to the resistance of the cylinder RM the same ratio as the length AB to V.　Q.E.D.

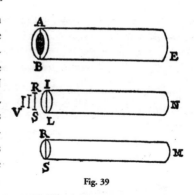

Fig. 39

END OF SECOND DAY.

[190]

THIRD DAY

CHANGE OF POSITION. [De Motu Locali]

My purpose is to set forth a very new science dealing with a very ancient subject. There is, in nature, perhaps nothing older than motion, concerning which the books

written by philosophers are neither few nor small; nevertheless I have discovered by experiment some properties of it which are worth knowing and which have not hitherto been either observed or demonstrated. Some superficial observations have been made, as, for instance, that the free motion [*naturalem motum*] of a heavy falling body is continuously accelerated;[27] but to just what extent this acceleration occurs has not yet been announced; for so far as I know, no one bas yet pointed out that the distances traversed, during equal intervals of time, by a body falling from rest, stand to one another in the same ratio as the odd numbers beginning with unity.[28]

It has been observed that missiles and projectiles describe a curved path of some sort; however no one has pointed out the fact that this path is a parabola. But this and other facts, not few in number or less worth knowing, I have succeeded in proving; and what I consider more important, there have been opened up to this vast and most excellent science, of which my work is merely the beginning, ways and means by which other minds more acute than mine will explore its remote corners.

This discussion is divided into three parts; the first part deals with motion which is steady or uniform; the second treats of motion as we find it accelerated in nature; the third deals with the so-called violent motions and with projectiles.

[191]

UNIFORM MOTION

In dealing with steady or uniform motion, we need a single definition which I give as follows:

DEFINITION

By steady or uniform motion, I mean one in which the distances traversed by the moving particle during any equal intervals of time, are themselves equal.

CAUTION

We must add to the old definition (which defined steady motion simply as one in which equal distances are traversed in equal times) the word "any," meaning by this, all equal intervals of time; for it may happen that the moving body will traverse equal distances during some equal intervals of time and yet the distances traversed during some small portion of these time-intervals may not be equal, even though the time-intervals be equal.

27. "Natural motion" of the author has here been translated into "free motion"-since this is the term used to-day to distinguish the "natural" from the "violent" motions of the Renaissance. *[Trans.]*
28. A theorem demonstrated below. *[Trans.]*

From the above definition, four axioms follow, namely:

AXIOM I

In the case of one and the same uniform motion, the distance traversed during a longer interval of time is greater than the distance traversed during a shorter interval of time.

AXIOM II

In the case of one and the same uniform motion, the time required to traverse a greater distance is longer than the time required for a less distance.

AXIOM III

In one and the same interval of time, the distance traversed at a greater speed is larger than the distance traversed at a less speed.

[192]

AXIOM IV

The speed required to traverse a longer distance is greater than that required to traverse a shorter distance during the same time-interval.

THEOREM I, PROPOSITION I

If a moving particle, carried uniformly at a constant speed, traverses two distances the time-intervals required are to each other in the ratio of these distances.

Let a particle move uniformly with constant speed through two distances AB, BC, and let the time required to traverse AB be represented by DE; the time required

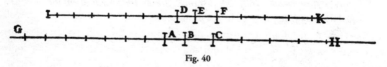

Fig. 40

to traverse BC, by EF; then I say that the distance AB is to the distance BC as the time DE is to the time EF.

Let the distances and times be extended on both sides towards G, H and I, K; let AG be divided into any number whatever of spaces each equal to AB, and in like manner lay off in DI exactly the same number of time-intervals each equal to DE. Again lay off in CH any number whatever of distances each equal to BC; and in FK exactly the same number of time-intervals each equal to EF; then will the distance BG and the time EI be equal and arbitrary multiples of the distance BA and the time ED; and likewise the distance HB and the time KE are equal and arbitrary multiples of the distance CB and the time FE.

And since DE is the time required to traverse AB, the whole time EI will be required for the whole distance BG, and when the motion is uniform there will be in EI as many time-intervals each equal to DE as there are distances in BG each equal to BA; and likewise it follows that KE represents the time required to traverse HB.

Since, however, the motion is uniform, it follows that if the distance GB is equal to the distance BH, then must also the time IE be equal to the time EK; and if GB is greater than BH, then also IE will be greater than EK; and if less, less.[29] There are then four quantities, the first AB, the second BC, the third DE, and the fourth EF; the time IE and the distance GB are arbitrary multiples of the first and the third, namely of the distance AB and the time DE.

[193]

But it has been proved that *both* of these latter quantities are either equal to, greater than, or less than the time EK and the space BH, which are arbitrary multiples of the second and the fourth. Therefore the first is to the second, namely the distance AB is to the distance BC, as the third is to the fourth, namely the time DE is to the time EF.

Q.E.D.

THEOREM II, PROPOSITION II

If a moving particle traverses two distances in equal intervals of time, these distances will bear to each other the same ratio as the speeds. And conversely if the distances are as the speeds then the times are equal.

Referring to Fig. 40, let AB and BC represent the two distances traversed in equal time-intervals, the distance AB for instance with the velocity DE, and the distance BC with the velocity EF. Then, I say, the distance AB is to the distance BC as the velocity DE is to the velocity EF. For if equal multiples of both distances and speeds be taken,

<hr>

29. The method here employed by Galileo is that of Euclid as set forth in the famous 5th Definition of the Fifth Book of his *Elements*, for which see *art. Geometry* Ency. Brit. 11th Ed. p. 683. *[Trans.]*

as above, namely, GB and IE of AB and DE respectively, and in like manner HB and KE of BC and EF, then one may infer, in the same manner as above, that the multiples GB and IE are either less than, equal to, or greater than equal multiples of BH and EK. Hence the theorem is established.

THEOREM III, PROPOSITION III

In the case of unequal speeds, the time-intervals required to traverse a given space are to each other inversely as the speeds.

Let the larger of the two unequal speeds be indicated by A; the smaller, by B; and let the motion corresponding to both traverse the given space CD. Then I say the time required to traverse the distance CD at speed A is to the time required to traverse the same distance at speed B,

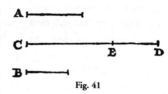

Fig. 41

as the speed B is to the speed A. For let CD be to CE as A is to B; then, from the preceding, it follows that the time required to complete the distance CD at speed A is the same as

[194]

the time necessary to complete CE at speed B; but the time needed to traverse the distance CE at speed B is to the time required to traverse the distance CD at the same speed as CE is to CD; therefore the time in which CD is covered at speed A is to the time in which CD is covered at speed B as CE is to CD, that is, as speed B is to speed A.   Q. E. D.

THEOREM IV, PROPOSITION IV

If two particles are carried with uniform motion, but each with a different speed, the distances covered by them during unequal intervals of time bear to each other the

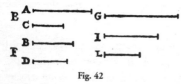

Fig. 42

compound ratio of the speeds and time intervals.

Let the two particles which are carried with uniform motion be E and F and let the ratio of the speed of the body E be to that of the body F as A is to B; but let the ratio of the time consumed by the motion of E be to the time consumed by the motion of F as C is to D. Then, I say, that the distance covered by E, with speed A in time C, bears to

518

the space traversed by F with speed B in time D a ratio which is the product of the ratio of the speed A to the speed B by the ratio of the time C to the time D. For if G is the distance traversed by E at speed A during the time-interval C, and if G is to I as the speed A is to the speed B; and if also the time-interval C is to the time-interval D as I is to L, then it follows that I is the distance traversed by F in the same time that G is traversed by E since G is to I in the same ratio as the speed A to the speed B. And since I is to L in the same ratio as the time-intervals C and D, if I is the distance traversed by F during the interval C, then L will be the distance traversed by F during the interval D at the speed B.

But the ratio of G to L is the product of the ratios G to I and I to L, that is, of the ratios of the speed A to the speed B and of the time-interval C to the time-interval D.   Q. E. D.

[195]

THEOREM V, PROPOSITION V

If two particles are moved at a uniform rate, but with unequal speeds, through unequal distances, then the ratio of the time-intervals occupied will be the product of the ratio of the distances by the inverse ratio of the speeds.

Let the two moving particles be denoted by A and B, and let the speed of A be to the speed of B in the ratio of V to T; in like manner let the distances traversed be in the ratio of S to R; then I say that the ratio of the time-interval during which the motion of A occurs to the time-interval occupied by the motion of B is the product of the ratio of the speed T to the speed V by the ratio of the distance S to the distance R.

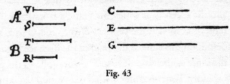

Fig. 43

Let C be the time-interval occupied by the motion of A, and let the time-interval C bear to a time-interval E the same ratio as the speed T to the speed V.

And since C is the time-interval during which A, with speed V, traverses the distance S and since T, the speed of B, is to the speed V, as the time-interval C is to the time-interval E, then E will be the time required by the particle B to traverse the distance S. If now we let the time-interval E be to the time-interval G as the distance S is to the distance R, then it follows that G is the time required by B to traverse the space R. Since the ratio of C to G is the product of the ratios C to E and E to G (while also

the ratio of C to E is the inverse ratio of the speeds of A and B respectively, i. e., the ratio of T to V); and since the ratio of E to G is the same as that of the distances S and R respectively, the proposition is proved.

[196]

THEOREM VI, PROPOSITION VI

If two particles are carried at a uniform rate, the ratio of their speeds will be the product of the ratio of the distances traversed by the inverse ratio of the time-intervals occupied.

Let A and B be the two particles which move at a uniform rate; and let the respective distances traversed by them have the ratio of V to T, but let the time-intervals be as S to R. Then I say the speed of A will bear to the speed of B a ratio which is the product

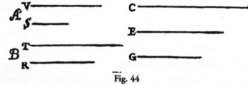

Fig. 44

of the ratio of the distance V to the distance T and the time-interval R to the time-interval S.

Let C be the speed at which A traverses the distance V during the time-interval S; and let the speed C bear the same ratio to another speed E as V bears to T; then E will be the speed at which B traverses the distance T during the time-interval S. If now the speed E is to another speed G as the time-interval R is to the time-interval S, then G will be the speed at which the particle B traverses the distance T during the time-interval R. Thus we have the speed C at which the particle A covers the distance V during the time S and also the speed G at which the particle B traverses the distance T during the time R. The ratio of C to G is the product of the ratio C to E and E to G; the ratio of C to E is by definition the same as the ratio of the distance V to distance T; and the ratio of E to G is the same as the ratio of R to S. Hence follows the proposition.

Salv. The preceding is what our Author has written concerning uniform motion. We pass now to a new and more discriminating consideration of naturally accelerated motion, such as that generally experienced by heavy falling bodies; following is the title and introduction.

[197]

NATURALLY ACCELERATED MOTION

The properties belonging to uniform motion have been discussed in the preceding section; but accelerated motion remains to be considered.

And first of all it seems desirable to find and explain a definition best fitting natural phenomena. For anyone may invent an arbitrary type of motion and discuss its properties; thus, for instance, some have imagined helices and conchoids as described by certain motions which are not met with in nature, and have very commendably established the properties which these curves possess in virtue of their definitions; but we have decided to consider the phenomena of bodies falling with an acceleration such as actually occurs in nature and to make this definition of accelerated motion exhibit the essential features of observed accelerated motions. And this, at last, after repeated efforts we trust we have succeeded in doing. In this belief we are confirmed mainly by the consideration that experimental results are seen to agree with and exactly correspond with those properties which have been, one after another, demonstrated by us. Finally, in the investigation of naturally accelerated motion we were led, by hand as it were, in following the habit and custom of nature herself, in all her various other processes, to employ only those means which are most common, simple and easy.

For I think no one believes that swimming or flying can be accomplished in a manner simpler or easier than that instinctively employed by fishes and birds.

When, therefore, I observe a stone initially at rest falling from an elevated position and continually acquiring new increments of speed, why should I not believe that such increases take place in a manner which is exceedingly simple and rather obvious to everybody? If now we examine the matter carefully we find no addition or increment more simple than that which repeats itself always in the same manner. This we readily understand when we consider the intimate relationship between time and motion; for just as uniformity of motion is defined by and conceived through equal times and equal spaces (thus we call a motion uniform when equal distances are traversed during equal time-intervals), so also we may, in a similar manner, through equal time-intervals, conceive additions of speed as taking place without complication; thus we may picture to our mind a motion as uniformly and continuously accelerated when, during any equal intervals of time whatever, equal increments of speed are given to it.

[198]

Thus if any equal intervals of time whatever have elapsed, counting from the time at which the moving body left its position of rest and began to descend, the amount of speed acquired during the first two time-intervals will be double that acquired during the first time-interval alone; so the amount added during three of these time-intervals will be treble; and that in four, quadruple that of the first time-interval. To put the matter more clearly, if a body were to continue its motion with the same speed which it had acquired during the first time-interval and were to retain this same uniform speed, then its motion would be twice as slow as that which it would have if its velocity had been acquired during *two* time-intervals.

And thus, it seems, we shall not be far wrong if we put the increment of speed as proportional to the increment of time; hence the definition of motion which we are about to discuss may be stated as follows: A motion is said to be uniformly accelerated, when starting from rest, it acquires, during equal time-intervals, equal increments of speed.

Sagr. Although I can offer no rational objection to this or indeed to any other definition, devised by any author whomsoever, since all definitions are arbitrary, I may nevertheless without offense be allowed to doubt whether such a definition as the above, established in an abstract manner, corresponds to and describes that kind of accelerated motion which we meet in nature in the case of freely falling bodies. And since the Author apparently maintains that the motion described in his definition is that of freely falling bodies, I would like to clear my mind of certain difficulties in order that I may later apply myself more earnestly to the propositions and their demonstrations.

Salv. It is well that you and Simplicio raise these difficulties. They are, I imagine, the same which occurred to me when I first saw this treatise, and which were removed either by discussion with the Author himself, or by turning the matter over in my own mind.

[199]

Sagr. When I think of a heavy body falling from rest, that is, starting with zero speed and gaining speed in proportion to the time from the beginning of the motion; such a motion as would, for instance, in eight beats of the pulse acquire eight degrees of speed; having at the end of the fourth beat acquired four degrees; at the end of the second, two; at the end of the first, one: and since time is divisible without limit, it follows from all these considerations that if the earlier speed of a body is less than its present speed in a constant ratio, then there is no degree of speed however small (or, one may say, no degree of slowness however great) with which we may not find this body travelling after starting from infinite slowness, i.e., from rest. So that if that speed which it had at the end of the fourth beat was such that, if kept uniform, the body would traverse two miles in an hour, and if keeping the speed which it had at the end of the second beat, it would traverse one mile an hour, we must infer that, as the instant of starting is more and more nearly approached, the body moves so slowly that, if it kept on moving at this rate, it would not traverse a mile in an hour, or in a day, or in a year or in a thousand years; indeed, it would not traverse a span in an even greater time; a phenomenon which baffles the imagination, while our senses show us that a heavy falling body suddenly acquires great speed.

Salv. This is one of the difficulties which I also at the beginning, experienced, but which I shortly afterwards removed; and the removal was effected by the very experiment which creates the difficulty for you. You say the experiment appears to show that immediately after a heavy body starts from rest it acquires a very considerable speed: and I say that the same experiment makes clear the fact that the initial motions of a falling body, no matter how heavy, are very slow and gentle. Place a heavy body upon a yielding material, and leave it there without any pressure except that owing to its own weight; it is clear that if one lifts this body a cubit or two and allows it to fall upon the same material, it will, with this impulse, exert a new and greater pressure than that caused by its mere weight; and this effect is brought about by the [weight of the] falling body together with the velocity acquired during the fall, an effect which will be greater and greater according to the height of the fall, that is according as the velocity of the falling body becomes greater. From the quality and intensity of the blow we are thus enabled to accurately estimate the speed of a falling body.

[200]

But tell me, gentlemen, is it not true that if a block be allowed to fall upon a stake from a height of four cubits and drives it into the earth, say, four finger-breadths, that coming from a height of two cubits it will drive the stake a much less distance, and from the height of one cubit a still less distance; and finally if the block be lifted only one finger-breadth how much more will it accomplish than if merely laid on top of the stake without percussion? Certainly very little. If it be lifted only the thickness of a leaf, the effect will be altogether imperceptible. And since the effect of the blow depends upon the velocity of this striking body, can any one doubt the motion is very slow and the speed more than small whenever the effect [of the blow] is imperceptible? See now the power of truth; the same experiment which at first glance seemed to show one thing, when more carefully examined, assures us of the contrary.

But without depending upon the above experiment, which is doubtless very conclusive, it seems to me that it ought not to be difficult to establish such a fact by reasoning alone. Imagine a heavy stone held in the air at rest; the support is removed and the stone set free; then since it is heavier than the air it begins to fall, and not with uniform motion but slowly at the beginning and with a continuously accelerated motion. Now since velocity can be increased and diminished without limit, what reason is there to believe that such a moving body starting with infinite slowness, that is, from rest, immediately acquires a speed of ten degrees rather than one of four, or of two, or of one, or of a half, or of a hundredth; or, indeed, of any of the infinite number of small values [of speed]? Pray listen. I hardly think you will refuse to grant that the gain of speed of the stone falling from rest follows the same sequence as the diminution and

loss of this same speed when, by some impelling force, the stone is thrown to its former elevation: but even if you do not grant this, I do not see how you can doubt that the ascending stone, diminishing in speed, must before coming to rest pass through every possible degree of slowness.

Simp. But if the number of degrees of greater and greater slowness is limitless, they will never be all exhausted, therefore such an ascending heavy body will never reach rest, but will continue to move without limit always at a slower rate; but this is not the observed fact.

[201]

Salv. This would happen, Simplicio, if the moving body were to maintain its speed for any length of time at each degree of velocity; but it merely passes each point without delaying more than an instant: and since each time-interval however small may be divided into an infinite number of instants, these will always be sufficient [in number] to correspond to the infinite degrees of diminished velocity.

That such a heavy rising body does not remain for any length of time at any given degree of velocity is evident from the following: because if, some time-interval having been assigned, the body moves with the same speed in the last as in the first instant of that time-interval, it could from this second degree of elevation be in like manner raised through an equal height, just as it was transferred from the first elevation to the second, and by the same reasoning would pass from the second to the third and would finally continue in uniform motion forever.

Sagr. From these considerations it appears to me that we may obtain a proper solution of the problem discussed by philosophers, namely, what causes the acceleration in the natural motion of heavy bodies? Since, as it seems to me, the force [virtù] impressed by the agent projecting the body upwards diminishes continuously, this force, so long as it was greater than the contrary force of gravitation, impelled the body upwards; when the two are in equilibrium the body ceases to rise and passes through the state of rest in which the impressed impetus [impeto] is not destroyed, but only its excess over the weight of the body has been consumed—the excess which caused the body to rise. Then as the diminution of the outside impetus [impeto] continues, and gravitation gains the upper hand, the fall begins, but slowly at first on account of the opposing impetus [virtù impressa], a large portion of which still remains in the body; but as this continues to diminish it also continues to be more and more overcome by gravity, hence the continuous acceleration of motion.

Simp. The idea is clever, yet more subtle than sound; for even if the argument were conclusive, it would explain only the case in which a natural motion is preceded by a violent motion, in which there still remains active a portion of the external force [virtù

esterna]; but where there is no such remaining portion and the body starts from an antecedent state of rest, the cogency of the whole argument fails.

Sagr. I believe that you are mistaken and that this distinction between cases which you make is superfluous or rather non-existent. But, tell me, cannot a projectile receive from the projector either a large or a small force [*virtù*] such as will throw it to a height of a hundred cubits, and even twenty or four or one?

[202]

Simp. Undoubtedly, yes.

Sagr. So therefore this impressed force [*virtù impressa*] may exceed the resistance of gravity so slightly as to raise it only a finger-breadth; and finally the force [*virtù*] of the projector may be just large enough to exactly balance the resistance of gravity so that the body is not lifted at all but merely sustained. When one holds a stone in his hand does he do anything but give it a force impelling [*virtù impellente*] it upwards equal to the power [*facoltà*] of gravity drawing it downwards? And do you not continuously impress this force [*virtù*] upon the stone as long as you hold it in the hand? Does it perhaps diminish with the time during which one holds the stone?

And what does it matter whether this support which prevents the stone from falling is furnished by one's hand or by a table or by a rope from which it hangs? Certainly nothing at all. You must conclude, therefore, Simplicio, that it makes no difference whatever whether the fall of the stone is preceded by a period of rest which is long, short, or instantaneous provided only the fall does not take place so long as the stone is acted upon by a force [*virtù*] opposed to its weight and sufficient to hold it at rest.

Salv. The present does not seem to be the proper time to investigate the cause of the acceleration of natural motion concerning which various opinions have been expressed by various philosophers, some explaining it by attraction to the center, others to repulsion between the very small parts of the body, while still others attribute it to a certain stress in the surrounding medium which closes in behind the falling body and drives it from one of its positions to another. Now, all these fantasies, and others too, ought to be examined; but it is not really worth while. At present it is the purpose of our Author merely to investigate and to demonstrate some of the properties of accelerated motion (whatever the cause of this acceleration may be)—meaning thereby a motion, such that the momentum of its velocity [*i momenti della sua velocità*] goes on increasing after departure from rest, in simple proportionality to the time, which is the same as saying that in equal time-intervals the body receives equal increments of velocity; and if we find the properties [of accelerated motion] which will be demonstrated later are realized in freely falling and accelerated bodies, we may conclude that the

assumed definition includes such a motion of falling bodies and that their speed [*accelerazione*] goes on increasing as the time and the duration of the motion.

[203]

Sagr. So far as I see at present, the definition might have been put a little more clearly perhaps without changing the fundamental idea, namely, uniformly accelerated motion is such that its speed increases in proportion to the space traversed; so that, for example, the speed acquired by a body in falling four cubits would be double that acquired in falling two cubits and this latter speed would be double that acquired in the first cubit. Because there is no doubt but that a heavy body falling from the height of six cubits has, and strikes with, a momentum [*impeto*] double that it had at the end of three cubits, triple that which it had at the end of one.

Salv. It is very comforting to me to have had such a companion in error; and moreover let me tell you that your proposition seems so highly probable that our Author himself admitted, when I advanced this opinion to him, that he had for some time shared the same fallacy. But what most surprised me was to see two propositions so inherently probable that they commanded the assent of everyone to whom they were presented, proven in a few simple words to be not only false, but impossible.

Simp. I am one of those who accept the proposition, and believe that a falling body acquires force [*vires*] in its descent, its velocity increasing in proportion to the space, and that the momentum [*momento*] of the falling body is doubled when it falls from a doubled height; these propositions, it appears to me, ought to be conceded without hesitation or controversy.

Salv. And yet they are as false and impossible as that motion should be completed instantaneously; and here is a very clear demonstration of it. If the velocities are in proportion to the spaces traversed, or to be traversed, then these spaces are traversed in equal intervals of time; if, therefore, the velocity with which the falling body traverses a space of eight feet were double that with which it covered the first four feet (just as the one distance is double the other) then the time-intervals required for these passages would be equal. But for one and the same body to fall eight feet and four feet in the same time is possible only in the case of instantaneous [discontinuous] motion; but observation shows us that the motion of a falling body occupies time, and less of it in covering a distance of four feet than of eight feet; therefore it is not true that its velocity increases in proportion to the space.

[204]

The falsity of the other proposition may be shown with equal clearness. For if we consider a single striking body the difference of momentum in its blows can depend

only upon difference of velocity; for if the striking body falling from a double height were to deliver a blow of double momentum, it would be necessary for this body to strike with a doubled velocity; but with this doubled speed it would traverse a doubled space in the same time-interval; observation however shows that the time required for fall from the greater height is longer.

Sagr. You present these recondite matters with too much evidence and ease; this great facility makes them less appreciated than they would be had they been presented in a more abstruse manner. For, in my opinion, people esteem more lightly that knowledge which they acquire with so little labor than that acquired through long and obscure discussion.

Salv. If those who demonstrate with brevity and clearness the fallacy of many popular beliefs were treated with contempt instead of gratitude the injury would be quite bearable; but on the other hand it is very unpleasant and annoying to see men, who claim to be peers of anyone in a certain field of study, take for granted certain conclusions which later are quickly and easily shown by another to be false. I do not describe such a feeling as one of envy, which usually degenerates into hatred and anger against those who discover such fallacies; I would call it a strong desire to maintain old errors, rather than accept newly discovered truths. This desire at times induces them to unite against these truths, although at heart believing in them, merely for the purpose of lowering the esteem in which certain others are held by the unthinking crowd. Indeed, I have heard from our Academician many such fallacies held as true but easily refutable; some of these I have in mind.

[205]

Sagr. You must not withhold them from us, but, at the proper time, tell us about them even though an extra session be necessary. But now, continuing the thread of our talk, it would seem that up to the present we have established the definition of uniformly accelerated motion which is expressed as follows:

> A motion is said to be equally or uniformly accelerated when, starting from rest, its momentum (celeritatis momenta) receives equal increments in equal times.

Salv. This definition established, the Author makes a single assumption, namely,

> The speeds acquired by one and the same body moving down planes of different inclinations are equal when the heights of these planes are equal.

By the height of an inclined plane we mean the perpendicular let fall from the upper end of the plane upon the horizontal line drawn through the lower end of the same plane. Thus, to illustrate, let the line AB be horizontal, and let the planes CA and CD be inclined to it; then the Author calls the perpendicular CB the "height" of the planes CA and CD; he supposes that the speeds acquired by one and the same body, descending along the planes CA and CD to the terminal points A and D are equal since the heights of these planes are the same, CB; and also it must be understood that this speed is that which would be acquired by the same body falling from C to B.

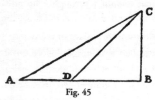

Fig. 45

Sagr. Your assumption appears to me so reasonable that it ought to be conceded without question, provided of course there are no chance or outside resistances, and that the planes are hard and smooth, and that the figure of the moving body is perfectly round, so that neither plane nor moving body is rough. All resistance and opposition having been removed, my reason tells me at once that a heavy and perfectly round ball descending along the lines CA, CD, CB would reach the terminal points A, D, B, with equal momenta [*impeti eguali*].

[206]

Salv. Your words are very plausible; but I hope by experiment to increase the probability to an extent which shall be little short of a rigid demonstration.

Imagine this page to represent a vertical wall, with a nail driven into it; and from the nail let there be suspended a lead bullet of one or two ounces by means of a fine vertical thread, AB, say from four to six feet long, on this wall draw a horizontal line DC, at right angles to the vertical thread AB, which hangs about two finger-breadths in front of the wall. Now bring the thread AB with the attached ball into the position AC and set it free; first it will be observed to descend along the arc CBD, to pass the point B, and to travel along the

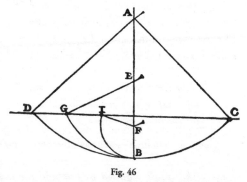

Fig. 46

arc BD, till it almost reaches the horizontal CD, a slight shortage being caused by the resistance of the air and the string; from this we may rightly infer that the ball in its

528

descent through the arc CB acquired a momentum [*impeto*] on reaching B, which was just sufficient to carry it through a similar arc BD to the same height. Having repeated this experiment many times, let us now drive a nail into the wall close to the perpendicular AB, say at E or F, so that it projects out some five or six finger-breadths in order that the thread, again carrying the bullet through the are CB, may strike upon the nail E when the bullet reaches B, and thus compel it to traverse the are BG, described about E as center. From this we can see what can be done by the same momentum [*impeto*] which previously starting at the same point B carried the same body through the arc BD to the horizontal CD. Now, gentlemen, you will observe with pleasure that the ball swings to the point G in the horizontal, and you would see the same thing happen if the obstacle were placed at some lower point, say at F, about which the ball would describe the arc BI, the rise of the ball always terminating exactly on the line CD. But when the nail is placed so low that the remainder of the thread below it will not reach to the height CD (which would happen if the nail were placed

[207]

nearer B than to the intersection of AB with the horizontal CD) then the thread leaps over the nail and twists itself about it.

This experiment leaves no room for doubt as to the truth of our supposition; for since the two arcs CB and DB are equal and similarly placed, the momentum [*momento*] acquired by the fall through the arc CB is the same as that gained by fall through the arc DB; but the momentum [*momento*] acquired at B, owing to fall through CB, is able to lift the same body [*mobile*] through the arc BD; therefore, the momentum acquired in the fall BD is equal to that which lifts the same body through the same arc from B to D; so, in general, every momentum acquired by fall through an arc is equal to that which can lift the same body through the same arc. But all these momenta [*momenti*] which cause a rise through the arcs BD, BG, and BI are equal, since they are produced by the same momentum, gained by fall through CB, as experiment shows. Therefore all the momenta gained by fall through the arcs DB, GB, IB are equal.

Sagr. The argument seems to me so conclusive and the experiment so well adapted to establish the hypothesis that we may, indeed, consider it as demonstrated.

Salv. I do not wish, Sagredo, that we trouble ourselves too much about this matter, since we are going to apply this principle mainly in motions which occur on plane surfaces, and not upon curved, along which acceleration varies in a manner greatly different from that which we have assumed for planes.

So that, although the above experiment shows us that the descent of the moving body through the arc CB confers upon it momentum [*momento*] just sufficient to

carry it to the same height through any of the arcs BD, BG, BI, we are not able, by similar means, to show that the event would be identical in the case of a perfectly round ball descending along planes whose inclinations are respectively the same as the chords of these arcs. It seems likely, on the other hand, that, since these planes form angles at the point B, they will present an obstacle to the ball which has descended along the chord CB, and starts to rise along the chord BD, BG, BI.

In striking these planes some of its momentum [*impeto*] will be lost and it will not be able to rise to the height of the line CD; but this obstacle, which interferes with the experiment, once removed, it is clear that the momentum [*impeto*] (which gains in

[208]

strength with descent) will be able to carry the body to the same height. Let us then, for the present, take this as a postulate, the absolute truth of which will be established when we find that the inferences from it correspond to and agree perfectly with experiment. The author having assumed this single principle passes next to the propositions which he clearly demonstrates; the first of these is as follows:

THEOREM I, PROPOSITION I

The time in which any space is traversed by a body starting from rest and uniformly accelerated is equal to the time in which that same space would be traversed by the same body moving at a uniform speed whose value is the mean of the highest speed and the speed just before acceleration began.

Let us represent by the line AB the time in which the space CD is traversed by a body which starts from rest at C and is uniformly accelerated; let the final and highest value of the speed gained during the interval AB be represented by the line EB drawn at right angles to AB; draw the line AE, then all lines drawn from equidistant points on AB and parallel to BE will represent the increasing values of the speed, beginning with the instant A. Let the point F bisect the line EB; draw FG parallel to BA, and GA parallel to FB, thus forming a parallelogram AGFB which will be equal in area to the triangle AEB, since the side GF bisects the side AE at the point I; for if the parallel lines in the triangle AEB are extended to GI, then the sum of all the parallels contained in the quadrilateral is equal to the sum of those

Fig. 47

530

contained in the triangle AEB; for those in the triangle IEF are equal to those contained in the triangle GIA, while those included in the trapezium AIFB are common. Since each and every instant of time in the time-interval AB has its corresponding point on the line AB, from which points parallels drawn in and limited by the triangle AEB represent the increasing values of the growing velocity, and since parallels contained within the rectangle represent the values of a speed which is not increasing, but constant, it appears, in like manner, that the momenta [*momenta*] assumed by the moving body may also be represented, in the case of the accelerated motion, by the increasing parallels of the triangle AEB, and, in the case of the uniform motion, by the

[209]

parallels of the rectangle GB. For, what the momenta may lack in the first part of the accelerated motion (the deficiency of the momenta being represented by the parallels of the triangle AGI) is made up by the momenta represented by the parallels of the triangle IEF.

Hence it is clear that equal spaces will be traversed in equal times by two bodies, one of which, starting from rest, moves with a uniform acceleration, while the momentum of the other, moving with uniform speed, is one-half its maximum momentum under accelerated motion.

Q.E.D.

THEOREM II, PROPOSITION II

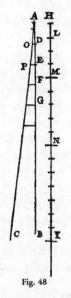

Fig. 48

The spaces described by a body falling from rest with a uniformly accelerated motion are to each other as the squares of the time-intervals employed in traversing these distances.

Let the time beginning with any instant A be represented by the straight line AB in which are taken any two time-intervals AD and AE. Let HI represent the distance through which the body, starting from rest at H, falls with uniform acceleration. If HL represents the space traversed during the time-interval AD, and HM that covered during the interval AE, then the space MH stands to the space LH in a ratio which is the square of the ratio of the time AE to the time AD; or we may say simply that the distances HM and HL are related as the squares of AE and AD.

Draw the line AC making any angle whatever with the line AB; and from the points D and E, draw the parallel lines DO and EP; of these two lines, DO represents the greatest velocity attained during the interval AD, while EP represents the maximum velocity acquired during the interval AE. But it has just been proved that so far as distances traversed are concerned it is precisely the same whether a body falls from rest with a uniform acceleration or whether it falls during an equal time-interval with a constant speed which is one-half the maximum speed attained during the accelerated motion. It follows therefore that the distances HM and HL are the same as would be traversed, during the time-intervals AE and AD, by uniform velocities equal to one-half those represented by DO and EP respectively. If, therefore, one can show that the distances HM and HL are in the same ratio as the squares of the time-intervals AE and AD, our proposition will be proven.

[210]

But in the fourth proposition of the first book [see above] it has been shown that the spaces traversed by two particles in uniform motion bear to one another a ratio which is equal to the product of the ratio of the velocities by the ratio of the times. But in this case the ratio of the velocities is the same as the ratio of the time-intervals (for the ratio of AE to AD is the same as that of 1/2 EP to 1/2 DO or of EP to DO). Hence the ratio of the spaces traversed is the same as the squared ratio of the time intervals. Q.E.D.

Evidently then the ratio of the distances is the square of the ratio of the final velocities, that is, of the lines EP and DO, since these are to each other as AE to AD.

COROLLARY I

Hence it is clear that if we take any equal intervals of time whatever, counting from the beginning of the motion, such as AD, DE, EF, FG, in which the spaces HL, LM, MN, NI are traversed, these spaces will bear to one another the same ratio as the series of odd numbers, 1, 3, 5, 7; for this is the ratio of the differences of the squares of the lines [which represent time], differences which exceed one another by equal amounts, this excess being equal to the smallest line [viz. the one representing a single time-interval]: or we may say [that this is the ratio] of the differences of the squares of the natural numbers beginning with unity.

While, therefore, during equal intervals of time the velocities increase as the natural numbers, the increments in the distances traversed during these equal time-intervals are to one another as the odd numbers beginning with unity.

Sagr. Please suspend the discussion for a moment since there just occurs to me an idea which I want to illustrate by means of a diagram in order that it may be clearer both to you and to me.

Let the line AI represent the lapse of time measured from the initial instant A; through A draw the straight line AF making A any angle whatever; join the terminal points I and F; divide the time AI in half at C; draw CB parallel to IF. Let us consider CB as the maximum value of the velocity which increases from zero at the beginning, in simple proportionality to the intercepts on the triangle ABC of lines drawn parallel to BC; or what is the same thing, let us suppose the velocity to increase in proportion to the time; then I admit without question, in view of the preceding argument, that the space described by a body falling in the aforesaid manner will be equal to the space traversed by the same body during the same length of time travelling with a uniform speed equal, to EC, the

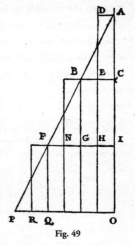

Fig. 49

[211]

half of BC. Further let us imagine that the body has fallen with accelerated motion so that, at the instant C, it has the velocity BC. It is clear that if the body continued to descend with the same speed BC, without acceleration, it would in the next time-interval CI traverse double the distance covered during the interval AC, with the uniform speed EC which is half of BC; but since the falling body acquires equal increments of speed during equal increments of time, it follows that the velocity BC, during the next time-interval CI will be increased by an amount represented by the parallels of the triangle BFG which is equal to the triangle ABC. If, then, one adds to the velocity GI half of the velocity FG, the highest speed acquired by the accelerated motion and determined by the parallels of the triangle BFG, he will have the uniform velocity with which the same space would have been described in the time CI; and since this speed IN is three times as great as EC it follows that the space described during the interval a is three times as great as that described during the interval AC. Let us imagine the motion extended over another equal time-interval IO, and the triangle extended to APO; it is then evident that if the motion continues during the interval IO, at the constant rate IF acquired by acceleration during the time AI, the space traversed during the interval IO will be four times that traversed during the first interval AC, because the speed IF is four times the speed EC. But if we enlarge our triangle so as to include FPQ which is equal to ABC, still assuming the acceleration to be constant, we shall add to the uniform speed an increment RQ, equal to EC; then the value of the equivalent uniform speed during the time-interval IO will be five times that during the first time-interval AC; therefore the space traversed will be quintuple that during the first interval

AC. It is thus evident by simple computation that a moving body starting from rest and acquiring velocity at a rate proportional to the time, will, during equal intervals of time, traverse distances which are related to each other as the odd numbers beginning with unity, 1, 3, 5;[30] or considering the total space traversed, that covered in double time will be quadruple that covered during unit time; in

[212]

triple time, the space is nine times as great as in unit time. And in general the spaces traversed are in the duplicate ratio of the times, i. e., in the ratio of the squares of the times.

Simp. In truth, I find more pleasure in this simple and clear argument of Sagredo than in the Author's demonstration which to me appears rather obscure; so that I am convinced that matters are as described, once having accepted the definition of uniformly accelerated motion. But as to whether this acceleration is that which one meets in nature in the case of falling bodies, I am still doubtful; and it seems to me, not only for my own sake but also for all those who think as I do, that this would be the proper moment to introduce one of those experiments-and there are many of them, I understand-which illustrate in several ways the conclusions reached.

Salv. The request which you, as a man of science, make, is a very reasonable one; for this is the custom-and properly so-in those sciences where mathematical demonstrations are applied to natural phenomena, as is seen in the case of perspective, astronomy, mechanics, music, and others where the principles, once established by well-chosen experiments, become the foundations of the entire superstructure. I hope therefore it will not appear to be a waste of time if we discuss at considerable length this first and most fundamental question upon which hinge numerous consequences of which we have in this book only a small number, placed there by the Author, who has done so much to open a pathway hitherto closed to minds of speculative turn. So far as experiments go they have not been neglected by the Author; and often, in his company, I have attempted in tile following manner to assure myself that the acceleration actually experienced by falling bodies is that above described.

[213]

A piece of wooden moulding or scantling, about 12 cubits long, half a cubit wide, and three finger-breadths thick, was taken; on its edge was cut a channel a little more

30. As illustrating the greater elegance and brevity of modern analytical methods, one may obtain the result of Prop. II directly from the fundamental equation

$s = 1/2g(t2^2 - t2^1) = g/2(t2 + t1)(t2 - t1)$

where g is the acceleration of gravity and s, the space traversed between the instants $t1$ and $t2=1$, say one second, then $s=g/2(t2+t1)$ where $t2+t1$, must always be an odd number, seeing that it is the sum of two consecutive terms in the series of natural numbers. *[Trans.]*

than one finger in breadth; having made this groove very straight, smooth, and polished, and having lined it with parchment, also as smooth and polished as possible, we rolled along it a hard, smooth, and very round bronze ball. Having placed this board in a sloping position, by lifting one end some one or two cubits above the other, we rolled the ball, as I was just saying, along the channel, noting, in a manner presently to be described, the time required to make the descent. We repeated this experiment more than once in order to measure the time with an accuracy such that the deviation between two observations never exceeded one-tenth of a pulse-beat. Having performed this operation and having assured ourselves of its reliability, we now rolled the ball only one-quarter the length of the channel; and having measured the time of its descent, we found it precisely one-half of the former. Next we tried other distances, comparing the time for the whole length with that for the half, or with that for two-thirds, or three-fourths, or indeed for any fraction; in such experiments, repeated a full hundred times, we always found that the spaces traversed were to each other as the squares of the times, and this was true for all inclinations of the plane, i. e., of the channel, along which we rolled the ball. We also observed that the times of descent, for various inclinations of the plane, bore to one another precisely that ratio which, as we shall see later, the Author had predicted and demonstrated for them.

For the measurement of time, we employed a large vessel of water placed in an elevated position; to the bottom of this vessel was soldered a pipe of small diameter giving a thin jet of water, which we collected in a small glass during the time of each descent, whether for the whole length of the channel or for a part of its length; the water thus collected was weighed, after each descent, on a very accurate balance; the differences and ratios of these weights gave us the differences and ratios of the times, and this with such accuracy that although the operation was repeated many, many times, there was no appreciable discrepancy in the results.

Simp. I would like to have been present at these experiments; but feeling confidence in the care with which you performed them, and in the fidelity with which you relate them, I am satisfied and accept them as true and valid.

Salv. Then we can proceed without discussion.

[214]

COROLLARY II

Secondly, it follows that, starting from any initial point, if we take any two distances, traversed in any time-intervals whatsoever, these time-intervals bear to one another the same ratio as one of the distances to the mean proportional of the two distances.

For if we take two distances ST and SY measured from the initial point S, the mean proportional of which is SX, the time of fall through ST is to the time of fall through SY as ST is to SX; or one may say the time of fall through SY is to the time of fall through ST as SY is to SX. Now since it has been shown that the spaces traversed are in the same ratio as the squares of the times; and since, moreover, the ratio of the space SY to the space ST is the square of the ratio SY to SX, it follows that the ratio of the times of fall through SY and ST is the ratio of the respective distances SY and SX.

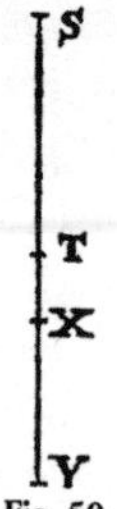

Fig. 50

SCHOLIUM

The above corollary has been proven for the case of vertical fall; but it holds also for planes inclined at any angle; for it is to be assumed that along these planes the velocity increases in the same ratio, that is, in proportion to the time, or, if you prefer, as the series of natural numbers.[31]

Salv. Here, Sagredo, I should like, if it be not too tedious to Simplicio, to interrupt for a moment the present discussion in order to make some additions on the basis of what has already been proved and of what mechanical principles we have already learned from our Academician. This addition I make for the better establishment on logical and experimental grounds, of the principle which we have above considered; and what is more important, for the purpose of deriving it geometrically, after first demonstrating a single lemma which is fundamental in the science of motion [*impeti*].

Sagr. If the advance which you propose to make is such as will confirm and fully establish these sciences of motion, I will gladly devote to it any length of time. Indeed,

[215]

I shall not only be glad to have you proceed, but I beg of you at once to satisfy the curiosity which you have awakened in me concerning your proposition; and I think that Simplicio is of the same mind.

Simp. Quite right.

Salv. Since then I have your permission, let us first of all consider this notable fact, that the momenta or speeds [*i momenti o le velocità*] of one and the same moving body vary with the inclination of the plane.

The speed reaches a maximum along a vertical direction, and for other directions diminishes as the plane diverges from the vertical. Therefore the impetus, ability, energy, [*l'impeto, il talento l'energia*] or, one might say, the momentum [*il momento*] of descent of the moving body is diminished by the plane upon which it is supported and along which it rolls.

---

31. The dialogue which intervenes between this Scholium and the following theorem was elaborated by Viviani, at the suggestion of Galileo. See *National Edition*, viii, 23. [*Trans.*]

For the sake of greater clearness erect the line AB perpendicular to the horizontal AC; next draw AD, AE, AF, etc., at different inclinations to the horizontal. Then I say that all the momentum of the falling body is along the vertical and is a maximum when it falls in that direction; the momentum is less along DA and still less along EA, and even less yet along the more inclined plane FA. Finally on the horizontal plane the momentum vanishes altogether; the body finds itself in a condition of indifference as to motion or rest; has no inherent tendency to move in any direction, and offers no resistance to being set in motion. For just as a heavy body or system of bodies cannot of itself move upwards, or recede from the common center [*comun centro*] toward which all heavy things tend, so it is

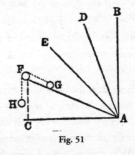

Fig. 51

impossible for any body of its own accord to assume any motion other than one which carries it nearer to the aforesaid common center. Hence, along the horizontal, by which we understand a surface, every point of which is equidistant from this same common center, the body will have no momentum whatever.

[216]

This change of momentum being clear, it is here necessary for me to explain something which our Academician wrote when in Padua, embodying it in a treatise on mechanics prepared solely for the use of his students, and proving it at length and conclusively when considering the origin and nature of that marvellous machine, the screw. What he proved is the manner in which the momentum [*impeto*] varies with the inclination of the plane, as for instance that of the plane FA, one end of which is elevated through a vertical distance FC. This direction FC is that along which the momentum of a heavy body becomes a maximum; let us discover what ratio this momentum bears to that of the same body moving along the inclined plane FA. This ratio, I say, is the inverse of that of the aforesaid lengths. Such is the lemma preceding the theorem which I hope to demonstrate a little later.

It is clear that the impelling force [*impeto*] acting on a body in descent is equal to the resistance or least force [*resistenza o forza minima*] sufficient to bold it at rest. In order to measure this force and resistance [*forza e resistenza*] I propose to use the weight of another body. Let us place upon the plane FA a body G connected to the weight H by means of a cord passing over the point F; then the body H will ascend or descend, along the perpendicular, the same distance which the body G ascends or descends along the inclined plane FA; but this distance will not be equal to the rise or fall of G along the vertical in which direction alone G, as other bodies, exerts its force

[*resistenza*]. This is clear. For if we consider the motion of the body G, from A to F in the triangle AFC to be made up of a horizontal component AC and a vertical component CF, and remember that this body experiences no resistance to motion along the horizontal (because by such a motion the body neither gains nor loses distance from the common center of heavy things) it follows that resistance is met only in consequence of the body rising through the vertical distance CF.

[217]

Since then the body G in moving from A to F offers resistance only in so far as it rises through the vertical distance CF, while the other body H must fall vertically through the entire distance FA, and since this ratio is maintained whether the motion be large or small, the two bodies being inextensibly connected, we are able to assert positively that, in case of equilibrium (bodies at rest) the momenta, the velocities, or their tendency to motion [*propensioni al moto*], i. e., the spaces which would be traversed by them in equal times, must be in the inverse ratio to their weights. This is what has been demonstrated in every case of mechanical motion.[32] So that, in order to hold the weight G at rest, one must give H a weight smaller in the same ratio as the distance CF is smaller than FA. If we do this, FA:FC=weight G:weight H; then equilibrium will occur, that is, the weights H and G will have the some impelling forces [*momenti eguali*], and the two bodies will come to rest.

And since we are agreed that the impetus, energy, momentum or tendency to motion of a moving body is as great as the force or least resistance [*forza o resistenza minima*] sufficient to stop it, and since we have found that the weight H is capable of preventing motion in the weight G it follows that the less weight H whose entire force [*momento totale*] is along the perpendicular, FC, will be an exact measure of the component of force [*momento parziale*] which the larger weight G exerts along the plane FA. But the measure of the total force [*total momento*] on the body G is its own weight, since to prevent its fall it is only necessary to balance it with an equal weight, provided this second weight be free to move vertically; therefore the component of the force [*momento parziale*] on G along the inclined plane FA will bear to the maximum and total force on this same body G along the perpendicular FC the same ratio as the weight H to the weight G. This ratio is, by construction, the same which the height, FC, of the inclined plane bears to the length FA. We have here the lemma which I proposed to demonstrate and which, as you will see, has been assumed by our Author in the second part of the sixth proposition of the present treatise.

Sagr. From what you have shown thus far, it appears to me that one might infer, arguing *ex aequali con la proportione perturbata*, that the tendencies [*momenti*] of one

32. A near approach to the principle of virtual work enunciated by John Bernoulli in 1717. [*Trans.*]

and the same body to move along planes differently inclined, but having the same vertical height, as FA and FI, are to each other inversely as the lengths of the planes.

[218]

Salv. Perfectly right. This point established, I pass to the demonstration of the following theorem:

If a body falls freely along smooth planes inclined at any angle whatsoever, but of the same height, the speeds with which it reaches the bottom are the same.

First we must recall the fact that on a plane of any inclination whatever a body starting from rest gains speed or momentum [*la quantitá dell'impeto*] in direct proportion to the time, in agreement with the definition of naturally accelerated motion given by the Author. Hence, as he has shown in the preceding proposition, the distances traversed are proportional to the squares of the times and therefore to the squares of the speeds, the speed relations are here the same as in the motion first studied [i. e., *vertical motion*], since in each case the gain of speed is proportional to the time.

Let AB be an inclined plane whose height above the level BC is AC. As we have seen above the force impelling [*l'impeto*] a body to fall along the vertical AC is to the force which drives the same body along the inclined plane AB as AB is to AC. On the incline AB , lay off AD a third proportional to AB and AC; then the force producing motion along AC is to that along AB (i. e., along AD) as the length AC is to the length AD. And therefore the body will traverse the space AD, along the incline AB, in the same time which it would occupy in falling the vertical distance AC, (since the forces [*momenti*] are in the same ratio as these distances); also the speed

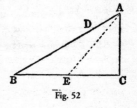

Fig. 52

at C is to the speed at D as the distance AC is to the distance AD. But, according to the definition of accelerated motion, the speed at B is to the speed of the same body at D as the time required to traverse AB is to the time required for AD; and, according to the last corollary of the second proposition, the time of passing through the distance AB bears to the time of passing through AD the same ratio as the distance AC (a mean proportional between AB and AD) to AD. Accordingly the two speeds at B and C each bear to the speed at D the same ratio, namely, that of the distances AC and AD; hence they are equal. This is the theorem which I set out to prove.

From the above we are better able to demonstrate the following third preposition of the Author in which he employs the following principle, namely, the time required to traverse an inclined plane is to that required to fall through the vertical height of the plane in the same ratio as the length of the plane to its height.

[219]

For, according to the second corollary of the second reposition, if BA represents the time required to pass over the distance BA, the time required to pass the distance AD will be a mean proportional between these two distances and will be represented by the line AC; but if AC represents the time needed to traverse AD it will also represent the time required to fall through the distance AC, since the distances AC and AD are traversed in equal times; consequently if AB represents the time required for AB then AC will represent the time required for AC. Hence the times required to traverse AB and AC are to each other as the distances AB and AC.

In like manner it can be shown that the time required to fall through AC is to the time required for any other incline AE as the length AC is to the length AE; therefore, *ex aequali*, the time of fall along the incline AB is to that along AE as the distance AB is to the distance AE, etc.[33]

One might by application of this same theorem, as Sagredo will readily see, immediately demonstrate the sixth proposition of the Author; but let us here end this digression which Sagredo has perhaps found rather tedious, though I consider it quite important or the theory of motion.

Sagr. On the contrary it has given me great satisfaction, and indeed I find it necessary for a complete grasp of this principle.

Salv. I will now resume the reading of the text.

[220]

THEOREM III, PROPOSITION III

If one and the same body, starting from rest, falls along an inclined plane and also along a vertical, each having the same height, the times of descent will be to each other as the lengths of the inclined plane and the vertical.

Let AC be the inclined plane and AB the perpendicular, each having the same vertical height above the horizontal, namely, BA; then I say, the time of descent of one and

33 Putting this argument in a modern and evident notation, one has AC=1/2 gt2c and AD = 1/2 AC/AB gt2/d If now AC^2=AB.AD, it follows at once that t_d=t_c. [Trans.] Q.D.E.

the same body along the plane AC bears a ratio to the time of fall along the perpendicular AB, which is the same as the ratio of the length AC to the AB.

[221]

Let DG, EI and LF be any lines parallel to the horizontal CB; then it follows from what has preceded that a body starting from A will acquire the same speed at the point G as at D, since in each case the vertical fall is the same; in like manner the speeds at I and E will be the same; so also those at L and F. And in general the speeds at the two extremities of any parallel drawn from any point on AB to the corresponding point on AC will be equal.

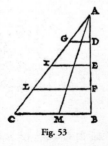

Fig. 53

[222]

Thus the two distances AC and AB are traversed at the same speed. But it has already been proved that if two distances are traversed by a body moving with equal speeds, then the ratio of the times of descent will be the ratio of the distances themselves; therefore, the time of descent along AC is to that along AB as the length of the plane AC is to the vertical distance AB.   Q.E.D.

[223]

Sagr. It seems to me that the above could have been proved clearly and briefly on the basis of a proposition already demonstrated, namely, that the distance traversed in the case of accelerated motion along AC or AB is the same as that covered by a uniform speed whose value is one-half the maximum speed, CB; the two distances AC and

[224]

AB having been traversed at the same uniform speed it is evident, from Proposition I, that the times of descent will be to each other as the distances.

COROLLARY

Hence we may infer that the times of descent along planes having different inclinations, but the same vertical height stand to one another in the same ratio as the lengths of the planes. For consider any plane AM extending from A to the horizontal CB; then it may be demonstrated in the same manner that the time of descent along AM is to the time along AB as the distance AM is to AB; but since the time along AB is to that along AC as the length AB is to the length AC, it follows, *ex æquali*, that as AM is to AC so is the time along AM to the time along AC.

THEOREM IV, PROPOSITION IV

The times of descent along planes of the same length but of different inclinations are to each other in the inverse ratio of the square roots of their heights.

From a single point B draw the planes BA and BC, having the same length but different inclinations; let AE and CD be horizontal lines drawn to meet the perpendicular BD; and let BE represent the height of the plane AB, and BD the height of BC; also

[225]

let BI be a mean proportional to BD and BE; then the ratio of BD to BI is equal to the square root of the ratio of BD to BE. Now, I say, the ratio of the times of descent along BA and BC is the ratio of BD to BI; so that the time of descent along BA is related to the height of the other plane BC, namely BD as the time along BC is related to the height BI. Now it must be proved that the time of descent along BA is to that along BC as the length BD is to the length BI.

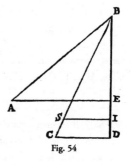

Fig. 54

Draw IS parallel to DC; and since it has been shown that the time of fall along BA is to that along the vertical BE as BA is to BE; and also that the time along BE is to that along BD as BE is to BI; and likewise that the time along BD is to that along BC as BD is to BC, or as BI to BS; it follows, *ex æquali*, that the time along BA is to that along BC as BA to BS, or BC to BS. However, BC is to BS as BD is to BI; hence follows our proposition.

THEOREM V, PROPOSITION V

The times of descent along planes of different length, slope and height bear to one another a ratio which is equal to the product of the ratio of the lengths by the square root of the inverse ratio of their heights.

Draw the planes AB and AC, having different inclinations, lengths, and heights. My theorem then is that the ratio of the time of descent along AC to that along AB is equal to the product of the ratio of AC to AB by the square root of the inverse ratio of their heights.

For let AD be a perpendicular to which are drawn the horizontal lines BG and CD; also let AL be a mean proportional to the heights AG and AD; from the point L draw a horizontal line meeting AC in F; accordingly AF will be a mean proportional between AC and AE. Now since the time of descent along AC is to that along AE as the length AF is to AE; and since the time along AE is to that along AB as AE is to AB, it is clear that the time along AC is to that along AB as AF is to AB.

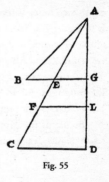

Fig. 55

[226]

Thus it remains to be shown that the ratio of AF to AB is equal to the product of the ratio of AC to AB by the ratio of AG to AL, which is the inverse ratio of the square roots of the heights DA and GA. Now it is evident that, if we consider the line AC in connection with AF and AB, the ratio of AF to AC is the same as that of AL to AD, or AG to AL which is the square root of the ratio of the heights AG and AD; but the ratio of AC to AB is the ratio of the lengths themselves. Hence follows the theorem.

THEOREM VI, PROPOSITION VI

If from the highest or lowest point in a vertical circle there be drawn any inclined planes meeting the circumference the times of descent along these chords are each equal to the other.

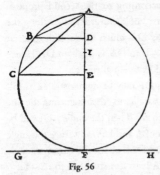

Fig. 56

On the horizontal line GH construct a vertical circle. From its lowest point—the point of tangency with the horizontal—draw the diameter FA and from the highest point, A, draw inclined planes to B and C, any points whatever on the circumference; then the times of descent along these are equal. Draw BD and CE perpendicular to the diameter; make AI a mean proportional between the heights of the planes, AE and AD; and since the, rectangles FA.AE and FA.AD are respectively equal to the squares of AC and AB, while the rectangle FA.AE is to the rectangle FA.AD as AE is to AD, it follows that the square of AC is to the square of AB as the length AE is to the length AD. But since the length AE is to AD as the square of AI is to the square of AD, it follows that the squares on the lines AC and AB

are to each other as the squares on the lines AI and AD, and hence also the length AC is to the length AB as AI is to AD. But it has previously been demonstrated that the ratio of the time of descent along AC to that along AB is equal to the product of the two ratios AC to AB and AD to AI; but this last ratio is the same as that of AB to AC. Therefore the ratio of the time of descent along AC to that along AB is the product of the two ratios, AC to AB and AB to AC. The ratio of these times is therefore unity. Hence follows our proposition.

By use of the principles of mechanics [*ex mechanicis*] one may obtain the same result, namely, that a falling body will require equal times to traverse the distances CA

[227]

and DA, indicated in the following figure. Lay off BA equal to DA, and let fall the perpendiculars BE and DF; it follows from the principles of mechanics that the component of the momentum [*momentum ponderis*] acting along the inclined plane ABC is

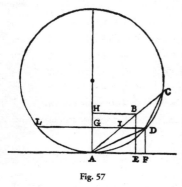

Fig. 57

to the total momentum [i. e., the momentum of the body falling freely] as BE is to BA; in like manner the momentum along the plane AD is to its total momentum [i. e., the momentum of the body falling freely] as DF is to DA, or to BA. Therefore the momentum of this same weight along the plane DA is to that along the plane ABC as the length DF is to the length BE; for this reason, this same weight will in equal times according to the second proposition of the first book, traverse spaces along the planes CA and DA which are to each other as

the lengths BE and DF. But it can be shown that CA is to DA as BE is to DF. Hence the falling body will traverse the two paths CA and DA in equal times.

Moreover the fact that CA is to DA as BE is to DF may be demonstrated as follows: Join C and D; through D, draw the line DGL parallel to AF and cutting the line AC in I; through B draw the line BH, also parallel to AF. Then the angle ADI will be equal to the angle DCA, since they subtend equal arcs LA and DA, and since the angle DAC is common, the sides of the triangles, CAD and DAI, about the common angle will be proportional to each other; accordingly as CA is to DA so is DA to IA, that is as BA is to IA, or as HA is to GA, that is as BE is to DF.   Q.E.D.

The same proposition may be more easily demonstrated as follows: On the horizontal line AB draw a circle whose diameter DC is vertical. From the upper end of this diameter draw any inclined plane, DF, extending to meet the circumference; then,

I say, a body will occupy the same time in falling along the plane DF as along the diameter DC. For draw FG parallel to AB and perpendicular to DC; join FC; and since the time of fall along DC is to that along DG as the mean proportional between CD and

[228]

GD is to GD itself; and since also DF is a mean proportional between DC and DG, the angle DFC inscribed in a semicircle being a right-angle, and FG being perpendicular to DC, it follows that the time of fall along DC is to that along DG as the length FD is to GD. But it has already been demonstrated that the time of descent along DF is to that along DG as the length DF is to DG; hence the times of descent along DF and DC each bear to the time of fall along DG the same ratio; consequently they are equal.

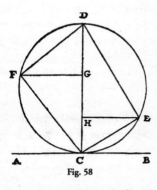

Fig. 58

In like manner it may be shown that if one draws the chord CE from the lower end of the diameter, also the line EH parallel to the horizon, and joins the points E and D, the time of descent along EC, will be the same as that along the diameter, DC.

COROLLARY I

From this it follows that the times of descent along all chords drawn through either C or D are equal one to another.

COROLLARY II

It also follows that, if from any one point there be drawn a vertical line and an inclined one along which the time of descent is the same, the inclined line will be a chord of a semicircle of which the vertical line is the diameter.

COROLLARY III

Moreover the times of descent along inclined planes will be equal when the vertical heights of equal lengths of these planes are to each other as the lengths of the planes themselves; thus it is clear that the times of descent along CA and DA, in the figure just before the last, are equal, provided the vertical height of AB (AB being equal to AD), namely, BE, is to the vertical height DF as CA is to DA.

Sagr. Please allow me to interrupt the lecture for a moment in order that I may clear up an idea which just occurs to me; one which, if it involve no fallacy, suggests at

[229]

least a freakish and interesting circumstance, such as often occurs in nature and in the realm of necessary consequences.

If, from any point fixed in a horizontal plane, straight lines be drawn extending indefinitely in all directions, and if we imagine a point to move along each of these lines with constant speed, all starting from the fixed point at the same instant and moving with equal speeds, then it is clear that all of these moving points will lie upon the circumference of a circle which grows larger and larger, always having the aforesaid fixed point as its center; this circle spreads out in precisely the same manner as the little waves do in the case of a pebble allowed to drop into quiet water, where the impact of the stone starts the motion in all directions, while the point of impact remains the center of these ever-expanding circular waves. But imagine a vertical plane from the highest point of which are drawn lines inclined at every angle and extending indefinitely; imagine also that heavy particles descend along these lines each with a naturally accelerated motion and each with a speed appropriate to the inclination of its line.

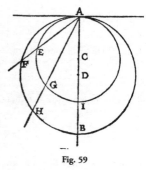

Fig. 59

If these moving particles are always visible, what will be the locus of their positions at any instant? Now the answer to this question surprises me, for I am led by the preceding theorems to believe that these particles will always lie upon the circumference of a single circle, ever increasing in size as the particles recede farther and farther from the point at which their motion began. To be more definite, let A be the fixed point from which are drawn the lines AF and AH inclined at any angle whatsoever. On the perpendicular AB take any two points C and D about which, as centers, circles are described passing through the point A, and cutting the inclined lines at the points F, H, B, E, G, I. From the preceding theorems it is clear that, if particles start, at the same instant, from A and descend along these lines, when one is at E another will be at G and another at I; at a later instant they will be found simultaneously at F, H and B; these, and indeed an infinite number of other particles travelling along an infinite

[230]

number of different slopes will at successive instants always lie upon a single ever-expanding circle. The two kinds of motion occurring in nature give rise therefore to two infinite series of circles, at once resembling and differing from each other; the one

takes its rise in the center of an infinite number of concentric circles; the other has its origin in the contact, at their highest points, of an infinite number of eccentric circles; the former are produced by motions which are equal and uniform; the latter by motions which are neither uniform nor equal among themselves, but which vary from one to another according to the slope.

Further, if from the two points chosen as origins of motion, we draw lines not only along horizontal and vertical planes but in all directions then just as in the former cases, beginning at a single point ever-expanding circles are produced, so in the latter case an infinite number of spheres are produced about a single point, or rather a single sphere which expands in size without limit; and this in two ways, one with the origin at the center, the other on the surface of the spheres.

Salv. The idea is really beautiful and worthy of the clever mind of Sagredo.

Simp. As for me, I understand in a general way how the two kinds of natural motions give rise to the circles and spheres; and yet as to the production of circles by accelerated motion and its proof, I am not entirely clear; but the fact that one can take the origin of motion either at the inmost center or at the very top of the sphere leads one to think that there may be some great mystery hidden in these true and wonderful results, a mystery related to the creation of the universe (which is said to be spherical in shape), and related also to the seat of the first cause [*prima causa*].

Salv. I have no hesitation in agreeing with you. But profound considerations of this kind belong to a higher science than ours [*a più alte dottrine che la nostre*]. We must be satisfied to belong to that class of less worthy workmen who procure from the quarry the marble out of which, later, the gifted sculptor produces those masterpieces which lay hidden in this rough and shapeless exterior. Now, if you please, let us proceed.

[231]

THEOREM VII, PROPOSITION VII

If the heights of two inclined planes are to each other in the same ratio as the squares of their lengths, bodies starting from rest will traverse these planes in equal times.

Take two planes of different lengths and different inclinations, AE and AB, whose heights are AF and AD: let AF be to AD as the square of AE is to the square of AB; then, I say, that a body, starting from rest at A, will traverse the planes AE and AB in equal times. From the vertical line, draw the horizontal parallel lines EF and DB, the latter cutting AE at G. Since $FA:DA=DV:EA^2:BA^2$, and since $FA:DA=EA:GA$, it follows that $EA:GA=EA^2:BA^2$. Hence BA is a mean proportional between FA and GA.

Now since the time of descent along AB bears to the time along AG the same ratio which AB bears to AG and since also the time of descent along AG is to the time along AE as AG is to a mean proportional between AG and AE, that is, to AB, it follows, *ex æquali*, that the time along AB is to the time along AE as AB is to itself. Therefore the times are equal.

Q.E.D.

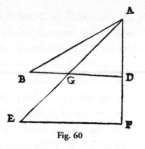

Fig. 60

THEOREM VIII, PROPOSITION VIII

The times of descent along all inclined planes which intersect one and the same vertical circle, either at its highest or lowest point, are equal to the time of fall along the vertical diameter; for those planes which fall short of this diameter the times are shorter; for planes which cut this diameter, the times are longer.

Let AB be the vertical diameter of a circle which touches the horizontal plane. It has already been proven that the times of descent along planes drawn from either end, A or B, to the circumference are equal. In order to show that the time of descent along

[232]

the plane DF which falls short of the diameter is shorter we may draw the plane DB which is both longer and

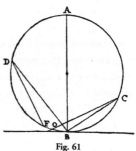

Fig. 61

less steeply inclined than DF; whence it follows that the time along DF is less than that along DB and consequently along AB. In like manner, it is shown that the time of descent along CO which cuts the diameter is greater: for it is both longer and less steeply inclined than CB. Hence follows the theorem.

THEOREM IX, PROPOSITION IX

If from any point on a horizontal line two planes, inclined at any angle, are drawn, and if they are cut by a line which makes with them angles alternately equal to the angles between these planes and the horizontal, then the times required to traverse those portions of the plane cut off by the aforesaid line are equal.

Through the point C on the horizontal line X, draw two planes CD and CE inclined at any angle whatever: at any point in the line CD lay off the angle CDF equal to the angle XCE; let the line DF cut CE at F so that the angles CDF and CFD are alternately equal to XCE and LCD; then, I say, the times of descent over CD and CF are equal. Now since the angle CDF is equal to the angle XCE by construction, it is evident that the angle CFD must be equal to the angle DCL. For if the common angle DCF be subtracted from the three angles of the triangle CDF, together equal to two right angles, (to which are also equal all the

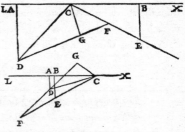

Fig. 62

angles which can be described about the point C on the lower side of the line LX) there remain in the triangle two angles, CDF and CFD, equal to the two angles XCE and LCD; but, by hypothesis, the angles CDF and XCE are equal; hence the remaining angle CFD is equal to the remainder DCL. Take CE equal to CD; from the points D and E draw DA and EB perpendicular to the horizontal line XL; and from the point C draw CG perpendicular to DF. Now since the angle CDG is equal to the angle ECB and since DGC and CBE are right angles, it follows that the triangles CDG and CBE are equiangular; consequently DC:CG=CE:EB. But DC is equal to CE, and therefore CG is equal to EB.

[233]

Since also the angles at C and at A, in the triangle DAC, are equal to the angles at F and G in the triangle CGF, we have CD:DA=FC:CG and, *permutando*, DC:CF=DA:CG=DA:BE. Thus the ratio of the heights of the equal planes CD and CE is the same as the ratio of the lengths DC and CF. Therefore, by Corollary I of Prop. VI, the times of descent along these planes will be equal. Q.E.D.

An alternative proof is the following: Draw FS perpendicular to the horizontal line AS. Then, since the triangle CSF is similar to the triangle DGC, we have SF:FC=GC:CD; and since the triangle CFG is similar to the triangle DCA, we have FC:CG=CD:DA. Hence, *ex aequali*, SF:CG=CG:DA. Therefore CG is a mean proportional between SF and DA, while

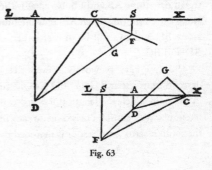

Fig. 63

DA:SF=DA^2:CG^2. Again since the triangle ACD is similar to the triangle CGF, we have DA:DC=GC:CF and, *permutando*, DA:CG=DC:CF: also DA^2:CG^2=DC^2:CF^2. But it has been shown that DA^2:CG^2=DA:FS. Therefore DC^2:CF^2=DA:FS. Hence from the above Prop. VII, since the heights DA and FS of the planes CD and CF are to each other as the squares of the lengths of the planes, it follows that the times of descent along these planes will be equal.

THEOREM X, PROPOSITION X

The times of descent along inclined planes of the same height, but of different slope, are to each other as the lengths of these planes; and this is true whether the motion starts from rest or whether it is preceded by a fall from a constant height.

Let the paths of descent be along ABC and ABD to the horizontal plane DC so that the falls along BD and BC are preceded by the fall along AB; then, I say, that the time of descent along BD is to the time of descent along BC as the length BD is to BC. Draw the horizontal line AF and extend DB until it cuts this line at F; let FE be a mean

[234]

proportional between DF and FB; draw EO parallel to DC; then AO will be a mean proportional between CA and AB. If now we represent the time of fall along AB by the length AB, then the time of descent along FB will be represented by the distance FB; so also the time of fall through the entire distance AC will be represented by the mean proportional AO: and for the entire distance FD by FE. Hence the time of fall along the remainder, BC, will be represented by BO, and that along the remainder, BD, by BE; but since BE:BO=BD:BC, it follows, if we allow the bodies to fall first along AB and FB, or, what is the same thing, along the common stretch AB, that the times of descent along BD and BC will be to each other as the lengths BD and BC.

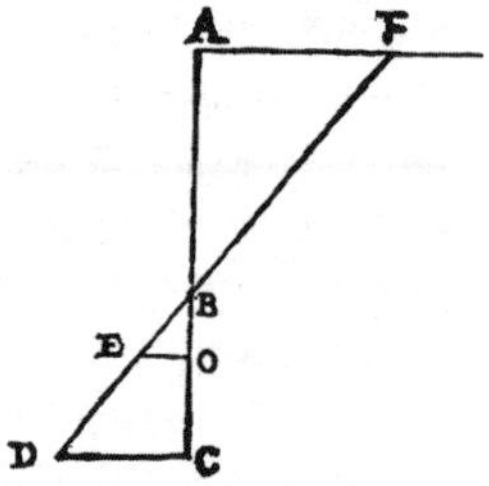

Fig. 64

But we have previously proven that the time of descent, from rest at B, along BD is to the time along BC in the ratio which the length BD bears to BC. Hence the times of descent along different planes of constant height are to each other as the lengths of these planes, whether the motion starts from rest or is preceded by a fall from a constant height.　Q.E.D.

THEOREM XI, PROPOSITION XI

If a plane be divided into any two parts and if motion along it starts from rest, then the time of descent along the first part is to the time of descent along the remainder as the length of this first part is to the excess of a mean proportional between this first part and the entire length over this first part.

Let the fall take place, from rest at A, through the entire distance AB which is divided at any point C; also let AF be a mean proportional between the entire length BA and the first part AC; then CF will denote the excess of the mean proportional FA over the first part AC. Now, I say, the time of descent along AC will be to the time of subsequent fall through CB as the length AC is to CF. This is evident, because the time along AC is to the time along the entire distance AB as AC is to the mean proportional AF. Therefore, *dividendo*, the time along AC will be to the time along the remainder CB as AC is to CF. If we agree to represent the time along AC by the length AC then the time along CB will be represented by CF.

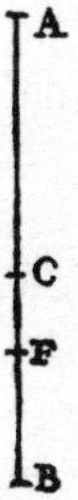

Fig. 65

Q.E.D.

[235]

In case the motion is not along the straight line ACB but along the broken line ACD to the horizontal line BD, and if from F we draw the horizontal line FE, it may in like manner be proved that the time along AC is to the time along the inclined line CD as AC is to CE. For the time along AC is to the time along CB as AC is to CF; but it has already been shown time along CB, after the fall through the distance AC, is to the time along CD, after descent through the same distance AC, as CB is to CD, or, as CF is to CE; therefore, *ex æquali*, the time along AC will be to the time along CD as the length AC is to the length CE.

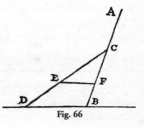

Fig. 66

THEOREM XII, PROPOSITION XII

If a vertical plane and any inclined plane are limited by two horizontals, and if we take mean proportionals between the lengths of these planes and those portions of them which lie between their point

of intersection and the upper horizontal, then the time of fall along the perpendicular bears to the time required to traverse the upper part of the perpendicular plus the time required to traverse the lower part of the intersecting plane the same ratio which the entire length of the vertical bears to a length which is the sum of the mean proportional on the vertical plus the excess of the entire length of the inclined plane over its mean proportional.

Let AF and CD be two horizontal planes limiting the vertical plane AC and the inclined plane DF; let the two last-mentioned planes intersect at B. Let AR be a mean proportional between the entire vertical AC and its upper part AB; and let FS be a mean proportional between FD and its upper part FB. Then, I say, the time of fall along the entire vertical path AC bears to the time of fall along its

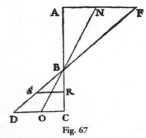

Fig. 67

upper portion AB plus the time of fall along the lower part of the inclined plane, namely, BD, the same ratio which the length AC bears to the mean proportional on the vertical, namely, AR, plus the length SD which is the excess of the entire plane DF over its mean proportional FS.

Join the points R and S giving a horizontal line RS. Now since the time of fall through the entire distance AC is to the time along the portion AB as CA is to the mean proportional AR it

follows that, if we agree to represent the time of fall through AC by the distance AC, the time of fall through the distance AB will be represented by AR; and the time of descent through the remainder, BC, will be represented by RC. But, if the time along AC is taken to be equal to the length AC, then the time along FD will be equal to the distance FD; and we may likewise infer that the time of descent

[236]

along BD, when preceded by a fall along FB or AB, is numerically equal to the distance DS. Therefore the time required to fall along the path AC is equal to AR plus RC; while the time of descent along the broken line ABD will be equal to AR plus SD.

Q.E.D.

The same thing is true if, in place of a vertical plane, one takes any other plane, as for instance NO; the method of proof is also the same.

PROBLEM I, PROPOSITION XIII

Given a perpendicular line of limited length, it is required to find a plane having a vertical height equal to the perpendicular and so inclined that a body, having fallen from rest along the perpendicular, will make its descent along the inclined plane in the same time which it occupied in falling through the given perpendicular.

Let AB denote the given perpendicular: prolong this line to C making BC equal to AB, and draw the horizontal lines CE and AG. It is required to draw a plane from B to the horizontal line CE such that after a body starting from rest at A has fallen through the distance AB, it will complete its path along this plane in an equal time.

Lay off CD equal to BC, and draw the line BD. Construct the line BE equal to the sum of BD and DC; then, I say, BE is the required plane. Prolong EB till it intersects the horizontal AG at G. Let GF be a mean proportional between GE and GB; then EF:FB=EG:GF, and EF^2:FB^2=EG^2:GF^2=EG:GB. But EG is twice GB; hence the square of EF is twice the square of FB; so also is the square of DB twice the square of BC. Consequently EF:FB=DB:BC,

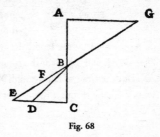

Fig. 68

and *componendo et permutando* EB:DB+BC=BF:BC. But EB=DB+BC; hence BF=BC=BA. If we agree that the length AB shall represent the time of fall along the line AB, then GB will represent the time of descent along GB, and GF the time along the entire distance GE; therefore BF will represent the time of descent along the difference of these paths, namely, BE, after fall from G or from A.   Q.E.F.

[237]

PROBLEM II, PROPOSITION XIV

Given an inclined plane and a perpendicular passing through it, to find a length on the upper part of the perpendicular through which a body will fall from rest in the same time which is required to traverse the inclined plane after fall through the vertical distance just determined.

Let AC be the inclined plane and DB the perpendicular. It is required to find on the vertical AD a length which will be traversed by a body, falling from rest, in the same time which is needed by the same body to traverse the plane AC after the aforesaid fall.

Draw the horizontal CB; lay off AE such that BA+2AC:AC=AC:AE, and lay off AR such that BA:AC=EA:AR. From R draw RX perpendicular to DB; then, I say, X is the

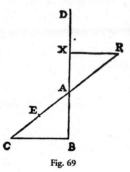

Fig. 69

point sought. For since BA+2AC:AC=AC:AE, it follows, *dividendo*, that BA+AC:AC=CE:AE. And since BA:AC=EA:AR, we have, *componendo*, BA+AC:AC=ER:RA. But BA+AC:AC=CE:AE, hence CE:EA=ER:RA= sum of the antecedents: sum of the consequents =CR:RE. Thus RE is seen to be a mean proportional between CR and RA. Moreover since it has been assumed that BA:AC=EA:AR, and since by similar triangles we have BA:AC=XA:AR, it follows that EA:AR=XA:AR. Hence EA and XA are equal. But if we agree that the time of fall through RA shall be represented by the length RA, then the time of fall along RC will be represented by the length RE which is a mean proportional between RA and RC; likewise AE will represent the time of descent along AC after descent along RA or along AX. But the time of fall through XA is represented by the length XA, while RA represents the time through RA. But it has been shown that XA and AE are equal.   Q.E.F.

[238]

PROBLEM III, PROPOSITION XV

Given a vertical line and a plane inclined to it, it is required to find a length on the vertical line below its point of intersection which will be traversed in the same time as the inclined plane, each of these motions having been preceded by a fall through the given vertical line.

Let AB represent the vertical line and BC the inclined plane; it is required to find a length on the perpendicular below its point of intersection, which after a fall from A will be traversed in the same time which is needed for BC after an identical fall from A. Draw the horizontal AD, intersecting the prolongation of CB at D; let DE be a

[239]

mean proportional between CD and DB; lay off BF equal to BE; also let AG be a third proportional to BA and AF. Then, I say, BG is the distance which a body, after falling through AB, will traverse in the same time which is needed for the

plane BC after the same preliminary fall. For if we assume that the time of fall along AB is represented by AB, then the time for DB will be represented by DB. And since DE is a mean proportional between BD and DC, this same DE will represent the time of descent along the entire distance DC while BE will represent the time required for the difference of these paths, namely, BC, provided in each case the fall is from rest at D or at A. In like manner we may infer that BF represents the time of descent through the distance BG after the same preliminary fall; but BF is equal to BE. Hence the problem is solved.

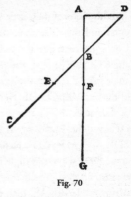

Fig. 70

THEOREM XIII, PROPOSITION XVI

If a limited inclined plane and a limited vertical line are drawn from the same point, and if the time required for a body, starting from rest, to traverse each of these is the same, then a body falling from any higher altitude will traverse the inclined plane in less time than is required for the vertical line.

Let EB be the vertical line and CE the inclined plane, both starting from the common point E, and both traversed in equal times by a body starting from rest at E; extend the vertical line upwards to any point A, from which falling bodies are allowed to start. Then, I say that, after the fall through AE, the inclined plane EC will be traversed in less time than the perpendicular EB. Join CB, draw the horizontal AD, and prolong CE backwards until it meets the latter in D; let DF be a mean proportional between CD and DE while AG is made a mean proportional between BA and AE. Draw FG and DG; then since the times of descent along EC and EB, starting from rest at E, are equal, it follows, according to Corollary II of Proposition VI that the angle at C is a right angle; but the angle at A is also a right angle

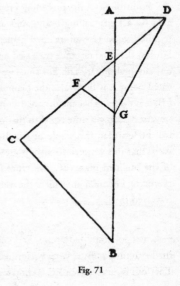

Fig. 71

and the angles at the vertex E are equal; hence the triangles AED and CEB are equiangular and the sides about the equal angles are proportional; hence BE:EC=DE:EA. Consequently the rectangle BE.EA is equal to the rectangle CE.ED; and since the rectangle CD.DE exceeds the rectangle CE.ED by the square of ED, and since the rectangle BA.AE exceeds the rectangle BE.EA by the square of EA, it follows that the excess of the rectangle CD.DE over the rectangle BA.AE, or what is the same thing, the excess of the square of FD over the square of AG, will be equal to the excess of the square of DE over the square of AE, which excess is equal to the square of AD. Therefore $FD^2=GA^2+AD^2=GD^2$. Hence DF is equal to DG, and the angle DGF is equal to the angle DFG while the angle EGF is less than the angle EFG, and the opposite side EF is less than the opposite side EG. If now we agree to represent the time of fall through AE by the length AE, then the time along DE will be represented by DE. And since AG is a mean proportional between BA and AE, it follows that AG will represent the time of fall through the total distance AB, and the difference EG will represent the time of fall, from rest at A, through the difference of path EB.

In like manner EF represents the time of descent along EC, starting from rest at D or falling from rest at A. But it has been shown that EF is less than EG; hence follows the theorem.

COROLLARY

From this and the preceding proposition, it is clear that the vertical distance covered by a freely falling body, after a preliminary fall, and during the time-interval required to traverse an inclined plane, is greater than the length of the inclined plane, but less than the distance traversed on the inclined plane during an equal time, without any preliminary fall. For since we have just shown that bodies falling from an elevated point A will traverse the plane EC in Fig. 71 in a shorter time than the vertical EB, it is evident that the distance along EB which will be traversed during a time equal to that of descent along EC will be less than the whole of EB. But now in order to show that this vertical distance is greater than the length of the inclined plane EC, we reproduce Fig. 70 of the preceding theorem in which the vertical length BG is traversed in

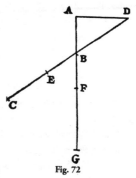

Fig. 72

[240]

the same time as BC after a preliminary fall through AB. That BG is greater than BC is shown as follows: since BE

and FB are equal while BA is less than BD, it follows that FB will bear to BA a greater ratio than EB bears to BD; and, *componendo*, FA will bear to BA a greater ratio than ED to DB; but FA:AB=GF:FB (since AF is a mean proportional between BA and AG) and in like manner ED:BD=CE:EB. Hence GB bears to BF a greater ratio than CB bears to BE; therefore GB is greater than BC.

PROBLEM IV, PROPOSITION XVII

Given a vertical line and an inclined plane, it is required to lay off a distance along the given plane which will be traversed by a body, after fall along the perpendicular, in the same time-interval which is needed for this body to fall from rest through the given perpendicular.

Let AB be the vertical line and BE the inclined plane. The problem is to determine on BE a distance such that a body, after falling through AB, will traverse it in a time equal to that required to traverse the perpendicular AB itself, starting from, rest.

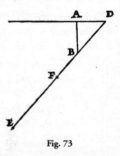

Draw the horizontal AD and extend the plane until it meets this line in D. Lay off FB equal to BA; and choose the point E such that BD:FD=DF:DE. Then, I say, the time of descent along BE, after fall through AB, is equal to the time of fall, from rest at A, through AB. For, if we assume that the length AB represents the tune of fall through AB, then the time of fall through DB will be represented by the time DB; and since BD:FD=DF:DE, it follows that DF will represent the time of descent along the entire plane DE while BF represents the time through the portion BE starting from rest at D; but the time of descent along BE after the preliminary descent along

Fig. 73

DB is the same as that after a preliminary fall through AB. Hence the time of descent along BE after AB will be BF which of course is equal to the time of fall through AB from rest at A. Q.E.F.

[241]

PROBLEM V, PROPOSITION XVIII

Given the distance through which a body will fall vertically from rest during a given time-interval, and given also a smaller time-interval, it is required to locate another [equal] vertical distance which the body will traverse during this given smaller time-interval.

557

Let the vertical line be drawn through A, and on this line lay off the distance AB which is traversed by a body falling from rest at A, during a time which may also be represented by AB. Draw the horizontal line CBE, and on it lay off BC to represent the given interval of time which is shorter than AB. It is required to locate, in the perpendicular above mentioned, a distance which is equal to AB and which will be descended in a time equal to BC. Join the points A and C; then, since BC<BA, it follows that the angle BAC<angle BCA. Construct the angle CAE equal to BCA and let E be the point where AE intersects the horizontal line; draw ED at right angles to AE, cutting the vertical at D; lay off DF equal to BA. Then, I say, that FD is that portion of the vertical which a body starting from rest at A will traverse during the assigned time-interval BC. For, if in the right-angled triangle AED a perpendicular be drawn from the right-angle at E to the side AD, then AE will be a mean proportional between DA and AB while BE will be a mean proportional between BD and BA, or between FA and AB (seeing that FA is equal to DB); and since it has been agreed to represent the time of fall through AB by the distance AB, it follows that AE, or EC, will represent the time of fall through the entire distance AD, while EB will represent the time through AF. Consequently the remainder BC will represent the time of fall through the remaining distance FD.   Q.E.F.

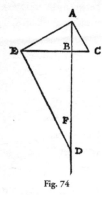

Fig. 74

[242]

PROBLEM VI, PROPOSITION XIX

Given the distance through which a body falls in a vertical line from rest and given also the time of fall, it is required to find the time in which the same body will, later, traverse an equal distance chosen anywhere in the same vertical line.

On the vertical line AB, lay off AC equal to the distance fallen from rest at A, also locate at random an equal distance DB. Let the time of fall through AC be represented by the length AC. It is required to find the time necessary to traverse DB after fall from rest at A. About the entire length AB describe the semicircle

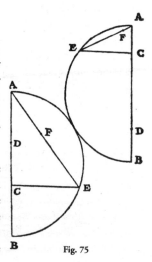

Fig. 75

AEB; from C draw CE perpendicular to AB; join the points A and E; the line AE will be longer than EC; lay off EF equal to EC. Then, I say, the difference FA will represent the time required for fall through DB. For since AE is a mean proportional between BA and AC and since AC represents the time of fall through AC, it follows that AE will represent the time through the entire distance AB. And since CE is a mean proportional between DA and AC (seeing that DA=BC) it follows that CE, that is, EF, will represent the time of fall through AD. Hence the difference AF will represent the time of fall through the difference DB.   Q.E.D.

COROLLARY

Hence it is inferred that if the time of fall from rest through any given distance is represented by that distance itself, then the time of fall, after the given distance has been increased by a certain amount, will be represented by the excess of the mean proportional between the increased distance and the original distance over the mean proportional between the original distance and the increment. Thus, for instance, if we agree that AB represents the time of fall, from rest at A, through the distance AB, and that AS is the increment, the time required to traverse AB, after fall through SA, will be the excess of the mean proportional between SB and BA over the mean proportional between BA and AS.

[243]

PROBLEM VII, PROPOSITION XX

Given any distance whatever and a portion of it laid off from the point at which motion begins, it is required to find another portion which lies at the other end of the distance and which is traversed in the same time as the first given portion.

Let the given distance be CB and let CD be that part of it which is laid off from the beginning of motion. It is required to find another part, at the end B, which is traversed in the same time as the assigned portion CD. Let BA be a mean proportional between BC and CD; also let CE be a third proportional to BC and CA. Then, I say, EB will be the distance which, after fall from C, will be traversed in the same time as CD itself. For if we agree that CB shall represent the time through the entire distance CB, then BA (which, of course, is a mean proportional between BC and CD) will represent the time along CD; and since CA is a mean proportional between BC and CE, it follows that CA will be the time through CE; but the total length

Fig. 76

CB represents the time through the total is distance CB. Therefore the difference BA will be the time along the difference of distances, EB, after falling from C; but this same BA was the time of fall through CD. Consequently the distances CD and EB are traversed, from rest at A, in equal times. Q.E.F.

[244.]

THEOREM XIV, PROPOSITION XXI

If, on the path of a body falling vertically from rest, one lays off a portion which is traversed in any time you please and whose upper terminus coincides with the point where the motion begins, and if this fall is followed by a motion deflected along any inclined plane, then the space traversed along the inclined plane, during a time-interval equal to that occupied in the previous vertical fall, will be greater than twice, and less than three times, the length of the vertical fall.

Fig. 77

Let AB be a vertical line drawn downwards from the horizontal line AE, and let it represent the path of a body falling from rest at A; choose any portion AC of this path. Through C draw any inclined plane, CG, along which the motion is

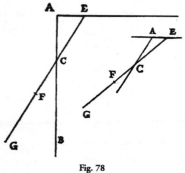

Fig. 78

continued after fall through AC. Then, I say, that the distance traversed along this plane CG, during the time-interval equal to that of the fall through AC, is more than twice, but less than three times, this same distance AC. Let us lay off CF equal to AC, and extend the plane GC until it meets the horizontal in E; choose G such that CE:EF=EF:EG. If now we assume that the time of fall along AC is represented by the length AC, then CE will represent the time of descent along CE, while CF, or CA, will

represent the time of descent along CG. It now remains to be shown that the distance CG is more than twice., and less than three times, the distance CA itself. Since CE:EF=EF:EG, it follows that CE:EF=CF:FG; but EC<EF; therefore CF will be less than FG and GC will be more than twice FC, or AC. Again since FE<2EC (for EC is greater than CA, or CF), we have GF less than twice FC, and also GC less than three times CF, or CA. Q.E.D.

This proposition may be stated in a more general form; since what has been proven for the case of a vertical and inclined plane holds equally well in the case of motion along a plane of any inclination followed by motion along any plane of less steepness, as can be seen from the adjoining figure. The method of proof is the same.

[245]

PROBLEM VIII, PROPOSITION XXII

Given two unequal time-intervals, also the distance through which a body will fall along a vertical line, from rest, during the shorter of these intervals, it is required to pass through the highest point of this vertical line a plane so inclined that the time of descent along it will be equal to the longer of the given intervals.

Let A represent the longer and B the shorter of the two unequal time-intervals, also let CD represent the length of the vertical fall, from rest, during the time B.

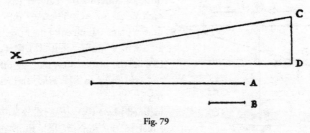

Fig. 79

It is required to pass through the point C a plane of such a slope that it will be traversed in the time A.

Draw from the point C to the horizontal a line CX of such a length that B:A=CD:CX. It is clear that CX is the plane along which a body will descend in the given time A. For it has been shown that the time of descent along an inclined plane bears to the time of fall through its vertical height the same ratio which the length of the plane bears to its vertical height. Therefore the time along CX is to the time along CD as the length CX is to the length CD, that is, as the time-interval A is to the time-interval B: but B is the time required to traverse the vertical distance, CD, starting from rest; therefore A is the time required for descent along the plane CX.

PROBLEM IX, PROPOSITION XXIII

Given the time employed by a body in falling through a certain dis-
tance along a vertical line, it is required to pass through the lower ter-
minus of this vertical fall, a plane so inclined that this body will, after
its vertical fall, traverse on this plane, during a time-interval equal
to that of the vertical fall, a distance equal to any assigned

[246]

distance, provided this assigned distance is more than twice and less
than three times, the vertical fall.

Let AS be any vertical line, and let AC denote both the length of the vertical fall, from rest at A, and also the time required for this fall. Let IR be a distance more than twice and less than three times, AC. It is required to pass a plane through the point C so inclined that a body, after fall through AC, will, during the time AC, traverse a

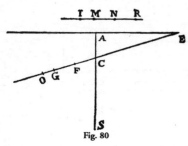

Fig. 80

distance equal to IR. Lay off RN and NM each equal to AC. Through the point C, draw a plane CE meeting the horizontal, AE, at such a point that IM:MN=AC:CE. Extend the plane to O, and lay off CF, FG and GO equal to RN, NM, and MI respectively. Then, I say, the time along the inclined plane CO, after fall through AC, is equal to the time of fall, from rest at A, through AC. For since OG:GF=FC:CE, it

follows, *componendo*, that OF:FG=OF:FC=FE:EC, and since an antecedent is to its consequent as the sum of the antecedents is to the sum of the consequents, we have OE:EF=EF:EC. Thus EF is a mean proportional between OE and EC. Having agreed to represent the time of fall through AC by the length AC it follows that EC will represent the time along EC, and EF the time along the entire distance EO, while the difference CF will represent the time along the difference CO; but CF=CA; therefore the problem is solved. For the time CA is the time of fall, from rest at A, through CA while CF (which is equal to CA) is the time required to traverse CO after descent along EC or after fall through AC.   Q.E.F.

It is to be remarked also that the same solution holds if the antecedent motion takes place, not along a vertical, but along an inclined plane. This case is illustrated in

[247]

the following figure where the antecedent motion is along the inclined plane AS underneath the horizontal AE. The proof is identical with the preceding.

SCHOLIUM

On careful attention, it will be clear that, the nearer the given line IR approaches to three times the length AC, the nearer the inclined plane, CO, along which the second

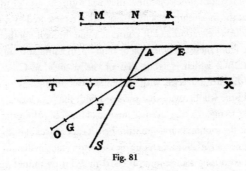

Fig. 81

motion takes place, approaches the perpendicular along which the space traverse during the time AC, will be three times the distance AC. For if. IR be taken nearly equal to three times AC, then IM will be almost equal to MN; and since, by construction, IM:MN=AC:CE, it follows that CE is but little greater than CA: consequently the point E will lie near the point A, and the lines CO and CS, forming a very acute angle, will almost coincide. But, on the other band, if the given line, IR, be only the least bit longer than twice AC, the line IM will be very short; from which it follows that AC will be very small in comparison with CE which is now so long that it almost coincides with the horizontal line drawn through C. Hence we can infer that, if, after descent along the inclined plane AC of the adjoining figure, the motion is continued along a horizontal line, such as CT, the distance traversed by a body, during a time equal to the time of fall through AC, will be exactly twice the distance AC. The argument here employed is the same as the preceding. For it is clear, since OE:EF=EF:EC, that FC measures the time of descent along CO. But, if the horizontal line TC which is twice as long as CA, be divided into two equal parts at V then this line must be extended indefinitely in the direction of X before it will intersect the line AE produced; and accordingly the ratio of the infinite length TX to the infinite length VX is the same as the ratio of the infinite distance VX to the infinite distance CX.

The same result may be obtained by another method of approach, namely, by returning to the same line of argument which was employed in the proof of the first proposition. Let us consider the triangle ABC, which, by lines drawn parallel to its

[248]

base, represents for us a velocity increasing in proportion to the time; if these lines are infinite in number, just as the points in the line AC are infinite or as the number of instants in any interval of time is infinite, they will form the area of the triangle. Let us now suppose that the maximum velocity attained-that represented by the line BC—to be continued, without acceleration and at constant value through another interval of time equal to the first. From these velocities will be built up, in a similar manner, the area of the parallelogram ADBC, which is twice that of the triangle ABC; accordingly the distance traversed with these velocities during any given interval of time will be twice that traversed with the velocities represented by the triangle during an equal interval of time. But along a horizontal plane the motion is uniform since here it experiences neither acceleration nor retardation; therefore we conclude that the distance CD traversed during a time-interval equal to AC is twice the distance AC; for the latter is covered by a motion, starting from rest and increasing in speed in proportion to the parallel lines in the tri-

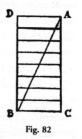

Fig. 82

angle, while the former is traversed by a motion represented by the parallel lines of the parallelogram which, being also infinite in number, yield an area twice that of the triangle.

Furthermore we may remark that any velocity once imparted to a moving body will be rigidly maintained as long as the external causes of acceleration or retardation are removed, a condition which is found only on horizontal planes; for in the case of planes which slope downwards there is already present a cause of acceleration, while on planes sloping upward there is retardation; from this it follows that motion along a horizontal plane is perpetual; for, if the velocity be uniform, it cannot be diminished or slackened, much less destroyed. Further, although any velocity which a body may have acquired through natural fall is permanently maintained so far as its own nature [*suapte natura*] is concerned, yet it must be remembered that if, after descent along a plane inclined downwards, the body is deflected to a plane inclined upward, there is already existing in this latter plane a cause of retardation; for in any such plane this same body is subject to a natural acceleration downwards. Accordingly we have here the superposition of two different states, namely, the velocity acquired during the preceding fall which if acting alone would carry the body at a uniform rate to infinity, and the velocity which results from a natural acceleration downwards common to all bodies. It seems altogether reasonable, therefore, if we wish to trace the future history of a body which has descended along some inclined

564

plane and has been deflected along some plane inclined upwards, for us to assume that the maximum speed acquired during descent is permanently maintained during the ascent. In the ascent, however, there

[249]

supervenes a natural inclination downwards, namely, a motion which, starting from rest, is accelerated at the usual rate. If perhaps this discussion is a little obscure, the following figure will help to make it clearer.

Let us suppose that the descent has been made along the downward sloping plane AB, from which the body is deflected so as to continue its motion along the upward sloping plane BC; and first let these planes be of equal length and placed so as to make equal angles with the horizontal line GH. Now it is well known that a body, starting from rest at A, and descending along AB, acquires a speed which is proportional to the time, which is a maximum at B, and which is maintained by the body so long as all causes of fresh acceleration or retardation are removed; the acceleration to which I refer is that to which the body would be subject if its motion were continued along the plane AB extended, while the retardation is that which the body would encounter if its motion were deflect- ed along the plane BC inclined upwards; but, upon the horizontal plane GH, the body would maintain a uniform velocity equal to that

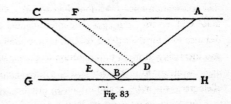

Fig. 83

which it had acquired at B after fall from A; moreover this velocity is such that, during an interval of time equal to the time of descent through AB, the body will traverse a horizontal distance equal to twice AB. Now let us imagine this same body to move with the same uniform speed along the plane BC so that here also during a time-interval equal to that of descent along AB, it will traverse along BC extended a distance twice AB; but let us suppose that, at the very instant the body begins its ascent it is subject- ed, by its very nature, to the same influences which surrounded it during its descent from A along AB, namely, it descends from rest under the same acceleration as that which was effective in AB, and it traverses, during an equal interval of time, the same distance along this second plane as it did along AB; it is clear that, by thus superpos- ing upon the body a uniform motion of ascent and an accelerated motion of descent, it will be carried along the plane BC as far as the point C where these two velocities become equal.

If now we assume any two points D and E, equally distant from the vertex B, we may then infer that the descent along BD takes place in the same time as the ascent

along BE. Draw DF parallel to BC; we know that, after descent along AD, the body will ascend along DF; or, if, on reaching D, the body is carried along the horizontal DE, it will reach E with the same momentum [*impetus*] with which it left D; hence from E the body will ascend as far as C, proving that the velocity at E is the same as that at D.

From this we may logically infer that a body which descends along any inclined plane and continues its motion along a plane inclined upwards will, on account of the

[250]

momentum acquired, ascend to an equal height above the horizontal; so that if the descent is along AB the body will be carried up the plane BC as far as the horizontal line ACD: and this is true whether the inclinations of the planes are the same or different, as in the case of

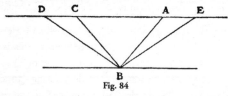
B
Fig. 84

the planes AB and BD. But by a previous postulate the speeds acquired by fall along variously inclined planes having the same vertical height are the same. If therefore the planes EB and BD have the same slope, the descent along EB will be able to drive the body

along BD as far as D; and since this propulsion comes from the speed acquired on reaching the point B, it follows that this speed at B is the same whether the body has made its descent along AB or EB. Evidently then the body will be carried up BD whether the descent has been made along AB or along EB. The time of ascent along BD is however greater than that along BC, just as the descent along EB occupies more time than that along AB; moreover it has been demonstrated that the ratio between the lengths of these times is the same as that between the lengths of the planes. We must next discover what ratio exists between the distances traversed in equal times along planes of different slope, but of the same elevation, that is, along planes which are included between the same parallel horizontal lines. This is done as follows:

THEOREM XV, PROPOSITION XXIV

Given two parallel horizontal planes and a vertical line connecting them; given also an inclined plane passing through the lower extremity of this vertical line; then, if a body fall freely along the vertical line and have its motion reflected along the inclined plane, the distance which it will traverse along this plane, during a time equal to that of the vertical fall, is greater than once but less than twice the vertical line.

Let BC and HG be the two horizontal planes, connected by the perpendicular AE; also let EB represent the inclined plane along which the motion takes place after the

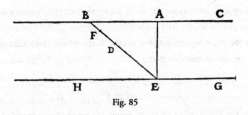

Fig. 85

body has fallen along AE and has been reflected from E towards B. Then, I say, that, during a time equal to that of fall along AE, the body will ascend the inclined plane through a distance which is greater than AE but less than twice AE. Lay off ED equal

[251]

to AE and choose F so that EB:BD=BD:BF. First we shall show that F is the point to which the moving body will be carried after reflection from E towards B during a time equal to that of fall along AE; and next we shall show that the distance EF is greater than EA but less than twice that quantity.

Let us agree to represent the time of fall along AE by the length AE, then the time of descent along BE, or what is the same thing, ascent along EB will be represented by the distance EB.

Now, since DB is a mean proportional between EB and BF, and since BE is the time of descent for the entire distance BE, it follows that BD will be the time of descent through BF, while the remainder DE will be the time of descent along the remainder FE. But the time of descent along the fall from rest at B is the same as the time of ascent from E to F after reflection from E with the speed acquired during fall either through AE or BE. Therefore DE represents the time occupied by the body in passing from E to F, after fall from A to E and after reflection along EB. But by construction ED is equal to AE. This concludes the first part of our demonstration.

Now since the whole of EB is to the whole of BD as the portion DB is to the portion BF, we have the whole of EB is to the whole of BD as the remainder ED is to the remainder DF; but EB>BD and hence ED>DF, and EF is less than twice DE or AE. Q.E.D.

The same is true when the initial motion occurs, not along a perpendicular, but upon an inclined plane: the proof is also the same provided the upward sloping plane is less steep, i. e., longer, than the downward sloping plane.

THEOREM XVI, PROPOSITION XXV

If descent along any inclined plane is followed by motion along a horizontal plane, the time of descent along the inclined plane bears to the time required to traverse any assigned length of the horizontal plane the same ratio which twice the length of the inclined plane bears to the given horizontal length.

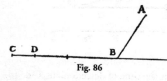
Fig. 86

Let CB be any horizontal line and AB an inclined plane; after descent along AB let the motion continue through the assigned horizontal distance BD. Then, I say, the time of descent along AB bears to the time spent in traversing BD the same ratio twice AB bears to BD. For, lay off BC equal to twice AB then it follows, from a previous proposition, that the time of descent along AB is equal to the time required to traverse BC; but the time along BC is to the time along DB as the length CB is to the length BD. Hence the time of

[252]

descent along AB is to the time along BD as twice the distance AB is to the distance BD.

Q.E.D.

PROBLEM X, PROPOSITION XXVI

Given a vertical height joining two horizontal parallel lines; given also a distance greater than once and less than twice this vertical height, it is required to pass through the foot of the given perpendicular an inclined plane such that, after fall through the given vertical height, a body whose motion is deflected along the plane will traverse the assigned distance in a time equal to the time of vertical fall.

Let AB be the vertical distance separating two parallel horizontal lines AO and BC; also let FE be greater than once and less than twice BA. The problem is to pass a plane through B, extending to the upper horizontal line, and such that a body, after having fallen from A to B, will, if its motion be deflected along the inclined plane, traverse a distance equal to EF in a time equal to that of fall along AB. Lay off ED equal to AB; then the remainder DF will be less than AB since the entire length EF is less than twice this quantity; also lay off DI equal to DF, and choose the point X such that EI:ID=DF:FX; from B, draw the plane BO equal in length to EX. Then, I say, that the

568

plane BO is the one along which, after fall through AB, a body will traverse the assigned distance FE in a time equal to the time of fall through AB. Lay off BR and RS equal to ED and DF respectively; then since EI:ID=DF:FX, we have, *componendo*, ED:DI=DX:XF=ED:DF=EX:XD=BO:OR=RO:OS. If we represent the time of fall

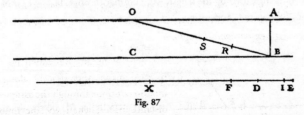

Fig. 87

along AB by the length AB, then OB will represent the time of descent along OB, and RO will stand for the time along OS, while the remainder BR will represent the time required for a body starting from rest at O to traverse the remaining distance SB. But the time of descent along SE starting from rest at O is equal to the time of ascent from B to S after fall through AB. Hence BO is that plane, passing through B, along which a body, after fall through AB, will traverse the distance BS, equal to the assigned distance EF, in the time-interval BR or BA.   Q.E.F.

[253]

THEOREM XVII, PROPOSITION XXVII

If a body descends along two inclined planes of different lengths but of the same vertical height, the distance which it will traverse, in the lower part of the longer plane, during a time-interval equal to that of descent over the shorter plane, is equal to the length of the shorter plane plus a portion of it to which the shorter plane bears the same ratio which the longer plane bears to the excess of the longer over the shorter plane.

Let AC be the longer plane, AB, the shorter, and AD the common elevation; on the lower part of AC lay off CE equal to AB. Choose F such that CA:AE=CA:CA-AB=CE:EF. Then, I say, that FC is that distance which will, after fall from A, be traversed during a time-interval equal to that required

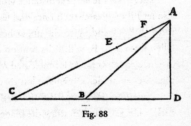

Fig. 88

for descent along AB. For since CA:AE=CE:EF, it follows that the remainder EA: the remainder AF=CA:AE. Therefore AE is a mean proportional between AC and AF. Accordingly if the length AB is employed to measure the time of fall along AB, then the distance AC will measure the time of descent through AC; but the time of descent through AF is measured by the length AE, and that through FC by EC. Now EC=AB; and hence follows the proposition.

[254]

PROBLEM XI, PROPOSITION XXVIII

Let AG be any horizontal line touching a circle; let AB be the diameter passing through the point of contact; and let AE and EB represent any two chords. The problem is to determine what ratio the time of fall through AB bears to the time of descent over both AE and EB. Extend BE till it meets the tangent at G, and draw AF so as to bisect the angle BAE. Then, I say, the time through AB is to the sum of the times along AE and EB as the length AE is to the sum of the lengths AE and EF. For since the angle FAB is equal to the angle FAE, while the angle EAG is equal to the angle ABF it fol-

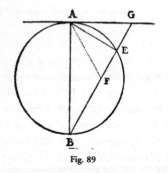
Fig. 89

lows that the entire angle GAF is equal to the sum of the angles FAB and ABF. But the angle GFA is also equal to the sum of these two angles. Hence the length GF is equal to the length GA; and since the rectangle BG.GE is equal to the square of GA, it will also be equal to the square of GF, or BG:GF=GF:GE. If now we agree to represent the time of descent along AE by the length AE, then the length GE will represent the time of descent along GE, while GF will stand for the time of descent through the entire distance GB; so also EF will denote the time through EB after fall from G or from A along AE. Consequently the time along AE, or AB, is to the time along AE and EB as the length AE is to AE+EF.   Q.E.D.

A shorter method is to lay off GF equal to GA, thus making GF a mean proportional between BG and GE. The rest of the proof is as above.

THEOREM XVIII, PROPOSITION XXIX

Given a limited horizontal line, at one end of which is erected a limited vertical line whose length is equal to one-half the given horizontal

line; then a body, falling through this given height and having its
motion deflected into a horizontal direction, will traverse the
given horizontal distance and vertical line in less time than
it will any other vertical distance plus the given horizontal
distance.

[255]

Let BC be the given distance in a horizontal plane; at the end B erect a perpendicular, on which lay off BA equal to half BC. Then, I say, that the time
required for a body, starting from rest
at A, to traverse the two distances, AB
and BC, is the least of all possible
times in which this same distance BC
together with a vertical portion,
whether greater or less than AB, can
be traversed.

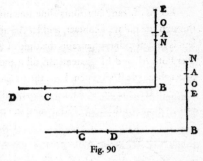

Fig. 90

Lay off EB greater than AB, as in
the first figure, and less than AB, as in
the second. It must be shown that the
time required to traverse the distance
EB plus BC is greater than that required for AB plus BC. Let us agree that the
length AB shall represent the time along AB, then the time occupied in traversing the horizontal portion BC will also be AB, seeing that BC=2AB; consequently the time required for both AB and BC will be twice AB. Choose the
point O such that EB:BO=BO:BA, then BO will represent the time of fall
through EB. Again lay off the horizontal distance BD equal to twice BE; whence
it is clear that BO represents the time along BD after fall through EB. Select a
point N such that DB:BC=EB:BA=OB:BN. Now since the horizontal motion is
uniform and since OB is the time occupied in traversing BD, after fall from E,
it follows that NB will be the time along BC after fall through the same height
EB. Hence it is clear that OB plus BN represents the time of traversing EB plus
BC; and since twice BA is the time along AB plus BC. it remains to be shown
that OB+BN>2BA.

But since EB:BO=BO:BA, it follows that EB:BA=OB^2:BA^2. Moreover since
EB:BA=OB:BN it follows that OB:BN=OB^2:BA^2. But OB:BN=(OB:BA)(BA:BN),
and therefore AB:BN=OB:BA, that is, BA is a mean proportional between BO and
BN. Consequently OB+BN>2BA.　　　　　　　　　　　　　　　　　Q.E.D.

[256]

THEOREM XIX, PROPOSITION XXX

A perpendicular is let fall from any point in a horizontal line; it is required to pass through any other point in this same horizontal line a plane which shall cut the perpendicular and along which a body will descend to the perpendicular in the shortest possible time. Such a plane will cut from the perpendicular a portion equal to the distance of the assumed point in the horizontal from the upper end of the perpendicular.

Let AC be any horizontal line and B any point in it from which is dropped the vertical line BD. Choose any point C in the horizontal line and lay off, on the vertical, the distance BE equal to BC; join C and E. Then, I say, that of all inclined planes that can be passed through C, cutting the perpendicular, CE is that one along which the descent to the perpendicular is accomplished in the shortest time. For, draw the plane CF cutting the vertical above E, and, the plane CG cutting the vertical below E; and draw IK, a parallel vertical line, touching at C a circle described with BC as radius. Let EK be drawn parallel to CF, and extended to meet the tangent, after cutting the circle at L. Now it is clear that the time of fall along LE is equal to the time along CE; but the time along KE is greater than along LE; therefore the time along KE is greater than along CE. But the time along KE is equal to the time along CF, since they have the

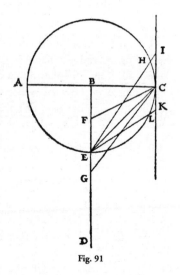

Fig. 91

same length and the same slope; and, in like manner, it follows that the planes CG and IE, having the same length and the same slope, will be traversed in equal times. Also, since HE<IE, the time along HE will be less than the time along IE. Therefore also the time along CE (equal to the time along HE), will be shorter than the time along IE. Q.E.D.

THEOREM XX, PROPOSITION XXXI

If a straight line is inclined at any angle to the horizontal and if, from any assigned point in the horizontal, a plane of quickest descent is to be drawn to the inclined line, that plane will be the one which bisects the angle contained between two lines drawn from

[257]

the given point, one perpendicular to the horizontal line, the other perpendicular to the inclined line.

Let CD be a line inclined at any angle to the horizontal AB; and from any assigned point A in the horizontal draw AC perpendicular to AB, and AE perpendicular to CD; draw FA so as to bisect the angle CAE. Then, I say, that of all the planes which can be drawn through the point A, cutting the line CD at any points whatsoever AF is the one of quickest descent [*in quo tempore omnium brevissimo fiat descensus*]. Draw FG parallel to AE; the alternate angles GFA and FAE will be equal; also the angle EAF is equal to the angle FAG. Therefore the sides GF and GA of the triangle FGA are equal. Accordingly if we describe a circle about G as center, with GA as radius, this circle will

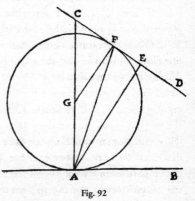

Fig. 92

pass through the point F, and will touch the horizontal at the point A and the inclined line at F; for GFC is a right angle, since GF and AE are parallel. It is clear therefore that all lines drawn from A to the inclined line, with the single exception of FA, will extend beyond the circumference of the circle, thus requiring more time to traverse any of them than is needed for FA.   Q.E.D.

LEMMA

If two circles one lying within the other are in contact, and if any straight line be drawn tangent to the inner circle, cutting the outer circle, and if three lines be drawn from the point at which the circles are in contact to three points on the tangential straight line, namely,

the point of tangency on the inner circle and the two points where the
straight line extended cuts the outer circle, then these three lines will
contain equal angles at the point of contact.

Let the two circles touch each other at the point A, the center of the smaller
being at B, the center of the larger at C. Draw the straight line FG touching the inner

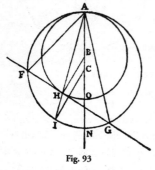

circle at H, and cutting the outer at the points F
and G; also draw the three lines AF, AH, and AG.
Then, I say, the angles contained by these lines,
FAH and GAH, are equal. Prolong AH to the circumference at I; from the centers of the circles,
draw BH and CI; join the centers B and C and
extend the line until it reaches the point of contact
at A and cuts the circles at the points O and N. But
now the lines BH and CI are parallel, because the
angles ICN and HBO are equal, each being twice
the angle IAN. And since BH, drawn from the cen-

Fig. 93

ter to the point of contact is perpendicular to FG, it follows that CI will also be perpendicular to FG and that the arc FI is equal to the arc IG· consequently the angle
FAI is equal to the angle IAG.   Q.E.D.

THEOREM XXI, PROPOSITION XXXII

If in a horizontal line any two points are chosen and if through one
of these points a line be drawn inclined towards the other, and if from
this other point a straight line is drawn to the inclined line in such a
direction that it cuts off from the inclined line a portion equal to the
distance between the two chosen points on the horizontal line, then
the time of descent along the line so drawn is less than along any other
straight line drawn from the same point to the same inclined line.
Along other lines which make equal angles on opposite side of this
line, the times of descent are the same.

Let A and B be any two points on a horizontal line: through B draw an inclined
straight line BC, and from B lay off a distance BD equal to BA; join the points A and
D. Then, I say, the time of descent along AD is less than along any other line drawn
from A to the inclined line BC. From the point A draw AE perpendicular to BA; and
from the point D draw DE perpendicular to BD, intersecting AE at E. Since in the
isosceles triangle ABD, we have the angles BAD and BDA equal, their complements

DAE and EDA are equal. Hence if, with E as center and EA as radius, we describe a circle it will pass through D and will touch the lines BA and BD at the points A and D. Now since A is the end of the vertical line AE, the descent along AD will occupy less time than along any other line drawn from the extremity A to the line BC and extending beyond the circumference of the circle; which concludes the first part of the proposition.

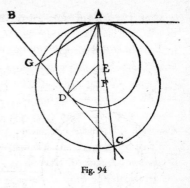

Fig. 94

[259]

If however, we prolong the perpendicular line AE, and choose any point F upon it, about which as center, we describe a circle of radius FA, this circle, AGC, will cut the tangent line in the points G and C. Draw the lines AG and AC which will according to the preceding lemma, deviate by equal angles from the median line AD. The time of descent along either of these lines is the same, since they start from the highest point A, and terminate on the circumference of the circle AGC.

PROBLEM XII, PROPOSITION XXXIII

Given a limited vertical line and an inclined plane of equal height, having a common upper terminal; it is required to find a point on the vertical line, extended upwards, from which a body will fall and, when deflected along the inclined plane, will traverse it in the same time-interval which is required for fall, from rest, through the given vertical height.

Let AB be the given limited vertical line and AC an inclined plane having the same altitude. It is required to find on the vertical BA, extended above A, a point from which a falling body will traverse the distance AC in the same time which is spent in falling, from rest at A, through the given vertical line AB. Draw the line DCE at right angles to AC, and lay off CD equal to AB; also join the points A and D; then the angle ADC will be greater than the angle CAD, since the side CA is greater than either AB or CD.

[260]

Make the angle DAE equal to the angle ADE, and draw EF perpendicular to AE; then EF will cut the inclined plane, extended both ways, at F. Lay off AI and AG each equal to CF; through G draw the horizontal line GH. Then, I say, H is the point sought.

For, if we agree to let the length AB represent the time of fall along the vertical AB, then AC will likewise represent the time of descent from rest at A, along AC; and since, in the right-angled triangle AEF, the line EC has been drawn from the right angle at E perpendicular to the base AF, it follows that AE will be a mean proportional between FA and AC, while CE will be a mean proportional between AC and CF, that is between CA and AI. Now, since AC represents the time of descent from A along AC, it follows that AE will be the time along the entire distance AF, and EC the time along AI. But since in the isosceles triangle AED the side EA is equal to the side

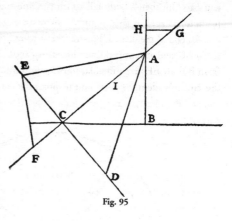

Fig. 95

ED it follows that ED will represent the time of fall along AF, while EC is the time of fall along AI. Therefore CD, that is AB, will represent the time of fall, from rest at A, along IF; which is the same as saying that AB is the time of fall, from G or from H, along AC.

E.F.

PROBLEM XIII, PROPOSITION XXXIV

Given a limited inclined plane and a vertical line having their highest point in common, it is required to find a point in the vertical line extended such that a body will fall from it and then traverse the inclined plane in the same time which is required to traverse the inclined plane alone starting from rest at the top of said plane.

Let AC and AB be an inclined plane and a vertical line respectively, having a common highest point at A. It is required to find a point in the vertical line, above A, such that a body, falling from it and afterwards having its motion directed along AB, will traverse both the assigned part of the vertical line and the plane AB in the same time

[261]

which is required for the plane AB alone, starting from rest at A. Draw BC a horizontal line and lay off AN equal to AC; choose the point L so that AB:BN=AL:LC, and lay off AI equal to AL; choose the point E such that CE, laid off on the vertical AC

576

produced, will be a third proportional to AC and BI. Then, I say, CE is the distance sought; so that, if the vertical line is extended above A and if a portion AX is laid off equal to CE, then a body falling from X will traverse both the distances, XA and AB, in the same time as that required, when starting from A, to traverse AB alone.

Draw XR parallel to BC and intersecting BA produced in R; next draw ED parallel to BC and meeting BA produced in D; on AD as diameter describe a semicircle; from B draw BF perpendicular to AD, and prolong it till it meets the circumference of the circle; evidently FB is a mean proportional between AB and BD, while FA is a

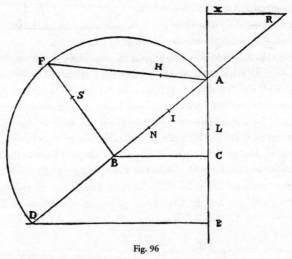

Fig. 96

mean proportional between DA and AB. Take BS equal to BI and FH equal to FB. Now since AB:BD=AC:CE and since BF is a mean proportional between AB and BD, while BI is a mean proportional between AC and CE, it follows that BA:AC=FB:BS, and since BA:AC=BA:BN=FB:BS we shall have, *convertendo*, BF:FS=AB:BN=AL:LC. Consequently the rectangle formed by FB and CL is equal to the rectangle whose sides are AL and SF; moreover, this rectangle AL.SF is the excess of the rectangle AL.FB, or

[262]

AI.BF, over the rectangle AI.BS, or AI.IB. But the rectangle FB.LC is the excess of the rectangle AC.BF over the rectangle AL.BF; and moreover the rectangle AC.BF is equal to the rectangle AB.BI since BA:AC=FB:BI; hence the excess of the rectangle AB.BI over the rectangle AI.BF, or AI.FH, is equal to the excess of the rectangle AI.FH over the rectangle AI.IB; therefore twice the rectangle AI.FH is equal to the sum of the

rectangles AB.BI and AI.IB, or $2AI.FH = 2AI.IB + \overline{BI}^2$. Add $\overline{AI}^2$ to each side, then $2AI.IB + \overline{BI}^2 + \overline{AI}^2 = \overline{AB}^2 = 2AI.FH = AI^2$. Again add $\overline{BF}^2$ to each side, then $AB^2 + BF^2 = \overline{AF}^2$ $= 2AI.FH + \overline{AI}^2 + \overline{BF}^2 = 2AI.FH + \overline{AI}^2 + \overline{FH}^2$. But $\overline{AF}^2 = 2AH.HF + \overline{AH}^2 + \overline{HF}^2$; and hence $2AI.FH + \overline{AI}^2 + \overline{FH}^2 = 2AH.HF + \overline{AH}^2 + \overline{HF}^2$. Subtracting $\overline{HF}^2$ from each side we have $2AI.FH + \overline{AI}^2 = 2AH.HF + \overline{AH}^2$. Since now FH is a factor common to both rectangles, it follows that AH is equal to AI; for if AH were either greater or smaller than AI, then the two rectangles AH.HF plus the square of HA would be either larger or smaller than the two rectangles AI.FH plus the square of IA, a result which is contrary to what we have just demonstrated.

If now we agree to represent the time of descent along AB by the length AB, then the time through AC will likewise be measured by AC; and IB, which is a mean proportional between AC and CE, will represent the time through CE, or XA, from rest at X. Now, since AF is a mean proportional between DA and AB, or between RB and AB, and since BF, which is equal to FH, is a mean proportional between AB and BD, that is between AB and AR, it follows, from a preceding proposition [Proposition XIX, corollary], that the difference AH represents the time of descent along AB either from rest at R or after fall from X, while the time of descent along AB, from rest at A, is measured by the length AB. But as has just been shown, the time of fall through XA is measured by IB, while the time of descent along AB, after fall, through RA or through XA, is IA. Therefore the time of descent through XA plus AB is measured by the length AB, which, of course, also measures the time of descent, from rest at A, along AB alone.   Q.E.F.

[263]

PROBLEM XIV, PROPOSITION XXXV

Given an inclined plane and a limited vertical line, it is required to find a distance on the inclined plane which a body, starting from rest, will traverse in the same time as that needed to traverse both the vertical and the inclined plane.

Let AB be the vertical line and BC the inclined plane. It is required to lay off on BC a distance which a body, starting from rest, will traverse in a time equal to that which is occupied by fall through the vertical AB and by descent of the plane. Draw the horizontal line AD, which intersects at E the prolongation of the inclined plane CB; lay off BF equal to BA, and about E as center, with EF as radius describe the circle FIG. Prolong FE until it intersects the circumference at G. Choose a point H such that GB:BF=BH:HF. Draw the line HI tangent to the circle at I. At B draw the line

BK perpendicular to FC, cutting the line EIL at L; also draw LM perpendicular to EL and cutting BC at M. Then, I say, BM is the distance which a body, starting from rest at B, will traverse in the same time which is required to descend from rest at A through both distances, AB and BM. Lay off EN equal to EL; then since GB:BF=BH:HF, we

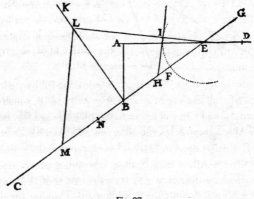

Fig. 97

shall have, *permutando*, GB:BH=BF:HF, and, *dividendo*, GH:BH=BH:HF,. Consequently the rectangle GH.HF is equal to the square on BH; but this same rectangle is also equal to the square on HI; therefore BH is equal to HI. Since, in the quadrilateral ILBH, the sides HB and HI are equal and since the angles at B and I are right angles, it follows that the sides BL and LI are also equal: but EI=EF; therefore the

[264]

total length LE, or NE, is equal to the sum of LB and EF. If we subtract the common part FF, the remainder FN will be equal to LB: but, by construction, FB=BA and, therefore, LB=AB+BN. If again we agree to represent the time of fall through AB by the length AB, then the time of descent along EB will be measured by EB; moreover since EN is a mean proportional between ME and EB it will represent the time of descent along the whole distance EM; therefore the difference of these distances, BM, will be traversed, after fall from EB, or AB, in a time which is represented by BN. But having already assumed the distance AB as a measure of the time of fall through AB, the time of descent along AB and BM is measured by AB+BN. Since EB measures the time of fall, from rest at E, along EB, the time from rest at B along BM will be the mean proportional between BE and BM, namely, BL. The time therefore for the path AB+BM, starting from rest at A is AB+BN; but the time for BM alone, starting from rest at B, is BL; and since it has already been shown that BL=AB+BN, the proposition follows.

Another and shorter proof is the following: Let BC be the inclined plane and BA the vertical; at B draw a perpendicular to EC, extending it both ways; lay off BH equal

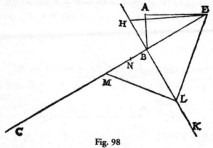

to the excess of BE over BA; make the angle HEL equal to the angle BHE; prolong EL until it cuts BK in L; at L draw LM perpendicular to EL and extend it till it meets BC in M; then, I say, BM is the portion of BC sought. For, since the angle MLE is a right angle, BL will be a mean proportional between MB and BE, while LE is a mean proportional between ME and BE; lay off EN equal

Fig. 98

to LE; then NE=EL=LH, and HB=NE-BL. But also HB=NE-(NB+BA); therefore BN+BA=BL. If now we assume the length EB as a measure of the time of descent along EB, the time of descent, from rest at B, along BM will be represented by BL; but, if the descent along BM is from rest at E or at A, then the time of descent will be measured by BN; and AB will measure the time along AB. Therefore the time required to traverse AB and BM, namely, the sum of the distances AB and BN, is equal to the time of descent, from rest at B, along BM alone.

Q.E.F.

[265]

LEMMA

Let DC be drawn perpendicular to the diameter BA; from the extremity B draw the line BED at random; draw the line FB. Then, I say, FB is a mean proportional between DB and BE. Join the points E and F. Through B, draw the tangent BG which will be parallel to CD. Now, since the angle DBG is equal to the angle FDB, and since the alternate angle of GBD is equal to EFB, it follows that the triangles FDB and FEB are similar and hence BD:BF=FB:BE.

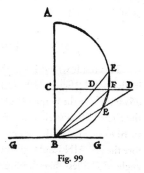

Fig. 99

LEMMA

Let AC be a line which is longer than DF, and let the ratio of AB to BC be greater than that of DE to EF. Then, I say, AB is greater than DE. For, if AB bears to BC a ratio greater than that of DE to EF, then DE will bear to some length shorter than EF,

the same ratio which AB bears to BC. Call this length EG; then since AB:BC=DE:EG, it follows, *componendo et convertendo*, that CA:AB=GD:DE. But since CA is greater than GD, it follows that BA is greater than DE.

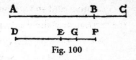

Fig. 100

LEMMA

Let ACIB be the quadrant of a circle; from B draw BE parallel to AC; about any point in the line BE describe a circle BOES, touching AB at B and

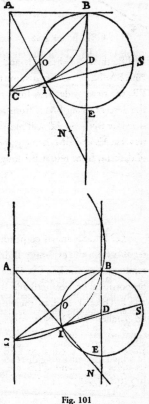

intersecting the circumference of the quadrant at I. Join the points C and B; draw the line CI, prolonging it to S. Then, I say, the line is always less than CO. Draw the line AI touching the circle BOE. Then, if the line DI be drawn, it will be equal to DB; but, since DB touches the quadrant, DI will also be tangent to it and will be at right angles to AI; thus AI touches the circle BOE at I. And since the angle AIC is greater than the angle ABC, subtending as it does a larger arc, it follows that the angle SIN is also greater than the angle ABC. Wherefore the arc IES is greater than the arc BO, and the line CS, being nearer the center, is longer that CB. Consequently CO is greater than CI, since SC:CB=OC:CI.

[266]

This result would be all the more marked if, as in the second figure, the arc BIC were less than a quadrant. For the perpendicular DB would then cut the circle CIB; and so also would DI which is equal to BD; the angle DIA would be obtuse and therefore the line AIN would cut the circle BIE. Since the angle ABC is less than the angle AIC, which is equal to SIN, and still less than the angle which the tangent at I would make with the line SI, it follows that the arc SEI is far greater than the arc BO; whence, etc.

Q.E.D.

Fig. 101

[267]

THEOREM XXII, PROPOSITION XXXVI

*If from the lowest point of a vertical circle, a chord is drawn sub-
tending an arc not greater than a quadrant, and if from the two ends
of this chord two other chords be drawn to any point on the arc, the
time of descent along the two latter chords will be shorter than along
the first, and shorter also, by the same amount, than along the lower
of these two latter chords.*

Let CBD be an arc, not exceeding a quadrant, taken from a vertical circle whose
point is C; let CD be the chord [*planum elevatum*] subtending this arc, and let there

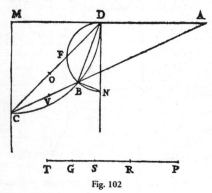

Fig. 102

be two other chords drawn from C and
D to any point B on the arc. Then, I
say, the time of descent along the two
chords [*plana*] DB and BC is shorter
than along DC alone, or along BC
alone, starting from rest at B. Through
the point D, draw the horizontal line
MDA cutting CB extended at A: draw
DN and MC at right angles to MD,
and BN at right angles to BD; about the
right-angled triangle DBN describe the
semicircle DFBN, cutting DC at F.
Choose the point O such that DO will

be a mean proportional between CD and DF; in like manner select V so that AV is a
mean proportional between CA and AB. Let the length PS represent the time of
descent along the whole distance DC or BC, both of which require the same time. Lay
off PR such that CD:DO=*time*PS.*time*PR. Then PR will represent the time in which
a body, starting from D, will traverse the distance DF, while RS will measure the time
in which the remaining distance, FC, will be traversed. But since PS is also the time of
descent, from rest at B, along BC, and if we choose T such that BC:CD=PS:PT then
PT will measure the time of descent from A to C, for we have already shown [Lemma]
that DC is a mean proportional between AC and CB. Finally choose the point G such
that CA:AV=PT:PG, then PG will be the time of descent from A to B, while GT will
be the residual time of descent along BC following descent from A to B. But, since the
diameter, DN, of the circle DFN is a vertical line, the chords DF and DB will be traversed in equal times; wherefore if one can prove that a body will traverse BC, after

582

descent along DB, in a shorter time than it will FC after descent along DF he will have proved the theorem. But a body descending from D along DB will traverse BC in the same time as if it had come from A along AB, seeing that the body acquires the same momentum in descending along DB as along AB.

[268]

Hence it remains only to show that descent along BC after AB is quicker than along FC after DF. But we have already shown that GT represents the time along BC after AB; also that RS measures the time along FC after DF. Accordingly it must be shown that RS is greater than GT, which may be done as follows: Since SP:PR=CD:DO, it follows, *invertendo et convertendo*, that RS:SP=OC:CD; also we have SP:PT=DC:CA. And since TP:PG=CA:AV, it follows, *invertendo*, that PT:TG=AC:CV, therefore, *ex æquali*, RS:GT=OC:CV. But, as we shall presently show, OC is greater than CV; hence the time RS is greater than the time GT, which was to be shown. Now, since [Lemma] CF is greater than CB and FD smaller than BA, it follows that CD:DF>CA:AB. But CD:DF=CO:OF, seeing that CD:DO=DO:DF; and CA:AB=CV^2:VB^2. Therefore CO:OF>CV:VB, and, according to the preceding lemma, CO>CV. Besides this it is clear that the time of descent along DC is to the time along DBC as DOC is to the sum of DO and CV.

[269]

SCHOLIUM

From the preceding it is possible to infer that the path of quickest descent [*lationem omnium velocissimam*] from one point to another is not the shortest path, namely, a straight line, but the are of a circle.[34] In the quadrant BAEC, having the side BC vertical, divide the arc AC into any number of equal parts, AD, DE, EF, FG, GC, and from C draw straight lines to the points A, D, E, F, G; draw also the straight lines AD, DE, EF, FG, GC. Evidently descent along the path ADC is quicker than along AC alone or along DC from rest at D. But a body, starting from rest at A, will traverse C more quickly than the path ADC; while, if it starts from rest at A, it will traverse the path DEC in a shorter time

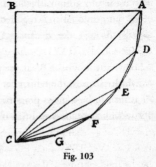

Fig. 103

34. It is well known that the first correct solution for the problem of quickest descent, under the condition of a constant force was given by John Bernoulli (1667-1748).*[Trans.]*

than DC alone. Hence descent along the three chords, ADEC, will take less time than along the two chords ADC. Similarly, following descent along ADE, the time required to traverse EFC is less than that needed for EC alone. Therefore descent is more rapid along the four chords ADEFC than along the three ADEC. And finally a body, after descent along ADEF, will traverse the two chords, FGC, more quickly than FC alone. Therefore, along the five chords, ADEFGC, descent will be more rapid than along the four, ADEFC. Consequently the nearer the inscribed polygon approaches a circle the shorter is the time required for descent from A to C.

What has been proven for the quadrant holds true also for smaller arcs; the reasoning is the same.

PROBLEM XV, PROPOSITION XXXVII

Given a limited vertical line and an inclined plane of equal altitude; it is required to find a distance on the inclined plane which is equal to the vertical line and which is traversed in an interval equal to the time of fall along the vertical line.

Let AB be the vertical line and AC the inclined plane. We must locate, on the inclined plane, a distance equal to the vertical line AB and which will be traversed by a body starting from rest at A in the same time needed for fall along the vertical line. Lay off AD equal to AB, and bisect the remainder DC at I. Choose the point E such that AC:CI=CI:AE and lay off DG equal to AE. Clearly EG is equal to AD, and also to AB. And further, I say that EG is that distance which will be traversed by a body, starting from rest at A, in the same time which is required for that body to fall through the distance AB. For since

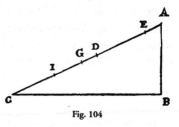

Fig. 104

AC:CI=CI:AE=ID:DG, we have, *convertendo*, CA:AI=DI:IG. And since the whole of CA is to the whole of AI as the portion CI is to the portion IG, it follows that the remainder IA is to the remainder AG as the whole of CA is to the whole of AI. Thus AI is seen to be a mean proportional between CA and AG, while CI is a mean proportional between CA and AE. If therefore the time of fall along AB is represented

[270]

by the length AB, the time along AC will be represented by AC, while CI, or ID, will measure the time along AE. Since AI is a mean proportional between CA and AG, and

since CA is a measure of the time along the entire distance AC, it follows that AI is the time along AG, and the difference IC is the time along the difference GC; but DI was the time along AE. Consequently the lengths DI and IC measure the times along AE and CG respectively. Therefore the remainder DA represents the time along EG, which of course is equal to the time along AB.   Q.E.F.

COROLLARY

From this it is clear that the distance sought is bounded at each end by portions of the inclined plane which are traversed in equal times.

PROBLEM XVI, PROPOSITION XXXVIII

Given two horizontal planes cut by a vertical line, it is required to find a point on the upper part of the vertical line from which bodies may fall to the horizontal planes and there, having their motion deflected into a horizontal direction, will, during an interval equal to the time of fall, traverse distances which bear to each other any assigned ratio of a smaller quantity to a larger.

Let CD and BE be the horizontal planes cut by the vertical ACB, and let the ratio of the smaller quantity to the larger be that of N to FG. It is required to find in the upper part of the vertical line, AB, a point from which a body falling to the plane CD and there having its motion deflected along this plane, will traverse, during an interval equal to its time of fall a distance such that if another body, falling from this same point to the plane BE, there have its motion deflected along this plane and continued during

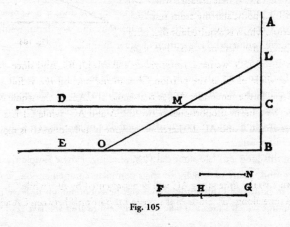

Fig. 105

585

an interval equal to its time of fall, will traverse a distance which bears to the former distance the ratio of FG to N. Lay off GH equal to N, and select the point L so that FH:HG=BC:CL. Then, I say, L is the point sought. For, if we lay off CM equal to twice CL, and draw the line LM cutting the plane BE at O, then BO will be equal to twice BL. And since FH:HG=BC:CL, we have, *componendo et convertendo*, HG:GF=N:GF=CL:LB=CM:BO. It is clear that, since CM is double the distance LC, the space CM is that which a body falling from L through LC will traverse in the plane CD; and, for the same reason, since BO is twice the distance BL, it is clear that BO is the distance which a body, after fall through LB, will traverse during an interval equal to the time of its fall through LB.

Q.E.F.

[271]

Sagr. Indeed, I think we may concede to our Academician, without flattery, his claim that in the principle [*principio*, i. e., accelerated motion] laid down in this treatise he has established a new science dealing with a very old subject. Observing with what ease and clearness he deduces from a single principle the proofs of so many theorems, I wonder not a little how such a question escaped the attention of Archimedes, Apollonius, Euclid and so many other mathematicians and illustrious philosophers, especially since so many ponderous tomes have been devoted to the subject of motion.

[272]

Salv. There is a fragment of Euclid which treats of motion, but in it there is no indication that he ever began to investigate the property of acceleration and the manner in which it varies with slope. So that we may say the door is now opened, for the first time, to a new method fraught with numerous and wonderful results which in future years will command the attention of other minds.

Sagr. I really believe that just as, for instance, the few properties of the circle proven by Euclid in the Third Book of his Elements lead to many others more recondite, so the principles which are set forth in this little treatise will, when taken up by speculative minds, lead to many another more remarkable result; and it is to be believed that it will be so on account of the nobility of the subject, which is superior to any other in nature.

During this long and laborious day, I have enjoyed these simple theorems more than their proofs, many of which, for their complete comprehension, would require more than an hour each; this study, if you will be good enough to leave the book in my hands, is one which I mean to take up at my leisure after we have read the remain-

ing portion which deals with the motion of projectiles; and this if agreeable to you we shall take up to-morrow.

Salv. I shall not fail to be with you.

END OF THE THIRD DAY.

[273]

FOURTH DAY

Salviati. Once more, Simplicio is here on time; so let us without delay take up the question of motion. The text of our Author is as follows:

THE MOTION OF PROJECTILES

In the preceding pages we have discussed the properties of uniform motion and of motion naturally accelerated along planes of all inclinations. I now propose to set forth those properties which belong to a body whose motion is compounded of two other motions, namely, one uniform and one naturally accelerated; these properties, well worth knowing, I propose to demonstrate in a rigid manner. This is the kind of motion seen in a moving projectile; its origin I conceive to be as follows:

Imagine any particle projected along a horizontal plane without friction; then we know, from what has been more fully explained in the preceding pages, that this particle will move along this same plane with a motion which is uniform and perpetual, provided the plane has no limits. But if the plane is limited and elevated, then the moving particle, which we imagine to be a heavy one, will on passing over the edge of the plane acquire, in addition to its previous uniform and perpetual motion, a downward propensity due to its own weight; so that the resulting motion which I call projection [*projectio*], is compounded of one which is uniform and horizontal and of another which is vertical and naturally accelerated. We now proceed to demonstrate some of its properties, the first of which is as follows:

[274]

THEOREM I, PROPOSITION I

A projectile which is carried by a uniform horizontal motion compounded with a naturally accelerated vertical motion describes a path which is a semi-parabola.

Sagr. Here, Salviati, it will be necessary to stop a little while for my sake and, I believe, also for the benefit of Simplicio; for it so happens that I have not gone very far in my study of Apollonius and am merely aware of the fact that he treats of the parabola and other conic sections, without an understanding of which I hardly think one will be able to follow the proof of other propositions depending upon them. Since even in this first beautiful theorem the author finds it necessary to prove that the path of a projectile is a parabola, and since, as I imagine, we shall have to deal with only this kind of curves, it will be absolutely necessary to have a thorough acquaintance, if not with all the properties which Apollonius has demonstrated for these figures, at least with those which are needed for the present treatment.

Salv. You are quite too modest, pretending ignorance of facts which not long ago you acknowledged as well known—I mean at the time when we were discussing the strength of materials and needed to use a certain theorem of Apollonius which gave you no trouble.

Sagr. I may have chanced to know it or may possibly have assumed it, so long as needed, for that discussion; but now when we have to follow all these demonstrations about such curves we ought not, as they say, to swallow it whole, and thus waste time and energy.

Simp. Now even though Sagredo is, as I believe, well equipped for all his needs, I do not understand even the elementary terms; for although our philosophers have treated the motion of projectiles, I do not recall their having described the path of a projectile except to state in a general way that it is always a curved line, unless the projection be vertically upwards. But if the little Euclid which I have learned since our previous discussion does not enable me to understand the demonstrations which are to follow, then I shall be obliged to accept the theorems on faith without fully comprehending them.

[275]

Salv. On the contrary, I desire that you should understand them from the Author himself, who, when he allowed me to see this work of his, was good enough to prove

588

for me two of the principal properties of the parabola because I did not happen to have at hand the books of Apollonius. These properties, which are the only ones we shall need in the present discussion, he proved in such a way that no prerequisite knowledge was required. These theorems are, indeed, given by Apollonius, but after many preceding ones, to follow which would take a long while. I wish to shorten our task by deriving the first property purely and simply from the mode of generation of the parabola and proving the second immediately from the first.

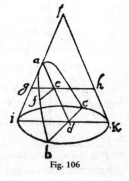

Fig. 106

Beginning now with the first, imagine a right cone, erected upon the base *ibkc* with apex at *l*. The section of this cone made by a plane drawn parallel to the side *lk* is the curve which is called a *parabola*. The base of this parabola *bc* cuts at right angles the diameter *ik* of the circle *ibkc*, and the axis *ad* is parallel to the side *lk*; now having taken any point in the curve *bfa* draw the straight line *fe* parallel to *bd*; then, I say, the square of *bd* is to the square of *fe* in the same ratio as the axis *ad* is to the portion *ae*. Through the point *e* pass a plane parallel to the circle *ibkc*, producing in the cone a circular section whose diameter is the line *geh*. Since *bd* is at right angles to *ik* in the circle *ibk*, the square of *bd* is equal to the rectangle formed by *id* and *dk*; so also in the upper circle which passes through the points *gfh* the square of *fe* is equal to the rectangle formed by *ge* and *eh*; hence the square of *bd* is to the square of *fe* as the rectangle *id.dk* is to the rectangle *ge.eh*. And since the line *ed* is parallel to *hk*, the line *eh*, being parallel to *dk*, is equal to it; therefore the rectangle *ge.eh* as *id* is to *id.dk* is to the rectangle *ge*, that is, as *da* is to *ae*; whence also the rectangle *id.dk* is to the rectangle *ge.eh*, that is, the square of *bd* is to the square of *fe*, as the axis *da* is to the portion *ae*.   Q.E.D.

[276]

The other proposition necessary for this discussion we demonstrate as follows. Let us draw a parabola whose axis *ca* is prolonged upwards to a point *d*; from any point *b* draw the line *bc* parallel to the base of the parabola; if now the point *d* is chosen so that *da* = *ca*, then, I say, the straight line drawn through the points *b* and *d* will be tangent to the parabola at *b*. For imagine, if possible, that this line cuts the parabola above or that its prolongation cuts it below, and through any point *g* in it draw the straight line *fge*. And since the square of *fe* is greater than the square of *ge*, the square of *fe* will bear a greater ratio to the square of *bc* than the square of *ge* to that of *bc*; and since, by the preceding proposition, the square of *fe* is to that of *bc* as the line *ea* is to *ca*, it follows that the line *ea* will bear to the line *ca* a greater ratio than the square of *ge* to that of

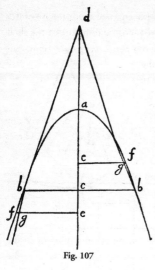

Fig. 107

bc, or, than the square of *ed* to that of *cd* (the sides of the triangles *deg* and *dcb* being proportional). But the line *ea* is to *ca*, or, *da*, in the same ratio as four times the rectangle *ea.ad* is to four times the square of *ad*, or, what is the same, the square of *cd*, since this is four times the square of *ad*; hence four times the rectangle *ea.ad* bears to the square of *cd* a greater ratio than the square of *ed* to the square of *cd*; but that would make four times the rectangle *ea.ad* greater than the square of *ed*; which is false, the fact being just the opposite, because the two portions *ea* and *ad* of the line *ed* are not equal. Therefore the line *db* touches the parabola without cutting it.   Q.E.D.

Simp. Your demonstration proceeds too rapidly and, it seems to me, you keep on assuming that all of Euclid's theorems are as familiar and available to me

[277]

as his first axioms, which is far from true. And now this fact which you spring upon us, that four times the rectangle *ea.ad* is less than the square of *de* because the two portions *ea* and *ad* of the line *de* are not equal brings me little composure of mind, but rather leaves me in suspense.

Salv. Indeed, all real mathematicians assume on the part of the reader perfect familiarity with at least the elements of Euclid; and here it is necessary in your case only to recall a proposition of the Second Book in which he proves that when a line is cut into equal and also into two unequal parts, the rectangle formed on the unequal parts is less than that formed on the equal (i. e, less than the square on half the line), by an amount which is the square of the difference between the equal and unequal segments. From this it is clear that the square of the whole line which is equal to four times the square of the half is greater than four times the rectangle of the unequal parts. In order to understand the following portions of this treatise it will be necessary to keep in mind the two elemental theorems from conic sections which we have just demonstrated; and these two theorems are indeed the only ones which the Author uses. We can now resume the text and see how he demonstrates his first proposition in which he shows that a body falling with a motion compounded of a uniform horizontal and a naturally accelerated [*naturale descendente*] one describes a semi-parabola.

Let us imagine an elevated horizontal line or plane *ab* along which a body moves with uniform speed from *a* to *b*. Suppose this plane to end abruptly at *b*; then at this

point the body will, on account of its weight, acquire also a natural motion downwards along the perpendicular *bn*. Draw the line *be* along the plane *ba* to represent the flow, or measure, of time; divide this line into a number of segments, *bc*, *cd*, *de*, representing equal intervals of time; from the points *b*, *c*, *d*, *e*, let

[278]

fall lines which are parallel to the perpendicular *bn*. On the first of these lay off any distance *ci*, on the second a distance four times as long, *df*; on the third, one nine times as long, *eh*; and so on, in proportion to the squares of *cb*, *db*, *eb*, or, we may say, in the squared ratio of these same lines. Accordingly we see that while the body moves from *b* to *c* with uniform speed, it also falls perpendicularly through the distance *ci*, and at the end of the time-interval *bc* finds itself at the point *i*. In like manner at the end of the time-interval *bd*, which is the double of *bc*, the vertical fall will be four times the

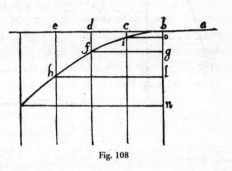

Fig. 108

first distance *ci*; for it has been shown in a previous discussion that the distance traversed by a freely falling body varies as the square of the time; in like manner the space *eh* traversed during the time *be* will be nine times *ci*; thus it is evident that the distances *eh*, *df*, *ci* will be to one another as the squares of the lines *be*, *bd*, *bc*. Now from the points *i*, *f*, *h* draw the straight lines *io*, *fg*, *hl* parallel to *be*; these lines *hl*, *fg*, *io* are equal to *eb*, *db* and *cb*, respectively; so also are the lines *bo*, *bg*, *bl* respectively equal to *ci*, *df*, and *eh*. The square of *hl* is to that of *fg* as the line *lb* is to *bg*; and the square of *fg* is to that of *io* as *gb* is to *bo*; therefore the points *i*, *f*, *h*, lie on one and the same parabola. In like manner it may be shown that, if we take equal time-intervals of any size whatever, and if we imagine the particle to be carried by a similar compound motion, the positions of this particle, at the ends of these time-intervals, will lie on one and the same parabola.   Q.E.D.

Salv. This conclusion follows from the converse of the first of the two propositions given above. For, having drawn a parabola through the points *b* and *h*, any other two points, *f* and *i*, not falling on the parabola must lie either within or without; consequently the line *fg* is either longer or shorter than the line which terminates on the parabola. Therefore the square of *hl* will not bear to the square of *fg* the same ratio as the line *lb* to *bg*, but a greater or smaller; the fact is, however, that the square of *hl* does

bear this same ratio to the square of *fg*. Hence the point *f* does lie on the parabola and so do all the others.

[279]

Sagr. One cannot deny that the argument is new, subtle and conclusive, resting as it does upon this hypothesis, namely, that the horizontal motion remains uniform, that the vertical motion continues to be accelerated downwards in proportion to the square of the time, and that such motions and velocities as these combine without altering, disturbing, or hindering each other,[35] so that as the motion proceeds the path of the projectile does not change into a different curve: but this, in my opinion, is impossible. For the axis of the parabola along which we imagine the natural motion of a falling body to take place stands perpendicular to a horizontal surface and ends at the center of the earth; and since the parabola deviates more and more from its axis no projectile can ever reach the center of the earth or, if it does, as seems necessary, then the path of the projectile must transform itself into some other curve very different from the parabola.

Simp. To these difficulties, I may add others. One of these is that we suppose the horizontal plane, which slopes neither up nor down, to be represented by a straight line as if each point on this line were equally distant from the center, which is not the case; for as one starts from the middle [of the line] and goes toward either end, he departs farther and farther from the center (of the earth] and is therefore constantly going uphill. Whence it follows that the motion cannot remain uniform through any distance whatever, but must continually diminish. Besides, I do not see how it is possible to avoid the resistance of the medium which must destroy the uniformity of the horizontal motion and change the law of acceleration of falling bodies. These various difficulties render it highly improbable that a result derived from such unreliable hypotheses should hold true in practice.

Salv. All these difficulties and objections which you urge are so well founded that it is impossible to remove them; and, as for me, I am ready to admit them all, which indeed I think our Author would also do. I grant that these conclusions proved in the abstract will be different when applied in the concrete and will be fallacious to this extent, that neither will the horizontal motion be uniform nor the natural acceleration be in the ratio assumed, nor the path of the projectile a parabola, etc. But, on the other hand, I ask you not to begrudge our Author that which other eminent men have assumed even if not strictly true. The authority of Archimedes alone will satisfy everybody. In his Mechanics and in his first quadrature of the parabola he takes for granted

35. A very near approach to Newton's Second Law of Motion. *[Trans.]*

that the beam of a balance or steelyard is a straight line, every point of which is equidistant from the common center of all heavy bodies, and that the cords by which heavy bodies are suspended are parallel to each other.

Some consider this assumption permissible because, in practice, our instruments and the distances involved are so small in comparison with the enormous distance from the center of the earth that we may consider a minute of arc on a great circle as a straight line, and may regard the perpendiculars let fall from its two extremities as parallel. For if in actual practice one had to consider such small quantities, it would be

[280]

necessary first of all to criticise the architects who presume, by use of a plumbline, to erect high towers with parallel sides. I may add that, in all their discussions, Archimedes and the others considered themselves as located at an infinite distance from the center of the earth, in which case their assumptions were not false, and therefore their conclusions were absolutely correct. When we wish to apply our proven conclusions to distances which, though finite, are very large, it is necessary for us to infer, on the basis of demonstrated truth, what correction is to be made for the fact that our distance from the center of the earth is not really infinite, but merely very great in comparison with the small dimensions of our apparatus. The largest of these will be the range of our projectiles—and even here we need consider only the artillery—which, however great, will never exceed four of those miles of which as many thousand separate us from the center of the earth; and since these paths terminate upon the surface of the earth only very slight changes can take place in their parabolic figure which, it is conceded, would be greatly altered if they terminated at the center of the earth.

As to the perturbation arising from the resistance of the medium this is more considerable and does not, on account of its manifold forms, submit to fixed laws and exact description. Thus if we consider only the resistance which the air offers to the motions studied by us, we shall see that it disturbs them all and disturbs them in an infinite variety of ways corresponding to the infinite variety in the form, weight, and velocity of the projectiles. For as to velocity, the greater this is, the greater will be the resistance offered by the air; a resistance which will be greater as the moving bodies become less dense [men gravi]. So that although the falling body ought to be displaced [andare accelerandosi] in proportion to the square of the duration of its motion, yet no matter how heavy the body, if it falls from a very considerable height, the resistance of the air will be such as to prevent any increase in speed and will render the motion uniform; and in proportion as the moving body is less dense [men grave] this uniformity will be so much the more quickly attained and after a shorter fall. Even horizontal

motion which, if no impediment were offered, would be uniform and constant is altered by the resistance of the air and finally ceases; and here again the less dense [*piu leggiero*] the body the quicker the process.

[281]

Of these properties [*accidenti*] of weight, of velocity, and also of form [*figura*], infinite in number, it is not possible to give any exact description; hence, in order to handle this matter in a scientific way, it is necessary to cut loose from these difficulties; and having discovered and demonstrated the theorems, in the case of no resistance., to use them and apply them with such limitations as experience will teach. And the advantage of this method will not be small; for the material and shape of the projectile may be chosen, as dense and round as possible, so that it will encounter the least resistance in the medium. Nor will the spaces and velocities in general be so great but that we shall be easily able to correct them with precision.

In the case of those projectiles which we use, made of dense [*grave*] material and round in shape, or of lighter material and cylindrical in shape, such as arrows, thrown from a sling or crossbow, the deviation from an exact parabolic path is quite insensible. Indeed, if you will allow me a little greater liberty, I can show you, by two experiments, that the dimensions of our apparatus are so small that these external and incidental resistances, among which that of the medium is the most considerable, are scarcely observable.

I now proceed to the consideration of motions through the air, since it is with these that we are now especially concerned; the resistance of the air exhibits itself in two ways: first by offering greater impedance to less dense than to very dense bodies, and secondly by offering greater resistance to a body in rapid motion than to the same body in slow motion.

Regarding the first of these, consider the case of two balls having the same dimensions, but one weighing ten or twelve times as much as the other; one, say, of lead, the other of oak, both allowed to fall from an elevation of 150 or 200 cubits.

Experiment shows that they will reach the earth with slight difference in speed, showing us that in both cases the retardation caused by the air is small; for if both balls start at the same moment and at the same elevation, and if the leaden one be slightly retarded and the wooden one greatly retarded, then the former ought to reach the earth a considerable distance in advance of the latter, since it is ten times as heavy. But this does not happen; indeed, the gain in distance of one over the other does not amount to the hundredth part of the entire fall. And in the case of a ball of stone weighing only a third or half as much as one of lead, the difference in their times of reaching the earth will be scarcely noticeable. Now since the speed [*impeto*] acquired by a leaden ball in

falling from a height of 200 cubits is so great that if the motion remained uniform the ball would, in an interval of time equal to that of the fall, traverse 400 cubits, and since this speed is so considerable in comparison with those which, by use of bows or other machines except fire arms, we are able to give to our projectiles, it follows that we may, without sensible error, regard as absolutely true those propositions which we are about to prove without considering the resistance of the medium.

[282]

Passing now to the second case, where we have to show that the resistance of the air for a rapidly moving body is not very much greater than for one moving slowly, ample proof is given by the following experiment. Attach to two threads of equal length—say four or five yards—two equal leaden balls and suspend them from the ceiling; now pull them aside from the perpendicular, the one through 80 or more degrees, the other through not more than four or five degrees; so that, when set free, the one falls, passes through the perpendicular, and describes large but slowly decreasing arcs of 160, 150, 140 degrees, etc.; the other swinging through small and also slowly diminishing arcs of 10, 8, 6, degrees, etc.

In the first place it must be remarked that one pendulum passes through its arcs of 180°, 160°, etc., in the same time that the other swings through its 10°, 8°, etc., from which it follows that the speed of the first ball is 16 and 18 times greater than that of the second. Accordingly, if the air offers more resistance to the high speed than to the low, the frequency of vibration in the large arcs of 180° or 160°, etc., ought to be less than in the small arcs of 10°, 8°, 4°, etc., and even less than in arcs of 2°, or 1°; but this prediction is not verified by experiment; because if two persons start to count the vibrations, the one the large, the other the small, they will discover that after counting tens and even hundreds they will not differ by a single vibration, not even by a fraction of one.

[283]

This observation justifies the two following propositions, namely, that vibrations of very large and very small amplitude all occupy the same time and that the resistance of the air does not affect motions of high speed more than those of low speed, contrary to the opinion hitherto generally entertained.

Sagr. On the contrary, since we cannot deny that the air hinders both of these motions, both becoming slower and finally vanishing, we have to admit that the retardation occurs in the same proportion in each case. But how? How, indeed, could the resistance offered to the one body be greater than that offered to the other except by the impartation of more momentum and speed [*impeto e velocità*] to the fast body than

to the slow? And if this is so the speed with which a body moves is at once the cause and measure [*cagione e misura*] of the resistance which it meets. Therefore, all motions, fast or slow, are hindered and diminished in the same proportion; a result, it seems to me, of no small importance.

Salv. We are able, therefore, in this second case to say that the errors, neglecting those which are accidental, in the results which we are about to demonstrate are small in the case of our machines where the velocities employed are mostly very great and the distances negligible in comparison with the semi-diameter of the earth or one of its great circles.

Simp. I would like to hear your reason for putting the projectiles of fire arms, i. e., those using powder, in a different class from the projectiles employed in bows, slings, and crossbows, on the ground of their not being equally subject to change and resistance from the air.

Salv. I am led to this view by the excessive and, so to speak, supernatural violence with which such projectiles are launched; for, indeed, it appears to me that without exaggeration one might say that the speed of a ball fired either from a musket or from a piece of ordnance is supernatural. For if such a ball be allowed to fall from some great elevation its speed will, owing to the resistance of the air, not go on increasing indefinitely; that which happens to bodies of small density in falling through short distances—I mean the reduction of their motion to uniformity— will also happen to a ball of iron or lead after it has fallen a few thousand cubits; this terminal or final speed [*terminata velocità*] is the maximum which such a heavy body can naturally acquire in falling through the air. This speed I estimate to be much smaller than that impressed upon the ball by the burning powder.

An appropriate experiment will serve to demonstrate this fact. From a height of one hundred or more cubits fire a gun [*archibuso*] loaded with a lead bullet, vertically downwards upon a stone pavement; with the same gun shoot against a similar stone from a distance of one or two cubits, and observe which of the two balls is the more flattened. Now if the ball which has come from the greater elevation is found to be the less flattened of the two, this will show that the air has hindered and diminished the speed initially imparted to the bullet by the powder, and that the air will not permit a bullet to acquire so great a speed, no matter from what height it falls; for if the speed impressed upon the ball by the fire does not exceed that acquired by it in falling freely [*naturalmente*] then its downward blow ought to be greater rather than less.

[284]

This experiment I have not performed, but I am of the opinion that a musket-ball or cannon-shot, falling from a height as great as you please, will not deliver so strong a

blow as it would if fired into a wall only a few cubits distant, i. e., at such a short range that the splitting or rending of the air will not be sufficient to rob the shot of that excess of supernatural violence given it by the powder.

The enormous momentum [*impeto*] of these violent shots may cause some deformation of the trajectory, making the beginning of the parabola flatter and less curved than the end; but, so far as our Author is concerned, this is a matter of small consequence in practical operations, the main one of which is the preparation of a table of ranges for shots of high elevation, giving the distance attained by the ball as a function of the angle of elevation; and since shots of this kind are fired from mortars [*mortari*] using small charges and imparting no supernatural momentum [*impeto sopranaturale*] they follow their prescribed paths very exactly.

But now let us proceed with the discussion in which the Author invites us to the study and investigation of the motion of a body [*impeto del mobile*] when that motion is compounded of two others; and first the case in which the two are uniform, the one horizontal, the other vertical.

[285]

THEOREM II, PROPOSITION II

When the motion of a body is the resultant of two uniform motions, one horizontal, the other perpendicular, the square of the resultant momentum is equal to the sum of the squares of the two component momenta.[36]

Let us imagine any body urged by two uniform motions and let *ab* represent the vertical displacement, while *bc* represents the displacement which, in the same interval of time, takes place in a horizontal direction. If then the distances *ab* and *bc* are traversed, during the same time-interval, with uniform motions the corresponding momenta will be to each

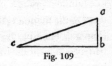

Fig. 109

other as the distances *ab* and *bc* are to each other; but the body which is urged by these two motions describes the diagonal *ac*; its momentum is proportional to *ac*. Also the square of *ac* is equal to the sum of the squares of *ab* and *bc*. Hence the square of the resultant momentum is equal to the sum of the squares of the two momenta *ab* and *bc*.   Q.E.D.

36. In the original this theorem reads as follows:
"*Si aliquod mobile duplici motu aequabili moveatur, nempe orizontali et perpendiculari, impetus seu momentum lationis ex utroque motu compositae erit potentia aequalis ambobus momentis priorum motuum.*"
    For the justification of this translation of the word "potentia" and of the use of the adjective "resultant" see below. [*Trans.*]
    37. See above. [*Trans.*]

Simp. At this point there is just one slight difficulty which needs to be cleared up; for it seems to me that the conclusion just reached contradicts a previous proposition[37] in which it is claimed that the speed [*impeto*] of a body coming from *a* to *b* is equal to that in coming from *a* to *c*; while now you conclude that the speed [*impeto*] at *c* is greater than that at *b*.

Salv. Both propositions, Simplicio, are true., yet there is a great difference between them. Here we are speaking of a body urged by a single motion which is the resultant of two uniform motions, while there we were speaking of two bodies each urged with naturally accelerated motions, one along the vertical *ab* the other along the inclined plane *ac*. Besides the time-intervals were there not supposed to be equal, that along the incline *ac* being greater than that along the vertical *ab*; but the motions of which we now speak, those along *ab*, *bc*, *ac*, are uniform and simultaneous.

[286]

Simp. Pardon me; I am satisfied; pray go on.

Salv. Our Author next undertakes to explain what happens when a body is urged by a motion compounded of one which is horizontal and uniform and of another which is vertical but naturally accelerated; from these two components results the path of a projectile, which is a parabola. The problem is to determine the speed [*impeto*] of the projectile at each point. With this purpose in view our Author sets forth as follows the manner, or rather the method, of measuring such speed [*impeto*] along the path which is taken by a heavy body starting from rest and falling with a naturally accelerated motion.

THEOREM III, PROPOSITION III

Let the motion take place along the line *ab*, starting from rest at *a*, and in this line choose any point *c*. Let *ac* represent the time, or the measure of the time, required for

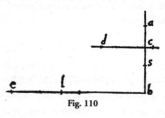

Fig. 110

the body to fall through the space *ac*; let *ac* also represent the velocity [*impetus seu momentum*] at *c* acquired by a fall through the distance *ac*. In the line *ab* select any other point *b*. The problem now is to determine the velocity at b acquired by a body in falling through the distance *ab* and to express this in terms of the velocity at *c*, the measure of which is the length *ac*. Take *as* a mean proportion-

al between *ac* and *ab*. We shall prove that the velocity at *b* is to that at *c* as the length *as* is to the length *ac*. Draw the horizontal line *cd*, having twice the length of *ac*, and

[287]

be, having twice the length of *ba*. It then follows, from the preceding theorems, that a body falling through the distance *ac*, and turned so as to move along the horizontal *cd* with a uniform speed equal to that acquired on reaching *c* will traverse the distance *cd* in the same interval of time as that required to fall with accelerated motion from *a* to *c*. Likewise *be* will be traversed in the same time as *ba*. But the time of descent through *ab* is *as*; hence the horizontal distance *be* is also traversed in the time *as*. Take a point *l* such that the time *as* is to the time *ac* as *be* is to *bl*; since the motion along *be* is uniform, the distance *bl*, if traversed with the speed [*momentum celeritatis*] acquired at *b*, will occupy the time *ac*; but in this same time-interval, *ac*, the distance *cd* is traversed with the speed acquired in *c*. Now two speeds are to each other as the distances traversed in equal intervals of time. Hence the speed at *c* is to the speed at *b* as *cd* is to *bl*. But since *dc* is to *be* as their halves, namely, as *ca* is to *ba*, and since *be* is to *bl* as *ba* is to *sa*; it follows that *dc* is to *bl* as *ca* is to *sa*. In other words, the speed at *c* is to that at *b* as *ca* is to *sa*, that is, as the time of fall through *ab*.

The method of measuring the speed of a body along the direction of its fall is thus clear; the speed is assumed to increase directly as the time.

But before we proceed further, since this discussion is to deal with the motion compounded of a uniform horizontal one and one accelerated vertically downwards— the path of a projectile, namely, a parabola-it is necessary that we define some common standard by which we may estimate the velocity, or momentum [*velocitatem, impetum seu momentum*] of both motions; and since from the innumerable uniform velocities one only, and that not selected at random, is to be compounded with a velocity acquired by naturally accelerated motion, I can think of no simpler way of selecting and measuring this than to assume another of the same kind.[38] For the sake of clearness, draw the vertical line *ac* to meet the horizontal line *bc*. *Ac* is the height and *bc* the amplitude of the semi-parabola *ab*, which is the resultant of the two motions, one that of a body falling from rest at *a*, through the distance *ac*, with naturally accelerated motion, the other a uniform motion along the horizontal *ad*.

[288]

The speed acquired at *c* by a fall through the distance *ac* is determined by the height *ac*; for the speed of a body falling from the same elevation is always one and the same; but along the horizontal one may give a body an infinite number of uniform speeds. However, in order that I may select one out of this multitude and separate it from the rest in a perfectly definite manner, I will extend the height *ca* upwards to *e* just as far

38. Galileo here proposes to employ as a standard of velocity the terminal speed of a body falling freely from a given height. [*Trans.*]

as is necessary and will call this distance *ae* the "sublimity." Imagine a body to fall from rest at *e*; it is clear that we may make its terminal speed at *a* the same as that with which

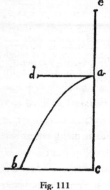

Fig. 111

the same body travels along the horizontal line *ad*; this speed will be such that, in the time of descent along *ea*, it will describe a horizontal distance twice the length of *ea*. This preliminary remark seems necessary.

The reader is reminded that above I have called the horizontal line *cb* the "amplitude" of the semi-parabola *ab*; the axis *ac* of this parabola, I have called its "altitude"; but the line *ea* the fall along which determines the horizontal speed I have called the "sublimity." These matters having been explained, I proceed with the demonstration.

Sagr. Allow me, please, to interrupt in order that I may point out the beautiful agreement between this thought of the Author and the views of Plato concerning the origin of the various uniform speeds with which the heavenly bodies revolve. The latter chanced upon the idea that a body could not pass from rest to any given speed and maintain it uniformly except by passing through all the degrees of speed intermediate between the given speed and rest. Plato thought that God, after having created the heavenly bodies, assigned them the proper and uniform speeds with which they were forever to revolve; and that He made them start from rest and move over definite distances under a natural and rectilinear acceleration such as governs the motion of terrestrial bodies. He added that once these bodies had gained their proper and permanent speed, their rectilinear motion was converted into a circular one, the only motion capable of maintaining uniformity, a motion in which the body revolves without either receding from or approaching its desired goal. This conception is truly worthy of Plato; and it is to be all the more highly prized since its underlying principles remained hidden until discovered by our Author who removed from them the mask and poetical dress and set forth the idea in correct historical perspective.

[289]

In view of the fact that astronomical science furnishes us such complete information concerning the size of the planetary orbits, the distances of these bodies from their centers of revolution, and their velocities, I cannot help thinking that our Author (to whom this idea of Plato was not unknown) bad some curiosity to discover whether or not a definite "sublimity" might be assigned to each planet, such that, if it were to start from rest at this particular height and to fall with naturally accelerated motion along a

straight line, and were later to change the speed thus acquired into uniform motion, the size of its orbit and its period of revolution would be those actually observed.

Salv. I think I remember his having told me that he once made the computation and found a satisfactory correspondence with observation. But he did not wish to speak of it, lest in view of the odium which his many new discoveries had already brought upon him, this might be adding fuel to the fire. But if any one desires such information he can obtain it for himself from the theory set forth in the present treatment.

We now proceed with the matter in hand, which is to prove:

PROBLEM I, PROPOSITION IV

To determine the momentum of a projectile at each particular point in its given parabolic path.

Let *bec* be the semi-parabola whose amplitude is *cd* and whose height is *db*, which latter extended upwards cuts the tangent of the parabola *ca* in *a*. Through the vertex draw the horizontal line *bi* parallel to *cd*. Now if the amplitude *cd* is equal to the entire height *da*, then *bi* will be equal to *ba* and also to *bd*; and if we take *ab* as the measure of the time required for fall through the distance *ab* and also of the momentum acquired at *b* in consequence of its fall from rest at *a*, then if we turn into a horizontal direction the momentum acquired by fall through *ab* [*impetum ab*] the space traversed in the same interval of time will be represented by *dc* which is twice *bi*. But a body which falls from rest at *b* along the line *bd* will during the same time-interval fall through the height of the parabola *bd*.

[290]

Hence a body falling from rest at *a*, turned into a horizontal direction with the speed *ab* will traverse a space equal to *dc*. Now if one superposes upon this motion a fall along *bd*, traversing the height *bd* while the parabola *bc* is described, then the momentum of the body at the terminal point *c* is the resultant of a uniform horizontal momentum, whose value is represented by *ab*, and of another momentum acquired by fall from *b* to the terminal point *d* or *c*; these two momenta are equal. If, therefore, we take *ab* to be the measure of one of these momenta, say, the uniform horizontal one, then *bi*, which is equal to *bd*, will represent the momentum acquired at *d* or *c*; and *ia* will represent the resultant of these two momenta, that is, the total momentum with which the projectile, travelling along the parabola, strikes at *c*.

With this in mind let us take any point on the parabola, say *e*, and determine the momentum with which the projectile passes that point. Draw the horizontal *ef* and take *bg* a mean proportional between *bd* and *bf*. Now since *ab*, or *bd*, is assumed to be

the measure of the time and of the momentum [*momentum velocitatis*] acquired by falling from rest at *b* through the distance *bd*, it follows that *bg* will measure the time

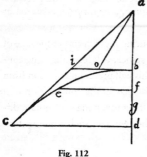

Fig. 112

and also the momentum [*impetus*] acquired at *f* by fall from *b*. If therefore we lay off *bo*, equal to *bg*, the diagonal line joining *a* and *o* will represent the momentum at the point *e*; because the length *ab* has been assumed to represent the momentum at *b* which, after diversion into a horizontal direction, remains constant; and because *bo* measures the momentum at *f* or *e*, acquired by fall, from rest at *b*, through the height *bf*. But the square of *ao* equals the sum of the squares of *ab* and *bo*. Hence the theorem sought.

Sagr. The manner in which you compound these different momenta to obtain their resultant strikes me as so novel that my mind is left in no small confusion. I do not refer to the composition of two uniform motions, even when unequal, and when one takes place along a horizontal, the other along a vertical direction; because in this case I am thoroughly convinced that the resultant is a motion whose square is equal to the sum of the squares of the two components. The confusion arises when one undertakes to compound a uniform horizontal motion with a vertical one which is naturally accelerated. I trust, therefore, we may pursue this discussion more at length.

[291]

Simp. And I need this even more than you since I am not yet as clear in my mind as I ought to be concerning those fundamental propositions upon which the others rest. Even in the case of the two uniform motions, one horizontal, the other perpendicular, I wish to understand better the manner in which you obtain the resultant from the components. Now, Salviati, you understand what we need and what we desire.

Salv. Your request is altogether reasonable and I will see whether my long consideration of these matters will enable me to make them clear to you. But you must excuse me if in the explanation I repeat many things already said by the Author.

Concerning motions and their velocities or momenta [*movimenti e lor velocità o impeti*] whether uniform or naturally accelerated, one cannot speak definitely until he has established a measure for such velocities and also for time. As for time we have the already widely adopted hours, first minutes and second minutes. So for velocities, just as for intervals of time, there is need of a common standard which shall be understood and accepted by everyone, and which shall be the same for all. As has already been stated, the Author considers the velocity of a freely falling body adapted to this purpose,

since this velocity increases according to the same law in all parts of the world; thus for instance the speed acquired by a leaden ball of a pound weight starting from rest and falling vertically through the height of, say, a spear's length is the same in all places; it is therefore excellently adapted for representing the momentum [*impeto*] acquired in the case of natural fall.

It still remains for us to discover a method of measuring momentum in the case of uniform motion in such a way that all who discuss the subject will form the same conception of its size and velocity [*grandezza e velocità*]. This will prevent one person from imagining it larger, another smaller, than it really is; so that in the composition of a given uniform motion with one which is accelerated different men may not obtain different values for the resultant. In order to determine and represent such a momentum

[292]

and particular speed [*impeto e velocità particolare*] our Author has found no better method than to use the momentum acquired by a body in naturally accelerated motion. The speed of a body which has in this manner acquired any momentum whatever will, when converted into uniform motion, retain precisely such a speed as, during a time-interval equal to that of the fall, will carry the body through a distance equal to twice that of the fall. But since this matter is one which is fundamental in our discussion it is well that we make it perfectly clear by means of some particular example.

Let us consider the speed and momentum acquired by a body falling through the height, say, of a spear [*picca*] as a standard which we may use in the measurement of other speeds and momenta as occasion demands; assume for instance that the time of such a fall is four seconds [*minuti secondi d'ora*]; now in order to measure the speed acquired from a fall through any other height, whether greater or less, one must not conclude that these speeds bear to one another the same ratio as the heights of fall; for instance, it is not true that a fall through four times a given height confers a speed four times as great as that acquired by descent through the given height; because the speed of a naturally accelerated motion does not vary in proportion to the time. As has been shown above, the ratio of the spaces is equal to the square of the ratio of the times.

If, then, as is often done for the sake of brevity, we take the same limited straight line as the measure of speed, and of the time, and also of the space traversed during that time, it follows that the duration of fall and the speed acquired by the same body in passing over any other distance, is not represented by this second distance, but by a mean proportional between the two distances. This I can better illustrate by an example. In the vertical line *ac*, lay off the portion *ab* to represent the distance traversed by a body falling freely with

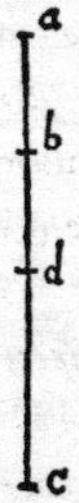

Fig. 113

accelerated motion: the time of fall may be represented by any limited straight line, but for the sake of brevity, we shall represent it by the same length *ab*; this length may also be employed as a measure of the momentum and speed acquired during the motion; in short, let *ab* be a measure of the various physical quantities which enter this discussion.

[293]

Having agreed arbitrarily upon *ab* as a measure of these three different quantities, namely, space, time, and momentum, our next task is to find the time required for fall through a given vertical distance *ac*, also the momentum acquired at the terminal point *c*, both of which are to be expressed in terms of the time and momentum represented by *ab*. These two required quantities are obtained by laying off *ad*, a mean proportional between *ab* and *ac*; in other words, the time of fall from *a* to *c* is represented by *ad* on the same scale on which we agreed that the time of fall from *a* to *b* should be represented by *ab*. In like manner we may say that the momentum [*impeto o grado di velocità*] acquired at *c* is related to that acquired at *b*, in the same manner that the line *ad* is related to *ab*, since the velocity varies direct as the time, a conclusion, which although employed as a postulate in Proposition III, is here amplified by the Author.

This point being clear and well-established we pass to the consideration of the momentum [*impeto*] in the case of two compound motions, one of which is compounded of a uniform horizontal and a uniform vertical motion, while the other is compounded of a uniform horizontal and a naturally accelerated vertical motion. If both components are uniform, and one at right angles to the other, we have already seen that the square of the resultant is

Fig. 114

obtained by adding the squares of the components as will be clear from the following illustration.

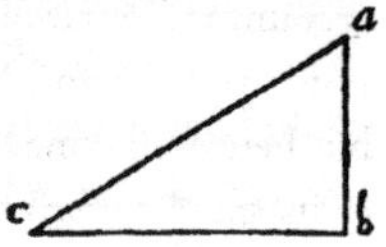

Let us imagine a body to move along the vertical *ab* with a uniform momentum [*impeto*] of 3, and on reaching *b* to move toward *c* with a momentum [*velocità ed impeto*] of 4, so that during the same time-interval it will traverse 3 cubits along the vertical and 4 along the horizontal. But a particle which moves with the resultant velocity [*velocità*] will, in the same time, traverse the diagonal *ac*, whose length is not 7 cubits—the sum of *ab* (3) and *bc* (4)—but 5, which is *in potenza* equal to the sum of 3 and 4, that is, the squares of 3 and 4 when added make 25, which is the square of *ac*, and is equal to the sum of the squares of *ab* and *bc*. Hence *ac* is represented by the side—or we may say the root—of a square whose area is 25, namely 5.

As a fixed and certain rule for obtaining the momentum which results from two

uniform momenta, one vertical, the other horizontal, we have therefore the following: take the square of each, add these together, and extract the square root of the sum, which will be the momentum resulting from the two. Thus, in the above example, the body which in virtue of its vertical motion would strike the horizontal plane with a momentum [*forza*] of 3, would owing to its horizontal motion alone strike at c with a momentum of 4; but if the body strikes with a momentum which is the resultant of these two, its blow will be that of a body moving with a momentum [*velocità e forza*] of 5; and such a blow will be the same at all points of the diagonal *ac*, since its components are always the same and never increase or diminish.

Let us now pass to the consideration of a uniform horizontal motion compounded with the vertical motion of a freely falling body starting from rest. It is at once clear that the diagonal which represents the motion compounded of these two is not a straight line, but, as has been demonstrated, a semi-parabola, in which the momentum [*impeto*] is always increasing because the speed [*velocità*] of the vertical component is always increasing. Wherefore, to determine the momentum [*impeto*] at any given point in the parabolic diagonal, it is necessary first to fix upon the uniform horizontal momentum [*impeto*] and then, treating the body as one falling freely, to find the vertical momentum at the given point; this latter can be determined only by taking into account the duration of fall, a consideration which does not enter into the composition of two uniform motions where the velocities and momenta are always the same; but here where one of the component motions has an initial value of zero and increases its speed [*velocità*] in direct proportion to the time, it follows that the time must determine the speed [*velocità*] at the assigned point. It only remains to obtain the momentum resulting from these two components (as in the case of uniform motions) by placing the square of the resultant equal to the sum of the squares of the two components. But here again it is better to illustrate by means of an example.

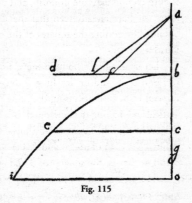

Fig. 115

On the vertical *ac* lay off any portion *ab* which we shall employ as a measure of the space traversed by a body falling freely along the perpendicular, likewise as a measure of the time and also of the speed [*grado di velocità*] or, we may say, of the momenta [*impeti*]. It is at once clear that if the momentum of a body at *b*, after having fallen

from rest at *a*, be diverted along the horizontal direction *bd*, with uniform motion, its speed will be such that, during the time-interval *ab*, it will traverse a distance which is represented by the line *bd* and which is twice as great as *ab*. Now choose a point *c*, such that *bc* shall be equal to *ab*, and through *c* draw the line *ce* equal and parallel to *bd*; through the points *b* and *e* draw the parabola *bei*. And since, during the time-interval *ab*, the horizontal distance *bd* or *ce*, double the length *ab*, is traversed with the momentum *ab*, and since during an equal time-interval the vertical distance *bc* is traversed, the body acquiring at *c* a momentum represented by the same horizontal, *bd*, it follows that during the time *ab* the body will pass from *b* to *e* along the parabola *be*, and will reach *e* with a momentum compounded of two momenta each equal to *ab*. And since one of these is horizontal and the other vertical, the square of the resultant momentum is equal to the sum of the squares of these two components, i. e., equal to twice either one of them.

Therefore, if we lay off the distance *bf*, equal to *ba*, and draw the diagonal *af*, it follows that the momentum [*impeto e percossa*] at *e* will exceed that of a body at *b* after having fallen from *a*, or what is the same thing, will exceed the horizontal momentum [*percossa dell'impeto*] along *bd*, in the ratio of *af* to *ab*.

Suppose now we choose for the height of fall a distance *bo* which is not equal to but greater than *ab*, and suppose that *bg* represents a mean proportional between *ba* and *bo*; then, still retaining *ba* as a measure of the distance fallen through, from rest at *a*, to *b*, also as a measure of the time and of the momentum which the falling body acquires at *b*, it follows that *bg* will be the measure of the time and also of the momentum which the body acquires in falling from *b* to *o*. Likewise just as the momentum *ab* during the time *ab* carried the body a distance along the horizontal equal to twice *ab*, so now, during the time-interval *bg*, the body will be carried in a horizontal direction through a distance which is greater in the ratio of *bg* to *ba*. Lay off *lb* equal to *bg* and draw the diagonal *al*, from which we have a quantity compounded of two velocities [*impeti*] one horizontal, the other vertical; these determine the parabola. The horizontal and uniform velocity is that acquired at *b* in falling from *a*; the other is that acquired at *o*, or, we may say, at *i*, by a body falling through the distance *bo*, during a time measured by the line *bg*, which line *bg* also represents the momentum of the body.

And in like manner we may, by taking a mean proportional between the two heights, determine the momentum [*impeto*] at the extreme end of the parabola where the height is less than the sublimity *ab*; this mean proportional is to be drawn along the

horizontal in place of *bf*, and also another diagonal in place of *af*, which diagonal will represent the momentum at the extreme end of the parabola.

To what has hitherto been said concerning the momenta, blows or shocks of projectiles, we must add another very important consideration; to determine the force and energy of the shock [*forza ed energia della percossa*] it is not sufficient to consider only the speed of the projectiles, but we must also take into account the nature and condition of the target which, in no small degree, determines the efficiency of the blow. First of all it is well known that the target suffers violence from the speed [*velocità*] of the projectile in proportion as it partly or entirely stops the motion; because if the blow falls upon an object which yields to the impulse [*velocità del percuziente*] without resistance such a blow will be of no effect; likewise when one attacks his enemy with a spear and overtakes him at an instant when he is fleeing with equal speed there will be no blow but merely a harmless touch. But if the shock falls upon an object which yields only in part then the blow will not have its full effect, but the damage will be in proportion to the excess of the speed of the projectile over that of the receding body; thus, for example, if the shot reaches the target with a speed of 10 while the latter recedes with a speed of 4, the momentum and shock [*impeto e percossa*] will be represented by 6. Finally the blow will be a maximum, in so far as the projectile is concerned, when the target does not recede at all but if possible completely resists and stops the motion of the projectile. I have said *in so far as the projectile is concerned* because if the target should approach the projectile the shock of collision [*colpo e l'incontro*] would be greater in proportion as the sum of the two speeds is greater than that of the projectile alone.

Moreover it is to be observed that the amount of yielding in the target depends not only upon the quality of the material, as regards hardness, whether it be of iron, lead, wool, etc., but also upon its position. If the position is such that the shot strikes

[297]

it at right angles, the momentum imparted by the blow [*impeto del colpo*] will be a maximum; but if the motion be oblique, that is to say slanting, the blow will be weaker; and more and more so in proportion to the obliquity; for, no matter how hard the material of the target thus situated, the entire momentum [*impeto e moto*] of the shot will not be spent and stopped; the projectile will slide by and will, to some extent, continue its motion along the surface of the opposing body.

All that has been said above concerning the amount of momentum in the projectile at the extremity of the parabola must be understood to refer to a blow received on a line at right angles to this parabola or along the tangent to the parabola at the given point; for, even though the motion has two components, one horizontal, the other

vertical, neither will the momentum along the horizontal nor that upon a plane perpendicular to the horizontal be a maximum, since each of these will be received obliquely.

Sagr. Your having mentioned these blows and shocks recalls to my mind a problem, or rather a question, in mechanics of which no author has given a solution or said anything which diminishes my astonishment or even partly relieves my mind.

My difficulty and surprise consist in not being able to see whence and upon what principle is derived the energy and immense force [*energia e forza immensa*] which makes its appearance in a blow; for instance we see the simple blow of a hammer, weighing not more than 8 or 10 lbs., overcoming resistances which, without a blow, would not yield to the weight of a body producing impetus by pressure alone, even though that body weighed many hundreds of pounds. I would. like to discover a method of measuring the force [*forza*] of such a percussion. I can hardly think it infinite, but incline rather to the view that it has its limit and can be counterbalanced and measured by other forces, such as weights, or by levers or screws or other mechanical instruments which are used to multiply forces in a manner which I satisfactorily understand.

Salv. You are not alone in your surprise at this effect or in obscurity as to the cause of this remarkable property. I studied this matter myself for a while in vain; but my confusion merely increased until finally meeting our Academician I received from him

[298]

great consolation. First he told me that he also had for a long time been groping in the dark; but later he said that, after having spent some thousands of hours in speculating and contemplating thereon, he had arrived at some notions which are far removed from our earlier ideas and which are remarkable for their novelty. And since now I know that you would gladly hear what these novel ideas are I shall not wait for you to ask but promise that, as soon as our discussion of projectiles is completed, I will explain all these fantasies, or if you please, vagaries, as far as I can recall them from the words of our Academician. In the meantime we proceed with the propositions of the author.

PROPOSITION V, PROBLEM

Having given a parabola, find the point, in its axis extended upwards, from which a particle must fall in order to describe this same parabola.

Let *ab* be the given parabola, *hb* its amplitude, and *he* its axis extended. The problem is to find the point *e* from which a body must fall in order that, after the momentum

which it acquires at *a* has been diverted into a horizontal direction, it will describe the parabola *ab*. Draw the horizontal *ag*, parallel to *bh*, and having laid off *af* equal to *ah*, draw the straight line *bf* which will be a tangent to the parabola at *b*, and will intersect the horizontal *ag* at *g*: choose *e* such that *ag* will be a mean proportional between *af* and *ae*. Now I say that *e* is the point above sought. That is, if a body falls from rest at this point *e*, and if the momentum acquired at the point *a* be diverted into a horizontal direction, and compounded with the momentum acquired at *h* in falling from rest at *a*, then the body will describe the parabola *ab*. For if we understand *ea* to be the

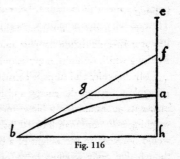

Fig. 116

measure of the time of fall from *e* to *a*, and also of the momentum acquired at *a*, then *ag* (which is a mean proportional between *ea* and *af*) will represent the time and momentum of fall from *f* to *a* or, what is the same thing, from *a* to *h*; and since a body falling from *e*, during the time *ea*, will, owing to the momentum acquired at *a*, traverse at uniform speed a horizontal distance which is twice *ea*, it follows that, the body will. if impelled by the same momentum, during the time-interval *ag* traverse a distance equal to twice *ag* which is the half of *bh*. This is true because, in the case of uniform motion, the spaces traversed vary directly as the times. And likewise if the

[299]

motion be vertical and start from rest, the body will describe the distance *ah* in the time *ag*. Hence the amplitude *bh* and the altitude *ah* are traversed by a body in the same time. Therefore the parabola *ab* will be described by a body falling from the sublimity of *e*.   Q.E.F.

COROLLARY

Hence it follows that half the base, or amplitude, of the semi-parabola (which is one-quarter of the entire amplitude) is a mean proportional between its altitude and the sublimity from which a falling body will describe this same parabola.

PROPOSITION VI, PROBLEM

Given the sublimity and the altitude of a parabola, to find its amplitude.

Let the line *ac*, in which lie the given altitude *cb* and sublimity *ab*, be perpendicular to the horizontal line *cd*. The problem is to find the amplitude, along the horizontal

cd, of the semi-parabola which is described with the sublimity *ba* and altitude *bc*. Lay off *cd* equal to twice the mean proportional between *cb* and *ba*. Then *cd* will be the amplitude sought, as is evident from the preceding proposition.

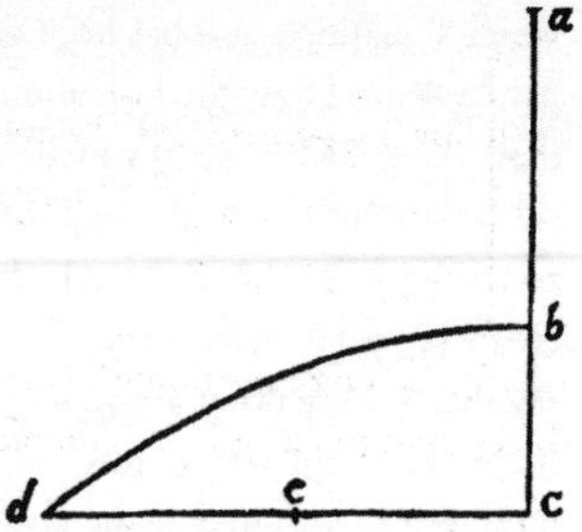

THEOREM. PROPOSITION VII

If projectiles describe semi-parabolas of the same amplitude, the momentum required to describe that one whose amplitude is double its attitude is less than that required for any other.

Fig. 117

Let *bd* be a semi-parabola whose amplitude *cd* is double its altitude *cb*; on its axis extended upwards lay off *ba* equal to its altitude *bc*. Draw the line *ad* which will be a tangent to the parabola at *d* and will cut the horizontal line *be* at the point *e*, making *be* equal to *bc* and also to *ba*. It is evident that this parabola will be described by a projectile whose uniform horizontal momentum is that which it would acquire at *b* in falling from rest at *a* and whose naturally accelerated vertical momentum is that of the body falling to *c*, from rest at *b*. From this it follows that the momentum at the terminal point *d*, compounded of these two, is represented by the diagonal *ae*, whose square is equal to the sum of the squares of the two components. Now let *gd* be any other parabola whatever having the same amplitude *cd*, but whose altitude *cg* is either greater or less than the altitude *bc*. Let *hd* be the tangent cutting the horizontal through

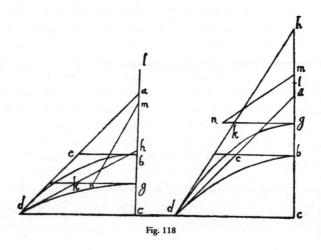

Fig. 118

g at *k*. Select a point *l* such that *hg*:*gk*=*gk*:*gl*. Then from a preceding proposition [V], it follows that *gl* will be the height from which a body must fall in order to describe the parabola *gd*.

Let *gm* be a mean proportional between *ab* and *gl*, then *gm* will [Prop. IV] represent the time and momentum acquired at *g* by a fall from *l*; for *ab* has been assumed as a measure of both time and momentum. Again let *gn* be a mean proportional between *bc* and *cg*; it will then represent the time and momentum which the body acquires at *c* in falling from *g*. If now we join *m* and *n*, this line *mn* will represent the momentum at *d* of the projectile traversing the parabola *dg*; which momentum is, I say, greater than that of the projectile travelling along the parabola *bd* whose measure was given by *ae*. For since *gn* has been taken as a mean proportional between *bc* and *gc*; and since *bc* is equal to *be* and also to *kg* (each of them being the half of *dc*) it follows that *cg*:*gn*=*gn*:*gk*, and as *cg* or (*hg*) is to *gk* so is $\overline{ng}^2$ to $\overline{gk}^2$: but by construction *hg*:*gk*=*gk*:*gl*. Hence $\overline{ng}^2$:$\overline{gk}^2$=*gk*:*gl*. But *gk*:*gl*=$\overline{gk}^2$:$\overline{gm}^2$, since *gm* is a mean proportional between *kg* and *gl*. Therefore the three squares *ng*, *kg*, *mg* form a continued proportion, $\overline{gn}^2$:$\overline{gk}^2$=$\overline{gk}^2$:$\overline{gm}^2$. And the sum of the two extremes which is equal to the square of *mn* is greater than twice the square of *gk*; but the square of *ae* is double the square of *gk*. Hence the square of *mn* is greater than the square of *ae* and the length *mn* is greater than the length *ae*.

Q.E.F.

301

COROLLARY

Conversely it is evident that less momentum will be required to send a projectile from the terminal point *d* along the parabola *bd* than along any other parabola having an elevation greater or less than that of the parabola *bd*, for which the tangent at *d* makes an angle of 45° with the horizontal. From which it follows that if projectiles are fired from the terminal point *d*, all having the same speed, but each having a different elevation, the maximum range, i. e., amplitude of the semi-parabola or of the entire parabola, will be obtained when the elevation is 45°: the other shots, fired at angles greater or less will have a shorter range.

Sagr. The force of rigid demonstrations such as occur only in mathematics fills me with wonder and delight. From accounts given by gunners, I was already aware of the fact that in the use of cannon and mortars, the maximum range, that is the one in which the shot goes farthest, is obtained when the elevation is 45° or, as they say, at the sixth point of the quadrant; but to understand why this happens far outweighs the mere information obtained by the testimony of others or even by repeated experiment.

Salv. What you say is very true. The knowledge of a single fact acquired through a discovery of its causes prepares the mind to understand and ascertain other facts without need of recourse to experiment, precisely as in the present case, where by argumentation alone the Author proves with certainty that the maximum range occurs when the elevation is 45°. He thus demonstrates what has perhaps never been observed in experience, namely, that of other shots those which exceed or fall short of 45° by equal amounts have equal ranges; so that if the balls have been fired one at an elevation of 7 points, the other at 5, they will strike the level at the same distance: the same is true if the shots are fired at 8 and at 4 points, at 9 and at 3, etc. Now let us hear the demonstration of this.

[302]

THEOREM. PROPOSITION VIII

The amplitudes of two parabolas described by projectiles fired with the same speed, but at angles of elevation which exceed and fall short of 45° by equal amounts, are equal to each other.

In the triangle *mcb* let the horizontal side *bc* and the vertical *cm*, which form a right angle at *c*, be equal to each other; then the angle *mbc* will be a semi-right angle; let the line *cm* be prolonged to *d*, such a point that the two angles at *b*, namely *mbe* and *mbd*, one above and the other below the diagonal *mb*, shall be equal. It is now to be proved that in the case of two parabolas described by two projectiles fired from *b* with the same speed, one at the angle of *ebc*, the other at the angle of *dbc*, their ampli-

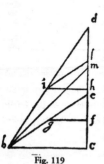

Fig. 119

tudes will be equal. Now since the external angle *bmc* is equal to the sum of the internal angles *mbd* and *dbm* we may also equate to them the angle *mbc*, but if we replace the angle *dbm* by *mbe*, then this same angle *mbc* is equal to the two *mbe* and *bdc*: and if we subtract from each side of this equation the angle *mbe*, we have the remainder *bdc* equal to the remainder *ebc*. Hence the two triangles *dcb* and *bce* are similar. Bisect the straight lines *dc* and *ec* in the points *h* and *f*: and draw the lines *hi* and *fg* parallel to the horizontal *cb*, and choose *l* such that *dh:hi=ih:hl*. Then the triangle *ihl* will be similar to *ihd*, and also to the triangle *egf*; and since *ih* and *gf* are equal, each being half of *bc*, it follows that *hl* is equal to *fe* and also to *fc*; and if we add to each of these the common part *fh*, it will be seen that *ch* is equal to *fl*.

Let us now imagine a parabola described through the points *h* and *b* whose altitude is *hc* and sublimity *hl*. Its amplitude will be *cb* which is double the length *hi* since

hi is a mean proportional between *dh* (or *ch*) and *hl*. The line *db* is tangent to the parabola at *b*, since *ch* is equal to *hd*. If again we imagine a parabola described through the points *f* and *b*, with a sublimity *fl* and altitude *fc*, of which the mean proportional is *fg*, or one-half of *cb*, then, as before, will *cb* be the amplitude and the line *eb* a tangent at *b*; for *ef* and *fc* are equal.

[303]

But the two angles *dbc* and *ebc*, the angles of elevation, differ by equal amounts from a 45° angle. Hence follows the proposition.

THEOREM. PROPOSITION IX

The amplitudes of two parabolas are equal when their altitudes and sublimities are inversely proportional.

Let the altitude *gf* of the parabola *fh* bear to the altitude *cb* of the parabola *bd* the same ratio which the sublimity *ba* bears to the sublimity *fe*; then I say the amplitude *hg* is equal to the amplitude *dc*. For since the first of these quantities, *gf*, bears to the second *cb* the same ratio which the third, *ba*, bears to the fourth *fe*, it follows that the area of the rectangle *gf.fe* is equal to that of the rectangle *cb.ba*; therefore squares which are equal to these rectangles are equal to each other. But [by Proposition VI] the square of half of *gh* is equal to the rectangle *gf.fe*; and the square of half of

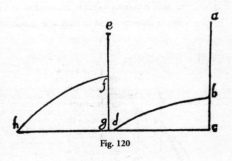

Fig. 120

cd is equal to the rectangle *cb.ba*. Therefore these squares and their sides and the doubles of their sides are equal. But these last are the amplitudes *gh* and *cd*. Hence follows the proposition.

LEMMA FOR THE FOLLOWING PROPOSITION

If a straight line be cut at any point whatever and mean proportionals between this line and each of its parts be taken, the sum of the

squares of these mean proportionals is equal to the square of the entire line.

Let the line *ab* be cut at *c*. Then I say that the square of the mean proportional between *ab* and *ac* plus the square of the mean proportional between *ab* and *cb* is equal to the square of the whole line *ab*. This is evident as soon as we describe a semicircle upon the entire line *ab*, erect a perpendicular *cd* at *c*, and draw *da* and *db*. For *da* is a mean proportional between *ab* and *ac* while *db* is a mean proportional between *ab* and *bc*: and since the angle *adb*, inscribed in a semicircle, is a right angle the sum of the squares of the lines *da* and *db* is equal to the square of the entire line *ab*. Hence follows the proposition.

Fig. 121

[304]

THEOREM. PROPOSITION X

The momentum [impetus seu momentum] acquired by a particle at the terminal point of any semi-parabola is equal to that which it would acquire in falling through a vertical distance equal to the sum of the sum of the sublimity and the altitude of the semi-parabola.[39]

Let *ab* be a semi-parabola having a sublimity *da* and an altitude *ac*, the sum of which is the perpendicular *dc*. Now I say the momentum of the particle at *b* is the same as that which it would acquire in falling freely from *d* to *c*. Let us take the length of *dc* itself as a measure of time and momentum, and lay off *cf* equal to the mean proportional between *cd* and *da*; also lay off *ce* a mean proportional between *cd* and *ca*. Now *cf* is the measure of the time and of the momentum acquired by fall, from rest at *d*, through the distance *da*; while *ce* is the time and momentum of fall, from rest at *a*, through the distance *ca*; also the diagonal *ef* will represent a momentum which is the resultant of these two, and is therefore the momentum at the terminal point of the parabola, *b*.

Fig. 122

And since *dc* has been cut at some point *a* and since *cf* and *ce* are mean proportionals between the whole of *cd* and its parts, *da* and *ac*, it follows, from the preceding lemma, that the sum of the squares of these mean proportionals is equal to the square of the whole: but the square of *ef* is also equal to the sum of these same squares; whence it follows that the line *ef* is equal to *dc*.

39. In modern mechanics this well-known theorem assumes the following form: *The speed of a projectile at any point is that produced by a fall from the directrix. [Trans.]*

Accordingly the momentum acquired at *c* by a particle in falling from *d* is the same as that acquired at *b* by a particle traversing the parabola *ab*.   Q.E.D.

COROLLARY

Hence it follows that, in the case of all parabolas where the sum of the sublimity and altitude is a constant, the momentum at the terminal point is a constant.

[305]

PROBLEM. PROPOSITION XI

Given the amplitude and the speed [impetus] at the terminal point
of a semi-parabola, to find its altitude.

Let the given speed be represented by the vertical line *ab*, and the amplitude by the horizontal line *bc*; it is required to find the sublimity of the semi-parabola whose terminal speed is *ab* and amplitude *bc*. From what precedes [Cor. Prop. V] it is clear

[306]

that half the amplitude *bc* is a mean proportional between the altitude and sublimity of the parabola of which the terminal speed is equal, in accordance with the preceding proposition, to the speed acquired by a body in falling from rest at *a* through the distance *ab*. Therefore the line *ba* must be cut at a point such that the rectangle formed by its two parts will be equal to the square of half *bc*, namely *bd*. Necessarily, therefore, *bd* must not exceed the half of *ba*; for of all the rectangles formed by parts of a straight line the one of greatest area is obtained when the line is divided into two equal parts. Let *e* be the middle point of the line *ab*; and now if *bd* be equal to *be* the problem is solved; for *be* will be the altitude and *ea* the sublimity of the parabola. (Incidentally we may observe a consequence already demonstrated, namely: of all parabolas described with any given terminal speed that for which the elevation is 45° will have the maximum amplitude.)

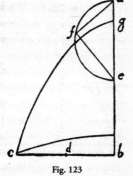

Fig. 123

But suppose that *bd* is less than half of *ba* which is to be divided in such a way that the rectangle upon its parts may be equal to the square of *bd*. Upon *ea* as diameter describe a semicircle *efa*, in which draw the chord *af*, equal to y: join *fe* and lay off the distance *eg* equal to *fe*.

Then the rectangle *bg.ga* plus the square of *eg* will be equal to the square of *ea*, and hence also to the sum of the squares of *af* and *fe*. If now we subtract the equal squares of *fe* and *ge* there remains the rectangle *bg.ga* equal to the square of *af*, that is, of *bd*, a line which is a mean proportional between *bg* and *ga*; from which it is evident that the semi-parabola whose amplitude is *bc* and whose terminal speed [*impetus*] is represented by *ba* has an altitude *bg* and a sublimity *ga*.

If however we lay off *bi* equal to *ga*, then *bi* will be the altitude of the semi-parabola *ic*, and *ia* will be its sublimity. From the preceding demonstration we are able to solve the following problem.

PROBLEM. PROPOSITION XII

To compute and tabulate the amplitudes of all semi-parabolas which are described by projectiles fired with the same initial speed [impetus].

From the foregoing it follows that, whenever the sum of the altitude and sublimity is a constant vertical height for any set of parabolas, these parabolas are described by projectiles having the same initial speed; all vertical heights thus obtained are therefore included between two parallel horizontal lines. Let *cb* represent a horizontal line and *ab* a vertical line of equal length; draw the diagonal *ac*; the angle *acb* will be one of 45°; let *d* be the middle point of the vertical line *ab*. Then the semi-parabola *dc* is the one which is determined by the sublimity *ad* and the altitude *db*, while its terminal speed at *c* is that which would be acquired at *b* by a particle falling from rest at *a*. If now *ag* be drawn parallel to *bc*, the sum of the attitude and sublimity for any other semi-parabola having the same terminal speed will, in the manner explained, be equal to the distance between the parallel lines *ag* and *bc*. Moreover, since it has already been shown that the amplitudes of two semi-parabolas are the same when their angles of elevation differ from 45° by like amounts, it follows that the same computation which is employed for the larger elevation will serve also for the smaller. Let us also assume 10000 as the greatest amplitude for a parabola whose angle of elevation is 45°; this then will be the length of the line *ba* and the amplitude of the semi-parabola *bc*. This number, 10000, is selected because in these

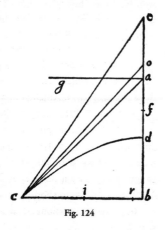

Fig. 124

calculations we employ a table of tangents in which this is the value of the tangent of 45°. And now, coming down to business, draw the straight line *ce* making an acute angle *ecb* greater than *acb*: the problem now is to draw the semi-parabola to which the line *ec* is a tangent and for which the sum of the sublimity and the altitude is the distance *ba*. Take the length of the tangent[40] *be* from the table of tangents, using the angle *bce* as an argument: let *f* be the middle point of *be*; next find a third proportional to *bf* and *bi* (the half of *bc*), which is of necessity greater than *fa*.[41] Call this *fo*. We have now discovered that, for the parabola inscribed in the triangle *ecb* having the tangent *ce* and the amplitude *cb*, the altitude is *bf* and the sublimity *fo*. But the total length of *bo* exceeds the distance

[307]

between the parallels *ag* and *cb*, while our problem was to keep it equal to this distance: for both the parabola sought and the parabola *dc* are described by projectiles fired from *c* with the same speed. Now since an infinite number of greater and smaller parabolas, similar to each other, may be described within the angle *bce* we must find another parabola which like *cd* has for the sum of its altitude and sublimity the height *ba*, equal to *bc*.

Therefore lay off *cr* so that, *ob:ba=bc:cr*; then *cr* will be the amplitude of a semi-parabola for which *bce* is the angle of elevation and for which the sum of the altitude and sublimity is the distance between the parallels *ga* and *cb*, as desired. The process is therefore as follows: One draws the tangent of the given angle *bce*; takes half of this tangent, and adds to it the quantity, *fo*, which is a third proportional to the half of this tangent and the half of *bc*; the desired amplitude *cr* is then found from the following proportion *ob:ba:bc:cr*. For example let the angle *ecb* be one of 50°; its tangent is 11918, half of which, namely *bf*, is 5959; half of *bc* is 5000; the third proportional of these halves is 4195, which added to *bf* gives the value 10154 for *bo*. Further, as *ob* is to *ab*, that is, as 10154 is to 10000, so is *bc*, or 10000 (each being the tangent of 45°) to *cr*, which is the amplitude sought and which has the value 9848, the maximum amplitude being *bc*, or 10000. The amplitudes of the entire parabolas are double these, namely, 19696 and 20000. This is also the amplitude of a parabola whose angle of elevation is 40°, since it deviates by an equal amount from one of 45°.

Sagr. In order to thoroughly understand this demonstration I need to be shown how the third proportional of *bf* and *bi* is, as the Author indicates, necessarily greater than *fa*.

---

40. The reader will observe that the word "tangent" is here used in a sense somewhat different from that of the preceding sentence. The "tangent *ec*" is a line which touches the parabola at *c*, but the "tangent *eb*" is the side of the right-angled triangle which lies opposite the angle *ecb*, a line whose length is proportional to the numerical value of the tangent of this angle. *[Trans.]*

41. This fact is demonstrated in the third paragraph below. *[Trans.]*

[308]

Salv. This result can, I think, be obtained as follows. The square of the mean proportional between two lines is equal to the rectangle formed by these two lines. Therefore the square of *bi* (or of *bd* which is equal to *bi*) must be equal to the rectangle formed by *fb* and the desired third proportional. This third proportional is necessarily greater than *fa* because the rectangle formed by *bf* and *fa* is less than the square of *bd* by an amount equal to the square of *df*, as shown in Euclid, II. 1. Besides it is to be observed that the point *f*, which is the middle point of the tangent *eb*, falls in general above *a* and only once at *a*; in which cases it is self-evident that the third proportional to the half of the tangent and to the sublimity *bi* lies wholly above *a*. But the Author has taken a case where it is not evident that the third proportional is always greater than *fa*, so that when laid off above the point *f* it extends beyond the parallel *ag*.

Now let us proceed. It will be worth while, by the use of this table, to compute another giving the altitudes of these semi-parabolas described by projectiles having the same initial speed. The construction is as follows:

[309]

AMPLITUDES OF SEMI-PARABOLAS DESCRIBED WITH THE SAME INITIAL SPEED.		AMPLITUDES OF SEMI-PARABOLAS DESCRIBED WITH THE SAME INITIAL SPEED.	
ANGLE OF ELEVATION	ANGLE OF ELEVATION	ANGLE OF ELEVATION	ANGLE OF ELEVATION
45° 10000		1° 3	46° 5173
46 9994	44°	2 13	47 5346
47 9976	43	3 28	48 5523
48 9945	42	4 50	49 5698
49 9902	41	5 76	50 5868
50 9848	40	6 108	51 6038
51 9782	39	7 150	52 6207
52 9704	38	8 194	53 6379
53 9612	37	9 245	54 6546
54 9511	36	10 302	55 6710
55 9396	35	11 365	56 6873
56 9272	34	12 432	57 7033
57 9136	33	13 506	58 7190
58 8989	32	14 585	59 7348
59 8829	31	15 670	60 7502
60 8659	30	16 760	61 7049
61 8481	29	17 855	62 7796
62 8290	28	18 955	63 7939
63 8090	27	19 1060	64 8078
64 7880	26	20 1170	65 8214

65	7660	25		21	1285	66	8346
66	7431	24		22	1402	67	8474
67	7191	23		23	1527	68	8597
68	6944	22		24	1685	69	8715
69	6692	21		25	1786	70	8830

AMPLITUDES OF SEMI-PARABOLAS DESCRIBED WITH THE SAME INITIAL SPEED.			ALTITUDES OF SEMI-PARABOLAS DESCRIBED WITH THE SAME INITIAL SPEED.		
ANGLE OF ELEVATION		ANGLE OF ELEVATION	ANGLE OF ELEVATION		ANGLE OF ELEVATION
70°	6428	20	26°	1922	71°	8940

ANGLE OF ELEVATION		ANGLE OF ELEVATION		ANGLE OF ELEVATION		ANGLE OF ELEVATION	
70°	6428	20		26°	1922	71°	8940
71	6157	19		27	2061	72	9045
72	5878	18		28	2204	73	9144
73	5592	17		29	2351	74	9240
74	5300	16		30	2499	75	9330
75	5000	15		31	2653	76	9415
76	4694	14		32	2810	77	9493
77	4383	13		33	2967	78	9567
78	4067	12		34	3128	79	9636
79	3746	11		35	3289	80	9698
80	3420	10		36	3456	81	9755
81	3090	9		37	3621	82	9806
82	2756	8		38	3793	83	9851
83	2419	7		39	3962	84	9890
84	2079	6		40	4132	85	9924
85	1736	5		41	4302	86	9951
86	1391	4		42	4477	87	9972
87	1044	3		43	4654	88	9987
88	698	2		44	4827	89	9998
89	349	1		45	5000	90	10000

[310]

PROBLEM. PROPOSITION XIII

From the amplitudes of semi-parabolas given in the preceding table to find the altitudes of each of the parabolas described with the same initial speed.

Let *bc* denote the given amplitude; and let *ob*, the sum of the altitude and sublimity, be the measure of the initial speed which is understood to remain constant. Next we must find and determine the altitude, which we shall accomplish by so dividing *ob* that the rectangle contained by its parts shall be equal to the square of half the amplitude, *bc*. Let *f* denote this point of division and *d* and *i* be the middle points of *ob* and *bc* respectively. The square of *ib* is equal to the rectangle *bf.fo*; but the square

of *do* is equal to the sum of the rectangle *bf.fo* and the square of *fd*. If, therefore, from the square of *do* we subtract the square of *bi* which is equal to the rectangle *bf.fo*, there will remain the square of *fd*. The altitude in question, *bf*, is now obtained by adding to this length, *fd*, the line *bd*. The process is then as follows: From the square of half of *bo* which is known, subtract the square of *bi* which as also known; take the square root of the remainder and add to it the known length *db*; then you have the required altitude, *bf*.

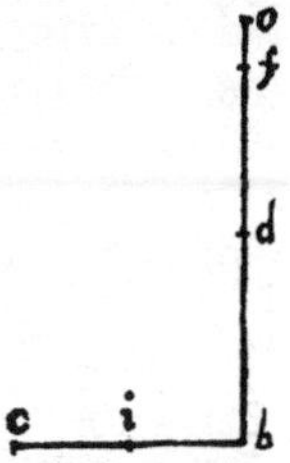

Fig. 125

Example. To find the altitude of a semi-parabola described with an angle of elevation of 55°. From the preceding table the amplitude is seen to be 9396, of which the half is 4698, and the square 22071204. When this is subtracted from the square of the half of *bo*, which is always 25,000,000, the remainder is 2928796, of which the square root is approximately 1710. Adding this to the half of *bo*, namely 5000, we have 6710 for the altitude of *bf*.

[311]

It will be worth while to add a third table giving the altitudes and sublimities for parabolas in which the amplitude is a constant.

Sagr. I shall be very glad to see this; for from it I shall learn the difference of speed and force [*degl' impeti e delle forze*] required to fire projectiles over the same range with what we call mortar shots. This difference will, I believe, vary greatly with the elevation so that if, for example, one wished to employ an elevation of 3° or 4°, or 87° or 88° and yet give the ball the same range which it had with an elevation of 45° (where we have shown the initial speed to be a minimum) the excess of force required will, I think, be very great.

Salv. You are quite right, sir; and you will find that in order to perform this operation completely, at all angles of elevation, you will have to make great strides toward an infinite speed. We pass now to the consideration of the table.

[312]

TABLE GIVING THE ALTITUDES AND SUBLIMITIES OF PARABOLAS OF CONSTANT AMPLITUDE, NAMELY 10000, COMPUTED FOR EACH DEGREE OF ELEVATION.

ANGLE OF ELEVATION	ALTITUDE	SUBLIMITY	ANGLE OF ELEVATION	ALTITUDE	SUBLIMITY
1°	87	286533	46°	5177	4828
2	175	142450	47	5363	4662
3	262	95802	48	5553	4502
4	349	71531	49	5752	4345
5	437	57142	50	5959	4196

6	525	47573	51	6174	4048
7	614	40716	52	6399	3906
8	702	35587	53	6635	3765
9	792	31565	54	6882	3632
10	881	28367	55	7141	3500
11	972	25720	56	7413	3372
12	1063	23518	57	7699	3247
13	1154	21701	58	8002	3123
14	1246	20056	59	8332	3004
15	1339	18663	60	8600	2887
16	1434	17405	61	9020	2771
17	1529	16355	62	9403	2658
18	1624	15389	63	9813	2547
19	1722	14522	64	10251	2438
20	1820	13736	65	10722	2331
21	1919	13024	66	11230	2226
22	2020	12376	67	11779	2122
23	2123	11778	68	12375	2020
24	2226	11230	69	13025	1919
25	2332	10722	70	13237	1819
26	2439	10253	71	14521	1721
27	2547	9814	72	15388	1624
28	2658	9404	73	16354	1528
29	2772	9020	74	17437	1433
30	2887	8659	75	18660	1339
31	3008	8336	76	20054	1246
32	3124	8001	77	21657	1154
33	3247	7699	78	23523	1062
34	3373	7413	79	25723	972
35	3501	7141	80	28356	881
36	3631	6882	81	31569	792
37	3768	6635	82	35577	702
38	3906	6395	83	40222	613
39	4049	6174	84	47572	525
40	4196	5959	85	57150	437
41	4346	5752	86	71503	349
42	4502	5553	87	95405	262
43	4662	5362	88	143181	174
44	4828	5177	89	286499	87
45	5000	5000	90	infinita	

[313]

PROPOSITION XIV

To find for each degree of elevation the altitudes and sublimities of parabolas of constant amplitude.

The problem is easily solved. For if we assume a constant amplitude of 10000, then half the tangent at any angle of elevation will be the altitude. Thus, to illustrate, a parabola having an angle of elevation of 30° and an amplitude of 10000, will have an altitude of 2887, which is approximately one-half the tangent. And now the altitude having been found, the sublimity is derived as follows. Since it has been proved that half the amplitude of a semi-parabola is the mean proportional between the altitude and sublimity, and since the altitude has already been found, and since the semi-amplitude is a constant, namely 5000, it follows that if we divide the square of the semi-amplitude by the altitude we shall obtain the sublimity sought. Thus in our example the altitude was found to be 2887: the square of 5000 is 25,000,000, which divided by 2887 gives the approximate value of the sublimity, namely 8659.

Salv. Here we see, first of all, how very true is the statement made above, that, for different angles of elevation, the greater the deviation from the mean, whether above or below, the greater the initial speed [*impeto e violenza*] required to carry the projectile over the same range. For since the speed is the resultant of two motions, namely, one horizontal and uniform, the other vertical and naturally accelerated; and since the sum of the altitude and sublimity represents this speed, it is seen from the preceding table that this sum is a minimum for an elevation of 45° where the altitude and sublimity are equal, namely, each 5000; and their sum 10000. But if we choose a greater elevation, say 50°, we shall find the altitude 5959, and the sublimity 4196, giving a sum of 10155; in like manner we shall find that this is precisely the value of the speed at 40° elevation, both angles deviating equally from the mean.

Secondly it is to be noted that, while equal speeds are required for each of two elevations that are equidistant from the mean, there is this curious alternation, namely, that the altitude and sublimity at the greater elevation correspond inversely to the sublimity and altitude at the lower elevation. Thus in the preceding example an elevation

[314]

of 50° gives an altitude of 5959 and a sublimity of 4196; while an elevation of 40° corresponds to an altitude of 4196 and a sublimity of 5959. And this holds true in general; but it is to be remembered that, in order to escape tedious calculations, no account has been taken of fractions which are of little moment in comparison with such large numbers.

Sagr. I note also in regard to the two components of the initial speed [*impeto*] that the higher the shot the less is the horizontal and the greater the vertical component; on the other band, at lower elevations where the shot reaches only a small height the horizontal component of the initial speed must be great. In the case of a projectile fired at an elevation of 90°, I quite understand that all the force [*forza*] in the world would

not be sufficient to make it deviate a single finger's breadth from the perpendicular and that it would necessarily fall back into its initial position; but in the case of zero elevation, when the shot is fired horizontally, I am not so certain that some force, less than infinite, would not carry the projectile some distance; thus not even a cannon can fire a shot in a perfectly horizontal direction, or as we say, point blank, that is, with no elevation at all. Here I admit there is some room for doubt. The fact I do not deny outright, because of another phenomenon apparently no less remarkable, but yet one for which I have conclusive evidence. This phenomenon is the impossibility of stretching a rope in such a way that it shall be at once straight and parallel to the horizon; the fact is that the cord always sags and bends and that no force is sufficient to stretch it perfectly straight.

Salv. In this case of the rope then, Sagredo, you cease to wonder at the phenomenon because you have its demonstration; but if we consider it with more care we may possibly discover some correspondence between the case of the gun and that of the string. The curvature of the path of the shot fired horizontally appears to result from two forces, one (that of the weapon) drives it horizontally and the other (its own weight) draws it vertically downward. So in stretching the rope you have the force which pulls it horizontally and its own weight which acts downwards. The circumstances in these two cases are, therefore, very similar. If then you attribute to the weight of the rope a power and energy [*possanza ed energia*] sufficient to oppose and overcome any stretching force, no matter how great, why deny this power to the bullet?

[315]

Besides I must tell you something which will both surprise and please you, namely, that a cord stretched more or less tightly assumes a curve which closely approximates the parabola. This similarity is clearly seen if you draw a parabolic curve on a vertical plane and then invert it so that the apex will lie at the bottom and the base remain horizontal; for, on hanging a chain below the base, one end attached to each extremity of the base, you will observe that, on slackening the chain more or less, it bends and fits itself to the parabola; and the coincidence is more exact in proportion as the parabola is drawn with less curvature or, so to speak, more stretched; so that in parabolas described with elevations less than 45° the chain fits its parabola almost perfectly.

Sagr. Then with a fine chain one would be able to quickly draw many parabolic lines upon a plane surface.

Salv. Certainly and with no small advantage as I shall show you later.

Simp. But before going further, I am anxious to be convinced at least of that proposition of which you say that there is a rigid demonstration; I refer to the statement that it is impossible by any force whatever to stretch a cord so that it will lie perfectly straight and horizontal.

Sagr. I will see if I can recall the demonstration; but in order to understand it, Simplicio, it will be necessary for you to take for granted concerning machines what is evident not alone from experiment but also from theoretical considerations, namely, that the velocity of a moving body [*velocità del movente*], even when its force [*forza*] is small, can overcome a very great resistance exerted by a slowly moving body, whenever the velocity of the moving body bears to that of the resisting body a greater ratio than the resistance [*resistenza*] of the resisting body to the force [*forza*] of the moving body.

[316]

Simp. This I know very well for it has been demonstrated by Aristotle in his *Questions in Mechanics*; it is also clearly seen in the lever and the steelyard where a counterpoise weighing not more than 4 pounds will lift a weight of 400 provided that the distance of the counterpoise from the axis about which the steelyard rotates be more than one hundred times as great as the distance between this axis and the point of support for the large weight. This is true because the counterpoise in its descent traverses a space more than one hundred times as great as that moved over by the large weight in the same time; in other words the small counterpoise moves with a velocity which is more than one hundred times as great as that of the large weight.

Sagr. You are quite right; you do not hesitate to admit that however small the force [*forza*] of the moving body it will overcome any resistance, however great, provided it gains more in velocity than it loses in force and weight [*vigore e gravità*]. Now let us return to the case of the cord. In the accompanying figure *ab* represents a line passing through two fixed points *a* and *b*; at the extremities of this line hang, as you see, two

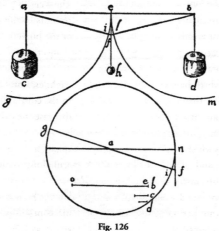

Fig. 126

large weights *c* and *d*, which stretch it with great force and keep it truly straight, seeing that it is merely a line without weight. Now I wish to remark that if from the middle point of this line, which we may call *e*, you suspend any small weight, say *h*, the

[317]

line *ab* will yield toward the point *f* and on account of its elongation will compel the two heavy weights *c* and *d* to rise. This I shall demonstrate as follows: with the points *a* and *b* as centers describe the two quadrants, *eig* and *elm*; now since the two semi-diameters *ai* and *bl* are equal to *ae* and *eb*, the remainders *fi* and *fl* are the excesses of the lines *af* and *fb* over *ae* and *eb*; they therefore determine the rise of the weights *c* and *d*, assuming of course that the weight *h* has taken the position *f*. But the weight *h* will take the position *f*, whenever the line *ef* which represents the descent of *h* bears to the line *fi*—that is, to the rise of the weights *c* and *d*—a ratio which is greater than the ratio of the weight of the two large bodies to that of the body *h*. Even when the weights of *c* and *d* are very great and that of *h* very small this will happen; for the excess of the weights *c* and *d* over the weight of *h* can never be so great but that the excess of the tangent *ef* over the segment *fi* may be proportionally greater. This may be proved as follows: Draw a circle of diameter *gai*; draw the line *bo* such that the ratio of its length to another length *c*, *c>d*, is the same as the ratio of the weights *c* and *d* to the weight *h*. Since *c>d*, the ratio of *bo* to *d* is greater than that of *bo* to *c*. Take *be* a third proportional to *ob* and *d*; prolong the diameter *gi* to a point *f* such that *gi:if=oe:eb*; and from the point *f* draw the tangent *fn*; then since we already have *oe:eb=gi:if*, we shall obtain, by compounding ratios, *ob:eb=gf:if*. But *d* is a mean proportional between *ob* and *be*; while *nf* is a mean proportional between *gf* and *fi*. Hence *nf* bears to *fi* the same ratio as that of *cb* to *d*, which is greater than that of the weights *c* and *d* to the weight *h*. Since then the descent, or velocity, of the weight *h* bears to the rise, or velocity, of the weights *c* and *d* a greater ratio than the weight of the bodies *c* and *d* bears to the weight of *h*, it is clear that the weight *h* will descend and the line *ab* will cease to be straight and horizontal.

And now this which happens in the case of a weightless cord *ab* when any small weight *h* is attached at the point *e*, happens also when the cord is made of ponderable matter but without any attached weight; because in this case the material of which the cord is composed functions as a suspended weight.

Simp. I am fully satisfied. So now Salviati can explain, as he promised, the advantage of such a chain and, afterwards, present the speculations of our Academician on the subject of impulsive forces [*forza della percossa*].

Salv. Let the preceding discussions suffice for to-day; the hour is already late and the time remaining will not permit us to clear up the subjects proposed; we may therefore postpone our meeting until another and more opportune occasion.

[318]

Sagr. I concur in your opinion, because after various conversations with intimate friends of our Academician I have concluded that this question of impulsive forces is very obscure, and I think that, up to the present, none of those who have treated this subject have been able to clear up its dark corners which lie almost beyond the reach of human imagination; among the various views which I have heard expressed one, strangely fantastic, remains in my memory, namely, that impulsive forces are indeterminate, if not infinite. Let us, therefore, await the convenience of Salviati. Meanwhile tell me what is this which follows the discussion of projectiles.

Salv. These are some theorems pertaining to the centers of gravity of solids, discovered by our Academician in his youth, and undertaken by him because he considered the treatment of Federigo Comandino to be somewhat incomplete. The propositions which you have before you would, he thought, meet the deficiencies of Comandino's book. The investigation was undertaken at the instance of the Illustrious Marquis Guid' Ubaldo Dal Monte, a very distinguished mathematician of his day, as is evidenced by his various publications. To this gentleman our Academician gave a copy of this work, hoping to extend the investigation to other solids not treated by Comandino. But a little later there chanced to fall into his hands the book of the great geometrician, Luca Valerio, where he found the subject treated so completely that he left off his own investigations, although the methods which he employed were quite different from those of Valerio.

Sagr. Please be good enough to leave this volume with me until our next meeting so that I may be able to read and study these propositions in the order in which they are written.

Salv. It is a pleasure to comply with your request and I only hope that the propositions will be of deep interest to you.

END OF FOURTH DAY.

APPENDIX

Containing some theorems, and their proofs, dealing with centers of gravity of solid bodies, written by the same Author at an earlier date.[42]

[FINIS]

---

42. Following the example of the National Edition, this *Appendix* which covers 18 pages of the Leyden Edition of 1638 is here omitted as being of minor interest. *[Trans.]*

Johannes Kepler

(1571-1630)

HIS LIFE AND WORK

If an award were ever given to the person in history who was most dedicated to the pursuit of absolute precision, the German astronomer Johannes Kepler might well be the recipient. Kepler was so obsessed with measurements that he even calculated his own gestational period to the minute—224 days, 9 hours, 53 minutes. (He had been born prematurely.) So it is no surprise that he toiled over his astronomical research to such a degree that he ultimately produced the most exact astronomical tables of his time, leading to the eventual acceptance of the sun-centered (heliocentric) theory of the planetary system.

Like Copernicus, whose work inspired him, Kepler was a deeply religious man. He viewed his continual study of universal properties as a fulfillment of his Christian duty to understand the very universe that God created. But unlike Copernicus, Kepler's life was anything but quiet and lacking in contrast. Always short of money, Kepler often resorted to publishing astrological calendars and horoscopes, which, ironically, gained him some local notoriety when their predictions turned out to be quite accurate. Kepler also suffered the early deaths of several of his children, as well as the indignity of having to defend in court his eccentric mother, Katherine, who had a reputation for practicing witchcraft and was nearly burned at the stake.

Kepler entered into a series of complex relationships, most notably with Tycho Brahe, the great naked-eye astronomical observer. Brahe dedicated years of his life to

recording and measuring celestial bodies, but he lacked the mathematical and analytical skills necessary to understand planetary motion. A man of wealth, Brahe hired Kepler to make sense of his observations of the orbit of Mars, which had perplexed astronomers for many years. Kepler painstakingly mapped Brahe's data on the motion of Mars to an ellipse, and this success lent mathematical credibility to the Copernican model of a sun-centered system. His discovery of elliptical orbits helped usher in a new era in astronomy. The motions of planets could now be predicted.

In spite of his achievements, Kepler never gained much wealth or prestige and was often forced to flee the countries where he sojourned because of religious upheaval and civil unrest. By the time he died at the age of fifty-nine in 1630 (while attempting to collect an overdue salary), Kepler had discovered three laws of planetary motion, which are still taught to students in physics classes in the twenty-first century. And it was Kepler's Third Law, not an apple, that led Isaac Newton to discover the law of gravitation.

Johannes Kepler was born on December 27, 1571, in the town of Weil der Stadt, in Württembung (now part of Germany). His father, Heinrich Kepler, was, according to Johannes, "an immoral, rough, and quarrelsome soldier" who deserted his family on several occasions to join up with mercenaries to battle a Protestant uprising in Holland. Heinrich is believed to have died somewhere in the Netherlands. The young Johannes lived with his mother, Katherine, in his grandfather's inn, where he was put to work at an early age waiting tables, despite his poor health. Kepler had nearsightedness as well as double vision, which was believed to have been caused by a near-fatal bout of smallpox; and he also suffered from abdominal problems and "crippled" fingers that limited his career potential choice, in the view of his family, to a life in the ministry.

"Bad-tempered" and "garrulous" were words Kepler used to describe his mother, Katherine, but he was aware from a young age that his father was the cause. Katherine herself had been raised by an aunt who practiced witchcraft and was burned at the stake. So it was no surprise to Kepler when his own mother faced similar charges later in her life. In 1577, Katherine showed her son the "great comet" that appeared in the sky that year, and Kepler later acknowledged that this shared moment with his mother had a lasting impact on his life. Despite a childhood filled with pain and anxiety, Kepler was obviously gifted, and he managed to procure a scholarship reserved for promising male children of limited means who lived in the German province of Swabia. He attended the German Schreibschule in Leonberg before transferring to a Latin school, which was instrumental in providing him with the Latin writing style he later employed in his work. Being frail and precocious, Kepler was beaten regularly by classmates, who considered him a know-it-all, and he soon turned to religious study as a way of escaping his predicament.

In 1587, Kepler enrolled at Tubingen University, where he studied theology and philosophy. He also established himself there as a serious student of mathematics and astronomy, and became an advocate of the controversial Copernican heliocentric theory. So public was young Kepler in his defense of the Copernican model of the universe that it was not uncommon for him to engage in public debate on the subject. Despite his main interest in theology, he was growing more and more intrigued by the mystical appeal of a heliocentric universe. Although he had intended to graduate from Tubingen in 1591 and join the university's theology faculty, a recommendation to a post in mathematics and astronomy at the Protestant school in Graz, Austria, proved irresistible. So, at the age of twenty-two, Kepler deserted a career in the ministry for the study of science. But he would never abandon his belief in God's role in the creation of the universe.

In the sixteenth century, the distinction between astronomy and astrology was fairly ambiguous. One of Kepler's duties as a mathematician in Graz was to compose an astrological calendar complete with predictions. This was a common practice at the time, and Kepler was clearly motivated by the extra money the job provided, but he could not have anticipated the public's reaction when his first calendar was published. He predicted an extraordinarily cold winter, as well as a Turkish incursion, and when both predictions came true, Kepler was triumphantly hailed as a prophet. Despite the clamor, he would never hold much respect for the work he did on the annual almanacs. He called astrology "the foolish little daughter of astronomy" and was equally dismissive of the public' s interest and the astrologer's intentions. "If ever astrologers are correct," he wrote, "it ought to be credited to luck." Still, Kepler never failed to turn to astrology whenever money became tight, which was a recurring theme in his life, and he did hold out hope of discovering some true science in astrology.

One day, while lecturing on geometry in Graz, Kepler experienced a sudden revelation that set him on a passionate journey and changed the course of his life. It was, he felt, the secret key to understanding the universe. On the blackboard, in front of the class, he drew an equilateral triangle within a circle, and another circle drawn within the triangle. It occurred to him that the ratio of the circles was indicative of the ratio of the orbits of Saturn and Jupiter. Inspired by this revelation, he assumed that all six planets known at the time were arranged around the sun in such a way that the geometric figures would fit perfectly between them. Initially he tested this hypothesis without success, using two-dimensional plane figures such as the pentagon, the square, and the triangle. He then returned to the Pythagorean solids, used by the ancient Greeks, who discovered that only five solids could be constructed from regular geometric figures. To Kepler, this explained why there could only be six planets (Mercury, Venus, Earth, Mars, Jupiter, and Saturn) with five spaces between them, and why these

spaces were not uniform. This geometric theory regarding planetary orbits and distances inspired Kepler to write *Mystery of the Cosmos* (*Mysterium Cosmographicum*), published in 1596. It took him about a year to write, and although the scheme was reasonably accurate, he was clearly very sure that his theories would ultimately bear out:

> *And how intense was my pleasure from this discovery can never be expressed in words. I no longer regretted the time wasted. Day and night I was consumed by the computing, to see whether this idea would agree with the Copernican orbits, or if my joy would be carried away by the wind. Within a few days everything worked, and I watched as one body after another fit precisely into its place among the planets.*

Kepler spent the rest of his life trying to obtain the mathematical proof and scientific observations that would justify his theories. *Mystery of the Cosmos* was the first decidedly Copernican work published since Copernicus' own *On the Revolutions*, and as a theologian and astronomer Kepler was determined to understand how and why God designed the universe. Advocating a heliocentric system had serious religious implications, but Kepler maintained that the sun's centrality was vital to God's design, as it kept the planets aligned and in motion. In this sense, Kepler broke with Copernicus' heliostatic system of a sun "near" the center and placed the sun directly in the center of the system.

Today, Kepler's polyhedra appear impracticable. But although the premise of *Mystery of the Cosmos* was erroneous, Kepler's conclusions were still astonishingly accurate and decisive, and were essential in shaping the course of modern science. When the book was published, Kepler sent a copy to Galileo, urging him to "believe and step forth," but the Italian astronomer rejected the work because of its apparent speculations. Tycho Brahe, on the other hand, was immediately intrigued. He viewed Kepler's work as new and exciting, and he wrote a detailed critique in the book's support. Reaction to *Mystery of the Cosmos*, Kepler would later write, changed the direction of his entire life.

In 1597, another event would change Kepler's life, as he fell in love with Barbara Müller, the first daughter of a wealthy mill owner. They married on April 27 of that year, under an unfavorable constellation, as Kepler would later note in his diary. Once again, his prophetic nature emerged as the relationship and the marriage dissolved. Their first two children died very young, and Kepler became distraught. He immersed himself in his work to distract himself from the pain, but his wife did not understand his pursuits. "Fat, confused, and simpleminded" was how he described her in his diary, though the marriage did last fourteen years, until her death in 1611 from typhus.

In September 1598, Kepler and other Lutherans in Graz were ordered to leave town by the Catholic archduke, who was bent on removing the Lutheran religion from Austria. After a visit to Tycho Brahe's Benatky Castle in Prague, Kepler was invited by the wealthy Danish astronomer to stay there and work on his research. Kepler was somewhat wary of Brahe, even before having met him. "My opinion of Tycho is this: he is superlatively rich, but he knows not how to make proper use of it, as is the case with most rich people," he wrote. "Therefore, one must try to wrest his riches from him."

If his relationship with his wife lacked complexity, Kepler more than made up for it when he entered into a working arrangement with the aristocratic Brahe. At first, Brahe treated the young Kepler as an assistant, carefully doling out assignments without giving him much access to detailed observational data. Kepler badly wanted to be regarded as an equal and given some independence, but the secretive Brahe wanted to use Kepler to establish his own model of the solar system—a non-Copernican model that Kepler did not support.

Kepler was immensely frustrated. Brahe had a wealth of observational data but lacked the mathematical tools to fully comprehend it. Finally, perhaps to pacify his restless assistant, Brahe assigned Kepler to study the orbit of Mars, which had confused the Danish astronomer for some time because it appeared to be the least circular. Kepler initially thought he could solve the problem in eight days, but the project turned out to take him eight years. Difficult as the research proved to be, it was not without its rewards, as the work led Kepler to discover that Mars's orbit precisely described an ellipse, as well as to formulate his first two "planetary laws," which he published in 1609 in *The New Astronomy* .

A year and a half into his working relationship with Brahe, the Danish astronomer became very ill at dinner and died a few days later of a bladder infection. Kepler took over the post of Imperial Mathematician and was now free to explore planetary theory without being constrained by the watchful eye of Tycho Brahe. Realizing an opportunity, Kepler immediately went after the Brahe data that he coveted before Brahe's heirs could take control of them. "I confess that when Tycho died," Kepler wrote later, "I quickly took advantage of the absence, or lack of circumspection, of the heirs, by taking the observations under my care, or perhaps usurping them." The result was Kepler's *Rudolphine Tables*, a compilation of the data from thirty years of Brahe's observations. To be fair, on his deathbed Brahe had urged Kepler to complete the tables; but Kepler did not frame the work according to any Tychonic hypothesis, as Brahe had hoped. Instead, Kepler used the data, which included calculations using logarithms he had developed himself, in predicting planetary positions. He was able to predict transits of the sun by Mercury and Venus, though he did not live long enough to witness them.

Kepler did not publish *Rudolphine Tables* until 1627, however, because the data he discovered constantly led him in new directions.

After Brahe's death, Kepler witnessed a nova, which later became known as "Kepler's nova," and he also experimented in optical theories. Though scientists and scholars view Kepler's optical work as minor in comparison with his accomplishments in astronomy and mathematics, the publication in 1611 of his book *Dioptrices,* changed the course of optics.

In 1605, Kepler announced his first law, the law of ellipses, which held that the planets move in ellipses with the sun at one focus. Earth, Kepler asserted, is closest to the sun in January and farthest from it in July as it travels along its elliptical orbit. His second law, the law of equal areas, maintained that the farther a planet sweeps out equal areas in equal times. Kepler demonstrated this by arguing that an imaginary line connecting any planet to the sun must sweep over equal areas in equal intervals of time. He published both laws in 1609 in his book *New Astronomy* (*Astronomia Nova*).

Yet despite his status as Imperial Mathematician and as a distinguished scientist whom Galileo sought out for an opinion on his new telescopic discoveries, Kepler was unable to secure for himself a comfortable existence. Religious upheaval in Prague jeopardized his new homeland, and in 1611 his wife and his favorite son died. Kepler was permitted, under exemption, to return to Linz, and in 1613 he married Susanna Reuttinger, a twenty-four-year-old orphan who would bear him seven children, only two of whom would survive to adulthood. It was at this time that Kepler's mother was accused of witchcraft, and in the midst of his own personal turmoil he was forced to defend her against the charge in order to prevent her being burned at the stake. Katherine was imprisoned and tortured, but her son managed to obtain an acquittal and she was released.

Because of these distractions, Kepler's return to Linz was not a productive time initially. Distraught, he turned his attention away from tables and began working on *Harmonies of the World* (*Harmonice Mundi*), a passionate work which Max Caspar, in his biography of Kepler, described as "a great cosmic vision, woven out of science, poetry, philosophy, theology, mysticism." Kepler finished *Harmonies of the World* on May 27, 1618. In this series of five books, he extended his theory of harmony to music, astrology, geometry, and astronomy. The series included his third law of planetary motion, the law that would inspire Isaac Newton some sixty years later, which maintained that the cubes of mean distances of the planets from the sun are proportional to the squares of their periods of revolution. In short, Kepler discovered how planets orbited, and in so doing paved the way for Newton to discover why.

Kepler believed he had discovered God's logic in designing the universe, and he was unable to hide his ecstasy. In Book 5 of *Harmonies of the World* he wrote:

I dare frankly to confess that I have stolen the golden vessels of the Egyptians to build a tabernacle for my God far from the bounds of Egypt. If you pardon me, I shall rejoice; if you reproach me, I shall endure. The die is cast, and I am writing the book, to be read either now or by posterity, it matters not. It can wait a century for a reader, as God himself has waited six thousand years for a witness.

The Thirty Years War, which beginning in 1618 decimated the Austrian and German lands, forced Kepler to leave Linz in 1626. He eventually settled in the town of Sagan, in Silesia. There he tried to finish what might best be described as a science fiction novel, which he had dabbled at for years, at some expense to his mother during her trial for witchcraft. *Dream of the Moon* (*Somnium seu astronomia lunari*), which features an interview with a knowing "demon" who explains how the protagonist could travel to the moon, was uncovered and presented as evidence during Katherine's trial. Kepler spent considerable energy defending the work as pure fiction and the demon as a mere literary device. The book was unique in that it was not only ahead of its time in terms of fantasy but also a treatise supporting Copernican theory.

In 1630, at the age of fifty-eight, Kepler once again found himself in financial straits. He set out for Regensburg, where he hoped to collect interest on some bonds in his possession as well as some money he was owed. However, a few days after his arrival he developed a fever, and died on November 15. Though he never achieved the mass renown of Galileo, Kepler produced a body of work that was extraordinarily useful to professional astronomers like Newton who immersed themselves in the details and accuracy of Kepler's science. Johannes Kepler was a man who preferred aesthetic harmony and order, and all that he discovered was inextricably linked with his vision of God. His epitaph, which he himself composed, reads: "I used to measure the heavens; now I shall measure the shadows of the earth. Although my soul was from heaven, the shadow of my body lies here."

HARMONIES
of the
WORLD
BOOK FIVE

Concerning the very perfect harmony of the celestial movements, and the genesis of eccentricities and the semidiameters, and as periodic times from the same.

After the model of the most correct astronomical doctrine of today, and the hypothesis not only of Copernicus but also of Tycho Brahe, whereof either hypotheses are today publicly accepted as most true, and the Ptolemaic as outmoded.

I commence a sacred discourse, a most true hymn to God the Founder, and I judge it to be piety, not to sacrifice many hecatombs of bulls to Him and to burn incense of innumerable perfumes and cassia, but first to learn myself, and afterwards to teach others too, how great He is in wisdom, how great in power, and of what sort in goodness. For to wish to adorn in every way possible the things that should receive adornment and to envy no thing its goods—this I put down as the sign of the greatest goodness, and in this respect I praise Him as good that in the heights of His wisdom He finds everything whereby each

thing may be adorned to the utmost and that He can do by His unconquerable power all that He has decreed.

Galen, *on the Use of Parts*. Book III

PROEM

[268] As regards that which I prophesied two and twenty years ago (especially that the five regular solids are found between the celestial spheres), as regards that of which I was firmly persuaded in my own mind before I had seen Ptolemy's *Harmonies*, as regards that which I promised my friends in the title of this fifth book before I was sure of the thing itself, that which, sixteen years ago, in a published statement, I insisted must be investigated, for the sake of which I spent the best part of my life in astronomical speculations, visited Tycho Brahe, [269] and took up residence at Prague: finally, as God the Best and Greatest, Who had inspired my mind and aroused my great desire, prolonged my life and strength of mind and furnished the other means through the liberality of the two Emperors and the nobles of this province of Austria-on-the-Anisana: after I had discharged my astronomical duties as much as sufficed, finally, I say, I brought it to light and found it to be truer than I had even hoped, and I discovered among the celestial movements the full nature of harmony, in its due measure, together with all its parts unfolded in Book III—not in that mode wherein I had conceived it in my mind (this is not last in my joy) but in a very different mode which is also very excellent and very perfect. There took place in this intervening time, wherein the very laborious reconstruction of the movements held me in suspense, an extraordinary augmentation of my desire and incentive for the job, a reading of the *Harmonies* of Ptolemy, which had been sent to me in manuscript by John George Herward, Chancellor of Bavaria, a very distinguished man and of a nature to advance philosophy and every type of learning. There, beyond my expectations and with the greatest wonder, I found approximately the whole third book given over to the same consideration of celestial harmony, fifteen hundred years ago. But indeed astronomy was far from being of age as yet; and Ptolemy, in an unfortunate attempt, could make others subject to despair, as being one who, like Scipio in Cicero, seemed to have recited a pleasant Pythagorean dream rather than to have aided philosophy. But both the crudeness of the ancient philosophy and this exact agreement in our meditations, down to the last hair, over an interval of fifteen centuries, greatly strengthened me in getting on with the job. For what need is there of many men? The very nature of things, in order to reveal herself to mankind, was at work in the different interpreters of different ages, and was the finger of God—to use the Hebrew expression; and here, in the minds of two men, who had wholly given themselves up to the contemplation of nature, there was the same conception as to the configuration of the world, although neither had

been the other's guide in taking this route. But now since the first light eight months
ago, since broad day three months ago, and since the sun of my wonderful speculation
has shone fully a very few days ago: nothing holds me back. I am free to give myself up
to the sacred madness, I am free to taunt mortals with the frank confession that I am
stealing the golden vessels of the Egyptians, in order to build of them a temple for my
God, far from the territory of Egypt. If you pardon me, I shall rejoice; if you are
enraged, I shall bear up. The die is cast, and I am writing the book—whether to be read
by my contemporaries or by posterity matters not. Let it await its reader for a hundred
years, if God Himself has been ready for His contemplator for six thousand years.

The chapters of this book are as follows:

1. Concerning the five regular solid figures.
2. On the kinship between them and the harmonic ratios.
3. Summary of astronomical doctrine necessary for speculation into the celestial
   harmonies.
4. In what things pertaining to the planetary movements the simple consonances
   have been expressed and that all those consonances which are present in song
   are found in the heavens.
5. That the clefs of the musical scale, or pitches of the system, and the genera of
   consonances, the major and the minor, are expressed in certain movements.
6. That the single musical Tones or Modes are somehow expressed by the single
   planets.
7. That the counterpoints or universal harmonies of all the planets can exist and
   be different from one another.
8. That four kinds of voice are expressed in the planets: soprano, contralto,
   tenor, and bass.
9. Demonstration that in order to secure this harmonic arrangement, those very
   planetary eccentricities which any planet has as its own, and no others, had to
   be set up.
10. Epilogue concerning the sun, by way of very fertile conjectures.

[270] Before taking up these questions, it is my wish to impress upon my readers
the very exhortation of Timaeus, a pagan philosopher, who was going to speak on the
same things: it should be learned by Christians with the greatest admiration, and
shame too, if they do not imitate him: Ἀλλ' ὦ Σώκρατεσ, τοῦτο γε δὴ πντεσ,
ὅσοι καὶ κατὰ βραχὺ σωφροσυνησ μετέουσιν, ἐπὶ πασῇ ὁρμῇ καὶ σμίκρου
καὶ μεγάλου πράγματοσ θεὸν ἀεί που καλοῦσιν. ἡμᾶς δὲ τοὺς περὶ τοῦ
πάντος λόγους ποιεῖσθαι πη μέλλοντας..., εἰ μὴ πανταπασι παραλλάτομεν,
ἀνάγκη θεούς τε καί θεὰς ἐπικαλουμενουσ εὔχεσθαι πάντα, κατὰ νοῦν
ἐκείνοισ μέν μάλιστα, ἑπομένως δέ ἡμῖν εἰπεῖν. *For truly, Socrates, since all who*

have the least particle of intelligence always invoke God whenever they enter upon any business, whether light or arduous; so too, unless we have clearly strayed away from all sound reason, we who intend to have a discussion concerning the universe must of necessity make our sacred wishes and pray to the Gods and Goddesses with one mind that we may say such things as will please and be acceptable to them in especial and, secondly, to you too.

1. CONCERNING THE FIVE REGULAR SOLID FIGURES

[271] It has been said in the second book how the regular plane figures are fitted together to form solids; there we spoke of the five regular solids, among others, on account of the plane figures. Nevertheless their number, five, was there demonstrated; and it was added why they were designated by the Platonists as the figures of the world, and to what element any solid was compared on account of what property. But now, in the anteroom of this book, I must speak again concerning these figures, on their own account, not on account of the planes, as much as suffices for the celestial harmonies; the reader will find the rest in the *Epitome of Astronomy*, Volume II, Book IV.

Accordingly, from the *Mysterium Cosmographicum*, let me here briefly inculcate the order of the five solids in the world, whereof three are primary and two secondary. For the *cube* (1) is the outmost and the most spacious, because firstborn and having the nature (*rationem*) of a *whole*, in the very form of its generation. There follows the *tetrahedron* (2), as if made a *part*, by cutting up the cube; nevertheless it is primary too, with a solid trilinear angle, like the cube. Within the tetrahedron is the *dodecahedron* (3), the last of primary figures, namely, like a solid composed of parts of a cube and similar parts of a tetrahedron, *i.e.*, of irregular tetrahedrons, wherewith the cube inside is roofed over. Next in order is the *icosahedron* (4) on account of its similarity, the last of the secondary figures and having a plurilinear solid angle. The *octahedron* (5) is inmost, which is similar to the cube and the first of the secondary figures and to which as inscriptile the first place is due, just as the first outside place is due to the cube as circumscriptile.

[272] However, there are as it were two noteworthy weddings of these figures, made from different classes: the males, the cube and the dodecahedron, among the primary; the females, the octahedron and the icosahedron, among the secondary, to which is added one as it were bachelor or hermaphrodite, the tetrahedron, because it is inscribed in itself, just as those female solids are inscribed in the males and are as it were subject to them, and have the signs of the feminine sex, opposite the masculine, namely, angles opposite planes. Moreover, just as the tetrahedron is the element, bowels, and as it were rib of the male cube, so the feminine octahedron is the element and part of the tetrahedron in another way; and thus the tetrahedron mediates in this marriage.

The main difference in these wedlocks or family relationships consists in the

following: the ratio of the cube is *rational*. For the tetrahedron is one third of the body of the cube, and the octahedron half of the tetrahedron, one sixth of the cube; while the ratio of the dodecahedron's wedding is *irrational (ineffabilis)* but *divine*.

The union of these two words commands the reader to be careful as to their significance. For the word *ineffabilis* here does not of itself denote any nobility, as elsewhere in theology and divine things, but denotes an inferior condition. For in geometry, as was said in the first book, there are many irrationals, which do not on that account participate in a divine proportion too. But you must look in the first book for what the divine ratio, or rather the divine section, is. For in other proportions there are four terms present; and three, in a continued proportion; but the divine requires a single relation of terms outside of that of the proportion itself, namely in such fashion that the two lesser terms, as parts make up the greater term, as a whole. Therefore, as much as is taken away from this wedding of the dodecahedron on account of its employing an irrational proportion,

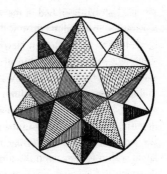

is added to it conversely, because its irrationality approaches the divine. This wedding also comprehends the solid star too, the generation whereof arises from the continuation of five planes of the dodecahedron till they all meet in a single point. See its generation in Book II.

Lastly, we must note the ratio of the spheres circumscribed around them to those inscribed in them: in the case of the tetrahedron it is rational, 100,000 : 33,333 or 3 : 1; in the wedding of the cube it is irrational, but the radius of the inscribed sphere is rational in square, and is itself the square root of one third the square on the radius (of the circumscribed sphere), namely 100,000 : 57,735; in the wedding of the dodecahedron, clearly irrational, 100,000 : 79,465; in the case of the star, 100,000 : 52,573, half the side of the icosahedron or half the distance between two rays.

2. ON THE KINSHIP BETWEEN THE HARMONIC RATIOS AND THE FIVE REGULAR FIGURES

[273] This kinship (*cognatio*) is various and manifold; but there are four degrees of kinship. For either the sign of kinship is taken from the outward form alone which the figures have, or else ratios which are the same as the harmonic arise in the construction of the side, or result from the figures already constructed, taken simply or together; or, lastly, they are either equal to or approximate the ratios of the spheres of the figure.

In the first degree, the ratios, where the character or greater term is 3, have kin-

ship with the triangular plane of the tetrahedron, octahedron, and icosahedron; but where the greater term is 4, with the square plane of the cube; where 5, with the pentagonal plane of the dodecahedron. This similitude on the part of the plane can also be extended to the smaller term of the ratio, so that wherever the number 3 is found as one term of the continued doubles, that ratio is held to be akin to the three figures first named: for example, 1 : 3 and 2 : 3 and 4 : 3 and 8 : 3, et cetera; but where the number is 5, that ratio is absolutely assigned to the wedding of the dodecahedron: for example, 2 : 5 and 4 : 5 and 8 : 5, and thus 3 : 5 and 3 : 10 and 6 : 5 and 12 : 5 and 24:5. The kinship will be less probable if the sum of the terms expresses this similitude, as in 2 : 3 the sum of the terms is equal to 5, as if to say that 2 : 3 is akin to the dodecahedron. The kinship on account of the outward form of the solid angle is similar: the solid angle is trilinear among the primary figures, quadrilinear in the octahedron, and quinquelinear in the icosahedron. And so if one term of the ratio participates in the number 3, the ratio will be connected with the primary bodies; but if in the number 4, with the octahedron; and finally, if in the number 5, with the icosahedron. But in the feminine solids this kinship is more apparent, because the characteristic figure latent within follows upon the form of the angle: the tetragon in the octahedron, the pentagon in the icosahedron; and so 3 : 5 would go to the sectioned icosahedron for both reasons.

The second degree of kinship, which is genetic, is to be conceived as follows: First, some harmonic ratios of numbers are akin to one wedding or family, namely, perfect ratios to the single family of the cube; conversely, there is the ratio which is never fully expressed in numbers and cannot be demonstrated by numbers in any other way, except by a long series of numbers gradually approaching it: this ratio is called *divine*, when it is perfect, and it rules in various ways throughout the dodecahedral wedding. Accordingly, the following consonances begin to shadow forth that ratio: 1:2 and 2: 3 and 2 : 3 and 5 : 8. For it exists most imperfectly in 1 : 2, more perfectly in 5 : 8, and still more perfectly if we add 5 and 8 to make 13 and take 8 as the numerator, if this ratio has not stopped being harmonic.

Further, in constructing the side of the figure, the diameter of the globe must be cut; and the octahedron demands its bisection, the cube and the tetrahedron its trisection, the dodecahedral wedding its quinquesection. Accordingly, the ratios between the figures are distributed according to the numbers which express those ratios. But the square on the diameter is cut too, or the square on the side of the figure is formed from a fixed part of the diameter. And then the squares on the sides are compared with the square on the diameter, and they constitute the following ratios: in the cube 1 : 3, in the tetrahedron 2 : 3, in the octahedron 1 : 2. Wherefore, if the two ratios are put together, the cubic and the tetrahedral will give 1 : 2; the cubic and the octahedral,

2 : 3; the octahedral and the tetrahedral, 3 : 4. The sides in the dodecahedral wedding are irrational.

Thirdly, the harmonic ratios follow in various ways upon the already constructed figures. For either the number of the sides of the plane is compared with the number of lines in the total figure; [274] and the following ratios arise: in the cube, 4:12 or 1:3; in the tetrahedron 3 : 6 or 1 : 2; in the octahedron 3 : 12 or 1 : 4; in the dodecahedron 5 : 30 or 1 : 6; in the icosahedron 3 : 30 or 1 : 10. Or else the number of sides of the plane is compared with the number of planes; then the cube gives 4 : 6 or 2 : 3, the tetrahedron 3 : 4, the octahedron 3 : 8, the dodecahedron 5 : 12, the icosahedron 3 : 20. Or else the number of sides or angles of the plane is compared with the number of solid angles, and the cube gives 4 : 8 or 1 : 2, the tetrahedron 3 : 4, the octahedron 3 : 6 or 1 : 2, the dodecahedron with its consort 5 : 20 or 3 : 12 (*i.e.*, 1 : 4). Or else the number of planes is compared with the number of solid angles, and the cubic wedding gives 6 : 8 or 3 : 4, the tetrahedron the ratio of equality, the dodecahedral wedding 12 : 20 or 3 : 5. Or else the number of all the sides is compared with the number of the solid angles, and the cube gives 8 : 12 or 2 : 3, the tetrahedron 4 : 6 or 2 : 3, and the octahedron 6 : 12 or 1 : 2, the dodecahedron 20 : 30 or 2 : 3, the icosahedron 12:30 or 2:5.

Moreover, the bodies too are compared with one another, if the tetrahedron is stowed away in the cube, the octahedron in the tetrahedron and cube, by geometrical inscription. The tetrahedron is one third of the cube, the octahedron half of the tetrahedron, one sixth of the cube, just as the octahedron, which is inscribed in the globe, is one sixth of the cube which circumscribes the globe. The ratios of the remaining bodies are irrational.

The fourth species or degree of kinship is more proper to this work: the ratio of the spheres inscribed in the figures to the spheres circumscribing them is sought, and what harmonic ratios approximate them is calculated. For only in the tetrahedron is the diameter of the inscribed sphere rational, namely, one third of the circumscribed sphere. But in the cubic wedding the ratio, which is single there, is as lines which are rational only in square. For the diameter of the inscribed sphere is to the diameter of the circumscribed sphere as the square root of the ratio 1 : 3. And if you compare the ratios with one another, the ratio of the tetrahedral spheres is the square of the ratio of the cubic spheres. In the dodecahedral wedding there is again a single ratio, but an irrational one, slightly greater than 4 : 5. Therefore the ratio of the spheres of the cube and octahedron is approximated by the following consonances: 1 : 2, as proximately greater, and 3 : 5, as proximately smaller. But the ratio of the dodecahedral spheres is approximated by the consonances 4 : 5 and 5 : 6, as proximately smaller, and 3 : 4 and 5 : 8, as proximately greater.

But if for certain reasons 1 : 2 and 1 : 3 are arrogated to the cube, the ratio of the spheres of the cube will be to the ratio of the spheres of the tetrahedron as the consonances 1 : 2 and 1 : 3, which have been ascribed to the cube, are to 1 : 4 and 1 : 9, which are to be assigned to the tetrahedron, if this proportion is to be used. For these ratios, too, are as the squares of those consonances. And because 1 : 9 is not harmonic, 1 : 8 the proximate ratio takes its place in the tetrahedron. But by this proportion approximately 4 : 5 and 3 : 4 will go with the dodecahedral wedding. For as the ratio of the spheres of the cube is approximately the cube of the ratio of the dodecahedral, so too the cubic consonances 1 : 2 and 2 : 3 are approximately the cubes of the consonances 4 : 5 and 3 : 4. For 4 : 5 cubed is 64 : 125, and 1 : 2 is 64 : 128. So 3 : 4 cubed is 27 : 64, and 1 : 3 is 27 : 81.

3. A SUMMARY OF ASTRONOMICAL DOCTRINE NECESSARY FOR SPECULATION INTO THE CELESTIAL HARMONIES

First of all, my readers should know that the ancient astronomical hypotheses of Ptolemy, in the fashion in which they have been unfolded in the *Theoricae* of Peurbach and by the other writers of epitomes, are to be completely removed from this discussion and cast out of [275] the mind. For they do not convey the true lay out of the bodies of the world and the polity of the movements.

Although I cannot do otherwise than to put solely Copernicus' opinion concerning the world in the place of those hypotheses and, if that were possible, to persuade everyone of it; but because the thing is still new among the mass of the intelligentsia (*apud vulgus studiosorum*), and the doctrine that the Earth is one of the planets and moves among the stars around a motionless sun sounds very absurd to the ears of most of them: therefore those who are shocked by the unfamiliarity of this opinion should know that these harmonical speculations are possible even with the hypotheses of Tycho Brahe—because that author holds, in common with Copernicus, everything else which pertains to the lay out of the bodies and the tempering of the movements, and transfers solely the Copernican annual movement of the Earth to the whole system of planetary spheres and to the sun, which occupies the centre of that system, in the opinion of both authors. For after this transference of movement it is nevertheless true that in Brahe the Earth occupies at any time the same place that Copernicus gives it, if not in the very vast and measureless region of the fixed stars, at least in the system of the planetary world. And accordingly, just as he who draws a circle on paper makes the writing-foot of the compass revolve, while he who fastens the paper or tablet to a turning lathe draws the same circle on the revolving tablet with the foot of the compass or stylus motionless; so too, in the case of Copernicus the Earth, by the real movement of its body, measures out a circle revolving midway between the circle of Mars on the

outside and that of Venus on the inside; but in the case of Tycho Brahe the whole planetary system (wherein among the rest the circles of Mars and Venus are found) revolves like a tablet on a lathe and applies to the motionless Earth, or to the stylus on the lathe, the midspace between the circles of Mars and Venus; and it comes about from this movement of the system that the Earth within it, although remaining motionless, marks out the same circle around the sun and midway between Mars and Venus, which in Copernicus it marks out by the real movement of its body while the system is at rest. Therefore, since harmonic speculation considers the eccentric movements of the planets, as if seen from the sun, you may easily understand that if any observer were stationed on a sun as much in motion as you please, nevertheless for him the Earth, although at rest (as a concession to Brahe), would seem to describe the annual circle midway between the planets and in an intermediate length of time. Wherefore, if there is any man of such feeble wit that he cannot grasp the movement of the Earth among the stars, nevertheless he can take pleasure in the most excellent spectacle of this most divine construction, if he applies to their image in the sun whatever he hears concerning the daily movements of the Earth in its eccentric—such an image as Tycho Brahe exhibits, with the Earth at rest.

And nevertheless the followers of the true Samian philosophy have no just cause to be jealous of sharing this delightful speculation with such persons, because their joy will be in many ways more perfect, as due to the consummate perfection of speculation, if they have accepted the immobility of the sun and the movement of the Earth.

Firstly [I], therefore, let my readers grasp that today it is absolutely certain among all astronomers that all the planets revolve around the sun, with the exception of the moon, which alone has the Earth as its centre: the magnitude of the moon's sphere or orbit is not great enough for it to be delineated in this diagram in a just ratio to the rest. Therefore, to the other five planets, a sixth, the Earth, is added, which traces a sixth circle around the sun, whether by its own proper movement with the sun at rest, or motionless itself and with the whole planetary system revolving.

Secondly [II]: It is also certain that all the planets are eccentric, *i.e.*, they change their distances from the sun, in such fashion that in one part of their circle they become farthest away from the sun, [276] and in the opposite part they come nearest to the sun. In the accompanying diagram three circles apiece have been drawn for the single planets: none of them indicate the eccentric route of the planet itself; but the mean circle, such as *BE* in the case of Mars, is equal to the eccentric orbit, with respect to its longer diameter. But the orbit itself, such as *AD*, touches *AF*, the upper of the three, in one place *A*, and the lower circle *CD*, in the opposite place *D*. The circle *GH* made with dots and described through the centre of the sun indicates the route of the sun according to Tycho Brahe. And if the sun moves on this route, then absolutely all the

points in this whole planetary system here depicted advance upon an equal route, each upon his own. And with one point of it (namely, the centre of the sun) stationed at one point of its circle, as here at the lowest, absolutely each and every point of the system will be stationed at the lowest part of its circle. However, on account of the smallness of the space the three circles of Venus unite in one, contrary to my intention.

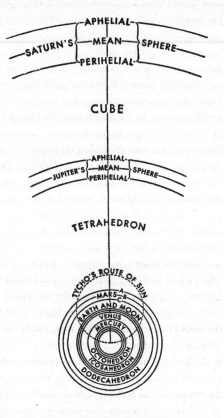

Thirdly [III]: Let the reader recall from my *Mysterium Cosmographicum*, which I published twenty-two years ago, that the number of the planets or circular routes around the sun was taken by the very wise Founder from the five regular solids, concerning which Euclid, so many ages ago, wrote his book which is called the *Elements* in that it is built up out of a series of propositions. But it has been made clear in the second book of this work that there cannot be more regular bodies, *i.e.*, that regular plane figures cannot fit together in a solid more than five times.

Fourthly [IV]: As regards the ratio of the planetary orbits, the ratio between two neighbouring planetary orbits is always of such a magnitude that it is easily apparent that each and every one of them approaches the single ratio of the spheres of one of the five regular solids, namely, that of the sphere circumscribing to the sphere inscribed in the figure. Nevertheless it is not wholly equal, as I once dared to promise concerning the final perfection of astronomy. For, after completing the demonstration of the intervals from Brahe's observations, I discovered the following: if the angles of the cube [277] are applied to the inmost circle of Saturn, the centres of the planes are approximately tangent to the middle circle of Jupiter; and if the angles of the tetrahedron are placed against the inmost circle of Jupiter, the centres of the planes of the tetrahedron

are approximately tangent to the outmost circle of Mars; thus if the angles of the octahedron are placed against any circle of Venus (for the total interval between the three has been very much reduced), the centres of the planes of the octahedron penetrate and descend deeply within the outmost circle of Mercury, but nonetheless do not reach as far as the middle circle of Mercury; and finally, closest of all to the ratios of the dodecahedral and icosahedral spheres—which ratios are equal to one another—are the ratios or intervals between the circles of Mars and the Earth, and the Earth and Venus; and those intervals are similarly equal, if we compute from the inmost circle of Mars to the middle circle of the Earth, but from the middle circle of the Earth to the middle circle of Venus. For the middle distance of the Earth is a mean proportional between the least distance of Mars and the middle distance of Venus. However, these two ratios between the planetary circles are still greater than the ratios of those two pairs of spheres in the figures, in such fashion that the centres of the dodecahedral planes are not tangent to the outmost circle of the Earth, and the centres of the icosahedral planes are not tangent to the outmost circle of Venus; nor, however, can this gap be filled by the semidiameter of the lunar sphere, by adding it, on the upper side, to the greatest distance of the Earth and subtracting it, on the lower, from the least distance of the same. But I find a certain other ratio of figures—namely, if I take the augmented dodecahedron, to which I have given the name of echinus, (as being fashioned from twelve quinquangular stars and thereby very close to the five regular solids), if I take it, I say, and place its twelve points in the inmost circle of Mars, then the sides of the pentagons, which are the bases of the single rays or points, touch the middle circle of Venus. In short: the cube and the octahedron, which are consorts, do not penetrate their planetary spheres at all; the dodecahedron and the icosahedron, which are consorts, do not wholly reach to theirs, the tetrahedron exactly touches both: in the first case there is falling short; in the second, excess; and in the third, equality, with respect to the planetary intervals.

Wherefore it is clear that the very ratios of the planetary intervals from the sun have not been taken from the regular solids alone. For the Creator, who is the very source of geometry and, as Plato wrote, "practices eternal geometry," does not stray from his own archetype. And indeed that very thing could be inferred from the fact that all the planets change their intervals throughout fixed periods of time, in such fashion that each has two marked intervals from the sun, a greatest and a least; and a fourfold comparison of the intervals from the sun is possible between two planets: the comparison can be made between either the greatest, or the least, or the contrary intervals most remote from one another, or the contrary intervals nearest together. In this way the comparisons made two by two between neighbouring planets are twenty in number, although on the contrary there are only five regular solids. But it is consonant that if the Creator had any concern for the ratio of the spheres in general, He would

also have had concern for the ratio which exists between the varying intervals of the single planets specifically and that the concern is the same in both cases and the one is bound up with the other. If we ponder that, we will comprehend that for setting up the diameters and eccentricities conjointly, there is need of more principles, outside of the five regular solids.

Fifthly [V]: To arrive at the movements between which the consonances have been set up, once more I impress upon the reader that in the *Commentaries on Mars* I have demonstrated from the sure observations of Brahe that daily arcs, which are equal in one and the same eccentric circle, are not traversed with equal speed; but that these dif-fering *delays in equal parts of the eccentric observe the ratio of their distances from the sun,* the source of movement; and conversely, that if equal times are assumed, namely, one natural day in both cases, the corresponding *true diurnal arcs* [278] *of one eccentric orbit have to one another the ratio which is the inverse of the ratio of the two distances from the sun.* Moreover, I demonstrated at the same time that *the planetary orbit is elliptical and the sun, the source of movement, is at one of the foci of this ellipse; and so, when the plan-et has completed a quarter of its total circuit from its aphelion, then it is exactly at its mean distance from the sun, midway between its greatest distance at the aphelion and its least at the perihelion.* But from these two axioms it results *that the diurnal mean movement of the planet in its eccentric is the same as the true diurnal arc of its eccentric at those moments wherein the planet is at the end of the quadrant of the eccentric measured from the aph-elion, although that true quadrant appears still smaller than the just quadrant.* Furthermore, it follows *that the sum of any two true diurnal eccentric arcs, one of which is at the same distance from the aphelion that the other is from the perihelion, is equal to the sum of the two mean diurnal arcs.* And as a consequence, *since the ratio of circles is the same as that of the diameters, the ratio of one mean diurnal arc to the sum of all the mean and equal arcs in the total circuit is the same as the ratio of the mean diurnal arc to the sum of all the true eccentric arcs, which are the same in number but unequal to one another.* And those things should first be known concerning the true diurnal arcs of the eccentric and the true movements, so that by means of them we may understand the movements which would be apparent if we were to suppose an eye at the sun.

Sixthly [VI]: But as regards the arcs which are apparent, as it were, from the sun, it is known even from the ancient astronomy that, among true movements which are equal to one another, that movement which is farther distant from the centre of the world (as being at the aphelion) will appear smaller to a beholder at that centre, but the movement which is nearer (as being at the perihelion) will similarly appear greater. Therefore, since moreover the true diurnal arcs at the near distance are still greater, on account of the faster movement, and still smaller at the distant aphelion, on account of the slowness of the movement, I demonstrated in the *Commentaries on Mars* that *the*

ratio of the apparent diurnal arcs of one eccentric circle is fairly exactly the inverse ratio of the squares of their distances from the sun. For example, if the planet one day when it is at a distance from the sun of 10 parts, in any measure whatsoever, but on the opposite day, when it is at the perihelion, of 9 similar parts: it is certain that from the sun its apparent progress at the aphelion will be to its apparent progress at the perihelion, as 81 : 100.

But that is true with these provisos: First, that the eccentric arcs should not be great, lest they partake of distinct distances which are very different—*i.e.*, lest the distances of their termini from the apsides cause a perceptible variation; second, that the eccentricity should not be very great, for the greater its eccentricity (*viz.*, the greater the arc becomes) the more the angle of its apparent movement increases beyond the measure of its approach to the sun, by Theorem 8 of Euclid's *Optics*; none the less in small arcs even a great distance is of no moment, as I have remarked in my *Optics*, Chapter 11. But there is another reason why I make that admonition. For the eccentric arcs around the mean anomalies are viewed obliquely from the centre of the sun. This obliquity subtracts from the magnitude of the apparent movement, since conversely the arcs around the apsides are presented directly to an eye stationed as it were at the sun. Therefore, when the eccentricity is very great, then the eccentricity takes away perceptibly from the ratio of the movements; if without any diminution we apply the mean diurnal movement to the mean distance, as if at the mean distance, it would appear to have the same magnitude which it does have—as will be apparent below in the case of Mercury. All these things are treated at greater length in Book V of the *Epitome of Copernican Astronomy*; but they have been mentioned here too because they have to do with the very terms of the celestial consonances, considered in themselves singly and separately.

Seventhly [VII]: If by chance anyone runs into those diurnal movements which are apparent [279] to those gazing not as it were from the sun but from the Earth, with which movements Book VI of the *Epitome of Copernican Astronomy* deals, he should know that their rationale is plainly not considered in this business. Nor should it be, since the Earth is not the source of the planetary movements, nor can it be, since with respect to deception of sight they degenerate not only into mere quiet or apparent stations but even into retrogradation, in which way a whole infinity of ratios is assigned to all the planets, simultaneously and equally. Therefore, in order that we may hold for certain what sort of ratios of their own are constituted by the single real eccentric orbits (although these too are still apparent, as it were to one looking from the sun, the source of movement), first we must remove from those movements of their own this image of the adventitious annual movement common to all five, whether it arises from the movement of the Earth itself, according to Copernicus, or from the annual movement

of the total system, according to Tycho Brahe, and the winnowed movements proper to each planet are to be presented to sight.

Eighthly [VIII]: So far we have dealt with the different delays or arcs of one and the same planet. Now we must also deal with the comparison of the movements of two planets. Here take note of the definitions of the terms which will be necessary for us. We give the name of *nearest apsides* of two planets to the perihelion of the upper and the aphelion of the lower, notwithstanding that they tend not towards the same region of the world but towards distinct and perhaps contrary regions. By *extreme movements* understand the slowest and the fastest of the whole planetary circuit; by *converging or converse extreme movements*, those which are at the nearest apsides of two planets—namely, at the perihelion of the upper planet and the aphelion of the lower; by *diverging or diverse*, those at the opposite apsides—namely, the aphelion of the upper and the perihelion of the lower. Therefore again, a certain part of my *Mysterium Cosmographicum*, which was suspended twenty-two years ago, because it was not yet clear, is to be completed and herein inserted. For after finding the true intervals of the spheres by the observations of Tycho Brahe and continuous labour and much time, at last, at last the right ratio of the periodic times to the spheres

> though it was late, looked to the unskilled man,
> yet looked to him, and, after much time, came,

and, if you want the exact time, was conceived mentally on the 8th of March in this year One Thousand Six Hundred and Eighteen but unfelicitously submitted to calculation and rejected as false, finally, summoned back on the 15th of May, with a fresh assault undertaken, outfought the darkness of my mind by the great proof afforded by my labor of seventeen years on Brahe's observations and meditation upon it uniting in one concord, in such fashion that I first believed I was dreaming and was presupposing the object of my search among the principles. But it is absolutely certain and exact that *the ratio which exists between the periodic times of any two planets is precisely the ratio of the $3/_2$th power of the mean distances,* i.e., *of the spheres themselves*; provided, however, that the arithmetic mean between both diameters of the elliptic orbit be slightly less than the longer diameter. And so if any one take the period, say, of the Earth, which is one year, and the period of Saturn, which is thirty years, and extract the cube roots of this ratio and then square the ensuing ratio by squaring the cube roots, he will have as his numerical products the most just ratio of the distances of the Earth and Saturn from the sun.[1] For the cube root of 1 is 1, and the square of it is 1; and the cube root of 30 is greater than 3, and therefore the square of it is greater than 9. And Saturn, at

1. For in the *Commentaries on Mars*, chapter 48, page 232, I have proved that this Arithmetic mean is either the diameter of the circle which is equal in length to the elliptic orbit, or else is very slightly less.

its mean distance from the sun, is slightly higher [280] than nine times the mean distance of the Earth from the sun. Further on, in Chapter 9, the use of this theorem will be necessary for the demonstration of the eccentricities.

Ninthly [IX]: If now you wish to measure with the same yardstick, so to speak, the true daily journeys of each planet through the ether, two ratios are to be compounded—the ratio of the true (not the apparent) diurnal arcs of the eccentric, and the ratio of the mean intervals of each planet from the sun (because that is the same as the ratio of the amplitude of the spheres), *i.e., the true diurnal arc of each planet is to be multiplied by the semidiameter of its sphere*: the products will be numbers fitted for investigating whether or not those journeys are in harmonic ratios.

Tenthly [X]: In order that you may truly know how great any one of these diurnal journeys appears to be to an eye stationed as it were at the sun, although this same thing can be got immediately from the astronomy, nevertheless it will also be manifest if you multiply the ratio of the journeys by the inverse ratio not of the mean, but of the true intervals which exist at any position on the eccentrics: *multiply the journey of the upper by the interval of the lower planet from the sun, and conversely multiply the journey of the lower by the interval of the upper from the sun.*

Eleventhly [XI]: And in the same way, if the apparent movements are given, at the aphelion of the one and at the perihelion of the other, or conversely or alternately, the ratios of the distances of the aphelion of the one to the perihelion of the other may be elicited. But where the mean movements must be known first, *viz.*, the inverse ratio of the periodic times, wherefrom the ratio of the spheres is elicited by Article VIII above: then *if the mean proportional between the apparent movement of either one of its mean movement be taken, this mean proportional is to the semidiameter of its sphere* (which is already known) *as the mean movement is to the distance or interval sought.* Let the periodic times of two planets be 27 and 8. Therefore the ratio of the mean diurnal movement of the one to the other is 8 : 27. Therefore the semidiameters of their spheres will be as 9 to 4. For the cube root of 27 is 3, that of 8 is 2, and the squares of these roots, 3 and 2, are 9 and 4. Now let the apparent aphelial movement of the one be 2 and the perihelial movement of the other $33^1/_3$. The mean proportionals between the mean movements 8 and 27 and these apparent ones will be 4 and 30. Therefore if the mean proportional 4 gives the mean distance of 9 to the planet, then the mean movement of 8 gives an aphelial distance 18, which corresponds to the apparent movement 2; and if the other mean proportional 30 gives the other planet a mean distance of 4, then its mean movement of 27 will give it a perihelial interval of $3^3/_5$. I say, therefore, that the aphelial distance of the former is to the perihelial distance of the latter as 18 to $3^3/_5$. Hence it is clear that if the consonances between the extreme movements of two planets are found and the periodic times are established for both, the extreme and the mean

distances are necessarily given, wherefore also the eccentricities.

Twelfthly [XII]: It is also possible, from the different extreme movements of one and the same planet, to find the *mean movement*. The mean movement is not exactly the arithmetic mean between the extreme movements, nor exactly the geometric mean, but it is as much less than the geometric mean as the geometric mean is less than the (arithmetic) mean between both means. Let the two extreme movements be 8 and 10: the mean movement will be less than 9, and also less than the square root of 80 by half the difference between 9 and the square root of 80. In this way, if the aphelial movement is 20 and the perihelial 24, the mean movement will be less than 22, even less than the square root of 480 by half the difference between that root and 22. There is use for this theorem in what follows.

[281] Thirteenthly [XIII]: From the foregoing the following proposition is demonstrated, which is going to be very necessary for us: Just as the ratio of the mean movements of two planets is the inverse ratio of the $3/2$th powers of the spheres, so the ratio of two apparent converging extreme movements always falls short of the ratio of the $3/2$th powers of the intervals corresponding to those extreme movements; and in what ratio the product of the two ratios of the corresponding intervals to the two mean intervals or to the semidiameters of the two spheres falls short of the ratio of the square roots of the spheres, in that ratio does the ratio of the two extreme converging movements exceed the ratio of the corresponding intervals; but if that compound ratio were to exceed the ratio of the square roots of the spheres, then the ratio of the converging movements would be less than the ratio of their intervals.[1]

Let the ratio of the spheres be $DH : AE$; let the ratio of the mean movements be $HI : EM$, the $3/2$th power of the inverse of the former. Let the least interval of the sphere of the first be CG; and the greatest interval of the sphere of the second be BF; and first let $DH : CG$ comp. $BF : AE$ be smaller than the $1/2$th power of $DH : AE$. And let GH be the apparent perihelial movement of the upper planet, and FL the aphelial of the lower, so that they are converging extreme movements.

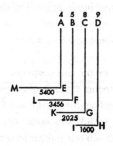

I say that

$$GK : FL = BF : CG$$
$$BF^{3/2} : CG^{3/2}.$$

For

$$HI : GK = CG^2 : DH^2$$

1. Kepler always measures the magnitude of a ratio from the greater term to the smaller, rather than from the antecedent to the consequent, as we do today. For example, as Kepler speaks, 2 : 3 is the same as 3 : 2, and 3 : 4 is greater than 7 : 8.—C. G. Wallis.

and
$$FL : EM = AE^2 : BF^2.$$
Hence
$$HI : GK \text{ comp. } FL : EM = CG^2 : DH^2 \text{ comp. } AE^2 : BF^2.$$
But
$$CG : DH \text{ comp. } AE : BF < AE^{1/2} : DH^{1/2}$$
by a fixed ratio of defect, as was assumed. Therefore too
$$HI : GK \text{ comp. } FL : EM \ AE^{2/2} : DH^{2/2}$$
$$AE : DH$$
by a ratio of defect which is the square of the former. But by number VIII
$$HI : EM = AE^{3/2} : DH^{3/2}.$$
Therefore let the ratio which is smaller by the total square of the ratio of defect be divided into the ratio of the $^3/_2$th powers; that is,
$$HI : EM \text{ comp. } GK : HI \text{ comp. } EM : FL \ AE^{1/2} : DH^{1/2}$$
by the excess squared. But
$$HI : EM \text{ comp. } GK : HI \text{ comp. } EM : FL = GK : FL.$$
Therefore
$$GK : FL \ AE^{1/2} : DH^{1/2}$$
by the excess squared. But
$$AE : DH = AE : BF \text{ comp. } BF : CG \text{ comp. } CG : DH.$$
And
$$CG : DH \text{ comp. } AE : BF \ AE^{1/2} : DH^{1/2}$$
by the simple defect. Therefore
$$BF : CG \ AE^{1/2} : DH^{1/2}$$
by the simple excess. But
$$GK : FL \ AE^{1/2} : DH^{1/2}$$
but by the excess squared. But the excess squared is greater than the simple excess. Therefore the ratio of the movements GK to FL is greater than the ratio of the corresponding intervals BF to CG.

In fully the same way, it is demonstrated even contrariwise that if the planets approach one another in G and F beyond the mean distances in H and E, in such fashion that the ratio of the mean distances $DH : AE$ becomes less than $DH^{1/2} : AE^{1/2}$, then the ratio of the movements $GK : FL$ becomes less than the ratio of the corresponding intervals $BF : CG$. For you need to do nothing more than to change the words *greater* to *less*, > to <, *excess* to *defect*, and conversely.

In suitable numbers, because the square root of $^4/_9$ is $^2/_3$; and $^5/_8$ is even greater than $^2/_3$ by the ratio of excess $^{15}/_{16}$; and the square of the ratio 8 : 9 [282] is the ratio 1600 : 2025, *i.e.*, 64 : 81; and the square of the ratio 4 : 5 is the ratio 3456 : 5400, *i.e.*,

16 : 25; and finally the $^3/_2$th power of the ratio 4 : 9 is the ratio 1600 : 5400, *i.e.*, 8 : 27: therefore too the ratio 2025 : 3456, *i.e.*, 75 : 128, is even greater than 5 : 8, *i.e.*, 75 : 120, by the same ratio of excess (*i.e.*, 120 : 128), 15 : 16; whence 2025 : 3456, the ratio of the converging movements, exceeds 5 : 8, the inverse ratio of the corresponding intervals, by as much as 5 : 8 exceeds 2 : 3, the square root of the ratio of the spheres. Or, what amounts to the same thing, the ratio of the two converging intervals is a mean between the ratio of the square roots of the spheres and the inverse ratio of the corresponding movements.

Moreover, from this you may understand that the ratio of the diverging movements is much greater than the ratio of the $^3/_2$th powers of the spheres, since the ratio of the $^3/_2$th powers is compounded with the squares of the ratio of the aphelial interval to the mean interval, and that of the mean to the perihelial.

4. In What Things Having to do with the Planetary Movements Have the Harmonic Consonances been Expressed by the Creator, and in What Way?

Accordingly, if the image of the retrogradation and stations is taken away and the proper movements of the planets in their real eccentric orbits are winnowed out, the following distinct things still remain in the planets: 1) The distances from the sun. 2) The periodic times. 3) The diurnal eccentric arcs. 4) The diurnal delays in those arcs. 5) The angles at the sun, and the diurnal area apparent to those as it were gazing from the sun. And again, all of these things, with the exception of the periodic times, are variable in the total circuit, most variable at the mean longitudes, but least at the extremes, when, turning away from one extreme longitude, they begin to return to the opposite. Hence when the planet is lowest and nearest to the sun and thereby delays the least in one degree of its eccentric, and conversely in one day traverses the greatest diurnal arc of its eccentric and appears fastest from the sun: then its movement remains for some time in this strength without perceptible variation, until, after passing the perihelion, the planet gradually begins to depart farther from the sun in a straight line; at that same time it delays longer in the degrees of its eccentric circle; or, if you consider the movement of one day, on the following day it goes forward less and appears even more slow from the sun until it has drawn close to the highest apsis and made its distance from the sun very great: for then longest of all does it delay in one degree of its eccentric; or on the contrary in one day it traverses its least arc and makes a much smaller apparent movement and the least of its total circuit.

Finally, all these things may be considered either as they exist in any one planet at different times or as they exist in different planets: whence, by the assumption of an infinite amount of time, all the affects of the circuit of one planet can concur in the

same moment of time with all the affects of the circuit of another planet and be compared, and then the total eccentrics, as compared with one another, have the same ratio as their semidiameters or mean intervals; but the arcs of two eccentrics, which are similar or designated by the same number (of degrees), nevertheless have their true lengths unequal in the ratio of their eccentrics. For example, one degree in the sphere of Saturn is approximately twice as long as one degree in the sphere of Jupiter. And conversely, the diurnal arcs of the eccentrics, as expressed in astronomical terms, do not exhibit the ratio of the true journeys which the globes complete in one day [283] through the ether, because the single units in the wider circle of the upper planet denote a quarter part of the journey, but in the narrower circle of the lower planet a smaller part.

Therefore let us take the second of the things which we have posited, namely, the periodic times of the planets, which comprehend the sums made up of all the delays—long, middling, short—in all the degrees of the total circuit. And we found that from antiquity down to us, the planets complete their periodic returns around the sun, as follows in the table:

			Therefore the mean diurnal moments		
	Days	Minutes of a day	Min.	Sec.	Thirds
Saturn	10,759	12	2	0	27
Jupiter	4,332	37	4	59	8
Mars	686	59	31	26	31
Earth with Moon	365	15	59	8	11
Venus	224	42	96	7	39
Mercury	87	58	245	32	25

Accordingly, in these periodic times there are no harmonic ratios, as is easily apparent, if the greater periods are continuously halved, and the smaller are continuously doubled, so that, by neglecting the intervals of an octave, we can investigate the intervals which exist within one octave.

	Saturn	Jupiter	Mars	Earth	Venus	Mercury	
	10,759D12'						
	5,379D36'	4,332D37'				87D58'	
Halves	2,689D48'	2,166D19'			224D42'	175D56'	Doubles
	1,344D54'	1,083D10'	686D59'	365D15'	449D24'	351D52'	
	672D27'	541D35'					

All the last numbers, as you see, are counter to harmonic ratios and seem, as it were, irrational. For let 687, the number of days of Mars, receive as its measure 120, which is the number of the division of the chord: according to this measure Saturn will have 117 for one sixteenth of its period, Jupiter less than 95 for one eighth of its peri-

od, the Earth less than 64, Venus more than 78 for twice its period, Mercury more than 61 for four times its period. These numbers do not make any harmonic ratio with 120, but their neighbouring numbers—60, 75, 80, and 96—do. And so, whereof Saturn has 120, Jupiter has approximately 97, the Earth more than 65, Venus more than 80, and Mercury less than 63. And whereof Jupiter has 120, the Earth has less than 81, Venus less than 100, Mercury less than 78. Likewise, whereof Venus has 120, the Earth has less than 98, Mercury more than 94. Finally, whereof the Earth has 120, Mercury has less than 116. But if the free choice of ratios had been effective here, consonances which are altogether perfect but not augmented or diminished would have been taken. Accordingly we find that God the Creator did not wish to introduce harmonic ratios between the sums of the delays added together to form the periodic times.

[284] And although it is a very probable conjecture (as relying on geometrical demonstrations and the doctrine concerning the causes of the planetary movements given in the *Commentaries on Mars*) that the bulks of the planetary bodies are in the ratio of the periodic times, so that the globe of Saturn is about thirty times greater than the globe of the Earth, Jupiter twelve times, Mars less than two, the Earth one and a half times greater than the globe of Venus and four times greater than the globe of Mercury: not therefore will even these ratios of bodies be harmonic.

But since God has established nothing without geometrical beauty, which was not bound by some other prior law of necessity, we easily infer that the periodic times have got their due lengths, and thereby the mobile bodies too have got their bulks, from something which is prior in the archetype, in order to express which thing these bulks and periods have been fashioned to this measure, as they seem disproportionate. But I have said that the periods are added up from the longest, the middling, and the slowest delays: accordingly geometrical fitnesses must be found either in these delays or in anything which may be prior to them in the mind of the Artisan. But the ratios of the delays are bound up with the ratios of the diurnal arcs, because the arcs have the inverse ratio of the delays. Again, we have said that the ratios of the delays and intervals of any one planet are the same. Then, as regards the single planets, there will be one and the same consideration of the following three: the arcs, the delays in equal arcs, and the distance of the arcs from the sun or the intervals. And because all these things are variable in the planets, there can be no doubt but that, if these things were allotted any geometrical beauty, then, by the sure design of the highest Artisan, they would have been received that at their extremes, at the aphelial and perihelial intervals, not at the mean intervals lying in between. For, given the ratios of the extreme intervals, there is no need of a plan to fit the intermediate ratios to a definite number. For they follow of themselves, by the necessity of planetary movement, from one extreme through all the intermediates to the other extreme.

Therefore the intervals are as follows, according to the very accurate observations of Tycho Brahe, by the method given in the *Commentaries on Mars* and investigated in very persevering study for seventeen years.

INTERVALS COMPARED WITH HARMONIC RATIOS[1]

Of Two Planets *Converging Diverging*				Of Single Planets
		Saturn's	aphelion 10,052. a.	More than a minor whole tone $^{10,000}/_{9,000}$
$^a/_b=^2/_1$,	$^b/_c=^5/_3$		perihelion 8,968. b.	Less than a major whole tone $^{10,000}/_{8,935}$
		Jupiter's	aphelion 5,451. c.	No concordant ratio but approximately
$^c/_f=^4/_1$,	$^d/_e=^3/_1$		perihelion 4,949. d.	11 : 10, a discordant or diminished 6 : 5.
		Mar's	aphelion 1,665. e.	Here 1662 : 1385 would be the consonance
$^e/_h=^5/_3$,	$^f/_g=^{17}/_{20}$		perihelion 1,382. f.	6 : 5, and 1665 : 1332 would be 5 : 4
$\frac{g}{k}=\frac{2}{1\frac{1}{2}}viz\frac{1000}{710}$,	$\frac{h}{i}=\frac{27}{20}$	Earth's	aphelion 1,018. g.	Here 1025 : 984 would be the diesis 24 : 25.
			perihelion 982. h.	Therefore it does not have the diesis.
		Venus'	aphelion 729 i.	Less than a sesquicomma.
$^i/_m=^{12}/_5$,	$^k/_i=^{243}/_{160}$		perihelion 719. k.	More than one third of a diesis.
		Mercury's	aphelion 470. l.	243 : 160, greater than a perfect fifth but
			perihelion 307. m.	less than a harmonic 8 : 5

[285] Therefore the extreme intervals of no one planet come near consonances except those of Mars and Mercury.

But if you compare the extreme intervals of different planets with one another, some harmonic light begins to shine. For the extreme diverging intervals of Saturn and Jupiter make slightly more than the octave; and the converging, a mean between the major and minor sixths. So the diverging extremes of Jupiter and Mars embrace approximately the double octave; and the converging, approximately the fifth and the

---

1. General Note: Throughout this text Kepler's *concinna* and *inconcinna* are translated as "concordant" and "discordant." *Concinna* is usually used by Kepler of all intervals whose ratios occur within the "natural system" or the just intonation of the scale. *Inconcinna* refers to all ratios that lie outside of this system of tuning. "Consonant" (*consonans*) and "dissonant" (*dissonans*) refer to qualities which can be applied to intervals within the musical system, in other words to "concords." "Harmony" (*harmonia*) is used sometimes in the sense of "concordance" and sometimes in the sense of "consonance."

Genus durum and *genus molle* are translated either as "major mode" and "minor mode," or as "major scale" and "minor scale," or as "major kind" and "minor kind" (of consonances). The use of *modus*, to refer to the ecclesiastical modes, occurs only in Chapter 6.

As our present musical terms do not apply strictly to the music of the sixteenth and seventeenth centuries, a brief explanation of terms here may be useful. This material is taken from Kepler's *Harmonies of the World*, Book III.

An octave system in the minor scale (*Systema octavae in cantu molli*)

		g	f	e	d	c	b	A	G
Ratios of string lengths:		72:	81:	90:	96:	108:	120:	128:	144

octave. But the diverging extremes of the Earth and Mars embrace somewhat more than the major sixth; the converging, an augmented fourth. In the next couple, the Earth and Venus, there is again the same augmented fourth between the converging extremes; but we lack any harmonic ratio between the diverging extremes: for it is less than the semi-octave (so to speak), *i.e.*, less than the square root of the ratio 2 : 1. Finally, between the diverging extremes of Venus and Mercury there is a ratio slightly less than the octave compounded with the minor third; between the converging there is a slightly augmented fifth.

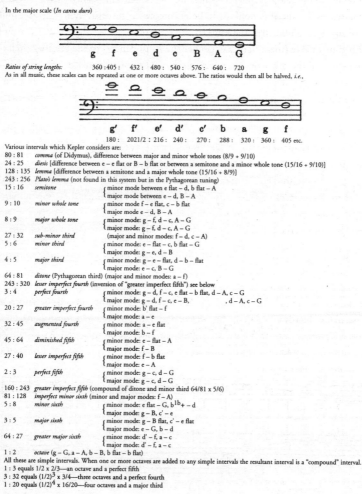

In the major scale (*In cantu duro*)

g f e d c B A G

Ratios of string lengths: 360 : 405 :   432 : 480 : 540 : 576 : 640 : 720
As in all music, these scales can be repeated at one or more octaves above. The ratios would then all be halved, *i.e.*,

g′   f′   e′   d′   c′   b   a   g   f

180 :   202½ : 216 : 240 : 270 : 288 : 320 : 360 : 405 etc.

Various intervals which Kepler considers are:

80 : 81	*comma* (of Didymus),	difference between major and minor whole tones (8/9 + 9/10)
24 : 25	*diesis*	[difference between e – e flat or B – b flat or between a semitone and a minor whole tone (15/16 + 9/10)]
128 : 135	*lemma*	[difference between a semitone and a major whole tone (15/16 + 8/9)]
243 : 256	*Plato's lemma*	(not found in this system but in the Pythagorean tuning)
15 : 16	*semitone*	minor mode between e flat – d, b flat – A major mode between e – d, B – A
9 : 10	*minor whole tone*	minor mode f – e flat, c – b flat major mode e – d, B – A
8 : 9	*major whole tone*	minor mode: g – f, d – c, A – G major mode: g – f, d – c, A – G
27 : 32	*sub-minor third*	(major and minor modes: f – d, c – A)
5 : 6	*minor third*	minor mode: e – flat – c, b flat – G major mode: g – e, d – B
4 : 5	*major third*	minor mode: g – e – flat, d – b – flat major mode: e – c, B – G
64 : 81	*ditone* (Pythagorean third)	(major and minor modes: a – f)
243 : 320	*lesser imperfect fourth*	(inversion of "greater imperfect fifth") see below
3 : 4	*perfect fourth*	minor mode: g – d, f – c, e flat – b flat, d – A, c – G major mode: g – d, f – c, e – B, , d – A, c – G
20 : 27	*greater imperfect fourth*	minor mode: b′ flat – f major mode: a – e
32 : 45	*augmented fourth*	minor mode: a – e flat major mode: b – f
45 : 64	*diminished fifth*	minor mode: e – flat – A major mode: f – B
27 : 40	*lesser imperfect fifth*	minor mode: f – b flat major mode: e – A
2 : 3	*perfect fifth*	minor mode: g – c, d – G major mode: g – c, d – G
160 : 243	*greater imperfect fifth*	(compound of ditone and minor third 64/81 x 5/6)
81 : 128	*imperfect minor sixth*	(minor and major modes: f – A)
5 : 8	*minor sixth*	minor mode: e flat – G, b¹b+ – d major mode: g – B, c′ – e
3 : 5	*major sixth*	minor mode: g – B flat, c′ – e flat major mode: e – G, b – d
64 : 27	*greater major sixth*	minor mode: d′ – f, a – c major mode: d′ – f, a – c
1 : 2	*octave*	(g – G, a – A, b – B, b flat – b flat)

All these are simple intervals. When one or more octaves are added to any simple intervals the resultant interval is a "compound" interval.
1 : 3 equals 1/2 x 2/3—an octave and a perfect fifth
3 : 32 equals $(1/2)^3$ x 3/4—three octaves and a perfect fourth
1 : 20 equals $(1/2)^4$ x 16/20—four octaves and a major third

Accordingly, although one interval was somewhat removed from harmonic ratios, this success was an invitation to advance further. Now my reasonings were as follows: First, in so far as these intervals are lengths without movement, they are not fittingly examined for harmonic ratios, because movement is more properly the subject of consonances, by reason of speed and slowness. Second, inasmuch as these same intervals are the diameters of the spheres, it is believable that the ratio of the five regular solids applied proportionally is more dominant in them, because the ratio of the geometrical solid bodies to the celestial spheres (which are everywhere either encompassed by celestial matter, as the ancients hold, or to be encompassed successively by the accumulation of many revolutions) is the same as the ratio of the plane figures which may be inscribed in a circle (these figures engender the consonances) to the celestial circles of movements and the other regions wherein the movements take place. Therefore, if we are looking for consonances, we should look for them not in these intervals in so far as they are the semidiameters of spheres but in them in so far as they are the measures of the movements, *i.e.*, in the movements themselves, rather. Absolutely no other than the mean intervals can be taken as the semidiameters of the spheres; but we are here dealing with the extreme intervals. Accordingly, we are not dealing with the intervals in respect to their spheres but in respect to their movements.

Accordingly, although for these reasons I had passed on to the comparison of the extreme movements, at first the ratios of the movements remained the same in magnitude as those which were previously the ratios of the intervals, only inverted. Wherefore too, certain ratios, which are discordant and foreign to harmonies, as before, have been found between the movements. But once again I judged that this happened to me deservedly, because I compared with one another eccentric arcs which

Concords: All intervals from diesis downward on above list.

Consonances: Minor and major thirds and sixths, perfect fourth, fifth, and octave.

"Adulterine" consonances: sub-minor third, ditone, lesser imperfect fourth and fifth, greater imperfect fourth and fifth, imperfect minor sixth, greater major sixth.

Dissonances: All other intervals.

Throughout this work Kepler, after the fashion of the theorists of his time, uses the ratios of string lengths rather than the ratios of vibrations as is usually done today. String lengths are, of course, inversely proportionate to the vibrations. That is, string lengths 4 : 5 are expressed in vibrations as 5 : 4. This accounts for the descending order of the scale, which follows the increasing numerical order. It is an interesting fact that Kepler's minor and major scales are inversions of each other and hence, when expressed in ratios of vibrations, are in the opposite order from those in ratios of string lengths:

Notes resulting from ratios of vibrations

72 : 81 : 90 : 96 : 108 : 120 : 128 : 144   360 : 405 : 432 : 480 : 540 : 576 : 640 : 720

Notes resulting from ratios of string lengths

An arbitrary pitch G is chosen to situate these ratios. This g or "gamma" was usually the lowest tone of the sixteenth-century musical gamut.

Elliott Carter, Jr.

are not expressed and numbered by a measure of the same magnitude but are numbered in degrees and minutes which are of diverse magnitude in diverse planets, nor do they from our place give the appearance of being as great as the number of each says, except only at the centre of the eccentric of each planet, which centre rests upon no body; and hence it is also unbelievable that there is any sense or natural instinct in that place in the world which is capable of perceiving this; or, rather, it was impossible, if I was comparing the eccentric arcs of different planets with respect to their appearance at their centres, which are different for different planets. But if diverse apparent magnitudes are compared with one another, they ought to be apparent in one place in the world in such a way that that which possesses the faculty of comparing them may be present in that place from which they are all apparent. Accordingly, I judged that the appearance of these eccentric arcs should be removed from the mind or else should be formed differently. But if I removed the appearance and applied my mind to the diurnal journeys of the planets, I saw that I had to employ the rule which I gave in Article IX of the preceding chapter. [286] Accordingly if the diurnal arcs of the eccentric are multiplied by the mean intervals of the spheres, the following journeys are produced:

		Diurnal movements	Mean intervals	Diurnal journeys
Saturn	at aphelion	1'53"	9510	1065
	at perihelion	2'7"		1208
Jupiter	at aphelion	4'44"	5200	1477
	at perihelion	5'15"		1638
Mars	at aphelion	28'44"	1524	2627
	at perihelion	34'34"		3161
Earth	at aphelion	58'60"	1000	3486
	at perihelion	60'13"		3613
Venus	at aphelion	95'29"	724	4149
	at perihelion	96'50"		4207
Mercury	at aphelion	201'0"	388	4680
	at perihelion	307'3"		7148

Thus Saturn traverses barely one seventh of the journey of Mercury; and hence, as Aristotle judged consonant with reason in Book II of *On the Heavens*, the planet which is nearer the sun always traverses a greater space than the planet which is farther away—as cannot hold in the ancient astronomy.

And indeed, if we weigh the thing fairly carefully, it will appear to be not very probable that the most wise Creator should have established harmonies between the

planetary journeys in especial. For if the ratios of the journeys are harmonic, all the other affects which the planets have will be necessitated and bound up with the journeys, so that there is no room elsewhere for establishing harmonies. But whose good will it be to have harmonies between the journeys, or who will perceive these harmonies? For there are two things which disclose to us harmonies in natural things: either light or sound: light apprehended through the eyes or hidden senses proportioned to the eyes, and sound through the ears. The mind seizes upon these forms and, whether by instinct (on which Book IV speaks profusely) or by astronomical or harmonic ratiocination, discerns the concordant from the discordant. Now there are no sounds in the heavens, nor is the movement so turbulent that any noise is made by the rubbing against the ether. Light remains. If light has to teach these things about the planetary journeys, it will teach either the eyes or a sensorium analogous to the eyes and situated in a definite place; and it seems that sense-perception must be present there in order that light of itself may immediately teach. Therefore there will be sense-perception in the total world, namely in order that the movements of all the planets may be presented to sense-perceptions at the same time. For that former route—from observations through the longest detours of geometry and arithmetic, through the ratios of the spheres and the other things which must be learned first, down to the journeys which have been exhibited—is too long for any natural instinct, for the sake of moving which it seems reasonable that the harmonies have been introduced.

Therefore with everything reduced to one view, I concluded rightly [287] that the true journeys of the planets through the ether should be dismissed, and that we should turn our eyes to the apparent diurnal arcs, according as they are all apparent from one definite and marked place in the world—namely, from the solar body itself, the source of movement of all the planets; and we must see, not how far away from the sun any one of the planets is, nor how much space it traverses in one day (for that is something for ratiocination and astronomy, not for instinct), but how great an angle the diurnal movement of each planet subtends in the solar body, or how great an arc it seems to traverse in one common circle described around the sun, such as the ecliptic, in order that these appearances, which were conveyed to the solar body by virtue of light, may be able to flow, together with the light, in a straight line into creatures, which are partakers of this instinct, as in Book IV we said the figure of the heavens flowed into the foetus by virtue of the rays.

Therefore, if you remove from the proper planetary movement the parallaxes of the annual orbit, which gives them the mere appearances of stations and retrogradations, Tycho's astronomy teaches that the diurnal movements of the planets in their orbits (which are apparent as it were to spectator at the sun) are as shown in the table on the opposite page.

Note that the great eccentricity of Mercury makes the ratio of the movements differ somewhat from the ratio of the square of the distances. For if you make the square of the ratio of 100, the mean distance, to 121, the aphelial distance, be the ratio of the aphelial movement to the mean movement of 245'32", then an aphelial movement of 167 will be produced; and if the square of the ratio of 100 to 79, the perihelial distance, be the ratio of the perihelial to the same mean movement, then the perihelial movement will become 393; and both cases are greater than I have here laid down, because the mean movement at the mean anomaly, viewed very obliquely, does not appear as great, *viz.*, not as great as 245'32", but about 5' less. Therefore, too, lesser aphelial and perihelial movements will be elicited. But the aphelial (appears) lesser and the perihelial greater, on account of theorem 8, Euclid's *Optics*, as I remarked in the preceding Chapter, Article VI.

Harmonies Between Two Planets	Apparent Diurnal Movements			Harmonies Between the Movements of Single Planets
Diverging Converging				
	Saturn	at aphelion	1'46" a.	1 : 48": 2'15" = 4 : 5,
		at perihelion	2'15" b.	major third
$a/_d = 1/_3,$ $b/_c = 1/_2$				
	Jupiter	at aphelion	4'30" c.	4'35": 5'30" = 5 : 6,
		at perihelion	5'30" d.	minor third
$c/_f = 1/_8,$ $d/_e = 5/_{24}$				
	Mars	at aphelion	26'14" e.	25'21" : 38'1" = 2 : 3,
		at perihelion	38'1" f.	the fifth
$e/_h = 5/_{12},$ $f/_g = 2/_3$				
	Earth	at aphelion	57'3" g.	57'28": 61'18" = 15 :
		at perihelion	61'18" h.	16, semitone
$g/_k = 3/_5,$ $h/_i = 5/_8$				
	Venus	at aphelion	94'50" i.	94'50": 98'47" = 24 :
		at perihelion	97'37" k.	25, diesis
$i/_m = 1/_4,$ $k/_l = 3/_5$				
	Mercury	at aphelion	164'0" l.	164'0" : 394'0" = 5 : 12,
		at perihelion	384'0" m.	octave and minor third

Accordingly, I could mentally presume, even from the ratios of the diurnal eccentric arcs given above, that there were harmonies and concordant intervals between these extreme apparent movements of the single planets, since I saw that everywhere there the square roots of harmonic ratios were dominant, but knew that the ratio of the apparent movements was the square of the ratio of the eccentric movements. But it is

possible by experience itself, or without any ratiocination to prove what is affirmed, as you see [288] in the preceding table. The ratios of the apparent movements of the single planets approach very close to harmonies, in such fashion that Saturn and Jupiter embrace slightly more than the major and minor thirds, Saturn with a ratio of excess of 53 : 54, and Jupiter with one of 54 : 55 or less, namely approximately a sesquicomma; the Earth, slightly more (namely 137 : 138, or barely a semicomma) than a semitone; Mars somewhat less (namely 29 : 30, which approaches 34 : 35 or 35 : 36) than a fifth; Mercury exceeds the octave by a minor third rather than a whole tone, *viz.*, it is about 38 : 39 (which is about two commas, *viz.*, 34 : 35 or 35 : 36) less than a whole tone. Venus alone falls short of any of the concords the diesis; for its ratio is between two and three commas, and it exceeds two thirds of a diesis, and is about 34:35 or 35:36, a diesis diminished by a comma.

The moon, too, comes into this consideration. For we find that its hourly apogeal movement in the quadratures, *viz.*, the slowest of all its movements, to be 26'26"; its perigeal movement in the syzygies, *viz.*, the fastest of all, 35'12", in which way the perfect fourth is formed very precisely. For one third of 26'26" is 8'49", the quadruple of which is 35'16". And note that the consonance of the perfect fourth is found nowhere else between the apparent movements; note also the analogy between the fourth in consonances and the quarter in the phases. And so the above things are found in the movements of the single planets.

But in the extreme movements of two planets compared with one another, the radiant sun of celestial harmonies immediately shines at first glance, whether you compare the diverging extreme movements or the converging. For the ratio between the diverging movements of Saturn and Jupiter is exactly the duple or octave; that between the diverging, slightly more than triple or the octave and the fifth. For one third of 5'30" is 1'50", although Saturn has 1'46" instead of that. Accordingly, the planetary movements will differ from a consonance by a diesis more or less, *viz.*, 26 : 27 or 27:28; and with less than one second acceding at Saturn's aphelion, the excess will be 34 : 35, as great as the ratio of the extreme movements of Venus. The diverging and converging movements of Jupiter and Mars are under the sway of the triple octave and the double octave and a third, but not perfectly. For one eighth of 38'1" is 4'45", although Jupiter has 4'30"; and between these numbers there is still a difference of 18 : 19, which is a mean between the semitone of 15 : 16 and the diesis of 24 : 25, namely, approximately a perfect lemma of 128 : 135.[1] Thus one fifth of 26'14" is 5'15", although Jupiter has 5'30"; accordingly in this case the quintuple ratio is diminished in the ratio of 21 : 22, the augment in the case of the other ratio, *viz.*, approximately a diesis of 24 : 25.

1. cf. Footnote to *Intervals Compared with Harmonic Ratios*, p. 186.

The consonance 5 : 24 comes nearer, which compounds a minor instead of a major third with the double octave. For one fifth of 5'30" is 1'6", which if multiplied by 24 makes 26'24", does not differ by more than a semicomma. Mars and the Earth have been allotted the least ratio, exactly the sesquialteral or perfect fifth: for one third of 57'3" is 19'1", the double of which is 38'2", which is Mars' very number, *viz.*, 38'11". They have also been allotted the greater ratio of 5 : 12, the octave and minor third, but more imperfectly. For one twelfth of 61'18" is 5'61/2," which if multiplied by 5 gives 25'33", although instead of that Mars has 26'14". Accordingly, there is a deficiency of a diminished diesis approximately, *viz.*, 35 : 36. But the Earth and Venus together have been allotted 3 : 5 as their greatest consonance and 5 : 8 as their least, the major and minor sixths, but again not perfectly. For one fifth of 97'37", which if multiplied by 3 gives 58'33", which is greater than the movement of the Earth in the ratio 34 : 35, which is approximately 35 : 36: by so much do the planetary ratios differ from the harmonic. Thus one eighth of 94'50" is 11'51"+, five times which is 59'16", which is approximately equal to the mean movement of the Earth. Wherefore here the planetary ratio is less than the harmonic [289] in the ratio of 29 : 30 or 30:31, which is again approximately 35 : 36, the diminished diesis; and thereby this least ratio of these planets approaches the consonance of the perfect fifth. For one third of 94'50" is 31'37", the double of which is 63'14", of which the 61'18" of the perihelial movement of the Earth falls short in the ratio of 31 : 32, so that the planetary ratio is exactly a mean between the neighbouring harmonic ratios. Finally, Venus and Mercury have been allotted the double octave as their greatest ratio and the major sixth as their least, but not absolute-perfectly. For one fourth of 384' is 96'0", although Venus has 94'50". Therefore the quadruple adds approximately one comma. Thus one fifth of 164' is 32'48", which if multiplied by 3 gives 98'24", although Venus has 97'37". Therefore the planetary ratio is diminished by about two thirds of a comma, *i.e.*, 126 : 127.

Accordingly the above consonances have been ascribed to the planets; nor is there any ratio from among the principal comparisons (*viz.*, of the converging and diverging extreme movements) which does not approach so nearly to some consonance that, if strings were tuned in that ratio, the ears would not easily discern their imperfection—with the exception of that one excess between Jupiter and Mars.

Moreover, it follows that we shall not stray far away from consonances if we compare the movements of the same field. For if Saturn's 4 : 5 comp. 53 : 54 are compounded with the intermediate 1 : 2, the product is 2 : 5 comp. 53 : 54, which exists between the aphelial movements of Saturn and Jupiter. Compound with that Jupiter's 5:6 comp. 54 : 55, and the product is 5 : 12 comp. 54 : 55, which exist between the perihelial movements of Saturn and Jupiter. Thus compound Jupiter's 5 : 6 comp. 54:55 with the intermediate ensuing ratio of 5 : 24 comp. 158 : 157, the product will be 1:6

comp. 36 : 35 between the aphelial movements. Compound the same 5 : 24 comp. 158 : 157 with Mars' 2 : 3 comp. 30 : 29, and the product will be 5 : 36 comp. 25 : 24 approximately, *i.e.*, 125 : 864 or about 1 : 7, between the perihelial movements. This ratio is still alone discordant. With 2 : 3 the third ratio among the intermediates, compound Mars' 2 : 3 less 29 : 30; the result will be 4 : 9 comp. 30 : 29, *i.e.*, 40 : 87, another discord between the aphelial movements. If instead of Mars' you compound the Earth's 15 : 16 comp. 137 : 138, you will make 5 : 8 comp. 137 : 138 between the perihelial movements. And if with the fourth of the intermediates, 5 : 8 comp. 31 : 30, or 2 : 3 comp. 31 : 32, you compound the Earth's 15 : 16 comp. 137 : 138, the product will be approximately 3 : 5 between the aphelial movements of the Earth and Venus. For one fifth of 94'50" is 18'58", the triple of which is 56'54", although the Earth has 57'3". If you compound Venus' 34 : 35 with the same ratio, the result will be 5 : 8 between the perihelial movements. For one eighth of 97'37" is 12'12"+ which if multiplied by 5 gives 61'1", although the Earth has 61'18". Finally, if with the last of the intermediate ratios, 3 : 5 comp. 126 : 127 you compound Venus' 34 : 35, the result is 3 : 5 comp. 24 : 25, and the interval, compounded of both, between the aphelial movements, is dissonant. But if you compound Mercury's 5 : 12 comp. 38 : 39, the double octave or 1 : 4 will be diminished by approximately a whole diesis, in proportion to the perihelial movements.

Accordingly, perfect consonances are found: between the converging movements of Saturn and Jupiter, the octave; between the converging movements of Jupiter and Mars, the octave and minor third approximately; between the converging movements of Mars and the Earth, the fifth; between their perihelial, the minor sixth; between the extreme converging movements of Venus and Mercury, the major sixth; between the diverging or even between the perihelial, the double octave: whence without any loss to an astronomy which has been built, most subtly of all, upon Brahe's observations, it seems that the residual very slight discrepancy can be discounted, especially in the movements of Venus and Mercury.

But you will note that where there is no perfect major consonance, as between Jupiter and Mars, there alone have I found the placing of the solid figure to be approximately perfect, since the perihelial distance of Jupiter is approximately three times the aphelial distance of Mars, in such fashion that this pair of planets strives after the perfect consonance in the intervals which it does not have in the movements.

[290] You will note, furthermore, that the major planetary ratio of Saturn and Jupiter exceeds the harmonic, *viz.*, the triple, by approximately the same quantity as belongs to Venus; and the common major ratio of the converging and diverging movements of Mars and the Earth are diminished by approximately the same. You will note thirdly that, roughly speaking, in the upper planets the consonances are established

between the converging movements, but in the lower planets, between movements in the same field. And note fourthly that between the aphelial movements of Saturn and the Earth there are approximately five octaves; for one thirty-second of 57'3" is 1'47", although the aphelial movement of Saturn is 1'46".

Furthermore, a great distinction exists between the consonances of the single planets which have been unfolded and the consonances of the planets in pairs. For the former cannot exist at the same moment of time, while the latter absolutely can; because the same planet, moving at its aphelion, cannot be at the same time at the opposite perihelion too, but of two planets one can be at its aphelion and the other at its perihelion at the same moment of time. And so the ratio of plain-song or monody, which we call choral music and which alone was known to the ancients,[1] to polyphony—called "figured song,";[2] the invention of the latest generations—is the same as the ratio of the consonances which the single planets designate to the consonances of the planets taken together. And so, further on, in Chapters 5 and 6, the single planets will be compared to the choral music of the ancients and its properties will be exhibited in the planetary movements. But in the following chapters, the planets taken together and the figured modern music will be shown to do similar things.

5. IN THE RATIOS OF THE PLANETARY MOVEMENTS WHICH ARE APPARENT AS IT WERE TO SPECTATORS AT THE SUN, HAVE BEEN EXPRESSED THE PITCHES OF THE SYSTEM, OR NOTES OF THE MUSICAL SCALE, AND THE MODES OF SONG (GENERA CANTUS), THE MAJOR AND THE MINOR[3]

Therefore by now I have proved by means of numbers gotten on one side from astronomy and on the other side from harmonies that, taken in every which way, harmonic ratios hold between these twelve termini or movements of the six planets revolving around the sun or that they approximate such ratios within an imperceptible part of least concord. But just as in Book III in the first chapter, we first built up the single harmonic consonances separately, and then we joined together all the consonances—as many as there were—in one common system or musical scale, or, rather, in one octave of them which embraces the rest in power, and by means of them we separated the others into their degrees or pitches (*loca*) and we did this in such a way that there would be a scale; so now also, after the discovery of the consonances (*harmoniis*) which

---

1. The choral music of the Greeks was monolinear, everyone singing the same melody together—E. C., Jr.

2. In plain-song all the time values of the notes were approximately equal, while in "figured song" time values of different lengths were indicated by the notes, which gave composers an opportunity both to regulate the way different contrapuntal parts joined together and to produce many expressive effects. Practically all melodies since this time are in "figured song" style.—E. C., Jr.

3. See note to *Intervals Compared with Harmonic Ratios.*

God Himself has embodied in the world, we must consequently see whether those single consonances stand so separate that they have no kinship with the rest, or whether all are in concord with one another. Notwithstanding it is easy to conclude, without any further inquiry, that those consonances were fitted together by the highest prudence in such fashion that they move one another about within one frame, so to speak, and do not jolt one another out of it; since indeed we see that in such a manifold comparison of the same terms there is no place where consonances do not occur. For unless in one scale all the consonances were fitted to all, it could easily have come about (and it has come about wherever necessity thus urges it) that many dissonances should exist. For example, if someone had set up a major sixth between the first and the second term, and likewise a major third between the second and the third term, without taking the first into account, then he would admit a dissonance and the discordant interval 12 : 25 between the first and third.

But come now, let us see whether that which we have already inferred by reasoning is really found in this way. [291] But let me premise some cautions, that we may be the less impeded in our progress. First, for the present, we must conceal those augments or diminutions which are less than a semitone; for we shall see later on what causes they have. Second, by continuous doubling or contrary halving of the movements, we shall bring everything within the range of one octave, on account of the sameness of consonance in all the octaves.

Accordingly the numbers wherein all the pitches or clefs (*loca seu claves*) of the octave system are expressed have been set out in a table in Book III, Chapter 7[1], *i.e.*,

1. The table is as follows:

Concordant Intervals	Lengths of Strings	In familiar notes	
	1080	High g	
Semitone			
Lemma	1152	f #	
	1215	f	
Semitone			
Diesis	1296	e	
	1350	e ♭	
Semitone			
Semitone	1440	d	
	1536	c #	
Lemma			
	1620	c	
Semitone			
Diesis	1728	b	
	1800	b ♭	
Semitone			
Semitone	1920	A	
	2048	G #	
Lemma			
	2160	Low G	

665

understand these numbers of the length of two strings. As a consequence, the speeds of the movements will be in the inverse ratios.

Now let the planetary movements be compared in terms of parts continuously halved. Therefore

Movement of Mercury	at perihelion,	7th subduple, or $1/128$,	3'0"
	at aphelion,	6th subduple, or $1/64$,	2'34"
Movement of Venus	at perihelion,	5th subduple, or $1/32$,	3'3"
	at aphelion,	5th subduple, or $1/32$,	2'58"
Movement of Earth	at perihelion,	5th subduple, or $1/32$,	1'55"
	at aphelion,	5th subduple, or $1/32$,	1'47"
Movement of Mars	at perihelion,	4th subduple, or $1/16$,	2'23"
	at aphelion,	3rd subduple, or $1/8$,	3'17"
Movement of Jupiter	at perihelion,	subduple, or $1/2$,	2'45"
	at aphelion,	subduple, or $1/2$	2'15"
Movement of Saturn	at perihelion,		2'15"
	at aphelion,		1'46"

Now the aphelial movement of Saturn at its slowest—*i.e.*, the slowest movement—marks *G*, the lowest pitch in the system with the number 1'46". Therefore the aphelial movement of the Earth will mark the same pitch, but five octaves higher, because its number is 1'47", and who wants to quarrel about one second in the aphelial movement of Saturn? But let us take it into account, nevertheless; the difference will not be greater than 106 : 107, which is less than a comma. If you add 27", one quarter of this 1'47", the sum will be 2'14", although the perihelial movement of Saturn has 2'15"; similarly the aphelial movement of Jupiter, but one octave higher. Accordingly, these two movements mark the note *b*, or else are very slightly higher. Take 36", one third of 1'47", and add it to the whole; you will get as a sum 2'23" for the note *c*; and here's the perihelion of Mars of the same magnitude but four octaves higher. To this same 1'47" add also 54", half of it, and the sum will be 2'41" for the note *d*; and here the perihelion of Jupiter is at hand, but one octave higher, for it occupies the nearest number, *viz.*, 2'45". If you add two thirds, *viz.*, 1'11", the sum will be 2'58"; and here's the aphelion of Venus at 2'58". Accordingly, it will mark the pitch or the note *e*, but five octaves higher. And the perihelial movement of Mercury, which is 3'0", does not exceed it by much but is seven octaves higher. Finally, divide the double of 1'47", *viz.*, 3'34", into nine parts and subtract one part of 24" from the whole; 3'10" will be left for the note *f*, which the 3'17" of the aphelial movement of Mars marks approximately but three octaves higher; and this number is slightly greater than the just number and approaches the note *f* sharp. For if one sixteenth of 3'34", *viz.*, $13\frac{1}{2}$", is subtracted from 3'34", then $3'20\frac{1}{2}$" is left, to which 3'17" is very near. And indeed in

music *f* sharp is often employed in place of *f*, as we can see everywhere.

Accordingly all the notes of the major scale (*cantus duri*) (except the note *a* which was not marked by harmonic division, in Book III, Chapter 2) are marked by all the extreme movements of the planets, except the perihelial movements of Venus and the Earth [292] and the aphelial movement of Mercury, whose number, 2'34", approaches the note *c* sharp. For subtract from the 2'41" of *d* one sixteenth or 10", and 2'30" remains for the note *c* sharp. Thus only the perihelial movement of Venus and the Earth are missing from this scale, as you may see in the table.

On the other hand, if the beginning of the scale is made at 2'15", the aphelial movement of Saturn, and we must express the note *G* in those degrees: then for the note *A* is 2'32", which closely approaches the aphelial movement of Mercury; for the note *b* flat, 2'42", which is approximately the perihelial movement of Jupiter, by the equipollence of octaves; for the note *c*, 3'0", approximately the perihelial movement of Mercury and Venus; for the note *d*, 3'23" and the aphelial movement of Mars is not much graver, *viz.*, 3'17", so that here the number is about as much less than its note as previously the number was greater than its note; for the note *e* flat, 3'36", which the aphelial movement of the Earth approximates; for the note *e*, 3'50", and the perihelial movement of the Earth is 3'49"; but the aphelial movement of Jupiter again

occupies *g*. In this way, all the notes except *f* are expressed within one octave of the minor scale by most of the aphelial and perihelial movements of the planets, especially by those which were previously omitted, as you see in the table.

Previously, however, *f* sharp was marked and *a* omitted; now *a* is marked, *f* sharp is omitted; for the harmonic division in Chapter 2 also omitted the note *f*.

Accordingly, the musical scale or system of one octave with all its pitches, by means of which natural song[1] is transposed in music, has been expressed in the heavens by a twofold way and in two as it were modes of song. There is this sole difference: in our harmonic sectionings both ways start together from one and the same terminus *G*; but here, in the planetary movements, that which was previously *b* now becomes *G* in the minor mode.

In the celestial movements, as follows:

| 2160 | 1920 | 1728 | 1620 | 1440 | 1296 | 1152 | 1080 | 960 | 864 | | 1728 | 1536 | 1440 | 1296 | 1152 | 1080 | 972 | 864 |

By harmonic sectionings, as follows:

| 2160 | 1920 | 1728 | 1620 | 1440 | 1296 | 1152 | 1080 | | 2160 | 1920 | 1800 | 1620 | 1440 | 1350 | 1215 | 1080 |

For as in music 2160 : 1800, or 6 : 5, so in that system which the heavens express, 1728 : 1440, namely, also 6 : 5; and so for most of the remaining, 2160 : 1800, 1620, 1440, 1350, 1080 as 1728 : 1440, 1296, 1152, 1080, 864.

Accordingly you won't wonder any more that a very excellent order of sounds or pitches in a musical system or scale has been set up by men, since you see that they are doing nothing else in this business except to play the apes of God the Creator and to act out, as it were, a certain drama of the ordination of the celestial movements.

But there still remains another way whereby we may understand the twofold musical scale in the heavens, where one and the same system but a twofold tuning (*tensio*) is embraced, one at the aphelial movement of Venus, the other at the perihelial, because the variety of movements of this planet is of the least magnitude, as being such as is comprehended within the magnitude of the diesis, the least concord. And the aphelial tuning (*tensio*), as above, has been given to the aphelial movements of Saturn, the Earth, Venus, and (relatively speaking) Jupiter, in *G, e, b*, but to the perihelial movements of Mars and (relatively speaking) Saturn and, as is apparent at first glance, to

1. Natural song: music in the basic major or minor system without accidentals. E. C., Jr.

those of Mercury, in *c*, *e*, and *b*. On the other hand, the perihelial tuning supplies a pitch even for the aphelial movements of Mars, Mercury, and (relatively speaking) Jupiter, but to the perihelial movements of Jupiter, Venus, and (relatively speaking) Saturn, and to a certain extent to that of the Earth and indubitably to that of Mercury too. For let us suppose that now not the aphelial movement of Venus but the 3'3" of the perihelial gets the pitch of *e*; it is approached very closely by the 3'0" of the perihelial movement of Mercury, through a double octave, at the end of Chapter 4. But if 18" or one tenth of this perihelial movement of Venus is subtracted, 2'45" remains, the perihelion of Jupiter, which occupies the pitch of *d*; and if one fifteenth or 12" is added, the sum will be 3'15", approximately the perihelion of Mars which occupies the pitch of *f*; and thus in *b*, the perihelial movement of Saturn and the aphelial movement of Jupiter have approximately the same tuning. But one eighth, or 23", if multiplied by 5, gives 1'55", which is the perihelial movement of the Earth; and, although it does not square with the foregoing in the same scale, as it does not give the interval 5 : 8 below *e* nor 24 : 25 above *G*, nevertheless if now the perihelial movement of Venus and so too the aphelial movement of Mercury, outside of the order, occupy the pitch *e*-flat instead of *e*, then there the perihelial movement of the Earth will occupy the pitch of *G*, and the aphelial movement of Mercury is in concord, because 1'1", or one third of 3'3", if multiplied by 5, gives 5'5", half of which, or 2'32", approximates the aphelion of Mercury, which in this extraordinary adjustment will occupy the pitch of *c*. Therefore, all these movements are of the same tuning with respect to one another; but the perihelial movement of Venus together with the three (or five) prior movements, *viz.*, in the same harmonic mode, divides the scale differently from the aphelial movement of the same in its tuning, *viz.*, in the major mode (*denere duro*). Moreover, the perihelial movement of Venus, together with the two posterior movements, divides the same scale differently, *viz.*, not into concords but merely into a different order of concords, namely one which belongs to the minor mode (*generis mollis*).

But it is sufficient to have laid before the eyes in this chapter what is the case casually, but it will be disclosed in Chapter 9 by the most lucid demonstrations why each and every one of these things was made in this fashion and what the causes were not merely of harmony but even of the very least discord.

6. IN THE EXTREME PLANETARY MOVEMENTS THE MUSICAL MODES OR TONES HAVE SOMEHOW BEEN EXPRESSED

[294] This follows from the aforesaid and there is no need of many words; for the single planets somehow mark the pitches of the system with their perihelial movement, in so far as it has been appointed to the single planets to traverse a certain fixed interval in the musical scale comprehended by the definite notes of it or the pitches of the

system, and beginning at that note or pitch of each planet which in the preceding chapter fell to the aphelial movement of that planet: *G* to Saturn and the Earth, *b* to Jupiter, which can be transposed higher to *G*, *f*-sharp to Mars, *e* to Venus, *a* to Mercury in the higher octave. See the single movements in the familiar terms of notes. They do not form articulately the intermediate positions, which you here see filled by notes, as they do the extremes, because they struggle from one extreme to the opposite not by leaps and intervals but by a continuum of tunings and actually traverse all the means (which are potentially infinite)—which cannot be expressed by me in any other way than by a continuous series of intermediate notes. Venus remains approximately in unison and does not equal even the least of the concordant intervals in the difference of its tension.

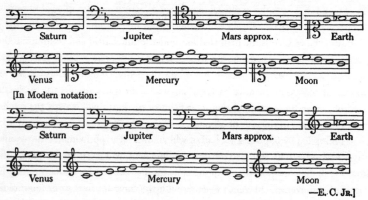

[In Modern notation:

—E. C. Jr.]

But the signature of two accidentals (flats) in a common staff and the formation of the skeletal outline of the octave by the inclusion of a definite concordant interval are a certain first beginning of the distinction of Tones or Modes (*modorum*). Therefore the musical Modes have been distributed among the planets. But I know that for the formation and determination of distinct Modes many things are requisite, which belong to human song, as containing (a) distinct (order of) intervals; and so I have used the word *somehow*.

But the harmonist will be free to choose his opinion as to which Mode each planet expresses as its own, since the extremes have been assigned to it here. From among the familiar Modes, I should give to Saturn the Seventh or Eighth, because if you place its key-note at *G*, the perihelial movement ascends to *b*; to Jupiter, the First or Second Mode, because its aphelial movement has been fitted to *G* and its perihelial movement arrives at *b* flat; to Mars, the Fifth or Sixth Mode, not only because Mars comprehends approximately the perfect fifth, which interval is common to all the Modes, but principally because when it is reduced with the others to a common system, it attains *c* with

its perihelial movement and touches *f* with its aphelial, which is the key-note of the Fifth or Sixth Mode or Tone; I should give the Third or Fourth Mode to the Earth, because its movement revolves within a semitone, while the first interval of those Modes is a semitone; but to Mercury will belong indifferently all the Modes or Tones on account of the greatness of its range; to Venus, clearly none on account of the smallness of its range; but on account of the common system the Third and Fourth Mode, because with reference to the other planets it occupies *e*. (The Earth sings MI, FA, MI so that you may infer even from the syllables that in this our domicile MIsery and FAmine obtain.)[1]

7. THE UNIVERSAL CONSONANCES OF ALL SIX PLANETS, LIKE COMMON FOUR-PART COUNTERPOINT, CAN EXIST

[295] But now, Urania, there is need for louder sound while I climb along the harmonic scale of the celestial movements to higher things where the true archetype of the fabric of the world is kept hidden. Follow after, ye modern musicians, and judge the thing according to your arts, which were unknown to antiquity. Nature, which is never not lavish of herself, after a lying-in of two thousand years, has finally brought you forth in these last generations, the first true images of the universe. By means of your concords of various voices, and through your ears, she has whispered to the human mind, the favorite daughter of God the Creator, how she exists in the innermost bosom.

(Shall I have committed a crime if I ask the single composers of this generation for some artistic motet instead of this epigraph? The Royal Psalter and the other Holy Books can supply a text suited for this. But alas for you! No more than six are in concord in the heavens. For the moon sings here monody separately, like a dog sitting on the Earth. Compose the melody; I, in order that the book may progress, promise that I will watch carefully over the six parts. To him who more properly expresses the celestial music described in this work, Clio will give a garland, and Urania will betroth Venus his bride.)

It has been unfolded above what harmonic ratios two neighbouring planets would embrace in their extreme movements. But it happens very rarely that two, especially the slowest, arrive at their extreme intervals at the same time; for example, the apsides of Saturn and Jupiter are about 81° apart. Accordingly, while this distance between them measures out the whole zodiac by definite twenty-year leaps,[2] eight hundred years pass by, and nonetheless the leap which concludes the eighth century, does not

---

1. See note on hexachordal system.
2. That is to say, since Saturn and Jupiter have one revolution with respect to one another every twenty years, they are 81° apart once every twenty years, while the end-positions of this 81° interval traverse the ecliptic in leaps, so to speak, and coincide with the apsides approximately once in eight hundred years. C. G. W.

carry precisely to the very apsides; and if it digresses much further, another eight hundred years must be awaited, that a more fortunate leap than that one may be sought; and the whole route must be repeated as many times as the measure of digression is contained in the length of one leap. Moreover, the other single pairs of planets have periods as that, although not so long. But meanwhile there occur also other consonances of two planets, between movements whereof not both are extremes but one or both are intermediate; and those consonances exist as it were in different tunings (*tensionibus*). For, because Saturn tends from *G* to *b*, and slightly further, and Jupiter from *b* to *d* and further; therefore between Jupiter and Saturn there can exist the following consonances, over and above the octave: the major and minor third and the perfect fourth, either one of the thirds through the tuning which maintains the amplitude of the remaining one, but the perfect fourth through the amplitude of a major whole tone. For there will be a perfect fourth not merely from *G* of Saturn to *cc* of Jupiter but also from *A* of Saturn to *dd* of Jupiter and through all the intermediates between the *G* and *A* of Saturn and the *cc* and *dd* of Jupiter. But the octave and the perfect fifth exist solely at the points of the apsides. But Mars, which got a greater interval as its own, received it in order that it should also make an octave with the upper planets through some amplitude of tuning. Mercury received an interval great enough for it to set up almost all the consonances with all the planets within one of its periods, which is not longer than the space of three months. On the other hand, the Earth, and Venus much more so, on account of the smallness of their intervals, limit the consonances, which they form not merely with the others but with one another in especial, to visible fewness. But if three planets are to concord in one harmony, many periodic returns are to be awaited; nevertheless there are many consonances, so that they may so much the more easily take place, while each nearest consonance follows after its neighbour, and very often threefold consonances are seen to exist between Mars, the Earth, and Mercury. But the consonances of four planets now begin to be scattered throughout centuries, and those of five planets throughout thousands of years.

But that all six should be in concord [296] has been fenced about by the longest intervals of time; and I do not know whether it is absolutely impossible for this to occur twice by precise evolving or whether that points to a certain beginning of time, from which every age of the world has flowed.

But if only one sextuple harmony can occur, or only one notable one among many, indubitably that could be taken as a sign of the Creation. Therefore we must ask, in exactly how many forms are the movements of all six planets reduced to one common harmony? The method of inquiry is as follows: let us begin with the Earth and Venus, because these two planets do not make more than two consonances and (wherein the cause of this thing is comprehended) by means of very short intensifications of the movements.

Therefore let us set up two, as it were, skeletal outlines of harmonies, each skeletal outline determined by the two extreme numbers wherewith the limits of the tunings are designated, and let us search out what fits in with them from the variety of movements granted to each planet.

HARMONIES OF ALL THE PLANETS, OR UNIVERSAL HARMONIES IN THE MAJOR MODE

In order that ♭ *may be in concord*			*At gravest Tuning*	*At most acute Tuning*	*[Modern notation*
☿	e⁷ b⁶ g⁶		380'20" 285'15" 228'12"	292'48" 234'16"	5 x 8va
♀	e⁶ e⁵		190'10" 95'5"	195'14" 97'37"	4 x 8va
☉	g⁴ b³		57'3" 35'39"	58'34" 36'36"	2 x 8va
♂	g³		28'32"	29'17"	8va
♃	b			4'34"	
♄	B G		2'14" 1'47"	1'49"	

E. C., Jr.]

In order that c *may be in concord*			*At gravest Tuning*	*At most acute Tuning*	*[Modern notation*
☿	e⁷ c⁷ g⁶		380'20" 204'16" 228'12"	312'21" 234'16"	5 x 8va
♀	e⁶ e⁵		190'10" 95'5"	195'14" 97'37"	4 x 8va
☉	g⁴ c⁴		57'3" 38'2"	58'34" 39'3"	☉ g⁴ b³

673

♂ g³		28'32"	29'17"	*8va*
♃ c¹		4'45"	4'53"	
♭ G		1'47"	1'49"	E. C., Jr.]

E. C., Jr.]

Saturn joins in this universal consonance with its aphelial movement, the Earth with its aphelial, Venus approximately with its aphelial; at highest tuning, Venus joins with its perihelial; at mean tuning, Saturn joins with its perihelial, Jupiter with its aphelial, Mercury with its perihelial. So Saturn can join in with two movements, Mars with two, Mercury with four. But with the rest remaining, the perihelial movement of Saturn and the aphelial of Jupiter are not allowed. But in their place, Mars joins in with perihelial movement.

The remaining planets join in with single movements, Mars alone with two, and Mercury with four.

[297] Accordingly, the second skeletal outline will be that wherein the other possible consonance, 5 : 8, exists between the Earth and Venus. Here one eighth of the 94'50"of the diurnal aphelial movement of Venus or 11'51"+, if multiplied by 5, equals the 59'16" of the movement of the Earth; and similar parts of the 97'37" of the perihelial movement of Venus are equal to the 61'1" of the movement of the Earth. Accordingly, the other planets are in concord in the following diurnal movements:

HARMONIES OF ALL THE PLANETS, OR UNIVERSAL HARMONIES
IN THE MINOR MODE

In order that ♭ may be in concord		At gravest Tuning	At most acute Tuning	[Modern notation
♄ eb⁷		379'20"		5 x 8va
bb⁷		284'32"	295'56"	
g⁶		237'4"	244'4"	
♀ eb⁶		189'40"	195'14"	4 x 8va
eb⁵		94'50"	97'37"	
☿ g⁴		59'16"	61'1"	2 x 8va
bb⁴		35'35"	36'37"	

674

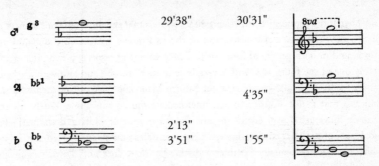

			At gravest	At most acute	
♂	g⁸		29'38"	30'31"	
♃	bb¹			4'35"	
			2'13"		
♄	G bb		3'51"	1'55"	

E. C., Jr.]

Here again, in the mean tuning Saturn joins in with its perihelial movement, Jupiter with its aphelial, Mercury with its perihelial. But at highest tuning approximately the perihelial movement of the Earth joins in.

In order that c *may be in concord* | At gravest Tuning | At most acute Tuning | [Modern notation

		At gravest Tuning	At most acute Tuning	
☿	eb⁷ c⁷ g⁶	379'20" 316'5" 237'4"	325'26" 244'4"	5 x 8va
♀	eb⁶ c⁶ eb⁵	189'40" 94'50"	195'14" 162'43" 97'37"	4 x 8va
♁	g⁴	59'16"	61'1"	2 x 8va
♂	g⁸	29'38"	30'31"	8va
♃	c¹	4'56"	5'5"	8va
♄	G	3'51"	1'55"	

E. C., Jr.]

And here, with the aphelial movement of Jupiter and the perihelial movement of Saturn removed, the aphelial movement of Mercury is practically admitted besides the perihelial. The rest remain.

Therefore astronomical experience bears witness that the universal consonances of all the movements can take place, and in the two modes (*generum*), the major and minor, and in both genera of form, or (if I may say so) in respect to two pitches and in any one of the four cases, with a certain latitude of tuning and also with a certain variety in the particular consonances of Saturn, Mars, and Mercury, of each with the rest; and that is not afforded by the intermediate movements alone, but by all the extreme movements too, except the aphelial movement of Mars and the perihelial movement of Jupiter; because since the former occupies *f* sharp; and the latter, *d* Venus, which occupies perpetually the intermediate *e* flat or *e*, does not allow those neighbouring dissonances in the universal consonance, as she would do if she had space to go beyond *e* or *e* flat. This difficulty is caused by the wedding of the Earth and Venus, or the male and the female. These two planets divide the kinds (*genera*) of consonances into the major and masculine and the minor and feminine, according as the one spouse has gratified the other—namely, either the Earth is in its aphelion, as if preserving [298] its marital dignity and performing works worthy of a man, with Venus removed and pushed away to her perihelion as to her distaff; or else the Earth has kindly allowed her to ascend into aphelion or the Earth itself has descended into its perihelion towards Venus and as it were, into her embrace, for the sake of pleasure, and has laid aside for a while its shield and arms and all the works befitting a man; for at that time the consonance is minor.

But if we command this contradictory Venus to keep quiet, *i.e.*, if we consider what the consonances not of all but merely of the five remaining planets can be, excluding the movement of Venus, the Earth still wanders around its g string and does not ascend a semitone above it. Accordingly *b*-flat, *b*, *c*, *d*, *e*-flat, and *e* can be in concord with *g*, whereupon, as you see, Jupiter, marking the d string with its perihelial movement, is brought in. Accordingly, the difficulty about Mars' aphelial movement remains. For the aphelial movement of the Earth, which occupies *g*, does not allow it on *f* sharp; but the perihelial movement, as was said above in Chapter V, is in discord with the aphelial movement of Mars by about half a diesis.

HARMONIES OF THE FIVE PLANETS, WITH VENUS LEFT OUT

Major mode (Genus durum)		At gravest Tuning	At most acute Tuning	[Modern notation
d⁷ b⁶ g⁶		342'18" 285'15" 228'12"	351'24" 292'48" 234'16"	5 x 8va
♀ in d⁶ discord e⁵		171'9" 95'5"	175'42" 97'37"	4 x 8va

	At gravest tuning	At most acute tuning	[Modern notation:
♂ g⁴ / b⁸	57'3" / 35'39"	58'34" / 36'36"	2 x 8va
g³	28'31"	29'17"	8va
♃ d¹ / b¹	5'21"	5'30" / 4'35"	
♭ B / G	2'13" / 1'47"		

E. C., Jr.]

Here at the most grave tuning, Saturn and the Earth join in with their aphelial movements; at the mean tuning, Saturn with its perihelial and Jupiter with its aphelial; at the most acute, Jupiter with its perihelial.

Minor mode (Genus molle)

	At gravest tuning	At most acute tuning	[Modern notation:
☿ d⁷ / b⁶ / g⁶	342'18" / 273'50" / 228'12"	351'24" / 280'57" / 234'16"	5 x 8va
♀ in d⁶ discord / e⁵	171'9" / 95'5"	175'42" / 97'37"	4 x 8va
♂ g⁴ / b³	57'3" / 34'14"	58'34" / 35'8"	2 x 8va
g³	28'31"	29'17"	8va
♃ d¹	5'21"	5'30"	
♭ B / G	2'8" / 1'47"	2'12" / 1'50"	

E. C., Jr.]

Here the aphelial movement of Jupiter is not allowed, but at the most acute tuning Saturn practically joins in with its perihelial movement.

But there can also exist the following harmony of the four planets, Saturn, Jupiter, Mars, and Mercury, wherein too the aphelial movement of Mars is present, but it is without latitude of tuning.

In order that b may be in concord   *[Modern notation:*

☿ d⁷ b⁶	335'50" 279'52"	
f♯⁶ d⁶	209'52" 167'55"	
♂ b³	34'59"	
f♯³	26'14"	
♃ d¹	5'15"	
♭ B	2'11"	

E. C., Jr.]

In order that a may

in concord
[Modern notation:

☿ d⁷ a⁶	
f♯⁶ d⁶	
♂ a³	
f♯³	

E. C., Jr.]

Accordingly the movements of the heavens are nothing except a certain everlasting polyphony (intelligible, not audible) with dissonant tunings, like certain syncopations or cadences (wherewith men imitate these natural dissonances), which tends towards fixed and prescribed clauses—the single clauses having six terms (like voices)—and which marks out and distinguishes the immensity of time with those notes.[1] Hence it

1. The comparison Kepler draws between the celestial harmonies and the polyphonic music of his time may be clarified by a simple example for four voices from—Palestrina, *O Crux:*

X   *Consonant harmonies*
Y   *Dissonant syncopations*
Z   *Resolutions of dissonances*

Cadence

As will be observed each of the few voices (as it would also be with the six to which Kepler refers) moves from one consonant chord to another while following a graceful melodic line. Sometimes bits of scales or passing tones are added to give a voice more melodic freedom expressiveness. For the same reason a voice may remain on the same note while the other voices change to a new chord. When this becomes a dissonance (called a syncopation) in the new chord it usually resolves by moving one step downward to a tone that is consonant with the other voices. As in this example each section or "clause" ends with a cadence.

E. C., Jr.

is no longer a surprise that man, the ape of his Creator, should finally have discovered the art of singing polyphonically (*per concentum*), which was unknown to the ancients, namely in order that he might play the everlastingness of all created time in some short part of an hour by means of an artistic concord of many voices and that he might to some extent taste the satisfaction of God the Workman with His own works, in that very sweet sense of delight elicited from this music which imitates God.

8. IN THE CELESTIAL HARMONIES WHICH PLANET SINGS SOPRANO, WHICH ALTO, WHICH TENOR, AND WHICH BASS?

Although these words are applied to human voices, while voices or sounds do not exist in the heavens, on account of the very great tranquillity of movements, and not even the subjects in which we find the consonances are comprehended under the true genus of movement, since we were considering the movements solely as apparent from the sun, and finally, although there is no such cause in the heavens, as in human singing, for requiring a definite number of voices in order to make consonance (for first there was the number of the six planets revolving around the sun, from the number of the five intervals taken from the regular figures, and then afterwards—in the order of nature, not of time—the congruence of the movements was settled): I do not know why but nevertheless this wonderful congruence with human song has such a strong effect upon me that I am compelled to pursue this part of the comparison, also, even without any solid natural cause. For those same properties which in Book III, [300] Chapter 16, custom ascribed to the bass and nature gave legal grounds for so doing are somehow possessed by Saturn and Jupiter in the heavens; and we find those of the tenor in Mars, those of the alto are present in the Earth and Venus, and those of the soprano are possessed by Mercury, if not with equality of intervals, at least proportionately. For howsoever in the following chapter the eccentricities of each planet are deduced from their proper causes and through those eccentricities the intervals proper to the movements of each, none the less there comes from that the following wonderful result (I do not know whether it is occasioned by the procurement and mere tempering of necessities): (1) as the bass is opposed to the alto, so there are two planets which have the nature of the alto, two that of the bass, just as in any Mode of song there is one (bass and one alto) on either side, while there are single representatives of the other single voices. (2) As the alto is practically supreme in a very narrow range (*in angustiis*) on account of necessary and natural causes unfolded in Book III, so the almost innermost planets, the Earth and Venus, have the narrowest intervals of movements, the Earth not much more than a semitone, Venus not even a diesis. (3) And as the tenor is free, but none the less progresses with moderation, so Mars alone—with

the single exception of Mercury—can make the greatest interval, namely a perfect fifth. (4) And as the bass makes harmonic leaps, so Saturn and Jupiter have intervals which are harmonic, and in relation to one another pass from the octave to the octave and perfect fifth. (5) And as the soprano is the freest, more than all the rest, and likewise the swiftest, so Mercury can traverse more than an octave in the shortest period. But this is altogether *per accidens*; now let us hear the reasons for the eccentricities.

9. THE GENESIS OF THE ECCENTRICITIES IN THE SINGLE PLANETS FROM THE PROCUREMENT OF THE CONSONANCES BETWEEN THEIR MOVEMENTS

Accordingly, since we see that the universal harmonies of all six planets cannot take place by chance, especially in the case of the extreme movements, all of which we see concur in the universal harmonies—except two, which concur in harmonies closest to the universal—and since much less can it happen by chance that all the pitches of the system of the octave (as set up in Book III) by means of harmonic divisions are designated by the extreme planetary movements, but least of all that the very subtle business of the distinction of the celestial consonances into two modes, the major and minor, should be the outcome of chance, without the special attention of the Artisan: accordingly it follows that the Creator, the source of all wisdom, the everlasting approver of order, the eternal and superexistent geyser of geometry and harmony, it follows, I say, that He, the Artisan of the celestial movements Himself, should have conjoined to the five regular solids the harmonic ratios arising from the regular plane figures, and out of both classes should have formed one most perfect archetype of the heavens: in order that in this archetype, as through the five regular solids the shapes of the spheres shine through on which the six planets are carried, so too through the consonances, which are generated from the plane figures, and deduced from them in Book III, the measures of the eccentricities in the single planets might be determined so as to proportion the movements of the planetary bodies; and in order that there should be one tempering together of the ratios and the consonances, and that the greater ratios of the spheres should yield somewhat to the lesser ratios of the eccentricities necessary for procuring the consonances, and conversely those in especial of the harmonic ratios which had a greater kinship with each solid figure should be adjusted to the planets—in so far as that could be effected by means of consonances. And in order that, finally, in that way both the ratios of the spheres and the eccentricities of the single planets might be born of the archetype simultaneously, while from the amplitude of the spheres and the bulk of the bodies the periodic times of the single planets might result.

[301] While I struggle to bring forth this process into the light of human intellect by means of the elementary form customary with geometers, may the Author of the

heavens be favourable, the Father of intellects, the Bestower of mortal senses, Himself immortal and superblessed, and may He prevent the darkness of our mind from bringing forth in this work anything unworthy of His Majesty, and may He effect that we, the imitators of God by the help of the Holy Ghost, should rival the perfection of His works in sanctity of life, for which He choose His church throughout the Earth and, by the blood of His Son, cleansed it from sins, and that we should keep at a distance all the discords of enmity, all contentions, rivalries, anger, quarrels, dissensions, sects, envy, provocations, and irritations arising through mocking speech and the other works of the flesh; and that along with myself, all who possess the spirit of Christ will not only desire but will also strive by deeds to express and make sure their calling, by spurning all crooked morals of all kinds which have been veiled and painted over with the cloak of zeal or of the love of truth or of singular erudition or modesty over against contentious teachers, or with any other showy garment. Holy Father, keep us safe in the concord of our love for one another, that we may be one, just as Thou art one with Thine Son, Our Lord, and with the Holy Ghost, and just as through the sweetest bonds of harmonies Thou hast made all Thy works one; and that from the bringing of Thy people into concord the body of Thy Church may be rebuilt up in the Earth, as Thou didst erect the heavens themselves out of harmonies.

PRIOR REASONS

I. Axiom. *It is reasonable that, wherever in general it could have been done, all possible harmonies were due to have been set up between the extreme movements of the planets taken singly and by twos, in order that that variety should adorn the world.*

II. Axiom. *The five intervals between the six spheres to some extent were due to correspond to the ratio of the geometrical spheres which inscribe and circumscribe the five regular solids, and in the same order which is natural to the figures.*

Concerning this, see Chapter 1 and the *Mysterium Cosmographicum* and the *Epitome of Copernican Astronomy*.

III. Proposition. *The intervals between the Earth and Mars, and between the Earth and Venus, were due to be least, in proportion to their spheres, and thereby approximately equal; middling and approximately equal between Saturn and Jupiter, and between Venus and Mercury; but greatest between Jupiter and Mars.*

For by Axiom II, the planets corresponding in position to the figures which make the least ratio of geometrical spheres ought likewise to make the least ratio; but those which correspond to the figures of middling ratio ought to make the greatest; and those which correspond to the figures of greatest ratio, the greatest. But the order holding between the figures of the dodecahedron and the icosahedron is the same as that between the pairs of planets, Mars and the Earth, and the Earth and Venus, and the

order of the cube and octahedron is the same as that of the pair Saturn and Jupiter and that of the pair Venus and Mercury; and, finally, the order of the tetrahedron is the same as that of the pair Jupiter and Mars (see Chapter 3). Therefore, the least ratio will hold between the planetary spheres first mentioned, while that between Saturn and Jupiter is approximately equal to that between Venus and Mercury; and, finally, the greatest between the spheres of Jupiter and Mars.

IV. Axiom. *All the planets ought to have their eccentricities diverse, no less than a movement in latitude, and in proportion to those eccentricities also their distances from the sun, the source of movement, diverse.*

As the essence of movement consists not in *being* but in *becoming*, so too the form or figure of the region which any planet traverses in its movement does not become solid immediately from the start but in the succession of time acquires at last not only length but also breadth and depth (its perfect ternary of dimensions); and, gradually, thus, by the interweaving and piling up of many circuits, the form of a concave sphere comes to be represented—just as out of the silk-worm's thread, by the interweaving and heaping together of many circles, the cocoon is built.

V. Proposition. *Two diverse consonances were to have been attributed to each pair of neighbouring planets.*

For, by Axiom IV, any planet has a longest and a shortest distance from the sun, wherefore, by Chapter 3, it will have both a slowest movement and a fastest. Therefore, there are two primary comparisons of the extreme movements, one of the diverging movements in the two planets, and the other of the converging. Now it is necessary that they be diverse from one another, because the ratio of the diverging movements will be greater, that of the converging, lesser. But, moreover, diverse consonances had to exist by way of diverse pairs of planets, so that this variety should make for the adornment of the world—by Axiom I—and also because the ratios of the intervals between two planets are diverse, by Proposition III. But to each definite ratio of the spheres there correspond harmonic ratios, in quantitative kinship, as has been demonstrated in Chapter 5 of this book.

VI. Proposition. *The two least consonances, 4 : 5 and 5 : 6, do not have a place between two planets.*

For

$$5 : 4 = 1{,}000 : 800$$

and

$$6 : 5 = 1{,}000 : 833.$$

But the spheres circumscribed around the dodecahedron and icosahedron have a greater ratio to the inscribed spheres than 1,000 : 795, etc., and these two ratios indicate the intervals between the nearest planetary spheres, or the least distances. For in

the other regular solids the spheres are farther distant from one another. But now the ratio of the movements is even greater than the ratios of the intervals, unless the ratio of the eccentricities to the spheres is vast—by Article XIII of Chapter 3. Therefore the least ratio of the movements is greater than 4 : 5 and 5 : 6. Accordingly, these consonances, being hindered by the regular solids, receive no place among the planets.

VII. Proposition. *The consonance of the perfect fourth can have no place between the converging movements of two planets, unless the ratios of the extreme movements proper to them are, if compounded, more than a perfect fifth.*

For let 3 : 4 be the ratio between the converging movements. And first, let there be no eccentricity, no ratio of movements proper to the single planets, but both the converging and the mean movements the same; then it follows that the corresponding intervals, which by this hypothesis will be the semidiameters of the spheres, constitute the $^2/_3$d power of this ratio, *viz.*, 4480 : 5424 (by Chapter 3). But this ratio is already less than the ratio of the spheres of any regular figure; and so the whole inner sphere would be cut by the regular planes of the figure inscribed in any outer sphere. But this is contrary to Axiom II.

Secondly, let there be some composition of the ratios between the extreme movements, and let the ratio of the converging movements be 3 : 4 or 75 : 100, but let the ratio of the corresponding intervals be 1,000 : 795, since no regular figure has a lesser ratio of spheres. And because the inverse ratio of the movements exceeds this ratio of the intervals by the excess 750 : 795, then if this excess is divided into the ratio 1,000 : 795, according to the doctrine of Chapter 3, the result will be 9434 : 7950, the square root of the ratio of the spheres. Therefore the square of this ratio, *viz.*, 8901 : 6320, *i.e.*, 10,000 : 7,100 is the ratio of the spheres. Divide this by 1000 : 795, the ratio of the converging intervals, the result will be 7100 : 7950, about a major whole tone. The compound of the two ratios which the mean movements have to the converging movements on either side must be at least so great, in order that the perfect fourth may be possible between the converging movements. Accordingly, the compound ratio of the diverging extreme intervals to the converging extreme intervals is about the square root of this ratio, *i.e.*, two tones, and again the converging intervals are the square of this, *i.e.*, more than a perfect fifth. Accordingly, if the compound of the proper movements of two neighbouring planets is less than a perfect fifth, a perfect fourth will not be possible between their converging movements.

VIII. Proposition. *The consonances 1 : 2 and 1 : 3, i.e., the octave and the octave plus a fifth were due to Saturn and Jupiter.*

For they are the first and highest of the planets and have obtained the first figure, the cube, by Chapter 1 of this book; and these consonances are first in the order of nature and are chief in the two families of figures, the bisectorial or tetragonal and the

triangular, by what has been said in Book I. But that which is chief, the octave 1 : 2, is approximately greater than the ratio of the spheres of the cube, [303] which is 1:√3; wherefore it is fitted to become the lesser ratio of the movements of the planets on the cube, by Chapter 3, Article XIII; and, as a consequence, 1 : 3 serves as the greater ratio.

But this is also the same as what follows: for if some consonance is to some ratio of the spheres of the figures, as the ratio of the movements apparent from the sun is to the ratio of the mean intervals, such a consonance will duly be attributed to the movements. But it is natural that the ratio of the diverging movements should be much greater than the ratio of the $^3/_2$th powers of the spheres, according to the end of Chapter 3, *i.e.*, it approaches the square of the ratio of the spheres; and moreover 1 : 3 is the square of the ratio of the spheres of the cube, which we call the ratio of 1 : √3. Therefore, the ratio of the diverging movements of Saturn and Jupiter is 1 : 3. (See above, Chapter 2, for many other kinships of these ratios with the cube.)

IX. Proposition. *The private ratios of the extreme movements of Saturn and Jupiter compounded were due to be approximately 2 : 3, a perfect fifth.*

This follows from the preceding; if the perihelial movement of Jupiter is triple the aphelial movement of Saturn, and conversely the aphelial movement of Jupiter is double the perihelial of Saturn, then 1 : 2 and 1 : 3 compounded inversely give 2 : 3.

X. Axiom. *When choice is free in other respects, the private ratio of movements, which is prior in nature or of a more excellent mode or even which is greater, is due to the higher planet.*

XI. Proposition. *The ratio of the aphelial movement of Saturn to the perihelial was due to be 4 : 5, a major third, but that of Jupiter's movements 5 : 6, a minor third.*

For as compounded together they are equivalent to 2 : 3; but 2 : 3 can be divided harmonically no other way than into 4 : 5 and 5 : 6. Accordingly God the composer of harmonies divided harmonically the consonance 2 : 3, (by Axiom I) and the harmonic part of it which is greater and of the more excellent major mode, as masculine, He gave to Saturn the greater and higher planet, and the lesser ratio 5 : 6 to the lower one, Jupiter (by Axiom X).

XII. Proposition. *The great consonance of 1 : 4, the double octave, was due to Venus and Mercury.*

For as the cube is the first of the primary figures, so the octahedron is the first of the secondary figures, by Chapter 1 of this book. And as the cube considered geometrically is outer and the octahedron is inner, *i.e.*, the latter can be inscribed in the former, so also in the world Saturn and Jupiter are the beginning of the upper and outer planets, or from the outside; and Mercury and Venus are the beginning of the inner planets, or from the inside, and the octahedron has been placed between their circuits: (see Chapter 3). Therefore, from among the consonances, one which is primary and

cognate to the octahedron is due to Venus and Mercury. Furthermore, from among the consonances, after 1 : 2 and 1 : 3, there follows in natural order 1 : 4; and that is cognate to 1 : 2, the consonance of the cube, because it has arisen from the same cut of figures, *viz.*, the tetragonal, and is commensurable with it, *viz.*, the double of it; while the octahedron is also akin to, and commensurable with the cube. Moreover, 1 : 4 is cognate to the octahedron for a special reason, on account of the number four being in that ratio, while a quadrangular figure lies concealed in the octahedron and the ratio of its spheres is said to be 1 : √2.

Accordingly the consonance 1 : 4 is a continued power of this ratio, in the ratio of the squares, *i.e.*, the 4th power of 1 : √2 (see Chapter 2). Therefore, 1 : 4 was due to Venus and Mercury. And because in the cube 1 : 2 has been made the smaller consonance of the two, since the outermost position is over against it, in the octahedron there will be 1 : 4, the greater consonance of the two, as the innermost position is over against it. But too, this is the reason why 1 : 4 has here been given as the greater consonance, not as the smaller.[1] For since the ratio of the spheres of the octahedron is the ratio of 1 : √3, then if it is postulated that the inscription of the octahedron among the planets is perfect (although it is not perfect, but penetrates Mercury's sphere to some extent—which is of advantage to us): accordingly, the ratio of the converging movements must be less than the $^3/_2$th powers of 1 : √3; but indeed 1 : 3 is plainly the square of the ratio 1 : √3 and is thus greater than the exact ratio; all the more then will 1 : 4 be greater than the exact ratio, as greater than 1 : 3. Therefore, not even the square root of 1 : 4 is allowed between the converging movements. Accordingly, 1 : 4 cannot be less than the octahedric; so it will be greater.

Further: 1 : 4 is akin to the octahedric square, where the ratio of the inscribed and circumscribed circles is 1 : √2, just as 1 : 3 is akin to the cube, where the ratio of the spheres is 1 : √3. For as 1 : 3 is a power of 1 : √3, *viz.*, its square, [304] so too here 1:4 is a power of 1 : √2, *viz.*, twice its square, *i.e.*, its quadruple power. Wherefore, if 1:3 was due to have been the greater consonance of the cube (by Proposition VII), accordingly 1 : 4 ought to become the greater consonance of its octahedron.

XIII. Proposition. *The greater consonance of approximately 1 : 8, the triple octave, and the smaller consonance of 5 : 24, the minor third and double octave, were due to the extreme movements of Jupiter and Mars.*

For the cube has obtained 1 : 2 and 1 : 3, while the ratio of the spheres of the tetrahedron, which is situated between Jupiter and Mars, called the triple ratio, is the square of the ratio of the spheres of the cube, which is called the ratio of 1 : √3. Therefore, it was proper that ratios of movements which are the squares of the cubic ratios should

---

1. *Smaller* (lesser) and *greater* consonances are equivalent to our modern "more closely spaced" and "more widely spaced" consonances. E. C., Jr.

be applied to the tetrahedron. But of the ratios 1 : 2 and 1 : 3 the following ratios are the squares: 1 : 4 and 1 : 9. But 1 : 9 is not harmonic, and 1 : 4 has already been used up in the octahedron. Accordingly, consonances neighbouring upon these ratios were to have been taken, by Axiom I. But the lesser ratio 1 : 8 and the greater 1 : 10 are the nearest. Choice between these ratios is determined by kinship with the tetrahedron, which has nothing in common with the pentagon, since 1 : 10 is of a pentagonal cut, but the tetrahedron has greater kinship with 1 : 8 for many reasons (see Chapter 2).

Further, the following also makes for 1 : 8: just as 1 : 3 is the greater consonance of the cube and 1 : 4 the greater consonance of the octahedron, because they are powers of the ratios between the spheres of the figures, so too 1 : 8 was due to be the greater consonance of the tetrahedron, because as its body is double that of the octahedron inscribed in it, as has been said in Chapter 1, so too the term 8 in the tetrahedral ratio is double the term 4 in the tetrahedral ratio.

Further, just as 1 : 2 the smaller consonance of the cube, is one octave, and 1 : 4, the greater consonance of the octahedron, is two octaves, so already 1 : 8, the greater consonance of the tetrahedron, was due to be three octaves. Moreover, more octaves were due to the tetrahedron than to the cube and octahedron, because, since the smaller tetrahedral consonance is necessarily greater than all the lesser consonances in the other figures (for the ratio of the tetrahedral spheres is greater than all the spheres of figures): too the greater tetrahedral consonance was due to exceed the greater consonances of the others in number of octaves. Finally, the triple of octave intervals has kinship with the triangular form of the tetrahedron, and has a certain perfection, as follows: every three is perfect; since even the octuple, the term (of the triple octave), is the first cubic number of perfect quantity, namely of three dimensions.

A greater consonance neighbouring upon 1 : 4 or 6 : 24 is 5 : 24, while a lesser is 6 : 20 or 3 : 10. But again 3 : 10 is of the pentagonal cut, which has nothing in common with the tetrahedron. But on account of the numbers 3 and 4 (from which the numbers 12, 24 arise) 5 : 24 has kinship with the tetrahedron. For we are here neglecting the other lesser terms, *viz.*, 5 and 3, because their lightest degree of kinship is with figures, as it is possible to see in Chapter 2. Moreover, the ratio of the spheres of the tetrahedron is triple; but the ratio of the converging intervals too ought to be approximately so great, by Axiom II. By Chapter 3, the ratio of the converging movements approaches the inverse ratio of the $3/_2$th powers of the intervals, but the $3/_2$th power of 3 : 1 is approximately 1000 : 193. Accordingly, whereof the aphelial movement of Mars is 1000, the (perihelial) of Jupiter will be slightly greater than 193 but much less than 333, which is one third of 1,000. Accordingly, not the consonance 10 : 3, *i.e.*, 1,000 : 333, but the consonance 24 : 5, *i.e.*, 1,000 : 208, takes place between the converging movements of Jupiter and Mars.

XIV. Proposition. *The private ratio of the extreme movements of Mars was due to be greater than 3 : 4, the perfect fourth, and approximately 18 : 25.*

For let there be the exact consonances 5 : 24 and 1 : 8 or 3 : 24, which are commonly attributed to Jupiter and Mars (Proposition XIII). Compound inversely 5 : 24, the lesser with 3 : 24, the greater; 3 : 5 results as the compound of both ratios. But the proper ratio of Jupiter alone has been found to be 5 : 6, in Proposition XI, above. Then compound this inversely with the composition 3 : 5, *i.e.*, compound 30 : 25 and 18:30; there results as the proper ratio of Mars 18 : 25, which is greater than 18 : 24 or 3 : 4. But it will become still greater, if, on account of the ensuing reasons, the common greater consonance 1 : 8 is increased.

XV. Proposition. *The consonances 2 : 3, the fifth; 5 : 8, the minor sixth; and 3 : 5, the major sixth were to have been distributed among the converging movements of Mars and the Earth, the Earth and Venus, Venus and Mercury, and in that order.*

For the dodecahedron and the icosahedron, the figures interspaced between Mars, the Earth, and Venus have the least ratio between their circumscribed and inscribed spheres. [305] Therefore from among possible consonances the least are due to them, as being cognate for this reason, and in order that Axiom II may have place. But the least consonances of all, *viz.*, 5 : 6 and 4 : 5, are not possible, by Proposition IV. Therefore, the nearest consonances greater than they, *viz.*, 3 : 4 or 2 : 3 or 5 : 8 or 3:5 are due to the said figures.

Again, the figure placed between Venus and Mercury, *viz.*, the octahedron, has the same ratio of its spheres as the cube. But by Proposition VII, the cube received the octave as the lesser consonance existing between the converging movements. Therefore, by proportionality, so great a consonance, *viz.*, 1 : 2, would be due to the octahedron as the lesser consonance, if no diversity intervened. But the following diversity intervenes: if compounded together, the private ratios of the single movements of the cubic planets, *viz.*, Saturn and Jupiter, did not amount to more than 2 : 3; while, if compounded, the ratios of the single movements of the octahedral planets, *viz.*, Venus and Mercury will amount to more than 2 : 3, as is apparent easily, as follows: For, as the proportion between the cube and octahedron would require if it were alone, let the lesser octahedral ratio be greater than the ratios here given, and thereby clearly as great as was the cubic ratio, *viz.*, 1 : 2; but the greater consonance was 1 : 4, by Proposition XII. Therefore if the lesser consonance 1 : 2 is divided into the one we have just laid down, 1 : 2, still remains as the compound of the proper movements of Venus and Mercury; but 1 : 2 is greater than 2 : 3 the compound of the proper movements of Saturn and Jupiter; and indeed a greater eccentricity follows upon this greater compound, by Chapter 3, but a lesser ratio of the converging movements follows upon the greater eccentricity, by the same Chapter 3. Wherefore by the addition of a greater

eccentricity to the proportion between the cube and the octahedron it comes about that a lesser ratio than 1 : 2 is also required between the converging movements of Venus and Mercury. Moreover, it was in keeping with Axiom I that, with the consonance of the octave given to the planets of the cube, another consonance which is very near (and by the earlier demonstration less than 1 : 2) should be joined to the planets of the octahedron. But 3 : 5 is proximately less than 1 : 2, and as the greatest of the three it was due to the figure having the greatest ratio of its spheres, *viz.*, the octahedron. Accordingly, the lesser ratios, 5 : 8 and 2 : 3 or 3 : 4, were left for the icosahedron and dodecahedron, the figures having a lesser ratio of their spheres.

But these remaining ratios have been distributed between the two remaining planets, as follows. For as, from among the figures, though of equal ratios between their spheres, the cube has received the consonance 1 : 2, while the octahedron the lesser consonance 3 : 5, in that the compound ratio of the private movements of Venus and Mercury exceeded the compound ratio of the private movements of Saturn and Jupiter; so also although the dodecahedron has the same ratio of its spheres as the icosahedron, a lesser ratio was due to it than to the icosahedron, but very close on account of a similar reason, *viz.*, because this figure is between the Earth and Mars, which had a great eccentricity in the foregoing. But Venus and Mercury, as we shall hear in the following, have the least eccentricities. But since the octahedron has 3 : 5, the icosahedron, whose species are in a lesser ratio, has the next slightly lesser, *viz.*, 5 : 8; accordingly, either 2 : 3, which remains, or 3 : 4 was left for the dodecahedron, but more likely 2:3, as being nearer to the icosahedral 5 : 8; since they are similar figures.

But 3 : 4 indeed was not possible. For although, in the foregoing, the private ratio of the extreme movements of Mars was great enough, yet the Earth—as has already been said and will be made clear in what follows—contributed its own ratio, which was too small for the compound ratio of both to exceed the perfect fifth. Accordingly, Proposition VII, 3 : 4 could not have place. And all the more so, because—as will follow in Proposition XVII—the ratio of the converging intervals was due to be greater than 1,000 : 795.

XVI. Proposition. *The private ratios of movements of Venus and Mercury, if compounded together, were due to make approximately 5 : 12.*

For divide the lesser harmonic ratio attributed in Proposition XV to this pair jointly into the greater of them, 1 : 4 or 3 : 12, by Proposition XII; there results 5 : 12, the compound ratio of the private movements of both. And so the private ratio of the extreme movements of Mercury alone is less than 5 : 12, the magnitude of the private movement of Venus. Understand this of these first reasons. For below, by the second reasons, through the addition of some variation to the joint consonances of both, it results that only the private ratio of Mercury is perfectly 5 : 12.

XVII. Proposition. *The consonance between the diverging movements of Venus and the Earth could not be less than 5 : 12.*

For in the private ratio of its movements Mars alone has received more than the perfect fourth and more than 18 : 25, by Proposition XIV. But their lesser consonance is the perfect fifth, [306] by Proposition XV. Accordingly, the ratio compounded of these two parts is 12 : 25. But its own private ratio is due to the Earth, by Axiom IV. Therefore, since the consonance of the diverging movements is made up out of the said three elements, it will be greater than 12 : 25. But the nearest consonance greater than 12 : 25, *i.e.*, 60 : 125, is 5 : 12, *viz.*, 60 : 144. Wherefore, if there is need of a consonance for this greater ratio of the two planets, by Axiom I, it cannot be less than 60:144 or 5 : 12.

Therefore up to now all the remaining pairs of planets have received their two consonances by necessary reasons; the pair of the Earth and Venus alone has as yet been allotted only one consonance, 5 : 8, by the axioms so far employed. Therefore, we must now take a new start and inquire into its remaining consonance, *viz.*, the greater, or the consonance of the diverging movements.

POSTERIOR REASONS

XVIII. Axiom. *The universal consonances of movements were to be constituted by a tempering of the six movements, especially in the case of the extreme movements.*

This is proved by Axiom I.

XIX. Axiom. *The universal consonances had to come out the same within a certain latitude of movements, namely, in order that they should occur the more frequently.*

For if they had been limited to indivisible points of the movements, it could have happened that they would never occur, or very rarely.

XX. Axiom. *As the most natural division of the kinds* (generum) *of consonances is into major and minor, as has been proved in Book 3, so the universal consonances of both kinds had to be procured between the extreme movements of the planets.*

XXI. Axiom. *Diverse species of both kinds of consonances had to be instituted, so that the beauty of the world might well be composed out of all possible forms of variety—and by means of the extreme movements, at least by means of some extreme movements.*

By Axiom I.

XXII. Proposition. *The extreme movements of the planets had to designate pitches or strings* (chordas) *of the octave system, or notes* (claves) *of as musical scale.*

For the genesis and comparison of consonances beginning from one common term has generated the musical scale, or the division of the octave into its pitches or tones (*sonos*), as has been proved in Book 3. Accordingly, since varied consonances between

the extremes of movements are required, by Axioms I, XX, and XXI, wherefore the real division of some celestial system or harmonic scale by the extremes of movements is required.

XXIII. Proposition. *It was necessary for there to be one pair of planets, between the movements of which no consonances could exist except the major sixth 3 : 5 and the minor sixth 5 : 8.*

For since the division into kinds of consonances was necessary, by Axiom XX, and by means of the extreme movements at the apsides, by XXII, because solely the extremes, *viz.*, the slowest and the fastest, need the determination of a manager and orderer, the intermediate tensions come of themselves, without any special care, with the passage of the planet from the slowest movement to the fastest: accordingly, this ordering could not take place otherwise than by having the diesis or 24 : 25 designated by the extremes of the two planetary movements, in that the kinds of consonances are distinguished by the diesis, as was unfolded in Book 3.

But the diesis is the difference either between two thirds, 4 : 5 and 5 : 6, or between two sixths, 3 : 5 and 5 : 8, or between those ratios increased by one or more octave intervals. But the two thirds, 4 : 5 and 5 : 6, did not have place between two planets, by Proposition VI, and neither the thirds nor the sixths increased by the interval of an octave have been found, except 5 : 12 in the pair of Mars and the Earth, and still not otherwise than along with the related 2 : 3, and so the intermediate ratios 5:8 and 3 : 5 and 1 : 2 were alike admitted. Therefore, it remains that the two sixths, 3:5 and 5 : 8, were to be given to one pair of planets. But too the sixths alone were to be granted to the variation of their movements, in such fashion that they would neither expand their term to the proximately greater interval of one octave, 1 : 2, [307] nor contract them to the narrows of the proximately lesser interval of the fifth, 2 : 3. For, although it is true that the same two planets, which make a perfect fifth with their extreme converging movements, can also make sixths and thus traverse the diesis too, still this would not smell of the singular providence of the Orderer of movements. For the diesis, the least interval—which is potentially latent in all the major intervals comprehended by the extreme movements—is itself at that time traversed by the intermediate movements varied by continuous tension, but it is not determined by their extremes, since the part is always less than the whole, *viz.*, the diesis than the greater interval 3 : 4 which exists between 2 : 3 and 1 : 2 and which whole would be here assumed to be determined by the extreme movements.

XXIV. Proposition. *The two planets which shift the kind (genus) of harmony, which is the difference between the private ratios of the extreme movements, ought to make a diesis, and the private ratio of one ought to be greater than a diesis, and they ought to make one of the sixths with their aphelial movements and the other with their perihelial.*

For, since the extremes of the movements make two consonances differing by a single diesis, that can take place in three ways. For either the movement of one planet will remain constant and the movement of the other will vary by a diesis, or both will vary by half a diesis and make 3 : 5, a major sixth, when the upper is at its aphelion and the lower in its perihelion, and when they move out of those intervals and advance towards one another, the upper into its perihelion and the lower into its aphelion, they make 5 : 8, a minor sixth; or, finally, one varies its movement from aphelion to perihelion more than the other does, and there is an excess of one diesis, and thus there is a major sixth between the two aphelia, and a minor sixth between the two perihelia. But the first way is not legitimate, for one of these planets would be without eccentricity, contrary to Axiom IV. The second way was less beautiful and less expedient; less beautiful, because less harmonic, for the private ratios of the movements of the two planets would have been out of tune (*inconcinnae*), for whatever is less than a diesis is out of tune; moreover it occasions one single planet to labour under this ill-concordant small difference—except that indeed it could not take place, because in this way the extreme movements would have wandered from the pitches of the system or the notes (*clavibus*) of the musical scale, contrary to Proposition XXII. Moreover, it would have been less expedient, because the sixths would have occurred only at those moments in which the planets would have been at the contrary apsides; there would have been no latitude within which these sixths and the universal consonances related to them could have occurred; accordingly, these universal consonances would have been very rare, with all the (*harmonic*) positions of the planets reduced to the narrow limits of definite and single points on their orbits, contrary to Axiom XIX. Accordingly, the third way remains: that both of the planets should vary their own private movements, but one more than the other, by one full diesis at the least.

XXV. Proposition. *The higher of the planets which shift the kind of harmony ought to have the ratio of its private movements less than a minor whole tone 9 : 10; while the lower, less than a semitone 15 : 16.*

For they will make 3 : 5 either with their aphelial movements or with their perihelial, by the foregoing proposition. Not with their perihelial, for then the ratio of their aphelial movements would be 5 : 8. Accordingly, the lower planet would have its private ratio one diesis more than the upper would, by the same foregoing proposition. But that is contrary to Axiom X. Accordingly, they make 3 : 5 with their aphelial movements, and with their perihelial 5 : 8, which is 24 : 25 less than the other. But if the aphelial movements make 3 : 5, a major sixth, therefore, the aphelial movement of the upper together with the perihelial of the lower will make more than a major sixth; for the lower planet will compound directly its full private ratio.

In the same way, if the perihelial movements make 5 : 8, a minor sixth, the perihelial

movement of the upper and the aphelial movement of the lower will make less than a minor sixth; for the lower planet will compound inversely its full private ratio. But if the private ratio of the lower equalled the semitone 15 : 16, then too a perfect fifth could occur over and above the sixths, because the minor sixth, diminished by a semitone, because the perfect fifth; but this is contrary to Proposition XXIII. Accordingly, the lower planet has less than a semitone in its own interval. And because the private ratio of the upper is one diesis greater than the private ratio of the lower, but the diesis compounded with the semitone makes 9 : 10 the minor whole tone.

XXVI. Proposition. *On the planets which shift the kind of harmony, the upper was due to have either a diesis squared, 576 : 625, i.e., approximately 12 : 13, as* [308] *the interval made by its extreme movements, or the semitone 15 : 16, or something intermediate differing by the comma 80 : 81 either from the former or the latter; while the lower planet, either the simple diesis 24 : 25, or the difference between a semitone and a diesis, which is 125 : 128, i.e., approximately 42 : 43; or, finally and similarly, something intermediate differing either from the former or from the latter by the comma 80 : 81, viz., the upper planet ought to make the diesis squared diminished by a comma, and the lower, the simple diesis diminished by a comma.*

For, by Proposition XXV, the private ratio of the upper ought to be greater than a diesis, but by the preceding proposition less than the (minor) whole tone 9 : 10. But indeed the upper planet ought to exceed the lower by one diesis, by Proposition XXIV. And harmonic beauty persuades us that, even if the private ratios of these planets cannot be harmonic, on account of their smallness, they should at least be from among the concordant (*ex concinnis*) if that is possible, by Axiom I. But there are only two concords less than 9 : 10, the (minor) whole tone, *viz.*, the semitone and the diesis; but they differ from one another not by the diesis but by some smaller interval, 125 : 128. Accordingly, the upper cannot have the semitone; nor the lower, the diesis; but either the upper will have the semitone 15 : 16, and the lower, 125 : 128, *i.e.*, 42 : 43; or else the lower will have the diesis 24 : 25, but the upper the diesis squared, approximately 12 : 13. But since the laws of both planets are equal, therefore, if the nature of the concordant had to be violated in their private ratios, it had to be violated equally in both, so that the difference between their private intervals could remain an exact diesis, which is necessary for distinguishing the kinds of consonances, by Proposition XXIV. But the nature of the concordant was then violated equally in both, if the interval whereby the private ratio of the upper planet fell short of the diesis squared and exceeded the semitone is the same interval whereby the private ratio of the lower planet fell short of a simple diesis and exceeded the interval 125 : 128.

Furthermore, this excess or defect was due to be the comma 80 : 81, because, once more, no other interval was designated by the harmonic ratios, and in order that the

comma might be expressed among the celestial movements as it is expressed in harmonics, namely, by the mere excess and defect of the intervals in respect to one another. For in harmonies the comma distinguishes between major and minor whole tones and does not appear in any other way.

It remains for us to inquire which ones of the intervals set forth are preferable—whether the diesis, the simple diesis for the lower planet and the diesis squared for the upper, or the semitone for the upper and 125 : 128 for the lower. And the dieses win by the following arguments: For although the semitone has been variously expressed in the musical scale, yet its allied ratio 125 : 128 has not been expressed. On the other hand, the diesis has been expressed variously and the diesis squared somehow, *viz.*, in the resolution of whole tones into dieses, semitones, and lemmas; for then, as has been said in Book III, Chapter 8, two dieses proximately succeed one another in two pitches. The other argument is that in the distinction into kinds, the laws of the diesis are proper but not at all those of the semitone. Accordingly, there had to be greater consideration of the diesis than of the semitone. It is inferred from everything that the private ratio of the upper planet ought to be 2916 : 3125 or approximately 14 : 15, and that of the lower, 243 : 250 or approximately 35 : 36.

It is asked whether the Highest Creative Wisdom has been occupied in making these tenuous little reckonings. I answer that it is possible that many reasons are hidden from me, but if the nature of harmony has not allowed weightier reasons—since we are dealing with ratios which descend below the magnitude of all concords—it is not absurd that God has followed even those reasons, wherever they appear tenuous, since He has ordained nothing without cause. It would be far more absurd to assert that God has taken at random these magnitudes below the limits prescribed for them, the minor whole tone; and it is not sufficient to say: He took them of that magnitude because He chose to do so. For in geometrical things, which are subject to free choice, God chose nothing without a geometrical cause of some sort, as is apparent in the edges of leaves, in the scales of fishes, in the skins of beasts and their spots and the order of the spots, and similar things.

XXVII. Proposition. *The ratio of movements of the Earth and Venus ought to have been greater than a major sixth between the aphelial movements; less than a minor sixth between the perihelial movements.*

By Axiom XX it was necessary to distinguish the kinds of consonances. But by Proposition XXIII that could not be done except through the sixths. Accordingly, since by Proposition XV the Earth and Venus, planets next to one another and icosahedral, had received the minor sixth, 5 : 8, it was necessary for the other sixth, 3 : 5, to be assigned to them, but not between the converging or diverging extremes, but between the extremes of the same field, one sixth [309] between the aphelial, and the other

between the perihelial, by Proposition XXIV. Furthermore, the consonance 3 : 5 is cognate to the icosahedron, since both are of the pentagonal cut. See Chapter 2.

Behold the reason why exact consonances are found between the aphelial and perihelial movements of these two planets, but not between the converging, as in the case of the upper planets.

XXVIII. Proposition. *The private ratio of movements fitting the Earth was approximately 14 : 15, Venus, approximately 35 : 36.*

For these two planets had to distinguish the kinds of consonances, by the preceding proposition; therefore, by Proposition XXVI, the Earth as the higher was due to receive the interval 2916 : 3125, *i.e.*, approximately 14 : 15, but Venus as the lower the interval 243 : 250, *i.e.*, approximately 35 : 36.

Behold the reason why these two planets have such small eccentricities and, in proportion to them, small intervals or private ratios of the extreme movements, although nevertheless the next higher planet, Mars, and the next lower, Mercury, have marked eccentricities and the greatest of all. And astronomy confirms the truth of this; for in Chapter 4 the Earth clearly had 14 : 15, but Venus 34 : 35, which astronomical certitude can barely discern from 35 : 36 in this planet.

XXIX. Proposition. *The greater consonance of the movements of Mars and the Earth, viz., that of the diverging movements, could not be from among the consonances greater than 5 : 12.*

Above, in Proposition XVII, it was not any one of the lesser ratios; but now it is not any one of the greater ratios either. For the other common or lesser consonance of these two planets is 2 : 3, when the private ratio of Mars, which by Proposition XIV exceeds 18 : 25, makes more than 12 : 25, *i.e.*, 60 : 125. Accordingly, compound the private ratio of the Earth 14 : 15, *i.e.*, 50 : 60, by the preceding proposition. The compound ratio is greater than 56 : 125, which is approximately 4 : 9, *viz.*, slightly greater than an octave and a major whole tone. But the next greater consonance than the octave and whole tone is 5 : 12, the octave and minor third.

Note that I do not say that this ratio is neither greater nor smaller than 5 : 12; but I say that if it is necessary for it to be harmonic, no other consonance will belong to it.

XXX. Proposition. *The private ratio of movements of Mercury was due to be greater than all the other private ratios.*

For by Proposition XVI the private movements of Venus and Mercury compounded together were due to make about 5 : 12. But the private ratio of Venus, taken separately, is only 243 : 250, *i.e.*, 1458 : 1500. But if it is compounded inversely with 5 : 12, *i.e.*, 625 : 1500, Mercury singly is left with 625 : 1458, which is greater than an octave and a major whole tone; although the private ratio of Mars, which is the greatest of all these among the remaining planets, is less than 2 : 3, *i.e.*, the perfect fifth.

And thereby the private ratios of Venus and Mercury, the lowest planets, if compounded together, are approximately equal to the compounded private ratios of the four higher planets, because, as will now be apparent immediately, the compounded private ratios of Saturn and Jupiter exceed 2 : 3; those of Mars fall somewhat short of 2 : 3: all compounded, 4 : 9, *i.e.*, 60 : 135. Compound the Earth's 14 : 15, *i.e.*, 56:60, the result will be 56 : 135, which is slightly greater than 5 : 12, which just now was the compound of the private ratios of Venus and Mercury. But this has not been sought for nor taken from any separate and singular archetype of beauty but comes of itself, by the necessity of the causes bound together by the consonances hitherto established.

XXXI. Proposition. *The aphelial movement of the Earth had to harmonize with the aphelial movement of Saturn, through some certain number of octaves.*

For, by Proposition XVIII, it was necessary for there to be universal consonances, wherefore also there had to be a consonance of Saturn with the Earth and Venus. But if one of the extreme movements of Saturn had harmonized with neither of the Earth's and Venus', this would have been less harmonic than if both of its extreme movements had harmonized with these planets, by Axiom I. Therefore both of Saturn's extreme movements had to harmonize, the aphelial with one of these two planets, the perihelial with the other, since nothing would hinder, as was the case with the first planet. Accordingly these consonances will be either identisonant[1] (*identisonae*) or diversisonant (*diversisonae*), *i.e.*, either of continued double proportion or of some other. But both of them cannot be of some other proportion, for between the terms 3 : 5 (which determine the greater consonance between the aphelial movements of the Earth and Venus, by Proposition XXVII) two harmonic means cannot be set up; for the sixth cannot be divided into three intervals (see Book III). Accordingly, Saturn could not, [310] by means of both its movements, make an octave with the harmonic means between 3 and 5; but in order that its movements should harmonize with the 3 of the Earth and the 5 of Venus, it is necessary that one of those terms should harmonize identically, or through a certain number of octaves, with the others, *viz.*, with one of the said planets. But since the identisonant consonances are more excellent, they had to be established between the more excellent extreme movements, *viz.*, between the aphelial, because too they have the position of a principle on account of the altitude of the planets and because the Earth and Venus claim as their private ratio somehow and as a prerogative the consonance 3 : 5, with which as their greater consonance we are now dealing. For although, by Proposition XXII, this consonance belongs to the perihelial movement of Venus and some intermediate movement of the Earth, yet the start is made at the extreme movements and the intermediate movements come after the beginnings.

1. "Identisonant consonances" are such as 3 : 5, 3 : 10, 3 : 20, etc.

Now, since on one side we have the aphelial movement of Saturn at its greatest altitude, on the other side the aphelial movement of the Earth rather than Venus is to be joined with it, because of these two planets which distinguish the kinds of harmony, the Earth, again, has the greater altitude. There is also another nearer cause: the posterior reasons—with which we are now dealing—take away from the prior reasons but only with respect to minima, and in harmonics that is with respect to all intervals less than concords. But by the prior reasons the aphelial movement not of Venus but of the Earth, will approximate the consonance of some number of octaves to be established with the aphelial movement of Saturn. For compound together, first, 4 : 5 the private ratio of Saturn's movements, *i.e.*, from the aphelion to the perihelial of Saturn (Proposition XI), secondly, the 1 : 2 of the converging movements of Saturn and Jupiter, *i.e.*, from the perihelion of Saturn to the aphelion of Jupiter (by Proposition VIII), thirdly, the 1 : 8 of the diverging movements of Jupiter and Mars, *i.e.*, from the aphelion of Jupiter to the perihelion of Mars (by Proposition XIV), fourthly, the 2 : 3 of the converging movements of Mars and the Earth, *i.e.*, from the perihelion of Mars to the aphelion of the Earth (by Proposition XV): you will find between the aphelion of Saturn and the perihelion of the Earth the compound ratio 1 : 30, which falls short of 1 : 32, or five octaves, by only 30 : 32, *i.e.*, 15 : 16 or a semitone. And so, if a semitone, divided into particles smaller than the least concord, is compounded with these four elements there will be a perfect consonance of five octaves between the aphelial movements of Saturn and the Earth, which have been set forth. But in order for the same aphelial movement of Saturn to make some number of octaves with the aphelial movement of Venus, it would have been necessary to snatch approximately a whole perfect fourth from the prior reasons; for if you compound 3 : 5, which exists between the aphelial movements of the Earth and Venus, with the ratio 1 : 30 compounded of the four prior elements, then as it were from the prior reasons, 1 : 50 is found between the aphelial movements of Saturn and Venus: This interval differs from 1 : 32, or five octaves, by 32 : 50, *i.e.*, 16 : 25, which is a perfect fifth and a diesis; and from six octaves, or 1 : 64, it differs by 50 : 64, *i.e.*, 25 : 32, or a perfect fourth minus a diesis. Accordingly, an indentisonant consonance was due to be established, not between the aphelial movements of Venus and Saturn but between those of Venus and the Earth, so that Saturn might keep a diversisonant consonance with Venus.

XXXII. Proposition. *In the universal consonances of planets of the minor scale the exact aphelial movement of Saturn could not harmonize precisely with the other planets.*

For the Earth by its aphelial movement does not concur in the universal consonance of the minor scale, because the aphelial movements of the Earth and Venus make the interval 3 : 5, which is of the major scale (by Proposition XVII). But by its aphelial movement Saturn makes an identisonant consonance with the aphelial movement of

the Earth (by Proposition XXXI). Therefore, neither does Saturn concur by its aphelial movement. Nevertheless, in place of the aphelial movement there follows some faster movement of Saturn, very near to the aphelial, and also in the minor scale—as was apparent in Chapter 7.

XXXIII. Proposition. *The major kind of consonances and musical scale is akin to the aphelial movements; the minor to the perihelial.*

For although a major consonance (*dura harmonia*) is set up not only between the aphelial movement of the Earth and the aphelial movement of Venus but also between the lower aphelial movements and the lower movements of Venus as far as its perihelion; and, conversely, there is a minor consonance not merely between the perihelial movement of Venus and the perihelial of the Earth but also between the higher movements of Venus as far as the aphelion and the higher movements of the Earth (by Propositions XX and XXIV). Accordingly, the major scale is designated properly only in the aphelial movements, the minor, only in the perihelial.

XXXIV. Proposition. *The major scale is more akin to the upper of the two planets, the minor, to the lower.*

[311] For, because the major scale is proper to the aphelial movements, the minor, to the perihelial (by the preceding proposition), while the aphelial are slower and graver than the perihelial; accordingly, the major scale is proper to the slower movements, the minor to the faster. But the upper of the two planets is more akin to the slow movements, the lower, to the fast, because slowness of the private movement always follows upon altitude in the world. Therefore, of two planets which adjust themselves to both modes, the upper is more akin to the major mode of the scale, the lower, to the minor. Further, the major scale employs the major intervals 4 : 5 and 3 : 5, and the minor, the minor ones, 5 : 6 and 5 : 8. But, moreover, the upper planet has both a greater sphere and slower, *i.e.*, greater movements and a lengthier circuit; but those things which agree greatly on both sides are rather closely united.

XXXV. Proposition. *Saturn and the Earth embrace the major scale more closely, Jupiter and Venus, the minor.*

For, first, the Earth, as compared with Venus and as designating both scales along with Venus, is the upper. Accordingly, by the preceding proposition, the Earth embraces the major scale chiefly; Venus, the minor. But with its aphelial movement Saturn harmonizes with the Earth's aphelial movement, through an octave (by Proposition XXXI): wherefore too (by Proposition XXXIII) Saturn embraces the major scale. Secondly, by the same proposition, Saturn by means of its aphelial movement nurtures more the major scale and (by Proposition XXXII) spits out the minor scale. Accordingly, it is more closely related to the major scale than to the minor, because the scales are properly designated by the extreme movements.

Now as regards Jupiter, in comparison with Saturn it is lower; therefore as the major scale is due to Saturn, so the minor is due to Jupiter, by the preceding proposition.

XXXVI. Proposition. *The perihelial movement of Jupiter had to concord with the perihelial movement of Venus in one scale but not also in the same consonance; and all the less so, with the perihelial movement of the Earth.*

For, because the minor scale chiefly was due to Jupiter, by the preceding proposition, while the perihelial movements are more akin to the minor scale (by Proposition XXX), accordingly, by its perihelial movement Jupiter had to designate the key of the minor scale, *viz.*, its definite pitch or key-note (*phthongum*). But too the perihelial movements of Venus and the Earth designate the same scale (by Proposition XXVIII); therefore the perihelial movement of Jupiter was to be associated with their perihelial movements in the same tuning, but it could not constitute a consonance with the perihelial movements of Venus. For, because (by Proposition VIII) it had to make about 1 : 3 with the aphelial movement of Saturn, *i.e.*, the note (*clavem*) d of that system, wherein the aphelial movement of Saturn strikes the note G, but the aphelial movement of Venus the note e: accordingly, it approached the note e within an interval of least consonance. For the least consonance is 5 : 6, but the interval between d and e is much smaller, *viz.*, 9 : 10, a whole tone. And although in the perihelial tension (*tensione*) Venus is raised from the d of the aphelial tension yet this elevation is less than a diesis, (by Proposition XXVIII). But the diesis (and hence any smaller interval) if compounded with a minor whole tone does not yet equal 5 : 6 the interval of least consonance. Accordingly, the perihelial movement of Jupiter could not observe 1 : 3 or thereabouts with the aphelial movement of Saturn and at the same time harmonize with Venus. Nor with the Earth. For if the perihelial movement of Jupiter had been adjusted to the key of the perihelial movement of Venus in the same tension in such fashion that below the quantity of least concord it should preserve with the aphelial movement of Saturn the interval 1 : 3, *viz.*, by differing from the perihelial movement of Venus by a minor whole tone, 9 : 10 or 36 : 40 (besides some octaves) towards the low. Now the perihelial movement of the Earth differs from the same perihelial movement of Venus by 5 : 8, *i.e.*, by 25 : 40. And so the perihelial movements of the Earth and Jupiter differ by 25 : 36, over and above some number of octaves. But that is not harmonic, because it is the square of 5 : 6, or a perfect fifth diminished by one diesis.

XXXVII. Proposition. *It was necessary for an interval equal to the interval of Venus to accede to the 2 : 3 of the compounded private consonances of Saturn and Jupiter and to 1 : 3 the great consonance common to them.*

For with its aphelial movement Venus assists in the proper designation of the major scale; with its perihelial, that of the minor scale, by Propositions XXVII and

XXXIII. But by its aphelial movement Saturn had to be in concord also with the major scale and thus with the aphelial movement of Venus, by Proposition XXXV, but Jupiter's perihelial with the perihelial of Venus, by the preceding proposition. Accordingly, as great as Venus makes its interval from aphelial to perihelial to be, so great an interval must also accede to that movement of Jupiter which makes 1 : 3 with the aphelial movement of Saturn—to the very perihelial movement of Jupiter. But the consonance of the converging movements of Jupiter and Saturn is precisely 1 : 2, by Proposition VIII. Accordingly, if the interval 1 : 2 is divided into the interval [312] greater than 1 : 3, there results, as the compound of the private ratios of both, something which is proportionately greater than 2 : 3.

Above, in Proposition XXVI, the private ratio of the movements of Venus was 243 : 250 or approximately 35 : 36; but in Chapter 4, between the aphelial movement of Saturn and the perihelial movement of Jupiter there was found a slightly greater excess beyond 1 : 3, *viz.*, between 26 : 27 and 27 : 28. But the quantity here prescribed is absolutely equalled, by the addition of a single second to the aphelial movement of Saturn, and I do not know whether astronomy can discern that difference.

XXXVIII. Proposition. *The increment 243 : 250 to 2 : 3, the compound of the private ratios of Saturn and Jupiter, which was up to now being established by the prior reasons, was to be distributed among the planets in such fashion that of it the comma 80 : 81 should accede to Saturn and the remainder, 19,683 : 20,000 or approximately 62 : 63, to Jupiter.*

It follows from Axiom XIX that this was to have been distributed between both planets so that each could with some latitude concur in the universal consonances of the scale akin to itself. But the interval 243 : 250 is smaller than all concords: accordingly no harmonic rules remain whereby it may be divided into two concordant parts, with the single exception of those of which there was need in the division of 24 : 25, the diesis, above in Proposition XXVI; namely, in order that it may be divided into the comma 80 : 81 (which is a primary one of those intervals which are subordinate to the concordant) and into the remainder 19,683 : 20,000, which is slightly greater than a comma, *viz.*, approximately 62 : 63. But not two but one comma had to be taken away, lest the parts should become too unequal, since the private ratios of Saturn and Jupiter are approximately equal (according to Axiom X extended even to concords and parts smaller than those) and also because the comma is determined by the intervals of the major whole tone and minor whole tone, not so two commas. Furthermore, to Saturn the higher and mightier planet was due not that part which was greater, although Saturn had the greater private consonance 4 : 5, but that one which is prior and more beautiful, *i.e.*, more harmonic. For in Axiom X the consideration of priority and harmonic perfection comes first, and the consideration of quantity comes last, because

there is no beauty in quantity of itself. Thus the movements of Saturn become 64 : 81, an adulterine[1] major third, as we have called them in Book III, Chapter 12, but those of Jupiter, 6,561 : 8,000.

I do not know whether it should be numbered among the causes of the addition of a comma to Saturn that the extreme intervals of Saturn can constitute the ratio 8:9, the major whole tone, or whether that resulted without further ado from the preceding causes of the movements. Accordingly, you here have, in place of a corollary, the reason why, above in Chapter 4, the intervals of Saturn were found to embrace approximately a major whole tone.

XXXIX. Proposition. *Saturn could not harmonize with its exact perihelial movement in the universal consonances of the planets of the major scale, nor Jupiter with its exact aphelial movement.*

For since the aphelial movement of Saturn had to harmonize exactly with the aphelial movements of the Earth and Venus (by Proposition XXXI), that movement of Saturn which is 4 : 5 or one major third faster than its aphelial will also harmonize with them. For the aphelial movements of the Earth and Venus make a major sixth, which, by the demonstrations of Book III, is divisible into a perfect fourth and a major third, therefore the movement of Saturn, which is still faster than this movement already harmonized but none the less below the magnitude of a concordant interval, will not exactly harmonize. But such a movement is Saturn's perihelial movement itself, because it differs from its aphelial movement by more than the interval 4 : 5, *viz.*, one comma or 80 : 81 more (which is less than the least concord), by Proposition XXXVIII. Accordingly the perihelial movement of Saturn does not exactly harmonize. But neither does the aphelial movement of Jupiter do so precisely. For while it does not harmonize precisely with the perihelial movement of Saturn, it harmonizes at a distance of a perfect octave (by Proposition VIII), wherefore, according to what has been said in Book III, it cannot precisely harmonize.

XL. Proposition. *It was necessary to add the lemma of Plato to 1 : 8, or the triple octave, the joint consonance of the diverging movements of Jupiter and Mars established by the prior reasons.*

For because, by Proposition XXXI, there had to be 1 : 32, *i.e.*, 12 : 384, between the aphelial movements of Saturn and the Earth, but there had to be 3:2, *i.e.*, 384:256, from the aphelion of the Earth to the perihelion of Mars [313] (by Proposition XV), and from the aphelion of Saturn to its perihelion, 4 : 5 or 12 : 15 with its increment (by Proposition XXXVII); finally, from the perihelion of Saturn to the aphelion of Jupiter 1 : 2 or 15 : 30 (by Proposition VIII); accordingly, there remains 30 : 256 from

---

1 See footnote to *Intervals Compared with Harmonic Ratios.*

the aphelion of Jupiter to the perihelion of Mars, by the subtraction of the increment of Saturn. But 30 : 256 exceeds 32 : 256 by the interval 30 : 32, *i.e.*, 15 : 16 or 240 : 256, which is a semitone. Accordingly, if the increment of Saturn, which (by Proposition XXXVIII) had to be 80 : 81, *i.e.*, 240 : 243, is compounded inversely with 240 : 243, the result is 243 : 256; but that is the lemma of Plato,[1] *viz.*, approximately 19 : 20, see Book III. Accordingly, Plato's lemma had to be compounded with the 1 : 8.

And so the great ratio of Jupiter and Mars, *viz.*, of the diverging movements, ought to be 243 : 2,048, which is somehow a mean between 243 : 2,187 and 243 : 1,944, *i.e.*, between 1 : 9 and 1 : 8, whereof proportionality required the first, above; and a nearer harmonic concord, the second.

XLI. Proposition. *The private ratio of the movements of Mars has necessarily been made the square of the harmonic ratio 5 : 6, viz., 25 : 86.*

For, because the ratio of the diverging movements of Jupiter and Mars had to be 243 : 2,048, *i.e.*, 729 : 6,144, by the preceding proposition, but that of the converging movements 5 : 24, *i.e.*, 1,280 : 6,144 (by Proposition XIII), therefore the compound of the private ratios of both was necessarily 729 : 1,280 or 72,900 : 128,000. But the private ratio of Jupiter alone had to be 6,561 : 8,000, *i.e.*, 104,976 : 128,000 (by Proposition XXVIII). Therefore, if the compound ratio of both is divided by this, the private ratio of Mars will be left as 72,900 : 104,976, *i.e.*, 25 : 36, the square root of which is 5 : 6.

In another fashion, as follows: There is 1 : 32 or 120 : 3,840 from the aphelial movement of Saturn to the aphelial movement of the Earth, but from that same movement to the perihelial of Jupiter there is 1 : 3 or 120 : 360, with its increment. But from this to the aphelial movement of Mars is 5 : 24 or 360 : 1,728. Accordingly, from the aphelial movement of Mars to the aphelial movement of the Earth, there remains 1,728 : 3,840 minus the increment of the ratio of the diverging movements of Saturn and Jupiter. But from the same aphelial movement of the Earth to the perihelial of Mars there is 3 : 2, *i.e.*, 3,840 : 2,500. Therefore between the aphelial and perihelial movements of Mars there remains the ratio 1,728 : 2,560, *i.e.*, 27 : 40 or 81 : 120, minus the said increment. But 81 : 120 is a comma less than 80 : 120 or 2 : 3. Therefore, if a comma is taken away from 2 : 3, and the said increment (which by Proposition XXXVIII is equal to the private ratio of Venus) is taken away too, the private ratio of Mars is left. But the private ratio of Venus is the diesis diminished by a comma, by Proposition XXVI. But the comma and the diesis diminished by a comma make a full diesis or 24 : 25. Therefore if you divide 2 : 3, *i.e.*, 24 : 36 by the diesis 24 : 25, Mars' private ratio of 25 : 36 is left, as before, the square root of which, or 5 : 6, goes to the intervals, by Chapter 3.

---

1. *Timaeus*, 36.

Behold again the reason why—above, in Chapter 4—the extreme intervals of Mars have been found to embrace the harmonic ratio 5 : 6.

XLII. Proposition. *The great ratio of Mars and the Earth, or the common ratio of the diverging movements, has been necessarily made to be 54 : 125, smaller than the consonance 5 : 12 established by the prior reasons.*

For the private ratio of Mars had to be a perfect fifth, from which a diesis has been taken away, by the preceding proposition. But the common or minor ratio of the converging movements of Mars and the Earth had to be a perfect fifth or 2 : 3, by Proposition XV. Finally, the private ratio of the Earth is the diesis squared, from which a comma is taken away, by Propositions XXVI and XXVIII. But out of these elements is compounded the major ratio or that of the diverging movements of Mars and the Earth—and it is two perfect fifths (or 4 : 9, *i.e.*, 108 : 243) plus a diesis diminished by a comma, *i.e.*, plus 243 : 250; namely, it is 108 : 250 or 54 : 125, *i.e.*, 608 : 1,500. But this is smaller than 625 : 1,500, *i.e.*, than 5 : 12, in the ratio 602 : 625, which is approximately 36 : 37, smaller than 625 : 1,5000, *i.e.*, than 5 : 12, in the ratio 602:625, which is approximately 36 : 37, smaller than the least concord.

XLIII. Proposition. *The aphelial movement of Mars could not harmonize in some universal consonance; nevertheless it was necessary for it to be in concord to some extent in the scale of the minor mode.*

For, because the perihelial movement of Jupiter has the pitch d of acute tuning in the minor mode, and the consonance 5 : 24 ought to have existed between that and the aphelial movement of Mars, therefore, the aphelial movement of Mars occupies the adulterine pitch of the same acute tuning. I say *adulterine* for, although in Book III, Chapter 12, the adulterine consonances were reviewed and deduced from the composition of systems, certain ones which exist in the simple natural system were omitted. [314] And so, after the line which ends 81 : 120, the reader may add: if you divide into it 4 : 5 or 32 : 40, there remains 27 : 32, the subminor sixth,[1] which exists between d and f or c and e^2 or a and c of even the simple octave. And in the ensuing table, the following should be in the first line; for 5 : 6 there is 27 : 32, which is deficient.

From that it is clear that in the natural system the true note (*clavem*) f, as regulated by my principles, constitutes a deficient or adulterine minor sixth with the note d. Accordingly since between the perihelial movement of Jupiter set up in the true note d and the aphelial movement of Mars there is a perfect minor sixth over and above the double octave, but not the diminished (by Proposition XIII), it follows that with its aphelial movement Mars designates the pitch which is one comma higher than the true note f; and so it will concord not absolutely but merely to a certain extent in this scale.

---

1. Here "sixth" (*sexta*) should probably be "third" (*tertia*). E. C., Jr.
2. C and e do not produce a subminor third in the "natural system." E. C., Jr.

But it does not enter into either the pure or the adulterine universal harmony. For the perihelial movement of Venus occupies the pitch of *e* in this tuning (*tensionem*). But there is dissonance between *e* and *f*, on account of their nearness. Therefore, Mars is in discord with the perihelial movement of one of the planets, *viz.*, Venus. But too it is in discord with the other movements of Venus; they are diminished by a comma less than a diesis: wherefore, since there is a semitone and a comma between the perihelial movement of Venus and the aphelial movement of Mercury, accordingly, between the aphelion of Venus and the aphelion of Mars there will be a semitone and a diesis (neglecting the octaves), *i.e.*, a minor whole tone, which is still a dissonant interval. Now the aphelial movement of Mars concords to that extent in the scale of the minor mode, but not in that of the major. For since the aphelial movement of Venus concords with the *e* of the major mode, while the aphelial movement of Mars (neglecting the octaves) has been made a minor whole tone higher than *e*, then necessarily the aphelial movement of Mars in this tuning would fall midway between *f* and *f* sharp and would make with *g* (which in this tuning would be occupied by the aphelial movement of the Earth) the plainly discordant interval 25 : 27, *viz.*, a major whole tone diminished by a diesis.

In the same way, it will be proved that the aphelial movement of Mars is also in discord with the movements of the Earth. For because it makes a semitone and comma with the perihelial movement of Venus, *i.e.*, 14 : 15 (by what has been said), but the perihelial movements of the Earth and Venus make a minor sixth 5 : 8 or 15 : 24 (by Proposition XXVII). Accordingly, the aphelial movement of Mars together with the perihelial movement of the Earth (the octaves added to it) will make 14 : 24 or 7 : 12, a discordant interval and one not harmonic, like 7 : 6. For any interval between 5 : 6 and 8 : 9 is dissonant and discordant, as 6 : 7 in this case. But no other movement of the Earth can harmonize with the aphelial movement of Mars. For it was said above that it makes the discordant interval 25 : 27 with the Earth (neglecting the octaves); but all from 6 : 7 or 24 : 28 to 25 : 27 are smaller than the least harmonic interval.

XLIV. Corollary. *Accordingly it is clear from the above Proposition XLIII concerning Jupiter and Mars, and from Proposition XXXIX concerning Saturn and Jupiter, and from Proposition XXXVI concerning Jupiter and the Earth, and from Proposition XXXII concerning Saturn, why—in Chapter 5, above—it was found that all the extreme movements of the planets had not been adjusted perfectly to one natural system or musical scale, and that all those which had been adjusted to a system of the same tuning did not distinguish the pitches (loca) of that system in a natural way or effect a purely natural succession of concordant intervals. For the reasons are prior whereby the single planets came into possession of their single consonances; those whereby all the planets, of the universal consonances; and finally, those whereby the universal consonances of the two modes, the major and the minor: when all those have been posited, an omniform adjustment to one natural system is pre-*

vented. But if those causes had not necessarily come first, there is no doubt that either one system and one tuning of it would have embraced the extreme movements of all the planets; or, if there was need of two systems for the two modes of song, the major and minor, the very order of the natural scale would have been expressed not merely in one mode, the major, but also in the remaining minor mode. Accordingly, here in Chapter 5, you have the promised causes of the discords through least intervals and intervals smaller than all concords.

XLV. Proposition. *It was necessary for an interval equal to the interval of Venus to be added to the common major consonance of Venus and Mercury, the double octave, and also the private consonance of Mercury, which were established above in Propositions XII and XIII by the prior reasons, [315] in order that the private ratio of Mercury should be a perfect 5 : 12 and that thus Mercury should with both its movements harmonize with the single perihelial movement of Venus.*

For, because the aphelial movement of Saturn, the highest and outmost planet, circumscribed around its regular solid, had to harmonize with the aphelial movement of the Earth, the highest movement of the Earth, which divides the classes of figures; it follows by the laws of opposites that the perihelial movement of Mercury as the innermost planet, inscribed in its figure, the lowest and nearest to the sun, should harmonize with the perihelial movement of the Earth, with the lowest movement of the Earth, the common boundary: the former in order to designate the major mode of consonances, the latter the minor mode, by Propositions XXXIII and XXXIV. But the perihelial movement of Venus had to harmonize with the perihelial movement of the Earth in the consonance 5 : 3, by Proposition XXVII; therefore too the perihelial movement of Mercury had to be tempered with the perihelial of Venus in one scale. But by Proposition XII the consonance of the diverging movements of Venus and Mercury was determined by the prior reasons to be 1 : 4; therefore, now by these posterior reasons it was to be adjusted by the accession of the total interval of Venus. Accordingly, not from further on, from the aphelion, but from the perihelion of Venus to the perihelion of Mercury there is a perfect double octave. But the consonance 3 : 5 of the converging movements is perfect, by Proposition XV. Accordingly if 1 : 4 is divided by 3 : 5, there remains to Mercury singly the private ratio 5 : 12, perfect too, but not further (by Proposition XVI, through the prior reasons) diminished by the private ratio of Venus.

Another reason. Just as only Saturn and Jupiter are touched nowhere on the outside by the dodecahedron and icosahedron wedded together, so only Mercury is untouched on the inside by these same solids, since they touch Mars on the inside, the Earth on both sides, and Venus on the outside. Accordingly, just as something equal to the private ratio of Venus has been added distributively to the private ratios of movements of Saturn and Jupiter, which are supported by the cube and tetrahedron; so now some-

thing as great was due to accede to the private ratio of solitary Mercury, which is comprehended by the associated figures of the cube and tetrahedron; because, as the octahedron, a single figure among the secondary figures, does the job of two among the primary, the cube and tetrahedron (concerning which see Chapter 1), so too among the lower planets there is one Mercury in place of two of the upper planets, *viz.*, Saturn and Jupiter.

Thirdly, just as the aphelial movement of the highest planet Saturn had to harmonize, in some number of octaves, *i.e.*, in the continued double ratio, 1 : 32, with the aphelial movement of the higher and nearer of the two planets which shift the mode of consonance (by Proposition XXXI); so, *vice versa*, the perihelial movement of the lowest planet Mercury, again through some number of octaves, *i.e.*, in the continued double ratio, 1 : 4, had to harmonize with the perihelial movement of the lower and similarly nearer of the two planets which shift the mode of consonance.

Fourthly, of the three upper planets, Saturn, Jupiter, and Mars, the single but extreme movements concord with the universal consonances; accordingly both extreme movements of the single lower planet, *viz.*, Mercury, had to concord with the same; for the middle planets, the Earth and Venus, had to shift the mode of consonances, by Propositions XXXIII and XXXIV.

Finally, in the three pairs of the upper planets perfect consonances have been found between the converging movements, but adjusted (*fermentatae*) consonances between the diverging movements and private ratios of the single planets; accordingly, in the two pairs of the lower planets, conversely, perfect consonances had to be found not between the converging movements chiefly, nor between the diverging, but between the movements of the same field. And because two perfect consonances were due to the Earth and Venus, therefore two perfect consonances were due to Venus and Mercury also. And the Earth and Venus had to receive as perfect a consonance between their aphelial movements as between their perihelial, because they had to shift the mode of their consonance; but Venus and Mercury, as not shifting the mode of their consonance, did not also require perfect consonances between both pairs, the aphelial movements and the perihelial; but there came in place of the perfect consonance of the aphelial movements, as being already adjusted the perfect consonance of the converging movements, so that just as Venus, the higher of the lower planets, has the least private ratio of all the private ratios of movements (by Proposition XXVI), and Mercury, the lower of the lower, has received the greatest ratio of all the private ratios of movements (by Proposition XXX), so too the private ratio of Venus should be the most imperfect of all the private ratios or the farthest removed from consonances, while the private ratio of Mercury should be most perfect of all the private ratios, *i.e.*, an absolute consonance without adjustment, and that finally the relations should be everywhere opposite.

For He Who is before the ages and on into the ages thus adorned the great things of His wisdom: nothing excessive, nothing defective, no room for any censure. How lovely are his works! All things, in twos, one [316] *against one, none lacking its opposite. He has strengthened the goods—adornment and propriety—of each and every one and established them in the best reasons, and who will be satiated seeing their glory?*

XLVI. Axiom. *If the interspacing of the solid figures between the planetary spheres is free and unhindered by the necessities of antecedent causes, then it ought to follow to perfection the proportionality of geometrical inscriptions and circumscriptions, and thereby the conditions of the ratio of the inscribed to the circumscribed spheres.*

For nothing is more reasonable than that physical inscription should exactly represent the geometrical, as the work, its pattern.

XLVII. Proposition. *If the inscription of the regular solids among the planets was free, the tetrahedron was due to touch with its angles precisely the perihelial sphere of Jupiter above it, and with centres of its planes precisely the aphelial sphere of Mars below it. But the cube and the octahedron, each placing its angles in the perihelial sphere of the planet above, were due to penetrate the sphere of the inside planet with the centres of their planes, in such fashion that those centres should turn within the aphelial and perihelial spheres: on the other hand, the dodecahedron and icosahedron, grazing with their angles the perihelial spheres of their planets on the outside, were due not quite to touch with the centres of their planes the aphelial spheres of their inner planets. Finally, the dodecahedral echinus, placing its angles in the perihelial sphere of Mars, was due to come very close to the aphelial sphere of Venus with the midpoints of its converted sides which interdistinguish two solid rays.*

For the tetrahedron is the middle one of the primary figures, both in genesis and in situation in the world; accordingly, it was due to remove equally both regions, that of Jupiter and that of Mars. And because the cube was above it and outside it, and the dodecahedron was below it and within it, therefore it was natural that their inscription should strive for the contrariety wherein the tetrahedron held a mean, and that the one of them should make an excessive inscription, and the other a defective, *viz.*, the one should somewhat penetrate the inner sphere, the other not touch it. And because the octahedron is cognate to the cube and has an equal ratio of spheres, but the icosahedron to the dodecahedron, accordingly, whatever the cube has of perfection of inscription, the same was due to the octahedron also, and whatever the dodecahedron, the same to the icosahedron too. And the situation of the octahedron's similar to the situation of the cube, but that of the icosahedron to the situation of the dodecahedron, because as the cube occupies the one limit to the outside, so the octahedron occupies the remaining limit to the inside of the world, but the dodecahedron and icosahedron are midway: accordingly even a similar inscription was proper, in the case of the dodecahedron, one penetrating the sphere of the inner planet, in that of the icosahedron, one falling short of it.

But the echinus, which represents the icosahedron with the apexes of its angles and the dodecahedron with the bases, was due to fill, embrace, or dispose both regions, that between Mars and the Earth with the dodecahedron as well as that between the Earth and Venus with the icosahedron. But the preceding axiom makes clear which of the opposites was due to which association. For the tetrahedron, which has a rational inscribed sphere, has been allotted the middle position among the primary figures and is surrounded on both sides by figures of incommensurable spheres, whereof the outer is the cube, the inner the dodecahedron, by Chapter 1 of this book. But this geometrical quality, *viz.*, the rationality of the inscribed sphere, represents in nature the perfect inscription of the planetary sphere. Accordingly, the cube and its allied figure have their inscribed spheres rational only in square, *i.e.*, in power alone; accordingly, they ought to represent a semiperfect inscription, where, even if not the extremity of the planetary sphere, yet at least something on the inside and rightfully a mean between the aphelial and perihelial spheres—if that is possible through other reasons—is touched by the centres of the planes of the figures. On the other hand, the dodecahedron and its allied figure have their inscribed spheres clearly irrational both in the length of the radius and in the square; accordingly, they ought to represent a clearly imperfect inscription and one touching absolutely nothing of the planetary sphere, *i.e.*, falling short and not reaching as far as the aphelial sphere of the planet with the centres of its planes.

Although the echinus is cognate to the dodecahedron and its allied figure, nevertheless it has a property similar to the tetrahedron. For the radius of the sphere inscribed in its inverted sides is indeed incommensurable with the radius of the circumscribed sphere, but it is, however, commensurable with the length of the distance between two neighbouring angles. And so the perfection of the commensurability of rays is approximately as great as in the tetrahedron; but elsewhere the imperfection is as great as in the [317] dodecahedron and its allied figure. Accordingly it is reasonable too that the physical inscription belonging to it should be neither absolutely tetrahedral nor absolutely dodecahedral but of an intermediate kind; in order that (because the tetrahedron was due to touch the extremity of the sphere with its planes, and the dodecahedron, to fall short of it by a definite interval) this wedge-shaped figure with the inverted sides should stand between the icosahedral space and the extremity of the inscribed sphere and should nearly touch this extremity—if nevertheless this figure was to be admitted into association with the remaining five, and if its laws could be allowed, with the laws of the others remaining. Nay, why do I say "could be allowed"? For they could not do without them. For if an inscription, which was loose and did not come into contact fitted the dodecahedron, what else could confine that indefinite

looseness within the limits of a fixed magnitude, except this subsidiary figure cognate to the dodecahedron and icosahedron, and which comes almost into contact with its inscribed sphere and does not fall short (if indeed it does fall short) any more than the tetrahedron exceeds and penetrates—with which magnitude we shall deal in the following.

This reason for the association of the echinus with the two cognate figures (*viz.*, in order that the ratio of the spheres of Mars and Venus, which they had left indefinite, should be made determinate) is rendered very probable by the fact that 1,000, the semidiameter of the sphere of the Earth, is found to be practically a mean proportional between the perihelial sphere of Mars and the aphelial sphere of Venus; as if the interval, which the echinus assigns to the cognate figures, has been divided between them as proportionally as possible.

XLVIII. Proposition. *The inscription of the regular solid figures between the planetary spheres was not the work of pure freedom; for with respect to very small magnitudes it was hindered by the consonances established between the extreme movements.*

For, by Axioms I and II, the ratio of the spheres of each figure was not due to be expressed immediately by itself, but by means of it the consonances most akin to the ratios of the spheres were first to be sought and adjusted to the extreme movements.

Then, in order that, by Axioms XVIII and XX, the universal consonances of the two modes could exist, it was necessary for the greater consonances of the single pairs to be readjusted somewhat, by means of the posterior reasons. Accordingly, in order that those things might stand, and be maintained by their own reasons, intervals were required which are somewhat discordant with those which arise from the perfect inscription of figures between the spheres, by the laws of movements unfolded in Chapter 3. In order that it be proved and made manifest how much is taken away from the single planets by the consonances established by their proper reasons; come, let us build up, out of them, the intervals of the planets from the sun, by a new form of calculation not previously tried by anyone.

Now there will be three heads to this inquiry: First, from the two extreme movements of each planet the similar extreme intervals between it and the sun will be investigated, and by means of them the radius of the sphere in those dimensions, of the extreme intervals, which are proper to each planet. Secondly, by means of the same extreme movements, in the same dimensions for all, the mean movements and their ratio will be investigated. Thirdly, by means of the ratio of the mean movements already disclosed, the ratio of the spheres or mean intervals and also one ratio of the

extreme intervals, will be investigated; and the ratio of the mean intervals will be compared with the ratios of the figures.

As regards the first: we must repeat, from Chapter 3, Article VI, that the ratio of the extreme movements is the inverse square of the ratio of the corresponding intervals from the sun. Accordingly, since the ratio of the squares is the square of the ratio of its sides, therefore, the numbers, whereby the extreme movements of the single planets are expressed, will be considered as squares and the extraction of their roots will give the extreme intervals, whereof it is easy to take the arithmetic mean as the semidiameter of the sphere and the eccentricity. Accordingly the consonances so far established have prescribed.

[318] *Planets Props.*	*Ratios of movements*	*The roots either prolonged or of their multiples*	*Therefore the semidiameter of the sphere*	*Eccen-tricity*	*In dimensions whereof the semidiameter of the sphere is 100,000*
Saturn by XXXVIII	64 : 81	80 : 90	85	5	5,882
Jupiter by XXXVIII	6,561 : 8,000	81,000 : 89,444	85,222	4,222	4,954
Mars by XLI	25 : 36	50 : 60	55	5	9,091
Earth by XXVIII	2,916 : 3,125	93,531 : 96,825	95,178	1,647	1,730
Venus by XXVIII	243 : 250	9,859 : 10,000	99,295	705	710
Mercury by XLV	5 : 12	63,250 : 98,000	80,625	17,375	21,551

For the second of the things proposed, we again have need of Chapter 3, Article XII, where it was shown that the number which expresses the movement which is as a mean in the ratio of the extremes is less than their arithmetic mean, also less than the geometric mean by half the difference between the geometric and arithmetic means. And because we are investigating all the mean movements in the same dimensions, therefore let all the ratios hitherto established between different twos and also all the private ratios of the single planets be set out in the measure of the least common divisible. Then let the means be sought: the arithmetic, by taking half the difference between the extreme movements of each planet, the geometric, by the multiplication of one extreme into the other and extracting the square root of the product; then by subtracting half the difference of the means from the geometric mean, let the number of the mean movement be constituted in the private dimensions of each planet, which can easily, by the rule of ratios, be converted into the common dimensions.

[319] Therefore, from the prescribed consonances, the ratio of the mean diurnal movements has been found, *viz.*, the ratio between the numbers of the degrees and minutes of each planet. It is easy to explore how closely that approaches to astronomy.

Harmonic ratios of two	Numbers of the extreme movements	Private ratios of the single planets	Continued means of the single planets		Halves of the differences	Number of the mean movement in dimensions	
			Arithmetic	Geometric		Private	Common
1	♄ 139,968	64 }					
			72.50	72.00	.25	71.75	156,917
1	♄ 177,147	81					
2	♃ 354,294	6,561 }					
			7,280.5	7,244.9	17.8	7,227.1	390,263
5	♃ 432,000	8,000					
24	♂ 2,073,600	25 }					
			30.50	30.00	.25	29.75	2,467,584
2	♂ 2,985,984	36					
32 3	♁ 4,478,976	2,916 }					
			3,020.500	3,018.692	.904	3,017.788	4,635,322
5 5	♁ 4,800,000	3,125					
5 8	♀ 7,464,960	243 }					
			246.500	246.475	.0125	246.4625	7,571,328
1 3	♀ 7,680,000	250					
5	☿ 12,800,000	5 }					
			8.500	7.746	.377	7.369	18,864,680
4	☿ 30,720,000	12					

The third head of things proposed requires Chapter 3, Article VIII. For when the ratio of the mean diurnal movements of the single planets has been found, it is possible to find the ratio of the spheres too. For the ratio of the mean movements is the $3/2$th power of the inverse ratio of the spheres. But, too, the ratio of the cube numbers is the $3/2$th power of the ratio of the squares of those same square roots, given in the table of Clavius, which he subjoined to his *Practical Geometry*. Wherefore, if the numbers of our mean movements (curtailed, if need be, of an equal number of ciphers) are sought among the cube numbers of that table, they will indicate on the left, under the heading of the squares, the numbers of the ratio of the spheres; then the eccentricities ascribed above to the single planets in the private ratio of the semidiameters of each may easily be converted by the rule of ratios into dimensions common to all, so that, by their addition to the semidiameters of the spheres and subtraction from them, the extreme intervals of the single planets from the sun may be established. Now we shall give to the semidiameter of the terrestrial sphere the round number 100,000, as is the practice in astronomy, and with the following design: because this number or its square

711

or its cube is always made up of mere ciphers; and so too we shall raise the mean movement of the Earth to the number 10,000,000,000 and by the rule of ratios make the number of the mean movement of any planet be to the number of the mean movement of the Earth, as 10,000,000,000 is to the new measurement. And so the business can be carried on with only five cube roots, by comparing those single cube roots with the one number of the Earth.

	Numbers of the mean movements						
	In the new dimensions found in inverse order among	Numbers of the ratio of the spheres found among	Semi-diameters	Eccentricities in dimensions Private as		Extreme intervals resulting	
In the original dimensions	the cubes	the squares	as above	above	Common	Aphelion	Perihelion
♄ 156,917	29,539,960	9,556	85	5	562	10,118	8,994
♃ 390,263	11,877,400	5,206	85,222	4,222	258	5,464	4,948
♂ 2,467,584	1,878,483	1,523	55	5	138	1,661	1,384
♁ 4,635,322	1,000,000	1,000	95,178	1,647	17	1,017	983
♀ 7,571,328	612,220	721	99,295	705	5	726	716
☿ 18,864,680	245,714	392	80,625	17,375	85	476	308

Accordingly, it is apparent in the last column what the numbers turn out to be whereby the converging intervals of two planets are expressed. All of them approach very near to those intervals, which I found from Brahe's observations. In Mercury alone is there some small difference. For astronomy is seen to give the following intervals to it: 470, 388, 306, all shorter. It seems that the reason for the dissonance may be referred either to the fewness of the observations or to the magnitude of the eccentricity. (See Chapter 3). But I hurry on to the end of the calculation.

For now it is easy to compare the ratio of the spheres of the figures with the ratio of the converging intervals.

[320] For if the semidiameter of the sphere circumscribed around the figure

which is commonly 100,000		Then the semidiameter of the sphere or circle becomes: inscribed in: instead of:			becomes:	Although by the consonances the interval is:	
In the cube	8,994	♄ (Saturn)	57,735		5,194	Mean	♃ 5,206
In the tetrahedron	4,948	♃ (Jupiter)	33,333		1,649	Aphelial	♂ 1,661
In the dodecahedron	1,384	♂ (Mars)	79,465		1,100	Aphelial	♁ 1,018
In the icosahedron	983	♁ (Earth)	79,465		781	Aphelial	♀ 726
In the echinus	1,384	♂ (Mars)	52,573		728	Aphelial	♀ 726
In the octahedron	716	♀ (Venus)	57,735		413	Mean	☿ 392

| In the square in the octahedron | 716 ♀ (Venus) | 70,711 | 506 | Aphelial | ☿ | 476 |
| | or 476 ☿ (Mercury) | 70,711 | 336 | Perihelial | ☿ | 308 |

That is to say, the planes of the cube extend down slightly below the middle circle of Jupiter; the octahedral planes, not quite to the middle circle of Mercury; the tetrahedral, slightly below the highest circle of Mars; the sides of the echinus, not quite to the highest circle of Venus; but the planes of the dodecahedron fall far short of the aphelial circle of the Earth; the planes of the icosahedron also fall short of the aphelial circle of Venus, and approximately proportionally; finally, the square in the octahedron is quite inept, and not unjustly, for what are plane figures doing among solids? Accordingly, you see that if the planetary intervals are deduced from the harmonic ratios of movements hitherto demonstrated, it is necessary that they turn out as great as these allow, but not as great as the laws of free inscription prescribed in Proposition XLV would require: because this κόσμοσ γεωμέτρικοσ (geometrical adornment) of perfect inscription was not fully in accordance with that other κόδμον ἁρμόνικον ἐν-δεχόμενον (possible harmonic adornment)—to use the words of Galen, taken from the epigraph to this Book V. So much was to be demonstrated by the calculation of numbers, for the elucidation of the prescribed proposition.

I do not hide that if I increase the consonance of the diverging movements of Venus and Mercury by the private ratio of the movements of Venus, and, as a consequence, diminish the private ratio of Mercury by the same, then by this process I produce the following intervals between Mercury and the sun: 469,388,307, which are very precisely represented by astronomy. But, in the first place, I cannot defend that diminishing by harmonic reasons. For the aphelial movement of Mercury will not square with that musical scale, nor in the planets which are opposite in the world is the planetary principle (*ratio*) of opposition of all conditions kept. Finally, the mean diurnal movement of Mercury becomes too great, and thereby the periodic time, which is the most certain fact in all astronomy, is shortened too much. And so I stay within the harmonic polity here employed and confirmed throughout the whole of Chapter 9. But none the less with this example I call you all forth, as many of you as have happened to read this book and are steeped in the mathematical disciplines and the knowledge of highest philosophy: work hard and either pluck up one of the consonances applied everywhere, interchange it with some other, and test whether or not you will come so near to the astronomy posited in Chapter 4, or else try by reasons whether or not you can build with the celestial movements something better and more expedient and destroy in part or in whole the layout applied by me. But let whatever pertains to the glory of Our Lord and Founder be equally permissible to you by way of this book, and up to this very hour I myself have taken the liberty of everywhere changing those

things which I was able to discover on earlier days and which were the conceptions of a sluggish care or hurrying ardour.

[321] XLIX. Envoi. *It was good that in the genesis of the intervals the solid figures should yield to the harmonic ratios, and the major consonances of two planets to the universal consonances of all, in so far as this was necessary.*

With good fortune we have arrived at 49, the square of 7; so that this may come as a kind of Sabbath, since the six solid eights of discourse concerning the construction of the heavens has gone before. Moreover, I have rightly made an *envoi* which could be placed first among the axioms: because God also, enjoying the works of His creation, "saw all things which He had made, and behold! they were very good."

There are two branches to the *envoi*: First, there is a demonstration concerning consonances in general, as follows: For where there is choice among different things which are not of equal weight, there the more excellent are to be put first and the more vile are to be detracted from, in so far as that is necessary, as the very word ὁ κόδμος, which signifies *adornment*, seems to argue. But inasmuch as life is more excellent than the body, the form than the material, by so much does harmonic adornment excel the geometrical.

For as life perfects the bodies of animate things, because they have been born for the exercise of life—as follows from the archetype of the world, which is the divine essence—so movement measures the regions assigned to the planets, each that of its own planet: because that region was assigned to the planet in order that it should move. But the five regular solids, by their very name, pertain to the intervals of the regions and to the number of them and the bodies; but the consonances to the movements. Again, as matter is diffuse and indefinite of itself, the form definite, unified, and determinant of the material, so too there are an infinite number of geometric ratios, but few consonances. For although among the geometrical ratios there are definite degrees of determinations, formation, and restriction, and no more than three can exist from the ascription of spheres to the regular solids; but nevertheless an accident common to all the rest follows upon even these geometrical ratios: an infinite possible section of magnitudes is presupposed, which those ratios whose terms are mutually incommensurable somehow involve in actuality too. But the harmonic ratios are all rational, the terms of all are commensurable and are taken from a definite and finite species of plane figures. But infinity of section represents the material, while commensurability or rationality of terms represents the form. Accordingly, as material desires the form, as the rough-hewn stone, of a just magnitude indeed, the form of a human body, so the geometric ratios of figures desire the consonances—not in order to fashion and form those consonances, but because this material squares better with this form, this quantity of stone with this statue, even this ratio of regular solids with this consonance—therefore in order so that

714

they are fashioned and formed more fully, the material by its form, the stone by the chisel into the form of an animate being; but the ratio of the spheres of the figure by its own, *i.e.*, the near and fitting, consonance.

The things which have been said up to now will become clearer from the history of my discoveries. Since I had fallen into this speculation twenty-four years ago, I first inquired whether the single planetary spheres are equal distances apart from one another (for the spheres are apart in Copernicus, and do not touch one another), that is to say, I recognized nothing more beautiful than the ratio of equality. But this ratio is without head or tail: for this material equality furnished no definite number of mobile bodies, no definite magnitude for the intervals. Accordingly, I meditated upon the similarity of the intervals to the spheres, *i.e.*, upon the proportionality. But the same complaint followed. For although to be sure, intervals which were altogether unequal were produced between the spheres, yet they were not unequally equal, as Copernicus wishes, and neither the magnitude of the ratio nor the number of the spheres was given. I passed on to the regular plane figures: [322] intervals were formed from them by the ascription of circles. I came to the five regular solids: here both the number of the bodies and approximately the true magnitude of the intervals was disclosed, in such fashion that I summoned to the perfection of astronomy the discrepancies remaining over and above. Astronomy was perfect these twenty years; and behold! there was still a discrepancy between the intervals and the regular solids, and the reasons for the distribution of unequal eccentricities among the planets were not disclosed. That is to say, in this house the world, I was asking not only why stones of a more elegant form but also what form would fit the stones, in my ignorance that the Sculptor had fashioned them in the very articulate image of an animated body. So, gradually, especially during these last three years, I came to the consonances and abandoned the regular solids in respect to minima, both because the consonances stood on the side of the form which the finishing touch would give, and the regular solids, on that of the material—which in the world is the number of bodies and the rough-hewn amplitude of the intervals—and also because the consonances gave the eccentricities, which the regular solids did not even promise—that is to say, the consonances made the nose, eyes, and remaining limbs a part of the statue, for which the regular solids had prescribed merely the outward magnitude of the rough-hewn mass.

Wherefore, just as neither the bodies of animate beings are made nor blocks of stone are usually made after the pure rule of some geometrical figure, but something is taken away from the outward spherical figure, however elegant it may be (although the just magnitude of the bulk remains), so that the body may be able to get the organs necessary for life, and the stone the image of the animate being; so too as the ratio which the regular solids had been going to prescribe for the planetary spheres is inferi-

or and looks only towards the body and material, it has to yield to the consonances, in so far as that was necessary in order for the consonances to be able to stand closely by and adorn the movement of the globes.

The other branch of the *envoi*, which concerns universal consonances, has a proof closely related to the first. (As a matter of fact, it was in part assumed above, in XVIII, among the Axioms.) For the finishing touch of perfection, as it were, is due rather to that which perfects the world more; and conversely that thing which occupies a second position is to be detracted from, if either is to be detracted from. But the universal harmony of all perfects the world more than the single twin consonances of different neighbouring twos. For harmony is a certain ratio of unity; accordingly the planets are more united, if they all are in concord together in one harmony, than if each two concord separately in two consonances. Wherefore, in the conflict of both, either one of the two single consonances of two planets was due to yield, so that the universal harmonies of all could stand. But the greater consonances, those of the diverging movements, were due to yield rather than the lesser, those of the converging movements. For if the divergent movements diverge, then they look not towards the planets of the given pair but towards other neighbouring planets, and if the converging movements converge, then the movements of one planet are converging toward the movement of the other, conversely: for example, in the pair Jupiter and Mars the aphelial movement of Jupiter verges toward Saturn, the perihelial of Mars towards the Earth: but the perihelial movement of Jupiter verges toward Mars, the aphelial of Mars toward Jupiter. Accordingly the consonance of the converging movements is more proper to Jupiter and Mars; the consonance of the diverging movements is somehow more foreign to Jupiter and Mars. But the ratio of union which brings together neighbouring planets by twos and twos is less disturbed if the consonance which is more foreign and more removed from them should be adjusted than if the private ratio should be, *viz.*, the one which exists between the more neighbouring movements of neighbouring planets. None the less this adjustment was not very great. For the proportionality has been found in which may stand the universal consonances of all the planets may exist (and these in two distinct modes), and in which (with a certain latitude of tuning merely equal to a comma) may also be embraced the single consonances of two neighbouring planets; the consonances of the converging movements in four pairs, perfect, of the aphelial movements in one pair, of the perihelial movements in two pairs, likewise perfect; the consonances of the diverging movements in four pairs, these, however, within the difference of one diesis (the very small interval by which the human voice [323] in figured song nearly always errs; the single consonance of Jupiter and Mars, this between the diesis and the semitone. Accordingly it is apparent that this mutual yielding is everywhere very good.)

Accordingly let this do for our *envoi* concerning the work of God the Creator. It now remains that at last, with my eyes and hands removed from the tablet of demonstrations and lifted up towards the heavens, I should pray, devout and supplicating, to the Father of lights: *O Thou Who dost by the light of nature promote in us the desire for the light of grace, that by its means Thou mayest transport us into the light of glory, I give thanks to Thee, O Lord Creator, Who hast delighted me with Thy makings and in the works of Thy hands have I exulted. Behold! now, I have completed the work of my profession, having employed as much power of mind as Thou didst give to me; to the men who are going to read those demonstrations I have made manifest the glory of Thy works, as much of its infinity as the narrows of my intellect could apprehend. My mind has been given over to philosophizing most correctly: if there is anything unworthy of Thy designs brought forth by me—a worm born and nourished in a wallowing place of sins—breathe into me also that which Thou dost wish men to know, that I may make the correction: If I have been allured into rashness by the wonderful beauty of Thy works, or if I have loved my own glory among men, while I am advancing in the work destined for Thy glory, be gentle and merciful and pardon me; and finally deign graciously to effect that these demonstrations give way to Thy glory and the salvation of souls and nowhere be an obstacle to that.*

10. EPILOGUE CONCERNING THE SUN, BY WAY OF CONJECTURE[1]

From the celestial music to the hearer, from the Muses to Apollo the leader of the Dance, from the six planets revolving and making consonances to the Sun at the centre of all the circuits, immovable in place but rotating into itself. For although the harmony is most absolute between the extreme planetary movements, not with respect to the true speeds through the ether but with respect to the angles which are formed by joining with the centre of the sun the termini of the diurnal arcs of the planetary orbits; while the harmony does not adorn the termini, *i.e.*, the single movements, in so far as they are considered in themselves but only in so far as by being taken together and compared with one another, they become the object of some mind; and although no object is ordained in vain, without the existence of some thing which may be moved by it, while those angles seem to presuppose some action similar to our eyesight or at least to that sense-perception whereby, in Book IV, the sublunary nature perceived the angles of rays formed by the planets on the Earth: still it is not easy for dwellers on the Earth to conjecture what sort of sight is present in the sun, what eyes there are, or what other instinct there is for perceiving those angles even without eyes and for evaluating the harmonies of the movements entering into the antechamber of the mind by whatever doorway, and finally what mind there is in the sun. None the less, however those

1. See Kepler's commentary on this epilogue in the *Epitome*, pages 10-11.

things may be, this composition of the six primary spheres around the sun, cherishing it with their perpetual revolutions and as it were adoring it (just as, separately, four moons accompany the globe of Jupiter, two Saturn, but a single moon by its circuit encompasses, cherishes, fosters the Earth and us its inhabitants, and ministers to us) and this special business of the harmonies, which is a most clear footprint of the highest providence over solar affairs, now being added to that consideration, [324] wrings from me the following confession: not only does light go out from the sun into the whole world, as from the focus or eye of the world, as life and heat from the heart, as every movement from the King and mover, but conversely also by royal law these returns, so to speak, of every lovely harmony are collected in the sun from every province in the world, nay, the forms of movements by twos flow together and are bound into one harmony by the work of some mind, and are as it were coined money from silver and gold bullion; finally, the curia, palace, and praetorium or throne-room of the whole realm of nature are in the sun, whatsoever chancellors, palatines, prefects the Creator has given to nature: for them, whether created immediately from the beginning or to be transported hither at some time, has He made ready those seats. For even this terrestrial adornment, with respect to its principal part, for quite a long while lacked the contemplators and enjoyers, for whom however it had been appointed; and those seats were empty. Accordingly the reflection struck my mind, what did the ancient Pythagoreans in Aristotle mean, who used to call the centre of the world (which they referred to as the "fire" but understood by that the sun) "the watchtower of Jupiter," Διος φυλακὴν; what, likewise, was the ancient interpreter pondering in his mind when he rendered the verse of the Psalm as: "He has placed His tabernacle in the sun."

But also I have recently fallen upon the hymn of Proclus the Platonic philosopher (of whom there has been much mention in the preceding books), which was composed to the Sun and filled full with venerable mysteries, if you excise that one κλῦθ (hear me) from it; although the ancient interpreter already cited has explained this to some extent, *viz.*, in invoking the sun, he understands Him Who has placed His tabernacle in the sun. For Proclus lived at a time in which it was a crime, for which the rulers of the world and the people itself inflicted all punishments, to profess Jesus of Nazareth, God Our Savior, and to contemn the gods of the pagan poets (under Constantine, Maxentius, and Julian the Apostate). Accordingly Proclus, who from his Platonic philosophy indeed, by the natural light of the mind, had caught a distant glimpse of the Son of God, that true light which lighteth every man coming into this world, and who already knew that divinity must never be sought with a superstitious mob in sensible things, nevertheless preferred to seem to look for God in the sun rather than in Christ a sensible man, in order that at the same time he might both deceive the pagans by

honoring verbally the Titan of the poets and devote himself to his philosophy, by drawing away both the pagans and the Christians from sensible beings, the pagans from the visible sun, the Christians from the Son of Mary, because, trusting too much to the natural light of reason, he spit out the mystery of the Incarnation; and finally that at the same time he might take over from them and adopt into his own philosophy whatever the Christians had which was most divine and especially consonant with Platonic philosophy.[1] And so the accusation of the teaching of the Gospel concerning Christ is laid against this hymn of Proclus, in its own matters: let that Titan keep as his private possessions χρῦσα ἡνία (golden reins) and ταμιεῖυν φαοῦς, μεσσατὶην, αἰθέρος ἕδρην, κοδμοῦ κραδιαῖον ἐριφεγγέα κυκλὸν (a treasury of light, a seat at the midpart of the ether, a radiant circle at the heart of the world), which visible aspect Copernicus too bestows upon him; let him even keep his παλιννοστοὺς διφρείς (cyclical chariot-drivings), although according to the ancient Pythagoreans he does not possess them but in their place τὸ κέντρον, Διὸς φυλακήν (the centre, the watchtower of Zeus)—which doctrine, misshapen by the forgetfulness of ages, as by a flood, was not recognized by their follower Proclus; let him also keep his γενεθλὴν Βλαστησασαν (offspring born) of himself, and whatever else is of nature; in turn, let the philosophy of Proclus yield to Christian doctrines, [325] let the sensible sun yield to the Son of Mary, the Son of God, Whom Proclus addresses under the name of the Titan ζωαρκεὸς, ὦ ἄνα, πηγὴς αὐτὸς ἔχων κλήδα (O lord, who dost hold the key of the life-supporting spring), and that πάντα τεῖς ἔπλήσας ἐλερσινοοῖο προνόιης (thou didst fulfill all things with thy mind-awakening foresight), and that immense power over the μοιρὰων (fates), and things which were read of in no philosophy before the promulgation of the Gospel,[2] the demons dreading him as their threatening scourge, the demons lying in ambush for souls, ὄφρα ὑφιτενοῦς λαθοῖντο πατρὸς περιφέγγεοσ αὐλής (in order that they might escape the notice of the light-filled hall of the lofty father); and who except the Word of the Father is that εἰκὼν παγγενεταο, θεοῦ, οὐφάευτος ἀπ' ἀρρήτου γενετῆρος παύσατο στοιχείων ὀρυμάγδος ἐπ ἀλληλοῖσιν ἰόντων (image of the all-begetting father, upon whose manifestation from an ineffable mother the sin of the elements changing into one another ceased), according to the following: *The Earth was unwrought and a chaotic mass, and darkness was upon the face of the abyss, and God divided the light from the darkness, the waters from the waters, the sea from the dry land; and: all things were made by the very Word.* Who except Jesus of Nazareth the Son of God, ψυχῶν ἀναγωγεύς (the shepherd of souls), to whom ἱκεσιὴ πολυδάκρυος (the prayer of a tearful suppliant)

---

1. It was the judgment of the ancients concerning his book *Metroace* that in it he set forth, not without divine rapture, his universal doctrine concerning God; and by the frequent tears of the author apparent in it all suspicion was removed from the hearers. None the less this same man wrote against the Christians eighteen epichiremata, to which John Philoponus opposed himself, reproaching Proclus with ignorance of Greek thought, which none the less he had undertaken to defend. That is to say, Proclus concealed those things which did not make for his own philosophy.

2. Nevertheless in Suidas some similar things are attributed to ancient Orpheus, nearly equal to Moses, as if his pupil; see too the hymns of Orpheus, on which Proclus wrote commentaries.

is to be offered, in order that He cleanse us from sins and wash us of the filth τῆς γενεθλῆς (of generation)—as if Proclus acknowledged the forms of original sin—and guard us from punishment and evil, πρηυνῶν θόον ὄμμα δικῆς (by making mild the quick eye of justice), namely, the wrath of the Father? And the other things we read of, which are as it were taken from the hymn of Zacharias (or, accordingly, was that hymn a part of the *Metroace?*) Αχλυν ἀποσκεδάσας ὀλεσίμβροτον ἰολοχεύτον (dispersing the poisonous, man-destroying mist), *viz.*, in order that He may give to souls living in darkness and the shadows of death the φάος ἁγνον (holy light) and ὄλβον ἀστυφελικτὸν ἀπ ἐυσεβίνἐρατείης (unshaken happiness from lovely piety); for that is to serve God in holiness and justice all our days.

Accordingly, let us separate out these and similar things and restore them to the doctrine of the Catholic Church to which they belong. But let us see what the principal reason is why there has been mention made of the hymn. For this same sun which ὕψοθεν ἁρμνίης ῥῦμα πλοῦσιον ἐξοτεύει (sluices the rich flow of harmony from on high)—so too Orpheus κόσμου τὸν ἐναρμόνιον δρόμον ἕλκων (making move the harmonious course of the world)—the same, concerning whose stock Phoebus about to rise κιθαρῇ ὑπὸ θέσκελα μελῶν εὐνάξει μεγὰ κῦμα βαρυφλσισβοῖο γενεθλῆς (sings marvellous things on his lyre and lulls to sleep the heavy-sounding surge of generation) and in whose dance Paean is the partner, πλήσας ἁρμονὶης παναπήμονος εὔρεα κόσμν (striking the wide sweep of innocent harmony)—him, I say, does Proclus at once salute in the first verse of the hymn as πῦρος νοεροῦ βασιλέα (king of intellectual fire). By that commencement, at the same time, he indicates what the Pythagoreans understood by the word of fire (so that it is surprising that the pupil should disagree with the masters in the position of the centre) and at the same time he transfers his whole hymn from the body of the sun and its quality and light, which are sensibles, to the intelligibles, and he has assigned to that πῦρ νοερὸς (intellectual fire) of his—perhaps the artisan fire of the Stoics—to that created God of Plato, that chief or self-ruling mind, a royal throne in the solar body, confounding into one the creature and Him through Whom all things have been created. But we Christians, who have been taught to make better distinctions, know that this eternal and uncreated "Word," Which was "with God" and Which is contained by no abode, although He is within all things, excluded by none, although He is outside of all things, took up into unity of person flesh out of the womb of the most glorious Virgin Mary, and, when the ministry of His flesh was finished, occupied as His royal abode the heavens, wherein by a certain excellence over and above the other parts of the world, *viz.*, through His glory and majesty, His celestial Father too is recognized to dwell, and has also promised to His faithful, mansions in that house of His Father: as for the remainder concerning that abode, we believe it superfluous to inquire into it too curiously or to forbid the

senses or natural reasons to investigate that which the eye has not seen nor the ear heard and into which the heart of man has not ascended; but we duly subordinate the created mind—of whatsoever excellence it may be—to Its Creator, and we introduce neither God-intelligences with Aristotle and the pagan philosophers nor armies of innumerable planetary spirits with the Magi, nor do we propose that they are either to be adored or summoned to intercourse with us by theurgic superstitions, for we have a careful fear of that; but we freely inquire by natural reasons what sort of thing each mind is, especially if in the heart of the world [326] there is any mind bound rather closely to the nature of things and performing the function of the soul of the world— or if also some intelligent creatures, of a nature different from human perchance do inhabit or will inhabit the globe thus animated (see my book *on the New Star*, Chapter 24, "On the Soul of the World and Some of Its Functions"). But if it is permissible, using the thread of analogy as a guide, to traverse the labyrinths of the mysteries of nature, not ineptly, I think, will someone have argued as follows: The relation of the six spheres to their common centre, thereby the centre of the whole world, is also the same as that of διανοὶα (discursive intellection) to νοῦς (intuitive intellection), according as those faculties are distinguished by Aristotle, Plato, Proclus, and the rest; and the relation of the single planets' revolutions in place around the sun to the ἀμετάθεδον (unvarying) rotation of the sun in the central space of the whole system (concerning which the sun-spots are evidence; this has been demonstrated in the *Commentaries on the Movement of Mars*) is the same as the relation of τὸ διανοητικὸν to τὸ νοερὸν, that of the manifold discourses of ratiocination to the most simple intellection of the mind. For as the sun rotating into itself moves all the planets by means of the form emitted from itself, so too—as the philosophers teach—mind, by understanding itself and in itself all things, stirs up ratiocinations, and by dispersing and unrolling its simplicity into them, makes everything to be understood. And the movements of the planets around the sun at their centre and the discourses of ratiocinations are so interwoven and bound together that, unless the Earth, our domicile, measured out the annual circle, midway between the other spheres—changing from place to place, from station to station—never would human ratiocination have worked its way to the true intervals of the planets and to the other things dependent from them, never would it have constituted astronomy. (See the *Optical Part of Astronomy*, Chapter 9.)

On the other hand, in a beautiful correspondence, simplicity of intellection follows upon the stillness of the sun at the centre of the world, in that hitherto we have always worked under the assumption that those solar harmonies of movements are defined neither by the diversity of regions nor by the amplitude of the expanses of the world. As a matter of fact, if any mind observes from the sun those harmonies, that mind is without the assistance afforded by the movement and diverse stations of his

abode, by means of which it may string together ratiocinations and discourse necessary for measuring out the planetary intervals. Accordingly, it compares the diurnal movements of each planet, not as they are in their own orbits but as they pass through the angles at the centre of the sun. And so if it has knowledge of the magnitude of the spheres, this knowledge must be present in it *a priori*, without any toil of ratiocination: but to what extent that is true of human mind and of sublunary nature has been made clear above, from Plato and Proclus.

Under these circumstances, it will not have been surprising if anyone who has been thoroughly warmed by taking a fairly liberal draft from that bowl of Pythagoras which Proclus gives to drink from in the very first verse of the hymn, and who has been made drowsy by the very sweet harmony of the dance of the planets begins to dream (by telling a story he may imitate Plato's Atlantis and, by dreaming, Cicero's Scipio): throughout the remaining globes, which follow after from place to place, there have been disseminated discursive or ratiocinative faculties, whereof that one ought assuredly to be judged the most excellent and absolute which is in the middle position among those globes, *viz.*, in man's earth, while there dwells in the sun simple intellect, $\pi\hat{\upsilon}\rho$ $\nu o\epsilon\rho\grave{o}\nu$, or $\nu o\hat{\upsilon}\varsigma$, the source, whatsoever it may be, of every harmony.

For if it was Tycho Brahe's opinion concerning that bare wilderness of globes that it does not exist fruitlessly in the world but is filled with inhabitants: with how much greater probability shall we make a conjecture as to God's works and designs even for the other globes, from that variety which we discern in this globe of the Earth. For He Who created the species which should inhabit the waters, beneath which however there is no room for the air [327] which living things draw in; Who sent birds supported on wings into the wilderness of the air; Who gave white bears and white wolves to the snowy regions of the North, and as food for the bears the whale, and for the wolves, birds' eggs; Who gave lions to the deserts of burning Libya and camels to the widespread plains of Syria, and to the lions an endurance of hunger, and to the camels an endurance of thirst: did He use up every art in the globe of the Earth so that He was unable, every goodness so that he did not wish, to adorn the other globes too with their fitting creatures, as either the long or short revolutions, or the nearness or removal of the sun, or the variety of eccentricities or the shine or darkness of the bodies, or the properties of the figures wherewith any region is supported persuaded?

Behold, as the generations of animals in this terrestrial globe have an image of the male in the dodecahedron, of the female in the icosahedron—whereof the dodecahedron rests on the terrestrial sphere from the outside and the icosahedron from the inside: what will we suppose the remaining globes to have, from the remaining figures? For whose good do four moons encircle Jupiter, two Saturn, as does this our moon this our domicile? But in the same way we shall ratiocinate concerning the globe of the sun

also, and we shall as it were incorporate conjectures drawn from the harmonies, *et cetera*—which are weighty of themselves—with other conjectures which are more on the side of the bodily, more suited for the apprehension of the vulgar. Is that globe empty and the others full, if everything else is in due correspondence? If as the Earth breathes forth clouds, so the sun black smoke? If as the Earth is moistened and grows under showers, so the sun shines with those combusted spots, while clear flamelets sparkle in its all fiery body. For whose use is all this equipment, if the globe is empty? Indeed, do not the senses themselves cry out that fiery bodies dwell here which are receptive of simple intellects, and that truly the sun is, if not the king, at least the queen πυρὸς νοεροῦ (of intellectual fire)?

Purposely I break off the dream and the very vast speculation, merely crying out with the royal Psalmist: *Great is our Lord and great His virtue and of His wisdom there is no number: praise Him, ye heavens, praise Him, ye sun, moon, and planets, use every sense for perceiving, every tongue for declaring your Creator. Praise Him, ye celestial harmonies, praise Him, ye judges of the harmonies uncovered* (and you before all, old happy Mastlin, for you used to animate these cares with words of hope): *and thou my soul, praise the Lord thy Creator, as long as I shall be: for out of Him and through Him and in Him are all things,* καὶ τὰ αἰσθητὰ καὶ τὰ νοερὰ (*both the sensible and the intelligible*); *for both whose whereof we are utterly ignorant and those which we know are the least part of them; because there is still more beyond. To Him be praise, honour, and glory, world without end. Amen.*

THE END

This work was completed on the 17th or 27th day of May, 1618; but Book V was reread (while the type was being set) on the 9th or 19th of February, 1619. At Linz, the capital of Austria—above the Enns.

Isaac Newton

(1642-1727)

HIS LIFE AND WORK

On February 5, 1676, Isaac Newton penned a letter to his bitter enemy, Robert Hooke, which contained the sentence, "If I have seen farther, it is by standing on the shoulders of giants." Often described as Newton's nod to the scientific discoveries of Copernicus, Galileo, and Kepler before him, it has become one of the most famous quotes in the history of science. Indeed, Newton did recognize the contributions of those men, some publicly and others in private writings. But in his letter to Hooke, Newton was referring to optical theories, specifically the study of the phenomena of thin plates, to which Hooke and Renè Descartes had made significant contributions.

Some scholars have interpreted the sentence as a thinly veiled insult to Hooke, whose crooked posture and short stature made him anything but a giant, especially in the eyes of the extremely vindictive Newton. Yet despite their feuds, Newton did appear to humbly acknowledge the noteworthy research in optics of both Hooke and Descartes, adopting a more conciliatory tone at the end of the letter.

Isaac Newton is considered the father of the study of infinitesimal calculus, mechanics and planetary motion, and the theory of light and color. But he secured his place in history by formulating gravitational force and defining the laws of motion and attraction in his landmark work, *Mathematical Principles of Natural Philosophy* (*Philosophiae Naturalis Principia Mathematica*) generally known as *Principia*. There

Newton fused the scientific contributions of Copernicus, Galileo, Kepler, and others into a dynamic new symphony. *Principia*, the first book on theoretical physics, is roundly regarded as the most important work in the history of science and the scientific foundation of the modern worldview.

Newton wrote the three books that form *Principia* in just eighteen months and, astonishingly, between severe emotional breakdowns—likely compounded by his competition with Hooke. He even went to such vindictive lengths as to remove from the book all references to Hooke's work, yet his hatred for his fellow scientist may have been the very inspiration for *Principia*.

The slightest criticism of his work, even if cloaked in lavish praise, often sent Newton into dark withdrawal for months or years. This trait revealed itself early in Newton's life and has led some to wonder what other questions Newton might have answered had he not been obsessed with settling personal feuds. Others have speculated that Newton's scientific discoveries and achievements were the result of his vindictive obsessions and might not have been possible had he been less arrogant.

As a young boy, Isaac Newton asked himself the questions that had long mystified humanity, and then went on to answer many of them. It was the beginning of a life full of discovery, despite some anguishing first steps. Isaac Newton was born in the English industrial town of Woolsthorpe, Lincolnshire, on Christmas Day of 1642, the same year in which Galileo died. His mother did not expect him to live long, as he was born very prematurely; he would later describe himself as having been so small at birth he could fit into a quart pot. Newton's yeoman father, also named Isaac, had died three month's earlier, and when Newton reached two years of age, his mother, Hannah Ayscough, remarried, wedding Barnabas Smith, a rich clergyman from North Witham.

Apparently there was no place in the new Smith family for the young Newton, and he was placed in the care of his grandmother, Margery Ayscough. The specter of this abandonment, coupled with the tragedy of never having known his father, haunted Newton for the rest of his life. He despised his stepfather; in journal entries for 1662 Newton, examining his sins, recalled "threatening my father and mother Smith to burne them and the house over them."

Much like his adulthood, Newton's childhood was filled with episodes of harsh, vindictive attacks, not only against perceived enemies but against friends and family as well. He also displayed the kind of curiosity early on that would define his life's achievements, taking an interest in mechanical models and architectural drawing. Newton spent countless hours building clocks, flaming kites, sundials, and miniature mills (powered by mice) as well as drawing elaborate sketches of animals and ships. At the age of five he attended schools at Skillington and Stoke but was considered one of the poorest students, receiving comments in teachers' reports such as "inattentive" and

"idle." Despite his curiosity and demonstrable passion for learning, he was unable to apply himself to schoolwork.

By the time Newton reached the age of ten, Barnabas Smith had passed away and Hannah had come into a considerable sum from Smith's estate. Isaac and his grandmother began living with Hannah, a half-brother, and two half-sisters. Because his work at school was uninspiring, Hannah decided that Isaac would be better off managing the farm and estate, and she pulled him out of the Free Grammar School in Grantham. Unfortunately for her, Newton had even less skill or interest in managing the family estate than he had in schoolwork. Hannah's brother, William, a clergyman, decided that it would be best for the family if the absent-minded Isaac returned to school to finish his education.

This time, Newton lived with the headmaster of the Free Grammar School, John Stokes, and he seemed to turn a corner in his education. One story has it that a blow to the head, administered by a schoolyard bully, somehow enlightened him, enabling the young Newton to reverse the negative course of his educational promise. Now demonstrating intellectual aptitude and curiosity, Newton began preparing for further study at a university. He decided to attend Trinity College, his uncle William's alma mater, at Cambridge University.

At Trinity, Newton became a subsizar, receiving an allowance toward the cost of his education in exchange for performing various chores such as waiting tables and cleaning rooms for the faculty. But by 1664 he was elected scholar, which status guaranteed him financial support and freed him from menial duties. When the university closed because of the bubonic plague in 1665, Newton retreated to Lincolnshire. In the eighteen months he spent at home during the plague he devoted himself to mechanics and mathematics, and began to concentrate on optics and gravitation. This "annus mirabilis" (miraculous year), as Newton called it, was one of the most productive and fruitful periods of his life. It is also around this time that an apple, according to legend, fell onto Newton's head, awakening him from a nap under a tree and spurring him on to define the laws of gravity. However far-fetched the tale, Newton himself wrote that a falling apple had "occasioned" his foray into gravitational contemplation, and he is believed to have performed his pendulum experiments then. "I was in the prime of my age for invention," Newton later recalled, "and minded Mathematicks and Philosophy more then at any time since."

When he returned to Cambridge, Newton studied the philosophy of Aristotle and Descartes, as well as the science of Thomas Hobbes and Robert Boyle. He was taken by the mechanics of Copernicus and Galileo's astronomy, in addition to Kepler's optics. Around this time, Newton began his prism experiments in light refraction and dispersion, possibly in his room at Trinity or at home in Woolsthorpe. A development at

the university that clearly had a profound influence on Newton's future —was the arrival of Isaac Barrow, who had been named the Lucasian Professor of Mathematics. Barrow recognized Newton's extraordinary mathematical talents, and when he resigned his professorship in 1669 to pursue theology he recommended the twenty-seven-year-old Newton as his replacement.

Newton's first studies as Lucasian Professor centered in the field of optics. He set out to prove that white light was composed of a mixture of various types of light, each producing a different color of the spectrum when refracted by a prism. His series of elaborate and precise experiments to prove that light was composed of minute particles drew the ire of scientists such as Hooke, who contended that light traveled in waves. Hooke challenged Newton to offer further proof of his eccentric optical theories. Newton's way of responding was one he did not outgrow as he matured. He withdrew, set out to humiliate Hooke at every opportunity, and refused to publish his book, Opticks , until after Hooke's death in 1703.

Early in his tenure as Lucasian Professor, Newton was well along in his study of pure mathematics, but he shared his work with very few of his colleagues. Already by 1666, he had discovered general methods of solving problems of curvature—what he termed "theories of fluxions and inverse fluxions." The discovery set off a dramatic feud with supporters of the German mathematician and philosopher Gottfried Wilhelm Leibniz, who more than a decade later published his findings on differential and integral calculus. Both men arrived at roughly the same mathematical principles, but Leibniz published his work before Newton. Newton's supporters claimed that Leibniz had seen the Lucasian Professor's papers years before, and a heated argument between the two camps, known as the Calculus Priority Dispute, did not end until Leibniz died in 1716. Newton's vicious attacks which often spilled over to touch on views about God and the universe, as well as his accusations of plagiarism left Leibniz impoverished and disgraced.

Most historians of science believe that the two men in fact arrived at their ideas independently and that the dispute was pointless. Newton's vitriolic aggression toward Leibniz took a physical and emotional toll on Newton as well. He soon found himself involved in another battle, this time over his theory of color and with the English Jesuits, and in 1678 he suffered a severe mental breakdown. The next year, his mother, Hannah passed away, and Newton began to distance himself from others. In secret, he delved into alchemy, a field widely regarded already in Newton's time as fruitless. This episode in the scientist's life has been a source of embarrassment to many Newton scholars. Only long after Newton died did it become apparent that his interest in chemical experiments was related to his later research in celestial mechanics and gravitation.

Newton had already begun forming theories about motion by 1666, but he was as yet unable to adequately explain the mechanics of circular motion. Some fifty years earlier, the German mathematician and astronomer Johannes Kepler had proposed three laws of planetary motion, which accurately described how the planets moved in relation to the sun, but he could not explain why the planets moved as they did. The closest Kepler came to understanding the forces involved was to say that the sun and the planets were "magnetically" related.

Newton set out to discover the cause of the planets' elliptical orbits. By applying his own law of centrifugal force to Kepler's third law of planetary motion (the law of harmonies) he deduced the inverse-square law, which states that the force of gravity between any two objects is inversely proportional to the square of the distance between the object's centers. Newton was thereby coming to recognize that gravitation is universal—that one and the same force causes an apple to fall to the ground and the moon to race around the earth. He then set out to test the inverse-square relation against known data. He accepted Galileo's estimate that the moon is sixty earth radii from the earth, but the inaccuracy of his own estimate of the earth's diameter made it impossible to complete the test to his satisfaction. Ironically, it was an exchange of letters in 1679 with his old adversary Hooke that renewed his interest in the problem. This time, he turned his attention to Kepler's second law, the law of equal areas, which Newton was able to prove held true because of centripetal force. Hooke, too, was attempting to explain the planetary orbits, and some of his letters on that account were of particular interest to Newton.

At an infamous gathering in 1684, three members of the Royal Society, Robert Hooke, Edmond Halley, and Christopher Wren, the noted architect of St. Paul's Cathedral, engaged in a heated discussion about the inverse-square relation governing the motions of the planets. In the early 1670s, the talk in the coffeehouses of London and other intellectual centers was that gravity emanated from the sun in all directions and fell off at a rate inverse to the square of the distance, thus becoming more and more diluted over the surface of the sphere as that surface expands. The 1684 meeting was, in effect, the birth of *Principia*. Hooke declared that he had derived from Kepler's law of ellipses the proof that gravity was an emanating force, but would withhold it from Halley and Wren until he was ready to make it public. Furious, Halley went to Cambridge, told Newton Hooke's claim, and proposed the following problem. "What would be the form of a planet's orbit about the sun if it were drawn towards the sun by a force that varied inversely as the square of the distance?" Newton's response was staggering. "It would be an ellipse," he answered immediately, and then told Halley that he had solved the problem four years earlier but had misplaced the proof in his office.

At Halley's request, Newton spent three months reconstituting and improving the proof. Then, in a burst of energy sustained for eighteen months, during which was so caught up in his work that he often forgot to eat, he further developed these ideas until their presentation filled three volumes. Newton chose to title the work *Philosophiae Naturalis Principia Mathematica*, in deliberate contrast with Descartes' *Principia Philosophiae*. The three books of Newton's *Principia* provided the link between Kepler's laws and the physical world. Halley reacted with "joy and amazement" to Newton's discoveries. To Halley, it seemed the Lucasian Professor had succeeded where all others had failed, and he personally financed publication of the massive work as a masterpiece and a gift to humanity.

Where Galileo had shown that objects were "pulled" toward the center of the earth, Newton was able to prove that this same force, gravity, affected the orbits of the planets. He was also familiar with Galileo's work on the motion of projectiles, and he asserted that the moon's orbit around the earth adhered to the same principles. Newton demonstrated that gravity explained and predict the moon's motions as well as the rising and falling of the tides on earth. Book 1 of *Principia* encompasses Newton's three laws of motion:

1. Every body perseveres in its state of resting, or uniformly moving in a right line, unless it is compelled to change that state by forces impressed upon it.
2. The change of motion is proportional to the motive force impressed; and is made in the direction of the right line in which that force is impressed.
3. To every action there is always opposed an equal reaction; or, the mutual actions of two bodies upon each other are always equal, and directed to contrary directions.

Book 2 began for Newton as something of an afterthought to Book 1; it was not included in the original outline of the work. It is essentially a treatise on fluid mechanics, and it allowed Newton room to display his mathematical ingenuity. Toward the end of the book, Newton concludes that the vortices invoked by Descartes to explain the motions of planets do not hold up to scrutiny, for the motions could be performed in free space without vortices. How that is so, Newton wrote, "may be understood by the first Book; and I shall now more fully treat of it in the following Book."

In Book 3, subtitled *System of the World*, by applying the laws of motion from book 1 to the physical world Newton concluded that "there is a power of gravity tending to all bodies, proportional to the several quantities of matter which they contain." He thus demonstrated that his law of universal gravitation could explain the motions of the six known planets, as well as moons, comets, equinoxes, and tides. The law states that all matter is mutually attracted with a force directly proportional to the product of their masses and inversely proportional to the square of the distance between them.

Newton, by a single set of laws, had united the earth with all that could be seen in the skies. In the first two "Rules of Reasoning" from Book 3, Newton wrote:

We are to admit no more causes of natural things than such as are both true and sufficient to explain their appearances. Therefore, to the same natural effects we must, as far as possible, assign the same causes.

It is the second rule that actually unifies heaven and earth. An Aristotelian would have asserted that heavenly motions and terrestrial motions are manifestly not the same natural effects and that Newton's second rule could not, therefore, be applied. Newton saw things otherwise.

Principia was moderately praised upon its publication in 1687, but only about five hundred copies of the first edition were printed. However, Newton's nemesis, Robert Hooke, had threatened to spoil any coronation Newton might have enjoyed. After Book 2 appeared, Hooke publicly claimed that the letters he had written in 1679 had provided scientific ideas that were vital to Newton's discoveries. His claims, though not without merit, were abhorrent to Newton, who vowed to delay or even abandon publication of Book 3. Newton ultimately relented and published the final book of *Principia* , but not before painstakingly removing from it every mention of Hooke's name.

Newton's hatred for Hooke consumed him for years afterward. In 1693, he suffered yet another nervous breakdown and retired from research. He withdrew from the Royal Society until Hooke's death in 1703, then was elected its president and reelected each year until his own death in 1727. He also withheld publication of *Opticks*, his important study of light and color that would become his most widely read work, until after Hooke was dead.

Newton began the eighteenth century in a government post as warden of the Royal Mint, where he utilized his work in alchemy to determine methods for reestablishing the integrity of the English currency. As president of the Royal Society, he continued to battle perceived enemies with inexorable determination, in particular carrying on his longstanding feud with Leibniz over their competing claims to have invented calculus. He was knighted by Queen Anne in 1705, and lived to see publication of the second and third editions of *Principia*.

Isaac Newton died in March 1727, after bouts of pulmonary inflammation and gout. As was his wish, Newton had no rival in the field of science. The man who apparently formed no romantic attachments with women (some historians have speculated on possible relationships with men, such as the Swiss natural philosopher Nicolas Fatio de Duillier) cannot, however, be accused of a lack of passion for his work. The poet Alexander Pope, a contemporary of Newton's, most elegantly described the great thinker's gift to humanity:

Nature and Nature's laws lay hid in night:
God said, "Let Newton be! and all was light."

For all the petty arguments and undeniable arrogance that marked his life, toward its end Isaac Newton was remarkably poignant in assessing his accomplishments: "I do not know how I may appear to the world, but to myself I seem to have been only like a boy, playing on the sea-shore, and diverting myself, in now and then finding a smoother pebble or prettier shell than ordinary, whilst the great ocean of truth lay all undiscovered before me."

THE
MATHEMATICAL PRINCIPLES
OF
NATURAL PHILOSOPHY.

DEFINITIONS.

DEFINITION I.

THE QUANTITY OF MATTER IS THE MEASURE OF THE SAME, ARISING FROM ITS DENSITY AND BULK CONJUNCTLY.

Thus air of a double density, in a double space, is quadruple in quantity; in a triple space, sextuple in quantity. The same thing is to be understood of snow, and fine dust or powders, that are condensed by compression or liquefaction; and of all bodies that are by any causes whatever differently condensed. I have no regard in this place to a medium, if any such there is, that freely pervades the interstices between the parts of bodies. It is this quantity that I mean hereafter everywhere under the name of body or mass, And the same is known by the weight of each body; for it is proportional to the weight, as I have found by experiments on pendulums, very accurately made, which shall be shewn hereafter.

DEFINITION II.

THE QUANTITY OF MOTION IS THE MEASURE OF THE SAME, ARISING FROM THE VELOCITY AND QUANTITY OF MATTER CONJUNCTLY.

The motion of the whole is the sum of the motions of all the parts; and therefore in a body double in quantity, with equal velocity, the motion is double; with twice the velocity, it is quadruple.

DEFINITION III.

THE *VIS INSITA*, OR INNATE FORCE OF MATTER, IS A POWER
OF RESISTING, BY WHICH EVERY BODY, AS MUCH AS IN IT
LIES, ENDEAVOURS TO PERSEVERE IN ITS PRESENT STATE,
WHETHER IT BE OF REST, OR OF MOVING UNIFORMLY
FORWARD IN A RIGHT LINE.

This force is ever proportional to the body whose force it is: and differs nothing from the inactivity of the mass, but in our manner of conceiving it. A body, from the inactivity of matter, is not without difficulty put out of its state of rest or motion. Upon which account, this *vis insita*, may, by a most significant name, be called *vis inertiæ*, or force of inactivity. But a body exerts this force only when another force, impressed upon it, endeavours to change its condition; and the exercise of this force may be considered both as resistance and impulse; it is resistance, in so far as the body, for maintaining its present state, withstands the force impressed; it is impulse, in so far as the body, by not easily giving way to the impressed force of another, endeavours to change the state of that other. Resistance is usually ascribed to bodies at rest, and impulse to those in motion; but motion and rest, as commonly conceived, are only relatively distinguished; nor are those bodies always truly at rest, which commonly are taken to be so.

DEFINITION IV.

AN IMPRESSED FORCE IS AN ACTION EXERTED UPON A BODY,
IN ORDER TO CHANGE ITS STATE, EITHER OF REST, OR OF
MOVING UNIFORMLY FORWARD IN A RIGHT LINE.

This force consists in the action only; and remains no longer in the body, when the action is over. For a body maintains every new state it acquires, by its *vis inertiæ* only. Impressed forces are of different origins as from percussion, from pressure, from centripetal force.

DEFINITION V.

A CENTRIPETAL FORCE IS THAT BY WHICH BODIES ARE
DRAWN OR IMPELLED, OR ANY WAY TEND, TOWARDS A POINT
AS TO A CENTRE.

Of this sort is gravity, by which bodies tend to the centre of the earth magnetism, by which iron tends to the loadstone; and that force, whatever it is, by which the planets are

perpetually drawn aside from the rectilinear motions, which otherwise they would pursue, and made to revolve in curvilinear orbits. A stone, whirled about in a sling, endeavours to recede from the hand that turns it; and by that endeavour, distends the sling, and that with so much the greater force, as it is revolved with the greater velocity, and as soon as ever it is let go, flies away. That force which opposes itself to this endeavour, and by which the sling perpetually draws back the stone towards the hand, and retains it in its orbit, because it is directed to the hand as the centre of the orbit, I call the centripetal force. And the same thing is to be understood of all bodies, revolved in any orbits. They all endeavour to recede from the centres of their orbits; and were it not for the opposition of a contrary force which restrains them to, and detains them in their orbits, which I therefore call centripetal, would fly off in right lines, with an uniform motion. A projectile, if it was not for the force of gravity, would not deviate towards the earth, but would go off from it in a right line, and that with an uniform motion, if the resistance of the air was taken away. It is by its gravity that it is drawn aside perpetually from its rectilinear course, and made to deviate towards the earth, more or less, according to the force of its gravity, and the velocity of its motion. The less its gravity is, for the quantity of its matter, or the greater the velocity with which it is projected, the less will it deviate from a rectilinear course, and the farther it will go. If a leaden hall, projected from the top of a mountain by the force of gunpowder with a given velocity, and in a direction parallel to the horizon, is carried in a curve line to the distance of two miles before it falls to the ground; the same, if the resistance of the air were taken away, with a double or decuple velocity, would fly twice or ten times as far. And by increasing the velocity, we may at pleasure increase the distance to which it might be projected, and diminish the curvature of the line, which it might describe, till at last it should fall at the distance of 10, 30, or 90 degrees, or even might go quite round the whole earth before it falls; or lastly, so that it might never fall to the earth, but go forward into the celestial spaces, and proceed in its motion *in infinitum*. And after the same manner that a projectile, by the force of gravity, may be made to revolve in an orbit, and go round the whole earth, the moon also, either by the force of gravity, if it is endued with gravity, or by any other force, that impels it towards the earth, may be perpetually drawn aside towards the earth, out of the rectilinear way, which by its innate force it would pursue; and would be made to revolve in the orbit which it now describes; nor could the moon without some such force, be retained in its orbit. If this force was too small, it would not sufficiently turn the moon out of a rectilinear course: if it was too great, it would turn it too much, and draw down the moon from its orbit towards the earth. It is necessary, that the force be of a just quantity, and it belongs to the mathematicians to find the force, that may serve exactly to retain a body in a given orbit, with a given velocity; and *vice versa*, to determine the curvilinear way,

into which a body projected from a given place, with a given velocity, may be made to deviate from its natural rectilinear way, by means of a given force.

The quantity of any centripetal force may be considered as of three kinds; absolute, accelerative, and motive.

DEFINITION VI.

THE ABSOLUTE QUANTITY OF A CENTRIPETAL FORCE IS THE
MEASURE OF THE SAME PROPORTIONAL TO THE EFFICACY OF
THE CAUSE THAT PROPAGATES IT FROM THE CENTRE,
THROUGH THE SPACES ROUND ABOUT.

Thus the magnetic force is greater in one load-stone and less in another according to their sizes and strength of intensity.

DEFINITION VII.

THE ACCELERATIVE QUANTITY OF A CENTRIPETAL FORCE IS
THE MEASURE OF THE SAME, PROPORTIONAL TO THE
VELOCITY WHICH IT GENERATES IN A GIVEN TIME.

Thus the force of the same load-stone is greater at a less distance, and less at a greater: also the force of gravity is greater in valleys, less on tops of exceeding high mountains; and yet less (as shall hereafter be shown), at greater distances from the body of the earth; but at equal distances, it is the same everywhere; because (taking away, or allowing for, the resistance of the air), it equally accelerates all falling bodies, whether heavy or light, great or small.

DEFINITION VIII.

THE MOTIVE QUANTITY OF A CENTRIPETAL FORCE, IS THE
MEASURE OF THE SAME, PROPORTIONAL TO THE MOTION
WHICH IT GENERATES IN A GIVEN TIME.

Thus the weight is greater in a greater body, less in a less body; and, in the same body, it is greater near to the earth, and less at remoter distances. This sort of quantity is the centripetency, or propension of the whole body towards the centre, or, as I may say, its weight; and it is always known by the quantity of an equal and contrary force just sufficient to hinder the descent of the body.

These quantities of forces, we may, for brevity's sake, call by the names of motive, accelerative, and absolute forces; and, for distinction's sake, consider them, with respect

to the bodies that tend to the centre; to the places of those bodies; and to the centre of force towards which they tend; that is to say, I refer the motive force to the body as an endeavour and propensity of the whole towards a centre, arising from the propensities of the several parts taken together; the accelerative force to the place of the body, as a certain power or energy diffused from the centre to all places around to move the bodies that are in them; and the absolute force to the centre, as endued with some cause, without which those motive forces would not be propagated through the spaces round about; whether that cause be some central body (such as is the load-stone, in the centre of the magnetic force, or the earth in the centre of the gravitating force), or anything else that does not yet appear. For I here design only to give a mathematical notion of those forces, without considering their physical causes and seats.

Wherefore the accelerative force will stand in the same relation to the motive, as celerity does to motion. For the quantity of motion arises from the celerity drawn into the quantity of matter; and the motive force arises from the accelerative force drawn into the same quantity of matter. For the sum of the actions of the accelerative force, upon the several articles of the body, is the motive force of the whole. Hence it is, that near the surface of the earth, where the accelerative gravity, or force productive of gravity, in all bodies is the same, the motive gravity or the weight is as the body: but if we should ascend to higher regions, where the accelerative gravity is less, the weight would be equally diminished, and would always be as the product of the body, by the accelerative gravity. So in those regions, where the accelerative gravity is diminished into one half, the weight of a body two or three times less, will be four or six times less.

I likewise call attractions and impulses, in the same sense, accelerative, and motive; and use the words attraction, impulse or propensity of any sort towards a centre, promiscuously, and indifferently, one for another; considering those forces not physically, but mathematically: wherefore, the reader is not to imagine, that by those words, I anywhere take upon me to define the kind, or the manner of any action, the causes or the physical reason thereof, or that I attribute form, in a true and physical sense, to certain centres (which are only mathematical points); when at any time I happen to speak of centres as attracting, or as endued with attractive powers.

SCHOLIUM.

Hitherto I have laid down the definitions of such words as are less known, and explained the sense in which I would have them to be understood in the following discourse. I do not define time, space, place and motion, as being well known to all. Only I must observe that the vulgar conceive those quantities under no other notions but from the relation they bear to sensible objects. And thence arise certain prejudices, for the removing of which, it will be convenient to distinguish them into absolute and relative, true and apparent, mathematical and common.

I. Absolute, true, and mathematical time, of itself, and from its own nature flows equably without regard to anything external, and by another name is called duration: relative, apparent, and common time, is some sensible and external (whether accurate or unequable) measure of duration by the means of motion, which is commonly used instead of true time; such as an hour, a day, a month, a year.

II. Absolute space, in its own nature, without regard to anything external, remains always similar and immovable. Relative space is some movable dimension or measure of the absolute spaces; which our senses determine by its position to bodies; and which is vulgarly taken for immovable space; such is the dimension of a subterraneous, an æreal, or celestial space, determined by its position in respect of the earth. Absolute and relative space, are the same in figure and magnitude; but they do not remain always numerically the same. For if the earth, for instance, moves, a space of our air, which relatively and in respect of the earth remains always the same, will at one time be one part of the absolute space into which the air panes; at another time it will be another part of the same, and so, absolutely understood, it will be perpetually mutable.

III. Place is a part of space which a body takes up, and is according to the space, either absolute or relative. I say, apart of space; not the situation, nor the external surface of the body. For the places of equal solids are always equal; but their superfices, by reason of their dissimilar figures, are often unequal. Positions properly have no quantity, nor are they so much the places themselves, as the properties of places. The motion of the whole is the same thing with the sum of the motions of the parts; that is, the translation of the whole, out of its place, is the same thing with the sum of the translations of the parts out of their places; and therefore the place of the whole is the same thing with the sum of the places of the parts, and for that reason, it is internal, and in the whole body.

IV. Absolute motion is the translation of a body from one absolute place into another; and relative motion, the translation from one relative place into another. Thus in a ship under sail, the relative place of a body is that part of the ship which the body possesses; or that part of its cavity which the body fills, and which therefore moves together with the ship: and relative rest is the continuance of the body in the same part of the ship, or of its cavity. But real, absolute rest, is the continuance of the body in the same part of that immovable space, in which the ship itself, its cavity, and all that it contains, is moved. Wherefore, if the earth is really at rest, the body, which relatively rests in the ship, will really and absolutely move with the same velocity which the ship has on the earth. But if the earth also moves, the true and absolute motion of the body will arise, partly from the true motion of the earth, in immovable space; partly from the relative motion of the ship on the earth; and if the body moves also relatively in the ship; its true motion will arise, partly from the true motion of the earth, in immovable

space, and partly from the relative motions as well of the ship on the earth, as of the body in the ship; and from these relative motions will arise the relative motion of the body on the earth. As if that part of the earth, where the ship is, was truly moved toward the east, with a velocity of 10010 parts; while the ship itself, with a fresh gale, and full sails, is carried towards the west, with a velocity expressed by 10 of those parts; but a sailor walks in the ship towards the east, with 1 part of the said velocity; then the sailor will be moved truly in immovable space towards the east, with a velocity of 10001 parts and relatively on the earth towards the west, with a velocity of 9 of those parts.

Absolute time, in astronomy, is distinguished from relative, by the equation or correction of the vulgar time. For the natural days are truly unequal, though they are commonly considered as equal, and used for a measure of time; astronomers correct this inequality for their more accurate deducing of the celestial motions. It may be that there is no such thing as an equable motion whereby time may be accurately measured. All motions may be accelerated and retarded, but the true, or equable progress of absolute time is liable to no change. The duration or perseverance of the existence of things remains the same, whether the motions are swift or slow, or none at all: and therefore it ought to be distinguished from what are only sensible measures thereof; and out of which we collect it, by means of the astronomical equation. The necessity of which equation, for determining the times of a phænomenon, is evinced as well from the experiments of the pendulum clock, as by eclipses of the satellites of *Jupiter*.

As the order of the parts of time is immutable, so also is the order of the parts of space. Suppose those parts to be moved out of their places, and they will be moved (if the expression may be allowed) out of themselves. For times and spaces are, as it were, the places as well of themselves as of all other things. All things are placed in time as to order of succession; and in space as to order of situation. It is from their essence or nature that they are places; and that the primary places of things should be moveable, is absurd. These are therefore the absolute places; and translations out of those places, are the only absolute motions.

But because the parts of space cannot be seen, or distinguished from one another by our senses, therefore in their stead we use sensible measures of them. For from the positions and distances of things from any body considered as immovable, we define all places; said then with respect to such places, we estimate all motions, considering bodies as transferred from some of those places into others. And so, instead of absolute places and motions, we use relative ones; and that without any inconvenience in common affairs; but in philosophical disquisitions, we ought to abstract from our senses, and consider things themselves, distinct from what are only sensible measures of them. For it may be that there is no body really at rest, to which the places and motions of others may be referred.

But we may distinguish rest and motion, absolute and relative, one from the other by their properties, causes and effects. It is a property of rest, that bodies really at rest do rest in respect to one another. And therefore as it is possible, that in the remote regions of the fixed stars, or perhaps far beyond them, there may be some body absolutely at rest; but impossible to know, from the position of bodies to one another in our regions whether any of these do keep the same position to that remote body; it follows that absolute rest cannot be determined from the position of bodies in our regions.

It is a property of motion, that the parts, which retain given positions to their wholes, do partake of the motions of those wholes. For all the parts of revolving bodies endeavour to recede from the axis of motion; and the impetus of bodies moving forward, arises from the joint impetus of all the parts. Therefore, if surrounding bodies are moved, those that are relatively at rest within them, will partake of their motion. Upon which account, the true and absolute motion of a body cannot be determined by the translation of it from those which only seem to rest; for the external bodies ought not only to appear at rest, but to be really at rest. For otherwise, all included bodies, beside their translation from near the surrounding ones, partake likewise of their true motions; and though that translation were not made they would not be really at rest, but only seem to be so. For the surrounding bodies stand in the like relation to the surrounded as the exterior part of a whole does to the interior, or as the shell does to the kernel; but, if the shell moves, the kernel will also move, as being part of the whole, without any removal from near the shell.

A property, near akin to the preceding, is this, that if a place is moved, whatever is placed therein moves along with it; and therefore a body, which is moved from a place in motion, partakes also of the motion of its place. Upon which account, all motions, from places in motion, are no other than parts of entire and absolute motions; and every entire motion is composed of the motion of the body out of its first place, and the motion of this place out of its place; and so on, until we come to some immovable place, as in the before-mentioned example of the sailor. Wherefore, entire and absolute motions can be not otherwise determined than by immovable places; and for that reason I did before refer those absolute motions to immovable places, but relative ones to movable places. Now no other places are immovable but those that, from infinity to infinity, do all retain the same given position one to another; and upon this account must ever remain unmoved; and do thereby constitute immovable space.

The causes by which true and relative motions are distinguished, one from the other, are the forces impressed upon bodies to generate motion. True motion is neither generated nor altered, but by some force impressed upon the body moved; but relative motion may be generated or altered without any force impressed upon the body. For it is sufficient only to impress some force on other bodies with which the former is

compared, that by their giving way, that relation may be changed, in which the relative rest or motion of this other body did consist. Again, true motion suffers always some change from any force impressed upon the moving body; but relative motion does not necessarily undergo any change by such forces. For if the same forces are likewise impressed on those other bodies, with which the comparison is made, that the relative position may be preserved, then that condition will be preserved in which the relative motion consists. And therefore any relative motion may be changed when the true motion remains unaltered, and the relative may be preserved when the true suffers some change. Upon which accounts, true motion does by no means consist in such relations.

The effects which distinguish absolute from relative motion are, the forces of receding from the axis of circular motion. For there are no such forces in a circular motion purely relative, but in a true and absolute circular motion, they are greater or less, according to the quantity of the motion. If a vessel, hung by a long cord, is so often turned about that the cord is strongly twisted, then filled with water, and held at rest together with the water; after, by the sudden action of another force, it is whirled about the contrary way, and while the cord is untwisting itself, the vessel continues for some time in this motion; the surface of the water will at first be plain, as before the vessel began to move; but the vessel, by gradually communicating its motion to the water, will make it begin sensibly to revolve, and recede by little and little from the middle, and ascend to the sides of the vessel, forming itself into a concave figure (as I have experienced), and the swifter the motion becomes, the higher will the water rise, till at last, performing its revolutions in the same times with the vessel, it becomes relatively at rest in it. This ascent of the water shows its endeavour to recede from the axis of its motion; and the true and absolute circular motion of the water, which is here directly contrary to the relative, discovers itself, and may be measured by this endeavour. At first, when the relative motion of the water in the vessel was greatest, it produced no endeavour to recede from the axis; the water showed no tendency to the circumference, nor any ascent towards the sides of the vessel, but remained of a plain surface, and therefore its true circular motion had not yet begun. But afterwards, when the relative motion of the water had decreased, the ascent thereof towards the sides of the vessel proved its endeavour to recede from the axis; and this endeavour showed the real circular motion of the water perpetually increasing' till it had acquired, its greatest quantity, when the water rested relatively in the vessel. And therefore this endeavour does not depend upon any translation of the water in respect of the ambient bodies, nor can true circular motion be defined by such translation. There is only one real circular motion of any one revolving body, corresponding to only one power of endeavouring to recede from its axis of motion, as its proper and adequate effect; but relative motions, in one and the same body, are innumerable, according to the various relations it bears to external

bodies, and like other relations, are altogether destitute of any real effect, any otherwise than they may perhaps partake of that one only true motion. And therefore in their system who suppose that our heavens, revolving below the sphere of the fixed stars, carry the planets along with them; the several parts of those heavens, and the planets, which are indeed relatively at rest in their heavens, do yet really move. For they change their position one to another (which never happens to bodies truly at rest), and being carried together with their heavens, partake of their motions, and as parts of revolving wholes, endeavour to recede from the axis of their motions.

Wherefore relative quantities are not the quantities themselves, whose names they bear, but those sensible measures of them (either accurate or inaccurate), which are commonly used instead of the measured quantities themselves. And if the meaning of words is to be determined by their use, then by the names time, space, place and motion, their measures are properly to be understood; and the expression will be unusual, and purely mathematical, if the measured quantities themselves are meant. Upon which account, they do strain the sacred writings, who there interpret those words for the measured quantities. Nor do those less defile the purity of mathematical and philosophical truths, who confound real quantities themselves with their relations and vulgar measures.

It is indeed a matter of great difficulty to discover, and effectually to distinguish, the true motions of particular bodies from the apparent; because the parts of that immovable space, in which those motions are performed, do by no means come under the observation of our senses. Yet the thing is not altogether desperate; for we have some arguments to guide us, partly from the apparent motions, which are the differences of the true motions; partly from the forces, which are the causes and effects of the true motions. For instance, if two globes, kept at a given distance one from the other by means of a cord that connects them, were revolved about their common centre of gravity, we might, from the tension of the cord, discover the endeavour of the globes to recede from the axis of their motion, and from thence we might compute the quantity of their circular motions. And then if any equal forces should be impressed at once on the alternate faces of the globes to augment or diminish their circular motions, from the increase or decrease of the tension of the cord, we might infer the increment or decrement of their motions; and thence would be found on what faces those forces ought to be impressed, that the motions of the globes might be most augmented; that is, we might discover their hindermost faces, or those which, in the circular motion, do follow. But the faces which follow being known, and consequently the opposite ones that precede, we should likewise know the determination of their motions. And thus we might find both the quantity and the determination of this circular motion, even in an immense vacuum, where there was nothing external or sensible with which the

globes could be compared. But now, if in that space some remote bodies were placed that kept always a given position one to another, as the fixed stars do in our regions, we could not indeed determine from the relative translation of the globes among those bodies, whether the motion did belong to the globes or to the bodies. But if we observed the cord, and found that its tension was that very tension which the motions of the globes required, we might conclude the motion to be in the globes, and the bodies to be at rest; and then, lastly, from the translation of the globes among the bodies, we should find the determination of their motions. But how we are to collect the true motions from their causes, effects, and apparent differences; and, *vice versa*, how from the motions, either true or apparent, we may come to the knowledge of their causes and effects, shall be explained more at large in the following tract For to this end it was that I composed it.

AXIOMS, OR LAWS OF MOTION.

LAW I.

EVERY BODY PERSEVERES IN ITS STATE OF REST, OR OF
UNIFORM MOTION IN A RIGHT LINE, UNLESS IT IS
COMPELLED TO CHANGE THAT STATE BY
FORCES IMPRESSED THEREON.

Projectiles persevere in their motions, so far as they are not retarded by the resistance of the air, or impelled downwards by the force of gravity A top, whose parts by their cohesion are perpetually drawn aside from rectilinear motions, does not cease its rotation, otherwise than as it is retarded by the air. The greater bodies of the planets and comets, meeting with less resistance in more free spaces, preserve their motions both progressive and circular for a much longer time.

LAW II.

THE ALTERATION OF MOTION IS EVER PROPORTIONAL TO
THE MOTIVE FORCE IMPRESSED; AND IS MADE IN THE
DIRECTION OF THE RIGHT LINE IN WHICH
THAT FORCE IS IMPRESSED.

If any force generates a motion, a double force will generate double the motion, a triple force triple the motion, whether that force be impressed altogether and at once, or gradually and successively. And this motion (being always directed the same way

with the generating force), if the body moved before, is added to or subducted from the former motion, according as they directly conspire with or are directly contrary to each other; or obliquely joined, when they are oblique, so as to produce a new motion compounded from the determination of both.

LAW III.

TO EVERY ACTION THERE IS ALWAYS OPPOSED AN EQUAL REACTION: OR THE MUTUAL ACTIONS OF TWO BODIES UPON EACH OTHER ARE ALWAYS EQUAL, AND DIRECTED TO CONTRARY PARTS.

Whatever draws or presses another is as much drawn or pressed by that other. If you press a stone with your finger, the finger is also pressed by the stone. If a horse draws a stone tied to a rope, the horse (if I may so say) will be equally drawn back towards the stone: for the distended rope, by the same endeavour to relax or unbend itself, will draw the horse as much towards the stone, as it does the stone towards the horse, and will obstruct the progress of the one as much as it advances that of the other.

If a body impinge upon another, and by its force change the motion of the other, that body also (because of the equality of the mutual pressure) will undergo an equal change, in its own motion, towards the contrary part. The changes made by these actions are equal, not in the velocities but in the motions of bodies; that is to say, if the bodies are not hindered by any other impediments. For, because the motions are equally changed, the changes of the velocities made towards contrary parts are reciprocally proportional to the bodies. This law takes place also in attractions, as will be proved in the next scholium.

COROLLARY I.

A BODY BY TWO FORCES CONJOINED WILL DESCRIBE THE DIAGONAL OF A PARALLELOGRAM, IN THE SAME TIME THAT IT WOULD DESCRIBE THE SIDES, BY THOSE FORCES APART.

If a body in a given time, by the force M impressed apart in the place A, should with an uniform motion be carried from A to B; and by the force N impressed apart in the same place, should be carried from A to C; complete the parallelogram ABCD, and, by both forces acting together, it will in the same time be carried in the diagonal from A to D. For since the force N acts in the direction of the line AC, parallel to BD, this force (by the second law) will not at all alter the

velocity generated by the other force M, by which the body is carried towards the line BD. The body therefore will arrive at the line BD in the same time, whether the force N be impressed or not; and therefore at the end of that time it will be found somewhere in the line BD. By the same argument, at the end of the same time it will be found somewhere in the line CD. Therefore it will be found in the point D, where both lines meet. But it will move in a right line from A to D, by Law I.

COROLLARY II.

AND HENCE IS EXPLAINED THE COMPOSITION OF ANY ONE DIRECT FORCE AD, OUT OF ANY TWO OBLIQUE FORCES AC AND CD; AND, ON THE CONTRARY, THE RESOLUTION OF ANY ONE DIRECT FORCE AD INTO TWO OBLIQUE FORCES AC AND CD: WHICH COMPOSITION AND RESOLUTION ARE ABUNDANTLY CONFIRMED FROM MECHANICS.

As if the unequal radii OM and ON drawn from the centre O of any wheel, should sustain the weights A and P by the cords MA and NP; and the forces of those weights to move the wheel were required. Through the centre O draw the right line KOL, meeting the cords perpendicularly in K and L; and from the centre O, with OL the greater of the distances OK and OL, describe a circle, meeting the cord MA in D: and drawing OD, make AC parallel and DC perpendicular thereto. Now, it being indifferent whether the points K, L, D, of the cords be fixed to the plane of the wheel or not, the weights will have the same effect whether they are suspended from the points K and L, or from D and L. Let the whole force of the weight A be represented by the line AD, and let it be resolved into the forces AC

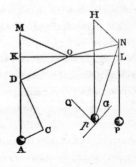

and CD; of which the force AC, drawing the radius OD directly from the centre, will have no effect to move the wheel: but the other force DC, drawing the radius DO perpendicularly, will have the same effect as if it drew perpendicularly the radius OL equal to OD; that is, it will have the same effect as the weight P, if that weight is to the weight A as the force DC is to the force DA; that is (because of the similar triangles ADC, DOK), as OK to OD or OL. Therefore the weights A and P, which are reciprocally as the radii OK and OL that lie in the same right line, will be equipollent, and so remain in equilibrio; which is the well known property of the balance, the lever, and the wheel. If either weight is greater than in this ratio, its force to move the wheel will be so much greater.

If the weight p, equal to the weight P, is partly suspended by the cord Np, partly sustained by the oblique plane pG; draw pH, NH, the former perpendicular to the horizon, the latter to the plane pG; and if the force of the weight p tending downwards is represented by the line pH, it may be resolved into the forces pN, HN. If there was any plane pQ, perpendicular to the cord pN, cutting the other plane pG in a line parallel to the horizon, and the weight p was supported only by those planes pQ, pG, it would press those planes perpendicularly with the forces pN, HN; to wit, the plane pQ with the force pN, and the plane pG with the force HN. And therefore if the plane pQ was taken away, so that the weight might stretch the cord, because the cord, now sustaining the weight, supplies the place of the plane that was removed, it will be strained by the same force N which pressed upon the plane before. Therefore, the tension of this oblique cord pN will be to that of the other perpendicular cord PN as pN to pH. And therefore if the weight p is to the weight A in a ratio compounded of the reciprocal ratio of the least distances of the cords PN, AM, from the centre of the wheel, and of the direct ratio of pH to pN, the weights will have the same effect towards moving the wheel and will therefore sustain each other; as any one may find by experiment.

But the weight p pressing upon those two oblique planes may be considered as a wedge between the two internal surfaces of a body split by it; and hence the forces of the wedge and the mallet may be determined; for because the force with which the weight p presses the plane pQ is to the force with which the same, whether by its own gravity, or by the blow of a mallet, is impelled in the direction of the line pH towards both the planes, as pH to pH; and to the force with which it presses the other plane pG, as pN to NH. And thus the force of the screw may be deduced from a like resolution of forces; it being no other than a wedge impelled with the force of a lever. Therefore the use of this Corollary spreads far and wide, and by that diffusive extent the truth thereof is farther confirmed. For on what has been said depends the whole doctrine of mechanics variously demonstrated by different authors. For from hence are easily deduced the forces of machines, which are compounded of wheels, pullies, levers, cords, and weights, ascending directly or obliquely, and other mechanical powers; as also the force of the tendons to move the bones of animals.

COROLLARY III.

THE QUANTITY OF MOTION, WHICH IS COLLECTED BY
TAKING THE SUM OF THE MOTIONS DIRECTED TOWARDS
THE SAME PARTS, AND THE DIFFERENCE OF THOSE THAT
ARE DIRECTED TO CONTRARY PARTS, SUFFERS NO CHANGE
FROM THE ACTION OF BODIES AMONG THEMSELVES.

For action and its opposite re-action are equal, by Law III, and therefore, by Law II, they produce in the motions equal changes towards opposite parts. Therefore if the motions are directed towards the same parts, whatever is added to the motion of the preceding body will be subducted from the motion of that which follows; so that the sum win be the same as before. If the bodies meet, with contrary motions, there will be an equal deduction from the motions of both; and therefore the difference of the motions directed towards opposite parts will remain the same.

Thus if a spherical body A with two parts of velocity is triple of a spherical body B which follows in the same right line with ten parts of velocity, the motion of A will be to that of B as 6 to 10. Suppose, then, their motions to be of 6 parts and of 10 parts, and the sum will be 16 parts. Therefore, upon the meeting of the bodies, if A acquire 3, 4, or 5 parts of motion, B will lose as many; and therefore after reflexion A will proceed with 9, 10, or 11 parts, and B with 7, 6, or 5 parts; the sum remaining always of 16 parts as before. If the body A acquire 9, 10, 11, or 12 parts of motion, and therefore after meeting proceed with 15, 16, 17, or 18 parts, the body B, losing so many parts as A has got, will either proceed with 1 part, having lost 9, or stop and remain at rest, as having lost its whole progressive motion of 10 parts; or it will go back with 1 part, having not only lost its whole motion, but (if I may so say) one part more; or it will go back with 2 parts, because a progressive motion of 12 parts is taken off. And so the sums of the conspiring motions 15+1 or 16+0, and the differences of the contrary motions 17-1 and 18-2, will always be equal to 16 parts, as they were before the meeting and reflexion of the bodies. But, the motions being known with which the bodies proceed after reflexion, the velocity of either will be also known, by taking the velocity after to the velocity before reflexion, as the motion after is to the motion before. As in the last case, where the motion of the body A was of 6 parts before reflexion and of 18 parts after, and the velocity was of 2 parts before reflexion, the velocity thereof after reflexion will be found to be of 6 parts; by saying, as the 6 parts of motion, before to 18 parts after, so are 2 parts of velocity before reflexion to 6 parts after.

But if the bodies are either not spherical, or, moving in different right lines, impinge obliquely one upon the other, and their motions after reflexion are required, in those cases we are first to determine the position of the plane that touches the concurring bodies in the point of concourse; then the motion of each body (by Corol. II) is to be resolved into two, one perpendicular to that plane, and the other parallel to it. This done, because the bodies act upon each other in the direction of a line perpendicular to this plane, the parallel motions are to be retained the same after reflexion as before; and to the perpendicular motions we are to assign equal changes towards the contrary parts; in such manner that the sum of the conspiring and the difference of the contrary motions may remain the same as before. From such kind of reflexions also

sometimes arise the circular motions of bodies about their own centres. But these are cases which I do not consider in what follows; and it would be too tedious to demonstrate every particular that relates to this subject.

COROLLARY IV.

THE COMMON CENTRE OF GRAVITY OF TWO OR MORE
BODIES DOES NOT ALTER ITS STATE OF MOTION OR REST BY
THE ACTIONS OF THE BODIES AMONG THEMSELVES: AND
THEREFORE THE COMMON CENTRE OF GRAVITY OF ALL
BODIES ACTING UPON EACH OTHER (EXCLUDING OUTWARD
ACTIONS AND IMPEDIMENTS) IS EITHER AT REST,
OR MOVES UNIFORMLY IN A RIGHT LINE.

For if two points proceed with an uniform motion in right lines, and their distance be divided in a given ratio, the dividing point will be either at rest, or proceed uniformly in a right line. This is demonstrated hereafter in Lem. XXIII and its Corol., when the points are moved in the same plane; and by a like way of arguing, it may be demonstrated when the points are not moved in the same plane. Therefore if any number of bodies move uniformly in right lines, the common centre of gravity of any two of them is either at rest, or proceeds uniformly in a right line; because the line which connects the centres of those two bodies so moving is divided at that common centre in a given ratio. In like manner the common centre of those two and that of a third body will be either at rest or moving uniformly in a right line because at that centre the distance between the common centre of the two bodies, and the centre of this last, is divided in a given ratio. In like manner the common centre of these three, and of a fourth body, is either at rest, or moves uniformly in a right line; because the distance between the common centre of the three bodies, and the centre of the fourth is there also divided in a given ratio, and so on ad infinitum. Therefore, in a system of bodies where there is neither any mutual action among themselves, nor any foreign force impressed upon them from without, and which consequently move uniformly in right lines, the common centre of gravity of them all is either at rest or moves uniformly forward in a right line.

Moreover, in a system of two bodies mutually acting upon each other, since the distances between their centres and the common centre of gravity of both are reciprocally as the bodies, the relative motions of those bodies, whether of approaching to or of receding from that centre, will be equal among themselves. Therefore since the changes which happen to motions are equal and directed to contrary parts, the common centre of those bodies, by their mutual action between themselves, is neither promoted nor

retarded, nor suffers any change as to its state of motion or rest. But in a system of several bodies, because the common centre of gravity of any two acting mutually upon each other suffers no change in its state by that action; and much less the common centre of gravity of the others with which that action does not intervene: but the distance between those two centres is divided by the common centre of gravity of all the bodies into parts reciprocally proportional to the total sums of those bodies whose centres they are: and therefore while those two centres retain their state of motion or rest, the common centre of all does also retain its state: it is manifest that the common centre of all never suffers any change in the state of its motion or rest from the actions of any two bodies between themselves. But in such a system all the actions of the bodies among themselves either happen between two bodies, or are composed of actions interchanged between some two bodies; and therefore they do never produce any alteration in the common centre of all as to its state of motion or rest. Wherefore since that centre, when the bodies do not act mutually one upon another, either is at rest or moves uniformly forward in some right line, it will, notwithstanding the mutual actions of the bodies among themselves, always persevere in its state, either of rest, or of proceeding uniformly in a right line, unless it is forced out of this state by the action of some power impressed from without upon the whole system. And therefore the same law takes place in a system consisting of many bodies as in one single body, with regard to their persevering in their state of motion or of rest. For the progressive motion, whether of one single body, or of a whole system of bodies, is always to be estimated from the motion of the centre of gravity.

COROLLARY V.

THE MOTIONS OF BODIES INCLUDED IN A GIVEN SPACE ARE THE SAME AMONG THEMSELVES, WHETHER THAT SPACE IS AT REST, OR MOVES UNIFORMLY FORWARDS IN A RIGHT LINE WITHOUT ANY CIRCULAR MOTION.

For the differences of the motions tending towards the same parts, and the sums of those that tend towards contrary parts, are, at first (by supposition), in both cases the same; and it is from those sums and differences that the collisions and impulses do arise with which the bodies mutually impinge one upon another. Wherefore (by Law II), the effects of those collisions will be equal in both cases; and therefore the mutual motions of the bodies among themselves in the one case will remain equal to the mutual motions of the bodies among themselves in the other. A clear proof of which we have from the experiment of a ship; where all motions happen after the same manner, whether the ship is at rest, or is carried uniformly forwards in a right line.

COROLLARY VI.

IF BODIES, ANY HOW MOVED AMONG THEMSELVES, ARE
URGED IN THE DIRECTION OF PARALLEL LINES BY EQUAL
ACCELERATIVE FORCES, THEY WILL ALL CONTINUE TO MOVE
AMONG THEMSELVES, AFTER THE SAME MANNER AS IF THEY
HAD BEEN URGED BY NO SUCH FORCES.

For these forces acting equally (with respect to the quantities of the bodies to be moved), and in the direction of parallel lines, will (by Law II) move all the bodies equally (as to velocity), and therefore will never produce any change in the positions or motions of the bodies among themselves.

SCHOLIUM.

Hitherto I have laid down such principles as have been received by mathematicians, and are confirmed by abundance of experiments. By the first two Laws and the first two Corollaries, Galileo discovered that the descent of bodies observed the duplicate ratio of the time, and that the motion of projectiles was in the curve of a parabola; experience agreeing with both, unless so far as these motions are a little retarded by 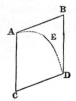 the resistance of the air. When a body is falling, the uniform force of its gravity acting equally, impresses, in equal particles of time, equal forces upon that body, and therefore generates equal velocities; and in the whole time impresses a whole force, and generates a whole velocity proportional to the time. And the spaces described in proportional times are as the velocities and the times conjunctly; that is, in a duplicate ratio of the times. And when a body is thrown upwards, its uniform gravity impresses forces and takes off velocities proportional to the times; and the times of ascending to the greatest heights are as the velocities to be taken off, and those heights are as the velocities and the times conjunctly, or in the duplicate ratio of the velocities. And if a body be projected in any direction, the motion arising from its projection is compounded with the motion arising from its gravity. As if the body A by its motion of projection alone could describe in a given time the right line AB, and with its motion of falling alone could describe in the same time the altitude AC; complete the paralellogram ABDC, and the body by that compounded motion will at the end of the time be found in the place D; and the curve line AED, which that body describes, will be a parabola, to which the right line AB will be a tangent in A; and whose ordinate BD will be as the square of the line AB. On the same Laws and Corollaries depend those things which have been demonstrated concerning the times

of the vibration of pendulums, and are confirmed by the daily experiments of pendulum clocks. By the same, together with the third Law, Sir Christ. Wren, Dr. Wallis, and Mr. Huygens, the greatest geometers of our times, did severally determine the rules of the congress and reflexion of hard bodies, and much about the same time communicated their discoveries to the Royal Society, exactly agreeing among themselves as to those rules. Dr. Wallis, indeed, was something more early in the publication; then followed Sir Christopher Wren, and, lastly, Mr. Huygens. But Sir Christopher Wren confirmed the truth of the thing before the Royal Society by the experiment of pendulums,

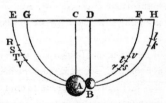

which Mr. Mariotte soon after thought fit to explain in a treatise entirely upon that subject. But to bring this experiment to an accurate agreement with the theory, we are to have a due regard as well to the resistance of the air as to the elastic force of the concurring bodies. Let the spherical bodies A, B be suspended by the parallel and equal strings AC, BD, from the centres C, D. About these centres, with those intervals, describe the semicircles EAF, GBH, bisected by the radii CA, DB. Bring the body A to any point R of the LAP, and (withdrawing the body B) let it go from thence, and after one oscillation suppose it to return to the point V: then RV will be the retardation arising from the resistance of the air. Of this RV let ST be a fourth part, situated in the middle, to wit, so as RS and TV may be equal, and RS may be to ST as 3 to 2 then will ST represent very nearly the retardation during the descent from S to A. Restore the body B to its place: and, supposing the body A to be let fall from the point S, the velocity thereof in the place of reflexion A, without sensible error, will be the same as if it had descended in vacuo from the point T. Upon which account this velocity may be represented by the chord of the arc TA. For it is a proposition well known to geometers, that the velocity of a pendulous body in the lowest point is as the chord of the arc which it has described in its descent. After reflexion, suppose the body A comes to the place s, and the body B to the place k. Withdraw the body B, and find the place v, from which if the body A, being let go, should after one oscillation return to the place r, st may be a fourth part of rv, so placed in the middle thereof as to leave rs equal to tv, and let the chord of the arc tA represent the velocity which the body A had in the place A immediately after reflexion. For t will be the true and correct place to which the body A should have ascended, if the resistance of the air had been taken off. In the same way we are to correct the place k to which the body B ascends, by finding the place l to which it should have ascended in vacuo. And thus everything may be subjected to experiment, in the same manner as if we were really placed in vacuo. These things being done, we are to take the product (if I may so say) of the body A, by the chord of

the arc TA (which represents its velocity), that we may have its motion in the place A immediately before reflexion; and then by the chord of the arc tA, that we may have its motion in the place A immediately after reflexion. And so we are to take the product of the body B by the chord of the arc Bl that we may have the motion of the same immediately after reflexion. And in like manner, when two bodies are let go together from different places, we are to find the motion of each, as well before as after reflexion; and then we may compare the motions between themselves, and collect the effects of the reflexion. Thus trying the thing with pendulums of ten feet, in unequal as well as equal bodies, and making the bodies to concur after a descent through large spaces, as of 8, 12, or 16 feet, I found always, without an error of 3 inches, that when the bodies concurred together directly, equal changes towards the contrary parts were produced in their motions, and, of consequence, that the action and reaction were always equal. As if the body A impinged upon the body B at rest with 9 parts of motion, and losing 7, proceeded after reflexion with 2, the body B was carried backwards with those 7 parts. If the bodies concurred with contrary motions, A with twelve parts of motion, and B with six, then if A receded with 2, B receded with 8; to wit, with a deduction of 14 parts of motion on each side. For from the motion of A subducting twelve parts, nothing will remain; but subducting 2 parts more, a motion will be generated of 2 parts towards the contrary way; and so, from the motion of the body B of 6 parts, subducting 14 parts, a motion is generated of 8 parts towards the contrary way. But if the bodies were made both to move towards the same way, A, the swifter, with 14 parts of motion, B, the slower, with 5, and after reflexion A went on with 5, B likewise went on with 14 parts; 9 parts being transferred from A to B. And so in other cases. By the congress and collision of bodies, the quantity of motion, collected from the sum of the motions directed towards the same way, or from the difference of those that were directed towards contrary ways, was never changed. For the error of an inch or two in measures may be easily ascribed to the difficulty of executing everything with accuracy. It was not easy to let go the two pendulums so exactly together that the bodies should impinge one upon the other in the lowermost place AB; nor to mark the places s, and k, to which the bodies ascended after congress. Nay, and some errors, too, might have happened from the unequal density of the parts of the pendulous bodies themselves, and from the irregularity of the texture proceeding from other causes.

But to prevent an objection that may perhaps be alledged against the rule, for the proof of which this experiment was made, as if this rule did suppose that the bodies were either absolutely hard, or at least perfectly elastic (whereas no such bodies are to be found in nature), I must add, that the experiments we have been describing, by no means depending upon that quality of hardness, do succeed as well in soft as in hard bodies. For if the rule is to be tried in bodies not perfectly hard, we are only to dimin-

ish the reflexion in such a certain proportion as the quantity of the elastic force requires. By the theory of Wren and Huygens, bodies absolutely hard return one from another with the same velocity with which they meet. But this may be affirmed with more certainty of bodies perfectly elastic. In bodies imperfectly elastic the velocity of the return is to be diminished together with the elastic force; because that force (except when the parts of bodies are bruised by their congress, or suffer some such extension as happens under the strokes of a hammer) is (as far as I can perceive) certain and determined, and makes the bodies to return one from the other with a relative velocity, which is in a given ratio to that relative velocity with which they met. This I tried in balls of wool, made up tightly, and strongly compressed. For, first, by letting go the pendulous bodies, and measuring their reflexion, I determined the quantity of their elastic force; and then, according to this force, estimated the reflexions that ought to happen in other cases of congress. And with this computation other experiments made afterwards did accordingly agree; the balls always receding one from the other with a relative velocity, which was to the relative velocity with which they met as about 5 to 9. Balls of steel returned with almost the same velocity: those of cork with a velocity something less; but in balls of glass the proportion was as about 15 to 16. And thus the third Law, so far as it regards percussions and reflexions, is proved by a theory exactly agreeing with experience.

In attractions, I briefly demonstrate the thing after this manner. Suppose an obstacle is interposed to hinder the congress of any two bodies A, B, mutually attracting one the other: then if either body, as A, is more attracted towards the other body B, than that other body B is towards the first body A, the obstacle will be more strongly urged by the pressure of the body A than by the pressure of the body B, and therefore will not remain in equilibrio: but the stronger pressure will prevail, and will make the system of the two bodies, together with the obstacle, to move directly towards the parts on which B lies; and in free spaces, to go forward in infinitum with a motion perpetually accelerated; which is absurd and contrary to the first Law. For, by the first Law, the system ought to persevere in its state of rest, or of moving uniformly forward in a right line; and therefore the bodies must equally press the obstacle, and be equally attracted one by the other. I made the experiment on the loadstone and iron. If these, placed apart in proper vessels, are made to float by one another in standing water, neither of them will propel the other; but, by being equally attracted, they will sustain each other's pressure, and rest at last in an equilibrium.

So the gravitation betwixt the earth and its parts is mutual. Let the earth FI be cut by any plane EG into two parts EGF and EGI, and their weights one towards the other will be mutually equal. For if by another plane HK, parallel to the former EG, the greater part EGI is cut into two parts EGKH and HKI, whereof HKI is equal to the

part EFG, first cut off, it is evident that the middle part EGKH, will have no propension by its proper weight towards either side, but will hang as it were, and rest in an equilibrium betwixt both. But the one extreme part HKI will with its whole weight bear upon and press the middle part towards the other extreme part EGF; and therefore the force with which EGI, the sum of the parts HKI and EGKH, tends towards the third part EGF, is equal to the weight of the part HKI, that is, to the weight of the third part EGF. And therefore the weights of the two parts EGI and EGF, one towards the other, are equal, as I was to prove.

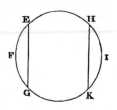

And indeed if those weights were not equal, the whole earth floating in the nonresisting æther would give way to the greater weight, and, retiring from it, would be carried off in infinitum.

And as those bodies are equipollent in the congress and reflexion, whose velocities are reciprocally as their innate forces, so in the use of mechanic instruments those agents are equipollent, and mutually sustain each the contrary pressure of the other, whose velocities, estimated according to the determination of the forces, are reciprocally as the forces.

So those weights are of equal force to move the arms of a balance; which during the play of the balance are reciprocally as their velocities upwards and downwards; that is, if the ascent or descent is direct, those weights are of equal force, which are reciprocally as the distances of the points at which they are suspended from the axis of the balance; but if they are turned aside by the interposition of oblique planes, or other obstacles, and made to ascend or descend obliquely, these bodies will be equipollent, which are reciprocally as the heights of their ascent and descent taken according to the perpendicular; and that on account of the determination of gravity downwards.

And in like manner in the pulley, or in a combination of pullies, the force of a hand drawing the rope directly, which is to the weight, whether ascending directly or obliquely, as the velocity of the perpendicular ascent of the weight to the velocity of the hand that draws the rope, will sustain the weight.

In clocks and such like instruments, made up from a combination of wheels, the contrary forces that promote and impede the motion of the wheels, if they are reciprocally as the velocities of the parts of the wheel on which they are impressed, will mutually sustain the one the other.

The force of the screw to press a body is to the force of the hand that turns the handles by which it is moved as the circular velocity of the handle in that part where it is impelled by the hand is to the progressive velocity of the screw towards the pressed body.

The forces by which the wedge presses or drives the two parts of the wood it cleaves are to the force of the mallet upon the wedge as the progress of the wedge in the direction of the force impressed upon it by the mallet is to the velocity with which the parts of the wood yield to the wedge, in the direction of lines perpendicular to the sides of the wedge. And the like account is to be given of all machines.

The power and use of machines consist only in this, that by diminishing the velocity we may augment the force, and the contrary: from whence in all sorts of proper machines, we have the solution of this problem; To move a given weight with a given power, or with a given force to overcome any other given resistance. For if machines are so contrived that the velocities of the agent and resistant are reciprocally as their forces, the agent will just sustain the resistant, but with a greater disparity of velocity will overcome it. So that if the disparity of velocities is so great as to overcome all that resistance which commonly arises either from the attrition of contiguous bodies as they slide by one another, or from the cohesion of continuous bodies that are to be separated, or from the weights of bodies to be raised, the excess of the force remaining, after all those resistances are overcome, will produce an acceleration of motion proportional thereto, as well in the parts of the machine as in the resisting body. But to treat of mechanics is not my present business. I was only willing to show by those examples the great extent and certainty of the third Law of motion. For if we estimate the action of the agent from its force and velocity conjunctly, and likewise the reaction of the impediment conjunctly from the velocities of its several parts, and from the forces of resistance arising from the attrition, cohesion, weight, and acceleration of those parts. the action and reaction in the use of all sorts of machines will be found always equal to one another. And so far as the action is propagated by the intervening instruments, and at last impressed upon the resisting body, the ultimate determination of the action will be always contrary to the determination of the reaction.

BOOK I.

OF THE MOTION OF BODIES.

SECTION I.

Of the method of first and last ratios of quantities, by the help whereof we demonstrate the propositions that follow.

LEMMA I.

QUANTITIES, AND THE RATIOS OF QUANTITIES, WHICH IN
ANY FINITE TIME CONVERGE CONTINUALLY TO EQUALITY,
AND BEFORE THE END OF THAT TIME APPROACH NEARER
THE ONE TO THE OTHER THAN BY ANY GIVEN DIFFERENCE,
BECOME ULTIMATELY EQUAL.

If you deny it, suppose them to be ultimately unequal, and let D be their ultimate difference. Therefore they cannot approach nearer to equality than by that given difference D; which is against the supposition.

LEMMA II.

IF IN ANY FIGURE AacE, TERMINATED BY THE RIGHT LINES
AA, AE, AND THE CURVE acE, THERE BE INSCRIBED ANY
NUMBER OF PARALLELOGRAMS AB, BC, CD,
&C., COMPREHENDED UNDER EQUAL BASES
AB, BC, CD, &C., AND THE SIDES, BB,
CC, DD, &C., PARALLEL TO ONE SIDE AA
OF THE FIGURE; AND THE PARALLELOGRAMS
AKBL, BLCM, CMDN, &C., ARE COMPLETED.
THEN IF THE BREADTH OF THOSE
PARALLELOGRAMS BE SUPPOSED TO BE
DIMINISHED, AND THEIR NUMBER TO BE
AUGMENTED IN INFINITUM; I SAY, THAT
THE ULTIMATE RATIOS WHICH THE INSCRIBED FIGURE
AKBLCMDD, THE CIRCUMSCRIBED FIGURE AALBMENDOE,
AND CURVILINEAR FIGURE AABCDE, WILL HAVE TO ONE
ANOTHER, ARE RATIOS OF EQUALITY.

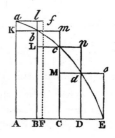

For the difference of the inscribed and circumscribed figures is the sum of the parallelograms Kl, Lm, Mn, Do, that is (from the equality of all their bases), the rectangle under one of their bases Kb and the sum of their altitudes Aa, that is, the rectangle ABla. But this rectangle, because its breadth AB is supposed diminished in infinitum, becomes less than any given space. And therefore (by Lem. I) the figures inscribed and circumscribed become ultimately equal one to the other; and much more will the intermediate curvilinear figure be ultimately equal to either.   Q.E.D.

LEMMA III.

The same ultimate ratios are also ratios of equality, when the breadths, AB, BC, DC, &c., of the parallelograms are unequal, and are all diminished in infinitum.

For suppose AF equal to the greatest breadth, and complete the parallelogram FAaf. This parallelogram will be greater than the difference of the inscribed and circumscribed figures; but, because its breadth AF is diminished in infinitum, it will become less than any given rectangle.   Q.E.D.

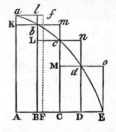

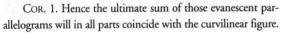

COR. 1. Hence the ultimate sum of those evanescent parallelograms will in all parts coincide with the curvilinear figure.

COR. 2. Much more will the rectilinear figure comprehended under the chords of the evanescent arcs ab, bc, cd, &c., ultimately coincide with the curvilinear figure.

COR. 3. And also the circumscribed rectilinear figure comprehended under the tangents of the same arcs.

COR. 4 And therefore these ultimate figures (as to their perimeters E) are not rectilinear, but curvilinear limits of rectilinear figures.

LEMMA IV.

IN TWO FIGURES AacE, PprT, YOU INSCRIBE (AS BEFORE) TWO RANKS OF PARALLELOGRAMS, AN EQUAL NUMBER IN EACH RANK, AND, WHEN THEIR BREADTHS ARE DIMINISHED IN INFINITRUM, THE ULTIMATE RATIOS OF THE PARALLELOGRAMS IN ONE FIGURE TO THOSE IN THE OTHER, EACH TO EACH RESPECTIVELY, ARE THE SAME; I SAY THAT THOSE TWO FIGURES AacE, PprT, ARE TO ONE ANOTHER IN THAT SAME RATIO.

For as the parallelograms in the one are severally to P the parallelograms in the other, so (by composition) is the sum of all in the one to the sum of all in the other; and so is the one figure to the other; because (by Lem. III) the former figure to the former sum, and the latter figure to the latter sum, are both in the ratio of equality.   Q.E.D.

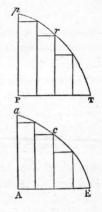

COR. Hence if two quantities of any kind are any how divided into an equal number of parts, and those parts, when their number is augmented, and their magnitude diminished in infinitum, have a given ratio one to the other, the first to the first, the second to the second, and so on in order, the whole quantities will be one to the other in that same given ratio. For if, in the figures of this Lemma, the parallelograms are taken one

to the other in the ratio of the parts, the sum of the parts will always be as the sum of the parallelograms; and therefore supposing the number of the parallelograms and parts to be augmented, and their magnitudes diminished in infinitum, those sums will be in the ultimate ratio of the parallelogram in the one figure to the correspondent parallelogram in the other; that is (by the supposition), in the ultimate ratio of any part of the one quantity to the correspondent part of the other.

LEMMA V.

IN SIMILAR FIGURES, ALL SORTS OF HOMOLOGOUS SIDES,
WHETHER CURVILINEAR OR RECTILINEAR, ARE PROPORTIONAL;
AND THE AREAS ARE IN THE DUPLICATE RATIO OF THE
HOMOLOGOUS SIDES.

LEMMA VI.

IF ANY ARC ACB, GIVEN IN POSITION
IS SUBTENDED BY ITS CHORD AB, AND
IN ANY POINT A, IN THE MIDDLE OF
THE CONTINUED
CURVATURE, IS TOUCHED BY A RIGHT
LINE AD, PRODUCED BOTH WAYS;
THEN IF THE POINTS A AND B
APPROACH ONE ANOTHER AND MEET, I
SAY THE ANGLE BAD, CONTAINED BETWEEN THE CHORD
AND THE TANGENT, WILL BE DIMINISHED IN INFINITUM,
AND ULTIMATELY WILL VANISH.

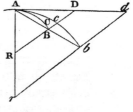

For if that angle does not vanish, the arc ACB will contain with the tangent AD an angle equal to a rectilinear angle; and therefore the curvature at the point A will not be continued, which is against the supposition.

LEMMA VII.

THE SAME THINGS BEING SUPPOSED, I SAY THAT THE ULTIMATE
RATIO OF THE ARC, CHORD, AND TANGENT, ANY ONE TO ANY
OTHER, IS THE RATIO OF EQUALITY.

For while the point B approaches towards the point A, consider always AB and AD as produced to the remote points b and d, and parallel to the secant BD draw bd: and let the arc Acb be always similar to the arc ACB. Then, supposing the points A and B to coincide, the angle dAb will vanish, by the preceding Lemma; and therefore the right

lines Ab, Ad (which are always finite), and the intermediate arc Acb, will coincide, and become equal among themselves. Wherefore, the rightlines AB, AD, and the intermediate arc ACB (which are always proportional to the former), will vanish, and ultimately acquire the ratio of equality. Q.E.D.

COR. 1. Whence if through B we draw BF parallel to the tangent, always cutting any right line AF passing through A in F, this line BF will be ultimately in the ratio of equality with the evanescent arc ACB; because, completing the parallelogram AFBD, it is always in a ratio of equality with AD.

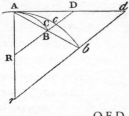

COR. 2. And if through B and A more right lines are drawn, as BE, BD, AF, AG, cutting the tangent AD and its parallel BF; the ultimate ratio of all the abscissas AD, AE, BF, BG, and of the chord and arc AB, any one to any other, will be the ratio of equality.

COR. 3. And therefore in all our reasoning about ultimate ratios, we may freely use any one of those lines for any other.

LEMMA VIII.

IF THE RIGHT LINES AR, BR, WITH THE ARC ACB, THE CHORD AB, AND THE TANGENT AD, CONSTITUTE THREE TRIANGLES RAB, RACB, RAD, AND THE POINTS A AND B APPROACH AND MEET: I SAY, THAT THE ULTIMATE FORM OF THESE EVANESCENT TRIANGLES IS THAT OF SIMILITUDE, AND THEIR ULTIMATE RATIO THAT OF EQUALITY.

For while the point B approaches towards the point A, consider always AB, AD, AR, as produced to the remote points b, d, and r, and rbd as drawn parallel to RD, and let the arc Acb be always similar to the arc ACB. Then supposing the points A and B to coincide, the angle bAd will vanish; and therefore the three triangles rAb, rAci, rAd(which are always finite), will coincide, and on that account become both similar and equal. And therefore the triangles RAB, RACB, RAD which are always similar and proportional to these, will ultimately become both similar and equal among themselves. Q.E.D.

COR. And hence in all reasonings about ultimate ratios, we may indifferently use any one of those triangles for any other.

LEMMA IX.

IF A RIGHT LINE AE. AND A CURVE LINE ABC, BOTH GIVEN
BY POSITION, CUT EACH OTHER IN A GIVEN ANGLE, A; AND
TO THAT RIGHT LINE, IN ANOTHER GIVEN ANGLE, BD, CE
ARE ORDINATELY APPLIED, MEETING THE CURVE IN B, C;
AND THE POINTS B AND C TOGETHER APPROACH TOWARDS
AND MEET IN THE POINT A: I SAY THAT THE AREAS OF THE
TRIANGLES ABD, ACE, WILL ULTIMATELY BE ONE TO THE
OTHER IN THE DUPLICATE RATIO OF THE SIDES.

For while the points B, C, approach towards the point A, suppose always AD to be produced to the remote points d and e, so as Ad, Ae may be proportional to AD, AE; and the ordinates db, ec, to be drawn parallel to the ordinates DB and EC, and meeting AB and AC produced in b and c. Let the curve Abc be similar to the curve

ABC, and draw the right line Ag so as to touch both curves in A, and cut the ordinates DB, EC, db ec, in F, G, f, g. Then, supposing the length Ae to remain the same, let the points B and C meet in the point A; and the angle cAg vanishing, the curvilinear areas Abd, Ace will coincide with the rectilinear areas Afd, A; and therefore (by Lem. V) will be one to the other in the duplicate ratio of the sides Ad, Ae. But the areas ABD, ACE are always proportional to these areas; and so the sides AD, AE are to these sides. And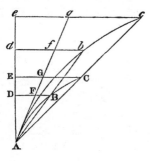

therefore the areas ABD, ACE are ultimately one to the other in the duplicate ratio of the sides AD, AE.   Q.E.D.

LEMMA X.

THE SPACES WHICH A BODY DESCRIBES BY ANY FINITE FORCE
URGING IT, WHETHER THAT FORCE IS DETERMINED AND
IMMUTABLE, OR IS CONTINUALLY AUGMENTED OR CONTINUALLY
DIMINISHED, ARE IN THE VERY BEGINNING OF THE MOTION ONE
TO THE OTHER IN THE DUPLICATE RATIO OF THE TIMES.

Let the times be represented by the lines AD, AE, and the velocities generated in those times by the ordinates DB, EC. The spaces described with these velocities will be as the areas ABD, ACE, described by those ordinates, that is, at the very beginning of the motion (by Lem. IX), in the duplicate ratio of the times AD, AE.   Q.E.D.

COR. 1. And hence one may easily infer, that the errors of bodies describing similar parts of similar figures in proportional times, are nearly as the squares of the times in which they are generated; if so be these errors are generated by any equal forces similarly applied to the bodies, and measured by the distances of the bodies from those places of the similar figures, at which, without the action of those form, the bodies would have arrived in those proportional times.

COR. 2. But the errors that are generated by proportional forces, similarly applied to the bodies at similar parts of the similar figures, are as the forces and the squares of the times conjunctly.

COR. 3. The same thing is to be understood of any spaces whatsoever described by bodies urged with different forces; all which, in the very beginning of the motion, are as the forces and the squares of the times conjunctly.

COR. 4. And therefore the forces are as the spaces described in the very beginning of the motion directly, and the squares of the times inversely.

COR. 5. And the squares of the times are as the spaces described directly, and the forces inversely.

SCHOLIUM.

If in comparing indetermined quantities of different sorts one with another, any one is said to be as any other directly or inversely, the meaning is, that the former is augmented or diminished in the same ratio with the latter, or with its reciprocal. And if any one is said to be as any other two or more directly or inversely, the meaning is, that the first is augmented or diminished in the ratio compounded of the ratios in which the others, or the reciprocals of the others, are augmented or diminished. As if A is said to be as B directly, and C directly, and D inversely, the meaning is, that A is augmented or diminished in the same ratio with $B \times C \times \frac{1}{D}$, that is to say, that A and $\frac{BC}{D}$ are one to the other in a given ratio.

LEMMA XI.

THE EVANESCENT SUBTENSE OF THE ANGLE OF CONTACT, IN ALL CURVES WHICH AT THE POINT OF CONTACT HAVE A FINITE CURVATURE, IS ULTIMATELY IN THE DUPLICATE RATIO OF THE SUBTENSE OF THE CONTERMINATE ARC.

CASE 1. Let AB be that arc, AD its tangent, BD the subtense of the angle of contact perpendicular on the tangent, AB the subtense of the arc. Draw BG perpendicular to the subtense AB, and AG to the tangent AD, meeting in G; then let the points D, B, and G, approach to the points d, b, and g, and suppose J to be the ultimate intersection

of the lines BG, AG, when the points D, B, have come to A. It is evident that the distance GJ may be less than any assignable. But (from the nature of the circles passing through q the points A, B, G, A, b, g,) AB^2= AG $\times$ BD, and A^2 = Ag $\times$ bd; and therefore the ratio of AB^2 to Ab^2 is compounded of the ratios of AG to Ag, and of Bd to bd. But because GJ may be assumed of less length than any assignable, the ratio of AG to Ag may be such as to differ from the ratio of equality by less than any assignable difference; and therefore the ratio of AB^2 to Ab^2 may be such as to differ from the ratio of BD to bd by less than any assignable difference. Therefore, by Lem. I, the ultimate ratio of AB^2 to Ab^2 is the same with the ultimate ratio of BD to bd.

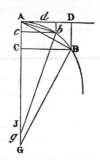

Q.E.D.

CASE 2. Now let BD be inclined to AD in any given angle, and the ultimate ratio of BD to bd will always be the same as before and therefore the same with the ratio of AB^2 to Ab^2. Q.E.D.

CASE. 3. And if we suppose the angle D not to be given, but that the right line BD converges to a given point, or is determined by any other condition whatever; nevertheless the angles D, d, being determined by the same law, will always draw nearer to equality, and approach nearer to each other than by any assigned difference, and therefore, by Lem. I, will at last be equal; and therefore the lines BD, bd are in the same ratio to each other as before. Q.E.D.

COR. 1. Therefore since the tangents AD, Ad, the arcs AB, Ab, and their sines, BC, bc, become ultimately equal to the chords AB, Ab, their squares will ultimately become as the subtenses BD, bd.

COR. 2. Their squares are also ultimately as the versed sines of the arcs, bisecting the chords, and converging to a given point. For those versed sines are as the subtenses BD, bd.

COR. 3. And therefore the versed sine is in the duplicate ratio of the time in which a body will describe the arc with a given velocity.

COR. 4. The rectilinear triangles ADB, Adb are ultimately in the triplicate ratio of the sides AD, Ad, and in a sesquiplicate ratio of the sides DB, db; as being in the ratio compounded of the sides AD to DB, and of Ad to db. So also the triangles ABC, Abc are ultimately in the triplicate ratio of the sides BC, bc. What I call the sesquiplicate ratio is the subduplicate of the triplicate, as being compounded of the simple and subduplicate ratio.

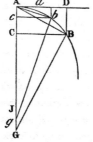

COR. 5. And because DB, *db* are ultimately parallel and in the duplicate ratio of the lines AD, A*d*, the ultimate curvilinear areas ADB, A*db* will be (by the nature of the parabola) two thirds of the rectilinear triangles ADB, A*db*; and the segments AB, A*b* will be one third of the same triangles. And thence those areas and those segments will be in the triplicate ratio as well of the tangents AD, A*d*, as of the chords and arcs AB, AB.

SCHOLIUM.

But we have all along supposed the angle of contact to be neither infinitely greater nor infinitely less than the angles of contact made by circles and their tangents; that is, that the curvature at the point A is neither infinitely small nor infinitely great, or that the interval AJ is of a finite magnitude. For DB may be taken as AD^3: in which case no circle can be drawn through the point A, between the tangent AD and the curve AB, and therefore the angle of contact will be infinitely less than those of circle. And by a like reasoning, if DB be made successfully as AD^4, AD^5 AD^6, AD^7 &c., we shall have a series of angles of contact, proceeding in infinitum, wherein every succeeding term is infinitely less than the preceding. And if DB be made successively as AD^2, AD$^{3/}_2$ AD$^{4/}_3$ AD$^{5/}_4$ AD$^{6/}_5$ AD^7, &c., we shall have another infinite series of angles of contact, the first of which is of the same sort with those of circles, the second infinitely greater, and every succeeding one infinitely greater than the preceding. But between any two of these angles another series of intermediate angles of contact may be interposed, proceeding both ways in infinitum, wherein every succeeding angle shall be infinitely greater or infinitely less than the preceding. As if between the terms AD^2 and AD^3 there were interposed the series AD$^{13/}_6$, AD$^{11/}_5$ AD$^{9/}_4$ AD$^{7/}_3$ AD$^{5/}_2$ AD$^{8/}_3$ AD$^{11/}_4$ AD$^{14/}_5$ AD$^{17/}_6$, &c. And again, between any two angles of this series, a new series of intermediate angles may be interposed, differing from one another by infinite intervals. Nor is nature confined to any bounds.

Those things which have been demonstrated of curve lines, and the superfices which they comprehend, may be easily applied to the curve superfices and contents of solids. These Lemmas are premised to avoid the tediousness of deducing perplexed demonstrations ad absurdum, according to the method of the ancient geometers. For demonstrations are more contracted by the method of indivisibles: but because the hypothesis of indivisibles seems somewhat harsh, and therefore that method is reckoned less geometrical, I chose rather to reduce the demonstrations of the following propositions to the first and last sums and ratios of nascent and evanescent quantities, that is, to the limits of those sums and ratios; and so to premise, as short as I could, the demonstrations of those limits. For hereby the same thing is performed as by the method of indivisibles; and now those principles being demonstrated, we may use

them with more safety. Therefore if hereafter I should happen to consider quantities as made up of particles, or should use little curve lines for right ones, I would not be understood to mean indivisibles, but evanescent divisible quantities; not the sums and ratios of determinate parts, but always the limits of sums and ratios; and that the force of such demonstrations always depends on the method laid down in the foregoing Lemmas.

Perhaps it may be objected, that there is no ultimate proportion, of evanescent quantities; because the proportion, before the quantities have vanished, is not the ultimate, and when they are vanished, is none. But by the same argument, it may be alledged, that a body arriving at a certain place, and there stopping, has no ultimate velocity: because the velocity, before the body comes to the place, is not its ultimate velocity; when it has arrived, is none. But the answer is easy; for by the ultimate velocity is meant that with which the body is moved, neither before it arrives at its last place and the motion ceases, nor after, but at the very instant it arrives; that is, that velocity with which the body arrives at its last place, and with which the motion ceases. And in like manner, by the ultimate ratio of evanescent quantities is to be understood the ratio of the quantities not before they vanish, nor afterwards, but with which they vanish. In like manner the first ratio of nascent quantities is that with which they begin to be. And the first or last sum is that with which they begin and cease to be (or to be augmented or diminished). There is a limit which the velocity at the end of the motion may attain, but not exceed. This is the ultimate velocity. And there is the like limit in all quantities and proportions that begin and cease to be. And since such limits are certain and definite, to determine the same is a problem strictly geometrical. But whatever is geometrical we may be allowed to use in determining and demonstrating any other thing that is likewise geometrical.

It may also be objected, that if the ultimate ratios of evanescent quantities are given, their ultimate magnitudes will be also given: and so all quantities will consist of indivisibles, which is contrary to what Euclid has demonstrated concerning incommensurables, in the 10th Book of his Elements. But this objection is founded on a false supposition. For those ultimate ratios with which quantities vanish are not truly the ratios of ultimate quantities, but limits towards which the ratios of quantities decreasing without limit do always converge; and to which they approach nearer than by any given difference, but never go beyond, nor in effect attain to, till the quantities are diminished in infinitum. This thing will appear more evident in quantities infinitely great. If two quantities, whose difference is given, be augmented in infinitum, the ultimate ratio of these quantities will be given, to wit, the ratio of equality; but it does not from thence follow, that the ultimate or greatest quantities themselves, whose ratio that is, will be given. Therefore if in what follows, for the sake of being

more easily understood, I should happen to mention quantities as least, or evanescent, or ultimate, you are not to suppose that quantities of any determinate magnitude are meant, but such as are conceived to be always diminished without end.

SECTION II.

OF THE INVENTION OF CENTRIPETAL FORCES.

PROPOSITION I THEOREM I.

THE AREAS, WHICH REVOLVING BODIES DESCRIBE BY RADII
DRAWN TO AN IMMOVABLE CENTRE OF FORCE DO LIE IN THE
SAME IMMOVABLE PLANES, AND ARE PROPORTIONAL TO THE
TIMES IN WHICH THEY ARE DESCRIBED.

For suppose the time to be divided into equal parts, and in the first part of that time let the body by its innate force describe the right line AB. In the second part of that time, the same would (by Law I.), if not hindered, proceed directly to *c*, along the line B*c* equal to AB; so that by the radii AS, BS, *c*S, drawn to the centre, the equal areas

ASB, BS*c*, would be described. But when the body is arrived at B, suppose that a centripetal force acts at once with a great impulse, and, turning aside the body from the right line B*c*, compels it afterwards to continue its motion along the right line BC. Draw *c*C parallel to BS meeting BC in C; and at the end of the second part of the time, the body (by Cor. I. of the Laws) will be found in C, in the same plane with the triangle ASB. Join SC, and, because SB and C*c* are parallel, the triangle SBC will be equal to the triangle SB, and therefore

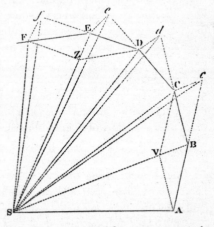

also to the triangle SAB. By the like argument, if the centripetal force acts successively in C, D, E, &c., and makes the body, in each single particle of time, to describe the right lines CD, DE, EF, &c., they will all lie in the same plane; and the triangle SCD will be equal to the triangle SBC, and SDE to SCD, and SEF to SDE. And therefore,

in equal times, equal areas are described in one immovable plane: and, by composition, any sums SADS, SAFS, of those areas, are one to the other as the times in which they are described. Now let the number of those triangles be augmented, and their breadth diminished in infinitum; and (by Cor. 4, Lem. III.) their ultimate perimeter ADF will be a curve line: and therefore the centripetal force, by which the body is perpetually drawn back from the tangent of this curve, will act continually; and any described areas SADS, SAFS, which are always proportional to the times of description, will, in this case also, be proportional to those times.   Q.E.D.

Cor. 1. The velocity of a body attracted towards an immovable centre, in spaces void of resistance, is reciprocally as the perpendicular let fall from that centre on the right line that touches the orbit. For the velocities in those places A, B, C, D, E, are as the bases AB, BC, CD, DE, EF, of equal triangles; and these bases are reciprocally as the perpendiculars let fall upon them.

Cor. 2. If the chords AB, BC of two arcs, successively described in equal times by the same body, in spaces void of resistance, are completed into a parallelogram ABCV, and the diagonal BV of this parallelogram, in the position which it ultimately acquires when those arcs are diminished in infinitum, is produced both ways, it will pass through the centre of force

Cor. 3. If the chords AB, BC, and DE, EF, of arcs described in equal times, in spaces void of resistance, are completed into the parallelograms ABCV, DEFZ: the forces in B and E are one to the other in the ultimate ratio of the diagonals BV, EZ, when those arcs are diminished in infinitum. For the motions BC and EF of the body (by Cor. 1 of the Laws) are compounded of the motions Bc, BV, and Ef, EZ: but BV and EZ, which are equal to Cc and Ff, in the demonstration of this Proposition, were generated by the impulses of the centripetal force in B and E, and are therefore proportional to those impulses.

Cor. 4. The forces by which bodies, in spaces void of resistance, are drawn back from rectilinear motions, and turned into curvilinear orbits, are one to another as the versed sines of arcs described in equal times; which versed sines tend to the centre of force, and bisect the chords when those axes are diminished to infinity. For such versed sines are the halves of the diagonals mentioned in Cor. 3.

Cor. 5. And therefore those forces are to the force of gravity as the said versed sines to the versed sines perpendicular to the horizon of those parabolic arcs which projectiles describe in the same time.

Cor. 6. And the same things do all hold good (by Cor. 5 of the Laws), when the planes in which the bodies are moved, together with the centres of force which are placed in those planes, are not at rest, but move uniformly forward in right lines.

PROPOSITION II. THEOREM II.

EVERY BODY THAT MOVES IN ANY CURVE LINE DESCRIBED
IN A PLANE, AND BY A RADIUS, DRAWN TO A POINT EITHER
IMMOVABLE, OR MOVING FORWARD WITH AN UNIFORM
RECTILINEAR MOTION, DESCRIBES ABOUT THAT POINT AREAS
PROPORTIONAL TO THE TIMES, IS URGED BY A CENTRIPETAL
FORCE DIRECTED TO THAT POINT.

CASE. 1. For every body that moves in a curve line, is (by Law 1) turned aside from its rectilinear course by the action of some force that impels it. And that force by which the body is turned off from its rectilinear course, and is made to describe, in equal times, the equal least triangles SAB, SBC, SCD, &c., about the immovable point S (by Prop. XL. Book 1, Elem. and Law II), acts in the place B, according to the direction of a line parallel to *c*C, that is, in the direction of the line BS, and in the place C, according to the direction of a line parallel to *d*D, that is, in the direction of the line CS, &c.; and therefore acts always in the direction of lines tending to the immovable point S.

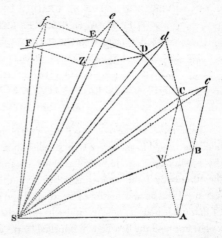

Q.E.D.

CASE 2. And (by Cor. 5 of the Laws) it is indifferent whether the superfices in which a body describes a curvilinear figure be quiescent, or moves together with the body, the figure described, and its point S, uniformly forward in right lines.

COR. 1. In non-resisting spaces or mediums, if the areas are not proportional to the times, the forces are not directed to the point in which the radii meet; but deviate therefrom in consequentia, or towards the parts to which the motion is directed, if the description of the areas is accelerated; but in antecedentia, if retarded.

COR. 2. And even in resisting mediums, if the description of the areas is accelerated, the directions of the forces deviate from the point in which the radii meet, towards the parts to which the motion tends.

SCHOLIUM.

A body may be urged by a centripetal force compounded of several forces; in which case the meaning of the Proposition is that the force which results out of all tends to the point S. But if any force acts perpetually in the direction of lines perpendicular to the described surface, this force will make the body to deviate from the plane of its motion: but will neither augment nor diminish the quantity of the described surface. and is therefore to be neglected in the composition of forces.

PROPOSITION III. THEOREM III.

EVERY BODY, THAT BY A RADIUS DRAWN TO THE CENTRE OF ANOTHER BODY, HOWSOEVER MOVED, DESCRIBES AREAS ABOUT THAT CENTRE PROPORTIONAL TO THE TIMES, IS URGED BY A FORCE COMPOUNDED OUT OF THE CENTRIPETAL FORCE TENDING TO THAT OTHER BODY, AND OF ALL THE ACCELERATIVE FORCE BY WHICH THAT OTHER BODY IS IMPELLED.

Let L represent the one, and T the other body; and (by Cor. 6 of the Laws) if both bodies are urged in the direction of parallel lines, by a new force equal and contrary to that by which the second body T is urged, the first body L will go on to describe about the other body T the same areas as before: but the force by which that other body T was urged will be now destroyed by an equal and contrary force; and therefore (by Law I.) that other body T, now left to itself, will either rest, or move uniformly forward in a right line: and the first body L impelled by the difference of the forces, that is, by the force remaining, will go on to describe about the other body T areas proportional to the times. And therefore (by Theor. II.) the difference of the forces is directed to the other body T as its centre.   Q.E.D

COR. 1. Hence if the one body L, by a radius drawn to the other body T, describes areas proportional to the times; and from the whole force, by which the first body L is urged (whether that force is simple, or, according to Cor. 2 of the Laws, compounded out of several forces), we subduct (by the same Cor.) that whole accelerative force by which the other body is urged; the whole remaining force by which the first body is urged will tend to the other body T, as its centre.

COR. 2. And, if these areas are proportional to the times nearly, the remaining force will tend to the other body T nearly.

COR. 3. And vice versa, if the remaining force tends nearly to the other body T, those areas will be nearly proportional to the times.

COR. 4. If the body L, by a radius drawn to the other body T, describes areas, which, compared with the times, are very unequal; and that other body T be either at rest, or moves uniformly forward in a right line: the action of the centripetal force tending to that other body T is either none at all, or it is mixed and compounded with very powerful actions of other forces: and the whole force compounded of them all, if they are many, is directed to another (immovable or moveable) centre. The same thing obtains, when the other body is moved by any motion whatsoever; provided that centripetal force is taken, which remains after subducting that whole force acting upon that other body T.

SCHOLIUM.

Because the equable description of areas indicates that a centre is respected by that force with which the body is most affected, and by which it is drawn back from its rectilinear motion, and retained in its orbit; why may we not be allowed, in the following discourse, to use the equable description of areas as an indication of a centre, about which all circular motion is performed in free spaces?

PROPOSITION IV. THEOREM IV.

THE CENTRIPETAL FORCES OF BODIES, WHICH BY EQUABLE
MOTIONS DESCRIBE DIFFERENT CIRCLES, TEND TO THE
CENTRES OF THE SAME CIRCLES; AND ARE ONE TO THE
OTHER AS THE SQUARES OF THE ARCS DESCRIBED IN EQUAL
TIMES APPLIED TO THE RADII OF THE CIRCLES.

These forces tend to the centres of the circles (by Prop. II, and Cor. 2, Prop. I.), and are one to another as the versed sines of the least arcs described in equal times (by Cor. 4, Prop. I.); that is, as the squares of the same arcs applied to the diameters of the circles (by Lem. VII.); and therefore since those arcs are as arcs described in any equal times, and the diameters are as the radii, the forces will be as the squares of any arcs described in the same time applied to the radii of the circles. Q.E.D.

COR. 1. Therefore, since those arcs are as the velocities of the bodies the centripetal forces are in a ratio compounded of the duplicate ratio of the velocities directly, and of the simple ratio of the radii inversely.

COR. 2. And since the periodic times are in a ratio compounded of the ratio of the radii directly, and the ratio of the velocities inversely, the centripetal forces are in a ratio compounded of the ratio of the radii directly, and the duplicate ratio of the periodic times inversely.

COR. 3. Whence if the periodic times are equal, and the velocities therefore as the radii, the centripetal forces will be also as the radii; and the contrary.

COR. 4. If the periodic times and the velocities are both in the subduplicate ratio of the radii, the centripetal forces will be equal among themselves; and the contrary.

COR. 5. If the periodic times are as the radii, and therefore the velocities equal, the centripetal forces will be reciprocally as the radii; and the contrary.

COR. 6. If the periodic times are in the sesquiplicate ratio of the radii, and therefore the velocities reciprocally in the subduplicate ratio of the radii, the centripetal forces will be in the duplicate ratio of the radii inversely; and the contrary.

COR. 7. And universally, if the periodic time is as any power R^n of the radius R, and therefore the velocity reciprocally as the power R^{n-1} of the radius, the centripetal force will be reciprocally as the power R^{2n-1} of the radius; and the contrary.

COR. 8. The same things all hold concerning the times, the velocities, and forces by which bodies describe the similar parts of any similar figures that have their centres in a similar position with those figures; as appears by applying the demonstration of the preceding cases to those. And the application is easy, by only substituting the equable description of areas in the place of equable motion, and using the distances of the bodies from the centres instead of the radii.

COR. 9. From the same demonstration it likewise follows, that the arc which a body, uniformly revolving in a circle by means of a given centripetal force, describes in any time, is a mean proportional between the diameter of the circle, and the space which the same body falling by the same given force would descend through in the same given time.

SCHOLIUM.

The case of the 6th Corollary obtains in the celestial bodies (as Sir Christopher Wren, Dr. Hooke, and Dr. Halley have severally observed); and therefore in what follows, I intend to treat more at large of those things which relate to centripetal force decreasing in a duplicate ratio of the distances from the centres.

Moreover, by means of the preceding Proposition and its Corollaries, we may discover the proportion of a centripetal force to any other known force, such as that of gravity. For if a body by means of its gravity revolves in a circle concentric to the earth, this gravity is the centripetal force of that body. But from the descent of heavy bodies, the time of one entire revolution, as well as the arc described in any given time, is given (by Cor. 9 of this Prop.). And by such propositions, Mr. Huygens, in his excellent book De Horologio Oscillatorio, has compared the force of gravity with the centrifugal forces of revolving bodies.

The preceding Proposition may be likewise demonstrated after this manner. In any circle suppose a polygon to be inscribed of any number of sides. And if a body, moved with a given velocity along the sides of the polygon, is reflected from the circle at the

several angular points, the force, with which at every reflection it strikes the circle, will be as its velocity: and therefore the sum of the forces, in a given time, will be as that velocity and the number of reflections conjunctly; that is (if the species of the polygon be given), as the length described in that given time, and increased or diminished in the ratio of the same length to the radius of the circle; that is, as the square of that length applied to the radius; and therefore the polygon, by having its sides diminished in infinitum, coincides with the circle, as the square of the arc described in a given time applied to the radius. This is the centrifugal force, with which the body impels the circle; and to which the contrary force, wherewith the circle continually repels the body towards the centre, is equal.

PROPOSITION V. PROBLEM I.

THERE BEING GIVEN, IN ANY PLACES, THE VELOCITY WITH
WHICH A BODY DESCRIBES A GIVEN FIGURE, BY MEANS OF
FORCES DIRECTED TO SOME COMMON CENTRE: TO FIND
THAT CENTRE.

Let the three right lines PT, TQV, VR touch the figure described in as many points, P, Q, R, and meet in T and V. On the tangents erect the perpendiculars PA, QB, RC, reciprocally proportional to the velocities of the body in the points P, Q, R, from which the perpendiculars were raised; that is, so that PA may be to QB as the velocity in Q to the velocity in P, and QB to RC as the velocity in R to the velocity in Q. Through the ends A, B, C, of the perpendiculars draw AD, DBE, EC, at right angles, meeting in D and E: and the right lines TD, VE produced, will meet in S, the centre required.

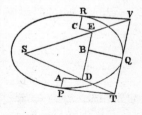

For the perpendiculars let fall from the centre S on the tangents PT, QT, are reciprocally as the velocities of the bodies in the points P and Q (by Cor. 1, Prop. I.), and therefore, by construction, as the perpendiculars AP, BQ directly; that is, as the perpendiculars let fall from the point D on the tangents. Whence it is easy to infer that the points S, D, T, are in one right line. And by the like argument the points S, E, V are also in one right line; and therefore the centre S is in the point where the right lines TD, VE meet.   Q.E.D.

PROPOSITION VI. THEOREM V.

IN A SPACE VOID OF RESISTANCE, IF A BODY REVOLVES IN
ANY ORBIT ABOUT AN IMMOVABLE CENTRE, AND IN THE

LEAST TIME DESCRIBES ANY ARC JUST THEN NASCENT; AND
THE VERSED SINE OF THAT ARC IS SUPPOSED TO BE DRAWN
BISECTING THE CHORD, AND PRODUCED PASSING THROUGH
THE CENTRE OF FORCE: THE CENTRIPETAL FORCE IN THE
MIDDLE OF THE ARC WILL BE AS THE VERSED SINE DIRECTLY
AND THE SQUARE OF THE TIME INVERSELY.

For the versed sine in a given time is as the force (by Cor. 4, Prop. 1); and augmenting the time in any ratio, because the arc will be augmented in the same ratio, the versed sine will be augmented in the duplicate of that ratio (by Cor. 2 and 3, Lem. XI.), and therefore is as the force and the square of the time. Subduct on both sides the duplicate ratio of the time, and the force will be as the versed sine directly, and the square of the time inversely. Q.E.D.

And the same thing may also be easily demonstrated by Corol. 4, Lem., X.

COR. 1. If a body P revolving about the centre S describes a curve line APQ, which a right line ZPR touches in any point P; and from any other point Q of the curve, QR is drawn parallel to the distance SP, meeting the tangent in R; and QT is drawn perpendicular to the distance SP; the centripetal force will be reciprocally as the solid $\frac{SP^2 \times QT^2}{QR}$, if the solid be taken of that magnitude which

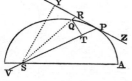

it ultimately acquires when the points P and Q coincide. For QR is equal to the versed sine of double the arc QP, whose middle is P : and double the triangle SQP, or SP × QT is proportional to the time in which that double arc is described; and therefore may be used for the exponent of the time.

COR. 2. By a like reasoning, the centripetal force is reciprocally as the solid $\frac{SY^2 \times QP^2}{QR}$; if SY is a perpendicular from the centre of force on PR the tangent of the orbit. For the rectangles SY × QP and SP × QT are equal.

COR. 3. If the orbit is either a circle, or touches or cuts a circle concentrically, that is, contains with a circle the least angle of contact or section, having the same curvature and the same radius of curvature at the point P; and if PV be a chord of this circle, drawn from the body through the centre of force; the centripetal force will be reciprocally as the solid SY² × PV– For PV is $\frac{QP^2}{QR}$.

COR. 4. The same things being supposed, the centripetal force is as the square of the velocity directly, and that chord inversely. For the velocity is reciprocally as the perpendicular SY, by Cor. 1. Prop. I.

COR. 5. Hence if any curvilinear figure APQ is given, and therein a point S is also given, to which a centripetal force is perpetually directed, that law of centripetal force may be found, by which the body P will be continually drawn back from a rectilinear course, and, being detained in the perimeter of that figure, will describe the same by a perpetual revolution. That is, we are to find, by computation, either the solid $\frac{SP^2 \times QT^2}{QR}$ or the solid $SY^2 \times PV$, reciprocally proportional to this force. Examples of this we shall give in the following Problems.

PROPOSITION VII. PROBLEM II.

IF A BODY REVOLVES IN THE CIRCUMFERENCE OF A CIRCLE; IT IS PROPOSED TO FIND THE LAW OF CENTRIPETAL FORCE DIRECTED TO ANY GIVEN POINT.

Let VQPA be the circumference of the circle; S the given point to which as to a centre the force tends; P the body moving in the circumference; Q the next place into which it is to move; and PRZ the tangent of the circle at the preceding L place. Through the point S draw the chord PV, and the diameter VA of the circle: join AP,

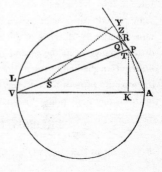

and draw QT perpendicular to SP, which produced, may meet the tangent PR in Z; and lastly, through the point Q, draw LR parallel to SP, meeting the circle in L, and the tangent PZ in R. And, because of the similar triangles ZQR, ZTP, VPA, we shall have RP², that is, QRL to QT² as AY² to PV². And therefore $\frac{QRL \times PV^2}{AV^2}$ is equal to QT². Multiply those equals by $\frac{SP^2 \times PV^3}{AV^2} = \frac{SP^2 \times QT^2}{QR}$, and the points P and Q coinciding, for RL write PV; then we shall have $\frac{SP^2 \times PV^3}{AV^2} = \frac{SP^2 \times QT^2}{QR}$. And therefore (by Cor 1 and 5,

Prop. VI.) the centripetal force is reciprocally as $\frac{SP^2 \times PV^3}{AV^2}$; that is (because AV² is given), reciprocally as the square of the distance or altitude SP, and the cube of the chord PV conjunctly. Q.E.I.

THE SAME OTHERWISE.

On the tangent PR produced let fall the perpendicular SY; and (because of the similar triangles SYP, VPA), we shall have AV to PV as SP to SY, and therefore $\frac{SP \times PV}{AV} = SY$, and $\frac{SP^2 \times PV^3}{AV^2} = SY^2 \times PV$. And therefore (by Corol. 3 and 5, Prop. VI), the

centripetal force is reciprocally as $\dfrac{SP^2 \times PV^3}{AV^2}$ that is (because AV is given), reciprocally as

$SP^2 \times PV^3$.
Q.E.I.

COR. 1. Hence if the given point S, to which the centripetal force always tends, is placed in the circumference of the circle, as at V, the centripetal force will be reciprocally as the quadrato-cube (or fifth power) of the altitude SP.

COR. 2. The force by which the body P in the circle APTV revolves about the centre of force S is to the force by which the same body P may revolve in the same circle, and in the same periodic time, about any other centre of force R, as $RP^2 \times SP$ to the cube of the right line SG, which from the first centre of force S is drawn parallel to the distance PR. of the body from the second centre of force R, meeting the tangent PG

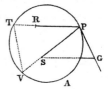

of the orbit in G. For by the construction of this Proposition, the former force is to the latter as $RP^2 \times PT^3$ to $SP^2 \times PV^3$; that is, as $SP \times RP^2$ to $\dfrac{SP^3 \times PV^3}{PT^3}$; or (because of the similar triangles PSG, TPV) to SG^3.

COR. 3. The force by which the body P in any orbit revolves about the centre of force S, is to the force by which the same body may revolve in the same orbit, and the same periodic time, about any other centre of force R, as the solid $SP \times RP^2$, contained under the distance of the body from the first centre of force S, and the square of its distance from the second centre of force R, to the cube of the right line SG, drawn from the first centre of the force S, parallel to the distance RP of the body from the second centre of force R, meeting the tangent PG of the orbit in G. For the force in this orbit at any point P is the same as in a circle of the same curvature.

PROPOSITION VIII. PROBLEM III.

IF A BODY MOVES IN THE SEMI-CIRCUMFERENCE PQA; IT IS
PROPOSED TO FIND THE LAW OF THE CENTRIPETAL FORCE
LENDING TO A POINT S, SO REMOTE, THAT ALL THE LINES
PS, RS DRAWN THERETO, MAY BE TAKEN FOR PARALLELS.

From C, the centre of the semi-circle, let the semi-diameter CA be drawn, cutting the parallels at right angles in M and N, and join CP. Because of the similar triangles CPM, PZT, and RZQ, we shall have CP^2 to PM^2 as PR^2 to QT^2; and, from the nature of the circle, PR^2 is equal to the rectangle $QR \times \overline{RN+QN}$ or, the points P, Q coinciding, to the rectangle $QR \times 2PM$. Therefore CP^2 is

to PM^2 as $QR \times 2PM$ to QT^2; and $\dfrac{QT^2}{QR} = \dfrac{2PM^3}{CP^2}$, and $\dfrac{QT^2 \times SP^2}{QR} = \dfrac{2PM^3 \times SP^2}{CP^2}$. And

774

therefore (by Corol. 1 and 5, Prop. VI.), the centripetal force is reciprocally as $\dfrac{2PM^3 \times SP^2}{CP^2}$; that is (neglecting the given ratio $\dfrac{2SP^2}{CP^2}$), reciprocally as PM^3.

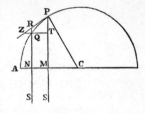

Q.E.I.

And the same thing is likewise easily inferred from the preceding Proposition.

SCHOLIUM

And by a like reasoning, a body will be moved in an ellipsis, or even in an hyperbola, or parabola, by a centripetal force which is reciprocally as the cube of the ordinate directed to an infinitely remote centre of force.

PROPOSITION IX. PROBLEM IV.

IF A BODY REVOLVES IN A SPIRAL PQS, CUTTING ALL THE
RADII SP, SQ, &C., IN A GIVEN ANGLE; IT IS PROPOSED TO
FIND THE LAW OF THE CENTRIPETAL FORCE TENDING TO
THE CENTRE OF THAT SPIRAL.

Suppose the indefinitely small angle PSQ to be given; because, then, all the angles are given, the figure SPRQT will be given in specie. Therefore the ratio $\dfrac{QT}{QR}$ is also given, $\dfrac{QT^2}{QR}$ is as QT, that is

(because the figure is given in specie), as SP. But if the angle PSQ is any way changed, the right line QR, subtending the angle of contact QPR (by

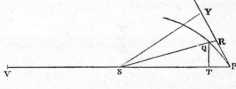

Lemma XI) will be changed in the duplicate ratio of PR or QT Therefore the ratio $\dfrac{QT^2}{QR}$ remains the same as before, that is, as SP. And $\dfrac{QT^2 \times SP^2}{QR}$ is as SP^3, and therefore (by Corol. 1 and 5, Prop. VI) the centripetal force is reciprocally as the cube of the distance SP. Q.E.I.

THE SAME OTHERWISE.

The perpendicular SY let fall upon the tangent, and the chord PV of the circle concentrically cutting the spiral, are in given ratios to the height SP; and therefore SP^3 is as $SY^2 \times PV$, that is (by Corel. 3 and 5, Prop. VI) reciprocally as the centripetal force.

·LEMMA XII.

ALL PARALLELOGRAMS CIRCUMSCRIBED ABOUT ANY CONJUGATE DIAMETERS OF A GIVEN ELLIPSIS OR HYPERBOLA ARE EQUAL AMONG THEMSELVES.

This is demonstrated by the writers on the conic sections.

PROPOSITION X. PROBLEM V.

IF A BODY REVOLVES IN AN ELLIPSIS; IT IS PROPOSED TO FIND THE LAW OF THE CENTRIPETAL FORCE TENDING TO THE CENTRE OF THE ELLIPSIS.

Suppose CA, CB to be semi-axes of the ellipsis; GP, DK, conjugate diameters; PF, QT perpendiculars to those diameters; Qv an ordinate to the diameter GP; and if the parallelogram QvPR be completed, then (by the properties of the conic sections) the rectangle PvG will be to Qv^2 as PC^2 to CD^2; and (because of the similar triangles QvT, PCF), Qv^2 to QT^2 as PC^2 to PF^2; and, by composition, the ratio of PvG to QT^2 is compounded of the ratio of PC^2 to CD^2, and of the ratio of PC^2 to PF^2, that is, vG to $\frac{QT^2}{Pv}$ as PC^2

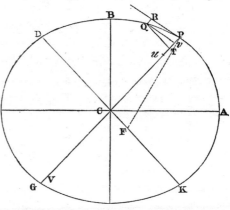

to $\frac{CD^2 \times PF^2}{PC^2}$. Put QR for Pv, and (by Lem. XII), BC $\times$ CA for CD $\times$ PF; also (the points P and Q coinciding) 2PC for vG; and multiplying the extremes and means together, we shall have $\frac{QT^2 \times PC^2}{QR}$ equal to $\frac{2BC^2 \times CA^2}{PC}$. Therefore (by Cor. 5, Prop. VI), the

centripetal force is reciprocally as $\frac{2BC^2 \times CA^2}{PC}$; that is (because $2BC^2 \times CA^2$ is given),

reciprocally as $\frac{1}{PC}$; that is, directly as the distance PC.   QEI.

THE SAME OTHERWISE.

In the right line PG on the other side of the point T, take the point u so that Tu may be equal to Tv; then take uV, such as shall be to vG as DC^2 to PC^2. And because

Qu^2 is to PuG as DC^2 to PC^2 (by the conic sections), we shall have $Qv^2 = Pv \times uV$. Add the rectangle uPv to both sides, and the square of the chord of the arc PQ will be equal to the rectangle VPv; and therefore a circle which touches the conic section in P, and passes through the point Q, will pass also through the point V. Now let the points P and Q meet, and the ratio of uV to vG, which is the same with the ratio of DC^2 to PC^2, will become the ratio of PV to PG, or PV to $2PC$; and therefore PV will be equal to $\frac{2DC^2}{PC}$. And therefore the force by which the body P revolves in the ellipsis will be

reciprocally as $\frac{2DC^2}{PC} \times PF2$ (by Cor. 3, Prop. VI); that is (because $2DC2 \times PF2$ is

given) directly as PC. Q.E.I.

COR. 1. And therefore the force is as the distance of the body from the centre of the ellipsis; and, vice versa, if the force is as the distance, the body will move in an ellipsis whose centre coincides with the centre of force, or perhaps in a circle into which the ellipsis may degenerate.

COR. 2. And the periodic times of the revolutions made in all ellipses whatsoever about the same centre will be equal. For those times in similar ellipses will be equal (by Corol. 3 and 8, Prop. IV); but in ellipses that have their greater axis common, they are one to another as the whole areas of the ellipses directly, and the parts of the areas described in the same time inversely; that is, as the lesser axes directly, and the velocities of the bodies in their principal vertices inversely; that is, as those lesser axes directly, and the ordinates to the same point of the common axes inversely; and therefore (because of the equality of the direct and inverse ratios) in the ratio of equality.

SCHOLIUM.

If the ellipsis, by having its centre removed to an infinite distance, degenerates into a parabola, the body will move in this parabola; and the force, now tending to a centre infinitely remote, will become equable. Which is Galileo's theorem. And if the parabolic section of the cone (by changing the inclination of the cutting plane to the cone) degenerates into an hyperbola, the body will move in the perimeter of this hyperbola, having its centripetal force changed into a centrifugal force. And in like manner as in the circle, or in the ellipsis, if the forces are directed to the centre of the figure placed in the abscissa, those forces by increasing or diminishing the ordinates in any given ratio, or even by changing the angle of the inclination of the ordinates to the abscissa, are always augmented or diminished in the ratio of the distances from the centre; provided the periodic times remain equal; so also in all figures whatsoever, if the ordinates are augmented or diminished in any given ratio, or their inclination is any way changed, the periodic time remaining the same, the forces directed to any centre

placed in the abscissa are in the several ordinates augmented or diminished in the ratio of the distances from the centre.

SECTION III.

OF THE MOTION OF BODIES IN ECCENTRIC CONIC SECTIONS.

PROPOSITION XL PROBLEM VI.

IF A BODY REVOLVES IN AN ELLIPSIS; IT IS REQUIRED TO FIND THE LAW OF THE CENTRIPETAL FORCE TENDING TO THE FOCUS OF THE ELLIPSIS.

Let S be the focus of the ellipsis. Draw SP cutting the diameter DK of the ellipsis in E, and the ordinate Qv in x; and complete the parallelogram QxPR. It is evident that EP is equal to the greater semi-axis AC: for drawing HI from the other focus H of the ellipsis parallel to EC, because CS, CH are equal, ES, EI will be also equal; so that EP is the half sum of PS, PI, that is (because of the parallels HI, PR, and the equal angles IPR, HPZ), of PS, PH, which taken together are equal to the whole axis 2AC. Draw QT perpendicular to SP, and putting L for the principal latus rectum of the ellipsis ($\frac{2BC^2}{AC}$), we shall have L $\times$ QR to L $\times$ Pv as QR to Pv, that is, as PE or AC to

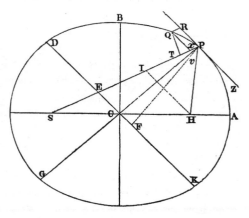

PC; and L $\times$ Pv to GvP as L to Gv; and GvP to Qv^2 as PC^2 to CD^2; and by (Corol. 2, Lem. VII) the points Q and P coinciding, Qv^2, is to Qx^2 in the ratio of equality; and Qx^2 or Qv^2 is to QT^2 as EP^2 to PF^2 that is, as CA^2 to PF^2, or (by Lem. XII) as CD^2 to CB^2. And compounding all those ratios together, we shall have L $\times$ QR to QT^2 as AC $\times$ L $\times$ PC^2 $\times$ CD^2, or 2CB^2 $\times$ PC^2 $\times$ CD^2 to PC $\times$ GvX CD^2 $\times$ CB^2, or as 2PC to Gv. But the points Q and P coinciding, 2PC and Gr are equal. And therefore the quantities L $\times$ QR and QT^2, proportional to these, will be also equal.

Let those equals be drawn into $\frac{SP^2}{QR}$ and $L \times SP^2$ will become equal to $\frac{SP^2 \times QT^2}{QR}$. And therefore (by Corol. 1 and 5, Prop. VI) the centripetal force is reciprocally as $L \times SP^2$, that is, reciprocally in the duplicate ratio of the distance SP. Q.E.I.

THE SAME OTHERWISE.

Since the force tending to the centre of the ellipsis, by which the body P may revolve in that ellipsis, is (by Corol. 1, Prop. X.) as the distance CP of the body from the centre C of the ellipsis; let CE be drawn parallel to the tangent PR of the ellipsis; and the force by which the same body P may revolve about any other point S of the ellipsis, if CE and PS intersect in E, will be as $\frac{PE^3}{SP^2}$ (by Cor. 3, Prop. VII.); that is, if the point S is the focus of the ellipsis, and therefore PE be given as SP^2 reciprocally. Q.E.I.

With the same brevity with which we reduced the fifth Problem to the parabola, and hyperbola, we might do the like here: but because of the dignity of the Problem and its use in what follows, I shall confirm the other cases by particular demonstrations.

PROPOSITION XII. PROBLEM VII.

SUPPOSE A BODY TO MOVE IN AN HYPERBOLA; IT IS REQUIRED TO FIND THE LAW OF THE CENTRIPETAL FORCE TENDING TO THE FOCUS OF THAT FIGURE.

Let CA, CB be the semi-axes of the hyperbola; PG, KD other conjugate diameters; PF a perpendicular to the diameter KD; and Qv an ordinate to the diameter GP. Draw SP cutting the diameter DK in E, and the ordinate Qv in x, and complete the parallelogram QRPx. It is evident that EP is equal to the semi-transverse axis AC; for drawing HI, from the other focus H of the hyperbola, parallel to EC, because CS, CH are equal, ES, EI will be also equal; so that EP is the half difference of PS, PI; that is (because of the parallels IH, PR, and the equal angles IPR HPZ), of PS, PH, the difference of which is equal to the whole axis 2AC. Draw QT perpendicular to SP; and putting L for the principal latus rectum of the hyperbola (that is, for $\frac{2BC^2}{AC}$), we shall have $L \times QR$ to $L \times Pv$ as QR to Pv, or Px to Pv, that is (because of the similar triangles Pxv, PEC), as PE to PC, or AC to PC. And $L \times Pv$ will be to $Gv \times Pv$ as L to Gv; and (by the properties of the conic sections) the rectangle GvP is to Qv^2 as PC^2 to CD^2; and by (Cor. 2, Lem. VII), Qv^2 to Qx^2, the points Q and P coinciding, becomes a ratio of equality; and Qx^2 or Qv^2 is to QT^2 as EP^2 to PF^2, that is, as CA^2 to PF^2, or (by Lem. XII.) as CD^2 to CB^2: and, compounding all those ratios together, we shall have $L \times QR$ to QT2 as $AC \times L \times PC^2 \times CD^2$, or $2CB^2 \times PC^2 \times CD^2$ to $PC \times Gv$

$\times$ CD^2 $\times$ CB^2, or as 2PC to Gv. But the points P and Q coinciding, 2PC and Gv are equal. And therefore the quantities L $\times$ QR and QT^2, proportional to them, will be also equal. Let those equals be drawn into $\frac{SP^2}{QR}$, and we shall have L $\times$ SP^2 equal to $\frac{SP^2 \times QT^2}{QR}$.

And therefore (by Cor. 1 and 5, Prop. VL) the centripetal force is reciprocally as L $\times$ SP^2, that is, reciprocally in the duplicate ratio of the distance SP.　　Q. E. I.

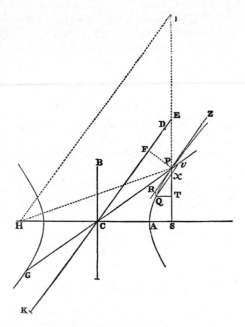

THE SAME OTHERWISE.

Find out the force tending from the centre C of the hyperbola. This will be proportional to the distance CP. But from thence (by Cor. 3, Prop. VII.) the force tending to the focus S will be as $\frac{P}{S}$ that is, because PE is given reciprocally as SP^2. Q.E.I.

And the same way may it be demonstrated, that the body having its centripetal changed into a centrifugal force, will move in the conjugate hyperbola.

LEMMA XIII.

THE LATUS RECTUM OF A PARABOLA BELONGING TO ANY VERTEX IS QUADRUPLE THE DISTANCE OF THAT VERTEX FROM THE FOCUS OF THE FIGURE.

This is demonstrated by the writers on the conic sections.

LEMMA XIV.

THE PERPENDICULAR, LET FALL FROM THE FOCUS OF A PARABOLA ON ITS TANGENT, IS A MEAN PROPORTIONAL

BETWEEN THE DISTANCES OF THE FOCUS FROM THE
POINT OF CONTACT, AND FROM THE PRINCIPAL
VERTEX OF THE FIGURE.

For, let AP be the parabola, S its focus, A its principal vertex, P the point of contact, PO an ordinate to the principal diameter, PM the tangent meeting the principal

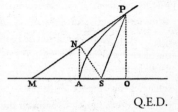

diameter in M, and SN the perpendicular from
the focus on the tangent: join AN, and because
of the equal lines MS and SP, MN and NP, MA
and AO, the right lines AN, OP, will be parallel;
and thence the triangle SAN will be right-angled
at A, and similar to the equal triangles SNM,
SNP; therefore PS is to SN as SN to SA.

Q.E.D.

COR. 1. PS^2 is to SN^2 as PS to SA.

COR. 2. And because SA is given, SN^2 will be as PS.

COR. 3. And the concourse of any tangent PM, with the right line SN, drawn
from the focus perpendicular on the tangent falls in the right line AN that touches the
parabola in the principal vertex.

PROPOSITION XIII. PROBLEM VIII.

IF A BODY MOVES IN THE PERIMETER OF A PARABOLA; IT IS
REQUIRED TO FIND THE LAW OF THE CENTRIPETAL FORCE
TENDING TO THE FOCUS OF THAT FIGURE.

Retaining the construction of the preceding Lemma, let P be the body in the
perimeter of the parabola; and from the place Q, into which it is next to succeed, draw
QR parallel and QT perpendicular to SP, as also Qv parallel to the tangent, and meet-

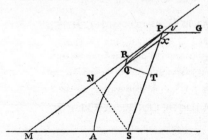

ing the diameter PG in v, and the distance SP in x. Now, because of the similar triangles Pxv, SPM, and of the equal
sides SP, SM of the one, the sides Px or
QR and Pv of the other will be also
equal. But (by the conic sections) the
square of the ordinate Qv is equal to the
rectangle under the latus rectum and the
segment Pv of the diameter; that is (by

Lem. XIII.), to the rectangle 4PS × Pv, or 4PS × QR; and the points P and Q coinciding, the ratio of Qv to Qx (by Cor. 2, Lem. VII.,) becomes a ratio of equality. And
therefore Qx^2, in this case, becomes equal to the rectangle 4PS × QR. But (because of

the similar triangles QxT, SPN), Q^2 is to QT^2 as PS^2 to SN^2, that is (by Cor. 1, Lem. XIV.), as PS to SA; that is, as 4PS $\times$ QR to 4SA $\times$ QR, and therefore (by Prop. IX. Lib. Y., Elem.) QT2 and 4SA $\times$ QR are equal. Multiply these equals by $\frac{SP^2}{QR}$ and $\frac{SP^2 \times QT^2}{QR}$ will become equal to SP^2 $\times$ 4SA: and therefore (by Cor. 1 and 5, Prop. VI.), the centripetal force is reciprocally as SP^2 $\times$ 4SA; that is, because 4SA is given, reciprocally in the duplicate ratio of the distance SP.   Q.E.D.

COR. 1. From the three last Propositions it follows, that if any body P goes from the place P with any velocity in the direction of any right line PR, and at the same time is urged by the action of a centripetal force that is reciprocally proportional to the square of the distance of the places from the centre, the body will move in one of the conic sections, having its focus in the centre of force; and the contrary. For the focus, the point of contact, and the position of the tangent, being given, a conic section may be described, which at that point shall have a given curvature. But the curvature is given from the centripetal force and velocity of the body being given; and two orbits, mutually touching one the other, cannot be described by the same centripetal force and the same velocity.

COR. 2. If the velocity with which the body goes from its place P in such, that in any infinitely small moment of time the lineola PR may be thereby described; and the centripetal force such as in the same time to move the same body through the space QR; the body will move in one of the conic sections, whose principal latus rectum is the quantity in its ultimate state, when the lineolæ PR, QR are diminished in infinitum. In these Corollaries I consider the circle as an ellipsis; and I except the case where the body descends to the centre in a right line.

PROPOSITION XIV. THEOREM VI

IF SEVERAL BODIES REVOLVE ABOUT ONE COMMON CENTRE, AND THE CENTRIPETAL FORCE IS RECIPROCALLY IN THE DUPLICATE RATIO OF THE DISTANCE OF PLACES FROM THE CENTRE; I SAY, THAT THE PRINCIPAL LATERA RECTA OF THEIR ORBITS ARE IN THE DUPLICATE RATIO OF THE AREAS, WHICH THE BODIES BY RADII DRAWN TO THE CENTRE DESCRIBE IN THE SAME TIME.

For (by Cor 2, Prop. XIII) the latus rectum L is equal to the quantity $\frac{QT^2}{QR}$ in its ultimate state when the points P and Q coincide. But the lineola QR in a given time is as the generating centripetal force; that is (by supposition), reciprocally as SP^2.

And therefore $\frac{QT^2}{QR}$ is as $QT^2 \times SP^2$; that is the latus

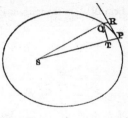

rectum L is in the duplicate ratio of the area $QT \times SP$. Q.E.D.

COR. Hence the whole area of the ellipsis, and the rectangle under the axes, which is proportional to it, is in the ratio compounded of the subduplicate ratio of the latus rectum, and the ratio of the periodic time. For the whole area is as the area $QT \times SP$, described in a given time, multiplied by the periodic time.

PROPOSITION XV. THEOREM VII.

THE SAME THINGS BEING SUPPOSED, I SAY, THAT THE PERIODIC TIMES IN ELLIPSES, ARE IN THE SESQUIPLICATE RATIO OF THEIR GREATER AXES.

For the lesser axis is a mean proportional between the greater axis and the latus rectum; and, therefore, the rectangle under the axes is in the ratio compounded of the subduplicate ratio of the latus rectum and the sesquiplicate ratio of the greater axis. But this rectangle (by Cor. 3, Prop. XIV) is in a ratio compounded of the subduplicate ratio of the latus rectum, and the ratio of the periodic time. Subduct from both sides the subduplicate ratio of the latus rectum, and there will remain the sesquiplicate ratio of the greater axis, equal to the ratio of the periodic time.   Q.E.D.

COR. Therefore the periodic times in ellipses are the same as in circles whose diameters are equal to the greater axes of the ellipses.

PROPOSITION XVI. THEOREM VIII

THE SAME THINGS BEING SUPPOSED, AND RIGHT LINES BEING, DRAWN TO THE BODIES THAT SHALL TOUCH THE ORBITS, AND PERPENDICULARS BEING LET FALL ON THOSE TANGENTS FROM THE COMMON FOCUS; I SAY, THAT THE VELOCITIES OF THE BODIES ARE IN A RATIO COMPOUNDED OF THE RATIO OF THE PERPENDICULARS INVERSELY, AND THE SUBDUPLICATE RATIO OF THE PRINCIPAL LATERA RECTA DIRECTLY.

From the focus S draw SY perpendicular to the tangent PR, and the velocity of the body P will be reciprocally in the subduplicate ratio of the quantity $\frac{SY^2}{L}$. For that velocity is as the infinitely small arc PQ described in a given moment of time that is (by

Lem. VII), as the tangent PR; that is (because of the proportionals PR to QT, and SP to SY), as $\frac{SP \times QT}{SY}$; or as SY reciprocally, and SP × QT directly; but SP × QT is as the area described in the given time, that is (by Prop. XIV), in the subduplicate ratio of the latus rectum.   Q.E.D.

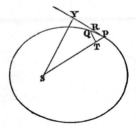

COR. 1. The principal latera recta are in a ratio compounded of the duplicate ratio of the perpendiculars and the duplicate ratio of the velocities.

COR. 2. The velocities of bodies, in their greatest and least distances from the common focus, are in the ratio compounded of the ratio of the distances inversely, and the subduplicate ratio of the principal latera recta directly. For those perpendiculars are now the distances.

COR. 3. And therefore the velocity in a conic section, at its greatest or least distance from the focus, is to the velocity in a circle, at the same distance from the centre, in the subduplicate ratio of the principal latus rectum to the double of that distance.

COR. 4. The velocities of the bodies revolving in ellipses, at their mean distances from the common focus, are the same as those of bodies revolving in circles, at the same distances; that is (by Cor. 6. Prop. IV), reciprocally in the subduplicate ratio of the distances. For the perpendiculars are now the lesser semi-axes, and these are as mean proportionals between the distances and the latera recta. Let this ratio inversely be compounded with the subduplicate ratio of the latera recta directly, and we shall have the subduplicate ratio of the distance inversely.

COR. 5. In the same figure, or even in different figures, whose principal latera recta are equal, the velocity of a body is reciprocally as the perpendicular let fall from the focus on the tangent.

COR. 6. In a, parabola, the velocity is reciprocally in the subduplicate ratio of the distance of the body from the focus of the figure; it is more variable in the ellipsis, and less in the hyperbola, than according to this ratio. For (by Cor. 2, Lem. XIV) the perpendicular let fall from the focus on the tangent of a parabola is in the subduplicate ratio of the distance. In the hyperbola the perpendicular is less variable; in the ellipsis more.

COR. 7. In a parabola, the velocity of a body at any distance from the focus is to the velocity of a body revolving in a circle, at the same distance from the centre, in the subduplicate ratio of the number 2 to 1; in the ellipsis it is less, and in the hyperbola greater, than according to this ratio. For (by Cor. 2 of this Prop.) the velocity at the vertex of a parabola is in this ratio, and (by Cor. 6 of this Prop. and Prop. IV) the same proportion holds in all distances. And hence, also, in a parabola, the velocity is everywhere equal to the velocity of a body revolving in a circle at half the distance; in the ellipsis it is less, and in the hyperbola greater.

Cor. 8. The velocity of a body revolving in any conic section is to the velocity of a body revolving in a circle, at the distance of half the principal latus rectum of the section, as that distance to the perpendicular let fall from the focus on the tangent of the section. This appears from Cor. 5.

Cor. 9. Wherefore since (by Cor. 6, Prop. IV), the velocity of a body revolving in this circle is to the velocity of another body revolving in any other circle reciprocally in the subduplicate ratio of the distances; therefore, ex æquo, the velocity of a body revolving in a conic section will be to the velocity of a body revolving in a circle at the same distance as a mean proportional between that common distance, and half the principal latus rectum of the section, to the perpendicular let fall from the common focus upon the tangent of the section.

PROPOSITION XVII. PROBLEM IX.

SUPPOSING THE CENTRIPETAL FORM TO BE RECIPROCALLY
PROPORTIONAL TO THE SQUARES OF THE DISTANCES OF
PLACES FROM THE CENTRE, AND THAT THE ABSOLUTE
QUANTITY OF THAT FORCE IS KNOWN; IT IS REQUIRED TO
DETERMINE THE LINE WHICH A BODY WILL DESCRIBE THAT
IS LET GO FROM A GIVEN PLACE WITH A GIVEN VELOCITY IN
THE DIRECTION OF A GIVEN RIGHT LINE.

Let the centripetal force tending to the point S be such as will make the body p revolve in any given orbit pq; and suppose the velocity of this body in the place p is known. Then from the place P suppose the body P to be let go with a given velocity in the direction of the line PR; but by virtue of a centripetal force to be immediately turned aside from that right line into the conic section PQ. This, the right line PR will therefore-touch in P. Suppose likewise that the right line pr touches the orbit pq in p; and if from S you suppose perpendiculars let fall on those tangents, the principal latus rectum of the conic section (by Cor. 1, Prop. XVI) will be to the principal latus rectum of that orbit in a ratio compounded of the duplicate ratio of the perpendiculars, and the duplicate ratio of the velocities; and is therefore given. Let this latus rectum be L; the focus S of the conic section is also given. Let the angle RPH be the complement of the angle RPS to two right; and the line PH, in which the other focus H is placed,

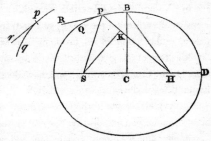

785

is given by position. Let fall SK perpendicular on PH, and erect the conjugate semi-axis BC; this done, we shall have $SP^2 - 2KPH + PH^2 = SH^2 - 4CH^2 = 4BH^2 - 4BC^2$ $= PH^2 - L \times \overline{SP+PI} = SP^2 + 2SPH + PH^2 - L \times \overline{P+PH}$. Add on both sides $2KPH - SP^2 - PH2 + L \times \overline{P+PH}$, and we shall have $L \times \overline{P+PH} = 2SPH + 2KPH$, or SP + PH to PH, as 2SP + 2KP to L. Whence PH is given both in length and position. That is, if the velocity of the body in P is such that the latus rectum L is less than 2SP + 2KP, PH will lie on the same side of the tangent PR with the line SP; and therefore the figure will be an ellipse, which from the given foci S, H, and the principal axis SP + PH, is given also. But if the velocity of the body is so great, that the latus rectum L becomes equal to 2SP + 2KP, the length PH will be infinite; and therefore, the figure will be a parabola, which has its axis SH parallel to the line PK, and is thence given. But if the body goes from its place P with a yet greater velocity, the length, PH is to be taken on the other side the tangent; and so the tangent passing between the foci, the figure will be an hyperbola having its principal axis equal to the difference of the lines SP and PH, and thence is given. For if the body, in these cases, revolves in a conic section so found; it is demonstrated in Prop. XI XII, and XIII, that the centripetal force will be reciprocally as the square of the distance of the body from the centre of force S; and therefore we have rightly determined the line PQ, which a body let go from a given place P with a given velocity, and in the direction of the right line PR given by position, would describe with such a force. Q.E.F.

COR. 1. Hence in every conic section, from the principal vertex D, the latus rectum L, and the focus S given, the other focus H is given, by taking DH to DS as the latus rectum to the difference between the latus rectum and 4DS. For the proportion, SP + PH to PH as 2SP + 2KP to L, becomes, in the case of this Corollary, DS + DH to DH as 4DS to L, and by division DS to DH as 4DS − L to L.

COR. 2. Whence if the velocity of a body in the principal vertex D is given, the orbit may be readily found; to wit, by taking its latus rectum to twice the distance DS, in the duplicate ratio of this given velocity to the velocity of a body revolving in a circle at the distance DS (by Cor. 3, Prop. XVI), and then taking DH to DS as the latus rectum to the difference between the latus rectum and 4DS.

COR. 3. Hence also if a body move in any conic section, and is forced out of its orbit by any impulse, you may discover the orbit in which it will afterwards pursue its course. For by compounding the proper motion of the body with that motion, which the impulse alone would generate, you will have the motion with which the body will go off from a given place of impulse in the direction of a right line given in position.

COR. 4. And if that body is continually disturbed by the action of some foreign force, we may nearly know its course, by collecting the changes which that force introduces in some points, and estimating the continual changes it will undergo in the intermediate places, from the analogy that appears in the progress of the series.

SCHOLIUM.

If a body P, by means of a centripetal force tending to any given point R, move in the perimeter of any given conic section whose centre is C; and the law of the centripetal force is required: draw CG parallel to the radius RP, and meeting the tangent PG of the orbit in G; and the force required (by Cor. 1, and Schol. Prop. X., and Cor. 3, Prop. VII.) will be as $\frac{CG^3}{RP^2}$.

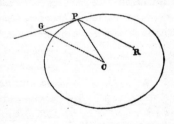

SECTION IV.

Of the finding of elliptic, parabolic, and hyperbolic orbits, from the focus given.

LEMMA XV.

IF FROM THE TWO FOCI S, H, OF ANY ELLIPSIS OR HYBERBOLA, WE DRAW TO ANY THIRD POINT V THE RIGHT LINES SV, HV, WHEREOF ONE HV IS EQUAL TO THE PRINCIPAL AXIS OF THE FIGURE, THAT IS, TO THE AXIS IN WHICH THE FOCI ARE SITUATED, THE OTHER, SV, IS BISECTED IN T BY THE PERPENDICULAR TR LET FALL UPON IT; THAT PERPENDICULAR TR WILL SOMEWHERE TOUCH THE CONIC SECTION : AND, VICE VERSA, IF IT DOES TOUCH IT, HV WILL BE EQUAL TO THE PRINCIPAL AXIS OF THE FIGURE.

For, let the perpendicular TR cut the right line HV, produced, if need be, in R; and join SR. Because TS, TV are equal, therefore the right lines SR, VR, as well as the angles TRS, TRV, will be also equal. Whence the point R will be in the conic section, and the perpendicular TR will touch the same; and the contrary.

Q.E.D.

PROPOSITION XVIII. PROBLEM X.

FROM A FOCUS AND THE PRINCIPAL AXES GIVEN, TO DESCRIBE ELLIPTIC AND HYPERBOLIC TRAJECTORIES, WHICH SHALL PASS THROUGH GIVEN POINTS, AND TOUCH RIGHT LINES GIVEN BY POSITION.

Let S be the common focus of the figures; AB the length of the principal axis of any trajectory; P a point through which the trajectory should pass; and TR a right line which it should touch. About the centre P, with the interval AB − SP, if the orbit is an ellipse, or AB + SP, if the orbit is an hyperbola, describe the circle HG. On the tan-

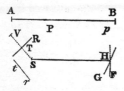

gent TR let fall the perpendicular ST, and produce the same to V, so that TV may be equal to ST; and about V as a centre with the interval. AB describe the circle FH. In this manner, whether two points P, p, are given, or two tangents TR, *tr*, or a point P and a tangent TR, we are to describe two circles. Let H be their common intersection, and from the foci S, H, with the given axis describe the trajectory: I say, the thing is done. For (because PH + SP in the ellipse, and PH − SP in the hyperbola, is equal to the axis) the described trajectory will pass through the point P, and (by the preceding Lemma) will touch the right line TR. And by the same argument it will either pass through the two points P, p, or touch the two right lines TR, *tr*.   Q.E.F.

PROPOSITION XIX PROBLEM XI.

ABOUT A GIVEN FOCUS, TO &SCRIBE A PARABOLIC TRAJEC-TORY, WHICH SHALL PASS THROUGH GIVEN POINTS, AND TOUCH RIGHT LINES GIVEN BY POSITION.

Let S be the focus, P a point, and TR a tangent of the trajectory to be described. About P as a centre, with the interval PS, describe the circle FG. From the focus let fall ST perpendicular on the tangent, and produce the same to V, so as TV may be equal to ST. After the same manner another circle *fg* is to be described, if another point *p* is

given; or another point *v* is to be found, if another tangent *tr* is given; then draw the right line IF, which shall touch the two cir-cles FG, *fg*, if two points P, p are given; or pass through the two points V, *v*, if two tangents TR, *tr*, are given: or touch the circle FG, and pass through the point V, if the point P and the tangent TR are given. On FI let fall the perpendicular SI, and bisect the same in K; and with the axis SK and principal vertex K describe a

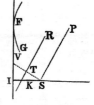

parabola: I say the thing is done. For this parabola (because SK is equal to IK, and SP to FP) will pass through the point P; and (by Cor. 3, Lem. XIV) because ST is equal to TV, and STR a right angle, it will touch the right line TR.   Q.E.F.

PROPOSITION XX. PROBLEM XII.

ABOUT A GIVEN FOCUS TO DESCRIBE ANY TRAJECTORY

GIVEN IN SPECIE WHICH SHALL PASS THROUGH GIVEN
POINTS, AND TOUCH RIGHT LINES GIVEN BY POSITION.

CASE 1. About the focus S it is required to describe a trajectory ABC, passing through two points B,C. Because the trajectory is given in specie, the ratio of the principal axis to the distance of the foci will be given. In that ratio take KB to BS, and LC to CS. About the centres B, C, with the intervals BK, CL, describe two circles; and on the right line KL, that touches the same in K and L, let fall the perpendicular SG; which out in A and *a*, so that GA may be to AS, and G*a* to *a*S, as KB to BS; and with

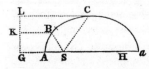

the axis A*a*, and vertices A, *a*, describe a trajectory: I say the thing is done. For let H be the other focus of the described figure, and seeing GA is to AS as G*a* to *a*S, then by division we shall have G*a*-GA, or A*a* to *a*S-AS, or SH in the same ratio, and therefore in the ratio which the principal axis of the figure to be described has to the distance of its foci; and therefore the described figure is of the same species with the figure which was to be described. And since KB to BS, and LC to CS, are in the same ratio, this figure will pass through the points B, C, as is manifest from the conic sections.

CASE 2. About the focus S it is required to describe a trajectory which shall somewhere touch two right lines TR, *tr*. From the focus on those tangents let fall the perpendiculars ST, S*t*, which produce to V, *v*, so that TV, *tv* may be equal to TS, *t*S. Bisect V*v* in O, and erect the indefinite perpendicular OH, and cut the right line VS infinitely produced in K and *k*, so that VK be to KS, and V*k* to *k*S, as the principal axis of the trajectory to be described is to the distance of its foci. On the diameter K*k* describe a

circle cutting OH in H; and with the foci S, H, and principal axis equal to VH, describe a trajectory: I say, the thing is done. For bisecting K*k* in X, and joining HX, HS, HV, H*v*, because VK is to KS as V*k* to *k*S; and by composition, as VK + V*k* to KS + *k*S; and by division, as V*k* – VK to *k*S – KS, that is, as 2VX to 2KX, and 2KX to 2SX, and therefore as VX to HX and HX to SX, the triangles

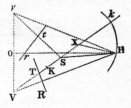

VXH, HXS will be similar; therefore VH will be to SH as VX to XH; and therefore as VK to KS. Wherefore VH, the principal axis of the described trajectory, has the same ratio to SH, the distance of the foci, as the principal axis of the trajectory which was to be described has to the distance of its foci; and is therefore of the same species. And seeing VH, *v*H are equal to the principal axis, and VS, *v*S are perpendicularly bisected by the right lines TR, tr, it is evident (by Lem. XV) that those right lines touch the described trajectory.   Q.E.F.

CASE. 3. About the focus S it is required to describe a trajectory, which shall touch

a right line TR in a given Point R. On the right line TR let fall the perpendicular ST, which produce to V, so that TV may be equal to ST; join VR, and out the right line VS indefinitely produced in K and *k*, so that VK may be to SK, and V*k* to S*k*, as the principal axis of the ellipsis to be described to the distance of its foci; and on the diam-

eter K*k* describing a circle, cut the right line VR produced in H; then with the foci S, H, and principal axis
equal to VH, describe a trajectory: I say, the thing is
done. For VH is to SH as VK to SK, and therefore as
the principal axis of the trajectory which was to be

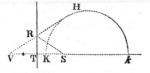

described to the distance of its foci (as appears from what we have demonstrated in Case 2); and therefore the described trajectory is of the same species with that which was to be described; but that the right line TR, by which the angle VRS is bisected, touches the trajectory in the point R, is certain from the properties of the conic sections.

Q.E.F.

CASE 4. About the focus S it is required to describe a trajectory APB that shall touch a right line TR, and pass through any given point P without the tangent, and shall be similar to the figure *apb*, described with the principal axis *ab*, and foci *s*, *h*. On the tangent TR let fall the perpendicular ST, which produce to V, so that TV may be equal to ST; and making the angles *hsq*, *shq*, equal to the angles VSP, SVP, about *q* as

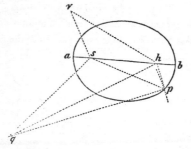

a centre, and with an interval which shall be to *ab* as SP to VS, describe a circle cutting the figure *apb* in *p*: join *sp*, and draw SH such that it may be to *sh* as SP is to *sp*, and may make the angle PSH equal to the angle *psh*, and the angle VSH equal to the angle *psq*. Then with the foci S, H, and principal axis AB, equal to the distance VH, describe a conic section: I say, the thing is done; for if *sv* is drawn so that it shall be to *sp* as *sh* is to *sq*,

and shall make the angle *vsp* equal to the angle *hsq*, and the angle *vsh* equal to the angle *psq*, the triangles *svh*, *spq*, will be similar, and therefore *vh* will be to *pq* as *sh* is to *sq*; that is (because of the similar triangles VSP, *hsq*, as VS is to SP, or as *ab* to *pq*. Wherefore *vh* and *ab* are equal. But, because of the similar triangles VSH, *vsh*, VH is to SH, or as *vh* to *sh*; that is, the axis of the conic section now described is to the distance of its foci as the axis *ab* to the distance of the foci *sh*; and therefore the figure now described is similar to the figure *aph*. But, because the

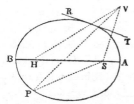

triangle PSH is similar to the triangle *psh*, this figure passes through the point P; and because VH is equal to its axis, and VS is perpendicularly bisected by the right line TR, the said figure touches the right line TR.   Q.E.F.

LEMMA XVI.

FROM THREE GIVEN POINTS TO DRAW TO A FOURTH POINT THAT IS NOT GIVEN THREE RIGHT LINES WHOSE DIFFERENCES SHALL BE EITHER GIVEN., OR NONE AT ALL.

CASE 1. Let the given points be A, B, C, and Z the fourth point which we are to find; because of the given difference of the lines AZ, BZ, the locus of the point Z will be an hyperbola whose foci are A and B, and whose principal axis is the given difference. Let that axis be MN. Taking PM to MA as MN is to AB, erect PR perpendicular to AB, and let fall ZR perpendicular to PR; then from the nature of the hyperbola, ZR will be to AZ as MN is to AB. And by the like argument, the locus of the point Z will be another hyperbola, whose foci are A, C, and whose principal axis is the difference between AZ and CZ; and QS a perpendicular on AC may be drawn, to which (QS) if from any point Z of this hyperbola a perpendicular ZS is let fall (this ZS), shall be to AZ as the difference between AZ and CZ is to AC. Wherefore the ratios of ZR and ZS to AZ are given, and consequently the ratio of ZR to ZS one to the

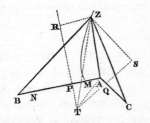

other; and therefore if the right lines RP, SQ, meet in T, and TZ and TA are drawn, the figure TRZS will be given in specie, and the right line TZ, in which the point Z is somewhere placed, will be given in position. There will be given also the right line TA, and the angle ATZ; and because the ratios of AZ and TZ to ZS are given, their ratio to each other is given also; and thence will be given likewise the triangle ATZ, whose vertex is the point Z.   Q.E.I.

CASE 2. If two of the three lines, for example AZ and BZ, are equal, draw the right line TZ so as to bisect the right line AB; then find the triangle ATZ as above.   Q.E.Ï.

CASE 3. If all the three are equal, the point Z will be placed in the centre of a circle that passes through the points A, B, C.   Q.E.I.

This problematic Lemma is likewise solved in Apollonius's Book of Tactions restored by Vieta.

PROPOSITION XXI. PROBLEM XIII.

ABOUT A GIVEN FOCUS TO DESCRIBE A TRAJECTORY THAT

SHALL PASS THROUGH GIVEN POINTS AND TOUCH RIGHT
LINES GIVEN BY POSITION.

Let the focus S, the point P, and the tangent TR be given, and suppose that the other focus H is to be found. On the tangent let fall the perpendicular ST, which produce to Y, so that TY may be equal to ST, and YH will be equal to the principal axis. Join SP, HP, and SP will be the difference between HP and the principal axis. After this

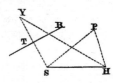

manner, if more tangents TR are given, or more points P, we shall always determine as many lines YH, or PH, drawn from the said points Y or P, to the focus H, which either shall be equal to the axes, or differ from the axes by given lengths SP; and therefore which shall either be equal among themselves, or shall have given differences; from whence (by the preceding Lemma), that other focus H is given. But having the foci and the length of the axis (which is either YH, or, if the trajectory be an ellipsis, PH + SP; or PH-SP, if it be an hyperbola), the trajectory is given.

Q.E.I.

SCHOLIUM.

When the trajectory is an hyperbola, I do not comprehend its conjugate hyperbola under the name of this trajectory. For a body going on with a continued motion can never pass out of one hyperbola into its conjugate hyperbola.

The case when three points are given is more readily solved thus. Let B, C, D, be the given points. Join BC, CD, and produce them to E, F, so as EB may be to EC as SB to SC; and FC to FD as SC to SD. On EF drawn and produced let fall the perpendiculars SG, BH, and in GS produced indefinitely take GA to AS, and G*a* to *a*S,

as HB is to BS; then A will be the vertex, and
A*a* the principal axis of the trajectory; which,
according as GA is greater than, equal to, or less
than AS, will be either an ellipsis, a parabola, or
an hyperbola; the point a in the first case falling
on the same side of the line GF as the point A;
in the second, going off to an infinite distance;
in the third, falling on the other side of the line
GF. For if on GF the perpendiculars CI, DK are

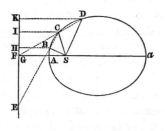

let fall, IC will be to HB as EC to EB; that is, as SC to SB: and by permutation, IC to SC as HB to SB, or as GA to SA. And, by the like argument, we may prove that KD is to SD in the same ratio. Wherefore the points B, C, D lie in a conic section described about the focus S, in such manner that all the right lines drawn from the focus S to the

several points of the section, and the perpendiculars let fall from the same points on the right line GF, are in that given ratio.

That excellent geometer M. De la Hire has solved this Problem much after the same way, in his Conics, Prop. XXV., Lib. VIII.

SECTION V.

How the orbits are to be found when neither focus is given.

LEMMA XVII.

IF FROM ANY POINT P OF A GIVEN CONIC SECTION, TO THE FOUR PRODUCED SIDES AB, CD, AC, DB, OF ANY TRAPEZIUM ABDC INSCRIBED IN THAT SECTION, AS MANY RIGHT LINES PQ, PR, PS, PT ARE DRAWN IN GIVEN ANGLES, EACH LINE TO EACH SIDE, THE RECTANGLE PQ × PR OF THOSE ON THE OPPOSITE SIDES AB, CD, WILL BE TO THE RECTANGLE PS × PT OF THOSE ON THE OTHER TWO OPPOSITE SIDES AC, BD, IN A GIVEN RATIO.

CASE 1. Let us suppose, first, that the lines drawn to one pair of opposite sides are parallel to either of the other sides; as PQ and PR to the side AC and PS and PT to the side AB. And farther, that one pair of the opposite sides, as AC and BD, are parallel betwixt themselves; then the right line which bisects those parallel sides will be one of the diameters of the conic section, and will likewise bisect RQ. Let O be the point in which RQ is bisected, and PO will be an ordinate to that diameter. Produce PO to K, so that OK may be equal to PO, and OK will be an ordinate on the other side of that diameter. Since, therefore, the points A, B, P and K are placed in the conic section, and PK 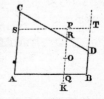 cuts AB in a given angle, the rectangle PQK (by Prop. XVII., XIX, XXI and XXIII, Book III., of Apollonius's Conics) will be to the rectangle AQB in a given ratio. But QK and PR are equal, as being the differences of the equal lines OK, OP, and OQ, OR; whence the rectangles PQK and PQ × PR are equal; and therefore the rectangle PQ × PR is to the rectangle A B, that *s*, to the rectangle PS × PT in a given ratio. Q.E.D.

CASE 2. Let us next suppose that the opposite sides AC and BD of the trapezium are not parallel. Draw B*d* parallel to AC, and meeting as well the right line ST in *t*, as the conic section in *d*. Join C*d* cutting PQ in *r*, and draw DM parallel to PQ, cutting C*d* in M, and AB in N. Then (because of the similar triangles BT*t*, DBIN),B*t* or PQ

is to T*t* as DN to NB. And so R*r* is to AQ or PS as DM to AN. Wherefore, by multiplying the antecedents by the antecedents, and the consequents by the consequents, as the rectangle PQ × R*r* is to the rectangle PS × T*t*, so will the rectangle NDM be to the rectangle ANB; and (by Case 1) so is the rectangle PQ × P*r* to the rectangle PS × P*t*; and by division, so is the rectangle PQ × PR to the rectangle PS × PT.

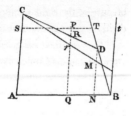

Q.E.D.

CASE 3. Let us suppose, lastly, the four lines PQ, PR, PS, PT, not to be parallel to

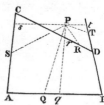

the sides AC, AB, but anyway inclined to them. In their place draw P*q*, P*r*, parallel to AC. and P*s*, P*t* parallel to AB; and because the angles of the triangles PQ*q*, PR*r*, PS*s*, PT*t* are given, the ratios of PQ to P*q*, PR to P*r*, PS to P*s*, PT to P*t* will be also given; and therefore the compounded ratios PQ × PR to P*qx* P*r*, and PS × PT to P*s* × P*t* are given. But from what we have demonstrated before, the ratio of P*q* × P*r* to P*s* × P*t* is given; and therefore also the ratio of PQ × PR to PS × PT. Q.E.D.

LEMMA XVIII.

THE SAME THINGS SUPPOSED, IF THE RECTANGLE
PQ × PR OF THE LINES DRAWN TO THE TWO
OPPOSITE SIDES OF THE TRAPEZIUM IS TO THE
RECTANGLE PS × PT OF THOSE DRAWN TO
THE OTHER TWO SIDES IN A GIVEN RATIO, THE
POINT P, FROM WHENCE THOSE LINES ARE DRAWN,
WILL BE PLACED IN A CONIC SECTION DESCRIBED
ABOUT THE TRAPEZIUM.

Conceive a conic section to be described passing through the points A, B, C, D, and any one of the infinite number of points P, as for example *p*; I say, the point P will be always placed in this section. If you deny the thing, join AP cutting this conic section somewhere else, if possible, than in P, as in *b*. Therefore if from those points *p* and *b*, in the given angles to the sides of the trapezium, we draw the right lines *pq*, *pr*, *ps*, *pt*, and *bk*, *bn*, *bf*, *bd*, we shall have, as *bk* × *bn* to *bf* × *bd* so (by Lem. XVII) *pq* × *pr* to *ps* × *pt*; and so (by supposition). PQ × PR to PS × PT. And

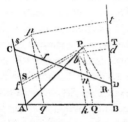

because of the similar trapezia bkAf, PQAS, as bk to bf, so PQ to PS. Wherefore by dividing the terms of the preceding proportion by the correspondent terms of this, we shall have bn to bd as PR to PT. And therefore the equiangular trapezia Dnbd, DRPT, are similar, and consequently their diagonals Db, DP do coincide. Wherefore b falls in the intersection of the right lines AP, DP, and consequently coincides with the point P. And therefore the point P, wherever it is taken, falls to be in the assigned conic section.

Q.E.D.

COR. Hence if three right lines PQ, PR, PS, are drawn from a common point P, to as many other right lines given in position, AB, CD, AC, each. to each, in as many angles respectively given, and the rectangle PQ × PR under any two of the lines drawn be to the square of the third PS in a given ratio; the point P, from which the right lines are drawn, will be placed in a conic section that touches the lines AB, CD in A and C; and the contrary. For the position of the three right lines AB, CD, AC remaining the same, let the line BD approach to and coincide with the line AC; then let the line PT come likewise to coincide with the line PS; and the rectangle PS × PT will become PS^2 and the right lines AB, CD, which before did cut the curve in the points A and B, C and D, can no longer cut, but only touch, the curve in those coinciding points.

SCHOLIUM.

In this Lemma, the name of conic section is to be understood in a large sense, comprehending as well the rectilinear section through the vertex of the cone, as the circular one parallel to the base. For if the point p happens to be in a right line, by which the points A and D, or C and B are joined, the conic section will be changed into two right lines, one of which is that right line upon which the point p falls, and the other is a right line that joins the other two of the four points. If the two opposite angles of the trapezium taken together are equal to two right angles, and if the four lines PQ, PR, PS, PT, are drawn to the sides thereof at right angles, or any other equal angles, and the rectangle PQ × PR under two of the lines drawn PQ and PR, is equal to the rectangle PS × PT under the other two

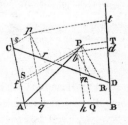

PS and PT, the conic section will become a circle. And the same thing will happen if the four lines are drawn in any angles, and the rectangle PQ × PR, under one pair of the lines drawn, is to the rectangle PS × PT under the other pair as the rectangle under the sines of the angles S, T, in which the two last lines PS, PT are drawn to the rectangle under the sines of the angles Q, R, in which the first two PQ, PR are drawn. In all other cases the locus of the point P will be one of the three figures which pass commonly by the name of the conic sections. But in room of the trapezium ABCD, we may substi-

tute a quadrilateral figure whose two opposite sides cross one another like diagonals. And one or two of the four points A, B, C, D may be supposed to be removed to an infinite distance, by which means the sides of the figure which converge to those points, will become parallel; and in this case the conic section will pass through the other points, and will go the same way as the parallels in infinitum.

LEMMA XIX.

TO FIND A POINT P FROM WHICH IF FOUR RIGHT LINES PQ, PR, PS, PT ARE DRAWN TO AS MANY OTHER RIGHT LINES AB, CD, AC, BD, GIVEN BY POSITION, EACH TO EACH, AT GIVEN ANGLES, THE RECTANGLE PQ × PR, UNDER ANY TWO OF THE LINES DRAWN, SHALL BE TO THE RECTANGLE PS × PT, UNDER THE OTHER TWO, IN A GIVEN RATIO.

Suppose the lines AB, CD, to which the two right lines PQ, PR, containing one of the rectangles, are drawn to meet two other lines, given by position, in the points A, B, C, D. From one of those, as A, draw any right line AH, in which you would find the point P. Let this cut the opposite lines BD, CD, in H and I; and, because all the angles of the figure are given, the ratio of PQ to PA, and PA to PS, and therefore of PQ to PS, will be also given. Subducting this ratio from the given ratio of PQ × PR to PS × PT, the ratio of PR to PT will be given; and adding the given ratios of PI to PR, and PT to PH, the ratio of PI to PH, and therefore the point P will be given.

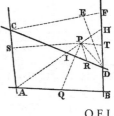

Q.E.I.

COR. 1. Hence also a tangent may be drawn to any point D of the locus of all the points P. For the chord PD, where the points P and D meet, that is, where AH is drawn through the point D, becomes a tangent. In which case the ultimate ratio of the evanescent lines IP and PH will be found as above. Therefore draw CF parallel to AD, meeting BD in F, and cut it in E in the same ultimate ratio, then DE will be the tangent; because CF and the evanescent IH are parallel, and similarly cut in E and P.

COR. 2. Hence also the locus of all the points P may be determined. Through any of the points A, B, C, D, as A, draw AE touching the locus, and through any other point B parallel to the tangent, draw BF meeting the locus in F; and find the point F by this Lemma. Bisect BF in G, and, drawing the indefinite line AG, this will be the position of the diameter to which BG and FG are ordinates. Let this AG meet the locus in H, and AH will be its diameter or latus transversum, to which the latus rectum will be as BG^2 to AG X, GH. If AG nowhere meets the locus, the line AH being infinite,

the locus will be a parabola; and its latus rectum corresponding to the diameter AG will be $\frac{BG^2}{AG}$. But if it does meet it anywhere, the locus will be an hyperbola, when the points A and H are placed on the same side the point G; and an ellipsis, if the point G falls between the points A and H; unless, perhaps, the angle AGB is a right angle, and at the same time BG^2 equal to the rectangle AGH, in which case the locus will be a circle.

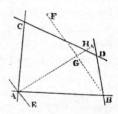

And so we have given in this Corollary a solution of that famous Problem of the ancients concerning four lines, begun by Euclid, and carried on by Apollonius; and this not an analytical calculus, but a geometrical composition, such as the ancients required.

LEMMA XX.

IF THE TWO OPPOSITE ANGULAR POINTS A AND P OF ANY PARALLELOGRAM ASPQ TOUCH ANY CONIC SECTION IN THE POINTS A AND P; AND THE SIDES AQ, AS OF ONE OF THOSE ANGLES, INDEFINITELY PRODUCED, MEET THE SAME CONIC SECTION IN B AND C; AND FROM THE POINTS OF CONCOURSE B AND C TO ANY FIFTH POINT D OF THE CONIC SECTION, TWO RIGHT LINES BD, CD ARE DRAWN MEETING THE TWO OTHER SIDES PS, PQ OF THE PARALLELOGRAM, INDEFINITELY PRODUCED IN T AND R; THE PARTS PR AND PT, CUT OFF FROM THE SIDES, WILL ALWAYS BE ONE TO THE OTHER IN A GIVEN RATIO. AND VICE VERSA, IF THOSE PARTS CUT OFF ARE ONE TO THE OTHER IN A GIVEN RATIO, THE LOCUS OF THE POINT D WILL BE A CONIC SECTION PASSING THROUGH THE FOUR POINTS A, B, C, P

CASE 1. Join BP, CP, and from the point D draw the two right lines DG, DE, of which the first DG shall be parallel to AB and meet PB, PQ, CA in H, I, G; and the other DE shall be parallel to AC, and meet PC, PS, AB, in F, K, E; and (by Lem. XVII)

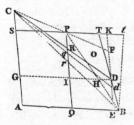

the rectangle DE × DF will be to the rectangle DG × DH in a given ratio. But PQ is to DE (or IQ) as PB to HB, and consequently as PT to DH; and by permutation PQ is to PT as DE to DH. Likewise PR is to DF as RC to DC, and therefore as (IG or) PS to DG; and by permutation PR is to PS as DF to DG; and, by compounding those ratios, the rectangle PQ × PR will

797

be to the rectangle PS × PT as the rectangle DE × DF is to the rectangle DG × DH and consequently in a given ratio. But PQ and PS are given, and therefore the ratio of PR to PT is given. Q.E.D.

CASE.. 2. But if PR and PT are supposed to be in a given ratio one to the other, then by going back again, by a like reasoning, it will follow that the rectangle DE × DF is to the rectangle DG × DH in a given ratio); and so the point D (by Lem. XVIII) will lie in a conic section passing through the points A, B, C, P, as its locus. Q.E.D.

COR. 1. Hence if we draw BC cutting PQ in r and in PT take Pt to Pr in the same ratio which PT has to PR; then B*t* will touch the conic section in the point B. For suppose the point D to coalesce with the point B, so that the chord BD vanishing, BT shall become a tangent, and CD and BT will coincide with CB and B*t*.

COR. 2. And, vice versa, if B*t* is a tangent, and the lines BD, CD meet in any point D of a conic section, PR will be to PT as Pr to Pt. And, on the contrary, if PR is to PT as Pr to P*t*, then BD and CD will meet in some point D of a conic section.

COR. 3. One conic section cannot cut another conic section in more than four points. For, if it is possible, let two conic sections pass through the five points A, B, C, P, O; and let the right line BD cut them in the points D, d, and the right line Cd cut the right line PQ in *q*. Therefore PR is to PT as Pq to PT: whence PR and P*q* are equal one to the other, against the supposition.

LEMMA XXI.

IF TWO MOVEABLE AND INDEFINITE RIGHT LINES BM, CM DRAWN THROUGH GIVEN POINTS B, C, AS POLES, DO BY THEIR POINT OF CONCOURSE M DESCRIBE A THIRD RIGHT LINE MN GIVEN BY POSITION; AND OTHER TWO INDEFINITE RIGHT LINES BD,CD ARE DRAWN, MAKING WITH THE FORMER TWO AT THOSE GIVEN POINTS B, C, GIVEN ANGLES, MBD, MCD : I SAY, THAT THOSE TWO RIGHT LINES BD, CD WILL BY THEIR POINT OF CONCOURSE D DESCRIBE A CONIC SECTION PASSING THROUGH THE POINTS B, C. AND, VICE VERSA, IF THE RIGHT LINES BD, CD DO BY THEIR POINT OF CONCOURSE D DESCRIBE A CONIC SECTION PASSING THROUGH THE GIVEN POINTS B, C, A, AND THE ANGLE DBM IS ALWAYS EQUAL TO THE GIVEN ANGLE ABC, AS WELL AS THE ANGLE DCM ALWAYS EQUAL TO THE GIVEN ANGLE; ACB, THE POINT M WILL LIE IN A RIGHT LINE GIVEN BY POSITION, AS ITS LOCUS.

For in the right line MN let a point N be given, and when the moveable point M falls on the immoveable point N, let the moveable point D fall on an immovable point P. Join CN, BN, CP BP and from the point P draw the right lines PT, PR meeting BD,

CD in T and R, and making the angle BPT equal to the given angle BNM, and the angle CPR equal to the given angle CNM. Wherefore since (by supposition) the angles MBD, NBP are equal, as also the angles MCD, NCP, take away the angles NBD and NCD that are common, and there will remain the angles NBM and PBT, NCM and PCR equal; and therefore the triangles NBM, PBT are similar, as also the triangles NCM, PCR. Wherefore PT is to NM as PB to NB; and PR to NM as PC to NC. But the points, B, C, N, P are immovable: wherefore PT and PR have a given ratio to NM and consequently a given ratio

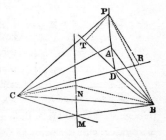

between themselves; and therefore, (by Lemma XX) the point D wherein the moveable right lines BT and CR perpetually concur, will be placed in a conic section passing through the points B, C, P. Q.E.D.

And, *vice versa*, if the moveable point D lies in a conic section passing through the given points B, C, A; and the angle DBM is always equal to the given angle ABC, and the angle DCM always equal to the given angle ACB, and when the point D falls successively on any two immovable points *p*, P, of the conic section, the moveable point M falls successively on two immovable points *n*, N. Through these points *n*, N, draw the right line *n*N: this line *n*N will be the perpetual locus of that moveable point M. For, if possible, let the point M be placed in any curve line. Therefore the point D will be placed in a conic section passing through the five points B, C, A, *p*, P, when the point M is perpetually placed in a curve line. But from what was demonstrated before, the point D will be also placed in a conic section passing through the same five points B, C, A, *p*, P, when the point M is perpetually placed in a right line. Wherefore the two conic sections will both pass through the same five points, against Corol. 3, Lem. XX. It is therefore absurd to suppose that the point M is placed in a curve line. Q.E.D.

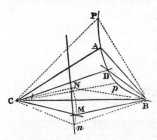

PROPOSITION XXII. PROBLEM XIV.

TO DESCRIBE A TRAJECTORY THAT SHALL PASS THROUGH FIVE GIVEN POINTS.

Let the five given points be A, B, C, P, D. From any one of them, as A, to any other two as B, C, which may be called the poles, draw the right lines AB, AC, and parallel to those the lines TPS, PRQ through the fourth point P. Then from the two

poles B, C, draw through the fifth point D two indefinite lines BDT, CRD, meeting with the last drawn lines TPS, PRQ (the former with the former, and the latter with

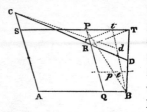

the latter) in T and R. Then drawing, the right line tr parallel to TR, cutting off from the right lines PT, PR, any segments Pt, Pr, proportional to PT, PR; and if through their extremities, t, r, and the poles B, C, the right lines Bt, Cr are drawn, meeting in d, that point d will be placed in the trajectory required. For (by Lem. XX) that point d is placed in a conic section passing through the four points A, B, C, P; and the lines Rr, Tt vanishing, the point d comes to coincide with the point D. Wherefore the conic section passes through the five points A, B, C, P, D.   Q.E.D.

THE SAME OTHERWISE.

Of the given points join any three, as A, B, C; and about two of them B, C, as poles, making the angles ABC, ACB of a given magnitude to revolve, apply the legs BA, CA, first to the point D, then to the point P, and mark the points M, N, in which the other legs BL, CL intersect each other in both cases.

Draw the indefinite right line MN, and let those moveable angles revolve about their poles B, C, in such manner that the intersection, which is now supposed to be m, of the legs BL, CL, or BM, CM, may always fall in that indefinite right line MN; and the intersection, which is now supposed to be d, of the legs BA CA, or BD, CD, will describe the trajectory required, PAD B.

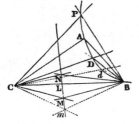

For (by Lem. XXI) the point d will be placed in a conic section passing through the points B, C; and when the point m comes to coincide with the points L, M, N, the point d will (by construction) come to coincide with the points A, D, P. Wherefore a conic section will be described that shall pass through the five points A, B, C, P, D.

Q.E.F.

COR. 1. Hence a right line may be readily drawn which shall be a tangent to the trajectory in any given point B. Let the point d come to coincide with the point B, and the right line Bd will become the tangent required.

COR. 2. Hence also may be found the centres, diameters, and latera recta of the trajectories, as in Cor. 2, Lem. XIX.

SCHOLIUM.

The former of these constructions will become something more simple by joining BP, and in that line, produced, if need be, taking B*p* to BP as PR is to PT; and through *p* draw the indefinite right line *pe* parallel to S PT, and in that line pe taking always *pe* equal to P*r*, and draw the right lines B*e*, C*r* to meet in *d*. For since

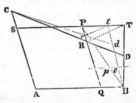

P*r* to P*t*, PR to PT, *p*B to PB, *pe* to P*t*, are all in the same ratio, *pe* and P*r* will be always be equal. After this manner the points of the trajectory are most readily found, unless you would rather describe the curve mechanically, as in the second construction.

PROPOSITION XXIII. PROBLEM XV.

To describe a trajectory that shall pass through four given points, and touch a right line given by position.

CASE. 1. Suppose that HB is the given tangent, B the point of contact, and C, D, P, the three other given points. Join BC, and draw PS parallel to BH, and PQ parallel to BC; complete the parallelogram BSPQ. Draw BD cutting SP in T, and CD cutting PQ in R. Lastly, draw any line tr parallel to TR, cutting off from PQ, PS, the segments P*r*, P*t* proportional to PR, PT respectively; and draw C*r*, B*t* their point of concourse *d* will (by Lem. XX) always fall on the trajectory to be described.

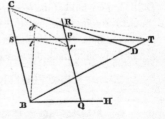

THE SAME OTHERWISE.

Let the angle CBH of a given magnitude revolve about the pole B, as also the rectilinear radius DC, both ways produced, about the pole C. Mark the points M, N, on which the leg BC of the angle cuts that radius when BH, the other leg thereof, meets the same radius in the points P and D. Then drawing the indefinite line MN, let that radius CP or CD and the leg BC of the angle perpetually meet in this line; and the point of concourse of the other leg BH with the radius will delineate the trajectory required.

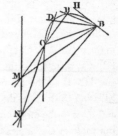

For if in the constructions of the preceding Problem the point A comes to a coincidence with the point B, the lines CA and CB will coincide, and the line AB, in its last situation, will become the tangent BH; and therefore the constructions there set down will become the same with the constructions here described. Wherefore the concourse of the leg BH with the radius will describe a conic section passing through the points C, D, P, and touching the line BH in the point B. Q.E.F.

CASE. 2. Suppose the four points B, C, D, P, given, being situated without the tangent HI. Join each two by the lines BD, CP meeting in G, and cutting the tangents in H and L. Cut the tangent in A in such manner that HA may be to IA as the rectangle under a mean proportional between CG and GP, and a mean proportional between

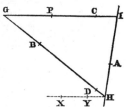

BH and HD is to a rectangle under a mean proportional between GD and GB, and a mean proportional between PI and IC, and A will be the point of contact. For if HX, a parallel to the right line PI, cuts the trajectory in any points X and Y, the point A (by the properties of the conic sections) will come to be so placed, that HA² will become to AI² in a ratio that is compounded out of the ratio of the rectangle XHY to the rectangle BHD, or of the rectangle CGP to the rectangle DGB; and the ratio of the rectangle BHD to the rectangle PIC. But after the point of contact A is found, the trajectory will be described as in the first Case. Q.E.F. But the point A may be taken either between or without the points H and I, upon which account a twofold trajectory may be described.

PROPOSITION XXIV. PROBLEM XVI.

To describe a trajectory that shall pass through three given points, and touch two right lines given by position.

Suppose HI, KL to be the given tangents, and B, C, D, the given points. Through any two of those points, as B, D, draw the indefinite right line BD meeting the tangents in the points H, K. Then likewise through any other two of these points, as C, D, draw the indefinite right line CD meeting the tangents in the points I, L. Cut the lines drawn in R and S, so that HR may be to KR as the mean proportional between BH and HD is to the mean proportional between BK and KD; and IS to LS as the mean proportional between CI and ID is to the mean proportional between CL and LD. But you may cut, at pleasure, either within or between the

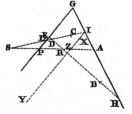

points K and H, I and L, or without them; then draw RS cutting the tangents in A and P, and A and P will be the points of contact. For if A and P are supposed to be the points of contact, situated anywhere else in the tangents, and through any of the points H, I, K, L, as I, situated in either tangent HI, a right line IY is drawn parallel to the other tangent KL, and meeting the curve in X and Y, and in that right line there be taken IZ equal to a mean proportional between IX and IY, the rectangle XIY or IZ², will (by the properties of the conic sections) be to LP² as the rectangle CID is to the rectangle CLD, that is (by the construction), as SI is to SL², and therefore IZ is to LP as SI to SL. Wherefore the points S, P, Z, are in one right line. Moreover, since the tangents meet in G, the rectangle XIY or IZ² will (by the properties of the conic sections) be to IA² as GP² is to GA², and consequently IZ will be to IA as GP to GA. Wherefore the points P, Z, A, lie in one right line, and therefore the points S, P, and A are in one right line. And the same argument will prove that the points R., P, and A are in one right line. Wherefore the points of contact A and P lie in the right line RS. But after these points are found, the trajectory may be described, as in the first Case of the preceding Problem. Q.E.F.

In this Proposition, and Case 2 of the foregoing, the constructions are the same, whether the right line XY cut the trajectory in X and Y, or not; neither do they depend upon that section. But the constructions being demonstrated where that right line does cut the trajectory, the constructions where it does not are also known; and therefore, for brevity's sake, I omit any farther demonstration of them.

LEMMA XXII.

To transform figures into other figures of the same kind.

Suppose that any figure HGI is to be transformed. Draw, at pleasure, two parallel lines AO, BL, cutting any third line AB, given by position, in A and B, and from any point G of the figure, draw out any right line GD, parallel to OA, till it meet the right line AB. Then from any given point O in the line OA, draw to the point D the right line OD, meeting BL in *d*; and from the point of concourse raise the right line *dg* containing any given angle with the right line BL, and having such ratio to O*d* as DG has to OD; and *g* will be the point in the new figure *hgi*, corresponding to the point G. And in like manner the several points of the first figure will give as many correspondent points of the new figure. If we therefore conceive the point G to be carried

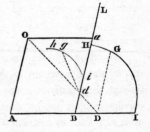

along by a continual motion through all the points of the first figure, the point g will be likewise carried along by a continual motion through all the points of the new figure, and describe the same. For distinction's sake, let us call DG the first ordinate, dg, the new ordinate, AD the first abscissa, ad the new abscissa; O the pole. OD the abscinding radius, OA the first ordinate radius, and Oa (by which the parallelogram OABa is completed) the new ordinate radius.

I say, then, that if the point G is placed in a right line given by position, the point g will be also placed in a right line given by position. If the point G is placed in a conic section, the point g, will be likewise placed in a conic section. And here I understand the circle as one of the conic sections. But farther,. if the point G is placed in a line of the third analytical order, the point g will also be placed in a line of the third order, and so on in curve lines of higher orders. The two lines in which the points G, g, are placed, will be always of the same analytical order. For as ad is to OA, so are Od to OD, dg, to DG, and AB to AD; and therefore AD is equal to $\frac{OA \times AB}{ad}$, and DG equal to $\frac{OA \times dg}{ad}$. Now if the point G is placed in a right line, and therefore, in any equation by which the relation between the abscissa AD and the ordinate GD is expressed, those indetermined lines AD and DG rise no higher than to one dimension, by writing this equation $\frac{OA \times AB}{ad}$ in place of AD, and $\frac{OA \times dg}{ad}$ in place of DG, a new equation will be produced, in which the new abscissa ad and new ordinate dg rise only to one dimension; and which therefore must denote a right line. But if AD and DG (or either of them) had risen to two dimensions in the first equation, ad and dg would likewise have risen to two dimensions in the second equation. And so on in three or more dimensions. The indetermined lines, ad, dg in the second equation, and AD, DG, in the first, will always rise to the same number of dimensions; and therefore the lines in which the points G, g, are placed are of the same analytical order.

I say farther, that if any right line touches the curve line in the first figure, the same right line transferred the same way with the curve into the new figure will touch that curve line in the new figure, and *vice versa*. For if any two points of the curve in the first figure are supposed to approach one the other till they come to coincide, the same points transferred will approach one the other till they come to coincide in the new figure; and therefore the right lines with which those points are joined will become together tangents of the curves in both figures. I might have given demonstrations of these assertions in a more geometrical form; but I study to be brief.

Wherefore if one rectilinear figure is to be transformed into another, we need only transfer the intersections of the right lines of which the first figure consists, and through the transferred intersections to draw right lines in the new figure. But if a curvilinear figure is to be transformed, we must transfer the points, the tangents, and other right lines, by means of which the curve line is defined. This Lemma is of use in

the solution of the more difficult Problems; for thereby we may transform the proposed figures, if they are intricate, into others that are more simple. Thus any right lines converging to a point are transformed into parallels, by taking for the first ordinate radius any right line that passes through the point of concourse of the converging lines, and that because their point of concourse is by this means made to go off *in infinitum*; and parallel lines are such as tend to a point infinitely remote. And after the problem is solved in the new figure, if by the inverse operations we transform the new into the first figure, we shall have the solution required.

This Lemma is also of use in the solution of solid problems. For as often as two conic sections occur, by the intersection of which a problem may be solved, any one of them may be transformed, if it is an hyperbola or a parabola, into an ellipsis, and then this ellipsis may be easily changed into a circle. So also a right line and a conic section, in the construction of plane problems, may be transformed into a right line and a circle.

PROPOSITION XXV. PROBLEM XVII.

TO DESCRIBE A TRAJECTORY THAT SHALL PASS THROUGH TWO GIVEN POINTS, AND TOUCH THREE RIGHT LINES GIVEN BY POSITION.

Through the concourse of any two of the tangents one with the other, and the concourse of the third tangent with the right line which passes through the two given points, draw an indefinite right line; and, taking this line for the first ordinate radius, transform the figure by the preceding Lemma into a new figure. In this figure those two tangents will become parallel to each other, and the third tangent will be parallel to the right line that passes through the two given points. Suppose *hi*, *kl* to be those

two parallel tangents, *ik* the third tangent, and *hl* a right line parallel thereto, passing through those points *a*, *b*, through which the conic section ought to pass in this new figure; and completing the parallelogram *hikl*, let the right lines *hi*, *ik*, *kl* be so cut in *c*, *d*, *e*, that *hc* may be to the square root of the rectangle *ahb*, *ic*, to *id*, and *ke* to *kd*, as the sum of the right lines *hi* and *kl* is to the sum of the three lines, the first where-of is the right line *ik*, and the other two are the square roots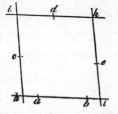

of the rectangles *ahb* and *alb*, and *c*, *d*, *e*, will be the points of contact. For by the properties of the conic sections, hc^2 to the rectangle *ahb*, and ic^2 to id^2, and ke^2 to kd^2, and el^2 to the rectangle *alb*, are all in the same ratio; and therefore *hc* to the square root of *ahb*, *ic* to *id*, *ke* to *kd*, and *el* to the square root of *alb*, are in the subduplicate of that ratio; and

by composition, in the given ratio of the sum of all the antecedents $hi + kl$, to the sum of all the consequents $\sqrt{ahb} + ik + \sqrt{alb}$. Wherefore from that given ratio we have the points of contact c, d, e, in the new figure. By the inverted operations of the last Lemma, let those points be transferred into the first figure, and the trajectory will be there described by Prob. XIV. Q.E.F. But according as the points *a*, *b*, fall between the points *k*, *l*, or without them, the points *c*, *d*, *e*, must be taken either between the points, *h*, *i*, *k*, *l*, or without them. If one of the points *a*, *b*, falls between the points *h*, *i*, and the other without the points *h*, *l*, the Problem is impossible.

PROPOSITION XXVI. PROBLEM XVIII.

TO DESCRIBE A TRAJECTORY THAT SHALL PASS THROUGH A GIVEN POINT, AND TOUCH FOUR RIGHT LINES GIVEN BY POSITION.

From the common intersections, of any two of the tangents to the common inter-section of the other two, draw an indefinite right line; and taking this line for the first ordinate radius, transform the figure (by Lem. XXII) into a new figure, and the two pairs of tangents, each of which before concurred in the first ordinate radius, will now become parallel. Let *hi* and *kl*, *ik* and *hl*, be those pairs of parallels completing the parallelogram *hikl*. And let *p* be the point in this new figure corresponding to the given point in the first figure. Through O the centre of the figure draw *pq*: and O*q* being equal to O*p*, *q* will be the other point through which the conic section must pass in this new figure. Let this point be transferred, by the inverse operation of Lem. XXII into the first figure, and there we shall have the two points through which the trajectory is to be described. But

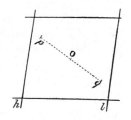

through those points that trajectory may be described by Prop. XVII.

LEMMA XXIII.

IF TWO RIGHT LINES, AS AC, BD GIVEN BY POSITION, AND TERMINATING IN GIVEN POINTS A, B ARE IN A GIVEN RATIO ONE TO THE OTHER, AND THE RIGHT LINE CD, BY WHICH THE INDETERMINED POINTS C, D ARE JOINED IS CUT IN K IN A GIVEN RATIO; I SAY, THAT THE POINT K WILL BE PLACED IN A RIGHT LINE GIVEN BY POSITION.

For let the right lines AC, BD meet in E, and in BE take BG to AE as BD is to AC, and let FD be always equal to the given line EG; and, by construction, EC will

be to GD, that is, to EF, as AC to BD, and therefore in a given ratio; and therefore the triangle EFC will be given in kind. Let CF be cut in L so as CL may be to CF in the ratio of CK to CD; and because that is a given ratio, the triangle EFL will be given in kind, and therefore the point L will be placed in the right line EL given by position. Join LK, and the triangles CLK, CFD will be similar; and because FD is a given line, and LK is to FD) in a given ratio, LK will be also given To this let EH be taken equal, and ELKH will be

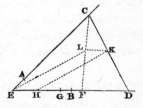

always a parallelogram. And therefore the point K is always placed in the side HK (given by position) of that parallelogram.   Q.E.D.

COR. Because the figure EFLC is given in kind, the three right lines EF, EL, and EC, that is, GD, HK, and EC, will have given ratios to each other.

LEMMA XXIV.

IF THREE RIGHT LINES, TWO WHEREOF ARE PARALLEL, AND
GIVEN BY POSITION, TOUCH ANY CONIC SECTION; I SAY,
THAT THE SEMI-DIAMETER OF THE SECTION WHICH IS
PARALLEL TO THOSE TWO IS A MEAN PROPORTIONAL
BETWEEN THE SEGMENTS OF THOSE TWO THAT ARE
INTERCEPTED BETWEEN THE POINTS OF CONTACT AND
THE THIRD TANGENT.

Let AF, GB be the two parallels touching the conic section ADB in A and B; EF the third right line touching the conic section in I, and meeting the two former tangents in F and G, and let CD be the semi-diameter of the figure parallel to those tangents; I say, that AF, CD, BG are continually proportional.

For if the conjugate diameters AB, DM meet the tangent FG in E and H, and cut one the other in C, and the parallelogram IKCL be completed; from the nature of the conic sections, EC will be to CA, as CA to CL; and so by division, EC – CA to CA – CL, or EA to AL; and by composition, EA to EA + AL or EL, as EC to EC+CA or EB; and therefore (because of

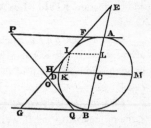

the similitude of the triangles EAF, EI, I, ECH, EBG) AF is to Ll as CH to BG. Likewise, from the nature of the conic sections, LI (or CK) is to CD as CD to CH; and therefore (*ex æquo perturbate*) AF is to CD as CD to BG.   Q.E.D.

COR. 1. Hence if two tangents FG, PQ meet two parallel tangents AF, BG in F and G, P and Q, and cut one the other in O; AF (*ex aquoperturbate*) will be to BQ as AP to BG, and by division, as FP to GQ, and therefore as FO to OG.

COR. 2. Whence also the two right lines PG, FQ drawn through the points P and G, F and Q, will meet in the right line ACB passing through the centre of the figure and the points of contact A, B.

LEMMA XXV.

IF FOUR SIDES OF A PARALLELOGRAM INDEFINITELY PRODUCED TOUCH ANY CONIC SECTION, AND ARE CUT BY A FIFTH TANGENT; I SAY, THAT, TAKING THOSE SEGMENTS OF ANY TWO CONTERMINOUS SIDES THAT TERMINATE IN OPPOSITE ANGLES OF THE PARALLELOGRAM, EITHER SEGMENT IS TO THE SIDE FROM WHICH IT IS CUT OFF AS THAT PART OF THE OTHER CONTERMINOUS SIDE WHICH IS INTERCEPTED BETWEEN THE POINT OF CONTACT AND THE THIRD SIDE IS TO THE OTHER SEGMENT.

Let the four sides ML, IK, KL, MI, of the parallelogram MLIK touch the conic section in A, B, C, D; and let the fifth tangent FQ out those sides in F, Q, H, and E; and taking the segments ME, KQ of the sides MI, KI, or the segments KH, MF of the sides KL, ML, I say, that ME is to MI as BK to KQ; and KH to KL as AM to MF. For, by Cor. 1 of the preceding Lemma, ME is to EI as (AM or) BK to BQ; and, by composition, ME is to MI as BK to KQ. Q.E.D. Also KH is to HL as (BK or) AM to AF; and by division, KH to KL as AM to MF. Q.E.D.

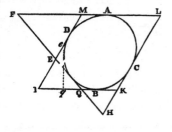

COR. 1. Hence if a parallelogram IKLM described about a given conic section is given, the rectangle KQ × ME, as also the rectangle KH × MF equal thereto, will be given. For, by reason of the similar triangles KQH, MFE, those rectangles are equal.

COR. 2. And if a sixth tangent *eq* is drawn meeting the tangents KI, MI in *q* and *e*, the rectangle KQ × ME will be equal to the rectangle K*q* × M*e*, and KQ will be to M*e* as K*q* to ME, and by division as Q*q* to E*e*.

COR. 3. Hence, also, if E*q*, *e*Q, are joined and bisected, and a right line is drawn through the points of bisection, this right line will pass through the centre of the conic

section. For since Q*q* is to E*e* as KQ to M*e*, the same right line will pass through the middle of all the lines E*q*, *e*Q, MK (by Lem, XXIII), and the middle point of the right line MK is the centre of the section.

PROPOSITION XXVII. PROBLEM XIX.

To describe a trajectory that may touch five right lines given by position.

Supposing ABG, BCF, GCD, FDE, EA to be the tangents given by position. Bisect in M and N, AF, BE, the diagonals of the quadrilateral figure ABPE contained under any four of them; and (by Cor. 3, Lem. XXV) the right line MN drawn through

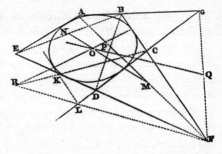

the points bisection will pass through the centre of the trajectory. Again, bisect in P and Q the diagonals (if I may so call them) BD, GF of the quadrilateral figure BGDF contained under any other four tangents, and the right line PQ drawn through the points of bisection will pass through the centre of the trajectory; and therefore the centre will be given in the

concourse of the bisecting lines. Suppose it to be O. Parallel to any tangent BC draw KL at such distance that the centre O may be placed in the middle between the parallels; this KL will touch the trajectory to be described. Let this cut any other two tangents GCD, FDE, in L and K. Through the points C and K, F and L, where the tangents not parallel, CL, FK meet the parallel tangents CF, KL, draw CK, FL meeting in R; and the right line OR drawn and produced, will cut the parallel tangents CF, KL, in the points of contact. This appears from Cor. 2, Lem. XXIV. And by the same method the other points of contact may be found, and then the trajectory may be described by Prob. XIV.   Q.E.F.

SCHOLIUM.

Under the preceding Propositions are comprehended those Problems wherein either the centres or asymptotes of the trajectories are given. For when points and tangents and the centre are given, as many other points and as many other tangents are given at an equal distance on the other side of the centre. And an asymptote is to be considered as a tangent, and its infinitely remote extremity (if we may say so) is a point of contact. Conceive the point of contact of any tangent removed *in infinitum*, and the

tangent will degenerate into an asymptote, and the constructions of the preceding Problems will be changed into the constructions of those Problems wherein the asymptote is given.

After the trajectory is described, we may find its axes and foci in this manner. In the construction and figure of Lem. XXI, let those legs BP, CP, of the moveable angles

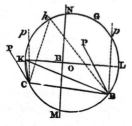

PBN, PCN, by the concourse of which the trajectory was described, be made parallel one to the other; and retaining that position, let them revolve about their poles B, C, in that figure. In the mean while let the other legs CN, BN, of those angles, by their concourse K or k, describe the circle BKGC. Let O be the centre of this circle; and from this centre upon the ruler MN, wherein those legs CN, BN did concur while the trajectory was described,

let fall the perpendicular OH meeting the circle in K and L. And when those other legs CK,BK meet in the point K that is nearest to the ruler, the first legs CP, BP will be parallel to the greater axis, and perpendicular on the lesser; and the contrary will happen if those legs meet in the remotest point L. Whence if the centre of the trajectory is given the axes will be given; and those being given, the foci will be readily found.

But the squares of the axes are one to the other as KH to LH, and thence it is easy to describe a trajectory given in kind through four given points. For if two of the given

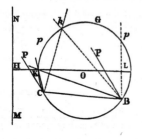

points are made the poles C, B, the third will give the moveable angles PCK, PBK; but those being given, the circle BGKC may be described. Then, because the trajectory is given in kind, the ratio of OH to OK, and therefore OH itself, will be given. About the centre O, with the interval OH, describe another circle, and the right line that touches this circle, and passes through the concourse of the legs CK, BK, when the first legs CP, BP meet in the fourth given point, will be the ruler MN, by means

of which the trajectory may be described. Whence also on the other hand a trapezium given in kind (excepting a few cases that are impossible) maybe inscribed in a given conic section.

There are also other Lemmas, by the help of which trajectories given in kind may be described through given points, and touching given lines. Of such a sort is this, that if a right line is drawn through any point given by position, that may cut a given conic section in two points, and the distance of the intersections is bisected, the point of bisection will touch another

conic section of the same kind with the former, and having its axes parallel to the axes of the former. But I hasten to things of greater use.

LEMMA XXVI.

TO PLACE THE THREE ANGLES OF A TRIANGLE, GIVEN BOTH IN KIND AND MAGNITUDE, IN RESPECT OF AS MANY RIGHT LINES GIVEN BY POSITION, PROVIDED THEY ARE NOT ALL PARALLEL AMONG THEMSELVES, IN SUCH MANNER THAT THE SEVERAL ANGLES MAY TOUCH THE SEVERAL LINES.

Three indefinite right lines AB, AC BC are given by position, and it is required so to place the triangle DEF that its angle D may touch the line AB, its angle E the line AC, and its angle F the line BC. Upon DE, DF, and EF, describe three segments of circles DRE, DGF; EMF, capable of angles equal to the angles BAC, ABC, ACB respectively. But those segments are to be described towards such sides of the lines DE, DF, EF, that the letters DRED may turn round about in the same order with the letters BACB; the letters DGFD in the same order with the letters ABCA; and the letters

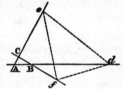

EMFE in the same order with the letters ACBA; then, completing those segments into entire circles let the two former circles cut one the other in G, and suppose P and Q to be their centres. Then joining GP, PQ, take Ga to AB as GP is to PQ; and about the centre G, with the interval Ga, describe a circle that may cut the first circle DGE in a. Join aD cutting the second circle DFG in b, as well as aE cutting the third circle EMF in c. Complete the figure ABCdef similar and equal to the figure abcDEF: I say, the thing is done.

For drawing Fc meeting aD in n, and joining aG, bG, QG, QD, PD, by construction the angle EaD is equal to the angle CAB, and the angle F equal to the angle

ACB; and therefore the triangle anc equiangular to the triangle ABC. Wherefore the angle anc or FnD is equal to the angle ABC, and consequently to the angle FbD; and therefore the point n falls on the point b. Moreover the angle GPQ, which is half the angle GPD at the centre, is equal to the angle GaD at the circumference; and the angle GQP, which is half the angle GQD at the centre, is equal to the complement to two right angles of the angle GbD at the circumference, and therefore equal to the angle

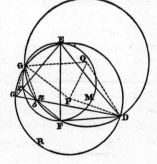

Gba. Upon which account the triangles GPQ, G*ab*, are similar, and G*a* is to *ab* as GP to PQ; that is (by, construction), as G*a* to AB. Wherefore *ab* and AB are equal; and consequently the triangles *abc*, ABC, which we have now proved to be similar, are also equal. And therefore since the angles D, E, F, of the triangle DEF do respectively touch the sides *ab*, *ac*, *bc* of the triangle *abc*, the figure ABC*def* may be completed similar and equal to the figure *abc*DEF, and by completing it the Problem will be solved. Q.E.F.

Cor. Hence a right line may be drawn whose parts given in length may be intercepted between three right lines given by position. Suppose the triangle DEF, by the access of its point D to the side EF, and by having the sides DE, DF placed *in directum* to be changed into a right line whose given part DE is to be interposed between the right lines AB, AC given by position; and its given part DF is to be interposed between the right lines AB, BC, given by position; then, by applying the preceding construction to this case, the Problem will be solved.

PROPOSITION XXVII. PROBLEM XX.

To describe a trajectory given both in kind and magnitude, given parts of which shall be interposed between three right lines given by position.

Suppose a trajectory is to be described that may be similar and equal to the curve line DEF, and may be cut by three right lines AB, AC, BC, given by position, into parts DE and EF, similar and equal to the given parts of this curve line.

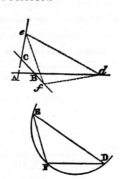

Draw the right lines DE, EF, DF: and place the angles D, E, F, of this triangle DEF, so as to touch those right lines given by position (by Lem. XXVI). Then about the triangle describe the trajectory, similar and equal to the curve DEF. Q.E.F.

LEMMA XXVII.

To describe a trapezium given in kind, the angles whereof may be so placed, in respect of four right lines given by position, that are neither all parallel among themselves, nor converge to one common point, that the several angles may touch the several lines.

Let the four right lines ABC, AD, BD, CE, be given by position; the first cutting the second in A, the third in B, and the fourth in C; and suppose a trapezium *fghi* is to be described that may be similar to the trapezium FGHI, and whose angle *f*, equal to the given angle F, may touch the right line ABC; and the other angles *g, h, i*, equal to the other given angles, G, H, I, may touch the other lines AD, BD, CE, respectively. Join FH, and upon FG, FH, FI describe as many segments of circles FSG, FTH, FVI, the first of which FSG may be capable of an angle equal to the angle BAD; the second FTH capable of an angle equal to the angle CBD; and the third FVI of an angle equal to the angle ACE. But the segments are to be described towards those sides of the lines FG, FH, FI, that the circular order of the letters FSGF may be the same as of the let-

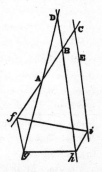

ters BADB, and that the letters FTHF may turn about in the same order as the letters CBDC and the letters FVIF in the same order as the letters ACEA. Complete the segments into entire circles, and let P be the centre of the first circle FSG, Q the centre of the second FTH. Join and produce both ways the line PQ, and in it take QR in the same ratio to PQ as BC has to AB. But QR is to be taken towards that side of the point Q, that the order of the letters P, Q, R may be the same as of the letters A, B, C; and about the centre R with the interval RF describe a fourth circle FN*c* cutting the third circle FVI in *c*. Join F*c* cutting the first circle in *a*,
and the second in *b*. Draw *a*G, *b*H, *c*I, and let the figure ABC*fghi* be made similar to the figure *abc*FGHI; and the trapezium *fghi* will be that which was required to be described.

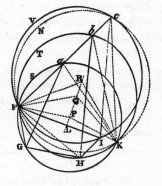

For let the two first circles FSG, FTH cut one the other in K; join PK, QK, RK, *a*K, *b*K, *c*K, and produce QP to L. The angles F*a*K, F*b*K, F*c*K at the circumferences are the halves of the angles FPK, FQK, FRK, at the centres, and therefore equal to LPK, LQK, LRK, the halves of those angles. Wherefore the figure PQRK is equiangular and sim-
ilar to the figure *abc*K, and consequently *ab* is to *bc* as PQ to QR, that is, as AB to BC. But by construction, the angles *f*A*g*, *f*B*h*, *f*C*i*, are equal to the angles F*a*G, F*b*H, F*c*I. And therefore the figure ABC may be completed similar to the figure *abc*FGHI. Which done a trapezium *fghi* will be constructed similar to the trapezium FGHI, and which by its angles *f, g, h, i* will touch the right lines ABC, AD, BD, CE.   Q.E.F.

Cor. Hence a right line may be drawn whose parts intercepted in a given order, between four right lines given by position, shall have a given proportion among themselves. Let the angles FGH, GHI, be so far increased that the right lines FG, GH, HI, may lie *in directum*; and by constructing the Problem in this case, a right line *fghi* will be drawn, whose parts *fg*, *gh*, *hi*, intercepted between the four right lines given by position, AB and AD, AD and BD, BD and CE, will be one to another as the lines FG, GH, HI, and will observe the same order among themselves. But the same thing may be more readily done in this manner.

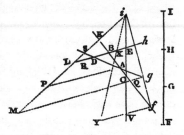

Produce AB to K and BD to L, so as BK may be to AB as HI to GH; and DL to BD as GI to FG; and join KL meeting the right line CE in *i*. Produce *i*L to M, so as LM may be to *i*L as GH to HI; then draw MQ parallel to LB, and meeting the right line AD in *g*, and join *gi* cutting AB, BD in *f*, *h*; I say, the thing, is done.

For let M*g* cut the right line AB in Q, and AD the right line KL in S, and draw AP parallel to BD, and meeting *i*L in P, and *g*M to L*h* (*g*; to *hi*, M*i* to L*i*, GI to HI, AK to BK) and AP to BL, will be in the same ratio. Cut DL in R, so as DL to RL may be in that same ratio; and because *g*S to *g*M, AS to AP, and DS to DL are proportional; therefore (*ex æquo*) as *g*S to L*h*, so will AS be to BL, and DS to RL; and mixtly, BL, – RL to L*h* – BL, as AS – DS to *g*S – AS. That is, BR is to B*h* as AD is to A*g*, and therefore as BD to *g*Q. And alternately BR is to BD as B*h* to *g*Q, or as *fh* to *fg*. But by construction the line BL was cut in D and R in the same ratio as the line FI in G and H; and therefore BR is to BD as FH to FG. Wherefore *fh* is to *fg* as FH to FG. Since, therefore, *gi* to *hi* likewise is as M*i* to L*i*, that is, as GI to HI, it is manifest that the lines FI, *fi*, are similarly out in G and H, *g* and *h*.   Q.E.F.

In the construction of this Corollary, after the line LK is drawn cutting CE in *i*, we may produce *i*E to V, so as EV may be to E*i* as FH to HI, and then draw V parallel to BD. It will come to the same, if about the centre *i* with an interval IH, we describe a circle cutting BD in X, and produce *i*X to Y so as *i*Y may be equal to IF, and then draw Y*f* parallel to BD.

Sir Christopher Wren and Dr. Wallis have long ago given other solutions of this Problem.

PROPOSITION XXIX. PROBLEM XXI

TO DESCRIBE A TRAJECTORY GIVEN IN KIND, THAT MAY BE CUT BY FOUR RIGHT LINES GIVEN BY POSITION, INTO PARTS GIVEN IN ORDER, KIND, AND PROPORTION.

Suppose a trajectory is to be described that may be similar to the curve line FGHI, and whose parts, similar and proportional to the parts FG, GH, HI of the other, may be intercepted between the right lines AB and AD, AD, and BD, BD and CE given by position, viz., the first between the first pair of those lines, the second between the second, and the third between the third. Draw the right lines FG, GH, HI, FI; and (by Lem. XXVII) describe a trapezium *fghi* that may be similar to the trapezium FGHI and whose angles *f, g, h, i,* may touch the right lines given by position AB, AD, BD, CE, severally according to their order. And then about this trapezium describe a trajectory, that trajectory will be similar to the curve line FGHI.

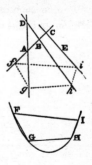

SCHOLIUM.

This problem may be likewise constructed in the following manner. Joining FG, GH, HI, FI, produce GF to V, and join FH, IG, and make the angles CAK, DAL equal to the angles FGH, VFH. Let AK, AL meet the right line BD in K and L, and thence draw KM, LN, of which let KM make the angle AKM equal to the angle GHI, and be itself to AK as HI is to GH; and let LN make the angle ALN equal to the angle FHI, and be itself to AL as HI to FH.

But AK, KM, AL, LN are to be drawn towards those sides of the lines AD, AK, AL, that the letters CAKMC, ALKA, DALND may be carried round in the same order as the letters FGHIF; and draw MN meeting the right line CE in *i*. Make the angle *i*EP equal to the angle IGF, and let PE be to E*i* as FG to GI; and through P draw PQ*f* that may with the right line ADE contain an angle PQE equal to the angle FIG, and may meet the right line AB in *f*, and join *fi*.

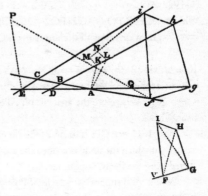

But PE and PQ are to be drawn towards those sides of the lines CE, PE, that the circular order of the letters PF P and PEQP may be the same as of the letters FGHIF; and if upon the line *fi*, in the same order of letters, and similar to the trapezium FGHI, a trapezium *fghi* is constructed, and a trajectory given in kind is circumscribed about it, the Problem will be solved.

So far concerning the finding of the orbits. It remains that we determine the motions of bodies in the orbits so found.

SECTION VI.

How the motions are to be found in given orbits.

PROPOSITION XXX. PROBLEM XXII.

TO FIND AT ANY ASSIGNED TIME THE PLACE OF A BODY MOVING IN A GIVEN PARABOLIC TRAJECTORY.

Let S be the focus, and A the principal vertex of the parabola; and suppose 4AS 3 M equal to the parabolic area to be out off APS, which either was described by the radius SP, since the body's departure from the vertex, or is to be described thereby before its arrival there. Now the quantity of that area to be cut off is known from the time which is proportional to it. Bisect AS in G, and erect the perpendicular GH equal to 3M, and a circle described about the centre H, with the interval HS, will cut the parabola in the place P required. For

letting fall PO perpendicular on the axis, and drawing PH, there will be AG2 + GH2 $(=\overline{HP2}= \overline{AO-AG}^2 + \overline{PO-GH}^2 = AO2 + PO2 - 2GAO - 2GH + PO +AG2+GH2.$ Whence 2GH 3 PO $(=AO2+PO2 - 2GAO) = AO2 + \frac{3}{4}$ PO2. For AO2 write AO 3 $\frac{AO-3AS}{6}$; then dividing all the terms by 3PO, and multiplying them by 2AS, we shall

have 4/3GH 3 AS $(= \frac{1}{6}AO\ 3\ PO + \frac{1}{2}AS\ 3\ PO = \frac{4AO-3SO}{6}$ 3 PO = $\overline{APO-SPO}$ 3 PO = to the area $^{APO-SPO})| $ = to the area APS. But GH was 3M, and therefore $^{4}/_{3}$GH 3 AS is 4AS × M Wherefore the area cut off APS is equal to the area that was to be cut off 4AS × M.   Q.E.D.

COR. 1. Hence GH is to AS as the time in which the body described the arc AP to the time in which the body described the arc between the vertex A and the perpendicular erected from the focus S upon the axis.

COR. 2. And supposing a circle ASP perpetually to pass through the moving body P, the velocity of the point H is to the velocity which the body had in the vertex A as 3 to S; and therefore in the same ratio is the line GH to the right line which the body, in the time of its moving from A to P, would describe with that velocity which it had in the vertex A.

COR. 3. Hence, also, on the other hand, the time may be found in which the body has described any assigned arc AP. Join AP, and on its middle point erect a perpendicular meeting the right line GH in H.

LEMMA XXVIII.

THERE IS NO OVAL FIGURE WHOSE AREA, CUT OF BY RIGHT LINES AT PLEASURE, CAN BE UNIVERSALLY FOUND BY MEANS OF EQUATIONS OF ANY NUMBER OF FINITE TERMS AND DIMENSIONS.

Suppose that within the oval any point is given, about which as a pole a right line is perpetually revolving with an uniform motion, while in that right line a moveable point going out from the pole moves always forward with a velocity proportional to the square of that right line with in the oval. By this motion that point will describe a spiral with infinite circumgyrations. Now if a portion of the area of the oval cut off by that right line could be found by a finite equation, the distance of the point from the pole, which is proportional to this area, might be found by the same equation, and therefore all the points of the spiral might be found by a finite equation also; and therefore the intersection of a right line given in position with the spiral might also be found by a finite equation. But every right line infinitely produced cuts a spiral in an infinite number of points; and the equation by which any one intersection of two lines is found at the same time exhibits all their intersections by as many roots, and therefore rises to as many dimensions as there are intersections. Because two circles mutually cut one another in two points, one of those intersections is not to be found but by an equation of two dimensions, by which the other intersection may be also found. Because there may be four intersections of two conic sections, any one of them is not to be found universally, but by an equation of four dimensions, by which they may be all found together. For if those intersections are severally sought, because the law and condition of all is the same, the calculus will be the same in every case, and therefore the conclusion always the same, which must therefore comprehend all those intersections at once within itself, and exhibit them all indifferently. Hence it is that the intersections of the conic sections with the curves of the third order, because they may amount to six, come out together by equations of six dimensions; and the intersections of two curves of the third order, because they may amount to nine, come out together by equations of nine dimensions. If this did not necessarily happen, we might reduce all solid to plane Problems, and those higher than solid to solid Problems. But here I speak of curves irreducible in power. For if the equation by which the curve is defined may be reduced to a lower power, the curve will not be one single curve, but composed of two, or more, whose intersections may be severally found by different calculusses. After the same manner the two intersections of right lines with the conic sections come out always by equations of two dimensions; the three intersections of right lines with the irreducible curves of the third order by equations of three dimensions; the four intersections of

right lines with the irreducible curves of the fourth order, by equations of four dimensions; and so on *in infinitum*. Wherefore the innumerable intersections of a right line with a spiral, since this is but one simple curve and not reducible to more curves, require equations infinite in number of dimensions and roots, by which they may be all exhibited together. For the law and calculus of all is the same. For if a perpendicular is let fall from the pole upon that intersecting right line, and that perpendicular together with the intersecting line revolves about the pole, the intersections of the spiral will mutually pass the one into the other; and that which was first or nearest, after one revolution, will be the second; after two, the third; and so on: nor will the equation in the mean time be changed but as the magnitudes of those quantities are changed, by which the position of the intersecting line is determined. Wherefore since those quantities after every revolution return to their first magnitudes, the equation will return to its first form; and consequently one and the same equation will exhibit all the intersections, and will therefore have an infinite number of roots, by which they may be all exhibited. And therefore the intersection of a right line with a spiral cannot be universally found by any finite equation; and of consequence there is no oval figure whose area, out off by right lines at pleasure, can be universally exhibited by any such equation.

By the same argument, if the interval of the pole and point by which the spiral is described is taken proportional to that part of the perimeter of the oval which is cut off, it may be proved that the length of the perimeter cannot be universally exhibited by any finite equation. But here I speak of ovals that are not touched by conjugate figures running out *in infinitum*.

COR. Hence the area of an ellipsis, described by a radius drawn from the focus to the moving body, is not to be found from the time given by a finite equation; and therefore cannot be determined by the description of curves geometrically rational. Those curves I call geometrically rational, all the points whereof may be determined by lengths that are definable by equations; that is, by the complicated ratios of lengths. Other curves (such as spirals, quadratrixes, and cycloids) I call geometrically irrational. For the lengths which are or are not as number to number (according to the tenth Book of Elements) are arithmetically rational or irrational. And therefore I cut off an area of an ellipsis proportional to the time in which it is described by a curve geometrically irrational, in the following manner.

PROPOSITION XXXI. PROBLEM XXIII.

TO FIND THE PLACE OF A BODY MOVING IN A GIVEN ELLIPTIC TRAJECTORY AT ANY ASSIGNED TIME.

Suppose A to be the principal vertex, S the focus, and O the centre of the ellipsis APB; and let P be the place of the body to be found. Produce OA to G so as OG may

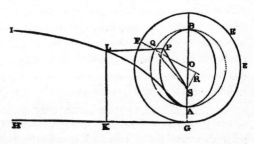

be to OA as OA to OS. Erect the perpendicular GH; and about the centre O, with the interval OG, describe the circle GEF; and on the ruler GH, as a base, suppose the wheel GEF to move forwards, revolving about its axis, and in the mean time by its point A describing the cycloid ALI. Which done, take GK to the perimeter GEFG of the wheel, in the ratio of the time in which the body proceeding from A described the arc AP, to the time of a whole revolution in the ellipsis. Erect the perpendicular KL meeting the cycloid in L; then LP drawn parallel to KG will meet the ellipsis in P, the required place of the body.

For about the centre O with the interval OA describe the semi-circle AQB, and let LP, produced, if need be, meet the arc AQ in Q, and join SQ, OQ. Let OQ meet the arc EFG in F, and upon OQ let fall the perpendicular SR. The area APS is as the area AQS, that is, as the difference between the sector OQA and the triangle OQS, or as the difference of the rectangles $\frac{1}{2}$OQ × AQ, and $\frac{1}{2}$OQ × SR, that is, because $\frac{1}{2}$OQ is given, as the difference between the arc AQ and the right line SR; and therefore (because of the equality of the given ratios SR to the sine of the arc AQ, OS to OA, OA to OG, AQ to GF; and by division, AQ – SR to GF – sine of the arc AQ) as GK, the difference between the arc GF and the sine of the arc AQ.   Q.E.D.

SCHOLIUM.

But since the description of this curve is difficult, a solution by approximation will be preferable. First, then, let there be found a certain angle B which may be to an angle

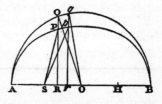

of 57,29578 degrees, which an arc equal to the radius subtends, as SH, the distance of the foci, to AB, the diameter of the ellipsis. Secondly, a certain length L, which may be to the radius in the same ratio inversely. And these being found, the Problem may be solved by the following analysis. By any construction (or even by conjecture), suppose we know P the place of the body near its true place p. Then letting fall on the axis of the ellipsis the ordinate PR from the proportion of the diameters of the ellipsis, the ordinate

RQ of the circumscribed circle AQB will be given; which ordinate is the sine of the angle AOQ, supposing AO to be the radius, and also cuts the ellipsis in P. It will be sufficient if that angle is found by a rude calculus in numbers near the truth. Suppose we also know the angle proportional to the time, that is, which is to four right angles as the time in which the body described the arc Ap, to the time of one revolution in the ellipsis. Let this angle be N. Then take an angle D, which may be to the angle B as the sine of the angle AOQ to the radius; and an angle E which may be to the angle N – AOQ + D as the length L to the same length L diminished by the cosine of the angle AOQ, when that angle is less than a right angle, or increased thereby when greater. In the next place, take an angle F that may be to the angle B as the sine of the angle AOQ + E to the radius, and an angle G, that may be to the angle N – AOQ – E + F as the length L to the same length L diminished by the cosine of the angle AOQ + E, when that angle is less than a right angle, or increased thereby when greater. For the third time take an angle H, that may be to the angle B as the sine of the angle AOQ + E + G to the radius; and an angle I to the angle N – AOQ – E – G + H, as the length L is to the same length L diminished by the cosine of the angle AOQ + E + G, when that angle is less than a right angle, or increased thereby when greater. And so we may proceed *in infinitum*. Lastly, take the angle AOq equal to the angle AOQ + E + G + I +, &c. and from its cosine Or and the ordinate pr, which is to its sine qr as the lesser axis of the ellipsis to the greater, we shall have p the correct place of the body. When the angle N – AOQ, + D happens to be negative, the sign + of the angle E must be every where changed into –, and the sign – into +. And the same thing is to be understood of the signs of the angles G and I, when the angles N – AOQ – E + F, and N – AOQ – E – G + H come out negative. But the infinite series AOQ + E + G + I +, &c. converges so very fast, that it will be scarcely ever needful to proceed beyond the second term E. And the calculus is founded upon this Theorem, that the area APS is as the difference between the arc AQ and the right line let fall from the focus S perpendicularly upon the radius OQ.

And by a calculus not unlike, the Problem is solved in the hyperbola. Let its centre be O, its vertex A, its focus S, and asymptote OK; and suppose the quantity of the area to be cut off is known, as being proportional to the time. Let that be A, and by conjecture suppose we know the position of a right line SP, that cuts off an area APS near the truth. Join OP, and from A and P to the asymptote draw AI, PK parallel to the other asymptote; and by the table of logarithms the area AIKP will be given, and equal thereto the area OPA, which subducted from the triangle OPS, will leave the area cut off APS. And by

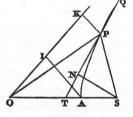

applying 2APS − 2A, or 2A − 2APS, the double difference of the area A that was to be cut off, and the area APS that is cut off, to the line SN that is let fall from the focus S, perpendicular upon the tangent TP, we shall have the length of the chord PQ. Which chord PQ is to be inscribed between A and P, if the area APS that is cut off be greater than the area, A that was to be cut off, but towards the contrary side of the point P, if otherwise: and the point Q will be the place of the body more accurately. And by repeating the computation the place may be found perpetually to greater and greater accuracy.

And by such computations we have a general analytical resolution of the Problem. But the particular calculus that follows is better fitted for astronomical purposes. Supposing AO, OB, OD, to be the semi-axis of the ellipsis, and L its latus rectum, and D the difference betwixt the lesser semi-axis OD, and $\frac{1}{2}$L, the half of the latus rectum: let an angle Y be found, whose sine may be to the radius as the rectangle under that difference D, and AO + OD the half sum of the axes to the square of the greater axis AB. Find also an angle Z, whose sine may be to the radius as the double rectangle under the distance of the foci

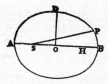

SH and that difference D to triple the square of half the greater semi-axis AO. Those angles being once found, the place of the body may be thus determined. Take the angle T proportional to the time in which the arc BP was described, or equal to what is called the mean motion; and an angle V the first equation of the mean motion to the angle Y, the greatest first equation, as the sine of double the angle T is to the radius; and an angle X, the second equation, to the angle Z, the second greatest equation, as the cube of the sine of the angle T is to the cube of the radius. Then take the angle BHP the mean motion equated equal to T + X + V, the sum of the angles T, V, X, if the angle T is less than a right angle; or equal to T + X − V, the difference of the same, if that angle T is greater than one and less than two right angles; and if HP meets the ellipsis in P, draw SP, and it will cut off the area BSP nearly proportional to the time.

This practice seems to be expeditious enough, because the angles V and X, taken in second minutes, if you please, being very small, it will be sufficient to find two or three of their first figures. But it is likewise sufficiently accurate to answer to the theory of the planet's motions. For even in the orbit of Mars, where the greatest equation of the centre amounts to ten degrees, the error will scarcely exceed one second. But when the angle of the mean motion equated BHP is found, the angle of the true motion BSP, and the distance SP, are readily had by the known methods.

And so far concerning the motion of bodies in curve lines. But it may also come to pass that a moving body shall ascend or descend in a right line; and I shall now go on to explain what belongs to such kind of motions.

SECTION VII.

Concerning the rectilinear ascent and descent of bodies.

PROPOSITION XXXII. PROBLEM XXIV.

SUPPOSING THAT THE CENTRIPETAL FORCE IS RECIPROCALLY
PROPORTIONAL TO THE SQUARE OF THE DISTANCE OF THE
PLACES FROM THE CENTRE; IT IS REQUIRED TO DEFINE
THE SPACES WHICH A BODY, FALLING DIRECTLY,
DESCRIBES IN GIVEN TIMES.

CASE. 1. If the body does not fall perpendicularly, it will (by Cor. I Prop. XIII)
describe some conic section whose focus is placed in the centre of force. Suppose that

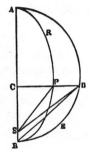

conic section to be ARPB and its focus S. And, first, if the figure
be an ellipsis, upon the greater axis thereof AB describe the semicircle ADB, and let the right line DPC pass through the falling
body, making right angles with the axis; and drawing DS, PS, the
area ASD will be proportional to the area ASP, and therefore also
to the time. The axis AB still remaining the same, let the breadth
of the ellipsis be perpetually diminished, and the area ASD will
always remain proportional to the time. Suppose that breadth to
be diminished *in infinitum*; and the orbit APB in that case coinciding with the axis AB, and the focus S with the extreme point of
the axis B, the body will descend. in the right line AC, and the
area ABD will become proportional to the time. Wherefore the space AC will be given
which the body describes in a given time by its perpendicular fall from the place A, if
the area ABD is taken proportional to the time, and from the point D the right line
DC is let fall perpendicularly on the right line AB.   Q.E.I.

CASE.. 2. If the figure RPB is an hyperbola, on the same principal diameter AB describe the rectangular hyperbola BED; and
because the areas CSP, CB P, SP B, are severally to the several
areas CSD, CBED, SDEB, in the given ratio of the heights CP,
CD, and the area. SP/B is proportional to the time in which the
body P will move through the arc P B, the area SDEB will be
also proportional to that time. Let the latus rectum of the hyperbola RPB be diminished *in infinitum*, the latus transversum
remaining the same; and the arc PB will come to coincide with
the right line CB, and the focus S, with the vertex B, and the

right line SD with the right line BD. And therefore the area BDEB will be proportional to the time in which the body C, by its perpendicular descent, describes the line CB. Q.E.I.

CASE. 3. And by the like argument, if the figure RPB is a parabola, and to the same principal vertex B another parabola BED is described, that may always remain given while the former parabola in whose perimeter the body P moves, by having its latus rectum diminished and reduced to nothing, comes to coincide with the line CB, the parabolic segment BDEB will be proportional to the time in which that body P or C will descend to the centre S or B.

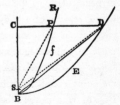

Q.E.I.

PROPOSITION XXXIII. THEOREM IX.

THE THINGS ABOVE FOUND BEING SUPPOSED, I SAY, THAT
THE VELOCITY OF A FALLING BODY IN ANY PLACE C IS TO
THE VELOCITY OF A BODY, DESCRIBING A CIRCLE ABOUT THE
CENTRE B AT THE DISTANCE BC, IN THE SUBDUPLICATE
RATIO OF AC, THE DISTANCE OF THE BODY FROM THE
REMOTER VERTEX A OF THE CIRCLE OR RECTANGULAR
HYPERBOLA, TO $^1/_2$AB, THE PRINCIPAL SEMI-DIAMETER
OF THE FIGURE.

Let AB, the common diameter of both figures RPB, DEB, be bisected in O; and draw the right line PT that may touch the figure RPB in P, and likewise cut that common diameter AB (produced, if need be) in T; and let SY be perpendicular to this line, and BQ to this diameter, and suppose the latus rectum of the figure RPB to be L. From Cor. 9, Prop. XVI, it is manifest that the velocity of a body, moving in the line RPB about the centre S, in any place P, is to the velocity of a body describing a circle about the same centre, at the distance SP, in the subduplicate ratio of the rectangle $^1/_2$L × SP to SY^2 For by the properties of the conic sections ACB is to CP^2 as 2AO to L, and therefore $\frac{2CP^2 \times AO}{ACB}$ is equal to L. Therefore those velocities are to each other in the

subduplicate ratio of $\frac{CP^2 \times AO \times SP}{ACB}$ to SY^2. Moreover, by the properties of the conic sections, CO is to BO as BO to TO and (by composition or division) as CB to BT. Whence (by division or composition) BO – or + CO will be to BO as CT to BT, that is, AC will be to AO as CP to BQ; and therefore $\frac{CP^2 \times AO \times SP}{ACB}$ is equal to $\frac{BQ^2 \times AC \times SP}{AO \times BC}$. Now suppose CP, the breadth of the figure RPB, to be diminished *in infinitum*, so as the point P may come to coincide with the point C, and the point S with the point B,

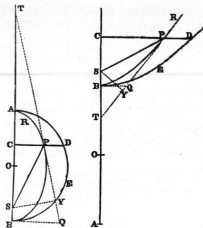

and the line SP with the line BC, and the line SY with the line BQ; and the velocity of the body now descending perpendicularly in the line CB will be to the velocity of a body describing a circle about the centre B, at the distance BC, in the subduplicate ratio of $\frac{BQ^2 \times AC \times SP}{AO \times BC}$ to SY^2, that is (neglecting the ratios of equality of SP to BC, and BQ^2 to SY^2), in the subduplicate ratio of AC to AO, or $\frac{1}{2}AB$. Q.E.D.

COR. 1. When the points B and S come to coincide, TC will become to TS as AC to AO.

COR. 2. A body revolving in any circle at a given distance from the centre, by its motion converted upwards, will ascend to double its distance from the centre.

PROPOSITION XXXIV. THEOREM X.

IF THE FIGURE BED IS A PARABOLA, I SAY, THAT THE VELOC-
ITY OF A FALLING BODY IN ANY PLACE C IS EQUAL TO THE
VELOCITY BY WHICH A BODY MAY UNIFORMLY DESCRIBE A
CIRCLE ABOUT THE CENTRE B AT HALF THE INTERVAL BC

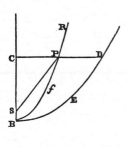

For (by Cor. 7, Prop. XVI) the velocity of a body describing a parabola RPB about the centre S, in any place P, is equal to the velocity of a body uniformly describing a circle about the same centre S at half the interval SP. Let the breadth CP of the parabola be diminished *in infinitum*, so as the parabolic arc P B may come to coincide with the right line CB, the centre S with the vertex B, and the interval SP with the interval BC, and the proposition will be manifest. Q.E.D.

PROPOSITION XXXV. THEOREM XI

THE SAME THINGS SUPPOSED, I SAY, THAT THE AREA OF THE
FIGURE DES, DESCRIBED BY THE INDEFINITE RADIUS SD, IS

EQUAL TO THE AREA WHICH A BODY WITH A RADIUS
EQUAL TO HALF THE LATUS RECTUM OF THE FIGURE DES,
BY UNIFORMLY REVOLVING ABOUT THE CENTRE S, MAY
DESCRIBE IN THE SAME TIME.

For suppose a body C in the smallest moment of time describes in falling the infinitely little line C*c*, while another body K, uniformly revolving about the centre S in the circle OK*k*, describes the arc K*k*. Erect the perpendiculars CD, *cd*,

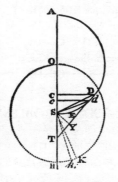 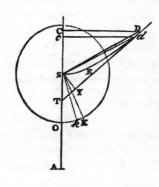

meeting the figure DES in D, *d*. Join SD, S*d*, SK S*k*, and draw D*d* meeting the axis AS in T, and thereon let fall the perpendicular SY.

CASE. 1. If the figure DES is a circle, or a rectangular hyperbola, bisect its transverse diameter AS in O, and SO will be half the latus rectum. And because TC is to TD as C*c* to D*d*, and TD to TS as CD to SY; *ex æquo* TC will be to TS as CD × C*c* to SY × D*d*. But (by Cor. 1, Prop. XXXIII) TC is to TS as AC to AO; to wit, if in the coalescence of the points D, *d*, the ultimate ratios of the lines are taken. Wherefore AC is to AO or SK as CD × C*c* to SY × D*d*. Farther, the velocity of the descending body in C is to the velocity of a body describing a circle about the centre S, at the interval SC, in the subduplicate ratio of AC to AO or SK (by Prop. XXXIII); and this velocity is to the velocity of a body describing the circle OK*k* in the subduplicate ratio of SK to SC (by Cor. 6, Prop IV); and, *ex æquo*, the first velocity to the last, that is, the little line C*c* to the arc K*k*, in the subduplicate ratio of AC to SC, that is, in the ratio of AC to CD. Wherefore CD × C is equal to AC × K*k*, and consequently AC to SK as AC × K*k* to SY × D*d*, and thence SK × K*k* equal to SY × D*d*, and $\frac{1}{2}$SK × K*k* equal to $\frac{1}{2}$SY × D*d*, that is, the area KS*k* equal to the area SD*d*. Therefore in every moment of time two equal particles, KS*k* and SD*d*, of areas are generated, which, if their magnitude is diminished, and their number increased *in infinitum*, obtain the ratio of equality, and consequently (by Cor. Lem. IV), the whole areas together generated are always equal. Q.E.D.

CASE. 2. But if the figure DES is a parabola, we shall find, as above, CD × C*c* to SY × D*d* as TC to TS, that is, as 2 to 1; and that therefore $\frac{1}{4}$CD × C*c* is equal to $\frac{1}{2}$SY × D*d*. But the velocity of the falling body in C is equal to the velocity with which a

circle may be uniformly described at the interval $\frac{1}{2}$ SC (by Prop. XXXIV). And this velocity to the velocity with which a circle may be described with the radius SK, that is, the little line C*c* to the arc K*k*, is (by Cor. 6, Prop. IV) in the subduplicate ratio of SK to $\frac{1}{2}$SC; that is, in the ratio of SK to $\frac{1}{2}$ CD. Wherefore $\frac{1}{2}$ SK × K*k* is equal to $\frac{1}{4}$ CD X. C*c*, and therefore equal to $\frac{1}{2}$ SY × D*d*; that is, the area KS*k* is equal to the area SD*d*, as above.   Q.E.D.

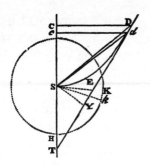

PROPOSITION XXXVI. PROBLEM XXV.

To determine the times of the descent of a body falling from a given place A.

Upon the diameter AS, the distance of the body from the centre at the beginning, describe the semi-circle ADS, as likewise the semi-circle OKH equal thereto, about the centre S. From any place C of the body erect the ordinate CD. Join SD, and make the sector OSK equal to the area ASD. It is evident (by Prop. XXXV) that the body in falling will describe the space AC in the same time in which another body, uniformly revolving about the centre S, may describe the arc OK.   Q.E.F.

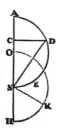

PROPOSITION XXXVII. PROBLEM XXVI.

To define the times of the ascent or descent of a body projected upwards or downwards from a given place.

Suppose the body to go off from the given place G, in the direction of the line GS, with any velocity. In the duplicate ratio of this velocity to the uniform velocity in a circle, with which the body may revolve about the centre S at the given interval SG, take GA to $\frac{1}{2}$AS. If that ratio is the same as of the number 2 to 1, the point A is infinitely remote; in which case a parabola is to be described with any latus rectum to the vertex S, and axis SG; as appears by Prop. XXXIV. But if that ratio is less or greater than the ratio of 2 to 1, in the former case a circle, in the latter a rectangular hyperbola, is to be described on the diameter SA; as appears by Prop. XXXIII. Then about the centre S, with an interval equal to half the latus rectum, describe the circle H K; and at the place G of the ascending or descending body, and at any other place C, erect the

826

perpendiculars GI, CD, meeting the conic section or circle in I and D. Then joining SI, SD, let the sectors HSK, HS*k* be made equal to the segments SEIS, SEDS, and (by Prop. XXXV) the body G will describe the space GC in the same time in which the body K may describe the arc K*k*.   Q.E.F.

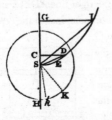
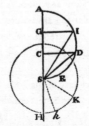
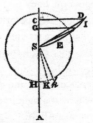

PROPOSITION XXXVIII. THEOREM XII.

SUPPOSING THAT THE CENTRIPETAL FORCE IS PROPORTIONAL TO THE ALTITUDE OR DISTANCE OF PLACES FROM THE CENTRE, I SAY, THAT THE TIMES AND VELOCITIES OF FALLING BODIES, AND THE SPACES WHICH THEY DESCRIBE, ARE RESPECTIVELY PROPORTIONAL TO THE ARCS, AND THE RIGHT AND VERSED SINES OF THE ARCS.

Suppose the body to fall from any place A in the right line AS; and about the centre of force S, with the interval AS, describe the quadrant of a circle AE; and let CD be the right sine of any arc AD; and the body A will in the time AD in falling describe the space AC, and in the place C will acquire the velocity CD.

This is demonstrated the same way from Prop. X, as Prop. XXXII was demonstrated from Prop. XI.

COR. 1. Hence the times are equal in which one body falling from the Place A arrives at the centre S, and another body revolving describes the quadrantal arc ADE.

COR. 2. Wherefore all the times are equal in which bodies falling from whatsoever places arrive at the centre. For all the periodic times of revolving bodies are equal (by Cor. 3, Prop. IV).

PROPOSITION XXXIX. PROBLEM XXVII.

SUPPOSING A CENTRIPETAL FORCE OF ANY KIND, AND GRANTING THE QUADRATURES OF CURVILINEAR FIGURES; IT IS REQUIRED TO FIND THE VELOCITY OF A BODY, ASCENDING

OR DESCENDING IN A RIGHT LINE, IN THE SEVERAL PLACES
THROUGH WHICH IT PASSES; AS ALSO THE TIME IN WHICH IT
WILL ARRIVE AT ANY PLACE: AND VICE VERSA.

Suppose the body E to fall from any place A in the right line ADEC; and from its place E imagine a perpendicular EG always erected proportional to the centripetal force in that place tending to the centre C; and let BFG be a curve line, the locus of the point G. And in the beginning of the motion suppose EG to coincide with the perpendicular AB; and the velocity of the body in any place E will be as a right line whose square is equal to the curvilinear area ABGE.   Q.E.I.

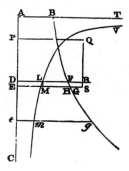

In EG take EM reciprocally proportional to a right line whose square is equal to the area ABGE, and let VLM be a curve line wherein the point M is always placed, and to which the right line AB produced is an asymptote; and the time in which the body in falling describes the line AE, will be as the curvilinear area ABTVME.   Q.E.I.

For in the right line AE let there be taken the very small line DE of a given length, and let DLF be the place of the line EMG, when the body was in D; and if the centripetal force be such, that a right line, whose square is equal to the area ABGE, is as the velocity of the descending body, the area itself will be as the square of that velocity; that is, if for the velocities in D and E we write V and V + I, the area ABFD will be as VV, and the area ABGE as VV + 2VI + II; and by division, the area DFGE as 2VI + II, and therefore $\frac{\text{DFGE}}{\text{DE}}$ will be as $\frac{2VI+II}{\text{DE}}$; that is, if we take the first ratios of those quantities when just nascent, the length DF is as the quantity $\frac{2VI}{\text{DE}}$, and therefore also as half that quantity $\frac{I+V}{\text{DE}}$. But the time in which the body in falling describes the very small line DE, is as that line directly and the velocity V inversely; and the force will be as the increment I of the velocity directly and the time inversely; and therefore if we take the first ratios when those quantities are just nascent, as $\frac{I+V}{\text{DE}}$ that is, as the length DF. Therefore a force proportional to DF or EG will cause the body to descend with a velocity that is as the right line whose square is equal to the area ABGE.   Q.E.D.

Moreover, since the time in which a very small line DE of a given length may be described is as the velocity inversely, and therefore also inversely as a right line whose square is equal to the area ABFD; and since the line DL, and by consequence the nascent area DLME, will be as the same right line inversely, the time will be as the area DLME, and the sum of all the times will be as the sum of all the areas; that is (by

Cor. Lem. IV), the whole time in which the line AE is described will be as the whole area ATVME.   Q.E.D.

COR. 1. Let P be the place from whence a body ought to fall, so as that, when urged by any known uniform centripetal force (such as gravity is vulgarly supposed to be), it may acquire in the place D a velocity equal to the velocity which another body, falling by any force whatever, both acquired in that place D. In the perpendicular DF let there be taken DR, which may be to DF as that uni-
form force to the other force in the place D. Complete the rectangle PDRQ, and cut off the area ABFD equal to that rectangle. Then A will be the place from whence the other body fell. For completing the rectangle DRSE, since the area ABFD is to the area DFGE as VV to 2VI, and therefore as 1/2 V to I, that is, as half the whole veloc-ity to the increment of the velocity of the body falling by the unequable force; and in like manner the area PGRD to the area DRSE as half the whole velocity to the incre-ment of the velocity of the body falling by the uniform

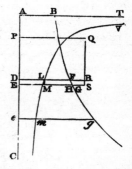

force; and since those increments (by reason of the equality of the nascent times) are as the generating forces, that is, as the ordinates DF, DR, and consequently as the nascent areas DFGE, DRSE; therefore, *ex æquo*, the whole areas ABFD, PQRD will be to one another as the halves of the whole velocities; and therefore, because the velocities are equal, they become equal also.

COR. 2. Whence if any body be projected either upwards or downwards with a given velocity from any place D, and there be given the law of centripetal force acting on it, its velocity will be found in any other place, as *e*, by erecting the ordinate *eg*, and taking that velocity to the velocity in the place D as a right line whose square is equal to the rectangle PQRD, either increased by the curvilinear area DF, if the place *e* is below the place D, or diminished by the same area DF, if it be higher, is to the right line whose square is equal to the rectangle PQRD alone.

COR. 3. The time is also known by erecting the ordinate *em* reciprocally propor-tional to the square root of PQRD + or − DF, and taking the time in which the body has described the line D to the time in which another body has fallen with an uniform force from P, and in falling arrived at D in the proportion of the curvilinear area DL to the rectangle 2PD × DL. For the time in which a body falling with an uniform force hath described the line PD, is to the time in which the same body has described the line PE in the subduplicate ratio of PD to PE: that is (the very small line DE being just nascent), in the ratio of PD to PD + 1/2 DE, or 2PD to 2PD + DE, and, by division, to the time in which the body hath described the small line DE, as 2PD to DE, and

therefore as the rectangle 2PD × DL to the area DLME; and the time in which both the bodies described the very small line DE is to the time in which the body moving unequably hath described the line D*e* as the area DLME to the area DL*me*; and, *ex æquo*, the first mentioned of these times is to the last as the rectangle 2PD × DL to the area DL*me*.

SECTION VIII.

Of the invention of orbits wherein bodies will revolve, being acted upon by any sort of centripetal force.

PROPOSITION XL. THEOREM XIII.

IF A BODY, ACTED UPON BY ANY CENTRIPETAL FORCE, IS ANY HOW MOVED, AND ANOTHER BODY ASCENDS OR DESCENDS IN A RIGHT LINE, AND THEIR VELOCITIES BE EQUAL IN ANY ONE CASE OF EQUAL ALTITUDES, THEIR VELOCITIES WILL BE ALSO EQUAL AT ALL EQUAL ALTITUDES.

Let a body descend from A through D and E, to the centre C; and let another body move from V in the curve line VIK*k*. From the centre C, with any distances, describe the concentric circles DI, EK, meeting the right line AC in D and E, and. the curve VIK in I and K. Draw IC meeting KE in N, and on IK let fall the perpendicular NT; and let the interval DE or IN between the circumferences of the circles be very small; and imagine the bodies in D and I to have equal velocities. Then because the distances CD and CI are equal, the centripetal forces in D and I will be also equal. Let those forces be expressed by the equal lineolæ DE and IN; and let the force IN (by Cor. 2 of the Laws of Motion) be resolved into two others, NT and IT. Then the force NT acting in the direction of the line NT perpendicular to the path ITK of the body will not at all affect or change the velocity of the body in that path, but only draw it aside from a rectilinear course, and make it deflect perpetually from the tangent of the orbit, and proceed in the curvilinear path ITK*k*. That whole force, therefore, will be spent in producing this effect; but the other force IT, acting in the direction of the course of the body, will be all employed in accelerating it, and in the least given time will produce an acceleration proportional to itself. Therefore the accelerations of the bodies in D and I, produced in equal times, are as the lines DE, IT (if we take the first ratios of the nascent lines DE, IN, IK, IT, NT); and in unequal times as those lines and the times conjunctly. But the

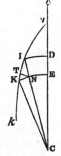

times in which DE and IK are described, are, by reason of the equal velocities (in D and I) as the spaces described DE and IK, and therefore the accelerations in the course of the bodies through the lines DE and IK are as DE and IT, and DE and IK conjunctly; that is, as the square of DE to the rectangle IT into IK. But the rectangle IT × IK is equal to the square of IN, that is, equal to the square of DE; and therefore the accelerations generated in the passage of the bodies from D and I to E and K are equal. Therefore the velocities of the bodies in E and K are also equal. and by the same reasoning they will always be found equal in any subsequent equal distances.   Q.E.D.

By the same reasoning, bodies of equal velocities and equal distances from the centre will be equally retarded in their ascent to equal distances.   Q.E.D.

COR. 1. Therefore if a body either oscillates by hanging to a string, or by any polished and perfectly smooth impediment is forced to move in a curve line; and another body ascends or descends in a right line, and their velocities be equal at any one equal altitude, their velocities will be also equal at all other equal altitudes. For by the string of the pendulous body, or by the impediment of a vessel perfectly smooth, the same thing will be effected as by the transverse force NT. The body is neither accelerated nor retarded by it, but only is obliged to leave its rectilinear course.

COR. 2. Suppose the quantity P to be the greatest distance from the centre to which a body can ascend, whether it be oscillating, or revolving in a trajectory, and so the same projected upwards from any point of a trajectory with the velocity it has in that point. Let the quantity A be the distance of the body from the centre in any other point of the orbit; and let the centripetal force be always as the power A^{n-1}, of the quantity A, the index of which power $n - 1$ is any number n diminished by unity. Then the velocity in every altitude A will be as $\sqrt{P^n - A^n}$, and therefore will be given. For by Prop. XXXIX, the velocity of a body ascending and descending in a right line is in that very ratio.

PROPOSITION XLI. PROBLEM XXVIII.

SUPPOSING A CENTRIPETAL FORCE OF ANY KIND, AND GRANTING THE QUADRATURES OF CURVILINEAR FIGURES, IT IS REQUIRED TO FIND AS WELL THE TRAJECTORIES IN WHICH BODIES WILL MOVE, AS THE TIMES OF THEIR MOTIONS IN THE TRAJECTORIES FOUND.

Let any centripetal force tend to the centre C, and let it be required to find the trajectory VIKk. Let there be given the circle VR, described from the centre C with any interval CV; and from the same centre describe any other circles ID, KE cutting the trajectory in I and K, and the right line CV in D and E. Then draw the right line

CNIX cutting the circles KE, VR in N and X, and the right line CKY meeting the circle VR in Y. Let the points I and K be indefinitely near; and let the body go on from V through I and K to *k*; and let the point A be the place from whence another body is to fall, so as in the place D to acquire a velocity equal to the velocity of the first body in I. And things remaining as in Prop. XXXIX, the lineola IK, described in the least given time will be as the velocity, and therefore as the right line whose square is equal to the area ABFD, and the triangle ICK proportional to the time will be

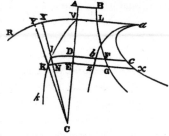

given, and therefore KN will be reciprocally as the altitude IC; that is (if there be given any quantity Q, and the altitude IC be called A), as $\frac{Q}{A}$. This quantity $\frac{Q}{A}$ call Z, and suppose the magnitude of Q to be such that in some case √ABFD may be to Z as IK to KN, and then in all cases √ABPD will be to Z as IK to KN, and ABFD to ZZ as IK² to KN², and by division ABFD − ZZ to ZZ as IN² to KN², and therefore $\sqrt{\overline{ABFD-ZZ}}$ to Z, or $\frac{Q}{A}$ as IN to KN; and therefore A × KN will be equal to $\frac{Q \times IN}{ABFD-ZZ}$. Therefore since YX × XC is to A × KN as CX², to AA, the rectangle XY × XCwill be equal to $\frac{Q \times IN \times CX^2}{AA\sqrt{ABFD-ZZ}}$. Therefore in the perpendicular DF let there be taken

continually D*b*, D*c* equal to $\frac{Q}{2\sqrt{ABFD-ZZ}}, \frac{Q \times CX^2}{2AA\sqrt{ABFD-ZZ}}$ respectively, and let the curve lines *ab*, *ac*, the foci of the points *b* and *c*, be described: and from the point V let the perpendicular V*a* be erected to the line AC, cutting off the curvilinear areas VD*ba*, VD*ca*, and let the ordinates E*z*, E*x*, be erected also. Then because the rectangle D*b* × IN or D*bz*E is equal to half the rectangle A × KN, or to the triangle ICK; and the rectangle D*c* × IN or D*cx*E is equal to half the rectangle YX × XC, or to the triangle XCY; that is, because the nascent particles D*bz*E, ICK of the areas VD*ba*, VIC are always equal; and the nascent particles D*cx*E, XCY of the areas VD*ca*, VCX are always equal; therefore the generated area VD*ca* will be equal to the generated area VIC, and therefore proportional to the time; and the generated area VD*ca* is equal to the generated sector VCX. If, therefore, any time be given during which the body has been moving from V, there will be also given the area proportional to it VD*ba*; and thence will be given the altitude of the body CD or CI; and the area VD*ca*, and the sector VCX equal thereto, together with its angle VCI. But the angle VCI, and the altitude CI being given, there is also given the place I, in which the body will be found at the end of that time.

COR. 1. Hence the greatest and least altitudes of the bodies, that is, the apsides of the trajectories, may be found very readily. For the apsides are those points in which a right line IC drawn through the centre falls perpendicularly upon the trajectory VIK; which comes to pass when the right lines IK and NK become equal; that is, when the area ABFD is equal to ZZ.

COR. 2. So also the angle KIN, in which the trajectory at any place cuts the line IC, may be readily found by the given altitude IC of the body: to wit, by making the sine of that angle to radius as KN to IK that is, as Z to the square root of the area ABFD.

COR. 3. If to the centre C, and the principal vertex V, there be described a conic section VRS; and from any point thereof, as R, there be drawn the tangent RT meeting the axis CV indefinitely produced in the point T; and then joining CR there be drawn the right line CP, equal to the abscissa CT, making an angle VCP proportional to the sector VCR: and if a centripetal force, reciprocally proportional to the cubes of

the distances of the places from the centre, tends to the centre C; and from the place V there sets out a body with a just velocity in the direction of a line perpendicular to the right line CV; that body will proceed in a trajectory VPQ, which the point P will always touch; and therefore if the conic section VRS be an hyperbola, the body will descend to the

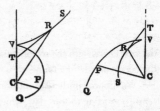

centre; but if it be an ellipsis, it will ascend perpetually, and go farther and farther off *in infinitum*. And, on the contrary, if a body endued with any velocity goes off from the place V, and according, as it begins either to descend obliquely to the centre, or ascends obliquely from it, the figure VRS be either an hyperbola or an ellipsis, the trajectory may be found by increasing or diminishing the angle VCP in a given ratio. And the centripetal force becoming centrifugal, the body will ascend obliquely in the trajectory VPQ, which is found by taking the angle VCP proportional to the elliptic sector VRC, and the length CP equal to the length CT, as before. All these things follow from the foregoing Proposition, by the quadrature of a certain curve, the invention of which, as being easy enough, for brevity's sake I omit.

PROPOSITION XLII. PROBLEM XXIX.

THE LAW OF CENTRIPETAL FORCE BEING GIVEN, IT IS REQUIRED TO FIND THE MOTION OF A BODY SETTING OUT FROM A GIVEN PLACE, WITH A GIVEN VELOCITY, IN THE DIRECTION OF A GIVEN RIGHT LINE.

Suppose the same things as in the three preceding propositions; and let the body go off from the place I in the direction of the little line, IK, with the same velocity as another body, by falling with an uniform centripetal force from the place P, may acquire in D; and let this uniform force be to the force with which the body is at first urged in I, as DR to DF. Let the body go on towards *k*; and about the centre C, with

the interval C*k*, describe the circle *ke*, meeting the right line PD in *e*, and let there be erected the lines *eg*, *ev*, *ew*, ordinately applied to the curves BF*g*, *abv*, *acw*. From the given rectangle PDRQ and the given law of centripetal force, by which the first body

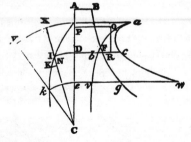

is acted on, the curve line BF*g* is also given, by the construction of Prop. XXVII, and its Cor. 1. Then from the given angle CIK is given the proportion of the nascent lines IK, KN; and thence, by the construction of Prob. XXVIII, there is given the quantity Q, with the curve lines *abv*, *acw*; and therefore, at the end of any time D*gve*, there is given both the altitude of the body C*e* or C*k*, and

the area D*cwe*, with the sector equal to it XC*y*, the angle IC*k*, and the place *k*, in which the body will then be found.   *Q.E.I.*

We suppose in these Propositions the centripetal force to vary in its recess from the centre according to some law, which any one may imagine at pleasure; but at equal distances from the centre to be everywhere the same.

I have hitherto considered the motions of bodies in immovable orbits. It remains now to add something concerning their motions in orbits which revolve round the centres of force.

SECTION IX.

Of the motion of bodies in moveable orbits; and of the motion of the apsides.

PROPOSITION XLIII. PROBLEM XXX.

IT IS REQUIRED TO MAKE A BODY MOVE IN A TRAJECTORY THAT REVOLVES ABOUT THE CENTRE OF FORCE IN THE SAME MANNER AS ANOTHER BODY IN THE SAME TRAJECTORY AT REST.

In the orbit VPK, given by position, let the body P revolve, proceeding from V towards K. From the centre C let there be continually drawn C*p*, equal to CP, making the angle VC*p* proportional to the angle VCP; and the area which the line C*p* describes will be to the area VCP, which the line CP describes at the same time, as the velocity of the describing line C*p* to the velocity of the describing line CP; that is, as the angle VC*p* to the angle VCP, therefore in a given ratio, and therefore proportional to the time. Since, then, the area described by the line C*p* in an immovable plane

is proportional to the time, it is manifest that a body, being acted upon by a just quantity of centripetal force may revolve with the point p in the curve line which the same

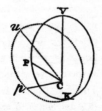

point p, by the method just now explained, may be made to describe an immovable plane. Make the angle VCu equal to the angle PCp, and the line Cu equal to CV, and the figure uCp equal to the figure VCP, and the body being always in the point p, will move in the perimeter of the revolving figure uCp, and will describe its (revolving) arc up in the same time that the other body P describes the similar and equal are VP in the quiescent

figure VPK. Find, then, by Cor. 5, Prop. VI., the centripetal force by which the body may be made to revolve in the curve line which the point p describes in an immovable plane, and the Problem will be solved. _Q.E.F._

PROPOSITION XLIV. THEOREM XIV.

THE DIFFERENCE OF THE FORCES, BY WHICH TWO BODIES
MAY BE MADE TO MOVE EQUALLY, ONE IN A QUIESCENT, THE
OTHER IN THE SAME ORBIT REVOLVING IT IN A TRIPLICATE
RATIO OF THEIR COMMON ALTITUDES INVERSELY.

Let the parts of the quiescent orbit VP, PK be similar and equal to the parts of the revolving orbit up, pk; and let the distance of the points P and K be supposed of the utmost smallness. Let fall a perpendicular kr from the point k to the right line pC, and produce it to m, so that mr may be to kr as the angle VCp to the angle VCP. Because

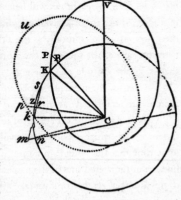

the altitudes of the bodies PC and pC, KC and kC, are always equal, it is manifest that the increments or decrements of the lines PC and pC are always equal; and therefore if each of the several motions of the bodies in the places P and p be resolved into two (by Cor. 2 of the Laws of Motion), one of which is directed towards the centre, or according to the lines PC, pC, and the other, transverse to the former, hath a direction perpendicular to the lines PC and C; the motions towards the centre will be equal, and the transverse motion of the body p will be to the transverse motion of the body P

as the angular motion of the line pC to the angular motion of the line PC; that is, as

the angle VCp to the angle VCP. Therefore, at the same time that the body P, by both its motions, comes to the point K, the body p, having an equal motion towards the centre, will be equally moved from p towards C; and therefore that time being expired, it will be found somewhere in the line mkr, which, passing through the point k, is perpendicular to the line pC; and by its transverse motion will acquire a distance from the line pC, that will be to the distance which the other body P acquires from the line PC as the transverse motion of the body p to the transverse motion of the other body P. Therefore since kr is equal to the distance which the body P acquires from the line PC, and mr is to kr as the angle VCp to the angle VCP, that is, as the transverse motion of the body p to the transverse motion of the body P, it is manifest that the body p, at the expiration of that time, will be found in the place m. These things will be so, if the bodies p and P are equally moved in the directions of the lines pC and PC, and are therefore urged with equal forces in those directions. But if we take an angle pCn that is to the angle pCk as the angle VCp to the angle VCP, and nC be equal to kC, in that case the body p at the expiration of the time will really be in n; and is therefore urged with a greater force than the body P, if the angle nCp is greater than the angle kCp, that is, if the orbit upk, move either *in consequentia*, or *in antecedentia*, with a celerity greater than the double of that with which the line CP moves *in consequentia*; and with a less force if the orbit moves slower *in antecedentia*. And the difference of the forces will be as the interval mn of the places through which the body would be carried by the action of that difference in that given space of time. About the centre C with the interval Cn or Ck suppose a circle described cutting the lines mr, mn produced in s and t, and the rectangle $mn \times mt$ will be equal to the rectangle $mk \times ms$, and therefore mn will be equal to $\frac{mk \times ms}{mt}$. But since the triangles pCk, pCn, in a given time, are of a given magnitude, kr, and mr, and their difference mk, and their sum ms, are reciprocally as the altitude pC, and therefore the rectangle $mk \times ms$ is reciprocally as the square of the altitude pC. But, moreover, mt is directly as $1/_2mt$, that is, as the altitude pC. These are the first ratios of the nascent lines; and hence $\frac{mk \, \acute{} \, ms}{mt}$, that is, the nascent lineola mn, and the difference of the forces proportional thereto, are reciprocally as the cube of the altitude pC.   Q.E.D.

COR. 1. Hence the difference of the forces in the places P and p, or K and k, is to the force with which a body may revolve with a circular motion from R to K, in the same time that the body P in an immovable orb describes the arc PK, as the nascent line mn to the versed sine of the nascent arc R, that is, as $\frac{mk \, \acute{} \, ms}{mt}$ to $\frac{rk^2}{2kC}$ or as $mk \times ms$ to the square of rk; that is, if we take given quantities F and G in the same ratio to one another as the angle VCP bears to the angle VCp, as GG – FF to FF. And, therefore, if from the centre C, with any distance CP or Cp, there be described a circular sector equal to the whole area VPC, which the body revolving, in an immovable orbit has by

a radius drawn to the centre described in any certain time, the difference of the forces, with which the body P revolves in an immovable orbit, and the body p in a movable orbit, will be to the centripetal force, with which another body by a radius drawn to the centre can uniformly describe that sector in the same time as the area VPC is described, as GG − FF to FF. For that sector and the area pCk are to one another as the times in which they are described.

COR. 2. If the orbit VPK be an ellipsis, having its focus C, and its highest apsis V, and we suppose the ellipsis upk similar and equal to it, so that pC may be always equal to PC, and the angle VCp be to the angle VCP in the given ratio of G, to F; and for the altitude PC or pC we put A, and 2R for the latus rectum of the ellipsis, the force with which a body may be made to revolve in a movable ellipsis will be

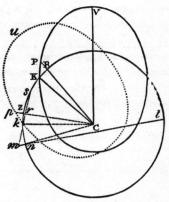

as $\frac{FF}{AA} + \frac{RGG-RFF}{A^3}$, and *vice versa*. Let the force with which a body may revolve in an immovable ellipsis be expressed by the quantity $\frac{FF}{AA}$, and the force in V will be $\frac{FF}{CV^2}$. But the force with which a body may revolve in

a circle at the distance CV, with the same velocity as a body revolving in an ellipsis has in V, is to the force with which a body revolving in an ellipsis is acted upon in the apsis V, as half the latus rectum of the ellipsis to the semi-diameter CV of the circle, and therefore is as $\frac{RFF}{CV^3}$; and the force which is to this, as GG − FF to FF, is as $\frac{RGG-RFF}{CV^3}$: and this force (by Cor. 1 of this Prop.) is the difference of the forces in V, with which the body P revolves in the immovable ellipsis VPK, and the body p in the movable ellipsis upk. Therefore since by this Prop. that difference at any other altitude A is to itself at the altitude CV as $\frac{1}{A^3}$ to $\frac{1}{CV^3}$ the same difference in every altitude A will be as

$\frac{RGG-RFF}{CV^3}$. Therefore to the force $\frac{FF}{AA}$, by which the body may revolve in an

immovable ellipsis VPK add the excess $\frac{RGG-RFF}{A^3}$, and the sum will be the whole force

$\frac{FF}{AA} + \frac{RGG-RFF}{A^3}$ by which a body may revolve in the same time in the movable ellipsis upk.

COR. 3. In the same manner it will be found, that, if the immovable orbit VPK be an ellipsis having its centre in the centre of the forces C, and there be supposed a movable ellipsis upk, similar, equal, and concentrical to it; and 2R be the principal latus rectum of that ellipsis, and 2T the latus transversum, or greater axis; and

the angle VCp be continually to the angle VCP as G to F; the forces with which bodies may revolve in the immovable and movable ellipsis, in equal times, will be as $\frac{FFA}{T^3}$ and $\frac{FFA}{T^3} + \frac{RGG - RFF}{A^3}$ respectively.

COR. 4. And universally, if the greatest altitude CV of the body be called T, and the radius of the curvature which the orbit VPK has in V, that is, the radius of a circle equally curve, be called R, and the centripetal force with which a body may revolve in any immovable trajectory VPK at the place V be called $\frac{VFF}{TT}$, and in other places P be indefinitely styled X; and the altitude CP be called A, and G – be taken to F in the given ratio of the angle VCp to the angle VCP; the centripetal force with which the same body will perform the same motions in the same time, in the same trajectory *upk* revolving with a circular motion, will be as the sum of the forces $X + \frac{VRGG - VRFF}{A^3}$.

COR. 5. Therefore the motion of a body in an immovable orbit being given, its angular motion round the centre of the forces may be increased or diminished in a given ratio; and thence new immovable orbits may be found in which bodies may revolve with new centripetal forces.

COR. 6. Therefore if there be erected the line VP of an indeterminate length, perpendicular to the line CV given by position, and CP be drawn, and Cp equal to it, making the angle VCp having a given ratio to the angle VCP, the force with which a body may revolve in the curve line V, which the point *p* is continually describing, will be reciprocally as the cube of the altitude Cp. For the body P, by its *vis inertiæ* alone, no other force impelling it, will proceed uniformly in the right line VP. Add, then, a force tending to the centre C reciprocally as the cube of the altitude CP or Cp, and (by what was just demonstrated)

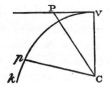

the body will deflect from the rectilinear motion into the curve line V*pk*. But this curve V*pk* is the same with the curve VPQ found in Cor. 3, Prop XLI, in which, I said, bodies attracted with such forces would ascend obliquely.

PROPOSITION XLV. PROBLEM XXXI.

TO FIND THE MOTION OF THE APSIDES IN ORBITS
APPROACHING VERY NEAR TO CIRCLES.

This problem is solved arithmetically by reducing the orbit, which a body revolving in a movable ellipsis (as in Cor. 2 and 3 of the above Prop.) describes in an immovable place, to the figure of the orbit whose apsides are required; and then seeking the apsides of the orbit which that body describes in an immovable plane. But orbits acquire the same figure, if the centripetal forces with which they are described, compared between themselves, are made proportional at equal altitudes. Let the point V be the

highest apsis, and write T for the greatest altitude CV, A for any other altitude CP or Cp, and $\times$ for the difference of the altitudes CV – CP; and the force with which a body moves in an ellipsis revolving about its focus C (as in Cor. 2), and which in Cor. 2 was as $\dfrac{FF}{AA} + \dfrac{RGG-RFF}{A^3}$,

that is as, $\dfrac{FFA+RGG-RFF}{A^3}$, by substituting T – $\times$ for A, will become as $\dfrac{RGG-RFF+TFF-FFX}{A^3}$.

In like manner any other centripetal force is to be reduced to a fraction whose denominator is A^3, and the numerators are to be made analogous by collating together the homologous terms. This will be made plainer by Examples.

Example 1. Let us suppose the centripetal force to be uniform, and therefore as $\dfrac{A^3}{A^3}$ or,

writing T - $\times$ for A in the numerator, as $\dfrac{T^3-3TTX+3TXX-X^3}{A^3}$. Then collating together the

correspondent terms of the numerators, that is, those that consist of given quantities, with those of given quantities, and those of quantities not given with those of quantities not given, it will become RGG – RFF + TFF to T^3 as – FFX to 3TTX + 3TXX – X^3, or as – FF to – 3TT + 3TX – XX. Now since the orbit is supposed extremely near to a circle, let it coincide with a circle; and because in that case R and T become equal, and $\times$ is infinitely diminished, the last ratios will be, as RGG to T^2, so – FF, to – 3TT, or as GG to TT, so FF to 3TT; and again, as GG to FF, so TT to 3TT, that is, as 1 to 3; and therefore G is to F, that is, the angle VCp to the angle VCP, as I to $\sqrt{3}$. Therefore since the body, in an immovable ellipsis, in descending from the upper to the lower apsis, describes an angle, if I may so speak, of 180 deg., the other body in a movable ellipsis, and therefore in the immovable orbit we are treating of, will in its descent from the upper to the lower apsis, describe an angle VCp of $\dfrac{T^3-3TTX+3TXX-X^3}{A^3}$ deg. And this comes to pass by reason of the likeness of this orbit which a body acted upon by an uniform centripetal force describes, and of that orbit which a body performing its circuits in a revolving ellipsis will describe in a quiescent plane. By this collation of the terms, these orbits are made similar; not universally, indeed, but then only when they approach very near to a circular figure. A body, therefore revolving with an uniform centripetal force in an orbit nearly circular, will always describe an angle of $\dfrac{T^3-3TTX+3TXX-X^3}{A^3}$

deg., or 103 deg., 55 m., 23 sec., at the centre; moving from the upper apsis to the lower apsis when it has once described that angle, and thence returning to the upper apsis when it has described that angle again; and so on *in infinitum*.

Exam. 2. Suppose the centripetal force to be as any power of the altitude A, as, for example, A^{n-3}, or $\dfrac{A^n}{A^3}$; where n - 3 and n signify any indices of powers whatever, whether integers or fractions, rational or surd, affirmative or negative. That numerator A^n or $\overline{T-X\,|}^n$ being reduced

to an indeterminate series by my method of converging series, will become $T^n - nXT^{n-1} + \frac{mn-n}{2}$; XXT^{n-2}, &c. And conferring these terms with the terms of the other numerator RGG – RFF + TFF – FFX, it becomes as RGG – RFF + TFF to T^n) so – FF to – $nT^{n-1} + \frac{nn-n}{2} XT^{n-2}$, &c. And taking the last ratios where the orbits approach to circles, it becomes as RGG to T^n, so – FF to – nT^{n-1}, or as GG to T^{n-1}, so FF to nT^{n-1}; and again, GG to FF, so T^{n-1} to nT^{n-1}, that is, as 1 to n.; and therefore G is to F, that is the angle VCp to the angle VCP, as 1 to $\sqrt{n}$. Therefore since the angle VCP, described in the descent of the body from the upper apsis to the lower apsis in an ellipsis, is of 180 deg., the angle VCp, described in the descent of the body from the upper apsis to the lower apsis in an orbit nearly circular which a body describes with a centripetal force proportional to the power A^{n-3}, will be equal to an angle of $\frac{180}{\sqrt{n}}$ deg., and this angle being repeated, the body will return from the lower to the upper apsis, and so on *in infinitum*. As if the centripetal force be as the distance of the body from the centre, that is, as A, or $\frac{A^4}{A^3}$ n will be equal to 4, and $\sqrt{n}$ equal to 2; and therefore the angle

between the upper and the lower apsis will be equal to $\frac{180}{2}$ deg., or 90 deg. Therefore the body having performed a fourth part of one revolution, will arrive at the lower apsis, and having performed another fourth part, will arrive at the upper apsis, and so on by turns *in infinitum*. This appears also from Prop. X. For a body acted on by this centripetal force will revolve in an immovable ellipsis, whose centre is the centre of force. If the centripetal force is reciprocally as the distance, that is, directly as $\frac{1}{A}$ or $\frac{A^2}{A^3}$,

n will be equal to 2; and therefore the angle between the upper and lower apsis will be $\frac{180}{\sqrt{2}}$ deg., or 127 deg., 16 min., 45 sec.; and therefore a body revolving with such a force, will by a perpetual repetition of this angle, move alternately from the upper to the lower and from the lower to the upper apsis for ever. So, also, if the centripetal force be reciprocally as the biquadrate root of the eleventh power of the altitude, that is, reciprocally as $A^{\frac{11}{4}}$, and, therefore, directly as $A^{\frac{11}{4}}$, or as $\frac{A\frac{1}{4}}{A^3}$ n will be equal to $\frac{1}{4}$, and

$\frac{180}{\sqrt{n}}$ deg. will be equal to 360 deg.; and therefore the body parting from the upper apsis, and from thence perpetually descending, will arrive at the lower apsis when it has completed one entire revolution; and thence ascending perpetually, when it has completed another entire revolution, it will arrive again at the upper apsis; and so alternately forever.

Exam. 3. Taking m and n for any indices of the powers of the altitude, and b and c for any given numbers, suppose the centripetal force to be as

$\frac{bA^m + cA^n}{A^3}$, that is, as $\frac{b \text{ into } \overline{T - X}^m + c \text{ into } \overline{T - X}^n}{A^3}$ or (by the method of converging series

above-mentioned) as $\frac{bT^m + cT^n - mbXT^{m-1} ncXT^{n-1} + \frac{mm-m}{2} bXXT^{m-2} + \frac{nn-n}{2} cXXT^{n-2}, \&c.}{A^3}$

and comparing the terms of the numerators, there will arise RGG − RFF + TFF to $bT^m + cT^n$ as − FF to $mbT^{m-1} - ncT^{n-1} + \frac{mm-m}{2} bXT^{m-2} + \frac{nn-n}{2} cXT^{n-2}$, &c. And taking the last ratios that arise when the orbits come to a circular form, there will come forth GG to $bT^{m-1} + cT^{n-1}$ as FF to $mbT^{m-1} + ncT^{n-1}$; and again, GG to FF as $bT^{m-1} + cT^{n-1}$ to $mbT^{n-1} + ncT^{n-1}$. This proportion, by expressing the greatest altitude CV or T arithmetically by unity, becomes, GG to FF as $b + c$ to $mb + nc$, and therefore as 1 to $\frac{mb + nc}{b + c}$. Whence G becomes to F, that is, the angle VC*p* to the

angle VCP, as 1 to $\sqrt{\frac{mb + nc}{b + c}}$. And therefore since the angle VCP between the upper and the lower apsis, in an immovable ellipsis, is of 180 deg., the angle VC*p* between the same apsides in an orbit which a body describes with a centripetal force, that is, as $\frac{bA^m + cA^n}{A^3}$, will be equal to an angle of 180 $\sqrt{\frac{b + c}{mb + nc}}$ deg. And by the same reasoning, if

the centripetal force be as $\frac{bA^m + cA^n}{A^3}$, the angle between the apsides will be found equal

to 180 $\sqrt{\frac{b - c}{mb - nc}}$ deg. After the same manner the Problem is solved in more difficult cases. The quantity to which the centripetal force is proportional must always be resolved into a converging series whose denominator is A^3. Then the given part of the numerator arising from that operation is to be supposed in the same ratio to that part of it which is not given, as the given part of this numerator RGG − RFF + TFF − FFX is to that part of the same numerator which is not given. And taking away the superfluous quantities, and writing unity for T, the proportion of G to F is obtained.

COR. 1. Hence if the centripetal force be as any power of the altitude, that power may be found from the motion of the apsides; and so contrariwise. That is, if the whole angular motion, with which the body returns to the same apsis, be to the angular motion of one revolution, or 360 deg., as any number as m to another as n, and the altitude called A; the force will be as the power $A^{\frac{nm}{mm} - 3}$ of the altitude A; the index of

which power is $\frac{nm}{mm} - 3$. This appears by the second example. Hence it is plain that the force in its recess from the centre cannot decrease in a greater then a triplicate ratio of the altitude. A body revolving with such a force, and parting from the apsis, if it once begins to descend, can never arrive at the lower apsis or least altitude, but will descend

to the centre, describing the curve line treated of in Cor. 3, Prop. XLI. But if it should, at its parting from the lower apsis, begin to ascend ever so little, it will ascend *in infinitum*, and never come to the upper apsis; but will describe the curve line spoken of in the same Cor. 3, and Cor. 6, Prop. XLIV. So that where the force in its recess from the centre decreases in a greater than a triplicate ratio of the altitude, the body at its parting from the apsis, will either descend to the centre, or ascend *in infinitum*, according as it descends or ascends at the beginning of its motion. But if the force in its recess from the centre either decreases in a less than a triplicate ratio of the altitude, or increases in any ratio of the altitude whatsoever, the body will never descend to the centre, but will at some time arrive at the lower apsis; and, on the contrary, if the body alternately ascending and descending from one apsis to another never comes to the centre, then either the force increases in the recess from the centre, or it decreases in a less than a triplicate ratio of the altitude; and the sooner the body returns from one apsis to another, the farther is the ratio of the forces from the triplicate ratio. As if the body should return to and from the upper apsis by an alternate descent and ascent in S revolutions, or in 4, or 2, or $1^1/_2$; that is, if m should be to n as 8, or 4, or 2, or $1^1/_2$ to 1, and therefore $\frac{nm}{mm} - 3$ be $1/_{64} - 3$ or $1/_{16} - 3$, or $1/_4 - 3$, or $4/_9 - 3$; then the force will be as $A^{1/64 - 3}$ 1 or $A^{1/16 - 3}$, or $A^{1/4 - 3}$, or $A^{4/9 - 3}$; that is, it will be reciprocally as $A^{3 - 1/64}$, or $A^{3 - 1/16}$, or $A^{3 - 1/4}$, or $A^{3 - 4/9}$. If the body after each revolution returns to the same apsis, and the apsis remains unmoved, then m will be to n as 1 to 1, and therefore $A^{\frac{nm}{mm} - 3}$ will be equal to A^{-2}, or $\frac{1}{AA}$; and therefore the decrease of the forces will be in a duplicate ratio of the altitude; as was demonstrated above. If the body in three fourth parts, or two thirds, or one third, or one fourth part of an entire revolution, return to the same apsis; m will be to n. as $3/_4$ or $2/_3$ or $1/_3$ or $1/_4$ to 1, and therefore $A^{\frac{nm}{mm} - 3}$ is equal to $A^{16/9 - 3}$, or $A^{9/4 - 3}$ or $A^{9 - 3}$, or $A^{16 - 3}$; and therefore the force is either reciprocally as $A^{11/9}$ or $A^{3/4}$, or directly as A^6 or A^{13}. Lastly if the body in its progress from the upper apsis to the same upper apsis again, goes over one entire revolution and three deg. more, and therefore that apsis in each revolution of the body moves three deg. *in consequentia*; then m will be to n as 363 deg. to 360 deg. or as 121 to 120, and therefore $A^{\frac{nm}{mm} - 3}$ will be equal to $A^{-\frac{29523}{14641}}$ and

therefore the centripetal force will be reciprocally as $A^{\frac{29523}{14641}}$ or reciprocally as very nearly. Therefore the centripetal force decreases in a ratio something greater than the duplicate; but approaching $59^3/_4$ times nearer to the duplicate than the triplicate.

COR. 2. Hence also if a body, urged by a centripetal force which is reciprocally as the square of the altitude, revolves in an ellipsis whose focus is in the centre of the

forces; and a new and foreign force should be added to or subducted from this centripetal force, the motion of the apsides arising from that foreign force may (by the third Example) be known; and so on the contrary. As if the force with which the body revolves in the ellipsis be as $\frac{1}{AA}$ and the foreign force subducted as cA, and therefore

the remaining force as $\frac{A - cA^4}{A^3}$; then (by the third Example) b will be equal to 1, m equal to 1, and n equal to 4; and therefore the angle of revolution between the apsides is equal to $180\sqrt{\frac{1-c}{1-4c}}$ deg. Suppose that foreign force to be 357.45 parts less than the

other force with which the body revolves in the ellipsis; that is, c to be $\frac{A - cA^4}{A^3}$; A or T

being equal to 1; and then $180\sqrt{\frac{1-c}{1-4c}}$ will be $180\sqrt{\frac{35645}{35345}}$ or 180.7623, that is, 180 deg., 45 min., 44 sec. Therefore the body, parting from the upper apsis, will arrive at the lower apsis with an angular motion of 180 deg., 45 min., 44 sec, and this angular motion being repeated, will return to the upper apsis; and therefore the upper apsis in each revolution will go forward 1 deg., 31 min., 28 sec. The apsis of the moon is about twice as swift.

So much for the motion of bodies in orbits whose planes pass through the centre of force. It now remains to determine those motions in eccentrical planes. For those authors who treat of the motion of heavy bodies used to consider the ascent and descent of such bodies, not only in a perpendicular direction, but at all degrees of obliquity upon any given planes; and for the same reason we are to consider in this place the motions of bodies tending to centres by means of any forces whatsoever, when those bodies move in eccentrical planes. These planes are supposed to be perfectly smooth and polished, so as not to retard the motion of the bodies in the least. Moreover, in these demonstrations, instead of the planes upon which those bodies roll or slide, and which are therefore tangent planes to the bodies, I shall use planes parallel to them, in which the centres of the bodies move, and by that motion describe orbits. And by the same method I afterwards determine the motions of bodies performed in curve superficies.

SECTION X.

Of the motion of bodies in given superficies, and of the reciprocal motion of funependulous bodies.

PROPOSITION XLVI. PROBLEM XXXII.

ANY KIND OF CENTRIPETAL FORCE BEING SUPPOSED, AND
THE CENTRE OF FORCE, AND ANY PLANE WHATSOEVER IN
WHICH THE BODY REVOLVES, BEING GIVEN, AND THE
QUADRATURES OF CURVILINEAR FIGURES BEING ALLOWED; IT
IS REQUIRED TO DETERMINE THE MOTION OF A BODY
GOING OFF FROM A GIVEN PLACE, WITH A GIVEN VELOCITY,
IN THE DIRECTION OF A GIVEN RIGHT LINE IN THAT PLANE.

Let S be the centre of force, SC the least distance of that centre from the given plane, P a body issuing from the place P in the direction of the right line PZ, Q the same body revolving in its trajectory, and PQR the trajectory itself which is required to be found, described in that given plane. Join CQ, QS, and if in QS we take SV proportional to the centripetal force with which the body is attracted towards the centre S, and draw VT parallel to CQ, and meeting SC in T; then will the force SV be resolved into two (by Cor. 2, of the Laws of Motion), the force ST, and the force TV; of which ST attracting the body in the direction of a line perpendicular to that plane, does not at all change its motion in that plane. But the action of the other force TV, coinciding with the position of the plane itself, attracts the body directly towards the given point C in that plane; and therefore causes the body to move in this plane in

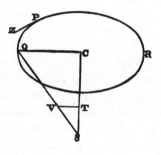

the same manner as if the force ST were taken away, and the body were to revolve in free space about the centre C by means of the force TV alone. But there being given the centripetal force TV with which the body Q revolves in free space about the given centre C, there is given (by Prop. XLII) the trajectory PQR which the body describes; the place Q, in which the body will be found at any given time; and, lastly, the velocity of the body in that place Q. And so è contra.   *Q.E.I.*

PROPOSITION XLVII. THEOREM XV.

SUPPOSING THE CENTRIPETAL FORCE TO BE PROPORTIONAL
TO THE DISTANCE OF THE BODY FROM THE CENTRE; ALL
BODIES REVOLVING IN ANY PLANES WHATSOEVER WILL
DESCRIBE ELLIPSES, AND COMPLETE THEIR REVOLUTIONS IN
EQUAL TIMES; AND THOSE WHICH MOVE IN RIGHT LINES

RUNNING BACKWARDS AND FORWARDS ALTERNATELY, WILL
COMPLETE THEIR SEVERAL PERIODS OF GOING AND
RETURNING IN THE SAME TIMES.

For letting all things stand as in the foregoing Proposition, the force SV, with which the body Q revolving in any plane PQR is attracted towards the centre S, is as the distance SQ; and therefore because SV and SQ, TV and CQ are proportional, the force TV with which the body is attracted towards the given point C in the plane of the orbit is as the distance CQ. Therefore the forces with which bodies found in the plane PQR are attracted towards the point C, are in proportion to the distances equal to the forces with which the same bodies are attracted every way towards the centre S; and therefore the bodies will move in the same times, and in the same figures, in any plane PQR about the point C as they would do in free spaces about the centre S; and therefore (by Cor. 2, Prop. X, and Cor. 2, Prop. XXXVIII.) they will in equal times either describe ellipses in that plane about the centre C, or move to and fro in right line passing through the centre C in that plane; completing the same periods of time in all cases. *Q.E.D.*

SCHOLIUM.

The ascent and descent of bodies in curve superficies has a near relation to these motions we have been speaking of. Imagine curve lines to be described on any plane, and to revolve about any given axes passing through the centre of force, and by that revolution to describe curve superficies; and that the bodies move in such sort that their centres may be always found in these superficies. If those bodies reciprocate to and fro with an oblique ascent and descent, their motions will be performed in planes passing through the axis, and therefore in the curve lines, by whose revolution those curve superficies were generated. In those cases, therefore, it will be sufficient to consider the motion in those curve lines.

PROPOSITION XLVIII. THEOREM XVI.

IF A WHEEL STANDS UPON THE OUTSIDE OF A GLOBE AT
RIGHT ANGLES THERETO, AND REVOLVING ABOUT ITS OWN
AXIS GOES FORWARD IN A GREAT CIRCLE, THE LENGTH OF
THE CURVILINEAR PATH WHICH ANY POINT, GIVEN IN THE
PERIMETER OF THE WHEEL, HATH DESCRIBED SINCE THE
TIME THAT IT TOUCHED THE GLOBE (WHICH CURVILINEAR
PATH WE MAY CALL THE CYCLOID OR EPICYCLOID), WILL BE
TO DOUBLE THE VERSED SINE OF HALF THE ARC WHICH

SINCE THAT TIME HAS TOUCHED THE GLOBE IN PASSING
OVER IT, AS THE SUM OF THE DIAMETERS OF THE GLOBE
AND THE WHEEL TO THE SEMI-DIAMETER OF THE GLOBE.

PROPOSITION XLIX. THEOREM XVII.

IF A WHEEL STAND UPON THE INSIDE OF A CONCAVE GLOBE
AT RIGHT ANGLES THERETO, AND REVOLVING ABOUT ITS
OWN AXIS GO FORWARD IN ONE OF THE GREAT CIRCLES OF
THE GLOBE, THE LENGTH OF THE CURVILINEAR PATH WHICH
ANY POINT, GIVEN IN THE PERIMETER OF THE WHEEL, HATH
DESCRIBED SINCE IT TOUCHED THE GLOBE, WILL BE TO THE
DOUBLE OF THE VERSED SINE OF HALF THE ARC WHICH IN
ALL THAT TIME HAS TOUCHED THE GLOBE IN PASSING OVER
IT, AS THE DIFFERENCE OF THE DIAMETERS OF THE GLOBE
AND THE WHEEL TO THE SEMI-DIAMETER OF THE GLOBE.

Let ABL be the globe, C its centre, BPV the wheel insisting thereon, E the centre
of the wheel, B the point of contact, and P the given point in the perimeter of the
wheel. Imagine this wheel to
proceed in the great circle
ABL from A through B
towards L, and in its progress
to revolve in such a manner
that the arcs AB, PB may be
always equal one to the other,
and the given point P in the
perimeter of the wheel may
describe in the mean time the
curvilinear path AP. Let AP be
the whole curvilinear path
described since the wheel
touched the globe in A, and
the length of this path AP will
be to twice the versed sine of
the are $^1/_2$PB as 2CE to CB.
For let the right line CE (produced if need be) meet the
wheel in V, and join CP, BP,

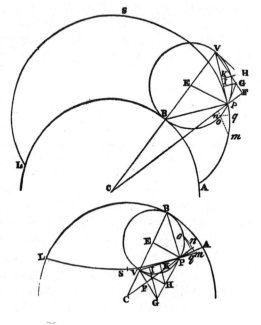

EP, VP; produce CP, and let fall thereon the perpendicular VF. Let PH, VH, meeting in H, touch the circle in P and V, and let PH cut VF in G, and to VP let fall the perpendiculars GI, HK. From the centre C with any interval let there be described the circle *nom*, cutting the right line CP in *n*, the perimeter of the wheel BP in *o*, and the curvilinear path AP in *m*; and from the centre V with the interval V*o* let there be described a circle cutting VP produced in *q*.

Because the wheel in its progress always revolves about the point of contact B, it is manifest that the right line BP is perpendicular to that curve line AP which the point P of the wheel describes, and therefore that the right line VP will touch this curve in the point P. Let the radius of the circle *nom* be gradually increased or diminished so that at last it become equal to the distance CP; and by reason of the similitude of the evanescent figure P*nomq*, and the figure PFGVI, the ultimate ratio of the evanescent lineolæ P*m*, P*n*, P*o*, P*q*, that is, the ratio of the momentary mutations of the curve AP, the right line CP, the circular arc BP, and the right line VP, will be the same as of the lines PV, PF, PG, PI, respectively. But since VF is perpendicular to CF, and VH to CV, and therefore the angles HVG, VCF equal; and the angle VHG (because the angles of the quadrilateral figure HVEP are right in V and P) is equal to the angle CEP, the triangles V HG, CEP will be similar; and thence it will come to pass that as EP is to CE so is HG to HV or HP, and so is KI to KP, and by composition or division as CB to CE so is PI to PK, and doubling the consequents as CB to 2CE so PI to PV, and so is P*q* to P*m*. Therefore the decrement of the line VP, that is, the increment of the line BV-VP to the increment of the curve line AP is in a given ratio of CB to 2CE, and therefore (by Cor. Lem. IV) the lengths BV – VP and AP, generated by those increments, are in the same ratio. But if BV be radius, VP is the cosine of the angle BVP or $\frac{1}{2}$BEP, and therefore BV – VP is the versed sine of the same angle, and therefore in this wheel, whose radius is $\frac{1}{2}$BV, BV – VP will be double the versed sine of the arc $\frac{1}{2}$BP. Therefore AP is to double the versed sine of the arc $\frac{1}{2}$BP as 2CE to CB.   *Q.E.D.*

The line AP in the former of these Propositions we shall name the cycloid without the globe, the other in the latter Proposition the cycloid within the globe, for distinction sake.

COR. 1. Hence if there be described the entire cycloid ASL, and the same be bisected in S, the length of the part PS will be to the length PV (which is the double of the sine of the angle VBP, when EB is radius) as 2CE to CB, and therefore in a given ratio.

COR. 2. And the length of the semi-perimeter of the cycloid AS will be equal to a right line which is to the diameter of the wheel BV as 2CE to CB.

PROPOSITION L. PROBLEM XXXIII.

TO CAUSE A PENDULOUS BODY TO OSCILLATE IN A GIVEN CYCLOID.

Let there be given within the globe QVS described with the centre C, the cycloid QRS, bisected in R, and meeting the superficies of the globe with its extreme points Q and S on either hand. Let there be drawn CR bisecting the arc QS in O, and let it be

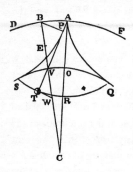

produced to A in such sort that CA may be to CO as CO to CR. About the centre C, with the interval CA, let there be described an exterior globe DAF; and within this globe, by a wheel whose diameter is AO, let there be described two semi-cycloids AQ, AS, touching the interior globe in Q and S, and meeting the exterior globe in A. From that point A, with a thread APT in length equal to the line AR, let the body T depend, and oscillate in such manner between the two semi-cycloids AQ, AS, that, as often as the pendulum parts from the perpendicular AR, the upper part of the thread AP may be applied to that semi-cycloid APS towards which the motion tends, and fold itself round that curve line, as if it were some solid obstacle, the remaining part of the same thread PT which has not yet touched the semi-cycloid continuing straight. Then will the weight T oscillate in the given cycloid QRS.

$Q.E.F.$

For let the thread PT meet the cycloid QRS in T, and the circle QOS in V, and let CV be drawn; and to the rectilinear part of the thread PT from the extreme points P and T let there be erected the perpendiculars BP, TW, meeting the right line CV in B and W. It is evident, from the construction and generation of the similar figures AS, SR, that those perpendiculars PB, TW, out off from CV the lengths VB, VW equal the diameters of the wheels OA, OR. Therefore TP is to VP (which is double the sine of the angle VBP when $1/2$BV is radius) as BW to BV, or AO +OR to AO, that is (since CA and CO, CO and CR, and by division AO and OR are proportional), as CA + CO to CA, or, if BV be bisected in E, as 2CE to CB. Therefore (by Cor. 1, Prop. XLIX), the length of the rectilinear part of the thread PT is always equal to the arc of the cycloid PS, and the whole thread APT is always equal to the half of the cycloid APS, that is (by Cor. 2, Prop. XLIX), to the length AR. And therefore contrariwise, if the string remain always equal to the length AR, the point T will always move in the given cycloid QRS.

$Q.E.D.$

COR. The string AR is equal to the semi-cycloid AS, and therefore has the same ratio to AC the semi-diameter of the exterior globe as the like semi-cycloid SR has to CO the semi-diameter of the interior globe.

PROPOSITION LI. THEOREM XVIII.

IF A CENTRIPETAL FORCE TENDING, ON ALL SIDES TO THE
CENTRE C OF A GLOBE, BE IN ALL PLACES AS THE DISTANCE
OF THE PLACE FROM THE CENTRE, AND BY THIS FORCE
ALONE ACTING, UPON IT, THE BODY T OSCILLATE (IN THE
MANNER ABOVE DESCRIBED) IN THE PERIMETER OF THE
CYCLOID QRS; I SAY, THAT ALL THE OSCILLATIONS, HOW
UNEQUAL SOEVER IN THEMSELVES, WILL BE PERFORMED
IN EQUAL TIMES.

For upon the tangent TW infinitely produced let fall the perpendicular CX, and join CT. Because the centripetal force with which the body T is impelled towards C is as the distance CT, let this (by Cor. 2, of the Laws) be resolved into the parts CX, TX, of which CX. impelling the body directly from P stretches the thread PT, and by the resistance the thread makes to it is totally employed, producing no other effect; but the other part TX, impelling the body transversely or towards X, directly accelerates the motion in the cycloid. Then it is plain that the acceleration of the body, proportional to this accelerating force, will be every moment as the length TX, that is (because CV,

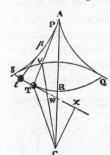

WV, and TX, TW proportional to them are given), as the length TW, that is (by Cor. 1, Prop. XLIX) as the length of the arc of the cycloid TR. If therefore two pendulums APT, Apt, be unequally drawn aside from the perpendicular AR, and let fall together, their accelerations will be always as the arcs to be described TR, tR. But the parts described at the beginning of the motion are as the accelerations, that is, as the wholes that are to be described at the beginning, and therefore the parts which remain to be described, and the subsequent accelerations proportional to those parts, are also as the wholes, and so on. Therefore the accelerations, and consequently the velocities generated, and the parts described with those velocities, and the parts to be described, are always as the wholes; and therefore the parts to be described preserving a given ratio to each other will vanish together, that is, the two bodies oscillating will arrive together at the perpendicular AR. And since on the other hand the ascent of the pendulums from the lowest place R through the same cycloidal arcs with a retrograde motion, is retarded in the several places they pass through by the same forces by which their descent was accelerated; it is plain that the velocities of their ascent and descent through the same arcs are equal, and consequently performed in equal times; and, therefore, since the two parts of the cycloid RS and RQ lying on either

side of the perpendicular are similar and equal, the two pendulums will perform as well the wholes as the halves of their oscillations in the same times. *Q.E.D.*

COR. The force with which the body T is accelerated or retarded in any place T of the cycloid, is to the whole weight of the same body in the highest place S or Q as the are of the cycloid TR is to the arc SR or QR.

PROPOSITION LII. PROBLEM XXXIV.

TO DEFINE THE VELOCITIES OF THE PENDULUMS IN THE
SEVERAL PLACES, AND THE TIMES IN WHICH BOTH THE
ENTIRE OSCILLATIONS, AND THE SEVERAL PARTS OF THEM
ARE PERFORMED.

About any centre G, with the interval GH equal to the arc of the cycloid RS, describe a semi-circle HKM bisected by the semi-diameter GK. And if a centripetal

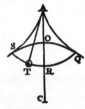

force proportional to the distance of the places from the centre tend to the centre G, and it be in the perimeter HIK equal to the centripetal force in the perimeter of the globe QOS tending towards its centre, and at the same time that the pendulum T is let fall from the highest place S, a body, as L, is let fall from H to G; then because the forces which act upon the bodies are equal at the beginning, and always proportional to the spaces to be described

TR, LG, and therefore if TR and LG are equal, are also equal in the places T and L, it is plain that those bodies describe at the beginning equal spaces ST, HL, and therefore are still acted upon equally, and continue to describe equal spaces. Therefore by Prop. XXXVIII, the time in which the body describes the arc ST is to the time of one oscillation, as the arc HI the time in which the body H arrives at L, to

the semi-periphery HKM, the time in which the body H will come to M. And the velocity of the pendulous body in the place T is to its velocity in the lowest place R, that is, the velocity of the body H in the place L to its velocity in the place G, or the momentary increment of the line HL, to the momentary increment of the line HG (the arcs HI, HK increasing with an equable flux) as the ordinate LI to the radius GK, or as $\sqrt{SR^2 - TR^2}$ to SR. Hence, since in unequal oscillations there are described in equal time arcs proportional to the entire arcs of the oscillations, there are obtained from the times given, both the velocities and the arcs described in all the oscillations universally. Which was first required.

Let now any pendulous bodies oscillate in different cycloids described within different globes, whose absolute forces are also different; and if the absolute force of any

globe QOS be called V, the accelerative force with which the pendulum is acted on in the circumference of this globe, when it begins to move directly towards its centre, will be as the distance of the pendulous body from that centre and the absolute force of the globe conjunctly, that is, as $CO \times V$. Therefore the lineola HY, which is as this accelerated force $CO \times V$, will be described in a given time; and if there be erected the perpendicular YZ meeting the circumference in Z, the nascent arc HZ will denote that given time. But that nascent arc HZ is in the subduplicate ratio of the rectangle GHY, and therefore as $\sqrt{GH \times CO \times V}$. Whence the time of an entire oscillation in the cycloid QRS (it being as the semi-periphery HKM, which denotes that entire oscillation, directly; and as the arc HZ which in like manner denotes a given time inversely) will be as GH directly and $\sqrt{GH \times CO \times V}$ inversely; that is, because GH and SR are equal, as

$$\sqrt{\frac{SR}{CO \times V}}$$ or (by Cor. Prop. L,) as $\sqrt{\frac{AR}{AC \times V}}$. Therefore the oscillations in all globes and cycloids, performed with what absolute forces soever, are in a ratio compounded of the subduplicate ratio of the length of the string directly, and the subduplicate ratio of the distance between the point of suspension and the centre of the globe inversely, and the subduplicate ratio of the absolute force of the globe inversely also. Q.E.I.

COR. 1. Hence also the times of oscillating, falling, and revolving bodies may be compared among themselves. For if the diameter of the wheel with which the cycloid is described within the globe is supposed equal to the semi-diameter of the globe, the cycloid will become a right line passing through the centre of the globe, and the oscillation will be changed into a descent and subsequent ascent in that right line. Whence there is given both the time of the descent from any place to the centre, and the time equal to it in which the body revolving uniformly about the centre of the globe at any distance describes an arc of a quadrant. For this time (by Case 2) is to the time of half the oscillation in any cycloid QRS as 1 to $\sqrt{\frac{AR}{AC \times V}}$.

COR. 2. Hence also follow what Sir *Christopher Wren* and M. *Huygens*, have discovered concerning the vulgar cycloid. For if the diameter of the globe be infinitely increased, its sphærical superficies will be changed into a plane, and the centripetal force will act uniformly in the direction of lines perpendicular to that plane, and this cycloid of our's will become the same with the common cycloid. But in that case the length of the arc of the cycloid between that plane and the describing point will become equal to four times the versed sine of half the arc of the wheel between the same plane and the describing point, as was discovered by Sir *Christopher Wren*. And a pendulum between two such cycloids will oscillate in a similar and equal cycloid in equal times, as M. *Huygens* demonstrated. The descent of heavy bodies also in the time of one oscillation will be the same as M. *Huygens* exhibited.

The propositions here demonstrated are adapted to the true constitution of the Earth, in so far as wheels moving in any of its great circles will describe, by the motions of nails fixed in their perimeters, cycloids without the globe; and pendulums, in mines and deep caverns of the Earth, must oscillate in cycloids within the globe, that those oscillations may be performed in equal times. For gravity (as will be shewn in the third book) decreases in its progress from the superficies of the Earth; upwards in a duplicate ratio of the distances from the centre of the Earth; downwards in a simple ratio of the same.

PROPOSITION LIII. PROBLEM XXXV.

GRANTING THE QUADRATURES OF CURVILINEAR FIGURES, IT IS REQUIRED TO FIND THE FORCES WITH WHICH BODIES MOVING IN GIVEN CURVE LINES MAY ALWAYS PERFORM THEIR OSCILLATIONS IN EQUAL TIMES.

Let the body T oscillate in any curve line STRQ, whose axis is AR passing through the centre of force C. Draw TX touching that curve in any place of the body T, and in that tangent TX take TY equal to the arc TR. The length of that arc is known from the common methods used for the quadratures of figures. From the point Y draw the right line YZ perpendicular to the tangent. Draw CT meeting that perpendicular in Z, and the centripetal force will be proportional to the right line TZ. *Q.E.I.*

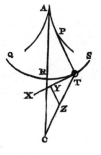

For if the force with which the body is attracted from T towards C be expressed by the right line TZ taken proportional to it, that force will he resolved into two forces TY, YZ, of which YZ drawing the body in the direction of the length of the thread PT, does not at all change its motion; whereas the other force TY directly accelerates or retards its motion in the curve STRQ. Wherefore since that force is as the space to be described TR, the accelerations or retardations of the body in describing two proportional parts (a greater and a less) of two oscillations, will be always as those parts, and therefore will cause those parts to be described together. But bodies which continually describe together parts proportional to the wholes, will describe the wholes together also. *Q.E.D.*

COR. 1. Hence if the body T, hanging by a rectilinear thread AT from the centre A, describe the circular arc STRQ, and in the mean time be acted on by any force tending downwards with parallel directions, which is to the uniform force of gravity as the arc TR to its sine TN, the times of the several oscillations will be equal. For because

TZ, AR are parallel, the triangles ATN, ZTY are similar; and therefore TZ will be to AT as TY to TN; that is, if the uniform force of gravity be expressed by the given length AT, the force TZ, by which the oscillations become isochronous, will be to the force of gravity AT, as the arc TR equal to TY is to TN the sine of that arc.

COR. 2. And therefore in clocks, if forces were impressed by some machine upon the pendulum which preserves the motion, and so compounded with the force of gravity that the whole force tending downwards should be always as a line produced by applying the rectangle under the arc TR and the radius AR to the sine TN, all the oscillations will become isochronous.

PROPOSITION LIV. PROBLEM XXXVI.

GRANTING THE QUADRATURES OF CURVILINEAR FIGURES, IT
IS REQUIRED TO FIND THE TIMES IN WHICH BODIES BY
MEANS OF ANY CENTRIPETAL FORCE WILL DESCEND OR
ASCEND IN ANY CURVE LINES DESCRIBED IN A PLANE PASSING
THROUGH THE CENTRE OF FORCE.

Let the body descend from any place S, and move in any curve ST R given in a plane passing through the centre of force C. Join CS, and let it be divided into innumerable equal parts, and let D*d* be one of those parts. From the centre C, with the intervals CD, C*d*, let the circles DT, *dt* be described, meeting the curve line ST*t*R in

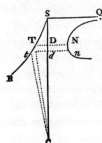

T and *t*. And because the law of centripetal force is given, and also the altitude CS from which the body at first fell, there will be given the velocity of the body in any other altitude CT (by Prop. XXXIX). But the time in which the body describes the lineola T*t* is as the length of that lineola, that is, as the secant of the angle *t*TC directly, and the velocity inversely. Let the ordinate DN, proportional to this time, be made perpendicular to the right line CS at the point D, and because D*d* is given, the rectangle D*d* × DN, that is, the area DN*nd*, will be proportional to the same time. Therefore if PN*n* be a curve line in which the point N is perpetually found, and its asymptote be the right line SQ standing upon the line CS at right angles, the area SQPND will be proportional to the time in which the body in its descent hath described the line ST; and therefore that area being found, the time is also given. *Q.E.I.*

PROPOSITION LV. THEOREM XIX.

IF A BODY MOVE IN ANY CURVE SUPERFICIES, WHOSE AXIS
PASSES THROUGH THE CENTRE OF FORCE, AND FROM THE
BODY A PERPENDICULAR BE LET FALL UPON THE AXIS; AND A
LINE PARALLEL AND EQUAL THERETO BE DRAWN FROM ANY
GIVEN POINT OF THE AXIS; I SAY, THAT THIS PARALLEL LINE
WILL DESCRIBE AN AREA PROPORTIONAL TO THE TIME.

Let BKL be a curve superficies, T a body revolving in it, STR a trajectory which
the body describes in the same, S the beginning of the trajectory, OMK the axis of the
curve superficies, TN a right line let fall perpendicularly from the body to the axis; OP
a line parallel and equal thereto drawn from the given point O in the axis; AP the
orthographic projection of the trajectory described by the point P in the plane AOP in
which the revolving line OP is found; A the beginning of that projection, answering to

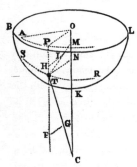

the point S; TC a right line drawn from the body to the
centre; TG a part thereof proportional to the centripetal force with which the body tends towards the
centre C; TM a right line perpendicular to the curve
superficies; T1 a part thereof proportional to the force
of pressure with which the body urges the superficies,
and therefore with which it is again repelled by the
superficies towards M; PTF a right line parallel to the
axis and passing through the body, and GF, IH right
lines let fall perpendicularly from the points G and I
upon that parallel PHTF. I say, now, that the area

AOP, described by the radius OP from the beginning of the motion, is proportional to
the time. For the force TG (by Cor. 2, of the Laws of Motion) is resolved into the forces
TF, FG; and the force TI into the forces TH, HI; but the forces TF, TH, acting in the
direction of the line PF perpendicular to the plane AOP, introduce no change in the
motion of the body but in a direction perpendicular to that plane. Therefore its
motion, so far as it has the same direction with the position of the plane, that is, the
motion of the point P, by which the projection AP of the trajectory is described in that
plane, is the same as if the forces TF, TH were taken away, and the body were acted on
by the forces FG, HI alone; that is, the same as if the body were to describe in the plane
AOP the curve AP by means of a centripetal force tending to the centre O, and equal
to the sum of the forces FG and HI. But with such a force as that (by Prop. 1) the area
AOP will be described proportional to the time.   *Q.E.D.*

COR. By the same reasoning, if a body, acted on by forces tending to two or more centres in any the same right line CO, should describe in a free space any curve line ST, the area AOP would be always proportional to the time.

PROPOSITION LVI. PROBLEM XXXVII.

GRANTING THE QUADRATURES OF CURVILINEAR FIGURES, AND SUPPOSING THAT THERE ARE GIVEN BOTH THE LAW OF CENTRIPETAL FORCE TENDING TO A GIVEN CENTRE, AND THE CURVE SUPERFICIES WHOSE AXIS PASSES THROUGH THAT CENTRE; IT IS REQUIRED TO FIND THE TRAJECTORY WHICH A BODY WILL DESCRIBE IN THAT SUPERFICIES, WHEN GOING OFF FROM A GIVEN PLACE WITH A GIVEN VELOCITY, AND IN A GIVEN DIRECTION IN THAT SUPERFICIES.

The last construction remaining, let the body T go from the given place S, in the direction of a line given by position, and turn into the trajectory sought STR, whose orthographic projection in the plane BDO is AP. And from the given velocity of the body in the altitude SC, its velocity in any other altitude TC will be also given. With that velocity, in a given moment of time, let the body describe the particle Tt of its trajectory, and let Pp be the projection of that particle described in the plane AOP. Join

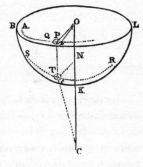

Op, and a little circle being described upon the curve superficies about the centre T with the interval Tt let the projection of that little circle in the plane AOP be the ellipsis pQ. And because the magnitude of that little circle Tt, and TN or PO its distance from the axis CO is also given, the ellipsis pQ will be given both in kind and magnitude, as also its position to the right line PO. And since the area POp is proportional to the time, and therefore given because the time is given, the angle POp will be given. And thence will be given p the common intersection of the ellipsis and the right line Op, together with the angle OPp, in which the projection APp of the trajectory cuts the line OP. But from thence (by conferring Prop. XLI, with its 2d Cor.) the manner of determining the curve APp easily appears. Then from the several points P of that projection erecting to the plane AOP, the perpendiculars PT meeting the curve superficies in T, there will be given the several points T of the trajectory. *Q.E.I.*

SECTION XI.

Of the motions of bodies tending to each other with centripetal forces.

I have hitherto been treating of the attractions of bodies towards an immovable centre; though very probably there is no such thing existent in nature. For attractions are made towards bodies, and the actions of the bodies attracted and attracting are always reciprocal and equal, by Law III; so that if there are two bodies, neither the attracted nor the attracting body is truly at rest, but both (by Cor. 4, of the Laws of Motion), being as it were mutually attracted, revolve about a common centre of gravity. And if there be more bodies, which are either attracted by one single one which is attracted by them again, or which all of them, attract each other mutually, these bodies will be so moved among themselves, as that their common centre of gravity will either be at rest, or move uniformly forward in a right line. I shall therefore at present go on to treat of the motion of bodies mutually attracting each other; considering the centripetal forces as attractions; though perhaps in a physical strictness they may more truly be called impulses. But these propositions are to be considered as purely mathematical; and therefore, laying aside all physical considerations, I make use of a familiar way of speaking, to make myself the more easily understood by a mathematical reader.

PROPOSITION LVII. THEOREM XX.

TWO BODIES ATTRACTING EACH OTHER MUTUALLY
DESCRIBE SIMILAR FIGURES ABOUT THEIR COMMON CENTRE
OF GRAVITY, AND ABOUT EACH OTHER MUTUALLY.

For the distances of the bodies from their common centre of gravity are reciprocally as the bodies; and therefore in a given ratio to each other; and thence, by composition of ratios, in a given ratio to the whole distance between the bodies. Now these distances revolve about their common term with an equable angular motion, because lying in the same right line they never change their inclination to each other mutually. But right lines that are in a given ratio to each other, and revolve about their terms with an equal angular motion, describe upon planes, which either rest with those terms, or move with any motion not angular, figures entirely similar round those terms. Therefore the figures described by the revolution of these distances are similar. *Q.E.D.*

PROPOSITION LVIII. THEOREM XXI.

IF TWO BODIES ATTRACT EACH OTHER MUTUALLY WITH FORCES
OF ANY KIND, AND IN THE MEAN TIME REVOLVE ABOUT THE

COMMON CENTRE OF GRAVITY; I SAY, THAT, BY THE SAME
FORCES, THERE MAY BE DESCRIBED ROUND EITHER BODY
UNMOVED A FIGURE SIMILAR AND EQUAL TO THE FIGURES
WHICH THE BODIES SO MOVING DESCRIBE ROUND EACH
OTHER MUTUALLY.

Let the bodies S and P revolve about their common centre of gravity C, proceeding from S to T, and from P to Q. From the given point s lot there be continually
drawn sp, sq, equal and parallel to SP, TQ; and the curve pqv, which the point p

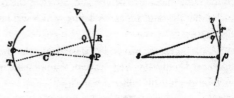

describes in its revolution round
the immovable point s, will be
similar and equal to the curves
which the bodies S and P describe
about each other mutually; and
therefore, by Theor. XX, similar
to the curves ST and PQV which the same bodies describe about their common centre of gravity C; and that because the proportions of the lines SC, CP, and SP or sp, to
each other, are given.

CASE. 1. The common centre of gravity C (by Cor. 4, of the Laws of Motion) is
either at rest, or moves uniformly in a right line. Let us first suppose it at rest, and in
s and p let there be placed two bodies, one immovable in s, the other movable in p, similar and equal to the bodies S and P. Then let the right lines PR and pr touch the curves
PQ and pq in P and p, and produce CQ and sq to R and r. And because the figures
CPRQ, $sprq$ are similar, RQ will be to rq as CP to sp, and therefore in a given ratio.
Hence if the force with which the body P is attracted towards the body S, and by consequence towards the intermediate point the centre C, were to the force with which the
body p is attracted towards the centre s, in the same given ratio, these forces would in
equal times attract the bodies from the tangents PR, pr to the arcs PQ, pq, through the
intervals proportional to them RQ, rq; and therefore this last force (tending to s) would
make the body p revolve in the curve pqv, which would become similar to the curve
PQV, in which the first force obliges the body P to revolve; and their revolutions would
be completed in the same times But because those forces are not to each other in the
ratio of CP to sp, but (by reason of the similarity and equality of the bodies S and s, P
and p and the equality of the distances SP, sp) mutually equal, the bodies in equal times
will be equally drawn from the tangents; and therefore that the body p may be attracted through the greater interval rq, there is required a greater time, which will be in the
subduplicate ratio of the intervals; because, by Lemma X, the spaces described at the
very beginning of the motion are in a duplicate ratio of the times. Suppose, then the
velocity of the body p to be to the velocity of the body P in a subduplicate ratio of the

distance *sp* to the distance CP, so that the arcs *pq*, PQ, which are in a simple proportion to each other, may be described in times that are in a subduplicate ratio of the distances; and the bodies P, *p*, always attracted by equal forces, will describe round the quiescent centres C and *s* similar figures PQV, *pqv*, the latter of which *pqv* is similar and equal to the figure which the body P describes round the movable body S.   *Q.E.D.*

CASE., 2. Suppose now that the common centre of gravity, together with the space in which the bodies are moved among themselves, proceeds uniformly in a right line; and (by Cor. 6, of the Laws of Motion) all the motions in this space will be performed in the same manner as before; and therefore the bodies will describe mutually about each other the same figures as before, which will be therefore similar and equal to the figure *pqv*.   *Q.E.D.*

COR. 1. Hence two bodies attracting each other with forces proportional to their distance, describe (by Prop. X) both round their common centre of gravity, and round each other mutually concentrical ellipses; and, *vice versa*, if such figures are described, the forces are proportional to the distances.

COR. 2. And two bodies, whose forces are reciprocally proportional to the square of their distance, describe (by Prop. XI, XII, XIII), both round their common centre of gravity, and round each other mutually, conic sections having their focus in the centre about which the figures are described. And, *vice versa*, if such figures are described, the centripetal forces are reciprocally proportional to the squares of the distance.

COR. 3. Any two bodies revolving round their common centre of gravity describe areas proportional to the times, by radii drawn both to that centre and to each other mutually.

PROPOSITION LIX. THEOREM XXII.

THE PERIODIC TIME OF TWO BODIES S AND P REVOLVING
ROUND THEIR COMMON CENTRE OF GRAVITY C, IS TO THE
PERIODIC TIME OF ONE OF THE BODIES P REVOLVING
ROUND THE OTHER S REMAINING UNMOVED, AND
DESCRIBING A FIGURE SIMILAR AND EQUAL TO THOSE
WHICH THE BODIES DESCRIBE ABOUT EACH OTHER
MUTUALLY, IN A SUBDUPLICATE RATIO OF THE OTHER BODY
S TO THE SUM OF THE BODIES S + P.

For, by the demonstration of the last Proposition, the times in which any similar arcs PQ and *pq* are described are in a subduplicate ratio of the distances CP and SP, or *sp*, that is, in a subduplicate ratio of the ody S to the sum of the bodies S + P. And by composition of ratios, the sums of the times in which all the similar arcs PQ and *pq*

are described, that is, the whole times in which the whole similar figures are described are in the same subduplicate ratio. *Q.E.D.*

PROPOSITION LX. THEOREM XXIII.

IF TWO BODIES S AND P, ATTRACTING EACH OTHER WITH FORCES RECIPROCALLY PROPORTIONAL TO THE SQUARES OF THEIR DISTANCE, REVOLVE ABOUT THEIR COMMON CENTRE OF GRAVITY; I SAY, THAT THE PRINCIPAL AXIS OF THE ELLIPSIS WHICH EITHER OF THE BODIES, AS P, DESCRIBES BY THIS MOTION ABOUT THE OTHER S, WILL BE TO THE PRINCIPAL AXIS OF THE ELLIPSIS, WHICH THE SAME BODY P MAY DESCRIBE IN THE SAME PERIODICAL TIME ABOUT THE OTHER BODY S QUIESCENT, AS THE SUM OF THE TWO BODIES S + P TO THE FIRST OF TWO MEAN PROPORTIONALS BETWEEN THAT SUM AND THE OTHER BODY S.

For if the ellipses described were equal to each other, their periodic times by the last Theorem would be in a subduplicate ratio of the body S to the sum of the bodies S + P. Let the periodic time in the latter ellipsis be diminished in that ratio, and the periodic times will become equal; but, by Prop. XV, the principal axis of the ellipsis will be diminished in a ratio sesquiplicate to the former ratio; that is, in a ratio to which the ratio of S to S + P is triplicate; and therefore that axis will be to the principal axis of the other ellipsis as the first of two mean proportionals between S + P and S to S + P. And inversely the principal axis of the ellipsis described about the movable body will be to the principal axis of that described round the immovable as S + P to the first of two mean proportionals between S + P and S. *Q.E.D.*

PROPOSITION LXI. THEOREM XXIV.

IF TWO BODIES ATTRACTING EACH OTHER WITH ANY KIND OF FORCES, AND NOT OTHERWISE AGITATED OR OBSTRUCTED, ARE MOVED IN ANY MANNER WHATSOEVER, THOSE MOTIONS WILL BE THE SAME AS IF THEY DID NOT AT ALL ATTRACT EACH OTHER MUTUALLY, BUT WERE BOTH ATTRACTED WITH THE SAME FORCES BY A THIRD BODY PLACED IN THEIR COMMON CENTRE OF GRAVITY; AND THE LAW OF THE ATTRACTING FORCES WILL BE THE SAME IN RESPECT OF THE DISTANCE OF THE BODIES FROM THE COMMON CENTRE, AS IN RESPECT OF THE DISTANCE BETWEEN THE TWO BODIES.

For those forces with which the bodies attract each other mutually, by tending to the bodies, tend also to the common centre of gravity lying directly between them; and therefore are the same as if they proceeded from in intermediate body. *Q.E.D.*

And because there is given the ratio of the distance of either body from that common centre to the distance between the two bodies, there is given, of course, the ratio of any power of one distance to the same power of the other distance; and also the ratio of any quantity derived in any manner from one of the distances compounded any how with given quantities, to another quantity derived in like manner from the other distance, and as many given quantities having that given ratio of the distances to the first Therefore if the force with which one body is attracted by another be directly or inversely as the distance of the bodies from each other, or as any power of that distance; or, lastly, as any quantity derived after any manner from that distance compounded with given quantities; then will the same force with which the same body is attracted to the common centre of gravity be in like manner directly or inversely as the distance of the attracted body from the common centre, or as any power of that distance; or, lastly, as a quantity derived in like sort from that distance compounded with analogous given quantities. That is, the law of attracting force will be the same with respect to both distances. *Q.E.D.*

PROPOSITION LXII. PROBLEM XXXVIII.

To determine the motions of two bodies which attract each other with forces reciprocally proportional to the squares of the distance between them, and are let full from given places.

The bodies, by the last Theorem, will be moved in the same manner as if they were attracted by a third placed in the common centre of their gravity : and by the hypothesis that centre will be quiescent at the beginning of their motion, and therefore (by Cor. 4, of the Laws of Motion) will be always quiescent. The motions of the bodies are therefore to be determined (by Prob. XXV) in the same manner as if they were impelled by forces tending to that centre; and then we shall have the motions of the bodies attracting each other mutually. *Q.E.I.*

PROPOSITION LXIII. PROBLEM XXXIX.

To determine the motions of two bodies attracting each other with forces reciprocally proportional to the squares of their distance, and going off from given places in given directions with given velocities.

The motions of the bodies at the beginning being given, there is given also the uniform motion of the common centre of gravity, and the motion of the space which moves along with this centre uniformly in a right line, and also the very first, or beginning motions of the bodies in respect of this space. Then (by Cor. 5, of the Laws, and the last Theorem) the subsequent motions will be performed in the same manner in that space, as if that space together with the common centre of gravity were at rest, and as if the bodies did not attract each other, but were attracted by a third body placed in that centre. The motion therefore in this movable space of each body going off from a given place, in a given direction, with a given velocity, and acted upon by a centripetal force tending to that centre, is to be determined by Prob. IX and XXVI, and at the same time will be obtained the motion of the other round the same centre. With this motion compound the uniform progressive motion of the entire system of the space and the bodies revolving in it, and there will be obtained the absolute motion of the bodies in immovable space. *Q.E.I.*

PROPOSITION LXIV. PROBLEM XL.

Supposing forces with which bodies mutually attract each other to increase in a simple ratio of their distances from the centres; it is required to find the motions of several bodies among themselves.

Suppose the first two bodies T and L to have their common centre of gravity in D. These, by Cor. 1, Theor. XXI, will describe ellipses having their centres in D, the magnitudes of which ellipses are known by Prob. V.

Let now a third body S attract the two former T and L with the accelerative forces ST, SL, and let it be attracted again by them. The force ST (by Cor. 2, of the Laws of Motion) is resolved into the forces SD, DT; and the force SL into the forces SD and DL. Now the forces DT, DL, which are as

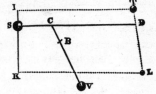

their sum TL, and therefore as the accelerative forces with which the bodies T and L attract each other mutually, added to the forces of the bodies T and L, the first to the first, and the last to the last, compose forces proportional to the distances DT and DL as before, but only greater than those former forces; and therefore (by Cor. 1, Prop. X, and Cor. 1, and 8, Prop. IV) they will cause those bodies to describe ellipses as before, but with a swifter motion. The remaining accelerative forces SD and DL by the motive forces SD × T and SD × L, which are as the bodies attracting those bodies equally and

in the direction of the lines TI, LK parallel to DS, do not at all change their situations with respect to one another, but cause them equally to approach to the line IK; which must be imagined drawn through the middle of the body S, and perpendicular to the line DS. But that approach to the line IK will be hindered by causing the system of the bodies T and L on one side, and the body S on the other, with proper velocities, to revolve round the common centre of gravity P. With such a motion the body S, because the sum of the motive forces SD × T and SD × L is proportional to the distance CS, tends to the centre C, will describe an ellipsis round the same centre C; and the point D, because the lines CS and CD are proportional, will describe a like ellipsis over against it. But the bodies T and L, attracted by the motive forces SD × T and SD × L, the first by the first, and the last by the last, equally and in the direction of the parallel lines TI and LK, as was said before, will (by Cor. 5 and 6, of the Laws of Motion) continue to describe their ellipses round the movable centre D, as before.   Q.E.I.

Let there be added a fourth body V, and, by the like reasoning, it will be demonstrated that this body and the point C will describe ellipses about the common centre of gravity B; the motions of the bodies T, L, and S round the centres D and C remaining the same as before; but accelerated. And by the same method one may add yet more bodies at pleasure.   Q.E.I.

This would be the case, though the bodies T and L attract each other mutually with accelerative forces either greater or less than those with which they attract the other bodies in proportion to their distance. Let all the mutual accelerative attractions be to each other as the distances multiplied into the attracting bodies; and from what has gone before it will easily be concluded that all the bodies will describe different ellipses with equal periodical times about their common centre of gravity B, in an immovable plane.   Q.E.I.

PROPOSITION LXV. THEOREM XXV.

Bodies, whose forces decrease in a duplicate ratio of their distances from their centres, may move among themselves in ellipses; and by radii drawn to the foci may describe areas proportional to the times very nearly.

In the last Proposition we demonstrated that case in which the motions will be performed exactly in ellipses. The more distant the law of the forces is from the law in that case, the more will the bodies disturb each other's motions; neither is it possible that bodies attracting each other mutually according to the law supposed in this Proposition should move exactly in ellipses, unless by keeping a certain proportion of

distances from each other. However, in the following cases the orbits will not much differ from ellipses.

CASE. 1. Imagine several lesser bodies to revolve about some very great one at different distances from it, and suppose absolute forces tending to every one of the bodies proportional to each. And because (by Cor. 4, of the Laws) the common centre of gravity of them all is either at rest, or moves uniformly forward in a right line, suppose the lesser bodies so small that the great body may be never at a sensible distance from that centre; and then the great body will, without any sensible error, be either at rest, or move uniformly forward in a right line; and the lesser will revolve about that great one in ellipses, and by radii drawn thereto will describe areas proportional to the times; if we except the errors that may be introduced by the receding of the great body from the common centre of gravity, or by the mutual actions of the lesser bodies upon each other. But the lesser bodies may be so far diminished, as that this recess and the mutual actions of the bodies on each other may become less than any assignable; and therefore so as that the orbits way become ellipses, and the areas answer to the times, without any error that is not less than any assignable.   Q.E.O.

CASE. 2. Let us imagine a system of lesser bodies revolving about a very great one in the manner just described, or any other system of two bodies revolving about each other to be moving uniformly forward in a right line, and in the mean time to be impelled sideways by the force of another vastly greater body situate at a great distance. And because the equal accelerative forces with which the bodies are impelled in parallel directions do not change the situation of the bodies with respect to each other, but only oblige the whole system to change its place while the parts still retain their motions among themselves, it is manifest that no change in those motions of the attracted bodies can arise from their attractions towards the greater, unless by the inequality of the accelerative attractions, or by the inclinations of the lines towards each other, in whose directions the attractions are made. Suppose, therefore, all the accelerative attractions made towards the great body to be among themselves as the squares of the distances reciprocally; and then, by increasing the distance of the great body till the differences of the right lines drawn from that to the others in respect of their length, and the inclinations of those lines to each other, be less than any given, the motions of the parts of the system will continue without errors that are not less than any given. And because, by the small distance of those parts from each other, the whole system is attracted as if it were but one body, it will therefore be moved by this attraction as if it were one body; that is, its centre of gravity will describe about the great body one of the conic sections (that is, a parabola or hyperbola when the attraction is but languid and an ellipsis when it is more vigorous); and by radii drawn thereto, it will describe areas proportional to the times, without any errors but those which arise from the distances of the parts, which are by the supposition exceedingly small, and may be diminished at pleasure.   Q.E.O.

By a like reasoning one may proceed to more compounded cases *in infinitum*.

COR. 1. In the second Case, the nearer the very great body approaches to the system of two or more revolving bodies, the greater will the perturbation be of the motions of the parts of the system among themselves; because the inclinations of the lines drawn from that great body to those parts become greater; and the inequality of the proportion is also greater.

COR. 2. But the perturbation will be greatest of all, if we suppose the accelerative attractions of the parts of the system towards the greatest body of all are not to each other reciprocally as the squares of the distances from that great body; especially if the inequality of this proportion be greater than the inequality of the proportion of the distances from the great body. For if the accelerative force, acting in parallel directions and equally, causes no perturbation in the motions of the parts of the system, it must of course, when it acts unequally, cause a perturbation somewhere, which will be greater or less as the inequality is greater or less. The excess of the greater impulses acting upon some bodies, and not acting upon others, must necessarily change their situation among themselves, And this perturbation, added to the perturbation arising from the inequality and inclination of the lines, makes the whole perturbation greater.

COR. 3. Hence if the parts of this system move in ellipses or circles without any remarkable perturbation, it is manifest that, if they are at all impelled by accelerative forces tending to any other bodies, the impulse is very weak, or else is impressed very near equally and in parallel directions upon all of them.

PROPOSITION LXVI. THEOREM XXVI.

IF THREE BODIES WHOSE FORCES DECREASE IN A DUPLICATE RATIO OF THE DISTANCES ATTRACT EACH OTHER MUTUALLY; AND THE ACCELERATIVE ATTRACTIONS OF ANY TWO TOWARDS THE THIRD BE BETWEEN THEMSELVES RECIPROCALLY AS THE SQUARES OF THE DISTANCES; AND THE TWO LEAST REVOLVE ABOUT THE GREATEST; I SAY, THAT THE INTERIOR OF THE TWO REVOLVING BODIES WILL, BY RADII DRAWN TO THE INNERMOST AND GREATEST, DESCRIBE ROUND THAT BODY AREAS MORE PROPORTIONAL TO THE TIMES, AND A FIGURE MORE APPROACHING TO THAT OF AN ELLIPSIS HAVING ITS FOCUS IN THE POINT OF CONCOURSE OF THE RADII, IF THAT GREAT BODY BE AGITATED BY THOSE ATTRACTIONS, THAN IT WOULD DO IF THAT REAL BODY WERE NOT ATTRACTED AT ALL BY THE LESSER, BUT REMAINED AT REST; OR THAN IT WOULD IF THAT GREAT

BODY WERE VERY MUCH MORE OR VERY MUCH LESS
ATTRACTED, OR VERY MUCH MORE OR VERY MUCH LESS
AGITATED, BY THE ATTRACTIONS.

This appears plainly enough from the demonstration of the second Corollary of
the foregoing Proposition; but it may be made out after this manner by a way of reasoning more distinct and more universally convincing.

CASE. 1. Let the lesser bodies P and S revolve in the same plane about the greatest
body T, the body P describing the interior orbit PAB, and S the exterior orbit ESE. Let
SK, be the mean distance of the bodies P and S; and let the accelerative attraction of
the body P towards S, at that mean distance, be expressed by that line SK. Make SL to
SK as the square of SK to the square of SP, and SL will be the accelerative attraction of

the body P towards S at any
distance SP. Join PT, and draw
LM parallel to it meeting ST in
M; and the attraction SL will
be resolved (by Cor. 2, of the
Laws of Motion) into the

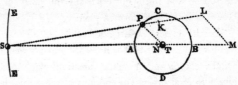

attractions SM, LM. And so the body P will be urged with a threefold accelerative
force. One of these forces tends towards T, and arises from the mutual attraction of the
bodies T and P. By this force alone the body P would describe round the body T, by
the radius PT, areas proportional to the times, and an ellipsis whose focus is in the centre of the body T; and this it would do whether the body T remained unmoved, or
whether it were agitated by that attraction. This appears from Prop. XI and Cor. 2 and
3 of Theor. XXI. The other force is that of the attraction LM, which, because it tends
from P to T, will be superadded to and coincide with the former force; and cause the
areas to be still proportional to the times, by Cor. 3, Theor. XXI. But because it is not
reciprocally proportional to the square of the distance PT, it will compose, when added
to the former, a force varying from that proportion; which variation will be the greater
by how much the proportion of this force to the former is greater, *caeteris paribus*.
Therefore, since by Prop. XI, and by Cor. 2, Theor. XXI, the force with which the ellipsis is described about the focus T ought to be directed to that focus, and to be reciprocally proportional to the square of the distance PT, that compounded force varying
from that proportion will make the orbit PAB vary from the figure of an ellipsis that
has its focus in the point T; and so much the more by how much the variation from
that proportion is greater; and by consequence by how much the proportion of the second force LM to the first force is greater, caerteris paribus. But now the third force SM,
attracting the body P in a direction parallel to ST, composes with the other force a new
force which is no longer directed from P to T; and which varies so much more from

this direction by how much the proportion of this third force to the other forces is greater, *caeteris paribus*; and therefore causes the body P to describe, by the radius TP, areas no longer proportional to the times; and therefore makes the variation from that proportionality so much greater by how much the proportion of this force to the others is greater. But this third force will increase the variation of the orbit PAB from the elliptical figure before-mentioned upon two accounts; first because that force is not directed from P to T; and, secondly, because it is not reciprocally proportional to the square of the distance PT. These things being premised, it is manifest that the areas are then most nearly proportional to the times, when that third force is the least possible, the rest preserving their former quantity; and that the orbit PAB does then approach nearest to the elliptical figure above-mentioned, when both the Second and third, but especially the third force, is the least possible; the first force remaining in its former quantity.

Let the accelerative attraction of the body T towards S be expressed by the line SN; then if the accelerative attractions SM and SN were equal, these, attracting the bodies T and P equally and in parallel directions would not at all change their situation with respect to each other. The motions of the bodies between themselves would be the same in that case as if those attractions did not act at all, by Cor. 6, of the Laws of Motion. And, by a like reasoning, if the attraction SN is less than the attraction SM, it will take away out of the attraction SM the part SN, so that there will remain only the part (of the attraction) MN to disturb the proportionality of the areas and times, and the elliptical figure of the orbit. And in like manner if the attraction SN be greater than the attraction SM, the perturbation of the orbit and proportion will be produced by the difference MN alone. After this manner the attraction SN reduces always the attraction SM to the attraction MN, the first and second attractions remaining perfectly unchanged; and therefore the areas and times come then nearest to proportionality, and the orbit PAB to the above-mentioned elliptical figure, when the attraction MN is either none, or the least that is possible; that is, when the accelerative attractions of the bodies P and T approach as near is possible to equality; that is, when the attraction SN is neither none at all, nor less than the least of all the attractions SM, but is, as it were, a mean between the greatest and least of all those attractions SM, that is, not much greater nor much less than the attraction SK.   *Q.E.D.*

CASE. 2. Let now the lesser bodies P, S, revolve about a greater T in different planes; and the force LM, acting in the direction of the line PT Situate in the plane of the orbit PAB, will have the same effect as before; neither will it draw the body P from the plane of its orbit. But the other force NM acting in the direction of a line parallel to ST (and which, therefore, when the body S is without the line of the nodes is inclined to the plane of the orbit PAB), besides the perturbation of the motion just

now spoken of as to longitude, introduces another perturbation also as to latitude, attracting the body P out of the plane of its orbit. And this perturbation, in any given situation of the bodies P and T to each other, will be as the generating force MN; and therefore becomes least when the force MN is least, that is (as was just now shown), where the attraction SN is not much greater nor much less than the attraction SK.   *Q.E.D.*

COR. 1. Hence it may be easily collected, that if several less bodies P, S, R, &c., revolve about a very great body T, the motion of the innermost revolving body P will be least disturbed by the attractions of the others, when the great body is as well attracted and agitated by the rest (according to the ratio of the accelerative forces) as the rest are by each other mutually.

COR. 2. In a system of three bodies, T, P, S, if the accelerative attractions of any two of them towards a third be to each other reciprocally as the squares of the distances, the body P, by the radius PT, will describe its area about the body T swifter near the conjunction A and the opposition B than it will near the quadratures C and D. For every force with which the body P is acted on and the body T is not, and which does not act in the direction of the line PT, does either accelerate or retard the description of the area, according as it is directed, whether *in consequentia* or *in antecedentia*. Such is the force NM. This force in the passage of the body P from O to A is directed *in consequentia* to its motion, and therefore accelerates it; then as far as D *in antecedentia*, and retards the motion; then *in consequentia* as far as B; and lastly *in antecedentia* as it moves from B to C.

COR. 3. And from the same reasoning it appears that the body P, *caeteris paribus*, moves more swiftly in the conjunction and opposition than in the quadratures.

COR. 4. The orbit of the body P, *caeteris paribus*, is more curve at the quadratures than at the conjunction and opposition. For the swifter bodies move, the less they deflect from a rectilinear path. And besides the force KL, or NM, at the conjunction and opposition, is contrary to the force with which the body T attracts the body P, and therefore diminishes that force; but the body P will deflect the less from a rectilinear path the less it is impelled towards the body T.

COR. 5. Hence the body P, *caeteris paribus*, goes farther from the body T at the quadratures than at the conjunction and opposition. This is said, however, supposing no regard had to the motion of eccentricity. For if the orbit of the body P be eccentrical, its eccentricity (as will be shewn presently by Cor. 9) will be greatest when the apsides are in

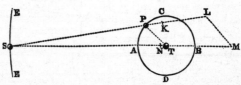

the syzygies; and thence it may sometimes come to pass that the body P, in its near approach to the farther apsis, may go farther from the body T at the syzygies than at the quadratures.

COR. 6. Because the centripetal force of the central body T, by which the body P is retained in its orbit, is increased at the quadratures by the addition caused by the force LM, and diminished at the syzygies by the subduction caused by the force KL, and, because the force KL is greater than LM, it is more diminished than increased; and, moreover, since that centripetal force (by Cor. 2, Prop. IV) is in a ratio compounded of the simple ratio of the radius TP directly, and the duplicate ratio of the periodical time inversely; it is plain that this compounded ratio is diminished by the action of the force KL; and therefore that the periodical time, supposing the radius of the orbit PT to remain the same, will be increased, and that in the subduplicate of that ratio in which the centripetal force is diminished; and, therefore, supposing this radius increased or diminished, the periodical time will be increased more or diminished less than in the sesquiplicate ratio of this radius, by Cor. 6, Prop. IV. If that force of the central body should gradually decay, the body P being less and less attracted would go farther and farther from the centre T; and, on the contrary, if it were increased, it would draw nearer to it. Therefore if the action of the distant body S, by which that force is diminished, were to increase and decrease by turns, the radius TP will be also increased and diminshed by turns; and the periodical time will be increased and diminished in a ratio compounded of the sesquiplicate ratio of the radius, and of the subduplicate of that ratio in which the centripetal force of the central body T is diminished or increased, by the increase or decrease of the action of the distant body S.

COR. 7. It also follows, from what was before laid down, that the axis of the ellipsis described by the body P, or the line of the apsides, does as to its angular motion go forwards and backwards by turns, but more forwards than backwards, and by the excess of its direct motion is in the whole carried forwards. For the force with which the body P is urged to the body T at the quadratures, where the force MN vanishes, is compounded of the force LM and the centripetal force with which the body T attracts the body P. The first force LM, if the distance PT be increased, is increased in nearly the same proportion with that distance, and the other force decreases in the duplicate ratio of the distance; and therefore the sum of these two forces decreases in a less than the duplicate ratio of the distance PT; and therefore, by Cor. 1, Prop. XLV, will make the line of the apsides, or, which is the same thing, the upper apsis, to go backward. But at

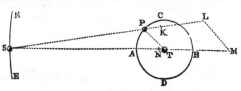

868

the conjunction and opposition the force with which the body P is urged towards the body T is the difference of the force KL, and of the force with which the body T attracts the body P; and that difference, because the force KL is very nearly increased in the ratio of the distance PT, decreases in more than the duplicate ratio of the distance PT; and therefore, by Cor. 1, Prop. XLV, causes the line of the apsides to go forwards. In the places between the syzygies and the quadratures, the motion of the line of the apsides depends upon both of these causes conjunctly, so that it either goes forwards or backwards in proportion to the excess of one of these causes above the other. Therefore since the force KL in the syzygies is almost twice as great as the force LM in the quadratures, the excess will be on the side of the force KL, and by consequence the line of the apsides will be carried forwards. The truth of this and the foregoing Corollary will be more easily understood by conceiving the system of the two bodies T and P to be surrounded on every side by several bodies S, S, S, &c., disposed about the orbit ESE. For by the actions of these bodies the action of the body T will be diminished on every side, and decrease in more than a duplicate ratio of the distance.

COR. 8. But since the progress or regress of the apsides depends upon the decrease of the centripetal force, that is, upon its being in a greater or less ratio than the duplicate ratio of the distance TP, in the passage of the body from the lower apsis to the upper; and upon a like increase in its return to the lower apsis again; and therefore becomes greatest where the proportion of the force at the upper apsis to the force at the lower apsis recedes farthest from the duplicate ratio of the distances inversely; it is plain, that, when the apsides are in the syzygies, they will, by reason of the subducting force KL or NM − LM, go forward more swiftly; and in the quadratures by the additional force LM go backward more slowly. Because the velocity of the progress or slowness of the regress is continued for a long time; this inequality becomes exceedingly great.

COR. 9. If a body is obliged, by a force reciprocally proportional to the square of its distance from any centre, to revolve in an ellipsis round that centre and afterwards in its descent from the upper apsis to the lower apsis, that force by a perpetual accession of new force is increased in more than a duplicate ratio of the diminished distance; it is manifest that the body being impelled always towards the centre by the perpetual accession of this new force, will incline more towards that centre than if it were urged by that force alone which decreases in a duplicate ratio of the diminished distance, and therefore will describe an orbit interior to that elliptical orbit, and at the lower apsis approaching nearer to the centre than before. Therefore the orbit by the accession of this new force will become more eccentrical. If now, while the body is returning from the lower to the upper apsis, it should decrease by the same degrees by which it increases before the body would return to its first distance; and therefore if the force decreases in

a yet greater ratio, the body, being now less attracted than before, will ascend to a still greater distance, and so the eccentricity of the orbit will be increased still more. Therefore if the ratio of the increase and decrease of the centripetal force be augmented each revolution, the eccentricity will be augmented also; and, on the contrary, if that ratio decrease, it will be diminished.

Now, therefore, in the system of the bodies T, P, S, when the apsides of the orbit PAB are in the quadratures, the ratio of that increase and decrease is least of all, and becomes greatest when the apsides are in the syzygies. If the apsides are placed in the quadratures, the ratio near the apsides is less, and near the syzygies greater, than the duplicate ratio of the distances; and from that greater ratio arises a direct motion of the line of the apsides, as was just now said. But if we consider the ratio of the whole increase or decrease in the progress between the apsides, this is less than the duplicate ratio of the distances. The force in the lower is to the force in the upper apsis in less than a duplicate ratio of the distance of the upper apsis from the focus of the ellipsis to the distance of the lower apsis from the same focus; and, contrariwise, when the apsides are placed in the syzygies, the force in the lower apsis is to the force in the upper apsis in a greater than a duplicate ratio of the distances. For the forces LM in the quadratures added to the forces of the body T compose forces in a less ratio; and the forces KL in the syzygies subducted from the forces of the body T, leave the forces in a greater ratio. Therefore the ratio of the whole increase and decrease in the passage between the apsides is least at the quadratures and greatest at the syzygies; and therefore in the passage of the apsides from the quadratures to the syzygies it is continually augmented, and increases the eccentricity of the ellipses; and in the passage from the syzygies to the quadratures it is perpetually decreasing, and diminishes the eccentricity.

COR. 10. That we may give an account of the errors as to latitude, let us suppose the plane of the orbit EST to remain immovable; and from the cause of the errors above explained, it is manifest, that, of the two forces NM, ML, which are the only and entire cause of them, the force ML acting always in the plane of the orbit PAB never disturbs the motions as to latitude; and that the force NM, when the nodes are in the syzygies, acting also in the same plane of the orbit, does not at that time affect those motions.

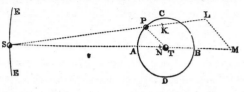

But when the nodes are in the quadratures, it disturbs them very much, and, attracting the body P perpetually out of the plane of its orbit, it diminishes the inclination of the plane in the passage of the body from the quadratures to the syzygies, and again increases the same in the passage from the syzygies to the quadratures. Hence it comes

to pass that when the body is in the syzygies, the inclination is then least of all, and returns to the first magnitude nearly, when the body arrives at the next node. But if the nodes are situate at the octants after the quadratures, that is, between C and A, D and B, it will appear, from what was just now shewn, that in the passage of the body P from either node to the ninetieth degree from thence, the inclination of the plane is perpetually diminished; then, in the passage through the next 45 degrees to the next quadrature, the inclination is increased; and afterwards, again, in its passage through another 45 degrees to the next node, it is diminished. Therefore the inclination is more diminished than increased, and is therefore always less in the subsequent node than in the preceding one, And, by a like reasoning, the inclination is more increased than diminished when the nodes are in the other octants between A and D, B and C. The inclination, therefore, is the greatest of all when the nodes are in the syzygies. In their passage from the syzygies to the quadratures the inclination is diminished at each appulse of the body to the nodes: and becomes least of all when the nodes are in the quadratures, and the body in the syzygies; then it increases by the same degrees by which it decreased before; and, when the nodes come to the next syzygies, returns to its former magnitude.

COR. 11. Because when the nodes are in the quadratures the body P is perpetually attracted from the plane of its orbit; and because this attraction is made towards S in its passage from the node C through the conjunction A to the node D; and to the contrary part in its passage from the node D through the opposition B to the node C; it is manifest that, in its motion from the node C, the body recedes continually from the former plane CD of its orbit till it comes to the next node; and therefore at that node, being now at its greatest distance from the first plane CD, it will pass through the plane of the orbit EST not in D, the other node of that plane, but in a point that lies nearer to the body S, which therefore becomes a new place of the node *in antecedentia* to its former place. And, by a like reasoning the nodes will continue to recede in their passage from this node to the next. The nodes, therefore, when situate in the quadratures, recede perpetually; and at the syzygies, where no perturbation can be produced in the motion as to latitude, are quiescent: in the intermediate places they partake of both conditions, and recede more slowly; and, therefore, being always either retrograde or stationary, they will be carried backwards, or *in antecedentia*, each revolution.

COR. 12. All the errors described in these corrollaries are a little greater at the conjunction of the bodies P, S, than at their opposition; because the generating forces NM and ML are greater.

COR. 13. And since the causes and proportions of the errors and variations mentioned in these Corollaries do not depend upon the magnitude of the body S, it follows that all things before demonstrated will happen, if the magnitude of the body S

be imagined so great as that the system of the two bodies P and T may revolve about it. And from this increase of the body S, and the consequent increase of its centripetal force, from which the errors of the body P arise, it will follow that all these errors, at equal distances, will be greater in this case, than in the other where the body S revolves about the system of the bodies P and T.

COR. 14. But since the forces NM, ML, when the body S is exceedingly distant, are very nearly as the force SK and the ratio PT to ST conjunctly; that is, if both the distance PT and the absolute force of the body S be given, as reciprocally; and since those forces NM, ML are the causes of all the errors and effects treated of in the foregoing Corollaries; it is manifest that all those effects, if the system of bodies T and P continue as before, and only the distance ST and the absolute force of the body S be changed, will be very nearly in a ratio compounded of the direct ratio of the absolute force of the body S, and the triplicate inverse ratio of the distance ST. Hence if the system of bodies T and P revolve about a distant body S, those forces NM, ML, and their effects, will be (by Cor. 2 and 6, Prop IV) reciprocally in a duplicate ratio of the periodical time. And thence, also, if the magnitude of the body S be proportional to its absolute force, those forces, NM, ML, and their effects, will be directly as the cube of the apparent diameter of the distant body S viewed from T, and so *vice versa*. For these ratios are the same as the compounded ratio above mentioned.

COR. 15. And because if the orbits ESE and PAB, retaining their figure, proportions, and inclination to each other, should alter their magnitude; and the forces of the bodies S and T should either remain, or be changed in any given ratio; these forces (that is, the force of the body T, which obliges the body P to deflect from a rectilinear course into the orbit PAB, and the force of the body S, which causes the body P to deviate from that orbit) would act always in the same manner, and in the same proportion; it follows, that all the effects will be similar and proportional, and the times of those effects proportional also; that is, that all the linear errors will be as the diameters of the orbits, the angular errors the same as before; and the times of similar linear errors, or equal angular errors, as the periodical times of the orbits.

COR. 16. Therefore if the figures of the orbits and their inclination to each other be given, and the magnitudes, forces, and distances of the bodies be any how changed, we may, from the errors and times of those errors in one case, collect very nearly the errors and times of the errors in any other case. But this may be done more expeditiously by the following method. The forces NM, ML, other things remaining unaltered, are as the radius TP; and their periodical effects (by Cor. 2, Lem. X) are as the forces and the square of the periodical time of the body P conjunctly. These are the linear errors of the body P; and hence the angular errors as they appear from the centre T (that is, the motion of the apsides and of the nodes, and all the apparent errors as to longitude

and latitude) are in each revolution of the body P as the square of the time of the revolution, very nearly. Let these ratios be compounded with the ratios in Cor. 14, and in any system of bodies T, P, S, where P revolves about T very near to it, and T revolves about S at a great distance, the angular errors of the body P, observed from the centre T, will be in each revolution of the body P as the square of the periodical time of the body P directly, and the square of the periodical time of the body T inversely. And therefore the mean motion of the line of the apsides will be in a given ratio to the mean motion of the nodes; and both those motions will be as the periodical time of the body P directly, and the square of the periodical time of the body T inversely. The increase or diminution of the eccentricity and inclination of the orbit PAB makes no sensible variation in the motions of the apsides and nodes, unless that increase or diminution be very great indeed.

COR. 17. Since the line LM becomes sometimes greater and sometimes less than the radius PT, let the mean quantity of the force LM be expressed by that radius PT; and then that mean force will be to the mean force SK or SN (which may be also expressed by ST) as the length PT to the length ST. But the mean force SN or ST, by which the body T is retained in the orbit it describes about S, is to the force with which

the body P is retained in its orbit about T in a ratio compounded of the ratio of the radius ST to the radius PT, and the duplicate ratio of the periodical time of the body P

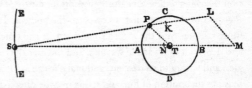

about T to the periodical time of the body T about S. And, *ex aequo*, the mean force LM is to the force by which the body P is retained in its orbit about T (or by which the same body P might revolve at the distance PT in the same periodical time about any immovable point T) in the same duplicate ratio of the periodical times. The periodical times therefore being given, together with the distance PT, the mean force LM is also given; and that force being given, there is given also the force MN, very nearly, by the analogy of the lines PT and MN.

COR. 18. By the same laws by which the body P revolves about the body T, let us suppose many fluid bodies to move round T at equal distances from it; and to be so numerous that they may all become contiguous to each other, so as to form a fluid annulus, or ring, of a round figure, and concentrical to the body T; and the several parts of this annulus, performing their motions by the same law as the body P, will draw nearer to the body T, and move swifter in the conjunction and opposition of themselves and the body S, than in the quadratures. And the nodes of this annulus, or its intersections with the plane of the orbit of the body S or T, will rest at the syzygies but

out of the syzygies they will be carried backward, or *in antecedentia* with the greatest swiftness in the quadratures, and more slowly in other places. The inclination of this annulus also will vary, and its axis will oscillate each revolution, and when the revolution is completed will return to its former situation, except only that it will be carried round a little by the praecession of the nodes.

COR. 19. Suppose now the sphaerical body T, consisting of some matter not fluid, to be enlarged, and to extend itself on every side as far as that annulus, and that a channel were cut all round its circumference containing water; and that this sphere revolves uniformly about its own axis in the same periodical time. This water being accelerated and retarded by turns (as in the last Corollary), will be swifter at the syzygies, and slower at the quadratures, than the surface of the globe, and so will ebb and flow in its channel after the manner of the sea. If the attraction of the body S were taken away, the water would acquire no motion of flux and reflux by revolving round the quiescent centre of the globe. The case is the same of a globe moving uniformly forwards in a right line, and in the mean time revolving about its centre (by Cor. 5 of the Laws of Motion), and of a globe uniformly attracted from its rectilinear course (by Cor. 6, of the same Laws). But let the body S come to act upon it, and by its unequable attraction the water will receive this new motion; for there will be a stronger attraction upon that part of the water that is nearest to the body, and a weaker upon that part which is more remote. And the force LM will attract the water downwards at the quadratures, and depress it as far as the syzygies; and the force KL will attract it upwards in the syzygies, and withhold its descent, and make it rise as far as the quadratures; except only in so far as the motion of flux and reflux may be directed by the channel of the water, and be a little retarded by friction.

COR. 20. If, now, the annulus becomes hard, and the globe is diminished, the motion of flux and reflux will cease; but the oscillating motion of the inclination and the praecession of the nodes will remain. Let the globe have the same axis with the annulus, and perform its revolutions in the same times, and at its surface touch the annulus within, and adhere to it; then the globe partaking of the motion of the annulus, this whole compages will oscillate, and the nodes will go backward, for the globe, as we shall show presently, is perfectly indifferent to the receiving of all impressions. The greatest

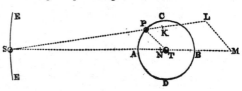

angle of the inclination of the annulus single is when the nodes are in the syzygies. Thence in the progress of the nodes to the quadratures, it endeavours to diminish its inclination, and by that endeavour impresses a motion upon the whole globe.

The globe retains this motion impressed, till the annulus by a contrary endeavour destroys that motion, and impresses a new motion in a contrary direction. And by this means the greatest motion of the decreasing inclination happens when the nodes are in the quadratures, and the least angle of inclination in the octants after the quadratures; and, again, the greatest motion of reclination happens when the nodes are in the syzygies; and the greatest angle of reclination in the octants following. And the case is the same of a globe without this annulus, if it be a little higher or a little denser in the equatorial than in the polar regions; for the excess of that matter in the regions near the equator supplies the place of the annulus. And though we should suppose the centripetal force of this globe to be any how increased, so that all its parts were to tend downwards, as the parts of our earth gravitate to the centre, yet the phaenomena of this and the preceding Corollary would scarce be altered; except that the places of the greatest and least height of the water will be different; for the water is now no longer sustained and kept in its orbit by its centrifugal force, but by the channel in which it flows. And, besides, the force LM attracts the water downwards most in the quadratures, and the force KL or NM – LM attracts it upwards most in the syzygies. And these forces conjoined cease to attract the water downwards, and begin to attract it upwards in the octants before the syzygies; and cease to attract the water upwards, and begin to attract the water downwards in the octants after the syzygies. And thence the greatest height of the water may happen about the octants after the syzygies; and the least height about the octants after the quadratures; excepting only so far as the motion of ascent or descent impressed by these forces may by the *vis insita* of the water continue a little longer, or be stopped a little sooner by impediments in its channel.

COR. 21. For the same reason that redundant matter in the equatorial regions of a globe causes the nodes to go backwards, and therefore by the increase of that matter that retrogradation is increased, by the diminution is diminished, and by the removal quite ceases: it follows, that, if more than that redundant matter be taken away, that is, if the globe be either, more depressed, or of a more rare consistence near the equator than near the poles, there will arise a motion of the nodes *in consequentia*.

COR. 22. And thence from the motion of the nodes is known the constitution of the globe. That is, if the globe retains unalterably the same poles, and the motion (of the nodes) be *in antecedentia*, there is a redundance of the matter near the equator; but if *in consequentia*, a deficiency. Suppose a uniform and exactly sphaerical globe to be first at rest in a free space; then by some impulse made obliquely upon its superficies to be driven from its place, and to receive a motion partly circular and partly right forward. Because this globe is perfectly indifferent to all the axes that pass through its centre, nor has a greater propensity to one axis or to one situation of the axis than to any other, it is manifest that by its own force it will never change its axis, or the inclination

of it. Let now this globe be impelled obliquely by a new impulse in the same part of its superficies as before and since the effect of an impulse is not at all changed by its coming sooner or later, it is manifest that these two impulses, successively impressed, will produce the same motion as if they were impressed at the same time; that is, the same motion as if the globe had been impelled by a simple force compounded of them both (by Cor. 2, of the Laws), that is, a simple motion about an axis of a given inclination. And the ease is the same if the second impulse were made upon any other place of the equator of the first motion; and also if the first impulse were made upon any place in the equator of the motion which would be generated by the second impulse alone; and therefore, also, when both impulses are made in any places whatsoever; for these impulses will generate the same circular motion as if they were impressed together, and at once, in the place of the intersections of the equators of those motions, which would be generated by each of them separately. Therefore, a homogeneous and perfect globe will not retain several distinct motions, but will unite all those that are impressed on it, and reduce them into one; revolving, as far as in it lies, always with a simple and uniform motion about one single given axis, with an inclination perpetually invariable. And the inclination of the axis, or the velocity of the rotation, will not be changed by centripetal force. For if the globe be supposed to be divided into two hemispheres, by any plane whatsoever passing through its own centre, and the centre to which the force is directed, that force will always urge each hemisphere equally; and therefore will not incline the globe any way as to its motion round its own axis. But let there be added any where between the pole and the equator a heap of new matter like a mountain, and this, by its perpetual endeavour to recede from the centre of its motion, will disturb the motion of the globe, and cause its poles to wander about its superficies, describing circles about themselves and their opposite points. Neither can this enormous evagation of the poles be corrected, unless by placing that mountain either in one of the poles; in which case, by Cor. 21, the nodes of the equator will go forwards; or in the equatorial regions, in which case, by Cor. 20, the nodes will go backwards; or, lastly, by adding on the other side of the axis anew quantity of matter, by which the mountain may be balanced in its motion; and then the nodes will either go forwards or backwards, as the mountain and this newly added matter happen to be nearer to the pole or to the equator.

PROPOSITION LXVII. THEOREM XXVII.

THE SAME LAWS OF ATTRACTION BEING SUPPOSED, I SAY, THAT THE EXTERIOR BODY S DOES, BY RADII DRAWN TO THE POINT O, THE COMMON CENTRE OF GRAVITY OF THE INTERIOR BODIES P AND T, DESCRIBE ROUND THAT CENTRE

AREAS MORE PROPORTIONAL TO THE TIMES, AND AN ORBIT
MORE APPROACHING TO THE FORM OF AN ELLIPSIS HAVING
ITS FOCUS IN THAT CENTRE, THAN IT CAN DESCRIBE ROUND
THE INNERMOST AND GREATEST BODY T BY RADII DRAWN
TO THAT BODY.

For the attractions of the body S towards T and P compose its absolute attraction, which is more directed towards
O, the common centre of gravity of the bodies T and P,
than it is to the greatest body T; and which is more in a
reciprocal proportion to the square of the distance SO,
than it is to the square of the distance ST; as will easily
appear by a little consideration.

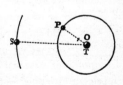

PROPOSITION LXVIII. THEOREM XXVIII.

THE SAME LAWS OF ATTRACTION SUPPOSED, I SAY, THAT
THE EXTERIOR BODY S WILL, BY RADII DRAWN TO O, THE
COMMON CENTRE OF GRAVITY OF THE INTERIOR BODIES P
AND T, DESCRIBE ROUND THAT CENTRE AREAS MORE
PROPORTIONAL TO THE TIMES, AND AN ORBIT MORE
APPROACHING TO THE FORM OF AN ELLIPSIS HAVING ITS
FOCUS IN THAT CENTRE, IF THE INNERMOST AND GREATEST
BODY BE AGITATED BY THESE ATTRACTIONS AS WELL AS THE
REST, THAN IT WOULD DO IF THAT BODY WERE EITHER AT
REST AS NOT ATTRACTED, OR WERE MUCH MORE OR MUCH
LESS ATTRACTED, OR MUCH MORE OR MUCH LESS AGITATED.

This may be demonstrated after the same manner as Prop. LXVI, but by a more
prolix reasoning, which I therefore pass over. It will be sufficient to consider it after
this manner. From the demonstration of the last Proposition it is plain that the centre, towards which the body S is urged by the two forces conjunctly, is very near to
the common centre of gravity of those two other bodies. If this centre were to coincide with that common centre, and moreover the common
centre of gravity of all the three bodies were at rest, the body
S on one side, and the common centre of gravity of the
other two bodies on the other side, would describe true
ellipses about that quiescent common centre. This appears

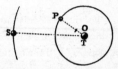

from Cor. 2, Prop LVIII, compared with what was demonstrated in Prop. LXIV, and
LXV. Now this accurate elliptical motion will be disturbed a little by the distance of

the centre of the two bodies from the centre towards which the third body S is attracted. Let there be added, moreover, a motion to the common centre of the three, and the perturbation will be increased yet more. Therefore the perturbation is least when the common centre of the three bodies is at rest; that is, when the innermost and greatest body T is attracted according to the same law as the rest are; and is always greatest when the common centre of the three, by the diminution of the motion of the body T, begins to be moved, and is more and more agitated.

COR. And hence if more lesser bodies revolve about the great one, it may easily be inferred that the orbits described will approach nearer to ellipses; and the descriptions of areas will be more nearly equable, if all the bodies mutually attract and agitate each other with accelerative forces that are as their absolute forces directly, and the squares of the distances inversely; and if the focus of each orbit be placed in the common centre of gravity of all the interior bodies (that is, if the focus of the first and innermost orbit be placed in the centre of gravity of the greatest and innermost body; the focus of the second orbit in the common centre of gravity of the two innermost bodies; the focus of the third orbit in the common centre of gravity of the three innermost; and so on), than if the innermost body were at rest, and was made the common focus of all the orbits.

PROPOSITION LXIX. THEOREM XXIX.

IN A SYSTEM OF SEVERAL BODIES A, B, C, D, &C., IF ANY ONE OF THOSE BODIES, AS A, ATTRACT ALL THE REST, B, C, D, &C., WITH ACCELERATIVE FORCES THAT ARE RECIPROCALLY AS THE SQUARES OF THE DISTANCES FROM THE ATTRACTING BODY; AND ANOTHER BODY, AS B, ATTRACTS ALSO THE REST, A, C, D, &C., WITH FORCES THAT ARE RECIPROCALLY AS THE SQUARES OF THE DISTANCES FROM THE ATTRACTING BODY; THE ABSOLUTE FORCES OF THE ATTRACTING BODIES A AND B WILL BE TO EACH OTHER AS THOSE VERY BODIES A AND B TO WHICH THOSE FORCES BELONG.

For the accelerative attractions of all the bodies B, C, D, towards A, are by the supposition equal to each other at equal distances; and in like manner the accelerative attractions of all the bodies towards B are also equal to each other at equal distances. But the absolute attractive force of the body A to the absolute attractive force of the body B as the accelerative attraction of all the bodies towards A to the accelerative attraction of all the bodies towards B at equal distances; and so is also the accelerative

attraction of the body B towards A to the accelerative attraction of the body K towards B. But the accelerative attraction of the body B towards A is to the accelerative attraction of the body A towards B as the mass of the body A to the mass of the body B; because the motive force which (by the 2d, 7th, and 8th Definition) are as the accelerative forces and the bodies attracted conjunctly are here equal to one another by the third Law. Therefore the absolute attractive force of the body A is to the absolute attractive force of the body B as the mass of the body A to the mass of the body B. *Q.E.D.*

COR. 1. Therefore if each of the bodies of the system A, 13, C, D, &c. does singly attract all the rest with accelerative forces that are reciprocally as the squares of the distances from the attracting body, the absolute forces of all those bodies will be to each other as the bodies themselves.

COR. 2. By a like reasoning, if each of the bodies of the system A, B, C, D, &c., does singly attract all the rest with accelerative forces, which are either reciprocally or directly in the ratio of any power whatever of the distances from the attracting body; or which are defined by the distances from each of the attracting bodies according to any common law; it is plain that the absolute forces of those bodies are as the bodies themselves.

COR. 3. In a system of bodies whose forces decrease in the duplicate ratio of the distances, if the lesser revolve about one very great one in ellipses, having their common focus in the centre of that great body, and of a figure exceedingly accurate; and moreover by radii drawn to that great body describe areas proportional to the times exactly; the absolute forces of those bodies to each other will be either accurately or very nearly in the ratio of the bodies. And so on the contrary. This appears from Cor. of Prop. XLVIII, compared with the first Corollary of this Prop.

SCHOLIUM.

These Propositions naturally lead us to the analogy there is between centripetal forces, and the central bodies to which those forces used to be directed; for it is reasonable to suppose that forces which are directed to bodies should depend upon the nature and quantity of those bodies, as we see they do in magnetical experiments. And when such cases occur, we are to compute the attractions of the bodies by assigning to each of their particles its proper force, and then collecting the sum of them all. I here use the word attraction in general for any endeavour, of what kind soever, made by bodies to approach to each other; whether that endeavour arise from the action of the bodies themselves, as tending mutually to or agitating each other by spirits emitted; or whether it arises from the action of the aether or of the air, or of any medium whatsoever, whether corporeal or incorporeal, any how impelling bodies placed therein towards each other.

In the same general sense I use the word impulse, not defining in this treatise the species or physical qualities of forces, but investigating the quantities and mathematical proportions of them; as I observed before in the Definitions. In mathematics we are to investigate the quantities of forces with their proportions consequent upon any conditions supposed; then, when we enter upon physics, we compare those proportions with the phenomena of Nature, that we may know what conditions of those forces answer to the several kinds of attractive bodies. And this preparation being made, we argue more safely concerning the physical species, causes, and proportions of the forces. Let us see, then, with what forces sphaerical bodies consisting of particles endued with attractive powers in the manner above spoken of must act mutually upon one another; and what kind of motions will follow from thence.

SECTION XII.

Of the attractive forces of sphaerical bodies.

PROPOSITION LXX. THEOREM XXX.

IF TO EVERY POINT OF A SPHAERICAL SURFACE THERE TEND
EQUAL CENTRIPETAL FORCES DECREASING IN THE DUPLICATE
RATIO OF THE DISTANCES FROM THOSE POINTS, I SAY, THAT
A CORPUSCLE PLACED WITHIN THAT SUPERFICIES WILL NOT
BE ATTRACTED BY THOSE FORCES ANY WAY.

Let HIKL, be that sphaerical superficies, and P a corpuscle placed within. Through P let there be drawn to this superficies to two lines HK, IL, intercepting very small arcs HI, KL; and because (by Cor. 3, Lem. VII) the triangles HPI, LPK are alike,

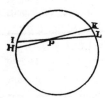

those arcs will be proportional to the distances HP LP; and any particles at HI and KL of the sphaerical superficies, terminated by right lines passing through P, will be in the duplicate ratio of those distances. Therefore the forces of these particles exerted upon the body P are equal between themselves. For the forces are as the particles directly, and the squares of the distances inversely. And these two ratios compose the ratio of equality. The attractions therefore, being made equally towards contrary parts, destroy each other. And by a like reasoning all the attractions through the whole sphaerical superficies are destroyed by contrary attractions. Therefore the body P will not be any way impelled by those attractions. *Q.E.D.*

PROPOSITION LXXI. THEOREM XXXI.

THE SAME THINGS SUPPOSED AS ABOVE, I SAY, THAT A CORPUSCLE PLACED WITHOUT THE SPHAERICAL SUPERFICIES IS ATTRACTED TOWARDS THE CENTRE OF THE SPHERE WITH A FORCE RECIPROCALLY PROPORTIONAL TO THE SQUARE OF ITS DISTANCE FROM THAT CENTRE.

Let AHKB, *ahkb*, be two equal sphaerical superficies described about the centre S, *s*; their diameters AB, *ab*; and let P and *p* be two corpuscles situate without the spheres in those diameters produced. Let there A be drawn from the corpuscles the lines PHK, PIL, *phk*, *pil*, cutting off from the great circles AHB, *ahb*, the equal arcs HK, *hk*, IL, *il*; and to those lines let fall the perpendiculars SD, *sd*, SE, *se*,

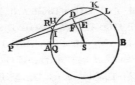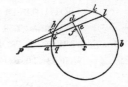

IR, *ir*; of which let SD, *sd*, cut PL, *pl*, in F and *f*. Let fall also to the diameters the perpendiculars IQ, *iq*. Let now the angles DPE, *dpe*, vanish; and because DS and *ds*, ES and *es* are equal, the lines PE, PF, and *pe*, *pf*, and the lineally DF, *df* may be taken for equal; because their last ratio, when the angles DPE, *dpe* vanish together, is the ratio of equality. These things then supposed, it will be, as PI to PF so is RI to DF, and as *pf* to *pi* so is *df* or DF to *ri*; and, *ex aequo*, as PI × *pf* to PF × *pi* so is RI to *ri*, that is (by Cor. 3, Lem. VII), so is the arc IH to the arc *ih*. Again, PI is to PS as IQ to SE, and *ps* to *pi* as *se* or SE to *iq*; and, *ex aequo*, PI × *ps* to PS × *pi* as IQ to *iq*. And compounding the ratios $PI^2 \times pf \times ps$ is to $pi^2 \times PF \times PS$, as IH × IQ to *ih* × *iq*; that is, as the circular superficies which is described by the arc IH, as the semi-circle AKB revolves about the diameter AB, is to the circular superficies described by the arc *ih* as the semi-circle *akb* revolves about the diameter *ab*. And the forces with which these superficies attract the corpuscles P and *p* in the direction of lines tending to those superficies are by the hypothesis as the superficies themselves directly, and the squares of the distances of the superficies from those corpuscles inversely; that is, as *pf* × *ps* to PF × PS. And these forces again are to the oblique parts of them which (by the resolution of forces as in Cor. 2, of the Laws) tend to the centres in the directions of the lines PS, *ps*, as PI to PQ, and *pi* to *pq*; that is (because of the like triangles PIQ and PSF, *piq* and *psf*), as PS to PF and *ps* to *pf*. Thence *ex aequo*, the attraction of the corpuscle P towards S is to the attraction of the corpuscle *p* towards *s* as $\frac{PF \times pf \times ps}{PS}$ is to $\frac{pf \times PF \times PS}{ps}$ that is, as

ps^2 to PS^2. And, by a like reasoning the forces with which the superficies described by

the revolution of the arcs KL, *kl* attract those corpuscles, will be as ps^2 to PS^2. And in the same ratio will be the forces of all the circular superficies into which each of the sphaerical superficies may be divided by taking *sd* always equal to SD, and *se* equal to SE. And therefore, by composition, the forces of the entire sphaerical superficies exerted upon those corpuscles will be in the same ratio.   *Q.E.D.*

PROPOSITION LXXII. THEOREM XXXII.

IF TO THE SEVERAL POINTS OF A SPHERE THERE TEND EQUAL CENTRIPETAL FORCES DECREASING IN A DUPLICATE RATIO OF THE DISTANCES FROM THOSE POINTS; AND THERE BE GIVEN BOTH THE DENSITY OF THE SPHERE AND THE RATIO OF THE DIAMETER OF THE SPHERE TO THE DISTANCE OF THE CORPUSCLE FROM ITS CENTRE; I SAY, THAT THE FORCE WITH WHICH THE CORPUSCLE IS ATTRACTED IS PROPORTIONAL TO THE SEMI-DIAMETER OF THE SPHERE.

For conceive two corpuscles to be severally attracted by two spheres, one by one, the other by the other, and their distances from the centres of the spheres to be proportional to the diameters of the spheres respectively , and the spheres to be resolved into like particles, disposed in a like situation to the corpuscles. Then the attractions of one corpuscle towards the several particles of one sphere will be to the attractions of the other towards as many analogous particles of the other sphere in a ratio compounded of the ratio of the particles, directly, and the duplicate ratio of the distances inversely. But the particles are as the spheres, that is, in a triplicate ratio of the diameters, and the distances are as the diameters; and the first ratio directly with the last ratio taken twice inversely, becomes the ratio of diameter to diameter.   *Q.E.D.*

COR. 1. Hence if corpuscles revolve in circles about spheres composed of matter equally attracting, and the distances from the centres of the spheres be proportional to their diameters, the periodic times will be equal.

COR. 2. And, *vice versa*, if the periodic times are equal, the distances will be proportional to the diameters. These two Corollaries appear from Cor. 3, Prop. IV.

COR. 3. If to the several points of any two solids whatever, of like figure and equal density, there tend equal centripetal forces decreasing in a duplicate ratio of the distances from those points, the forces, with which corpuscles placed in a like situation to those two solids will be attracted by them, will be to each other as the diameters of the solids.

PROPOSITION LXXIII. THEOREM XXXIII.

IF TO THE SEVERAL POINTS OF A GIVEN SPHERE THERE TEND
EQUAL CENTRIPETAL FORCES DECREASING IN A DUPLICATE
RATIO OF THE DISTANCES FROM THE POINTS; I SAY, THAT A
CORPUSCLE PLACED WITHIN THE SPHERE IS ATTRACTED BY A
FORCE PROPORTIONAL TO ITS DISTANCE FROM THE CENTRE.

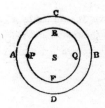

In the sphere ABCD, described about the centre S, let
there be placed the corpuscle P; and about the same centre S, with the interval SP, conceive described an interior
sphere PEQF. It is plain (by Prop. LXX) that the concentric sphaerical superficies, of which the difference AEBF
of the spheres is composed, have no effect at all upon the
body P, their attractions being destroyed by contrary
attractions. There remains, therefore, only the attraction of the interior sphere
PEQF. And (by Prop. LXXII) this is as the distance PS. *Q.E.D.*

SCHOLIUM.

By the superficies of which I here imagine the solids composed, I do not mean
superficies purely mathematical, but orbs so extremely thin that their thickness is as
nothing; that is, the evanescent orbs of which the sphere will at last consist, when the
number of the orbs is increased, and their thickness diminished without end. In like
manner, by the points of which lines, surfaces, and solids are said to be composed, are
to be understood equal particles, whose magnitude is perfectly inconsiderable.

PROPOSITION LXXIV. THEOREM XXXIV.

THE SAME THINGS SUPPOSED, I SAY, THAT A CORPUSCLE
SITUATE WITHOUT THE SPHERE IS ATTRACTED WITH A
FORCE RECIPROCALLY PROPORTIONAL TO THE SQUARE OF
ITS DISTANCE FROM THE CENTRE.

For suppose the sphere to be divided into innumerable concentric sphaerical
superficies, and the attractions of the corpuscle arising from the several superficies will be reciprocally proportional to the square of the distance of the corpuscle from the centre of the sphere (by Prop. LXXI). And, by composition, the sum
of those attractions, that is, the attraction of the corpuscle towards the entire
sphere, will be in the same ratio. *Q.E.D.*

COR. 1. Hence the attractions of homogeneous spheres at equal distances from the centres will be as the spheres themselves. For (by Prop. LXXII) if the distances be proportional to the diameters of the spheres, the forces will be as the diameters. Let the greater distance be diminished in that ratio; and the distances now being equal, the attraction will be increased in the duplicate of that ratio; and therefore will be to the other attraction in the triplicate of that ratio; that is, in the ratio of the spheres.

COR. 2. At any distances whatever the attractions are as the spheres applied to the squares of the distances.

COR. 3. If a corpuscle placed without an homogeneous sphere is attracted by a force reciprocally proportional to the square of its distance from the centre, and the sphere consists of attractive particles, the force of every particle will decrease in a duplicate ratio of the distance from each particle.

PROPOSITION LXXV. THEOREM XXXV.

IF TO THE SEVERAL POINTS OF A GIVEN SPHERE THERE TEND
EQUAL CENTRIPETAL FORCES DECREASING IN A DUPLICATE
RATIO OF THE DISTANCES FROM THE POINTS; I SAY, THAT
ANOTHER SIMILAR SPHERE WILL BE ATTRACTED BY IT WITH A
FORCE RECIPROCALLY PROPORTIONAL TO THE SQUARE OF
THE DISTANCE OF THE CENTRES.

For the attraction of every particle is reciprocally as the square of its distance from the centre of the attracting sphere (by Prop. LXXIV), and is therefore the same as if that whole attracting force issued from one single corpuscle placed in the centre of this sphere. But this attraction is as great as on the other hand the attraction of the same corpuscle would be, if that were itself attracted by the several particles of the attracted sphere with the same force with which they are attracted by it. But that attraction of the corpuscle would be (by Prop. LXXIV) reciprocally proportional to the square of its distance from the centre of the sphere; therefore the attraction of the sphere, equal thereto, is also in the same ratio.   Q.E.D.

COR. 1. The attractions of spheres towards other homogeneous spheres, are as the attracting spheres applied to the squares of the distances of their centres from the centres of those which they attract.

COR. 2. The case is the same when the attracted sphere does also attract. For the several points of the one attract the several points of the other with the same force with which they themselves are attracted by the others again; and therefore since in all attractions (by Law III) the attracted and attracting point are both equally acted on, the force will be doubled by their mutual attractions, the proportions remaining.

COR. 3. Those several truths demonstrated above concerning the motion of bodies about the focus of the conic sections will take place when an attracting sphere is placed in the focus, and the bodies move without the sphere.

COR. 4. Those things which were demonstrated before of the motion of bodies about the centre of the conic sections take place when the motions are performed within the sphere.

PROPOSITION LXXVI. THEOREM XXXVI.

IF SPHERES BE HOWEVER DISSIMILAR (AS TO DENSITY OF MATTER AND ATTRACTIVE FORCE) IN THE SAME RATIO ONWARD FROM THE CENTRE TO THE CIRCUMFERENCE; BUT EVERY WHERE SIMILAR, AT EVERY GIVEN DISTANCE FROM THE CENTRE, ON ALL SIDES ROUND ABOUT; AND THE ATTRACTIVE FORCE OF EVERY POINT DECREASES IN THE DUPLICATE RATIO OF THE DISTANCE OF THE BODY ATTRACTED; I SAY, THAT THE WHOLE FORCE WITH WHICH ONE OF THESE SPHERES ATTRACTS THE OTHER WILL BE RECIPROCALLY PROPORTIONAL TO THE SQUARE OF THE DISTANCE OF THE CENTRES.

Imagine several concentric similar spheres, AB, CD, EF, &c., the innermost of which added to the outermost may compose a matter more dense towards the centre, or subducted from them may leave the same more lax and rare. Then, by Prop. LXXV, these

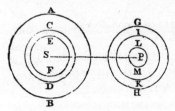

spheres will attract other similar concentric spheres GH, IK, LM, &c., each the other, with forces reciprocally proportional to the square of the distance SP. And, by composition or division, the sum of all those forces, or the excess of any of them above the others; that is, the entire force with which the whole sphere AB (composed of any concentric spheres or of their differences) will attract the whole sphere GH (composed of any concentric spheres or their differences) in the same ratio. Let the number of the concentric spheres be increased *in infinitum*, so that the density of the matter together with the attractive force may, in the progress from the circumference to the centre, increase or decrease according to any given law; and by the addition of matter not attractive, let the deficient density be supplied, that so the spheres may acquire any form desired; and the force with which one of these attracts the other will be still, by the former reasoning, in the same ratio of the square of the distance inversely. *Q.E.D.*

COR. 1. Hence if many spheres of this kind, similar in all respects, attract each other mutually, the accelerative attractions of each to each, at any equal distances of the centres, will be as the attracting spheres.

COR. 2. And at any unequal distances, as the attracting spheres applied to the squares of the distances between the centres.

COR. 3. The motive attractions, or the weights of the spheres towards one another, will be at equal distances of the centres as the attracting and attracted spheres conjunctly; that is, as the products arising from multiplying the spheres into each other.

COR. 4. And at unequal distances, as those products directly, and the squares of the distances between the centres inversely.

COR. 5. These proportions take place also when the attraction arises from the attractive virtue of both spheres mutually exerted upon each other. For the attraction is only doubled by the conjunction of the forces, the proportions remaining as before.

COR. 6. If spheres of this kind revolve about others at rest, each about each : and the distances between the centres of the quiescent and revolving bodies are proportional to the diameters of the quiescent bodies; the periodic times will be equal.

COR. 7. And, again, if the periodic times are equal, the distances will be proportional to the diameters.

COR. 8. All those truths above demonstrated, relating to the motions of bodies about the foci of conic sections, will take place when an attracting sphere, of any form and condition like that above described, is placed in the focus.

COR. 9. And also when the revolving bodies are also attracting spheres of any condition like that above described.

PROPOSITION LXXVII. THEOREM XXXVII.

IF TO THE SEVERAL POINTS OF SPHERES THERE TEND
CENTRIPETAL FORCES PROPORTIONAL LO THE DISTANCES
OF THE POINTS FROM THE ATTRACTED BODIES; I SAY,
THAT THE COMPOUNDED FORCE WITH WHICH TWO
SPHERES ATTRACT EACH OTHER MUTUALLY IS AS THE
DISTANCE BETWEEN THE CENTRES OF THE SPHERES.

CASE. 1. Let AEBF be a sphere; S its centre, P a corpuscle attracted : PASB the axis of the sphere passing through the centre of the corpuscle; EF, *ef* two planes cutting the sphere, and perpendicular to the axis, and equidistant, one on one side, the other on the other, from the centre of the sphere; G and *g* the intersections of the planes and the axis; and H any point in the plane EF. The centripetal force of the point H upon the corpuscle P, exerted in the direction of the line PH, is as the distance PH; and (by

Cor. 2, of the Laws) the same exerted in the direction of the line PG, or towards the centre S, is as the length PG. Therefore the force of all the points in the plane EF (that is, of that whole plane) by which the corpuscle P is attracted towards the centre S is as

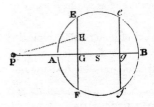

the distance PG multiplied by the number of those points, that is, as the solid contained under that plane EF and the distance PG. And in like manner the force of the plane *ef*, by which the corpuscle P is attracted towards the centre S, is as that plane drawn into its distance Pg, or as the equal plane EF drawn into that distance Pg; and the sum of the forces of both planes as the plane EF drawn into the sum of the distances PG + Pg, that is, as that plane drawn into twice the distance PS of the centre and the corpuscle; that is, as twice the plane EF drawn into the distance PS, or as the sum of the equal planes EF + *ef* drawn into the same distance. And, by a like reasoning, the forces of all the planes in the whole sphere, equi-distant on each side from the centre of the sphere, are as the sum of those planes drawn into the distance PS, that is, as the whole sphere and the distance PS conjunctly.   Q.E.D.

CASE. 2. Let now the corpuscle P attract the sphere AEBF. And, by the same reasoning, it will appear that the force with which the sphere is attracted is as the distance PS.   Q.E.D.

CASE. 3. Imagine another sphere composed of innumerable corpuscles P; and because the force with which every corpuscle is attracted is as the distance of the corpuscle from the centre of the first sphere, and as the same sphere conjunctly, and is therefore the same as if it all proceeded from a single corpuscle situate in the centre of the sphere, the entire force with which all the corpuscles in the second sphere are attracted, that is, with which that whole sphere is attracted, will be the same as if that sphere were attracted by a force issuing from a single corpuscle in the centre of the first sphere; and is therefore proportional to the distance between the centres of the spheres.

Q.E.D.

CASE. 4. Let the spheres attract each other mutually, and the force will be doubled, but the proportion will remain.   Q.E.D.

CASE. 5. Let the corpuscle *p* be placed within the sphere AEBF; and because the force of the Plane *ef* upon the corpuscle is as the solid contained under that plane and the distance *pg*, and the contrary force of the plane EF as the solid contained under that plane and the distance Pg; the force compounded of both will be as the difference of the solids, that is, as the sum of the equal planes drawn into half the difference

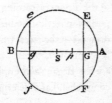

of the distances; that is, as that sum drawn into pS, the distance of the corpuscle from the centre of the sphere. And, by a like reasoning, the attraction of all the planes EF, *ef*, throughout the whole sphere, that is, the attraction of the whole sphere, is conjunctly as the sum of all the planes, or as the whole sphere, and as pS, the distance of the corpuscle from the centre of the sphere. *Q.E.D.*

CASE. 6. And if there be composed a new sphere out of innumerable corpuscles such as p, situate within the first sphere AEBF, it may be proved, as before, that the attraction, whether single of one sphere towards the other, or mutual of both towards each other, will be as the distance pS of the centres. *Q.E.D.*

PROPOSITION LXXVIII. THEOREM XXXVIII.

IF SPHERES IN THE PROGRESS FROM THE CENTRE TO THE CIRCUMFERENCE BE HOWEVER DISSIMILAR AND UNEQUABLE, BUT SIMILAR ON EVERY SIDE ROUND ABOUT AT ALL GIVEN DISTANCES FROM THE CENTRE; AND THE ATTRACTIVE FORCE OF EVERY POINT BE AS THE DISTANCE OF THE ATTRACTED BODY; I SAY, THAT THE ENTIRE FORCE WITH WHICH TWO SPHERES OF THIS KIND ATTRACT EACH OTHER MUTUALLY, IS PROPORTIONAL TO THE DISTANCE BETWEEN THE CENTRES OF THE SPHERES.

This is demonstrated from the foregoing Proposition, in the same manner as Proposition LXXVI was demonstrated from Proposition LXXV.

COR. Those things that were above demonstrated in Prop, X and LXIV, of the motion of bodies round the centres of conic sections, take place when all the attractions are made by the force of sphaerical bodies of the condition above described, and the attracted bodies are spheres of the same kind.

SCHOLIUM.

I have now explained the two principal cases of attractions; to wit, when the centripetal forces decrease in a duplicate ratio of the distances, or increase in a simple ratio of the distances, causing the bodies in both cases to revolve in conic sections, and composing sphaerical bodies whose centripetal forces observe the same law of increase or decrease in the recess from the centre as the forces of the particles themselves do; which is very remarkable. It would be tedious to run over the other cases, whose conclusions are less elegant and important, so particularly as I have done these. I choose rather to comprehend and determine them all by one general method as follows.

LEMMA XXIX.

IF ABOUT THE CENTRE S THERE BE DESCRIBED ANY CIRCLE AS
AEB, AND ABOUT THE CENTRE P THERE BE ALSO DESCRIBED
TWO CIRCLES EF, ef, CUTTING THE FIRST IN E AND e, AND
THE LINE PS IN F AND f; AND THERE BE LET FALL TO PS THE
PERPENDICULARS ED, ed; I SAY, THAT IF THE DISTANCE OF
THE ARCS EF, ef BE SUPPOSED TO BE INFINITELY DIMINISHED,
THE LAST RATIO OF THE LINE D*d* TO THE EVANESCENT LINE
F*f* IS THE SAME AS THAT OF THE LINE PE TO THE LINE PS.

For if the line P*e* cut the arc EF in *q*, and the right line E*e*, which coincides with the evanescent arc E*e*, be produced, and meet the right line PS in T; and there be let fall from S to PE the

perpendicular SG; then, because
of the like triangles DTE, *d*T*e*,
DES, it will be as D*d* to E*e* so DT
to TE, or DE to ES; and because
the triangles, E*eq*, ESG (by Lem.
VIII, and Cor. 3, Lem. VII) are
similar, it will be as E*e* to *eq* or F*f*
so ES to SG; and, *ex aequo*, as D*d*

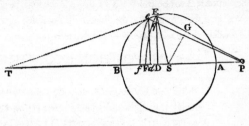

to F*f* so DE to SG; that is (because of the similar triangles PDE, PGS), so is PE to PS.   Q.E.D.

PROPOSITION LXXIX. THEOREM XXXIX.

SUPPOSE A SUPERFICIES AS EF*fe* TO HAVE ITS BREADTH
INFINITELY DIMINISHED, AND TO BE JUST VANISHING; AND
THAT THE SAME SUPERFICIES BY ITS REVOLUTION ROUND THE
AXIS PS DESCRIBES A SPHERICAL CONCAVO-CONVEX SOLID, TO
THE SEVERAL EQUAL PARTICLES OF WHICH THERE TEND EQUAL
CENTRIPETAL FORCES; I SAY, THAT THE FORCE WITH WHICH
THAT SOLID ATTRACTS A CORPUSCLE SITUATE IN P IS IN A
RATIO COMPOUNDED OF THE RATIO OF THE SOLID DE2 × F*f*
AND THE RATIO OF THE FORCE WITH WHICH THE GIVEN PARTICLE
IN THE PLANE F*f* WOULD ATTRACT THE SAME CORPUSCLE.

For if we consider, first, the force of the sphaerical superficies FE which is generated by the revolution of the arc FE, and is cut any where, as in *r*, by the line *de*, the annular part of the superficies generated by the revolution of the arc *r*E will be as the lineola D*d*, the radius of the sphere PE remainimg the same; as Archimedes has

demonstrated in his Book of the Sphere and Cylinder. And the force of this superficies exerted in the direction of the lines PE or Pr situate all round in the conical superficies, will be as this annular superficies itself; that is as the lineola Dd, or, which is the same, as the rectangle under the given radius PE of the sphere and the lineola Dd; but that force, exerted in the direction of the line PS tending to the centre S, will be less in the ratio PD to PE, and therefore will be as PD × Dd. Suppose now the line DF to be divided into innumerable little equal particles, each of which call Dd, and then the superficies FE will be divided into so many equal annuli, whose forces will be as the

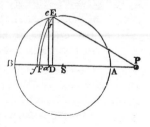

sum of all the rectangles PD × Dd, that is, as $\frac{1}{2}$PF^2 – $\frac{1}{2}$PD^2 and therefore as DE^2 × Ff. Let now the superficies FE be drawn into the altitude Ff; and the force of the solid EFfe exerted upon the corpuscle P will be as DE^2 × Ff; that is, if the force be given which any given particle as Ff exerts upon the corpuscle P at the distance PF. But if that force be not given, the force of the solid EFfe will be as the solid DE^2 × Ff and that force not given, conjunctly. *Q.E.D.*

PROPOSITION LXXX. THEOREM XL.

IF TO THE SEVERAL EQUAL PARTS OF A SPHERE ABE DESCRIBED ABOUT THE CENTRE S THERE TEND EQUAL CENTRIPETAL FORCES; AND FROM THE SEVERAL POINTS D IN THE AXIS OF THE SPHERE AB IN WHICH A CORPUSCLE, AS P, IS PLACED, THERE BE ERECTED THE PERPENDICULARS DE MEETING THE SPHERE IN E, AND IF IN THOSE PERPENDICULARS THE LENGTHS DN BE TAKEN AS THE QUANTITY $\frac{\mathrm{DE}^2 \times \mathrm{PS}}{\mathrm{PE}}$, AND AS THE FORCE WHICH A PARTICLE OF THE SPHERE

SITUATE IN THE AXIS EXERTS AT THE DISTANCE PE UPON THE CORPUSCLE P CONJUNCTLY; I SAY, THAT THE WHOLE FORCE WITH WHICH THE CORPUSCLE P IS ATTRACTED TOWARDS THE SPHERE IS AS THE AREA ANB, COMPREHENDED UNDER THE AXIS OF THE SPHERE AB, AND THE CURVE LINE ANB, THE LOCUS OF THE POINT N.

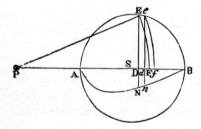

For supposing the construction in the last Lemma and Theorem to stand, conceive the axis of the sphere AB to be divided into innumerable equal particles Dd, and the whole sphere to be divided into so many sphaerical concavo-convex laminae EFfe; and erect the perpendicular dn. By the last

Theorem, the force with which the laminae EFfe attracts the corpuscle P is as $DE^2 \times$ Ff and the force of one particle exerted at the distance PE or PF, conjunctly. But (by the last Lemma) Dd is to Ff as PE to PS, and therefore Ff is equal to $\frac{PS \times Dd}{PE}$; and DE^2

x Ff is equal to $Dd \times \frac{DE^2 \times PS}{PE}$; and therefore the force of the laminae EFfe is as

$Dd \times \frac{DE^2 \times PS}{PE}$ and the force of a particle exerted at the distance PF conjunctly; that is, by the supposition, as DN $\times$ Dd, or as the evanescent area DNnd. Therefore the forces of all the laminae, exerted upon the corpuscle P are as all the areas DNnd, that is, the whole force of the sphere will be as the whole area ANB.   *Q.E.D.*

COR. 1. Hence if the centripetal force tending to the several particles remain always the same at all distances, and DN be made as $\frac{DE^2 \times PS}{PE}$, the whole force with which the

corpuscle is attracted by the sphere is as the area ANB.

COR. 2. If the centripetal force of the particles be reciprocally as the distance of the corpuscle attracted by it, and DN be made as $\frac{DE^2 \times PS}{PE^2}$, the force with which the

corpuscle P is attracted by the whole sphere will be as the area ANB.

COR. 3. If the centripetal force of the particles be reciprocally as the cube of the distance of the Corpuscle attracted by it, and DN be made as $\frac{DE^2 \times PS}{PE^4}$, the force with

which the corpuscle is attracted by the whole sphere will be as the area ANB.

COR. 4. And universally if the centripetal force tending to the several particles of the sphere be supposed to be reciprocally as the quantity V; and DN be made as $\frac{DE^2 \times PS}{PE \times V}$; the force with which a corpuscle is attracted by the whole sphere will be as

the area ANB.

PROPOSITION LXXXI. PROBLEM XLI.

THE THINGS REMAINING AS ABOVE, IT IS REQUIRED TO MEASURE THE AREA ANB.

From the point P let there be drawn the right line PH touching the sphere in H; and to the axis PAB, letting fall the perpendicular HI, bisect PI in L; and (by Prop. XII, Book II, Elem.) PE^2 is equal to $PS^2 + SE^2 + 2PSD$. But because the triangles SPH, SHI are alike, SE^2 or SH^2 is equal to the rectangle PSI. Therefore PE^2 is equal to the rectangle contained under PS and PS + SI + 2SD; that is, under PS and 2LS + 2SD; that is, under PS and 2LD. Moreover DE^2 is equal to $SE^2 - SD^2$, or $SE^2 - LS^2 + 2SLD - LD^2$,

that is, 2SLD – LD^2 – ALB. For LS^2 – SE^2 or LS^2 – SA^2 (by Prop. VI, Book II, Elem.) is equal to the rectangle ALB. Therefore if instead of DE^2, we write 2SLD – LD^2 – ALB, the quantity $\frac{DE^2 \times PS}{PE \times V}$, which (by Cor. 4 of the foregoing,

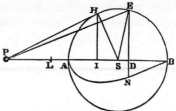

Prop.) is as the length of the ordinate DN, will now resolve itself into three parts $\frac{2SLD \times PS}{PE \times V} - \frac{LD^2 \times PS}{PE \times V} - \frac{ALB \times PS}{PE \times V}$ where if instead of V we

write the inverse ratio of the centripetal force, and instead of PE the mean proportional between PS and 2LD, those three parts will become ordinates to so many curve lines, whose areas are discovered by the common methods.   *Q.E.D.*

EXAMPLE 1. If the centripetal force tending to the several particles of the sphere be reciprocally as the distance; instead of V write PE the distance, then 2PS × LD for PE^2; and DN will become as $SL - \frac{1}{2}LD - \frac{ALB}{LD}$. Suppose DN equal to its double

$2SL - LD - \frac{ALB}{LD}$; and 2SL the given part of the ordinate drawn into the length AB will describe the rectangular area 2SL 3 AB; and the indefinite part LD, drawn perpendicularly into the same length with a continued motion, in such sort as in its motion one way or another it may either by increasing or decreasing remain always equal to the length LD, will describe the area $\frac{LB^2 - LA^2}{2}$, that is, the area SL × AB; which taken from the former area 2SL × AB, leaves the area SL × AB. But the third part $\frac{ALB}{LD}$ drawn after the same manner with a continued motion perpendicularly into the same length, will describe the area of an hyperbola, which subducted from the area SL × AB will leave ANB the area sought. Whence arises this construction of the Problem. At the points, L, A, B, erect the perpendiculars L*l*, A*a*, B*b*; making A*a* equal to LB, and B*b* equal to LA. Making L*l* and LB asymptotes, describe through the points *a*, *b*, the hyperbolic

curve *ab*. And the chord *ba* being drawn, will inclose the area *aba* equal to the area sought ANB.

EXAMPLE 2. If the centripetal force tending to the several particles of the sphere be reciprocally as the cube of the distance, or (which is the same thing) as that cube applied to any given plane; write $\frac{PE^3}{2AS^2}$ for V, and 2PS × LD for PE^2; and DN will become as $\frac{SL \times AS^2 AS^2 ALB \times AS^2}{PS \times LD 2PS 2PS \times LD^2}$ that is (because PS, AS, SI are continually

proportional), as $\frac{LSI}{LD} - \frac{1}{2}SI - \frac{ALB \times SI}{2LD^2}$. If we draw then these three parts into the length

AB, the first $\frac{LSI}{LD}$ will generate the area of an hyperbola; the second $\frac{1}{2}SI$ the area $\frac{1}{2}AB \times SI$; the third $\frac{ALB \times SI}{2LD^2}$ the area $\frac{ALB \times SI}{2LA}\frac{ALB \times SI}{2LB}$, that is,

$\frac{1}{2}AB \times SI$. From the first subduct the sum of the second and third, and there will remain ANB, the area sought. Whence arises this construction of the problem. At the points L, A, S, B, erect the perpendiculars Ll Aa Ss, Bb, of which suppose Ss equal to SI; and through the point s, to the asymptotes Ll, LB, describe the hyperbola asb meeting the perpendiculars Aa, Bb, in a and b; and the rectangle 2ASI, subducted from the hyberbolic area AasbB, will leave ANB the area sought.

EXAMPLE 3. If the centripetal force tending to the several particles of the spheres decrease in a quadruplicate ratio of the distance from the particles; write $\frac{PE^4}{2AS^3}$ for V, then $\sqrt{2PS+LD}$ for PE, and DN will become as

$\frac{SI^2 \times SL}{\sqrt{2SI}} \times \frac{1}{\sqrt{LD^3}} - \frac{SI^2}{2\sqrt{2SI}} \times \frac{1}{\sqrt{LD}} - \frac{SI^2 \times ALB}{2\sqrt{2SI}} \times \frac{1}{\sqrt{LD^5}}$. These three parts drawn into the length AB, produce so many areas, viz.

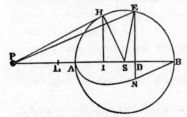

$\frac{2SI^2 \times SL}{\sqrt{2SI}}$ into $\frac{1}{\sqrt{LA}} - \frac{1}{\sqrt{LB}}$; $\frac{SI^2}{\sqrt{2SI}}$ into $\overline{\sqrt{LB} - \sqrt{LA}}$; and $\frac{SI^2 \times ALB}{3\sqrt{2SI}}$ into $\frac{1}{\sqrt{LA^3}} - \frac{1}{\sqrt{LB^3}}$. And these after due reduction come forth $\frac{2SI^2 \times SL}{LI}$, SI^2, and $SI^2 + \frac{2SI^3}{3LI}$. And these by subducting the last from the first, become $\frac{4SI^3}{3LI}$.

Therefore the entire force with which the corpuscle P is attracted towards the centre of the sphere is as s $\frac{SI^3}{PI}$, that is, reciprocally as $PS^3 \times PI$ Q.E.I.

By the same method one may determine the attraction of a corpuscle situate within the sphere, but more expeditiously by the following Theorem.

PROPOSITION LXXXII. THEOREM XLI.

IN A SPHERE DESCRIBED ABOUT THE CENTRE S WITH THE INTERVAL SA, IF THERE BE TAKEN SI, SA, SP CONTINUALLY PROPORTIONAL; I SAY, THAT THE ATTRACTION OF A CORPUSCLE WITHIN THE SPHERE

IN ANY PLACE I IS TO ITS ATTRACTION WITHOUT THE SPHERE IN THE
PLACE P IN A RATIO COMPOUNDED OF THE SUBDUPLICATE RATIO OF
IS, PS, THE DISTANCES FROM THE CENTRE, AND THE SUBDUPLICATE
RATIO OF THE CENTRIPETAL FORCES TENDING TO THE CENTRE IN
THOSE PLACES P AND I.

As if the centripetal forces of the particles of the sphere be reciprocally as the distances
of the corpuscle attracted by them; the force with
which the corpuscle situate in I is attracted by the
entire sphere will be to the force with which it is
attracted in P in a ratio compounded of the
subduplicate ratio of the distance SI to the distance SP, and the subduplicate ratio of the centripetal force in the place I arising from any particle in the centre to the centripetal force in the
place P arising from the same particle in the cen-

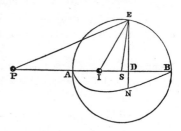

tre; that is, in the subduplicate ratio of the distances SI, SP to each other reciprocally. These
two subduplicate ratios compose the ratio of equality, and therefore the attractions in I and
P produced by the whole sphere are equal. By the like calculation, if the forces of the particles of the sphere are reciprocally in a duplicate ratio of the distances, it will be found that
the attraction in I is to the attraction in P as the distance SP to the semi-diameter SA of
the sphere. If those forces are reciprocally in a triplicate ratio of the distances, the attractions in I and P will be to each other as SP^2 to SA^2; if in a quadruplicate ratio, as SP^3 to
SA^3. Therefore since the attraction in P was found in this last case to be reciprocally as PS^3
$\times$ PI, the attraction in I will be reciprocally as $SA^3 \times$ PI, that is, because SA^3 is given reciprocally as PI. And the progression is the same *in infinitum*. The demonstration of this
Theorem is as follows:

The things remaining as above constructed, and a corpuscle being in any place P,
the ordinate DN was found to be as $\frac{DE^2 \times PS}{PE \times V}$. Therefore if IE be drawn, that ordinate
for any other place of the corpuscle, as I, will become (*mutatis mutandis*) as $\frac{DE^2 \times IS}{IE \times V}$.
Suppose the centripetal forces flowing from any point of the sphere, as E, to be to each
other at the distances IE and PE as PE^n to IE^n (where the number n denotes the index
of the powers of PE and IE), and those ordinates will become as $\frac{DE^2 \times PS}{PE \times PE^n}$ and $\frac{DE^2 \times IS}{IE \times IE^n}$
whose ratio to each other is as $PS \times IE \times IE^N$ to $IS \times PE \times PE^N$.
Because SI, SE, SP are in continued proportion, the triangles SPE, SEI are alike; and
thence IE is to PE as IS to SE or SA. For the ratio of IE to PE write the ratio of IS to
SA; and the ratio of the ordinates becomes that of $PS \times IE^N$ to $SA \times PE^N$. But the
ratio of PS to SA is subduplicate of that of the distances PS, SI; and the ratio of IE^N

to PE^n (because IE is to PE as IS to SA) is subduplicate of that of the forces at the distances PS, IS. Therefore the ordinates, and consequently the areas which the ordinates describe, and the attractions proportional to them, are in a ratio compounded of those subduplicate ratios.   *Q.E.D.*

PROPOSITION LXXXIII. PROBLEM XLII.

TO FIND THE FORCE WITH WHICH A CORPUSCLE PLACED IN THE CENTRE OF A SPHERE IS ATTRACTED TOWARDS ANY SEGMENT OF THAT SPHERE WHATSOEVER.

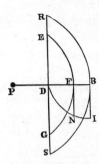

Let P be a body in the centre of that sphere and RBSD a segment thereof contained under the plane RDS, and the sphaerical superficies RBS. Let DB be cut in F by a sphaerical superficies EFG described from the centre P, and let the segment be divided into the parts BREFGS, FEDG. Let us suppose that segment to be not a purely mathematical but a physical superficies, having some, but a perfectly inconsiderable thickness. Let that thickness be called O, and (by what *Archimedes* has demonstrated) that superficies will be as PF × DF x 0. Let us suppose besides the attractive forces of the particles of the sphere to be reciprocally as that power of the distances, of which n is index; and the force with which the superficies EFG attracts the body P will be (by Prop. LXXIX) as $\frac{DE^2 \times O}{PF^n}$, that is, as $\frac{2DF \times O}{PF^{n}-1} - \frac{DF^2 \times O}{PF^n}$. Let the perpendicular FN drawn into O be proportional to this quantity; and the curvilinear area BDI, which the ordinate FN, drawn through the length DB with a continued motion will describe, will be as the whole force with which the whole segment RBSD attracts the body P.   *Q.E.I.*

PROPOSITION LXXXIV. PROBLEM XLIII.

TO FIND THE FORCE WITH WHICH A CORPUSCLE, PLACED WITHOUT THE CENTRE OF A SPHERE IN THE AXIS OF ANY SEGMENT, IS ATTRACTED BY THAT SEGMENT.

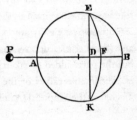

Let the body P placed in the axis ADB of the segment EBK be attracted by that segment. About the centre P, with the interval PE, let the sphaerical superficies EFK be described; and let it divide the segment into two parts EBKFE and EFKDE. Find the force of the first of those parts by Prop. LXXXI, and the force of the latter part by Prop. LXXXIII, and the sum of the forces will be the force of the whole segment EBKDE.   *Q.E.I.*

SCHOLIUM.

The attractions of sphaerical bodies being now explained, it comes next in order to treat of the laws of attraction in other bodies consisting in like manner of attractive particles; but to treat of them particularly is not necessary to my design. It will be sufficient to subjoin some general propositions relating to the forces of such bodies, and the motions thence arising, because the knowledge of these will be of some little use in philosophical inquiries.

SECTION XIII.

Of the attractive forces of bodies which are not of a sphaerical figure.

PROPOSITION. LXXXV. THEOREM XLII.

IF A BODY BE ATTRACTED BY ANOTHER, AND ITS ATTRACTION
BE VASTLY STRONGER WHEN IT IS CONTIGUOUS TO THE
ATTRACTING BODY THAN WHEN THEY ARE SEPARATED FROM
ONE ANOTHER BY A VERY SMALL INTERVAL; THE FORCES OF
THE PARTICLES OF THE ATTRACTING BODY DECREASE, IN
THE RECESS OF THE BODY ATTRACTED, IN MORE THAN A
DUPLICATE RATIO OF THE DISTANCE OF THE PARTICLES.

For if the forces decrease in a duplicate ratio of the distances from the particles, the attraction towards a sphaerical body being (by Prop. LXXIV) reciprocally as the square of the distance of the attracted body from the centre of the sphere, will not be sensibly increased by the contact, and it will be still less increased by it, if the attraction, in the recess of the body attracted, decreases in a still less proportion. The proposition, therefore, is evident concerning attractive spheres. And the case is the same of concave sphaerical orbs attracting external bodies. And much more does it appear in orbs that attract bodies placed within them, because there the attractions diffused through the cavities of those orbs are (by Prop. LXX) destroyed by contrary attractions, and therefore have no effect even in the Place of contact. Now if from these spheres and sphaerical orbs we take away any parts remote from the place of contact, and add new parts any where at pleasure, we may change the figures of the attractive bodies at pleasure; but the parts added or taken away, being remote from the place of contact, will cause no remarkable excess of the attraction arising from the contact of the two bodies. Therefore the proposition holds good in bodies of all figures. *Q.E.D.*

PROPOSITION LXXXVI. THEOREM XLIII.

IF THE FORCES OF THE PARTICLES OF WHICH AN
ATTRACTIVE BODY IS COMPOSED DECREASE, IN THE RECESS
OF THE ATTRACTIVE BODY, IN A TRIPLICATE OR MORE THAN
A TRIPLICATE RATIO OF THE DISTANCE FROM THE PARTICLES,
THE ATTRACTION WILL BE VASTLY STRONGER IN THE POINT
OF CONTACT THAN WHEN THE ATTRACTING AND ATTRACTED
BODIES ARE SEPARATED FROM EACH OTHER, THOUGH BY
NEVER SO SMALL AN INTERVAL.

For that the attraction is infinitely increased when the attracted corpuscle comes to touch an attracting sphere of this kind, appears, by the solution of Problem XLI, exhibited in the second and third Examples. The same will also appear (by comparing those Examples and Theorem XLI together) of attractions of bodies made towards concavo-convex orbs, whether the attracted bodies be placed without the orbs, or in the cavities within them. And by adding to or taking from those spheres and orbs any attractive matter any where without the place of contact, so that the attractive bodies may receive any assigned figure, the Proposition will hold good of all bodies universally. *Q.E.D.*

PROPOSITION LXXXVII. THEOREM XLIV.

IF TWO BODIES SIMILAR TO EACH OTHER, AND CONSISTING
OF MATTER EQUALLY ATTRACTIVE, ATTRACT SEPARATELY
TWO CORPUSCLES PROPORTIONAL TO THOSE BODIES, AND IN
A LIKE SITUATION TO THEM, THE ACCELERATIVE
ATTRACTIONS OF THE CORPUSCLES TOWARDS THE ENTIRE
BODIES WILL BE AS THE ACCELERATIVE ATTRACTIONS OF THE
CORPUSCLES TOWARDS PARTICLES OF THE BODIES PROPORTIONAL
TO THE WHOLES, AND ALIKE SITUATED IN THEM.

For if the bodies are divided into particles proportional to the wholes, and alike situated in there, it will be, as the attraction towards any particle of one of the bodies to the attraction towards the correspondent particle in the other body, so are the attractions towards the several particles of the first body, to the attractions towards the several correspondent particles of the other body; and, by composition, so is the attraction towards the first whole body to the attraction towards the second whole body. *Q.E.D.*

COR. 1. Therefore if, as the distances of the corpuscles attracted increase, the attractive forces of the particles decrease in the ratio of any power of the distances, the accelerative attractions towards the whole bodies will be as the bodies directly, and those powers of the distances inversely. As if the forces of the particles decrease in a duplicate ratio of the distances from the corpuscles attracted, and the bodies are as A^3 and B^3, and therefore both the cubic sides of the bodies, and the distance of the attracted corpuscles from the bodies, are as A and B; the accelerative attractions towards the bodies will be as $\frac{A^3}{A^2}$ and $\frac{B^3}{B^2}$ that is, as A and B the cubic sides of those bodies. If the forces of the particles decrease in a triplicate ratio of the distances from the attracted corpuscles, the accelerative attractions towards the whole bodies will be as $\frac{A^3}{A^3}$ and $\frac{B^3}{B^3}$ that is, equal. If the forces decrease in a quadruplicate ratio, the attractions towards the bodies will be as $\frac{A^3}{A^4}$ and $\frac{B^3}{B^4}$, that is, reciprocally as the cubic sides A and B. And so in other cases.

COR. 2. Hence, on the other hand, from the forces with which like bodies attract corpuscles similarly situated, may be collected the ratio of the decrease of the attractive forces of the particles as the attracted corpuscle recedes from them; if so be that decrease is directly or inversely in any ratio of the distances.

PROPOSITION LXXXVIII. THEOREM XLV.

IF THE ATTRACTIVE FORCES OF THE EQUAL PARTICLES OF ANY BODY
BE AS THE DISTANCE OF THE PLACES FROM THE PARTICLES, THE
FORCE OF THE WHOLE BODY WILL TEND TO ITS CENTRE OF
GRAVITY; AND WILL BE THE SAME WITH THE FORCE OF A GLOBE,
CONSISTING OF SIMILAR AND EQUAL MATTER, AND HAVING ITS
CENTRE IN THE CENTRE OF GRAVITY.

Let the particles A, B, of the body RSTV attract any corpuscle Z with forces which, supposing the particles to be equal between themselves, are as the distances AZ, BZ; but, if they are supposed unequal, are as those particles and their distances AZ, BZ, conjunctly, or (if I may so speak) as those particles drawn into their distances AZ, BZ respectively. And let those forces be expressed by the contents under A × AZ, and B × BZ. Join AB, and let it be cut in G, so that AG may be to BG as the particle B to the particle A; and G will be the common centre of gravity of the particles A and B. The force A × AZ will (by Cor. 2, of the Laws) be

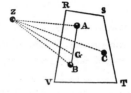

resolved into the forces A × GZ and A × AG; and the force B × RZ into the forces B × GZ and B × BG. Now the forces A × AG and B × BG, because A is proportional

to B, and BG to AG, are equal, and therefore having contrary directions destroy one another. There remain then the forces A × GZ and B × GZ. These tend from Z towards the centre G, and compose the force $\overline{A+B}$ × GZ; that is, the same force as if the attractive particles A and B were placed in their common centre of gravity G, composing there a little globe.

By the same reasoning, if there be added a third particle C, and the force of it be compounded with the force $\overline{A+B}$ × GZ tending to the centre G, the force thence arising will tend to the common centre of gravity of that globe in G and of the particle C; that is, to the common centre of gravity of the three particles A, B, C; and will be the same as if that globe and the particle C were placed in that common centre composing a greater globe there; and so we may go on *in infinitum*. Therefore the whole force of all the particles of any body whatever RSTV is the same as if that body, without removing its centre of gravity, were to put on the form of a globe.   *Q.E.D.*

COR. Hence the motion of the attracted body Z will be the same as if the attracting body RSTV were sphaerical and therefore if that attracting body be either at rest, or proceed uniformly in a right line, the body attracted will move in an ellipsis having its centre in the centre of gravity of the attracting body.

PROPOSITION LXXXIX. THEOREM XLVI.

IF THERE BE SEVERAL BODIES CONSISTING OF EQUAL PARTICLES
WHOSE FORCES ARE AS THE DISTANCES OF THE PLACES FROM EACH,
THE FORCE COMPOUNDED OF ALL THE FORCES BY WHICH ANY
CORPUSCLE IS ATTRACTED WILL TEND TO THE COMMON CENTRE OF
GRAVITY OF THE ATTRACTING BODIES; AND WILL BE THE SAME AS IF
THOSE ATTRACTING BODIES, PRESERVING THEIR COMMON CENTRE
OF GRAVITY, SHOULD UNITE THERE, AND BE FORMED INTO A GLOBE.

This is demonstrated after the same manner as the foregoing Proposition.

COR. Therefore the motion of the attracted body will be the same as if the attracting bodies, preserving their common centre of gravity, should unite there, and be formed into a globe. And, therefore, if the common centre of gravity of the attracting bodies be either at rest, or proceed uniformly in a right line, the attracted body will move in an ellipsis having its centre in the common centre of gravity of the attracting bodies.

PROPOSITION XC. PROBLEM XLIV.

IF TO THE SEVERAL POINTS OF ANY CIRCLE THERE TEND EQUAL
CENTRIPETAL FORCES, INCREASING OR DECREASING IN ANY RATIO

OF THE DISTANCES; IT IS REQUIRED TO FIND THE FORCE WITH
WHICH A CORPUSCLE IS ATTRACTED, THAT IS, SITUATE ANY WHERE
IN A RIGHT LINE WHICH STANDS AT RIGHT ANGLES TO THE PLANE
OF THE CIRCLE AT ITS CENTRE.

Suppose a circle to be described about the centre A with any interval AD in a plane to
which the right line AP is perpendicular; and let it be required to find the force with which

a corpuscle P is attracted towards the same. From any point E of
the circle, to the attracted corpuscle P, let there be drawn the right
line PE. In the right line PA take PF equal to PE, and make a perpendicular FK, erected at F, to be as the force with which the point
E attracts the corpuscle P. And let the curve line IKL be the locus
of the point K. Let that curve meet the plane of the circle in L. In
PA take PH equal to PD, and erect the perpendicular HI meeting
that curve in I; and the attraction of the corpuscle P towards the
circle will be as the area AHIL drawn into the altitude AP. *Q.E.I.*

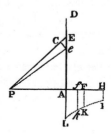

For let there be taken in AE a very small line E*e*. Join P*e*, and in PE, PA take PC, P*f*
equal to P*e*. And because the force, with which any point E of the annulus described about
the centre A with the interval AP in the aforesaid plane attracts to itself the body P, is supposed to be as FK; and, therefore, the force with which that point attracts the body P towards
A is as $\frac{AP \times FK}{PE}$; and the force with which the whole annulus attracts the body P towards A is

as the annulus and $\frac{AP \times FK}{PE}$ conjunctly; and that annulus also is as the rectangle under the
radius AE and the breadth E*e*, and this rectangle (because PE and AE, E*e* and CE are proportional) is equal to the rectangle PE × CE or PE × F*f*, the force with which that annulus
attracts the body P towards A will be as PE × F*f* and $\frac{AP \times FK}{PE}$ conjunctly; that is, as the

content under F*f* × FK × AP, or as the area FK*kf* drawn into AP. And therefore the sum of
the forces with which all the annuli, in the circle described about the centre A with the interval
AD, attract the body P towards A, is as the whole area AHIKL drawn into AP.   *Q.E.D.*

COR. 1. Hence if the forces of the points decrease in the duplicate ratio of the distances,
that is, if FK be as $\frac{1}{PF^2}$, and therefore the area AHIKL as $\frac{1}{PA} - \frac{1}{PH}$; the attraction of the cor-

puscle P towards the circle will be as $1 - \frac{PA}{PH}$; that is, as $\frac{AH}{PH}$.

COR. 2. And universally if the forces of the points at the distances D be reciprocally as
any power D^n of the distances; that is, if FK be as $\frac{1}{D^n}$ and therefore the area AHIKL as

$\frac{1}{PA^{n-1}} - \frac{1}{PH^{n-1}}$; the attraction of the corpuscle P towards the circle will be as $\frac{1}{PA^{n-2}} - \frac{PA}{PH^{n-1}}$.

COR. 3. And if the diameter of the circle be increased *in infinitum*, and the number n be greater than unity; the attraction of the corpuscle P towards the whole infinite plane will be reciprocally as PA^{n-2}, because the other term $\frac{PA}{PH^{n-1}}$ vanishes.

PROPOSITION XCI. PROBLEM XLV.

TO FIND THE ATTRACTION OF A CORPUSCLE SITUATE IN THE
AXIS OF A ROUND SOLID, TO WHOSE SEVERAL POINTS THERE
TEND EQUAL CENTRIPETAL FORCES DECREASING IN ANY
RATIO OF THE DISTANCES WHATSOEVER.

Let the corpuscle P, situate in the axis AB of the solid DECG, be attracted towards that solid. Let the solid be cut by any circle as RFS, perpendicular to the axis; and in its semi-

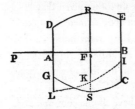

diameter FS, in any plane PALKB passing through the axis, let there be taken (by Prop. XC) the length FK proportional to the force with which the corpuscle P is attracted towards that circle. Let the locus of the point K be the curve line LKI, meeting the planes of the outermost circles AL and BI in L and I; and the attraction of the corpuscle P towards the solid will be as the area LABI. *Q.E.I.*

COR. 1. Hence if the solid be a cylinder described by the parallelogram ADEB revolved about the axis AB, and the centripetal forces tending to the several points be reciprocally as the squares of the distances from the points; the attraction of the corpuscle P towards this cylinder will be as AB – PE + PD. For the ordinate FK (by Cor. 1, Prop. XC) will be as $1-\frac{PF}{PR}$.

The part 1 of this quantity, drawn into the length AB, describes the area 1 x AB; and the other part $\frac{PF}{PR}$, drawn into the length PB describes the area 1 into $\overline{PE-AD}$ (as may be easily shewn from the quadrature of the curve LKI); and, in like manner, the same part drawn into the length PA describes the area 1 into $\overline{PD-AD}$, and drawn into AB, the difference of PB and PA, describes 1 into $\overline{PE-PD}$, the difference of the areas. From the first content 1 x AB take away the last content 1 into $\overline{PE-PD}$,

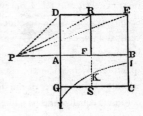

and there will remain the area LABI equal to 1 into $\overline{AB-PE+PD}$. Therefore the force, being proportional to this area, is as AB - PE + PD.

COR. 2. Hence also is known the force by which a spheroid AGBC attracts any body P situate externally in its axis AB. Let NKRM be a conic section whose ordinate ER perpendicular to PE may be always equal to the length of the line PD, continually drawn to the point D in which that ordinate cuts the spheroid From the vertices A, B, of the spheroid, let there be erected to its axis AB the perpendiculars AK, BM,

respectively equal to AP, BP, and therefore meeting the conic section in K and M; and join KM cutting off from it the segment KMRK. Let S be the centre of the spheroid, and SC its greatest semi-diameter; and the force with which the spheroid attracts the body P will be to the force with which a sphere described with the diameter AB attracts the same body as $\frac{AS \times CS^2 - PS \times KMRK}{PS^2 + CS^2 - AS^2}$ is to $\frac{AS^3}{3PS^2}$. And by a calculation

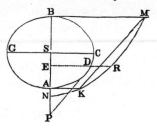

founded on the same principles may be found the forces of the segments of the spheroid.

COR. 3. If the corpuscle be placed within the spheroid and in its axis, the attraction will be as its distance from the centre. This may be easily collected from the following reasoning, whether the particle be in the axis or in any other given diameter. Let AGOF be an attracting spheroid, S its centre, and P the body attracted. Through the body P let there be drawn the semi-diameter SPA, and two right lines DE, FG meeting the spheroid in D and E, F and G; and let PCM, HLN be the superficies of two interior spheroids similar and concentrical to the exterior, the first of which passes through the body P, and cuts the right lines DE, FG in B and C; and the latter cuts the same right lines in H and I, K and L. Let the spheroids have all one common axis, and the parts of the right lines intercepted on both sides DP and BE, FP and CG, DH and IE, FK and LG, will be mutually equal; because the right lines DE, PB, and HI, are bisected in the same point, as are also the right lines FG, PC, and KL. Conceive now DPF, EPG to represent opposite cones described with the infinitely small vertical angles DPF, EPG, and the lines DH, EI to be

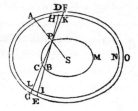

infinitely small also. Then the particles of the cones DHKF, GLIE, cut off by the spheroidical superficies, by reason of the equality of the lines DH and EI, will be to one another as the squares of the distances from the body P, and will therefore attract that corpuscle equally. And by a like reasoning if the spaces DPF, EGCB be divided into particles by the superficies of innumerable similar spheroids concentric to the former and having one common axis, all these particles will equally attract on both aides the body P towards contrary parts. Therefore the forces of the cone DPF, and of the conic segment EGCB, are equal, and by their contrariety destroy each other. And the case is the same of the forces of all the matter that lies without the interior spheroid PCBM. Therefore the body P is attracted by the interior spheroid PCBM alone, and therefore (by Cor. 3, Prop. LXXII) its attraction is to the force with which the body A is attracted by the whole spheroid AGOD as the distance PS to the distance AS.   Q.E.D.

PROPOSITION XCII. PROBLEM XLVI.

AN ATTRACTING BODY BEING GIVEN, IT IS REQUIRED TO FIND THE RATIO OF THE DECREASE OF THE CENTRIPETAL FORCES TENDING TO ITS SEVERAL POINTS.

The body given must be formed into a sphere, a cylinder, or some regular figure, whose law of attraction answering to any ratio of decrease may be found by Prop. LXXX, LXXXI, and XCI. Then, by experiments, the force of the attractions must be found at several distances, and the law of attraction towards the whole, made known by that means, will give the ratio of the decrease of the forces of the several parts; which was to be found.

PROPOSITION XCIII. THEOREM XLVII.

IF A SOLID BE PLANE ON ONE SIDE, AND INFINITELY EXTENDED ON ALL OTHER SIDES, AND CONSIST OF EQUAL PARTICLES EQUALLY ATTRACTIVE, WHOSE FORCES DECREASE, IN THE RECESS FROM THE SOLID, IN THE RATIO OF ANY POWER GREATER THAN THE SQUARE OF THE DISTANCES, AND A CORPUSCLE PLACED TOWARDS EITHER PART OF THE PLANE IS ATTRACTED BY THE FORCE OF THE WHOLE SOLID; I SAY, THAT THE ATTRACTIVE FORCE OF THE WHOLE SOLID, IN THE RECESS FROM ITS PLACE SUPERFICIES, WILL DECREASE IN THE RATIO OF A POWER WHOSE SIDE IS THE DISTANCE OF THE CORPUSCLE FROM THE PLANE, AND ITS INDEX LESS BY 3 THAN THE INDEX OF THE POWER OF THE DISTANCES.

CASE.: 1. Let LG*l* be the plane by which the solid is terminated. Let the solid lie on that hand of the plane that is towards I, and let it be resolved into innumerable planes *m*HM, *n*IN, *o*KO, &c., parallel to GL. And first let the attracted body C be placed without the solid. Let there be drawn

CGHI perpendicular to those innumerable planes, and let the attractive forces of the points of the solid decrease in the ratio of a power of the distances whose index is the number *n* not less than 3. Therefore (by Cor. 3, Prop. XC) the force with which any plane *m*HM attracts the point C reciprocally as CH^{n-2}. In the plane *m*HM take the length HM reciprocally proportional to CH^{n-2}, and that force will be as HM. In like manner in the several

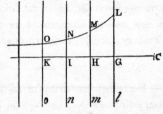

planes *l*GL, *n*IN, *o*KO, &c., take the lengths GL, IN, KO, &c., reciprocally proportional to CG^{n-2}, CI^{n-2}, CK^{n-2}, &c., and the forces of those planes will be as the lengths so taken, and therefore the sum of the forces as the sum of the lengths, that is, the force of the whole solid as the area GLOK

produced infinitely towards OK. But that area (by the known methods of quadratures) is reciprocally as CG^{n-3}, and therefore the force of the whole solid is reciprocally as CG^{n-3}. *Q.E.D.*

CASE. 2. Let the corpuscle C be now placed on that hand of the plane lGL that is within the solid, and take the distance CK equal to the distance CG. And the part of the solid LGlo$KO terminated by the parallel planes lGL, oKO, will attract the corpuscle C, situate in the middle, neither one way nor another, the contrary actions of the opposite points destroying one another by reason of their equality. Therefore

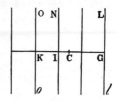

the corpuscle C is attracted by the force only of the solid situate beyond the plane OK. But this force (by Case 1) is reciprocally as CK^{n-3}, that is, (because CG, CK are equal) reciprocally as CK^{n-3}. *Q.E.D.*

COR. 1. Hence if the solid LGIN be terminated on each side by two infinite parallel planes LG, IN, its attractive force is known, subducting from the attractive force of the whole infinite solid LGKO the attractive force of the more distant part NIKO infinitely produced towards KO.

COR. 2. If the more distant part of this solid be rejected, because its attraction compared with the attraction of the nearer part is inconsiderable, the attraction of that nearer part will, as the distance increases, decrease nearly in the ratio of the power CG^{n-3}.

COR. 3. And hence if any finite body, plane on one side, attract a corpuscle situate over against the middle of that plane, and the distance between the corpuscle and the plane compared with the dimensions of the attracting body be extremely small; and the attracting body consist of homogeneous particles, whose attractive forces decrease in the ratio of any power of the distances greater than the quadruplicate; the attractive force of the whole body will decrease very nearly in the ratio of a power whose side is that very small distance, and the index less by 3 than the index of the former power. This assertion does not hold good, however, of a body consisting of particles whose attractive forces decrease in the ratio of the triplicate power of the distances; because, in that case, the attraction of the remoter part of the infinite body in the second Corollary is always infinitely greater than the attraction of the nearer part.

SCHOLIUM.

If a body is attracted perpendicularly towards a given plane, and from the law of attraction given, the motion of the body be required; the Problem will be solved by seeking (by Prop. XXXIX) the motion of the body descending in a right line towards that plane, and (by Cor. 2, of the Laws) compounding that motion with an uniform motion performed in the direction of lines parallel to that plane. And, on the contrary,

if there be required the law of the attraction tending towards the plane in perpendicular directions, by which the body may be caused to move in any given curve line, the Problem will be solved by working after the manner of the third Problem.

But the operations may be contracted by resolving the ordinates into converging series. As if to a base A the length B be ordinately applied in any given angle, and that length be as any power of the base $A\frac{m}{n}$; and there be sought the force with which a body, either attracted towards the base or driven from it in the direction of that ordinate, may be caused to move in the curve line which that ordinate always describes with its superior extremity; I suppose the base to be increased by a very small part O, and I resolve the ordinate $\overline{A+O}\frac{m}{n}$ into an infinite series

$A\frac{m}{n}+\frac{m}{n}OA\frac{m-n}{n}+\frac{mm-mn}{2nn}OOA\frac{m-2}{n}$ c&., and I suppose the force proportional to the term

of this series in which O is of two dimensions, that is, to the term $\frac{mm-mn}{2nn}OOA\frac{m-2}{n}$.

Therefore the force sought is as $\frac{mm-mn}{nn}A\frac{m-2n}{n}$, or, which is the same thing, as

$\frac{mm-mn}{nn}B\frac{m-2n}{m}$.

As if the ordinate describe a parabola, m being = 2, and n = 1, the force will be as the given quantity 2B°, and therefore is given. Therefore with a given force the body will move in a parabola, as *Galileo* has demonstrated. If the ordinate describe an hyperbola, m being = 0 − 1, and n = 1, the force will be as $2A^{-3}$ or $2B^3$; and therefore a force which is as the cube of the ordinate will cause the body to move in an hyperbola. But leaving this kind of propositions, I shall go on to some others relating to motion which I have not yet touched upon.

SECTION XIV.

Of the motion of very small bodies when agitated by centripetal forces tending to the several parts of any very great body.

PROPOSITION XCIV. THEOREM XLVIII.

IF TWO SIMILAR MEDIUMS BE SEPARATED FROM EACH OTHER BY A
SPACE TERMINATED ON BOTH SIDES BY PARALLEL PLANES, AND A
BODY IN ITS PASSAGE THROUGH THAT SPACE BE ATTRACTED OR
IMPELLED PERPENDICULARLY TOWARDS EITHER OF THOSE MEDIUMS,
AND NOT AGITATED OR HINDERED BY ANY OTHER FORCE; AND THE
ATTRACTION BE EVERY WHERE THE SAME AT EQUAL DISTANCES

FROM EITHER PLANE, TAKEN TOWARDS THE SAME HAND OF
THE PLANE; I SAY, THAT THE SINE OF INCIDENCE UPON
EITHER PLANE WILL BE TO THE SINE OF EMERGENCE FROM
THE OTHER PLANE IN A GIVEN RATIO.

CASE. 1. Let A*a* and B*b* be two parallel planes and let the body light upon the first, plane A*a* in the direction of the line GH, and in its whole passage through the intermediate space let it be attracted or impelled towards the medium of incidence, and by that action let it be made to describe a curve line HI, and let it emerge in the direction of the line IK. Let there be erected IM perpendicular to B*b* the plane of emergence, and meeting the line of incidence GH prolonged in M, and the plane of incidence A*a* in R; and let the line of emergence KI be produced and meet HM in L. About the centre L, with the interval LI, let a circle be described cutting both HM in P and Q, and MI produced in N; and, first, if the attraction or impulse be supposed uniform, the curve HI (by what *Galileo* has demonstrated) be a parabola, whose property is that of a rectangle under its given latus rectum and the line IM is equal to the square of HM; and moreover the line HM will be bisected in L. Whence if to MI there be let fall the perpendicular LO, MO, OR will be equal; and adding the equal lines ON, OI, the wholes MN, IR will be equal also. Therefore since IR is given, MN is also given, and the rectangle NMI is to the rectangle under the latus rectum and IM, that is, to HM2 in a given ratio. But the rectangle NMI is equal to the rectangle PMQ, that is, to

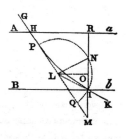

the difference of the squares ML2, and PL2 or LI2; and HM2 hath a given ratio to its fourth part ML2; therefore the ratio of ML2 – LI2 to ML2 is given, and by conversion the ratio of LI2 to ML2, and its subduplicate, the ratio of LI to ML. But in every triangle, as LMI, the sines of the angles are proportional to the opposite sides. Therefore the ratio of the sine of the angle of incidence LMR to the sine of the angle of emergence LIR is given.   *Q.E.D.*.

CASE. 2. Let now the body pass successively through several spaces terminated with parallel planes A*ab*B, B*bc*C, &c., and let it be acted on by a force which is uniform in each of them separately, but different in the different spaces; and by what was just demonstrated, the sine of the angle of incidence on the first plane A*a* is to the sine of emergence from the second plane B*b* in a

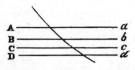

given ratio; and this sine of incidence upon the second plane B*b* will be to the sine of emergence from the third plane C*c* in a given ratio; and this sine to the sine of emergence from the fourth plane D*d* in a given ratio; and so on *in infinitum*; and, by equality, the sine of incidence on the first plane to the sine of emergence from the last plane in a given ratio. Let now the intervals of the planes be diminished, and their number be infinitely increased, so that the action of attraction or impulse,

exerted according to any assigned law, may become continual, and the ratio of the sine of incidence on the first plane to the sine of emergence from the last plane being all along given, will be given then also.   *Q.E.D.*

PROPOSITION XCV. THEOREM XLIX.

THE SAME THINGS BEING SUPPOSED, I SAY, THAT THE VELOCITY OF
THE BODY BEFORE ITS INCIDENCE IS TO ITS VELOCITY AFTER EMER-
GENCE AS THE SINE OF EMERGENCE TO THE SINE OF INCIDENCE.

Make AH and I*d* equal, and erect the perpendiculars AG, *d*K meeting the lines of incidence and emergence GH, IK, in G and K. In GH take TH equal to IK, and to the plane A*a* let fall a perpendicular T*v*. And (by Cor. 2 of the Laws of Motion) let the motion of the body be resolved into two, one perpendicular to the planes A*a*, B*b*, C*c*, &c, and another parallel to

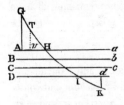

them. The force of attraction or impulse, acting in directions perpendicular to those planes, does not at all alter the motion in parallel directions; and therefore the body proceeding with this motion will in equal times go through those equal parallel intervals that lie between the line AG and the point H, and between the point I and the line *d*K; that is, they will describe the lines GH, IK in equal times. Therefore the velocity before incidence is to the velocity after emergence as GH to IK or TH, that is, as AH or Id to *v*H, that is (supposing TH or IK radius), as the sine of emergence to the sine of incidence.   *Q.E.D.*

PROPOSITION XCVI. THEOREM L.

THE SAME THINGS BEING SUPPOSED, AND THAT THE MOTION
BEFORE INCIDENCE IS SWIFTER THAN AFTERWARDS; I SAY, THAT IF
THE LINE OF INCIDENCE BE INCLINED CONTINUALLY, THE BODY
WILL BE AT LAST REFLECTED, AND THE ANGLE OF REFLEXION WILL
BE EQUAL TO THE ANGLE OF INCIDENCE.

For conceive the body passing between the parallel planes A*a*, B*b*, C*c*, &c., to describe parabolic arcs as above; and let those arcs be HP, PQ, QR, &c. And let the obliquity of the line of incidence GH to the first plane A*a* be such that the sine of incidence maybe to the radius of the circle whose sine it is, in the same ratio which the same sine of incidence hath to the sine of emergence from the plane D*d* into the space D*de*E; and because the sine of emergence is now become equal to radius, the angle of emergence will be a right one, and therefore the line of emergence will coincide with the plane D*d*. Let the body come to this plane in the point R; and because the line of

emergence coincides with that plane, it is manifest that the body can proceed no farther towards the plane E*e*. But neither can it proceed in the line of emergence R*d*;
because it is perpetually attracted or impelled towards the medium of incidence. It will
return, therefore, between the planes C*c*, D*d*,
describing an are of a parabola QR*q*, whose principal
vertex (by what *Galileo* has demonstrated) is in R,
cutting the plane C*c* in the same angle at *q*, that it

did before at Q; then going on in the parabolic arcs *qp*, *ph*, &c., similar and equal to
the former arcs QP, PH, &c., it will cut the rest of the planes in the same angles at *p*,
h, &c., as it did before in P, H, &c., and will emerge at last with the same obliquity at
h with which it first impinged on that plane at H. Conceive now the intervals of the
planes A*a*, B*b*, C*c*, D*d*, E*e*, &c., to be infinitely diminished, and the number infinitely increased, so that the action of attraction or impulse, exerted according to any
assigned law, may become continual; and, the angle of emergence remaining all along
equal to the angle of incidence, will be equal to the same also at last.   *Q.E.D.*

SCHOLIUM.

These attractions bear a great resemblance to the reflexions and refractions of
light made in a given ratio of the secants, as was discovered by *Snellius*; and consequently in a given ratio of the sines, as was exhibited by *Des Cartes*. For it is now certain from the phenomena of *Jupiter's* satellites, confirmed by the observations of different astronomers, that light is propagated in succession, and requires about seven
or eight minutes to travel from the sun to the earth. Moreover, the rays of light that
are in our air (as lately was discovered by *Grimaldus*, by the admission of light into
a dark room through a small hole, which I have also tried) in their passage near the
angles of bodies, whether transparent or opaque (such as the circular and rectangular edges of gold, silver and brass coins, or of knives, or broken pieces of stone or
glass), are bent or inflected round those bodies as if they were attracted to them; and
those rays which in
their passage come
nearest to the bodies
are the most inflected,
as if they were most
attracted; which thing
I myself have also care-

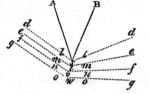
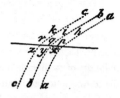

fully observed. And those which pass at greater distances are less inflected; and those
at still greater distances are a little inflected the contrary way, and form three fringes
of colours. In the figure s represents the edge of a knife, or any kind of wedge A*s*B;

and *gowog, fnunf, emtme, dlsld,* are rays inflected towards the knife in the arcs *owo, nvn, mtm, lsl;* which inflection is greater or less according to their distance from the knife. Now since this inflection of the rays is performed in the air without the knife, it follows that the rays which fall upon the knife are first inflected in the air before they touch the knife. And the case is the same of the rays falling upon glass. The refraction, therefore, is made not in the point of incidence, but gradually, by a continual inflection of the rays; which is done partly in the air before they touch the glass, partly (if I mistake not) within the glass, after they have entered it; as is represented in the rays *ckzc, biyb, ahxa,* falling upon *r, q, p,* and inflected between *k* and *z, i* and *y, h* and *x.* Therefore because of the analogy there is between the propagation of the rays of light and the motion of bodies, I thought it not amiss to add the following Propositions for optical uses; not at all considering the nature of the rays of light, or inquiring whether they are bodies or not; but only determining the trajectories of bodies which are extremely like the trajectories of the rays.

PROPOSITION XCVII. PROBLEM XLVII.

SUPPOSING THE SINE OF INCIDENCE UPON ANY SUPERFICIES TO BE
IN A GIVEN RATIO TO THE SINE OF EMERGENCE; AND THAT THE
INFLECTION OF THE PATHS OF THOSE BODIES NEAR THAT SUPERFI
CIES IS PERFORMED IN A VERY SHORT SPACE, WHICH MAY BE CON
SIDERED AS A POINT; IT IS REQUIRED TO DETERMINE SUCH A SUPER
FICIES AS MAY CAUSE ALL THE CORPUSCLES ISSUING FROM ANY ONE
GIVEN PLACE TO CONVERGE TO ANOTHER GIVEN PLACE.

Let A be the place from whence the corpuscles diverge; B the place to which they should converge; CDE the curve line which by its revolution round the axis AB describes the superficies sought; D, E, any two points of that curve; and EF, EG, perpendiculars let fall on the paths of the bodies AD, DB. Let the point D approach to and coalesce with the point E; and the ultimate ratio of the line DF by which AD is increased, to the line DG by which DB is diminished, will be the same as that of the sine of incidence to the sine of emergence. Therefore the ratio of the increment of the line AD to the decrement of the line DB is given; and therefore if in the axis AB there be taken any where the point C through which the curve CDE must pass, and CM the increment of AC be taken in that given ratio to CN the decrement of BC, and from the centres A, B, with the intervals AM, BN, there be described two circles cutting each other in D; that point D will touch the curve sought CDE, and, by touching it any where at pleasure, will determine that curve. *Q.E.I.*

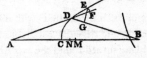

COR. 1. By causing the point A or B to go off sometimes *in infinitum*, and sometimes to move towards other parts of the point C, will be obtained all those figures which *Cartesius* has exhibited in his Optics and Geometry relating to refractions. The invention of which *Cartesius* having thought fit to conceal, is here laid open in this Proposition.

COR. 2. If a body lighting on any superficies CD in the direction of a right line AD, drawn according to any law, should emerge in the direction of another right line DK; and from the point C there be drawn curve lines CP, CQ, always perpendicular to AD, DK;

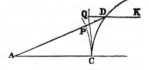

the increments of the lines PD, QD, and therefore the lines themselves PD, QD, generated by those increments, will be as the sines of incidence and emergence to each other, and *è contra*.

PROPOSITION XCVIII. PROBLEM XLVIII.

THE SAME THINGS SUPPOSED; IF ROUND THE AXIS AB ANY
ATTRACTIVE SUPERFICIES BE DESCRIBED AS CD, REGULAR OR
IRREGULAR, THROUGH WHICH THE BODIES ISSUING FROM
THE GIVEN PLACE A MUST PASS; IT IS REQUIRED TO FIND A
SECOND ATTRACTIVE SUPERFICIES EF, WHICH MAY MAKE
THOSE BODIES CONVERGE TO A GIVEN PLACE B.

Let a line joining AB cut the first superficies in C and the second in E, the point D being taken any how at pleasure. And supposing the sine of incidence on the first superficies to the sine of emergence from the same, and the sine of emergence from the second superficies to the sine of incidence on the same, to be as any given quantity M to another given quantity N; then produce AB to G, so that BG may be to CE as M − N to N; and AD to H, so that AH

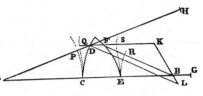

may be equal to AG; and DF to K, so that DK may be to DH as N to M. Join KB, and about the centre D with the interval DH describe a circle meeting KB produced in L, and draw BF parallel to DL; and the point F will touch the line EF, which, being turned round the axis AB, will describe the superficies sought.   *Q.E.F.*

For conceive the lines CP, CQ to be every where perpendicular to AD, DF, and the lines ER, ES to FB, FD respectively, and therefore QS to be always equal to CE; and (by Cor. 2, Prop. XCVII) PD will be to QD as M to N, and therefore as DL, to

DK, or FB to FK; and by division as DL − FB or PH − PD - FB to FD or FQ − QD; and by composition as PH − FB to FQ, that is (because PH and CG, QS and CE, are equal), as CE + BG − FR to CE − FS. But (because BG is to CE as M − N to N) it comes to pass also that CE + BG is to CE as M to N; and therefore, by division, FR is to FS as M to N; and therefore (by Cor. 2, Prop. XCVII) the superficies EF compels a body, falling upon it in the direction DF, to go on in the line FR to the place B.　　　*Q.E.D.*

SCHOLIUM.

In the same manner one may go on to three or more superficies. But of all figures the sphaerical is the most proper for optical uses. If the object glasses of telescopes were made of two glasses of a sphaerical figure, containing water between them, it is not unlikely that the errors of the refractions made in the extreme parts of the superficies of the glasses may be accurately enough corrected by the refractions of the water. Such object glasses are to be preferred before elliptic and hyperbolic glasses, not only because they may be formed with more ease and accuracy, but because the pencils of rays situate without the axis of the glass would be more accurately refracted by them. But the different refrangibility of different rays is the real obstacle that hinders optics from being made perfect by sphaerical or any other figures. Unless the errors thence arising can be corrected, all the labour spent in correcting the others is quite thrown away.

BOOK II.
OF THE MOTION OF BODIES.

SECTION I.

Of the motion of bodies that are resisted in the ratio of the velocity.

PROPOSITION I. THEOREM I.

IF A BODY IS RESISTED IN THE RATIO OF ITS VELOCITY, THE
MOTION LOST BY RESISTANCE IS AS THE SPACE GONE OVER
IN ITS MOTION.

For since the motion lost in each equal particle of time is as the velocity, that is, as the particle of space gone over, then, by composition, the motion lost in the whole time will be as the whole space gone over. *Q.E.D.*

COR. Therefore if the body, destitute of all gravity, move by its innate force only in free spaces, and there be given both its whole motion at the beginning, and also the motion remaining after some part of the way is gone over, there will be given also the whole space which the body can describe in an infinite time. For that space will be to the space now described as the whole motion at the beginning is to the part lost of that motion.

LEMMA I.

Quantities proportional to their differences are continually proportional.

Let A be to A — B asB to B — C and C to C — D, &c., and, by conversion, A will be to B as B to C and C to D, &c. *Q.E.D.*

PROPOSITION II. THEOREM II.

IF A BODY IS RESISTED IN THE RATIO OF ITS VELOCITY,
AND MOVES, BY ITS VIS INSITA, ONLY, THROUGH A SIMILAR
MEDIUM, AND THE TIMES BE TAKEN EQUAL, THE VELOCITIES
IN THE BEGINNING OF EACH OF THE TIMES ARE IN A
GEOMETRICAL PROGRESSION, AND THE SPACES DESCRIBED
IN EACH OF THE TIMES ARE AS THE VELOCITIES.

CASE 1. Let the time be divided into equal particles; and if at the very beginning of each particle we suppose the resistance to act with one single impulse which is as the velocity, the decrement of the velocity in each of the particles of time will be as the same velocity. Therefore the velocities are proportional to their differences, and therefore (by Lem. 1, Book II) continually proportional. Therefore if out of an equal number of particles there be compounded any equal portions of time, the velocities at the beginning of those times will be as terms in a continued progression, which are taken by intervals, omitting every where an equal number of intermediate terms. But the ratios of these terms are compounded of the equal ratios of the intermediate terms equally repeated, and therefore are equal. Therefore the velocities, being proportional to those terms, are in geometrical progression. Let those equal particles of time be diminished, and their number increased *in infinitum*, so that the impulse of resistance may become continual; and the velocities at the beginnings of equal times, always continually proportional, will be also in this case continually proportional.   *Q.E.D.*

CASE 2. And, by division, the differences of the velocities, that is, the parts of the velocities lost in each of the times, are as the wholes; but the spaces described in each of the times are as the lost parts of the velocities (by Prop. 1, Book 1), and therefore are also as the wholes.   *Q.E.D.*

COROL. Hence if to the rectangular asymptotes AC, CH, the hyperbola BG is described, and AB, DG be drawn perpendicular to the asymptote AC, and both the velocity of the body, and the resistance of the medium, at the very beginning of the motion, be expressed by any given line AC, and, after some time is elapsed, by the indefinite line DC; the time may be expressed by the area ABGD, and the space described in that time by the line AD. For if that area, by the motion of the point D, be uniformly increased in the same manner as the time, the right line DC will decrease in a geometrical ratio in the same manner as the velocity; and the parts of the right line AC, described in equal times, will decrease in the same ratio.

PROPOSITION III. PROBLEM I.

TO DEFINE THE MOTION OF A BODY WHICH, IN A SIMILAR MEDIUM, ASCENDS OR DESCENDS IN A RIGHT LINE, AND IS RESISTED IN THE RATIO OF ITS VELOCITY, AND ACTED UPON BY AN UNIFORM FORCE OF GRAVITY.

The body ascending, let the gravity be expounded by any given rectangle BACH; and the resistance of the medium, at the beginning of the ascent, by the rectangle BADE, taken on the contrary side of the right line AB. Through the point B, with the rectangular asymptotes AC, CH, describe an hyperbola, cutting the perpendiculars

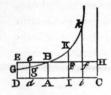

DE, *de*, in G, *g*; and the body ascending will in the time DG*gd* describe the space EG*ge*; in the time DGBA, the space of the whole ascent EGB; in the time ABKI, the space of descent BFK; and in the time IK*ki* the space of descent KF*fk*; and the velocities of the bodies (proportional to the resistance of the medium) in these periods of time will be ABED, AB*ed*, O, ABFI, AB*fi* respectively; and the greatest velocity which the body can acquire by descending will be BACH.

For let the rectangle BACH be resolved into innumerable rectangles A*k*, KI, L*m*, M*n*, &c., which shall be as the increments of the velocities produced in so many equal times; then will O, A*k*, A*l*, A*m*, A*n*, &c., be as the whole velocities, and therefore (by supposition) as the resistances of the medium in the beginning of each of the equal times. Make AC to AK, or ABHC to AB*k*K, as the force of gravity to the resistance in the beginning of the second time; then from the force of gravity subduct the resistances, and ABHC, K*k*HC, L*l*HC, M*m*HC, &c., will be as the absolute forces with which the body is acted upon in the beginning of, each of the times, and therefore (by Law I) as the increments of the velocities, that is, as the rectangles A*k*, K*l*, L*m*, M*n*, &c., and therefore (by Lem. 1, Book II) in a geometrical progression. Therefore, if

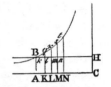

the right lines K*k*, L*l*, M*m*, N*n*, &c., are produced so as to meet the hyperbola in *q, r, s, t,* &c., the areas AB*q*K, K*qr*L, L*rs*M, M*st*N, &c., will be equal, and therefore analogous to the equal times and equal gravitating forces. But the area AB*q*K (by Corol. 3, Lem. VII and VIII, Book I) is to the area B*kq* as K*q* to ½*kq*, or AC to ½*AK*, that is, as the force of gravity to the resistance in the middle of the first time. And by the like reasoning, the areas *q*KL*r*, *r*LM*s*, *s*MN*t*, &c., are to the areas *qklr*, *rlms*, *smnt*, &c., as the gravitating forces to the resistances in the middle of the second, third, fourth time, and so on. Therefore since the equal areas BAK*q*, *q*KL*r*, *r*LM*s*, *s*MN*t*, &c., are analogous to the gravitating forces, the areas B*kq*, *qklr*, *rlms*, *smnt*, &c., will be analogous to the resistances in the middle of each of the times, that is (by supposition), to the velocities, and so to the spaces described. Take the sums of the analogous quantities, and the areas B*kq*, B*lr*, B*ms*, B*nt*, &c., will be analogous to the whole spaces described; and also the areas AB*q*K, AB*r*L, AB*s*M, AB*t*N, &c., to the times. Therefore the body, in descending, will in anytime AB*r*L describe the space B*lr*, and in the time L*rt*N the space *rlnt*.   Q.E.D. And the like demonstration holds in ascending motion.

COROL. 1. Therefore the greatest velocity that the body can acquire by filling is to the velocity acquired in any given time as the given force of gravity which perpetually acts upon it to the resisting force which opposes it at the end of that time.

COROL. 2. But the time being augmented in an arithmetical progression, the sum of that greatest velocity and the velocity in the ascent, and also their difference in the descent, decreases in a geometrical progression.

COROL. 3. Also the differences of the spaces, which are described in equal differences of the times, decrease in the same geometrical progression.

COROL. 4. The space described by the body is the difference of two spaces, whereof one is as the time taken from the beginning of the descent, and the other as the velocity; which [spaces] also at the beginning of the descent are equal among themselves.

PROPOSITION IV. PROBLEM II.

SUPPOSING, THE FORCE OF GRAVITY IN ANY SIMILAR MEDIUM TO BE UNIFORM, AND TO TEND PERPENDICULARLY TO THE PLANE OF THE HORIZON; TO DEFINE THE MOTION OF A PROJECTILE THERE-IN, WHICH SUFFERS RESISTANCE PROPORTIONAL TO ITS VELOCITY.

Let the projectile go from any place D in the direction of any right line DP, and let its velocity at the beginning of the motion be expounded by the length DP. From the point P let fall the perpendicular PC on the horizontal line DC, and cut DC in A, so that DA may be to AC as the resistance of the medium arising from the

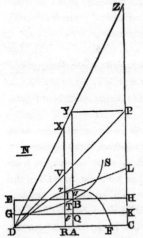

motion upwards at the beginning to the force of gravity; or (which comes to the same) so that the rectangle under DA and DP may be to that under AC and CP as the whole resistance at the beginning of the motion to the force of gravity. With the asymptotes DC, CP describe any hyperbola GTBS cutting the perpendiculars DG, AB in G and B; complete the parallelogram DGKC, and let its side GK out AB in Q. Take a line N in the same ratio to QB as DC is in to CP; and from any point R of the right line DC erect RT perpendicular to it, meeting the hyperbola in T, and the right lines EH, GK, DP in I, t, and V; in that perpendicular take Vr equal to $\frac{tGT}{N}$, or which is the same thing, take Rr

equal to $\frac{GTIE}{N}$; and the projectile in the time DRTG will arrive at the point r describing the curve line DraF, the locus of the point r, thence it will come to its greatest height a in the perpendicular

AB; and afterwards ever approach to the asymptote PC. And its velocity in any point r will be as the tangent rL to the curve.   *Q.E.I.*

For N is to QB as DC to CP or DR to RV, and therefore RV is equal to $\frac{DR \times QB}{N}$

and Rr (that is, RV — Vr, or $\frac{DR \times QB - tGT}{N}$) is equal to $\frac{DR \times AB - RDGT}{N}$. Now let the time be expounded by the area RDGT and (by Laws, Cor. 2), distinguish the motion of the body into two others, one of ascent, the other lateral. And since the resistance is as the motion, let that also be distinguished into two parts proportional and contrary to the parts of the motion : and therefore the length described by the lateral motion will be (by Prop. II, Book II) as the line DR, and the height (by Prop. III, Book II) as the area DR x AB — RDGT, that is, as the line Rr. But in the very beginning of the motion the area RDGT is equal to the rectangle DR x AQ, and therefore that line Rr (or $\frac{DR \times AB - DR \times AQ}{N}$) will then be to DR as AB — AQ or QB to N, that is, as CP to DC; and therefore as the motion upwards to the motion lengthwise at the beginning. Since, therefore, Rr is always as the height, and DR always as the length, and Rr is to DR at the beginning as the height to the length, it follows, that Rr is always to DR as the height to the length; and therefore that the body will move in the line DraF, which is the locus of the point r.   *Q.E.D.*

COR. 1. Therefore Rr is equal to $\frac{DR \times AB}{N} - \frac{RDGT}{N}$; and therefore if RT be produced to X so that RX may be equal to $\frac{DR \times AB}{N}$, that is, if the parallelogram ACPY be completed, and DY cutting CP in Z be drawn, and RT be produced till it meets DY in X; Xr will be equal to $\frac{RDGT}{N}$, and therefore proportional to the time.

COR. 2. Whence if innumerable lines CR, or, which is the same, innumerable lines ZX, be taken in a geometrical progression, there will be as many lines Xr in an arithmetical progression. And hence the curve DraF is easily delineated by the table of logarithms.

COR. 3. If a parabola be constructed to the vertex D, and the diameter DG produced downwards, and its latus rectum is to 2 DP as the whole resistance at the beginning of

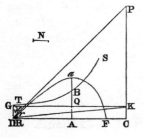

the motion to the gravitating force, the velocity with which the body ought to go from the place D, in the direction of the right line DP, so as in an uniform resisting medium to describe the curve DraF, will be the same as that with which it ought to go from the same place D in the direction of the same right line DP, so as to describe a parabola in a non-resisting medium. For the latus rectum of this parabola, at the

very beginning of the motion, is $\frac{DV^2}{Vr}$; and Vr is $\frac{tGT}{N}$ $_{or}$ $\frac{DR \times Tt}{2N}$ or $\frac{CK \times DR}{DC}$. But a right

line, which, if drawn, would touch the hyperbola GTS in G, is parallel to DK, and

therefore Tt is $\frac{QB \times DC}{CP}$, and N is $\frac{DR^2 \times CK \times CP}{2DC^2 \times QB}$. And therefore Vr is equal to $\frac{DV^2 \times CK \times CP}{2DP \times QB}$, that is (because DR and DC, DV and DP are proportionals), to

$\frac{DV^2}{Vr}$; and the latus rectum $\frac{2DP^2 \times QB}{CK \times CP}$ comes out $\frac{2DP^2 \times DA}{AC \times CP}$, that is (because QB and

CK, DA and AC are proportional), $\frac{2DP^2 \times DA}{AC \times CP}$, and therefore is to 2DP as DP x DA to CP x AC; that is, as the resistance to the gravity. *Q.E.D.*

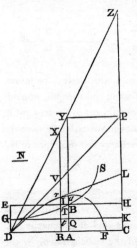

COR. 4. Hence if a body be projected from any place D with a given velocity, in the direction of a right line DP given by position, and the resistance of the medium, at the beginning of the motion, be given, the curve D*ra*F, which that body will describe, may be found. For the velocity being given, the latus rectum of the parabola is given, as is well known. And taking 2DP to that latus rectum, as the force of gravity to the resisting force, DP is also given. Then cutting DC in A, so that CP × AC may be to DP × DA in the same ratio of the gravity to the resistance, the point A will be given. And hence the curve D*ra*F is also given.

COR. 5. And, on the contrary, if the curve D*ra*F be given, there will be given both the velocity of the body and the resistance of the medium in each of the places *r*. For the ratio of CP × AC to DP × DA being given, there is given both the resistance of the medium at the beginning of the motion and the latus rectum of the parabola; and thence the velocity at the beginning of the motion is given also. Then from the length of the tangent L there is given both the velocity proportional to it, and the resistance proportional to the velocity in any place *r*.

COR. 6. But since the length 2DP is to the latus rectum of the parabola as the gravity to the resistance in D; and, from the velocity augmented, the resistance is augmented in the same ratio, but the latus rectum of the parabola is augmented in the duplicate of that ratio, it is plain that the length 2DP is augmented in that simple ratio only; and is therefore always proportional to the velocity; nor will it be augmented or diminished by the change of the angle CDP, unless the velocity be also changed.

COR. 7. Hence appears the method of determining the curve D*ra*F nearly from the phenomena, and thence collecting the resistance and velocity with which the body is projected. Let two similar and equal bodies be projected with the same velocity, from the place D, in different angles CDP, CD*p*; and let the places F, *f*, where they fall upon the horizontal plane DC, be known. Then taking any length for DP or D*p* suppose the resistance in D to be to the gravity in any ratio whatsoever, and let that ratio be expounded by any length SM. Then, by computation, from that assumed length DP, find the lengths DF, D*f*; and from the ratio, found by calculation, subduct the same ratio as found by experiment; and let the difference be expounded by the perpendicular MN. Repeat the same a second and a third time, by assuming always a new ratio SM of the resistance to the gravity, and collecting a new difference MN. Draw the affirmative differences on one side of the right line SM, and the negative on the other side; and through the points N, N, N, draw a regular curve NNN, cutting the right line SMMM in X,

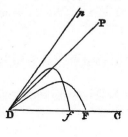

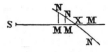

and SX will be the true ratio of the resistance to the gravity, which was to be found. From this ratio the length DF is to be collected by calculation; and a length, which is to the assumed length DP as the length DF known by experiment to the length DF just now found, will be the true length DP. This being known, you will have both the curve line D*ra*F which the body describes, and also the velocity and resistance of the body in each place.

SCHOLIUM.

But, yet, that the resistance of bodies is in the ratio of the velocity, is more a mathematical hypothesis than a physical one. In mediums void of all tenacity, the resistances made to bodies are in the duplicate ratio of the velocities. For by the action of a swifter body, a greater motion in proportion to a greater velocity is communicated to the same quantity of the medium in a less time; and in an equal time, by reason of a greater quantity of the disturbed medium, a motion is communicated in the duplicate ratio greater; and the resistance (by Law II and III) is as the motion communicated. Let us, therefore see what motions arise from this law of resistance.

SECTION II.

Of the motion of bodies that are resisted in the duplicate ratio of their velocities.

PROPOSITION V. THEOREM III.

IF A BODY IS RESISTED IN THE DUPLICATE RATIO OF ITS VELOCITY,
AND MOVES BY ITS INNATE FORCE ONLY THROUGH A SIMILAR
MEDIUM, AND THE TIMES BE TAKEN IN A GEOMETRICAL PROGRES-
SION, PROCEEDING FROM LESS TO GREATER TERMS: I SAY, THAT THE
VELOCITIES AT THE BEGINNING OF EACH OF THE TIMES ARE IN THE
SAME GEOMETRICAL PROGRESSION INVERSELY; AND THAT THE
SPACES ARE EQUAL, WHICH ARE DESCRIBED IN EACH OF THE TIMES.

For since the resistance of the medium is proportional to the square of the velocity, and the decrement of the velocity is proportional to the resistance: if the time be divided into innumerable equal particles, the squares of the velocities at the beginning of each of the times will be proportional to the differences of the same velocities. Let those particles of time be AK, KL, LM, &c., taken in the right line CD; and erect the perpendiculars AB, Kk, Ll, Mm, &c., meeting the hyperbola, BklmG, described with the centre C, and the rectangular asymptotes CD, CH, in B, k, l, m, &c.; then AB will be to Kk as CK to CA, and, by division, AB — Kk to Kk as AK to CA, and alternately, AB — Kk to AK as Kk to CA; and therefore as AB × Kk to AB × CA. Therefore since AK and AB × CA are given, AB — Kk will be as AB × Kk; and, lastly, when AB and Kk coincide, as. And, by the like reasoning, Kk – Ll, Ll – Mm, &c., will be as, Kk^2,

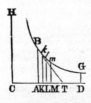

Ll^2, &c. Therefore the squares of the lines AB, Kk, Ll, Mm, &c., are as their differences; and, therefore, since the squares of the velocities were shewn above to be as their differences, the progression of both will be alike. This being demonstrated it follows also that the areas described by these lines are in a like progression with the spaces described by these velocities. Therefore if the velocity at the beginning of the first time AK be expounded by the line AB, and the velocity at the beginning of the second time KL by the line Kk, and the length described in the first time by the area AKkB, all the following velocities will be expounded by the following lines Ll, Mm, &c., and the lengths described, by the areas Kl, Lm, &c. And, by composition, if the whole time be expounded by AM, the sum of its parts, the whole length described will be expounded by AMmB the sum of its parts. Now conceive the time AM to be divided into the parts AK, KL, LM, &c so that CA, CK, CL, CM, &c. may be in a geometrical progression; and those parts will be in the same progression, and the velocities AB, Kk, Ll, Mm, &c., will be in the same progression inversely, and the spaces described Ak, Kl, Lm, &c., will be equal.   Q.E.D.

COR. 1. Hence it appears, that if the time be expounded by any part AD of the asymptote, and the velocity in the beginning of the time by the ordinate AB, the velocity at the end of the time will be expounded by the ordinate DG; and the whole space described by the adjacent hyperbolic area ABGD; and the space which any body can describe in the same time AD, with the first velocity AB, in a non-resisting medium, by the rectangle AB × AD.

COR. 2. Hence the space described in a resisting medium is given, by taking it to the space described with the uniform velocity AB in a non-resisting medium, as the hyperbolic area ABGD to the rectangle AB × AD.

COR. 3. The resistance of the medium is also given, by making it equal, in the very beginning of the motion, to an uniform centripetal force, which could generate, in a body falling through a non-resisting medium, the velocity AB in the time AC. For if BT be drawn touching the hyperbola in B, and meeting the asymptote in T, the right line AT will be equal to AC, and will express the time in which the first resistance, uniformly continued, may take away the whole velocity AB.

COR. 4. And thence is also given the proportion of this resistance to the force of gravity, or any other given centripetal force.

COR. 5. And, *vice versa*, if there is given the proportion of the resistance to any given centripetal force, the time AC is also given, in which a centripetal force equal to the resistance may generate any velocity as AB; and thence is given the point B, through which the hyperbola, having CH, CD for its asymptotes, is to be described; as also the space ABGD, which a body, by beginning its motion with that velocity AB, can describe in any time AD, in a similar resisting medium.

PROPOSITION VI. THEOREM IV.

HOMOGENEOUS AND EQUAL SPHERICAL BODIES, OPPOSED BY
RESISTANCES THAT ARE IN THE DUPLICATE RATIO OF THE
VELOCITIES, AND MOVING ON BY THEIR INNATE FORCE ONLY,
WILL, IN TIMES WHICH ARE RECIPROCALLY AS THE VELOCITIES
AT THE BEGINNING, DESCRIBE EQUAL SPACES, AND LOSE
PARTS OF THEIR VELOCITIES PROPORTIONAL TO THE WHOLES.

To the rectangular asymptotes CD, CH describe any hyperbola B*b*E*e*, cutting the perpendiculars AB, *ab*, DE, *de* in B, *b*, E, *e*; let the initial velocities be expounded by the perpendiculars AB, DE, and the times by the lines A*a*, D*d*. Therefore as A*a* is to D*d*, so (by the hypothe-

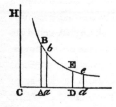

sis) is DE to AB, and so (from the nature of the hyperbola) is CA to CD; and, by composition, so is C*a* to C*d*. Therefore the areas AB*ba*, DE*ed*, that is, the spaces described, are equal among themselves, and the first velocities AB, DE are proportional to the last *ab*, *de*; and therefore, by division, proportional to the parts of the velocities lost, AB — *ab*, DE — *de*.   *Q.E.D.*

PROPOSITION VII. THEOREM V.

IF SPHERICAL BODIES ARE RESISTED IN THE DUPLICATE
RATIO OF THEIR VELOCITIES, IN TIMES WHICH ARE AS THE
FIRST MOTION DIRECTLY, AND THE FIRST RESISTANCES

INVERSELY, THEY WILL LOSE PARTS OF THEIR MOTIONS PROPORTIONAL TO THE WHOLES, AND WILL DESCRIBE SPACES PROPORTIONAL TO THOSE TIMES AND THE FIRST VELOCITIES CONJUNCTLY.

For the parts of the motions lost are as the resistances and times conjunctly. Therefore, that those parts may be proportional to the wholes, the resistance and time conjunctly ought to be as the motion. Therefore the time will be as the motion directly and the resistance inversely. Wherefore the particles of the times being taken in that ratio, the bodies will ,always lose parts of their motions proportional to the wholes, and therefore will retain velocities always proportional to their first velocities. And because of the given ratio of the velocities, they will always describe spaces which are as the first velocities and the times conjunctly.   *Q.E.D.*

COR. 1. Therefore if bodies equally swift are resisted in a duplicate ratio of their diameters, homogeneous globes moving with any velocities whatsoever, by describing spaces proportional to their diameters, will lose parts of their motions proportional to the wholes. For the motion of each globe will be as its velocity and mass conjunctly, that is, as the velocity and the cube of its diameter; the resistance (by supposition) will be as the square of the diameter and the square of the velocity conjunctly; and the time (by this proposition) is in the former ratio directly, and in the latter inversely, that is, as the diameter directly and the velocity inversely; and therefore the space, which is proportional to the time and velocity is as the diameter.

COR. 2. If bodies equally swift are resisted in a sesquiplicate ratio of their diameters, homogeneous globes, moving with any velocities whatsoever, by describing spaces that are in a sesquiplicate ratio of the diameters, will lose parts of their motions proportional to the wholes.

COR. 3. And universally, if equally swift bodies are resisted in the ratio of any power of the diameters, the spaces, in which homogeneous globes, moving with any velocity whatsoever, will lose parts of their motions proportional to the wholes, will be as the cubes of the diameters applied to that power. Let those diameters be D and E; and if the resistances, where the velocities are supposed equal, are as D^n and E^n; the spaces in which the globes, moving with any velocities whatsoever, will lose parts of their motions proportional to the wholes, will be as D^{3-n} and E^{3-n}. And therefore homogeneous globes, in describing spaces proportional to and, will retain their velocities in the same ratio to one another as at the beginning.

COR. 4. Now if the globes are not homogeneous, the space described by the denser globe must be augmented in the ratio of the density. For the motion, with an equal velocity, is greater in the ratio of the density, and the time (by

this Prop.) is augmented in the ratio of motion directly, and the space described in the ratio of the time.

COR. 5. And if the globes move in different mediums, the space, in a medium which, *caeteris paribus*, resists the most, must be diminished in the ratio of the greater resistance. For the time (by this Prop.) will be diminished in the ratio of the augmented resistance, and the space in the ratio of the time.

LEMMA II.

THE MOMENT OF ANY *GENITUM* IS EQUAL TO THE MOMENTS OF EACH OF THE GENERATING SIDES DRAWN INTO THE INDICES OF THE POWERS OF THOSE SIDES, AND INTO THEIR CO-EFFICIENTS CONTINUALLY.

I call any quantity a *genitum* which is not made by addition or subduction of divers parts, but is generated or produced in arithmetic by the multiplication, division, or extraction of the root of any terms whatsoever; in geometry by the invention of contents and sides, or of the extremes and means of proportionals. Quantities of this kind are products, quotients, roots, rectangles, squares, cubes, square and cubic sides, and the like. These quantities I here consider as variable and indetermined, and increasing or decreasing, as it were, by a perpetual motion or flux; and I understand their momentaneous increments or decrements by the name of moments; so that the increments may be esteemed as added or affirmative moments; and the decrements as subducted or negative ones. But take care not to look upon finite particles as such. Finite particles are not moments, but the very quantities generated by the moments. We are to conceive them as the just nascent principles of finite magnitudes. Nor do we in this Lemma regard the magnitude of the moments, but their first proportion, as nascent. It will be the same thing, if, instead of moments, we use either the velocities of the increments and decrements (which may also be called the motions, mutations, and fluxions of quantities), or any finite quantities proportional to those velocities. The co-efficient of any generating side is the quantity which arises by applying the genitum to that side.

Wherefore the sense of the Lemma is, that if the moments of any quantities A, B, C, &c., increasing or decreasing by a perpetual flux, or the velocities of the mutations which are proportional to them, be called $a, b, c,$ &c., the moment or mutation of the generated rectangle AB will be aB + bA; the moment of the generated content ABC will be aBC + bAC + cAB; and the moments of the generated powers A^2, A^3, A^4, $A^{\frac{1}{2}}$, $A^{\frac{3}{2}}$, $A^{\frac{1}{3}}$, $A^{\frac{2}{3}}$, A^{-1}, A^{-2}, $A^{-\frac{1}{2}}$, will be $2a$A, $3a$A^2, $4a$A^3, $\frac{1}{2}a$A$^{-\frac{1}{2}}$, $\frac{3}{2}a$A$^{\frac{1}{2}}$, $\frac{1}{3}a$A$^{-\frac{2}{3}}$, $\frac{2}{3}a$A$^{-\frac{1}{3}}$,

$-a$A^{-2}, $-2a$A^{-3}, $-\frac{1}{2}a$A$^{-\frac{3}{2}}$ respectively; and, in general, that the moment of any power $A^{\frac{n}{m}}$,

will be $\frac{n}{m}aA\frac{n-m}{m}$. Also, that the moment of the generated quantity A^2B will be $2aAB$ $+bA^2$; the moment of the generated quantity $A^3B^4C^2$ will be $3aA^2B^4C^2 + 4bA^3B^3C^2$ $+ 2cA^3B^4C$; and the moment of the generated quantity $\frac{A^3}{B^2}$ or A^3B^{-2} will be $3aA^2B^{-2}$ $- sbA^3B^{-3}$; and so on. The Lemma is thus demostrated.

CASE 1. Any rectangle, as AB, augmented by a perpetual flux, when, as yet, there wanted of the sides A and B half their moments $\frac{1}{2}a$ and $\frac{1}{2}b$, was $A-\frac{1}{2}a$ into $B-\frac{1}{2}b$, or

$AB-\frac{1}{2}aB-\frac{1}{2}bA+\frac{1}{4}ab$; but as soon as the sides A and B are augmented by the other half

moments, the rectangle becomes $A+\frac{1}{2}a$ into $B+\frac{1}{2}b$, or $AB+\frac{1}{2}aB+\frac{1}{2}bA+\frac{1}{4}ab$. From this rectangle subduct the former rectangle, and there will remain the excess $aB + bA$. Therefore with the whole increments a and b of the sides, the increment $aB + bA$ of the rectangle is generated. Q.E.D.

CASE 2. Suppose AB always equal to G, and then the moment of the content ABC or GC (by Case 1) will be, $gC + cG$, that is (putting AB and $aB + BA$ for G and g), $aBC + bAC + cAB$. And the reasoning is the same for contents under ever so many sides. Q.E.D.

CASE 3. Suppose the sides A, B, and C, to be always equal among themselves; and the moment $aB + bA$, of A^2, that is, of the rectangle AB, will be $2aA$; and the moment $aBC + bAC + cAB$ of A^3, that is, of the content ABC, will be $3aA^2$. And by the same reasoning the moment of any power A^n is naA^{n-1}. Q.E.D.

CASE 4. Therefore since $\frac{1}{A}$ into A is 1, the moment of $\frac{1}{A}$ drawn into A, together with $\frac{1}{A}$

drawn into a, will be the moment of 1, that is, nothing. Therefore the moment of $\frac{1}{A}$, or of A^{-1}, is

$\frac{-a}{A^2}$. And generally since $\frac{1}{A^n}$ into A^n is 1, the moment of $\frac{1}{A^n}$ drawn into A^n together with $\frac{1}{A^n}$ into

naA^{n-1} will be nothing. And, therefore, the moment o $\frac{1}{A^n}$ or A^{-n} will $-\frac{na}{A^{n+1}}$. QE.D.

CASE 5. And since $A^{\frac{1}{2}}$ into $A^{\frac{1}{2}}$ is A, the moment of $A^{\frac{1}{2}}$ drawn into $2A^{\frac{1}{3}}$ will be a (by Case 3);

and, therefore, the moment of $A^{\frac{1}{2}}$ will be $\frac{a}{2A^{\frac{1}{2}}}$ or $\frac{1}{2}aA-\frac{1}{2}$. And, generally, putting $A^{\frac{m}{n}}$ equal to B,

then A^m will be equal to B^n, and therefore maA^{m-1} equal to nbB^{n-1}, and maA^{-1} equal to nbB^{-1}, or

$nbA-\frac{m}{n}$; and therefore $AB-\frac{1}{2}aB-\frac{1}{2}bA+\frac{1}{4}ab$ is equal to b, that is, equal to the moment $A^{\frac{m}{n}}$.

Q.E.D.

CASE 6. Therefore the moment of any generated quantity $A^m B^n$ is the moment of A^m drawn into B^n, together with the moment of B^n drawn into A^m, that is, $ma A^{m-1} B^n + nb B^{n-1}$; and that whether the indices m and n of the powers be whole numbers or fractions, affirmative or negative. And the reasoning is the same for contents under more powers.
Q.E.D.

COR. 1. Hence in quantities continually proportional, if one term is given, the moments of the rest of the terms will be as the same terms multiplied by the number of intervals between them and the given term. Let A, B, C, D, E, F, be continually proportional; then if the term C is given, the moments of the rest of the terms will be among themselves as — 2A, — B, D, 2E, 3F.

COR. 2. And if in four proportionals the two means are given, the moments of the extremes will be as those extremes. The same is to be understood of the sides of any given rectangle.

COR. 3. And if the sum or difference of two squares is given, the moments of the sides will be reciprocally as the sides.

SCHOLIUM.

In a letter of mine to Mr. J. *Collins*, dated *December* 10, 1672, having described a method of tangents, which I suspected to be the same with *Slusius's* method, which at that time was not made public, I subjoined these words: *This is one particular, or rather a Corollary, of a general method, which extends itself, without any troublesome calculation, not only to the drawing of tangents to any curve lines, whether geometrical or mechanical, or any how respecting right lines or other curves, but also to the resolving other abstruser kinds of problems about the crookedness, areas, lengths, centres of gravity of curves, &c.; nor is it (as* Hudden's *method de Maximis & Minimis) limited to equations which are free from surd quantities. This method I have interwoven with that other of working in equations, by reducing them to infinite series.* So far that letter. And these last words relate to a treatise I composed on that subject in the year 1671. The foundation of that general method is contained in the preceding Lemma.

PROPOSITION VIII. THEOREM VI.

IF A BODY IN AN UNIFORM MEDIUM, BEING UNIFORMLY
ACTED UPON BY THE FORCE OF GRAVITY, ASCENDS OR
DESCENDS IN A RIGHT LINE; AND THE WHOLE SPACE
DESCRIBED BE DISTINGUISHED INTO EQUAL PARTS, AND IN
THE BEGINNING, OF EACH OF THE PARTS (BY ADDING OR
SUBDUCTING THE RESISTING FORCE OF THE MEDIUM TO

OR FROM THE FORCE OF GRAVITY, WHEN THE BODY
ASCENDS OR DESCENDS) YOU COLLECT THE ABSOLUTE
FORCES; I SAY, THAT THOSE ABSOLUTE FORCES ARE IN A
GEOMETRICAL PROGRESSION.

For let the force of gravity be expounded by the given line AC; the force of resistance by the indefinite line AK; the absolute force in the descent of the body by the difference KC; the velocity of the body by a line AP, which shall be a mean proportional between AK and AC, and therefore in a subduplicate ratio of the resistance; the increment of the resistance made in a given particle of time by the lineola KL, and the contemporaneous increment of the velocity by the lineola PQ; and with the centre C,

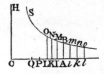

and rectangular asymptotes CA, CH, describe any hyperbola BNS meeting the erected perpendiculars AB, KN, LO in B, N and O. Because AK is as AP^2, the moment KL of the one will be as the moment 2APQ of the other, that is, as AP × KC; for the increment PQ of the velocity is (by Law II) proportional to the generating force KC. Let the ratio of KL be compounded with the ratio KN, and the rectangle KL × KN will become as AP × KC × KN; that is (because the rectangle KC × KN is given), as AP. But the ultimate ratio of the hyperbolic area KNOL to the rectangle KL × KN becomes, when the points K and L coincide, the ratio of equality. Therefore that hyperbolic evanescent area is as AP. Therefore the whole hyperbolic area ABOL is composed of particles KNOL which are always proportional to the velocity AP; and therefore is itself proportional to the space described with that velocity. Let that area be now divided into equal parts, as ABMI, IMNK, KNOL, &c., and the absolute forces AC, IC, KC, LC, &c., will be in a geometrical progression. Q.E.D. And by a like reasoning, in the ascent of the body, taking, on the contrary side of the point A, the equal areas AB*mi*, *imnk*, *knol*, &c., it will appear that the absolute forces AC, *i*C, *k*C, *l*C, &c., are continually proportional. Therefore if all the spaces in the ascent and descent are taken equal, all the absolute forces *l*C, *k*C, *i*C, AC, IC, KC, LC, &c., will be continually proportional. Q.E.D.

COR. 1. Hence if the space described be expounded by the hyperbolic area ABNK, the force of gravity, the velocity of the body, and the resistance of the medium, may be expounded by the lines AC, AP, and AK respectively; and *vice versa*.

COR. 2. And the greatest velocity which the body can ever acquire in an infinite descent will be expounded by the line A.C.

COR. 3. Therefore if the resistance of the medium answering to any given velocity be known, the greatest velocity will be found, by taking it to that given velocity in a ratio subduplicate of the ratio which the force of gravity bears to that known resistance of the medium.

PROPOSITION IX. THEOREM VII.

Supposing what is above demonstrated, I say, that if the
tangents of angles of the sector of a circle, and of an
hyperbola, be taken proportional to the velocities, the
radius being of a fit magnitude, all the time of the ascent
to the highest place will be as the sector of the circle,
and all the time of descending from the highest place as
the sector of the hyperbola.

To the right line AC, which expresses the force of gravity, let AD be drawn perpendicular
and equal. From the centre D with the semi-diameter
AD describe as well the quadrant AtE of a circle, as the
rectangular hyperbola AVZ, whose axis is AK, principal vertex A, and asymptote DC. Let Dp, DP be
drawn; and the circular sector AtD will be as all the
time of the ascent to the highest place; and the hyperbolic sector ATD as all the time of descent from the
highest place; if so be that the tangents Ap, AP of those
sectors be as the velocities.

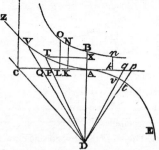

CASE 1. Draw Dvq cutting off the moments or
least particles tDv and qDp, described in the same time, of the sector ADt and of the triangle
ADp. Since those particles (because of the common angle D) are in a duplicate ratio of the sides,
the particle tDv will be as $\frac{qDp \times tD^2}{pD^2}$, that is (because tD is given), as $\frac{qDp}{pD^2}$. But pD^2 is AD^2 + Ap^2,

that is AD^2 + AD × Ak, or AD × Ck; and qDp is $\frac{1}{2}$AD × pq. Therefore tDv, the particle of the

sector, is as $\frac{pq}{Ck}$; that is, as the least decrement pq of the velocity directly, and the force Ck which
diminishes the velocity, inversely; and therefore as the particle of time answering to the
decrement of the velocity. And, by composition, the sum of all the particles tDv in the sector ADt will be as the sum of the particles of time answering to each of the lost particles pq
of the decreasing velocity Ap, till that velocity, being diminished into nothing, vanishes; that
is, the whole sector ADt is as the whole time of ascent to the highest place.   Q.E.D.

CASE 2. Draw DQV cutting off the least particles TDV and PDQ of the sector
DAV, and of the triangle DAQ; and these particles will be to each other as DT^2 to DP^2,
that is (if TX and AP are parallel), as DX^2 to DA^2 or TX^2 to AP^2; and, by division, as
DX^2 — TX^2 to DA^2 — AP^2. But, from the nature of the hyperbola, DX^2 — TX^2 is
AD^2; and, by the supposition, AP^2 is AD × AK. Therefore the particles are to each other as

AD^2 to $AD^2 - AD \times AK$; that is, as AD to $AD - AK$ or AC to CK: and therefore the particle TDV of the sector is $\frac{PDQ \times AC}{CK}$; and CK therefore (because AC and AD are given) as $\frac{PQ}{CK}$; that is, as the increment of the velocity directly, and as the force generating the increment inversely; and therefore as the particle of the time answering to the increment. And, by composition, the sum of the particles of time, in which all the particles PQ of the velocity AP are generated, will be as the sum of the particles of the sector ATD; that is, the whole time will be as the whole sector. Q.E.D.

COR. 1. Hence if AB be equal to a fourth part of AC, the space which a body will describe by falling in any time will be to the space which the body could describe, by moving uniformly on in the same time with its greatest velocity AC, as the area ABNK, which expresses the space described in falling to the area ATD, which expresses the time. For since AC is to AP as AP to AK, then (by Cor. 1, Lem. II, of this Book) LK is to PQ as 2AK to AP, that is, as 2AP to AC, and thence LK is to $\frac{1}{2}$PQ is AP to $\frac{1}{4}$AC

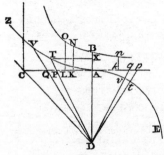

or AB; and KN is to AC or AD as AB to CK; and therefore, *ex aequo*, LKNO to DPQ as AP to CK. But DPQ was to DTV as CK to AC. Therefore, *ex aequo*, LKNO is to DTV as AP to AC; that is, as the velocity of the falling body to the greatest velocity which the body by falling can acquire. Since, therefore, the moments LKNO and DTV of the areas ABNK and ATD are as the velocities, all the parts of those areas generated in the same time will be as the spaces described in the same time; and therefore the whole areas ABNK and ADT, generated from the beginning, will be as the whole spaces described from the beginning of the descent. Q.E.D.

COR. 2. The same is true also of the space described in the ascent. That is to say, that all that space is to the space described in the same time, with the uniform velocity AC, as the area AB*nk* is to the sector AD*t*.

COR. 3. The velocity of the body, falling in the time ATD, is to the velocity which it would acquire in the same time in a non-resisting space, as the triangle APD to the hyperbolic sector ATD. For the velocity in a non-resisting medium would be as the time ATD, and in a resisting medium is as AP, that is, as the triangle APD. And those velocities, at the beginning of the descent, are equal among themselves, as well as those areas ATD, APD.

COR. 4. By the same argument, the velocity in the ascent is to the velocity with which the body in the same time, in a non-resisting space, would lose

all its motion of ascent, as the triangle A*p*D to the circular sector A*t*D; or as the right line A*p* to the arc A*t*.

COR. 5. Therefore the time in which a body, by falling in a resisting medium, would acquire the velocity AP, is to the time in which it would acquire its greatest velocity AC, by falling in a non-resisting space, as the sector ADT to the triangle ADC: and the time in which it would lose its velocity A*p*, by ascending in a resisting medium, is to the time in which it would lose the same velocity by ascending in a non-resisting space, as the arc A*t* to its tangent A*p*.

COR. 6. Hence from the given time there is given the space described in the ascent or descent. For the greatest velocity of a body descending *in infinitum* is given (by Corol. 2 and 3, Theor. VI, of this Book); and thence the time is given in which a body would acquire that velocity by falling in a non-misting space. And taking the sector ADT or AD*t* to the triangle ADC in the ratio of the given time to the time just now found, there will be given both the velocity AP or A*p*, and the area ABNK or AB*nk*, which is to the sector ADT, or AD*t*, as the space sought to the space which would, in the given time, be uniformly described with that greatest velocity found just before.

COR. 7. And by going backward, from the given space of ascent or descent AB*nk* or ABNK, there will be given the time AD*t* or ADT.

PROPOSITION X. PROBLEM III.

SUPPOSE THE UNIFORM FORCE OF GRAVITY TO TEND DIRECTLY
TO THE PLANE OF THE HORIZON, AND THE RESISTANCE TO BE AS
THE DENSITY OF THE MEDIUM AND THE SQUARE OF THE
VELOCITY CONJUNCTLY: IT IS PROPOSED TO FIND THE DENSITY
OF THE MEDIUM IN EACH PLACE, WHICH SHALL MAKE THE BODY
MOVE IN ANY GIVEN CURVE LINE; THE VELOCITY OF THE BODY
AND THE RESISTANCE OF THE MEDIUM IN EACH PLACE.

Let PQ be a plane perpendicular to the plane of the scheme itself; PFHQ a curve line meeting that plane in the points P and Q; G, H, I, K four places of the body going

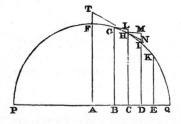

on in this curve from F to Q; and GB, HC, ID, KE four parallel ordinates let fall from these points to the horizon, and standing on the horizontal line PQ at the points B, C, D, E; and let the distances BC, CD, DE, of the ordinates be equal among themselves. From the points G and H let the right lines GL, HN, be drawn touching the curve in G and H, and meeting the ordi-

nates CH, DI, produced upwards, in L and N: and complete the parallelogram

HCDM. And the times in which the body describes the arcs GH, HI, will be in a subduplicate ratio of the altitudes LH, NI, which the bodies would describe in those times, by falling from the tangents; and the velocities will be as the lengths described GH, HI directly, and the times inversely. Let the times be expounded by T and t, and the velocities by $\frac{GH}{T}$ and $\frac{HI}{t}$; and the decrement of the velocity produced in the time

t will be expounded by $\frac{GH}{T} - \frac{HI}{t}$. This decrement arises from the resistance which retards the body, and from the gravity which accelerates it. Gravity, in a falling body, which in its fall describes the space NI, produces a velocity with which it would be able to describe twice that space in the same time, as *Galileo* has demonstrated; that is, the velocity $\frac{2NI}{t}$: but if the body describes the arc HI, it augments that arc only by the

length HI - HN or $\frac{MI \times NI}{HI}$; and therefore generates only the velocity $\frac{2MI \times NI}{t \times HI}$. Let this velocity be added to the before mentioned decrement, and we shall have the decrement of the velocity arising from the resistance alone, that is, $\frac{GH}{T} - \frac{HI}{t} + \frac{2MI \times NI}{t \times HI}$. Therefore

since, in the same time, the action of gravity generates, in a falling body, the velocity

$\frac{2NI}{t}$, the resistance will be to the gravity as $\frac{GH}{T} - \frac{HI}{t} + \frac{2MI \times NI}{t \times HI}$ to $\frac{2NI}{t}$ or as

$\frac{t \times GH}{T} - HI + \frac{2MI \times NI}{HI}$ to 2NI.

Now for the abscissas CB, CD, CE, put — o, o, $2o$. For the ordinate CH put P; and for MI put any series $Qo + Ro^2 + So^3 +$, &c. And all the terms of the series after the first, that is, $Ro^2 + So^3 +$, &c., will be NI; and the ordinates DI, EK, and BG will

be P — Qo — Ro^2 — So^3 —, &c., P — $2Qo$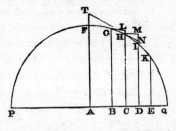
— $4Ro^2$ — $8So^3$ -, &c., and P + Qo — Ro^2
+So^3 —, &c., respectively. And by squaring the
differences of the ordinates BG — CH and CH
- DI, and to the squares thence produced
adding the squares of BC and CD themselves,
you will have $oo + QQoo — 2QRo^3 +$,&c., and
$oo + QQoo + 2QRo^3 +$,&c., the squares of the
arcs GH,HI; whose roots $o\sqrt{1+QQ} - \frac{QRoo}{\sqrt{1+QQ}}$, and $\sqrt{1+QQ} - \frac{QRoo}{\sqrt{1+QQ}}$ are the arcs GH and

HI. Moreover, if from the ordinate CH there be subducted half the sum of the ordinates BG and DI, and from the ordinate DI there be subducted half the sum of the ordinates CH and EK, there will remain Roo and $Roo + 3So^3$, the versed sines of the arcs GI and

929

HK. And these are proportional to the lineolae LH and NI, and therefore in the duplicate ratio of the infinitely small times T and t: and thence the ratio $\frac{t}{T}$ is $\sqrt{\frac{R+3So}{R}}$ or

$\frac{R+\frac{3}{2}So}{R}$ and $\frac{t \times GH}{T} - HI + \frac{2MI \times NI}{HI}$, by substituting the values of $\frac{t}{T}$, GH, HI, MI and

NI just found, becomes $\frac{3Soo}{2R}\sqrt{1+QQ}$. And since 2NI is 2R$oo$, the resistance will be now

to the gravity as $\frac{3Soo}{2R}\sqrt{1+QQ}$ to 2Roo, that is, as $3S\sqrt{1+Q\zeta}$ to 4RR.

And the velocity will be such, that a body going off therewith from any place H, in the direction of the tangent HN, would describe, *in vacuo*, a parabola whose diameter is HC, and its latus rectum $\frac{HN^2}{NI}$ or $\frac{1+QQ}{R}$.

And the resistance is as the density of the medium and the square of the velocity conjunctly; and therefore the density of the medium is as the resistance directly, and the square of the velocity inversely; that is, as $\frac{3S\sqrt{1+QQ}}{4RR}$ directly and $\frac{1+QQ}{R}$ inversely;

that is, as $\frac{S}{R\sqrt{1+QQ}}$.   Q.E.I.

COR. 1. If the tangent HN be produced both ways, so as to meet any ordinate AF in T $\frac{HT}{AC}$ will be equal to $\sqrt{1+QQ}$, and therefore in what has gone before may be put for

$\sqrt{1+QQ}$. By this means the resistance will be to the gravity as 3S x HT to 4RR x AC;

the velocity will be $\frac{HT}{AC\sqrt{R}}$, and the density of the medium will be as $\frac{S \times AC}{R \times HT}$.

COR. 2. And hence, if the curve line PFHQ be defined by the relation between the base or abscissa AC and the ordinate CH, as is usual, and the value of the ordinate be resolved into a converging series, the Problem will be expeditiously solved by the first terms of the series; as in the following examples.

EXAMPLE 1. Let the line PFHQ be a semi-circle described upon the diameter PQ, to find the density of the medium that shall make a projectile move in that line.

Bisect the diameter PQ in A; and call AQ, n; AC, a; CH, e; and CD, o; then DI^2 or $AQ^2 - AD^2 = nn - aa - 2ao - oo$, or $ee - 2ao - oo$; and the root being extracted by our method, will give $DI = e - \frac{ao}{e} - \frac{oo}{2e} - \frac{aaoo}{2e^3} - \frac{ao^3}{2e^3} - \frac{a^3o^3}{2e^5} -$, &c. Here put nn for $es + aa$, and DI

will become $= e - \frac{ao}{e} - \frac{nnoo}{2e^3} - \frac{anno^3}{2e^5} -$, &c.

Such series I distinguish into successive terms after this manner: I call that the first

term in which the infinitely small quantity o is not found; the second, in which that quantity is of one dimension only; the third, in which it arises to two dimensions; the fourth, in which it is of three; and so *ad infinitum*. And the first term, which here is e,

will always denote the length of the ordinate CH, standing at the beginning of the indefinite quantity o. The second term, which here is $\frac{ao}{e}$, will denote the difference between CH and DN; that is, the lineola, MN which is cut off by completing the parallelogram HCDM; and therefore always determines the position of the tangent HN; as, in this case, by taking MN to

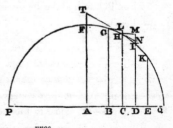

HM as $\frac{ao}{e}$ to o, or a to e. The third term, which here is $\frac{nnoo}{2e^3}$, will represent the lineola IN, which lies between the tangent and the curve; and therefore determines the angle of contact IHN, or the curvature which the curve line has in H. If that lineola IN is of a finite magnitude, it will be expressed by the third term, together with those that follow *in infinitum*. But if that lineola be diminished *in infinitum*, the terms following become infinitely less than the third term, and therefore may be neglected. The fourth term determines the variation of the curvature; the fifth, the variation of the variation; and so on. Whence, by the way, appears no contemptible use of these series in the solution of problems that depend upon tangents, and the curvature of curves.

Now compare the series $e - \frac{ao}{e} - \frac{nnoo}{2e^3} - \frac{anno^3}{2e^5}$ &c., with the series $P - Qo - Roo - So^3 -$ &c., and for P, Q, R and S, put e, $\frac{a}{e}$, $\frac{nn}{2e^3}$, $\frac{ann}{2e^5}$, and for $\sqrt{1+QQ}$ put $\sqrt{1 + \frac{aa}{ee}}$ or $\frac{n}{e}$; and the density of the medium will come out as $\frac{a}{ne}$; that is (because n is given), as $\frac{a}{e}$ or $\frac{AC}{CH}$, that is, as that length of the tangent HT, which is terminated at the semi-diameter AF standing perpendicularly on PQ: and the resistance will be to the gravity as $3a$ to $2n$, that is, as 3AC to the diameter PQ of the circle; and the velocity will be as $\sqrt{CH}$. Therefore if the body goes from the place F, with a due velocity, in the direction of a line parallel to PQ, and the density of the medium in each of the places H is as the length of the tangent HT, and the resistance also in any place H is to the force of gravity as 3AC to PQ, that body will describe the quadrant FHQ of a circle. Q.E.I.

But if the same body should go from the plus P, in the direction of a line perpendicular to PQ, and should begin to move in an arc of the semicircle PFQ, we must take AC or a on the contrary side of the centre A; and therefore its sign must be changed, and we must put $- a$ for $+ a$. Then the density of the medium would come out as $\cdot \frac{a}{e}$. But nature does not admit of a negative density, that is, a density which accelerates the motion of bodies; and therefore it cannot naturally come to pass that a body by ascending

from P should describe the quadrant PF of a circle. To produce such an effect, a body ought to be accelerated by an impelling medium, and not impeded by a resisting one.

EXAMPLE 2. Let the line PFQ be a parabola, having its axis AF perpendicular to the horizon PQ, to find the density of the medium, which will make a projectile move in that line.

From the nature of the parabola, the rectangle PDQ is equal to the rectangle under the ordinate DI and some given right line; that is, if that right line be called b; PC, a; PQ, c; CH, e; and CD, o; the rectangle $a + o$ into $c - a - o$ or $ac - aa - 2ao + cc - oo$, is equal to the rectangle b into DI, and therefore DI is equal to $\frac{ac - aa}{b} + \frac{c - 2a}{b}o - \frac{oo}{b}$. Now the second term $\frac{c - 2a}{b}o$ of this series is to be put for Qo, and the third term $\frac{oo}{b}$ for Roo. But since there are no more terms, the co-efficient S of the fourth term will vanish; and therefore the quantity $\frac{S}{R\sqrt{1 + QQ}}$, to which the density of the medium is proportional, will be nothing. Therefore, where the medium is of no density, the projectile will move in a parabola; as *Galileo* hath heretofore demonstrated.   Q.E.I.

EXAMPLE 3. Let the line AGK be an hyperbola, having its asymptote NX perpendicular to the horizontal plane AK, to find the density of the medium that will make a projectile move in that line.

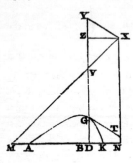

Let MX be the other asymptote, meeting the ordinate DG produced in V; and from the nature of the hyperbola, the rectangle of XV into VG will be given. There is also given the ratio of DN to VX, and therefore the rectangle of DN into VG is given. Let that be bb: and, completing the parallelogram DNXZ, let BN be called a; BD, o, NX, c; and let the given ratio of VZ to ZX or DN be $\frac{m}{n}$. Then DN will be equal to $a - o$, VG equal to $\frac{bb}{a - o}$, VZ equal to $\frac{m}{n} \times \overline{a - o}$

and GD or NX - VZ - VG equal to $c - \frac{m}{n}a + \frac{m}{n}o - \frac{bb}{a - o}$. Let the term $\frac{bb}{a - o}$ be resolved into the converging series $\frac{bb}{a} + \frac{bb}{aa}o + \frac{bb}{a^3}oo + \frac{bb}{a^4}o^3$, &c., and GD will become equal to $c - \frac{m}{n}a - \frac{bb}{a} + \frac{m}{n}o - \frac{bb}{aa}o - \frac{bb}{a^3}o^2 - \frac{bb}{a^4}o^3$ &c. The second term $\frac{m}{n}o - \frac{bb}{aa}o$ of this series is to be used

for Qo; the third $\frac{bb}{a^3}o^2$, with its sign changed for Ro^2; and the fourth $\frac{bb}{a^4}o^3$, with its sign

changed also for So^3, and their coefficients $\frac{m}{n} - \frac{bb}{aa}$, $\frac{bb}{a^3}$ and $\frac{bb}{a^4}$ are to be put for Q, R, and S in the former rule. Which being done, the density of the medium will come out as $\overline{\frac{bb}{a^3}\sqrt{1 + \frac{mm}{nn} - \frac{2mbb}{naa} + \frac{b^4}{a^4}}}$ or $\overline{\frac{1}{\sqrt{aa + \frac{mm}{nn}aa - \frac{2mbb}{n} + \frac{b^4}{aa}}}}$, that is, if in VZ you take VY equal to

VG, as $\frac{1}{XY}$. For aa and $\frac{XY^2}{VG}$ are the squares of XZ and ZY. But the ratio of the resistance to gravity is found to be that of 3XY to 2YG; and the velocity is that with which the body would describe a parabola, whose vertex is G, diameter DG, latus rectum $\frac{XY^2}{VG}$.

Suppose, therefore, that the densities of the medium in each of the places G are reciprocally as the distances XY, and that the resistance in any place G is to the gravity as 3XY to 2YG; and a body let go from the place A, with a due velocity, will describe that hyperbola AGK. Q.E.I.

EXAMPLE 4. Suppose, indefinitely, the line AGK to be an hyperbola described with the centre X, and the asymptotes MX, NX, so that, having constructed the rectangle XZDN, whose side ZD cuts the hyperbola in G and its asymptote in V, VG may be reciprocally as any power DN^n of the line ZX or DN, whose index is the number n: to find the density of the inedium in whichs projected body will describe this curve.

For BN, BD, NX, put A, O, C, respectively, and let VZ be to XZ or DN as d to e, and VG be equal to $\frac{bb}{DN^n}$;

then DN will be equal to A – O, $VG = \frac{bb}{A-O^n}$,

$VZ = \frac{d}{e}\overline{A - O}$, and GD or NX - VZ - VG equal to

$C - \frac{d}{e}A + \frac{d}{e}O - \frac{bb}{A-O^n}$. Let the term $\overline{A-O^n}$ be resolved into an infinite series

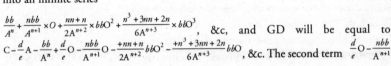

$\frac{bb}{A^n} + \frac{nbb}{A^{n+1}} \times O + \frac{nn+n}{2A^{n+2}} \times bbO^2 + \frac{n^3 + 3nn + 2n}{6A^{n+3}} \times bbO^3$, &c, and GD will be equal to $C - \frac{d}{e}A - \frac{bb}{A^n} + \frac{d}{e}O - \frac{nbb}{A^{n+1}}O - \frac{+nn+n}{2A^{n+2}}bbO^2 - \frac{+n^3 + 3nn + 2n}{6A^{n+3}}bbO$, &c. The second term $\frac{d}{e}O - \frac{nbb}{A^{n+1}}$

of this series is to be used for Qo, the third $\frac{nn+n}{2A^{n+2}}bbO^2$ for Roo, the fourth $\frac{n^3 + 3nn + 2n}{6A^{n+3}}bbO^3$ for So^3. And thence the density of the medium $\overline{R\sqrt{1 + QC}}$ in

any place G, will be $\dfrac{\overline{3\sqrt{A^2}+\dfrac{dd}{ee}A^2-\dfrac{2dnbb}{eA^n}A+\dfrac{nnb^4}{A^{2n}}}^{n+2}}{}$, and therefore if in VZ you take VY

equal to $n \times$ VG, that density is reciprocally as XY. For A^2 and $\dfrac{dd}{ee}A^2-\dfrac{2dnbb}{eA^n}A+\dfrac{nnb^4}{A^{2n}}$ are the squares of XZ and ZY. But the resistance in the same place G is to the force of gravity as $3S \times \dfrac{XY}{A}$ to 4RR, that is, as XY to $\dfrac{2nn+2n}{n+2}$VG.

And the velocity there is the same wherewith the projected body would move in a parabola, whose vertex is G, diameter GD, and latus rectum $\dfrac{1+QR}{R}$ or $\dfrac{2XY^2}{nn+n \times VC}$. Q.E.I.

SCHOLIUM.

In the same manner that the density of the medium comes out to be as $\dfrac{S \times AC}{R \times HT}$ in Cor. 1, if the resistance is put as any power V^n of the velocity V, the density of the medium will come out to be as $\dfrac{S}{4-n} \times \overline{\dfrac{AC}{HT}}^{n-1}$. And therefore if a curve can be found, such that the ratio of $\dfrac{\dfrac{S}{4-n}}{R^{\frac{2}{}}}$ to $\overline{\dfrac{HT}{AC}}^{n-1}$, or of $\dfrac{S^2}{R^{4-n}}$ to $\overline{1+QQ}^{n-1}$ may be given; the body, in an

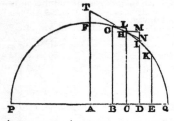

uniform medium, whose resistance is as the power V^n of the velocity V, will move in this curve. But let us return to more simple curves.

Because there can be no motion in a parabola except in a non-resisting medium, but in the hyperbolas here described it is produced by a perpetual resistance; it is evident that the line which a projectile describes in an uniformly resisting medium approaches nearer to these hyperbolas than to a parabola. That line is certainly of the hyperbolic kind, but about the vertex it is more distant from the asymptotes, and in the parts remote from the vertex draws nearer to them than these hyperbolas here described. The difference, however, is not so great between the one and the other but that these latter may be commodiously enough used in practice instead of the former. And perhaps these may prove more useful than an hyperbola that is more accurate, and at the same time more compounded. They may be made use of, then, in this manner.

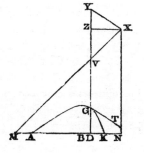

934

Complete the parallelogram XYGT, and the right line GT will touch the hyperbola in G, and therefore the density of the medium in G is reciprocally as the tangent GT, and the velocity there as $v\dfrac{GT^2}{GV}$ and the resistance

is to the force of gravity as $\dfrac{2nn+2n}{n+2} \times GV$.

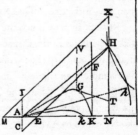

Therefore if a body projected from the place A, in the direction of the right line AH, describes the hyperbola AGK and AH produced meets the asymptote NX in H, and AI drawn parallel to it meets the other asymptote MX in I; the density of the medium in A will be reciprocally as AH, and the velocity of the body as $\sqrt{\dfrac{AH^2}{AI}}$, and the resistance to the force of

gravity as AH to $\dfrac{AH^2}{AI}$. Hence the following rules are deduced.

RULE 1. If the density of the medium at A, and the velocity with which the body is projected remain the same, and the angle NAH be changed; the lengths AH, AI, HX will remain. Therefore if those lengths, in any one case, are found, the hyperbola may afterwards be easily determined from any given angle NAH.

RULE 2. If the angle NAH, and the density of the medium at A, remain the same, and the velocity with which the body is projected be changed, the length AH will continue the same; and AI will be changed in a duplicate ratio of the velocity reciprocally.

RULE 3. If the angle NAH, the velocity of the body at A, and the accelerative gravity remain the same, and the proportion of the resistance at A to the motive gravity be augmented in any ratio; the proportion of AH to AI will be augmented in the same ratio, the latus rectum of the above mentioned parabola remaining the same, and also the length $\dfrac{AH^2}{AI}$ proportional to it; and therefore AH will be diminished in the

same ratio, and AI will be diminished in the duplicate of that ratio. But the proportion of the resistance to the weight is augmented, when either the specific gravity is made less, the magnitude remaining equal, or when the density of the medium is made greater, or when, by diminishing the magnitude, the resistance becomes diminished in a less ratio than the weight.

RULE 4. Because the density of the medium is greater near the vertex of the hyperbola than it is in the place A, that a mean density may be preserved, the ratio of the least of the tangents GT to the tangent AH ought to be found, and the density in A augmented in a ratio a little greater than that of half the sum of those tangents to the least of the tangents GT.

RULE 5. If the lengths AH, AI are given, and the figure AGK is to be described, produce HN to X, so that HX may be to AI as $n + 1$ to 1; and with the centre X, and the asymptotes MX, NX, describe an hyperbola through the point A, such that AI may be to any of the lines VG as XV^n to XI^n.

RULE 6. By how much the greater the number it is, so much the more accurate are these hyperbolas in the ascent of the body from A, and less accurate in its descent to K; and the contrary. The conic hyperbola, keeps a mean ratio between these, and is more simple than the rest. Therefore if the hyperbola be of this kind, and you are to find the point K, where the projected body falls upon any right line AN passing through the point A, let AN produced meet the asymptotes MX, NX in M and N, and take NK equal to AM.

RULE 7. And hence appears an expeditious method of determining this hyperbola from the phenomena. Let two similar and equal bodies be projected with the same velocity, in different angles HAK, hAk, and let them fall upon the plane of the horizon in K and k; and note the proportion of AK to Ak. Let it be as d to e. Then erecting a perpendicular AI of any length, assume any how the length AH or Ah, and thence graphically, or by scale and compass, collect the lengths AK, Ak (by Rule 6). If the ratio of AK to Ak be the same with that of d to e, the length of AH was rightly assumed. If not, take on the indefinite right line SM, the length SM equal to the assumed AH; and

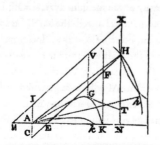
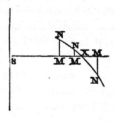

erect a perpendicular MN equal to the difference $\frac{AK}{Ak} - \frac{d}{e}$ of the ratios drawn into any

given right line. By the like method, from several assumed lengths AH, you may find several points N; and draw through them all a regular curve NNXN, cutting the right line SMMM in X. Lastly, assume AR equal to the abscissa SX, and thence find again the length AK; and the lengths, which are to the assumed length AI, and this last AH, as the length AK known by experiment, to the length AK last found, will be the true lengths AI and AH, which were to be found. But these being given, there will be given also the resisting force of the medium in the place A, it being to the force of gravity as

AH to $^4/_3$AI. Let the density of the medium be increased by Rule 4, and if the resisting force just found be increased in the same ratio, it will become still more accurate.

RULE 8. The lengths AH, HX being found; let there be now required the position of the line AH, according to which a projectile thrown with that given velocity shall fall upon any point K. At the points A and K, erect the lines AC, KF perpendicular to the horizon; whereof let AC be drawn downwards, and be equal to AI or $^1/_2$HX. With the asymptotes AK, KF, describe an hyperbola, whose conjugate shall pass through the point C; and from the centre A, with the interval AH, describe a circle cutting that

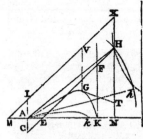

hyperbola in the point H; then the projectile thrown in the direction of the right line AH will fall upon the point K. Q.E.I. For the point H, because of the given length AH, must be somewhere in the circumference of the described circle. Draw CH meeting AK and KF in E and F; and because CH, MX are parallel, and AC, AI equal, AE will be equal to AM, and therefore also equal to KN. But CE is to AE as FH to KN, and therefore CE and FH are equal. Therefore the point

H falls upon the hyperbolic curve described with the asymptotes AK, KF whose conjugate passes through the point C; and is therefore found in the common intersection of this hyperbolic curve and the circumference of the described circle. Q.E.D. It is to be observed that this operation is the same, whether the right line AKN be parallel to the horizon, or inclined thereto in any angle; and that from two intersections H, h, there arise two angles NAH, NAh; and that in mechanical practice it is sufficient once to describe a circle, then to apply a ruler CH, of an indeterminate length, so to the point C, that its part FH, intercepted between the circle and the right line FK, may be equal to its part CE placed between the point C and the right line AK.

What has been said of hyperbolas may be easily applied to parabolas. For if a parabola be represented by XAGK, touched by a right line XV in the vertex X, and the

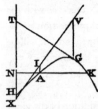

ordinates IA, VG be as any powers XI^n, XV^n, of the abscissas XI, XV; draw XT, GT, AH, whereof let XT be parallel to VG, and let GT, AH touch the parabola in G and A: and a body projected from any place A, in the direction of the right line AH, with a due velocity, will describe this parabola, if the density of the medium in each of the places G be reciprocally as the tangent GT. In that case the velocity in G will be the same as would cause a body, moving in a non-resisting space, to describe a conic parabola, having G for its vertex, VG produced downwards for its diameter, and $\dfrac{2GT^2}{nn-n \times VG}$ for its

latus rectum. And the resisting force in G will be to the force of gravity as GT to

$\frac{2nn-2n}{n-2}$ VG. Therefore if NAK represent an horizontal line, and both the density of the medium at A, and the velocity with which the body is projected, remaining the same, the angle NAH be any how altered, the lengths AH, AI, HX will remain; and thence will be given the vertex X of the parabola, and the position of the right line XI; and by taking VG to IA as XV^n to XI^n, there will be given all the points G of the parabola, through which the projectile will pass.

SECTION III.

Of the motions of bodies which are resisted partly in the ratio of the velocities, and partly in the duplicate of the same ratio.

PROPOSITION XI. THEOREM VIII.

IF A BODY BE RESISTED PARTLY IN THE RATIO AND PARTLY IN
THE DUPLICATE RATIO OF ITS VELOCITY, AND MOVES IN A
SIMILAR MEDIUM BY ITS INNATE FORCE ONLY; AND THE
TIMES BE TAKEN IN ARITHMETICAL PROGRESSION; THEN
QUANTITIES RECIPROCALLY PROPORTIONAL TO THE VELOCITIES,
INCREASED BY A CERTAIN GIVEN QUANTITY,
WILL BE IN GEOMETRICAL PROGRESSION.

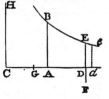

With the centre C, and the rectangular asymptotes CAD*d* and CH, describe an hyperbola BE*e*, and let AB, DE, *de*, be parallel to the asymptote CH. In the asymptote CD let A, G be given points; and if the time be expounded by the hyperbolic area ABED uniformly increasing, I say, that the velocity may be expressed by the length DF, whose reciprocal GD, together with the given line CG, compose the length CD increasing in a geometrical progression.

For let the areola DE*ed* be the least given increment of the time, and D*d* will be reciprocally as DE, and therefore directly as CD. Therefore the decrement of $\frac{1}{GD}$, which (by Lem. II, Book II) is $\frac{Dd}{GD^2}$, will be also as $\frac{CD}{GD^2}$ or $\frac{CG+GD}{GD^2}$, that is, as $\frac{1}{GD}+\frac{CG}{GD^2}$. Therefore the time ABED uniformly increasing by the addition of the given particles ED*de*, it follows that $\frac{1}{GD}$ decreases in the same ratio with the velocity. For the decrement of the velocity is as the resistance, that is (by the supposition), as the sum

of two quantities, whereof one is as the velocity, and the other as the square of the velocity; and the decrement of $\frac{1}{GD}$ is as the sum of the quantities $\frac{1}{GD}$ and $\frac{CG}{GD^2}$, whereof the first is $\frac{1}{GD}$ itself, and the last $\frac{CG}{GD^2}$ is as $\frac{1}{GD^2}$: therefore $\frac{1}{GD}$ is as the velocity, the decrements of both being analogous. And if the quantity GD reciprocally proportional to $\frac{1}{GD}$, be augmented by the given quantity CG; the sum CD, the time ABED uniformly increasing, will increase in a geometrical progression. Q.E.D.

COR. 1. Therefore, if, having the points A and G given, the time be expounded by the hyperbolic area ABED, the velocity may be expounded by $\frac{1}{GD}$ the reciprocal of GD.

COR. 2. And by taking GA to GD as the reciprocal of the velocity at the beginning to the reciprocal of the velocity at the end of any time ABED, the point G will be found. And that point being found the velocity may be found from any other time given.

PROPOSITION XII. THEOREM IX.

THE SAME THINGS BEING SUPPOSED, I SAY, THAT IF THE
SPACES DESCRIBED ARE TAKEN IN ARITHMETICAL PROGRESSION,
THE VELOCITIES AUGMENTED BY A CERTAIN GIVEN
QUANTITY WILL BE IN GEOMETRICAL PROGRESSION.

In the asymptote CD let there be given the point R, and, erecting the perpendicular RS meeting the hyperbola in S, let the space described be expounded by the hyperbolic area RSED; and the velocity will be as the length GD, which, together with the given line CG, composes a length CD decreasing in a geometrical progression, while the space RSED increases in an arithmetical progression.

For, because the increment ED*de* of the space is given, the lineola D*d*, which is the decrement of GD, will be reciprocally as ED, and therefore directly as CD; that is, as the sum of the same GD and the given length CG. But the decrement of the velocity, in a time reciprocally proportional thereto, in which the given particle of space D*de*E is described, is as the resistance and the time conjunctly, that is, directly as the sum of two quantities, whereof one is as the velocity, the other as the square of the velocity, and inversely as the velocity; and therefore directly as the sum of two quantities, one of which is given, the other is as the velocity. Therefore the decrement both of the velocity and the line GD is as a given quantity and a decreasing quantity conjunctly; and, because the decrements are analogous, the decreasing quantities will always be analogous; viz., the velocity, and the line GD. Q.E.D.

COR. 1. If the velocity be expounded by the length GD, the space described will be as the hyperbolic area DESR.

COR. 2. And if the point R be assumed any how, the point G will be found, by taking GR to GD as the velocity at the beginning to the velocity after any space RSED is described. The point G being given, the space is given from the given velocity: and the contrary.

COR. 3. Whence since (by Prop. XI) the velocity is given from the given time, and (by this Prop.) the space is given from the given velocity; the space will be given from the given time: and the contrary.

PROPOSITION XIII. THEOREM X.

SUPPOSING THAT A BODY ATTRACTED DOWNWARDS BY AN
UNIFORM GRAVITY ASCENDS OR DESCENDS IN A RIGHT
LINE; AND THAT THE SAME IS RESISTED PARTLY IN THE
RATIO OF ITS VELOCITY, AND PARTLY IN THE DUPLICATE
RATIO THEREOF: I SAY, THAT, IF RIGHT LINES PARALLEL TO
THE DIAMETERS OF A CIRCLE AND AN HYPERBOLA BE
DRAWN THROUGH THE ENDS OF THE CONJUGATE
DIAMETERS, AND THE VELOCITIES BE AS SOME SEGMENTS
OF THOSE PARALLELS DRAWN FROM A GIVEN POINT, THE
TIMES WILL BE AS THE SECTORS OF THE AREAS CUT OF BY
RIGHT LINES DRAWN FROM THE CENTRE TO THE ENDS OF
THE SEGMENTS; AND THE CONTRARY.

CASE 1. Suppose first that the body is ascending, and from the centre D, with any semi-diameter DB, describe a quadrant BETF of a circle, and through the end B of the semi-diameter DB draw the indefinite line BAP, parallel to the semi-diameter DF. In that line let there be given the point A, and take the segment AP proportional to the velocity. And since one part of the resistance is as the velocity, and another part as the square of the velocity, let the whole resistance be as $AP^2 - 2BAP$. Join DA, DP, cutting the circle in E and T, and let the gravity be expounded by DA^2, so that the 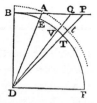 gravity shall be to the resistance in P as DA^2 to $AP^2 + 2BAP$; and the time of the whole ascent will be as the sector EDT of the circle.

For draw DVQ, cutting off the moment PQ of the velocity AP, and the moment DTV of the sector DET answering to a given moment of time; and that decrement PQ of the velocity will be as the sum of the forces of gravity DA^2 and of resistance

$AP^2 + 2BAP$, that is (by Prop. XII, Book II, Elem.), as DP^2. Then the area DPQ, which is proportional to PQ, is as DP^2, and the area DTV, which is to the area DPQ as DT^2 to DP^2, is as the given quantity DT^2. Therefore the area EDT decreases uniformly according to the rate of the future time, by subduction of given particles DTV, and is therefore proportional to the time of the whole ascent.   Q.E.D.

CASE 2. If the velocity in the ascent of the body be expounded by the length AP as before, and the resistance be made as $AP^2 + 2BAP$, and if the force of gravity be less than can be expressed by DA^2; take BD of such a length, that $AB^2 - BD^2$ may be proportional to the gravity, and let DF be perpendicular and equal to DB, and through the vertex F describe the hyperbola FTVE, whose conjugate semi-diameters are DB and DF, and which cuts DA in E, and DP, DQ in T and V; and the time of the whole ascent will be as the hyperbolic sector TDE.

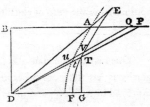

For the decrement PQ of the velocity, produced in a given particle of time, is as the sum of the resistance $AP^2 + 2BAP$ and of the gravity $AB^2 - BD^2$, that is, as $BP^2 - BD^2$. But the area DTV is to the area DPQ as DT^2 to DP^2; and, therefore, if GT be drawn perpendicular to DF, as GT^2 or $GD^2 - GF^2$ to, and as GD^2 to BP^2, and, by division, as DF^2 to $BP^2 - BD^2$. Therefore since the area DPQ is as PQ, that is, as $BP^2 - BD^2$, the area DTV will be as the given quantity DF^2. Therefore the area EDT decreases uniformly in each of the equal particles of time, by the subduction of so many given particles DTV, and therefore is proportional to the time.   Q.E.D.

CASE 3. Let AP be the velocity in the descent of the body, and AP^2 the force of resistance, and $BD^2 - AB^2$ the force of gravity, the angle DBA being a right one. And if with the centre D, and the principal vertex B, there be described a rectangular hyperbola BETV cutting DA, DP, and DQ produced in E, T, and V; the sector DET of this hyperbola will be as the whole time of descent.

For the increment PQ of the velocity, and the area DPQ proportional to it, is as the excess of the gravity above the resistance, that is, as $BD^2 - AB^2 - 2BAP - AP^2$ or $BD^2 - BP^2$. And the area DTV is to the area DPQ as DT^2 to DP^2; and therefore as GT^2 or $GD^2 - BD^2$ to BP^2, and as GD^2 to BD^2, and, by division, as BD^2 to $BD^2 - BP^2$. Therefore since the area DPQ is as $BD^2 - BP^2$, the area DTV will be as the given quantity BD^2. Therefore the area EDT increases uniformly in the several equal particles of time by the addition of as many given particles DTV, and therefore is proportional to the time of the descent.   Q.E.D.

COR. If with the centre D and the semi-diameter DA there be drawn through the vertex A an arc A*t* similar to the arc ET, and similarly subtending the angle ADT, the velocity AP will be to the velocity which the body in the time EDT, in a non-resisting space, can lose in its ascent, or acquire in its descent, as the area of the triangle DAP to the area of the sector DA*t*; and therefore is given from the time given. For the velocity in a non-resisting medium is proportional to the time, and therefore to this sector; in a resisting medium, it is as the triangle; and in both mediums, where it is least, it approaches to the ratio of equality, as the sector and triangle do.

SCHOLIUM.

One may demonstrate also that case in the ascent of the body, where the force of gravity is less than can be expressed by DA^2 or $AB^2 + DB^2$, and greater than can be expressed by $AB^2 - DB^2$, and must be expressed by AB^2. But I hasten to other things.

PROPOSITION XIV. THEOREM XI.

THE SAME THINGS BEING SUPPOSED, I SAY, THAT THE SPACE DESCRIBED IN THE ASCENT OR DESCENT IS AS THE DIFFERENCE OF THE AREA BY WHICH THE TIME IS EXPRESSED, AND OF SOME OTHER AREA WHICH IS AUGMENTED OR DIMINISHED IN AN ARITHMETICAL PROGRESSION; IF THE FORCES COMPOUNDED OF THE RESISTANCE AND THE GRAVITY BE TAKEN IN A GEOMETRICAL PROGRESSION.

Take AC (in these three figures) proportional to the gravity, and AK to the resistance; but take them on the same side of the point A, if the body is descending, otherwise on the contrary. Erect A*b*, which make to DB as DB^2 to 4BAC: and to the rectangular asymptotes CK, CH, describe the hyperbola *b*N; and, erecting KN perpendicular to CK, the area A*b*NK will be augmented or diminished in an arithmetical progression, while the forces CK are taken in a geometrical progression. I say, therefore, that the distance of the body from its greatest altitude is as the excess of the area A*b*NK above the area DET.

For since AK is as the resistance, that is, as $AP^2 \times 2BAP$; assume any given quantity

Z, and put AK equal to $\dfrac{AP^2 + 2BAP}{Z}$; then (by Lem. II of this Book) the moment KL of

AK will be equal to $\dfrac{2APQ + 2BA \times PQ}{Z}$ or $\dfrac{2BPQ}{Z}$, and the moment KLON of the area

A*b*NK will be equal to $\dfrac{2BPQ \times LO}{Z}$ or $\dfrac{BPQ \times BD^3}{2Z \times CK \times AB}$.

CASE 1. Now if the body ascends, and the gravity be as $AB^2 + BD^2$, BET being a

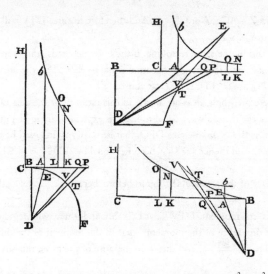

circle, the line AC, which is proportional to the gravity, will be $\dfrac{AB^2 + BD^2}{Z}$, and DP^2 or $AP^2 + 2BAP + AB^2 + BD^2$ will be $AK \times Z + AC \times Z$ or $CK \times Z$; and therefore the area DTV will be to the area DPQ as DT^2 or DB^2 to $CK \times Z$.

CASE 2. If the body ascends, and the gravity be as $AB^2 - BD^2$, the line AC will be

$\dfrac{AB^2 - BD^2}{Z}$, and DT^2 will be to DP^2 as DF^2 or DB^2 to $BP^2 - BD^2$ or $AP^2 + 2BAP + AB^2 - BD^2$,

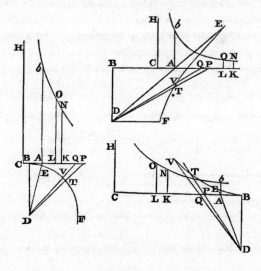

943

that is, to AK x Z + AC x Z or CK x Z. And therefore the area DTV will be to the area DPQ as DB^2 to CK x Z.

CASE 3. And by the same reasoning, if the body descends, and therefore the gravity is as $BD^2 - AB^2$, and the line AC becomes equal to $\frac{BD^2 - AB^2}{Z}$; the area DTV will be to the area DPQ as DB^2 to CK × Z: as above.

Since, therefore, these areas are always in this ratio, if for the area DTV, by which the moment of the time, always equal to itself, is expressed, there be put any determinate rectangle, as BD × m, the area DPQ, that is, $^1/_2$BD × PQ, will be to BD × m as CK × Z to BD^2. And thence $PQ \times BD^3$ becomes equal to 2BD × m × CK × Z, and the moment KLON of the area AbNK, found before, becomes $\frac{BP \times BD \times m}{AB}$. From the area DET subduct its moment DTV or BD x m, and there will remain $\frac{AP \times BD \times m}{AB}$. Therefore the difference of the moments, that is, the moment of the difference of the areas, is equal to $\frac{AP \times BD \times m}{AB}$; and therefore (because of the given quantity $\frac{BD \times m}{AB}$) as

the velocity AP; that is, as the moment of the space which the body describes in its ascent or descent. And therefore the difference of the areas, and that space, in creasing or decreasing by proportional moments, and beginning together or vanishing together, are proportional. Q.E.D.

COR. If the length, which arises by applying the area DET to the line BD, be called M; and another length V be taken in that ratio to the length M which the line DA has to the line DE; the space which a body, in a resisting medium, describes in its whole ascent or descent, will be to the space which a body, in a non-resisting medium, falling from rest, can describe in the same time, as the difference of the aforesaid areas to BL; and therefore is given from the time given. For the space in a non-resisting medium is in a duplicate ratio of the time, or as V^2; and, because BD and AB are given, as

$\frac{BD \times V^2}{AB}$. This area is equal to the area $\frac{DA^2 \times BD \times M^2}{DE^2 \times AB}$ and the moment of M is m; and therefore the moment of this area is $\frac{DA^2 \times BD \times 2M \times m}{DE^2 \times AB}$. But this moment is to the moment of the difference of the aforesaid areas DET and AbNK, viz., to $\frac{AP \times BD \times m}{AB}$, as $\frac{DA^2 \times BD \times M}{DE^2}$ to $\frac{1}{2}$BD × AP, or as $\frac{DA^2}{DE^2}$ into DET to DAP; and, therefore, when the areas DET and DAP are least, in the ratio of equality. Therefore the area $\frac{BD \times V^2}{AB}$ and the

difference of the areas DET and A*b*NK, when all these areas are least, have equal moments; and are therefore equal. Therefore since the velocities, and therefore also the spaces in both mediums described together, in the beginning of the descent, or the end of the ascent, approach to equality, and therefore us then one to another as the area $\frac{BD \times V^2}{AB}$, and the difference of the areas DET and A*b*NK; and moreover since the space,

in a non-resisting medium, is perpetually as $\frac{BD \times V^2}{AB}$, and the space, in a resisting medium, is perpetually as the difference of the areas DET and A*b*NK; it necessarily follows, that the spaces, in both mediums, described in any equal times, are one to another as that area $\frac{BD \times V^2}{AB}$ and the difference of the areas DET and A*b*NK. Q.E.D.

SCHOLIUM.

The resistance of spherical bodies in fluids arises partly from the tenacity, partly from the attrition, and partly from the density of the medium. And that part of the resistance which arises from the density of the fluid as I said, in a duplicate ratio of the velocity; the other part, which arises from the tenacity of the fluid, is uniform, or as the moment of the time; and, therefore, we might now proceed to the motion of bodies, which are resisted partly by an uniform force, or in the ratio of the moments of the time, and partly in the duplicate ratio of the velocity. But it is sufficient to have cleared the way to this speculation in Prop. VIII and IX foregoing, and their Corollaries. For in those Propositions, instead of the uniform resistance made to an ascending body arising from its gravity, one may substitute the uniform resistance which arises from the tenacity of the medium, when the body moves by its *vis insita* alone; and when the body ascends in a right line, add this uniform resistance to the force of gravity, and subduct it when the body descends in a right line. One might also go on to the motion of bodies which are resisted in part uniformly, in part in the ratio of the velocity, and in part in the duplicate ratio of the same velocity. And I have opened a way to this in Prop. XIII and XIV foregoing, in which the uniform resistance arising from the tenacity of the medium may be substituted for the force of gravity, or be compounded with it as before. But I hasten to other things.

SECTION IV.

Of the circular motion of bodies in resisting mediums.

LEMMA III.

LET PQR BE A SPIRAL CUTTING ALL THE RADII SP, SQ, SR, &C., IN EQUAL ANGLES. DRAW THE RIGHT LINE PT TOUCHING THE SPIRAL

IN ANY POINT P, AND CUTTING, THE RADIUS SQ IN T; DRAW PO, QO PERPENDICULAR TO THE SPIRAL, AND MEETING IN O, AND JOIN SO. I SAY, THAT IF THE POINTS P AND Q APPROACH AND COINCIDE, THE ANGLE PSO WILL BECOME A RIGHT ANGLE, AND THE ULTI-MATE RATIO OF THE RECTANGLE TQ × 2PS TO PQ² WILL BE THE RATIO OF EQUALITY.

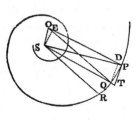

For from the right angles OPQ, OQR, subduct the equal angles SPQ, SQR, and there will remain the equal angles OPS, OQS. Therefore a circle which passes through the points OSP will pass also through the point Q. Let the points P and Q coincide, and this circle will touch the spiral in the place of coincidence PQ, and will therefore cut the right line OP perpendicularly. Therefore OP will become a diameter of this circle, and the angle OSP, being in a semi-circle, becomes a right one. Q.E.D.

Draw QD, SE perpendicular to OP, and the ultimate ratios of the lines will be as follows: TQ to PD as TS or PS to PE, or 2PO to 2PS; and PD to PQ as PQ to 2PO; and, *ex aequo perturbatè*, to TQ to PQ as PQ to 2PS. Whence PQ² becomes equal to TQ x 2PS. Q.E.D.

PROPOSITION XV. THEOREM XII.

IF THE DENSITY OF A MEDIUM IN EACH PLACE THEREOF BE RECIPRO-CALLY AS THE DISTANCE OF THE PLACES FROM AN IMMOVABLE CEN-TRE AND, THE CENTRIPETAL FORCE BE IN THE DUPLICATE RATIO OF THE DENSITY; I SAY, THAT A BODY MAY REVOLVE IN A SPIRAL WHICH CUTS ALL THE RADII DRAWN FROM THAT CENTRE IN A GIVEN ANGLE.

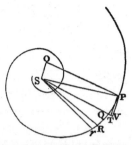

Suppose every thing to be as in the foregoing Lemma, and produce SQ to V so that SV may be equal to SP. In any time let a body, in a resisting medium, describe the least arc PQ, and in double the time the least arc PR; and the decrements of those arcs arising from the resistance, or their differences from the arcs which would be described in a non-resisting medium in the same times, will be to each other as the squares of the times in which they are generated; therefore the decrement of the arc PQ is the fourth part of the decrement of the arc PR. Whence also if the area QSr be taken equal to the area PSQ, the decrement of the arc PQ will be equal

to half the lineola Rr; and therefore the force of resistance and the centripetal force are to each other as the lineola $\frac{1}{2}$Rr and TQ which they generate in the same time. Because

the centripetal force with which the body is urged in P is reciprocally as SP^2, and (by Lem. X, Book I) the lineola TQ, which is generated by that force, is in a ratio compounded of the ratio of this force and the duplicate ratio of the time in which the arc PQ is described (for in this case I neglect the resistance, as being infinitely less than the centripetal force), it follows that $TQ \times SP^2$, that is (by the last Lemma), $\frac{1}{2}PQ^2 \times SP$, will be in a duplicate ratio of the time, and therefore the time is as $\sqrt{PQ} \times SP$; and the velocity of the body, with which the arc PQ is described in that time, as $\frac{PQ}{PQ \times \sqrt{SP}}$ or $\frac{1}{\sqrt{SP}}$, that is, in the subduplicate ratio of SP reciprocally.

And, by a like reasoning, the velocity with which the arc QR is described, is in the subduplicate ratio of SQ reciprocally. Now those arcs PQ and QR are as the describing velocities to each other; that is, in the subduplicate ratio of SQ to SP, or as SQ to $\sqrt{SP \times SQ}$; and, because of the equal angles SPQ, SQr, and the equal areas PSQ, QSr, the arc PQ is to the arc Qr as SQ to SP. Take the differences of the proportional consequents, and the arc PQ will be to the arc Rr as SQ to $SP - \sqrt{SP \times SQ}$ or $\mathcal{2}$. For the

points P and Q coinciding, the ultimate ratio of $SP - \sqrt{SP \times SQ}$ to $\frac{1}{2}VQ$ is the ratio of equality. Because the decrement of the arc PQ arising from the resistance, or its double Rr, is as the resistance and the square of the time conjunctly, the resistance will be as $\frac{Rr}{PQ^2 \times SP}$. But PQ was to R$r$ as SQ to $\frac{1}{2}VQ$, and thence $\frac{Rr}{PQ^2 \times SP}$ becomes as

$\frac{\frac{1}{2}VQ}{PQ \times SP \times SQ}$, or as $\frac{\frac{1}{2}OS}{OP \times SP^2}$. For the points P and Q coinciding, SP and SQ coincide also, and the angle PVQ becomes a right one; and, because of the similar triangles PVQ, PSO, PQ becomes to $\frac{1}{2}VQ$ as OP to $\frac{1}{2}OS$. Therefore $\frac{OS}{OP \times SP^2}$ is as the

resistance, that is, in the ratio of the density of the medium in P and the duplicate ratio of the velocity conjunctly. Subduct the duplicate ratio of the velocity, namely, the ratio $\frac{1}{SP}$, and there will remain the density of the medium in P, as $\frac{OS}{OP \times SP}$. Let the spiral be

given, and, because of the given ratio of OS to OP, the density of the medium in P will be as $\frac{1}{SP}$. Therefore in a medium whose density is reciprocally as SP the distance from the centre, a body will revolve in this spiral. Q.E.D.

COR. 1. The velocity in any place P, is always the same wherewith a body in a non-resisting medium with the same centripetal force would revolve in a circle, at the same distance SP from the centre.

Cor. 2. The density of the medium, if the distance SP be given, is as $\frac{OS}{SP}$, but if that distance is not given, as $\frac{OS}{OP \times SP}$. And thence a spiral may be fitted to any density of the medium.

Cor. 3. The force of the resistance in any place P is to the centripetal force in the same place as $\frac{1}{2}$ OS to OP. For those forces are to each other as $\frac{1}{2}$ Rr and TQ, or as

$$\frac{\frac{1}{4}VQ \times PC}{SQ} \quad \text{and} \quad \frac{\frac{1}{2}PQ^2}{SP}, \text{ that is, as } \frac{1}{2} \text{ VQ and PQ, or } \frac{1}{2} \text{ OS and OP. The spiral therefore}$$

being given, there is given the proportion of the resistance to the centripetal force; and, *vice versa*, from that proportion given the spiral is given.

Cor. 4. Therefore the body cannot revolve in this spiral, except where the force of resistance is less than half the centripetal force. Let the resistance be made equal to half the centripetal force, and the spiral will coincide with the right line PS, and in that right line the body will descend to the centre with a velocity that is to the velocity, with which it was proved before, in the case of the parabola (Theor. X, Book I), the descent would be made in a non-resisting medium, in the subduplicate ratio of unity to the number two. And the times of the descent will be here reciprocally as the velocities, and therefore given.

Cor. 5. And because at equal distances from the centre the velocity is the same in the spiral PQR as it is in the right line SP, and the length of the spiral is to the length of the right line PS in a given ratio, namely, in the ratio of OP to OS; the time of the descent in the spiral will be to the time of the descent in the right line SP in the same given ratio, and therefore given.

Cor. 6. If from the centre S, with any two given intervals, two circles are described; and these circles remaining, the angle which the spiral makes with the radius PS be any how changed; the number of revolu-tions which the body can complete in the space between

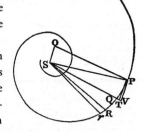

the circumferences of those circles, going round in the spiral from one circumference to another, will be as $\frac{PS}{OS}$, or as the tangent of the angle which the spiral makes with the radius PS; and the time of the same revolutions will be as $\frac{OP}{OS}$, that is, as the secant of the same angle, or reciprocally as the density of the medium.

Cor. 7. If a body, in a medium whose density is reciprocally as the distances of places from the centre, revolves in any curve AEB about that centre, and cuts the first

radius AS in the same angle in B as it did before in A, and that with a velocity that shall be to its first velocity in A reciprocally in a subduplicate ratio of the distances from the

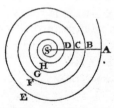

centre (that is, as AS to a mean proportional between AS and BS) that body will continue to describe innumerable similar revolutions BFC, CGD, &c., and by its intersections will distinguish the radius AS into parts AS, BS, CS, DS, &c., that are continually proportional. But the times of the revolutions will be as the perimeters of the orbits AEB, BFC, CGD, &c., directly, and the velocities at the beginnings A, B, C of those orbits inversely; that is as $\frac{3}{AS^2}$, $\frac{3}{BS^2}$, $\frac{3}{CS^2}$. And the whole time in

which the body will arrive at the centre, will be to the time of the first revolution as the sum of all the continued proportionals $\frac{3}{AS^2}$, $\frac{3}{BS^2}$, $\frac{3}{CS^2}$, going on *ad infinitum*, to the

first term $\frac{3}{AS^2}$; that is, as the first term $\frac{3}{AS^2}$ to the difference of the two first $\frac{3}{AS^2} - \frac{3}{BS^2}$

or as $\frac{2}{3}$ AS to AB very nearly. Whence the whole time may be easily found.

COR. 8. From hence also may be deduced, near enough, the motions of bodies in mediums whose density is either uniform, or observes any other assigned law. From the centre S, with intervals SA, SB, SC, &c., continually proportional, describe as many circles; and suppose the time of the revolutions between the perimeters of any two of those circles, in the medium whereof we treated, to be to the time of the revolutions between the same in the medium proposed as the mean density of the proposed medium between those circles to the mean density of the medium whereof we treated, between the same circles, nearly: and that the secant of the angle in which the spiral above determined, in the medium whereof we treated, cuts the radius AS, is in the same ratio to the secant of the angle in which the new spiral, in the proposed medium, cuts the same radius: and also that the number of all the revolutions between the same two circles is nearly as the tangents of those angles. If this be done every where between every two circles, the motion will be continued through all the circles. And by this means one may without difficulty conceive at what rate and in what time bodies ought to revolve in any regular medium.

COR. 9. And although these motions becoming eccentrical should be performed in spirals approaching to an oval figure, yet, conceiving the several revolutions of those spirals to be at the same distances from each other, and to approach to the centre by the same degrees as the spiral above described, we may also understand how the motions of bodies may be performed in spirals of that kind.

PROPOSITION XVI. THEOREM XIII.

IF THE DENSITY OF THE MEDIUM IN EACH OF THE PLACES BE RECIP-
ROCALLY AS THE DISTANCE OF THE PLACES FROM THE IMMOVEABLE
CENTRE, AND THE CENTRIPETAL FORCE BE RECIPROCALLY AS ANY
POWER OF THE SAME DISTANCE, I SAY, THAT THE BODY MAY
REVOLVE IN A SPIRAL INTERSECTING ALL THE RADII DRAWN FROM
THAT CENTRE IN A GIVEN ANGLE.

This is demonstrated in the same manner as the foregoing Proposition. For if the centripetal force in P be reciprocally as any power SP^{n+1} of the distance SP whose index is $n + 1$; it will be collected, as above, that the time in which the body describes any arc PQ, will be as $PQ \times PS^{\frac{1}{2}n}$; and the resistance in P as $\dfrac{Rr}{PQ^2 \times SP^n}$

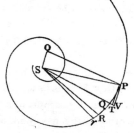

, or as $\dfrac{\overline{1 - \frac{1}{2}n} \times VQ}{PQ \times SP^n \times SQ}$, and therefore as $\dfrac{\overline{1 - \frac{1}{2}n} \times OS}{OP \times SP^{n+1}}$, that is,

(because $\dfrac{\overline{1 - \frac{1}{2}n} \times OS}{OP}$ is a given quantity), reciprocally as SP^{n+1}. And therefore, since the velocity is reciprocally as $SP^{\frac{1}{2}n}$, the density in P will be reciprocally as SP.

COR. 1. The resistance is to the centripetal force as $\overline{1 - \frac{1}{2}n} \times OS$ to OP.

COR. 2. If the centripetal force be reciprocally as SP^3, $1 - \frac{1}{2}n$ will be = 0; and therefore the resistance and density of the medium will be nothing, as in Prop. IX, Book I.

COR. 3. If the centripetal force be reciprocally as any power of the radius SP, whose index is greater than the number 3, the affirmative resistance will be changed into a negative.

SCHOLIUM.

This Proposition and the former, which relate to mediums of unequal density, are to be understood of the motion of bodies that are so small, that the greater density of the medium on one side of the body above that on the other is not to be considered. I suppose also the resistance, *caeteris paribus*, to be proportional to its density. Whence, in mediums whose force of resistance is not as the density, the density must be so much augmented or diminished, that either the excess of the resistance maybe taken away, or the defect supplied.

PROPOSITION XVII. PROBLEM IV.

To find the centripetal force and the resisting force of the medium, by which a body, the law of the velocity being given, shall revolve in a given spiral.

Let that spiral be PQR. From the velocity, with which the body goes over the very small arc PQ, the time will be given; and from the altitude TQ, which is as the centripetal force, and the square of the time, that force will be given. Then from the difference RS*r* of the areas PSQ and QSR described in equal particles of time, the retardation of the body will be given; and from the retardation will be found the resisting force and density of the medium.

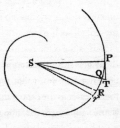

PROPOSITION XVIII. PROBLEM V.

The law of centripetal force being given, to find the density of the medium in each of the places thereof, by which a body may describe a given spiral.

From the centripetal force the velocity in each place must be found; then from the retardation of the velocity the density of the medium is found, as in the foregoing Proposition.

But I have explained the method of managing these Problems in the tenth Proposition and second Lemma, of this Book; and will no longer detain the reader in these perplexed disquisitions. I shall now add some things relating to the forces of progressive bodies, and to the density and resistance of those mediums in which the motions hitherto treated of, and those akin to them, are performed.

SECTION V.

Of the density and compression of fluids; and of hydrostatics.

THE DEFINITION OF A FLUID.

A fluid is any body whose parts yield to any force impressed on it, by yielding, are easily moved among themselves.

PROPOSITION XIX. THEOREM XIV.

All the parts of a homogeneous and unmoved fluid included in any unmoved vessel, and compressed on every

SIDE (SETTING ASIDE THE CONSIDERATION OF CONDENSATION, GRAVITY, AND ALL CENTRIPETAL FORCES), WILL BE EQUALLY PRESSED ON EVERY SIDE, AND REMAIN IN THEIR PLACES WITHOUT ANY MOTION ARISING FROM THAT PRESSURE.

CASE 1. Let a fluid be included in the spherical vessel ABC, and uniformly compressed on every side: I say, that no part of it will be moved by that pressure. For if, any part, as D, be moved, all such parts at the same distance from the centre on every side must necessarily be moved at the same time by a like motion; because the pressure of them all is similar and equal; and all other motion is excluded that does not arise from that pressure. But if these parts come all of them nearer to the centre, the fluid must be condensed towards the centre, contrary to the supposition. If they recede from it, the fluid must be condensed towards the circumference; which is also contrary to the supposition. Neither can they move in any one direction retaining their distance from the centre, because for the same reason, they may move in a contrary direction; but the same part cannot be moved contrary ways at the same time. Therefore no part of the fluid will be moved from its place.

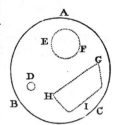

Q.E.D.

CASE 2. I say now, that all the spherical parts of this fluid are equally pressed on every side. For let EF be a spherical part of the fluid; if this be not pressed equally on every side, augment the lesser pressure till it be pressed equally on every side; and its parts (by Case 1) will remain in their places. But before the increase of the pressure, they would remain in their places (by Case 1); and by the addition of a new pressure they will be moved, by the definition of a fluid, from those places. Now these two conclusions contradict each other. Therefore it was false to say that the sphere EF was not pressed equally on every side.

Q.E.D.

CASE 3. I say besides, that different spherical parts have equal pressures. For the contiguous spherical parts press each other mutually and equally in the point of contact (by Law III). But (by Case 2) they are pressed on every side with the same force. Therefore any two spherical parts not contiguous, since an intermediate spherical part can touch both, will be pressed with the same force.

Q.E.D.

CASE: 4. 1 say now, that all the parts of the fluid are every where pressed equally. For any two parts may be touched by spherical parts in any points whatever; and there they will equally press those spherical parts (by Case 3), and are reciprocally equally pressed by them (by Law III).

Q.E.D.

CASE 5. Since, therefore, any part GHI of the fluid is inclosed by the rest of the fluid as in a vessel, and is equally pressed on every side; and also its parts equally press one another, and are at rest among themselves; it is manifest that all the parts of any fluid

as GHI, which is pressed equally on every side, do press each other mutually and equally, and are at rest among themselves.   Q.E.D.

CASE 6. Therefore if that fluid be included in a vessel of a yielding substance, or that is not rigid, and be not equally pressed on every side, the same will give way to a stronger pressure, by the Definition of fluidity.

CASE 7. And therefore, in an inflexible or rigid vessel, a fluid will not sustain a stronger pressure on one side than on the other, but will give way to it, and that in a moment of time; because the rigid side of the vessel does not follow the yielding liquor. But the fluid, by thus yielding, will press against the opposite side, and so the pressure will tend on every side to equality. And because the fluid, as soon as it endeavours to recede from the part that is most pressed, is withstood by the resistance of the vessel on the opposite side, the pressure will on every side be reduced to equality, in a moment of time, without any local motion : and from thence the parts of the fluid (by Case 5) will press each other mutually and equally, and be at rest among themselves. Q.E.D.

COR. Whence neither will a motion of the parts of the fluid among themselves be changed by a pressure communicated to the external superficies, except so far as either the figure of the superficies may be somewhere altered, or that all the parts of the fluid, by pressing one another more intensely or remissly, may slide with more or less difficulty among themselves.

PROPOSITION XX. THEOREM XV.

IF ALL THE PARTS OF A SPHERICAL FLUID, HOMOGENEOUS AT EQUAL
DISTANCES FROM THE CENTRE, LYING ON A SPHERICAL CONCENTRIC
BOTTOM, GRAVITATE TOWARDS THE CENTRE OF THE WHOLE, THE
BOTTOM WILL SUSTAIN THE WEIGHT OF A CYLINDER, WHOSE BASE IS
EQUAL TO THE SUPERFICIES OF THE BOTTOM, AND WHOSE ALTI-
TUDE IS THE SAME WITH THAT OF THE INCUMBENT FLUID.

Let DHM be the superficies of the bottom, and AEI the upper superficies of the fluid. Let the fluid be distinguished into concentric orbs of equal thickness, by the innumerable spherical superficies BFK, CGL: and conceive the force of gravity to act only in the upper superficies of every orb, and the actions to be equal on the equal parts of all the superficies. Therefore the upper superficies AE is pressed by the single force of its own gravity, by which all the parts of the upper orb, and the second superficies BFK, will (by Prop. XIX), according to its measure, be equally pressed. The second superficies BFK is pressed likewise by the force of its own gravity, which, added to the former force, makes the pressure double. The third superficies CGL is, according to its measure, acted on by this pressure and the force of its own gravity besides, which makes

its pressure triple. And in like manner the fourth superficies receives a quadruple pressure, the fifth superficies a quintuple, and so on. Therefore the pressure acting on every superficies is not as the solid quantity of the incumbent fluid, but as the number of the orbs reaching to the upper surface of the fluid; and is equal to the gravity of the lowest orb multiplied by the number of orbs: that is, to the gravity of a solid whose ultimate ratio to the cylinder abovementioned (when the number of the orbs is increased and their thickness diminished, *ad infinitum*, so that the action of gravity from the lowest superficies to the uppermost may become continued) is the ratio of equality. Therefore the lowest superficies sustains the weight of the cylinder above determined. Q.E.D. And

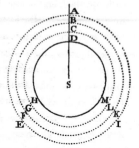

by a like reasoning the Proposition will be evident, where the gravity of the fluid decreases in any assigned ratio of the distance from the centre, and also where the fluid is more rare above and denser below. Q.E.D.

COR. 1. Therefore the bottom is not pressed by the whole weight of the incumbent fluid, but only sustains that part of it which is described in the Proposition; the rest of the weight being sustained archwise by the spherical figure of the fluid.

COR. 2. The quantity of the pressure is the same always at equal distances from the centre, whether the superficies pressed be parallel to the horizon, or perpendicular, or oblique; or whether the fluid, continued upwards from the compressed superficies, rises perpendicularly in a rectilinear direction, or creeps obliquely through crooked cavities and canals, whether those passages be regular or irregular, wide or narrow. That the pressure is not altered by any of these circumstances, may be collected by applying the demonstration of this Theorem to the several cases of fluids.

COR. 3. From the same demonstration it may also be collected (by Prop. XIX), that the parts of a heavy fluid acquire no motion among themselves by the pressure of the incumbent weight, except that motion which arises from condensation.

COR. 4. And therefore if another body of the same specific gravity, incapable of condensation, be immersed in this fluid, it will acquire no motion by the pressure of the incumbent weight: it will neither descend nor ascend, nor change its figure. If it be spherical, it will remain so, notwithstanding the pressure; if it be square, it will remain square; and that, whether it be soft or fluid; whether it swims freely in the fluid, or lies at the bottom. For any internal part of a fluid is in the same state with the submersed body; and the case of all submersed bodies that have the same magnitude, figure, and specific gravity, is alike. If a submersed body, retaining its weight, should dissolve and put on the form of a fluid, this body, if before it would have ascended, descended, or

from any pressure assume a new figure, would now likewise ascend, descend, or put on a new figure; and that, because its gravity and the other causes of its motion remain. But (by Case 5, Prop. XIX) it would now be at rest, and retain its figure. Therefore also in the former case.

COR. 5. Therefore a body that is specifically heavier than a fluid contiguous to it will sink; and that which is specifically lighter will ascend, and attain so much motion and change of figure as that excess or defect of gravity is able to produce. For that excess or defect is the same thing as an impulse, by which a body, otherwise in equilibrium with the parts of the fluid, is acted on; and may be compared with the excess or defect of a weight in one of the scales of a balance.

COR. 6. Therefore bodies placed in fluids have a twofold gravity the one true and absolute, the other apparent, vulgar, and comparative. Absolute gravity is the whole force with which the body tends downwards; relative and vulgar gravity is the excess of gravity with which the body tends downwards more than the ambient fluid. By the first kind of gravity the parts of all fluids and bodies gravitate in their proper places; and therefore their weights taken together compose the weight of the whole. For the whole taken together is heavy, as may be experienced in vessels full of liquor; and the weight of the whole is equal to the weights of all the parts, and is therefore composed of them. By the other kind of gravity bodies do not gravitate in their places; that is, compared with one another, they do not preponderate, but, hindering one another's endeavours to descend, remain in their proper places, as if they were not heavy. Those things which are in the air, and do not preponderate, are commonly looked on as not heavy. Those which do preponderate are commonly reckoned heavy, in as much as they are not sustained by the weight of the air. The common weights are nothing else but the excess of the true weights above the weight of the air. Hence also, vulgarly, those things are called light which are less heavy, and, by yielding to the preponderating air, mount upwards. But these are only comparatively light and not truly so, because they descend *in vacuo*. Thus, in water, bodies which; by their greater or less gravity, descend or ascend, are comparatively and apparently heavy or light; and their comparative and apparent gravity or levity is the excess or defect by which their true gravity either exceeds the gravity of the water or is exceeded by it. But those things which neither by preponderating descend, nor, by yielding to the preponderating fluid, ascend, although by their true weight they do increase the weight of the whole, yet comparatively, and in the sense of the vulgar, they do not gravitate in the water. For these cases are alike demonstrated.

COR. 7. These things which have been demonstrated concerning gravity take place in any other centripetal forces.

COR. 8. Therefore if the medium in which any body moves be acted on either by its own gravity, or by any other centripetal force, and the body be urged more powerfully by the same force; the difference of the forces is that very motive force, which, in the foregoing Propositions, I have considered as a centripetal force. But if the body be more lightly urged by that force, the difference of the forces becomes a centrifugal force, and is to be considered as such.

COR. 9. But since fluids by pressing the included bodies do not change their external figures, it appears also (by Cor. Prop. XIX) that they will not change the situation of their internal parts in relation to one another; and therefore if animals were immersed therein, and that all sensation did arise from the motion of their parts, the fluid will neither hurt the immersed bodies, nor excite any sensation, unless so far as those bodies may be condensed by the compression. And the case is the same of any system of bodies encompassed with a compressing fluid. All the parts of the system will be agitated with the same motions as if they were placed in a vacuum, and would only retain their comparative gravity; unless so far as the fluid may somewhat resist their motions, or be requisite to conglutinate them by compression.

PROPOSITION XXI. THEOREM XVI.

LET THE DENSITY OF ANY FLUID BE PROPORTIONAL TO THE COMPRESSION, AND ITS PARTS BE ATTRACTED DOWNWARDS BY A CENTRIPETAL FORCE RECIPROCALLY PROPORTIONAL TO THE DISTANCES FROM THE CENTRE: I SAY, THAT, IF THOSE DISTANCES BE TAKEN CONTINUALLY PROPORTIONAL, THE DENSITIES OF THE FLUID AT THE SAME DISTANCES WILL BE ALSO CONTINUALLY PROPORTIONAL.

Let ATV denote the spherical bottom of the fluid, S the centre, SA, SB, SC, SD, SE, SF, &c., distances continually proportional. Erect the perpendiculars AH, BI, CK, DL, EM, FN, &c., which shall be as the densities of the medium in the places A, B, C, D, E, F; and the specific gravities in those places will be as $\frac{AH}{AS}$, $\frac{BI}{BS}$, $\frac{CK}{CS}$, &c., or, which is all one, as $\frac{AH}{AB}$, $\frac{BI}{BC}$, $\frac{CK}{CD}$. Suppose, first, these gravities to be uniformly continued from A to B, from B to C, from C to D, &c., the decrements in the points B, C, D, &c., being taken by steps. And these gravities drawn into the altitudes AB, BC, CD, &c., will give the pressures AH, BI, CK, &c., by which the bottom ATV is acted on (by Theor. XV). Therefore the particle A sustains all the pressures AH, BI, CK, DL, &c., proceeding *in infinitum*; and the particle B sustains the pressures of all but the first AH; and the particle C all but the two first AH, BI; and so on: and therefore the density AH of the first particle A is to the density BI of the second particle B as the

sum of all AH + BI + CK + DL, *in infinitum*, to the sum of all BI + CK + DL, &c. And BI the density of the second particle B is to CK the density of the third C, as the sum of all BI + CK + DL, &c., to the sum of all CK + DL, &c. Therefore these sums

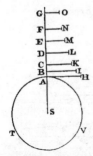

are proportional to their differences AH, BI, CK, &c., and therefore continually proportional (by Lem. I of this Book); and therefore the differences AH, BI, CK, &c., proportional to the sums, are also continually proportional. Wherefore since the densities in the places A, B, C, &c., are as AH, BI, CK, &c., they will also be continually proportional. Proceed intermissively, and, *ex aequo*, at the distances SA, SC, SE, continually proportional, the densities AH, CK, EM will be continually proportional. And by the same reasoning, at any distances SA, SD, SG, continually proportional, the densities AH, DL, GO, will be continually proportional. Let

now the points A, B, C, D, E, &c., coincide, so that the progression of the specific gravities from the bottom A to the top of the fluid may be made continual; and at any distances SA, SD, SG, continually proportional, the densities AH, DL, GO, being all along continually proportional, will still remain continually proportional. Q.E.D.

Cor. Hence if the density of the fluid in two places, as A and E, be given, its density in any other place Q may be collected. With the centre S, and the rectangular asymptotes SQ, SX, describe an hyperbola cutting the perpendiculars AH, EM, QT in *a*, *e*, and *q*, as also the perpendiculars HX, MY, TZ, let fall upon the asypmtote SX, in *h*, *m*, and *t*. Make the area

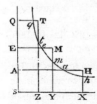

YmtZ to the given area YmhX as the given area EeqQ to the given area EeaA; and the line Zt produced will cut off the line Qt proportional to the density. For if the lines SA, SE, SQ are continually proportional, the areas EeqQ, EeaA will be equal, and thence the areas YmtZ, XhmY, proportional to them, will be also equal: and the lines SX, SY, SZ, that is, AH, EM, QT continually proportional, as they ought to be. And if the lines SA, SE, SQ, obtain any other order in the series

of continued proportionals, the lines AH, EM, QT, because of the proportional hyperbolic areas, will obtain the same order in another series of quantities continually proportional.

PROPOSITION XXII. THEOREM XVII.

Let the density of any fluid be proportional to the compression, and its parts be attracted downwards by a gravitation reciprocally proportional to the squares of the distances from the centre : I say, that if the distances be taken in harmonic progression, the densities of the fluid at those distances will be in a geometrical progression.

Let S denote the centre, and SA, SB, SC, SD, SE, the distances in F geometrical progression. Erect the perpendiculars AH, BI, CK, &c., which shall be as the densities of the fluid in the places A, B, C, D, E, &c., and the specific gravities thereof in those places will be as $\frac{AH}{SA^2}$, $\frac{BI}{SB^2}$, $\frac{CK}{SC^2}$, &c. Suppose these gravities to be uniformly

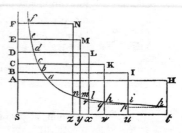

continued, the first from A to B, the second from B to C, the third from C to D, &c. And these drawn into the altitudes AD, BC, CD, DE, &c., or, which is the same thing, into the distances SA, SB, SC, &c., proportional to those altitudes, will give $\frac{AH}{SA}$, $\frac{BI}{SB}$, $\frac{CK}{SC}$, &c., the exponents of the pressures. Therefore since the densities are as the sums of those pressures, the differences AH - BI, BI - CK, &c., of the densities will be as the differences of those sums $\frac{AH}{SA}$, $\frac{BI}{SB}$, $\frac{CK}{SC}$, &c. With the centre S, and the asymptotes SA, Sx, describe any hyperbola, cutting the perpendiculars AH, BI, CK, &c., in a, b, c, &c., and the perpendiculars Ht, Iu, Kw, let fall upon the asymptote Sx, in h, i, k; and the differences of the densities tu, uw, &c, will be as $\frac{AH}{SA}$, $\frac{BI}{SB}$, &c. And the rectangles tu x th, uw x ui, &c., or tp, uq, &c., as $\frac{AH \times th}{SA}$, $\frac{BI \times ui}{SB}$ &c., that is, as Aa, Bb, &c. For, by the nature of the hyperbola, SA is to AH or St as th to Aa, and therefore $\frac{AH \times th}{SA}$ is equal to Aa. And, by a like reasoning, $\frac{BI \times ui}{SB}$ is equal to Bb,&c. But Aa, Bb, Cc, &c., are continually proportional, and therefore proportional to their differences Aa − Bb, Bb − Cc, &c., therefore the rectangles tp, ug, &c., are proportional to those differences; as also the sums of the rectangles tp + uq, or tp + uq + wr to the sums of the differences Aa − Cc or Aa − Dd. Suppose several of these terms and the sum of all the differences, as Aa − Ff, will be proportional to the sum of all the rectangles, as zthn. Increase the number of terms, and diminish the distances of the points A, B, C; &c., in infinitum, and those rectangles will become equal to the hyperbolic area zthn, and therefore the difference Aa − Ff is proportional to this area. Take now any distances, as SA, SD, SF, in harmonic progression, and the differences Aa − Dd, Dd − Ff will be equal; and therefore the areas thlx, xlnz, proportional to those differences will be equal among themselves, and the densities St, Sx, Sz, that is, AH, DI, FN, continually proportional. Q.E.D.

COR. Hence if any two densities of the fluid, as AH and BI, be given, the area thiu, answering to their difference tu, will be given; and thence the density FN will be found at any height SF, by talking the area thnz to that given area thiu as the difference Aa - Ff to the difference Aa - Bb.

958

SCHOLIUM.

By a like reasoning it may be proved, that if the gravity of the particles of a fluid be diminished in a triplicate ratio of the distances from the centre; and the reciprocals of the squares of the distances SA, SB, SC, &c., (namely, $\frac{SA^3}{SA^2}, \frac{SA^3}{SB^2}, \frac{SA^3}{SC^2}$) be taken in an

arithmetical progression, the densities AH, BI, CK, &c., will be in a geometrical progression. And if the gravity be diminished in a quadruplicate ratio of the distances, and the reciprocals of the cubes of the distances (as $\frac{SA^4}{SA^3}, \frac{SA^4}{SB^3}, \frac{SA^4}{SC^3}$, &c.,) be taken in

arithmetical progression, the densities AH, BI, CK, &c., will be in geometrical progression. And so *in infinitum*. Again; if the gravity of the particles of the fluid be the same at all distances, and the distances be in arithmetical progression, the densities will be in a geometrical progression as Dr. Halley has found. If the gravity be as the distance, and the squares of the distances be in arithmetical progression, the densities will be in geometrical progression. And so *in infinitum*. These things will be so, when the density of the fluid condensed by compression is as the force of compression; or, which is the same thing, when the space possessed by the fluid is reciprocally as this force. Other laws of condensation may be supposed, as that the cube of the compressing force may be as the biquadrate of the density; or the triplicate ratio of the force the same with the quadruplicate ratio of the density : in which case, if the gravity be reciprocally as the square of the distance from the centre, the density will be reciprocally as the cube of the distance. Suppose that the cube of the compressing force be as the quadrato-cube of the density; and if the gravity be reciprocally as the square of the distance, the density will be reciprocally in a sesquiplicate ratio of the distance. Suppose the compressing force to be in a duplicate ratio of the density, and the gravity reciprocally in a duplicate ratio of the distance, and the density will be reciprocally as the distance. To run over all the cases that might be offered would be tedious. But as to our own air, this is certain from experiment, that its density is either accurately, or very nearly at least, as the compressing force; and therefore the density of the air in the atmosphere of the earth is as the weight of the whole incumbent air, that is, as the height of the mercury in the barometer.

PROPOSITION XXIII. THEOREM XVIII.

IF A FLUID BE COMPOSED OF PARTICLES MUTUALLY FLYING
EACH OTHER, AND THE DENSITY BE AS THE COMPRESSION, THE

CENTRIFUGAL FORCES OF THE PARTICLES WILL BE
RECIPROCALLY PROPORTIONAL TO THE DISTANCES OF THEIR
CENTRES. AND, VICE VERSA, PARTICLES FLYING EACH OTHER,
WITH FORCES THAT ARE RECIPROCALLY PROPORTIONAL TO THE
DISTANCES OF THEIR CENTRES, COMPOSE AN ELASTIC FLUID,
WHOSE DENSITY IS AS THE COMPRESSION.

Let the fluid be supposed to be included in a cubic space ACE, and then to be reduced by compression into a lesser cubic space *ace*; and the distances of the particles retaining a like situation with respect to each other in both the spaces, will be as the sides AB, *ab* of the cubes; and the densities of the mediums will be recip-rocally as the containing spaces AB^3, ab^3. In the plane side of the greater cube ABCD take the square DP equal to the plane side *db* of the lesser cube : and, by the supposition, the pressure with which the square DP urges the inclosed fluid will be to the pressure with which that square *db* urges the inclosed fluid as the densities of the medi-ums are to each other, that is, as ab^3 to AB^3. But the pressure with which the square DB urges the included fluid is to the pressure with which the square DP urges the same fluid as the square DB to the square DP, that is, as AB^2 to ab^2. Therefore, *ex aequo*, the pressure

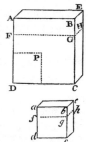

with which the square DB urges the fluid is to the pressure with which the square *db* urges the fluid as *ab* to AB. Let the planes FGH, *fgh*, be drawn through the middles of the two cubes, and divide the fluid into two parts. These parts will press each other mutu-ally with the same forces with which they are themselves pressed by the planes AC, *ac*, that is, in the proportion of *ab* to AB: and therefore the centrifugal forces by which these pressures are sustained are in the same ratio. The number of the particles being equal, and the situation alike, in both cubes, the forces which all the particles exert, according to the planes FGH, *fgh*, upon all, are as the forces which each exerts on each. Therefore the forces which each exerts on each, according to the plane FGH in the greater cube, are to the forces which each exerts on each, according to the plane *fgh* in the lesser cube, as *ab* to AB, that is, reciprocally as the distances of the particles from each other. Q.E.D.

And, *vice versa*, if the forces of the single particles are reciprocally as the distances, that is, reciprocally as the sides of the cubes AB, *ab*; the sums of the forces will be in the same ratio, and the pressures of the sides DB, *db* as the sum of the forces; and the pressure of the square DP to the pressure of the side DB as ab^2 to AB^2. And, *ex aequo*, the pressure of the square DP to the pressure of the side *db* as ab^3 to AB^3; that is, the force of compression in the one to the force of compression in the other as the density in the former to the density in the latter. Q.E.D.

SCHOLIUM.

By a like reasoning, if the centrifugal forces of the particles are reciprocally in the duplicate ratio of the distances between the centres, the cubes of the compressing forces will be as the biquadrates of the densities. If the centrifugal forces be reciprocally in the triplicate or quadruplicate ratio of the distances, the cubes of the compressing forces will be as the quadrato-cubes, or cubo-cubes of the densities. And universally, if D be put for the distance, and E for the density of the compressed fluid, and the centrifugal forces be reciprocally as any power D^n of the distance, whose index is the number n, the compressing forces will be as the cube roots of the power E^{n+2}, whose index is the number $n + 2$; and the contrary. All these things are to be understood of particles whose centrifugal forces terminate in those particles that are next them, or are diffused not much further. We have an example of this in magnetical bodies. Their attractive virtue is terminated nearly in bodies of their own kind that are next them. The virtue of the magnet is contracted by the interposition of an iron plate, and is almost terminated at it: for bodies further off are not attracted by the magnet so much as by the iron plate. If in this manner particles repel others of their own kind that lie next them, but do not exert their virtue on the more remote, particles of this kind will compose such fluids as are treated of in this Proposition. If the virtue of any particle diffuse itself every way *in infinitum*, there will be required a greater force to produce an equal condensation of a greater quantity of the fluid. But whether elastic fluids do really consist of particles so repelling each other, is a physical question. We have here demonstrated mathematically the property of fluids consisting of particles of this kind, that hence philosophers may take occasion to discuss that question.

SECTION VI.

Of the motion and resistance of funependulous bodies.

PROPOSITION XXIV. THEOREM XIX.

THE QUANTITIES OF MATTER IN FUNEPENDULOUS BODIES, WHOSE CENTRES OF OSCILLATION, ARE EQUALLY DISTANT FROM THE CENTRE OF SUSPENSION, ARE IN A RATIO COMPOUNDED OF THE RATIO OF THE WEIGHTS AND THE DUPLICATE RATIO OF THE TIMES OF THE OSCILLATIONS IN VACUO.

For the velocity which a given force can generate in a given matter in a given time is as the force and the time directly, and the matter inversely. The greater the force or the time is, or the less the matter, the greater velocity will be generated. This is manifest from the second Law of Motion. Now if pendulums are of the same length, the

motive forces in places equally distant from the perpendicular are as the weights: and therefore if two bodies by oscillating describe equal arcs, and those arcs are divided into equal parts; since the times in which the bodies describe each of the correspondent parts of the arcs are as the times of the whole oscillations, the velocities in the correspondent parts of the oscillations will be to each other as the motive forces and the whole times of the oscillations directly, and the quantities of matter reciprocally: and therefore the quantities of matter are as the forces and the times of the oscillations directly and the velocities reciprocally. But the velocities reciprocally are as the times, and therefore the times directly and the velocities reciprocally are as the squares of the times; and therefore the quantities of matter are as the motive forces and the squares of the times, that is, as the weights and the squares of the times.   Q.E.D.

COR. 1. Therefore if the times are equal, the quantities of matter in each of the bodies are as the weights.

COR. 2. If the weights are equal, the quantities of matter will be as the squares of the times.

COR. 3. If the quantities of matter are equal, the weights will be reciprocally as the squares of the times.

COR. 4. Whence since the squares of the times, *caeteris paribus*, are as the lengths of the pendulums, therefore if both the times and quantities of matter are equal, the weights will be as the lengths of the pendulums.

COR. 5. And universally, the quantity of matter in the pendulous body is as the weight and the square of the time directly, and the length of the pendulum inversely.

COR. 6. But in a non-resisting medium, the quantity of matter in the pendulous body is as the comparative weight and the square of the time directly, and the length of the pendulum inversely. For the comparative weight is the motive force of the body in any heavy medium, as was shewn above; and therefore does the same thing in such a non-resisting medium as the absolute weight does in a vacuum.

COR. 7. And hence appears a method both of comparing bodies one among another, as to the quantity of matter in each; and of comparing the weights of the same body in different places, to know the variation of its gravity. And by experiments made with the greatest accuracy, I have always found the quantity of matter in bodies to be proportional to their weight.

PROPOSITION XXV. THEOREM XX.

FUNEPENDULOUS BODIES THAT ARE, IN ANY MEDIUM,
RESISTED IN THE RATIO OF THE MOMENTS OF TIME, AND
FUNEPENDULOUS BODIES THAT MOVE IN A NON-RESISTING
MEDIUM OF THE SAME SPECIFIC GRAVITY, PERFORM THEIR
OSCILLATIONS IN A CYCLOID IN THE SAME TIME, AND
DESCRIBE PROPORTIONAL PARTS OF ARCS TOGETHER.

Let AB be an arc of a cycloid, which a body D, by vibrating in a non-resisting medium, shall describe in any time. Bisect that arc in C, so that C may be the lowest point thereof; and the accelerative force with which the body is urged in any place D, or d or E will be as the length of the arc CD, or Cd, or CE. Let that force be expressed by that same arc; and since the resistance is as the moment of the time, and therefore given, let it be expressed by the given part CO of the cycloidal arc, and take the arc Od

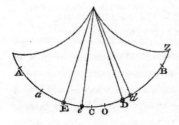

in the same ratio to the arc CD that the arc OB has to the arc CB: and the force with which the body in d is urged in a resisting medium, being the excess of the force Cd above the resistance CO, will be expressed by the arc Od, and will therefore be to the force with which the body D is urged in a non-resisting medium in the place D, as the arc Od to the arc CD; and therefore

also in the place B, as the arc OB to the arc CB. Therefore if two bodies D, d go from the place B, and are urged by these forces; since the forces at the beginning are as the arc CB and OB, the first velocities and arcs first described will be in the same ratio. Let those area be BD and Bd, and the remaining arcs CD, Od, will be in the same ratio. Therefore the forces, being proportional to those arcs CD, Od, will remain in the same ratio as at the beginning, and therefore the bodies will continue describing together arcs in the same ratio. Therefore the forces and velocities and the remaining arcs CD; Od, will be always as the whole arcs CB, OB, and therefore those remaining arcs will be described together. Therefore the two bodies D and d will arrive together at the places C and O; that which moves in the non-resisting medium, at the place C, and the other, in the resisting medium, at the place O. Now since the velocities in C and O are as the arcs CB, OB, the arcs which the bodies describe when they go farther will be in the same ratio. Let those arcs be CE and Oe. The force with which the body D in a non-resisting medium is retarded in E is as CE, and the force with which the body d in the resisting medium is retarded in e, is as the sum of the force Ce and the resistance CO, that is, as Oe; and therefore the forces with which the bodies are retarded are as the arcs CB, OB, proportional to the arcs CE, Oe; and therefore the velocities, retarded in that given ratio, remain in the same given ratio. Therefore the velocities and the arcs described with those velocities are always to each other in that given ratio of the arcs CB and OB; and therefore if the entire arcs AB, aB are taken in the same ratio, the bodies D and d will describe those arcs together, and in the places A and a will lose all their motion together. Therefore the whole oscillations are isochronal, or are performed in equal times; and any parts of the arcs, as BD, Bd, or BE, Be, that are described together, are proportional to the whole arcs BA, Ba.   Q.E.D.

COR. Therefore the swiftest motion in a resisting medium does not fall upon the lowest point C, but is found in that point O, in which the whole arc described B*a* is bisected. And the body, proceeding from thence to *a*, is retarded at the same rate with which it was accelerated before in its descent from B to O.

PROPOSITION XXVI. THEOREM XXI.

FUNEPENDULOUS BODIES, THAT ARE RESISTED IN THE RATIO OF THE VELOCITY, HAVE THEIR OSCILLATIONS IN A CYCLOID ISOCHRONAL.

For if two bodies, equally distant from their centres of suspension, describe, in oscillating, unequal arcs, and the velocities in the correspondent parts of the arcs be to each other as the whole arcs; the resistances, proportional to the velocities, will be also to each other as the same arcs. Therefore if these resistances be subducted from or added to the motive forces arising from gravity which are as the same arcs, the differences or sums will be to each other in the same ratio of the arcs; and since the increments and decrements of the velocities are as these differences or sums, the velocities will be always as the whole arcs; therefore if the velocities are in any one case as the whole arcs, they will remain always in the same ratio. But at the beginning of the motion, when the bodies begin to descend and describe those arcs, the forces, which at that time are proportional to the arcs, will generate velocities proportional to the arcs. Therefore the velocities will be always as the whole arcs to be described, and therefore those arcs will be described in the same time.   Q.E.D.

PROPOSITION XXVII. THEOREM XXII.

IF FUNEPENDULOUS BODIES ARE RESISTED IN THE DUPLICATE RATIO OF THEIR VELOCITIES, THE DIFFERENCES BETWEEN THE TIMES OF THE OSCILLATIONS IN A RESISTING MEDIUM, AND THE TIMES OF THE OSCILLATIONS IN A NON-RESISTING MEDIUM OF THE SAME SPECIFIC GRAVITY, WILL BE PROPORTIONAL TO THE ARCS DESCRIBED IN OSCILLATING NEARLY.

For let equal pendulums in a resisting medium describe the unequal arcs A, B; and the resistance of the body in the arc A will be to the resistance of the body in the correspondent part of the arc B in the duplicate ratio of the velocities, that is, as AA to BB nearly. If the resistance in the arc B were to the resistance in the arc A as AB to AA, the times in the arcs A and B would be

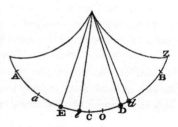

equal (by the last Prop.) Therefore the resistance AA in the arc A, or AB in the arc B, causes the excess of the time in the arc A above the time in a non-resisting medium; and the resistance BB causes the excess of the time in the arc B above the time in a non-resisting medium. But those excesses are as the efficient forces AB and BB nearly, that is, as the arcs A and B.   Q.E.D.

COR. 1. Hence from the times of the oscillations in unequal arcs in a resisting medium, may be known the times of the oscillations in a non-resisting medium of the same specific gravity. For the difference of the times will be to the excess of the time in the lesser arc above the time in a non-resisting medium as the difference of the arcs to the lesser arc.

COR. 2. The shorter oscillations are more isochronal, and very short ones are performed nearly in the same times as in a non-resisting medium. But the times of those which are performed in greater arcs are a little greater, because the resistance in the descent of the body, by which the time is prolonged, is greater, in proportion to the length described in the descent than the resistance in the subsequent ascent, by which the time is contracted. But the time of the oscillations, both short and long, seems to be prolonged in some measure by the motion of the medium. For retarded bodies are resisted somewhat less in proportion to the velocity, and accelerated bodies somewhat more than those that proceed uniformly forwards; because the medium, by the motion it has received from the bodies, going forwards the same way with them, is more agitated in the former case, and less in the latter; and so conspires more or less with the bodies moved. Therefore it resists the pendulums in their descent more, and in their ascent less, than in proportion to the velocity; and these two causes concurring prolong the time.

PROPOSITION XXVIII. THEOREM XXIII.

IF A FUNEPENDULOUS BODY, OSCILLATING IN A CYCLOID, BE
RESISTED IN THE RATIO OF THE MOMENTS OF THE TIME, ITS
RESISTANCE WILL BE TO THE FORCE OF GRAVITY AS THE EXCESS
OF THE ARC DESCRIBED IN THE WHOLE DESCENT ABOVE THE
ARC DESCRIBED IN THE SUBSEQUENT ASCENT TO TWICE THE
LENGTH OF THE PENDULUM.

Let BC represent the arc described in the descent, Ca the arc described in the ascent, and Aa the difference of the axes : and things remaining as they were constructed and demonstrated in Prop. XXV, the force with which the oscillating body is urged in any place D will be to the force of resistance as the arc CD to the arc CO,

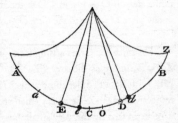

which is half of that difference A*a*. Therefore the force with which the oscillating body is urged at the beginning or the highest point of the cycloid, that is, the force of gravity, will be to the resistance as the arc of the cycloid, between that highest point and lowest point C, is to the arc M; that is (doubling those arcs), as the whole cycloidal arc, or twice the length of the pendulum, to the arc A*a*.   Q.E.D.

PROPOSITION XXIX. PROBLEM VI.

SUPPOSING THAT A BODY OSCILLATING IN A CYCLOID IS
RESISTED IN A DUPLICATE RATIO OF THE VELOCITY: TO FIND
THE RESISTANCE IN EACH PLACE.

Let B*a* be an arc described in one entire oscillation, C the lowest point of the cycloid, and CZ half the whole cycloidal arc, equal to the length of the pendulum; and

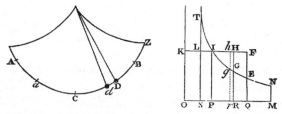

let it be required to find the resistance of the body in any place D. Cut the indefinite right line OQ in the points O, S, P, Q, so that (erecting the perpendiculars OK, ST, PI, QE, and with the centre O, and the aysmptotes OK, OQ, describing the hyperbola TIGE cutting the perpendiculars ST, PI, QE in T, I, and E, and through the point I drawing KF, parallel to the asymptote OQ, meeting the asymptote OK in K, and the perpendiculars ST and QE in L and F) the hyperbolic area PIEQ may be to the hyperbolic area PITS as the arc BC, described in the descent of the body, to the arc C*a* described in the ascent; and that the area IEF may be to the area ILT as OQ to OS. Then with the perpendicular MN cut off the hyperbolic area PINM, and let that area be to the hyperbolic area PIEQ as the arc CZ to the arc BC described in the descent. And if the perpendicular. RG cut off the hyperbolic area PIGR, which shall be to the area PIEQ as any arc CD to the arc BC described in the whole descent, the resistance in any place D will be to the force of gravity as the area $\frac{OR}{OQ}$ IEF – IGH to the area PINM.

For since the forces arising from gravity with which the body is urged in the places Z, B, D, *a*, are as the arcs CZ, CB, CD, C*a* and those arcs are as the areas PINM, PIFQ, PIGR, PITS; let those areas be the exponents both of the arcs and of the forces respectively. Let D*d* be a very small space described by the body in its descent: and let it be expressed by the very small area RG*gr* comprehended between the parallels RG,

rg; and produce rg to h, so that GHhg and RGgr may be the contemporaneous decrements of the areas IGH, PIGR. And the increment GH$hg - \dfrac{Rr}{OQ}$ IEF, or $Rr \times HG - \dfrac{Rr}{OQ}$ IEF, of the area $\dfrac{OR}{OQ}$ IEF $-$ IGH will be to the decrement RGgr, or $Rr \times RG$, of the area PIGR, as HG $- \dfrac{IEF}{OQ}$ to RG; and therefore as $OR \times HG - \dfrac{OR}{OQ}$ IEF to OR x GR or OP x PI, that is (because of the equal quantities OR x HG, OR x HR - OR x GR, ORHK - OPIK, PIHR and PIGR + IGH), as PIGR $+$ IGH $- \dfrac{OR}{OQ}$ IEF to OPIK. Therefore if the area $\dfrac{OR}{OQ}$ IEF $-$ IGH be called Y, and RGgr the decrement of the area PIGR be given, the increment of the area Y will be as PIGR $-$ Y.

Then if V represent the force arising from the gravity, proportional to the arc CD to be described, by which the body is acted upon in D, and R be put for the resistance, V - R will be the whole force with which the body is urged in D. Therefore the increment of the velocity is as V - R and the particle of time in which it is generated conjunctly. But the velocity itself is as the contemporaneous increment of the space described directly and the same particle of time inversely. Therefore, since the resistance is, by the supposition, as the square of the velocity, the increment of the resistance will (by Lem. II) be as the velocity and the increment of the velocity conjunctly, that is, as the moment of the space and V - R conjunctly; and, therefore, if the moment of the space be given, as V - R; that is, if for the force V we put its exponent PIGR, and the resistance R be expressed by any other area Z, as PIGR - Z.

Therefore the area PIGR uniformly decreasing by the subduction of given moments, the area Y increases in proportion of PIGR - Y, and the area Z in proportion of PIGR - Z. And therefore if the areas Y and Z begin together, and at the beginning are equal, these, by the addition of equal moments, will continue to be equal; and in like manner decreasing by equal moments, will vanish together. And, *vice versa*, if they together begin and vanish, they will have equal moments and be always equal; and that, because if the resistance Z be augmented, the velocity together with the arc Ca, described in the ascent of the body, will be diminished; and the point in which all the motion together with the resistance ceases coming nearer to the point C, the resistance vanishes sooner than the area Y. And the contrary will happen when the resistance is diminished.

Now the area Z begins and ends where the resistance is nothing, that is, at the beginning of the motion where the arc CD is equal to the arc CB, and the right line RG falls upon the right line QE; and at the end of the motion where the arc CD is equal to the arc Ca, and RG falls upon the right line ST. And the area Y or $\dfrac{OR}{OQ}$ IEF $-$ IGH

begins and ends also where the resistance is nothing, and therefore where $\frac{OR}{OQ}$IEF and IGH are equal; that is (by the construction), where the right line RG falls successively

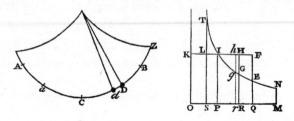

upon the right lines QE and ST. Therefore those areas begin and vanish together, and are therefore always equal. Therefore the area $\frac{OR}{OQ}$IEF–IGH is equal to the area Z, by

which the resistance is expressed, and therefore is to the area, PINM, by which the gravity is expressed, as the resistance to the gravity.   Q.E.D.

COR. 1. Therefore the resistance in the lowest place C is to the force of gravity as the area $\frac{OP}{OQ}$IEF to the area PINM.

COR. 2. But it becomes greatest where the area PIHR is to the area IEF as OR to OQ. For in that case its moment (that is, PIGR - Y) becomes nothing.

COR. 3. Hence also may be known the velocity in each place, as being in the subduplicate ratio of the resistance, and at the beginning of the motion equal to the velocity of the body oscillating in the same cycloid without any resistance.

However, by reason of the difficulty of the calculation by which the resistance and the velocity are found by this Proposition, we have thought fit to subjoin the Proposition following.

PROPOSITION XXX. THEOREM XXIV.

IF A RIGHT LINE AB BE EQUAL TO THE ARC OF A CYCLOID
WHICH AN OSCILLATING BODY DESCRIBES, AND AT EACH OF
ITS POINTS D THE PERPENDICULARS DK BE ERECTED,
WHICH SHALL BE TO THE LENGTH OF THE PENDULUM AS
THE RESISTANCE OF THE BODY IN THE CORRESPONDING
POINTS OF THE ARC TO THE FORCE OF GRAVITY; I SAY, THAT
THE DIFFERENCE BETWEEN THE ARC DESCRIBED IN THE
WHOLE DESCENT AND THE ARC DESCRIBED IN THE WHOLE

SUBSEQUENT ASCENT DRAWN INTO HALF THE SUM OF THE
SAME ARCS WILL BE EQUAL TO THE AREA BK*a* WHICH ALL
THOSE PERPENDICULARS TAKE UP.

Let the arc of the cycloid, described in one entire oscillation, be expressed by the right

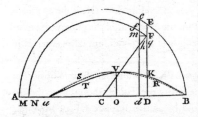

line *a*B, equal to it, and the arc which would
have been described *in vacuo* by the length AB.
Bisect AB in C, and the point C will represent
the lowest point of the cycloid, and CD will be
as the force arising from gravity, with which the
body in D is urged in the direction of the tangent of the cycloid, and will have the same ratio
to the length of the pendulum as the force in D

has to the force of gravity. Let that force, therefore, be expressed by that length CD, and the
force of gravity by the length of the pendulum; and if in DE you take DK in the same ratio
to the length of the pendulum as the resistance has to the gravity, DK will lie the exponent
of the resistance. From the centre C with the interval CA or CB describe a semi-circle BE*e*A.
Let the body describe, in the least time, the space D*d*; and, erecting the perpendiculars DE,
de, meeting the circumference in E and *e*, they will be as the velocities which the body
descending *in vacuo* from the point B would acquire in the places D and *d*. This appears by
Prop. LII, Book I. Let therefore, these velocities be expressed by those perpendiculars DE, *de*;
and let DF be the velocity which it acquires in D by falling from B in the resisting medium.
And if from the centre C with the interval CF we describe the circle F*f*M meeting the right
lines *de* and AB in *f* and M, then M will be the place to which it would thenceforward, without farther resistance, ascend, and *df* the velocity it would acquire in *d*. Whence, also, if F*g*
represents the moment of the velocity which the body D, in describing the least space D*d*,
loses by the resistance of the medium; and CN be taken equal to C*g*; then will N be the place
to which the body, if it met no farther resistance, would thenceforward ascend, and MN will
be the decrement of the ascent arising from the loss of that velocity. Draw F*m* perpendicular to *df*, and the decrement F*g* of the velocity DF generated by the resistance DK will be to
the increment *fm* of the same velocity, generated by the force CD, as the generating force DK
to the generating force CD. But because of the similar triangles F*mf*, F*hg*, FDC, *fm* is to F*m*
or D*d* as CD to DF; and, *ex aequo*, F*g* to D*d* as DK to DF. Also F*h* is to F*g*, as DF to CF;
and, *ex aequo perturbate*, F*h* or MN to D*d* as DK to CF or CM; and therefore the sum of
all the MN × CM will be equal to the sum of all the D*d* × DK. At the moveable point M
suppose always a rectangular ordinate erected equal to the indeterminate CM, which by a
continual motion is drawn into the whole length A*a*; and the trapezium described by that
motion, or its equal, the rectangle A*a* × $\frac{1}{2}$ *a*B, will be equal to the sum of all the MN x CM,
and therefore to the sum of all the D*d* x DK, that is, to the area BKVT*a*.   Q.E.D.

COR. Hence from the law of resistance, and the difference Aa of the arcs Ca, CB, may be collected the proportion of the resistance to the gravity nearly.

For if the resistance DK be uniform, the figure BKTa will be a rectangle under Ba and DK; and thence the rectangle under $\frac{1}{2}$ Ba and Aa will be equal to the rectangle under Ba and DK, and DK will be equal to $\frac{1}{2}$ Aa. Wherefore since DK is the exponent of the resistance, and the length of the pendulum the exponent of the gravity, the resistance will be to the gravity as $\frac{1}{2}$ Aa to the length of the pendulum; altogether as in Prop. XXVIII is demonstrated.

If the resistance be as the velocity, the figure BKTa will be nearly an ellipsis. For if a body, in a non-resisting medium, by one entire oscillation, should describe the length BA, the velocity in any place D would be as the ordinate DE of the circle described on the diameter AB. Therefore since Ba in the resisting medium, and BA in the non-resisting one, are described nearly in the same times; and therefore the velocities in each of the points of Ba are to the velocities in the correspondent points of the length BA nearly as Ba is to BA, the velocity in the point D in the resisting medium will be as the ordinate of the circle or ellipsis described upon the diameter Ba; and therefore the figure BKVTa will be nearly an ellipsis. Since the resistance is supposed proportional to the velocity, let OV be the exponent of the resistance in the middle point O; and an ellipsis BRVSa described with the centre O, and the semi-axes OB, OV, will be nearly equal to the figure BKVTa, and to its equal the rectangle Aa x BO. Therefore Aa x BO is to OV x BO as the area of this ellipsis to OV x BO; that is, Aa is to OV as the area of the semi-circle to the square of the radius, or as 11 to 7 nearly; and, therefore, $\frac{7}{11}$ Aa is to the length of the pendulum as the resistance of the oscillating body in O to its gravity.

Now if the resistance DK be in the duplicate ratio of the velocity, the figure BKVTa will be almost a parabola having V for its vertex and OV for its axis, and therefore will be nearly equal to the rectangle under $\frac{2}{3}$ Ba and OV. Therefore the rectangle under $\frac{1}{2}$ Ba and Aa is equal to the rectangle $\frac{2}{3}$ Ba x OV, and therefore OV is equal to $\frac{3}{4}$ Aa; and therefore the resistance in O made to the oscillating body is to its gravity as $\frac{3}{4}$ Aa to the length of the pendulum.

And I take these conclusions to be accurate enough for practical uses. For since an ellipsis or parabola BRVSa falls in with the figure BKVTa in the middle point V, that figure, if greater towards the part BRV or VSa than the other, is less towards the contrary part, and is therefore nearly equal to it.

PROPOSITION XXXI. THEOREM XXV.

IF THE RESISTANCE MADE TO AN OSCILLATING BODY IN
EACH OF THE PROPORTIONAL PARTS OF THE ARCS DESCRIBED
BE AUGMENTED OR DIMINISHED IN A GIVEN RATIO, THE
DIFFERENCE BETWEEN THE ARE DESCRIBED IN THE DESCENT
AND THE ARC DESCRIBED IN THE SUBSEQUENT ASCENT WILL
BE AUGMENTED OR DIMINISHED IN THE SAME RATIO.

For that difference arises from the retardation of the pendulum by the resistance of the
medium, and therefore is as the whole retardation and the retarding resistance proportional thereto. In the foregoing Proposition the rectangle under the right line $\frac{1}{2}$ aB and the difference Aa of the

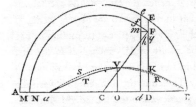

arcs CB, Ca, was equal to the area BKTa. And that area, if the length aB remains, is augmented or diminished in the ratio of the ordinates DK; that is, in the ratio of the resistance and is therefore as the length aB and the resistance conjunctly. And therefore the rectangle under Aa and $\frac{1}{2}$ aB is as aB and the resistance conjunctly, and therefore Aa is as the resistance.   Q.E.D.

COR. 1. Hence if the resistance be as the velocity, the difference of the arcs in the same medium will be as the whole arc described: and the contrary.

COR. 2. If the resistance be in the duplicate ratio of the velocity, that difference will be in the duplicate ratio of the whole arc: and the contrary.

COR. 3. And universally, if the resistance be in the triplicate or any other ratio of the velocity, the difference will be in the same ratio of the velocity: and the contrary.

COR. 4. If the resistance be partly in the simple ratio of the velocity, and partly in the duplicate ratio of the same, the difference will be partly in the ratio of the whole arc, and partly in the duplicate ratio of it: and the contrary. So that the law and ratio of the resistance will be the same for the velocity as the law and ratio, of that difference for the length of the arc.

COR. 5. And therefore if a pendulum describe successively unequal arcs, and we can find the ratio of the increment or decrement of this difference for the length of the arc described, there will be had also the ratio of the increment or decrement of the resistance for a greater or less velocity.

GENERAL SCHOLIUM.

From these propositions we may find the resistance of mediums by pendulums oscillating therein. I found the resistance of the air by the following experiments. I suspended

a wooden globe or ball weighing $57^7/_{22}$ ounces troy, its diameter $6^7/_8$ *London* inches, by a fine thread on a firm hook, so that the distance between the hook and the centre of oscillation of the globe was $10^1/_2$ feet. I marked on the thread a point 10 feet and 1 inch distant from the centre of suspension; and even with that point I placed a ruler divided into inches, by the help whereof I observed the lengths of the arcs described by the pendulum. Then I numbered the oscillations in which the globe would lose $^1/_8$ part of its motion. If the pendulum was drawn aside from the perpendicular to the distance of 2 inches, and thence let go, so that in its whole descent it described an arc of 2 inches, and in the first whole oscillation, compounded of the descent and subsequent ascent, an arc of almost 4 inches, the same in 164 oscillations lost $^1/_8$ part of its motion, so as in its last ascent to describe an arc of $1^3/_4$ inches. If in the first descent it described an arc of 4 inches, it lost $^1/_8$ part of its motion in 121 oscillations, so as in its last ascent to describe an arc of $3^1/_2$ inches. If in the first descent it described an arc of 8, 16, 32, or 64 inches, it lost $^1/_8$ part of its motion in 69, $35^1/_2$, $18^1/_2$, $9^2/_3$ oscillations, respectively. Therefore the difference between the arcs described in the first descent and the last ascent was in the 1st, 2d, 3d, 4th, 5th, 6th cases, $^1/_4$, $^1/_2$, 1, 2, 4, 8 inches respectively. Divide those differences by the number of oscillations in each case, and in one mean oscillation, wherein an arc of $3^3/_4$, $7^1/_2$, 15, 30, 60, 120 inches was described, the difference of the arcs described in the descent and subsequent ascent will be $^1/_{656}$, $^1/_{242}$, $^1/_{69}$, $^4/_{71}$, $^8/_{37}$, $^{24}/_{29}$ parts of an inch, respectively. But these differences in the greater oscillations are in the duplicate ratio of the arcs described nearly, but in lesser oscillations something greater than in that ratio; and therefore (by Cor. 2, Prop. XXXI of this Book) the resistance of the globe, when it moves very swift, is in the duplicate ratio of the velocity, nearly; and when it moves slowly, somewhat greater than in that ratio.

Now let V represent the greatest velocity in any oscillation, and let A, B, and C be given quantities, and let us suppose the difference of the arcs to be $AV + BV^2 + CV^2$. Since the greatest velocities are in the cycloid as $^1/_2$ the arcs described in oscillating, and in the circle as $^1/_2$ the chords of those arcs; and therefore in equal arcs are greater in the cycloid than in the circle in the ratio of $^1/_2$ the arcs to their chords; but the times in the circle are greater than in the cycloid, in a reciprocal ratio of the velocity; it is plain that the differences of the arcs (which are as the resistance and the square of the time conjunctly) are nearly the same in both curves: for in the cycloid those differences must be on the one hand augmented, with the resistance, in about the duplicate ratio of the arc to the chord, because of the velocity augmented in the simple ratio of the same; and on the other hand diminished, with the square of the time in the same duplicate ratio. Therefore to reduce these observations to the cycloid, we must take the same differences of the arcs as were observed in the circle, and suppose the greatest velocities analogous

to the half, or the whole arcs, that is, to the numbers $^1/_2$, 1, 2, 4, 8, 16. Therefore in the 2d, 4th, and 6th cases, put 1, 4, and 16 for V; and the difference of the arcs in the 2d case will become $\dfrac{\frac{1}{2}}{121} = A + B + C$; in the 4th case, $\dfrac{\frac{2}{1}}{35^2} = 4A + 8B + 16C$;

in the 6th case, $\dfrac{\frac{8}{2}}{9^3} = 16A + 64B + 256C$. These equations reduced give A =

0,0000916, B = 0,0010847, and C = 0,0029558. Therefore the difference of the arcs is as $0,0000916,\ V\ =\ 0,0010847V^{\frac{3}{2}}\ +\ 0,0029558V^2$: and therefore since (by Cor. Prop.

XXX, applied to this case) the resistance of the globe in the middle of the arc described in oscillating, where the velocity is V, is to its weight as $\dfrac{7}{11}AV + \dfrac{7}{10}BV^{\frac{3}{2}} + \dfrac{3}{4}CV^2$ to

the length of the pendulum, if for A, B, and C you put the numbers found, the resistance of the globe will be to its weight as $0,0000583V + 0,0007593V^{\frac{3}{2}} + 0,0022169V^2$ to

the length of the pendulum between the centre of suspension and the ruler, that is, to 121 inches. Therefore since V in the second case represents 1, in the 4th case 4, and in the 6th case 16, the resistance will be to the weight of the globe, in the 2d case, as 0,0030345 to 121; in the 4th, as 0,041748 to 121; in the 6th, as 0,61705 to 121.

The arc, which the point marked in the thread described in the 6th case, was of $121 - \dfrac{8}{9\frac{2}{3}}$, or 119 $^5/_{29}$ inches. And therefore since the radius was 121 inches, and the

length of the pendulum between the point of suspension and the centre of the globe was 126 inches, the arc which the centre of the globe described was 124 $^3/_{31}$ inches. Because the greatest velocity of the oscillating body, by reason of the resistance of the air, does not fall on the lowest point of the arc described, but near the middle place of the whole arc, this velocity will be nearly the same as if the globe in its whole descent in a non-resisting medium should describe 62 $^3/_{62}$ inches, the half of that arc, and that in a cycloid, to which we have above reduced the motion of the pendulum; and therefore that velocity will be equal to that which the globe would acquire by falling perpendicularly from a height equal to the versed sine of that arc. But that versed sine in the cycloid is to that are 62 $^3/_{62}$ as the same arc to twice the length of the pendulum 252, and therefore equal to 15,278 inches. Therefore the velocity of the pendulum is the same which a body would acquire by falling, and in its fall describing a space of 15,278 inches. Therefore with such a velocity the globe meets with a resistance which is to its weight as 0,61705 to 121, or (if we take that part only of the resistance which is in the duplicate ratio of the velocity) as 0,56752 to 121.

I found, by an hydrostatical experiment, that the weight of this wooden globe was to the weight of a globe of water of the same magnitude as 55 to 97: and therefore since 121 is to 213,4 in the same ratio, the resistance made to this globe of water, moving forwards with the above-mentioned velocity, will be to its weight as 0,56752 to 213,4, that is, as 1 to 376 $^1/_{50}$. Whence since the weight of a globe of water, in the time in which the globe with a velocity uniformly continued describes a length of 30,556 inches, will generate all that velocity in the falling globe, it is manifest that the force of resistance uniformly continued in the same time will take away a velocity, which will be less than the other in the ratio of 1 to 376$^1/_{50}$, that is, the $\frac{1}{376\frac{1}{50}}$ part of the whole velocity.

And therefore in the time that the globe, with the same velocity uniformly continued, would describe the length of its semi-diameter, or 3$^7/_{16}$ inches, it would lose the $^1/_{3342}$ part of its motion.

I also counted the oscillations in which the pendulum lost $^1/_4$ part of its motion. In the following table the upper numbers denote the length of the arc described in the first descent, expressed in inches and parts of an inch; the middle numbers denote the length of the arc described in the last ascent; and in the lowest place are the numbers of the oscillations. I give an account of this experiment, as being more accurate than that in which only $^1/_8$ part of the motion was lost. I leave the calculation to such as are disposed to make it.

First descent	2	4	8	16	32	64
Last ascent	1 $^1/_2$	3	6	12	24	48
Numb. of oscill.	374	272	162 $^1/_2$	83 $^1/_3$	41 $^2/_3$	22 $^2/_3$

I afterward suspended a leaden globe of 2 inches in diameter, weighing 26 $^1/_4$ ounces troy by the same thread, so that between the centre of the globe and the point of suspension there was an interval of 10 $^1/_2$ feet, and I counted the oscillations in which a given part of the motion was lost. The first of the following tables exhibits the number of oscillations in which $^1/_8$ part of the whole motion was lost; the second the number of oscillations in which there was lost $^1/_4$ part of the same.

First descent	1	2	4	8	16	32	64
Last ascent	$^7/_8$	$^7/_4$	3 $^1/_2$	7	14	28	56
Numb. of oscill.	226	228	193	140	90 $^1/_2$	53	30
First descent	1	2	4	8	16	32	64
Last ascent	$^3/_4$	1 $^1/_2$	3	6	12	24	48
Numb. of oscill.	510	518	420	318	204	121	70

Selecting in the first table the 3d, 5th, and 7th observations, and expressing the greatest velocities in these observations particularly by the numbers 1, 4, 16 respectively,

and generally by the quantity V as above there will come out in the 3d observation $\frac{1}{2}$ =A + B + C, in the 5th observation $\frac{2}{90\frac{1}{2}}$ =4A + 8B + 16C, in the 7th

observation 8/30 = 16A + 64B + 256C. These equations reduced give A = 0,001414, B = 0,000297, C = 0,000879. And thence the resistance of the globe moving with the velocity V will be to its weight 261, ounces in the same ratio as $0,0009V + 0,000208V\frac{2}{3} + 0,000659V^2$ to 121 inches, the length of the pendulum. And if we regard that part only of the resistance which is in the duplicate ratio of the velocity, it will be to the weight of the globe as $0,000659V^2$ to 121 inches. But this part of the resistance in the first experiment was to the weight of the wooden globe of $57^7/_{22}$ ounces as $0,002217V^2$ to 121; and thence the resistance of the wooden globe is to the resistance of the leaden one (their velocities being equal) as $57^7/_{22}$ into 0,002217 to $26^1/_4$ into 0,000659, that is, as $7^1/_3$ to 1. The diameters of the two globes were $6^7/_8$ and 2 inches, and the squares of these are to each other as $47^1/_4$, and 4, or $11^{13}/_{16}$ and 1, nearly. Therefore the resistances of these equally swift globes were in less than a duplicate ratio of the diameters. But we have not yet considered the resistance of the thread, which was certainly very considerable, and ought to be subducted from the resistance of the pendulums here found. I could not determine this accurately, but I found it greater than a third part of the whole resistance of the lesser pendulum; and thence I gathered that the resistances of the globes, when the resistance of the thread is subducted, are nearly in the duplicate ratio of their diameters. For the ratio of $7^1/_3$ - $1/_3$ to 1 - $1/_3$, or $10^1/_2$ to 1 is not very different from the duplicate ratio of the diameters $11^{13}/_{16}$ to 1.

Since the resistance of the thread is of less moment in greater globes, I tried the experiment also with a globe whose diameter was, $18^3/_4$ inches. The length of the pendulum between the point of suspension and the centre of oscillation was $122^1/_2$ inches, and between the point of suspension and the knot in the thread $109^1/_2$ inches. The arc described by the knot at the first descent of the pendulum was 32 inches. The arc described by the same knot in the last ascent after five oscillations was 28 inches. The sum of the arcs, or the whole arc described in one mean oscillation, was 60 inches. The difference of the arcs 4 inches. The $1/_{10}$ part of this, or the difference between the descent and ascent in one mean oscillation, is $2/_5$ of an inch. Then as the radius 109 $1/_2$ to the radius $122^1/_2$, so is the whole arc of 60 inches described by the knot in one mean oscillation to the whole arc of $67^1/_8$ inches described by the centre of the globe in one mean oscillation; and so is the difference $2/_5$ to a new difference 0,4475. If the length of the arc described were to remain, and the length of the pendulum should be augmented in the ratio of 126 to $122^1/_2$, the time of the oscillation would be augmented, and the velocity of the pendulum would be diminished in the subduplicate of

that ratio; so that the difference 0,4475 of the arcs described in the descent and subsequent ascent would remain. Then if the arc described be augmented in the ratio of $124^3/_{31}$ to $67^1/_8$, that difference 0,4475 would be augmented in the duplicate of that ratio, and so would become 1,5295. These things would be so upon the supposition that the resistance of the pendulum were in the duplicate ratio of the velocity. Therefore if the pendulum describe the whole arc of $124^3/_{31}$ inches, and its length between the point of suspension and the centre of oscillation be 126 inches, the difference of the arcs described in the descent and subsequent ascent would be 1,5295 inches. And this difference multiplied into the weight of the pendulous globe, which was 208 ounces, produces 318,136. Again; in the pendulum above-mentioned, made of a wooden globe, when its centre of oscillation, being 126 inches from the point of suspension, described the whole arc of $124^3/_{31}$ inches, the difference of the arcs described in the descent and ascent was $^{126}/_{121}$ into $\dfrac{8}{9\frac{2}{3}}$. This multiplied into the weight of the globe, which was

$5^7/_{22}$ ounces, produces 49,396. But I multiply these differences into the weights of the globes, in order to find their resistances. For the differences arise from the resistances, and are as the resistances directly and the weights inversely. Therefore the resistances are as the numbers 318,136 and 49,396. But that part of the resistance of the lesser globe which is in the duplicate ratio of the velocity, was to the whole resistance as 0,56752 to 0,61675, that is, as 45,453 to 49,396; whereas that part of the resistance of the greater globe is almost equal to its whole resistance; and so those parts are nearly as 318,136 and 45,453, that is, as 7 and 1. But the diameters of the globes are $18^3/_4$ and $6\,^7/_8$; and their squares $351^9/_{16}$ and $47^{17}/_{64}$ are as 7,438 and 1, that is, as the resistances of the globes 7 and 1, nearly. The difference of these ratios is scarce greater than may arise from the resistance of the thread. Therefore those parts of the resistances which are, when the globes are equal, as the squares of the velocities, are also, when the velocities are equal, as the squares of the diameters of the globes.

But the greatest of the globes I used in these experiments was not perfectly spherical, and therefore in this calculation I have, for brevity's sake, neglected some little niceties; being not very solicitous for an accurate calculus in an experiment that was not very accurate. So that I could wish that these experiments were tried again with other globes, of a larger size, more in number, and more accurately formed; since the demonstration of a vacuum depends thereon. If the globes be taken in a geometrical proportion, as suppose whose diameters are 4, 8, 16, 32 inches; one may collect from the progression observed in the experiments what would happen if the globes were still larger.

In order to compare the resistances of different fluids with each other, I made the following trials. I procured a wooden vessel 4 feet long, 1 foot broad, and 1 foot high.

This vessel, being uncovered, I filled with spring water, and, having immersed pendulums therein, I made them oscillate in the water. And I found that a leaden globe weighing 166 $1/_6$, ounces, and in diameter 3 $5/_8$ inches, moved therein as it is set down in the following table; the length of the pendulum from the point of suspension to a certain point marked in the thread being 126 inches, and to the centre of oscillation 134 $3/_8$ inches.

The arc described in the first descent, by a point marked in the thread was inches.	64	32	16	8	4	2	1	$1/_2$	$1/_4$
The arc described in the last ascent was inches.	48	24	12	6	3	$1 1/_2$	$3/_4$	$3/_8$	$3/_{16}$
The difference of the arcs, proportional to the motion lost, was inches.	16	8	4	2	1	$1/_2$	$1/_4$	$1/_8$	$1/_{16}$
The number of the oscillations in water.		$29/_{60}$	$1 1/_3$	3	7	$11 1/_4$	$12 2/_3$	$13 1/_3$	
The number of the oscillations in air.	$85 1/_2$	287	535						

In the experiments of the 4th column there were equal motions lost in 535 oscillations made in the air, and $1 1/_5$ in water. The oscillations in the air were indeed a little swifter than those in the water. But if the oscillations in the water were accelerated in such a ratio that the motions of the pendulums might be equally swift in both mediums, there would be still the same number $1 1/_5$ is of oscillations in the water, and by these the same quantity of motion would be lost as before; because the resistance if increased, and the square of the time diminished in the same duplicate ratio. The pendulums, therefore, being of equal velocities, there were equal motions lost in 535 oscillations in the air, and $1 1/_5$ in the water; and therefore the resistance of the pendulum in the water is to its resistance in the air as 535 to $1 1/_5$. This is the proportion of the whole resistances in the case of the 4th column.

Now let $AV + CV^2$ represent the difference of the arcs described in the descent and subsequent ascent by the globe moving in air with the greatest velocity V; and since the greatest velocity is in the case of the 4th column to the greatest velocity in the case of the 1st column as 1 to 8; and that difference of the arcs in the case of the 4th column to the difference in the case of the 1st column as $2/_{535}$ to $\dfrac{16}{85\frac{1}{2}}$, or as $85 1/_2$ to 4280; put in these cases 1 and 8 for the velocities, and $85 1/_2$ and 4280 for the differences of the arcs, and A + C will be = $85 1/_2$ and 8A + 64C = 4280 or A + 8C = 535; and then by reducing these equations, there will come out 7C = $449 1/_2$ and C = $64 3/_{14}$ and A = $21 2/_7$; and therefore the resistance, which is as $7/_{11}AV + 3/_4CV^2$, will become as $13 6/_{11}$ V + $48 9/_{56} V^2$. Therefore in the case of the 4th column, where the velocity was 1, the

whole resistance is to its part proportional to the square of the velocity as $13 \, ^6/_{11}V + 48 \, ^9/_{56} \, V^2$ or $61 \, ^{13}/_{17}$ to $48 \, ^9/_{56}$; and therefore the resistance of the pendulum in water is to that part of the resistance in air, which is proportional to the square of the velocity, and which in swift motions is the only part that deserves consideration, as $61^{13}/_{17}$ to $48^9/_{56}$ and 535 to $1^1/_5$ conjunctly, that is, as 571 to 1. If the whole thread of the pendulum oscillating in the water had been immersed, its resistance would have been still greater; so that the resistance of the pendulum oscillating in the water, that is, that part which is proportional to the square of the velocity, and which only needs to be considered in swift bodies, is to the resistance of the same whole pendulum, oscillating in air with the same velocity, as about 950 to 1, that is as, the density of water to the density of air, nearly.

In this calculation we ought also to have taken in that part of the resistance of the pendulum in the water which was as the square of the velocity; but I found (which will perhaps seem strange) that the resistance in the water was augmented in more than a duplicate ratio of the velocity. In searching after the cause, I thought upon this, that the vessel was too narrow for the magnitude of, the pendulous globe, and by its narrowness obstructed the motion of the water as it yielded to the oscillating globe. For when, I immersed a pendulous globe, whose diameter was one inch only, the resistance was augmented nearly in a duplicate ratio of the velocity. I tried this by making a pendulum of two globes, of which the lesser and lower oscillated in the water, and the greater and higher was fastened to the thread just above the water, and, by oscillating in the air, assisted the motion of the pendulum, and continued it longer. The experiments made by this contrivance proved according to the following table.

Arc descr. in first descent	16	8	4	2	1	$^1/_2$	$^1/_4$
Arc descr. in last ascent	12	6	3	$1^1/_2$	$^3/_4$	$^3/_8$	$^3/_{16}$
Diff. of arcs, proport. to motion lost	4	2	1	$^1/_2$	$^1/_4$	$^1/_8$	$^1/_{16}$
Number of oscillations	$3 \, ^3/_8$	$6 \, ^1/_2$	$12^1/_{12}$	$21^1/_5$	34	53	$62^1/_5$

In comparing the resistances of the mediums with each other, I also caused iron pendulums to oscillate in quicksilver. The length of the iron wire was about 3 feet, and the diameter of the pendulous globe about $^1/_3$ of an inch. To the wire, just above the quicksilver, there was fixed another leaden globe of a bigness sufficient to continue the motion of the pendulum for some time. Then a vessel, that would hold about 3 pounds of quicksilver, was filled by turns with quicksilver and common water, that, by making the pendulum oscillate successively in these two different fluids, I might find the proportion of their resistances; and the resistance of the quicksilver proved to be to the resistance of water as about 13 or 14 to 1; that is, as the density of quicksilver to the density of water. When I made use of a pendulous globe something bigger, as of one

whose diameter was about $1/2$ or $2/3$ of an inch, the resistance of the quicksilver proved to be to the resistance of the water as about 12 or 10 to 1. But the former experiment is more to be relied on, because in the latter the vessel was too narrow in proportion to the magnitude of the immersed globe; for the vessel ought to have been enlarged together with the globe. I intended to have repeated these experiments with larger vessels, and in melted metals, and other liquors both cold and hot; but I had not leisure to try all: and besides, from what is already described, it appears sufficiently that the resistance of bodies moving swiftly is nearly proportional to the densities of the fluids in which they move. I do not say accurately; for more tensious fluids, of equal density, will undoubtedly resist more than those that are more liquid; as cold oil more than warm, warm oil more than rainwater, and water more than spirit of wine. But in liquors, which are sensibly fluid enough, as in air, in salt and fresh water, in spirit of wine, of turpentine, and salts, in oil cleared of its faeces by distillation and warmed, in oil of vitriol, and in mercury, and melted metals, and any other suchlike, that are fluid enough to retain for some time the motion impressed upon them by the agitation of the vessel, and which being poured out are easily resolved into drops, I doubt not but the rule already laid down may be accurate enough, especially if the experiments be made with larger pendulous bodies and more swiftly moved.

Lastly, since it is the opinion of some that there is a certain ethereal medium extremely rare and subtile, which freely pervades the pores of all bodies; and from such a medium, so pervading the pores of bodies, some resistance must needs arise; in order to try whether the resistance, which we experience in bodies in motion, be made upon their outward superficies only, or whether their internal parts meet with any considerable resistance upon their superficies, I thought of the following experiment. I suspended a round deal box by a thread 11 feet long on a steel hook, by means of a ring of the same metal, so as to make a pendulum of the aforesaid length. The hook had a sharp hollow edge on its upper part, so that the upper arc of the ring pressing on the edge might move the more freely; and the thread was fastened to the lower arc of the ring. The pendulum being thus prepared, I drew it aside from the perpendicular to the distance of about 6 feet, and that in a plane perpendicular to the edge of the hook, lest the ring, while the pendulum oscillated, should slide to and fro on the edge of the hook: for the point of suspension, in which the ring touches the hook, ought to remain immovable. I therefore accurately noted the place to which the pendulum was brought, and letting it go, I marked three other places, to which it returned at the end of the 1st, 2d, and 3d oscillation. This I often repeated, that I might find those places as accurately as possible. Then I filled the box with lead and other heavy metals that were near at hand. But, first, I weighed the box when empty, and that part of the thread that went round it, and half the remaining part, extended between the hook and the suspended

box; for the thread so extended always acts upon the pendulum, when drawn aside from the perpendicular, with half its weight. To this weight I added the weight of the air contained in the box. And this whole weight was about $1/78$ of the weight of the box when filled with the metals. Then because the box when full of the metals, by extending the thread with its weight, increased the length of the pendulum, I shortened the thread so as to make the length of the pendulum, when oscillating, the same as before. Then drawing aside the pendulum to the place first marked, and letting it go, I reckoned about 77 oscillations before the box returned to the second mark, and as many afterwards before it came to the third mark, and as many after that before it came to the fourth mark. From whence I conclude that the whole resistance of the box, when full, had not a greater proportion to the resistance of the box, when empty, than 78 to 77. For if their resistances were equal, the box, when full, by reason of its *vis insita*, which was 78 times greater than the *vis insita* of the same when empty, ought to have continued its oscillating motion so much the longer, and therefore to have returned to those marks at the end of 78 oscillations. But it returned to them at the end of 77 oscillations.

Let, therefore, A represent the resistance of the box upon its external superficies, and B the resistance of the empty box on its internal superficies; and if the resistances to the internal parts of bodies equally swift be as the matter, or the number of particles that are resisted, then 78B will be the resistance made to the internal parts of the box, when full; and therefore the whole resistance A + B of the empty box will be to the whole resistance A + 78B of the full box as 77 to 78, and, by division, A + B to 77B as 77 to 1; and thence A + B to B as 77 x 77 to 1, and, by division again, A to B as 5928 to 1. Therefore the resistance of the empty box in its internal parts will be above 5000 times less than the resistance on its external superficies. This reasoning depends upon the supposition that the greater resistance of the full box arises not from any other latent cause, but only from the action of some subtile fluid upon the included metal.

This experiment is related by memory, the paper being lost in which I had described it; so that I have been obliged to omit some fractional parts, which are slipt out of my memory; and I have no leisure to try it again. The first time I made it, the hook being weak, the full box was retarded sooner. The cause I found to be, that the hook was not strong enough to bear the weight of the box; so that, as it oscillated to and fro, the hook was bent sometimes this and sometimes that way. I therefore procured a hook of sufficient strength, so that the point of suspension might remain unmoved, and then all things happened as is above described.

SECTION VII.

Of the motion of fluids, and the resistance made to projected bodies.

PROPOSITION XXXII. THEOREM XXVI.

SUPPOSE TWO SIMILAR SYSTEMS OF BODIES CONSISTING OF AN EQUAL NUMBER OF PARTICLES, AND LET THE CORRESPONDENT PARTICLES BE SIMILAR AND PROPORTIONAL, EACH IN ONE SYSTEM TO EACH IN THE OTHER, AND HAVE A LIKE SITUATION AMONG THEMSELVES, AND THE SAME GIVEN RATIO OF DENSITY TO EACH OTHER; AND LET THEM BEGIN TO MOVE AMONG THEMSELVES IN PROPORTIONAL TIMES, AND WITH LIKE MOTIONS (THAT IS, THOSE IN ONE SYSTEM AMONG ONE ANOTHER, AND THOSE IN THE OTHER AMONG ONE ANOTHER). AND IF THE PARTICLES THAT ARE IN THE SAME SYSTEM DO NOT TOUCH ONE ANOTHER, EXCEPT IN THE MOMENTS OF REFLEXION; NOR ATTRACT, NOR REPEL EACH OTHER, EXCEPT WITH ACCELERATIVE FORCES THAT ARE AS THE DIAMETERS OF THE CORRESPONDENT PARTICLES INVERSELY, AND THE SQUARES OF THE VELOCITIES DIRECTLY; I SAY, THAT THE PARTICLES OF THOSE SYSTEMS WILL CONTINUE TO MOVE AMONG THEMSELVES WITH LIKE MOTIONS AND IN PROPORTIONAL TIMES.

Like bodies in like situations are said to be moved among themselves with like motions and in proportional times, when their situations at the end of those times are always found alike in respect of each other; as suppose we compare the particles in one system with the correspondent particles in the other. Hence the times will be proportional, in which similar and proportional parts of similar figures will be described by correspondent particles. Therefore if we suppose two systems of this kind, the correspondent particles, by reason of the similitude of the motions at their beginning, will continue to be moved with like motions, so long as they move without meeting one another; for if, they are acted on by no forces, they will go on uniformly in right lines, by the 1st Law. But if they do agitate one another with some certain forces, and those forces are as the diameters of the correspondent particles inversely and the squares of the velocities directly, then, because the particles are in like situations, and their forces are proportional, the whole forces with which correspondent particles are agitated, and which are compounded of each of the agitating forces (by Corol. 2 of the Laws), will

have like directions, and have the same effect as if they respected centres placed alike among the particles; and those whole forces will be to each other as the several forces which compose them, that is, as the diameters of the correspondent particles inversely, and the squares of the velocities directly: and therefore will came, correspondent particles to continue to describe like figures. These things will be so (by Cor. 1 and 8, Prop. IV., Book I), if those centres are at rest; but if they are moved, yet, by reason of the similitude of the translations, their situations among the particles of the system will remain similar, so that the changes introduced into the figures described by the particles will still be similar. So that the motions of correspondent and similar particles will continue similar till their first meeting with each other; and thence will arise similar collisions, and similar reflexions: which will again beget similar motions of the particles among themselves (by what was just now shewn), till they mutually fall upon one another again, and so on *ad infinitum.*

COR. 1. Hence if any two bodies, which are similar and in like situations to the correspondent particles of the systems, begin to move amongst them in like manner and in proportional times, and their magnitudes and densities be to each other as the magnitudes and densities of the corresponding particles, these bodies will continue to be moved in like manner and in proportional times; for the case of the greater parts of both systems and of the particles is the very same.

COR . 2. And if all the similar and similarly situated parts of both systems be at rest among themselves; and two of them, which are greater than the rest, and mutually correspondent in both systems, begin to move in lines alike posited, with any similar motion whatsoever, they will excite similar motions in the rest of the parts of the systems, and will continue to move among those parts in like manner and in proportional times; and will therefore describe spaces proportional to their diameters.

PROPOSITION XXXIII. THEOREM XXVII.

THE SAME THINGS BEING SUPPOSED, I SAY, THAT THE GREATER
PARTS OF THE SYSTEMS ARE RESISTED IN A RATIO COMPOUNDED
OF THE DUPLICATE RATIO OF THEIR VELOCITIES, AND THE
DUPLICATE RATIO OF THEIR DIAMETERS, AND THE SIMPLE
RATIO OF THE DENSITY OF THE PARTS OF THE SYSTEMS.

For the resistance arises partly from the centripetal or centrifugal forces with which the particles of the system mutually act on each other, partly from the collisions and reflexions of the particles and the greater parts. The resistances of the first kind are to each other as the whole motive forces from which they arise, that is, as the whole accelerative forces and the quantities of matter in corresponding parts; that is (by the sup-

position), as the squares of the velocities directly, and the distances of the corresponding particles inversely, and the quantities of matter in the correspondent parts directly: and therefore since the distances of the particles in one system are to the correspondent distances of the particles of the others as the diameter of one particle or part in the former system to the diameter of the correspondent particle or part in the other, and since the quantities of matter are as the densities of the parts and the cubes of the diameters; the resistances are to each other as the squares of the velocities and the squares of the diameters and the densities of the parts of the systems. Q.E.D. The resistances of the latter sort are as the number of correspondent reflexions and the forces of those reflexions conjunctly; but the number of the reflexions are to each other as the velocities of the corresponding parts directly and the spaces between their reflexions inversely. And the forces of the reflexions are as the velocities and the magnitudes and the densities of the corresponding parts conjunctly; that is, as the velocities and the cubes of the diameters and the densities of the parts. And, joining all these ratios, the resistances of the corresponding parts are to each other as the squares of the velocities and the squares of the diameters and the densities of the parts conjunctly.   Q.E.D.

COR. 1. Therefore if those systems are two elastic fluids, like our air, and their parts are at rest among themselves; and two similar bodies proportional in magnitude and density to the parts of the fluids, and similarly situated among those parts, be any how projected in the direction of lines similarly posited; and the accelerative forces with, which the particles of the fluids mutually act upon each other are as the diameters of the bodies projected inversely and the squares of their velocities directly; those bodies will excite similar motions in the fluids in proportional times, and will describe similar spaces and proportional to their diameters.

COR. 2. Therefore in the same fluid a projected body that moves swiftly meets with a resistance that is, in the duplicate ratio of its velocity, nearly. For if the forces with which distant particles act mutually upon one another should be augmented in the duplicate ratio of the velocity, the projected body would be resisted in the same duplicate ratio accurately; and therefore in a medium, whose parts when at a distance do not act mutually with any force on one another, the resistance is in the duplicate ratio of the velocity accurately. Let there be, therefore, three mediums A, B, C, consisting of similar and equal parts regularly disposed at equal distances. Let the parts of the mediums A and B recede from each other with forces that are among themselves as T and V; and let the parts of the medium C be entirely destitute of any such forces. And if four equal bodies D, E, F, G, move in these mediums, the two first D and E in the two first A and B, and the other two F and G in the third C; and if the velocity of the body D be to the velocity of the body E, and the velocity of the body F to the velocity of the body G, in the subduplicate ratio of the force T to the force V: the resistance of the

body D to the resistance of the body E, and the resistance of the body F to the resistance of the body G, will be in the duplicate ratio of the velocities; and therefore the resistance of the body D will be to the resistance of the body F as the resistance of the body E to the resistance of the body G. Let the bodies D and F be equally swift, as also the bodies E and G; and, augmenting the velocities of the bodies D and F in any ratio, and diminishing the forces of the particles of the medium B in the duplicate of the same ratio, the medium B will approach to the form and condition of the medium C at pleasure; and therefore the resistances of the equal and equally swift bodies E and G in these mediums will perpetually approach to equality so that their difference will at last become less than any given. Therefore since the resistances of the bodies D and F are to each other as the resistances of the bodies E and G, those will also in like manner approach to the ratio of equality. Therefore the bodies D and F, when they move with very great swiftness, meet with resistances very nearly equal; and therefore since the resistance of the body F is in a duplicate ratio of the velocity, the resistance of the body D will be nearly in the same ratio.

COR. 3. The resistance of a body moving very swift in an elastic fluid is almost the same as if the parts of the fluid were destitute of their centrifugal forces, and did not fly from each other; if so be that the elasticity of the fluid arise from the centrifugal forces of the particles, and the velocity be so great as not to allow the particles time enough to act.

COR. 4. Therefore, since the resistances of similar and equally swift bodies, in a medium whose distant parts do not fly from each other, are as the squares of the diameters, the resistances made to bodies moving with very great and equal velocities in an elastic fluid will be as the squares of the diameters, nearly.

COR. 5. And since similar, equal, and equally swift bodies, moving through mediums of the same density, whose particles do not fly from each other mutually, will strike against an equal quantity of matter in equal times, whether the particles of which the medium consists be more and smaller, or fewer and greater, and therefore impress on that matter an equal quantity of motion, and in return (by the 3d Law of Motion) suffer an equal re-action from the same, that is, are equally resisted; it is manifest, also, that in elastic fluids of the same density, when the bodies move with extreme swiftness, their resistances are nearly equal, whether the fluids consist of gross parts, or of parts ever so subtile. For the resistance of projectiles moving with exceedingly great celerities is not much diminished by the subtilty of the medium.

COR. 6. All these things are so in fluids whose elastic force takes its rise from the centrifugal forces of the particles. But if that force arise from some other cause, as from the expansion of the particles after the manner of wool, or the boughs of trees, or any other cause, by which the particles are hindered from moving freely among

themselves, the resistance, by reason of the lesser fluidity of the medium, will be greater than in the Corollaries above.

PROPOSITION XXXIV THEOREM XXVIII.

IF IN A RARE MEDIUM, CONSISTING OF EQUAL PARTICLES
FREELY DISPOSED AT EQUAL DISTANCES FROM EACH OTHER,
A GLOBE AND A CYLINDER DESCRIBED ON EQUAL DIAMETERS
MOVE WITH EQUAL VELOCITIES IN THE DIRECTION OF THE
AXIS OF THE CYLINDER, THE RESISTANCE OF THE GLOBE
WILL BE BUT HALF SO GREAT AS THAT OF THE CYLINDER.

For since the action of the medium upon the body is the same (by Cor. 5 of the Laws) whether the body move in a quiescent medium, or whether the particles of the medium impinge with the same velocity upon the quiescent body, let us consider the body as if it were quiescent, and see with what force it would be impelled by the moving medium. Let, therefore, ABKI represent a spherical body described from the centre C with the semi-diameter CA, and let the particles of the medium impinge with a given velocity upon that spherical body in the directions of right lines parallel to AC; and let FB be one of those right lines. In FB take LB equal to the semi-diameter CB, and draw BD touching the sphere in B. Upon KC and BD let fall the perpendiculars BE, LD; and

the force with which a particle of the medium, impinging on the globe obliquely in the direction FB, would strike the globe in B, will be to the force with which the same particle, meeting the cylinder ONGQ described about the globe with the axis ACI, would strike it perpendicularly in b, as LD to LB, or BE to BC. Again; the efficacy of this force to move the globe, according to the direction of its incidence FB or AC, is to the effi-

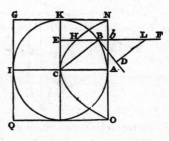

cacy of the same to move the globe, according to the direction of its determination, that is, in the direction of the right line BC in which it impels the globe directly, as BE to BC. And, joining these ratios, the efficacy of a particle, falling upon the globe obliquely in the direction of the right line FB, to move the globe in the direction of its incidence, is to the efficacy of the same particle falling in the same line perpendicularly on the cylinder, to move it in the same direction, as BE2 to BC2. Therefore if in bE, which is perpendicular to the circular base of the cylinder NAO, and equal to the radius AC, we take bH equal to BE2/CB; then bH will be to bE as the effect of the particle upon the globe to the effect of the particle upon the cylinder. And therefore

the solid which is formed by all the right lines bH will be to the solid formed by all the right lines bE as the effect of all the particles upon the globe to the effect of all the particles upon the cylinder. But the former of these solids is a paraboloid whose vertex is C, its axis CA, and *latus rectum* CA, and the latter solid is a cylinder circumscribing the paraboloid; and it is known that a paraboloid is half its circumscribed cylinder. Therefore the whole force of the medium upon the globe is half of the entire force of the same upon the cylinder. And therefore if the particles of the medium are at rest, and the cylinder and globe move with equal velocities, the resistance of the globe will be half the resistance of the cylinder.

Q.E.D.

SCHOLIUM.

By the same method other figures may be compared together as to their resistance; and those may be found which are most apt to continue their motions in resisting mediums. As if upon the circular base CEBH from the centre O, with the radius OC, and the altitude OD, one would construct a frustum CBGF of a cone, which should meet with less resistance than any other frustum constructed with the same base and altitude, and forwards towards D in the direction of its axis: bisect the altitude OD in Q, and produce OQ to S, so that QS may be equal to QC, and S will be the vertex of the cone whose frustum is sought.

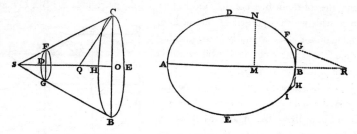

Whence, by the bye, since the angle CSB is always acute, it follows, that, if the solid ADBE be generated by the convolution of an elliptical or oval figure ADBE about its axis AB, and the generating figure be touched by three right lines FG, GH, HI, in the points F, B, and I, so that GH shall be perpendicular to the axis in the point of contact B, and FG, HI maybe inclined to GH in the angles FGB, BHI of 135 degrees: the solid arising from the convolution of the figure ADFGHIE about the same axis AB will be less resisted than the former solid; if so be that both move forward in the direction of their axis AB, and that the extremity B of each go foremost. Which Proposition I conceive may be of use in the building of ships.

If the figure DNFG be such a curve, that if, from any point thereof, as N, the perpendicular NM be let fall on the axis AB, and from the given point G there be drawn

the right line GR parallel to a right line touching the figure in N, and cutting the axis produced in R, MN becomes to GR as GR^3 to $4BR \times GB^2$, the solid described by the revolution of this figure about its axis AB, moving in the before-mentioned rare medium from A towards B, will be less resisted than any other circular solid whatsoever, described of the same length and breadth.

PROPOSITION XXXV. PROBLEM VII.

IF A RARE MEDIUM CONSIST OF VERY SMALL QUIESCENT PARTICLES OF EQUAL MAGNITUDES, AND FREELY DISPOSED AT EQUAL DISTANCES FRONT ONE ANOTHER: TO FIND THE RESISTANCE OF A GLOBE MOVING UNIFORMLY FORWARD IN THIS MEDIUM.

CASE 1. Let a cylinder described with the same diameter and altitude be conceived to go forward with the same velocity in the direction of its axis through the same medium; and let us suppose that the particles of the medium, on which the globe or cylinder falls, fly back with as great a force of reflexion as possible. Then since the resistance of the globe (by the last Proposition) is but half the resistance of the cylinder, and since the globe is to the cylinder as 2 to 3, and since the cylinder by falling perpendicularly on the particles, and reflecting them with the utmost force, communicates to them a velocity double to its own; it follows that the cylinder, in moving forward uniformly half the length of its axis, will communicate a motion to the particles which is to the whole motion of the cylinder as the density of the medium to the density of the cylinder; and that the globe, in the time it describes one length of its diameter in moving uniformly forward, will communicate the same motion to the particles; and in the time that it describes two thirds of its diameter, will communicate a motion to the particles which is to the whole motion of the globe as the density of the medium to; the density of the globe. And therefore the globe meets with a resistance, which is to the force by which its whole motion may be either taken away or generated in the time in which it describes two thirds of its diameter moving uniformly forward, as the density of the medium to the density of the globe.

CASE 2. Let us suppose that the particles of the medium incident on the globe or cylinder are not reflected; and then the cylinder falling perpendicularly on the particles will communicate its own simple velocity to them, and therefore meets a resistance but half so great as in the former case, and the globe also meets with a resistance but half so great.

CASE 3. Let us suppose the particles of the medium to fly back from the globe with a force which is neither the greatest, nor yet none at all, but with a certain mean force;

then the resistance of the globe will be in the same mean ratio between the resistance in the first case and the resistance in the second. Q.E.I.

COR. 1. Hence if the globe and the particles are infinitely bard, and destitute, of all elastic force, and therefore of all force of reflexion; the resistance of the globe will be to the force by which its whole motion may be destroyed or generated, in the time that the globe describes four third parts of its diameter, as the density of the medium to the density of the globe.

COR. 2. The resistance of the globe, *caeteris paribus*, is in the duplicate ratio of the velocity.

COR. 3. The resistance of the globe, *caeteris paribus*, is in the duplicate ratio of the diameter.

COR. 4. The resistance of the globe is, *caeteris paribus*, as the density of the medium.

COR. 5. The resistance of the globe is in a ratio compounded of the duplicate ratio of the velocity, and the duplicate ratio of the diameter, and the ratio of the density of the medium.

COR. 6. The motion of the globe and its resistance maybe thus expounded. Let AB be the time in which the globe may, by its resistance uniformly continued, lose its whole motion. Erect AD, BC perpendicular to AB. Let BC be that whole motion, and through the point C, the asymptotes being AD, AB, describe the hyperbola CF. Produce AB to any point E. Erect the perpendicular EF meeting the hyperbola in F. Complete the parallelogram CBEG, and draw AF meeting BC in H. Then if the globe in any time BE, with its first motion BC uniformly continued, describes in a non-resisting medium the space CBEG expounded by the area of the parallelogram, the same in a resisting medium will describe the space CBEF expounded by the area of the

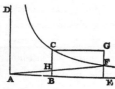

hyperbola; and its motion at the end of that time, will be expounded by EF, the ordinate of the hyperbola, there being lost of its motion the part FG. And its resistance at the end of the same time will be expounded by the length BH, there being lost of its resistance the part CH. All these things appear by Cor. 1 and 3, Prop. V., Book II.

COR. 7. Hence if the globe in the time T by the resistance R uniformly continued lose its whole motion M, the same globe in the time *t* in a resisting medium, wherein the resistance R decreases in a duplicate ratio of the velocity, will lose out of its motion M the part $\frac{tM}{T+t}$, the part $\frac{TM}{T+t}$ remaining; and will describe a space

which is to the space described in the same time *t*, with the uniform motion M, as the

logarithm of the number $\frac{T+t}{T}$ multiplied by the number 2,302585092994 is to

the number t because the hyperbolic area BCFE is to the rectangle BCGH in that proportion.

SCHOLIUM.

I have exhibited in this Proposition the resistance and retardation of spherical projectiles in mediums that are not continued, and shewn that this resistance is to the force by which the whole motion of the globe may be destroyed or produced in the time in which the globe can describe two thirds of its diameter, with a velocity uniformly continued, as the density of the medium to the density of the globe, if so be the globe and the particles of the medium be perfectly elastic, and are endued with the utmost force of reflexion; and that this force, where the globe and particles of the medium are infinitely hard and void of any reflecting force, is diminished one half. But in continued mediums, as water, hot oil, and quicksilver, the globe as it passes through them does not immediately strike against all the particles of the fluid that generate the resistance made to it, but presses only the particles that lie next to it, which press the particles beyond, which press other particles, and so on; and in these mediums the resistance is diminished one other half. A globe in these extremely fluid mediums meets with a resistance that is to the force by which its whole motion may be destroyed or generated in the time wherein it can describe, with that motion uniformly continued, eight third parts of its diameter, as the density of the medium to the density of the globe. This I shall endeavour to shew in what follows.

PROPOSITION XXXVI. PROBLEM VIII.

To DEFINE THE MOTION OF WATER RUNNING OUT OF A CYLINDRICAL VESSEL THROUGH A HOLE MADE AT THE BOTTOM.

Let ACDB be a cylindrical vessel, AB the mouth of it, CD the bottom parallel to the horizon, EF a circular hole in the middle of the bottom, G the centre of the hole, and GH the axis of the cylinder perpendicular to the horizon. And suppose a cylinder of ice APQB to be of the same breadth with the cavity of the vessel, and to have the same axis, and to descend perpetually with an uniform motion, and that its parts, as soon as they touch the superficies AB, dissolve into water, and flow down by their weight into the vessel, and in their fall compose the cataract or column of water ABNFEM, passing through the hole EF, and filling up the same exactly. Let the uniform velocity of the descending ice and of the contiguous water in the circle AB be that which the water would acquire by falling through the space IH; and let IH and HG lie in the same right

line; and through the point I let there be drawn the right line KL parallel to the horizon, and meeting the ice on both the sides thereof in K and L. Then the velocity of the water running out at the hole EF will be the same that it would acquire by falling from I through the space IG. Therefore, by *Galileo's* Theorems, IG will be to IH in the duplicate ratio of the velocity of the water that runs out at the hole to the velocity of the water in the circle AB, that is, in the duplicate ratio of the circle AB to the circle EF: those circles being reciprocally as the velocities of the water which in the same time and in equal quantities passes severally through each of them, and completely fills them both. We are now considering the velocity with which the water tends to the plane of the horizon. But the motion parallel to the same, by which the parts of the falling water approach to each

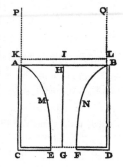

other, is not here taken notice of; since it is neither produced by gravity, nor at all changes the motion perpendicular to the horizon which the gravity produces. We suppose, indeed, that the parts of the water cohere a little, that by their cohesion they may in falling approach to each other with motions parallel to the horizon in order to form one single cataract, and to prevent their being divided into several: but the motion parallel to the horizon arising from this cohesion does not come under our present consideration.

CASE 1. Conceive now the whole cavity in the vessel, which encompasses the falling water ABNPEM, to be full of ice, so that the water may pass through the ice as through a funnel. Then if the water pass very near to the ice only, without touching it; or, which is the same thing, if by reason of the perfect smoothness of the surface of the ice, the water, though touching it, glides over it with the utmost freedom, and without the least resistance; the water will run through the hole EF with the same velocity as before, and the whole weight of the column of water ABNFEM will be all taken up as before in forcing out the water, and the bottom of the vessel will sustain the weight of the ice encompassing that column.

Let now the ice in the vessel dissolve into water; yet will the efflux of the water remain, as to its velocity, the same as before. It will not be less, because the ice now dissolved will endeavour to descend; it will not be greater, because the ice, now become water, cannot descend without hindering the descent of other water equal to its own descent. The same force ought always to generate the same velocity in the effluent water.

But the hole at the bottom of the vessel, by reason of the oblique motions of the particles of the effluent water, must be a little greater than before. For now the particles

of the water do not all of them pass through the hole perpendicularly, but, flowing down on all parts from the sides of the vessel, and converging towards the hole, pass through it with oblique motions; and in tending downwards meet in a stream whose diameter is a little smaller below the hole than at the hole itself; its diameter being to the diameter of the hole as 5 to 6, or as $5^1/_2$ to $6^1/_2$ very nearly, if I took the measures of those diameters right. I procured a very thin flat plate, having a hole pierced in the middle, the diameter of the circular hole being $^5/_8$ parts of an inch. And that the stream of running waters might not be accelerated in falling, and by that acceleration become narrower, I fixed this plate not to the bottom, but to the side of the vessel, so as to make the water go out in the direction of a line parallel to the horizon. Then, when the vessel was full of water, I opened the hole to let it run out; and the diameter of the stream, measured with great accuracy at the distance of about half an inch from the hole, was $^{21}/_{46}$ of an inch. Therefore the diameter of this circular hole was to the diameter of the stream very nearly as 25 to 21. So that the water in passing through the hole converges on all sides, and, after it has run out of the vessel, becomes smaller by converging in that manner, and by becoming smaller is accelerated till it comes to the distance of half an inch from the hole, and at that distance flows in a smaller stream and with greater celerity than in the hole itself, and this in the ratio of 25×25 to 21×21, or 17 to 12, very nearly; that is, in about the subduplicate ratio of 2 to 1. Now it is certain from experiments, that the quantity of water running out in a given time through a circular hole made in the bottom of a vessel is equal to the quantity, which, flowing with the aforesaid velocity, would run out in the same time through another circular hole, whose diameter is to the diameter of the former as 21 to 25. And therefore that running water in passing through the hole itself has a velocity downwards equal to that which a heavy body would acquire in falling through half the height of the stagnant water in the vessel, nearly. But, then, after it has run out, it is still accelerated by converging, till it arrives at a distance from the hole that is nearly equal to its diameter, and acquires a velocity greater than the other in about the subduplicate ratio of 2 to 1; which velocity a heavy body would nearly acquire by falling through the whole height of the stagnant water in the vessel.

Therefore in what follows let the diameter of the stream be represented by that lesser hole which we called EF. And imagine another plane VW above the hole EF, and parallel to the plane thereof, to be placed at a distance equal to the diameter of the same hole, and to be pierced through with a greater hole ST, of such a magnitude that a stream which will exactly fill the lower hole EF way pass through it; the diameter of which hole will therefore be to the diameter of the lower hole as 25 to 21, nearly. By this means the water will run perpendicularly out at the lower hole; and the quantity of the water running out will be, according to the magnitude of this last hole, the same,

very nearly, which the solution of the Problem requires. The space included between the two planes and the falling stream may be considered as the bottom of the vessel. But, to make the solution more simple and mathematical, it is better to take the lower plane alone for the bottom of the vessel, and to suppose that the water which flowed through the ice as through a funnel, and ran out of the vessel through the hole EF made in the lower plane, preserves its motion continually, and that the ice continues at

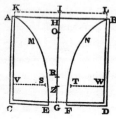

rest. Therefore in what follows let ST be the diameter of a circular hole described from the centre Z, and let the stream run out of the vessel through that hole, when the water in the vessel is all fluid. And let EF be the diameter of the hole, which the stream, in falling through, exactly fills up, whether the water runs out of the vessel by that upper hole ST, or flows through the middle of the ice in the vessel, as through a funnel. And let the diameter of the upper hole ST be to the diameter of the lower EF as about 25 to 21, and let the perpendicular distance between the planes of the holes be equal to the diameter of the lesser hole EF. Then the velocity of the water downwards, in running out of the vessel through the hole ST, will be in that hole the same that a body may acquire by falling from half the height IZ; and the velocity of both the falling streams will be in the hole EF, the same which a body would acquire by falling from the whole height IG.

CASE 2. If the hole EF be not in the middle of the bottom of the vessel, but in some other part thereof, the water will still run out with the same velocity as before, if the magnitude of the hole be the same. For though an heavy body takes a longer time in descending to the same depth, by an oblique line, than by a perpendicular line, yet in both cases it acquires in its descent the same velocity; as *Galileo* has demonstrated.

CASE 3. The velocity of the water is the same when it runs out through a hole in the side of the vessel. For if the hole be small, so that the interval between the superficies AB and KL may vanish as to sense, and the stream of water horizontally issuing out may form a parabolic figure; from the *latus rectum* of this parabola may be collected, that the velocity of the effluent water is that which a body may acquire by falling the height IG or HG of the stagnant water in the vessel. For, by making an experiment, I found that if the height of the stagnant water above the hole were 20 inches, and the height of the hole above a plane parallel to the horizon were also 20 inches, a stream of water springing out from thence would fall upon the plane, at the distance of 37 inches, very nearly, from a perpendicular let fall upon that plane from the hole. For without resistance the stream would have fallen upon the plane at the distance of 40 inches, the *latus rectum* of the parabolic stream being 80 inches.

CASE 4. If the effluent water tend upward, it will still issue forth with the same velocity. For the small stream of water springing upward, ascends with a perpendicular motion to GH or GI, the height of the stagnant water in the vessel; excepting in so far as its ascent is hindered a little by the resistance of the air; and therefore it springs out with the same velocity that it would acquire in falling from that height. Every particle of the stagnant water is equally pressed on all sides (by Prop. XIX., Book II), and, yielding to the pressure, tends always with an equal force, whether it descends through the hole in the bottom of the vessel, or gushes out in an horizontal direction through a hole in the side, or passes into a canal, and springs up from thence through a little hole made in the upper part of the canal. And it may not only be collected from reasoning, but is manifest also from the well-known experiments just mentioned, that the velocity with which the water runs out is the very same that is assigned in this Proposition.

CASE 5. The velocity of the effluent water is the same, whether the figure of the hole be circular, or square, or triangular, or any other figure equal to the circular; for the velocity of the effluent water does not depend upon the figure of the hole, but arises from its depth below the plane KL.

CASE 6. If the lower part of the vessel ABDC be immersed into stagnant water, and

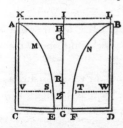

the height of the stagnant water above the bottom of the vessel be GR, the velocity with which the water that is in the vessel will run out at the hole EF into the stagnant water will be the same which the water would acquire by falling, from the height IR; for the weight of all the water in the vessel that is below the superficies of the stagnant water will be sustained in equilibrio by the weight of the stagnant water, and therefore does not at all accelerate the

motion of the descending water in the vessel. This case will also appear by experiments, measuring the times in which the water will run out.

COR. 1. Hence if CA the depth of the water be produced to K, so that AK may be to CK in the duplicate ratio of the area of a hole made in any part of the bottom to the area of the circle AB, the velocity of the effluent water will be equal to the velocity which the water would acquire by falling from the height KC.

COR. 2. And the force with which the whole motion of the effluent water may be generated is equal to the weight of a cylindric column of water, whose base is the hole EF, and its altitude 2GI or 2CK. For the effluent water, in the time it becomes equal to this column, may acquire, by falling by its own weight from the height GI, a velocity equal to that with which it runs out.

COR. 3. The weight of all the water in the vessel ABDC is to that part of the weight which is employed in forcing out the water as the sum of the circles AB and EF

to twice the circle EF. For let IO be a mean proportional between IH and IG, and the water running out at the hole EF will, in the time that a drop falling from I would describe the altitude IG, become equal to a cylinder whose base is the circle EF and its altitude 2IG, that is, to a cylinder whose base is the circle AB, and whose altitude is 2IO. For the circle EF is to the circle AB in the subduplicate ratio of the altitude IH to the altitude IG; that is, in the simple ratio of the mean proportional IO to the altitude IG. Moreover, in the time that a drop falling from I can describe the altitude IH, the water that runs out will have become equal to a cylinder whose base is the circle AB, and its altitude 2IH; and in the time that a drop falling from I through H to G describes HG, the difference of the altitudes, the effluent water, that is, the water contained within the solid ABNFEM, will be equal to the difference of the cylinders, that is, to a cylinder whose base is AB, and its altitude 2HO. And therefore all the water contained in the vessel ABDC is to the whole falling water contained in the said solid ABNFEM as HG to 2HO, that is, as HO + OG to 2HO, or IH + IO to 2IH. But the weight of all the water in the solid ABNFEM is employed in forcing out the water; and therefore the weight of all the water in the vessel is to that part of the weight that is employed in forcing out the water as IH + IO to 2IH, and therefore as the sum of the circles EF and AB to twice the circle EF.

COR. 4. And hence the weight of all the water in the vessel ABDC is to the other part of the weight which is sustained by the bottom of the vessel as the sum of the circles AB and EF to the difference of the same circles.

COR. 5. And that part of the weight which the bottom of the vessel sustains is to the other part of the weight employed in forcing out the water as the difference of the circles AB and EF to twice the lesser circle EP, or as the area of the bottom to twice the hole.

COR. 6. That part of the weight which presses upon the bottom is to the whole weight of the water perpendicularly incumbent thereon as the circle AB to the sum of the circles AB and EF, or as the Circle AB to the excess of twice the circle AB above the area of the bottom. For that part of the weight which presses upon the bottom is to the weight of the whole water in the vessel as the difference of the circles AB and EF to the sum of the same circles (by Cor. 4); and the weight of the whole water in the vessel is to the weight of the whole water perpendicularly incumbent on the bottom as the circle AB to the difference of the circles AB and EF. Therefore, *ex aequo perturbatè*;, that part of the weight which presses upon the bottom is to the weight of the whole water perpendicularly incumbent thereon as the circle AB to the sum of the circles AB and EF, or the excess of twice the circle AB above the bottom.

COR. 7. If in the middle of the hole EF there be placed the little circle PQ described about the centre G, and parallel to the horizon, the weight of water which

that little circle sustains is greater than the weight of a third part of a cylinder of water whose base is that little circle and its height GH. For let ABNFEM be the cataract or column of falling water whose axis is GH, as above, and let all the water, whose fluidity is not requisite for the ready and quick descent of the water, be supposed to be congealed, as well round about the cataract, as above the little circle. And let PHQ be the column of water congealed above the little circle, whose vertex is H, and its altitude GH. And suppose this cataract to fall with its whole weight downwards, and not in the least to lie against or to press PHQ, but to glide freely by it without any friction, unless, perhaps, just at the very vertex of the ice, where the cataract at the beginning of its fill may tend to a concave figure. And as the congealed water AMEC, BNFD, lying round the cataract, is convex in its internal superficies AME, BNF, towards the falling cataract, so this column PHQ will be convex towards the cataract also, and will therefore be greater than a cone whose base is that little circle PQ and its altitude GH; that is, greater than a third part of a cylinder described with the same base and altitude. Now that little circle sustains the weight of this column, that is, a weight greater than the weight of the cone, or a third part of the cylinder.

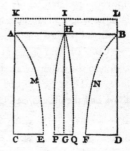

COR. 8. The weight of water which the circle PQ, when very small, sustains, seems to be less than the weight of two thirds of a cylinder of water whose base is that little circle, and its altitude HG. For, things standing as above supposed, imagine the half of a spheroid described whose base is that little circle, and its semi-axis or altitude HG. This figure will be equal to two thirds of that cylinder, and will comprehend within it the column of congealed water PHQ, the weight of which is sustained by that little circle. For though the motion of the water tends directly downwards, the external superficies of that column must yet meet the base PQ in an angle somewhat acute, because the water in its fall is perpetually accelerated, and by reason of that acceleration become narrower. Therefore, since that angle is less than a right one, this column in the lower parts thereof will lie within the hemi-spheroid. In the upper parts also it will be acute or pointed; because to make it otherwise, the horizontal motion of the water must be at the vertex infinitely more swift than its motion towards the horizon. And the less this circle PQ is, the more acute will the vertex of this column be; and the circle being diminished *in infinitum*, the angle PHQ will be diminished *in infinitum*, and therefore the column will lie within the hemi-spheroid. Therefore that column is less than that hemi-spheroid, or than two-third parts of the cylinder whose base is that little circle, and its altitude GH. Now the little circle sustains a force of water equal to the weight

of this column, the weight of the ambient water being employed in causing its efflux out at the hole.

COR. 9. The weight of water which the little circle PQ sustains, when it is very small, is very nearly equal to the weight of a cylinder of water whose base is that little circle, and its altitude $1/2$GH; for this weight is an arithmetical mean between the weights of the cone and the hemi-spheroid above mentioned. But if that little circle be not very small, but on the contrary increased till it be equal to the hole EF, it will sustain the weight of all the water lying perpendicularly above it, that is, the weight of a cylinder of water whose base is that little circle, and its altitude GH.

COR. 10. And (as far as I can judge) the weight which this little circle sustains is always to the weight of a cylinder of water whose base is that little circle, and its altitude $1/2$GH, as EF^2 to $EF^2 - 1/2PQ^2$, or as the circle EF to the excess of this circle above half the little circle PQ, very nearly.

LEMMA IV.

IF A CYLINDER MOVE UNIFORMLY FORWARD IN THE DIRECTION
OF ITS LENGTH, THE RESISTANCE MADE THERETO IS NOT AT
ALL CHANGED BY AUGMENTING OR DIMINISHING THAT
LENGTH; AND, IS THEREFORE THE SAME WITH THE RESISTANCE
OF A CIRCLE, DESCRIBED WITH THE SAME DIAMETER, AND
MOVING, FORWARD WITH THE SAME VELOCITY IN THE DIRECTION
OF A RIGHT LINE PERPENDICULAR TO ITS PLANE.

For the sides are not at all opposed to the motion; and a cylinder becomes a circle when its length is diminished *in infinitum.*

PROPOSITION XXXVII. THEOREM XXIX.

IF A CYLINDER MOVE UNIFORMLY FORWARD IN A COMPRESSED,
INFINITE, AND NON-ELASTIC FLUID, IN THE DIRECTION OF ITS
LENGTH, THE RESISTANCE ARISING FROM THE MAGNITUDE OF ITS
TRANSVERSE SECTION IS TO THE FORCE BY WHICH ITS
WHOLE MOTION MAY BE DESTROYED OR GENERATED, IN THE
TIME THAT IT MOVES FOUR TIMES ITS LENGTH, AS THE
DENSITY OF THE MEDIUM TO THE DENSITY OF THE
CYLINDER, NEARLY.

For let the vessel ABDC touch the surface of stagnant water with its bottom CD, and let the water run out of this vessel into the stagnant water through the cylindric canal EFTS perpendicular to the horizon; and let the little circle PQ be placed parallel

to the horizon any where in the middle of the canal; and produce CA to K, so that AK may be to CK in the duplicate of the ratio, which the excess of the orifice of the canal EF above the little circle PQ bears to the circle AB. Then it is manifest (by Case 5, Case 6, and Cor. 1, Prop. XXXVI) that the velocity of the water passing through the annular space between the little circle and the sides of the vessel will be the very same which the water would acquire by falling, and in its fall describing the altitude KC or IG.

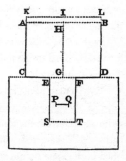

And (by Cor. 10, Prop. XXXVI) if the breadth of the vessel be infinite, so that the lineola HI may vanish, and the altitudes IG, HG become equal; the force of the water that flows down and presses upon the circle will be to the weight of a cylinder whose base is that little circle, and the altitude $\frac{1}{2}$IG, as EF^2 to EF^2 − $\frac{1}{2}$PQ^2, very nearly. For the force of the water flowing downward uniformly through the whole canal will be the same upon the little circle PQ in whatsoever part of the canal it be placed.

Let now the orifices of the canal EF, ST be closed, and let the little circle ascend in the fluid compressed on every side, and by its ascent let it oblige the water that lies above it to descend through the annular space between the little circle and the sides of the canal. Then will the velocity of the ascending little circle be to the velocity of the descending water as the difference of the circles EF and PQ is to the circle PQ; and the velocity of the ascending little circle will lie to the sum of the velocities, that is, to the relative velocity of the descending water with which it passes by the little circle in its ascent, as the difference of the circles EF and PQ to the circle EF, or as EF^2 − PQ^2 to EF^2. Let that relative velocity be equal to the velocity with which it was shewn above that the water would pass through the annular space, if the circle were to remain unmoved, that is, to the velocity which the water would acquire by falling, and in its fall describing the altitude IG; and the force of the water upon the ascending circle will be the same as before (by Cor. 5, of the Laws of Motion); that is, the resistance of the ascending little circle will be to the weight of a cylinder of water whose base is that little circle, and its altitude $\frac{1}{2}$IG, as EF^2 to EF^2 − $\frac{1}{2}$PQ^2, nearly. But the velocity of the little circle will be to the velocity which the water acquires by falling, and in its fall describing the altitude IG, as EF^2 − PQ^2 to EF^2.

Let the breadth of the canal be increased *in infinitum*; and the ratios between EF^2 − PQ^2 and EF^2 − $\frac{1}{2}$PQ^2, will become at last ratios of equality. And therefore the velocity of the little circle will now be the same which the water would acquire in falling, and in its fall describing the altitude IG; and the resistance will become equal to the

weight of a cylinder whose base is that little circle, and its altitude half the altitude IG, from which the cylinder must fall to acquire the velocity of the ascending circle; and with this velocity the cylinder in the time of its fall will describe four times its length. But the resistance of the cylinder moving forward with this velocity in the direction of its length is the same with the resistance of the little circle (by Lem. IV), and is therefore nearly equal to the force by which its motion may be generated while it describes four times its length.

If the length of the cylinder be augmented or diminished, its motion, and the time in which it describes four times its length, will be augmented or diminished in the same ratio, and therefore the force by which the motion so increased or diminished, may be destroyed or generated, will continue the same; because the time is increased or diminished in the same proportion; and therefore that force remains still equal to the resistance of the cylinder, because (by Lem. IV) that resistance will also remain the same.

If the density of the cylinder be augmented or diminished, its motion, and the force by which its motion may be generated or destroyed in the same time, will be augmented or diminished in the same ratio. Therefore the resistance of any cylinder whatsoever will be to the force by which its whole motion may be generated or destroyed, in the time during which it moves four times its length, as the density of the medium to the density of the cylinder, nearly. Q.E.D.

A fluid must be compressed to become continued; it must be continued and nonelastic, that all the pressure arising from its compression may be propagated in an instant; and so, acting equally upon all parts of the body moved, may produce no change of the resistance. The pressure arising from the motion of the body is spent in generating a motion in the parts of the fluid, and this creates the resistance. But the pressure arising from the compression of the fluid, be it ever so forcible, if it be propagated in an instant, generates no motion in the parts of a continued fluid, produces no change at all of motion therein; and therefore neither augments nor lessens the resistance. This is certain, that the action of the fluid arising from the compression cannot be stronger on the hinder parts of the body moved than on its fore parts, and therefore cannot lessen the resistance described in this proposition. And if its propagation be infinitely swifter than the motion of the body pressed, it will not be stronger on the fore parts than on the hinder parts. But that action will be infinitely swifter, and propagated in an instant, if the fluid be continued and nonelastic.

COR. 1. The resistances, made to cylinders going uniformly forward in the direction of their lengths through continued infinite mediums are in a ratio compounded of the duplicate ratio of the velocities and the duplicate ratio of the diameters, and the ratio of the density of the mediums.

COR. 2. If the breadth of the canal be not infinitely increased but the cylinder go forward in the direction of its length through an included quiescent medium, its axis all the while coinciding with the axis of the canal, its resistance will be to the force by which its whole motion, in the time in which it describes four times its length, may be generated or destroyed, in a ratio compounded of the ratio of EF^2 to $EF^2 - \frac{1}{2}PQ^2$ once, and the ratio of EF^2 to $EF^2 - PQ^2$ twice, and the ratio of the density of the medium to the density of the cylinder.

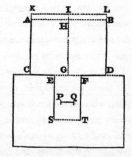

COR. 3. The same thing supposed, and that a length L is to the quadruple of the length of the cylinder in a ratio compounded of the ratio $EF^2 - \frac{1}{2}PQ^2$ to EF^2 once, and the ratio of $EF^2 - PQ^2$ to EF^2 twice; the resistance of the cylinder will be to the force by which its whole motion, in the time during which it describes the length L, may be destroyed or generated, as the density of the medium to the density of the cylinder.

SCHOLIUM.

In this proposition we have investigated that resistance alone which arises from the magnitude of the transverse section of the cylinder, neglecting that part of the same which may arise from the obliquity of the motions. For as, in Case 1, of Prop. XXXVI., the obliquity of the motions with which the parts of the water in the vessel converged on every side to the hole EF hindered the efflux of the water through the hole, so, in this Proposition, the obliquity of the motions, with which the parts of the water, pressed by the antecedent extremity of the cylinder, yield to the pressure, and diverge on all sides, retards their passage through the places that lie round that antecedent extremity, toward the hinder parts of the cylinder, and causes the fluid to be moved to a greater distance; which increases the resistance, and that in the same ratio almost in which it diminished the efflux of the water out of the vessel, that is, in the duplicate ratio of 25 to 21, nearly. And as, in Case 1, of that Proposition, we made the parts of the water pass through the hole EF perpendicularly and in the greatest plenty, by supposing all the water in the vessel lying round the cataract to be frozen, and that part of the water whose motion was oblique and useless to remain without motion, so in this Proposition, that the obliquity of the motions may be taken away, and the parts of the water may give the freest passage to the cylinder, by yielding to it with the most direct and quick motion possible, so that only so much resistance may remain as arises from the magnitude of the transverse section, and which is incapable of diminution, unless by diminishing the diameter of the cylinder; we must conceive those parts of the fluid

whose motions are oblique and useless, and produce resistance, to be at rest among themselves at both extremities of the cylinder, and there to cohere and be joined to the cylinder.

Let ABCD be a rectangle, and let AE and BE be two parabolic arcs, described with the axis AB, and with a *latus rectum* that is to the space HG, which must be described by the cylinder in falling, in order to acquire the velocity with which it moves, as HG to $^1/_2$ AB. Let CF and DF be two other parabolic arcs described with the axis CD, and a *latus rectum* quadruple of the former; and by the convolution of the figure about the axis EF let there be generated a solid, whose middle part ABDO is the cylinder we are here speaking of, and whose extreme parts ABE and CDF contain the parts of the fluid at

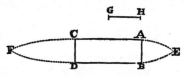

rest among themselves, and concreted into two hard bodies, adhering to the cylinder at each end like a head and tail. Then if this solid EACFDB move in the direction of the length of its axis FE toward the parts beyond E, the resistance will be the same which we have here determined in this Proposition, nearly; that is, it will have the same ratio to the force with which the whole motion of the cylinder may be destroyed or generated, in the time that it is describing the length 4AC with that motion uniformly continued, as the density of the fluid has to the density of the cylinder, nearly. And (by Cor. 7, Prop. XXXVI) the resistance must be to this force in the ratio of 2 to 3, at the least.

LEMMA V.

IF A CYLINDER, A SPHERE, AND A SPHEROID, OF EQUAL BREADTHS
BE PLACED SUCCESSIVELY IN THE MIDDLE OF A CYLINDRIC CANAL,
SO THAT THEIR AXES MAY COINCIDE WITH THE AXIS OF THE
CANAL, THESE BODIES WILL EQUALLY HINDER THE PASSAGE OF
THE WATER THROUGH THE CANAL.

For the spaces lying between the sides of the canal, and the cylinder, sphere, and spheroid, through which the water passes, are equal; and the water will pass equally through equal spaces.

This is true, upon the supposition that all the water above the cylinder, sphere, or spheroid, whose fluidity is not necessary to make the passage of the water the quickest possible, is congealed, as was explained above in Cor. 7, Prop. XXXVI.

LEMMA VI.

THE SAME SUPPOSITION REMAINING, THE FORE-MENTIONED
BODIES ARE EQUALLY ACTED ON BY THE WATER FLOWING
THROUGH THE CANAL.

This appears by Lem. V and the third Law. For the water and the bodies act upon each other mutually and equally.

LEMMA VII.

IF THE WATER BE AT REST IN THE CANAL, AND THESE BODIES
MOVE WITH EQUAL VELOCITY AND THE CONTRARY WAY
THROUGH THE CANAL, THEIR RESISTANCES WILL BE EQUAL
AMONG THEMSELVES.

This appears from the last Lemma, for the relative motions remain the same among themselves.

SCHOLIUM.

The case is the same of all convex and round bodies, whose axes coincide with the axis of the canal. Some difference may arise from a greater or less friction; but in these *Lemmata* we suppose the bodies to be perfectly smooth, and the medium to be void of all tenacity and friction; and that those parts of the fluid which by their oblique and superfluous motions may disturb, hinder, and retard the flux of the water through the canal, are at rest among themselves; being fixed like water by frost, and adhering to the fore and hinder parts of the bodies in the manner explained in the Scholium of the last Proposition; for in what follows we consider the very least resistance that round bodies described with the greatest given transverse sections can possibly meet with.

Bodies swimming upon fluids, when they move straight forward, cause the fluid to ascend at their fore parts and subside at their hinder parts, especially if they are of an obtuse figure; and thence they meet with a little more resistance than if they were acute at the head and tail. And bodies moving in elastic fluids, if they are obtuse behind and before, condense the fluid a little more at their fore parts, and relax the same at their hinder parts; and therefore meet also with a little more resistance than if they were acute at the head and tail. But in these Lemmas and Propositions we are not treating of elastic, but non-elastic fluids; net of bodies floating on the surface of the fluid, but deeply immersed therein. And when the resistance of bodies in non-elastic fluids is once known, we may then augment this resistance a little in elastic fluids, as our air; and in the surfaces of stagnating fluids, as lakes and seas.

PROPOSITION XXXVIII. THEOREM XXX.

IF A GLOBE MOVE UNIFORMLY FORWARD IN A COMPRESSED,
INFINITE, AND NONELASTIC FLUID, ITS RESISTANCE IS TO
THE FORCE BY WHICH ITS WHOLE MOTION MAY BE

DESTROYED OR GENERATED, IN THE TIME THAT IT
DESCRIBES EIGHT THIRD PARTS OF ITS DIAMETER, AS THE
DENSITY OF THE FLUID TO THE DENSITY OF THE GLOBE,
VERY NEARLY.

For the globe is to its circumscribed cylinder as two to three; and therefore the force which can destroy all the motion of the cylinder, while the same cylinder is describing the length of four of its diameters, will destroy all the motion of the globe, while the globe is describing two thirds of this length, that is, eight third parts of its own diameter. Now the resistance of the cylinder is to this force very nearly as the density of the fluid to the density of the cylinder or globe (by Prop. XXXVII), and the resistance of the globe is equal to the resistance of the cylinder (by Lem. V, VI, and VII).   Q.E.D.

COR. 1. The resistances of globes in infinite compressed mediums are in a ratio compounded of the duplicate ratio of the velocity, and the duplicate ratio of the diameter, and the ratio of the density of the mediums.

COR. 2. The greatest velocity, with which a globe can descend by its comparative weight through a resisting fluid, is the same which it may acquire by falling with the same weight, and without any resistance, and in its fall describing a space that is, to four third parts of its diameter as the density of the globe to the density of the fluid. For the globe in the time of its fall, moving with the velocity acquired in falling, will describe a space that will be to eight third parts of its diameter as the density of the globe to the density of the fluid; and the force of its weight which generates this motion will be to the force that can generate the same motion, in the time that the globe describes eight third parts of its diameter, with the same velocity as the density of the fluid to the density of the globe; and therefore (by this Proposition) the force of weight will be equal to the force of resistance, and therefore cannot accelerate the globe.

COR. 3. If there be given both the density of the globe and its velocity at the beginning of the motion, and the density of the compressed quiescent fluid in which the globe moves, there is given at any time both the velocity of the globe and its resistance, and the space described by it (by Cor. 7, Prop. XXXV).

COR. 4. A globe moving in a compressed quiescent fluid of the same density with itself will lose half its motion before it can describe the length of two of its diameters (by the same Cor. 7).

PROPOSITION XXXIX. THEOREM XXXI.

IF A GLOBE MOVE UNIFORMLY FORWARD THROUGH A FLUID
INCLOSED AND COMPRESSED IN A CYLINDRIC CANAL, ITS

RESISTANCE IS TO THE FORCE BY WHICH ITS WHOLE MOTION
MAY BE GENERATED OR DESTROYED, IN THE TIME IN WHICH
IT DESCRIBES EIGHT THIRD PARTS OF ITS DIAMETER, IN A
RATIO COMPOUNDED OF THE RATIO OF LIKE ORIFICE OF
THE CANAL TO THE EXCESS OF THAT ORIFICE ABOVE HALF
THE GREATEST CIRCLE OF THE GLOBE; AND THE DUPLICATE
RATIO OF THE ORIFICE OF THE CANAL TO THE EXCESS OF
THAT ORIFICE ABOVE THE GREATEST CIRCLE OF THE GLOBE;
AND THE RATIO OF THE DENSITY OF THE FLUID TO THE
DENSITY OF THE GLOBE, NEARLY.

This appears by Cor. 2, Prop. XXXVII, and the demonstration proceeds in the same manner as in the foregoing Proposition.

SCHOLIUM.

In the last two Propositions we suppose (as was done before in Lem. V) that all the water which precedes the globe, and whose fluidity increases the resistance of the same, is congealed. Now if that water becomes fluid, it will somewhat increase the resistance. But in these Propositions that increase is so small, that it may be neglected, because the convex superficies of the globe produces the very same effect almost as the congelation of the water.

PROPOSITION XL. PROBLEM IX.

TO FIND BY PHENOMENA THE RESISTANCE OF A GLOBE MOVING THROUGH A PERFECTLY FLUID COMPRESSED MEDIUM.

Let A be the weight of the globe *in vacuo*, B its weight in the resisting medium, D the diameter of the globe, F a space which is to $^4/_3$D as the density of the globe to the density of the medium, that is, as A to A − B, G the time in which the globe falling with the weight B without resistance describes the space F, and H the velocity which the body acquires by that fall. Then H will be the greatest velocity with which the globe can possibly descend with the weight B in the resisting medium, by Cor. 2, Prop XXXVIII; and the resistance which the globe meets with, when descending with that velocity, will be equal to its weight B; and the resistance it meets with in any other velocity will be to the weight B in the duplicate ratio of that velocity to the greatest velocity H, by Cor. 1, Prop. XXXVIII.

This is the resistance that arises from the inactivity of the matter of the fluid. That resistance which arises from the elasticity, tenacity, and friction of its parts, may be thus investigated.

Let the globe be let fall so that it may descend in the fluid by the weight B; and let P be the time of falling, and let that time be expressed in seconds, if the time G be given in seconds. Find the absolute number N agreeing to the logarithm 0,4342944819 $2P/_G$ and let L be the logarithm of the number $\frac{N+1}{N}$: and the velocity acquired in falling will be $\frac{N-1}{N+1}$ H, and the height described will be $\frac{2PF}{G}$ - 1,3862943611F + 4,605170186LF. If the fluid be of a sufficient depth, we may neglect the term 4,605170186LF; and $\frac{2PF}{G}$ - 1,3862943611F will be the altitude described, nearly. These things appear by Prop. IX, Book II, and its Corollaries, and are true upon this supposition, that the globe meets with no other resistance but that which arises from the inactivity of matter. Now if it really meet with any resistance of another kind, the descent will be slower, and from the quantity of that retardation will be known the quantity of this new resistance.

That the velocity and descent of a body falling in a fluid might more easily be known, I have composed the following table; the first column of which denotes the times of descent; the second shews the velocities acquired in falling, the greatest velocity being 100000000; the third exhibits the spaces described by falling in those times, 2F being the space which the body describes in the time G with the greatest velocity; and the fourth gives the spaces described with the greatest velocity in the same times.

The numbers in the fourth column are $2P/_G$, and by subducting the number 1,3862944 - 4,6051702L, are found the numbers in the third column; and these numbers must be multiplied by the space F to obtain the spaces described in falling. A fifth column is added to all these, containing the spaces described in the same times by a body falling *in vacuo* with the force of B its comparative weight.

The Times P.	Velocities of the body falling in the fluid.	The space described in falling in the fluid.	The spaces described with the greatest motion.	The spaces described by falling *in vacuo*.
0,001G	99999$^{29}/_{30}$	0,000001F	0,002F	0,00001F
0,01G	999967	0,0001F	0,02F	0,0001F
0,1G	9966799	0,0099834F	0,2F	0,01F
0,2G	19737532	0,0397361F	0,4F	0,04F
0,3G	29131261	0,0886815F	0,6F	0,09F
0,4G	37994896	0,1559070F	0,8F	0,16F
0,5G	46211716	0,2402290F	1,0F	0,25F
0,6G	53704957	0,3402706F	1,2F	0,36F
0,7G	60436778	0,4545405F	1,4F	0,19F
0,8G	66403677	0,5815071F	1,6F	0,64F
0,9G	71629787	0,7196609F	1,8F	0,81F
1G	76159416	0,8675617F	2F	1F
2G	96402758	2,6500055F	4F	4F
3G	99505475	4,6186570F	6F	9F
4G	99932930	6,6143765F	8F	16F
5G	99990920	8,6137964F	10F	25F
6G	99998771	10,6137179F	12F	36F
7G	99999834	12,6137073F	14F	49F
8G	99999980	14,6137059F	16F	64F
9G	39999997	16,6137057F	18F	81F
10G	99999999$^{3}/_{5}$	18,6137056F	20F	100F

SCHOLIUM.

In order to investigate the resistances of fluids from experiments, I procured a square wooden vessel, whose length and breadth on the inside was 9 inches *English* measure, and its depth 9 feet $^1/_2$; this I filled with rain-water: and having provided globes made up of wax, and lead included therein, I noted the times of the descents of these globes, the height through which they descended being 112 inches. A solid cubic foot of *English* measure contains 76 pounds *troy* weight of rain water; and a solid inch contains $^{19}/_{36}$ ounces *troy* weight, or $253^1/_3$ grains; and a globe of water of one inch in diameter contains 132,645 grains in air, or 132,8 grains *in vacuo*; and any other globe will be as the excess of its weight *in vacuo* above its weight in water.

EXPER. 1. A globe whose weight was $156^1/_4$ grains in air, and 77 grains in water, described the whole height of 112 inches in 4 seconds. And, upon repeating the experiment, the globe spent again the very same time of 4 seconds in falling.

The weight of this globe *in vacuo* is $156^{13}/_{38}$ grains; and the excess of this weight above the weight of the globe in water is $79^{13}/_{38}$ grains. Hence the diameter of the globe appears to be 0,84224 parts of an inch. Then it will be, as that excess to the weight of the globe *in vacuo*, so is the density of the water to the density of the globe; and so is $^8/_3$ parts of the diameter of the globe (viz. 2,24597 inches) to the space 2F, which will be therefore 4,4256 inches. Now a globe falling *in vacuo* with its whole weight of $156^{13}/_{38}$ grains in one second of time will describe $193^1/_3$ inches; and falling in water in the same time with the weight of 77 grains without resistance, will describe 95,219 inches; and in the time G, which is to one second of time in the subduplicate ratio of the space F, or of 2,2128 inches to 95,219 inches, will describe 2,2128 inches, and will acquire the greatest velocity H with which it is capable of descending in water. Therefore the time G is 0",15244. And in this time G, with that greatest velocity H, the globe will describe the space 2F, which is 4,4256 inches; and therefore in 4 seconds will describe a space of 116,1245 inches. Subduct the space 1,3862944F, or 3,0676 inches, and there will remain a space of 113,0569 inches, which the globe falling through water in a very wide vessel will describe in 4 seconds. But this space, by reason of the narrowness of the wooden vessel before mentioned, ought to be diminished in a ratio compounded of the subduplicate ratio of the orifice of the vessel to the excess of this orifice above half a great circle of the globe, and of the simple ratio of the same orifice to its excess above a great circle of the globe, that is, in a ratio of 1 to 0,9914. This done, we have a space of 112,08 inches, which a globe falling through the water in this wooden vessel in 4 seconds of time ought nearly to describe by this theory; but it described 112 inches by the experiment.

EXPER. 2. Three equal globes, whose weights were severally $76^1/_3$ grains in air, and $5^1/_{16}$ grains in water, were let fall successively; and every one fell through the water in 15 seconds of time, describing in its fall a height of 112 inches.

By computation, the weight of each globe *in vacuo* is $76^5/_{12}$ grains; the excess of this weight above the weight in water is 71 grains $^{17}/_{48}$; the diameter of the globe 0,81296 of an inch; $^8/_3$ parts of this diameter 2,16789 inches; the space 2F is 2,3217 inches; the space which a globe of $5^1/_{16}$ grains in weight would describe in one second without resistance, 12,808 inches, and the time G0",301056. Therefore the globe, with the greatest velocity it is capable of receiving from a weight of $5^1/_{16}$ grains in its descent through water, will describe in the time 0",301056 the space of 2,3217 inches; and in 15 seconds the space 115,678 inches. Subduct the space 1,3862944F, or 1,609 inches, and there remains the space 114,069 inches; which therefore the falling globe ought to describe in the same time, if the vessel were very wide. But because our vessel was narrow, the space ought to be diminished by about 0,895 of an inch. And so the space will remain 113,174 inches, which a globe falling in this vessel ought nearly to describe in 15 seconds, by the theory. But by the experiment it described 112 inches. The difference is not sensible.

EXPER. 3. Three equal globes, whose weights were severally 121 grains in air, and 1 grain in water, were successively let fall; and they fell through the water in the times 46", 47", and 50", describing a height of 112 inches.

By the theory, these globes ought to have fallen in about 40". Now whether their falling more slowly were occasioned from hence, that in slow motions the resistance arising from the force of inactivity does really bear a less proportion to the resistance arising from other causes; or whether it is to be attributed to little bubbles that might chance to stick to the globes, or to the rarefaction of the wax by the warmth of the weather, or of the hand that let them fall; or, lastly, whether it proceeded from some insensible errors in weighing the globes in the water, I am not certain. Therefore the weight of the globe in water should be of several grains, that the experiment may be certain, and to be depended on.

EXPER. 4. I began the foregoing experiments to investigate the resistances of fluids, before I was acquainted with the theory laid down in the Propositions immediately preceding. Afterward, in order to examine the theory after it was discovered, I procured a wooden vessel, whose breadth on the inside was 82 inches, and its depth 15 feet and $^1/_3$. Then I made four globes of wax, with lead included, each of which weighed $139^1/_4$ grains in air, and $7^1/_8$ grains in water. These I let fall, measuring the times of their falling in the water with a pendulum oscillating to half seconds. The globes were cold, and had remained so some time, both when they were weighed and when they were let fall; because warmth rarefies the wax, and by rarefying it diminishes the weight of the globe

in the water; and wax, when rarefied, is not instantly reduced by cold to its former density. Before they were let fall, they were totally immersed under water, lest, by the weight of any part of them that might chance to be above the water, their descent should be accelerated in its beginning. Then, when after their immersion they were perfectly at rest, they were let go with the greatest care, that they might not receive any impulse from the hand that let them down. And they fell successively in the times of $47^1/_2$, $48^1/_2$, 50, and 51 oscillations, describing a height of 15 feet and 2 inches. But the weather was now a little colder than when the globes were weighed, and therefore I repeated the experiment another day; and then the globes fell in the times of 49, $49^1/_2$, 50, and 53; and at a third trial in the times of $49^1/_2$, 50, 51, and 53 oscillations. And by making the experiment several times over, I found that the globes fell mostly in the times of $49^1/_2$ and 50 oscillations. When they fell slower, I suspect them to have been retarded by striking against the sides of the vessel.

Now, computing from the theory, the weight of the globe *in vacuo* is $139^2/_5$ grains; the excess of this weight above the weight of the globe in water $132^{11}/_{46}$ grains; the diameter of the globe 0,99868 of an inch; $^8/_3$ parts of the diameter 2,66315 inches; the space 2F 2,8066 inches; the space which a globe weighing $7^1/_8$ grains falling without resistance describes in a second of time 9,88164 inches; and the time G0",376843. Therefore the, globe with the greatest velocity with which it is capable of descending through the water by the force of a weight of $7^1/_8$ grains, will in the time 0",376843 describe a space of 2,8066 inches, and in one second of time a space of 7,44766 inches, and in the time 25", or in 50 oscillations, the space 186,1915 inches. Subduct the space 1,366294F, or 1,9454 inches, and there will remain the space 184,2461 inches which the globe will describe in that time in a very wide vessel. Because our vessel was narrow, let this space be diminished in a ratio compounded of the subduplicate ratio of the orifice of the vessel to the excess of this orifice above half a great circle of the globe, and of the simple ratio of the same orifice to its excess above a great circle of the globe; and we shall have the space of 181,86 inches, which the globe ought by the theory to describe in this vessel in the time of 50 oscillations, nearly. But it described the space of 182 inches, by experiment, in $49^1/_2$ or 50 oscillations.

EXPER. 5. Four globes weighing $154^3/_8$ grains in air, and $21^1/_2$ grains in water, being let fall several times, fell in the times of $28^1/_2$, 29, $29^1/_2$, and 30, and sometimes of 31, 32, and 33 oscillations, describing a height of 15 feet and 2 inches.

They ought by the theory to have fallen in the time of 29 oscillations, nearly.

EXPER. 6. Five globes, weighing $212^3/_8$ grains in air, and $79^1/_2$ in water, being several times let fall, fell in the times of 15, $15^1/_2$, 16, 17, and 18 oscillations, describing a height of 15 feet and 2 inches.

By the theory they ought to have fallen in the time of 15 oscillations, nearly.

EXPER. 7. Four globes, weighing $293^3/_8$ grains in air, and $35^7/_8$ grains in water, being let fall several times, fell in the times of $29^1/_2$, 30, $30^1/_2$, 31, 32, and 33 oscillations, describing a height of 15 feet and 1 inch and $^1/_2$.

By the theory they ought to have fallen in the time of 28 oscillations, nearly.

In searching for the cause that occasioned these globes of the same weight and magnitude to fall, some swifter and some slower, I hit upon this; that the globes, when they were first let go and began to fall, oscillated about their centres; that side which chanced to be the heavier descending first, and producing an oscillating motion. Now by oscillating thus, the globe communicates a greater motion to the water than if it descended without any oscillations; and by this communication loses part of its own motion with which it should descend; and therefore as this oscillation is greater or lesser it will be more or less retarded. Besides, the globe always recedes from that side of itself which is descending in the oscillation, and by so receding comes nearer to the sides of the vessel, so as even to strike against them sometimes. And the heavier the globes are, the stronger this oscillation is; and the greater they are, the more is the water agitated by it. Therefore to diminish this oscillation of the globes, I made new ones of lead and wax, sticking the lead in one side of the globe very near its surface; and I let fall the globe in such a manner, that, as near as possible, the heavier side might be lowest at the beginning of the descent. By this means the oscillations became much less than before, and the times in which the globes fell were not so unequal: as in the following experiments.

EXPER. 8. Four globes weighing 139 grains in air, and $6^1/_2$ in water, were let fall several times, and fell mostly in the time of 51 oscillations, never in more than 52, or in fewer than 50, describing a height of 182 inches.

By the theory they ought to fall in about the time of 52 oscillations.

EXPER. 9. Four globes weighing $273^1/_4$ grains in air, and 140 $^3/_4$ in water, being several times let fall, fell in never fewer than 12, and never more than 13 oscillations, describing a height of 182 inches.

These globes by the theory ought to have fallen in the time of $11^1/_3$ oscillations, nearly.

EXPER. 10. Four globes, weighing 384 grains in air, and $119^1/_2$ in water, being let fall several times, fell in the times of $17^3/_4$, 18, $18^1/_2$, and 19 oscillations describing a height of $181^1/_2$ inches. And when they fell in the time of 19 oscillations, I sometimes heard them hit against the sides of the vessel before they reached the bottom.

By the theory they ought to have fallen in the time of $15^5/_9$ oscillations, nearly.

EXPER. 11. Three equal globes, weighing 48 grains in the air, and $3^{29}/_{32}$ in water, being several times let fall, fell in the times of $43^1/_2$, 44, $44^1/_2$, 45, and

46 oscillations, and mostly in 44 and 45, describing a height of $182^1/_2$ inches, nearly.

By the theory they ought to have fallen in the time of 46 oscillations and $5/_9$, nearly.

EXPER. 12. Three equal globes, weighing 141 grains in air, and $4\,^3/_8$ in water, being let fall several times, fell in the times of 61, 62, 63, 64, and 65 oscillations, describing a space of 162 inches.

And by the theory they ought to have fallen in $64^1/_2$ oscillations, nearly.

From these experiments it is manifest, that when the globes fell slowly, as in the second, fourth, fifth, eighth, eleventh, and twelfth experiments, the times of falling are rightly exhibited by the theory; but when the globes fell more swiftly, as in the sixth, ninth, and tenth experiments, the resistance was somewhat greater than in the duplicate ratio of the velocity. For the globes in falling oscillate a little; and this oscillation, in those globes that are light and fall slowly, soon ceases by the weakness of the motion; but in greater and heavier globes, the motion being strong, it continues longer, and is not to be checked by the ambient water till after several oscillations. Besides, the more swiftly the globes move, the less are they pressed by the fluid at their hinder parts; and if the velocity be perpetually increased, they will at last leave an empty space behind them, unless the compression of the fluid be increased at the same time. For the compression of the fluid ought to be increased (by Prop. XXXII and XXXIII) in the duplicate ratio of the velocity, in order to preserve the resistance in the same duplicate ratio. But because this is not done, the globes that move swiftly are not so much pressed at their hinder parts as the others; and by the defect of this pressure it comes to pass that their resistance is a little greater than in a duplicate ratio of their velocity.

So that the theory agrees with the phaenomena of bodies falling in water It remains that we examine the phaenomena of bodies falling in air.

EXPER. 13. From the top of St. *Paul's* Church in *London*, in *June* 1710, there were let fall together two glass globes, one full of quicksilver, the other of air; and in their fall they described a height of 220 *English* feet. A wooden table was suspended upon iron hinges on one side, and the other side of the same was supported by a wooden pin. The two globes lying upon this table were let fall together by pulling out the pin by means of an iron wire reaching from thence quite down to the ground; so that, the pin being removed, the table, which had then no support but the iron hinges, fell downward, and turning round upon the hinges, gave leave to the globes to drop off from it. At the same instant, with the same pull of the iron wire that took out the pin, a pendulum oscillating to seconds was let go, and began to oscillate. The diameters and weights of the globes, and their times of falling, are exhibited in the following table.

The globes filled with mercury.			The globes full of air.		
Weights.	Diameters.	Times in falling.	Weights.	Diameters.	Times in falling.
908 grains	0,8 of an inch	4"	510 grains	5,1 inches	8" 1/2
983	0,8	4 -	642	5,2	8
866	0,8	4	599	5,1	8
747	0,75	4 +	515	5,0	8 1/4
808	0,75	4	483	5,0	8 1/2
784	0,75	4 +	641	5,2	8

But the times observed must be corrected; for the globes of mercury (by *Galileo's* theory), in 4 seconds of time, will describe 257 *English* feet, and 220 feet in only 3"42". So that the wooden table, when the pin was taken out, did not turn upon its hinges so quickly as it ought to have done; and the slowness of that revolution hindered the descent of the globes at the beginning. For the globes lay about the middle of the table, and indeed were rather nearer to the axis upon which it turned than to the pin. And hence the times of falling were prolonged about 18"'; and therefore ought to be corrected by subducting that excess, especially in the larger globes, which, by reason of the largeness of their diameters, lay longer upon the revolving table than the others. This being done, the times in which the six larger globes fell will come forth 8" 12"', 7" 42"', 7" 42"', 7" 57"', 8" 12"' and 7" 42"'.

Therefore the fifth in order among the globes that were full of air being 5 inches in diameter, and 483 grains in weight, fell in 8" 12"', describing a space of 220 feet. The weight of a bulk of water equal to this globe is 16600 grains; and the weight of an equal bulk of air is 16600/860 grains, or $19^3/_{10}$ grains; and therefore the weight of the globe *in vacuo* is $502^3/_{10}$ grains; and this weight is to the weight of a bulk of air equal to the globe as $502^3/_{10}$ to $19^3/_{10}$; and so is 2F to $^8/_3$ of the diameter of the globe, that is, to $13^1/_3$ inches. Whence 2F becomes 28 feet 11 inches. A globe, falling *in vacuo* with its whole weight of $502^3/_{10}$ grains, will in one second of time describe $193^1/_3$ inches as above; and with the weight of 483 grains will describe 185,905 inches; and with that weight 483 grains *in vacuo* will describe the space F, or 14 feet $5^1/_2$ inches, in the time of 57"' 58", and acquire the greatest velocity it is capable of descending with in the air. With this velocity the globe in 8" 12"' of time will describe 245 feet and $5^1/_3$ inches. Subduct 1,3863F, or 20 feet and $^1/_2$ an inch, and there remain 225 feet 5 inches. This space, therefore, the falling globe ought by this theory to describe in 8" 12"'. But by the experiment it described a space of 220 feet The difference is insensible.

By like calculations applied to the other globes full of air, I composed the following table.

The weight of the globes.	The diameters.	The times of falling from a height of 220 feet.	The space which they would describe by the theory.		The excesses.	
510 grains	5,1 inches	8" 12"'	226 feet	11 inch.	6 feet	11 inch.
642	5,2	7 42	230	9	10	9
599	5,1	7 42	227	10	7	0
515	5	7 57	224	5	4	5
483	5	8 12	225	5	5	5
641	5,2	7 42	230	7	10	7

EXPER. 14. *Anno* 1719, in the month of *July*, Dr. *Desaguliers* made some experiments of this kind again, by forming hogs' bladders into spherical orbs; which was done by means of a concave wooden sphere, which the bladders, being wetted well first, were put into. After that being blown full of air, they were obliged to fill up the spherical cavity that contained them; and then, when dry, were taken out. These were let fall from the lantern on the top of the cupola of the same church, namely, from a height of 272 feet; and at the same moment of time there was let fall a leaden globe, whose weight was about 2 pounds *troy* weight. And in the mean time some persons standing in the upper part of the church where the globes were let fall observed the whole times of falling; and others standing on the ground observed the differences of the times between the fall of the leaden weight and the fall of the bladder. The times were measured by pendulums oscillating to half seconds. And one of those that stood upon the ground had a machine vibrating four times in one second; and another had another machine accurately made with a pendulum vibrating four times in a second also. One of those also who stood at the top of the church had a like machine; and these instruments were so contrived, that their motions could be stopped or renewed at pleasure. Now the leaden globe fell in about four seconds and $1/4$ of time; and from the addition of this time to the difference of time above spoken of, was collected the whole time in which the bladder was falling. The times which the five bladders spent in falling, after the leaden globe had reached the ground, were, the first time, $14^3/_4$", $12^3/_4$", $14^5/_8$", $17^3/_4$", and $16^7/_8$"; and the second time, $14^1/_2$", $14^1/_4$", 14", 19", and $16^3/_4$". Add to these 4", the time in which the leaden globe was falling, and the whole times in which the five bladders fell were, the first time, 19", 17", $18^7/_8$", 22", and $21^1/_8$"; and the second time, $18^3/_4$", $18^1/_2$", $18^1/_4$", $23^1/_4$", and 21". The times observed at the top of the church were, the first time, $19^3/_8$", $17^1/_4$", $18^3/_4$", $22^1/_8$", and 21 $5/_8$"; and the second time, 19", $18^5/_8$", $18^3/_8$", 24", and $21^1/_4$". But the bladders did not always fall directly down, but sometimes fluttered a little in the air, and waved to and fro, as they were descending. And by these motions the times of their falling were prolonged, and increased by half a second sometimes, and sometimes by a whole second. The second and fourth bladder fell most directly the first time, and the

first and third the second time. The fifth bladder was wrinkled, and by its wrinkles was a little retarded. I found their diameters by their circumferences measured with a very fine thread wound about them twice. In the following table I have compared the experiments with the theory; making the density of air to be to the density of rain-water as 1 to 860, and computing the spaces which by the theory the globes ought to describe in falling.

The weights of the bladders.	The diameters	The times of falling from a height of 272 feet	The spaces which by the theory ought to have been described in those times.		The difference between the theory and the experiments.	
128 grains	5,28 inches	19"	271 feet	11 inches	- 0 ft.	1 in.
156	5,19	17	272	0 1/2	+ 0	0 1/2
137 1/2	5,3	18	272	7	+ 0	7
97 1/2	5,26	22	277	4	+ 5	4
99 1/8	5	21 1/8	282	0	+ 10	0

Our theory, therefore, exhibits rightly, within a very little, all the resistance that globes moving either in air or in water meet with; which appears to be proportional to the densities of the fluids in globes of equal velocities and magnitudes.

In the Scholium subjoined to the sixth Section, we shewed, by experiments of pendulums, that the resistances of equal and equally swift globes moving in air, water, and quicksilver, are as the densities of the fluids. We here prove the same more accurately by experiments of bodies falling in air and water. For pendulums at each oscillation excite a motion in the fluid always contrary to the motion of the pendulum in its return; and the resistance arising from this motion, as also the resistance of the thread by which the pendulum is suspended, makes the whole resistance of a pendulum greater than the resistance deduced from the experiments of falling bodies. For by the experiments of pendulums described in that Scholium, a globe of the same density as water in describing the length of its semi-diameter in air would lose the $1/3342$ part of its motion. But by the theory delivered in this seventh Section, and confirmed by experiments of falling bodies, the same globe in describing the same length would lose only a part of its motion equal to $1/4586$, supposing the density of water to be to the density of air as 860 to 1. Therefore the resistances were found greater by the experiments of pendulums (for the reasons just mentioned) than by the experiments of falling globes; and that in the ratio of about 4 to 3. But yet since the resistances of pendulums oscillating, in air, water, and quicksilver, are alike increased by like causes, the proportion of the resistances in these mediums will be rightly enough exhibited by the experiments of pendulums, as well as by the experiments of falling bodies. And from all this it may be concluded, that the resistances of bodies, moving in any fluids whatsoever, though of the most extreme fluidity, are, *caeteris paribus*, as the densities of the fluids.

These things being thus established, we may now determine what part of its motion any globe projected in any fluid whatsoever would nearly lose in a given time. Let D be the diameter of the globe, and V its velocity at the beginning of its motion, and T the time in which a globe with the velocity V can describe *in vacuo* a space that is, to the space $^8/_3$D as the density of the globe to the density of the fluid; and the globe projected in that fluid will, in any other time t lose the part $\dfrac{tV}{T+t}$, the part $\dfrac{TV}{T+t}$ remaining; and will describe a space, which will be to that described in the same time *in vacuo* with the uniform velocity V, as the logarithm of the number $\dfrac{T+t}{T}$ multiplied by the number 2,302585093 is the number $^t/_T$ by Cor. 7, Prop. XXXV. In slow motions the resistance may be a little less, because the figure of a globe is more adapted to motion than the figure of a cylinder described with the same diameter. In swift motions the resistance may be a little greater, because the elasticity and compression of the fluid do not increase in the duplicate ratio of the velocity. But these little niceties I take no notice of.

And though air, water, quicksilver, and the like fluids, by the division of their parts *in infinitum*, should be subtilized, and become mediums infinitely fluid, nevertheless, the resistance they would make to projected globes would be the same. For the resistance considered in the preceding Propositions arises from the inactivity of the matter; and the inactivity of matter is essential to bodies, and always proportional to the quantity of matter. By the division of the parts of the fluid the resistance arising from the tenacity and friction of the parts may be indeed diminished; but the quantity of matter will not be at all diminished by this division; and if the quantity of matter be the same, its force of inactivity will be the same; and therefore the resistance here spoken of will be the same, as being always proportional to that force. To diminish this resistance, the quantity of matter in the spaces through which the bodies move must be diminished; and therefore the celestial spaces, through which the globes of the planets and comets are perpetually passing towards all parts, with the utmost freedom, and without the least sensible diminution of their motion, must be utterly void of any corporeal fluid, excepting, perhaps, some extremely rare vapours and the rays of light.

Projectiles excite a motion in fluids as they pass through them, and this motion arises from the excess of the pressure of the fluid at the fore parts of the projectile above the pressure of the same at the hinder parts; and cannot be less in mediums infinitely fluid than it is in air, water, and quicksilver, in proportion to the density of matter in each. Now this excess of pressure does, in proportion to its quantity, not only excite a motion in the fluid, but also acts upon the projectile so as to retard its motion; and therefore the resistance in every fluid is as the motion excited by the projectile in the fluid; and cannot be less in the most subtile rather in proportion to the density of that

aether, than it is in air, water, and quicksilver, in proportion to the densities of those fluids.

SECTION VIII.

Of motion propagated through fluids.

PROPOSITION XLI. THEOREM XXXII.

A PRESSURE IS NOT PROPAGATED THROUGH A FLUID IN
RECTILINEAR DIRECTIONS UNLESS WHERE THE PARTICLES
OF THE FLUID LIE IN A RIGHT LINE.

If the particles *a*, *b*, *c*, *d*, *e*, lie in a right line, the pressure may be indeed directly propagated from *a* to *e*; but then the particle *e* will urge the obliquely posited particles *f* and *g* obliquely, and those particles *f* and *g* will not sustain this pressure, unless they be supported by the particles *h* and *k* lying beyond them; but the particles that support

them are also pressed by them; and those particles cannot sustain that pressure, without being supported by, and pressing upon, those particles that lie still farther, as *l* and *m*, and so on *in infinitum*. Therefore the pressure, as soon as it is propagated to particles that lie out of right lines, begins to deflect towards one hand and the other, and will be propagated obliquely *in infinitum*; and after it has begun to be propagated obliquely, if it reaches more distant particles lying out of the light line, it will deflect again on each hand and this it will do as often as it lights on particles that do not lie exactly in a right line.   Q.E.D.

COR. If any part of a pressure, propagated through a fluid from a given point, be intercepted by any obstacle, the remaining part, which is not intercepted, will deflect into the spaces behind the obstacle. This may be demonstrated also after the following manner. Let a pressure be propagated from the point A towards any part, and, if it be possible, in rectilinear directions; and the obstacle NBCK being perforated in BC, let all the pressure be intercepted but the coniform part APQ passing through the circular hole BC. Let the cone APQ be divided into frustums by the transverse planes, *de*, *fg*, *hi*. Then while the cone ABC, propagating the pressure, urges the conic frustum *degf* beyond it on the superficies *de*, and this frustum urges the next frustum *fgih* on the superficies *fg*, and that frustum urges a third frustum, and so *in infinitum*; it is manifest (by the third Law) that the first frustum *defg* is, by the re-action of the second frustum *fghi*, as much urged and pressed on the superficies *fg*, as it urges and presses that second frustum. Therefore the frustum *degf* is compressed on both sides, that is,

between the cone *Ade* and the frustum *fhig*; and therefore (by Case 6, Prop. XIX) cannot preserve its figure, unless it be compressed with the same force on all sides.

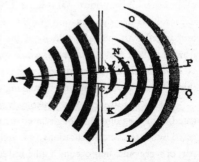

Therefore with the same force with which it is pressed on the superficies *de*, *fg*, it will endeavour to break forth at the sides *df*, *eg*; and there (being not in the least tenacious or hard, but perfectly fluid) it will run out, expanding itself, unless there be an ambient fluid opposing that endeavour. Therefore, by the effort it makes to run out, it will press the ambient fluid, at its sides *df*, *eg*, with the same force that it does the frustum *fghi*; and therefore, the pressure will be propagated as much from the sides *df*, *eg*, into the spaces NO, KL this way and that way, as it is propagated from the superficies *fg*, towards PQ.   Q.E.D.

PROPOSITION XLII. THEOREM XXXIII.

ALL MOTION PROPAGATED THROUGH A FLUID DIVERGES FROM A RECTILINEAR PROGRESS INTO THE UNMOVED SPACES.

CASE 1. Let a motion be propagated from the point A through the hole BC, and, if it be possible, let it proceed in the conic space BCQP according to right lines diverging from the point A. And let us first suppose this motion to be that of waves in the surface of standing water; and let *de*, *fg*, *hi*, *kl*, &c., be the tops of the several waves, divided from each other by as any intermediate

valleys or hollows. Then, because the water in the ridges of the waves is higher than in the unmoved parts of the fluid KL, NO, it will run down from off the tops of those ridges, *e*, *g*, *i*, *l*, &c., *d*, *f*, *h*, *k*, &c., this way and that way towards KL and NO; and because the water is more depressed in the hollows of the waves than in the unmoved parts of the fluid KL, NO, it will run down into those hollows out of those unmoved parts. By the first deflux the ridges of the waves will dilate themselves this way and that way, and be propagated towards KL and NO. And because the motion of the waves from A towards PQ is carried on by a

continual deflux from the ridges of the waves into the hollows next to them, and therefore cannot be swifter than in proportion to the celerity of the descent; and the descent of the water on each side towards KL and NO must be performed with the same velocity; it follows that the dilatation of the waves on each side towards KL and NO will be propagated with the game velocity as the waves themselves go forward with directly from A to PQ, And therefore the whole space this way and that way towards KL and NO will be filled by the dilated waves *rfgr*, *shis*, *tkit*, *vmnv*, &c. Q.E.D. That these things are so, anyone may find by making the experiment in still water.

CASE 2. Let us suppose that *de*, *fg*, *hi*, *kl*, *mn*, represent pulses successively propagated from the point A through an elastic medium. Conceive the pulses to be propagated by successive condensations and rarefactions of the medium, so that the densest part of every pulse may occupy a spherical superficies described about the centre A, and that equal intervals intervene between the successive pulses. Let the lines *de*, *fg*, *hi*, *kl*, &c., represent the densest parts of the pulses, propagated through the hole BC; and because the medium is denser there than in the spaces on either side towards KL and NO, it will dilate itself as well towards those spaces KL, NO, on each hand, as towards the rare intervals between the pulses; and thence the medium, becoming always more rare next the intervals, and more dense next the pulses, will partake of their motion. And because the progressive motion of the pulses arises from the perpetual relaxation of the denser parts towards the antecedent rare intervals; and since the pulses will relax themselves on each hand towards the quiescent parts of the medium KL, NO, with very near the same celerity; therefore the pulses will dilate themselves on all sides into the unmoved parts KL, NO, with almost the same celerity with which they are propagated directly from the centre A; and therefore will fill up the whole space KLON. Q.E.D. And we find the same by experience also in sounds which are heard through a mountain interposed; and, if they come into F, chamber through the window, dilate themselves into all the parts of the room, and are heard in every corner; and not as reflected from the opposite walls, but directly propagated from the window, as far as our sense can judge.

CASE 3. Let us suppose, lastly, that a motion of any kind is propagated from A through the hole BC. Then since the cause of this propagation is that the parts of the medium that are near the centre A disturb and agitate those which lie farther from it; and since the parts which are urged are fluid, and therefore recede every way towards those spaces where they are less pressed, they will by consequence recede towards all the parts of the quiescent medium; as well to the parts on each band, as KL and NO, as to those right before, as PQ; and by this means all the motion, as soon as it has passed through the hole BC, will begin to dilate itself, and from thence, as from its principle and centre, will be propagated directly every way. Q.E.D.

PROPOSITION XLIII. THEOREM XXXIV.

EVERY TREMULOUS BODY IN AN ELASTIC MEDIUM PROPAGATES THE MOTION OF THE PULSES ON EVERY SIDE RIGHT FORWARD, BUT IN A NON-ELASTIC MEDIUM EXCITES A CIRCULAR MOTION.

CASE 1. The parts of the tremulous body, alternately going and returning, do in going urge and drive before them those parts of the medium that lie nearest, and by that impulse compress and condense them; and in returning suffer those compressed parts to recede again, and expand themselves. Therefore the parts of the medium that lie nearest to the tremulous body move to and fro by turns, in like manner as the parts of the tremulous body itself do; and for the same cause that the parts of this body agitate these parts of the medium, these parts, being agitated by like tremors, will in their turn agitate others next to themselves; and these others, agitated in like manner, will agitate those that lie beyond them, and so on *in infinitum*. And in the same manner as the first parts of the medium were condensed in going, and relaxed in returning, so will the other parts be condensed every time they go, and expand themselves every time they return. And therefore they will not be all going and all returning at the same instant (for in that case they would always preserve determined distances from each other, and there could be no alternate condensation and rarefaction); but since, in the places where they are condensed, they approach to, and, in the places where they are rarefied, recede from each other, therefore some of them will be going while others are returning; and so on *in infinitum*. The parts so going, and in their going condensed, are pulses, by reason of the progressive motion with which they strike obstacles in their way; and therefore the successive pulses produced by a tremulous body will be propagated in rectilinear directions; and that at nearly equal distances from each other, because of the equal intervals of time in which the body, by its several tremors produces the several pulses. And though the parts of the tremulous body go and return in some certain and determinate direction, yet the pulses propagated from thence through the medium will dilate themselves towards the sides, by the foregoing Proposition; and will be propagated on all sides from that tremulous body, as from a common centre, in superficies nearly spherical and concentrical. An example of this we have in waves excited by shaking a finger in water, which proceed not only forward and backward agreeably to the motion of the finger, but spread themselves in the manner of concentrical circles all round the finger, and are propagated on every side. For the gravity of the water supplies the place of elastic force.

CASE 2. If the medium be not elastic, then, because its parts cannot be condensed by the pressure arising from the vibrating parts of the tremulous body, the motion will

be propagated in an instant towards the parts where the medium yields most easily, that is, to the parts which the tremulous body would otherwise leave vacuous behind it. The case is the same with that of a body projected in any medium whatever. A medium yielding to projectiles does not recede *in infinitum*, but with a circular motion comes round to the spaces which the body leaves behind it. Therefore as often as a tremulous body tends to any part, the medium yielding to it comes round in a circle to the parts which the body leaves; and as often as the body returns to the first place, the medium will be driven from the place it came round to, and return to its original place. And though the tremulous body be not firm and hard, but every way flexible, yet if it continue of a given magnitude, since it cannot impel the medium by its tremors any where without yielding to it somewhere else, the medium receding from the parts of the body where it is pressed will always come round in a circle to the parts that yield to it. Q.E.D.

COR. It is a mistake, therefore, to think, as some have done, that the agitation of the parts of flame conduces to the propagation of a pressure in rectilinear directions through an ambient medium. A pressure of that kind must be derived not from the agitation only of the parts of flame, but from the dilatation of the whole.

PROPOSITION XLIV. THEOREM XXXV.

IF WATER ASCEND AND DESCEND ALTERNATELY IN THE
ERECTED LEGS KL, MN, OF A CANAL OR PIPE; AND A
PENDULUM BE CONSTRUCTED WHOSE LENGTH BETWEEN
THE POINT OF SUSPENSION AND THE CENTRE OF OSCILLATION
IS EQUAL TO HALF THE LENGTH OF THE WATER IN THE
CANAL; I SAY, THAT THE WATER WILL ASCEND AND DESCEND
IN THE SAME TIMES IN WHICH THE PENDULUM OSCILLATES.

I measure the length of the water along the axes of the canal and its legs, and make it equal to the sum of those axes; and take no notice of the resistance of the water arising from its attrition by the sides of the canal. Let, therefore, AB, CD, represent the mean height of the water in both legs; and when the water in the leg KL ascends to the height EF, the water will descend in the leg MN to the height GH. Let P be a pendulous body, VP the thread, V the point of suspension, RPQS the cycloid which the pendulum describes, P its lowest point, PQ an arc equal to the height AE. The force with which the motion of the water is accelerated and retarded alternately is the excess of the weight of the water in one leg above the weight in the other; and, therefore, when the water in the leg KL ascends to EF, and in the other leg descends to GH, that force is double the weight of the water EABF, and therefore is to the weight of the whole

water as AE or PQ to VP or PR. The force also with which the body P is accelerated or retarded in any place, as Q, of a cycloid, is (by Cor. Prop. LI) to its whole weight as its distance PQ from the lowest place P to the length PR of the cycloid. Therefore the

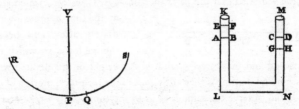

motive forces of the water and pendulum, describing the equal spaces AE, PQ, are as the weights to be moved; and therefore if the water and pendulum are quiescent at first, those forces will move them in equal times, and will cause them to go and return together with a reciprocal motion. Q.E.D.

COR. 1. Therefore the reciprocations of the water in ascending and descending are all performed in equal times, whether the motion be more or less intense or remiss.

COR. 2. If the length of the whole water in the canal be of $6^1/_9$ feet of *French* measure, the water will descend in one second of time, and will ascend in another second, and so on by turns *in infinitum*, for a pendulum of $3^1/_{18}$ such feet in length will oscillate in one second of time.

COR. 3. But if the length of the water be increased or diminished, the time of the reciprocation will be increased or diminished in the subduplicate ratio of the length.

PROPOSITION XLV. THEOREM XXXVI.

THE VELOCITY OF WAVES IS IN THE SUBDUPLICATE RATIO OF THE BREADTHS.

This follows from the construction of the following Proposition.

PROPOSITION XLVI. PROBLEM X.

TO FIND THE VELOCITY OF WAVES.

Let a pendulum be constructed, whose length between the point of suspension and the centre of oscillation is equal to the breadth of the waves and in the time that the pendulum will perform one single oscillation the waves will advance forward nearly a space equal to their breadth.

That which I call the breadth of the waves is the transverse measure lying between the deepest part of the hollows, or the tops of the ridges. Let ABCDEF represent the

surface of stagnant water ascending and descending in successive waves; and let A, C, E, &c., be the tops of the waves; and let B, D, F, &c., be the intermediate hollows. Because the motion of the waves is carried on by the successive ascent and descent of the water, so that the parts thereof, as A, C, E, &c., which are highest

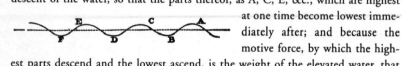

at one time become lowest immediately after; and because the motive force, by which the highest parts descend and the lowest ascend, is the weight of the elevated water, that alternate ascent and descent will be analogous to the reciprocal motion of the water in the canal, and observe the same laws as to the times of its ascent and descent; and therefore (by Prop. XLIV) if the distances between the highest places of the waves A, C, E, and the lowest B, D, F, be equal to twice the length of any pendulum, the highest parts A, C, E, will become the lowest in the time of one oscillation, and in the time of another oscillation will ascend again. Therefore between the passage of each wave, the time of two oscillations will intervene; that is, the wave will describe its breadth in the time that pendulum will oscillate twice; but a pendulum of four times that length, and which therefore is equal to the breadth of the waves, will just oscillate once in that time.   Q.E.D.

COR. 1. Therefore waves, whose breadth is equal to $3^1/_{18}$ *French* feet, will advance through a space equal to their breadth in one second of time; and therefore in one minute will go over a space of $183^1/_3$ feet; and in an hour a space of 11000 feet, nearly.

COR. 2. And the velocity of greater or less waves will be augmented or diminished in the subduplicate ratio of their breadth.

These things are true upon the supposition that the parts of water ascend or descend in a right line; but, in truth, that ascent and descent is rather performed in a circle; and therefore I propose the time defined by this Proposition as only near the truth.

PROPOSITION XLVII. THEOREM XXXVII.

IF PULSES ARE PROPAGATED THROUGH A FLUID, THE SEVERAL PARTICLES OF THE FLUID, GOING AND RETURNING WITH THE SHORTEST RECIPROCAL MOTION, ARE ALWAYS ACCELERATED OR RETARDED ACCORDING TO THE LAW OF THE OSCILLATING PENDULUM.

Let AB, BC, CD, &c., represent equal distances of successive pulses; ABC the line of direction of the motion of the successive pulses propagated from A to B; E, F, G

three physical points of the quiescent medium situate in the right line AC at equal distances from each other; E*e*, F*f*, G*g* equal spaces of extreme shortness, through which those points go and return with a reciprocal motion in each vibration; ϵ, ϕ, γ any intermediate places of the same points; EF, FG physical lineolae, or linear parts of the medium lying between those points, and successively transferred into the places $\epsilon\phi$, $\phi\gamma$, and *ef*, *fg*. Let there be drawn the right line PS equal to the right line E*e*. Bisect the same in O, and from the centre O, with the interval OP, describe the circle SIP*i*. Let the whole time of one vibration; with its proportional parts, be expounded by the whole circumference of this circle and its parts, in such sort, that, when any time PH or PHS*h* is completed, if there be let fall to PS the perpendicular HL or *hl*, and there be taken Eϵ equal to PL or P*l*, the physical point E may be found in ϵ. A point, as E, moving according to this law with a reciprocal motion, in its going from E through ϵ to *e*, and returning again through ϵ to E, will perform its several vibrations with the degrees of acceleration and retardation with those of an oscillating pendulum. We are now to prove that the several physical points of the medium will be agitated with such a kind of motion. Let us suppose, then, that a medium hath such a motion excited in it from any cause whatsoever, and consider what will follow from thence.

In the circumference PHS*h* let there be taken the equal arcs, HI, IK, or *hi*, *ik*, having the same ratio to the whole circumference as the equal right lines EF, FG have to BC, the whole interval of the pulses. Let fell the perpendiculars IM, KN, or *im*, *kn*; then because the points E, F, G are successively agitated with like motions, and perform their entire vibrations composed of their going and return, while the pulse is transferred from B to C; if PH or PHS*h* be the time elapsed since the beginning of the motion of the point E, then will PI or PHS*i* be the time elapsed since the beginning of the motion of the point F, and PK or PHS*k* the time elapsed since the beginning of the motion of the point G; and therefore Eϵ, Fϕ, Gγ, will be respectively equal to PL, PM, PN, while the points are going, and to P*l*, P*m*, P*n*, when the points are returning. Therefore $\epsilon\gamma$ or EG + Gγ - Eϵ will, when the points are going, be equal to EG - LN and in their return equal to EG + *ln*. But $\epsilon\gamma$ is the breadth or expansion of the part EG of the medium in the place $\epsilon\gamma$; and therefore the expansion of that part in its going is to its mean expansion as EG - LN to EG; and in its return, as EG + *ln* or EG + LN to EG.

Therefore since LN is to KH as IN to the radius OP, and KH to EG as the circumference PHS*h*P to BC; that is, if we put V for the radius of a circle whose circumference is equal to BC the interval of the pulses, as OP to V; and, *ex aequo*, LN to EG as IM to V; the expansion of the part EG, or of the physical point F in the place $\epsilon\gamma$, to the mean expansion of the same part in its first place EG, will be as V - IM to V in going, and as V + *im* to V in its return. Hence the elastic force of the point F in the place $\epsilon\gamma$ to its mean elastic force in the place EG is as $\frac{1}{V-IM}$ to $1/V$ in its going, and

as $\frac{1}{V+im}$ to $1/V$ in its return. And by the same reasoning the elastic forces of the physical points E and G in going are as $\frac{1}{V-HL}$ and $\frac{1}{V-KN}$ to $1/V$; and the difference of the forces to the mean elastic force of the medium as $\frac{HL-KN}{VV-V\times HL-V\times KN+HI\times KN}$ to $1/V$; that is, as $\frac{HL-KN}{VV}$ to $1/V$, or as HL

– KN to V; if we suppose (by reason of the very short extent of the vibrations) HL and KN to be indefinitely less than the quantity V. Therefore since the quantity V is given, the difference of the form is as HL – KN; that is (because HL – KN is proportional to HK, and OM to OI or OP; and because HK and OP are given) as OM; that is, if F*f* be bisected in Ω, as $\Omega\phi$. And for the same reason the difference of the elastic forces of the physical points ϵ and γ, in the return of the physical lineola $\epsilon\gamma$, is as $\Omega\phi$. But that difference (that is, the excess of the elastic force of the point ϵ above the elastic force of the point γ) is the very force by which the intervening physical lineola $\epsilon\gamma$ of the medium is accelerated in going, and retarded in returning; and therefore the accelerative force of the physical lineola $\epsilon\gamma$ is as its distance from Ω, the middle place of the vibration, Therefore (by Prop. XXXVIII, Book I) the time is rightly expounded by the arc PI; and the linear part of the medium $\epsilon\gamma$ is moved according to the law above-mentioned, that is, according to the law of a pendulum oscillating; and the case is the same of all the linear parts of which the whole medium is compounded.   Q.E.D.

COR. Hence it appears that the number of the pulses propagated is the same with the number of the vibrations of the tremulous body, and is not multiplied in their progress. For the physical lineola $\epsilon\gamma$ as soon as it returns to its first place is at rest; neither will it move again, unless it receives a new motion either from the impulse of the tremulous body, or of the pulses propagated from that body. As soon, therefore, as the pulses cease to be propagated from the tremulous body, it will return to a state of rest, and move no more.

PROPOSITION XLVIII. THEOREM XXXVIII.

THE VELOCITIES OF PULSES PROPAGATED IN AN ELASTIC FLUID ARE IN A RATIO COMPOUNDED OF THE SUBDUPLICATE RATIO OF

THE ELASTIC FORCE DIRECTLY, AND THE SUBDUPLICATE RATIO OF THE DENSITY INVERSELY; SUPPOSING THE ELASTIC FORCE OF THE FLUID TO BE PROPORTIONAL TO ITS CONDENSATION.

CASE 1. If the mediums be homogeneous, and the distances of the pulses in those mediums be equal amongst themselves, but the motion in one medium is more intense than in the other, the contractions and dilatations of the correspondent parts will be as those motions; not that this proportion is perfectly accurate. However, if the contractions and dilatations are not exceedingly intense, the error will not be sensible; and therefore this proportion may be considered as physically exact. Now the motive elastic forces are as the contractions and dilatations; and the velocities generated in the same time in equal parts are as the forces. Therefore equal and corresponding parts of corresponding pulses will go and return together, through spaces proportional to their contractions and dilatations, with velocities that are as those spaces; and therefore the pulses, which in the time of one going and returning advance forward a space equal to their breadth, and are always succeeding into the places of the pulses that immediately go before them, will, by reason of the equality of the distances, go forward in both mediums with equal velocity.

CASE 2. If the distances of the pulses or their lengths are greater in one medium than in another, let us suppose that the correspondent parts describe spaces, in going and returning, each time proportional to the breadths of the pulses; then will their contractions and dilatations be equal; and therefore if the mediums are homogeneous, the motive elastic form, which agitate them with a reciprocal motion, will be equal also. Now the matter to be moved by these forces is as the breadth of the pulses; and the space through which they move every time they go and return is in the same ratio. And, moreover, the time of one going and returning is in a ratio compounded of the subduplicate ratio of the matter, and the subduplicate ratio of the space; and therefore is as the space. But the pulses advance a space equal to their breadths in the times of going once and returning once; that is, they go over spaces proportional to the times, and therefore are equally swift.

CASE 3. And, therefore in mediums of equal density and elastic force all the pulses are equally swift. Now if the density or the elastic force of the medium were augmented, then, because the motive force is increased in the ratio of the elastic force, and the matter to be moved is increased in the ratio of the density, the time which is necessary for producing the same motion as before will be increased in the subduplicate ratio of the density, and will be diminished in the subduplicate ratio of the elastic force. And therefore the velocity of the pulses will be in a ratio compounded of the subduplicate ratio of the density of the medium inversely, and the subduplicate ratio of the elastic force directly.   Q.E.D.

This Proposition will be made more clear from the construction of the following Problem.

PROPOSITION XLIX. PROBLEM XI.

THE DENSITY AND ELASTIC FORCE OF A MEDIUM BEING GIVEN, TO FIND THE VELOCITY OF THE PULSES.

Suppose the medium to be pressed by an incumbent weight after the manner of our air; and let A be the height of a homogeneous medium, whose weight is equal to the incumbent weight, and whose density is the same with the density of the compressed medium in which the pulses are propagated. Suppose a pendulum to be constructed whose length between the point of suspension and the centre of oscillation is A: and in the time in which that pendulum will perform one entire oscillation composed of its going and returning, the pulse will be propagated right onwards through a space equal to the circumference of a circle described with the radius A.

For, letting those things stand which were constructed in Prop. XLVII, if any physical line, as EF, describing the space PS in each vibration, be acted on in the extremities P and S of every going and return that it makes by an elastic force that is equal to its weight, it will perform its several vibrations in the time in which the same might oscillate in a cycloid whose whole perimeter is equal to the length PS; and that because equal forces will impel equal corpuscles through equal spares in the same or equal times. Therefore since the times of the oscillations are in the subduplicate ratio of the lengths of the pendulums, and the length of the pendulum is equal to half the arc of the whole cycloid, the time of one vibration would be to the time of the oscillation of a pendulum whose length is A in the subduplicate ratio of the length $^1/_2$PS or PO to the length A. But the elastic force with which the physical lineola EG is urged, when it is found in its extreme places P, S, was (in the demonstration of Prop. XLVII) to its whole elastic force as HL – KN to V, that is (since the point K now falls upon P), as HK to V: and all that force, or which is the same thing, the incumbent weight by which the lineola EG is compressed, is to the weight of the lineola as the altitude A of the incumbent weight to EG the length of the lineola; and therefore, *ex aequo*, the force with which the lineola EG is urged in the places P and S is to the weight of that lineola as HK × A to V × EG; or as PO x A to VV; because HK was to EG as PO to V. Therefore since the times in which equal bodies are impelled through equal spaces are reciprocally in the subduplicate ratio of the forces, the time of one vibration, produced by the action of that elastic force, will be to the time of a vibration, produced by the impulse of the weight in a subduplicate ratio of VV to PO × A, and therefore to the time of the oscillation of a pendulum whose length is A in the subduplicate ratio of VV to PO × A, and

the subduplicate ratio of PO to A conjunctly; that is, in the entire ratio of V to A. But in the time of one vibration composed of the going and returning of the pendulum, the pulse will be propagated right onward through a space equal to its breadth BC. Therefore the time in which a, pulse runs over the space BC is to the time of one oscillation composed of the going and returning of the pendulum as V to A, that is, as BC too the circumference of a circle whose radius is A. But the time in which the pulse will run over the space BC is to the time in which it will run over a length equal to that circumference in the same ratio; and therefore in the time of such an oscillation the pulse will run over a length equal to that circumference.   Q.E.D.

COR. 1. The velocity of the pulses is equal to that which heavy bodies acquire by falling with an equally accelerated motion, and in their fall describing half the altitude A. For the pulse will, in the time of this fall, supposing it to move with the velocity acquired by that fall, run over a space that will be equal to the whole altitude A; and therefore in the time of one oscillation composed of one going and return, will go over a space equal to the circumference of a circle described with the radius A; for the time of the fall is to the time of oscillation as the radius of a circle to its circumference.

COR. 2. Therefore since that altitude A is as the elastic force of the fluid directly, and the density of the same inversely, the velocity of the pulses will be in a ratio compounded of the subduplicate ratio of the density inversely, and the subduplicate ratio of the elastic force directly.

PROPOSITION L. PROBLEM XII.

TO FIND THE DISTANCES OF THE PULSES.

Let the number of the vibrations of the body, by whose tremor the pulses are produced, be found to any given time. By that number divide the space which a pulse can go over in the same time, and the part found will be the breadth of one pulse.   Q.E.I.

SCHOLIUM.

The last Propositions respect the motions of light and sounds; for since light is propagated in right lines, it is certain that it cannot consist in action alone (by Prop. XLI and XLII). As to sounds, since they arise from tremulous bodies, they can be nothing else but pulses of the air propagated through it (by Prop. XLIII); and this is confirmed by the tremors which sounds, if they be loud and deep, excite in the bodies near them, as we experience in the sound of drums; for quick and short tremors are less easily excited. But it is well known that any sounds, falling upon strings in unison with the sonorous bodies, excite tremors in those strings. This is also confirmed from the velocity of sounds; for since the specific gravities of rain-water and quicksilver are to one another as about 1 to $13^2/_3$, and when the mercury in the barometer is at the height of 30 inches of our measure, the specific gravities of the air and of rain-water are to one another as about 1 to 870, therefore the specific gravity of air and quicksilver are to each other as 1 to 11890. Therefore when the height of the quicksilver is at 30 inches, a height of uniform air, whose weight would be sufficient to compress our air to the density we find it to be of, must be equal to 356700 inches, or 29725 feet of our measure; and this is that very height of the medium, which I have called A in the construction of the foregoing Proposition. A circle whose radius is 29725 feet is 186768 feet in circumference. And since a pendulum $39^1/_5$ inches in length completes one oscillation, composed of its going and return, in two seconds of time, as is commonly known, it follows that a pendulum 29725 feet, or 356700 inches in length will perform a like oscillation in $190^3/_4$ seconds. Therefore in that time a sound will go right onwards 186768 feet, and therefore in one second 979 feet.

But in this computation we have made no allowance for the crassitude of the solid particles of the air, by which the sound is propagated instantaneously. Because the weight of air is to the weight of water as 1 to 870, and because salts are almost twice as dense as water; if the particles of air are supposed to be of near the same density as those of water or salt, and the rarity of the air arises from the intervals of the particles; the diameter of one particle of air will be to the interval between the centres of the particles as 1 to about 9 or 10, and to the interval between the particles themselves as 1 to 8 or 9. Therefore to 979 feet, which, according to the above calculation, a sound will advance forward in one second of time, we may add $979/_9$, or about 109 feet, to compensate for the crassitude of the particles of the air: and then a sound will go forward about 1088 feet in one second of time.

Moreover, the vapours floating in the air being of another spring, and a different tone, will hardly, if at all, partake of the motion of the true air in which the sounds are propagated. Now if these vapours remain unmoved that motion will be propagated the

swifter through the true air alone, and that in the subduplicate ratio of the defect of the matter. So if the atmosphere consist of ten parts of true air and one part of vapours, the motion of sounds will be swifter in the subduplicate ratio of 11 to 10, or very nearly in the entire ratio of 21 to 20, than if it were propagated through eleven parts of true air: and therefore the motion of sounds above discovered must be increased in that ratio. By this means the sound will pass through 1142 feet in one second of time.

These things will be found true in spring and autumn, when the air is rarefied by the gentle warmth of those seasons, and by that means its elastic force becomes somewhat more intense. But in winter, when the air is condensed by the cold, and its elastic force is somewhat remitted, the motion of sounds will be slower in a subduplicate ratio of the density; and, on the other hand, swifter in the summer.

Now by experiments it actually appears that sounds do really advance in one second of time about 1142 feet of *English* measure, or 1070 feet of *French* measure.

The velocity of sounds being known, the intervals of the pulses are known also. For M. *Sauveur*, by some experiments that he made, found that an open pipe about five *Paris* feet in length gives a sound of the same tone with a viol-string that vibrates a hundred times in one second. Therefore there are near 100 pulses in a space of 1070 *Paris* feet, which a sound runs over in a second of time; and therefore one pulse fills up a space of about $10^7/_{10}$ *Paris* feet, that is, about twice the length of the pipe. From whence it is probable that the breadths of the pulses, in all sounds made in open pipes, are equal to twice the length of the pipes.

Moreover, from the Corollary of Prop. XLVII appears the reason why the sounds immediately cease with the motion of the sonorous body, and why they are heard no longer when we are at a great distance from the sonorous bodies than when we are very near them. And besides, from the foregoing principles, it plainly appears how it comes to pass that sounds are so mightily increased in speaking-trumpets; for all reciprocal motion uses to be increased by the generating cause at each return. And in tubes hindering the dilatation of the sounds, the motion decays more slowly, and recurs more forcibly; and therefore is the more increased by the new motion impressed at each return. And these are the principal phaenomena of sounds.

SECTION IX.

Of the circular motion of fluids.

HYPOTHESIS.

THE RESISTANCE ARISING FROM THE WANT OF LUBRICITY IN
THE PARTS OF A FLUID, IS, CÆTERIS PARIBUS, PROPORTIONAL TO

THE VELOCITY WITH WHICH THE PARTS OF THE FLUID ARE
SEPARATED FROM EACH OTHER.

PROPOSITION LI. THEOREM XXXIX.

IF A SOLID CYLINDER INFINITELY LONG, IN AN UNIFORM
AND INFINITE FLUID, REVOLVE WITH AN UNIFORM MOTION
ABOUT AN AXIS GIVEN IN POSITION, AND THE FLUID BE
FORCED ROUND BY ONLY THIS IMPULSE OF THE CYLINDER,
AND EVERY PART OF THE FLUID PERSEVERE UNIFORMLY IN
ITS MOTION; I SAY, THAT THE PERIODIC TIMES OF THE
PARTS OF THE FLUID ARE AS THEIR DISTANCES FROM THE
AXIS OF THE CYLINDER.

Let AFL be a cylinder turning uniformly about the axis S, and let the concentric circles BGM, CHN, DIO, EKP, &c., divide the fluid into innumerable concentric cylindric solid orbs of the same thickness. Then, because the fluid is homogeneous, the impressions which the contiguous orbs make upon each other mutually will be (by the Hypothesis) as their translations from each other, and as the contiguous superficies

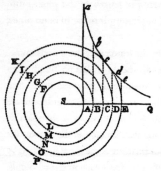

upon which the impressions are made. If the impression made upon any orb be greater or less on its concave than on its convex side, the stronger impression will prevail, and will either accelerate or retard the motion of the orb, according as it agrees with, or is contrary to, the motion of the same. Therefore, that every orb may persevere uniformly in its motion, the impressions made on both sides must be equal and their directions contrary. Therefore since the impressions are as the contiguous superficies, and as their translations from one another, the translations will be inversely as the superficies, that is, inversely as the distances of the superficies from the axis. But the differences of the angular motions about the axis are as those translations applied to the distances, or as the translations directly and the distances inversely; that is, joining these ratios together, as the squares of the distances inversely. Therefore if there be erected the lines Aa, Bb, Cc, Dd, Ee, &c., perpendicular to the several parts of the infinite right line SABCDEQ, and reciprocally proportional to the squares of SA, SB, SC, SD, SE, &c., and through the extremities of those perpendiculars there be supposed to pass an hyperbolic curve, the sums of the differences, that is, the whole angular motions, will be as the correspondent sums of

the lines A*a*, B*b*, C*c*, D*d*, E*e*, that is (if to constitute a medium uniformly fluid the number of the orbs be increased and their breadth diminished in, infinitum), as the hyperbolic areas A*a*Q, B*b*Q, C*c*Q, D*d*Q, E*e*Q, &c., analogous to the sums; and the times, reciprocally proportional to the angular motions, will be also reciprocally proportional to those areas. Therefore the periodic time of any particle as D, is reciprocally as the area D*d*Q, that is (as appears from the known methods of quadratures of curves), directly as the distance SD.   Q.E.D.

COR. 1. Hence the angular motions of the particles of the fluid are reciprocally as their distances from the axis of the cylinder, and the absolute velocities are equal.

COR. 2. If a fluid be contained in a cylindric vessel of an infinite length, and contain another cylinder within, and both the cylinders revolve about one common axis, and the times of their revolutions be as their semi-diameters, and every part of the fluid perseveres in its motion, the periodic times of the several parts will be as the distances from the axis of the cylinders.

COR. 3. If there be added or taken away any common quantity of angular motion from the cylinder and fluid moving in this manner; yet because this new motion will not alter the mutual attrition of the parts of the fluid, the motion of the parts among themselves will not be changed; for the translations of the parts from one another depend upon the attrition. Any part will persevere in that motion, which, by the attrition made on both sides with contrary directions, is no more accelerated than it is retarded.

COR. 4. Therefore if there be taken away from this whole system of the cylinders and the fluid all the angular motion of the outward cylinder, we shall have the motion of the fluid in a quiescent cylinder.

COR. 5. Therefore if the fluid and outward cylinder are at rest, and the inward cylinder revolve uniformly, there will be communicated a circular motion to the fluid, which will be propagated by degrees through the whole fluid; and will go on continually increasing, till such time as the several parts of the fluid acquire the motion determined in Cor. 4.

COR. 6. And because the fluid endeavours to propagate its motion still farther, its impulse will carry the outmost cylinder also about with it, unless the cylinder be violently detained; and accelerate its motion till the periodic times of both cylinders become equal among themselves. But if the outward cylinder be violently detained, it will make an effort to retard the motion of the fluid; and unless the inward cylinder preserve that motion by means of some external force impressed thereon, it will make it cease by degrees.

All these things will be found true by making the experiment in deep standing water.

PROPOSITION LII. THEOREM XL.

IF A SOLID SPHERE, IN AN UNIFORM AND INFINITE FLUID,
REVOLVES ABOUT AN AXIS GIVEN IN POSITION WITH AN
UNIFORM MOTION, AND THE FLUID BE FORCED ROUND BY
ONLY THIS IMPULSE OF THE SPHERE; AND EVERY PART OF
THE FLUID PERSEVERES UNIFORMLY IN ITS MOTION; I SAY,
THAT THE PERIODIC TIMES OF THE PARTS OF THE FLUID
ARE AS THE SQUARES OF THEIR DISTANCES FROM THE
CENTRE OF THE SPHERE.

CASE 1. Let AFL be a sphere turning uniformly about the axis S, and let the concentric circles BGM, CHN, DIO, EKP, &c., divide the fluid into innumerable concentric orbs of the same thickness. Suppose those orbs to be solid; and, because the fluid is homogeneous, the impressions which the contiguous orbs make one upon another will be (by the supposition) as their translations from one another, and the contiguous

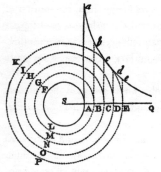

superficies upon which the impressions are made. If the impression upon any orb be greater or less upon its concave than upon its convex side, the more forcible impression will prevail, and will either accelerate or retard the velocity of the orb, according as it is directed with a conspiring or contrary motion to that of the orb. Therefore that every orb may persevere uniformly in its motion, it is necessary that the impressions made upon both sides of the orb should be equal, and have contrary directions. Therefore since the impressions are as the contiguous superficies, and as their translations from one another, the translations will be inversely as the superficies, that is, inversely as the squares of the distances of the superficies from the centre. But the differences of the angular motions about the axis are as those translations applied to the distances, or as the translations directly and the distances inversely; that is: by compounding those ratios, as the cubes of the distances inversely, Therefore if upon the several parts of the infinite right line SABCDEQ there be erected the perpendiculars Aa, Bb, Cc, Dd, Ee, &c., reciprocally proportional to the cubes of SA, SB, SC, SD, SE, &c., the sums of the differences, that is, the whole angular motions will be as the corresponding sums of the lines Aa, Bb, Cc, Dd, Ee, &c., that is (if to constitute an uniformly fluid medium the number of the orbs be increased and their thickness diminished *in infinitum*), as the hyperbolic areas AaQ, BbQ, CcQ,

DdQ, EeQ, &c., analogous to the sums; and the periodic times being reciprocally proportional to the angular motions, will be also reciprocally proportional to those areas. Therefore the periodic time of any orb DIO is reciprocally as the area DdQ, that is (by the known methods of quadratures), directly as the square of the distance SD. Which was first to be demonstrated.

CASE 2. From the centre of the sphere let there be drawn a great number of indefinite right lines, making given angles with the axis, exceeding one another by equal differences; and, by these lines revolving about the axis, conceive the orbs to be cut into innumerable annuli; then will every annulus have four annuli contiguous to it, that is, one on its inside, one on its outside, and two on each hand. Now each of these annuli cannot be impelled equally and with contrary directions by the attrition of the interior and exterior annuli, unless the motion be communicated according to the law which we demonstrated in Case 1. This appears from that demonstration. And therefore any series of annuli, taken in any right line extending itself *in infinitum* from the globe, will move according to the law of Case 1, except we should imagine it hindered by the attrition of the annuli on each side of it. But now in a motion, according to this law, no such is, and therefore cannot be, any obstacle to the motions persevering according to that law. If annuli at equal distances from the centre revolve either more swiftly or more slowly near the poles than near the ecliptic, they will be accelerated if slow, and retarded if swift, by their mutual attrition; and so the periodic times will continually approach to equality, according to the law of Case 1. Therefore this attrition will not at all hinder the motion from going on according to the law of Case 1, and therefore that law will take place; that is, the periodic times of the several annuli will be as the squares of their distances from the centre of the globe. Which was to be demonstrated in the second place.

CASE 3. Let now every annulus be divided by transverse sections into innumerable particles constituting a substance absolutely and uniformly fluid; and because these sections do not at all respect the law of circular motion, but only serve to produce a fluid substance, the law of circular motion will continue the same as before. All the very small annuli will either not at all change their asperity and force of mutual attrition upon account of these sections, or else they will change the same equally. Therefore the proportion of the causes remaining the same, the proportion of the effects will remain the same also; that is, the proportion of the motions and the periodic times. Q.E.D. But now as the circular motion, and the centrifugal force thence arising, is greater at the ecliptic than at the poles. There must be some cause operating to retain the several particles in their circles; otherwise the matter that is at the ecliptic will always recede from the centre, and come round about to the poles by the outside of the vortex, and from thence return by the axis to the ecliptic with a perpetual circulation.

COR. 1. Hence the angular motions of the parts of the fluid about the axis of the globe are reciprocally as the squares of the distances from the centre of the globe, and the absolute velocities are reciprocally as the same squares applied to the distances from the axis.

COR. 2. If a globe revolve with a uniform motion about an axis of a given position in a similar and infinite quiescent fluid with an uniform motion, it will communicate a whirling motion to the fluid like that of a vortex, and that motion will by degrees be propagated onward *in infinitum*; and this motion will be increased continually in every part of the fluid, till the periodical times of the several parts become as the squares of the distances from the centre of the globe.

COR. 3. Because the inward parts of the vortex are by reason of their greater velocity continually pressing upon and driving forward the external parts, and by that action are perpetually communicating motion to them, and at the same time those exterior parts communicate the same quantity of motion to those that lie still beyond them, and by this action preserve the quantity of their motion continually unchanged, it is plain that the motion is perpetually transferred from the centre to the circumference of the vortex, till it is quite swallowed up and lost in the boundless extent of that circumference. The matter between any two spherical superficies concentrical to the vortex will never be accelerated; because that matter will be always transferring the motion it receives from the matter nearer the centre to that matter which lies nearer the circumference.

COR. 4. Therefore, in order to continue a vortex in the same state of motion, some active principle is required from which the globe may receive continually the same quantity of motion which it is always communicating to the matter of the vortex. Without such a principle it will undoubtedly come to pass that the globe and the inward parts of the vortex, being always propagating their motion to the outward parts, and not receiving any new motion, will gradually move slower and slower, and at last be carried round no longer.

COR. 5. If another globe should be swimming in the same vortex at a certain distance from its centre, and in the mean time by some force revolve constantly about an axis of a given inclination, the motion of this globe will drive the fluid round after the manner of a vortex; and at first this new and small vortex will revolve with its globe about the centre of the other; and in the mean time its motion will creep on farther and farther, and by degrees be propagated *in infinitum*, after the manner of the first vortex. And for the same reason that the globe of the new vortex was carried about before by the motion of the other vortex, the globe of this other will be carried about by the motion of this new vortex, so that the two globes will revolve about some intermediate point, and by reason of that circular motion mutually fly from each other,

unless some force restrains them. Afterward, if the constantly impressed forces, by which the globes persevere in their motions, should cease, and every thing be left to act according to the laws of mechanics, the motion of the globes will languish by degrees (for the reason assigned in Cor. 3 and 4) and the vortices at last will quite stand still.

COR. 6. If several, globes in given places should constantly revolve with determined velocities about axes given in position, there would arise front them as many vortices going on *in infinitum*. For upon the same account that any one globe propagates its motion *in infinitum*, each globe apart will propagate its own motion *in infinitum* also; so that every part of the infinite fluid will be agitated with a motion resulting from the actions of all the globes. Therefore the vortices will not be confined by any certain limits, but by degrees run mutually into each other; and by the mutual actions of the vortices on each other, the globes will be perpetually moved from their places, as was shewn in the last Corollary; neither can they possibly keep any certain position among themselves, unless some force restrains them. But if those forces, which are constantly impressed upon the globes to continue these motions, should cease, the matter (for the reason assigned in Cor. 3 and 4) will gradually stop, and cease to move in vortices.

COR. 7. If a similar fluid be inclosed in a spherical vessel, and, by the uniform rotation of a globe in its centre, is driven round in a vortex; and the globe and vessel revolve the same way about the same axis, and their periodical times be as the squares of the semi-diameters; the parts of the fluid will not go on in their motions without acceleration or retardation, till their periodical times are as the squares of their distances from the centre of the vortex. No constitution of a vortex can be permanent but this.

COR. 8. If the vessel, the inclosed fluid, and the globe, retain this motion, and revolve besides with a common angular motion about any given axis, because the mutual attrition of the parts of the fluid is not changed by this motion, the motions of the parts among each other will not be changed; for the translations of the parts among themselves depend upon this attrition. Any part will persevere in that motion in which its attrition on one side retards it just as much an its attrition on the other side accelerates it.

COR. 9. Therefore if the vessel be quiescent, and the motion of the globe be given, the motion of the fluid will be given. For conceive a plane to pass through the axis of the globe, and to revolve with a contrary motion; and suppose the sum of the time of this revolution and of the revolution of the globe to be to the time of the revolution of the globe as the square of the semi-diameter of the vessel to the square of the semi-diameter of the globe; and the periodic times of the parts of the fluid in respect of this plane will be as the squares of their distances from the centre of the globe.

COR. 10. Therefore if the vessel move about the same axis with the globe, or with a given velocity about a different one, the motion of the fluid will be given. For if from the whole system we take away the angular motion of the vessel, all the motions will remain the same among themselves as before, by Cor. 8, and those motions will be given by Cor. 9.

COR. 11. If the vessel and the fluid are quiescent, and the globe revolves with an uniform motion, that motion will be propagated by degrees through the whole fluid to the vessel, and the vessel will be carried round by it, unless violently detained; and the fluid and the vessel will be continually accelerated till their periodic times become equal to the periodic times of the globe. If the vessel be either withheld by some force, or revolve with any constant and uniform motion, the medium will come by little and little to the state of motion defined in Cor. 8, 9, 10, nor will it ever persevere in any other state. But if then the forces, by which the globe and vessel revolve with certain motions, should cease, and the whole system be left to act according to the mechanical laws, the vessel and globe, by means of the intervening fluid, will act upon each other, and will continue to propagate their motions through the fluid to each other, till their periodic times become equal among themselves, and the whole system revolves together like one solid body.

SCHOLIUM.

In all these reasonings I suppose the fluid to consist of matter of uniform density and fluidity; I mean, that the fluid is such, that a globe placed any where therein may propagate with the same motion of its own, at distances from itself continually equal, similar and equal motions in the fluid in the same interval of time. The matter by its circular motion endeavours to recede from the axis of the vortex, and therefore presses all the matter that lies beyond. This pressure makes the attrition greater, and the separation of the parts more difficult; and by consequence diminishes the fluidity of the matter. Again; if the parts of the fluid are in any one place denser or larger than in the others, the fluidity will be less in that place, because there are fewer superficies where the parts can be separated from each other. In these cases I suppose the defeat of the fluidity to be supplied by the smoothness or softness of the parts, or some other condition; otherwise the matter where it is less fluid will cohere more, and be more sluggish, and therefore will receive the motion more slowly, and propagate it farther than agrees with the ratio above assigned. If the vessel be not spherical, the particles will move in lines not circular, but answering to the figure of the vessel; and the periodic times will be nearly as the squares of the mean distances from the centre. In the parts between the centre and the circumference the motions will be slower where the spaces are wide, and swifter where narrow; but yet the particles win not tend to the circumference at all the

more for their greater swiftness; for they then describe arcs of less curvity, and the conatus of receding from the centre is as much diminished by the diminution of this curvature as it is augmented by the increase of the velocity. As they go out of narrow into wide spaces, they recede a little farther from the centre, but in doing so are retarded; and when they come out of wide into narrow spaces, they are again accelerated; and so each particle is retarded and accelerated by turns for ever. These things win come to pass in a rigid vessel; for the state of vortices in an infinite fluid is known by Cor. 6 of this Proposition.

I have endeavoured in this Proposition to investigate the properties of vortices, that I might find whether the celestial phaenomena can be explained by them; for the phaenomenon is this, that the periodic times of the planets revolving about Jupiter are in the sesquiplicate ratio of their distances from Jupiter's centre; and the same rule obtains also among the planets that revolve about the sun. And these rules obtain also with the greatest accuracy, as far as has been yet discovered by astronomical observation. Therefore if those planets are carried round in vortices revolving about Jupiter and the sun, the vortices must revolve according to that law. But here we found the periodic times of the parts of the vortex to be in the duplicate ratio of the distances from the centre of motion; and this ratio cannot be diminished and reduced to the sesquiplicate, unless either the matter of the vortex be more fluid the farther it is from the centre, or the resistance arising from the want of lubricity in the parts of the fluid should, as the velocity with which the parts of the fluid are separated goes on increasing, be augmented with it in a greater ratio than that in which the velocity increases. But neither of these suppositions seem reasonable. The more gross and less fluid parts will tend to the circumference, unless they are heavy towards the centre. And though, for the sake of demonstration, I proposed, at the beginning of this Section, an Hypothesis that the resistance is proportional to the velocity, nevertheless, it is in truth probable that the resistance is in a less ratio than that of the velocity; which granted, the periodic times of the parts of the vortex will be in a greater than the duplicate ratio of the distances from its centre. If, as some think, the vortices move more swiftly near the centre, then slower to a certain limit, then again swifter near the circumference, certainly neither the sesquiplicate, nor any other certain and determinate ratio, can obtain in them. Let philosophers then see how that phaenomenon of the sesquiplicate ratio can be accounted for by vortices.

PROPOSITION LIII. THEOREM XLI.

BODIES CARRIED ABOUT IN A VORTEX, AND RETURNING IN
THE SAME ORB, ARE OF THE SAME DENSITY WITH THE
VORTEX, AND ARE MOVED ACCORDING TO THE SAME LAW

WITH THE PARTS OF THE VORTEX, AS TO VELOCITY AND
DIRECTION OF MOTION.

For if any small part of the vortex, whose particles or physical points preserve a given situation among each other, be supposed to be congealed, this particle will move according to the same law as before, since no change is made either in its density, *vis insita*, or figure. And again; if a congealed or solid part of the vortex be of the same density with the rest of the vortex, and be resolved into a fluid, this will move according to the same law as before, except in so far as its particles, now become fluid, may be moved among themselves. Neglect, therefore, the motion of the particles among themselves as not at all concerning the progressive motion of the whole, and the motion of the whole will be the same as before. But this motion will be the same with the motion of other parts of the vortex at equal distances from the centre; because the solid, now resolved into a fluid, is become perfectly like to the other parts of the vortex. Therefore a solid, if it be of the same density with the matter of the vortex, will move with the same motion as the parts thereof, being relatively at rest in the matter that surrounds it. If it be more dense, it will endeavour more than before to recede from the centre; and therefore overcoming that force of the vortex, by which, being, as it were, kept in equilibrio, it was retained in its orbit, it will recede from the centre, and in its revolution describe a spiral, returning no longer into the same orbit. And, by the same argument, if it be more rare, it will approach to the centre. Therefore it can never continually go round in the same orbit, unless it be of the same density with the fluid. But we have shewn in that case that it would revolve according to the same law with those parts of the fluid that are at the same or equal distances from the centre of the vortex.

COR. 1. Therefore a solid revolving in a vortex, and continually going round in the same orbit, is relatively quiescent in the fluid that carries it.

COR. 2. And if the vortex be of an uniform density, the same body may revolve at any distance from the centre of the vortex.

SCHOLIUM.

Hence it is manifest that the planets are not carried round in corporeal vortices; for, according to the *Copernican* hypothesis, the planets going round the sun revolve in ellipses, having the sun in their common focus and by radii drawn to the sun describe areas proportional to the times. But now the parts of a vortex can never revolve with such a motion. Let AD, BE, CF, represent three orbits described about the sun S, of which let the utmost circle CF be concentric to the sun; and let the aphelia of the two innermost be A, B; and their perihelia D, E. Therefore a body revolving in the orb CF,

describing, by a radius drawn to the sun, areas proportional to the times, will move with an uniform motion. And, according to the laws of astronomy, the body revolving in the orb BE will move slower in its aphelion B, and swifter in its perihelion E; whereas, according to the laws of mechanics, the matter of the vortex ought to move more swiftly in the narrow space between A and C than in the wide space between D and F; that is, more swiftly in the aphelion than in the perihelion. Now these two conclusions contradict each other. So at the beginning of the sign of Virgo, where the aphelion, of Mars is at present, the distance between the orbits of Mars and Venus is to the distance between the same orbits, at the beginning of the sign of Pisces, as about 3 to 2; and therefore the matter of the vortex between those orbits ought to be swifter at the beginning of Pisces than at the beginning of

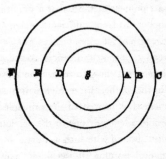

Virgo in the ratio of 3 to 2; for the narrower the space is through which the same quantity of matter passes in the same time of one revolution, the greater will be the velocity with which it passes through it. Therefore if the earth being relatively at rest in this celestial matter should be carried round by it, and revolve together with it about the sun, the velocity of the earth at the beginning of Pisces would be to its velocity at the beginning of Virgo in a sesquialteral ratio. Therefore the sun's apparent diurnal motion at the beginning of Virgo ought to be above 70 minutes, and at the beginning of Pisces less than 48 minutes; whereas, on the contrary, that apparent motion of the sun is really greater at the beginning of Pisces than at the beginning of Virgo, as experience testifies; and therefore the earth is swifter at the beginning of Virgo than at the beginning of Pisces; so that the hypothesis of vortices is utterly irreconcileable with astronomical phaenomena, and rather serves to perplex than explain the heavenly motions. How these motions are performed in free spaces without vortices, may be understood by the first Book; and I shall now more fully treat of it in the following Book.

BOOK III.

In the preceding Books I have laid down the principles of philosophy, principles not philosophical, but mathematical; such, to wit, as we may build our reasonings upon in philosophical inquiries. These principles are the laws and conditions of certain motions, and powers or forces, which chiefly have respect to philosophy; but, lest they should have appeared of themselves dry and barren, I have illustrated them here and

there with some philosophical scholiums, giving an account of such things as are of more general nature, and which philosophy seems chiefly to be founded on; such as the density and the resistance of bodies, spaces void of all bodies, and the motion of light and sounds. It remains that, from the same principles, I now demonstrate the frame of the System of the World. Upon this subject I had, indeed, composed the third Book in a popular method, that it might be read by many; but afterward, considering that such as had not sufficiently entered into the principles could not easily discern the strength of the consequences, nor lay aside the prejudices to which they had been many years accustomed, therefore, to prevent the disputes which might be raised upon such accounts, I chose to reduce the substance of this Book into the form of Propositions (in the mathematical way), which should be read by those only who had first made themselves masters of the principles established in the preceding Books: not that I would advise any one to the previous study of every Proposition of those Books; for they abound with such as might cost too much time, even to readers of good mathematical learning. It is enough if one carefully reads the Definitions, the Laws of Motion, and the first three Sections of the first Book, he may then pass on to this Book, and consult such of the remaining Propositions of the first two Books, as the references in this, and his occasions, shall require.

RULES OF REASONING IN PHILOSOPHY.

RULE I.

WE ARE TO ADMIT NO MORE CAUSES OF NATURAL THINGS
THAN SUCH AS ARE BOTH TRUE AND SUFFICIENT TO EXPLAIN
THEIR APPEARANCES.

To this purpose the philosophers say that Nature does nothing in vain, and more is in vain when less will serve; for Nature is pleased with simplicity, and affects not the pomp of superfluous causes.

RULE II.

THEREFORE TO THE SAME NATURAL EFFECTS WE MUST, AS
FAR AS POSSIBLE, ASSIGN THE SAME CAUSES.

As to respiration in a man and in a beast; the descent of stones in Europe and in *America*; the light of our culinary fire and of the sun; the reflection of light in the earth, and in the planets.

RULE III.

THE QUALITIES OF BODIES, WHICH ADMIT NEITHER INTENSION NOR REMISSION OF DEGREES, AND WHICH ARE FOUND TO BELONG TO ALL BODIES WITHIN THE REACH OF OUR EXPERIMENTS, ARE TO BE ESTEEMED THE UNIVERSAL QUALITIES OF ALL BODIES WHATSOEVER.

For since the qualities of bodies are only known to us by experiments, we are to hold for universal all such as universally agree with experiments; and such as are not liable to diminution can never be quite taken away. We are certainly not to relinquish the evidence of experiments for the sake of dreams and vain fictions of our own devising; nor are we to recede from the analogy of Nature, which uses to be simple, and always consonant to itself. We no other way know the extension of bodies than by our senses, nor do these reach it in all bodies; but because we perceive extension in all that are sensible, therefore we ascribe it universally to all others also. That abundance of bodies are hard, we learn by experience; and because the hardness of the whole arises from the hardness of the parts, we therefore justly infer the hardness of the undivided particles not only of the bodies we feel but of all others. That all bodies are impenetrable, we gather not from reason, but from sensation. The bodies which we handle we find impenetrable, and thence conclude impenetrability to be an universal property of all bodies whatsoever. That all bodies are moveable, and endowed with certain powers (which we call the *vires inertiae*) of persevering in their motion, or in their rest, we only infer from the like properties observed in the bodies which we have seen. The extension, hardness, impenetrability, mobility, and *vis inertia* of the whole, result from the extension, hardness, impenetrability, mobility, and *vires inertia* of the parts; and thence we conclude the least particles of all bodies to be also all extended, and hard and impenetrable, and moveable, and endowed with their proper *vires inertiae*. And this is the foundation of all philosophy. Moreover, that the divided but contiguous particles of bodies may be separated from one another, is matter of observation; and, in the particles that remain undivided, our minds are able to distinguish yet lesser parts, as is mathematically demonstrated. But whether the parts so distinguished, and not yet divided, may, by the powers of Nature, be actually divided and separated from one another, we cannot certainly determine. Yet, had we the proof of but one experiment that any undivided particle, in breaking a hard and solid body, suffered a division, we might by virtue of this rule conclude that the undivided as well as the divided particles may be divided and actually separated to infinity.

Lastly, if it universally appears, by experiments and astronomical observations, that all bodies about the earth gravitate towards the earth, and that in proportion to the

quantity of matter which they severally contain, that the moon likewise, according to the quantity of its matter, gravitates towards the earth; that, on the other hand, our sea gravitates towards the moon; and all the planets mutually one towards another; and the comets in like manner towards the sun; we must in consequence of this rule, universally allow that all bodies whatsoever are endowed with a principle of mutual gravitation. For the argument from the appearances concludes with more force for the universal gravitation of all bodies than for their impenetrability; of which, among those in the celestial regions, we have no experiments, nor any manner of observation. Not that I affirm gravity to be essential to bodies: by their *vis insita* I mean nothing but their *vis inertiae*. This is immutable. Their gravity is diminished as they recede from the earth.

RULE IV.

In experimental philosophy we are to look upon propositions collected by general induction from phaenomena as accurately or very nearly true, notwithstanding any contrary hypotheses that may be imagined, till such time as other phaenomena occur, by which they may either be made more accurate, or liable to exceptions.

This rule we must follow, that the argument of induction may not be evaded by hypotheses.

PHAENOMENA, OR APPEARANCES.

PHAENOMENON I.

That the circumjovial planets, by radii drawn to Jupiter's centre, describe areas proportional to the times of description; and that their periodic times, the fixed stars being at rest, are in the sesquiplicate proportion of their distances from its centre.

This we know from astronomical observations. For the orbits of these planets differ but insensibly from circles concentric to Jupiter; and their motions in those circles are found to be uniform. And all astronomers agree that their periodic times are in the sesquiplicate proportion of the semi-diameters of their orbits; and so it manifestly appears from the following table.

The periodic times of the satellites of Jupiter.
1^d. 18^h. 27'. 34". 3^d. 13^h. 13' 42". 7^d. 42' 36".   16^d. 16^b. 32' 9".
The distances of the satellites from Jupiter's centre.

From the observations of	1	2	3	4	
Borelli	5$^2/_3$	8$^2/_3$	14	24$^2/_3$	} *semi-diameter of*
Townly *by the Microm*	5,52	8,78	13,47	24,72	*Jupiter.*
Cassini *by the Telescope.*	5	8	13	23	
Cassini *by the eclip. of the satel*	5^2/3	9	14$^{23}/_{60}$	25$^3/_{10}$	
From the periodic times	5,667	9,017	14,384	25,299	

Mr. *Pound* has determined, by the help of excellent micrometers, the diameters of Jupiter and the elongation of its satellites after the following manner. The greatest heliocentric elongation of the fourth satellite from Jupiter's centre was taken with a micrometer in a 15 feet telescope, and at the mean distance of Jupiter from the earth was found about 8' 16". The elongation of the third satellite was taken with a micrometer in a telescope of 123 feet, and at the same distance of Jupiter from the earth was found 4' 42". The greatest elongations of the other satellites, at the same distance of Jupiter from the earth, are found from the periodic times to be 2' 56" 47''', and 1' 51" 6'''.

The diameter of Jupiter taken with the micrometer in a 123 feet telescope several times, and reduced to Jupiter's mean distance from the earth, proved always less than 40", never less than 38", generally 39". This diameter in shorter telescopes is 40", or 41"; for Jupiter's light is a little dilated by the unequal refrangibility of the rays, and this dilatation bears a less ratio to the diameter of Jupiter in the longer and more perfect telescopes than in those which are shorter and less perfect. The times in which two satellites, the first and the third, passed over Jupiter's body, were observed, from the beginning of the ingress to the beginning of the egress, and from the complete ingress to the complete egress, with the long telescope. And from the transit of the first satellite, the diameter of Jupiter at its mean distance from the earth came forth 37$^1/_8$", and from the transit of the third 37$^3/_8$". There was observed also the time in which the shadow of the first satellite passed over Jupiter's body, and thence the diameter of Jupiter at its mean distance from the earth came out about 37". Let us suppose its diameter to be 37$^1/_4$" very nearly, and then the greatest elongations of the first, second, third, and fourth satellite will be respectively equal to 5,965, 9,494, 15,141, and 26,63 semi-diameters of Jupiter.

PHAENOMENON II.

THAT THE CIRCUMSATURNAL PLANETS, BY RADII DRAWN TO SATURN'S CENTRE, DESCRIBE AREAS PROPORTIONAL TO THE TIMES OF DESCRIPTION; AND THAT THEIR PERIODIC TIMES, THE FIXED STARS BEING AT REST, ARE IN THE SESQUIPLICATE PROPORTION OF THEIR DISTANCES FROM ITS CENTRE.

For, as *Cassini* from his own observations has determined, their distances from Saturn's centre and their periodic times are as follow.

The periodic times of the satellites of Saturn.
1^d. 21^h. 18' 27". 2^d. 17^h. 41' 22". 4^d. 12^h. 25' 12". 15^d. 22^h. 41' 14". 79^d. 7^h. 48' 00".

The distances of the satellites from Saturn's centre, in semi-diameters of its ring.

From observations	$1^{19}/_{20}$	$2^1/_2$	$3^1/_2$	8	24
From the periodic times	1,93	2,47	3,45	8	23,35

The greatest elongation of the fourth satellite from Saturn's centre is commonly determined from the observations to be eight of these semi-diameters very nearly. But the greatest elongation of this satellite from Saturn's centre, when taken with an excellent micrometer in Mr. *Huygens'* telescope of 123 feet, appeared to be eight semi-diameters and $^7/_{19}$ of a semi-diameter. And from this observation and the periodic times the distances of the satellites from Saturn's centre in semi-diameters of the ring are 2,1, 2,69, 3,75, 8,7, and 25,35. The diameter of Saturn observed in the same telescope was found to be to the diameter of the ring as 3 to 7; and the diameter of the ring, *May* 29-29, 1719, was found to be 43"; and thence the diameter of the ring when Saturn is at its mean distance from the earth is 42", and the diameter of Saturn 18". These things appear so in very long and excellent telescopes, because in such telescopes the apparent magnitudes of the heavenly bodies bear a greater proportion to the dilatation of light in the extremities of those bodies than in shorter telescopes.

If we, then, reject all the spurious light, the diameter of Saturn will not amount to more than 16".

PHAENOMENON III.

THAT THE FIVE PRIMARY PLANETS, MERCURY, VENUS, MARS, JUPITER, AND SATURN, WITH THEIR SEVERAL ORBITS, ENCOMPASS THE SUN.

That Mercury and Venus revolve about the sun, is evident from their moon-like appearances. When they shine out with a full face, they are, in respect of us, beyond or above the sun; when they appear half full, they are about the same height on one side or other of the sun; when horned, they are below or between us and the sun; and they are sometimes, when directly under, seen like spots traversing the sun's disk. That Mars surrounds the sun, is as plain from its full face when near its conjunction with the sun, and from the gibbous figure which it shews in its quadratures. And the same thing is demonstrable of Jupiter and Saturn, from their appearing full in all situations; for the shadows of their satellites that appear sometimes upon their disks make it plain that the light they shine with is not their own, but borrowed from the sun.

PHAENOMENON IV.

THAT THE FIXED STARS BEING AT REST, THE PERIODIC TIMES OF THE FIVE PRIMARY PLANETS, AND (WHETHER OF THE SUN ABOUT THE EARTH, OR) OF THE EARTH ABOUT THE SUN, ARE IN THE SESQUIPLICATE PROPORTION OF THEIR MEAN DISTANCES FROM THE SUN.

This proportion, first observed by *Kepler*, is now received by all astronomers; for the periodic times are the same, and the dimensions of the orbits are the same, whether the sun revolves about the earth, or the earth about the sun. And as to the measures of the periodic times, all astronomers are agreed about them. But for the dimensions of the orbits, *Kepler* and *Bullialdus*, above all others, have determined them from observations with the greatest accuracy; and the mean distances corresponding to the periodic times differ but insensibly from those which they have assigned, and for the most part fall in between them; as we may see from the following table.

The periodic times with respect to the fixed stars, of the planets and earth revolving about the sun, in days and decimal parts of a day.

♄	♃	♂	☉	♀	☿
10759,275	4332,514	686,9785	365,2565	224,6176	87,9692

The mean distances of the planets and of the earth from the sun.

	♄	♃	♂
According to Kepler	951000.	519650.	152350.
" to Bullialdus	954198.	522520.	152350.
" to the periodic times	954006.	520096.	152369

	♂	♀	☿
According to Kepler	100000.	72400.	38806
" to Bullialdus	100000.	72398.	38585
" to the periodic times	100000.	72333.	38710.

As to Mercury and Venus, there can be no doubt about their distances from the sun; for they are determined by the elongations of those planet's from the sun; and for the distances of the superior planets, all dispute is cut off by the eclipses of the satellites of Jupiter. For by those eclipses the position of the shadow which Jupiter projects is determined; whence we have the heliocentric longitude of Jupiter. And from its heliocentric and geocentric longitudes compared together, we determine its distance.

PHAENOMENON V.

THEN THE PRIMARY PLANETS, BY RADII DRAWN TO THE
EARTH, DESCRIBE AREAS NO WISE PROPORTIONAL TO THE
TIMES; BUT THAT THE AREAS WHICH THEY DESCRIBE BY
RADII DRAWN TO THE SUN ARE PROPORTIONAL TO THE
TIMES OF DESCRIPTION.

For to the earth they appear sometimes direct, sometimes stationary, nay, and sometimes retrograde. But from the sun they are always seen direct, and to proceed with a motion nearly uniform, that is to say, a little swifter in the perihelion and a little slower in the aphelion distances, so as to maintain an equality in the description of the areas. This a noted proposition among astronomers, and particularly demonstrable in Jupiter, from the eclipses of his satellites; by the help of which eclipses, as we have said, the heliocentric longitudes of that planet, and its distances from the sun, are determined.

PHAENOMENON VI.

THAT THE MOON, BY A RADIUS DRAWN TO THE EARTH'S
CENTRE, DESCRIBES AN AREA PROPORTIONAL TO THE TIME
OF DESCRIPTION.

This we gather from the apparent motion of the moon, compared with its apparent diameter, it is true that the motion of the moon is a little disturbed by the action of the sun: but in laying down these Phaenomena, I neglect those small and inconsiderable errors.

PROPOSITIONS.

PROPOSITION I. THEOREM I.

THAT THE FORCES, BY WHICH THE CIRCUMJOVIAL PLANETS ARE
CONTINUALLY DRAWN OF FROM RECTILINEAR MOTIONS, AND
RETAINED IN THEIR PROPER ORBITS, TEND TO JUPITER'S CENTRE;
AND ARE RECIPROCALLY AS THE SQUARES OF THE DISTANCES OF
THE PLACES OF THOSE PLANETS FROM THAT CENTRE.

The former part of this Proposition appears from Phaen. I, and Prop. II or III,
Book I; the latter from Phaen. I, and Cor. 6, Prop. IV, of the same Book.

The same thing we are to understand of the planets which encompass
Saturn, by Phaen. II.

PROPOSITION II. THEOREM II.

THAT THE FORCES, BY WHICH THE PRIMARY PLANETS ARE
CONTINUALLY DRAWN OFF FROM RECTILINEAR MOTIONS, AND
RETAINED IN THEIR PROPER ORBITS, TEND TO THE SUN; AND
ARE RECIPROCALLY AS THE SQUARES OF THE DISTANCES OF
THE PLACES OF THOSE PLANETS FROM THE SUN'S CENTRE.

The former part of the Proposition is manifest from Phaen. V, and Prop. II, Book
I; the latter from Phaen. IV, and Cor. 6, Prop. IV, of the same Book. But this part of
the Proposition is, with great accuracy, demonstrable from the quiescence of the aphelion points; for a very small aberration from the reciprocal duplicate proportion would
(by Cor. 1, Prop. XLV, Book I) produce a motion of the apsides sensible enough in
every single revolution, and in many of them enormously great.

PROPOSITION III. THEOREM III.

THAT THE FORCE BY WHICH THE MOON IS RETAINED IN ITS ORBIT
LEADS TO THE EARTH; AND IS RECIPROCALLY AS THE SQUARE OF
THE DISTANCE OF ITS PLACE FROM THE EARTH'S CENTRE.

The former part of the Proposition is evident from Phaen. VI, and Prop. II or III,
Book I; the latter from the very slow motion of the moon's apogee; which in every sin-
gle revolution amounting but to 30° 31' *in consequentia*, may be neglected. For (by
Cor. 1, Prop. XLV, Book I) it appears, that, if the distance of the moon from the earth's

centre is to the semi-diameter of the earth as D to 1, the force from which such a motion will result, is reciprocally as $D^2 \, {}^4/_{243}$, i.e., reciprocally as the power of D, whose exponent is $2^4/_{243}$ that is to say, in the proportion of the distance something greater than reciprocally duplicate, but which comes $59^3/_4$ times nearer to the duplicate than to the triplicate proportion. But in regard that this motion is owing to the action of the sun (as we shall afterwards shew), it is here to be neglected. The action of the sun, attracting the moon from the earth, is nearly as the moon's distance from the earth; and therefore (by what we have shewed in Cor. 2, Prop. XLV, Book I) is to the centripetal force of the moon as 2 to 357,45, or nearly so; that is, as 1 to $178^{29}/_{40}$. And if we neglect so inconsiderable a force of the sun, the remaining force, by which the moon is retained in its orb, will be reciprocally as D^2. This will yet more fully appear from comparing this force with the force of gravity, as is done in the next Proposition.

COR. If we augment the mean centripetal force by which the moon is retained in its orb, first in the proportion of $177^{29}/_{40}$ to $178^{29}/_{40}$, and then in the duplicate proportion of the semi-diameter of the earth to the mean distance of the centres of the moon and earth, we shall have the centripetal force of the moon at the surface of the earth; supposing this force, in descending to the earth's surface, continually to increase in the reciprocal duplicate proportion of the height.

PROPOSITION IV. THEOREM IV.

THAT THE MOON GRAVITATES TOWARDS THE EARTH, AND BY THE FORCE OF GRAVITY IS CONTINUALLY DRAWN OF FROM A RECTILINEAR MOTION, AND RETAINED IN ITS ORBIT.

The mean distance of the moon from the earth in the syzygies in semi-diameters of the earth, is, according to *Ptolemy* and most astronomers, 59; according to *Vendelin* and *Huygens*, 60; to *Copernicus*, $60^1/_3$; to *Street*, $60^2/_5$; and to *Tycho*, $56^1/_2$. But *Tycho*, and all that follow his tables of refraction, making the refractions of the sun and moon (altogether against the nature of light) to exceed the refractions of the fixed stars, and that by four or five minutes *near the horizon*, did thereby increase the moon's *horizontal* parallax by a like number of minutes, that is, by a twelfth or fifteenth part of the whole parallax. Correct this error, and the distance will become about $60^1/_2$ semi-diameters of the earth, near to what others have assigned. Let us assume the mean distance of 60 diameters in the syzygies; and suppose one revolution of the moon, in respect of the fixed stars, to be completed in $27^d. \ 7^h. \ 43'$, as astronomers have determined; and the circumference of the earth to amount to 123249600 *Paris* feet, as the French have found by mensuration. And now if we imagine the moon, deprived of all motion, to be let go, so as to descend towards the earth with the impulse of all that

force by which (by Cor. Prop. III) it is retained in its orb, it will in the space of one minute of time, describe in its fall $15^1/_{12}$ *Paris* feet. This we gather by a calculus, founded either upon Prop. XXXVI, Book I, or (which comes to the same thing) upon Cor. 9, Prop. IV, of the same Book. For the versed sine of that arc, which the moon, in the space of one minute of time, would by its mean motion describe at the distance of 60 semi-diameters of the earth, is nearly $15^1/_{12}$ *Paris* feet, or more accurately 15 feet, 1 inch, and 1 line $^4/_9$. Wherefore, since that force, in approaching to the earth, increases in the reciprocal duplicate proportion of the distance, and, upon that account, at the surface of the earth, is 60×60 times greater than at the moon, a body in our regions, falling with that force, ought in the space of one minute of time, to describe $60 \times 60 \times 15^1/_{12}$ *Paris* feet; and, in the space of one second of time, to describe $15^1/_{12}$ of those feet; or more accurately 15 feet, 1 inch, and 1 line $^4/_9$. And with this very force we actually find that bodies here upon earth do really descend; for a pendulum oscillating seconds in the latitude of *Paris* will be 3 *Paris* feet, and 8 lines 12 in length, as Mr. *Huygens* has observed. And the space which a heavy body describes by falling in one second of time is to half the length of this pendulum in the duplicate ratio of the circumference of a circle to its diameter (as Mr. *Huygens* has also shewn), and is therefore 15 *Paris* feet, 1 inch, 1 line $^7/_9$. And therefore the force by which the moon is retained in its orbit becomes, at the very surface of the earth, equal to the force of gravity which we observe in heavy bodies there. And therefore (by Rule I and II) the force by which the moon is retained in its orbit is that very same force which we commonly call gravity; for, were gravity another force different from that, then bodies descending to the earth with the joint impulse of both forces would fall with a double velocity, and in the space of one second of time would describe $30^1/_6$ *Paris* feet; altogether against experience.

This calculus is founded on the hypothesis of the earth's standing still for if both earth and moon move about the sun, and at the same time about their common centre of gravity, the distance of the centres of the moon and earth from one another will be $60^1/_2$ semi-diameters of the earth; as may be found by a computation from Prop. LX, Book I.

SCHOLIUM.

The demonstration of this Proposition may be more diffusely explained after the following manner. Suppose several moons to revolve about the earth, as in the system of Jupiter or Saturn; the periodic times of these moons (by the argument of induction) would observe the same law which *Kepler* found to obtain among the planets; and therefore their centripetal forces would be reciprocally as the squares of the distances from the centre of the earth, by Prop. I, of this Book. Now if the lowest of these were very small, and were so near the earth as almost to touch the tops of the highest mountains, the

centripetal force thereof, retaining it in its orb, would be very nearly equal to the weights of any *terrestrial* bodies that should be found upon the tops of those mountains, as may be known by the foregoing computation. Therefore if the same little moon should be deserted by its centrifugal force that carries it through its orb, and so be disabled from going onward therein, it would descend to the earth; and that with the same velocity as heavy bodies do actually fall with upon the tops of those very mountains; because of the equality of the forces that oblige them both to descend. And if the force by which that lowest moon would descend were different from gravity, and if that moon were to gravitate towards the earth, as we find terrestrial bodies do upon the tops of mountains, it would then descend with twice the velocity, as being impelled by both these form conspiring together. Therefore since both these forces, that is, the gravity of heavy bodies, and the centripetal forces of the moons, respect the centre of the earth, and are similar and equal between themselves, they will (by Rule I and II) have one and the same cause. And therefore the force which retains the moon in its orbit is that very force which we commonly call gravity; because otherwise this little moon at the top of a mountain must either be without gravity, or fall twice as swiftly as heavy bodies are wont to do.

PROPOSITION V. THEOREM V.

THAT THE CIRCUMJOVIAL PLANETS GRAVITATE TOWARDS JUPITER; THE CIRCUMSATURNAL TOWARDS SATURN; THE CIRCUMSOLAR TOWARDS THE SUN; AND BY THE FORCES OF THEIR GRAVITY ARE DRAWN OFF FROM RECTILINEAR MOTIONS, AND RETAINED IN CURVILINEAR ORBITS.

For the revolutions of the circumjovial planets about Jupiter, of the circumsaturnal about Saturn, and of Mercury and Venus, and the other circumsolar planets, about the sun, are appearances of the same sort with the revolution of the moon about the earth; and therefore, by Rule II, must be owing to the same sort of causes; especially since it has been demonstrated, that the forces upon which those revolutions depend tend to the centres of Jupiter, of Saturn, and of the sun; and that those forces, in receding from Jupiter, from Saturn, and from the sun, decrease in the same proportion, and according to the same low, as the force of gravity does in receding from the earth.

COR. 1. There is, therefore, a power of gravity tending to all the planets; for, doubtless, Venus, Mercury, and the rest, are bodies of the same sort with Jupiter and Saturn. And since all attraction (by Law III) is mutual, Jupiter will therefore gravitate towards all his own satellites, Saturn towards his, the earth towards the moon, and the sun towards all the primary planets.

COR. 2. The force of gravity which tends to any one planet is reciprocally as the square of the distance of places from that planet's centre.

COR. 3. All the planets do mutually gravitate towards one another, by Cor. 1 and 2. And hence it is that Jupiter and Saturn, when near their conjunction, by their mutual attractions sensibly disturb each other's motions. So the sun disturbs the motions of the moon; and both sun and moon disturb our sea, as we shall hereafter explain.

SCHOLIUM.

The force which retains the celestial bodies in their orbits has been hitherto called centripetal force; but it being now made plain that it can be no other than a gravitating force, we shall hereafter call it gravity. For the cause of that centripetal force which retains the moon in its orbit will extend itself to all the planets, by Rule I, II, and IV.

PROPOSITION VI. THEOREM VI.

THAT ALL BODIES GRAVITATE TOWARDS EVERY PLANET; AND THAT THE WEIGHTS OF BODIES TOWARDS ANY THE *SAME* PLANET, AT EQUAL DISTANCES FROM THE CENTRE OF THE PLANET, ARE PROPORTIONAL TO THE QUANTITIES OF MATTER WHICH THEY SEVERALLY CONTAIN.

It has been, now of a long time, observed by others, that all sorts of heavy bodies (allowance being made for the inequality of retardation which they suffer from a small power of resistance in the air) descend to the earth *from equal heights* in equal times; and that equality of times we may distinguish to a great accuracy, by the help of pendulums. I tried the thing in gold, silver, lead, glass, sand, common salt, wood, water, and wheat. I provided two wooden boxes, round and equal: I filled the one with wood, and suspended an equal weight of gold (as exactly as I could) in the centre of oscillation of the other. The boxes hanging by equal threads of 11 feet made a couple of pendulum perfectly equal in weight and figure, and equally receiving the resistance of the air. And, placing the one by the other, I observed them to play together forward and backward, for a long time, with equal vibrations. And therefore the quantity of matter in the gold (by Cor. 1 and 6, Prop. XXIV, Book II) was to the quantity of matter in the wood as the action of the motive force (or *vis motrix*) upon all the gold to the action of the same upon all the wood; that is, as the weight of the one to the weight of the other: and the like happened in the other bodies. By these experiments, in bodies of the same weight, I could manifestly have discovered a difference of matter less than the thousandth part of the whole, had any such been. But, without all doubt, the nature of gravity towards the planets is the same as towards the earth. For, should we imagine

our terrestrial bodies removed to the orb of the moon, and there, together with the moon, deprived of all motion, to be let go, so as to fall together towards the earth, it is certain, from what we have demonstrated before, that, in equal times, they would describe equal spaces with the moon, and of consequence are to the moon, in quantity of matter, as their weights to its weight. Moreover, since the satellites of Jupiter perform their revolutions in times which observe the sesquiplicate proportion of their distances from Jupiter's centre, their accelerative gravities towards Jupiter will be reciprocally as the squares of their distances from Jupiter's centre; that is, equal, at equal distances. And, therefore, these satellites, if supposed to fall *towards Jupiter* from equal heights, would describe equal spaces in equal times, in like manner as heavy bodies do on our earth. And, by the same argument, if the circumsolar planets were supposed to be let fall at equal distances from the sun, they would, in their descent towards the sun, describe equal spaces in equal times. But forces which equally accelerate unequal bodies must be as those bodies: that is to say, the weights of the planets *towards the sun* must be as their quantities of matter. Further, that the weights of Jupiter and of his satellites towards the sun are proportional to the several quantities of their matter, appears from the exceedingly regular motions of the satellites (by Cor. 3, Prop. LXV, Book I). For if some of those bodies were more strongly attracted to the sun in proportion to their quantity of matter than others, the motions of the satellites would be disturbed by that inequality of attraction (by Cor. 2, Prop. LXV, Book I). If, at equal distances from the sun, any satellite, in proportion to the quantity of its matter, did gravitate towards the sun with a force greater than Jupiter in proportion to his, according to any given proportion, suppose of d to e; then the distance between the centres of the sun and of the satellite's orbit would be always greater than the distance between the centres of the sun and of Jupiter nearly in the subduplicate of that proportion: as by some computations I have found. And if the satellite did gravitate towards the sun with a force, lesser in the proportion of e to d, the distance of the centre of the satellite's orb from the sun would be less than the distance of the centre of Jupiter from the sun in the subduplicate of the same proportion. Therefore if, at equal distances from the sun, the accelerative gravity of any satellite towards the sun were greater or less than the accelerative gravity of Jupiter towards the sun but by one $^1/_{1000}$ part of the whole gravity, the distance of the centre of the satellite's orbit from the sun would be greater or less than the distance of Jupiter from the sun by one $^1/_{2000}$ part of the whole distance; that is, by a fifth part of the distance of the utmost satellite from the centre of Jupiter; an eccentricity of the orbit which would be very sensible. But the orbits of the satellites are concentric to Jupiter, and therefore the accelerative gravities of Jupiter, and of all its satellites towards the sun, are equal among themselves. And by the same argument, the weights of Saturn and of his satellites towards the sun, at equal distances

from the sun, are as their several quantities of matter; and the weights of the moon and of the earth towards the sun are either none, or accurately proportional to the masses of matter which they contain. But some they are, by Cor. 1 and 3, Prop. V.

But further; the weights of all the parts of every planet towards any other planet are one to another as the matter in the several parts; for if some parts did gravitate more, others less, than for the quantity of their matter, then the whole planet, according to the sort of parts with which it most abounds, would gravitate more or less than in proportion to the quantity of matter in the whole. Nor is it of any moment whether these parts are external or internal; for if, for example, we should imagine the terrestrial bodies with us to be raised up to the orb of the moon, to be there compared with its body: if the weights of such bodies were to the weights of the external parts of the moon as the quantities of matter in the one and in the other respectively; but to the weights of the internal parts in a greater or less proportion, then likewise the weights of those bodies would be to the weight of the whole moon in a greater or less proportion; against what we have shewed above.

COR. 1. Hence the weights of bodies do not depend upon their forms and textures; for if the weights could be altered with the forms, they would be greater or less, according to the variety of forms, in equal matter; altogether against experience,

COR. 2. Universally, all bodies about the earth gravitate towards the earth; and the weights of all, at equal distances from the earth's centre, are as the quantities of matter which they severally contain. This is the quality of all bodies within the reach of our experiments; and therefore (by Rule III) to be affirmed of all bodies whatsoever. If the *aether*, or any other body, were either altogether void of gravity, or were to gravitate lest in proportion to its quantity of matter, then, because (according to *Aristotle, Des Cartes*, and others) there is no difference betwixt that and other bodies but in *mere* form of matter, by a successive change from form to form, it might be changed at last into a body of the same condition with those which gravitate most in proportion to their quantity of matter; and, on the other hand, the heaviest bodies, acquiring the first form of that body, might by degrees quite lose their gravity. And therefore the weights would depend upon the forms of bodies, and with those forms might be changed: contrary to what was proved in the preceding Corollary.

COR. 3. All spaces are not equally full; for if all spaces were equally full, then the specific gravity of the fluid which fills the region of the air, on account of the extreme density of the matter, would fall nothing short of the specific gravity of quicksilver, or gold, or any other the most dense body; and, therefore, neither gold, nor any other body, could descend in air; for bodies do not descend in fluids, unless they are specifically heavier than the fluids. And if the quantity of matter in a given space can, by any rarefaction, be diminished, what should hinder a diminution to infinity?

Cor. 4. If all the solid particles of all bodies are of the same density, nor can be rarefied without pores, a void, space, or vacuum must be granted.

By bodies of the same density, I mean those whose *vires inertiae* are in the proportion of their bulks.

Cor. 5. The power of gravity is of a different nature from the power of magnetism; for the magnetic attraction is not as the matter attracted. Some bodies are attracted more by the magnet; others less; most bodies not at all. The power of magnetism in one and the same body may be increased and diminished; and is sometimes far stronger, for the quantity of matter, than the power of gravity; and in receding from the magnet decreases not in the duplicate but almost in the triplicate proportion of the distance, as nearly as I could judge from some rude observations.

PROPOSITION VII. THEOREM VII.

That there is a power of gravity tending to all bodies, proportional to the several quantities of matter which they contain.

That all the planets mutually gravitate one towards another, we have proved before; as well as that the force of gravity towards every one of them, considered apart, is reciprocally as the square of the distance of places from the centre of the planet. And thence (by Prop. LXIX, Book I, and its Corollaries) it follows, that the gravity tending towards all the planets is proportional to the matter which they contain.

Moreover, since all the parts of any planet A gravitate towards any other planet B; and the gravity of every part is to the gravity of the whole as the matter of the part to the matter of the whole; and (by Law III) to every action corresponds an equal re-action; therefore the planet B will, on the other hand, gravitate towards all the parts of the planet A; and its gravity towards any one part will be to the gravity towards the whole as the matter of the part to the matter of the whole. Q.E.D.

Cor. 1. Therefore the force of gravity towards any whole planet arises from, and is compounded of, the forces of gravity towards all its parts. Magnetic and electric attractions afford us examples of this; for all attraction towards the whole arises from the attractions towards the several parts. The thing may be easily understood in gravity, if we consider a greater planet, as formed of a number of lesser planets, meeting together in one globe; for *hence it would appear that* the force of the whole must arise from the forces of the component parts. If it is objected, that, according to this law, all bodies with us must mutually gravitate one towards another, whereas no such gravitation any where appears, I answer, that since the gravitation towards these bodies is to the gravitation towards the whole earth as these bodies are to the whole earth, the gravitation towards them must be far less than to fall under the observation of our senses.

COR. 2. The force of gravity towards the several equal particles of any body is reciprocally as the square of the distance of places from the particles; as appears from Cor. 3, Prop. LXXIV, Book I.

PROPOSITION VIII. THEOREM VIII.

IN TWO SPHERES MUTUALLY GRAVITATING, EACH TOWARDS
THE OTHER, IF THE MATTER IN PLANES ON ALL SIDES
ROUND ABOUT AND EQUI-DISTANT FROM THE CENTRES IS
SIMILAR, THE WEIGHT OF EITHER SPHERE TOWARDS THE
OTHER WILL BE RECIPROCALLY AS THE SQUARE OF THE
DISTANCE BETWEEN THEIR CENTRES.

After I had found that the force of gravity towards a whole planet did arise from and was compounded of the forces of gravity towards all its parts, and towards every one part was in the reciprocal proportion of the squares of the distances from the part, I was yet in doubt whether that reciprocal duplicate proportion did accurately hold, or but nearly so, in the total force compounded of so many partial ones; for it might be that the proportion which accurately enough took place in greater distances should be wide of the truth near the surface of the planet, where the distances of the particles are unequal, and their situation dissimilar. But by the help of Prop. LXXV and LXXVI, Book I, and their Corollaries, I was at last satisfied of the truth of the Proposition, as it now lies before us.

COR. 1. Hence we may find and compare together the weights of bodies towards different planets; for the weights of bodies revolving in circles about planets are (by Cor. 2, Prop. IV, Book I) as the diameters of the circles directly, and the squares of their periodic times reciprocally; and their weights at the surfaces of the planets, or at any other distances from their centres, are (by this Prop.) greater or less in the reciprocal duplicate proportion of the distances. Thus from the periodic times of Venus, revolving about the sun, in 224^d. $16\,^3/_4{}^h$, of the utmost circumjovial satellite revolving about Jupiter, in 16^d. $16\,^8/_{15}{}^h$.; of the Huygenian satellite about Saturn in 15^d. $22\,^2/_3{}^h$. and of the moon about the earth in 27^d. 7^h. $43'$; compared with the mean distance of Venus from the sun, and with the greatest heliocentric elongations of the outmost circumjovial satellite from Jupiter's centre, $8'\,16''$; of the Huygenian satellite from the centre of Saturn, $3'\,4''$; and of the moon from the earth, $10'\,33''$: by computation I found that the weight of equal bodies, at equal distances from the centres of the sun, of Jupiter, of Saturn, and of the earth, towards the sun, Jupiter, Saturn, and the earth, were one to another, as 1, $^1/_{1067}$, and $^1/_{169282}$, respectively. Then because as the distances are increased or diminished, the weights are diminished or increased in a duplicate ratio,

the weights of equal bodies towards the sun, Jupiter, Saturn, and the earth, at the distances 10000, 997, 791, and 109 from their centres, that is, at their very superficies, will be as 10000, 943, 529, and 435 respectively. How much the weights of bodies are at the superficies of the moon, will be shewn hereafter.

COR. 2. Hence likewise we discover the quantity of matter in the several planets; for their quantities of matter are as the forces of gravity at equal distances from their centres; that is, in the sun, Jupiter, Saturn, and the earth, as 1, $1/1067$, and $1/169282$ respectively. If the parallax of the sun be taken greater or less than 10" 30''', the quantity of matter in the earth must be augmented or diminished in the triplicate of that proportion.

COR. 3. Hence also we find the densities of the planets; for (by Prop. LXXII, Book I) the weights of equal and similar bodies towards similar spheres are, at the surfaces of those spheres, as the diameters of the spheres; and therefore the densities of dissimilar spheres are as those weights applied to the diameters of the spheres. But the true diameters of the Sun, Jupiter, Saturn, and the earth, were one to another as 10000, 997, 791, and 109; and the weights towards the same as 10000, 943, 529, and 435 respectively; and therefore their densities are as 100, 94,1, 67, and 400. The density of the earth, which comes out by this computation, does not depend upon the parallax of the sun, but is determined by the parallax of the moon, and therefore is here truly defined. The sun, therefore, is a little denser than Jupiter, and Jupiter than Saturn, and the earth four times denser than the sun; for the sun, by its great heat, is kept in a sort of a rarefied state. The moon is denser than the earth, as shall appear afterward.

COR. 4. The smaller the planets are, they are, *caeteris paribus*, of so much the greater density; for so the powers of gravity on their several surfaces come nearer to equality. They are likewise, *caeteris paribus*, of the greater density, as they are nearer to the sun. So Jupiter is more dense than Saturn, and the earth than Jupiter; for the planets were to be placed at different distances from the sun, that, according to their degrees of density, they might enjoy a greater or less proportion to the sun's heat. Our water, if it were removed as far as the orb of Saturn, would be converted into ice, and in the orb of Mercury would quickly fly away in vapour; for the light of the sun, to which its heat is proportional, is seven times denser in the orb of Mercury than with us: and by the thermometer I have found that a sevenfold heat of our summer sun will make water boil. Nor are we to doubt that the matter of Mercury is adapted to its heat, and is therefore more dense than the matter of our earth; since, in a denser matter, the operations of Nature require a stronger heat.

PROPOSITION IX THEOREM IX.

THAT THE FORCE OF GRAVITY, CONSIDERED DOWNWARD FROM THE SURFACE OF THE PLANETS, DECREASES NEARLY IN THE PROPORTION OF THE DISTANCES FROM THEIR CENTRES.

If the matter of the planet were of an uniform density, this Proposition would be accurately true (by Prop. LXXIII. Book I). The error, therefore, can be no greater than what may arise from the inequality of the density.

PROPOSITION X. THEOREM X.

THAT THE MOTIONS OF THE PLANETS IN THE HEAVENS MAY SUBSIST AN EXCEEDINGLY LONG TIME.

In the Scholium of Prop. XL, Book II, I have shewed that a globe of water frozen into ice, and moving freely in our air, in the time that it would describe the length of its semi-diameter, would lose by the resistance of the air $1/4386$ part of its motion; and the same proportion holds nearly in all globes, how great soever, and moved with whatever velocity. But that our globe of earth is of greater density than it would be if the whole consisted of water only, I thus make out. If the whole consisted of water only, whatever was of less density than water, because of its less specific gravity, would emerge and float above. And upon this account, if a globe of terrestrial matter, covered on all sides with water, was less dense than water, it would emerge somewhere; and, the subsiding water falling back, would be gathered to the opposite side. And such is the condition of our earth, which in a great measure is covered with seas. The earth, if it was not for its greater density, would emerge from the seas, and, according to its degree of levity, would be raised more or less above their surface, the water of the seas flowing backward to the opposite side. By the same argument, the spots of the sun, which float upon the lucid matter thereof, are lighter than that matter; and, however the planets have been formed while they were yet in fluid masses, all the heavier matter subsided to the centre. Since, therefore, the common matter of our earth on the surface thereof is about twice as heavy as water, and a little lower, in mines, is found about three, or four, or even five times more heavy, it is probable that the quantity of the whole matter of the earth may be five or six times greater than if it consisted all of water; especially since I have before shewed that the earth is about four times more dense than Jupiter. If, therefore, Jupiter is a little more dense than water, in the space of thirty days, in which that planet describes the length of 459 of its semi-diameters, it would, in a medium of the same density with our air, lose almost a tenth part of its motion. But since the resistance of mediums decreases in proportion to their weight or density, so that water, which is $13^3/_5$ times lighter than quicksilver, resists less in that proportion; and air, which is 860 times lighter than water, resists less in the same proportion; therefore in the heavens, where the weight of the medium in which the planets move is immensely diminished, the resistance will almost vanish.

It is shewn in the Scholium of Prop. XXII, Book II, that at the height of 200 miles above the earth the air is more rare than it is at the superficies of the earth in the ratio of 30 to 0,0000000000003998, or as 75000000000000 to 1 nearly. And hence the planet Jupiter, revolving in a medium of the same density with that superior air, would not lose by the resistance of the medium the 1000000th part of its motion in 1000000 years. In the spaces near the earth the resistance is produced only by the air, exhalations, and vapours. When these are carefully exhausted by the air-pump from under the receiver, heavy bodies fall within the receiver with perfect freedom, and without the least sensible resistance; gold itself, and the lightest down, let fall together, will descend with equal velocity; and though they fall through a space of four, six, and eight feet, they will come to the bottom at the same time; as appears from experiments. And therefore the celestial regions being perfectly void of air and exhalations, the planets and comets meeting no sensible resistance in those spaces will continue their motions through them for an immense tract of time.

HYPOTHESIS I.

THAT THE CENTRE OF THE SYSTEM OF THE WORLD IS IMMOVABLE.

This is acknowledged by all, while some contend that the earth, others that the sun, is fixed in that centre. Let us see what may from hence follow.

PROPOSITION XI. THEOREM XI.

THAT THE COMMON CENTRE OF GRAVITY OF THE EARTH, THE SUN, AND ALL THE PLANETS, IS IMMOVABLE.

For (by Cor. 4 of the Laws) that centre either is at rest, or moves uniformly forward in a right line; but if that centre moved, the centre of the worb:1 would move also, against the Hypothesis.

PROPOSITION XII. THEOREM XII.

THAT THE SUN IS AGITATED BY A PERPETUAL MOTION, BUT NEVER RECEDES FAR FROM THE COMMON CENTRE OF GRAVITY OF ALL THE PLANETS.

For since (by Cor. 2, Prop. VIII) the quantity of matter in the sun is to the quantity of matter in Jupiter as 1067 to 1; and the distance of Jupiter from the sun is to the semi-diameter of the sun in a proportion but a small matter greater, the common centre of gravity of Jupiter and the sun will fall upon a point a little without the surface of

the sun. By the same argument, since the quantity of matter in the sun is to the quantity of matter in Saturn as 3021 to 1, and the distance of Saturn from the sun is to the semi-diameter of the sun in a proportion but a small matter less, the common centre of gravity of Saturn and the sun will fall upon a point a little within the surface of the sun. And, pursuing the principles of this computation, we should find that though the earth and all the planets were placed on one side of the sun, the distance of the common centre of gravity of all from the centre of the sun would scarcely amount to one diameter of the sun. In other cases, the distances of those centres are always less; and therefore, since that centre of gravity is in perpetual rest, the sun, according to the various positions of the planets must perpetually be moved every way, but will never recede far from that centre.

COR. Hence the common centre of gravity of the earth, the sun, and all the planets, is to be esteemed the centre of the world; for since the earth, the sun, and all the planets, mutually gravitate one towards another, and are therefore, according to their powers of gravity, in perpetual agitation, as the Laws of Motion require, it is plain that their moveable centres cannot be taken for the immovable centre of the world. If that body were to be placed in the centre, towards which other bodies gravitate most (according to common opinion), that privilege ought to be allowed to the sun; but since the sun itself is moved, a fixed point is to be chosen from which the centre of the sun recedes least, and from which it would recede yet less if the body of the sun were denser and greater, and therefore less apt to be moved.

PROPOSITION XIII. THEOREM XIII.

THE PLANETS MOVE IN ELLIPSES WHICH HAVE THEIR COMMON FOCUS IN THE CENTRE OF THE SUN; AND, BY RADII DRAWN TO THAT CENTRE, THEY DESCRIBE AREAS PROPORTIONAL TO THE TIMES OF DESCRIPTION.

We have discoursed above of these motions from the Phaenomena. Now that we know the principles on which they depend, from those principles we deduce the motions of the heavens à priori. Because the weights of the planets towards the sun are reciprocally as the squares of their distances from the sun's centre, if the sun was at rest, and the other planets did not mutually act one upon another, their orbits would be ellipses, having the sun in their common focus; and they would describe areas proportional to the times of description, by Prop. I and XI, and Cor. 1, Prop. XIII, Book I. But the mutual actions of the planets one upon another are so very small, that they may be neglected; and by Prop. LXVI, Book I, they less disturb the motions of the planets around the sun in motion than if those motions were performed about the sun at rest.

It is true, that the action of Jupiter upon Saturn is not to be neglected; for the force of gravity towards Jupiter is to the force of gravity towards the sun (at equal distances, Cor. 2, Prop. VIII) as 1 to 1067; and therefore in the conjunction of Jupiter and Saturn, because the distance of Saturn from Jupiter is to the distance of Saturn from the sun almost as 4 to 9, the gravity of Saturn towards Jupiter will be to the gravity of Saturn towards the sun as 81 to 16 x 1067; or, as 1 to about 211. And hence arises a perturbation of the orb of Saturn in every conjunction of this planet with Jupiter, so sensible, that astronomers are puzzled with it. As the planet is differently situated in these conjunctions, its eccentricity is sometimes augmented, sometimes diminished; its aphelion is sometimes carried forward, sometimes backward, and its mean motion is by turns accelerated and retarded; yet the whole error in its motion about the sun, though arising from so great a force, may be almost avoided (except in the mean motion) by placing the lower focus of its orbit in the common centre of gravity of Jupiter and the sun (according to Prop. LXVII, Book I), and therefore that error, when it is greatest, scarcely exceeds two minutes; and the greatest error in the mean motion scarcely exceeds two minutes yearly. But in the conjunction of Jupiter and Saturn, the accelerative forces of gravity of the sun towards Saturn, of Jupiter towards Saturn, and of Jupiter towards the sun, are almost as 16, 81, and $\frac{16 \times 81 \times 3021}{25}$ or 156609; and

therefore the difference of the form of gravity of the sun towards Saturn, and of Jupiter towards Saturn, is to the force of gravity of Jupiter towards the sun as 65 to 156609, or as 1 to 2409. But the greatest power of Saturn to disturb the motion of Jupiter is proportional to this difference; and therefore the perturbation of the orbit of Jupiter is much less than that of Saturn's. The perturbations of the other orbits are yet far less, except that the orbit of the earth is sensibly disturbed by the moon. The common centre of gravity of the earth and moon moves in an ellipsis about the sun in the focus thereof, and, by a radius drawn to the sun, describes areas proportional to the times of description. But the earth in the mean time by a menstrual motion is revolved about this common centre.

PROPOSITION XIV. THEOREM XIV.

The aphelions and nodes of the orbits of the planets are fixed.

The aphelions are immovable by Prop. XI, Book I; and so are the planes of the orbits, by Prop. I of the same Book. And if the planes are fixed, the nodes must be so too. It is true, that some inequalities may arise from the mutual actions of the planets and comets in their revolutions; but these will be so small, that they may be here passed by.

COR. 1. The fixed stars are immovable, seeing they keep the same position to the aphelions and nodes of the planets.

COR. 2. And since these stars are liable to no sensible parallax from the annual motion of the earth, they can have no force, because of their immense distance, to produce any sensible effect in our system. Not to mention that the fixed stars, every where promiscuously dispersed in the heavens, by their contrary attractions destroy their mutual actions, by Prop. LXX, Book I.

SCHOLIUM.

Since the planets near the sun (viz. Mercury, Venus, the, Earth, and Mars) are so small that they can act with but little force upon each other, therefore their aphelions and nodes must be fixed, excepting in so far as they are disturbed by the actions of Jupiter and Saturn, and other higher bodies. And hence we may find, by the theory of gravity, that their aphelions move a little *in consequentia*, in respect of the fixed stars, and that in the sesquiplicate proportion of their several distances from the sun. So that if the aphelion of Mars, in the space of a hundred years, is carried 33' 20" *in consequentia*, in respect of the fixed stars, the aphelions of the Earth, of Venus, and of Mercury, will in a hundred years be carried forwards 17' 40", 10' 53", and 4' 16", respectively. But these motions are so inconsiderable, that we have neglected them in this Proposition.

PROPOSITION XV. PROBLEM I.

TO FIND THE PRINCIPAL DIAMETERS OF THE ORBITS OF THE PLANETS.

They are to be taken in the sub-sesquiplicate proportion of the periodic times, by Prop. XV, Book I, and then to be severally augmented in the proportion of the sum of the masses of matter in the sun and each planet to the first of two mean proportionals betwixt that sum and the quantity of matter in the sun, by Prop. LX, Book I.

PROPOSITION XVI. PROBLEM II.

TO FIND THE ECCENTRICITIES AND APHELIONS OF THE PLANETS.

This Problem is resolved by Prop. XVIII, Book I.

PROPOSITION XVII. THEOREM XV.

THAT THE DIURNAL MOTIONS OF THE PLANETS ARE UNIFORM, AND THAT THE LIBRATION OF THE MOON ARISES FROM ITS DIURNAL MOTION.

The Proposition is proved from the first Law of Motion, and Cor. 22, Prop. LXVI, Book I. Jupiter, with respect to the fixed stars, revolves in 91^h. 56'; Mars in 24^h. 39'; Venus in about 23^h.; the Earth in 23^h. 56'; the Sun in 25$^1/_2$, days, and the moon in 27 days, 7 hours, 43'. These things appear by the Phænomena. The spots in the sun's body return to the same situation on the sun's disk, with respect to the earth, in 27$^1/_2$ days; and therefore with respect to the fixed stars the sun revolves in about 25$^1/_2$ days. But because the lunar day, arising from its uniform revolution about its axis, is menstrual, *that is, equal to the time of its periodic revolution in its orb*, therefore the same face of the moon will be always nearly turned to the upper focus of its orb; but, as the situation of that focus requires, will deviate a little to one side and to the other from the earth in the lower focus; and this is the libration in longitude; for the libration in latitude arises from the moon's latitude, and the inclination of its axis to the plant of the ecliptic. This theory of the libration of the moon, Mr. *N. Mercator*, in his Astronomy, published at the beginning of the year 1676, explained more fully out of the letters I sent him. The utmost satellite of Saturn seems to revolve about its axis with a motion like this of the moon, respecting Saturn continually with the same face; for in its revolution round Saturn, as often as it comes to the eastern part of its orbit, it is scarcely visible, and generally quite disappears; which is like to be occasioned by some spots in that part of its body, which is then turned towards the earth, as M. *Cassini* has observed. So also the utmost satellite of Jupiter seems to revolve about its axis with a like motion, because in that part of its body which is turned from Jupiter it has a spot, which always appears as if it were in Jupiter's own body, whenever the satellite passes between Jupiter and our eye.

PROPOSITION XVIII. THEOREM XVI.

That the axes of the planets are less than the diameters drawn perpendicular to the axes.

The equal gravitation of the parts on all sides would give a spherical figure to the planets, if it was not for their diurnal revolution in a circle. By that circular motion it comes to pass that the parts receding from the axis endeavour to ascend about the equator; and therefore if the matter is in a fluid state, by its ascent towards the equator it will enlarge the diameters there, and by its descent towards the poles it will shorten the axis. So the diameter of Jupiter (by the concurring observations of astronomers) is found shorter betwixt pole and pole than from east to west. And, by the same argument, if our earth was not higher about the equator than at the poles, the seas would subside about the poles, and, rising towards the equator, would lay all things there under water.

PROPOSITION XIX. PROBLEM III.

TO FIND THE PROPORTION OF THE AXIS OF A PLANET TO THE DIAMETER, PERPENDICULAR THERETO.

Our countryman, Mr. *Norwood*, measuring a distance of 905751 feet of *London* measure between *London* and *York*, in 1635, and observing the difference of latitudes to be 2° 28', determined the measure of one degree to be 367196 feet of *London* measure, that is 57300 *Paris* toises. M. *Picart*, measuring an arc of one degree, and 22' 55" of the meridian between *Amiens* and *Malvoisine*, found an arc of one degree to be 57060 *Paris* toises. M. *Cassini*, the father, measured the distance upon the meridian from the town of *Collioure* in *Roussillon* to the Observatory of *Paris*; and his son added the distance from the Observatory to the Citadel of *Dunkirk*. The whole distance was $486156^1/_2$ toises and the difference of the latitudes of *Callioure* and *Dunkirk* was 8 degrees, and 31' $11^5/_6$". Hence an arc of one degree appears to be 57061 *Paris* toises. And from these measures we conclude that the circumference of the earth is 123249600, and its semi-diameter 19615800 *Paris* feet, upon the supposition that the earth is of a spherical figure.

In the latitude of *Paris* a heavy body falling in a second of time describes 15 *Paris* feet, 1 inch, $1^7/_8$ line, as above, that is, 2173 lines$^7/_9$. The weight of the body is diminished by the weight of the ambient air. Let us suppose the weight lost thereby to be $^1/_{11000}$ part of the whole weight; then that heavy body falling *in vacuo* will describe a height of 2174 lines in one second of time.

A body in every sidereal day of 23^h. 56' 4" uniformly revolving in a circle at the distance of 19615800 feet from the centre, in one second of time describes an arc of 1433,46 feet; the versed sine of which is 0,05236561 feet, or 7,54064 lines. And therefore the force with which bodies descend in the latitude of *Paris* is to the centrifugal force of bodies in the equator arising from the diurnal motion of the earth as 2174 to 7,54064.

The centrifugal force of bodies in the equator is to the centrifugal force with which bodies recede directly from the earth in the latitude of *Paris* 48° 50' 10" in the duplicate proportion of the radius to the cosine of the latitude, that is, as 7,54064 to 3,267. Add this force to the force with which bodies descend by their weight in the latitude of *Paris*, and a body, in the latitude of *Paris*, falling by its whole undiminished force of gravity, in the time of one second, will describe 2177,267 lines, or 15 *Paris* feet, 1 inch, and 5,267 lines. And the total force of gravity in that latitude will be to the centrifugal force of bodies in the equator of the earth as 2177,267 to 7,54064, or as 289 to 1.

Wherefore if APBQ represent the figure of the earth, now no longer spherical, but generated by the rotation of an ellipsis about its lesser axis PQ; and ACQ*qca* a canal

full of water, reaching from the pole Qq to the centre Cc, and thence rising to the equator Aa; the weight of the water in the leg of the canal ACca will be to the weight of water in the other leg QCcq as 299 to 288, because the centrifugal force arising from the circular motion sustains and takes off one of the 289 parts of the weight (in the one leg), and the weight of 288 in the other sustains the rest. But by computation (from

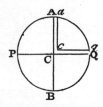

Cor. 2, Prop. XCI, Book I) I find, that, if the matter of the earth was all uniform, and without any motion, and its axis PQ were to the diameter AB as 100 to 101, the force of gravity in the place Q towards the earth would be to the force of gravity in the same place Q towards a sphere described about the centre C with the radius PC, or QC, as 126 to 125. And, by the same argument, the force of gravity in the place A towards the spheroid generated by the rotation of the ellipsis APBQ about the axis AB is to the force of gravity in the same place A, towards the sphere described about the centre C with the radius AC, as 125 to 126. But the force of gravity in the place A towards the earth is a mean proportional betwixt the forces of gravity towards the spheroid and this sphere; because the sphere, by having its diameter PQ diminished in the proportion of 101 to 100, is transformed into the figure of the earth; and this figure, by having a third diameter perpendicular to the two diameters AB and PQ diminished in the same proportion, is converted into the said spheroid; and the force of gravity in A, in either case, is diminished nearly in the same proportion. Therefore the force of gravity in A towards the sphere described about the centre C with the radius AC, is to the force of gravity in A towards the earth as 126 to 125 $^1/_2$. And the force of gravity in the place Q towards the sphere described about the centre C with the radius QC, is to the force of gravity in the place A towards the sphere described about the centre C, with the radius AC, in the proportion of the diameters (by Prop. LXXII, Book I), that is, as 100 to 101. If, therefore, we compound those three proportions 126 to 125, 126 to 125$^1/_2$, and 100 to 101, into one, the force of gravity in the place Q towards the earth will be to the force of gravity in the place A towards the earth as 126 × 126 × 100 to 125 × 125$^1/_2$ × 101; or as 501 to 500.

Now since (by Cor. 3, Prop. XCI, Book I) the force of gravity in either leg of the canal ACca, or QCcq, is as the distance of the places from the centre of the earth, if those legs are conceived to be divided by transverse, parallel, and equidistant surfaces, into parts proportional to the wholes, the weights of any number of parts in the one leg ACca will be to the weights of the same number of parts in the other leg as their magnitudes and the accelerative forces of their gravity conjunctly, that is, as 101 to 100, and 500 to 501, or as 505 to 501. And therefore if the centrifugal force of every part in the leg ACca, arising from the diurnal motion, was to the weight of the same

part as 4 to 505, so that from the weight of every part, conceived to be divided into 505 parts, the centrifugal force might take off four of those parts, the weights would remain equal in each leg, and therefore the fluid would rest in an equilibrium. But the centrifugal force of every part is to the weight of the same part as 1 to 289; that is, the centrifugal force, which should be $4/505$ parts of the weight, is only $1/289$ part thereof. And, therefore, I say, by the rule of proportion, that if the centrifugal force $4/505$ make the height of the water in the leg AC*ca* to exceed the height of the water in the leg QC*cq* by one $1/100$ part of its whole height, the centrifugal force $1/289$ will make the excess of the height in the leg AC*ca* only $1/289$ part of the height of the water in the other leg QC*cq*; and therefore the diameter of the earth at the equator, is to its diameter from pole to pole as 230 to 229. And since the mean semi-diameter of the earth, according to *Picart's* mensuration, is 19615800 *Paris* feet, or 3923,16 miles. (reckoning 5000 feet to a mile), the earth will be higher at the equator than at the poles by 85472 feet, or $17 1/16$ miles. And its height at the equator will be about 19658600 feet, and at the poles 19573000 feet.

If, the density and periodic time of the diurnal revolution remaining the same, the planet was greater or less than the earth, the proportion of the centrifugal force to that of gravity, and therefore also of the diameter betwixt the poles to the diameter at the equator, would likewise remain the same. But if the diurnal motion was accelerated or retarded in any proportion, the centrifugal force would be augmented or diminished nearly in the same duplicate proportion; and therefore the difference of the diameters will be increased or diminished in the same duplicate ratio very nearly. And if the density of the planet was augmented or diminished in any proportion, the force of gravity tending towards it would also be augmented or diminished in the same proportion: and the difference of the diameters contrariwise would be diminished in proportion as the force of gravity is augmented, and augmented in proportion as the force of gravity is diminished. Wherefore, since the earth, in respect of the fixed stars, revolves in 23^h. 56', but Jupiter in 9^h. 56', and the squares of their periodic times are as 29 to 5, and their densities as 400 to $94 1/2$, the difference of the diameters of Jupiter will be to its lesser diameter as $\frac{29}{5} \times \frac{400}{94 \frac{1}{2}} \times \frac{1}{225}$ to 1, or as 1 to $9 1/2$, nearly. Therefore the diameter

of Jupiter from east to west is to its diameter from pole to pole nearly as $10 1/3$ to $9 1/3$. Therefore since its greatest diameter is 37", its lesser diameter lying between the poles will be 33" 25". Add thereto about 3" for the irregular refraction of light, and the apparent diameters of this planet will become 40" and 36" 25"'; which are to each other as $11 1/6$ to $10 1/6$, very nearly. These things are so upon the supposition that the body of Jupiter is uniformly dense. But now if its body be denser towards the plane of the equator than towards the poles, its diameters may be to each other as 12 to 11, or 13 to 12, or perhaps as 14 to 13.

And *Cassini* observed in the year 1691, that the diameter of Jupiter reaching from east to west is greater by about a fifteenth part than the other diameter. Mr. *Pound* with his 123 feet telescope, and an excellent micrometer, measured the diameters of Jupiter in the year 1719, and found them as follow.

The Times.			Greatest diam.	Lesser diam.	The diam. to each other.
	Day.	Hours.	Parts	Parts	
January	28	6	13,40	12,28	As 12 to 11
March	6	7	13,12	12,20	$13^3/_4$ to $12^3/_4$
March	9	7	13,12	12,08	$12^2/_3$ to $11^2/_3$
April	9	9	12,32	11,48	14 1/2 to 13 1/2

So that the theory agrees with the phaenomena; for the planets are more heated by the sun's rays towards their equators, and therefore are a little more condensed by that heat than towards their poles.

Moreover, that there is a diminution of gravity occasioned by the diurnal rotation of the earth, and therefore the earth rises higher there than it does at the poles (supposing that its matter is uniformly dense), will appear by the experiments of pendulums related under the following Proposition.

PROPOSITION XX. PROBLEM IV.

To find and compare together the weights of bodies in the different regions of our earth.

Because the weights of the unequal legs of the canal of water ACQ*qca* are equal; and the weights of the parts proportional to the whole legs, and alike situated in them, are one to another as the weights of the wholes, and therefore equal betwixt themselves; the weights of equal parts, and alike situated in the legs, will be reciprocally as the legs, that is, reciprocally as 230 to 229. And the case is the same in all homogeneous equal bodies alike situated in the legs of the canal. Their weights are reciprocally as the legs, that is, reciprocally as the distances of the bodies from the centre of the earth. Therefore if the bodies are situated in the uppermost parts of the canals, or on the surface of the earth, their weights will be one to another reciprocally as their distances from the centre. And, by the same argument, the weights in all other places round the whole surface of the earth are reciprocally as the distances of the places from the centre; and, therefore, in the hypothesis of the earth's being a spheroid are given in proportion.

Whence arises this Theorem, that the increase of weight in passing from the equator to the poles is nearly as the versed sine of double the latitude; or, which comes to the same thing, as the square of the right sine of the latitude; and the area of the degrees of latitude in the meridian increase nearly in the same proportion. And, therefore, since the latitude of *Paris* is 48° 50', that of places under the equator 00° 00', and that of places under the poles 90°; and the versed sines of double those arcs are 11334,00000 and 20000, the radius being 10000; and the force of gravity at the pole is to the force of gravity at the equator as 230 to 229; and the excess of the force of gravity at the pole to the force of gravity at the equator as 1 to 229; the excess of the force of gravity in the latitude of *Paris* will be to the force of gravity at the equator as $1 \times {}^{11334}/_{20000}$ to 229, or as 5667 to 2290000, And therefore the whole forces of gravity in those places will be one to the other as 2295667 to 2290000. Wherefore since the lengths of pendulums vibrating in equal times are as the forces of gravity, and in the latitude of *Paris*, the length of a pendulum vibrating seconds is 3 *Paris* feet, and $8^{1}/_{2}$ lines, or rather because of the weight of the air, 63 lines, the length of a pendulum vibrating in the same time under the equator will be shorter by 1,087 lines. And by a like calculus the following table is made.

Latitude of the place.	Length of the pendulum		Measure of one degree in the meridian.
Deg.	Feet.	Lines.	Toises.
0	3	. 7,468	56637
5	3	. 7,482	56642
10	3	. 7,526	56659
15	3	. 7,596	56687
20	3	. 7,692	66724
25	3	. 7,812	56769
30	3	. 7,948	56823
35	3	. 8,099	56882
40	3	. 8,261	56945
1	3	. 8,294	56958
2	3	. 8,327	56971
3	3	. 8,361	56984
4	3	. 8,394	56997
45	3	. 8,428	57010
6	3	. 8,461	57022
7	3	. 8,494	57035
8	3	. 8,528	57048
9	3	. 8,561	57061
50	3	. 8,594	57074
55	3	. 8,756	57137
60	3	. 8,907	57196
65	3	. 9,044	57250
70	3	. 9,162	57295
75	3	. 9,258	57332
80	3	. 9,329	57360
85	3	. 9,372	57377
90	3	. 9,387	57382

By this table, therefore, it appears that the inequality of degrees is so small, that the figure of the earth, in geographical matters, may be considered as spherical; especially if the earth be a little denser towards the plane of the equator than towards the poles.

Now several astronomers, sent into remote countries to make astronomical observations, have found that pendulum clocks do accordingly move slower near the equator than in our climates. And, first of all, in the year 1672, M. *Richer* took notice of it in the island of *Cayenne*; for when, in the month of *August*, he was observing the transits of the fixed stars over the meridian, he found his clock to go slower than it ought in respect of the mean motion of the sun at the rate of 2' 28" a day. Therefore, fitting up a simple pendulum to vibrate in seconds, which were measured by an excellent clock, he observed the length of that simple pendulum; and this he did over and over every week for ten months together. And upon his return to *France*, comparing the length of that pendulum with the length of the pendulum at *Paris* (which was 3 *Paris* feet and $8^3/_5$ lines), he found it shorter by $1^1/_4$ line.

Afterwards, our friend Dr. *Halley*, about the year 1677, arriving at the island of St. *Helena*, found his pendulum clock to go slower there than at *London* without marking the difference. But he shortened the rod of his clock by more than the $^1/_8$ of an inch, or $1^1/_2$ line; and to effect this, because the length of the screw at the lower end of the rod was not sufficient, he interposed a wooden ring betwixt the nut and the ball.

Then, in the year 1682, M. *Varin* and M. *des Hayes* found the length of a simple pendulum vibrating in seconds at the Royal Observatory of *Paris* to be 3 feet and $8^5/_9$ lines. And by the same method in the island of *Goree*, they found the length of an isochronal pendulum to be 3 feet and $6^5/_9$ lines, differing from the former by two lines. And in the same year, going to the islands of *Guadaloupe* and *Martinico*, they found that the length of an isochronal pendulum in those islands was 3 feet and $6^1/_2$ lines.

After this, M. *Couplet*, the son, in the month of *July* 1697, at the Royal Observatory of *Paris*, so fitted his pendulum clock to the mean motion of the sun, that for a considerable time together the clock agreed with the motion of the sun. In *November* following, upon his arrival at *Lisbon*, he found his clock to go slower than before at the rate of 2' 13", in 24 hours. And next *March* coming to *Paraiba*, he found his clock to go slower than at *Paris*, and at the rate 4' 12" in 24 hours; and he affirms, that the pendulum vibrating in seconds was shorter at *Lisbon* by 21 lines, and at *Paraiba* by 32 lines, than at *Paris*. He had done better to have reckoned those differences $1^1/_3$ and $2^5/_9$ for these differences correspond to the differences of the times 2' 13" and 4' 12". But this gentleman's observations are so gross, that we cannot confide in them.

In the following years, 1699, and 1700, M. *des Hayes*, making another voyage to *America*, determined that in the island of *Cayenne* and *Granada* the length of the pendulum vibrating in seconds was a small matter less than 3 feet and $6^1/_2$ lines; that in the island of St. *Christophers* it was 3 feet and $6^3/_4$, lines; and in the island of St. *Domingo* 3 feet and 7 lines.

And in the year 1704, P. Feuillé, at *Puerto Bello* in *America*, found that the length of the pendulum vibrating in seconds was 3 *Paris* feet, and only 5 $^7/_{12}$ lines, that is, almost 3 lines shorter than at *Paris*, but the observation was faulty. For afterward, going to the island of *Martinico*, he found the length of the isochronal pendulum there 3 *Paris* feet and $5^{10}/_{12}$ lines.

Now the latitude of *Paraiba* is 6° 38' south; that of *Puerto Bello* 9° 33' north; and the latitudes of the islands *Cayenne*, *Goree*, *Gaudaloupe*, *Martinico*, *Granada*, St. *Christophers*, and St. *Domingo*, are respectively 4° 55', 14° 40", 15° 00', 14° 44', 12° 06', 17° 191, and 19° 48', north. And the excesses of the length of the pendulum at *Paris* above the lengths of the isochronal pendulums observed in those latitudes are a little greater than by the table of the lengths of the pendulum before computed. And therefore the earth is a little higher under the equator than by the preceding calculus, and a little denser at the centre than in mines near the surface, unless, perhaps, the heats of the torrid zone have a little extended the length of the pendulums.

For M. *Picart* has observed, that a rod of iron, which in frosty weather in the winter season was one foot long, when heated by fire, was lengthened into one foot and $^1/_4$ line. Afterward M. *de la Hire* found that a rod of iron, which in the like winter season was 6 feet long, when exposed to the heat of the summer sun, was extended into 6 feet and $^2/_3$ line. In the former case the heat was greater than in the latter; but in the latter it was greater than the heat of the external parts of a human body; for metals exposed to the summer sun acquire a very considerable degree of heat. But the rod of a pendulum clock is never exposed to the heat of the summer sun, nor ever acquires a heat equal to that of the external parts of a human body; and, therefore, though the 3 feet rod of a pendulum clock will indeed be a little longer in the summer than in the winter season, yet the difference will scarcely amount to $^1/_4$ line. Therefore the total difference of the lengths of isochronal pendulums in different climates cannot be ascribed to the difference of heat; nor indeed to the mistakes of the French astronomers. For although there is not a perfect agreement betwixt their observations, yet the errors are so small that they may be neglected; and in this they all agree, that isochronal pendulums are shorter under the equator than at the Royal Observatory of *Paris*, by a difference not less than $1^1/_4$ line, nor greater than $2^2/_3$ lines. By the observations of M. *Richer*, in the island of Cayenne, the difference was $1^1/_4$ line. That difference being corrected by those of M. *des Hayes*, becomes $1^1/_2$ line or $1^3/_4$ line. By the

less accurate observations of others, the same was made about two lines. And this disagreement might arise partly from the errors of the observations, partly from the dissimilitude of the internal parts of the earth, and the height of mountains; partly from the different heats of the air.

I take an iron rod of 3 feet long to be shorter by a sixth part of one line in winter time with us here in *England* than in the summer. Because of the great heats under the equator, subduct this quantity from the difference of one line and a quarter observed by M. *Richer*, and there will remain one line $^1/_{12}$, which agrees very well with $1^{87}/_{1000}$ line collected, by the theory a little before. M. *Richer* repeated his observations, made in the island of *Cayenne*, every week for ten months together, and compared the lengths of the pendulum which he had there noted in the iron rods with the lengths thereof which he observed in *France*. This diligence and care seems to have been wanting to the other observers. If this gentleman's observations are to be depended on, the earth is higher under the equator than at the poles, and that by an excess of about 17 miles; as appeared above by the theory.

PROPOSITION XXI. THEOREM XVII.

THAT THE EQUINOCTIAL POINTS GO BACKWARD, AND THAT THE AXIS OF THE EARTH, BY A NUTATION IN EVERY ANNUAL REVOLUTION, TWICE VIBRATES TOWARDS THE ECLIPTIC, AND AS OFTEN RETURNS TO ITS FORMER POSITION.

The proposition appears from Cor. 20, Prop. LXVI, Book I; but that motion of nutation must be very small, and, indeed, scarcely perceptible.

PROPOSITION XXII. THEOREM XVIII.

THAT ALL THE MOTIONS OF THE MOON, AND ALL THE INEQUALITIES OF THOSE MOTIONS, FOLLOW FROM THE PRINCIPLES WHICH WE HAVE LAID DOWN.

That the greater planets, while they are carried about the sun, may in the mean time carry other lesser planets, revolving about them; and that those lesser planets must move in ellipses which have their foci in the centres of the greater, appears from Prop. LXV, Book I. But then their motions will be several ways disturbed by the action of the sun, and they will suffer such inequalities as are observed in our moon. Thus our moon (by Cor. 2, 3, 4, and 5, Prop. LXVI, Book I) moves faster, and, by a radius drawn to the earth, describes an area greater for the time, and has its orbit less curved, and therefore approaches nearer to the earth in the syzygies than in the quadratures, excepting

in so far as these effects are hindered by the motion of eccentricity; for (by Cor. 9, Prop. LXVI, Book I) the eccentricity is greatest when the apogeon of the moon is in the syzygies, and least when the same is in the quadratures; and upon this account the perigeon moon is swifter, and nearer to us, but the apogeon moon slower, and farther from us, in the syzygies than in the quadratures. Moreover, the apogee goes forward, and the nodes backward; and this is done not with a regular but an unequal motion. For (by Cor. 7 and 8, Prop. LXVI, Book I) the apogee goes more swiftly forward in its syzygies, more slowly backward in its quadratures; and, by the excess of its progress above its regress, advances yearly *in consequentia*. But, contrariwise, the nodes (by Cor. 11, Prop. LXVI, Book I) are quiescent in their syzygies, and go fastest back in their quadratures. Farther, the greatest latitude of the moon (by Cor. 10, Prop. LXVI, Book I) is greater in the quadratures of the moon than in its syzygies. And (by Cor. 6, Prop. LXVI, Book I) the mean motion of the moon is slower in the perihelion of the earth than in its aphelion. And these are the principal inequalities (of the moon) taken notice of by astronomers.

But there are yet other inequalities not observed by former astronomers, by which the motions of the moon are so disturbed, that to this day we have not been able to bring them under any certain rule. For the velocities or horary motions of the apogee and nodes of the moon, and their equations, as well as the difference betwixt the greatest eccentricity in the syzygies, and the least eccentricity in the quadratures, and that inequality which we call the variation, are (by Cor. 14, Prop. LXVI, Book I) in the course of the year augmented and diminished in the triplicate proportion of the sun's apparent diameter. And besides (by Cor. I and 2, Lem. 10, and Cor. 16, Prop. LXVI, Book I) the variation is augmented and diminished nearly in the duplicate proportion of the time between the quadratures. But in astronomical calculations, this inequality is commonly thrown into and confounded with the equation of the moon's centre.

PROPOSITION XXIII. PROBLEM V.

TO DERIVE THE UNEQUAL MOTIONS OF THE SATELLITES OF JUPITER AND SATURN FROM THE MOTIONS OF OUR MOON.

From the motions of our moon we deduce the corresponding motions of the moons or satellites of Jupiter in this manner, by Cor. 16, Prop. LXVI, Book I. The mean motion of the nodes of the outmost satellite of Jupiter is to the mean motion of the nodes of our moon in a proportion compounded of the duplicate proportion of the periodic times of the earth about the sun to the periodic times of Jupiter about the sun, and the simple proportion of the periodic time of the satellite about Jupiter to the periodic time of our moon about the earth; and, therefore, those nodes, in the space of a

hundred years, are carried 8° 24' backward, or *in antecedentia*. The mean motions of the nodes of the inner satellites are to the mean motion of the nodes of the outmost as their periodic times to the periodic time of the former, by the same Corollary, and are thence given. And the motion of the apsis of every satellite *in consequentia* is to the motion of its nodes *in antecedentia* as the motion of the apogee of our moon to the motion of its nodes (by the same Corollary), and is thence given. But the motions of the apsides thus found must be diminished in the proportion of 5 to 9, or of about 1 to 2, on account of a cause which I cannot here descend to explain. The greatest equations of the nodes, and of the apsis of every satellite, are to the greatest equations of the nodes, and apogee of our moon respectively, as the motions of the nodes and apsides of the satellites, in the time of one revolution of the former equations, to the motions of the nodes and apogee of our moon, in the time of one revolution of the latter equations. The variation of a satellite seen from Jupiter is to the variation of our moon in the same proportion as the whole motions of their nodes respectively during the times in which the satellite and our moon (after parting from) are revolved (again) to the sun, by the same Corollary; and therefore in the outmost satellite the variation does not exceed 5" 12"'.

PROPOSITION XXIV. THEOREM XIX.

THAT THE FLUX AND REFLUX OF THE SEA ARISE FROM THE ACTIONS OF THE SUN AND MOON.

By Cor. 19 and 20, Prop. LXVI, Book I, it appears that the waters of the sea ought twice to rise and twice to fall every day, as well lunar as solar; and that the greatest height of the waters in the open and deep seas ought to follow the appulse of the luminaries to the meridian of the place by a less interval than 6 hours; as happens in all that eastern tract of the *Atlantic* and *AEthiopic* seas between *France* and the *Cape of Good Hope*; and on the coasts of *Chili* and *Peru* in the South Sea; in all which shores the flood falls out about the second, third, or fourth hour, unless where the motion propagated from the deep ocean is by the shallowness of the channels, through which it passes to some particular places, retarded to the fifth, sixth, or seventh hour, and even later. The hours I reckon from the appulse of each luminary to the meridian of the place, as well under as above the horizon; and by the hours of the lunar day I understand the 24th parts of that time which the moon, by its apparent diurnal motion, employs to come about again to the meridian of the place which it left the day before. The force of the sun or moon in raising the sea is greatest in the appulse of the luminary to the meridian of the place; but the force impressed upon the sea at that time continues a little while after the impression, and is afterwards increased by a new though less force still

acting upon it. This makes the sea rise higher and higher, till this new force becoming too weak to raise it any more, the sea rises to its greatest height. And this will come to pass, perhaps, in one or two hours, but more frequently near the shores in about three hours, or even more, where the sea is shallow.

The two luminaries excite two motions, which will not appear distinctly, but between them will arise one mixed motion compounded out of both. In the conjunction or opposition of the luminaries their forces will be conjoined, and bring on the greatest flood and ebb. In the quadratures the sun will raise the waters which the moon depresses, and depress the waters which the moon raises, and from the difference of their forces the smallest of all tides will follow. And because (as experience tells us) the force of the moon is greater than that of the sun, the greatest height of the waters will happen about the third lunar hour. Out of the syzygies and quadratures, the greatest tide, which by the single force of the moon ought to fall out at the third lunar hour, and by the single force of the sun at the third solar hour, by the compounded forces of both must fall out in an intermediate time that approaches nearer to the third hour of the moon than to that of the sun. And, therefore, while the moon is passing from the syzygies to the quadratures, during which time the 3d hour of the sun precedes the 3d hour of the moon, the greatest height of the waters will also precede the 2d hour of the moon, and that, by the greatest interval, a little after the octants of the moon; and, by like intervals, the greatest tide will follow the 3d lunar hour, while the moon is passing from the quadratures to the syzygies. Thus, it happens in the open sea; for in the mouths of rivers the greater tides come later to their height.

But the effects of the luminaries depend upon their distances from the earth; for when they are less distant, their effects are greater, and when more distant, their effects are less, and that in the triplicate proportion of their apparent diameter. Therefore it is that the sun, in the winter time, being then in its perigee, has a greater effect, and makes the tides in the syzygies something greater, and those in the quadratures something less than in the summer season; and every month the moon, while in the perigee, raises greater tides than at the distance of 15 days before or after, when it is in its apogee. Whence it comes to pass that two highest tides do not follow one the other in two immediately succeeding syzygies.

The effect of either luminary doth likewise depend upon its declination or distance from the equator; for if the luminary was placed at the pole, it would constantly attract all the parts of the waters without any intension or remission of its action, and could cause no reciprocation of motion. And, therefore, as the luminaries decline from the equator towards either pole, they will, by degrees, lose their force, and on this account will excite lesser tides in the solstitial than in the equinoctial syzygies. But in the solstitial quadratures they will raise greater tides than in the quadratures about the

equinoxes; because the force of the moon, then situated in the equator, most exceeds the force of the sun. Therefore the greatest tides fall out in those syzygies, and the least in those quadratures, which happen about the time of both equinoxes: and the greatest tide in the syzygies is always succeeded by the least tide in the quadratures, as we find by experience. But, because the sun is less distant from the earth in winter than in summer, it comes to pass that the greatest and least tides more frequently appear before than after the vernal equinox, and more frequently after than before the autumnal.

Moreover, the effects of the luminaries depend upon the latitudes of places. Let ApEP represent the earth covered with deep waters; C its centre; P, p its poles; AE the equator; P any place without the equator; Ff the parallel of the place; Dd the correspondent parallel on the other side of the equator; L the place of the moon three hours before; H the place of the earth directly under it; h the opposite place; K, k the places

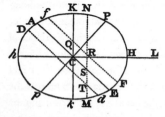

at 90 degrees distance; CH, Ch, the greatest heights of the sea from the centre of the earth; and CK, Ck, its least heights: and if with the axes Hh, Kk, an ellipsis is described, and by the revolution of that ellipsis about its longer axis Hh a spheroid HPKhpk is formed, this spheroid will nearly represent the figure of the sea; and CF, Cf, CD, Cd, will represent the heights of the sea in the places Ff, Dd.

But farther; in the said revolution of the ellipsis any point N describes the circle NM cutting the parallels Ff, Dd, in any places RT, and the equator AE in S; CN will represent the height of the sea in all those places R, S, T, situated in this circle. Wherefore, in the diurnal revolution of any place F, the greatest flood will be in F, at the third hour after the appulse of the moon to the meridian above the horizon; and afterwards the greatest ebb in Q, at the third hour after the setting of the moon; and then the greatest flooding, at the third hour after the appulse of the moon to the meridian under the horizon; and, lastly, the greatest ebb in Q, at the third hour after the rising of the moon; and the latter flood in f will be less than the preceding flood in F. For the whole sea is divided into two hemispherical floods, one in the hemisphere KHk on the north side, the other in the opposite hemisphere Khk; which we may therefore call the northern and the southern floods. These floods, being always opposite the one to the other, come by turns to the meridians of all places, after an interval of 12 lunar hours. And seeing the northern countries partake more of the northern flood, and the southern countries more of the southern flood, thence arise tides, alternately greater and less in all places without the equator, in which the luminaries rise and set. But the greatest tide will happen when the moon declines towards the vertex of the place, about the third hour after the appulse of the moon to the meridian above the horizon; and when the

moon changes its declination to the other side of the equator, that which was the greater tide will be changed into a lesser. And the greatest difference of the floods will fall out about the times of the solstices; especially if the ascending node of the moon is about the first of Aries. So it is found by experience that the morning tides in winter exceed those of the evening, and the evening tides in summer exceed those of the morning; at *Plymouth* by the height of one foot, but at *Bristol* by the height of 15 inches, according to the observations of *Collepress* and *Sturmy*.

But the motions which we have been describing suffer some alteration from that force of reciprocation, which the waters, being once moved, retain a little while *by their vis insita*. Whence it comes to pass that the tides may continue for some time, though the actions of the luminaries should cease. This power of retaining the impressed motion lessens the difference of the alternate tides, and makes those tides which immediately succeed after the syzygies greater, and those which follow next after the quadratures less. And hence it is that the alternate tides at *Plymouth* and *Bristol* do not differ much more one from the other than by the height of a foot or 15 inches, and that the greatest tides of all at those ports are not the first but the third after the syzygies. And, besides, all the motions are retarded in their passage through shallow channels, so that the greatest tides of all, in some straits and mouths of rivers, are the fourth or even the fifth after the syzygies.

Farther, it may happen that the tide may be propagated from the ocean through different channels towards the same port, and may pass quicker through some channels than through others; in which case the same tide, divided into two or more succeeding one another, may compound new motions of different kinds. Let us suppose two equal tides flowing towards the same port from different places, the one preceding the other by 6 hours; and suppose the first tide to happen at the third hour of the appulse of the moon to the meridian of the port. If the moon at the time of the appulse to the meridian was in the equator, every 6 hours alternately there would arise equal floods, which, meeting with as many equal ebbs, would so balance one the other, that for that day, the water would stagnate and remain quiet. If the moon then declined from the equator, the tides in the ocean would be alternately greater and less, as was said; and from thence two greater and two lesser tides would be alternately propagated towards that port. But the two greater floods would make the greatest height of the waters to fall out in the middle time betwixt both; and the greater and lesser floods would make the waters to rise to a mean height in the middle time between them, and in the middle time between the two lesser floods the waters would rise to their least height. Thus in the space of 24 hours the waters would come, not twice, as commonly, but once only to their greatest, and once only to their least height; and their greatest height, if the moon declined towards the elevated pole, would happen at the 6th or

30th hour after the appulse of the moon to the meridian; and when the moon changed its declination, this flood would be changed into an ebb. An example of all which Dr. *Halley* has given us, from the observations of seamen in the port of *Batsham*, in the kingdom of *Tounquin*, in the latitude of 20° 50' north. In that port, on the day which follows after the passage of the moon over the equator, the waters stagnate: when the moon declines to the north, they begin to flow and ebb, not twice, as in other ports, but once only every day; and the flood happens at the setting, and the greatest ebb at the rising of the moon. This tide increases with the declination of the moon till the 7th or 8th day; then for the 7 or 8 days following it decreases at the same rate as it had increased before, and ceases when the moon changes its declination, crossing over the equator to the south. After which the flood is immediately changed into an ebb; and thenceforth the ebb happens at the setting and the flood at the rising of the moon; till the moon, again passing the equator, changes its declination. There are two inlets to this port and the neighboring channels, one from the seas of *China*, between the continent and the island of *Leuconia*; the other from the *Indian* sea, between the continent and the island of *Borneo*. But whether there be really two tides propagated through the said channels, one from the *Indian* sea in the space of 12 hours, and one from the sea of *China* in the space of 6 hours, which therefore happening at the 3d and 9th lunar hours, by being compounded together, produce those motions; or whether there be any other circumstances in the state of those seas, I leave to be determined by observations on the neighbouring shores.

Thus I have explained the causes of the motions of the moon and of the sea. Now it is fit to subjoin something concerning the quantity of those motions.

PROPOSITION XXV PROBLEM VI.

To find the forces with which the sun disturbs the motions of the moon.

Let S represent the sun, T the earth, P the moon, CADB the moon's orbit. In SP take SK equal to ST; and let SL be to SK in the duplicate proportion of SK to SP: draw LM parallel to PT; and if ST or SK is supposed to represent the accelerated force of gravity of the earth towards the sun, SL will represent the accelerative force of gravity of the moon towards the sun. But that force is compounded of the parts SM and LM, of

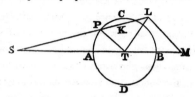

which the force LM, and that part of SM which is represented by TM, disturb the motion of the moon, as we have shewn in Prop. LXVI, Book I, and its Corollaries.

Forasmuch as the earth and moon are revolved about their common centre of gravity, the motion of the earth about that centre will be also disturbed by the like forces; but we may consider the sums both of the forces and of the motions as in the moon, and represent the sum of the forces by the lines TM and ML, which are analogous to them both. The force ML (in its mean quantity) is to the centripetal force by which the moon may be retained in its orbit revolving about the earth at rest, at the distance PT, in the duplicate proportion of the periodic time of the moon about the earth to the periodic time of the earth about the sun (by Cor. 17, Prop. LXVI, Book I); that is, in the duplicate proportion of $27^d.7^h.43'$ to $365^d. 6^h. 9'$; or as 1000 to 178725; or as 1 to $178^{29}/_{40}$. But in the 4th Prop. of this Book we found, that, if both earth and moon were revolved about their common centre of gravity, the mean distance of the one from the other would be nearly $60^1/_2$ mean semi-diameters of the earth; and the force by which the moon maybe kept revolving in its orbit about the earth in rest at the distance PT of $60^1/_2$ semi-diameters of the earth, is to the force by which it may be revolved in the same time, at the distance of 60 semi-diameters, as $60^1/_2$ to 60: and this force is to the force of gravity with us very nearly as 1 to 60 x 60. Therefore the mean force ML is to the force of gravity on the surface of our earth as $1 \times 60^1/_2$ to $60 \times 60 \times 60 \times 178^1/_2$, or as 1 to 638092,6; whence by the proportion of the lines TM, ML, the force TM is also given; and these are the forces with which the sun disturbs the motions of the moon.   Q.E.I.

PROPOSITION XXVI. PROBLEM VII.

TO FIND THE HORARY INCREMENT OF THE AREA WHICH THE MOON, BY A RADIUS DRAWN TO THE EARTH, DESCRIBES IN A CIRCULAR ORBIT.

We have above shewn that the area which the moon describes by a radius drawn to the earth is proportional to the time of description, excepting in so far as the moon's motion is disturbed by the action of the sun; and here we propose to investigate the inequality of the moment, or horary increment *of that area or motion so disturbed*. To render the calculus more easy, we shall suppose the orbit of the moon to be circular, and neglect all inequalities but that only which is now under consideration; and, because of the immense distance of the sun, we shall farther suppose that the lines SP and ST are parallel. By this means, the force LM will be always reduced to its mean quantity TP, as well as the force TM to its mean quantity 3PK. These forces (by Cor. 2 of the Laws of Motion) compose the force TL; and this force, by letting fall the perpendicular LE upon the radius TP, if resolved into the forces TE, EL; of which the force TE, acting constantly in the direction of the radius TP, neither accelerates nor retards

the description of the area TPC made by that radius TP; but EL, *acting on the radius* TP in a perpendicular direction, accelerates or retards *the description of the area* in proportion as it accelerates or retards the moon. That acceleration of the moon, in its passage from the quadrature C to the conjunction A, is in every moment of time as the *generating* accelerative force EL, that is, as $\frac{3PK \times TK}{TP}$. Let the time be represented by the TP mean motion of the moon, or (which comes to the same thing) by the angle CTP,

or even by the arc CP. At right angles upon CT erect CG equal to CT; and, supposing the quadrantal arc AC to be divided into an infinite number of equal parts P*p*, &c., these *parts* may represent the like *infinite* number of the equal parts of time. Let fall *pk* perpendicular on CT, and draw TG meeting with KP, *kp* produced in F and *f*,

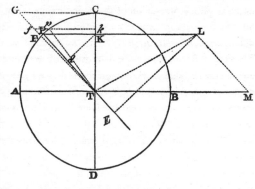

then will FK be equal to TK, and K*k* be to PK as P*p* to T*p*, that is, in a given proportion; and therefore FK x K*k*, or the area FK*kf*, will be as $\frac{3PK \times TK}{TP}$, that is, as

EL; and compounding, the whole area GCKF will be as the sum of all the forces EL impressed upon the moon in the whole time CP; and therefore also as the velocity generated by that sum, that is, as the acceleration of the description of the area CTP, or as the increment of the moment *thereof*. The force by which the moon may in its periodic time CADB of 27^d. 7^h. 43' be retained revolving about the earth in rest at the distance TP, would cause a body falling in the time CT to describe the length CT, and at the same time to acquire a velocity equal to that with which the moon is moved in its orbit. This appears from Cor. 9, Prop. IV., Book I. But since K*d*, drawn perpendicular on TP, is *but* a third part of EL, and *equal* to the half of TP, or ML, in the octants, the force EL in the octants, where it is greatest, will exceed the force ML in the proportion of 3 to 2; and therefore will be to that force by which the moon in its periodic time may be retained revolving about the earth at rest as 100 to $^2/_3$ x 17872$^1/_2$, or 11915; and in the time CT will generate a velocity equal to 100/11915 parts of the velocity of the moon; but in the time CPA will generate a greater velocity in the proportion of CA to CT or TP. Let the greatest force EL in the octants be represented by the area FK x K*k*, or by the rectangle $^1/_2$ TP $\times$ P*p*, which is equal thereto; and the velocity which that greatest force can generate in any time CP will be to the velocity which any other

lesser force EL can generate in the same time as the rectangle $1/_2$TP x CP to the area KCGF; but the velocities generated in the whole time CPA will be one to the other as the rectangle $1/_2$TP x CA to the triangle TCG, or as the quadrantal arc CA to the radius TP; and therefore the latter velocity generated in the whole time will be $^{100}/_{11915}$ parts of the velocity of the moon. To this velocity of the moon, which is proportional to the mean moment of the area (supposing this mean moment to be represented by the number 11915), we add and subtract the half of the other velocity; the sum 11915 + 50, or 11965, will represent the greatest moment of the area in the syzygy A; and the difference 11915 − 50, or 11865, the least moment thereof in the quadratures. Therefore the areas which in equal times are described in the syzygies and quadratures are one to the other as 11965 to 11865. And if to the least moment 11865 we add a moment which shall be to 100, the difference of the two former moments, as the trapezium FKCG to the triangle TCG, or, which comes to the same thing, as the square of the sine PK to the square of the radius TP (that is, as Pd to TP), the sum will represent the moment of the area when the moon is in any intermediate place P.

But these things take place only in the hypothesis that the sun and the earth are at rest, and that the synodical revolution of the moon is finished in 27^d. 7^h. 43'. But since the moon's synodical period is really 29^d. 12^h. 44', the increments of the moments must be enlarged in the same proportion as the time is, that is, in the proportion of 1080853 to 1000000. Upon which account, the whole increment, which was $^{100}/_{11915}$ parts of the mean moment, will now become $^{100}/_{11023}$ parts thereof; and therefore the moment of the area in the quadrature of the moon will be to the moment thereof in the syzygy as 11023 − 50 to 11023 + 50; or as 10973 to 11073; and to the moment thereof, when the moon is in any intermediate place P, as 10973 to 10973 + Pd; that is, supposing TP = 100.

The area, therefore, which the moon, by a radius drawn to the earth, describes in the several little equal parts of time, is nearly as the sum A the number 219,46, and the versed sine of the double distance of the moon from the nearest quadrature, considered in a circle which hath unity for its radius. Thus it is when the variation in the octants is in its mean quantity. But if the variation there is greater or less, that versed sine must be augmented or diminished in the same proportion.

PROPOSITION XXVII. PROBLEM VIII.

FROM THE HORARY MOTION OF THE MOON TO FIND ITS DISTANCE FROM THE EARTH.

The area which the moon, by a radius drawn to the earth, describes in every moment of time, is as the horary motion of the moon and the square of the distance

of the moon from the earth conjunctly. And therefore the distance of the moon from the earth is in a proportion compounded of the subduplicate proportion of the area directly, and the subduplicate proportion of the horary motion inversely.   Q.E.I.

COR. 1. Hence the apparent diameter of the moon is given; for it is reciprocally as the distance of the moon from the earth. Let astronomers try how accurately this rule agrees with the phænomena.

COR. 2. Hence also the orbit of the moon may be more exactly defined from the phænomena than hitherto could be done.

PROPOSITION XXVIII. PROBLEM IX.

TO FIND THE DIAMETERS OF THE ORBIT, IN WHICH, WITHOUT ECCENTRICITY, THE MOON WOULD MOVE.

The curvature of the orbit which a body describes, if attracted in lines perpendicular to the orbit, is as the force of attraction directly, and the square of the velocity inversely. I estimate the curvatures of lines compared one with another according to the evanescent proportion of the sines or tangents of their angles of contact to equal radii, supposing those radii to be infinitely diminished. But the attraction of the moon towards the earth in the syzygies is the excess of its gravity towards the earth above the force of the sun 2PK (see Fig. Prop. XXV), by which force the accelerative gravity of the moon towards the sun exceeds the accelerative gravity of the earth towards the sun, or is exceeded by it. But in the quadratures that attraction is the sum of the gravity of the moon towards the earth, and the sun's force KT, by which the moon is attracted towards the earth. And these attractions, putting N for $\frac{AT+CT}{2}$ are nearly as

$$\frac{178725}{AT^2} - \frac{2000}{CT \times N} \quad \text{and} \quad \frac{178725}{CT^2} + \frac{1000}{AT \times N}$$, or as 178725N x CT^2 – 2000AT^2 x CT, and 178725N x AT^2 + 1000CT^2 x AT. For if the accelerative gravity of the moon towards the earth be represented by the number 178725, the mean force ML, which in the quadratures is PT or TK, and draws the moon towards the earth, will be 1000, and the mean force TM in the syzygies will be 3000; from which, if we subtract the mean force ML, there will remain 2000, the force by which the moon in the syzygies is drawn from the earth, and which we above called 2PK. But the velocity of the moon in the syzygies A and B is to its velocity in the quadratures C and D as CT to AT, and the moment of the area, which the moon by a radius drawn to the earth describes in the syzygies, to the moment of that area *described* in the quadratures conjunctly; that is, as 11073CT to 10973AT. Take this ratio twice inversely, and the former ratio once directly, and the curvature of the orb of the moon in the syzygies will be to the curvature thereof in the quadratures as 120406729 × 178725AT^2 × CT^2 ×

N – 120406729 × 2000AT^4 × CT to 122611329 × 178725AT^2 × CT^2 × N + 122611329 x 1000CT^4 × AT, that is, as 2151969AT × CT × N – 24081AT^3 to 2191371AT × CT × N + 12261CT^3.

Because the figure of the moon's orbit is unknown, let us, in its stead, assume the ellipsis DBCA, in the centre of which we suppose the earth to be situated, and the greater axis DC to lie between the quadratures as the lesser AB between the syzygies. But since the plane of this ellipsis is revolved about the earth by an angular motion, and the orbit, whose curvature we now examine, should be described in a plane void of such motion, we are to consider the figure which the moon, while it is revolved in

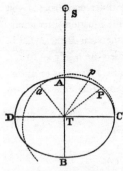

that ellipsis, describes in this plane, that is to say, the figure C*pa*, the several points *p* of which are found by assuming any point P in the ellipsis, which may represent the place of the moon, and drawing T*p* equal to TP in such manner that the angle PT*p* may be equal to the apparent motion of the sun from the time of the last quadrature in C; or (which comes to the same thing) that the angle CT*p* may be to the angle CTP as the time of the synodic revolution of the moon to the time of the periodic revolution thereof, or as 29^d. 12^h. 44' to 27^d. 7^h. 43'.

If, therefore, in this proportion we take the angle CT*a* to the right angle CTA, and make T*a* of equal length with TA, we shall have *a* the lower and C the upper apsis of this orbit C*pa*. But, by computation, I find that the difference betwixt the curvature of this orbit C*pa* at the vertex *a*, and the curvature of a circle described about the centre T with the interval TA, is to the difference between the curvature of the ellipsis at the vertex A, and the curvature of the same circle, in the duplicate proportion of the angle CTP to the angle CT*p*; and that the curvature of the ellipsis in A is to the curvature of that circle in the duplicate proportion of TA to TC; and the curvature of that circle to the curvature of a circle described about the centre T with the interval TC as TC to TA; but that the curvature of this last arch is to the curvature of the ellipsis in C in the duplicate proportion of TA to TC; and that the difference betwixt the curvature of the ellipsis in the vertex C, and the curvature of this last circle, is to the difference betwixt the curvature of the figure C*pa*, at the vertex C, and the curvature of this same last circle, in the duplicate proportion of the angle CT*p* to the angle CTP; all which proportions are easily drawn from the sines of the angles of contact, and of the differences of those angles. But, by comparing those proportions together, we find the curvature of the figure C*pa* at *a* to be to its curvature at C as AT^3 – $^{16824}/_{10000}$ CT^{2}AT to CT^3 + $^{16824}/_{10000}$ AT^2 x CT; where the number $^{16824}/_{10000}$ represents the difference of the squares of the angles CTP and CT*p*, applied to the

square of the lesser angle CTP; or (which is all one) the difference of the squares of the times 27^d. 7^h. 43', and 29^d. 12^h. 44', applied to the square of the time 27^d. 7^h. 43'.

Since, therefore, a represents the syzygy of the moon, and C its quadrature, the proportion now found must be the same with that proportion of the curvature of the moon's orb in the syzygies to the curvature thereof in the quadratures, which we found above. Therefore, in order to find the proportion of CT to AT, let us multiply the extremes and the means, and the terms which come out, applied to AT x CT, become $2062,79CT^4 - 2151969N \times CT^3 + 368676N \times AT \times CT^2 + 36342 AT^2 \times CT^2 - 362047N \times AT^2 \times CT + 2191371N \times AT^3 + 4051,4AT^4 = 0$. Now if for the half sum N of the terms AT and CT we put 1, and x for their half difference, then CT will be = $1 + x$, and AT = $1 - x$. And substituting those values in the equation, after resolving thereof, we shall find x = 0,00719; and from thence, the semi-diameter CT = 1,00719, and the semi-diameter AT = 0,99281, which numbers are nearly as 70$^1/_{24}$, and 69$^1/_{24}$. Therefore the moon's distance from the earth in the syzygies is to its distance in the quadratures (setting aside the consideration of eccentricity) as 69$^1/_{24}$ to 70$^1/_{24}$; or, in round numbers, as 69 to 70.

PROPOSITION XXIX. PROBLEM X.

To find the variation of the moon.

This inequality is owing partly to the elliptic figure of the moon's orbit, partly to the inequality of the moments of the area which the moon by a radius drawn to the earth describes. If the moon P revolved in the ellipsis DBCA about the earth quiescent in the centre of the ellipsis, and by the radius TP, drawn to the earth, described the area CTP, proportional to the time *of description*; and the greatest semi-diameter CT of the ellipsis was to the least TA as 70 to 69; the tangent of the angle CTP would be to the tangent of the angle of the mean motion, computed from the quadrature C, as the semi-diameter TA of the ellipsis to its semi-diameter TC, or as 69 to 70. But the description of the area CTP, as the moon advances from the quadrature to the syzygy, ought to be in such manner accelerated, that the moment of the area in the moon's syzygy may be to the moment thereof in its quadrature as 11073 to 10973; and that the excess of the moment in any intermediate place P above the moment in the quadrature may be as the square of the sine of the angle CTP; which we may effect with accuracy enough, if we diminish the tangent of the angle CTP in the subduplicate proportion of the number 10973 to the number 11073, that is, in proportion of the number 68,6877 to the number 69. Upon which account the tangent of the angle CTP will now be to the tangent of the mean motion as 68,6877 to 70; and the angle CTP in the octants, where the mean motion is 45°, will be found 44° 27' 28", which subtracted from 45°, the

angle of the mean motion, leaves the greatest variation 32' 32". Thus it would be, if
the moon, in passing from the quadrature to the syzygy, described an angle CTA of 90
degrees only. But because of the motion of the earth, by which the sun is apparently
transferred *in consequentia*, the moon, before it overtakes the sun, describes an angle
CT*a*, greater than a right angle, in the proportion of the time of the synodic revolution of the moon to the time of its periodic revolution, that is, in the proportion of
$29^d. 12^h. 44'$, to $27^d. 7^h. 43'$. Whence it comes to pass that all the angles about the
centre T are dilated in the same proportion; and the greatest variation, which otherwise would be *but* 32' 32", now augmented in the said proportion, becomes 35' 10".

And this is its magnitude in the mean distance of the sun from the earth, neglecting the differences which may arise from the curvature of the *orbis magnus*, and the
stronger action of the sun upon the moon when horned and new, than when gibbous
and full. In other distances of the sun from the earth, the greatest variation is in a proportion compounded of the duplicate proportion of the time of the synodic revolution
of the moon (the time of the year being given) directly, and the triplicate proportion
of the distance of the sun from the earth inversely. And, therefore, in the apogee of the
sun, the greatest variation is 33' 14", and in its perigee 37' 11", if the eccentricity of
the sun is to the transverse semi-diameter of the *orbis magnus* as $16^{15}/_{16}$ to 1000.

Hitherto we have investigated the variation in an orb not eccentric, in which,
to wit, the moon in its octants is always in its mean distance from the earth. If
the moon, on account of its eccentricity, is more or less removed from the earth
than if placed in this orb, the variation may be something greater, or something
less, than according to this rule. But I leave the excess or defect to the determination of astronomers from the phaenomena.

PROPOSITION XXX. PROBLEM XI.

To find the horary motion of the nodes of the
moon in a circular orbit.

Let S represent the sun, T the earth, P the moon, NP*n* the orbit of the moon, N*pn*
the orthographic projection of the orbit upon the plane of the ecliptic; N, *n* the nodes,
*n*TN*m* the line of the nodes produced indefinitely; PI, PK perpendiculars upon the
lines ST, Q*q*; P*p* a perpendicular upon the plane of the ecliptic; A, B the moon's syzygies in the plane of the ecliptic; AZ a perpendicular let fall upon N*n*, the line of the
nodes; Q, *q* the quadratures of the moon in the plane of the ecliptic, and *p*K a perpendicular on the line Q*q* lying between the quadratures. The force of the sun to disturb the motion of the moon (by Prop. XXV) is twofold, one proportional to the line
LA the other to the line MT, in the scheme of that Proposition; and the moon by the

former force is drawn towards the earth, by the latter towards the sun, in a direction parallel to the right line ST joining the earth and the sun. The former force LM acts in the direction of the plane of the moon's orbit, and therefore makes no change upon the situation thereof, and is upon that account to be neglected: the latter force MT, by which the plane of the moon's orbit is disturbed, is the same with the force 3PK or 3IT. And this force (by Prop. XXV) is to the force by which the moon may, in its periodic

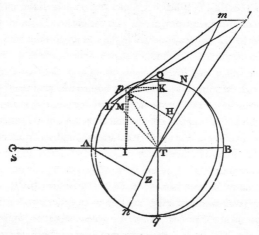

time, be uniformly revolved in a circle about the earth at rest, as 3IT to the radius of the circle multiplied by the number 178,725, or as IT to the radius thereof multiplied by 59,575. But in this calculus, and all that follows, I consider all the lines drawn from the moon to the sun as parallel to the line which joins the earth and the sun; because what inclination there is almost as much diminishes all effects in some cases as it augments them in others; and we are now inquiring after the mean motions of the nodes, neglecting such niceties as are of no moment, and would only serve to render the calculus more perplexed.

Now suppose PM to represent an arc which the moon describes in the least moment of time, and ML a little line, the half of which the moon, by the impulse of the said force 3IT, would describe in the same time; and joining PL, MP, let them be produced to *m* and *l*, where they cut the plane of the ecliptic, and upon T*m* let fall the perpendicular PH. Now, since the right line ML is parallel to the plane of the ecliptic, and therefore can never meet with the right line *ml* which lies in that plane, and yet both those right lines lie in one common plane LMP*ml*, they will be parallel, and upon that account the triangles LMP, *lm*P will be similar. And seeing MP*m* lies in the plane of the orbit, in which the moon did move while in the place P, the point *m* will fall

upon the line Nn, which passes through the nodes N, n, of that orbit. And because the force by which the half of the little line LM is generated, if the whole had been together, and at once impressed in the point P, would have generated that whole line, and caused the moon to move in the arc whose chord is LP; that is to say, would have transferred the moon from the plane MPmT into the plane LPlT; therefore the angular motion of the nodes generated by that force will be equal to the angle mTl. But ml is to mP as ML to MP; and since MP, because of the time given, is also given, ml will be as the rectangle ML x mP, that is, as the rectangle IT x mP. And if Tml is a right angle, the angle mTl will be as $\frac{ml}{\text{T}m}$ and therefore as $\frac{\text{IT} \times \text{P}m}{\text{T}m}$, that is (because Tm and mP,

TP and PH are proportional), as $\frac{\text{IT} \times \text{PH}}{\text{TP}}$; and, therefore, because TP is given, as IT x PH. But if the angle Tml or STN is oblique, the angle mTl will be yet less, in proportion of the sine of the angle STN to the radius, or AZ to AT. And therefore the velocity of the nodes is as IT x PH x AZ, or as the solid content of the sines of the three angles TPI, PTN, and STN.

If these are right angles, as happens when the nodes are in the quadratures, and the moon in the syzygy, the little line ml will be removed to an infinite distance, and the angle mTl will become equal to the angle mPl. But in this case the angle mPl is to the angle PTM, which the moon in the same time by its apparent motion describes about the earth, as I to 59,575. For the angle mPl is equal to the angle LPM, that is, to the angle of the moon's deflexion from a rectilinear path; which angle, if the gravity of the moon should have then ceased, the said force of the sun 3IT would by itself have generated in that given time; and the angle PTM is equal to the angle of the moon's deflexion from a rectilinear path; which angle, if the force of the sun 3IT should have then ceased, the force alone by which the moon is retained in its orbit would have generated in the same time. And these forces (as we have above shewn) are the one to the other as 1 to 59,575. Since, therefore, the mean horary motion of the moon (in respect of the fixed stars) is 32' 56" 27"' 12$\frac{1}{2}$ ^{iv}, the horary motion of the node in this case will be 33" 10"' 33^{iv} 12^{v}. But in other cases the horary motion will be to 33" 10"' 33^{iv} 12^{v} as the solid content of the sines of the three angles TPI, PTN, and. STN (or of the distances of the moon from the quadrature, of the moon from the node, and of the node from the sun) to the cube of the radius. And as often as the sine of any angle is changed from positive to negative, and from negative to positive, so often must the regressive be changed into a progressive, and the progressive into a regressive motion. Whence it comes to pass that the nodes are progressive as often as the moon happens to be placed between either quadrature, and the node nearest to that quadrature. In other cases they are regressive, and by the excess of the regress above the progress, they are monthly transferred *in antecedentia*.

COR. 1. Hence if from P and M, the extreme points of a least arc PM, on the line Q*q* joining the quadratures we let fall the perpendiculars PK M*k*, and produce the same till they out the line of the nodes N*n* in D and *d*, the horary motion of the nodes will be as the area MPD*d*, and the square of the line AZ conjunctly. For let PK, PH, and AZ, be the three said sines, viz. PK the sine of the distance of the moon from the quadrature, PH the sine of the distance of the moon from the node, and AZ the sine

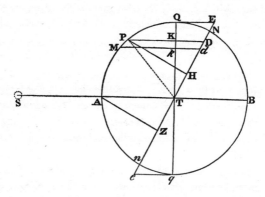

of the distance of the node from the sun; and the velocity of the node will be as the solid content of PK x PH x AZ. But PT is to PK as PM to K*k*; and, therefore, because PT and PTA are given, K*k* will be as PK. Likewise AT is to PD as AZ to PH, and therefore PH is as the rectangle PD x AZ; and, by compounding those proportions, PK x PH is as the solid content K*k* x PD x AZ, and PK x PH x AZ as K*k* x PD x AZ²; that is, as the area PD*d*M and AZ² conjunctly.   Q.E.D.

COR. 2. In any given position of the nodes their mean horary motion is half their horary motion in the moon's syzygies; and therefore is to 16" 35"' 16^{iv}. 36^{v}. as the square of the sine of the distance of the nodes from the syzygies to the square of the radius, or as AZ² to AT². For if the moon, by an uniform motion, describes the semi-circle QA*q*, the sum of all the areas PD*d*M, during the time of the moon's passage from Q to M, will make up the area QMD*Æ*, terminating at the tangent QE of the circle; and by the time that the moon has arrived at the point *n*, that sum will make up the whole area EQA*n* described by the line PD: but when the moon proceeds from *n* to *q*, the line PD will fall without the circle, and describe the area *nqe*, terminating at the tangent *qe* of the circle, which area, because the nodes were before regressive, but are now progressive, must be subducted from the former area, and, being itself equal to the area QEN, will leave the semi-circle NQA*n*. While, therefore, the moon describes a semi-circle, the sum of all the areas PD*d*M will be the area of that semi-circle; and while the moon describes a complete circle, the sum of those areas will be the area of

the whole circle. But the area PDdM, when the moon is in the syzygies, is the rectangle of the arc PM into the radius PT; and the sum of all the areas, *every one* equal to this area, in the time that the moon describes a complete circle, is the rectangle of the whole circumference into the radius of the circle; and this rectangle, being double the area of the circle, will be double the quantity of the former sum

If, therefore, the nodes went on with that velocity uniformly continued which they acquire in the moon's syzygies, they would describe a space double of that which they describe in fact; and, therefore, the mean motion, by which, if uniformly continued, they would describe the same space with that which they do in fact describe by an unequal motion, is *but* one-half of that motion which they are possessed of in the moon's syzygies. Wherefore since their greatest horary motion, if the nodes are in the quadratures, is 33" 10"' 33^{iv}. 12^{v}. their mean horary motion in this case will be 16" 35"' 16^{iv}. 36^{v}. And seeing the horary motion of the nodes is every where as AZ^2 and the area PDdM conjunctly, and, therefore, in the moon's syzygies, the horary motion of the nodes is as AZ^3 and the area PDdM conjunctly, that is (because the area PDdM described in the syzygies is given), as AZ^2, therefore the mean motion also will be as AZ^2; and, therefore, when the nodes are without the quadratures, this motion will be to 16" 35"' 16^{iv}. 36^{v}. as AZ^2 to AT^2.   Q.E.D.

PROPOSITION XXXI. PROBLEM XII.

TO FIND THE HORARY MOTION OF THE NODES OF THE MOON IN AN ELLIPTIC ORBIT.

Let Q$pmaq$ represent an ellipsis described with the greater axis Qq, and the lesser axis ab; QAqB a circle circumscribed; T the earth in the common centre of both; S the sun; p the moon moving in this ellipsis; and pm an arc which it describes in the least moment of time; N and n the nodes joined by the line Nn; pK and mk perpendiculars upon the axis Qq, produced both ways till they meet the circle in P and M, and the line of the nodes in D and d. And if the moon, by a radius drawn to the earth, describes an area proportional to the time *of description*, the horary motion of the node in the ellipsis will be as the area pDdm and AZ^2 conjunctly.

For let PF touch the circle in P, and produced meet TN in F; and pf touch the ellipsis in p, and produced meet the same TN in f, and both tangents concur in the axis TQ at Y. And let ML represent the space which the moon, by the impulse of the above-mentioned force 3IT or 3PK, would describe with a transverse motion, in the meantime while revolving in the circle it describes the arc PM; and ml denote the space which the moon revolving in the ellipsis would describe in the same time by the impulse of the same force 3IT or 3PK; and let LP and lp be produced till they meet

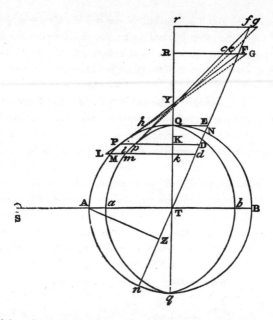

the plane of the ecliptic in G and *g*, and FG and *fg* be joined, of which FG produced may cut *pf*, *pg*, and TGL, in *c*, *e*, and R respectively; and *fg* produced may cut TQ in *r*. Because the force 3IT or 3PK in the circle is to the force 3IT or 3*p*K in the ellipsis as PK to *p*K, or as AT to *a*T, the space ML generated by the former force will be to the space *ml* generated by the latter as PK to *p*K; that is, because of the similar figures PYK*p* and FYR*c*, as FR to *c*R. But (because of the similar triangles PLM, PGF) ML is to FG as PL to PG, that is (on account of the parallels L*k*, PK, GR), as *pl* to *pe*, that is (because of the similar triangles *plm*, *cpe*), as *lm* to *ce*; and inversely as LM is to *lm*, or as PR is to *c*R, so is FG to *ce*. And therefore if *fg* was to *ce* as *fy* to *c*Y, that is, as *fr* to *c*R (that is, as *fr* to FR and FR to *c*R conjunctly, that is, as *f*T to FT, and FG to *ce* conjunctly), because the ratio of FG to *ce*, expunged on both sides, leaves the ratios *fg* to FG and *f*T to FT, *fg* would be to FG as *f*T to FT; and, therefore, the angles which FG and *fg*, would subtend at the earth T would be equal to each other. But these angles (by what we have shewn in the preceding Proposition) are the motions of the nodes, while the moon describes in the circle the arc PM, in the ellipsis the arc *pm*; and therefore the motions of the nodes in the circle and in the ellipsis would be equal to each other. Thus, I say, it would be, if *fg* was to *ce* as *fy* to *c*Y, that is, if *fg* was equal to $\frac{ce \times fY}{cY}$. But because of the similar triangles *fgp*, *cep*, *fg* is to *ce* as *fp* to *cp*; and therefore

fg is equal to $\frac{ce \times fp}{cp}$; and therefore the angle which fg subtends in fact is to the former angle which FG subtends, that is to say, the motion of the nodes in the ellipsis is to the motion of the same in the circle as this fg or $\frac{ce \times fp}{cp}$ to the force fg

or $\frac{ce \times fY}{cY}$, that is, as $fp \times cY$ to $fY \times cp$, or as fp to fY, and cY to cp; that is, if ph parallel to TN meet FP in h, as Fh to FY and FY to FP; that is, as Fh to FP or Dp to DP, and therefore as the area $Dpmd$ to the area DPMd. And, therefore, seeing (by Corol. 1, Prop. XXX) the latter area and AZ^2 conjunctly are proportional to the horary motion of the nodes in the circle, the former area and AZ^2 conjunctly will be proportional to the horary motion of the nodes in the ellipsis.   Q.E.D.

COR. Since, therefore, in any given position of the nodes, the sum of all the areas $pDdm$, in the time while the moon is carried from the quadrature to any place m, is the area mpQEd terminated at the tangent of the ellipsis QE; and the sum of all those areas, in one entire revolution, is the area of the whole ellipsis; the mean motion of the nodes in the ellipsis will be to the mean motion of the nodes in the circle as the ellipsis to the circle; that is, as Ta to TA, or 69 to 70. And, therefore, since (by Corol 2, Prop. XXX) the mean horary motion of the nodes in the circle is to 16" 35"' 16^{iv}. 36^v. as AZ^2 to AT^2, if we take the angle 16" 2"' 3^{iv}. 30^v. to the angle 16" 35"' 16^{iv}. 36^v. as 69 to 70, the mean horary motion of the nodes in the ellipsis will be to 16" 21"' 31^{iv}. 30^v. as AZ^2 to AT^2; that is, as the square of the sine of the distance of the node from the sun to the square of the radius.

But the moon, by a radius drawn to the earth, describes the area in the syzygies with a greater velocity than it does that in the quadratures, and upon that account the time is contracted in the syzygies, and prolonged in the quadratures; and together with the time the motion of the nodes is likewise augmented or diminished. But the moment of the area in the quadrature of the moon was to the moment thereof in the syzygies as 10973 to 11073; and therefore the mean moment in the octants is to the excess in the syzygies, and to the defect in the quadratures, as 11023, the half sum of those numbers, to their half difference 50. Wherefore since the time of the moon in the several little equal parts of its orbit is reciprocally as its velocity, the mean time in the octants will be to the excess of the time in the quadratures, and to the defect of the time in the syzygies arising from this cause, nearly as 11023 to 50. But, reckoning from the quadratures to the syzygies, I find that the excess of the moments of the area, in the several places above the least moment in the quadratures, is nearly as the square of the sine of the moon's distance from the quadratures; and therefore the difference betwixt the moment in any place, and the mean moment in the octants, is as the difference betwixt the square of the sine of the moon's distance from the quadratures, and the

square of the sine of 45 degrees, or half the square of the radius; and the increment of the time in the several places between the octants and quadratures, and the decrement thereof between the octants and syzygies, is in the same proportion. But the motion of the nodes, while the moon describes the several little equal parts of its orbit, is accelerated or retarded in the duplicate proportion of the time; for that motion, while the moon describes PM, is (*caeteris paribus*) as ML, and ML is in the duplicate proportion of the time. Wherefore the motion of the nodes in the syzygies, in the time while the moon describes given little parts of its orbit, is diminished in the duplicate proportion of the number 11073 to the number 11023; and the decrement is to the remaining motion as 100 to 10973; but to the whole motion as 100 to 11073 nearly. But the decrement in the places between the octants and syzygies, and the increment in the places between the octants and quadratures, is to this decrement nearly as the whole motion in these places to the whole motion in the syzygies, and the difference betwixt the square of the sine of the moon's distance from the quadrature, and the half square of the radius, to the half square of the radius conjunctly. Wherefore, if the nodes are in the quadratures, and we take two places, one on one side, one on the other, equally distant from the octant and other two distant by the same interval, one from the syzygy, the other from the quadrature, and from the decrements of the motions in the two places between the syzygy and octant we subtract the increments of the motions in the two other places between the octant and the quadrature, the remaining decrement will be equal to the decrement in the syzygy, as will easily appear by computation; and therefore the mean decrement, which ought to be subducted from the mean motion of the nodes, is the fourth part of the decrement in the syzygy. The whole horary motion of the nodes in the syzygies (when the moon by a radius drawn to the earth was supposed to describe an area proportional to the time) was $32'' \, 42''' \, 7^{iv}$. And we have shewn that the decrement of the motion of the nodes, in the time while the moon, now moving with greater velocity, describes the same space, was to this motion as 100 to 11073; and therefore this decrement is $17''' \, 43^{iv}. \, 11^v$. The fourth part of which $4'''$ $25^{iv}. \, 48^v$. subtracted from the mean horary motion above found, $16'' \, 21''' \, 3^{iv}. \, 30^v$. leaves $16'' \, 16''' \, 37^{iv}. \, 42^v$. their correct mean horary motion.

If the nodes are without the quadratures, and two places are considered, one on one side, one on the other, equally distant from the syzygies, the sum of the motions of the nodes, when the moon is in those places, will be to the sum of their motions, when the moon is in the same places and the nodes in the quadratures, as AZ^2 to AT^2. And the decrements of the motions arising from the causes but now explained will be mutually as the motions themselves, and therefore the remaining motions will be mutually betwixt themselves as AZ^2 to AT^2; and the mean motions will be as the remaining motions. And, therefore, in any given position of the nodes, their correct

mean horary motion is to 16" 16''' 37^{iv}. 42^{v}. as AZ^2 to AT^2; that is, as the square of the sine of the distance of the nodes from the syzygies to the square of the radius.

PROPOSITION XXXII. PROBLEM XIII.

TO FIND THE MEAN MOTION OF THE NODES OF THE MOON.

The yearly mean motion is the sum of all the mean horary motions throughout the course of the year. Suppose that the node is in N, and that, after every hour is elapsed, it is drawn back again to its former place; so that, notwithstanding its proper motion, it may constantly remain in the same situation with respect to the fixed stars; while in the mean time the sun S, by the motion of the earth, is seen to leave the node, and to proceed till it completes its apparent annual course by an uniform motion. Let Aa rep-

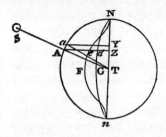

resent a given least arc, which the right line TS always drawn to the sun, by its intersection with the circle NAn, describes in the least given moment of time; and the mean horary motion (from what we have above shewn) will be as AZ^2, that is (because AZ and ZY are proportional), as the rectangle of AZ into ZY, that is, as the area AZYa; and the sum of all the mean horary motions from the beginning will be as the sum of all the areas aYZA, that is, as the area NAZ. But the greatest AZYa is equal to the rectangle of the arc Aa, into the radius of the circle; and therefore the sum of all these rectangles in the whole circle will be to the like sum of all the greatest rectangles as the area of the whole circle to the rectangle of the whole circumference into the radius, that is, as 1 to 2. But the horary motion corresponding to that greatest rectangle was 16" 16''' 37^{iv}. 42^{v}. and this motion in the complete course of the sidereal year, 365^d. 6^h. 9', amounts to 39° 38' 7" 50''', and therefore the half thereof, 19° 49' 3" 55''', is the mean motion of the nodes corresponding to the whole circle. And the motion of the nodes, in the time while the sun is carried from N to A, is to 19° 49' 3" 55''' as the area NAZ to the whole circle.

Thus it would be if the node was after every hour drawn back again to its former place, that so, after a complete revolution, the sun at the year's end would be found again in the same node which it had left when the year begun. But, because of the motion of the node in the mean time, the sun most needs meet the node sooner; and now it remains that we compute the abbreviation of the time. Since, then, the sun, in the coarse of the year, travels 360 degrees, and the node in the same time by its greatest motion would be carried 39° 39' 7" 50''', or 39,6355 degrees; and the mean motion

of the node in any place N is to its mean motion in its quadratures as AZ^2 to AT^2; the motion of the sun will be to the motion of the node in N as $360AT^2$ to $39,6355AZ^2$; that is, as $9,0927646AT^2$ to AZ^2. Wherefore if we suppose the circumference NAn of the whole circle to be divided into little equal parts, such as Aa, the time in which the sun would describe the little arc Aa, if the circle was quiescent, will be to the time of which it would describe the same arc, supposing the circle together with the nodes to be revolved about the centre T, reciprocally as $9,0827646AT^2$ to $9,0827646AT^2 + AZ^3$; for the time is reciprocally as the velocity with which the little arc is described, and this velocity is the sum of the velocities of both sun and node. If, therefore, the sector NTA represent the time in which the sun by itself, without the motion of the node, would describe the arc NA, and the indefinitely small part ATa of the sector represent the little moment of the time in which it would describe the least arc An; and (letting fall aY perpendicular upon Nn) if in AZ we take dZ of such length that the rectangle of dZ into ZY may be to the least part ATa of the sector as AZ^2 to $9,0827646AT^2 + AZ^2$, that is to say, that dZ may be to $^1/_2$ AZ as AT^2 to $9,0827646AT^2 + AZ^2$; the rectangle of dZ into ZY will represent the decrement of the time arising from the motion of the node, while the arc Aa is described; and if the curve NdGn is the locus where the point d is always found, the curvilinear area NdZ will be as the whole decrement *of time* while the whole arc NA is described; and, therefore, the excess of the sector NAT above the area NdZ will be as the whole time. But because the motion of the node in a less time is less in proportion of the time, the area AaYZ must also be diminished in the same proportion; which may be done by taking in AZ the line eZ of such length, that it may be to the length of AZ as AZ^2 to $9,0827646AT^2 + AZ^2$; for so the rectangle of eZ into ZY will be to the area AZYa as the decrement of the time in which the arc Aa is described to the whole time in which it would have been described, if the node had been quiescent; and, therefore, that rectangle will be as the decrement of the motion of the node. And if the curve NeFn is the locus of the point e, the whole area NaZ, which is the sum of all the decrements *of that motion*, will be as the whole decrement *thereof* during the time in which the arc AN is described; and the remaining area NAe will be as the remaining motion, which is the true motion of the node, during the time in which the whole arc NA is described by the joint motions of both sun and node. Now the area of the semi-circle is to the area of the figure NeFn found by the method of infinite series nearly as 793 to 60. But the motion corresponding *or proportional* to the whole circle was 19° 49' 3" 55'''; and therefore the motion corresponding to double the figure NeFn is 1° 29' 58" 2''', which taken from the former motion leaves 18° 19' 5" 53''', the whole motion of the node with respect to the fixed stars in the interval between two of its conjunctions with the sun; and this motion subducted from the annual motion of the sun 360°, leaves 341° 40' 54" 7''', the motion of the sun in the

interval between the same conjunctions. But as this motion is to the annual motion 360°, so is the motion of the node but just now found 18° 19' 5" 53''' to its annual motion, which will therefore be 19° 18' 1" 23'''; and this is the mean motion of the nodes in the sidereal year. By astronomical tables, it is 19° 21' 21" 50'''. The difference is less than $1/300$ part of the whole motion, and seems to arise from the eccentricity of the moon's orbit, and its inclination to the plane of the ecliptic. By the eccentricity of this orbit the motion of the nodes is too much accelerated; and, on the other hand, by the inclination of the orbit, the motion of the nodes is something retarded, and reduced to its just velocity.

PROPOSITION XXXIII. PROBLEM XIV.

To find the true motion of the nodes of the moon.

In the time which is as the area NTA–NdZ (in the preceding Fig.) that motion is as the area NAe, and is thence given; but because the calculus is too difficult, it will be better to use the following construction of the Problem. About the centre C, with any interval CD, describe the circle BEFD; produce DC to A so as AB may be to AC as the mean motion to half the mean true motion when the nodes are in their quadratures (that is, as 19° 18' 1" 23''' to 19° 49' 3" 55'''; and therefore BC to AC as the difference of those motions 0° 31' 2" 32''' to the latter motion 19° 49' 3" 55''', that is, as 1 to 38 $3/10$). Then through the point D draw the indefinite line Gg, touching the cir-

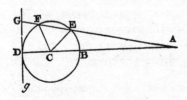

cle in D; and if we take the angle BCE, or BCF, equal to the double distance of the sun from the place of the node, as found by the mean motion, and drawing AE or AF cutting the perpendicular DG in G, we take another angle which shall be to the whole motion of the node in the interval between its syzygies (that is, to 9° 11' 3") as the tangent DG to the whole circumference of the circle BED, and add this last angle (for which the angle DAG may be used) to the mean motion of the nodes, while they are passing from the quadratures to the syzygies, and subtract it from their mean motion while they are passing from the syzygies to the quadratures, we shall have their true motion; for the true motion so found will nearly agree with the true motion which comes out from assuming the times as the area NTA – NdZ, and the motion of the node as the area NAe; as whoever will please to examine and make the computations will find: and this is the semi-menstrual equation of the motion of the nodes. But there is also a menstrual equation, but which is by no means necessary for finding of the moon's latitude; for since the variation of the inclination of the

moon's orbit to the plane of the ecliptic is liable to a twofold inequality, the one semi-menstrual, the other menstrual, the menstrual inequality *of this variation*, and the menstrual equation of the nodes, so moderate and correct each other, that in computing the latitude of the moon both may be neglected.

COR. From this and the preceding Prop. it appears that the nodes are quiescent in their syzygies, but regressive in their quadratures, by an hourly motion of 16" 19"' 26^{iv}.; and that the equation of the motion of the nodes in the octants is 1° 30'; all which exactly agree with the phaenomena of the heavens.

SCHOLIUM.

Mr. *Machin*, Astron., Prof. Gresh., and Dr. *Henry Pemberton*, separately found out the motion of the nodes by a different method. Mention has been made of this method in another place. Their several papers, both of which I have seen, contained two Propositions, and exactly agreed with each other in both of them. Mr. *Machin's* paper coming first to my hands, I shall here insert it.

OF THE MOTION OF THE MOON'S NODES.

"PROPOSITION I.

THE MEAN MOTION OF THE SUN FROM THE NODE IS
DEFINED BY A GEOMETRIC MEAN PROPORTIONAL BETWEEN
THE MEAN MOTION OF THE SUN AND THAT MEAN MOTION
WITH WHICH THE SUN RECEDES WITH THE GREATEST
SWIFTNESS FROM THE NODE IN THE QUADRATURES.

"Let T be the earth's place, N*n* the line of the moon's nodes at any given time, KTM a perpendicular thereto, TA a right line revolving about the centre with the same angular velocity with which the sun and the node recede from one another, in such sort that the angle between the quiescent right line N*n* and the revolving line TA may be always equal to the distance of the places of the sun and node. Now if any right line TK be divided into parts TS and SK, and those parts be taken as the mean horary motion of the sun to the mean horary motion of the node in the quadratures, and there be taken the right line TH, a mean proportional between the part TS and the whole TK, this right line will be proportional to the sun's mean motion from the node.

"For let there be described the circle NK*n*M from the centre T and with the radius TK, and about the same centre, with the semi-axis TH and TN, let there be described an ellipsis NH*n*L; and in the time in which the sun recedes from the node through the

arc Na, if there be drawn the right line Tba, the area of the sector NTa will be the exponent of the sum of the motions of the sun and node in the same time. Let, therefore, the extremely small arc aA be that which the right line Tba, revolving according to the aforesaid law, will uniformly describe in a given particle of time, and the extremely small sector TAa will be as the sum of the velocities with which the sun and node are carried two different ways in that time. Now the sun's velocity is almost uniform, its inequality being so small as scarcely to produce the least inequality in the mean motion of the nodes. The other part of this sum, namely, the mean quantity of the velocity of the node, is increased in the recess from the syzygies in a duplicate ratio of the sine of its distance from the sun (by Cor. Prop. XXXI, of this Book), and, being greatest in its quadratures with the sun in K, is in the same ratio to the sun's velocity as SK to TS, that is, as (the difference of the squares of TK and TH, or) the rectangle KHM to TH^2. But the ellipsis NBH divides the sector ATa, the exponent of the sum of these two velocities, into two parts ABba and BTb, proportional to the velocities. For produce BT to the circle in β, and from the point B let fall upon the greater axis

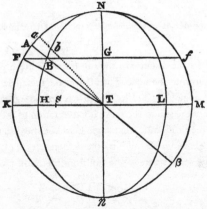

the perpendicular BG, which being produced both ways may meet the circle in the points F and f; and because the space ABba is to the sector TBb as the rectangle ABβ to BT^2 (that rectangle being equal to the difference of the squares of TA and TB, because the right line Aβ is equally cut in T, and unequally in B), therefore when the space ABba is the greatest of all in K, this ratio will be the same as the ratio of the rectangle KHM to HT^2. But the greatest mean velocity of the node was shewn above to be in that very ratio to the velocity of the sun; and therefore in the quadratures the sector ATa is divided into parts proportional to the velocities. And because the rectangle KHM is to HT^2 as FBf to BG^2, and the rectangle ABβ is equal to the rectangle FBf, therefore the little area ABba, where it is greatest, is to the remaining sector TBb as the

rectangle ABβ to BG². But the ratio of these little areas always was as the rectangle AB(
to BT²; and therefore the little area AB*ba* in the place A is less than its correspondent
little area in the quadratures in the duplicate ratio of BG to BT, that is, in the duplicate ratio of the sine of the sun's distance from the node. And therefore the sum of all
the little areas AB*ba*, to wit, the space ABN, will be as the motion of the node in the
time in which the sun hath been going over the arc NA since he left the node; and the
remaining space, namely, the elliptic sector NTB, will be as the sun's mean motion in
the same time. And because the mean annual motion of the node is that motion which
it performs in the time that the sun completes one period of its course, the mean
motion of the mode from the sun will be to the mean motion of the sun itself as the
area of the circle to the area of the ellipsis; that is, as the right line TK to the right line
TH, which is a mean proportional between TK and TS; or, which comes to the same
as the mean proportional TH to the right line TS.

"PROPOSITION II.

THE MEAN MOTION OF THE MOONS NODES BEING GIVEN,
TO FIND THEIR TRUE MOTION.

"Let the angle A be the distance of the sun from the mean place of the node, or
the sun's mean motion from the node. Then if we take the angle B, whose tangent is
to the tangent of the angle A as TH to TK, that is, in the sub-duplicate ratio of the
mean horary motion of the sun to the mean horary motion of the sun from the node,
when the node is in the quadrature, that angle B will be the distance of the sun from
the node's true place. For join FT, and, by the demonstration of the last Proportion,
the angle FTN will be the distance of the sun from the mean place of the node, and

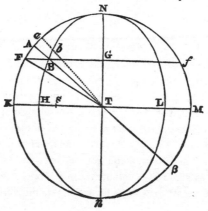

the angle ATN the distance from the true place, and the tangents of these angles are between themselves as TK to TH.

"COR. Hence the angle FTA is the equation of the moon's nodes; and the sine of this angle, where it is greatest in the octants, is to the radius as KH to TK + TH. But the sine of this equation in any other place A is to the greatest sine as the sine of the sums of the angles FTN + ATN to the radius; that is, nearly as the sine of double the distance of the sun from the mean place of the node (namely, 2FTN) to the radius.

"SCHOLIUM.

"If the mean horary motion of the nodes in the quadratures be $16" \ 16''' \ 37^{iv}. \ 42^{v}$. that is, in a whole sidereal year, $39° \ 38' \ 7" \ 50'''$, TH will be to TK in the subduplicate ratio of the number 9,0827646 to the number 10,0827646, that is, as 18,6524761 to 19,6524761. And, therefore, TH is to HK as 18,6524761 to 1; that is, as the motion of the sun in a sidereal year to the mean motion of the node $19° \ 18' \ 1" \ 23^{2}/_{3}'''$.

"But if the mean motion of the moon's nodes in 20 Julian years is $386° \ 50' \ 15"$, as is collected from the observations made use of in the theory of the moon, the mean motion of the nodes in one sidereal year will be $19° \ 20' \ 31" \ 58'''$. and TH will be to HK as $360°$ to $19° \ 20' \ 31" \ 69'''$; that is, as 18,61214 to 1: and from hence the mean horary motion of the nodes in the quadratures will come out $16" \ 18''' \ 48^{iv}$. And the greatest equation of the nodes in the octants will be $1° \ 29' \ 57"$."

PROPOSITION XXXIV. PROBLEM XV.

TO FIND THE HORARY VARIATION OF THE INCLINATION OF THE MOON'S ORBIT TO THE PLANE OF THE ECLIPTIC.

Let A and a represent the syzygies; Q and q the quadratures; N and n the nodes; P the place of the moon in its orbit; p the orthographic projection of that place upon the plane of the ecliptic; and mTl the momentaneous motion of the nodes as above. If upon Tm we let fall the perpendicular PG, and joining pG we produce it till it meet Tl in g, and join also Pg, the angle PGp will be the inclination of the moon's orbit to the plane of the ecliptic when the moon is in P; and the angle Pgp will be the inclination of the same after a small moment of time is elapsed; and therefore the angle GPg will be the momentaneous variation of the inclination. But this angle GPg is to the angle GTg as TG to PG and Pp to PG conjunctly. And, therefore, if for the moment of time we assume an hour, since the angle GTg (by Prop. XXX) is to the angle $33" \ 10''' \ 33^{iv}$. as IT x PG x AZ to AT^3, the angle GPg, (or the horary variation of the inclination) will be to the angle $33"$ $10''' \ 33^{iv}$. as IT $\times$ AZ $\times$ TG $\times \frac{Pp}{PG}$ to AT^3. Q.E.I.

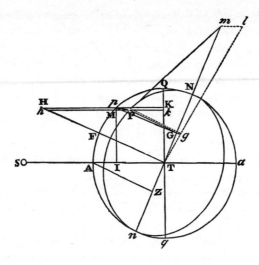

And thus it would be if the moon was uniformly revolved in a circular orbit. But if the orbit is elliptical, the mean motion of the nodes will he diminished in proportion of the lesser axis to the greater, as we have shewn above; and the variation of the inclination will be also diminished in the same proportion.

COR. 1. Upon Nn erect the perpendicular TF, and let pM be the horary motion of the moon in the plane of the ecliptic; upon QT let fall the perpendiculars pK, Mk, and produce them till they meet TF in H and h; then IT will be to AT as Kk to Mp; and TG to Hp as TZ to AT; and, therefore, IT x TG will be equal to $\frac{Kk \times Hp \times TZ}{Mp}$, that is, equal to the area HpMh multiplied into the ratio $\frac{TZ}{Mp}$: and therefore the horary variation of the inclination will be to 33" 10"' 33^{iv}. as the area HpMh multiplied into AZ x $\frac{TZ}{Mp}$ x $\frac{Pp}{PG}$ to AT³.

COR. 2. And, therefore, if the earth and nodes were after every hour drawn back from their new and instantly restored to their old places, so as their situation might continue given for a whole periodic month together, the whole variation of the inclination during that mouth would be to 33" 10"' 33^{iv}. as the aggregate of all the areas HpMh, generated in the time of one revolution of the point p (with due regard in summing to their proper signs + –), multiplied into AZ x TZ x $\frac{Pp}{PG}$ to Mp x AT³;

that is, as the whole circle QAqa multiplied into AZ x TZ x $\frac{Pp}{PG}$ to Mp x AT³, that is,

as the circumference QAqa multiplied into AZ x TZ x $\frac{Pp}{PG}$ to 2Mp x AT².

COR. 3. And, therefore, in a given position of the nodes, the mean horary variation, from which, if uniformly continued through the whole month, that menstrual variation might be generated, is to $33'' 10''' 33^{iv}$. as $AZ \times TZ \times \frac{Pp}{PG}$ to $2AT^2$, or as Pp

$\times \frac{AZ \times TZ}{\frac{1}{2}AT}$ to $PG \times 4AT$; that is (because Pp is to PG as the sine of the aforesaid inclination

to the radius, and $\frac{AZ \times TZ}{\frac{1}{2}AT}$ to $4AT$ as the sine of double the angle ATn to four times the radius), as the sine of the same inclination multiplied into the sine of double the distance of the nodes from the sun to four times the square of the radius.

COR. 4. Seeing the horary variation of the inclination, when the nodes are in the quadratures, is (by this Prop.) to the angle $33'' 10''' 33^{iv}$. as $IT \times AZ \times TG \times \frac{Pp}{PG}$ to AT^3,

that is, as $\frac{IT \times TG}{\frac{1}{2}AT} \times \frac{Pp}{PG}$ to $2AT$, that is, as the sine of double the distance of the moon

from the quadratures multiplied into $\frac{Pp}{PG}$ to twice the radius, the sum of all the horary variations during the time that the moon, in this situation of the nodes, passes from the quadrature to the syzygy (that is, in the space of $177 \, ^1/_6$ hours) will be to the sum of as many angles $33'' 10''' 33^{iv}$. or $5878''$, as the sum of all the sines of double the distance of the moon from the quadratures multiplied into $\frac{Pp}{PG}$ to the sum of as many

diameters; that is, as the diameter multiplied $\frac{Pp}{PG}$ to the circumference; that is, if the

inclination be $5° 1'$, as $7 \times {}^{874}/_{10000}$ to 22, or as 278 to 10000. And, therefore, the whole variation, composed out of the sum of all the horary variations in the aforesaid time, is $163''$, or $2' 43''$.

PROPOSITION XXXV. PROBLEM XVI.

TO A GIVEN TIME TO FIND THE INCLINATION OF THE MOON'S ORBIT TO THE PLANE OF THE ECLIPTIC.

Let AD be the sine of the greatest inclination, and AB the sine of the least. Bisect BD in C; and round the centre 0, with the interval BC, describe the circle BGD. In AC take CE in the same proportion to EB as EB to twice BA. And if to the time given we set off the angle AEG equal to double the distance of the nodes from the

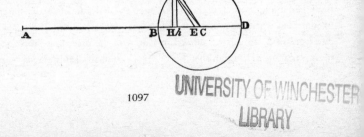

quadratures, and upon AD let fall the perpendicular GH, AH will be the sine of the inclination required.

For GE² is equal to GH² + HE² = BHD + HE² = HBD + HE² – BH² = HBD + BE² – 2BH x BE = BE² + 2EC x BH = 2EC x AB + 2EC x BH = 2EC x AH; wherefore since 2EC is given, GE² will be as AH. Now let AEg represent double the distance of the nodes from the quadratures, in a given moment of time after, and the arc Gg, on account of the given angle GEg, will be as the distance GE. But Hh is to Gg as GH to GC, and, therefore, Hh is as the rectangle GH x Gg, or GH x GE, that is, as $\frac{GH}{GE}$ x

GE², or $\frac{GH}{GE}$ x AH; that is, as AH and the sine of the angle AEG conjunctly. If, therefore, in any one case, AH be the sine of inclination, it will increase by the same increments as the sine of inclination doth, by Cor. 3 of the preceding Prop. and therefore win always continue equal to that sine. But when the point G falls upon either point B or D, AH is equal to this sine, and therefore remains always equal thereto.   Q.E.D.

In this demonstration I have supposed that the angle BEG, representing double the distance of the nodes from the quadratures, increaseth uniformly; for I cannot descend to every minute circumstance of inequality. Now suppose that BEG is a right angle, and that Gg is in this case the horary increment of double the distance of the nodes from the sun; then, by Cor. 3 of the last Prop. the horary variation of the inclination in the same case will be to 33" 10"' 33^{iv}. as the rectangle of AH, the sine of the inclination, into the sine of the right angle BEG, double the distance of the nodes from the sun, to four times the square of the radius; that is, as AH, the sine of the mean inclination, to four times the radius; that is, seeing the mean inclination is about 5° 8$^{1}/_{2}$, as its sine 896 to 40000, the quadruple of the radius, or as 224 to 10000. But the whole variation corresponding to BD, the difference of the sines, is to this horary variation as the diameter BD to the arc Gg, that is, conjunctly as the diameter BD to the semi-circumference BGD, and as the time of 2079 $^{7}/_{10}$ hours, in which the node proceeds from the quadratures to the syzygies, to one hour, that is, as 7 to 11, and 2079$^{7}/_{10}$ to 1. Wherefore, compounding all these proportions, we shall have the whole variation BD to 33" 10"' 33^{iv}. as 224 x 7 x 2079$^{7}/_{10}$ to 110000, that is, as 29645 to 1000; and from thence that variation BD will come out 16' 23$^{1}/_{2}$".

And this is the greatest variation of the inclination, abstracting from the situation of the moon in its orbit; for if the nodes are in the syzygies, the inclination suffers no change from the various positions of the moon. But if the nodes are in the quadratures, the inclination is less when the moon is in the syzygies than when it is in the quadratures by a difference of 2' 43", as we shewed in Cor. 4 of the preceding Prop.; and the whole mean variation BD, diminished by 1' 21$^{1}/_{2}$", the half of this excess, becomes 15' 2",

when the moon is in the quadratures; and increased by the same, becomes 17' 45" when the moon is in the syzygies. If, therefore, the moon be in the syzygies, the whole variation in the passage of the nodes from the quadratures to the syzygies will be 17'45"; and, therefore, if the inclination be 5° 17' 20", when the nodes are in the syzygies, it will be 4° 59' 35" when the nodes are in the quadratures and the moon in the syzygies. The truth of all which is confirmed by observations.

Now if the inclination of the orbit should be required when the moon is in the syzygies, and the nodes any where between them and the quadratures, let AB be to AD as the sine of 4° 59' 35" to the sine of 5° 17' 20", and take the angle AEG equal to double the distance of the nodes from the quadratures; and AH will be the sine of the inclination desired, To this inclination of the orbit the inclination of the same is equal, when the moon is 900 distant from the nodes. In other situations of the moon, this menstrual inequality, to which the variation of the inclination is obnoxious in the calculus of the moon's latitude, is balanced, and in a manner took off, by the menstrual inequality of the motion of the nodes (as we said before), and therefore may be neglected in the computation of the said latitude.

SCHOLIUM.

By these computations of the lunar motions I was willing to shew that by the theory of gravity the motions of the moon could be calculated from their physical causes. By the same theory I moreover found that the annual equation of the mean motion of the moon arises from the various dilatation which the orbit of the moon suffers from the action of the sun according to Cor. 6, Prop. LXVI, Book I. The force of this action is greater in the perigeon sun, and dilates the moon's orbit; in the apogeon sun it is less, and permits the orbit to be again contracted. The moon moves slower in the dilated and faster in the contracted orbit; and the annual equation, by which this inequality is regulated, vanishes in the apogee and perigee of the sun. In the mean distance of the sun from the earth it arises to about 11' 50"; in other distances of the sun it is proportional to the equation of the sun's centre, and is added to the mean motion of the moon, while the earth is passing from its aphelion to its perihelion, and subducted while the earth is in the opposite semi-circle. Taking for the radius of the *orbis magnus* 1000, and $16^7/_8$ for the earth's eccentricity, this equation, when of the greatest magnitude, by the theory of gravity comes out 11' 49". But the eccentricity of the earth seems to be something greater, and with the eccentricity this equation will be augmented in the same proportion. Suppose the eccentricity $16^{11}/_{12}$, and the greatest equation will be 11' 51".

Farther; I found that the apogee and nodes of the moon move faster in the perihelion of the earth, where the force of the sun's action is greater, than in the aphelion

thereof, and that in the reciprocal triplicate proportion of the earth's distance from the sun; and hence arise annual equations of those motions proportional to the equation of the sun's centre. Now the motion of the sun is in the reciprocal duplicate proportion of the earth's distance from the sun; and the greatest equation of the centre which this inequality generates is 1° 56' 20", corresponding to the above mentioned eccentricity of the sun, $16^{11}/_{12}$. But if the motion of the sun had been in the reciprocal triplicate proportion of the distance, this inequality would have generated the greatest equation 2° 54' 30"; and therefore the greatest equations which the inequalities of the motions of the moon's apogee and nodes do generate are to 2° 54' 30" as the mean diurnal motion of the moon's apogee and the mean diurnal motion of its nodes are to the mean diurnal motion of the sun. Whence the greatest equation of the mean motion of the apogee comes out 19' 43", and the greatest equation of the mean motion of the nodes 9' 24". The former equation is added, and the latter subducted, while the earth is passing from its perihelion to its aphelion, and contrariwise when the earth is in the opposite semi-circle.

By the theory of gravity I likewise found that the action of the sun upon the moon is something greater when the transverse diameter of the moon's orbit passeth through the sun than when the same is perpendicular upon the line which joins the earth and the sun; and therefore the moon's orbit is something larger in the former than in the latter case. And hence arises another equation of the moon's mean motion, depending upon the situation of the moon's apogee in respect of the sun, which is in its greatest quantity when the moon's apogee is in the octants of the sun, and vanishes when the apogee arrives at the quadratures or syzygies; and it, is added to the mean motion while the moon's apogee is passing from the quadrature of the sun to the syzygy, and subducted while the apogee is passing from the syzygy to the quadrature. This equation, which I shall call the semiannual, when greatest in the octants of the apogee, arises to about 3' 45", so far as I could collect from the phaenomena: and this is its quantity in the mean distance of the sun from the earth, But it is increased and diminished in the reciprocal triplicate proportion of the sun's distance, and therefore is nearly 3' 34" when that distance is greatest, and 3' 56" when least. But when the moon's apogee is without the octants, it becomes less, and is to its greatest quantity as the sine of double the distance of the moon's apogee from the nearest syzygy or quadrature to the radius.

By the same theory of gravity, the action of the sun upon the moon is something greater when the line of the moon's nodes passes through the sun than when it is at right angles with the line which joins the sun and the earth; and hence arises another equation of the moon's mean motion, which I shall call the second semi-annual; and this is greatest when the nodes are in the octants of the sun, and vanishes when they

are in the syzygies or quadratures; and in other positions of the nodes is proportional to the sine of double the distance of either node from the nearest syzygy or quadrature. And it is added to the mean motion of the moon, if the run is *in antecedentia*, to the node which is nearest to him, and subducted if *in consequentia*, and in the octants, where it is of the greatest magnitude, it arises to 47" in the mean distance of the sun from the earth, as I find from the theory of gravity. In other distances of the sun, this equation, greatest in the octants of the nodes, is reciprocally as the cube of the sun's distance from the earth; and therefore in the sun's perigee it comes to about 49", and in its apogee to about 45".

By the same theory of gravity, the moon's apogee goes forward at the greatest rate when it is either in conjunction with or in opposition to the sun, but in its quadratures with the sun it goes backward; and the eccentricity comes, in the former case, to its greatest quantity; in the latter to its least, by Cor. 7, 8, and 9, Prop. LXVI, Book I. And those inequalities, by the Corollaries we have named, are very great, and generate the principal which I call the semiannual equation of the apogee; and this semi-annual equation in its greatest quantity comes to about 12° 18', as nearly as I could collect from the phenomena. Our countryman, *Horrox*, was the first who advanced the theory of the moon's mov-

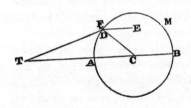

ing in an ellipsis about the earth placed in its lower focus. Dr. *Halley* improved the notion, by putting the centre of the ellipsis in an epicycle whose centre is uniformly revolved about the earth; and from the motion in this epicycle the mentioned inequalities in the progress and regress of the apogee, and in the quantity of eccentricity, do arise. Suppose the mean distance of the moon from the earth to be divided into 100000 parts, and let T represent the earth, and TC the moon's mean eccentricity of 5505 such parts. Produce TC to B, so as CB may be the sine of the greatest semi-annual equation 12° 18' to the radius TC; and the circle BDA described about the centre C, with the interval CB, will be the epicycle spoken of, in which the centre of the moon's orbit is placed, and revolved according to the order of the letters BDA. Set off the angle BCD equal to twice the annual argument, or twice the distance of the sun's true place from the place of the moon's apogee once equated, and CTD will be the semi-annual equation of the moon's apogee, and TD the eccentricity of its orbit, tending to the place of the apogee now twice equated. But, having the moon's mean motion, the place of its apogee, and its eccentricity, as well as the longer axis of its orbit 200000, from these *data* the true place of the moon in its orbit, together with its distance from the earth, may be determined by the methods commonly known.

In the perihelion of the earth, where the force of the sun is greatest, the centre of the moon's orbit moves faster about the centre C than in the aphelion, and that in the reciprocal triplicate proportion of the sun's distance from the earth. But, because the equation of the sun's centre is included in the annual argument, the centre of the moon's orbit moves faster in its epicycle BDA, in the reciprocal duplicate proportion of the sun's distance from the earth. Therefore, that it may move yet faster in the reciprocal simple proportion of the distance, suppose that from D, the centre of the orbit, a right line DE is drawn, tending towards the moon's apogee once equated, that is, parallel to TC; and set off the angle EDF equal to the excess of the aforesaid annual argument above the distance of the moon's apogee from the sun's perigee *in consequentia*, or, which comes to the same thing, take the angle CDF equal to the complement of the sun's true anomaly to 360°; and let DF be to DC as twice the eccentricity of the *orbis magnus* to the sun's mean distance from the earth, and the sun's mean diurnal motion from the moon's apogee to the sun's mean diurnal motion from its own apogee conjunctly, that is, as $33^7/_8$ to 1000, and 52' 27" 16"' to 59' 8" 10"' conjunctly, or as 3 to 100; and imagine the centre of the moon's orbit placed in the point F to be revolved in an epicycle whose centre is D, and radius DF, while the point D moves in the circumference of the circle DABD: for by this means the centre of the moon's orbit comes to describe a certain curve line about the centre Q with a velocity which will be almost reciprocally as the cube of the sun's distance from the earth, as it ought to be.

The calculus of this motion is difficult, but may be rendered more easy by the following approximation. Assuming, as above, the moon's mean distance from the earth of 100000 parts, and the eccentricity TC of 5505 such parts, the line CB or CD will be found $1172^3/_4$ and DF $35^1/_5$, of those parts; and this line DF at the distance TC subtends the angle at the earth, which the removal of the centre of the orbit from the place D to the place F generates in the motion of this centre; and double this line DF in a parallel position, at the distance of the upper focus of the moon's orbit from the earth, subtends at the earth the same angle as DF did before, which that removal generates in the motion of this upper focus; but at the distance of the moon from the earth this double line 2DF at the upper focus, in a parallel position to the first line DF, subtends an angle at the moon, which the said removal generates in the motion of the moon, which angle may be therefore called the second equation of the moon's centre; and this equation, in the mean distance of the moon from the earth, is nearly as the sine of the angle which that line DP contains with the line drawn from the point F to the moon, and when in its greatest quantity amounts to 2' 25". But the angle which the line DF contains with the line drawn from the point F to the moon is found either by subtracting the angle EDP from the mean anomaly of the moon, or by adding the distance of the moon from the sun to the distance of the moons apogee from the

apogee of the sun; and as the radius to the sine of the angle thus found, so is 2' 25" to the second equation of the centre: to be added, if the forementioned sum be less than a semi-circle; to be subducted, if greater. And from the moon's place in its orbit thus corrected, its longitude may be found in the syzygies of the luminaries.

The atmosphere of the earth to the height of 35 or 40 miles refracts the sun's light. This refraction scatters and spreads the light over the earth's shadow; and the dissipated light near the limits of the shadow dilates the shadow. Upon which account, to the diameter of the shadow, as it comes out by the parallax, I add 1 or $1^1/_3$ minute in lunar eclipses.

But the theory of the moon ought to be examined and proved from the phaenomena, first in the syzygies, then in the quadratures, and last of all in the octants: and whoever pleases to undertake the work will find it not amiss to assume the following mean motions of the sun and moon at the Royal Observatory of *Greenwich*, to the last day of *December* at noon, *anno* 1700, O.S. viz. The mean motion of the sun ♉ 20° 43' 40", and of its apogee ♋ 7° 44' 30"; the mean motion of the moon ♒ 15° 21' 00"; of its apogee, ♓ 8° 20' 00"; and of its ascending node ♌ 27° 24' 20"; and the difference of meridians betwixt the Observatory at *Greenwich* and the Royal Observatory at *Paris*, 0^h. 9' 20": but the mean motion of the moon and of its apogee are not yet obtained with sufficient accuracy.

PROPOSITION XXXVI. PROBLEM XVII.

To find the force of the sun to move the sea.

The sun's force ML or PT to disturb the motions of the moon, was (by Prop. XXV.) in the moon's quadratures, to the force of gravity with us, as 1 to 638092,6; and the force TM – LM or 2PK in the moon's syzygies is double that quantity. But, descending to the surface of the earth, these forces are diminished in proportion of the distances from the Centre of the earth, that is, in the proportion of $60^1/_2$ to 1; and therefore the former force on the earth's surface is to the force of gravity as 1 to 38604600; and by this force the sea is depressed in such places as are 90 degrees distant from the sun, But by the other force, which is twice as great, the sea is raised not only in the places directly under the sun, but in those also which are directly opposed to it; and the sum of these forces is to the force of gravity as 1 to 12868200. And because the same force excites the same motion, whether it depresses the waters in those places which are 90 degrees distant from the sun, or raises them in the places which are directly under and directly opposed to the sun, the aforesaid sum will be the total force of the sun to disturb the sea, and will have the same effect as if the whole was employed in raising the sea in the places directly under and directly opposed to the sun, and did not act at all in the places which are 90 degrees removed from the sun.

And this is the force of the sun to disturb the sea in any given place, where the sun is at the same time both vertical, and in its mean distance from the earth. In other positions of the sun, its force to raise the sea is as the versed sine of double its altitude above the horizon of the place directly, and the cube of the distance from the earth reciprocally.

COR. Since the centrifugal force of the parts of the earth, arising from the earth's diurnal motion, which is to the force of gravity as 1 to 289, raises the waters under the equator to a height exceeding that under the poles by 85472 *Paris* feet, as above, in Prop. XIX., the force of the sun, which we have now shewed to be to the force of gravity as 1 to 12868200, and therefore is to that centrifugal force as 289 to 12868200, or as 1 to 44527, will be able to raise the waters in the places directly under and directly opposed to the sun to a height exceeding that in the places which are 90 degrees removed from the sun only by one *Paris* foot and $113^1/_{30}$ inches; for this measure is to the measure of 85472 feet as 1 to 44527.

PROPOSITION XXXVII. PROBLEM XVIII.

TO FIND THE FORCE OF THE MOON TO MOVE THE SEA.

The force of the moon to move the sea is to be deduced from its proportion to the force of the sun, and this proportion is to be collected from the proportion of the motions of the sea, which are the effects of those forces. Before the mouth of the river *Avon*, three miles below *Bristol*, the height of the ascent of the water in the vernal and autumnal syzygies of the luminaries (by the observations of *Samuel Sturmy*) amounts to about 45 feet, but in the quadratures to 25 only. The former of those heights arises from the sum of the aforesaid forces, the latter from their difference. If, therefore, S and L are supposed to represent respectively the forces of the sun and moon while they are in the equator, as well as in their mean distances from the earth, we shall have L + S to L − S as 45 to 25, or as 9 to 5.

At *Plymouth* (by the observations of *Samuel Colepress*) the tide in its mean height rises to about 16 feet, and in the spring and autumn the height thereof in the syzygies may exceed that in the quadratures by more than 7 or 8 feet. Suppose the greatest difference of those heights to be 9 feet, and L + S will be to L − S as $20^1/_2$ to $11^1/_2$, or as 41 to 23; a proportion that agrees well enough with the former. But because of the great tide at *Bristol*, we are rather to depend upon the observations of *Sturmy*; and, therefore, till we procure something that is more certain, we shall use the proportion of 9 to 5.

But because of the reciprocal motions of the waters, the greatest tides do not happen at the times of the syzygies of the luminaries, but, as we have said before, are the

third in order after the syzygies; or (reckoning from the syzygies) follow next after the third appulse of the moon to the meridian of the place after the syzygies; or rather (as *Sturmy* observes) are the third after the day of the new or full moon, or rather nearly after the twelfth hour from the new or full moon, and therefore fall nearly upon the forty-third hour after the new or full of the moon. But in this port they fall out about the seventh hour after the appulse of the moon to the meridian of the place; and therefore follow next after the appulse of the moon to the meridian, when the moon is distant from the sun, or from opposition with the sun by about 18 or 19 degrees *in consequentia*. So the summer and winter seasons come not to their height in the solstices themselves, but when the sun is advanced beyond the solstices by about a tenth part of its whole course, that is, by about 36 or 37 degrees. In like manner, the greatest tide is raised after the appulse of the moon to the meridian of the place, when the moon has passed by the sun, *or the opposition thereof,* by about a tenth part of the whole motion from *one greatest* tide to *the next following greatest* tide. Suppose that distance about $18^1/_2$ degrees; and the sun's force in this distance of the moon from the syzygies and quadratures will be of less moment to augment and diminish that part of the motion of the sea which proceeds from the motion of the moon than in the syzygies and quadratures themselves in the proportion of the radius to the co-sine of double this distance, or of an angle of 37 degrees; that is, in proportion of 10000000 to 7986355; and, therefore, in the preceding analogy, in place of S we must put 0,7986355S.

But farther; the force of the moon in the quadratures must be diminished, on account of its declination from the equator; for the moon in those quadratures, or rather in $18^1/_2$ degrees past the quadratures, declines from the equator by about 23° 13'; and the force of either luminary to move the sea is diminished as it declines from the equator nearly in the duplicate proportion of the co-sine of the declination; and therefore the force of the moon in those quadratures is only 0,8570327L; whence we have L + 0,7986355S to 0,8570327L – 0,7986355S as 9 to 5.

Farther yet; the diameters of the orbit in which the moon should move, setting aside the consideration of eccentricity, are one to the other as 66 to 70; and therefore the moon's distance from the earth in the syzygies is to its distance in the quadratures, *caeteris paribus,* as 69 to 70; and its distances, when $18^1/_2$, degrees advanced beyond the syzygies, where the greatest tide was excited, and when $18^1/_2$ degrees passed by the quadratures, where the least tide was produced, are to its mean distance as 69,098747 and 69,897345 to $69^1/_2$. But the force of the moon to move the sea is in the reciprocal triplicate proportion of its distance; and therefore its forces, in the greatest and least of those distances, are to its force in its mean distance as 0,9830427 and 1,017522 to 1. From whence we have 1,017522L x 0,7986355S to 0,9830427 x 0,8570327L – 0,7986355S as 9 to 5; and S to L as 1 to 4,4815. Wherefore since the force of the sun is to the force of gravity as 1 to 12868200, the moon's force will be to the force of gravity as 1 to 2671400.

COR. 1. Since the waters excited by the sun's force rise to the height of a foot and $11^{1}/_{30}$ inches, the moon's force will raise the same to the height of 8 feet and $7^{5}/_{22}$ inches; and the joint forces of both will raise the same to the height of 10 $^{1}/_{2}$ feet; and when the moon is in its perigee to the height of $12^{1}/_{2}$ feet, and more, especially when the wird sets the same way as the tide. And a force of that quantity is abundantly sufficient to excite all the motions of the sea, and agrees well with the proportion of those motions; for in such seas as lie free and open from east to west, as in the *Pacific* sea, and in those tracts of the *Atlantic* and *AEthiopic* seas which lie without the tropics, the waters commonly rise to 6, 9, 12, or 15 feet; but in the *Pacific* sea, which is of a greater depth, as well as of a larger extent, the tides are said to be greater than in the *Atlantic* and *AEthiopic* seas; for to have a full tide raised, an extent of sea from east to west is required of no less than 90 degrees. In the *AEthiopic* sea, the waters rise to a less height within the tropics than in the temperate zones, because of the narrowness of the sea between *Africa* and the southern parts of *America*. In the middle of the open sea the waters cannot rise without falling together, and at the same time, upon both the eastern and western shores, when, notwithstanding, in our narrow seas, they ought to fall on those shores by alternate turns; upon which account there is commonly but a small flood and ebb in such islands as lie far distant from the continent. On the contrary, in some ports, where to fill and empty the bays alternately the waters are with great violence forced in and out through shallow channels, the flood and ebb must be greater than ordinary; as at *Plymouth* and *Chepstow Bridge* in *England*, at the mountains of St. *Michael*, and the town of *Auranches*, in *Normandy*, and at *Cambaia* and *Pegu* in the *East Indies*. In these places the sea is hurried in and out with such violence, as sometimes to lay the shores under water, sometimes to leave them dry for many miles. Nor is this force of the influx and efflux to be broke till it has raised and depressed the waters to 30, 40, or 50 feet and above. And a like account is to be given of long and shallow channels or straits, such as the *Magellanic* straits, and those channels which environ *England*. The tide in such ports and straits, by the violence of the influx and efflux, is augmented above measure. But on such shores as lie towards the deep and open sea with a steep descent, where the waters may freely rise and fall without that precipitation of influx and efflux, the proportion of the tides agrees with the forces of the sun and moon.

COR. 2. Since the moon's force to move the sea is to the force of gravity as 1 to 2871400, it is evident that this force is far less than to appear sensibly in statical or hydrostatical experiments, or even in those of pendulums. It is in the tides only that this force shews itself by any sensible effect.

COR. 3. Because the force of the moon to move the sea is to the like force of the sun as 4,4815 to 1, and the forces (by Cor. 14, Prop. LXVI, Book I) are as the densities

of the bodies of the sun and moon and the cubes of their apparent diameters conjunctly, the density of the moon will be to the density of the *sun* as 4,4815 to 1 directly, and the cube of the moon's diameter to the cube of the sun's diameter inversely; that is (seeing the mean apparent diameters of the moon and sun are 31' 16 $1/2$", and 32' 12"), as 4891 to 1000. But the density of the sun was to the density of the earth as 1000 to 4000; and therefore the density of the moon is to the density of the earth as 4891 to 4000, or as 11 to 9. Therefore the body of the moon is more dense and more earthly than the earth itself.

Cor. 4. And since the true diameter of the moon (from the observations of astronomers) is to the true diameter of the earth as 100 to 365, the mass of matter in the moon will be to the mass of matter in the earth as 1 to 39,788.

Cor. 5. And the accelerative gravity on the surface of the moon will be about three times less than the accelerative gravity on the surface of the earth.

Cor. 6. And the distance of the moon's centre from the centre of the earth will be to the distance of the moon's centre from the common centre, of gravity of the earth and moon as 40,788 to 39,788.

Cor. 7. And the mean distance of the centre of the moon from the centre of the earth will be (in the moon's octants) nearly 60$2$, of the greatest semi-diameters of the earth; for the greatest semi-diameter of the earth was 19658600 *Paris* feet, and the mean distance of the centres of the earth and moon, consisting of 60$1/2$ such semi-diameters, is equal to 1187379440 feet. And this distance (by the preceding Cor.) is to the distance of the moon's centre, from the common centre of gravity of the earth and moon as 40,788 to 39,788; which latter distance, therefore, is 1158268534 feet. And since the moon, in respect of the fixed stars, performs its revolution in 27^d. 7^h. 43$4/9$', the versed sine of that angle which the moon in a minute of time describes is 12752341 to the radius 1000,000000,000000; and as the radius is to this versed sine, so are 1158268534 feet to 14,7706353 feet. The moon, therefore, falling towards the earth by that force which retains it in its orbit, would in one minute of time describe 14,7706353 feet; and if we augment this force in the proportion of 178$29/40$ to 177$29/40$, we shall have the total force of gravity at the orbit of the moon, by Cor. Prop. III; and the moon falling by this force, in one minute of time would describe 14,8538067 feet. And at the 60th part of the distance of the moon from the earth's centre, that is, at the distance of 197896573 feet from the centre of the earth, a body falling by its weight, would, in one second of time, likewise describe 14,8538067 feet. And, therefore, at the distance of 19615800, which compose one mean semi-diameter of the earth, a heavy body would describe in falling 15,11175, or 15 feet, 1 inch, and 4$1/11$ lines, in the same time. This will be the descent of bodies in the latitude of 45 degrees. And by the foregoing table, to be found under Prop.

XX, the descent in the latitude of *Paris* will be a little greater by an excess of about $2/3$ parts of a line. Therefore, by this computation, heavy bodies in the latitude of *Paris* falling *in vacuo* will describe 15 *Paris* feet, 1 inch, $4^{25}/_{33}$ lines, very nearly, in one second of time. And if the gravity be, diminished by taking away a quantity equal to the centrifugal force arising in that latitude from the earth's diurnal motion, heavy bodies falling there will describe in one second of time 15 feet, 1 inch, and $1^1/_2$ line. And with this velocity heavy bodies do really fall in the latitude of *Paris*, as we have shewn above in Prop. IV and XIX.

Cor. 8. The mean distance of the centres of the earth and moon in the syzygies of the moon is equal to 60 of the greatest semi-diameters of the earth, subducting only about one 30th part of a semi-diameter: and in the moon's quadratures the mean distance of the same centres is $60^5/_6$ such semi-diameters of the earth; for these two distances are to the mean distance of the moon in the octants as 69 and 70 to $69^1/_2$, by Prop. XXVIII.

Cor. 9. The mean distance of the centres of the earth and moon in the syzygies of the moon is 60 mean semi-diameters of the earth, and a 10th part of one semi-diameter; and in the moon's quadratures the mean distance of the same centres is 61 mean semi-diameters of the earth, subducting one 30th part of one semi-diameter.

Cor. 10. In the moon's syzygies its mean horizontal parallax in the latitudes of 0, 30, 38, 45, 52, 60, 90 degrees is 57' 20", 57' 16", 57' 14", 57' 12", 57' 10", 57' 8", 57' 4", respectively.

In these computations I do not consider the magnetic attraction of the earth, whose quantity is very small and unknown: if this quantity should ever be found out, and the measures of degrees upon the meridian, the lengths of isochronous pendulums in different parallels, the laws of the motions of the sea, and the moon's parallax, with the apparent diameters of the sun and moon, should be more exactly determined from phaenomena: we should then be enabled to bring this calculation to a greater accuracy.

PROPOSITION XXXVIII. PROBLEM XIX.

To find the figure of the moon's body.

If the moon's body were fluid like our sea, the force of the earth to raise that fluid in the nearest and remotest parts would be to the force of the moon by which our sea is raised in the places under and opposite to the moon as the accelerative gravity of the moon towards the earth to the accelerative gravity of the earth towards the moon, and the diameter of the moon to the diameter of the earth conjunctly; that is, as 39,788 to 1, and 100 to 365 conjunctly, or as 1081 to 100. Wherefore, since our sea, by the force of the moon, is raised to $8^3/_5$ feet, the lunar fluid would be raised by the force of the

earth to 93 feet; and upon this account the figure of the moon would be a spheroid, whose greatest diameter produced would pass through the centre of the earth, and exceed the diameters perpendicular thereto by 186 feet. Such a figure, therefore, the moon affects, and must have put on from the beginning. Q.E.I.

COR. Hence it is that the same face of the moon always respects the earth; nor can the body of the moon possibly rest in any other position, but would return always by a libratory motion to this situation; but those librations, however, must be exceedingly slow, because of the weakness of the forces which excite them; so that the face of the moon, which should be always obverted to the earth, may, for the reason assigned in Prop. XVII. be turned towards the other focus of the moon's orbit, without being immediately drawn back, and converted again towards the earth.

LEMMA I.

If APEp *represent the earth uniformly dense, marked with the centre* C, *the poles* P, p, *and the equator* AE; *and if about the centre* C, *with the radius* CP, *we suppose the sphere* Pape *to be described, and* QR *to denote the plane an which a right line, drawn from the centre of the sun to the centre of the earth, insists at right angles; and further suppose that the several particles of the whole exterior earth* PapAPepE, *without the height of the said sphere, endeavour to recede towards this side and that side from the plane* QR, *every particle by a force proportional to its distance from that plane; I say, in the first place, that the whole force and efficacy of all the particles that are situate in* AE, *the circle of the equator, and disposed uniformly without the globe, encompassing, the same after the manner of a ring, to wheel the earth about its centre, is to the whole force and efficacy of as many particles in that point* A *of the equator which is at the greatest distance from the plane* QR, *to wheel the earth about its centre with a like circular motion, as 1 to 2. And that circular motion will be performed about an axis lying in the common section of the equator and the plane* QR.

For let there be described from the centre K, with the diameter IL, the semi-circle INL. Suppose the semi-circumference INL to be divided into innumerable equal parts,

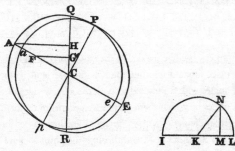

and from the several parts N to the diameter IL let fall the sines NM. Then the sum of the squares of all the sines NM will be equal to the sums of the squares of the sines KM, and both sums together will be equal to the sums of the squares of as many semi-diameters KN; and therefore the sum of the squares of all the sines NM will be but half so great as the sum of the squares of as many semi-diameters KN.

Suppose now the circumference of the circle AE to be divided into the like number of little equal parts, and from every such part P a perpendicular FG to be let fall upon the plane QR, as well as the perpendicular AH from the point A. Then the force by which the particle F recedes from the plane QR will (by supposition) be as that perpendicular PG; and this force multiplied by the distance CG will represent the power of the particle F to turn the earth round its centre. And, therefore, the power of a particle in the place F will be to the power of a particle in the place A as FG x GC to AH x HC; that is, as FC^2 to AC^2: and therefore the whole power of all the particles F, in their proper places F, will be to the power of the like number of particles in the place A as the sum of all the FC^2 to the sum of all the AC^2, that is (by what we have demonstrated before), as 1 to 2.   Q.E.D.

And because the action of those particles is exerted in the direction of lines perpendicularly receding from the plane QR, and that equally from each side of this plane, they will wheel about the circumference of the circle of the equator, together with the adherent body of the earth, round an axis which lies as well in the plane QR as in that of the equator.

LEMMA II.

THE SAME THINGS STILL SUPPOSED, I SAY, IN THE SECOND
PLACE, THAT THE TOTAL FORCE OR POWER OF ALL THE
PARTICLES SITUATED EVERY WHERE ABOUT THE SPHERE TO
TURN THE EARTH ABOUT THE SAID AXIS IS TO THE WHOLE
FORCE OF THE LIKE NUMBER OF PARTICLES, UNIFORMLY
DISPOSED ROUND THE WHOLE CIRCUMFERENCE OF THE
EQUATOR AE IN THE FASHION OF A RING, TO TURN THE WHOLE
EARTH ABOUT WITH THE LIKE CIRCULAR MOTION, AS 2 TO 5.

For let IK be any lesser circle parallel to the equator AE, and let L*l* be any two equal particles in this circle, situated without the sphere P*ape*; and if upon the plane QR, which, is at right angles with a radius drawn to the sun, we let fall the perpendiculars LM, *lm*, the total forces by which these particles recede from the plane QR will be proportional to the perpendiculars LM, *lm*. Let the right line L*l* be drawn parallel to the plane P*ape*, and bisect the same in X; and through the point X draw N*n* parallel to the

plane QR, and meeting the perpendiculars LM, lm, in N and n; and upon the plane QR let fall the perpendicular XY. And the contrary forces of the particles L and l to wheel about the earth contrariwise are as LM x MC, and lm x mC; that is, as LN x MC

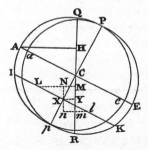

+ NM x MC, and ln x mC − nm x mC; or LN x MC + NM x MC, and LN x mC − NM x mC, and LN x Mm − NM x $\overline{\text{MC} + n}$, the difference of the two, is the force of both taken together to turn the earth round. The affirmative part of this difference LN x Mm, or 2LN x NX, is to 2AH x HC, the force of two particles of the same size situated in A, as LX^2 to AC^2; and the negative part $\overline{\text{MC} + n}$, or 2XY x CY, is to 2AH x HC, the force of the same two particles situated in A, as CX^2 to AC^2. And therefore the difference of the parts, that is, the force of the two particles L and l, taken together, to wheel the earth about, is to the force of two particles, equal to the former and situated in the place A, to turn in like manner the earth round, as $LX^2 - CX^2$ to AC^2. But if the circumference IK of the circle IK is supposed to be divided into an infinite number of little equal parts L, all the LX^2 will be to the like number of IX^2 as 1 to 2 (by Lem. 1); and to the same number of AC^2 as IX^2 to $2AC^2$; and the same number of CX^2 to as many AC^2 as $2CX^2$ to $2AC^2$. Wherefore the united forces of all the particles in the circumference of the circle IK are to the joint forces of as many particles in the place A as $IX^2 - 2CX^2$ to $2AC^2$; and therefore (by Lem. 1) to the united form of as many particles in the circumference of the circle AE as $IX^2 - 2CX^2$ to AC^2.

Now if Pp, the diameter of the sphere, is conceived to be divided into an infinite number of equal parts, upon which a like number of circles IK are supposed to insist, the matter in the circumference of every circle IK will be as IX^2; and therefore the force of that matter to turn the earth about will be as IX^2 into $IX^2 - 2CX^2$: and the force of the same matter, if it was situated in the circumference of the circle AE, would be as IX^2 into AC^2. And therefore the force of all the particles of the whole matter situated without the sphere in the circumferences of all the circles is to the force of the like number of particles situated in the circumference of the greatest circle AE as all the IX^2 into $IX^2 - 2CX^2$ to as many IX^2 into AC^2; that is, as all the $AC^2 - CX^2$ into $AC^2 - 3CX^2$ to as many $AC^2 - CX^2$ into AC^2; that is, as all the $AC^4 - 4AC^2$ x $CX^2 + 3CX^4$ to as many $AC^4 - AC^2$ x CX^2; that is, as the whole fluent quantity, whose fluxion is $AC^4 - 4AC^2$ x $CX^2 + 3CX^4$, to the whole fluent quantity, whose fluxion is $AC^4 - AC^2$ x CX^2; and, therefore, by the method of fluxions, as AC^4 x $CX - \frac{4}{3}AC^2$ x $CX^3 + \frac{3}{5}CX^5$ to AC^4 x $CX - \frac{1}{3}AC^2$ x CX^3; that is, if for CX we write the whole Cp, or AC, as $\frac{4}{15}AC^5$ to $\frac{2}{3}AC^5$; that is, as 2 to 5. Q.E.D.

LEMMA III.

The same things still supposed, I say, in the third place, that the motion of the whole earth about the axis above-named arising from the motions of all the particles, will be to the motion of the aforesaid ring about the same axis in a proportion compounded of the proportion of the matter in the earth to the matter in the ring; and the proportion of three squares of the quadrantal arc of any circle to two squares of its diameter, that is, in the proportion of the matter to the matter, and of the number 925275 to the number 1000000.

For the motion of a cylinder revolved about its quiescent axis is to the motion of the inscribed sphere revolved together with it as any four equal squares to three circles inscribed in three of those squares; and the motion of this cylinder is to the motion of an exceedingly thin ring surrounding both sphere and cylinder in their common contact as double the matter in the cylinder to triple the matter in the ring; and this motion of the ring, uniformly continued about the axis of the cylinder, is to the uniform motion of the same about its own diameter performed in the same periodic time as the circumference of a circle to double its diameter.

HYPOTHESIS II.

If the other parts of the earth were taken away, and the remaining ring was carried alone about the sun in the orbit of the earth by the annual motion, while by the diurnal motion it was in the mean time revolved about its own axis inclined to the plane of the ecliptic by an angle of $23^1/_2$, degrees, the motion of the equinoctial points would be the same, whether the ring were fluid, or whether it consisted of a hard and rigid matter.

PROPOSITION XXXIX. PROBLEM XX.

To find the precession of the equinoxes.

The middle horary motion of the moon's nodes in a circular orbit, when the nodes are in the quadratures, was 16" 35'" 16^{iv}. 36^v.; the half of which, 8" 17'" 38^{iv}. 18^v. (for

the reasons above explained) is the mean horary motion of the nodes in such an orbit, which motion in a whole sidereal year becomes 20° 11' 46". Because, therefore, the nodes of the moon in such an orbit would be yearly transferred 20° 11' 46" *in antecedentia*; and, if there were more moons, the motion of the nodes of every one (by Cor. 16, Prop. LXVI, Book I) would be as its periodic time; if upon the surface of the earth a moon was revolved in the time of a sidereal day, the annual motion of the nodes of this moon would be to 20° 11' 46" as 23^h. 56', the sidereal day, to 27^d. 7^h. 43', the periodic time of our moon, that is, as 1436 to 39343. And the same thing would happen to the nodes of a ring of moons encompassing the earth, whether these moons did not mutually touch each the other, or whether they were molten, said formed into a continued ring, or whether that ring should become rigid and inflexible.

Let us, then, suppose that this ring is in quantity of matter equal to the whole exterior earth P*ap*AP*ep*E, which lies without the sphere P*ape* (see fig. Lem. II); and because this sphere is to that exterior earth as aC^2 to AC^2 − aC^2, that is (seeing PC or aC the least semi-diameter of the earth is to AC the greatest semi-diameter of the same as 229 to 230), as 52441 to 459; if this ring encompassed the earth round the equator, and both together were revolved about the diameter of the ring, the motion of the ring (by Lem. III) would be to the motion of the inner sphere as 459 to 52441 and 1000000 to 925275 conjunctly, that is, as 4590 to 485223; and therefore the motion of the ring would be to the sum of the motions of both ring and sphere as 4590 to 489813. Wherefore if the ring adheres to the sphere, and communicates its motion to the sphere, by which its nodes or equinoctial points recede, the motion remaining in the ring will be to its former motion as 4590 to 489813; upon which account the motion of the equinoctial points will be diminished in the same proportion. Wherefore the annual motion of the equinoctial points of the body, composed of both ring and sphere, will be to the motion 20° 11' 46" as 1436 to 39343 and 4590 to 489813 conjunctly, that is, as 100 to 292369. But the forces by which the nodes of a number of moons (as we explained above), and therefore by which the equinoctial points of the ring recede (that is, the forces 3IT, in fig. Prop. XXX), are in the several particles as the distances of those particles from the plane QR; and by these forces the particles recede from that plane: and therefore (by Lem. II) if the matter of the ring was spread all over the surface of the sphere, after the fashion of the figure P*ap*AP*ep*E, in order to make up that exterior part of the earth, the total force or power of all the particles to wheel about the earth round any diameter of the equator, and therefore to move the equinoctial points, would become less than before in the proportion of 2 to 5. Wherefore the annual regress of the equinoxes now would be to 20° 11' 46" as 10 to 73092; that is, would be 9" 56"' 50^{iv}.

But because the plane of the equator is inclined to that of the ecliptic, this motion is to be diminished in the proportion of the sine 91706 (which is the co-sine of $23^1/_2$ deg.) to the radius 100000; and the remaining motion will now be 9" 7'" 20^{iv}. which is the annual precession of the equinoxes arising from the force of the sun.

But the force of the moon to move the sea, was to the force of the sun nearly as 4,4815 to 1; and the force of the moon to move the equinoxes is to that of the sun in the same proportion. Where the annual precession of the equinoxes proceeding from the force of the moon comes out 40" 52'" 52^{iv}. and the total annual precession arising from the united forces of both will be 50" 00'" 12^{iv}. the quantity of which motion agrees with the phænomena; for the precession of the equinoxes, by astronomical observations, is about 50" yearly.

If the height of the earth at the equator exceeds its height at the poles by more than $17^1/_6$ miles, the matter thereof will be more rare near the surface than at the centre; and the precession of the equinoxes will be augmented by the excess of height, and diminished by the greater rarity.

And now we have described the system of the sun, the earth, moon, and planets, it remains that we add something about the comets.

LEMMA IV.

THAT THE COMETS ARE HIGHER THAN THE MOON, AND IN THE REGIONS OF THE PLANETS.

As the comets were placed by astronomers above the moon, because they were found to have no diurnal parallax, so their annual parallax is a convincing proof of their descending into the regions of the planets; for all the comets which move in a direct course according to the order of the signs, about the end of their appearance become

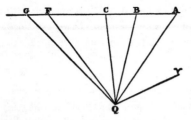

more than ordinarily slow or retrograde, if the earth is between them and the sun; and more than ordinarily swift, if the earth is approaching to a heliocentric opposition with them; where, on the other hand, those which move against the order of the signs, towards the end of their appearance appear swifter than they ought to be, if the earth is between them and the sun; and slower, and perhaps retrograde, if the earth is in the other side of its orbit. And these appearances proceed chiefly from the diverse situations which the earth acquires in the course of its motion, after the same manner as it happens to the planets, which appear sometimes retrograde, sometimes more slowly,

and sometimes more swiftly, progressive according as the motion of the earth falls in with that of the planet, or is directed the contrary way. If the earth move the same way with the comet, but, by an angular motion about the sun, so much swifter that right lines drawn from the earth to the comet converge towards the parts beyond the comet, the comet seen from the earth, because of its slower motion, will appear retrograde; and even if the earth is slower than the comet, the motion of the earth being subducted, the motion of the comet will at least appear retarded; but if the earth tends the contrary way to that of the comet, the motion of the comet will from thence appear accelerated; and from this apparent acceleration, or retardation, or regressive motion, the distance of the comet may be inferred in this manner. Let ♈QA, ♈QB, ♈QC, be three observed longitudes of the comet about the time of its first appearing, and ♈QF its last observed longitude before its disappearing. Draw the right line ABC, whose parts AB, BC, intercepted between the right lines QA and QB, QB and QC, may be one to the other as the two times between the three first observations. Produce AC to G, so as AG may be to AB as the time between the first and last observation to the time between the first and second; and join QG. Now if the comet

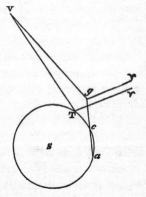

did move uniformly in a right line, and the earth either stood still, or was likewise carried forwards in a right line by an uniform motion, the angle ♈QG would be the longitude of the comet at the time of the last observation. The angle, therefore, FQG, which is the difference of the longitude, proceeds from the inequality of the motions of the comet and the earth; and this angle, if the earth and comet move contrary ways, is added to the angle ♈QG, and accelerates the apparent motion of the comet; but if the comet move the same way with the earth, it is subtracted, and either retards the motion of the comet, or perhaps renders it retrograde, as we have but now explained. This angle, therefore, proceeding chiefly from the motion of the earth, is justly to be esteemed the parallax of the comet; neglecting, to wit, some little increment or decrement that may arise from the unequal motion of the comet in its orbit; and from this parallax we thus deduce the distance of the comet. Let S represent the sun, acT the *orbis magnus*, a the earth's place in the first observation, c the place of the earth in the third observation, T the place of the earth in the last observation, and T♈ a right line drawn to the beginning of Aries. Set off the angle ♈TV equal to the angle ♈QF, that is, equal to the longitude of the comet at the time when the earth is in T; join ac, and produce it to g, so as ag may be to ac as AG to AC; and g will be the place at which the earth would have arrived in the time of the last observation, if

it had continued to move uniformly in the right line ac. Wherefore, if we draw gϒ parallel to Tϒ, and make the angle ϒgV equal to the angle ϒQG, this angle ϒgV will be equal to the longitude of the comet seen from the place g, and the angle TVg will be the parallax which arises from the earth's being transferred from the place g into the place T; and therefore V will be the place of the comet in the plane of the ecliptic. And this place V is commonly lower than the orb of Jupiter.

The same thing may be deduced from the incurvation of the way of the comets; for these bodies move almost in great circles, while their velocity is great; but about the end of their course, when that part of their apparent motion which arises from the parallax bears a greater proportion to their whole apparent motion, they commonly deviate from those circles, and when the earth goes to one side, they deviate to the other; and this deflexion, because of its corresponding with the motion of the earth, must arise chiefly from the parallax; and the quantity thereof is so considerable, as, by my computation, to place the disappearing comets a good deal lower than Jupiter. Whence it follows that when they approach nearer to us in their perigees and perihelions they often descend below the orbs of Mars and the inferior planets.

The near approach of the comets is farther confirmed from the light of their heads; for the light of a celestial body, illuminated by the sun, and receding to remote parts, is diminished in the quadruplicate proportion of the distance; to wit, in one duplicate proportion, on account of the increase of the distance from the sun, and in another duplicate proportion, on account of the decrease of the apparent diameter. Wherefore if both the quantity of light and the apparent diameter of a comet are given, its distance will be also given, by taking the distance of the comet to the distance of a planet in the direct proportion of their diameters and the reciprocal subduplicate proportion of their lights. Thus, in the comet of the year 1682, Mr. *Flamsted* observed with a telescope of 16 feet, and measured with a micrometer, the least diameter of its head, 2' 00"; but the nucleus or star in the middle of the head scarcely amounted to the tenth part of this measure; and therefore its diameter was only 11" or 12"; but in the light and splendor of its head it surpassed that of the comet in the year 1680, and might be compared with the stars of the first or second magnitude. Let us suppose that Saturn with its ring was about four times more lucid; and because the light of the ring was almost equal to the light of the globe within, and the apparent diameter of the globe is about 21", and therefore the united light of both globe and ring would be equal to the light of a globe whose diameter is 30", it follows that the distance of the comet was to the distance of Saturn as 1 to $\sqrt{4}$ inversely, and 12" to 30 directly; that is, as 24 to 30, or 4 to 5. Again; the comet in the month of *April* 1665, as *Hevelius* informs us, excelled almost all the fixed stars in splendor, and even Saturn itself, as being of a much more vivid colour; for this comet was more lucid than that other which had appeared about the end of the

preceding year, and had been compared to the stars of the first magnitude. The diameter of its head was about 6'; but the nucleus, compared with the planets by means of a telescope, was plainly less than Jupiter; and sometimes judged less, sometimes judged equal, to the globe of Saturn within the ring. Since, then, the diameters of the heads of the comets seldom exceed 8' or 12', and the diameter of the nucleus or central star is but about a tenth or perhaps fifteenth part of the diameter of the head, it appears that these stars are generally of about the same apparent magnitude with the planets. But in regard that their light may be often compared with the light of Saturn, yea, and sometimes exceeds it, it is evident that all comets in their perihelions must either be placed below or not far above Saturn; and they are much mistaken who remove them almost as far as the fixed stars; for if it was so, the comets could receive no more light from our sun than our planets do from the fixed stars.

So far we have gone, without considering the obscuration which comets suffer from that plenty of thick smoke which encompasseth their heads, and through which the heads always shew dull, as through a cloud; for by how much the more a body is obscured by this smoke, by so much the more near it must be allowed to come to the sun, that it may vie with the planets in the quantity of light which it reflects. Whence it is probable that the comets descend far below the orb of Saturn, as we proved before from their parallax. But, above all, the thing is evinced from their tails, which must be owing either to the sun's light reflected by a smoke arising from them, and dispersing itself through the aether, or to the light of their own heads. In the former case, we must shorten the distance of the comets, lest we be obliged to allow that the smoke arising from their heads is propagated through such a vast extent of space, and with such a velocity and expansion as will seem altogether incredible; in the latter ease, the whole light of both head and tail is to be ascribed to the central nucleus. But, then, if we suppose all this light to be united and condensed within the disk of the nucleus, certainly the nucleus will by far exceed Jupiter itself in splendor, especially when it emits a very large and lucid tail. If, therefore, under a less apparent diameter, it reflects more light, it must be much more illuminated by the sun, and therefore much nearer to it; and the same argument will bring down the heads of comets sometimes within the orb of Venus, viz., when, being hid under the sun's rays, they emit such huge and splendid tails, like beams of fire, as sometimes they do; for if all that light was supposed to be gathered together into one star, it would sometimes exceed not one Venus only, but a great many such united into one.

Lastly; the same thing is inferred from the light of the heads, which increases in the recess of the comets from the earth towards the sun, and decreases in their return from the sun towards the earth; for so the comet of the year 1665 (by the observations of *Hevelius*), from the time that it was first seen, was always losing of its apparent

motion, and therefore had already passed its perigee; but yet the splendor of its head was daily increasing, till, being hid under the sun's rays, the comet ceased to appear. The comet of the year 1683 (by the observations of the same *Hevelius*), about the end of *July*, when it first appeared, moved at a very slow rate, advancing only about 40 or 45 minutes in its orb in a day's time; but from that time its diurnal motion was continually upon the increase, till *September* 4, when it arose to about 5 degrees; and therefore, in all this interval of time, the comet was approaching to the earth. Which is likewise proved from the diameter of its head, measured with a micrometer: for, *August* 6, *Hevelius* found it only 6' 05", including the coma, which, *September* 2, he observed to be 9' 07", and therefore its head appeared far less about the beginning than towards the end of the motion; though about the beginning, because nearer to the sun, it appeared far more lucid than towards the end, as the same *Hevelius* declares. Wherefore in all this interval of time, on account of its recess from the sun, it decreases in splendor, notwithstanding its access towards the earth. The comet of the year 1618, about the middle of *December*, and that of the year 1680, about the end of the same month, did both move with their greatest velocity, and were therefore then in their perigees; but the greatest splendor of their heads was seen two weeks before, when they had just got clear of the sun's rays; and the greatest splendor of their tails a little more early, when yet nearer to the sun. The head of the former comet (according to the observations of *Cysatus*), *December* 1, appeared greater than the stars of the first magnitude; and, *December* 16 (then in the perigee), it was but little diminished in magnitude, but in the splendor and brightness of its light a great deal. *January* 7, *Kepler*, being uncertain about the head, left off observing. *December* 12, the head of the latter comet was seen and observed by Mr. *Flamsted*, when but 9 degrees distant from the sun; which is scarcely to be done in a star of the third magnitude. *December* 15 and 17, it appeared as a star of the third magnitude, its lustre being diminished by the brightness of the clouds near the setting sun. *December* 26, when it moved with the greatest velocity, being almost in its perigee, it was less than the month of *Pegasus*, a star of the third magnitude. *January* 3, it appeared as a star of the fourth. *January* 9, as one of the fifth. *January* 13, it was hid by the splendor of the moon, then in her increase. *January* 25, it was scarcely equal to the stars of the seventh magnitude. If we compare equal intervals of time on one side and on the other from the perigee, we shall find that the head of the comet, which at both intervals of time was far, but yet equally, removed from the earth, and should have therefore shone with equal splendor, appeared brightest on the side of the perigee towards the sun, and disappeared on the other. Therefore, from the great difference of light in the one situation and in the other, we conclude the great vicinity of the sun and cornet in the former; for the light of comets uses to be regular, and to appear greatest when the heads move fastest, and are therefore in their perigees; excepting in so far as it is increased by their nearness to the sun.

COR. 1. Therefore the comets shine by the sun's light, which they reflect.

COR. 2. From what has been said, we may likewise understand why comets are so frequently seen in that hemisphere in which the sun is, and so seldom in the other. If they were visible in the regions far above Saturn, they would appear more frequently in the parts opposite to the sun; for such as were in those parts would be nearer to the earth, whereas the presence of the sun must obscure and hide those that appear in the hemisphere in which he is. Yet, looking over the history of comets, I find that four or five times more have been seen in the hemisphere towards the sun than in the opposite hemisphere; besides, without doubt, not a few, which have been hid by the light of the sun: for comets descending into our parts neither emit tails nor are so well illuminated by the sun, as to discover themselves to our naked eyes, until they are come nearer to us than Jupiter. But the far greater part of that spherical space, which is described about the sun with so small an interval, lies on that side of the earth which regards the sun; and the comets in that greater part are commonly more strongly illuminated, as being for the most part nearer to the sun.

COR. 3. Hence also it is evident that the celestial spaces are void of resistance; for though the comets are carried in oblique paths, and sometimes contrary to the course of the planets, yet they move every way with the greatest freedom, and preserve their motions for an exceeding long time, even where contrary to the course of the planets. I am out in my judgment if they are not a sort of planets revolving in orbits returning into themselves with a perpetual motion; for, as to what some writers contend, that they are no other than meteors, led into this opinion by the perpetual changes that happen to their heads, it seems to have no foundation; for the heads of comets are encompassed with huge atmospheres, and the lowermost parts of these atmospheres must be the densest; and therefore it is in the clouds only, not in the bodies of the comets themselves, that these changes are seen. Thus the earth, if it was viewed from the planets, would, without all doubt, shine by the light of its cloud and the solid body would scarcely appear through the surrounding clouds. Thus also the belts of Jupiter are formed in the clouds of that planet, for they change their position one to another, and the solid body of Jupiter is hardly to be seen through them; and much more must the bodies of comets be hid under their atmospheres, which are both deeper and thicker.

PROPOSITION XL. THEOREM XX.

THAT THE COMETS MOVE IN SOME OF THE CONIC
SECTIONS, HAVING THEIR FOCI IN THE CENTRE OF
THE SUN; AND BY RADII DRAWN TO THE SUN DESCRIBE
AREAS PROPORTIONAL TO THE TIMES.

This proposition appears from Cor. 1, Prop. XIII, Book I, compared with Prop. VIII, XII, and XIII, Book III.

COR. 1. Hence if comets are revolved in orbits returning into themselves, those orbits will be ellipses; and their periodic times be to the periodic times of the planets in the sesquiplicate proportion of their principal axes. And therefore the comets, which for the most part of their course are higher than the planets, and upon that account describe orbits with greater axes, will require a longer time to finish their revolutions. Thus if the axis of a comet's orbit was four times greater than the axis of the orbit of Saturn, the time of the revolution of the comet would be to the time of the revolution of Saturn, that is, to 30 years, as $4\sqrt{4}$ (or 8) to 1, and would therefore be 240 years.

COR. 2. But their orbits will be so near to parabolas, that parabolas may be used for them without sensible error.

COR. 3. And, therefore, by Cor. 7, Prop. XVI, Book I, the velocity of every comet will always be to the velocity of any planet, supposed to be revolved at the same distance in a circle about the sun, nearly in the subduplicate proportion of double the distance of the planet from the centre of the sun to the distance of the comet from the sun's centre, very nearly. Let us suppose the radius of the *orbis magnus*, or the greatest semi-diameter of the ellipsis which the earth describes, to consist of 100000000 parts; and then the earth by its mean diurnal motion will describe 1720212 of those parts, and $71675^1/_2$, by its horary motion. And therefore the comet, at the same mean distance of the earth from the sun, with a velocity which is to the velocity of the earth as $\sqrt{2}$ to 1, would by its diurnal motion describe 2432747 parts, and $101364^1/_2$ parts by its horary motion. But at greater or less distances both the diurnal and horary motion will be to this diurnal and horary motion in the reciprocal subduplicate proportion of the distances, and is therefore given.

COR. 4. Wherefore if the *latus rectum* of the parabola is quadruple of the radius of the *orbis magnus*, and the square of that radius is supposed to consist of 100000000 parts, the area which the comet will daily describe by a radius drawn to the sun will be $1216373^1/_2$ parts, and the horary area will be $50682^1/_4$ parts. But, if the *latus rectum* is greater or less in any proportion, the diurnal and horary area will be less or greater in the subduplicate of the same proportion reciprocally.

LEMMA V.

TO FIND A CURVE LINE OF THE PARABOLIC KIND WHICH SHALL PASS THROUGH ANY GIVEN NUMBER OF POINTS.

Let those points be A, B, C, D, E, F, &c., and from the same to any right line HN, given in position, let fall as many perpendiculars AH, BI, CK, DL, EM, FN, &c.

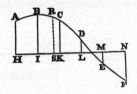

b	$2b$	$3b$	$4b$	$5b$
	c	$2c$	$3c$	$4c$
	d	$2d$	$3d$	
		e	$2e$	
		f		

CASE 1. If HI, IK, KL, &c., the intervals of the points H, I, K, L, M &c., are equal, take b, $2b$, $3b$, $4b$, $5b$, &c., the first differences of the perpendiculars AH, BI, CK, &c.; their second differences c, $2c$, $3c$, $4c$, &c.; their third, d, $2d$, $3d$, &c, that is to say, so as AH – BI may be = b, BI – CK = $2b$, CK – DL = $3b$, DL + EM = $4b$, – EM + FN = $5b$, &c.; then $b - 2b = c$, &c., and so on to the last difference, which is here f. Then, erecting any perpendicular RS, which may be considered as an ordinate of the curve required, in order to find the length of this ordinate, suppose the intervals HI, IK, KL, LK &c, to be units, and let AH = a, - HS = p, $\frac{1}{2}p$ into – IS = q, $\frac{1}{3}q$ into + SK = r, $\frac{1}{4}r$ into + SL = s, $\frac{1}{5}s$ into + SM = t; proceeding, to wit, to ME, the last perpendicular but one, and prefixing negative signs before the terms HS, IS, &c., which lie from S towards A; and affirmative signs before the terms SK, SL, &c., which lie on the other side of the point S; and, observing well the signs, RS will be = $a + bp + rq + dr + es + ft$, + &c.

CASE 2. But if HI, IK, &c., the intervals of the points H, I, K, L, &c., are unequal, take b, $2b$, $3b$, $4b$, $5b$, &c., the first differences of the perpendiculars AH, BI, CK, &c., divided by the intervals between those perpendiculars; c, $2c$, $3c$, $4c$, &c., their second differences, divided by the intervals between every two; d, $2d$, $3d$, &c., their third differences, divided by the intervals between every three; e, $2e$, &c., their fourth differences, divided by the intervals between every four; and so forth; that is, in such manner, that b may be $\frac{AH - BI}{HI}$, $2b = \frac{BI - CK}{IK}$, $3b = \frac{CK - DL}{KL}$, &c., then $c = \frac{b - 2b}{HK}$, $2c = \frac{2b - 3b}{IL}$, $3c = \frac{3b - 4b}{KM}$, &c., then $d = \frac{c - 2c}{HL}$, $2d = \frac{2c - 3c}{IM}$, &c. And those differences being found, let AH be = a, – HS = p, p into – IS = q, q into + SK = r, r into + SL = s, s into + SM = t; proceeding, to wit, to ME, the last perpendicular but one; and the ordinate RS will be = $a + bp + cq + dr + es + ft$, + &c.

COR. Hence the areas of all curves may be nearly found; for if some number of points of the curve to be squared are found, and a parabola be supposed to be drawn through those points, the area of this parabola will be nearly the same with the area of the curvilinear figure proposed to be squared: but the parabola can be always squared geometrically by methods vulgarly known.

LEMMA VI.

CERTAIN OBSERVED PLACES OF A COMET BEING GIVEN, TO FIND THE PLACE OF THE SAME TO ANY INTERMEDIATE GIVEN TIME.

Let HI, IK, KL, LM (in the preceding Fig.), represent the times between the observations; HA, IB, KC, LD, ME, five observed longitudes of the comet; and HS the given time between the first observation and the longitude required. Then if a regular curve ABCDE is supposed to be drawn through the points A, B, C, D, E, and the ordinate RS is found out by the preceding lemma, RS will be the longitude required.

After the same method, from five observed latitudes, we may find the latitude to a given time.

If the differences of the observed longitudes are small, suppose of 4 or 5 degrees, three or four observations will be sufficient to find a new longitude and latitude; but if the differences are greater, as of 10 or 20 degrees, five observations ought to be used.

LEMMA VII.

THROUGH A GIVEN POINT P TO DRAW A RIGHT LINE BC, WHOSE PARTS PB, PC, CUT OFF BY TWO RIGHT LINES AB, AC, GIVEN IN POSITION, MAY BE ONE TO THE OTHER IN A GIVEN PROPORTION.

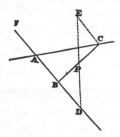

From the given point P suppose any right line PD to be drawn to either of the right lines given, as AB; and produce the same towards AC, the other given right line, as far as r, so as PE may be to PD in the given proportion. Let EC be parallel to AD. Draw CPB, and PC will be to PB as PE to PD.   Q.E.F.

LEMMA VIII.

LET ABC BE A PARABOLA, HAVING ITS FOCUS IN S. BY THE CHORD AC BISECTED IN I CUT OFF THE SEGMENT ABCI, WHOSE DIAMETER IS Iμ AND VERTEX μ. IN Iμ PRODUCED TAKE μO EQUAL TO ONE HALF OF Iμ. JOIN OS, AND PRODUCE IT TO ξ, SO AS Sξ MAY BE EQUAL TO 2SO. NOW,

SUPPOSING A COMET TO REVOLVE IN THE ARC CBA, DRAW
ξB, CUTTING AC IN E; I SAY, THE POINT E WILL CUT OFF
FROM THE CHORD AC THE SEGMENT AE, NEARLY PROPOR-
TIONAL TO THE TIME.

For, if we join EO, cutting the parabolic arc ABC in Y, and draw μX touching the
same arc in the vertex μ, and meeting EO in X, the curvilinear area AEXμA will be to
the curvilinear area ACYμA as AE to AC; and, therefore, since the triangle ASE is to
the triangle ASC in the same proportion, the whole area ASEXμA will be to the whole

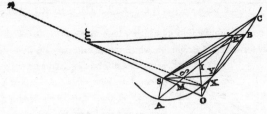

area ASCYμA as AE to AC. But, because ξO is to SO as 3 to 1, and EO to XO in the
same proportion, SK will be parallel to EB; and, therefore, joining BX, the triangle
SEB will be equal to the triangle XEB. Wherefore if to the area ASEXμA we add the
triangle EXB, and from the sum subduct the triangle SEB, there will remain the area
ASBXμA, equal to the area ASEXμA, and therefore in proportion to the area ASCYμA
as AE to AC. But the area ASBYμA is nearly equal to the area ASBXμA; and this area
ASBYμA is to the area ASCYμA as the time of description of the arc AB to the time
of description of the whole arc AC; and, therefore, AE is to AC nearly in the proportion of the times.   Q.E.D.

COR. When the point B falls upon the vertex μ of the parabola, AE is to AC accurately in the proportion of the times.

SCHOLIUM.

If we join μξ cutting AC in δ, and in it take ξn in proportion to μB as 27MI to
16Mμ, and draw Bn, this Bn will cut the chord AC, in the proportion of the times,
more accurately than before; but the point n is to be taken beyond or on this side the
point ξ, according as the point B is more or less distant from the principal vertex of
the parabola than the point μ.

LEMMA IX.

THE RIGHT LINES Iμ AND μM, AND THE LENGTH $\frac{AI^2}{4S\mu}$ ARE
EQUAL AMONG THEMSELVES.

For $4S\mu$ is the *latus rectum* of the parabola belonging to the vertex μ.

LEMMA X.

PRODUCE $S\mu$ TO N AND P, SO AS μN MAY BE ONE THIRD OF μI, AND SP MAY BE TO SN AS SN TO $S\mu$; AND IN THE TIME THAT A COMET WOULD DESCRIBE THE ARC $A\mu$C, IF IT WAS SUPPOSED TO MOVE ALWAYS FORWARDS WITH THE VELOCITY WHICH IT HATH IN A HEIGHT EQUAL TO SP, IT WOULD DESCRIBE A LENGTH EQUAL TO THE CHORD AC.

For if the comet with the velocity which it hath in it was in the said time supposed to move uniformly forward in the right line which touches the parabola in μ, the area which it

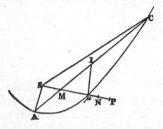

would describe by a radius drawn to the point S would be equal to the parabolic area $ASC\mu A$; and therefore the space contained under the length described in the tangent and the length $S\mu$ would be to the space contained under the lengths AC and SM as the area $ASC\mu A$ to the triangle ASC, that is, as SN to SM. Wherefore AC is to the length described in the tangent as $S\mu$ to SN. But since the velocity of the comet in the height SP (by Cor. 6, Prop. XVI., Book I) is to the velocity of the same in the height $S\mu$ in the reciprocal subduplicate proportion of SP to $S\mu$, that is, in the proportion of $S\mu$ to SN, the length described with this velocity will be to the length in the same time described in the tangent as $S\mu$ to SN, Wherefore since AC, and the length described with this new velocity, are in the same proportion to the length described in the tangent, they must be equal betwixt themselves.   Q.E.D.

COR. Therefore a comet, with that velocity which it both in the height $S\mu + {}^2/_3 I\mu$, would in the same time describe the chord AC nearly.

LEMMA XI.

IF A COMET VOID OF ALL MOTION WAS LET FALL FROM THE HEIGHT SN, OR $S\mu + {}^1/_3 I\mu$, TOWARDS THE SUN, AND WAS STILL IMPELLED TO THE SUN BY THE SAME FORCE UNIFORM-LY CONTINUED BY WHICH IT WAS IMPELLED AT FIRST, THE SAME, IN ONE HALF OF THAT TIME IN WHICH IT MIGHT DESCRIBE THE ARC AC IN ITS OWN ORBIT, WOULD IN DESCENDING DESCRIBE A SPACE EQUAL TO THE LENGTH $I\mu$.

For in the same time that the comet would require to describe the parabolic arc AC, it would (by the last Lemma), with that velocity which it hath in the height SP, describe the chord AC: and, therefore (by Cor. 7, Prop. XVI, Book I), if it was in the

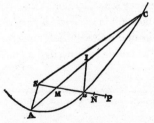

same time supposed to revolve by the force of its own gravity in a circle whose semi-diameter was SP, it would describe an arc of that circle, the length of which would be to the chord of the parabolic arc AC in the subduplicate proportion of 1 to 2. Wherefore if with that weight, which in the height SP it hath towards the sun, it should fall from that height towards the sun, it would

(by Cor. 9, Prop. XVI, Book I) in half the said time describe a space equal to the square of half the said chord applied to quadruple the height SP, that it would describe the space $\frac{AI^2}{4SP}$. But since the weight of the comet towards the sun in the height SN is

to the weight of the same towards the sun in the height SP as SP to Sμ, the comet, by the weight which it both in the height SN, in falling from that height towards the sun, would in the same time describe the space $\frac{AI^2}{4S\mu}$; that is, a space equal to the

length Iμ or μM.   Q.E.D.

PROPOSITION XLI. PROBLEM XXI.

FROM THREE OBSERVATIONS GIVEN TO DETERMINE THE ORBIT OF A COMET MOVING IN A PARABOLA.

This being a Problem of very great difficulty, I tried many methods of resolving it; and several of those Problems, the composition whereof I have given in the first Book, tended to this purpose. But afterwards I contrived the following solution, which is something more simple.

Select three observations distant one from another by intervals of time nearly equal; but let that interval of time in which the comet moves more slowly be somewhat

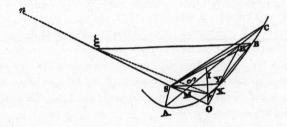

greater than the other; so, to wit, that the difference of the times may be to the sum of the times as the sum of the times to about 600 days; or that the point E may fall upon M nearly, and may err therefrom rather towards I than towards A. If such direct observations are not at hand, a new place of the comet must be found, by Lem. VI.

Let S represent the sun; T, *t*, τ, three places of the earth in the *orbis magnus*; TA, *t*B, τC, three observed longitudes of the comet; V the time between the first observation and the second; W the time between the second and the third, X the length which

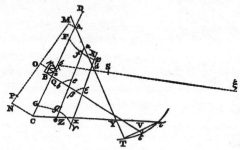

in the whole time V + W the comet might describe with that velocity which it hath in the mean distance of the earth from the sun, which length is to be found by Cor. 3, Prop. XL, Book III; and *t*V a perpendicular upon the chord Tτ. In the mean observed longitude *t*B take at pleasure the point B, for the place of the comet in the plane of the ecliptic; and from thence, towards the sun S, draw the line BE, which may be to the perpendicular *t*V as the content under SB and S*t*² to the cube of the hypothenuse of the right angled triangle, whose sides are SB, and the tangent of the latitude of the comet in the second observation to the radius *t*B. And through the point F, (by Lemma VII) draw the right line AEC, whose parts AE and EC, terminating in the right lines TA and τC, may be one to the other as the times V and W: then A and C will be nearly the places of the comet in the place of the ecliptic in the first and third observations, if B was its place rightly assumed in the second.

Upon AC, bisected in I, erect the perpendicular I*i*. Through B draw the obscure line B*i* parallel to AC. Join the obscure line S*i*, cutting AC in λ, and complete the parallelogram *i*I λμ. Take Iσ equal to 3Iλ; and through the sun S draw the obscure line σξ equal to 3Sσ + 3*i*λ. Then, cancelling the letters A, E, C, I, from the point B towards the point ξ, draw the new obscure line BE, which may be to the former BE in the duplicate proportion of the distance BS to the quantity Sμ + ¹/₃ *i*λ. And through the point E draw again the right line AEC by the same rule as before; that is, so as its parts AE and EC may be one to the other as the times V and W between the observations. Thus A and C will be the places of the comet more accurately.

Upon AC, bisected in I, erect the perpendiculars AM, CN, IO, of which AM and CN may be the tangents of the latitudes in the first and third observations, to the radii TA and τC. Join MN, cutting IO in O. Draw the rectangular parallelogram *i*Iλμ, as before. In IA produced take ID equal to Sμ + $^2/_3$*i*λ. Then in MN, towards N, take MP, which may be to the above found length X in the subduplicate proportion of the mean distance of the earth from the sun (or of the semi-diameter of the *orbis magnus*) to the distance OD. If the point P fall upon the point N; A, B, and C, will be three places of the comet, through which its orbit is to be described in the plane of the ecliptic. But if the point P falls not upon the point N, in the right line AC take CG equal to NP, so as the points G and P may lie on the same side of the line NC.

By the same method as the points E, A, C, G, were found from the assumed point B, from other points *b* and ß assumed at pleasure, find out the new points *e*, *a*, *c*, *g*; and ε, α, κ, γ. Then through G, *g*, and γ, draw the circumference of a circle G*g*γ, cutting the right line τC in Z: and Z will be one place of the comet in the plane of the ecliptic. And in AC, *ac*, ακ, taking AF, *af*, αφ, equal respectively to CG, *cg*, κγ; through the points F, *f*, and φ, draw the circumference of a circle F*f*φ, cutting the right line AT in X; and the point X will be another place of the comet in the plane of the ecliptic. And at the points X and Z, erecting the tangents of the latitudes of the comet to the radii TX and τZ, two places of the comet in its own orbit will be determined. Lastly, if (by Prop. XIX, Book I) to the focus S a parabola is described passing through those two places, this parabola will be the orbit of the comet. Q.E.I.

The demonstration of this construction follows from the preceding Lemmas, because the right line AC is cut in E in the proportion of the times, by Lem. VII., as it ought to be, by Lem. VIII.; and BE, by Lem. XI., is a portion of the right line BS or Bξ in the plane of the ecliptic, intercepted between the arc ABC and the chord AEC; and MP (by Cor. Lem. X.) is the length of the chord of that arc, which the comet should describe in its proper orbit between the first and third observation, and therefore is equal to MN, providing B is a true place of the comet in the plane of the ecliptic.

But it will be convenient to assume the points B, *b*, β, not at random, but nearly true. If the angle AQ*t*, at which the projection of the orbit in the plane of the ecliptic cuts the right line *t*B, is rudely known, at that angle with B*t* draw the obscure line AC, which may be to $^4/_3$Tτ in the subduplicate proportion of SQ to S*t*; and, drawing the right line SEB so as its part EB may be equal to the length V*t*, the point B will be determined, which we are to use for the first time. Then, cancelling the right line AC, and drawing anew AC according to the preceding construction, and, moreover, finding the length MP, in *t*B take the point *b*, by this rule, that, if TA and τC intersect each other in Y, the distance Y*b* may be to the distance YB in a proportion compounded of the

proportion of MP to MN, and the subduplicate proportion of SB to Sb. And by the same method you may find the third point β, if you please to repeat the operation the third time; but if this method is followed, two operations generally will be sufficient; for if the distance Bb happens to be very small, after the points F, f, and G, g, are found, draw the right lines Ff and Gg, and they will out TA and τC in the points required, X and Z.

EXAMPLE.

Let the comet of the year 1680 be proposed. The following table shews the motion thereof, as observed by *Flamsted*, and calculated afterwards by him from his observations, and corrected by Dr. *Halley* from the same observations.

	Time.		Sun's	Comet's	
	Appar.	True	Longitude.	Longitude.	Lat. N.
	h. "	h. ' "	o ' "	o ' "	o ' "
1680, *Dec.* 12	4.46	4.46. 0	♉ 1.51.23	♉ 6.32.30	8.28. 0
21	6.32 1/2	6.36.59	11.06.44	♒ 5.08.12	21.42.13
24	6.12	6.17.52	14.09.26	18.49.23	25.23. 5
26	5.14	5.20.44	16.09.22	28.24.13	27.00.52
29	7.55	8.03.02	19.19.42	♓ 13.10.41	28.09.58
30	8.02	8.10.26	20.21.09	17.38.20	28.11.53
1681, *Jan.* 5	5.51	6.01.38	26.22.18	♈ 8.48.53	26.15. 7
9	6.49	7.00.53	♒ 0.29.02	18.44.04	24.11.56
10	5.54	6.06.10	1.27.43	20.40.50	23.43.52
13	6.56	7.08.55	4.33.20	25.59.48	22.17.28
25	7.44	7.58.42	16.45.36	♉ 9.35. 0	17.51.11
30	8.07	8.21.53	21.49.58	13.19.51	16.42.18
Feb. 2	6.20	6.34.51	24.46.59	15.13.53	16.04. 1
5	6.50	7.04.41	27.49.51	16.59.06	15.27. 3

To these you may add some observations of mine.

	Ap.	Comet's	
	Time	Longitude	Lat. N.
	h. '	o ' "	o ' "
1681, *Feb.* 25	8.30	♉ 26.18.35	12.46.46
27	8.15	27.04.30	12.36.12
Mar. 1	11.0	27.52.42	12.23.40
2	8.0	28.12.48	12.19.38
5	11.30	29.18. 0	12.03.16
7	9.30	♊ 0. 4. 0	11.57. 0
9	8.30	0.43. 4	11.45.52

These observations were made by a telescope of 7 feet, with a micrometer and threads placed in the focus of the telescope; by which instruments we determined the positions both of the fixed stars among themselves, and of the comet in respect of the fixed stars. Let A represent the star of the fourth magnitude in the left heel of *Perseus* (*Bayer*'s o), B the following star of the third magnitude in the left foot (*Bayer*'s ζ), C a star of the sixth magnitude (*Bayer*'s n) in the heel of the same foot, and D, E, F, G, H, I, K, L, M, N, O, Z, α, β, γ, δ, other smaller stars in the same foot; and let p, P, Q, R, S, T, V, X, represent the places of the comet in the observations above set down; and, reckoning the distance AB of $80^7/_{12}$ parts, AC was $52^1/_4$ of those parts; BC, $58^5/_6$; AD, $57^5/_{12}$; BD, $82^6/_{11}$; CD, $23\ ^2/_3$; AE, $29^4/_7$; CE, $57^1/_2$; DE, $49^{11}/_{12}$; AI, $27^7/_{12}$; BI, $52^1/_6$; CI, $36^7/_{12}$; DI, $53^5/_{11}$; AK, $38^2/_3$; BK, 43; CK, $31^5/_9$; FK, 29; FB, 23; FC, $36^1/_4$; AH, $18^6/_7$; DH, $50^7/_8$; BN, $46^5/_{12}$; CN, $31^1/_3$; BL, $45^5/_{12}$; NL, $31^5/_7$. HO was to HI as 7 to 6, and, produced, did pass between the stars D and E, so as the distance of the star D from this right line was $1/_6$CD. LM was to LN as 2 to 9, and, produced, did pass through the star H. Thus were the positions of the fixed stars determined in respect of one another.

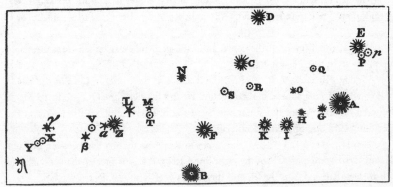

Mr. *Pound* has since observed a second time the positions of these fixed stars amongst themselves, and collected their longitudes and latitudes according to the following table.

The fixed stars.	Their Longitudes		Latitude North.	The fixed stars.	Their Longitudes		Latitude North.
		° ′ ″				° ′ ″	
A	♉	26.41.50	12. 8.36	L	♉	29.33.34	12. 7.48
B		28.40.23	11.17.54	M		29.18.51	12. 7.20
C		27.58.30	12.40.25	N		28.48.29	12.31. 9
E		26.27.17	12.52. 7	Z		29.44.48	11.57.13
F		28,28.37	11.52.22	a		29.52. 3	11.55.48
G		26.56. 8	12. 4.58	ß	♊	0. 8.23	11.48.56
H		27.11.45	12. 2. 1	?		0.40.10	11.55.18
I		27.25. 2	11.53.11	d		1. 3.20	11.30.42
K		27.42. 7	11.53.26				

The positions of the comet to these fixed stars were observed to be as follow:

Friday, *February* 25, O.S. at $8^1/_2$^h. P.M. the distance of the comet in p from the star E was less than $^3/_{13}$AE, and greater than $^1/_5$AE, and therefore nearly equal to $^3/_{14}$AE; and the angle ApE was a little obtuse, but almost right. For from A, letting fall a perpendicular on pE, the distance of the comet from that perpendicular was $^1/_5 p$E.

The same night, at $9^1/_2$^h., the distance of the comet in P from the star E was greater than $\frac{1}{4\frac{1}{2}}$AE, and less than $\frac{1}{5\frac{1}{4}}$AE, and therefore nearly equal to $\frac{1}{4\frac{2}{8}}$ of AE, or

$^8/_{39}$AE. But the distance of the comet from the perpendicular let fall from the star A upon the right line PE was $^4/_5$PE.

Sunday, *February* 27, $8^1/_4$^h. P.M. the distance of the comet in Q from the star O was equal to the distance of the stars O and H; and the right line QO produced passed between the stars K and B. I could not, by reason of intervening clouds, determine the position of the star to greater accuracy.

Tuesday, *March* 1, 11^h. P.M. the comet in R lay exactly in a line between the stars K and C, so as the part CR of the right line CRK was a little greater than $^1/_3$CK, and a little less than $^1/_3$CK + $^1/_8$CR, and therefore = $^1/_3$CK + $^1/_{16}$CR, or $^{16}/_{45}$CK.

Wednesday, *March* 2, 8^h. P.M. the distance of the comet in S from the star C was nearly $^4/_9$FC; the distance of the star F from the right line CS produced was $^1/_{24}$FC; and the distance of the star B from the same right line was five times greater than the distance of the star F; and the right line NS produced passed between the stars H and I five or six times nearer to the star H than to the star I.

Saturday, *March* 5, $11^1/_2$h. P.M. when the comet was in T, the right line MT was equal to $^1/_2$ML, and the right line LT produced passed between B and F four or five times nearer to F than to B, cutting off from BF a fifth or sixth part thereof towards F: and MT produced passed on the outside of the space BF towards the star B four times nearer to the star B than to the star F. M was a very small

star, scarcely to be seen by the telescope; but the star L was greater, and of about the eighth magnitude.

Monday, March 7, $9^1/_2$h. P.M. the comet being in V, the right line Vα produced did pass between B and F, cutting off, from BF towards F, $^1/_{10}$ of BF, and was to the right line Vβ as 5 to 4. And the distance of the comet from the right line αβ was $^1/_2$ Vβ.

Wednesday, March 9, $8^1/_2$h. P.M. the comet being in X, the right line γX was equal to $^1/_4$γδ; and the perpendicular let fall from the star δ upon the right γX was $^2/_3$ of γδ.

The same night, at 12^h. the comet being in Y, the right line γY was equal to $^1/_3$ of γδ, or a little less, as perhaps $^5/_{16}$ of γδ; and a perpendicular let fall from the star δ on the right line γY was equal to about $^1/_6$ or $^1/_7$γδ. But the comet being then extremely near the horizon, was scarcely discernible, and therefore its place could not be determined with that certainty as in the foregoing observations.

From these observations, by constructions of figures and calculations, I deduced the longitudes and latitudes of the comet; and Mr. Pound, by correcting the places of the fixed stars, hath determined more correctly the places of the comet, which correct places are set down above. Though my micrometer was none of the best, yet the errors in longitude and latitude (as derived from my observations) scarcely exceed one minute. The comet (according to my observations), about the end of its motion, began to decline sensibly towards the north, from the parallel which it described about the end of *February*.

Now, in order to determine the orbit of the comet out of the observations above described, I selected those three which *Flamsted* made, *Dec.* 21, *Jan.* 5, and *Jan.* 25; from which I found S*t* of 9842,1 parts, and V*t* of 455, such as the semi-diameter of the *orbis magnus* contains 10000. Then for the first observation, assuming *t*B of 5657 of those parts, I found SB 9747, BE for the first time 412, Sμ 9503, *i*λ 413, BE for the second time 421, OD 10186, X 8528,4, PM 8450, MN 8475, NP 25; from whence, by the second operation, I collected the distance *tb* 5640; and by this operation I at last deduced the distances TX 4775 and τZ 11322. From which, limiting the orbit, I found its descending node in ♋, and ascending node in ♉ 1° 53'; the inclination of its plane to the plane of the ecliptic 61° 20 $^1/_3$, the vertex thereof (or the perihelion of the comet) distant from the node 8° 39', and in ♐ 27° 43', with latitude 7° 34' south; its *latus rectum* 236,8; and the diurnal area described by a radius drawn to the sun 93585, supposing the square of the semi-diameter of the *orbis magnus* 100000000; that the comet in this orbit moved directly according to the order of the signs, and on *Dec.* 8d. 00h. 04' P.M. was in the vertex or perihelion of its orbit. All which I determined by scale and compass, and the chords of angles, taken from the table of natural sines, in a pretty large figure, in which, to wit, the radius of the *orbis magnus* (consisting of 10000 parts) was equal to $16^1/_3$ inches of an English foot.

Lastly, in order to discover whether the comet did truly move in the orbit so determined, I investigated its places in this orbit partly by arithmetical operations, and partly by scale and compass, to the times of some of the observations, as may be seen in the following table:—

	Dist. from sun.	Longitude computed.	Latitude computed.	Longitude observed.	Latitude observed.	Dif. Lo.	Dif. Lat.
Dec. 12	2792	♉ 6°.32'	8°18¹/₂	♉ 6°.31¹/₂	8°.26	+ 1	− 7¹/₂
29	8403	♓ 13.13²/₅	28.00	♓ 13.11³/₄	28.10¹/₁₂	+ 2	− 10 ¹/₁₂
Feb. 5	16669	♉ 17.00	15.29²/₃	♉ 16.59⁷/₈	15.27²/₅	+ 0	+ 2¹/₄
Mar. 5	21737	29.19³/₄	12. 4	29.20⁶/₇	12.3 ¹/₂	- 1	+ ¹/₂

But afterwards Dr. *Halley* did determine the orbit to a greater accuracy by an arithmetical calculus than could be done by linear descriptions; and, retaining the place of the nodes in ♋ and ♑ 1° 53', and the inclination of the plane of the orbit to the ecliptic 61° 20¹/₃', as well as the time of the comet's being *in perihelio*, *Dec.* 8ᵈ. 00ʰ. 04', he found the distance of the perihelion from the ascending node measured in the comet's orbit 9° 20', and the *latus rectum* of the parabola 2430 parts, supposing the mean distance of the sun from the earth to be 100000 parts; and from these data, by an accurate arithmetical calculus, he computed the places of the comet to the times of the observations as follows:—

True Time.		Dist. from the sun.	Longitude computed		Latitude computed	Errors in Long.	Lat.
	d. h. ' "			o ' "	o ' "	' "	' "
Dec.	12. 4. 46.	28028	♉	6.29.25	8.26. 0 bor.	− 3. 5	− 2. 0
	21. 6. 37.	61076	♒	5. 6.30	21.43.20	− 1.42	+ 1. 7
	24. 6. 18.	70008		18.48.20	25.22.40	− 1.3	− 0.25
	26. 5. 20.	75576		28.22.45	27. 1.36	− 1.28	+ 0.44
	29. 8. 3.	84021	♓	13.12.40	28.10.10	+ 1.59	+ 0.12
	30. 8. 10.	86661		17.40. 5	28.11.20	+ 1.45	− 0.33
Jan.	5. 6. 1.¹/₂	101440	♈	8.49.49	26.15.15	+ 0.56	+ 0. 8
	9. 7. 0.	110959		18.44.36	24.12.54	+ 0.32	+ 0.58
	10. 6. 6.	113162		20.41. 0	23.44.10	+ 0.10	+ 0.18
	13. 7. 9.	120000		26. 0.21	22.17.30	+ 0.33	+ 0. 2
	25. 7. 59.	145370	♉	9.33.40	17.57.55	− 1.10	+ 1.25
	30. 8. 22.	155303		13.17.41	16.42. 7	− 2.10	− 0.11
Feb.	2. 6. 35	160951		15.11.11	16. 4.15	− 2.42	+ 0.14
	5. 7. 4.¹/₂	166686		16.58.55	15.29.13	− 0.41	+ 2. 0
	25. 8. 4	202570		26.15.46	12.48. 0	− 2.49	+ 1.10
Mar.	5. 11. 3	216205		29.18.35	12. 5.40	+ 0.35	+ 2.14

This comet also appeared in the *November* before, and at *Coburg*, in Saxony, was observed by Mr. *Gottfried Kirch*, on the 4th of that month, on the 6th and 11th O.S.; from its positions to the nearest fixed stars observed with sufficient accuracy, sometimes with a two feet, and sometimes with a ten feet telescope; from the difference of longitudes of *Coburg* and *London*, 11°; and from the plates of the fixed stairs observed by Mr. *Pound*, Dr. *Halley* has determined the places of the comet as follows:—

Nov. 3, 17^h. 2', apparent time at *London*, the comet was in a ♌ deg. 51', with 1 deg. 17' 45" latitude north.

November 5, 15^h. 58' the comet was in ♍ 3° 23', with 1° 6' north lat.

November 10, 16^h. 31', the comet was equally distant from two stars in ♌, which are σ and τ in *Bayer*; but it had not quite touched the right line that joins them, but was very little distant from it. In *Flamsted*'s catalogue this star σ was then in ♍ 14° 15', with 1 deg. 41' lat. north nearly, and τ in ♍ 17° 3$^1/_2$' with 0 deg. 34' lat. south; and the middle point between those stars was ♍ 15° 39$^1/_2$', with 0° 33$^1/_2$' lat. north. Let the distance of the comet from that right line be about 10' or 12'; and the difference of the longitude of the comet and that middle point will be 7'; and the difference of the latitude nearly 7$^1/_2$'; and thence it follows that the comet was in ♍ 15° 32', with about 26' lat. north.

The first observation from the position of the comet with respect to certain small fixed stars had all the exactness that could be desired; the second also was accurate enough. In the third observation, which was the least accurate, there might be an error of 6 or 7 minutes, but hardly greater. The longitude of the comet, as found in the first and most accurate observation, being computed in the aforesaid parabolic orbit, comes out ♌ 29° 30' 22", its latitude north 1° 25' 7", and its distance from the sun 115546.

Moreover, Dr. *Halley*, observing that a remarkable comet had appeared four times at equal intervals of 575 years (that is, in the mouth of *September* after *Julius Caesar* was killed; *An. Chr.* 531, in the consulate of *Lampadius* and *Orestes*; *An. Chr.* 1106, in the month of *February*; and at the end of the year 1680; and that with a long and remarkable tail, except when it was seen after *Caesar*'s death, at which time, by reason of the inconvenient situation of the earth, the tail was not so conspicuous), set himself to find out an elliptic orbit whose greater axis should be 1382957 parts, the mean distance of the earth from the sun containing 10000 such; in which orbit a comet might revolve in 575 years; and, placing the ascending node in ♋ 2° 2', the inclination of the plane of the orbit to the plane of the ecliptic in an angle of 61° 6' 48", the perihelion of the comet in this plane in ♐ 22° 44' 25", the equal time of the perihelion *December* 7^d. 23^h. 9', the distance of the perihelion from the ascending node in the plane of the ecliptic 9° 17' 35", and its conjugate axis 18481,2, he computed the motions of the

comet in this elliptic orbit. The places of the comet, as deduced from the observations, and as arising from computation made in this orbit, may be seen in the following table.

True time.		Longitude observed	Latitude North obs.	Longitude comp.	Latitude computed	Errors in Long.	Lat.
	d. h. '	*o ' "*	*o ' "*	*o ' "*	*o ' "*	*' "*	*' '*
Nov.	3.16.47	♌ 29.51. 0	1.17.45	♌ 29.51.22	1.17.32 N	+ 0.22	− 0.13
	5.15.37	♍ 3.23. 0	1. 6. 0	♍ 3.24.32	1. 6. 9	+ 1.32	+ 0. 9
	10.16.18	15.32. 0	0.27. 0	15.33. 2	0.25. 7	+ 1. 2	−1.53
	16.17.00			♎ 8.16.45	0.53. 7 S		
	18.21.34			18.52.15	1.26.54		
	20.17. 0			28.10.36	1.53.35		
	23.17. 5			♏ 13.22.42	2.29. 0		
Dec.	12. 4.46	♉ 6.32.30	8.28. 0	♉ 6.31.20	8.29. 6 N	− 1.10	+ 1. 6
	21. 6.37	♒ 5. 8.12	21.42.13	♒ 5. 6.14	91.44.42	− 1.58	+ 2.29
	24. 6.18	18.49.23	25.23. 5	18.47.30	25.23.35	− 1.53	+ 0.30
	26. 5.21	28.24.13	27. 0.52	28.21.42	27. 2. 1	− 2.31	+ 1. 9
	29. 8. 3	♓ 13.10.41	28. 9.58	♓ 13.11.14	28.10.38	+ 0.33	+ 0.40
	30. 8.10	17.38. 0	28.11.53	17.38.27	28.11.37	+ 0. 7	− 0.16
Jan.	5. 6. 1½	♈ 8.48.53	26.15. 7	♈ 8.48.51	26.14.57	− 0. 2	− 0.10
	9. 7. 1	18.44. 4	24.11.56	18.43.51	24.12.17	− 0.13	+ 0.21
	10. 6. 6	20.40. 5	23.43.32	20.40.23	23.43.25	− 0.27	− 0. 7
	13. 7. 9	25.59.48	22.17.28	26. 0. 8	22.16.32	+ 0.20	− 0.56
	25. 7.59	♉ 9.35. 0	17.56.30	♉ 9.34.11	17.56. 6	− 0.49	− 0.24
	30. 8.22	13.19.51	16.42.18	13.18.28	16.40. 5	− 1.23	− 2.13
Feb.	2. 6.35	15.13.53	16. 4. 1	15.11.59	16. 2.17	− 1.54	− 1.54
	5. 7.4½	16.59. 6	15.27. 3	16.59.17	15.27. 0	+ 0.11	− 0. 3
	25. 8.41	26.18.35	12.46.46	26.16.59	12.45.22	− 1.36	− 1.24
Mar.	1.11.10	27.52.42	12.23.40	27.51.47	12.22.28	− 0.55	− 1.12
	5.11.39	29.18. 0	12. 3.16	29.20.11	12. 2.50	+ 2.11	− 0.26
	9. 8.38	♊ 0.43. 4	11.45.52	♊ 0.42.43	11.45.35	− 0.21	−0.17

The observations of this comet from the beginning to the end agree as perfectly with the motion of the comet in the orbit just now described as the motions of the planets do with the theories from whence they are calculated; and by this agreement plainly evince that it was one and the same comet that appeared all that time, and also that the orbit of that comet is here rightly defined.

In the foregoing table we have emitted the observations of *Nov.* 16, 18, 20, and 23, as not sufficiently accurate, for at those times several persons had observed the comet. *Nov.* 17, O.S. *Ponthaeus* and his companions, at 6ʰ. in the morning at *Rome* (that is, 5ʰ. 10' at *London*), by threads directed to the fixed stars, observed the comet in ♎ 8° 30', with latitude 0° 40' south. Their observations may be seen in a treatise which *Ponthaeus* published concerning this comet. *Cellius*, who was present, and communicated his observations in a letter to *Cassini* saw the comet at the same hour in ♎ 8° 30', with latitude 0° 30' south. It was likewise seen by *Galletius* at the

same hour at *Avignon* (that is, at 5^h. 42' morning at *London*) in ♎ 8° 30' 8° without latitude. But by the theory the comet was at that time in ♎ 8° 16' 45", and its latitude was 0° 53' 7" south.

Nov. 18, at 6^h. 30' in the morning at *Rome* (that is, at 5^h. 40' at *London*), *Ponthæus* observed the comet in ♎ 13° 30', with latitude 1° 20' south; and *Cellius* in ♎ 13° 60' with latitude 1° 00' south. But at 5^h. 30' in the morning at *Avignon*, *Galletius* saw it in ♎ 13° 00', with latitude 1° 00' south. In the University of *La Fleche*, in *France*, at 5^h. in the morning (that is, at 5^h. 9' at *London*), it was seen by *P. Ango*, in the middle between two small stars, one of which is the middle of the three which lie in a right line in the southern hand of Virgo, *Bayer*'s ψ; and the other is the outmost of the wing, *Bayer*'s θ. Whence the comet was then in ♎ 12° 46' with latitude 50' south. And I was informed by Dr. *Halley*, that on the same day at *Boston* in *New England*, in the latitude of 42^1/$_2$ deg. at 5^h. in the morning (that is, at 9^h. 44' in the morning at *London*), the comet was seen near ♎ 14°, with latitude 1° 30' south.

Nov. 19, at 4^1/$_2$^{h}. at *Cambridge*, the comet (by the observation of a young man) was distant from *Spica* ♍ about 2° towards the north west. Now the spike was at that time in ♎ 19° 23' 47", with latitude 2° 1' 59" south. The same day, at 5^h. in the morning at *Boston* in *New England*, the comet was distant from *Spica* ♍ 1°, with the difference of 40' in latitude. The same day, in the island of *Jamaica*, it was about 1° distant from *Spica* ♍. The same day, Mr. *Arthur Storer*, at the river *Patuxent*, near *Hunting Creek*, in *Maryland*, in the confines of *Virginia*, in lat. 38^1/$_2$ ° at 5 in the morning (that is, at 10^h. at *London*), saw the comet above *Spica* ♍, and very nearly joined with it, the distance between them being about ^{3}/$_4$ of one deg. And from these observations compared, I conclude, that at 9^h. 44' at *London* the comet was in ♎ 18° 50', with about 1° 25' latitude south. Now by the theory the comet was at that time in ♎ 18° 52' 15", with 1° 26' 54" lat. south.

Nov. 20, *Montenari*, professor of astronomy at *Padua*, at 6^h. in the morning at *Venice* (that is, 5^h. 10' at *London*), saw the comet in ♎ 23°, with latitude 1° 30' south. The same day, at *Boston*, it was distant from *Spica* ♍ by about 4° of longitude east, and therefore was in ♎ 23° 24' nearly.

Nov. 21, *Ponthæus* and his companions, at 7^1/$_4$^{h}. in the morning, observed the comet in ♎ 27° 50', with latitude 1° 16' south; *Cellius*, in ♎ 28°; *P. Ango* at 5^h. in the morning, in ♎ 27° 45'; *Montenari* in ♎ 27° 51'. The same day, in the island of *Jamaica*, it was seen near the beginning of ♏, and of about the same latitude with *Spica* ♍, that is, 2° 2'. The same day, at 5^h. morning, at *Ballasore*, in the *East Indies* (that is, at 11^h. 20' of the night preceding at *London*), the distance of the comet from ♍ was taken 7° 35' to the east. It was in a right line between the spike and the balance, and therefore was then in ♎ 26° 58', with about 1° 11' lat. south; and after 5^h. 40' (that is,

at 5ʰ. morning at *London*), it was in ♎ 28° 12'. with 1° 16' lat. south. Now by the theory the comet was then in ♎ 28° 10' 36", with 1° 53' 35" lat. south.

Nov. 22, the comet was seen by *Montenari* in ♍ 2° 33'; but at *Boston* in *New England*, it was found in about ♍ 3°, and with almost the same latitude as before, that is, 1° 30'. The same day, at 5ʰ. morning at *Ballasore*, the comet was observed in ♍ 1° 50'; and therefore at 5ʰ. morning at *London*, the comet was in ♍ 3° 5' nearly. The same day, at 6¹/₂ʰ. in the morning at *London*, Dr. *Hook* observed it in about ♍ 3° 30', and that in the right line which passeth through *Spica* ♍ and *Cor Leonis*; not, indeed, exactly, but deviating a little from that line towards the north. *Montenari* likewise observed, that this day, and some days after, a right line drawn from the comet through *Spica* passed by the south side of *Cor Leonis* at a very small distance therefrom. The right line through *Cor Leonis* and *Spica* ♍ did cut the ecliptic in ♍ 3° 46' at an angle of 2° 51'; and if the comet had been in this line and in ♍ 3°, its latitude would have been 2° 26'; but since *Hook* and *Montenari* agree that the comet was at some small distance from this line towards the north, its latitude must have been something less. On the 20th, by the observation of *Montenari*, its latitude was almost the same with that of *Spica* ♍, that is, about 1° 30'. But by the agreement of *Hook*, *Montenari*, and *Ango*, the latitude was continually increasing, and therefore must now, on the 22d, be sensibly greater than 1° 30'; and, taking a mean between the extreme limits but now stated, 2° 26' and 1° 30', the latitude will be about 1° 58'. *Hook* and *Montenari* agree that the tail of the comet was directed towards *Spica* ♍, declining a little from that star towards the south according to *Hook*; but towards the north according to *Montenari*; and, therefore, that declination was scarcely sensible; and the tail, lying nearly parallel to the equator, deviated a little from the opposition of the sun towards the north.

Nov. 23, O.S. at 5ʰ. morning, at Nuremberg (that is, at 4¹/₂ʰ. at *London*.), Mr. *Zimmerman*, saw the comet in ♍ 8° 8', with 2° 31' south lat. its place being collected by taking its distances from fixed stars.

Nov. 24, before sun-rising, the comet was seen by *Montenari* in ♍ 12° 52' on the north side of the right line through *Cor Leonis* and *Spica* ♍, and therefore its latitude was something less than 2° 38'; and since the latitude, as we said, by the concurring observations of *Montenari*, *Ango*, and *Hook*, was continually increasing; therefore, it was now, on the 24th, something greater than 1° 58'; and, taking the mean quantity, may be reckoned 2° 18', without any considerable error. *Ponthaeus* and *Galletius* will have it that the latitude was now decreasing; and *Cellius*, and the observer in *New England*, that it continued the same, viz., of about 1°, or 1¹/₂°. The observations of *Ponthæus* and *Cellius* are more rude, especially those which were made by taking the azimuths and altitudes; as are also the observations of *Galletius*. Those are better which were made by taking the position of the comet to the fixed stars by *Montenari*, *Hook*,

Ango, and the observer in *New England*, and sometimes by *Ponthaeus* and *Cellius*. The same day, at 5ʰ. morning, at *Ballasore*, the comet was observed in ♍ 11° 45'; and, therefore, at 5ʰ. morning, at *London*, was in ♍ 13° nearly. And, by the theory, the comet was at that time in ♍ 13° 22' 42".

Nov. 25, before sunrise, *Montenari* observed the comet in ♍ 17³/₄ nearly; and *Cellius* observed at the same time that the comet was in a right line between the bright star in the right thigh of Virgo and the southern scale of Libra; and this right line cuts the comet's way in ♍ 18° 36'. And, by the theory, the comet was in ♍ 18¹/₃° nearly.

From all this it is plain that these observations agree with the theory, so far as they agree with one another; and by this agreement it is made clear that it was one and the same comet that appeared all the time from *Nov.* 4 to *Mar.* 9. The path of this comet did twice cut the plane of the ecliptic, and therefore was not a right line. It did cut the ecliptic not in opposite parts of the heavens, but in the end of Virgo and beginning of Capricorn, including an arc of about 98°; and therefore the way of the comet did very much deviate from the path of a great circle; for in the month of *Nov.* it declined at least 3° from the ecliptic towards the south; and in the month of *Dec.* following it declined 29° from the ecliptic towards the north; the two parts of the orbit in which the comet descended towards the sun, and ascended again from the sun, declining one from the other by an apparent angle of above 30°, as observed by *Montenari*. This comet travelled over 9 signs, to wit, from the last deg. of ♌ to the beginning of ♐, beside the sign of ♌, through which it passed before it began to be seen; and there is no other theory by which a comet can go over so great a part of the heavens with a regular motion. The motion of this comet was very unequable; for about the 20th of *Nov.* it described about 5° a day. Then its motion being retarded between *Nov.* 26 and *Dec.* 12, to wit, in the space of 15¹/₂ days, it described only 40°. But the motion thereof being afterwards accelerated, it described near 5° a day, till its motion began to be again retarded. And the theory which justly corresponds with a motion so unequable, and through so great a part of the heavens, which observes the same laws with the theory of the planets, and which accurately agrees with accurate astronomical observations, cannot be otherwise than true.

And, thinking it would not be improper, I have given a true representation of the orbit which this comet described, and of the tail which it emitted in several places, in the annexed figure; protracted in the plane of the trajectory. In this scheme ABC represents the trajectory of the comet, D the sun DE the axis of the trajectory, DF the line of the nodes, GH the intersection of the sphere of the *orbis magnus* with the plane of the trajectory, I the place of the comet *Nov.* 4, *Ann.* 1680; K the place of the same *Nov.* 11; L the place of the same *Nov.* 19; M its place *Dec.* 12; N its place *Dec.* 21; O its

place *Dec.* 29; P its place *Jan.* 5 following; Q its place *Jan.* 25; R its place *Feb.* 5; S its place *Feb.* 25; T its place *March* 5; and V its place *March* 9. In determining the length of the tail, I made the following observations.

Nov. 4 and 6, the tail did not appear; *Nov.* 11, the tail just begun to shew itself, but did not appear above $1/2$ deg. long through a 10 feet telescope; *Nov.* 17, the tail was seen by *Ponthaeus* more than 15° long; *Nov.* 18, in *New-England*, the tail appeared 30° long, and directly opposite to the sun, extending itself to the planet Mars, which was then in ♍, 9° 54'; *Nov.* 19. in *Maryland*, the tail was found 15° or 20° long; *Dec.* 10 (by the observation of Mr. *Flamsted*), the tail passed through the middle of the distance intercepted between the tail of the Serpent of *Ophiuchus* and the star δ in the south wing of *Aquila*, and did terminate near the stars A, ω, *b*, in *Bayer*'s tables. Therefore the end of the tail was in ♉ $19^1/_2$°, with latitude about $34^1/_4$° north; *Dec.* 11, it ascended to the head of *Sagitta* (*Bayer*'s α, β), terminating in ♉ 26° 43', with latitude 38° 34' north; *Dec,* 12, it passed through the middle of *Sagitta*, nor did it reach much farther; terminating in ♒ 4°, with latitude $42^1/_2$° north nearly. But these things are to be understood of the length of the brighter part of the tail; for with a more faint light, observed, too, perhaps, in a serener sky, at *Rome, Dec.* 12, 5^h. 40', by the observation of *Ponthaeus*, the tail arose to 10° above the rump of the Swan, and the side thereof towards the west and towards, the north was 45' distant from this star. But about that time the tail was 3° broad towards the upper end; and therefore the middle thereof was 2° 15' distant from that star towards the south, and the upper end was ♓ in 22°, with latitude 61° north; and thence the tail was about 70° long; *Dec.* 21, it extended almost to *Cassiopeia*'s chair, equally distant from β and from *Schedir*, so as its distance from either of the two was equal to the distance of the one from the other, and therefore did terminate in ♈ 24°, with latitude $47^1/_2$°; *Dec.* 29, it reached to a contact with *Scheat* on its left, and exactly filled up the space between the two stars in the northern foot of *Andromeda*, being 54° in length; and therefore terminated in ♉ 19°, with 35° of latitude; *Jan.* 5, it touched the star p in the breast of *Andromeda* on its right side, and the star (of the girdle on its left; and, according to our observations, was 40° long; but it was curved, and the convex side thereof lay to the south; and near the head of the comet it made an angle of 4° with the circle which passed through the sun and the comet's head; but towards the other end it was inclined to that circle in an angle of about 10° or 11°; and the chord of the tail contained with that circle an angle of 8°. *Jan.* 13, the tail terminated between *Alamech* and *Algol*, with a light that was sensible enough; but with a faint light it ended over against the star κ in *Perseus*'s side. The distance of the end of the tail from the circle passing through the sun and the comet was 3° 50'; and the inclination of the chord of the tail to that circle was $8^1/_2$°. *Jan.* 25 and 26, it shone with a faint light to the length, of 6° or 7°; and for a night or two after,

when there was a very clear sky, it extended to the length of 12°, or something more, with a light that was very faint and very hardly to be seen; but the axis thereof was exactly directed to the bright star in the eastern shoulder of Auriga, and therefore deviated from the opposition of the sun towards the north by an angle of 10°. Lastly, *Feb.* 10, with a telescope I observed the tail 2° long; for that fainter light which I spoke of did not appear through the glasses. But *Ponthæus* writes, that, on *Feb.* 7, he saw the tail 12° long. *Feb.* 25, the comet was without a tail, and so continued till it disappeared.

Now if one reflects upon the orbit described, and duly considers the other appearances of this comet, he will be easily satisfied that the bodies of comets are solid, compact, fixed, and durable, like the bodies of the planets; for if they were nothing else but the vapours or exhalations of the earth, of the sun, and other planets, this comet, in its passage by the neighbourhood of the sun, would have been immediately dissipated; for the heat of the sun is as the density of its rays, that is, reciprocally as the square of the distance of the places from the sun. Therefore, since on *Dec.* 8, when the comet was in its perihelion, the distance thereof from the centre of the sun was to the distance of the earth from the same as about 6 to 1000, the sun's heat on the comet was at that time to the heat of the summer sun with us as 1000000 to 36, or as 28000 to 1. But the heat of boiling water is about 3 times greater than the heat which dry earth acquires from the summer-sun, as I have tried; and the heat of red-hot iron (if my conjecture is right) is about three or four times greater than the heat of boiling water. And therefore the heat which dry earth on the comet, while in its perihelion, might have conceived from the rays of the sun, was about 2000 times greater than the heat of red-hot iron. But by so fierce a heat, vapours and exhalations, and every volatile matter, must have been immediately consumed and dissipated.

This comet, therefore, must have conceived an immense heat from the sun, and retained that heat for an exceeding long time; for a globe of iron of an inch in diameter, exposed red-hot to the open air, will scarcely lose all its heat in an hour's time; but a greater globe would retain its heat longer in the proportion of its diameter, because the surface (in proportion to which it is cooled by the contact of the ambient air) is in that proportion less in respect of the quantity of the included hot matter; and therefore a globe of red hot iron equal to our earth, that is, about 40000000 feet in diameter, would scarcely cool in an equal number of days, or in above 50000 years. But I suspect that the duration of heat may, on account of some latent causes, increase in a yet less proportion than that of the diameter; and I should be glad that the true proportion was investigated by experiments.

It is farther to be observed, that the comet in the month of *December*, just after it had been heated by the sun, did emit a much longer tail, and much more splendid, than in the month of *November* before, when it had not yet arrived at its perihelion;

and, universally, the greatest and most fulgent tails always arise from comets immediately after their passing by the neighbourhood of the sun. Therefore the heat received by the comet conduces to the greatness of the tail: from whence, I think I may infer, that the tail is nothing else but a very fine vapour, which the head or nucleus of the comet emits by its heat.

But we have had three several opinions about the tails of comets; for some will have it that they are nothing else but the beams of the sun's light transmitted through the comets' heads, which they suppose to be transparent; others, that they proceed from the refraction which light suffers in passing from the comet's head to the earth: and, lastly, others, that they are a sort of clouds or vapour constantly rising from the comets' heads, and tending towards the parts opposite to the sun. The first is the opinion of such as are yet unacquainted with optics; for the beams of the sun are seen in a darkened room only in consequence of the light that is reflected, from them by the little particles of dust and smoke which are always flying about in the air; and, for that reason, in air impregnated with thick smoke, those beams appear with great brightness, and move the sense vigorously; in a yet finer air they appear more faint, and are less easily discerned; but in the heavens, where there is no matter to reflect the light, they can never be seen at all. Light is not seen as it is in the beam, but as it is thence reflected to our eyes; for vision can be no otherwise produced than by rays falling upon the eyes; and, therefore, there must be some reflecting matter in those parts where the tails of the comets are seen: for otherwise, since all the celestial spaces are equally illuminated by the sun's light, no part of the heavens could appear with more splendor than another. The second opinion is liable to many difficulties. The tails of comets are never seen variegated with those colours which commonly are inseparable from refraction; and the distinct transmission of the light of the fixed stars and planets to us is a demonstration that the aether or celestial medium is not endowed with any refractive power: for as to what is alleged, that the fixed stars have been sometimes seen by the Egyptians environed with a *Coma* or *Capitlitium*, because that has but rarely happened, it is rather to be ascribed to a casual refraction of clouds; and so the radiation and scintillation of the fixed stars to the refractions both of the eyes and air; for upon laying a telescope to the eye, those radiations and scintillations immediately disappear. By the tremulous agitation of the air and ascending vapours, it happens that the rays of light are alternately turned aside from the narrow space of the pupil of the eye; but no such thing can have place in the much wider aperture of the object-glass of a telescope; and hence it is that a scintillation is occasioned in the former case, which ceases in the latter; and this cessation in the latter case is a demonstration of the regular transmission of light through the heavens, without any sensible refraction. But, to obviate an objection that may be made from the appearing of no tail in such comets as shine but with a faint light, as if

the secondary rays were then too weak to affect the eyes, and for that reason it is that the tails of the fixed stars do not appear, we are to consider, that by the means of telescopes the light of the fixed stars may be augmented above an hundred fold, and yet no tails are seen; that the light of the planets is yet more copious without any tail; but that comets are seen sometimes with huge tails, when the light of their heads is but faint and dull. For so it happened in the comet of the year 1680, when in the month of *December* it was scarcely equal in light to the stars of the second magnitude, and yet emitted a notable tail, extending to the length of 40°, 50°, 60°, or 70°, and upwards; and afterwards, on the 27th and 28th of *January*, when the head appeared but as a star of the 7th magnitude, yet the tail (as we said above), with a light that was sensible enough, though faint, was stretched out to 6 or 7 degrees in length, and with a languishing light that was more difficultly Been, even to 12°, and upwards. But on the 9th and 10th of *February*, when to the naked eye the head appeared no more, through a telescope I viewed the tail of 2° in length. But farther; if the tail was owing to the refraction of the celestial matter, and did deviate from the opposition of the sun, according to the figure of the heavens, that deviation in the same places of the heavens should be always directed towards the same parts. But the comet of the year 1680, *December* 28^d. 8$^{1/2}$^{h}. P.M. at *London*, was seen in $\mathcal{H}$ 8° 41', with latitude north 28° 6'; while the sun was in $\mathfrak{d}$18° 26'. And the comet of the year 1577, *December* 29^d. was in $\mathcal{H}$ 8° 41', with latitude north 28° 40', and the sun, as before, in about $\mathfrak{d}$ 18° 26'. In both cases the situation of the earth was the same, and the comet appeared in the same place of the heavens; yet in the former case the tail of the comet (as well by my observations as by the observations of others) deviated from the opposition of the sun towards the north by an angle of 4$^1/_2$ degrees; whereas in the latter there was (according to the observations of *Tycho*) a deviation of 21 degrees towards the south. The refraction, therefore, of the heavens being thus disproved, it remains that the *phaenomena* of the tails of comets must be derived from some reflecting matter.

And that the tails of comets do arise from their heads, and tend towards the parts opposite to the sun, is farther confirmed from the laws which the tails observe. As that, lying in the planes of the comets' orbits which pass through the sun, they constantly deviate from the opposition of the sun towards the parts which the comets' heads in their progress along these orbits have left. That to a spectator, placed in those planes, they appear in the parts directly opposite to the sun; but, as the spectator recedes from those planes, their deviation begins to appear, and daily becomes greater. That the deviation, *caeteris paribus*, appears less when the tail is more oblique to the orbit of the comet, as well as when the head of the comet approaches nearer to the sun, especially if the angle of deviation is estimated near the head of the comet. That the tails which have no deviation appear straight, but the tails which deviate are likewise bended into

a certain curvature. That this curvature is greater when the deviation is greater; and is more sensible when the tail, *caeteris paribus*, is longer; for in the shorter tails the curvature is hardly to be perceived. That the angle of deviation is less near the comet's head, but greater towards the other end of the tail; and that because the convex side of the tail regards the parts from which the deviation is made, and which lie in a right line drawn out infinitely from the sun through the comet's head. And that the tails that are long and broad, and shine with a stronger light, appear more resplendent and more exactly defined on the convex than on the concave side. Upon which accounts it is plain that the *phaenomena* of the tails of comets depend upon the motions of their head

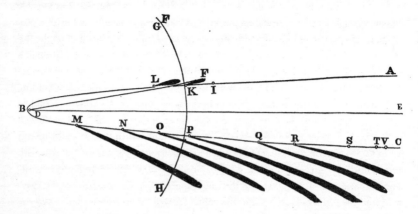

and by no means upon the places of the heavens in which their heads are seen; and that, therefore, the tails of comets do not proceed from the refraction of the heavens, but from their own heads, which furnish the matter that forms the tail. For, as in our air, the smoke of a heated body ascends either perpendicularly if the body is at rest, or obliquely if the body is moved obliquely, so in the heavens, where all bodies gravitate towards the sun, smoke and vapour must (as we have already said) ascend from the sun, and either rise perpendicularly if the smoking body is at rest, or obliquely if the body, in all the progress of its motion, is always leaving those places from which the upper or higher parts of the vapour had risen before; and that obliquity will be least where the vapour ascends with most velocity, to wit, near the smoking body, when that is near the sun. But, because the obliquity varies, the column of vapour will be incurvated; and because the vapour in the preceding sides is something more recent, that is, has ascended something more late; from the body, it will therefore be something more dense on that aide, and must on that account reflect more light, as well as be better defined. I add nothing concerning the sudden uncertain agitation of the tails of comets, and their

irregular figures, which authors sometimes describe, because they may arise from the mutations of our air, and the motions of our clouds, in part obscuring those tails; or, perhaps, from parts of the *Via Lactea*, which might have been confounded with and mistaken for parts of the tails of the comets as they passed by.

But that the atmospheres of comets may furnish a supply of vapours great enough to fill so immense spaces, we may easily understand from the rarity of our own air; for the air near the surface of our earth possesses a space 850 times greater than water of the same weight; and therefore a cylinder of air 850 feet high is of equal weight with a cylinder of water of the same breadth, and but one foot high. But a cylinder of air reaching to the top of the atmosphere is of equal weight with a cylinder of water about 33 feet high: and, therefore, if from the whole cylinder of air the lower part of 850 feet high is taken away, the remaining upper part will be of equal weight with a cylinder of water 32 feet high: and from thence (and by the hypothesis, confirmed by many experiments, that the compression of air is as the weight of the incumbent atmosphere, and that the force of gravity is reciprocally as the square of the distance from the centre of the earth) raising a calculus, by Cor. Prop. XXII, Book II, I found, that, at the height of one semi-diameter of the earth, reckoned from the earth's surface, the air is more rare than with us in a far greater proportion than of the whole space within the orb of Saturn to a spherical space of one inch in diameter; and therefore if a sphere of our air of but one inch in thickness was equally rarefied with the air at the height of one semi-diameter of the earth from the earth's surface, it would fill all the regions of the planets to the orb of Saturn, and far beyond it. Wherefore since the air at greater distances is immensely rarefied, and the *coma* or atmosphere of comets is ordinarily about ten times higher, reckoning from their centres, than the surface of the nucleus, and the tails rise yet higher, they must therefore be exceedingly rare; and though, on account of the much thicker atmospheres of comets, and the great gravitation of their bodies towards the sun, as well as of the particles of their air and vapours mutually one towards another, it may happen that the air in the celestial spaces and in the tails of comets is not so vastly rarefied, yet from this computation it is plain that a very small quantity of air and vapour is abundantly sufficient to produce all the appearances of the tails of comets; for that they are, indeed, of a very notable rarity appears from the shining of the stars through them. The atmosphere of the earth, illuminated by the sun's light, though but of a few miles in thickness, quite obscures and extinguishes the light not only of all the stars, but even of the moon itself; whereas the smallest stars are seen to shine through the immense thickness of the tails of comets, likewise illuminated by the sun, without the least diminution of their splendor. Nor is the brightness of the tails of most comets ordinarily greater than that of our air, an inch or two in thickness, reflecting in a darkened room the light of the sun-beams let in by a hole of the window-shutter.

And we may pretty nearly determine the time spent during the ascent of the vapour from the comet's head to the extremity of the tail, by drawing a right line from the extremity of the tail to the sun, and marking the place where that right line intersects the comet's orbit; for the vapour that is now in the extremity of the tail, if it has ascended in a right line from the sun, must have begun to rise from the head at the time when the head was in the point of intersection. It is trite, the vapour does not rise in a right line from the sun, but, retaining the motion which it had from the comet before its ascent, and compounding that motion with its motion of ascent, arises obliquely; and, therefore, the solution of the Problem will be more exact, if we draw the line which intersects the orbit parallel to the length of the tail; or rather (because of the curvilinear motion of the comet) diverging a little from the line or length of the tail. And by means of this principle I found that the vapour which, *January* 25, was in the extremity of the tail, had begun to rise from the head before *December* 11, and therefore had spent in its whole ascent 45 days; but that the whole tail which appeared on *December* 10 had finished its ascent in the space of the two days then elapsed from the time of the comet's being in its perihelion. The vapour, therefore, about the beginning and in the neighbourhood of the sun rose with the greatest velocity, and afterwards continued to ascend with a motion constantly retarded by its own gravity; and the higher it ascended, the more it added to the length of the tail; and while the tail continued to be seen, it was made up of almost all that vapour which had risen since the time of the comet's being in its perihelion; nor did that part of the vapour which had risen first, and which formed the extremity of the tail, cease to appear, till its too great distance, as well from the sun, from which it received its light, as from our eyes, rendered it invisible. Whence also it is that the tails of other comets which are short do not rise from their heads with a swift and continued motion, and soon after disappear, but are permanent and lasting columns of vapours and exhalations, which, ascending from the heads with a slow motion of many days, and partaking of the motion of the heads which they had from the beginning, continue to go along together with them through the heavens. From whence again we have another argument proving the celestial spaces to be free, and without resistance, since in them not only the solid bodies of the planets and comets, but also the extremely rare vapours of comets' tails, maintain their rapid motions with great freedom, and for an exceeding long time.

Kepler ascribes the ascent of the tails of the comets to the atmospheres of their heads; and their direction towards the parts opposite to the sun to the action of the rays of light carrying along with them the matter of the comets' tails; and without any great incongruity we may suppose, that, in so free spaces, so fine a matter as that of the aether may yield to the action of the rays of the sun's light, though those rays are not able sensibly to move the gross substances in our parts, which are clogged with so palpable a

resistance. Another author thinks that there may be a sort of particles of matter endowed with a principle of levity, as well as others are with a power of gravity; that the matter of the tails of comets may be of the former sort, and that its ascent from the sun may be owing to its levity; but, considering that the gravity of terrestrial bodies is as the matter of the bodies, and therefore can be neither more nor less in the same quantity of matter, I am inclined to believe that this ascent may rather proceed from the rarefaction of the matter of the comets' tails. The ascent of smoke in a chimney is owing to the impulse of the air with which it is entangled. The air rarefied by heat ascends, because its specific gravity is diminished, and in its ascent carries along with it the smoke with which it is engaged; and why may not the tail of a comet rise from the sun after the same manner? For the sun's rays do not act upon the mediums which they pervade otherwise than by reflection and refraction; and those reflecting particles heated by this action, heat the matter of the æther which is involved with them. That matter is rarefied by the heat which it acquires, and because, by this rarefaction, the specific gravity with which it tended towards the sun before is diminished, it will ascend therefrom, and carry along with it the reflecting particles of which the tail of the comet is composed. But the ascent of the vapours is further promoted by their circumgyration about the sun, in consequence whereof they endeavour to recede from the sun, while the sun's atmosphere and the other matter of the heavens are either altogether quiescent, or are only moved with a slower circumgyration derived from the rotation of the sun. And these are the causes of the ascent of the tails of the comets in the neighbourhood of the sun, where their orbits are bent into a greater curvature, and the comets themselves are plunged into the danger and therefore heavier parts of the sun's atmosphere: upon which account they do then emit tails of an huge length; for the tails which then arise, retaining their own proper motion, and in the mean time gravitating towards the sun, must be revolved in ellipses about the sun in like manner as the heads are and by that motion must always accompany the heads, and freely adhere to them. For the gravitation of the vapours towards the sun can no more force the tails to abandon the heads, and descend to the sun, than the gravitation of the heads can oblige them to fall from the tails. They must by their common gravity either fall together towards the sun, or be retarded together in their common ascent therefrom; and, therefore (whether from the causes already described, or from any others), the tails and heads of comets may easily acquire and freely retain any position one to the other, without disturbance or impediment from that common gravitation.

The tails, therefore, that rise in the perihelion positions of the comets will go along with their heads into far remote parts, and together with the heads will either return again from thence to us, after a long course of years, or rather will be there rarefied, and by degrees quite vanish away; for afterwards, in the descent of the heads towards the sun,

new short tails will be emitted from the heads with a slow motion; and those tails by degrees will be augmented immensely, especially in such comets as in their perihelion distances descend as low as the sun's atmosphere; for all vapour in those free spaces is in a perpetual state of rarefaction and dilatation; and from hence it is that the tails of all comets are broader at their upper extremity than near their heads. And it is not unlikely but that the vapour, thus perpetually rarefied and dilated, may be at last dissipated and scattered through the whole heavens, and by little and little be attracted towards the planets by its gravity, and mixed with their atmosphere; for as the seas are absolutely necessary to the constitution of our earth, that from them, the sun, by its heat, may exhale a sufficient quantity of vapours, which, being gathered together into clouds, may drop down in rain, for watering of the earth, and for the production and nourishment of vegetables; or, being condensed with cold on the tops of mountains (as some philosophers with reason judge), may run down in springs and rivers; so for the conservation of the seas, and fluids of the planets, comets seem to be required that, from their exhalations and vapours condensed, the wastes of the planetary fluids spent upon vegetation and putrefaction, and converted into dry earth, may be continually supplied and made up; for all vegetables entirely derive their growths from fluids, and afterwards, in great measure, are turned into dry earth by putrefaction; and a sort of slime is always found to settle at the bottom of putrefied fluids; and hence it is that the bulk of the solid earth is continually increased; and the fluids, if they are not supplied from without, must be in a continual decrease, and quite fail at last. I suspect, moreover, that it is chiefly from the comets that spirit comes, which is indeed the smallest, but the most subtle and useful part of our air, and so much required to sustain the life of all things with us.

The atmospheres of comets, in their descent towards the sun, by running out into the tails, are spent and diminished, and become narrower, at least on that side which regards the sun; and in receding from the sun, when they less run out into the tails, they are again enlarged, if *Hevelius* has justly marked their appearances. But they are seen least of all just after they have been most heated by the sun, and on that account then emit the longest and most resplendent tails; and, perhaps, at the same time, the nuclei are environed with a denser and blacker smoke in the lowermost parts of their atmosphere; for smoke that is raised by a great and intense heat is commonly the denser and blacker. This the head of that comet which we have been describing, at equal distances both from the sun and from the earth, appeared darker after it had passed by its perihelion than it did before; for in the month of *December* it was commonly compared with the stars of the third magnitude, but in *November* with those of the first or second; and such as saw both appearances have described the first as of another and greater comet than the second. For, *November* 19, this comet appeared to a young man at *Cambridge*, though with a pale and dull light, yet equal to *Spica Virginis*; and at that

time it shone with greater brightness than it did afterwards. And *Montenari, November* 20, st. vet. observed it larger than the stars of the first magnitude, its tail being then 2 degrees long. And Mr. *Storer* (by letters which have come into my hands) writes that in the month of *December*, when the tail appeared of the greatest bulk and splendor, the head was but small, and far less than that which was seen in the month of *November* before sun-rising; and, conjecturing at the cause of the appearance, he judged it to proceed from there being a greater quantity of matter in the head at first, which was afterwards gradually spent.

And, which farther makes for the same purpose, I find, that the heads of other comets, which did put forth tails of the greatest bulk and splendour, have appeared but obscure and small. For in *Brazil, March* 5,1668, 7ʰ. P.M., St. N. P. *Valentinus Estancius* saw a comet near the horizon, and towards the south west, with a head so small as scarcely to be discerned, but with a tail above measure splendid, so that the reflection thereof from the sea was easily seen by those who stood upon the shore; and it looked like a fiery beam extended 23° in length from the west to south, almost parallel to the horizon. But this excessive splendor continued only three days, decreasing apace afterwards; and while the splendor was decreasing, the bulk of the tail increased: whence in *Portugal* it is said to have taken up one quarter of the heavens, that is, 45 degrees, extending from west to east with a very notable splendor, though the whole tail was not seen in those parts, because the head was always hid under the horizon: and from the increase of the bulk and decrease of the splendor of the tail, it appear that the head was then in its recess from the sun, and had been very near to it in its perihelion, as the comet of 1680 was. And we read, in the *Saxon* Chronicle, of a like comet appearing in the year 1106, *the star whereof was small and obscure* (as that of 1690), *but the splendour of its tail was very bright, and like a huge fiery beam stretched out in a direction between the east and north*, as *Hevelius* has it also from *Simeon*, the monk of *Durham*. This comet appeared in the beginning of *February*, about the evening, and towards the south west part of heaven; from whence, and from the position of the tail, we infer that the head was near the sun. *Matthew Paris* says, *It was distant from the sun by about a cubit, from three of the clock* (rather six) *till nine, putting forth a long tail*. Such also was that most resplendent comet described by *Aristotle*, lib. 1, *Meteor*. 6. *The head whereof could not be seen, because it had set before the sun, or at least was hid under the sun's rays; but next day it was seen as well as might be; for, having left the sun but a very little way, it set immediately after it. And the scattered light of the head, obscured by the too great splendour* (of the tail) *did not yet appear. But afterwards* (as *Aristotle* says) *when the splendour* (of the tail) *was now diminished* (the head of), *the comet recovered its native brightness; and the splendour* (of its tail) *reached now to a third part of the heavens* (that is, to 60°). *This appearance was in the winter season* (an. 4, Olymp. 101), *and, rising to* Orion's

girdle, it there vanished away. It is true that the comet of 1618, which came out direct-ly from under the sun's rays with a very large tail, seemed to equal, if not to exceed, the stars of the first magnitude; but, then, abundance of other comets have appeared yet greater than this, that put forth shorter tails; some of which are said to have appeared as big as Jupiter, others as big as Venus, or even as the moon.

We have said, that comets are a sort of planets revolved in very eccentric orbits about the sun; and as, in the planets which are without tails, those are commonly less which are revolved in lesser orbits, and nearer to the sun, so in comets it is probable that those which in their perihelion approach nearer to the sun are generally of less magnitude, that they may not agitate the sun too much by their attractions. But as to the transverse diameters of their orbits, and the periodic times of their revolutions, I leave them to be determined by comparing comets together which after long intervals of time return again in the same orbit. In the mean time, the following Proposition may give some light in that inquiry.

PROPOSITION XLII. PROBLEM XXII.

TO CORRECT A COMET'S TRAJECTORY FOUND AS ABOVE.

OPERATION 1. Assume that position of the plane of the trajectory which was deter-mined according to the preceding proposition; and select three places of the comet, deduced from very accurate observations, and at great distances one from the other. Then suppose A to represent the time between the first observation and the second, and B the time between the second and the third; but it will be convenient that in one of those times the comet be in its perigeon, or at least not far from it. From those apparent places find, by trigonometric operations; the three true places of the comet in that assumed plane of the trajectory; then through the places found, and about the cen-tre of the sun as the focus, describe a conic section by arithmetical operations, accord-ing to Prop. XXI., Book I. Let the areas of this figure which are terminated by radii drawn from the sun to the places found be D and E; to wit, D the area between the first observation and the second, and E the area between the second and third; and let T represent the whole time in which the whole area D + E should be described with the velocity of the comet found by Prop. XVI., Book I.

OPER. 2. Retaining the inclination of the plane of the trajectory to the plane of the ecliptic, let the longitude of the nodes of the plane of the trajectory be increased by the addition of 20 or 30 minutes, which call P. Then from the aforesaid three observed places of the comet let the three true places be found (as before) in this new plane; as also the orbit passing through those places, and the two areas of the same described between the two observations, which call *d* and *e*; and let *t* be the whole time in which the whole area *d* + *e* should be described.

OPER. 3. Retaining the longitude of the nodes in the first operation, let the inclination of the plane of the trajectory to the plane of the ecliptic decreased by adding thereto 20' or 30', which call Q. Then from the aforesaid three observed apparent places of the comet let the three true places be found in this new plane, as well as the orbit passing through them, and the two areas of the same described between the observation, which call δ and ϵ; and let τ be the whole time in which the whole area $\delta + \delta$ should be described.

Then taking C to 1 as A to B; and G to 1 as D to E; and g to 1 as d to e; and γ to 1 as δ to ϵ; let S be the true time between the first observation and the third; and, observing well the signs + and -, let such numbers m and n be found out as will make $2G - 2C$, $= mG - mg + nG - n\tau$; and $2T - 2S = mT - mt + nT - n\tau$. And if, in the first operation, I represents the inclination of the plane of the trajectory to the plane of the ecliptic, and K the longitude of either node, then $I + nQ$ will be the true inclination of the plane of the trajectory to the plane of the ecliptic, and $K + mP$ the true longitude of the node. And, lastly, if in the first, second, and third operations, the quantities R, r, and (, represent the parameters of the trajectory, and the quantities $\frac{1}{L}, \frac{1}{l}, \frac{1}{\lambda}$, the

transverse diameters of the same, then $R + mr - mR + n\rho - nR$ will be the true parameter, and $\frac{1}{L + ml - mL + n\lambda - nL}$ will be the true transverse diameter of the trajectory which the comet describes; and from the transverse diameter given the periodic time of the comet is also given. Q.E.I. But the periodic times of the revolutions of comets, and the transverse diameters of their orbits, cannot be accurately enough determined but by comparing comets together which appear at different times. If, after equal intervals of time, several comets are found to have described the same orbit, we may thence conclude that they are all but one and the same comet revolved in the same orbit; and then from the times of their revolutions the transverse diameters of their orbits will be given, and from those diameters the elliptic orbits themselves will be determined.

To this purpose the trajectories of many comets ought to be computed, supposing those trajectories to be parabolic; for such trajectories will always nearly agree with the *phaenomena*, as appears not only from the parabolic trajectory of the comet of the year 1680, which I compared above with the observations, but likewise from that of the notable comet which appeared in the year 1664 and 1665, and was observed by *Hevelius*, who, from his own observations, calculated the longitudes and latitudes thereof, though with little accuracy. But from the same observations Dr. *Halley* did again compute its places; and from those new places determined its trajectory, finding its ascending node in ♊ 21° 13' 55"; the inclination of the orbit to the plane of the ecliptic 21° 18' 40"; the distance of its perihelion from the node, estimated in the comet's orbit, 49° 27' 30", its perihelion in ♌ 8° 40' 30", with heliocentric latitude

south 16° 01' 45"; the comet to have been in its perihelion *November* 21^d. 11^h. 52' P.M. equal time at *London*, or 13^h. 8' at *Dantzick*, O.S.; and that the *latus rectum* of the parabola was 410286 such parts as the sun's mean distance from the earth is supposed to contain 100000. And how nearly the places of the comet computed in this orbit agree with the observations, will appear from the annexed table, calculated by Dr. *Halley*.

Appar. Time at Dantzick.	The observed Distances of the Comet from		The observed Places.			The Places computed in the Orb.
d. h. '		° ' "			° ' "	° ' "
December 3 18.29 ½	The Lion's heart The Virgin's spike	46.24.20 22.52.10	Long. Lat. S.	♎	7.01.00 21.39. 0	♎ 7. 1.29 21.38.50
4.18. 1½	The Lion's heart The Virgin's spike	46. 2.45 23.52.40	Long. Lat. S.	♎	6.15. 0 22.24. 0	♎ 6.16. 5 22.24. 0
7.17.48	The Lion's heart The Virgin's spike	44.48. 0 27.56.40	Long. Lat. S.	♎	3. 6. 0 25.22. 0	♍ 3. 7.33 25.21.40
17.14.43	The Lion's heart Orion's right shoulder	63.15.15 45.43.30	Long. Lat. S.	♌	2.56. 0 49.25. 0	♌ 2.56. 0 49.25. 0
19. 9.25	Procyon Bright star of Whale's jaw	35.13.50 52.56. 0	Long. Lat. S.	♊	28.40.30 45.48. 0	♊ 28.43. 0 45.46. 0
20. 9. 53½	Procyon Bright star of Whale's jaw	40.49. 0 40.04. 0	Long. Lat. S	♊	13.03. 0 39.54. 0	♊ 13. 5. 0 39.53.40
21. 9. 9½	Orion's right shoulder Bright star of Whale's jaw	26.21.25 29.28. 0	Long. Lat. S.	♊	2.16. 0 33.41. 0	♊ 2.18.30 33.39.40
22. 9. 0	Orion's right shoulder Bright star of Whale's jaw	29.47. 0 20.29.30	Long. Lat. S.	♉	24.24. 0 27.45. 0	♉ 24.27. 0 27.46. 0
26. 7.58	The bright star of Aries Aldebaran	23.20. 0 26.44. 0	Long. Lat. S.	♉	9. 0. 0 12.36. 0	♉ 9. 2.28 12.34.13
27. 6.45	The bright star of Aries Aldebaran	20.45. 0 28.10. 0	Long. Lat. S.	♉	7. 5.40 10.23. 0	♉ 7. 8.45 10.23.13
28. 7.39	The bright star of Aries Palilicium	18.29. 0 29.37. 0	Long. Lat. S.	♉	5.24.45 8.22.50	♉ 5.27.52 8.23.37
31. 6.45	Andromeda's girdle Palilicium	30.48.10 32.53.30	Long. Lat. S.	♉	2. 7.40 4.13. 0	♉ 2. 8.20 4.16.25
Jan. 1665 7. 7.37½	Andromeda's girdle Palilicium	25.11. 0 37.12.25	Long. Lat. N.	♈	28.24.47 0.54. 0	♈ 28.24. 0 0.53. 0
13. 7. 0	Andromeda's head Palilicium.	28. 7.10 38.55.20	Long. Lat. N.	♈	27. 6.54 3. 6.50	♈ 27. 6.39 3. 7.40
24. 7.29	Andromeda's girdle Palilicium	20.32.15 40. 5. 0	Long. Lat. N.	♈	26.29.15 5.25.50	♈ 26.28.50 5.26. 0
Feb. 7. 8.37			Long. Lat. N.	♈	27. 4.46 7. 3.29	♈ 27.24.55 7. 3.15
22. 8.46			Long. Lat. N.	♈	28.29.46 8.12.36	♈ 28.29.58 8.10.25
March. 1. 8.16			Long. Lat. N.	♈	29.18 15 8.36.26	♈ 29.18.20 8.36.12
7. 8.37			Long. Lat. N.	♉	0. 2.48 8.56.30	♉ 0. 2.42 8.56.56

In *February*, the beginning of the year 1665, the first star of Aries, which I shall hereafter call γ, was in ♈ 28° 30' 15", with 7° 8' 58" north lat.; the second star of Aries was in ♈ 29° 17' 18", with 8° 28' 16" north lat.; and another star of the seventh magnitude, which I call, was in ♈ 28° 24' 45", with 8° 28' 13" north lat. The comet *Feb.* 7ᵈ. 7ʰ. 30' at *Paris* (that is, *Feb.* 7ᵈ. 8ʰ. 37' at *Dantzick*) O.S. made a triangle with those stars γ and A, which was right-angled in γ; and the distance of the comet from the star γ was equal to the distance of the stars γ and A, that is, 1° 19' 46" of a great circle; and therefore in the parallel of the latitude of the star γ it was 1° 20' 26". Therefore if from the longitude of the star γ there be subducted the longitude 1° 20' 26", there will remain the longitude of the comet ♈ 27° 9' 49". M. *Auzout*, from this observation of his, placed the comet in ♈ 27° 0', nearly; and, by the scheme in which Dr. *Hooke* delineated its motion, it was then in ♈ 26° 59' 24". I place it in ♈ 27° 4' 46", taking the middle between the two extremes.

From the same observations, M. *Auzout* made the latitude of the comet at that time 7° and 4' or 5' to the north; but he had done better to have made it 7° 3' 29", the difference of the latitudes of the comet and the star γ being equal to the difference of the longitude of the stars γ and A.

February 22ᵈ. 7ʰ. 30' at *London*, that is, *February* 22ᵈ. 8ʰ. 46' at *Dantzick*, the distance of the comet from the star A, according to Dr. *Hooke's* observation, as was delineated by himself in a scheme, and also by the observations of M. *Auzout*, delineated in like manner by M. *Petit*, was a fifth part of the distance between the star A and the first star of Aries, or 15° 57'; and the distance of the comet from a right line joining the star A and the first of Aries was a fourth part of the same fifth part, that is, 4'; and therefore the comet was in ♈ 28° 29' 46", with 8° 12' 36" north lat.

March 1, 7ʰ. 0' at *London*, that is, *March* 1, 8ʰ. 16' at *Dantzick*, the comet was observed near the second star in Aries, the distance between them being to the distance between the first and second stars in Aries, that is, to 1° 33', as 4 to 45 according to Dr. *Hooke*, or as 2 to 23 according to M. Gottignies. And, therefore, the distance of the comet from the second star in Aries was 8' 16" according to Dr. *Hooke*, or 8' 5" according to M. Gottignies; or, taking a mean between both, 8' 10". But, according to M. Gottignies, the comet had gone beyond the second star of Aries about a fourth or a fifth part of the space that it commonly went over in a day, to wit, about 1' 35" (in which he agrees very well with M. *Auzout*); or, according to Dr. *Hooke*, not quite so much, as perhaps only 1'. Wherefore if to the longitude of the first star in Aries we add 1', and 8' 10" to its latitude, we shall have the longitude of the comet ♈ 29° 18', with 8° 36' 26" north lat.

March 7, 7ʰ. 30' at *Paris* (that is, *March* 7, 8ʰ. 37' at *Dantzick*), from the observations of M. *Auzout*, the distance of the comet from the second star in Aries was equal

to the distance of that star from the star A, that is, 52' 29"; and the difference of the longitude of the comet and the second star in Aries was 45' or 46', or, taking a mean quantity, 45' 30"; and therefore the comet was in ♈ 0° 2' 48". From the scheme of the observations of M. *Auzout*, constructed by M. *Petit*, *Hevelius* collected the latitude of the comet 8° 54'. But the engraver did not rightly trace the curvature of the comet's way towards the end of the motion; and *Hevelius*, in the scheme of M. *Auzout*'s observations which he constructed himself, corrected this irregular curvature, and so made the latitude of the comet 8° 55' 30". And, by farther correcting this irregularity, the latitude may become 8° 56', or 8° 57'.

This comet was also seen *March* 9, and at that time its place must have been in ♉ 0° 18', with 9° 3$^1/_2$' north lat. nearly.

This comet appeared three months together, in which space of time it travelled over almost six signs, and in one of the days thereof described almost 20 deg. Its course did very much deviate from a great circle, bending towards the north, and its motion towards the end from retrograde became direct; and, notwithstanding, its course was so uncommon, yet by the table it appears that the theory, from beginning to end, agrees with the observations no less accurately than the theories of the planets usually do with the observations of them; but we are to subduct about 2' when the comet was swiftest, which we may effect by taking off 12', from the angle between the ascending node and the perihelion, or by making that angle 49° 27, 18". The annual parallax of both these comets (this and the preceding) was very conspicuous, and by its quantity demonstrates the annual motion of the earth in the *orbis magnus*.

This theory is likewise confirmed by the motion of that comet, which in the year 1683 appeared retrograde, in an orbit whose plane contained almost a right angle with the plane of the ecliptic, and whose ascending node (by the computation of Dr. *Halley*) was in ♍ 23° 23'; the inclination of its orbit to the ecliptic 83° 11'; its perihelion in ♊ 25° 29' 30", its perihelion distance from the sun 56020 of such parts as the radius of the *orbis magnus* contains 100000; and the time of its perihelion *July* 2^d. 3^h. 50'. And the places thereof, computed by Dr. *Halley* in this orbit, are compared with the places of the same observed by Mr. *Flamsted*, in the following table:—

1683 Eq. time.	Sun's place	Comet's Long. com.	Lat. Nor. comput.	Comet's Long. obs'd	Lat. Nor. observed	Diff. Long.	Diff. Lat.
d. h. '	° ' "	° ' "	° ' "	° ' "	° ' "	' "	' "
July 13.12.55	♌ 1.02.30	♋ 13.05.42	29.29.13	♋ 13. 6.42	29.29.20	+ 1.00	+ 0.07
15.11.15	2.53.12	11.37.48	29.34. 0	11.39.43	29.34.50	+ 1.55	+ 0.50
17.10.20	4.45.45	10. 7. 6	29.33.30	10. 8.40	29.34. 0	+ 1.34	+ 0.30
23.13.40	10.38.21	5.10.27	28.51.42	5.11.30	28.50.28	+ 1.03	- 1.14
25.14. 5	12.35.28	3.27.53	24.24.47	3.27. 0	28.23.40	- 0.53	- 1. 7
31. 9.42	18.09.22	♊ 27.55. 3	26.22.52	♊ 27.54.24	26.22.25	- 0.39	- 0.27
31.14.55	18.21.53	27.41. 7	26.16.57	27.41. 8	26.14.50	+ 0. 1	- 2. 7
Aug. 2.14 56	20.17.16	25.29.32	25.16.19	25.28.46	25.17.28	- 0.46	+ 1. 9
4.10.49	22.02.50	23.18.20	24.10.49	23.16.55	24.12.19	- 1.25	+ 1.30
6.10. 9	21.16.45	20.42.21	22.47. 5	20.40.32	22.49. 5	- 1.51	+ 2. 0
9.10.26	26.50.52	16. 7.57	20. 6.37	16. 5.55	20. 6.10	- 2. 2	- 0.27
15.14. 1	♍ 2.47.13	3.30.48	11.37.33	3.26.18	11.32. 1	- 4.30	- 5.32
16.15.10	3.48. 2	0.43. 7	9.34.16	0.41.55	9.34.13	- 1.12	- 0. 3
18.16.44	6.45.33	♉ 24.52.53	5.11.15 South.	♉ 24.49. 5	5. 9.11 South	-3.48	- 2. 4
22.14.44	9.35.49	11. 7.14	5.16.58	11.07.12	5.16.58	- 0. 2	- 0. 3
23.15.52	10.36 48	7. 2.18	8.17. 9	7. 1.17	8.16.41	- 1. 1	- 0.28
26.16. 2	♈ 13.31.20	24.45.31	16.38. 0	♈ 24.44.00	16.38.20	- 1.31	+ 0.20

This theory is yet farther confirmed by the motion of that retrograde comet which appeared in the year 1682. The ascending node of this (by Dr. *Halley's* computation) was in ♉ 21° 16' 30"; the inclination of its orbit to the plane of the ecliptic 17° 56' 00"; its perihelion in ♒ 2° 52' 50"; its perihelion distance from the sun 58328 parts, of which the radius of the *orbis magnus* contains 100000; the equal time of the comet's being in its perihelion *Sept.* 4^d. 7^h. 39'. And its places, collected from Mr. *Flamsted's* observations, are compared with its places computed from our theory in the following table:

1682 App. time.	Sun's place	Comet's Lon. comp.	Lat. Nor comp.	Com. Long. observed.	Lat. Nor observ.	Diff. Long.	Diff. Lat.
d. h. '	° ' "	° ' "	° ' "	° ' "	° ' "	' "	' "
Aug 19.16.38	♍ 7. 0. 7	♌ 18.14.28	25.50. 7	♌ 18.14.40	25.49.55.	- 0.12	+ 0.12
20.15.38	7.55.52	24.46.23	26.14.42	24.46.22	26.12.52	+ 0. 1	+ 1.50
21. 8.21	8.36.14	29.37.15	26.20. 3	29.38.02	26.17.37	- 0.47	+ 2,26
22. 8. 8	9.33.55	♍ 6.29.53	26. 8.42	♍ 6.30. 3	26. 7.12	- 0.10	+ 1.30
29.08.20	16.22.40	♎ 12.37.54	18.37.47	♎ 12.37.49	18.34. 5	+ 0. 5	+ 3.42
30. 7.46	17.19.41	15.36. 1	17.26.43	15.35.18	17.27.17	+ 0.43	- 0.34
Sept. 1. 7.33	19.16. 9	20.30.53	15.13. 0	20.27. 4	15. 9.49	+ 3.49	+ 3.11
4. 7.22	22.11.28	25.42. 0	12.23.48	25.40.58	12.22. 0	+ 1. 2	+ 1.48
5. 7.32	23.10.29	27. 0.46	11.33.08	26.59.24	11.33.51	+ 1.22	- 0.43
8. 7.16	26. 5.58	29.58.44	9.26.46	29.58.45	9.26.43	- 0. 1	+ 0. 3
9. 7.26	27. 5. 9	♍ 0.44.10	8.49.10	♍ 0.44. 4	8.48.25	+ 0. 6	+ 0.45

This theory is also confirmed by the retrograde motion of the comet that appeared in the year 1723. The ascending node of this comet (according to the computation of Mr. *Bradley*, Savilian Professor of Astronomy at *Oxford*) was in ♈ 14° 16'. The inclination of the orbit to the plane of the ecliptic 49° 59'. Its perihelion was in ♉ 12° 15' 20". Its perihelion distance from the sun 998651 parts, of which the radius of the *orbis*

magnus contains 1000000, and the equal time of its perihelion *September* 16^d. 16^h. 10'. The places of this comet computed in this orbit by Mr. *Bradley*, and compared with the places observed by himself, his uncle Mr. *Pound*, and Dr. *Halley*, may be seen in the following table.

1723 Eq. time.		Comet's Long. obs.	Lat. Nor obs.	Comet's Lon. com.	Lat. Nor comp.	Diff. Lon.	Diff. Lat.
	d. h. '	o ' "	o ' "	o ' "	o ' "	' "	' "
Oct. 9. 8. 5	♒	7.22.15	5. 2. 0	♒ 7.21.26	5. 2.47	+ 49	– 47
10. 6.21		6.41.12	7.44.13	6.41.42	7.43.18	– 50	+ 55
12. 7.22		5.39.58	11.55. 0	5.40.19	11.54.55	– 21	+ 5
14. 8.57		4.59.49	14.43.50	5. 0.37	14.44. 1	– 48	– 11
15. 6.35		4.47.41	15.40.51	4.47.45	15.40.55	– 4	– 4
21. 6.22		4. 2.32	19.41.49	4. 2.21	19.42. 3	+ 11	– 14
22. 6.24		3.59. 2	20. 8.12	3.59.10	20. 8.17	– 8	– 5
24. 8. 2		3.55.29	20.55 18	3.55.11	20.55. 9	+ 18	+ 9
29. 8.56		3.56.17	22.20.27	3.56.42	22.20.10	– 26	+ 17
30. 6.20		3.58. 9	22.32.28	3.58.17	22.32.12	– 8	+ 16
Nov. 5. 5.53		4.16.30	23.38.33	4.16.23	23.38. 7	+ 7	+ 26
8. 7. 6		4.29.36	24. 4.30	4.29.54	24. 4.40	– 18	– 10
14. 6.20		5. 2.16	24.48.46	5. 2.51	24.48.16	– 35	+ 30
20. 7.45		5.42.20	25.24.45	5.43.13	25.25.17	– 53	– 32
Dec. 7. 6.45		8. 4.13	26.54.18	8. 3.55	26.53.42	+ 18	+ 36

From these examples it is abundantly evident that the motions of comets are no less accurately represented by our theory than the motions of the planets commonly are by the theories of them; and, therefore, by means of this theory, we may enumerate the orbits of comets, and so discover the periodic time of a comet's revolution in any orbit; whence, at last, we shall have the transverse diameters of their elliptic orbits and their aphelion distances.

That retrograde comet which appeared in the year 1607 described an orbit whose ascending node (according to Dr. *Halley*'s computation) was in ♉ 20° 21'; and the inclination of the plane of the orbit to the plane of the ecliptic 17° 2'; whose perihelion was in ♒ 2° 16'; and its perihelion distance from the sun 58680 of such parts as the radius of the *orbis magnus* contains 100000; and the comet was in its perihelion *October* 16^d. 3^h. 50'; which orbit agrees very nearly with the orbit of the comet which was seen in 1682. If these were not two different comets, but one and the same, that comet will finish one revolution in the space of 75 years; and the greater axis of its orbit will be to the greater axis of the *orbis magnus* as $\sqrt[3]{75 \times 75}$ to 1, or as 1778 to 100, nearly. And the aphelion distance of this comet from the sun will be to the mean distance of the earth from the sun as about 35 to 1; from which data it will be no hard matter to determine the elliptic orbit of this comet. But these things are to be supposed on condition, that, after the space of 75 years, the same comet shall return again in the same orbit. The other comets seem to ascend to greater heights, and to require a longer time to perform their revolutions.

But, because of the great number of comets, of the great distance of their aphelions from the sun, and of the slowness of their motions in the aphelions, they will, by their mutual gravitations, disturb each other; so that their eccentricities and the times of their revolutions will be sometimes a little increased, and sometimes diminished. Therefore we are not to expect that the same comet will return exactly in the same orbit, and in the same periodic times: it will be sufficient if we find the changes no greater than may arise from the causes just spoken of.

And hence a reason may be assigned why comets are not comprehended within the limits of a zodiac, as the planets are; but, being confined to no bounds, are with various motions dispersed all over the heavens; namely, to this purpose, that in their aphelions, where their motions are exceedingly slow, receding to greater distances one from another, they may suffer less disturbance from their mutual gravitations: and hence it is that the comets which descend the lowest, and therefore move the slowest in their aphelions, ought also to ascend the highest.

The comet which appeared in the year 1680 was in its perihelion less distant from the sun than by a sixth part of the sun's diameter; and because of its extreme velocity in that proximity to the sun, and some density of the sun's atmosphere, it must have suffered some resistance and retardation; and therefore, being attracted something nearer to the sun in every revolution, will at last fall down upon the body of the sun. Nay, in its aphelion, where it moves the slowest, it may sometimes happen to be yet farther retarded by the attractions of other comets, and in consequence of this retardation descend to the sun. So fixed stars, that have been gradually wasted by the light and vapours emitted from them for a long time, may be recruited by comets that fall upon them; and from this fresh supply of new fuel those old stars, acquiring new splendor, may pass for new stars. Of this kind are such fixed stars as appear on a sudden, and shine with a wonderful brightness at first, and afterwards vanish by little and little. Such was that star which appeared in *Cassiopeia's* chair; which *Cornelius Gemma* did not see upon the 8th of *November*, 1572, though he was observing that part of the heavens upon that very night, and the sky was perfectly serene; but the next night (*November* 9) be saw it shining much brighter than any of the fixed stars, and scarcely inferior to veil us in splendor. *Tycho Brahe* saw it upon the 11th of the same month, when it shone with the greatest lustre; and from that time he observed it to decay by little and little; and in 16 months' time it entirely disappeared. In the month of *November*, when it first appeared, its light was equal to that of *Venus*. In the month of *December* its light was a little diminished, and was now become equal to that of *Jupiter*. In *January* 1573 it was less than *Jupiter*, and greater than *Sirius*; and about the end of *February* and the beginning of *March* became equal to that star. In the months of *April* and *May* it was equal to a star of the second magnitude; in *June*, *July*, and *August*, to a

star of the third magnitude; in *September*, *October*, and *November*, to those of the fourth magnitude; in *December* and *January* 1574 to those of the fifth; in *February* to those of the sixth magnitude; and in *March* it entirely vanished. Its colour at the beginning was clear, bright, and inclining to white; afterwards it, turned a little yellow; and in *March* 1573 it became ruddy like *Mars* or *Aldebaran*: in *May* it turned to a kind of dusky whiteness, like that we observe in *Saturn*; and that colour it retained ever after, but growing always more and more obscure. Such also was the star in the right foot of *Serpentarius*, which *Kepler*'s scholars first observed *September* 30, O.S. 1604, with a light exceeding that of *Jupiter*, though the night before it was not to be seen; and from that time it decreased by little and little, and in 15 or 16 months entirely disappeared. Such a new star appearing with an unusual splendor is said to have moved *Hipparchus* to observe, and make a catalogue of, the fixed stars. As to those fixed stars that appear and disappear by turns and increase slowly and by degrees, and, scarcely ever exceed the stars of the third magnitude, they seem to be of another kind, which revolve about their axes, and, having a light and a dark side, shew those two different sides by turns. The vapours which arise from the sun, the fixed stars, and the tails of the comets, may meet at last with, and fall into, the atmospheres of the planets by their gravity, and there be condensed and turned into water and humid spirits; and from thence, by a slow heat, pass gradually into the form of salts, and sulphurs, and tinctures, and mud, and clay, and sand, and stones, and coral, and other terrestrial substances.

GENERAL SCHOLIUM.

The hypothesis of vortices is pressed with many difficulties. That every planet by a radius drawn to the sun may describe areas proportional to the times of description, the periodic times of the several parts of the vortices should observe the duplicate proportion of their distances from the sun; but that the periodic times of the planets may obtain the sesquiplicate proportion of their distances from the sun, the periodic times of the parts of the vortex ought to be in the: sesquiplicate proportion of their distances. That the smaller vortices may maintain their lesser revolutions about *Saturn*, *Jupiter*, and other planets, and swim quietly and undisturbed in the greater vortex of the sun, the periodic times of the parts of the sun's vortex should be equal; but the rotation of the sun and planets about their axes, which ought to correspond with the motions of their vortices, recede far from all these proportions. The motions of the comets are exceedingly regular, are governed by the same laws with the motions of the planets, and can by no means be accounted for by the hypothesis of vortices, for comets are carried with very eccentric motions through all parts of the heavens indifferently, with a freedom that is incompatible with the notion of a vortex.

Bodies projected in our air suffer no resistance but from the air. Withdraw the air, as is done in Mr. *Boyle*'s vacuum, and the resistance ceases; for in this void a bit of fine down and a piece of solid gold descend with equal velocity. And the parity of reason must take place in the celestial spaces above the earth's atmosphere; in which spaces, where there is no air to resist their motions, all bodies will move with the greatest freedom; and the planets and comets will constantly pursue their revolutions in orbits given in kind and position, according to the laws above explained; but though these bodies may, indeed, persevere in their orbits by the mere laws of gravity, yet they could by no means have at first derived the regular position of the orbits themselves from those laws.

The six primary planets are revolved about the sun in circles concentric with the sun, and with motions directed towards the same parts, and almost in the same plane. Ten moons are revolved about the earth, Jupiter and Saturn, in circles concentric with them, with the same direction of motion, and nearly in the planes of the orbits of those planets; but it is not to be conceived that mere mechanical causes could give birth to so many regular motions, since the comets range over all parts of the heavens in very eccentric orbits; for by that kind of motion they pass easily through the orbs of the planets, and with great rapidity; and in their aphelions, where they move the slowest, and we detained the longest, they recede to the greatest distances from each other, and thence suffer the least disturbance from their mutual attractions. This most beautiful system of the sun, planets, and comets, could only proceed from the counsel and dominion of an intelligent and powerful Being. And if the fixed stars are the centres of other like systems, these being formed by the like wise counsel, must be all subject to the dominion of One; especially since the light of the fixed stars is of the same nature with the light of the sun, and from every system light passes into all the other systems: and lest the systems of the fixed stars should, by their gravity, fall on each other mutually, he hath placed thou systems at immense distances one from another.

This Being governs all things, not as the soul of the world, but as Lord over all; and on account of his dominion he is wont to be called *Lord God* παιτοκράτωρ, or *Universal Ruler*; for *God* is a relative word, and has a respect to servants; and *Deity* is the dominion of God not over his own body, as those imagine who fancy God to be the soul of the world, but over servants. The Supreme God is a Being eternal, infinite, absolutely perfect; but a being, however perfect, without dominion, cannot be said to be Lord God; for we say, my God, your God, the God of *Israel*, the God of Gods, and Lord of Lords; but we do not say, my Eternal, year Eternal the Eternal of *Israel*, the Eternal of Gods; we do not say,

my Infinite, or my Perfect: these are titles which have no respect to servants. The word *God*[1] usually signifies *Lord*; but every lord is not a God. It is the dominion of a spiritual being which constitutes a God: a true, supreme, or imaginary dominion makes a true, supreme, or imaginary God. And from his true dominion it follows that the true God is a living, intelligent, and powerful Being; and, from his other perfections, that he is supreme, or most perfect. He is eternal and infinite, omnipotent and omniscient; that is, his duration reaches from eternity to eternity; his presence from infinity to infinity; he governs all things and knows all things that are or can be done. He is not eternity or infinity, but eternal and infinite; he is not duration or space, but he endures and is present. He endures for ever, and is every wherē present; and by existing always and every where, he constitutes duration and space. Since every particle of space is *always*, and every indivisible moment of duration is *every where*, certainly the Maker and Lord of all things cannot be *never* and *no where*. Every soul that has perception is, though in different times and in different organs of sense and motion, still the same indivisible person. There are given successive parts in duration, coexistent parts in space, but neither the one nor the other in the person of a man, or his thinking principle; and much less can they be found in the thinking substance of God. Every man, so far as he is a thing that has perception, is one and the same man during his whole life, in all and each of his organs of sense. God is the same God, always and every where. He is omnipresent not *virtually* only, but also *substantially*; for virtue cannot subsist without substance. In him[2] are all things contained and moved; yet neither affects the other: God suffers nothing from the motion of bodies; bodies find no resistance from the omnipresence of God. It is allowed by all that the Supreme God exists necessarily; and by the same necessity he exists *always* and *every where*. Whence also he is all similar, all eye, all ear, all brain, all arm, all power to perceive, to understand, and to act; but in a manner not at all human, in a manner not at all corporeal, in a manner utterly unknown to us. As a blind man has no idea of colours, so have we no idea of the manner by which the all-wise God perceives and understands all things; He is utterly void of all body and bodily figures and can therefore neither be seen, nor heard, nor touched; nor ought he to be worshipped under the representation of any corporeal thing. We have ideas of his attributes; but what the real substance of any thing is we know not. In bodies, we see only their figures and colours, we hear only the sounds, we touch only their outward surfaces, we smell only the smells, and taste the savours;

---

[1] Dr. Pocock derives the Latin word *Deus* from the Arabic *du* (in the oblique case *di*), which signifies *Lord*. And in this same princes are called *gods*, *Psal.* lxxxii. ver. 6; and *John* x. ver. 35. And *Moses* is called a *god* to his brother *Aaron*, and a god to *Pharaoh* (*Exod.* iv. ver. 16; and vii. ver. 1). And in the same sense the souls of dead princes were formerly, by the Heathens, called *gods*, but falsely, became of their want of dominion.

[2] This was the opinion of the Ancients. So *Pythagoras*, in *Cicer. de Nat. Deor.* lib. i. *Thales, Anaxagoras, Virgil, Georg.* lib. iv. ver. 220; and *AEneid,* lib. vi. ver. 721. *Philo Allegor,* at the beginning of lib. i. *Aratus,* in his *Phaenom.* at the beginning. So also the sacred writers; as *St. Paul, Acts,* xvii. ver. 27, 28. St. *John's* Gosp. chap. xiv. ver. 2. *Moses,* in *Deut.* iv. ver. 39; and x. ver. 14. *David, Psal.* cxxxix. ver. 7, 8, 9. *Solomon,* 1 *Kings,* viii. ver. 27. *Job,* xxii. ver. 12, 13, 14. *Jeremiah,* xxiii. ver. 23, 24. The Idolaters supposed the sun, moon, and stars, the souls of man, and other parts of the world, to be parts of the Supreme God, and therefore, to be worshipped; but erroneously.

but their inward substances are not to be known either by our, senses, or by any reflex act of our minds: much less, then, have we any idea of the substance of God. We know him only by his most wise and excellent contrivances of things, and final causes; we admire him for his perfections; but we reverence and adore him on account of his dominion: for we adore him as his servants; and a god without dominion, providence, and final causes, is nothing else but Fate and Nature. Blind metaphysical necessity, which is certainly the same always and every where, could produce no variety of things. All that diversity of natural things which we find suited to different times and places could arise from nothing but the ideas and will of a Being necessarily existing. But, by way of allegory, God is said to see, to speak, to laugh, to love, to hate, to desire, to give, to receive, to rejoice, to be angry, to fight, to frame, to work, to build; for our notions of God are taken from the ways of mankind by a certain similitude, which, though not perfect, has some likeness, however. And thus much concerning God; to discourse of whom from the appearances of things, does certainly belong to Natural Philosophy.

Hitherto we have explained the phaenomena of the heavens and of our sea by the power of gravity, but have not yet assigned the cause of this power. This is certain, that it must proceed from a cause that penetrates to the very centres of the sun and planets, without suffering the least diminution of its force; that operates not according to the quantity of the surfaces of the particles upon which it acts (as mechanical causes use to do), but according to the quantity of the solid matter which they contain, and propagates its virtue on all sides to immense distances, decreasing always in the duplicate proportion of the distances. Gravitation towards the sun is made up out of the gravitations towards the several particles of which the body of the sun is composed; and in receding from the sun decreases accurately in the duplicate proportion of the distances as far as the orb of Saturn, as evidently appears from the quiescence of the aphelions of the planets; nay, and even to the remotest aphelions of the comets, if those aphelions are also quiescent, But hitherto I have not been able to discover the cause of those properties of gravity from phaenomena, and I frame no hypotheses; for whatever is not deduced from the phaenomena, is to be called an hypothesis; and hypotheses, whether metaphysical or physical, whether of occult qualities or mechanical, have no place in experimental philosophy. In this philosophy particular propositions are inferred from the phaenomena, and afterwards rendered general by induction. Thus it was that the impenetrability, the mobility, and the impulsive force of bodies, and the laws of motion and of gravitation, were discovered, And to us it is enough that gravity does really exist, and act according to the laws which we have explained, and abundantly serves to account for all the motions of the celestial bodies, and of our sun.

And now we might add something concerning a certain most subtle Spirit which pervades and lies hid in all gross bodies; by the force and action of which Spirit the particles of bodies mutually attract one another at near distances, and cohere, if contiguous; and electric bodies operate to greater distance, as well repelling as attracting the neighbouring corpuscles; and light is emitted, reflected, refracted, inflected, and heats bodies; and all sensation is excited, and the members of animal bodies move at the command of the will, namely, by the vibrations of this Spirit, mutually propagated along the solid filaments of the nerves, from the outward organs of sense to the brain, and from the brain into the muscles. But these are things that cannot be explained in few words, nor are we furnished with that sufficiency of experiments which is required to an accurate determination and demonstration of the laws by which this electric and elastic Spirit operates.

END OF THE MATHEMATICAL PRINCIPLES.

Albert Einstein

(1879-1955)

HIS LIFE AND WORK

Genius isn't always immediately recognized. Although Albert Einstein would become the greatest theoretical physicist who ever lived, when he was in grade school in Germany his headmaster told his father, "He'll never make a success of anything." When Einstein was in his mid-twenties, he couldn't find a decent teaching job even though he had graduated from the Federal Polytechnic School in Zurich as a teacher of mathematics and physics. So he gave up hope of obtaining a university position and applied for temporary work in Bern. With the help of a classmate's father, Einstein managed to secure a civil-service post as an examiner in the Swiss patent office. He worked six days a week, earning $600 a year. That's how he supported himself while working toward his doctorate in physics at the University of Zurich.

In 1903, Einstein married his Serbian sweetheart, Mileva Maric, and the couple moved into a one-bedroom flat in Bern. Two years later, she bore him a son, Hans Albert. The period surrounding Hans's birth was probably the happiest time in Einstein's life. Neighbors later recalled seeing the young father absentmindedly pushing a baby carriage down the city streets. From time to time, Einstein would reach into the carriage and remove a pad of paper on which to jot down notes to himself. It seems likely that the notepad in the baby's stroller contained some of the formulas and equations that led to the theory of relativity and the development of the atomic bomb.

During these early years at the patent office, Einstein spent most of his spare time studying theoretical physics. He composed a series of four seminal scientific papers,

which set forth some of the most momentous ideas in the long history of the quest to comprehend the universe. Space and time would never be looked at the same way again. Einstein's work won him the Nobel Prize in Physics in 1921, as well as much popular acclaim.

As Einstein pondered the workings of the universe, he received flashes of understanding that were too deep for words. "These thoughts did not come in any verbal formulation," Einstein was once quoted as saying. "I rarely think in words at all. A thought comes, and I may try to express it in words afterward."

Einstein eventually settled in the United States, where he publicly championed such causes as Zionism and nuclear disarmament. But he maintained his passion for physics. Right up until his death in 1955, Einstein kept seeking a unified field theory that would link the phenomena of gravitation and electromagnetism in one set of equations. It is a tribute to Einstein's vision that physicists today continue to seek a grand unification of physical theory. Einstein revolutionized scientific thinking in the twentieth century and beyond.

Albert Einstein was born at Ulm, in the former German state of Wüettemberg, on March 14, 1879, and grew up in Munich. He was the only son of Hermann Einstein and Pauline Koch. His father and uncle owned an electrotechnical plant. The family considered Albert a slow learner because he had difficulty with language. (It is now thought that he may have been dyslexic.) Legend has it that when Hermann asked the headmaster of his son's school about the best profession for Albert, the man replied, "It doesn't matter. He'll never make a success of anything."

Einstein did not do well in school. He didn't like the regimentation, and he suffered from being one of the few Jewish children in a Catholic school. This experience as an outsider was one that would repeat itself many times in his life.

One of Einstein's early loves was science. He remembered his father's showing him a pocket compass when he was around five years old, and marveling that the needle always pointed north, even if the case was spun. In that moment, Einstein recalled, he "felt something deeply hidden had to be behind things."

Another of his early loves was music. Around the age of six, Einstein began studying the violin. It did not come naturally to him; but when after several years he recognized the mathematical structure of music, the violin became a lifelong passion—although his talent was never a match for his enthusiasm.

When Einstein was ten his family enrolled him in the Luitpold Gymnasium, which is where, according to scholars, he developed a suspicion of authority. This trait served Einstein well later in life as a scientist. His habit of skepticism made it easy for him to question many long-standing scientific assumptions.

In 1895, Einstein attempted to skip high school by passing an entrance examination to the Federal Polytechnic School in Zurich, where he hoped to pursue a degree in electrical engineering. This is what he wrote about his ambitions at the time:

> *"If I were to have the good fortune to pass my examinations, I would go to Zurich. I would stay there for four years in order to study mathematics and physics. I imagine myself becoming a teacher in those branches of the natural sciences, choosing the theoretical part of them. Here are the reasons which lead me to this plan. Above all, it is my disposition for abstract and mathematical thought, and my lack of imagination and practical ability.*

Einstein failed the arts portion of the exam and so was denied admission to the polytechnic. His family instead sent him to secondary school at Aarau, in Switzerland, hoping that it would earn him a second chance to enter the Zurich school. It did, and Einstein graduated from the polytechnic in 1900. At about that time he fell in love with Mileva Maric, and in 1901 she gave birth, out of wedlock, to their first child, a daughter named Lieserl. Very little is known for certain about Lieserl, but it appears that she either was born with a crippling condition or fell very ill as an infant, then was put up for adoption, and died at about two years of age. Einstein and Maric married in 1903.

The year Hans was born, 1905, was a miracle year for Einstein. Somehow he managed to handle the demands of fatherhood and a full-time job and still publish four epochal scientific papers, all without benefit of the resources that an academic appointment might have provided.

In the spring of that year, Einstein submitted three papers to the German periodical *Annals of Physics* (*Annalen der Physik*). The three appeared together in the journal's volume 17. Einstein characterized the first paper, on the light quantum, as "very revolutionary." In it, he examined the phenomenon of the quantum (the fundamental unit of energy) discovered by the German physicist Max Planck. Einstein explained the photoelectric effect, which holds that for each electron emitted, a specific amount of energy is released. This is the quantum effect that states that energy is emitted in fixed amounts that can be expressed only as whole integers. This theory formed the basis for a great deal of quantum mechanics. Einstein suggested that light be considered a collection of independent particles of energy, but remarkably, he offered no experimental data. He simply argued hypothetically for the existence of these "light quantum" for aesthetic reasons.

Initially, physicists were hesitant to endorse Einstein's theory. It was too great a departure from scientifically accepted ideas of the time, and far beyond anything

Planck had discovered. It was this first paper, titled "On a Heuristic View concerning the Production and Transformation of Light"—not his work on relativity—that won Einstein the Nobel Prize in Physics in 1921.

In his second paper, "On a New Determination of Molecular Dimensions"—which Einstein wrote as his doctoral dissertation—and his third, "On the Movement of Small Particles Suspended in Stationary Liquids Required by the Molecular-Kinetic Theory of Heat," Einstein proposed a method to determine the size and motion of atoms. He also explained Brownian motion, a phenomenon described by the British botanist Robert Brown after studying the erratic movement of pollen suspended in fluid. Einstein asserted that this movement was caused by impacts between atoms and molecules. At the time, the very existence of atoms was still a subject of scientific debate, so there could be no underestimating the importance of these two papers. Einstein had confirmed the atomic theory of matter.

In the last of his 1905 papers, entitled "On the Electrodynamics of Moving Bodies," Einstein presented what became known as the special theory of relativity. The paper reads more like an essay than a scientific communication. Entirely theoretical, it contains no notes or bibliographic citations. Einstein wrote this 9,000-word treatise in just five weeks, yet historians of science consider it every bit as comprehensive and revolutionary as Isaac Newton's *Principia*.

What Newton had done for our understanding of gravity, Einstein had done for our view of time and space, managing in the process to overthrow the Newtonian conception of time. Newton had declared that "absolute, true, and mathematical time, of itself and from its own nature, flows equably without relation to anything external." Einstein held that all observers should measure the same speed for light, regardless of how fast they themselves are moving. Einstein also asserted that the mass of an object is not unchangeable but rather increases with the object's velocity. Experiments later proved that a small particle of matter, when accelerated to 86 percent of the speed of light, has twice as much mass as it does at rest.

Another consequence of relativity is that the relation between energy and mass may be expressed mathematically, which Einstein did in the famous equation $E=mc^2$. This expression—that energy is equivalent to mass times the square of the speed of light—led physicists to understand that even miniscule amounts of matter have the potential to yield enormous amounts of energy. Completely converting to energy just a part of the mass of a few atoms would, then, result in a colossal explosion. Thus did Einstein's modest-looking equation lead scientists to consider the consequences of splitting the atom (nuclear fission) and, at the urging of governments, to develop the atomic bomb. In 1909, Einstein was appointed professor of theoretical physics at the University of Zurich, and three years later he fulfilled his ambition to return to the

Federal Polytechnic School as a full professor. Other prestigious academic appointments and directorships followed. Throughout, he continued to work on his theory of gravity as well as his general theory of relativity. But as his professional status continued to rise, his marriage and health began to deteriorate. He and Mileva began divorce proceedings in 1914, the same year he accepted a professorship at the University of Berlin. When he later fell ill, his cousin Elsa nursed him back to health, and around 1919 they were married.

Where the special theory of relativity radically altered concepts of time and mass, the general theory of relativity changed our concept of space. Newton had written that "absolute space, in its own nature, without relation to anything external, remains always similar and immovable." Newtonian space is Euclidean, infinite, and unbounded. Its geometric structure is completely independent of the physical matter occupying it. In it, all bodies gravitate toward one another without having any effect on the structure of space. In stark contrast, Einstein's general theory of relativity asserts that not only does a body's gravitational mass act on other bodies, it also influences the structure of space. If a body is massive enough, it induces space to curve around it. In such a region, light appears to bend.

In 1919, Sir Arthur Eddington sought evidence to test the general theory. Eddington organized two expeditions, one to Brazil and the other to West Africa, to observe the light from stars as it passed near a massive body—the sun—during a total solar eclipse on May 29. Under normal circumstances such observations would be impossible, as the weak light from distant stars would be blotted out by daylight, but during the eclipse such light would briefly be visible.

In September, Einstein received a telegram from Hendrik Lorentz, a fellow physicist and close friend. It read: "Eddington found star displacement at rim of Sun, preliminary measurements between nine-tenths of a second and twice that value." Eddington's data were in keeping with the displacement predicted by the special relativity theory. His photographs from Brazil seemed to show the light from known stars in a different position in the sky during the eclipse than they were at nighttime, when their light did not pass near the sun. The theory of general relativity had been confirmed, forever changing the course of physics. Years later, when a student of Einstein's asked how he would have reacted had the theory been disproved, Einstein replied, "Then I would have felt sorry for the dear Lord. The theory is correct."

Confirmation of general relativity made Einstein world-famous. In 1921, he was elected a member of the British Royal Society. Honorary degrees and awards greeted him at every city he visited. In 1927, he began developing the foundation of quantum mechanics with the Danish physicist Niels Bohr, even as he continued to pursue his dream of a unified field theory. His travels in the United States led to his appointment

in 1932 as a professor of mathematics and theoretical physics at the Institute for Advanced Study in Princeton, New Jersey.

A year later, he settled permanently in Princeton after the ruling Nazi party in Germany began a campaign against "Jewish science." Einstein's property was confiscated, and he was deprived of German citizenship and positions in German universities. Until then, Einstein had considered himself a pacifist. But when Hitler turned Germany into a military power in Europe, Einstein came to believe that the use of force against Germany was justified. In 1939, at the dawn of World War II, Einstein became concerned that the Germans might be developing the capability to build an atomic bomb—a weapon made possible by his own research and for which he therefore felt a responsibility. He sent a letter to President Franklin D. Roosevelt warning of such a possibility and urging that the United States undertake nuclear research. The letter, composed by his friend and fellow scientist Leo Szilard, became the impetus for the formation of the Manhattan Project, which produced the world's first atomic weapons. In 1944, Einstein put a handwritten copy of his 1905 paper on special relativity up for auction and donated the proceeds—six million dollars—to the Allied war effort.

After the war, Einstein continued to involve himself with causes and issues that concerned him. In November 1952, having shown strong support for Zionism for many years, he was asked to accept the presidency of Israel. He respectfully declined, saying that he was not suited for the position. In April 1955, only one week before his death, Einstein composed a letter to the philosopher Bertrand Russell in which he agreed to sign his name to a manifesto urging all nations to abandon nuclear weapons.

Einstein died of heart failure on April 18, 1955. Throughout his life, he had sought to understand the mysteries of the cosmos by probing it with his thought rather than relying on his senses. "The truth of a theory is in your mind," he once said, "not in your eyes."

THE PRINCIPLE OF RELATIVITY

Translated by W. Perrett and G. B. Jeffery

ON THE ELECTRODYNAMICS OF MOVING BODIES

It is known that Maxwell's electrodynamics—as usually understood at the present time—when applied to moving bodies, leads to asymmetries which do not appear to be inherent in the phenomena. Take, for example, the reciprocal electrodynamic action of a magnet and a conductor. The observable phenomenon here depends only on the relative motion of the conductor and the magnet, whereas the customary view draws a sharp distinction between the two cases in which either the one or the other of these bodies is in motion. For if the magnet is in motion and the conductor at rest, there arises in the neighbourhood of the magnet an electric field with a certain definite energy, producing a current at the places where parts of the conductor are situated. But if the magnet is stationary and the conductor in motion, no electric field arises in the neighbourhood of the magnet. In the conductor, however, we find an electromotive force, to which in itself there is no corresponding energy, but which gives rise—assuming equality of relative motion in the two cases discussed—to electric currents of the same path and intensity as those produced by the electric form in the former case.

Examples of this sort, together with the unsuccessful attempts to discover any motion of the earth relatively to the "light medium," suggest that the phenomena of electrodynamics as well as of mechanics possess no properties corresponding to the idea of absolute rest. They suggest rather that, as has already been shown to the first order of small quantities, the same laws of electrodynamics and optics will be valid for all frames of reference for which the equations of mechanics hold good.[1] We will raise this conjecture (the purport of which will hereafter be called the "Principle of Relativity") to the status of a postulate, and also introduce another postulate, which is only apparently

---

1. The preceding memoir by Lorentz was not at this time known to the author.

irreconcilable with the former, namely, that light is always propagated in empty space with a definite velocity c which is independent of the state of motion of the emitting body. These two postulates suffice for the attainment of a simple and consistent theory of the electrodynamics of moving bodies based on Maxwell's theory for stationary bodies. The introduction of a "luminiferous ether" will prove to be superfluous inasmuch as the view here to be developed will not require an "absolutely stationary space" provided with special properties, nor assign a velocity-vector to a point of the empty space in which electromagnetic processes take place.

The theory to be developed is based—like all electrodynamics—on the kinematics of the rigid body, since the assertions of any such theory have to do with the relationships between rigid bodies (systems of co-ordinates), clocks, and electromagnetic processes. Insufficient consideration of this circumstance lies at the root of the difficulties which the electrodynamics of moving bodies at present encounters.

I. KINEMATICAL PART

§ 1. DEFINITION OF SIMULTANEITY

Let us take a system of co-ordinates in which the equations of Newtonian mechanics hold good.[1] In order to render our presentation more precise and to distinguish this system of co-ordinates verbally from others which will be introduced hereafter, we call it the "stationary system."

If a material point is at rest relatively to this system of co-ordinates, its position can be defined relatively thereto by the employment of rigid standards of measurement and the methods of Euclidean geometry, and can be expressed in Cartesian co-ordinates.

If we wish to describe the *motion* of a material point, we give the values of its co-ordinates as functions of the time. Now we must bear carefully in mind that a mathematical description of this kind has no physical meaning unless we are quite clear as to what we understand by "time." We have to take into account that all our judgments in which time plays a part are always judgments of *simultaneous events*. If, for instance, I say, "That train arrives here at 7 o'clock," I mean something like this: "The pointing of the small hand of my watch to 7 and the arrival of the train are simultaneous events."[2]

It might appear possible to overcome all the difficulties attending the definition of "time" by substituting "the position of the small hand of my watch" for "time." And in fact such a definition is satisfactory when we are concerned with defining a time

---

1. I.e. to the first approximation.
2. We shall not here discuss the inexactitude which lurks in the concept of simultaneity of two events at approximately the same place, which can only be removed by an abstraction.

exclusively for the place where the watch is located; but it is no longer satisfactory when we have to connect in time series of events occurring at different places, or—what comes to the same thing—to evaluate the times of events occurring at places remote from the watch.

We might, of course, content ourselves with time values determined by an observer stationed together with the watch at the origin of the co-ordinates, and co-ordinating the corresponding positions of the hands with light signals, given out by every event to be timed, and reaching him through empty space. But this co-ordination has the disadvantage that it is not independent of the standpoint of the observer with the watch or clock, as we know from experience. We arrive at a much more practical determination along the following line of thought.

If at the point A of space there is a clock, an observer at A can determine the time values of events in the immediate proximity of A by finding the positions of the hands which are simultaneous with these events. If there is at the point B of space another clock in all respects resembling the one at A, it is possible for an observer at B to determine the time values of events in the immediate neighbourhood of B. But it is not possible without further assumption to compare, in respect of time, an event at A with an event at B. We have so far defined only an "A time" and a "B time." We have not defined a common "time" for A and B, for the latter cannot be defined at all unless we establish *by definition* that the "time" required by light to travel from A to B equals the "time" it requires to travel from B to A. Let a ray of light start at the "A time" t_A from A towards B, let it at the "B time" t_B be reflected at B in the direction of A, and arrive again at A at the "A time" t'_A.

In accordance with definition the two clocks synchronize if

$$t_B - t_A = t'_A - t_B.$$

We assume that this definition of synchronism is free from contradictions, and possible for any number of points; and that the following relations are universally valid:

1. If the clock at B synchronizes with the clock at A, the clock at A synchronizes with the clock at B.

2. If the clock at A synchronizes with the clock at B and also with the clock at C, the clocks at B and C also synchronize with each other.

Thus with the help of certain imaginary physical experiments we have settled what is to be understood by synchronous stationary clocks located at different places, and have evidently obtained a definition of "simultaneous" or "synchronous," and of "time." The "time" of an event is that which is given simultaneously with the event by a stationary clock located at the place of the event, this clock being synchronous, and

indeed synchronous for all time determinations, with a specified stationary clock.

In agreement with experience we further assume the quantity

$$\frac{2AB}{t'_A - t_A} = c$$

to be a universal constant—the velocity of light in empty space.

It is essential to have time defined by means of stationary clocks in the stationary system, and the time now defined being appropriate to the stationary system we call it "the time of the stationary system."

§ 2. ON THE RELATIVITY OF LENGTHS AND TIMES

The following reflexions are based on the principle of relativity and on the principle of the constancy of the velocity of light. These two principles we define as follows:—

1. The laws by which the states of physical systems undergo change are not affected, whether these changes of state be referred to the one or the other of two systems of co-ordinates in uniform translatory motion.

2. Any ray of light moves in the "stationary" system of co-ordinates with the determined velocity c, whether the ray be emitted by a stationary or by a moving body. Hence

$$\text{velocity} = \frac{\text{light path}}{\text{time interval}}$$

where time interval is to be taken in the sense of the definition in § 1.

Let there be given a stationary rigid rod; and let its length be l as measured by a measuring-rod which is also stationary. We now imagine the axis of the rod lying along the axis of x of the stationary system of co-ordinates, and that a uniform motion of parallel translation with velocity v along the axis of x in the direction of increasing x is then imparted to the rod. We now inquire as to the length of the moving rod, and imagine its length to be ascertained by the following two operations:—

(a) The observer moves together with the given measuring-rod and the rod to be measured, and measures the length of the rod directly by superposing the measuring-rod, in just the same way as if all three were at rest.

(b) By means of stationary clocks set up in the stationary system and synchronizing in accordance with § 1, the observer ascertains at what points of the stationary system the two ends of the rod to be measured are located at a definite time. The distance between these two points, measured by the measuring-rod already employed, which in this case is at rest, is also a length which may be designated "the length of the rod."

In accordance with the principle of relativity the length to be discovered by the operation (a)—we will call it "the length of the rod in the moving system"—must be

equal to the length l of the stationary rod.

The length to be discovered by the operation (b) we will call "the length of the (moving) rod in the stationary system." This we shall determine on the basis of our two principles, and we shall find that it differs from l.

Current kinematics tacitly assumes that the lengths determined by these two operations are precisely equal, or in other words, that a moving rigid body at the epoch t may in geometrical respects be perfectly represented by *the same* body *at rest* in a definite position.

We imagine further that at the two ends A and B of the rod, clocks are placed which synchronize with the clocks of the stationary system, that is to say that their indications correspond at any instant to the "time of the stationary system" at the places where they happen to be. These clocks are therefore "synchronous in the stationary system."

We imagine further that with each clock there is a moving observer, and that these observers apply to both clocks the criterion established in § 1 for the synchronization of two clocks. Let a ray of light depart from A at the time[1] t_A, let it be reflected at B at the time t_B, and reach A again at the time t'_A. Taking into consideration the principle of the constancy of the velocity of light we find that

$$t_B - t_A = \frac{r_{AB}}{c - v} \quad \text{and} \quad t'_A - t_B = \frac{r_{AB}}{c + v}$$

where r_{AB} denotes the length of the moving rod—measured in the stationary system. Observers moving with the moving rod would thus find that the two clocks were not synchronous, while observers in the stationary system would declare the clocks to be synchronous.

So we see that we cannot attach any *absolute* signification to the concept of simultaneity, but that two events which, viewed from a system of co-ordinates, are simultaneous, can no longer be looked upon as simultaneous events when envisaged from a system which is in motion relatively to that system.

§ 3. THEORY OF THE TRANSFORMATION OF CO-ORDINATES AND TIMES FROM A STATIONARY SYSTEM TO ANOTHER SYSTEM IN UNIFORM MOTION OF TRANSLATION RELATIVELY TO THE FORMER

Let us in "stationary" space take two systems of co-ordinates, i.e. two systems, each of three rigid material lines, perpendicular to one another, and issuing from a point. Let the axes of X of the two systems coincide, and their axes of Y and Z respectively be parallel. Let each system be provided with a rigid measuring-rod and a number of

1. "Time" here denotes "time of the stationary" and also "position of hands of the moving clock situated at the place, under discussion."

clocks, and let the two measuring-rods, and likewise all the clocks of the two systems, be in all respects alike.

Now to the origin of one of the two systems (k) let a constant velocity v be imparted in the direction of the increasing x of the other stationary system (K), and let this velocity be communicated to the axes of the co-ordinates, the relevant measuring-rod, and the clocks. To any time of the stationary system K there then will correspond a definite position of the axes of the moving system, and from reasons of symmetry we are entitled to assume that the motion of k may be such that the axes of the moving system are at the time t (this "t" always denotes a time of the stationary system) parallel to the axes of the stationary system.

We now imagine space to be measured from the stationary system K by means of the stationary measuring-rod, and also from the moving system k by means of the measuring-rod moving with it; and that we thus obtain the co-ordinates x, y, z, and ξ, η, ζ respectively. Further, let the time t of the stationary system be determined for all points thereof at which there are clocks by means of light signals in the manner indicated in § 1; similarly let the time τ of the moving system be determined for all points of the moving system at which there are clocks at rest relatively to that system by applying the method, given in § 1, of light signals between the points at which the latter clocks are located.

To any system of values x, y, z, t, which completely defines the place and time of an event in the stationary system, there belongs a system of values ξ, η, ζ, τ, determining that event relatively to the system k, and our task is now to find the system of equations connecting these quantities.

In the first place it is clear that the equations must be *linear* on account of the properties of homogeneity which we attribute to space and time.

If we place $x' = x - vt$, it is clear that a point at rest in the system k must have a system of values x', y, z, independent of time. We first define τ as a function of x', y, z, and t. To do this we have to express in equations that τ is nothing else than the summary of the data of clocks at rest in system k, which have been synchronized according to the rule given in § 1.

From the origin of system k let a ray be emitted at the time τ_0 along the X-axis to x', and at the time τ_1 be reflected thence to the origin of the co-ordinates, arriving there at the time τ_2; we then must have $\frac{1}{2}(\tau_0 + \tau_2) = \tau_1$, or, by inserting the arguments of the function τ and applying the principle of the constancy of the velocity of light in the stationary system:

$$\frac{1}{2}\left[\tau(0,0,0,t) + \tau\left(0,0,0,t + \frac{x'}{c-v} + \frac{x'}{c+v}\right)\right] = \tau\left(x',0,0,t + \frac{x'}{c-v}\right)$$

Hence, if x' be chosen infinitesimally small,

$$\frac{1}{2}\left(\frac{1}{c-v}+\frac{1}{c+v}\right)\frac{\partial \tau}{\partial t}=\frac{\partial \tau}{\partial x'}+\frac{1}{c-v}\frac{\partial \tau}{\partial t}$$

or

$$\frac{\partial \tau}{\partial x'}+\frac{v}{c^2-v^2}\frac{\partial \tau}{\partial t}=0$$

It is to be noted that instead of the origin of the co-ordinates we might have chosen any other point for the point of origin of the ray, and the equation just obtained is therefore valid for all values of x', y, z.

An analogous consideration—applied to the axes of Y and Z—it being borne in mind that light is always propagated along these axes, when viewed from the stationary system, with the velocity $\sqrt{(c^2-v^2)}$, gives us

$$\frac{\partial \tau}{\partial y}=0, \frac{\partial \tau}{\partial z}=0.$$

Since τ is a linear function, it follows from these equations that

$$\tau = a\left(t-\frac{v}{c^2-v^2}x'\right)$$

where a is a function $\phi(v)$ at present unknown, and where for brevity it is assumed that at the origin of k, $\tau = 0$, when $t = 0$.

With the help of this result we easily determine the quantities ξ, η, ζ by expressing in equations that light (as required by the principle of the constancy of the velocity of light, in combination with the principle of relativity) is also propagated with velocity c when measured in the moving system. For a ray of light emitted at the time $\tau = 0$ in the direction of the increasing ξ

$$\xi = c\tau \quad \text{or} \quad \xi = ac\left(t-\frac{v}{c^2-v^2}x'\right).$$

But the ray moves relatively to the initial point of k, when measured in the stationary system, with the velocity $c - v$, so that

$$\frac{x'}{c-v}=t$$

If we insert this value of t in the equation for ξ, we obtain

$$\xi = a\frac{c^2}{c^2-v^2}x'.$$

In an analogous manner we find, by considering rays moving along the two other axes, that

$$\eta = c\tau = ac\left(t-\frac{v}{c^2-v^2}x'\right)$$

when

$$\frac{y}{\sqrt{\left(c^2 - v^2\right)}} = t, x' = 0$$

Thus

$$\eta = a\frac{c}{\sqrt{\left(c^2 - v^2\right)}}y \quad \text{and} \quad \zeta = a\frac{c}{\sqrt{\left(c^2 - v^2\right)}}z$$

Substituting for x' its value, we obtain

$$\tau = \phi(v)\beta\left(t - vx/c^2\right),$$
$$\xi = \phi(v)\beta(x - vt),$$
$$\eta = \phi(v)y,$$
$$\zeta = \phi(v)z$$

where

$$\beta = \frac{1}{\sqrt{\left(1 - v^2/c^2\right)}},$$

and ϕ is an as yet unknown function of v. If no assumption whatever be made as to the initial position of the moving system and as to the zero point of τ, an additive constant is to be placed on the right side of each of these equations.

We now have to prove that any ray of light, measured in the moving system, is propagated with the velocity c, if, as we have assumed, this is the case in the stationary system; for we have not as yet furnished the proof that the principle of the constancy of the velocity of light is compatible with the principle of relativity.

At the time $t = \tau = 0$, when the origin of the co-ordinates is common to the two systems, let a spherical wave be emitted therefrom, and be propagated with the velocity c in system K. If (x, y, z) be a point just attained by this wave, then

$$x^2 + y^2 + z^2 = c^2t^2.$$

Transforming this equation with the aid of our equations of transformation we obtain after a simple calculation

$$\xi^2 + \eta^2 + \zeta^2 = c^2\tau^2$$

The wave under consideration is therefore no less a spherical wave with velocity of propagation c when viewed in the moving system. This shows that our two fundamental principles are compatible.[1]

In the equations of transformation which have been developed there enters an unknown function ϕ of v, which we will now determine.

1. The equations of the Lorentz transformation may be more simply deduced directly from the condition that in virtue of these equations the relation $x^2 + y^2 + z^2 = c^2t^2$ shall have as its consequence the second relation $\xi^2 + \eta^2 + \zeta^2 = c^2\tau^2$.

For this purpose we introduce a third system of co-ordinates K', which relatively to the system k is in a state of parallel translatory motion parallel to the axis of X, such that the origin of co-ordinates of system k moves with velocity $-v$ on the axis of X. At the time $t = 0$ let all three origins coincide, and when $t = x = y = z = 0$ let the time t' of the system K be zero. We call the co-ordinates, measured in the system K', x', y', z', and by a twofold application of our equations of transformation we obtain

$$
\begin{aligned}
t' &= \phi(-v)\beta(-v)(\tau + v\xi/c^2) &= \phi(v)\phi(-v)t, \\
x' &= \phi(-v)\beta(-v)(\xi + v\tau) &= \phi(v)\phi(-v)x, \\
y' &= \phi(-v)\eta &= \phi(v)\phi(-v)y, \\
z' &= \phi(-v)\zeta &= \phi(v)\phi(-v)z,
\end{aligned}
$$

Since the relations between x', y', z' and x, y, z do not contain the time t, the systems K and K' are at rest with respect to one another, and it is clear that the transformation from K to K' must be the identical transformation. Thus

$$\phi(v)\phi(-v) = 1$$

We now inquire into the signification of $\phi(v)$. We give our attention to that part of the axis of Y of system k which lies between $\xi = 0, \eta = 0, \zeta = 0$ and $\xi = 0, \eta = l, \zeta = 0$. This part of the axis of Y is a rod moving perpendicularly to its axis with velocity v relatively to system K. Its ends possess in K the co-ordinates

$$x_1 = vt, y_1 = \frac{l}{\phi(v)}, z_1 = 0$$

and

$$x_2 = vt, y_2 = z_2 = 0$$

The length of the rod measured in K is therefore $l/\phi(v)$; and this gives us the meaning of the function $\phi(v)$. From reasons of symmetry it is now evident that the length of a given rod moving perpendicularly to its axis, measured in the stationary system, must depend only on the velocity and not on the direction and the sense of the motion. The length of the moving rod measured in the stationary system does not change, therefore, if v and $-v$ are interchanged. Hence follows that $l/\phi(v) = l/\phi(-v)$, or

$$\phi(v) = \phi(-v)$$

It follows from this relation and the one previously found that $\phi(v) = 1$, so that the transformation equations which have been found become

$$
\begin{aligned}
\tau &= \beta\left(t - vx/c^2\right), \\
\xi &= \beta(x - vt), \\
\eta &= y, \\
\zeta &= z
\end{aligned}
$$

where

$$\beta = 1 / \sqrt{(1 - v^2 / c^2)}$$

§ 4. PHYSICAL MEANING OF THE EQUATIONS OBTAINED IN RESPECT TO MOVING RIGID BODIES AND MOVING CLOCKS

We envisage a rigid sphere[1] of radius R, at rest relatively to the moving system k, and with its centre at the origin of co-ordinates of k. The equation of the surface of this sphere moving relatively to the system K with velocity v is

$$\xi^2 + \eta^2 + \zeta^2 =$$

The equation of this surface expressed in x, y, z at the time $t = 0$ is

$$\frac{x^2}{\left(\sqrt{(1 - v^2 / c^2)}\right)^2} + y^2 + z^2 = R^2$$

A rigid body which, measured in a state of rest, has the form of a sphere, therefore has in a state of motion—viewed from the stationary system—the form of an ellipsoid of revolution with the axes

$$R\sqrt{(1 - v^2/c^2)}, R, R$$

Thus, whereas the Y and Z dimensions of the sphere (and therefore of every rigid body of no matter what form) do not appear modified by the motion, the X dimension appears shortened in the ratio $1:\sqrt{(1 - v^2/c^2)}$, i.e. the greater the value of v, the greater the shortening. For $v = c$ all moving objects—viewed from the "stationary" system—shrivel up into plain figures. For velocities greater than that of light our deliberations become meaningless; we shall, however, find in what follows, that the velocity of light in our theory plays the part, physically, of an infinitely great velocity.

It is clear that the same results hold good of bodies at rest in the "stationary" system, viewed from a system in uniform motion.

Further, we imagine one of the clocks which are qualified to mark the time t when at rest relatively to the stationary system, and the time τ when at rest relatively to the moving system, to be located at the origin of the co-ordinates of k, and so adjusted that it marks the time τ. What is the rate of this clock, when viewed from the stationary system?

Between the quantities x, t, and τ, which refer to the position of the clock, we have, evidently, $x = vt$ and

1. That is, a body possessing spherical form when examined at rest.

$$\tau = \frac{1}{\sqrt{\left(1 - v^2/c^2\right)}}\left(t - vx/c^2\right)$$

Therefore,

$$\tau = t\sqrt{\left(1 - v^2/c^2\right)} = t - (1 - \sqrt{\left(1 - v^2/c^2\right)})t$$

whence it follows that the time marked by the clock (viewed in the stationary system) is slow by $1 - \sqrt{(1 - v^2/c^2)}$ seconds per second, or—neglecting magnitudes of fourth and higher order—by $\frac{1}{2}v^2c^2$.

From this there ensues the following peculiar consequence. If at the points A and B of K there are stationary clocks which, viewed in the stationary system, are synchronous; and if the clock at A is moved with the velocity v along the line AB to B, then on its arrival at B the two clocks no longer synchronize, but the clock moved from A to B lags behind the other which has remained at B by $\frac{1}{2}v^2c^2$ (up to magnitudes of fourth and higher order), t being the time occupied in the journey from A to B.

It is at once apparent that this result still holds good if the clock moves from A to B in any polygonal line, and also when the points A and B coincide.

If we assume that the result proved for a polygonal line is also valid for a continuously curved line, we arrive at this result: If one of two synchronous clocks at A is moved in a closed curve with constant velocity until it returns to A, the journey lasting t seconds, then by the clock which has remained at rest the travelled clock on its arrival at A will be $\frac{1}{2}v^2c^2$ second slow. Thence we conclude that a balance-clock[1] at the equator must go more slowly, by a very small amount, than a precisely similar clock situated at one of the poles under otherwise identical conditions.

§ 5. THE COMPOSITION OF VELOCITIES

In the system k moving along the axis of X of the system K with velocity v, let a point move in accordance with the equations

$$\xi = w_\xi \tau, \eta = w_\eta \tau, \zeta = 0$$

where w_ξ and w_η denote constants.

Required: the motion of the point relatively to the system K. If with the help of the equations of transformation developed in § 3 we introduce the quantities x, y, z, t into the equations of motion of the point, we obtain

---

1. Not a pendulum-clock, which is physically a system to which the Earth belongs. This case had to be excluded.

$$x = \frac{w_\xi + v}{1 + v w_\xi / c^2} w,$$

$$y = \frac{\sqrt{\left(1 - v^2 / c^2\right)}}{1 + v w_\xi / c^2} w_\eta t,$$

$$z = 0$$

Thus the law of the parallelogram of velocities is valid according to our theory only to a first approximation. We set

$$V^2 = \left(\frac{dx}{dt}\right)^2 + \left(\frac{dy}{dt}\right)^2$$

$$w^2 = w_\xi{}^2 + w_\eta{}^2,$$

$$a = \tan^{-1} w_y / w_x,$$

a is then to be looked upon as the angle between the velocities v and w. After a simple calculation we obtain

$$V = \frac{\sqrt{\left[\left(v^2 + w^2 + 2vw \cos a\right) - \left(vw \sin a / c^2\right)^2\right]}}{1 + vw \cos a / c^2}$$

It is worthy of remark that v and w enter into the expression for the resultant velocity in a symmetrical manner. If w also has the direction of the axis of X, we get

$$V = \frac{v + w}{1 + vw / c^2}$$

It follows from this equation that from a composition of two velocities which are less than c, there always results a velocity less than c. For if we set $v = c - \kappa$, $w = c - \lambda$, κ and λ being positive and less than c, then

$$V = c \frac{2c - \kappa - \lambda}{2c - \kappa - \lambda + \kappa\lambda/c} < c$$

It follows, further, that the velocity of light c cannot be altered by composition with a velocity less than that of light. For this case we obtain

$$V = \frac{c + w}{1 + w / c} = c$$

We might also have obtained the formula for V, for the case when v and w have the same direction, by compounding two transformations in accordance with § 3. If in addition to the systems K and k figuring in § 3 we introduce still another system of co-ordinates k' moving parallel to k, its initial point moving on the axis of X with the velocity w, we obtain equations between the quantities x, y, z, t and the corresponding

quantities of k', which differ from the equations found in § 3 only in that the place of "v" is taken by the quantity

$$\frac{v + w}{1 + vw / c^2}$$

from which we see that such parallel transformations—necessarily—form a group.

We have now deduced the requisite laws of the theory of kinematics corresponding to our two principles, and we proceed to show their application to electrodynamics.

II. ELECTRODYNAMICAL PART

§ 6. TRANSFORMATION OF THE MAXWELL-HERTZ EQUATIONS FOR EMPTY SPACE. ON THE NATURE OF THE ELECTROMOTIVE FORCES OCCURRING IN A MAGNETIC FIELD DURING MOTION

Let the Maxwell-Hertz equations for empty space hold good for the stationary system K, so that we have

$$\frac{1}{c}\frac{\partial X}{\partial x} = \frac{\partial N}{\partial y} - \frac{\partial M}{\partial z}, \quad \frac{1}{c}\frac{\partial L}{\partial t} = \frac{\partial Y}{\partial z} - \frac{\partial Z}{\partial y},$$

$$\frac{1}{c}\frac{\partial Y}{\partial t} = \frac{\partial L}{\partial z} - \frac{\partial N}{\partial x}, \quad \frac{1}{c}\frac{\partial M}{\partial t} = \frac{\partial Z}{\partial x} - \frac{\partial X}{\partial z},$$

$$\frac{1}{c}\frac{\partial Z}{\partial t} = \frac{\partial M}{\partial x} - \frac{\partial L}{\partial y}, \quad \frac{1}{c}\frac{\partial N}{\partial t} = \frac{\partial X}{\partial y} - \frac{\partial Y}{\partial x}$$

where (X, Y, Z) denotes the vector of the electric force, and (L, M, N) that of the magnetic force.

If we apply to these equations the transformation developed in § 3, by referring the electromagnetic processes to the system of co-ordinates there introduced, moving with the velocity v, we obtain the equations

$$\frac{1}{c}\frac{\partial X}{\partial \tau} = \frac{\partial}{\partial \tau}\left\{\beta\left(N - \frac{v}{c}Y\right)\right\} - \frac{\partial}{\partial \zeta}\left\{\beta\left(M + \frac{v}{c}Z\right)\right\},$$

$$\frac{1}{c}\frac{\partial}{\partial \tau}\left\{\beta\left(Y - \frac{v}{c}N\right)\right\} = \frac{\partial L}{\partial \xi} \qquad\qquad - \frac{\partial}{\partial \zeta}\left\{\beta\left(N - \frac{v}{c}Y\right)\right\},$$

$$\frac{1}{c}\frac{\partial}{\partial \tau}\left\{\beta\left(Z + \frac{v}{c}M\right)\right\} = \frac{\partial}{\partial \xi}\left\{\beta\left(M + \frac{v}{c}Z\right)\right\} - \frac{\partial L}{\partial \eta},$$

$$\frac{1}{c}\frac{\partial L}{\partial \tau} = \frac{\partial}{\partial \zeta}\left\{\beta\left(Y - \frac{v}{c}N\right)\right\} - \frac{\partial}{\partial \eta}\left\{\beta\left(Z + \frac{v}{c}M\right)\right\},$$

$$\frac{1}{c}\frac{\partial}{\partial \tau}\left\{\beta\left(M + \frac{v}{c}Z\right)\right\} = \frac{\partial}{\partial \xi}\left\{\beta\left(Z + \frac{v}{c}M\right)\right\} - \frac{\partial X}{\partial \zeta},$$

$$\frac{1}{c}\frac{\partial}{\partial \tau}\left\{\beta\left(N + \frac{v}{c}Y\right)\right\} = \frac{\partial X}{\partial \eta} \qquad\qquad - \frac{\partial}{\partial \xi}\left\{\beta\left(Y + \frac{v}{c}N\right)\right\},$$

where

$$b = 1/\sqrt{(1 - v^2/c^2)}$$

Now the principle of relativity requires that if the Maxwell-Hertz equations for empty space hold good in system K, they also hold good in system k; that is to say that the vectors of the electric and the magnetic force—(x', Y', Z') and (L', M', N')—of the moving system k, which are defined by their ponderomotive effects on electric or magnetic masses respectively, satisfy the following equations:—

$$\frac{1}{c}\frac{\partial X'}{\partial \tau} = \frac{\partial N'}{\partial \eta} - \frac{\partial M'}{\partial \zeta}, \quad \frac{1}{c}\frac{\partial L'}{\partial \tau} = \frac{\partial Y'}{\partial \zeta} - \frac{\partial Z'}{\partial \eta},$$

$$\frac{1}{c}\frac{\partial Y'}{\partial \tau} = \frac{\partial L'}{\partial \zeta} - \frac{\partial N'}{\partial \xi}, \quad \frac{1}{c}\frac{\partial M'}{\partial \tau} = \frac{\partial Z'}{\partial \xi} - \frac{\partial X'}{\partial \zeta},$$

$$\frac{1}{c}\frac{\partial Z'}{\partial \tau} = \frac{\partial M'}{\partial \xi} - \frac{\partial L'}{\partial \eta}, \quad \frac{1}{c}\frac{\partial N'}{\partial \tau} = \frac{\partial X'}{\partial \eta} - \frac{\partial Y'}{\partial \xi}.$$

Evidently the two systems of equations found for system k must express exactly the same thing, since both systems of equations are equivalent to the Maxwell-Hertz equations for system K. Since, further, the equations of the two systems agree, with the exception of the symbols for the vectors, it follows that the functions occurring in the systems of equations at corresponding places must agree, with the exception of a factor $\psi(v)$, which is common for all functions of the one system of equations, and is independent of ξ, η, ζ and τ but depends upon v. Thus we have the relations

$$X' = \psi(v)X, \qquad\qquad L' = \psi(v)L,$$

$$Y' = \psi(v)\beta\left(Y - \frac{v}{c}N\right), \quad M' = \psi(v)\beta\left(M + \frac{v}{c}Z\right),$$

$$Z' = \psi(v)\beta\left(Z - \frac{v}{c}M\right), \quad N' = \psi(v)\beta\left(N + \frac{v}{c}Y\right).$$

If we now form the reciprocal of this system of equations, firstly by solving the equations just obtained, and secondly by applying the equations to the inverse transformation (from k to K), which is characterized by the velocity $-v$, it follows, when we consider that the two systems of equations thus obtained must be identical, that $\psi(v)\,\psi(-v) = 1$. Further, from reasons of symmetry[1] $\psi(v) = \psi(-v)$, and therefore

$$\psi(v) = 1,$$

and our equations assume the form

$$X' = X, \qquad\qquad L' = L$$
$$Y' = \beta\left(Y - \frac{v}{c}N\right), \quad M' = \beta\left(M + \frac{v}{c}Z\right),$$
$$Z' = \beta\left(Z - \frac{v}{c}M\right), \quad N' = \beta\left(N - \frac{v}{c}Y\right).$$

As to the interpretation of these equations we make the following remarks: Let a point charge of electricity have the magnitude "one" when measured in the stationary system K, i.e. let it when at rest in the stationary system exert a force of one dyne upon an equal quantity of electricity at a distance of one cm. By the principle of relativity this electric charge is also of the magnitude "one" when measured in the moving system. If this quantity of electricity is at rest relatively to the stationary system, then by definition the vector (X, Y, Z) is equal to the force acting upon it. If the quantity of electricity is at rest relatively to the moving system (at least at the relevant instant), then the force acting upon it, measured in the moving system, is equal to the vector (x', Y', Z'). Consequently the first three equations above allow themselves to be clothed in words in the two following ways:

1. If a unit electric point charge is in motion in an electromagnetic field, there acts upon it, in addition to the electric force, an "electromotive force" which, if we neglect the terms multiplied by the second and higher powers of v/c, is equal to the vector-product of the velocity of the charge and the magnetic force, divided by the velocity of light. (Old manner of expression.)

2. If a unit electric point charge is in motion in an electromagnetic field, the force acting upon it is equal to the electric force which is present at the locality of the charge, and which we ascertain by transformation of the field to a system of co-ordinates at rest relatively to the electrical charge. (New manner of expression.)

The analogy holds with "magnetomotive forces." We see that electromotive force plays in the developed theory merely the part of an auxiliary concept, which owes its introduction to the circumstance that electric and magnetic forces do not exist independently of the state of motion of the system of co-ordinates.

1. If, for example, X = Y = Z = L = M = O, and N ≠ O, then from reasons of symmetry it is clear that when v changes sign without changing its numerical value, Y' must also change sign without changing its numerical value.

Furthermore it is clear that the asymmetry mentioned in the introduction as arising when we consider the currents produced by the relative motion of a magnet and a conductor, now disappears. Moreover, questions as to the "seat" of electrodynamic electromotive forces (unipolar machines) now have no point.

§ 7. THEORY OF DOPPLER'S PRINCIPLE AND OF ABERRATION

In the system K, very far from the origin of co-ordinates, let there be a source of electrodynamic waves, which in a part of space containing the origin of co-ordinates may be represented to a sufficient degree of approximation by the equations

$$X = X_0 \sin \Phi, \quad L = L_0 \sin \Phi,$$
$$Y = Y_0 \sin \Phi, \quad M = M_0 \sin \Phi,$$
$$Z = Z_0 \sin \Phi, \quad N = N_0 \sin \Phi,$$

where

$$\Phi = \omega \left\{ t - \frac{1}{c} \left(lx + my + nz \right) \right\}.$$

Here (X_0, Y_0, Z_0) and (L_0, M_0, N_0) are the vectors defining the amplitude of the wave-train, and l, m, n the direction-cosines of the wave-normals. We wish to know the constitution of these waves, when they are examined by an observer at rest in the moving system k.

Applying the equations of transformation found in § 6 for electric and magnetic forces, and those found in § 3 for the co-ordinates and the time, we obtain directly

$$X' = X_0 \sin \Phi', \qquad\qquad L' = L_0 \sin \Phi',$$
$$Y' = \beta \left(Y_0 - v N_0 / c \right) \sin \Phi', \quad M' = \beta \left(M_0 + v Z_0 / c \right) \sin \Phi',$$
$$Z' = \beta \left(Z_0 + v M_0 / c \right) \sin \Phi', \quad N = \beta \left(N_0 - v Y_0 / c \right) \sin \Phi',$$
$$\Phi' = \omega' \left\{ \tau - \frac{1}{c} \left(l' \xi + m' \eta + n' \zeta \right) \right\}$$

where

$$\omega' = \omega \beta \left(1 - l v / c \right),$$
$$l' = \frac{l - v / c}{1 - l v / c},$$
$$m' = \frac{m}{\beta \left(1 - l v / c \right)},$$
$$n' = \frac{n}{\beta \left(1 - l v / c \right)}.$$

From the equation for ω' it follows that if an observer is moving with velocity v relatively to an infinitely distant source of light of frequency ν, in such a way that the connecting line "source—observer" makes the angle ϕ with the velocity of the observer

referred to a system of co-ordinates, which is at rest relatively to the source of light, the frequency ν' of the light perceived by the observer is given by the equation

$$\nu' = \nu \frac{1 - \cos\phi \cdot v/c}{\sqrt{\left(1 - v^2/c^2\right)}}.$$

This is Doppler's principle for any velocities whatever. When $\phi = 0$ the equation assumes the perspicuous form

$$\nu' = \nu \sqrt{\frac{1 - v/c}{1 + v/c}}.$$

We see that, in contrast with the customary view, when $v = -c$, $\nu' = \infty$.

If we call the angle between the wave-normal (direction of the ray) in the moving system and the connecting line "source—observer" ϕ', the equation for l assumes the form

$$\cos\phi' = \frac{\cos\phi - v/c}{1 - \cos\phi \cdot v/c}.$$

This equation expresses the law of aberration in its most general form. If $\phi = {}^1/_2\pi$, the equation becomes simply

$$\cos \phi' = - v/c.$$

We still have to find the amplitude of the waves, as it appears in the moving system. If we call the amplitude of the electric or magnetic force A or A' respectively, accordingly as it is measured in the stationary system or in the moving system, we obtain

$$A'^2 = A^2 \frac{\left(1 - \cos\phi \cdot v/c\right)^2}{1 + v^2/c^2}.$$

which equation, if $\phi = 0$, simplifies into

$$A'^2 = A^2 \frac{1 - v/c}{1 + v/c}.$$

It follows from these results that to an observer approaching a source of light with the velocity c, this source of light must appear of infinite intensity.

§ 8. Transformation of the Energy of Light Rays. Theory of the Pressure of Radiation Exerted on Perfect Reflectors

Since $A^2/8\pi$ equals the energy of light per unit of volume, we have to regard $A'^2/8\pi$, by the principle of relativity, as the energy of light in the moving system. Thus A'^2/A^2 would be the ratio of the "measured in motion" to the "measured at rest" energy

of a given light complex, if the volume of a light complex were the same, whether measured in K or in k. But this is not the case. If l, m, n are the direction-cosines of the wave-normals of the light in the stationary system, no energy passes through the surface elements of a spherical surface moving with the velocity of light:

$$(x - lct)^2 + (y - mct)^2 + (z - nct)^2 = R^2.$$

We may therefore say that this surface permanently encloses the same light complex. We inquire as to the quantity of energy enclosed by this surface, viewed in system k, that is, as to the energy of the light complex relatively to the system k.

The spherical surface—viewed in the moving system—is an ellipsoidal surface, the equation for which, at the time $\tau = 0$, is

$$(\beta\xi - l\,\beta\xi\,v/c)^2 + (\eta - m\,\beta\xi\,v/c)^2 + (\zeta - n\,\beta\xi\,v/c)^2 = R^2$$

If S is the volume of the sphere, and S' that of this ellipsoid, then by a simple calculation

$$\frac{S'}{S} = \frac{\sqrt{1 - v^2/c^2}}{1 - \cos\phi \cdot v/c}.$$

Thus, if we call the light-energy enclosed by this surface E when it is measured in the stationary system, and E' when measured in the moving system, we obtain

$$\frac{E'}{E} = \frac{A^2 S'}{A^2 S} = \frac{1 - \cos\phi \cdot v/c}{\sqrt{\left(1 - v^2/c^2\right)}},$$

and this formula, when $\phi = 0$, simplifies into

$$\frac{E'}{E} = \sqrt{\frac{1 - v/c}{1 + v/c}}.$$

It is remarkable that the energy and the frequency of a light complex vary with the state of motion of the observer in accordance with the same law.

Now let the co-ordinate plane $\xi = 0$ be a perfectly reflecting surface, at which the plane waves considered in § 7 are reflected. We seek for the pressure of light exerted on the reflecting surface, and for the direction, frequency, and intensity of the light after reflexion.

Let the incidental light be defined by the quantities A, $\cos\phi$, ν (referred to system K). Viewed from k the corresponding quantities are

$$A' = A \frac{1 - \cos\phi \cdot v/c}{\sqrt{\left(1 - v^2/c^2\right)}},$$

$$\cos\phi' = \frac{\cos\phi - v/c}{1 - \cos\phi \cdot v/c},$$

$$v' = v \frac{1 - \cos\phi \cdot v/c}{\sqrt{\left(1 - v^2/c^2\right)}}.$$

For the reflected light, referring the process to system k, we obtain

$$A'' = A'$$
$$\cos\phi'' = -\cos\phi'$$
$$v'' = v'$$

Finally, by transforming back to the stationary system K, we obtain for the reflected light

$$A''' = A'' \frac{1 + \cos\phi'' \cdot v/c}{\sqrt{\left(1 - v^2/c^2\right)}} = A \frac{1 - 2\cos\phi \cdot v/c + v^2/c^2}{1 - v^2/c^2},$$

$$\cos\phi''' = \frac{\cos\phi'' + v/c}{1 + \cos\phi'' \cdot v/c} = -\frac{\left(1 + v^2/c^2\right)\cos\phi - 2v/c}{1 - 2\cos\phi \cdot v/c + v^2/c^2}$$

$$v''' = v'' \frac{1 + \cos\phi'' v/c}{\sqrt{\left(1 - v^2/c^2\right)}} = v \frac{1 - 2\cos\phi \cdot v/c + v^2/c^2}{1 - v^2/c^2}.$$

The energy (measured in the stationary system) which is incident upon unit area of the mirror in unit time is evidently $A^2(c\cos\phi - v)/8\pi$. The energy leaving the unit of surface of the mirror in the unit of time is $A'''^2(-c\cos\phi''' + v)/8\pi$. The difference of these two expressions is, by the principle of energy, the work done by the pressure of light in the unit of time. If we set down this work as equal to the product Pv, where P is the pressure of light, we obtain

$$P = 2 \cdot \frac{A^2}{8\pi} \frac{\left(\cos\phi - v/c\right)^2}{1 - v^2/c^2}.$$

In agreement with experiment and with other theories, we obtain to a first approximation

$$P = 2 \cdot \frac{A^2}{8\pi} \cos^2\phi.$$

All problems in the optics of moving bodies can be solved by the method here employed. What is essential is that the electric and magnetic force of the light which is influenced by a moving body be transformed into a system of co-ordinates; at rest relatively to the body. By this means all problems in the optics of moving bodies will be reduced to a series of problems in the optics of stationary bodies.

§ 9. TRANSFORMATION OF THE MAXWELL-HERTZ EQUATIONS WHEN CONVECTION-CURRENTS ARE TAKEN INTO ACCOUNT

We start from the equations

$$\frac{1}{c}\left\{\frac{\partial X}{\partial t} + u_x\rho\right\} = \frac{\partial N}{\partial y} - \frac{\partial M}{\partial z}, \quad \frac{1}{c}\frac{\partial L}{\partial t} = \frac{\partial Y}{\partial z} - \frac{\partial Z}{\partial y},$$

$$\frac{1}{c}\left\{\frac{\partial Y}{\partial t} + u_y\rho\right\} = \frac{\partial L}{\partial z} - \frac{\partial N}{\partial x}, \quad \frac{1}{c}\frac{\partial M}{\partial t} = \frac{\partial Z}{\partial x} - \frac{\partial X}{\partial z},$$

$$\frac{1}{c}\left\{\frac{\partial Z}{\partial t} + u_z\rho\right\} = \frac{\partial M}{\partial x} - \frac{\partial L}{\partial y}, \quad \frac{1}{c}\frac{\partial N}{\partial t} = \frac{\partial X}{\partial y} - \frac{\partial Y}{\partial x},$$

where

$$\rho = \frac{\partial X}{\partial x} + \frac{\partial Y}{\partial y} + \frac{\partial Z}{\partial z}$$

denotes 4π times the density of electricity, and (u_x, u_y, u_z) the velocity-vector of the charge. If we imagine the electric charges to be invariably coupled to small rigid bodies (ions, electrons), these equations are the electromagnetic basis of the Lorentzian electrodynamics and optics of moving bodies.

Let these equations be valid in the system K, and transform them, with the assistance of the equations of transformation given in §§ 3 and 6, to the system k. We then obtain the equations

$$\frac{1}{c}\left\{\frac{\partial X'}{\partial \tau} + u_\xi\rho'\right\} = \frac{\partial N'}{\partial \eta} - \frac{\partial M'}{\partial \zeta}, \quad \frac{1}{c}\frac{\partial L'}{\partial \tau} = \frac{\partial Y'}{\partial \zeta} - \frac{\partial Z'}{\partial \eta},$$

$$\frac{1}{c}\left\{\frac{\partial Y'}{\partial \tau} + u_\eta\rho'\right\} = \frac{\partial L'}{\partial \zeta} - \frac{\partial N'}{\partial \xi}, \quad \frac{1}{c}\frac{\partial M'}{\partial \tau} = \frac{\partial Z'}{\partial \xi} - \frac{\partial X'}{\partial \zeta},$$

$$\frac{1}{c}\left\{\frac{\partial Z'}{\partial \tau} + u_\zeta\rho'\right\} = \frac{\partial M'}{\partial \xi} - \frac{\partial L'}{\partial \eta}, \quad \frac{1}{c}\frac{\partial N'}{\partial \tau} = \frac{\partial X'}{\partial \eta} - \frac{\partial Y'}{\partial \xi},$$

where

$$u_\xi = \frac{u_x - v}{1 - u_x v/c^2}$$

$$u_\eta = \frac{u_y}{\beta\left(1 - u_x v/c^2\right)}$$

$$u_\zeta = \frac{u_z}{\beta\left(1 - u_x v/c^2\right)},$$

and

$$\rho' = \frac{\partial X'}{\partial \xi} + \frac{\partial Y'}{\partial \eta} + \frac{\partial Z'}{\partial \zeta}$$
$$= \beta\left(1 - u_x v / c^2\right)\rho.$$

Since—as follows from the theorem of addition of velocities (§ 5)—the vector (u_ξ, u_η, u_ζ) is nothing else than the velocity of the electric charge, measured in the system k, we have the proof that, on the basis of our kinematical principles, the electrodynamic foundation of Lorentz's theory of the electrodynamics of moving bodies is in agreement with the principle of relativity.

In addition I may briefly remark that the following important law may easily be deduced from the developed equations: If an electrically charged body is in motion anywhere in space without altering its charge when regarded from a system of co-ordinates moving with the body, its charge also remains—when regarded from the "stationary" system K—constant.

§ 10. DYNAMICS OF THE SLOWLY ACCELERATED ELECTRON

Let there be in motion in an electromagnetic field an electrically charged particle (in the sequel called an "electron"), for the law of motion of which we assume as follows:

If the electron is at rest at a given epoch, the motion of the electron ensues in the next instant of time according to the equations

$$m\frac{d^2 x}{dt^2} = \varepsilon X$$

$$m\frac{d^2 y}{dt^2} = \varepsilon Y$$

$$m\frac{d^2 z}{dt^2} = \varepsilon Z$$

where x, y, z denote the co-ordinates of the electron, and m the mass of the electron, as long as its motion is slow.

Now, secondly, let the velocity of the electron at a given epoch be v. We seek the law of motion of the electron in the immediately ensuing instants of time.

Without affecting the general character of our considerations, we may and will assume that the electron, at the moment when we give it our attention, is at the origin of the co-ordinates, and moves with the velocity v along the axis of X of the system K. It is then clear that at the given moment ($t = 0$) the electron is at rest relatively to a system of co-ordinates which is in parallel motion with velocity v along the axis of X.

From the above assumption, in combination with the principle of relativity, it is clear that in the immediately ensuing time (for small values of t) the electron, viewed from the system k, moves in accordance with the equations

$$m \frac{d^2 \xi}{dt^2} = \varepsilon X',$$

$$m \frac{d^2 \eta}{dt^2} = \varepsilon Y',$$

$$m \frac{d^2 \zeta}{dt^2} = \varepsilon Z',$$

in which the symbols ξ, η, ζ, τ, x', Y', Z' refer to the system k. If, further, we decide that when $t = x = y = z = 0$ then $\tau = \xi = \eta = \zeta = 0$, the transformation equations of §§ 3 and 6 hold good, so that we have

$$\xi = \beta(x - vt), \eta = y, \zeta = z, \tau = \beta(t - vx/c^2)$$

$$x' = X, Y' = \beta(Y - v N/c), Z' = \beta(Z + v M/c).$$

With the help of these equations we transform the above equations of motion from system k to system K, and obtain

$$\left. \begin{array}{l} \dfrac{d^2 x}{dt^2} = \dfrac{\varepsilon}{m\beta^3} X \\[2mm] \dfrac{d^2 y}{dt^2} = \dfrac{\varepsilon}{m\beta} \left(Y - \dfrac{v}{c} N \right) \\[2mm] \dfrac{d^2 z}{dt^2} = \dfrac{\varepsilon}{m\beta} \left(Z + \dfrac{v}{c} M \right) \end{array} \right\} \ \ldots (A)$$

Taking the ordinary point of view we now inquire as to the "longitudinal" and the "transverse" mass of the moving electron. We write the equations (A) in the form

$$m\beta^3 \frac{d^2 x}{dt^2} = \varepsilon X = \varepsilon X',$$

$$m\beta^2 \frac{d^2 y}{dt^2} = \varepsilon\beta \left(Y - \frac{v}{c} N \right) = \varepsilon Y',$$

$$m\beta^2 \frac{d^2 z}{dt^2} = \varepsilon\beta \left(Z + \frac{v}{c} M \right) = \varepsilon Z',$$

and remark firstly that ϵx', ϵY', ϵZ' are the components of the ponderomotive force acting upon the electron, and are so indeed as viewed in a system moving at the moment with the electron, with the same velocity as the electron. (This force might be measured, for example, by a spring balance at rest in the last-mentioned system.) Now if we call this force simply "the force acting upon the electron,"[1] and maintain the equation—mass x acceleration = force—and if we also decide that the accelerations are to be measured in the stationary system K, we derive from the above equations

1. The definition of force here given is not advantageous, as was first shown by M. Planck. It is more to the point to define force in such a way that the laws of momentum and energy assume the simplest form.

$$\text{Longitudinal mass} = \frac{m}{\left(\sqrt{1 - v^2/c^2}\right)^3}.$$

$$\text{Transverse mass} = \frac{m}{1 - v^2/c^2}.$$

With a different definition of force and acceleration we should naturally obtain other values for the masses. This shows us that in comparing different theories of the motion of the electron we must proceed very cautiously.

We remark that these results as to the mass are also valid for ponderable material points, because a ponderable material point can be made into an electron (in our sense of the word) by the addition of an electric charge, *no matter how small*.

We will now determine the kinetic energy of the electron. If an electron moves from rest at the origin of co-ordinates of the system K along the axis of X under the action of an electrostatic force X, it is clear that the energy withdrawn from the electrostatic field has the value $\int \varepsilon X \, dx$. As the electron is to be slowly accelerated, and consequently may not give off any energy in the form of radiation, the energy withdrawn from the electrostatic field must be put down as equal to the energy of motion W of the electron. Bearing in mind that during the whole process of motion which we are considering, the first of the equations (A) applies, we therefore obtain

$$W = \int \varepsilon X \, dx = m \int_0^v \beta^3 v \, dv$$

$$= mc^2 \left\{ \frac{1}{\sqrt{1 - v^2/c^2}} - 1 \right\}$$

Thus, when $v = c$, W becomes infinite. Velocities greater than that of light have—as in our previous results—no possibility of existence.

This expression for the kinetic energy must also, by virtue of the argument stated above, apply to ponderable masses as well.

We will now enumerate the properties of the motion of the electron which result from the system of equations (A), and are accessible to experiment.

1. From the second equation of the system (A) it follows that an electric force Y and a magnetic force N have an equally strong deflective action on an electron moving with the velocity v, when $Y = Nv/c$. Thus we see that it is possible by our theory to determine the velocity of the electron from the ratio of the magnetic power of deflexion A_m to the electric power of deflexion A_e, for any velocity, by applying the law

$$\frac{A_m}{A_e} = \frac{v}{c}.$$

This relationship may be tested experimentally, since the velocity of the electron can be directly measured, e.g. by means of rapidly oscillating electric and magnetic fields.

2. From the deduction for the kinetic energy of the electron it follows that between the potential difference, P, traversed and the acquired velocity v of the electron there must be the relationship

$$P = \int X dx = \frac{m}{\varepsilon} c^2 \left\{ \frac{1}{\sqrt{1 - v^2/c^2}} - \right.$$

3. We calculate the radius of curvature of the path of the electron when a magnetic force N is present (as the only deflective force), acting perpendicularly to the velocity of the electron. From the second of the equations (A) we obtain

$$-\frac{d^2 y}{dt^2} = \frac{v^2}{R} = \frac{\varepsilon}{m} \frac{v}{c} N \sqrt{1 - \frac{v^2}{c^2}}$$

or

$$R = \frac{mc^2}{\varepsilon} \cdot \frac{v/c}{\sqrt{\left(1 - v^2/c^2\right)}} \cdot \frac{1}{N}$$

These three relationships are a complete expression for the laws according to which, by the theory here advanced, the electron must move.

In conclusion I wish to say that in working at the problem here dealt with I have had the loyal assistance of my friend and colleague M. Besso, and that I am indebted to him for several valuable suggestions.

DOES THE INERTIA OF A BODY DEPEND UPON ITS ENERGY-CONTENT?

Translated from "Ist die Trägheit eins Körpers von seinem Energiegehalt abhängig?" Annalen der Physik, 17, 1905.

The results of the previous investigation lead to a very interesting conclusion, which is here to be deduced.

I based that investigation on the Maxwell-Hertz equations for empty space, together with the Maxwellian expression for the electromagnetic energy of space, and in addition the principle that:

The laws by which the states of physical system alter are independent of the alternative, to which of two systems of co-ordinates, in uniform motion of parallel translation relatively to each other, these alterations of state are referred (principle of relativity).

With these principles[1] as my basis I deduced *inter alia* the following result (§ 8):

Let a system of plane waves of light, referred to the system of co-ordinates (x, y, z), possess the energy l; let the direction of the ray (the wave-normal) make an angle ϕ with the axis of x of the system. If we introduce a new system of co-ordinates (ξ, η, ζ) moving in uniform parallel translation with respect to the system (x, y, z), and having its origin of co-ordinates in motion along the axis of x with the velocity v, then this quantity of light—measured in the system (ξ, η, ζ)—possesses the energy

$$l^* = l \frac{1 - \frac{v}{c}\cos\phi}{\sqrt{1 - v^2/c^2}}$$

where c denotes the velocity of light. We shall make use of this result in what follows.

Let there be a stationary body in the system (x, y, z), and let its energy—referred to the system (x, y, z)—be E_0. Let the energy of the body relative to the system (ξ, η, ζ) moving as above with the velocity v, be H_0.

Let this body send out, in a direction making an angle ϕ with the axis of x, plane waves of light, of energy $1/2 L$ measured relatively to (x, y, z), and simultaneously an equal quantity of light in the opposite direction. Meanwhile the body remains at rest with respect to the system (x, y, z). The principle of energy must apply to this process, and in fact (by the principle of relativity) with respect to both systems of co-ordinates. If we call the energy of the body after the emission of light E_1 or H_1 respectively, measured relatively to the system (x, y, z) or (ξ, η, ζ) respectively, then by employing the relation given above we obtain

$$E_0 = E_1 + \frac{1}{2}L + \frac{1}{2}L,$$

$$H_0 = H_1 + \frac{1}{2}L\frac{1 - \frac{v}{c}\cos\phi}{\sqrt{1 - v^2/c^2}} + \frac{1}{2}L\frac{1 + \frac{v}{c}\cos\phi}{\sqrt{1 - v^2/c^2}}$$

$$= H_1 + \frac{L}{\sqrt{1 - v^2/c^2}}$$

By subtraction we obtain from these equations

---

1. The principle of the constancy of the velocity of light is of course contained in Maxwell's equations.

$$H_0 - E_0 - (H_1 - E_1) = L\left\{\frac{1}{\sqrt{1 - v^2/c^2}} - 1\right\}.$$

The two differences of the form H – E occurring in this expression have simple physical significations. H and E are energy values of the same body referred to two systems of co-ordinates which are in motion relatively to each other, the body being at rest in one of the two systems (system (x, y, z)). Thus it is clear that the difference H – E can differ from the kinetic energy K of the body, with respect to the other system (ξ, η, ζ), only by an additive constant C, which depends on the choice of the arbitrary additive constants of the energies H and E. Thus we may place

$$H_0 - E_0 = K_0 + C,$$

$$H_1 - E_1 = K_1 + C,$$

since C does not change during the emission of light. So we have

$$K_0 - K_1 = L\left\{\frac{1}{\sqrt{1 - v^2/c^2}} - 1\right\}.$$

The kinetic energy of the body with respect to (ξ, η, ζ) diminishes as a result of the emission of light, and the amount of diminution is independent of the properties of the body. Moreover, the difference $K_0 - K_1$, like the kinetic energy of the electron (§ 10), depends on the velocity.

Neglecting magnitudes of fourth and higher orders we may place

$$K_0 - K_1 = \frac{1}{2}\frac{L}{c^2}v^2.$$

From this equation it directly follows that:

If a body gives off the energy L in the form of radiation, its mass diminishes by L/c^2. The fact that the energy withdrawn from the body becomes energy of radiation evidently makes no difference, so that we are led to the more general conclusion that:

The mass of a body is a measure of its energy-content; if the energy changes by L, the mass changes in the same sense by $L/9 \times 10^{20}$, the energy being measured in ergs, and the mass in grammes.

It is not impossible that with bodies whose energy-content is variable to a high degree (e.g. with radium salts) the theory may be successfully put to the test.

If the theory corresponds to the facts, radiation conveys inertia between the emitting and absorbing bodies.

ON THE INFLUENCE OF GRAVITATION ON THE PROPAGATION OF LIGHT

Translated from "Über den Einfluss der Schwerkraft auf die Ausbreitung des Lichtes," Annalen der Physik, 35, 1911.

In a memoir published four years ago[1] I tried to answer the question whether the propagation of light is influenced by gravitation. I return to this theme, because my previous presentation of the subject does not satisfy me, and for a stronger reason, because I now see that one of the most important consequences of my former treatment is capable of being tested experimentally. For it follows from the theory here to be brought forward, that rays of light, passing close to the sun, are deflected by its gravitational field, so that the angular distance between the sun and a fixed star appearing near to it is apparently increased by nearly a second of arc.

In the course of these reflexions further results are yielded which relate to gravitation. But as the exposition of the entire group of considerations would be rather difficult to follow, only a few quite elementary reflexions will be given in the following pages, from which the reader will readily be able to inform himself as to the suppositions of the theory and its line of thought. The relations here deduced, even if the theoretical foundation is sound, are valid only to a first approximation.

§ 1. A HYPOTHESIS AS TO THE PHYSICAL NATURE OF THE GRAVITATIONAL FIELD

In a homogeneous gravitational field (acceleration of gravity γ) let there be a stationary system of co-ordinates K, orientated so that the lines of force of the gravitational field run in the negative direction of the axis of z. In a space free of gravitational fields let there be a second system of co-ordinates K', moving with uniform acceleration (γ) in the positive direction of its axis of z. To avoid unnecessary complications, let us for the present disregard the theory of relativity, and regard both systems from the customary point of view of kinematics, and the movements occurring in them from that of ordinary mechanics.

Relatively to K, as well as relatively to K', material points which are not subjected to the action of other material points, move in keeping with the equations

1. A. Einstein, Jahrbuch für Radioakt. und Elektronik, 4, 1907

$$\frac{d^2 x}{dt^2} = 0, \frac{d^2 y}{dt^2} = 0, \frac{d^2 z}{dt^2} = -$$

For the accelerated system K' this follows directly from Galileo's principle, but for the system K, at rest in a homogeneous gravitational field, from the experience that all bodies in such a field are equally and uniformly accelerated. This experience, of the equal falling of all bodies in the gravitational field, is one of the most universal which the observation of nature has yielded; but in spite of that the law has not found any place in the foundations of our edifice of the physical universe.

But we arrive at a very satisfactory interpretation of this law of experience, if we assume that the systems K and K' are physically exactly equivalent, that is, if we assume that we may just as well regard the system K as being in a space free from gravitational fields, if we then regard K as uniformly accelerated. This assumption of exact physical equivalence makes it impossible for us to speak of the absolute acceleration of the system of reference, just as the usual theory of relativity forbids us to talk of the absolute velocity of a system;[1] and it makes the equal falling of all bodies in a gravitational field seem a matter of course.

As long as we restrict ourselves to purely mechanical processes in the realm where Newton's mechanics holds sway, we are certain of the equivalence of the systems K and K'. But this view of ours will not have any deeper significance unless the systems K and K' are equivalent with respect to all physical processes, that is, unless the laws of nature with respect to K are in entire agreement with those with respect to K'. By assuming this to be so, we arrive at a principle which, if it is really true, has great heuristic importance. For by theoretical consideration of processes which take place relatively to a system of reference with uniform acceleration, we obtain information as to the career of processes in a homogeneous gravitational field. We shall now show, first of all, from the standpoint of the ordinary theory of relativity, what degree of probability is inherent in our hypothesis.

§ 2. ON THE GRAVITATION OF ENERGY

One result yielded by the theory of relativity is that the inertia mass of a body increases with the energy it contains; if the increase of energy amounts to E, the increase in inertia mass is equal to E/c^2, when c denotes the velocity of light. Now is there an increase of gravitating mass corresponding to this increase of inertia mass? If not, then a body would fall in the same gravitational field with varying acceleration according to the energy it contained. That highly satisfactory result of the theory of

---

1. Of course we cannot replace any arbitrary gravitational field by a state of motion of the system without a gravitational field, any more than, by a transformation of relativity, we can transform all points of a medium in any kind of motion to rest.

relativity by which the law of the conservation of mass is merged in the law of conservation of energy could not be maintained, because it would compel us to abandon the law of the conservation of mass in its old form for inertia mass, and maintain it for gravitating mass.

But this must be regarded as very improbable. On the other hand, the usual theory of relativity does not provide us with any argument from which to infer that the weight of a body depends on the energy contained in it. But we shall show that our hypothesis of the equivalence of the systems K and K' gives us gravitation of energy as a necessary consequence.

Let the two material systems S_1 and S_2, provided with instruments of measurement, be situated on the z-axis of K at the distance h from each other,[1] so that the gravitation potential in S_2 is greater than that in S_1 by γh. Let a definite quantity of energy E be emitted from S_2 towards S_1. Let the quantities of energy in S_1 and S_2 be measured by contrivances which—brought to one place in the system z and there compared—shall be perfectly alike. As to the process of this conveyance of energy by radiation we can make no *a priori* assertion, because we do not know the influence of the gravitational field on the radiation and the measuring instruments in S_1 and S_2.

But by our postulate of the equivalence of K and K' we are able, in place of the system K in a homogeneous gravitational field, to set the gravitation-free system K', which moves with uniform acceleration in the direction of positive z, and with the z-axis of which the material systems S_1 and S_2 are rigidly connected.

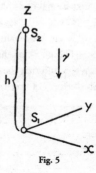

Fig. 5

We judge of the process of the transference of energy by radiation from S_2 to S_1 from a system K_0, which is to be free from acceleration. At the moment when the radiation energy E_2 is emitted from S_2 toward S_1, let the velocity of K' relatively to K_0 be zero. The radiation will arrive at S_1 when the time h/c has elapsed (to a first approximation). But at this moment the velocity of S_1 relatively to K_0 is $\gamma h/c = v$. Therefore by the ordinary theory of relativity the radiation arriving at S_1 does not possess the energy E_2, but a greater energy E_1, which is related to E_2, to a first approximation by the equation[2]

$$E_1 = E_2\left(1 + \frac{v}{c}\right) = E_2\left(1 + \gamma\frac{h}{c^2}\right) \qquad (1)$$

By our assumption exactly the same relation holds if the same process takes place in the system K, which is not accelerated, but is provided with a gravitational field. In this case we may replace γh by the potential Φ of the gravitation vector in S_2, if the arbitrary constant of Φ in S_1 is equated to zero. We then have the equation

1. The dimensions of S1 and S2, are regarded as infinitely small in comparison with h.
2. See above.

$$E_1 = E_2 + \frac{E_2}{c^2}\Phi \qquad (1a)$$

This equation expresses the law of energy for the process under observation. The energy E_1 arriving at S_1 is greater than the energy E_2, measured by the same means, which was emitted in S_2, the excess being the potential energy of the mass E_2/c^2 in the gravitational field. It thus proves that for the fulfilment of the principle of energy we have to ascribe to the energy E, before its emission in S_2, a potential energy due to gravity, which corresponds to the gravitational mass E/c^2. Our assumption of the equivalence of K and K' thus removes the difficulty mentioned at the beginning of this paragraph which is left unsolved by the ordinary theory of relativity.

The meaning of this result is shown particularly clearly if we consider the following cycle of operations:

1. The energy E, as measured in S_2, is emitted in the form of radiation in S_2 towards S_1, where, by the result just obtained, the energy $E(1 + \gamma h/c^2)$, as measured in S_1, is absorbed.

2. A body W of mass M is lowered from S_2 to S_1, work $M\gamma h$ being done in the process.

3. The energy E is transferred from S_1 to the body W while W is in S_1. Let the gravitational mass M be thereby changed so that it acquires the value M'.

4. Let W be again raised to S_2, work $M'\gamma h$ being done in the process.

5. Let E be transferred from W back to S_2.

The effect of this cycle is simply that S_1 has undergone the increase of energy $E\gamma h/c^2$, and that the quantity of energy $M'\gamma k - M\gamma h$ has been conveyed to the system in the form of mechanical work. By the principle of energy, we must therefore have

$$E\gamma \frac{h}{c^2} = M'\gamma h - M\gamma h,$$

or

$$M' - M = E/c^2 \ldots (1b)$$

The increase in gravitational mass is thus equal to E/c^2, and therefore equal to the increase in inertia mass as given by the theory of relativity.

The result emerges still more directly from the equivalence of the systems K and K', according to which the gravitational mass in respect of K is exactly equal to the inertia mass in respect of K'; energy must therefore possess a gravitational mass which is equal to its inertia mass. If a mass M_0 be suspended on a spring balance in the system K', the balance will indicate the apparent weight $M_0\gamma$ on account of the inertia of M_0. If the quantity of energy E be transferred to M_0, the spring balance, by the law of the inertia of energy, will indicate $(M_0 + E/c^2)\gamma$. By reason of our fundamental assumption

exactly the same thing must occur when the experiment is repeated in the system K, that is, in the gravitational field.

3. TIME AND THE VELOCITY OF LIGHT IN THE GRAVITATIONAL FIELD

If the radiation emitted in the uniformly accelerated system K' in S_2 toward S_1 had the frequency ν_2 relatively to the clock in S_2, then, relatively to S_1, at its arrival in S_1 it no longer has the frequency ν_2, relatively to an identical clock in S_1, but a greater frequency ν_1, such that to a first approximation

$$\nu_1 = \nu_2\left(1 + \gamma\frac{h}{c^2}\right). \tag{2}$$

For if we again introduce the unaccelerated system of reference K_0, relatively to which, at the time of the emission of light, K' has no velocity, then S_1, at the time of arrival of the radiation at S_1, has, relatively to K_0, the velocity $\gamma h/c$, from which, by Doppler's principle, the relation as given results immediately.

In agreement with our assumption of the equivalence of the systems K' and K, this equation also holds for the stationary system of co-ordinates K, provided with a uniform gravitational field, if in it the transference by radiation takes place as described. It follows, then, that a ray of light emitted in S_2 with a definite gravitational potential, and possessing at its emission the frequency ν_2—compared with a clock in S_2—will, at its arrival in S_1, possess a different frequency ν_1—measured by an identical clock in S_1. For γh we substitute the gravitational potential Φ of S_2—that of S_1 being taken as zero—and assume that the relation which we have deduced for the homogeneous gravitational field also holds for other forms of field. Then

$$\nu_1 = \nu_2\left(1 + \frac{\Phi}{c^2}\right) \tag{2a}$$

This result (which by our deduction is valid to a first approximation) permits, in the first place, of the following application. Let ν_0 be the vibration-number of an elementary light-generator, measured by a delicate clock at the same place. Let us imagine them both at a place on the surface of the Sun (where our S_2 is located). Of the light there emitted, a portion reaches the Earth (S_1), where we measure the frequency of the arriving light with a clock U in all respects resembling the one just mentioned. Then by (2a),

$$\nu = \nu_0\left(1 + \frac{\Phi}{c^2}\right)$$

where Φ is the (negative) difference of gravitational potential between the surface of the Sun and the Earth. Thus according to our view the spectral lines of sunlight, as compared with the corresponding spectral lines of terrestrial sources of light, must be

somewhat displaced toward the red, in fact by the relative amount

$$\frac{\nu_0 - \nu}{\nu_0} = -\frac{\Phi}{c^2} = 2.10^{-6}$$

If the conditions under which the solar bands arise were exactly known, this shifting would be susceptible of measurement. But as other influences (pressure, temperature) affect the position of the centres of the spectral lines, it is difficult to discover whether the inferred influence of the gravitational potential really exists.[1]

On a superficial consideration equation (2), or (2a), respectively, seems to assert an absurdity. If there is constant transmission of light from S_2 to S_1, how can any other number of periods per second arrive in S_1 than is emitted in S_2? But the answer is simple. We cannot regard ν_2 or respectively ν_1 simply as frequencies (as the number of periods per second) since we have not yet determined the time in system K. What ν_2 denotes is the number of periods with reference to the time-unit of the clock U in S_2, while ν_1 denotes the number of periods per second with reference to the identical clock in S_1. Nothing compels us to assume that the clocks U in different gravitation potentials must be regarded as going at the same rate. On the contrary, we must certainly define the time in K in such a way that the number of wave crests and troughs between S_2 and S_1 is independent of the absolute value of time; for the process under observation is by nature a stationary one. If we did not satisfy this condition, we should arrive at a definition of time by the application of which time would merge explicitly into the laws of nature, and this would certainly be unnatural and unpractical. Therefore the two clocks in S_1 and S_2 do not both give the "time" correctly. If we measure time in S_1 with the clock U, then we must measure time in S_2 with a clock which goes $1 + \Phi/c^2$ times more slowly than the clock U when compared with U at one and the same place. For when measured by such a clock the frequency of the ray of light which is considered above is at its emission in S_2

$$\nu_0\left(1 + \frac{\Phi}{c^2}\right)$$

and is therefore, by (2a), equal to the frequency ν_1 of the same ray of light on its arrival in S_1.

This has a consequence which is of fundamental importance for our theory. For if we measure the velocity of light at different places in the accelerated, gravitation-free system K', employing clocks U of identical constitution, we obtain the same magnitude at all these places. The same holds good, by our fundamental assumption, for the system K as well. But from what has just been said we must use clocks of unlike

1. L. F. Jewell (Journ. de Phys., 6, 1897, p. 84) and particularly Ch. Fabry and H. Boisson (Comptes rendus, 148, 1909, pp. 688-690) have actually found such displacements of fine spectral lines toward the red end of the spectrum, of the order of magnitude here calculated, but have ascribed them to an effect of pressure in the absorbing layer.

constitution, for measuring time at places with differing gravitation potential. For measuring time at a place which, relatively to the origin of the co-ordinates, has the gravitation potential Φ, we must employ a clock which—when removed to the origin of co-ordinates—goes $(1 + \Phi/c^2)$ times more slowly than the clock used for measuring time at the origin of co-ordinates. If we call the velocity of light at the origin of co-ordinates c_0, then the velocity of light c at a place with the gravitation potential Φ will be given by the relation

$$c = c_0\left(1 + \frac{\Phi}{c^2}\right) \tag{3}$$

The principle of the constancy of the velocity of light holds good according to this theory in a different form from that which usually underlies the ordinary theory of relativity.

4. BENDING OF LIGHT-RAYS IN THE GRAVITATIONAL FIELD

From the proposition which has just been proved, that the velocity of light in the gravitational field is a function of the place, we may easily infer, by means of Huyghens's principle, that light-rays propagated across a gravitational field undergo deflexion. For let E be a wave front of a plane light-wave at the time t, and let P_1 and P_2 be two points in that plane at

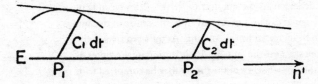

unit distance from each other. P_1 and P_2 lie in the plane of the paper, which is chosen so that the differential coefficient of Φ, taken in the direction of the normal to the plane, vanishes, and therefore also that of c. We obtain the corresponding wave front at time $t + dt$, or, rather, its line of section with the plane of the paper, by describing circles round the points P_1 and P_2 with radii $c_1 dt$ and $c_2 dt$ respectively, where c_1 and c_2 denote the velocity of light at the points P_1 and P_2 respectively, and by drawing the tangent to these circles. The angle through which the light-ray is deflected in the path $c dt$ is therefore

$$\left(c_1 - c_2\right)dt = -\frac{\partial c}{\partial n'}dt,$$

if we calculate the angle positively when the ray is bent toward the side of increasing n'. The angle of deflexion per unit of path of the light-ray is thus

$$-\frac{1}{c}\frac{\partial c}{\partial n'} \text{ or by (3) } -\frac{1}{c^2}\frac{\partial \Phi}{\partial n'}$$

Finally, we obtain for the deflexion which a light-ray experiences toward the side n' on any path (s) the expression

$$a = -\frac{1}{c^2}\int\frac{\partial \Phi}{\partial n'}\,ds \qquad (4)$$

We might have obtained the same result by directly considering the propagation of a ray of light in the uniformly accelerated system K', and transferring the result to the system K, and thence to the case of a gravitational field of any form.

By equation (4) a ray of light passing along by a heavenly body suffers a deflexion to the side of the diminishing gravitational potential, that is, on the side directed toward the heavenly body, of the magnitude

$$a = -\frac{1}{c^2}\int_{\theta=-\frac{1}{2}\pi}^{\theta=\frac{1}{2}\pi}\frac{kM}{r^2}\cos\theta\,ds = 2\frac{kM}{r^2\Delta}$$

where k denotes the constant of gravitation, M the mass of the heavenly body, Δ the distance of the ray from the centre of the body. A ray of light going past the Sun would accordingly undergo deflexion to the amount of $4\cdot10^{-6}$ = .83 seconds of arc. The angular distance of the star from the centre of the Sun appears to be increased by this amount. As the fixed stars in the parts of the sky near the Sun are visible during total eclipses of the Sun, this consequence of the theory may be compared with experience. With the planet Jupiter the displacement to be expected reaches to about $1/100$ of the amount given. It would be a most desirable thing if astronomers would take up the question here raised. For apart from any theory there is the question whether it is possible with the equipment at present available to detect an influence of gravitational fields on the propagation of light.

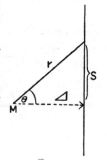

THE FOUNDATION OF THE GENERAL THEORY OF RELATIVITY

Translated from "Die Grundlage der allgemeinen Relativitätstheorie," Annalen der Physik, 49, 1916.

A. FUNDAMENTAL CONSIDERATIONS ON THE POSTULATE OF RELATIVITY

§ 1. OBSERVATIONS ON THE SPECIAL THEORY OF RELATIVITY

The special theory of relativity is based on the following postulate, which is also satisfied by the mechanics of Galileo and Newton.

If a system of co-ordinates K is chosen so that, in relation to it, physical laws hold good in their simplest form, the *same* laws also hold good in relation to any other system of co-ordinates K' moving in uniform translation relatively to K. This postulate we call the "special principle of relativity." The word "special" is meant to intimate that the principle is restricted to the case when K' has a motion of uniform translation relatively to K, but that the equivalence of K' and K does not extend to the case of non-uniform motion of K' relatively to K.

Thus the special theory of relativity does not depart from classical mechanics through the postulate of relativity, but through the postulate of the constancy of the velocity of light *in vacuo*, from which, in combination with the special principle of relativity, there follow, in the well-known way, the relativity of simultaneity, the Lorentzian transformation, and the related laws for the behaviour of moving bodies and clocks.

The modification to which the special theory of relativity has subjected the theory of space and time is indeed far-reaching, but one important point has remained unaffected. For the laws of geometry, even according to the special theory of relativity, are to be interpreted directly as laws relating to the possible relative positions of solid bodies at rest; and, in a more general way, the laws of kinematics are to be interpreted as laws which describe the relations of measuring bodies and clocks. To two selected material points of a stationary rigid body there always corresponds a distance of quite definite length, which is independent of the locality and orientation of the body, and is also independent of the time. To two selected positions of the hands of a clock at rest relatively to the privileged system of reference there always corresponds an interval of time of a definite length, which is independent of place and time. We shall soon see that the general theory of relativity cannot adhere to this simple physical interpretation of space and time.

§ 2. THE NEED FOR AN EXTENSION OF THE POSTULATE OF RELATIVITY

In classical mechanics, and no less in the special theory of relativity, there is an inherent epistemological defect which was, perhaps for the first time, clearly pointed

out by Ernst Mach. We will elucidate it by the following example: Two fluid bodies of the same size and nature hover freely in space at so great a distance from each other and from all other masses that only those gravitational forces need be taken into account which arise from the interaction of different parts of the same body. Let the distance between the two bodies be invariable, and in neither of the bodies let there be any relative movements of the parts with respect to one another. But let either mass, as judged by an observer at rest relatively to the other mass, rotate with constant angular velocity about the line joining the masses. This is a verifiable relative motion of the two bodies. Now let us imagine that each of the bodies has been surveyed by means of measuring instruments at rest relatively to itself, and let the surface of S_1 prove to be a sphere, and that of S_2 an ellipsoid of revolution. Thereupon we put the question— What is the reason for this difference in the two bodies? No answer can be admitted as epistemologically satisfactory,[1] unless the reason given is an *observable fact of experience*. The law of causality has not the significance of a statement as to the world of experience, except when *observable facts* ultimately appear as causes and effects.

Newtonian mechanics does not give a satisfactory answer to this question. It pronounces as follows: The laws of mechanics apply to the space R_1, in respect to which the body S_1 is at rest, but not to the space R_2 in respect to which the body S_2 is at rest. But the privileged space R_1 of Galileo, thus introduced, is a merely *factitious* cause, and not a thing that can be observed. It is therefore clear that Newton's mechanics does not really satisfy the requirement of causality in the case under consideration, but only apparently does so, since it makes the factitious cause R_1 responsible for the observable difference in the bodies S_1 and S_2.

The only satisfactory answer must be that the physical system consisting of S_1 and S_2 reveals within itself no imaginable cause to which the differing behaviour of S_1 and S_2 can be referred. The cause must therefore lie *outside* this system. We have to take it that the general laws of motion, which in particular determine the shapes of S_1 and S_2, must be such that the mechanical behaviour of S_1 and S_2 is partly conditioned, in quite essential respects, by distant masses which we have not included in the system under consideration. These distant masses and their motions relative to S_1 and S_2 must then be regarded as the seat of the causes (which must be susceptible to observation) of the different behaviour of our two bodies S_1 and S_2. They take over the rôle of the factitious cause R_1. Of all imaginable spaces R_1, R_2, etc., in any kind of motion relatively to one another, there is none which we may look upon as privileged *a priori* without reviving the above-mentioned epistemological objection. *The laws of physics must be of such a nature that they apply to systems of reference in any kind of motion.* Along this road we arrive at an extension of the postulate of relativity.

1. Of course an answer may be satisfactory from the point of view of epistemology, and yet be unsound physically, if it is in conflict with other experiences.

In addition to this weighty argument from the theory of knowledge, there is a well-known physical fact which favours an extension of the theory of relativity. Let K be a Galilean system of reference, i.e. a system relatively to which (at least in the four-dimensional region under consideration) a mass, sufficiently distant from other masses, is moving with uniform motion in a straight line. Let K' be a second system of reference which is moving relatively to K in *uniformly accelerated* translation. Then, relatively to K', a mass sufficiently distant from other masses would have an accelerated motion such that its acceleration and direction of acceleration are independent of the material composition and physical state of the mass.

Does this permit an observer at rest relatively to K' to infer that he is on a "really" accelerated system of reference? The answer is in the negative; for the above-mentioned relation of freely movable masses to K' may be interpreted equally well in the following way. The system of reference K' is unaccelerated, but the space-time territory in question is under the sway of a gravitational field, which generates the accelerated motion of the bodies relatively to K'.

This view is made possible for us by the teaching of experience as to the existence of a field of force, namely, the gravitational field, which possesses the remarkable property of imparting the same acceleration to all bodies.[1] The mechanical behaviour of bodies relatively to K' is the same as presents itself to experience in the case of systems which we are wont to regard as "stationary" or as "privileged." Therefore, from the physical standpoint, the assumption readily suggests itself that the systems K and K' may both with equal right be looked upon as "stationary," that is to say, they have an equal title as systems of reference for the physical description of phenomena.

It will be seen from these reflexions that in pursuing the general theory of relativity we shall be led to a theory of gravitation, since we are able to "produce" a gravitational field merely by changing the system of co-ordinates. It will also be obvious that the principle of the constancy of the velocity of light *in vacuo* must be modified, since we easily recognize that the path of a ray of light with respect to K' must in general be curvilinear, if with respect to K light is propagated in a straight line with a definite constant velocity.

§ 3. The Space-Time Continuum. Requirement of General Co-Variance for the Equations Expressing General Laws of Nature

In classical mechanics, as well as in the special theory of relativity, the co-ordinates of space and time have a direct physical meaning. To say that a point-event has the X_1 co-ordinate x_1 means that the projection of the point-event on the axis of X_1, determined by rigid rods and in accordance with the rules of Euclidean geometry, is obtained by measuring off a given rod (the unit of length) x_1 times from the origin of

---

1. Eötvös has proved experimentally that the gravitational field has this property in great accuracy.

co-ordinates along the axis of X_1. To say that a point-event has the X_4 co-ordinates x_4 = t, means that a standard clock, made to measure time in a definite unit period, and which is stationary relatively to the system of co-ordinates and practically coincident in space with the point-event,[1] will have measured off x_4 = t periods at the occurrence of the event.

This view of space and time has always been in the minds of physicists, even if, as a rule, they have been unconscious of it. This is clear from the part which these concepts play in physical measurements; it must also have underlain the reader's reflexions on the preceding paragraph (§ 2) for him to connect any meaning with what he there read. But we shall now show that we must put it aside and replace it by a more general view, in order to be able to carry through the postulate of general relativity, if the special theory of relativity applies to the special case of the absence of a gravitational field.

In a space which is free of gravitational fields we introduce a Galilean system of reference K (x, y, z, t), and also a system of co-ordinates K' (x', y', z', t') in uniform rotation relatively to K. Let the origins of both systems, as well as their axes of Z, permanently coincide. We shall show that for a space-time measurement in the system K' the above definition of the physical meaning of lengths and times cannot be maintained. For reasons of symmetry it is clear that a circle around the origin in the X, Y plane of K may at the same time be regarded as a circle in the X', Y' plane of K'. We suppose that the circumference and diameter of this circle have been measured with a unit measure infinitely small compared with the radius, and that we have the quotient of the two results. If this experiment were performed with a measuring-rod at rest relatively to the Galilean system K, the quotient would be π. With a measuring-rod at rest relatively to K', the quotient would be greater than π. This is readily understood if we envisage the whole process of measuring from the "stationary" system K, and take into consideration that the measuring-rod applied to the periphery undergoes a Lorentzian contraction, while the one applied along the radius does not. Hence Euclidean geometry does not apply to K'. The notion of co-ordinates defined above, which presupposes the validity of Euclidean geometry, therefore breaks down in relation to the system K'. So, too, we are unable to introduce a time corresponding to physical requirements in K', indicated by clocks at rest, relatively to K'. To convince ourselves of this impossibility, let us imagine two clocks of identical constitution placed, one at the origin of co-ordinates, and the other at the circumference of the circle, and both envisaged from the "stationary" system K. By a familiar result of the special theory of relativity, the clock at the circumference—judged from K—goes more slowly than the other, because the former is in motion and the latter at rest. An observer at the common origin of co-ordinates, capable of observing the clock at the circumference by means of light, would therefore

1. We assume the possibility of verifying "simultaneity" for events immediately proximate in space, or—to speak more precisely—for immediate proximity or coincidence in space-time, without giving a definition of this fundamental concept.

see it lagging behind the clock beside him. As he will not make up his mind to let the velocity of light along the path in question depend explicitly on the time, he will interpret his observations as showing that the clock at the circumference "really" goes more slowly than the clock at the origin. So he will be obliged to define time in such a way that the rate of a clock depends upon where the clock may be.

We therefore reach this result:—In the general theory of relativity, space and time cannot be defined in such a way that differences of the spatial co-ordinates can be directly measured by the unit measuring-rod, or differences in the time co-ordinate by a standard clock.

The method hitherto employed for laying co-ordinates into the space-time continuum in a definite manner thus breaks down, and there seems to be no other way which would allow us to adapt systems of co-ordinates to the four-dimensional universe so that we might expect from their application a particularly simple formulation of the laws of nature. So there is nothing for it but to regard all imaginable systems of co-ordinates, on principle, as equally suitable for the description of nature. This comes to requiring that:—

The general laws of nature are to be expressed by equations which hold good for all systems of co-ordinates, that is, are co-variant with respect to any substitutions whatever (generally co-variant).

It is clear that a physical theory which satisfies this postulate will also be suitable for the general postulate of relativity. For the sum of *all* substitutions in any case includes those which correspond to all relative motions of three-dimensional systems of co-ordinates. That this requirement of general co-variance, which takes away from space and time the last remnant of physical objectivity, is a natural one, will be seen from the following reflexion. All our space-time verifications invariably amount to a determination of space-time coincidences. If, for example, events consisted merely in the motion of material points, then ultimately nothing would be observable but the meetings of two or more of these points. Moreover, the results of our measurings are nothing but verifications of such meetings of the material points of our measuring instruments with other material points, coincidences between the hands of a clock and points on the clock dial, and observed point-events happening at the same place at the same time.

The introduction of a system of reference serves no other purpose than to facilitate the description of the totality of such coincidences. We allot to the universe four space-time variables x_1, x_2, x_3, x_4 in such a way that for every point-event there is a corresponding system of values of the variables $x_1 \ldots x_4$. To two coincident point-events there corresponds one system of values of the variables $x_1 \ldots x_4$, i.e. coincidence is characterized by the identity of the co-ordinates. If, in place of the variables $x_1 \ldots x_4$, we

introduce functions of them, x'_1, x'_2, x'_3, x'_4, as a new system of co-ordinates, so that the systems of values are made to correspond to one another without ambiguity, the equality of all four co-ordinates in the new system will also serve as an expression for the space-time coincidence of the two point-events. As all our physical experience can be ultimately reduced to such coincidences, there is no immediate reason for preferring certain systems of co-ordinates to others, that is to say, we arrive at the requirement of general co-variance.

§ 4. THE RELATION OF THE FOUR CO-ORDINATES TO MEASUREMENT IN SPACE AND TIME

It is not my purpose in this discussion to represent the general theory of relativity as a system that is as simple and logical as possible, and with the minimum number of axioms; but my main object is to develop this theory in such a way that the reader will feel that the path we have entered upon is psychologically the natural one, and that the underlying assumptions will seem to have the highest possible degree of security. With this aim in view let it now be granted that:

For infinitely small four-dimensional regions the theory of relativity in the restricted sense is appropriate, if the co-ordinates are suitably chosen.

For this purpose we must choose the acceleration of the infinitely small ("local") system of co-ordinates so that no gravitational field occurs; this is possible for an infinitely small region. Let X_1, X_2, X_3, be the co-ordinates of space, and X_4 the appertaining co-ordinate of time measured in the appropriate unit.[1] If a rigid rod is imagined to be given as the unit measure, the co-ordinates, with a given orientation of the system of co-ordinates, have a direct physical meaning in the sense of the special theory of relativity. By the special theory of relativity the expression

$$ds^2 = -dX_1^2 - dX_2^2 - dX_3^2 + dX_4^2 \tag{1}$$

then has a value which is independent of the orientation of the local system of co-ordinates, and is ascertainable by measurements of space and time. The magnitude of the linear element pertaining to points of the four-dimensional continuum in infinite proximity, we call ds. If the ds belonging to the element $dX_1 \ldots dX_4$ is positive, we follow Minkowski in calling it time-like; if it is negative, we call it space-like.

To the "linear element" in question, or to the two infinitely proximate point-events, there will also correspond definite differentials $dx_1 \ldots dx_4$ of the four-dimensional co-ordinates of any chosen system of reference. If this system, as well as the "local" system, is given for the region under consideration, the $dX\nu$ will allow themselves to be represented here by definite linear homogeneous expressions of the $dx\sigma$:—

1. The unit of time is to be chosen so that the velocity of light *in vacuo* as measured in the "local" system of co-ordinates is to be equal to unity.

$$dX_v = \sum_\sigma a_{v\sigma} dx_\sigma \tag{2}$$

Inserting these expressions in (1), we obtain

$$ds^2 = \sum_{\tau\sigma} g_{\sigma\tau} dx_\sigma dx_\tau \tag{3}$$

where the $g_{\sigma\tau}$ will be functions of the x_σ. These can no longer be dependent on the orientation and the state of motion of the "local" system of co-ordinates, for ds^2 is a quantity ascertainable by rod-clock measurement of point-events infinitely proximate in space-time, and defined independently of any particular choice of co-ordinates. The $g_{\sigma\tau}$ are to be chosen here so that $g_{\sigma\tau} = g_{\tau\sigma}$; the summation is to extend over all values of σ and τ, so that the sum consists of 4 x 4 terms, of which twelve are equal in pairs.

The case of the ordinary theory of relativity arises out of the case here considered, if it is possible, by reason of the particular relations of the $g_{\sigma\tau}$ in a finite region, to choose the system of reference in the finite region in such a way that the $g_{\sigma\tau}$ assume the constant values

$$\left.\begin{matrix} -1 & 0 & 0 & 0 \\ 0 & -1 & 0 & 0 \\ 0 & 0 & -1 & 0 \\ 0 & 0 & 0 & +1 \end{matrix}\right\} \tag{4}$$

We shall find hereafter that the choice of such co-ordinates is, in general, not possible for a finite region.

From the considerations of § 2 and § 3 it follows that the quantities $g\tau\sigma$ are to be regarded from the physical standpoint as the quantities which describe the gravitational field in relation to the chosen system of reference. For, if we now assume the special theory of relativity to apply to a certain four-dimensional region with the co-ordinates properly chosen, then the $g_{\sigma\tau}$ have the values given in (4). A free material point then moves, relatively to this system, with uniform motion in a straight line. Then if we introduce new space-time co-ordinates x_1, x_2, x_3, x_4, by means of any substitution we choose, the $g^{\sigma\tau}$ in this new system will no longer be constants, but functions of space and time. At the same time the motion of the free material point will present itself in the new co-ordinates as a curvilinear non-uniform motion, and the law of this motion will be independent of the nature of the moving particle. We shall therefore interpret this motion as a motion under the influence of a gravitational field. We thus find the occurrence of a gravitational field connected with a space-time variability of the g_σ. So, too, in the general case, when we are no longer able by a suitable choice of co-ordinates to apply the special theory of relativity to a finite region, we shall hold fast to the view

that the $g_{\sigma\tau}$ describe the gravitational field.

Thus, according to the general theory of relativity, gravitation occupies an exceptional position with regard to other forces, particularly the electromagnetic forces, since the ten functions representing the gravitational field at the same time define the metrical properties of the space measured.

B. MATHEMATICAL AIDS TO THE FORMULATION OF GENERALLY COVARIANT EQUATIONS

Having seen in the foregoing that the general postulate of relativity leads to the requirement that the equations of physics shall be covariant in the face of any substitution of the co-ordinates $x_1 \ldots x_4$, we have to consider how such generally covariant equations can be found. We now turn to this purely mathematical task, and we shall find that in its solution a fundamental rôle is played by the invariant ds given in equation (3), which, borrowing from Gauss's theory of surfaces, we have called the "linear element."

The fundamental idea of this general theory of covariants is the following:—Let certain things ("tensors") be defined with respect to any system of co-ordinates by a number of functions of the co-ordinates, called the "components" of the tensor. There are then certain rules by which these components can be calculated for a new system of co-ordinates, if they are known for the original system of co-ordinates, and if the transformation connecting the two systems is known. The things hereafter called tensors are further characterized by the fact that the equations of transformation for their components are linear and homogeneous. Accordingly, all the components in the new system vanish, if they all vanish in the original system. If, therefore, a law of nature is expressed by equating all the components of a tensor to zero, it is generally covariant. By examining the laws of the formation of tensors, we acquire the means of formulating generally covariant laws.

§ 5. CONTRAVARIANT AND COVARIANT FOUR-VECTORS

Contravariant Four-vectors.—The linear element is defined by the four "components" dx_ν, for which the law of transformation is expressed by the equation

$$dx'_\sigma = \sum_\nu \frac{\partial x'_\sigma}{\partial x_\nu} dx_\nu$$

(5)

The dx'_σ are expressed as linear and homogeneous functions of the dx_ν. Hence we may look upon these co-ordinate differentials as the components of a "tensor" of the particular kind which we call a contravariant four-vector. Any thing which is defined relatively to the system of co-ordinates by four quantities A^ν, and which is transformed by the same law

$$A'^\sigma = \sum_\nu \frac{\partial x'_\sigma}{\partial x_\nu} A^\nu, \tag{5a}$$

we also call a contravariant four-vector. From (5a) it follows at once that the sums $A^\sigma \pm B^\sigma$ are also components of a four-vector, if A^σ and B^σ are such. Corresponding relations hold for all "tensors" subsequently to be introduced. (Rule for the addition and subtraction of tensors.)

Covariant Four-vectors.—We call four quantities $A\nu$ the components of a covariant four-vector, if for any arbitrary choice of the contravariant four-vector B^ν

$$\sum_\nu A_\nu B^\nu = \text{Invariant} \tag{6}$$

The law of transformation of a covariant four-vector follows from this definition. For if we replace B^ν on the right-hand side of the equation

$$\sum_\sigma A'_\sigma B'^\sigma = \sum_\nu A_\nu B^\nu$$

by the expression resulting from the inversion of (5a),

$$\sum_\sigma \frac{\partial x_\nu}{\partial x'_\sigma} B'^\sigma,$$

we obtain

$$\sum_\sigma B'^\sigma \sum_\nu \frac{\partial x_\nu}{\partial x'_\sigma} A_\nu = \sum_\sigma B'^\sigma A'_\sigma.$$

Since this equation is true for arbitrary values of the B'^σ, it follows that the law of transformation is

$$A'_\sigma = \sum_\nu \frac{\partial x_\nu}{\partial x'_\sigma} A_\nu. \tag{7}$$

Note on a Simplified Way of Writing the Expressions.—A glance at the equations of this paragraph shows that there is always a summation with respect to the indices which occur twice under a sign of summation (e.g. the index ν in (5)), and only with respect to indices which occur twice. It is therefore possible, without loss of clearness, to omit the sign of summation. In its place we introduce the convention:—If an index occurs twice in one term of an expression, it is always to be summed unless the contrary is expressly stated.

The difference between covariant and contravariant four-vectors lies in the law of transformation ((7) or (5) respectively). Both forms are tensors in the sense of the general remark above. Therein lies their importance. Following Ricci and Levi-Civita, we

denote the contravariant character by placing the index above, the covariant by placing it below.

§ 6. Tensors of the Second and Higher Ranks

Contravariant Tensors.—If we form all the sixteen products of the components $A^{\mu\nu}$ and B^ν of two contravariant four-vectors

$$A^{\mu\nu} = A^\mu B^\nu \tag{8}$$

then by (8) and (5a) $A^{\mu\nu}$ satisfies the law of transformation

$$A^{\sigma\tau} = \frac{\partial x'_\sigma}{\partial x_\mu} \frac{\partial x'_\tau}{\partial x_\nu} A^{\mu\nu} \tag{9}$$

We call a thing which is described relatively to any system of reference by sixteen quantities, satisfying the law of transformation (9), a contravariant tensor of the second rank. Not every such tensor allows itself to be formed in accordance with (8) from two four-vectors, but it is easily shown that any given sixteen $A^{\mu\nu}$ can be represented as the sums of the $A^\mu B^\nu$ of four appropriately selected pairs of four-vectors. Hence we can prove nearly all the laws which apply to the tensor of the second rank defined by (9) in the simplest manner by demonstrating them for the special tensors of the type (8).

Contravariant Tensors of Any Rank.—It is clear that, on the lines of (8) and (9), contravariant tensors of the third and higher ranks may also be defined with 4^3 components, and so on. In the same way it follows from (8) and (9) that the contravariant four-vector may be taken in this sense as a contravariant tensor of the first rank.

Covariant Tensors.—On the other hand, if we take the sixteen products $A_{\mu\nu}$, of two covariant four-vectors A_μ and B_ν,

$$A_{\mu\nu} = A_\mu B_\nu, \tag{10}$$

the law of transformation for these is

$$A'_{\sigma\tau} = \frac{\partial x_\mu}{\partial x'_\sigma} \frac{\partial x_\nu}{\partial x'_\tau} A_{\mu\nu} \tag{11}$$

This law of transformation defines the covariant tensor of the second rank. All our previous remarks on contravariant tensors apply equally to covariant tensors.

NOTE.—It is convenient to treat the scalar (or invariant) both as a contravariant and a covariant tensor of zero rank.

Mixed Tensors.—We may also define a tensor of the second rank of the type

$$A^\nu_\mu = A_\mu B^\nu$$

which is covariant with respect to the index μ, and contravariant with respect to the

index ν. Its law of transformation is

$$A^\tau_\sigma = \frac{\partial x'_\tau}{\partial x_\nu} \frac{\partial x_\mu}{\partial x'_\sigma} A^\nu_\mu \tag{13}$$

Naturally there are mixed tensors with any number of indices of covariant character, and any number of indices of contravariant character. Covariant and contravariant tensors may be looked upon as special cases of mixed tensors.

Symmetrical Tensors.—A contravariant, or a covariant tensor, of the second or higher rank is said to be symmetrical if two components, which are obtained the one from the other by the interchange of two indices, are equal. The tensor $A^{\mu\nu}$, or the tensor $A_{\mu\nu}$, is thus symmetrical if for any combination of the indices μ, ν,

$$A^{\mu\nu} = A^{\nu\mu}, \tag{14}$$

or respectively,

$$A_{\mu\nu} = A_{\nu\mu} \tag{14a}$$

It has to be proved that the symmetry thus defined is a property which is independent of the system of reference. It follows in fact from (9), when (14) is taken into consideration, that

$$A'^{\sigma\tau} = \frac{\partial x'_\sigma}{\partial x_\mu} \frac{\partial x'_\tau}{\partial x_\nu} A^{\mu\nu} = \frac{\partial x'_\sigma}{\partial x_\mu} \frac{\partial x'_\tau}{\partial x_\nu} A^{\nu\mu} = \frac{\partial x'_\sigma}{\partial x_\nu} \frac{\partial x'_\tau}{\partial x_\mu} A^{\mu\nu} = A'^{\tau\sigma}$$

The last equation but one depends upon the interchange of the summation indices μ and ν, i.e. merely on a change of notation.

Antisymmetrical Tensors.—A contravariant or a covariant tensor of the second, third, or fourth rank is said to be antisymmetrical if two components, which are obtained the one from the other by the interchange of two indices, are equal and of opposite sign. The tensor $A^{\mu\nu}$, or the tensor $A_{\mu\nu}$, is therefore antisymmetrical, if always

$$A^{\mu\nu} = -A^{\nu\mu}, \tag{15}$$

or respectively,

$$A_{\mu\nu} = -A_{\nu\mu} \tag{15a}$$

Of the sixteen components $A^{\mu\nu}$, the four components $A^{\mu\mu}$ vanish; the rest are equal and of opposite sign in pairs, so that there are only six components numerically different (a six-vector). Similarly we see that the antisymmetrical tensor of the third rank $A^{\mu\nu\sigma}$ has only four numerically different components, while the antisymmetrical tensor $A^{\mu\nu\sigma\tau}$ has only one. There are no antisymmetrical tensors of higher rank than the fourth in a continuum of four dimensions.

§ 7. MULTIPLICATION OF TENSORS

Outer Multiplication of Tensors.—We obtain from the components of a tensor of rank n and of a tensor of rank m the components of a tensor of rank $n + m$ by multiplying each component of the one tensor by each component of the other. Thus, for example, the tensors T arise out of the tensors A and B of different kinds,

$$T_{\mu\nu\sigma} = A_{\mu\nu}B_\sigma,$$
$$T^{\mu\nu\sigma\tau} = A^{\mu\nu}B^{\sigma\tau},$$
$$T^{\sigma\tau}_{\mu\nu} = A_{\mu\nu}B^{\sigma\nu}.$$

The proof of the tensor character of T is given directly by the representations (8), (10), (12), or by the laws of transformation (9), (11), (13). The equations (8), (10), (12) are themselves examples of outer multiplication of tensors of the first rank.

"Contraction" of a Mixed Tensor.—From any mixed tensor we may form a tensor whose rank is less by two, by equating an index of covariant with one of contravariant character, and summing with respect to this index ("contraction"). Thus, for example, from the mixed tensor of the fourth rank $A^{\sigma\tau}_{\mu\nu}$, we obtain the mixed tensor of the second rank,

$$A^{\tau}_{\nu} = A^{\mu\tau}_{\mu\nu}\left(= \sum_{\mu} A^{\mu\tau}_{\mu\nu}\right),$$

and from this, by a second contraction, the tensor of zero rank,

$$A = A^{\nu}_{\nu} = A^{\mu\nu}_{\mu\nu}.$$

The proof that the result of contraction really possesses the tensor character is given either by the representation of a tensor according to the generalization of (12) in combination with (6), or by the generalization of (13).

Inner and Mixed Multiplication of Tensors.—These consist in a combination of outer multiplication with contraction.

Examples.—From the covariant tensor of the second rank $A_{\mu\nu}$ and the contravariant tensor of the first rank B^σ we form by outer multiplication the mixed tensor

$$D^{\sigma}_{\mu\nu} = A_{\mu\nu}B^{\sigma}.$$

On contraction with respect to the indices ν and σ, we obtain the covariant four-vector

$$D_\mu = D^{\nu}_{\mu\nu} = A_{\mu\nu}B^{\nu}.$$

This we call the inner product of the tensors $A_{\mu\nu}$ and B^σ. Analogously we form from the tensors $A_{\mu\nu}$, and $B^{\sigma\tau}$, by outer multiplication and double contraction, the inner

product $A_{\mu\nu}B^{\mu\nu}$. By outer multiplication and one contraction, we obtain from $A\mu\nu$ and $B^{\sigma\tau}$ the mixed tensor of the second rank $D^{\tau}_{\mu} = A_{\mu\nu}B^{\nu\tau}$. This operation may be aptly characterized as a mixed one, being "outer" with respect to the indices μ and τ, and "inner" with respect to the indices ν and σ.

We now prove a proposition which is often useful as evidence of tensor character. From what has just been explained, $A_{\mu\nu}B^{\mu\nu}$ is a scalar if $A_{\mu\nu}$, and $B^{\sigma\tau}$ are tensors. But we may also make the following assertion: If $A_{\mu\nu}B^{\mu\nu}$ is a scalar *for any choice of the tensor* $B^{\mu\nu}$, then $A_{\mu\nu}$ has tensor character. For, by hypothesis, for any substitution,

$$A'_{\sigma\tau}B'^{\sigma\tau} = A'_{\mu\nu}B'^{\mu\nu}.$$

But by an inversion of (9)

$$B^{\mu\nu} = \frac{\partial x_{\mu}}{\partial x'_{\sigma}}\frac{\partial x_{\nu}}{\partial x'_{\tau}}B'^{\sigma\tau}.$$

This, inserted in the above equation, gives

$$\left(A'_{\sigma\tau} - \frac{\partial x_{\mu}}{\partial x'_{\sigma}}\frac{\partial x_{\nu}}{\partial x'_{\tau}}A_{\mu\nu}\right)B'^{\sigma\tau} = 0.$$

This can only be satisfied for arbitrary values of $B'^{\sigma\tau}$ if the bracket vanishes. The result then follows by equation (11). This rule applies correspondingly to tensors of any rank and character, and the proof is analogous in all cases.

The rule may also be demonstrated in this form: If B^{μ} and C^{ν} are any vectors, and if, for all values of these, the inner product $A_{\mu\nu}B^{\mu}C^{\nu}$ is a scalar, then $A_{\mu\nu}$ is a covariant tensor. This latter proposition also holds good even if only the more special assertion is correct, that with any choice of the four-vector B^{μ} the inner product $A_{\mu\nu}B^{\mu}B^{\nu}$ is a scalar, if in addition it is known that $A_{\mu\nu}$ satisfies the condition of symmetry $A_{\mu\nu} = A_{\nu\mu}$. For by the method given above we prove the tensor character of $(A_{\mu\nu} + A_{\nu\mu})$, and from this the tensor character of $A_{\mu\nu}$ follows on account of symmetry. This also can be easily generalized to the case of covariant and contravariant tensors of any rank.

Finally, there follows from what has been proved, this law, which may also be generalized for any tensors: If for any choice of the four-vector B^{ν} the quantities $A_{\mu\nu}B^{\nu}$ form a tensor of the first rank, then $A_{\mu\nu}$ is a tensor of the second rank. For, if C^{μ} is any four-vector, then on account of the tensor character of $A_{\mu\nu}B^{\nu}$, the inner product $A_{\mu\nu}B^{\nu}C^{\mu}$ is a scalar for any choice of the two four-vectors B^{ν} and C^{μ}. From which the proposition follows.

§ 8. SOME ASPECTS OF THE FUNDAMENTAL TENSOR $g_{\mu\nu}$

The Covariant Fundamental Tensor.—In the invariant expression for the square of the linear element,

$$ds^2 = g_{\mu\nu}dx_\mu dx_\nu,$$

the part played by the dx_μ is that of a contravariant vector which may be chosen at will. Since further, $g_{\mu\nu} = g_{\nu\mu}$, it follows from the considerations of the preceding paragraph that $g_{\mu\nu}$ is a covariant tensor of the second rank. We call it the "fundamental tensor." In what follows we deduce some properties of this tensor which, it is true, apply to any tensor of the second rank. But as the fundamental tensor plays a special part in our theory, which has its physical basis in the peculiar effects of gravitation, it so happens that the relations to be developed are of importance to us only in the case of the fundamental tensor.

The Contravariant Fundamental Tensor.—If in the determinant formed by the elements $g_{\mu\nu}$, we take the co-factor of each of the $g_{\mu\nu}$ and divide it by the determinant $g = |g_{\mu\nu}|$, we obtain certain quantities $g^{\mu\nu}(= g^{\nu\mu})$ which, as we shall demonstrate, form a contravariant tensor.

By a known property of determinants

$$g_{\mu\sigma}g^{\nu\sigma} = \delta_\mu^\nu \tag{16}$$

where the symbol δ_μ^ν denotes 1 or 0, according as $\mu = \nu$ or $\mu \neq \nu$.

Instead of the above expression for ds^2 we may thus write

$$g_{\mu\sigma}\delta_\nu^\sigma dx_\mu dx_\nu$$

or, by (16)

$$g_{\mu\sigma}g_{\nu\tau}g^{\sigma\tau}dx_\mu dx_\nu$$

But, by the multiplication rules of the preceding paragraphs, the quantities

$$d\xi_\sigma = g_{\mu\sigma}dx_\mu$$

form a covariant four-vector, and in fact an arbitrary vector, since the dx_μ are arbitrary. By introducing this into our expression we obtain

$$ds^2 = g^{\sigma\tau}d\xi_\sigma d\xi_\tau$$

Since this, with the arbitrary choice of the vector $d\xi\sigma$, is a scalar, and $g^{\sigma\tau}$ by its definition is symmetrical in the indices σ and τ, it follows from the results of the preceding paragraph that $g^{\sigma\tau}$ is a contravariant tensor.

It further follows from (16) that δ_μ is also a tensor, which we may call the mixed fundamental tensor.

The Determinant of the Fundamental Tensor.—By the rule for the multiplication of determinants

$$\left| g_{\mu\alpha} g^{\alpha\nu} \right| = \left| g_{\mu\alpha} \right| \times \left| g^{\alpha\nu} \right|$$

On the other hand

$$\left| g_{\mu\alpha} g^{\alpha\nu} \right| = \left| \delta_\mu^\nu \right| = 1.$$

It therefore follows that

$$\left| g_{\mu\nu} \right| \times \left| g^{\mu\nu} \right| = 1 \tag{17}$$

The Volume Scalar.—We seek first the law of transformation of the determinant $g = \left| g_{\mu\nu} \right|$. In accordance with (11)

$$g' = \left| \frac{\partial x_\mu}{\partial x'_\sigma} \frac{\partial x}{\partial x'_\tau} g_{\mu\nu} \right|.$$

Hence, by a double application of the rule for the multiplication of determinants, it follows that

$$g' = \left| \frac{\partial x_\mu}{\partial x'_\sigma} \right| \cdot \left| \frac{\partial x_\nu}{\partial x'_\tau} \right| \cdot \left| g_{\mu\nu} \right| = \left| \frac{\partial x_\mu}{\partial x'_\sigma} \right|^2 g,$$

or

$$\sqrt{g'} = \left| \frac{\partial x_\mu}{\partial x'_\sigma} \right| \cdot \sqrt{g}.$$

On the other hand, the law of transformation of the element of volume

$$d\tau = \int dx_1 dx_2 dx_3 dx_4$$

is, in accordance with the theorem of Jacobi,

$$d\tau' = \left| \frac{\partial x'_\sigma}{\partial x_\mu} \right| d\tau.$$

By multiplication of the last two equations, we obtain

$$\sqrt{g'} d\tau' = \sqrt{g} d\tau \tag{18}.$$

Instead of $\sqrt{g}$, we introduce in what follows the quantity $\sqrt{-g}$, which is always real on account of the hyperbolic character of the space-time continuum. The invariant $\sqrt{-g} \, d\tau$ is equal to the magnitude of the four-dimensional element of volume in the "local" system of reference, as measured with rigid rods and clocks in the sense of the special theory of relativity.

Note on the Character of the Space-time Continuum.—Our assumption that the special theory of relativity can always be applied to an infinitely small region, implies that ds^2 can always be expressed in accordance with (1) by means of real quantities $dX_1 \ldots dX_4$. If we denote by $d\tau_0$ the "natural" element of volume dX_1, dX_2, dX_3, dX_4, then

$$d\tau_0 = \sqrt{-g}\, d\tau \tag{18a}$$

If $\sqrt{-g}$ were to vanish at a point of the four-dimensional continuum, it would mean that at this point an infinitely small "natural" volume would correspond to a finite volume in the co-ordinates. Let us assume that this is never the case. Then g cannot change sign. We will assume that, in the sense of the special theory of relativity, g always has a finite negative value. This is a hypothesis as to the physical nature of the continuum under consideration, and at the same time a convention as to the choice of co-ordinates.

But if $-g$ is always finite and positive, it is natural to settle the choice of co-ordinates *a posteriori* in such a way that this quantity is always equal to unity. We shall see later that by such a restriction of the choice of co-ordinates it is possible to achieve an important simplification of the laws of nature.

In place of (18), we then have simply $d\tau' = d\tau$, from which, in view of Jacobi's theorem, it follows that

$$\left| \frac{\partial x'_\sigma}{\partial x_\mu} \right| = 1 \tag{19}$$

Thus, with this choice of co-ordinates, only substitutions for which the determinant is unity are permissible.

But it would be erroneous to believe that this step indicates a partial abandonment of the general postulate of relativity. We do not ask "What are the laws of nature which are covariant in face of all substitutions for which the determinant is unity?" but our question is "What are the generally covariant laws of nature?" It is not until we have formulated these that we simplify their expression by a particular choice of the system of reference.

The Formation of New Tensors by Means of the Fundamental Tensor.—Inner, outer, and mixed multiplication of a tensor by the fundamental tensor give tensors of different character and rank. For example,

$$A^\mu = g^{\mu\sigma} A_\sigma,$$
$$A = g_{\mu\nu} A^{\mu\nu}.$$

The following forms may be specially noted:—

$$A^{\mu\nu} = g^{\mu\alpha} g^{\nu\beta} A_{\alpha\beta},$$
$$A_{\mu\nu} = g_{\mu\alpha} g_{\nu\beta} A^{\alpha\beta}$$

(the "complements" of covariant and contravariant tensors respectively), and

$$B_{\mu\nu} = g_{\mu\nu} g^{\alpha\beta} A_{\alpha\beta}$$

1216

We call $B_{\mu\nu}$ the reduced tensor associated with $A_{\mu\nu}$. Similarly,

$$B^{\mu\nu} = g^{\mu\nu}g_{\alpha\beta}A^{\alpha\beta}.$$

It may be noted that $g^{\mu\nu}$ is nothing more then the complement of $g_{\mu\nu}$, since

$$g^{\mu\alpha}g^{\nu\beta}g_{\alpha\beta} = g^{\mu\alpha}\delta^{\nu}_{\alpha} = g^{\mu\nu}.$$

§ 9. THE EQUATION OF THE GEODETIC LINE. THE MOTION OF A PARTICLE

As the linear element ds is defined independently of the system of co-ordinates, the line drawn between two points P and P' of the four-dimensional continuum in such a way that $\int ds$ is stationary—a geodetic line—has a meaning which also is independent of the choice of co-ordinates. Its equation is

$$\delta \int_P^{P'} ds = 0 \tag{20}$$

Carrying out the variation in the usual way, we obtain from this equation four differential equations which define the geodetic line; this operation will be inserted here for the sake of completeness. Let λ be a function of the co-ordinates x_ν, and let this define a family of surfaces which intersect the required geodetic line as well as all the lines in immediate proximity to it which are drawn through the points P and P'. Any such line may then be supposed to be given by expressing its co-ordinates x_ν as functions of λ. Let the symbol δ indicate the transition from a point of the required geodetic to the point corresponding to the same λ on a neighbouring line. Then for (20) we may substitute

$$\left. \begin{aligned} \int_{\lambda_1}^{\lambda_2} \delta w d\lambda &= 0 \\ w^2 &= g_{\mu\nu}\frac{dx_\mu}{d\lambda}\frac{dx_\nu}{d\lambda} \end{aligned} \right\} \tag{20a}$$

But since

$$\delta w = \frac{1}{w}\left\{\frac{1}{2}\frac{\partial g_{\mu\nu}}{\partial x_\sigma}\frac{dx_\mu}{d\lambda}\frac{dx_\nu}{d\lambda}\delta x_\sigma + g_{\mu\nu}\frac{dx_\mu}{d\lambda}\delta\left(\frac{dx_\nu}{d\lambda}\right)\right\},$$

and

$$\delta\left(\frac{dx_\nu}{d\lambda}\right) = \frac{d}{d\lambda}(\delta x_\nu),$$

we obtain from (20a), after a partial integration,

$$\int_{\lambda_1}^{\lambda_2} \kappa_\sigma \delta x_\sigma d\lambda = 0,$$

where

$$\kappa_\sigma = -\frac{d}{d\lambda}\left\{\frac{g_{\mu\nu}}{w}\frac{dx_\mu}{d\lambda}\right\} - \frac{1}{2w}\frac{\partial g_{\mu\nu}}{\partial x_\sigma}\frac{dx_\mu}{d\lambda}\frac{dx_\nu}{d\lambda} \qquad (20b)$$

Since the values of δx_σ are arbitrary, it follows from this that

$$\kappa_\sigma = 0 \qquad (20c)$$

are the equations of the geodetic line.

If ds does not vanish along the geodetic line we may choose the "length of the arc" s, measured along the geodetic line, for the parameter λ. Then $w = 1$, and in place of (20c) we obtain

$$g_{\mu\nu}\frac{d^2 x_\mu}{ds^2} + \frac{\partial g_{\mu\nu}}{\partial x_\sigma}\frac{dx_\sigma}{ds}\frac{dx_\mu}{ds} - \frac{1}{2}\frac{\partial g_{\mu\nu}}{\partial x_\sigma}\frac{dx_\mu}{ds}\frac{dx_\nu}{ds} = 0$$

or, by a mere change of notation,

$$g_{\alpha\sigma}\frac{d^2 x_\alpha}{ds^2} + [\mu\nu, \sigma]\frac{\partial x_\mu}{\partial s}\frac{dx_\nu}{ds} = 0 \qquad (20d)$$

where, following Christoffel, we have written

$$[\mu\nu, \sigma] = \frac{1}{2}\left(\frac{\partial g_{\mu\sigma}}{\partial x_\nu} + \frac{\partial g_{\nu\sigma}}{\partial x_\mu} - \frac{\partial g_{\mu\nu}}{\partial x_\sigma}\right) \qquad (21)$$

Finally, if we multiply (20d) by $g^{\sigma\tau}$ (outer multiplication with respect to τ, inner with respect to σ), we obtain the equations of the geodetic line in the form

$$\frac{d^2 x_\tau}{ds^2} + \{\mu\nu, \tau\}\frac{dx_\mu}{ds}\frac{dx_\nu}{ds} = 0 \qquad (22)$$

where, following Christoffel, we have set

$$\{\mu\nu, \tau\} = g^{\tau\alpha}[\mu\nu, \alpha] \qquad (23)$$

§ 10. THE FORMATION OF TENSORS BY DIFFERENTIATION

With the help of the equation of the geodetic line we can now easily deduce the laws by which new tensors can be formed from old by differentiation. By this means we are able for the first time to formulate generally covariant differential equations. We reach this goal by repeated application of the following simple law:

If in our continuum a curve is given, the points of which are specified by the arcual distance s measured from a fixed point on the curve, and if, further, ϕ is an invariant function of space, then $d\phi/ds$ is also an invariant. The proof lies in this, that ds is an invariant as well as $d\phi$.

As

$$\frac{d\phi}{ds} = \frac{\partial \phi}{\partial x_\mu}\frac{dx_\mu}{ds}$$

therefore

$$\psi = \frac{\partial \phi}{\partial x_\mu} \frac{dx_\mu}{ds}$$

is also an invariant, and an invariant for all curves starting from a point of the continuum, that is, for any choice of the vector dx_μ. Hence it immediately follows that

$$A_\mu = \frac{\partial \phi}{\partial x_\mu} \qquad (24)$$

is a covariant four-vector—the "gradient" of ϕ.

According to our rule, the differential quotient

$$\chi = \frac{\partial \psi}{\partial s}$$

taken on a curve, is similarly an invariant. Inserting the value of ψ, we obtain in the first place

$$\chi = \frac{\partial^2 \phi}{\partial x_\mu \partial x_\nu} \frac{dx_\mu}{ds} \frac{dx_\nu}{ds} + \frac{\partial \phi}{\partial x_\mu} \frac{d^2 x_\mu}{ds^2}$$

The existence of a tensor cannot be deduced from this forthwith. But if we may take the curve along which we have differentiated to be a geodetic, we obtain on substitution for $d^2 x_\nu / ds^2$ from (22),

$$\chi = \left(\frac{\partial^2 \phi}{\partial x_\mu \partial x_\nu} - \{\mu\nu, \tau\} \frac{\partial \phi}{\partial x_\tau} \right) \frac{dx_\mu}{ds} \frac{dx_\nu}{ds}.$$

Since we may interchange the order of the differentiations, and since by (23) and (21) $\{\mu\nu, \tau\}$ is symmetrical in μ and ν, it follows that the expression in brackets is symmetrical in μ and ν. Since a geodetic line can be drawn in any direction from a point of the continuum, and therefore dx_μ / ds is a four-vector with the ratio of its components arbitrary, it follows from the results of § 7 that

$$A_{\mu\nu} = \frac{\partial^2 \phi}{\partial x_\mu \partial x_\nu} - \{\mu\nu, \tau\} \frac{\partial \phi}{\partial x_\tau} \qquad (25)$$

is a covariant tensor of the second rank. We have therefore come to this result: from the covariant tensor of the first rank

$$A_\mu = \frac{\partial \phi}{\partial x_\mu}$$

we can, by differentiation, form a covariant tensor of the second rank

$$A_{\mu\nu} = \frac{\partial A_\mu}{\partial x_\nu} - \{\mu\nu, \tau\} A_\tau \qquad (26)$$

We call the tensor $A_{\mu\nu}$ the "extension" (covariant derivative) of the tensor A_μ. In the first place we can readily show that the operation leads to a tensor, even if the vector A_μ cannot be represented as a gradient. To see this, we first observe that

$$\psi \frac{\partial \phi}{\partial x_\mu}$$

is a covariant vector, if ψ and ϕ are scalars. The sum of four such terms

$$S_\mu = \psi^{(1)} \frac{\phi \partial^{(1)}}{\partial x_\mu} + . + . + \psi^{(4)} \frac{\partial \phi^{(4)}}{\partial x_\mu},$$

is also a covariant vector, if $\psi^{(1)}$, $\phi^{(1)} \ldots \psi^{(4)}$, $\phi^{(4)}$ are scalars. But it is clear that any covariant vector can be represented in the form S_μ. For, if A_μ is a vector whose components are any given functions of the x_ν, we have only to put (in terms of the selected system of co-ordinates)

$$\psi^{(1)} = A_1, \quad \phi^{(1)} = x_1,$$
$$\psi^{(2)} = A_2, \quad \phi^{(2)} = x_2,$$
$$\psi^{(3)} = A_3, \quad \phi^{(3)} = x_3,$$
$$\psi^{(4)} = A_4, \quad \phi^{(4)} = x_4,$$

in order to ensure that S_μ shall be equal to A_μ.

Therefore, in order to demonstrate that $A_{\mu\nu}$ is a tensor if *any* covariant vector is inserted on the right-hand side for A_μ, we only need show that this is so for the vector S_μ. But for this latter purpose it is sufficient, as a glance at the right-hand side of (26) teaches us, to furnish the proof for the case

$$A_\mu = \psi \frac{\partial \phi}{\partial x_\mu}.$$

Now the right-hand side of (25) multiplied by ψ,

$$\psi \frac{\partial^2 \phi}{\partial x_\mu \partial x_\nu} - \left\{ \mu\nu, \tau \right\} \psi \frac{\partial \phi}{\partial x_\tau}$$

is a tensor. Similarly

$$\frac{\partial \psi}{\partial x_\mu} \frac{\partial \phi}{\partial x_\nu}$$

being the outer product of two vectors, is a tensor. By addition, there follows the tensor character of

$$\frac{\partial}{\partial x_\nu} \left(\psi \frac{\partial \phi}{\partial x_\mu} \right) - \left\{ \mu\nu, \tau \right\} \left(\psi \frac{\partial \phi}{\partial x_\tau} \right).$$

As a glance at (26) will show, this completes the demonstration for the vector

$$\psi \frac{\partial \phi}{\partial x_\mu}$$

and consequently, from what has already been proved, for any vector A_μ.

By means of the extension of the vector, we may easily define the "extension" of a covariant tensor of any rank. This operation is a generalization of the extension of a vector. We restrict ourselves to the case of a tensor of the second rank, since this suffices to give a clear idea of the law of formation.

As has already been observed, any covariant tensor of the second rank can be represented[1] as the sum of tensors of the type $A_\mu B_\nu$. It will therefore be sufficient to deduce the expression for the extension of a tensor of this special type. By (26) the expressions

$$\frac{\partial A_\mu}{\partial x_\sigma} - \{\sigma\mu, \tau\} A_\tau,$$

$$\frac{\partial B_\nu}{\partial x_\sigma} - \{\sigma\nu, \tau\} B_\tau,$$

are tensors. On outer multiplication of the first by B_ν, and of the second by A_μ, we obtain in each case a tensor of the third rank. By adding these, we have the tensor of the third rank

$$A_{\mu\nu\sigma} = \frac{\partial A_{\mu\nu}}{\partial x_\sigma} - \{\sigma\mu, \tau\} A_{\tau\nu} - \{\sigma\nu, \tau\} A_{\mu\tau} \tag{27}$$

where we have put $A_{\mu\nu} = A_\mu B_\nu$. As the right-hand side of (27) is linear and homogeneous in the $A_{\mu\nu}$ and their first derivatives, this law of formation leads to a tensor, not only in the case of a tensor of the type $A_\mu B_\nu$, but also in the case of a sum of such tensors, i.e. in the case of any covariant tensor of the second rank. We call $A_{\mu\nu\sigma}$ the extension of the tensor $A_{\mu\nu}$.

It is clear that (26) and (24) concern only special cases of extension (the extension of the tensors of rank one and zero respectively).

In general, all special laws of formation of tensors are included in (27) in combination with the multiplication of tensors.

§ 11. SOME CASES OF SPECIAL IMPORTANCE

The Fundamental Tensor.—We will first prove some lemmas which will be useful hereafter. By the rule for the differentiation of determinants

$$dg = g^{\mu\nu} g\, dg_{\mu\nu} = g_{\mu\nu} g\, dg^{\mu\nu} \tag{28}$$

The last member is obtained from the last but one, if we bear in mind that $g_{\mu\nu} dg^{\mu'\nu} = \delta_\mu^{\mu'}$, so that $g_{\mu\nu} g^{\mu\nu} = 4$, and consequently

$$g_{\mu\nu} dg^{\mu\nu} + g^{\mu\nu} dg_{\mu\nu} = 0.$$

From (28), it follows that

---

1. By outer multiplication of the vector with arbitrary components $A_{11}, A_{12}, A_{13}, A_{14}$ by the vector with components 1, 0, 0, 0, we produce a tensor with components

$$\begin{array}{cccc} A_{11} & A_{12} & A_{13} & A_{14} \\ 0 & 0 & 0 & 0 \\ 0 & 0 & 0 & 0 \\ 0 & 0 & 0 & 0 \end{array}$$

By the addition of four tensors of this type, we obtain the tensor $A_{\mu\nu}$ with any assigned components.

$$\frac{1}{\sqrt{-g}}\frac{\partial\sqrt{-g}}{\partial x_\sigma}=\frac{1}{2}\frac{\partial\log(-g)}{\partial x_\sigma}=\frac{1}{2}g^{\mu\nu}\frac{\partial g_{\mu\nu}}{\partial x_\sigma}=\frac{1}{2}g_{\mu\nu}\frac{\partial g^{\mu\nu}}{\partial x_\sigma} \qquad (29)$$

Further, from $g_{\mu\sigma}g^{\nu\sigma}=\delta^\nu_\mu$, it follows on differentiation that

$$\left.\begin{array}{l}g_{\mu\sigma}dg^{\nu\sigma}=-g^{\nu\sigma}dg_{\mu\sigma}\\[2mm] g_{\mu\sigma}\dfrac{\partial g^{\nu\sigma}}{\partial x_\lambda}=-g^{\nu\sigma}d\dfrac{\partial g_{\mu\sigma}}{\partial x_\lambda}\end{array}\right\} \qquad (30)$$

From these, by mixed multiplication by $g^{\sigma\tau}$ and $g_{\nu\lambda}$ respectively, and a change of notation for the indices, we have

$$\left.\begin{array}{l}dg^{\mu\nu}=-g^{\mu\alpha}g^{\nu\beta}dg_{\alpha\beta}\\[2mm] \dfrac{\partial g^{\mu\nu}}{\partial x_\sigma}=-g^{\mu\alpha}g^{\nu\beta}\dfrac{\partial g_{\alpha\beta}}{\partial x_\sigma}\end{array}\right\} \qquad (31)$$

and

$$\left.\begin{array}{l}dg_{\mu\nu}=-g_{\mu\alpha}g_{\nu\beta}dg^{\alpha\beta}\\[2mm] \dfrac{\partial g_{\mu\nu}}{\partial x_\sigma}=-g_{\mu\alpha}g_{\nu\beta}\dfrac{\partial g^{\alpha\beta}}{\partial x_\sigma}\end{array}\right\} \qquad (32)$$

The relation (31) admits of a transformation, of which we also have frequently to make use. From (21)

$$\frac{\partial g_{\alpha\beta}}{\partial x_\sigma}=[\alpha\sigma,\beta]+[\beta\sigma,\alpha] \qquad (33)$$

Inserting this in the second formula of (31), we obtain, in view of (23)

$$\frac{\partial g^{\mu\nu}}{\partial x_\sigma}=-g^{\mu\tau}\{\tau\sigma,\nu\}-g^{\nu\tau}\{\tau\sigma,\mu\} \qquad (34)$$

Substituting the right-hand side of (34) in (29), we have

$$\frac{1}{\sqrt{-g}}\frac{\partial\sqrt{-g}}{\partial x_\sigma}=\{\mu\sigma,\mu\} \qquad (29a)$$

The "Divergence" of a Contravariant Vector.—If we take the inner product of (26) by the contravariant fundamental tensor $g^{\mu\nu}$, the right-hand side, after a transformation of the first term, assumes the form

$$\frac{\partial}{\partial x_\nu}\left(g^{\mu\nu}A_\mu\right)-A_\mu\frac{\partial g^{\mu\nu}}{\partial x_\nu}-\frac{1}{2}\left(\frac{\partial g_{\mu\alpha}}{\partial x_\nu}+\frac{\partial g_{\nu\alpha}}{\partial x_\mu}-\frac{\partial g_{\mu\nu}}{\partial x_\alpha}\right)g^{\mu\nu}A_\tau.$$

In accordance with (31) and (29), the last term of this expression may be written

$$\frac{1}{2}\frac{\partial g^{\tau\nu}}{\partial x_\nu}A_\tau+\frac{1}{2}\frac{\partial g^{\tau\mu}}{\partial x_\mu}A_\tau+\frac{1}{\sqrt{-g}}\frac{\partial\sqrt{-g}}{\partial x_\sigma}g^{\mu\nu}A_\tau.$$

As the symbols of the indices of summation are immaterial, the first two terms of this expression cancel the second of the one above. If we then write $g^{\mu\nu}A_\mu=A^\nu$, so that A^ν like A_μ is an arbitrary vector, we finally obtain

$$\Phi = \frac{1}{\sqrt{-g}} \frac{\partial}{\partial x_\nu} \left(\sqrt{-g}\, A^\nu \right) \tag{35}$$

This scalar is the *divergence* of the contravariant vector A^ν.

The "Curl" of a Covariant Vector.—The second term in (26) is symmetrical in the indices μ and ν. Therefore $A_{\mu\nu} - A_{\nu\mu}$ is a particularly simply constructed antisymmetrical tensor. We obtain

$$B_{\mu\nu} = \frac{\partial A_\mu}{\partial x_\nu} - \frac{\partial A_\nu}{\partial x_\mu} \tag{36}$$

Antisymmetrical Extension of a Six-vector.—Applying (27) to an antisymmetrical tensor of the second rank $A_{\mu\nu}$, forming in addition the two equations which arise through cyclic permutations of the indices, and adding these three equations, we obtain the tensor of the third rank

$$B_{\mu\nu\sigma} = A_{\mu\nu\sigma} + A_{\nu\sigma\mu} + A_{\sigma\mu\nu} = \frac{\partial A_{\mu\nu}}{\partial x_\sigma} + \frac{\partial A_{\nu\sigma}}{\partial x_\mu} + \frac{\partial A_{\sigma\mu}}{\partial x_\nu} \tag{37}$$

which it is easy to prove is antisymmetrical.

The Divergence of a Six-vector.—Taking the mixed product of (27) by $g^{\mu\alpha}g^{\nu\beta}$, we also obtain a tensor. The first term on the right-hand side of (27) may be written in the form

$$\frac{\partial}{\partial x_\sigma}\left(g^{\mu\alpha}g^{\nu\beta}A_{\mu\nu} \right) - g^{\mu\alpha}\frac{\partial g^{\nu\beta}}{\partial x_\sigma}A_{\mu\nu} - g^{\nu\beta}\frac{\partial g^{\mu\alpha}}{\partial x_\sigma}A_{\mu\nu}.$$

If we write $A_\sigma^{\alpha\beta}$ for $g^{\mu\alpha}g^{\nu\beta}A_{\mu\nu\sigma}$ and $A^{\alpha\beta}$ for $g^{\mu\alpha}g^{\nu\beta}A_{\mu\nu}$, and in the transformed first term replace

$$\frac{\partial g^{\nu\beta}}{\partial x_\sigma} \text{ and } \frac{\partial g^{\mu\alpha}}{\partial x_\sigma}$$

by their values as given by (34), there results from the right-hand side of (27) an expression consisting of seven terms, of which four cancel, and there remains

$$A_\sigma^{\alpha\beta} = \frac{\partial A^{\alpha\beta}}{\partial x_\sigma} + \left\{ \sigma\gamma,\, \alpha \right\} A^{\gamma\beta} + \left\{ \sigma\gamma,\, \beta \right\} A^{\alpha\gamma} \tag{38}$$

This is the expression for the extension of a contravariant tensor of the second rank, and corresponding expressions for the extension of contravariant tensors of higher and lower rank may also be formed.

We note that in an analogous way we may also form the extension of a mixed tensor:—

$$A_{\mu\sigma}^\alpha = \frac{\partial A_\mu^\alpha}{\partial x_\sigma} + \left\{ \sigma\mu,\, \tau \right\} A_\tau^\alpha + \left\{ \sigma\tau,\, \alpha \right\} A_\mu^\tau \tag{39}$$

On contracting (38) with respect to the indices β and σ (inner multiplication by δ_β^σ), we obtain the vector

$$A^\alpha = \frac{\partial A^{\alpha\beta}}{\partial x_\beta} + \{\beta\gamma, \beta\} A^{\alpha\gamma} + \{\beta\gamma, \alpha\} A^{\gamma\beta}.$$

On account of the symmetry of $\{\beta\gamma, \alpha\}$ with respect to the indices β and γ, the third term on the right-hand side vanishes, if $A^{\alpha\beta}$ is, as we will assume, an antisymmetrical tensor. The second term allows itself to be transformed in accordance with (29a). Thus we obtain

$$A^\alpha = \frac{1}{\sqrt{-g}} \frac{\partial\left(\sqrt{-g}\, A^{\alpha\beta}\right)}{\partial x_\beta} \qquad (40)$$

This is the expression for the divergence of a contravariant six-vector.

The Divergence of a Mixed Tensor of the Second Rank.—Contracting (39) with respect to the indices α and σ, and taking (29a) into consideration, we obtain

$$\sqrt{-g}\, A_\mu = \frac{\partial\left(\sqrt{-g}\, A_\mu^\sigma\right)}{\partial x_\sigma} - \{\sigma\mu, \tau\} \sqrt{-g}\, A_\tau^\sigma \qquad (41)$$

If we introduce the contravariant tensor $A^{\rho\sigma} = g^{\rho\tau} A_\tau^\sigma$ in the last term, it assumes the form

$$-[\sigma\mu, \rho] \sqrt{-g}\, A^{\rho\sigma}.$$

If, further, the tensor $A^{\rho\sigma}$ is symmetrical, this reduces to

$$-\frac{1}{2} \sqrt{-g} \frac{\partial g_{\rho\sigma}}{\partial x_\mu} A^{\rho\sigma}.$$

Had we introduced, instead of $A^{\rho\sigma}$, the covariant tensor $A_{\rho\sigma} = g_{\rho\alpha} g_{\sigma\beta} A^{\alpha\beta}$, which is also symmetrical, the last term, by virtue of (31), would assume the form

$$\frac{1}{2} \sqrt{-g} \frac{\partial g^{\rho\sigma}}{\partial x_\mu} A_{\rho\sigma}.$$

In the case of symmetry in question, (41) may therefore be replaced by the two forms

$$\sqrt{-g}\, A_\mu = \frac{\partial\left(\sqrt{-g}\, A_\mu^\sigma\right)}{\partial x_\sigma} - \frac{1}{2} \frac{\partial g^{\rho\sigma}}{\partial x_\mu} \sqrt{-g}\, A^{\rho\sigma} \qquad (41a)$$

$$\sqrt{-g}\, A_\mu = \frac{\partial\left(\sqrt{-g}\, A_\mu^\sigma\right)}{\partial x_\sigma} + \frac{1}{2} \frac{\partial g^{\rho\sigma}}{\partial x_\mu} \sqrt{-g}\, A_{\rho\sigma} \qquad (41b)$$

which we have to employ later on.

§ 12. THE RIEMANN-CHRISTOFFEL TENSOR

We now seek the tensor which can be obtained from the fundamental tensor *alone*, by differentiation. At first sight the solution seems obvious. We place the fundamental

tensor of the $g_{\mu\nu}$ in (27) instead of any given tensor $A_{\mu\nu}$, and thus have a new tensor, namely, the extension of the fundamental tensor. But we easily convince ourselves that this extension vanishes identically. We reach our goal, however, in the following way. In (27) place

$$A_{\mu\nu} = \frac{\partial A_{\mu}}{\partial x_{\nu}} - \{\mu\nu, \rho\} A_{\rho},$$

i.e. the extension of the four-vector A_{μ}. Then (with a somewhat different naming of the indices) we get the tensor of the third rank

$$A_{\mu\sigma\tau} = \frac{\partial^2 A_{\mu}}{\partial x_{\sigma}\partial x_{\tau}} - \{\mu\sigma, \rho\}\frac{\partial A_{\rho}}{\partial x_{\tau}} - \{\mu\tau, \rho\}\frac{\partial A_{\rho}}{\partial x_{\sigma}} - \{\sigma\tau, \rho\}\frac{\partial A_{\mu}}{\partial x_{\rho}}$$
$$+ \left[-\frac{\partial}{\partial x_{\tau}}\{\mu\sigma, \rho\} + \{\mu\tau, \alpha\}\{\alpha\sigma, \rho\} + \{\sigma\tau, \alpha\}\{\alpha\mu, \rho\} \right] A_{\rho}.$$

This expression suggests forming the tensor $A_{\mu\sigma\tau} - A_{\mu\tau\sigma}$. For, if we do so, the following terms of the expression for $A_{\mu\sigma\tau}$ cancel those of $A_{\mu\tau\sigma}$, the first, the fourth, and the member corresponding to the last term in square brackets; because all these are symmetrical in σ and τ. The same holds good for the sum of the second and third terms. Thus we obtain

$$A_{\mu\sigma\tau} - A_{\mu\tau\sigma} = B^{\rho}_{\mu\sigma\tau} A_{\rho} \qquad (42)$$

where

$$B^{\rho}_{\mu\sigma\tau} = -\frac{\partial}{\partial x_{\tau}}\{\mu\sigma, \rho\} + \frac{\partial}{\partial x_{\sigma}}\{\mu\tau, \rho\} - \{\mu\sigma, \alpha\}\{\alpha\tau, \rho\} + \{\mu\tau, \alpha\}\{\alpha\sigma, \rho\} \qquad (43)$$

The essential feature of the result is that on the right side of (42) the A_{ρ} occur alone, without their derivatives. From the tensor character of $A_{\mu\sigma\tau} - A_{\mu\tau\sigma}$ in conjunction with the fact that A_{ρ} is an arbitrary vector, it follows, by reason of § 7, that $B^{\rho}_{\mu\sigma\tau}$ is a tensor (the Riemann-Christoffel tensor).

The mathematical importance of this tensor is as follows: If the continuum is of such a nature that there is a co-ordinate system with reference to which the $g_{\mu\nu}$ are constants, then all the $B^{\rho}_{\mu\sigma\tau}$ vanish. If we choose any new system of co-ordinates in place of the original ones, the $g_{\mu\nu}$ referred thereto will not be constants, but in consequence of its tensor nature, the transformed components of $B^{\rho}_{\mu\sigma\tau}$ will still vanish in the new system. Thus the vanishing of the Riemann tensor is a necessary condition that, by an appropriate choice of the system of reference, the $g_{\mu\nu}$ may be constants. In our problem this corresponds to the case in which,[1] with a suitable choice of the system of reference, the special theory of relativity holds good for a *finite* region of the continuum.

Contracting (43) with respect to the indices τ and ρ we obtain the covariant tensor of second rank

1. The mathematicians have proved that this is also a *sufficient* condition.

where $\qquad\qquad\qquad\qquad\qquad\qquad\qquad\qquad\qquad\qquad\qquad\quad$ (44)

$$\left.\begin{array}{l} G_{\mu\nu} = B^{\rho}_{\mu\nu\rho} = R_{\mu\nu} + S_{\mu\nu} \\[6pt] R_{\mu\nu} = -\dfrac{\partial}{\partial x_{\alpha}}\{\mu\nu,\,\alpha\} + \{\mu\alpha,\,\beta\}\{\nu\beta,\,\alpha\} \\[6pt] S_{\mu\nu} = \dfrac{\partial^2 \log\sqrt{-g}}{\partial x_{\mu}\,\partial x_{\nu}} - \{\mu\nu,\,\alpha\}\dfrac{\partial \log\sqrt{-g}}{\partial x_{\alpha}} \end{array}\right\}$$

Note on the Choice of Co-ordinates.—It has already been observed in § 8, in connexion with equation (18a), that the choice of co-ordinates may with advantage be made so that $\sqrt{-g} = 1$. A glance at the equations obtained in the last two sections shows that by such a choice the laws of formation of tensors undergo an important simplification. This applies particularly to $G_{\mu\nu}$, the tensor just developed, which plays a fundamental part in the theory to be set forth. For this specialization of the choice of coordinates brings about the vanishing of $S_{\mu\nu}$, so that the tensor $G_{\mu\nu}$ reduces to $R_{\mu\nu}$.

On this account I shall hereafter give all relations in the simplified form which this specialization of the choice of co-ordinates brings with it. It will then be an easy matter to revert to the *generally* covariant equations, if this seems desirable in a special case.

C. THEORY OF THE GRAVITATIONAL FIELD

§ 13. EQUATIONS OF MOTION OF A MATERIAL POINT IN THE GRAVITATIONAL FIELD. EXPRESSION FOR THE FIELD-COMPONENTS OF GRAVITATION

A freely movable body not subjected to external forces moves, according to the special theory of relativity, in a straight line and uniformly. This is also the case, according to the general theory of relativity, for a part of four-dimensional space in which the system of co-ordinates K_0, may be, and is, so chosen that they have the special constant values given in (4).

If we consider precisely this movement from any chosen system of co-ordinates K_1, the body, observed from K_1, moves, according to the considerations in § 2, in a gravitational field. The law of motion with respect to K_1 results without difficulty from the following consideration. With respect to K_0 the law of motion corresponds to a four-dimensional straight line, i.e. to a geodetic line. Now since the geodetic line is defined independently of the system of reference, its equations will also be the equation of motion of the material point with respect to K_1. If we set

$$\Gamma^{\tau}_{\mu\nu} = -\{\mu\nu,\,\tau\}$$ (45)

the equation of the motion of the point with respect to K_1, becomes

$$\frac{d^2 x_\tau}{ds^2} = \Gamma_{\mu\nu}^\tau \frac{dx_\mu}{ds} \frac{dx_\nu}{ds} \qquad (46)$$

We now make the assumption, which readily suggests itself, that this covariant system of equations also defines the motion of the point in the gravitational field in the case when there is no system of reference K_0, with respect to which the special theory of relativity holds good in a finite region. We have all the more justification for this assumption as (46) contains only *first* derivatives of the $g_{\mu\nu}$, between which even in the special case of the existence of K_0, no relations subsist.[1]

If the $\Gamma_{\mu\nu}^\tau$ vanish, then the point moves uniformly in a straight line. These quantities therefore condition the deviation of the motion from uniformity. They are the components of the gravitational field.

§ 14. THE FIELD EQUATIONS OF GRAVITATION IN THE ABSENCE OF MATTER

We make a distinction hereafter between "gravitational field" and "matter" in this way, that we denote everything but the gravitational field as "matter." Our use of the word therefore includes not only matter in the ordinary sense, but the electromagnetic field as well.

Our next task is to find the field equations of gravitation in the absence of matter. Here we again apply the method employed in the preceding paragraph in formulating the equations of motion of the material point. A special case in which the required equations must in any case be satisfied is that of the special theory of relativity, in which the $g_{\mu\nu}$, have certain constant values. Let this be the case in a certain finite space in relation to a definite system of co-ordinates K_0. Relatively to this system all the components of the Riemann tensor $B_{\mu\nu\tau}^\rho$, defined in (43), vanish. For the space under consideration they then vanish, also in any other system of co-ordinates.

Thus the required equations of the matter-free gravitational field must in any case be satisfied if all $B_{\mu\sigma\tau}^\rho$ vanish. But this condition goes too far. For it is clear that, e.g., the gravitational field generated by a material point in its environment certainly cannot be "transformed away" by any choice of the system of co-ordinates, i.e. it cannot be transformed to the case of constant $g_{\mu\nu}$.

This prompts us to require for the matter-free gravitational field that the symmetrical tensor $G_{\mu\nu}$, derived from the tensor $B_{\mu\nu\tau}^\rho$, shall vanish. Thus we obtain ten equations for the ten quantities $g_{\mu\nu}$, which are satisfied in the special case of the vanishing of all $B_{\mu\nu\tau}^\rho$. With the choice which we have made of a system of co-ordinates, and taking (44) into consideration, the equations for the matter-free field are

1. It is only between the second (and first) derivatives that, by § 12, the relations $B_{\mu\sigma\tau}^\rho = 0$ subsist.

$$\left.\begin{array}{r} \dfrac{\partial \Gamma^{\alpha}_{\mu\nu}}{\partial x_{\alpha}} + \Gamma^{\alpha}_{\mu\beta}\Gamma^{\beta}_{\nu\alpha} = 0 \\[2mm] \sqrt{-g} = 1 \end{array}\right\} \qquad (47)$$

It must be pointed out that there is only a minimum of arbitrariness in the choice of these equations. For besides $G_{\mu\nu}$ there is no tensor of second rank which is formed from the $g_{\mu\nu}$ and its derivatives, contains no derivations higher than second, and is linear in these derivatives.[1]

These equations, which proceed, by the method of pure mathematics, from the requirement of the general theory of relativity, give us, in combination with the equations of motion (46), to a first approximation Newton's law of attraction, and to a second approximation the explanation of the motion of the perihelion of the planet Mercury discovered by Leverrier (as it remains after corrections for perturbation have been made). These facts must, in my opinion, be taken as a convincing proof of the correctness of the theory.

§ 15. THE HAMILTONIAN FUNCTION FOR THE GRAVITATIONAL FIELD. LAWS OF MOMENTUM AND ENERGY

To show that the field equations correspond to the laws of momentum and energy, it is most convenient to write them in the following Hamiltonian form:

$$\left.\begin{array}{l} \delta \int H \, d\tau = 0 \\[1mm] H = g^{\mu\nu}\Gamma^{\alpha}_{\mu\beta}\Gamma^{\beta}_{\nu\alpha} \\[1mm] \sqrt{-g} = 1 \end{array}\right\} \qquad (47a)$$

where, on the boundary of the finite four-dimensional region of integration which we have in view, the variations vanish.

We first have to show that the form (47a) is equivalent to the equations (47). For this purpose we regard H as a function of the $g^{\mu\nu}$ and the $g^{\mu\nu}_{\sigma}\left(= \partial g^{\mu\nu} / \partial x_{\sigma}\right)$. Then in the first place

$$\delta H = \Gamma^{\alpha}_{\mu\beta}\Gamma^{\beta}_{\nu\alpha}\delta g^{\mu\nu} + 2g^{\mu\nu}\Gamma^{\alpha}_{\mu\beta}\delta\Gamma^{\beta}_{\nu\alpha}$$
$$= -\Gamma^{\alpha}_{\mu\beta}\Gamma^{\beta}_{\nu\alpha}\delta g^{\mu\nu} + 2\Gamma^{\alpha}_{\mu\beta}\delta\left(g^{\mu\nu}\Gamma^{\beta}_{\nu\alpha}\right).$$

But

$$\delta\left(g^{\mu\nu}\Gamma^{\beta}_{\nu\alpha}\right) = -\frac{1}{2}\delta\left[g^{\mu\nu}g^{\beta\nu}\left(\frac{\partial g_{\nu\lambda}}{\partial x_{\alpha}} + \frac{\partial g_{\alpha\lambda}}{\partial x_{\nu}} - \frac{\partial g_{\alpha\nu}}{\partial x_{\lambda}}\right)\right].$$

The terms arising from the last two terms in round brackets are of different sign, and result from each other (since the denomination of the summation indices is immaterial)

1. Properly speaking, this can be affirmed only of the tensor

$$G_{\mu\nu} + \lambda g_{\mu\nu}g^{\alpha\beta}G_{\alpha\beta}$$

where λ is a constant. If, however, we set this tensor = 0, we come back again to the equations $G_{\mu\nu} = 0$.

through interchange of the indices μ and β. They cancel each other in the expression for δH, because they are multiplied by the quantity $\Gamma_{\mu\beta}^{\alpha}$, which is symmetrical with respect to the indices μ and β. Thus there remains only the first term in round brackets to be considered, so that, taking (31) into account, we obtain

$$\delta H = \Gamma_{\mu\beta}^{\alpha}\Gamma_{\nu\alpha}^{\beta}\delta g^{\mu\nu} + \Gamma_{\mu\beta}^{\alpha}\delta g_{\alpha}^{\mu\beta}$$

Thus

$$\left.\begin{array}{l} \dfrac{\partial H}{\partial g^{\mu\nu}} = -\Gamma_{\mu\beta}^{\alpha}\Gamma_{\nu\alpha}^{\beta} \\[3mm] \dfrac{\partial H}{\partial g_{\sigma}^{\mu\nu}} = \Gamma_{\mu\nu}^{\sigma} \end{array}\right\} \tag{48}$$

Carrying out the variation in (47a), we get in the first place

$$\frac{\partial}{\partial x_{\alpha}}\left(\frac{\partial H}{\partial g_{\alpha}^{\mu\nu}}\right) - \frac{\partial H}{\partial g^{\mu\nu}} = 0, \tag{47b}$$

which, on account of (48), agrees with (47), as was to be proved.

If we multiply (47b) by $g_{\sigma}^{\mu\nu}$, then because

$$\frac{\partial g_{\sigma}^{\mu\nu}}{\partial x_{\alpha}} = \frac{\partial g_{\alpha}^{\mu\nu}}{\partial x_{\sigma}}$$

and, consequently,

$$g_{\sigma}^{\mu\nu}\frac{\partial}{\partial x_{\alpha}}\left(\frac{\partial H}{\partial g_{\alpha}^{\mu\nu}}\right) = \frac{\partial}{\partial x_{\alpha}}\left(g_{\sigma}^{\mu\nu}\frac{\partial H}{\partial g_{\alpha}^{\mu\nu}}\right) - \frac{\partial H}{\partial g_{\alpha}^{\mu\nu}}$$

we obtain the equation

$$\frac{\partial}{\partial x_{\alpha}}\left(g_{\sigma}^{\mu\nu}\frac{\partial H}{\partial g_{\alpha}^{\mu\nu}}\right) - \frac{\partial H}{\partial x_{\sigma}} = 0$$

or[1]

$$\left.\begin{array}{l} \dfrac{\partial t_{\sigma}^{\alpha}}{\partial x_{\alpha}} = 0 \\[3mm] -d\kappa t_{\sigma}^{\alpha} = g_{\sigma}^{\mu\nu}\dfrac{\partial H}{\partial g_{\alpha}^{\mu\nu}} - \delta_{\sigma}^{\alpha}H \end{array}\right\} \tag{49}$$

where, on account of (48), the second equation of (47), and (34)

$$\kappa t_{\sigma}^{\alpha} = \frac{1}{2}\delta_{\sigma}^{\alpha}g^{\mu\nu}\Gamma_{\mu\beta}^{\lambda}\Gamma_{\nu\lambda}^{\beta} - g^{\mu\nu}\Gamma_{\mu\beta}^{\alpha}\Gamma_{\nu\sigma}^{\beta} \tag{50}$$

It is to be noticed that t_{σ}^{α} is not a tensor; on the other hand (49) applies to all systems of co-ordinates for which $\sqrt{-g} = 1$. This equation expresses the law of conservation of momentum and of energy for the gravitational field. Actually the integration of this equation over a three-dimensional volume V yields the four equations

---

1. The reason for the introduction of the factor -2κ will be apparent later.

$$\frac{d}{dx_4}\int t_\sigma^4 dV = \int \left(l t_\sigma^1 + m t_\sigma^2 + n t_\sigma^3\right) dS \tag{49a}$$

where l, m, n denote the direction-cosines of direction of the inward drawn normal at the element dS of the bounding surface (in the sense of Euclidean geometry). We recognize in this the expression of the laws of conservation in their usual form. The quantities t_σ^α we call the "energy components" of the gravitational field.

I will now give equations (47) in a third form, which is particularly useful for a vivid grasp of our subject. By multiplication of the field equations (47) by $g^{\nu\sigma}$ these are obtained in the "mixed" form. Note that

$$g^{\nu\sigma}\frac{\partial\Gamma_{\mu\nu}^\alpha}{\partial x_\alpha}\left(g^{\nu\sigma}\Gamma_{\mu\nu}^\alpha\right) - \frac{\partial g^{\nu\sigma}}{\partial x_\alpha}\Gamma_{\mu\nu}^\alpha$$

which quantity, by reason of (34), is equal to

$$\frac{\partial}{\partial x_\alpha}\left(g^{\nu\sigma}\Gamma_{\mu\nu}^\alpha\right) - g^{\nu\beta}\Gamma_{\alpha\beta}^\sigma\Gamma_{\mu\nu}^\alpha - g^{\sigma\beta}\Gamma_{\alpha\beta}^\nu\Gamma_{\mu\nu}^\alpha,$$

or (with different symbols for the summation indices)

$$\frac{\partial}{\partial x_\alpha}\left(g^{\sigma\beta}\Gamma_{l\beta}^\alpha\right) - g^{\gamma\delta}\Gamma_{\gamma\beta}^\sigma\Gamma_{\delta\mu}^\beta - g^{\nu\sigma}\Gamma_{\mu\beta}^\alpha\Gamma_{\nu\alpha}^\beta.$$

The third term of this expression cancels with the one arising from the second term of the field equations (47); using relation (50), the second term may be written

$$\kappa\left(t_\mu^\sigma - \frac{1}{2}\delta_\mu^\sigma t\right)$$

where $t = t_\alpha^\alpha$. Thus instead of equations (47) we obtain

$$\left.\begin{array}{c}\frac{\partial}{\partial x_\alpha}\left(g^{\sigma\beta}\Gamma_{l\beta}^\alpha\right) = -\kappa\left(t_\mu^\sigma - \delta_\mu^\sigma t\right)\\[2mm]\sqrt{-g} = 1\end{array}\right\} \tag{51}$$

§ 16. THE GENERAL FORM OF THE FIELD EQUATIONS OF GRAVITATION

The field equations for matter-free space formulated in § 15 are to be compared with the field equation

$$\nabla^2\phi = 0$$

of Newton's theory. We require the equation corresponding to Poisson's equation

$$\nabla^2\phi = 4\pi\lambda\rho,$$

where ρ denotes the density of matter.

The special theory of relativity has led to the conclusion that inert mass is nothing

more or less than energy, which finds its complete mathematical expression in a symmetrical tensor of second rank, the energy-tensor. Thus in the general theory of relativity we must introduce a corresponding energy-tensor of matter T_σ^α, which, like the energy-components t_σ [equations (49) and (50)] of the gravitational field, will have mixed character, but will pertain to a symmetrical covariant tensor.[1]

The system of equation (51) shows how this energy-tensor (corresponding to the density ρ in Poisson's equation) is to be introduced into the field equations of gravitation. For if we consider a complete system (e.g. the solar system), the total mass of the system, and therefore its total gravitating action as well, will depend on the total energy of the system, and therefore on the ponderable energy together with the gravitational energy. This will allow itself to be expressed by introducing into (51), in place of the energy-components of the gravitational field alone, the sums $t_\mu^\sigma + T_\mu^\sigma$ of the energy-components of matter and of gravitational field. Thus instead of (51) we obtain the tensor equation

$$\left.\begin{array}{c} \dfrac{\partial}{\partial x_\alpha}\left(g^{\sigma\beta}T_{\mu\beta}^\alpha\right) = -\kappa\left[\left(t_\mu^\sigma + T_\mu^\sigma\right) - \dfrac{1}{2}\delta_\mu^\sigma\left(t + T\right)\right] \\ \sqrt{-g} = 1 \end{array}\right\} \tag{52}$$

where we have set $T = T_\mu^\mu$ (Laue's scalar). These are the required general field equations of gravitation in mixed form. Working back from these, we have in place of (47)

$$\left.\begin{array}{c} \dfrac{\partial}{\partial x_\alpha}\Gamma_{\mu\nu}^\alpha + \Gamma_{\mu\beta}^\alpha\Gamma_{\mu\alpha}^\beta = -\kappa\left(T_{\mu\nu} - \dfrac{1}{2}g_{\mu\nu}T\right), \\ \sqrt{-g} = 1 \end{array}\right\} \tag{53}$$

It must be admitted that this introduction of the energy-tensor of matter is not justified by the relativity postulate alone. For this reason we have here deduced it from the requirement that the energy of the gravitational field shall act gravitatively in the same way as any other kind of energy. But the strongest reason for the choice of these equations lies in their consequence, that the equations of conservation of momentum and energy, corresponding exactly to equations (49) and (49a), hold good for the components of the total energy. This will be shown in § 17.

§ 17. THE LAWS OF CONSERVATION IN THE GENERAL CASE

Equation (52) may readily be transformed so that the second term on the right-hand side vanishes. Contract (52) with respect to the indices μ and σ, and after multiplying the resulting equation by $\frac{1}{2}\delta_\mu^\sigma$, subtract it from equation (52). This gives

1. $g_{\alpha\tau}T_\mu^\sigma = T_{\sigma\tau}$ and $g_{\sigma\beta}T_\sigma^\alpha = T^{\sigma\beta}$ are to be symmetrical tensors.

$$\frac{\partial}{\partial x_\alpha}\left(g^{\sigma\beta}\Gamma^\alpha_{\mu\beta} - \frac{1}{2}\delta^\sigma_\mu g^{\lambda\beta}\Gamma^\alpha_{\lambda\beta}\right) = -\kappa\left(t^\sigma_\mu + T^\sigma_\mu\right) \qquad (52a)$$

On this equation we perform the operation $\partial/\partial x_\sigma$. We have

$$\frac{\partial^2}{\partial x_\alpha \partial x_\sigma}\left(g^\sigma \Gamma^\alpha_{\beta\mu}\right) = -\frac{1}{2}\frac{\partial^2}{\partial x_\alpha \partial x_\sigma}\left[g^{\sigma\beta}g^{\alpha\lambda}\left(\frac{\partial g_{\mu\lambda}}{\partial x_\beta} + \frac{\partial g_{\beta\lambda}}{\partial x_\mu} - \frac{\partial g_{\mu\beta}}{\partial x_\lambda}\right)\right].$$

The first and third terms of the round brackets yield contributions which cancel one another, as may be seen by interchanging, in the contribution of the third term, the summation indices α and σ on the one hand, and β and λ on the other. The second term may be re-modelled by (31), so that we have

$$\frac{\partial^2}{\partial x_\sim \partial x_\sim}\left(g^{\sigma\beta}\Gamma^\alpha_{\mu\beta}\right) = \cdot \qquad (54)$$

The second term on the left-hand side of (52a) yields in the first place

$$-\frac{1}{2}\frac{\partial^2}{\partial x_\alpha \partial x_\mu}\left(g^{\lambda\beta}\Gamma^\alpha_{\lambda\beta}\right)$$

or

$$\frac{1}{4}\frac{\partial^2}{\partial x_\alpha \partial x_\mu}\left[g^{\lambda\beta}g^{\alpha\delta}\left(\frac{\partial g_{\delta\lambda}}{\partial x_\beta} + \frac{\partial g_{\delta\beta}}{\partial x_\lambda} - \frac{\partial g_{\lambda\beta}}{\partial x_\delta}\right)\right].$$

With the choice of co-ordinates which we have made, the term deriving from the last term in round brackets disappears by reason of (29). The other two may be combined, and together, by (31), they give

$$-\frac{1}{2}\frac{\partial^3 g^{\alpha\beta}}{\partial x_\alpha \partial x_\beta \partial x_\mu},$$

so that in consideration of (54), we have the identity

$$\frac{\partial^2}{\partial x_\alpha \partial x_\sigma}\left(g^{\rho\beta}\Gamma_{\mu\beta} - \frac{1}{2}\delta^\delta_\mu g^{\lambda\beta}\Gamma^\alpha_{\lambda\beta}\right) = 0 \qquad (55)$$

From (55) and (52a), it follows that

$$\frac{\partial\left(t^\sigma_\mu + T^\sigma_\mu\right)}{\partial x_\sigma} = 0 \qquad (56)$$

Thus it results from our field equations of gravitation that the laws of conservation of momentum and energy are satisfied. This may be seen most easily from the consideration which leads to equation (49a); except that here, instead of the energy components t^σ of the gravitational field, we have to introduce the totality of the energy components of matter and gravitational field.

§ 18. THE LAWS OF MOMENTUM AND ENERGY FOR
MATTER, AS A CONSEQUENCE OF THE FIELD EQUATIONS

Multiplying (53) by $\partial g^{\mu\nu}/\partial x_\sigma$, we obtain, by the method adopted in § 15, in view of the vanishing of

$$g_{\mu\nu}\frac{\partial g^{\mu\nu}}{\partial x_\sigma},$$

the equation

$$\frac{\partial t_\sigma^\alpha}{\partial x_\alpha}+\frac{1}{2}\frac{\partial g^{\mu\nu}}{\partial x_\sigma}T_{\mu\nu}=0,$$

or, in view of (56),

$$\frac{\partial T_\sigma^\alpha}{\partial x_\alpha}+\frac{1}{2}\frac{\partial g^{\mu\nu}}{\partial x_\sigma}T_{\mu\nu}=0 \qquad (57)$$

Comparison with (41b) shows that with the choice of system of co-ordinates which we have made, this equation predicates nothing more or less than the vanishing of divergence of the material energy-tensor. Physically, the occurrence of the second term on the left-hand side shows that laws of conservation of momentum and energy do not apply in the strict sense for matter alone, or else that they apply only when the $g^{\mu\nu}$ are constant, i.e. when the field intensities of gravitation vanish. This second term is an expression for momentum, and for energy, as transferred per unit of volume and time from the gravitational field to matter. This is brought out still more clearly by re-writing (57) in the sense of (41) as

$$\frac{\partial T_\sigma^\alpha}{\partial x_\alpha}=\Gamma_{\alpha\sigma}^\beta T_\beta^\alpha \qquad (57a)$$

The right side expresses the energetic effect of the gravitational field on matter.

Thus the field equations of gravitation contain four conditions which govern the course of material phenomena. They give the equations of material phenomena completely, if the latter is capable of being characterized by four differential equations independent of one another.[1]

D. MATERIAL PHENOMENA

The mathematical aids developed in part B enable us forthwith to generalize the physical laws of matter (hydrodynamics, Maxwell's electrodynamics), as they are formulated in the special theory of relativity, so that they will fit in with the general theory of relativity. When this is done, the general principle of relativity does not indeed afford us a further limitation of possibilities; but it makes us acquainted with the influence of the gravitational field on all processes, without our having to introduce any new hypothesis whatever.

---

1. On this question cf. H. Hilbert, Nachr. d. K. Gesellsch. d. Wiss. zu Göttingen, Math.-phys. Klasse, 1915, p. 3.

Hence it comes about that it is not necessary to introduce definite assumptions as to the physical nature of matter (in the narrower sense). In particular it may remain an open question whether the theory of the electromagnetic field in conjunction with that of the gravitational field furnishes a sufficient basis for the theory of matter or not. The general postulate of relativity is unable on principle to tell us anything about this. It must remain to be seen, during the working out of the theory, whether electromagnetics and the doctrine of gravitation are able in collaboration to perform what the former by itself is unable to do.

§ 19. EULER'S EQUATIONS FOR A FRICTIONLESS ADIABATIC FLUID

Let p and ρ be two scalars, the former of which we call the "pressure," the latter the "density" of a fluid; and let an equation subsist between them. Let the contravariant symmetrical tensor

$$T^{\alpha\beta} = -g^{\alpha\beta}p + \rho\frac{dx_\alpha}{ds}\frac{dx_\beta}{ds} \tag{58}$$

be the contravariant energy-tensor of the fluid. To it belongs the covariant tensor

$$T_{\mu\nu} = -g_{\mu\nu}p + g_{\mu\alpha}g_{\nu\beta}\frac{dx_\alpha}{ds}\frac{dx_\beta}{ds}\rho, \tag{58a}$$

as well as the mixed tensor[1]

$$T_\sigma^\alpha = -\delta_\sigma^\alpha p + g_{\sigma\beta}\frac{dx_\alpha}{ds}\frac{dx_\beta}{ds}\rho \tag{58b}$$

Inserting the right-hand side of (58b) in (57a), we obtain the Eulerian hydrodynamical equations of the general theory of relativity. They give, in theory, a complete solution of the problem of motion, since the four equations (57a), together with the given equation between p and ρ, and the equation

$$g_{\alpha\beta}\frac{dx_\alpha}{ds}\frac{dx_\beta}{ds} = 1,$$

are sufficient, $g_{\alpha\beta}$ being given, to define the six unknowns

$$p, \rho, \frac{dx_1}{ds}, \frac{dx_2}{ds}, \frac{dx_3}{ds}, \frac{dx_4}{ds}.$$

If the $g_{\mu\nu}$ are also unknown, the equations (53) are brought in. These are eleven equations for defining the ten functions $g_{\mu\nu}$, so that these functions appear over-defined. We must remember, however, that the equations (57a) are already contained in the equations (53), so that the latter represent only seven independent equations. There is good reason for this lack of definition, in that the wide freedom of the choice of coordinates

---

1. For an observer using a system of reference in the sense of the special theory of relativity for an infinitely small region, and moving with it, the density of energy T_4^4 equals $\rho - p$. This gives the definition of ρ. Thus ρ is not constant for an incompressible fluid.

causes the problem to remain mathematically undefined to such a degree that three of the functions of space may be chosen at will.[1]

§ 20. MAXWELL'S ELECTROMAGNETIC FIELD EQUATIONS FOR FREE SPACE

Let ϕ_ν be the components of a covariant vector—the electromagnetic potential vector. From them we form, in accordance with (36), the components $F_{\rho\sigma}$ of the covariant six-vector of the electromagnetic field, in accordance with the system of equations

$$F_{\rho\sigma} - \frac{\partial \phi_\rho}{\partial x_\sigma} - \frac{'}{'} \qquad (59)$$

It follows from (59) that the system of equations

$$\frac{\partial F_{\rho\sigma}}{\partial x_\tau} + \frac{\partial F_{\sigma\tau}}{\partial x_\rho} + \frac{\partial F_{\tau\rho}}{\partial x_\sigma} = 0 \qquad (60)$$

is satisfied, its left side being, by (37), an antisymmetrical tensor of the third rank. System (60) thus contains essentially four equations which are written out as follows:—

$$\left. \begin{array}{l} \dfrac{\partial F_{23}}{\partial x_4} + \dfrac{\partial F_{34}}{\partial x_2} + \dfrac{\partial F_{42}}{\partial x_3} = 0 \\[2mm] \dfrac{\partial F_{34}}{\partial x_1} + \dfrac{\partial F_{41}}{\partial x_3} + \dfrac{\partial F_{13}}{\partial x_4} = 0 \\[2mm] \dfrac{\partial F_{41}}{\partial x_2} + \dfrac{\partial F_{12}}{\partial x_4} + \dfrac{\partial F_{24}}{\partial x_1} = 0 \\[2mm] \dfrac{\partial F_{12}}{\partial x_3} + \dfrac{\partial F_{23}}{\partial x_1} + \dfrac{\partial F_{31}}{\partial x_2} = 0 \end{array} \right\} \qquad (60a)$$

This system corresponds to the second of Maxwell's systems of equations. We recognize this at once by setting

$$\left. \begin{array}{ll} F_{23} = H_x, & F_{14} = E_x \\ F_{31} = H_y, & F_{24} = E_y \\ F_{12} = H_z, & F_{34} = E_z \end{array} \right\} \qquad (61)$$

Then in place of (60a) we may set, in the usual notation of three-dimensional vector analysis,

$$\left. \begin{array}{l} -\dfrac{\partial H}{\partial t} = \operatorname{curl} E \\[2mm] \operatorname{div} H = 0 \end{array} \right\} \qquad (60b)$$

We obtain Maxwell's first system by generalizing the form given by Minkowski. We introduce the contravariant six-vector associated with $F^{\alpha\beta}$

---

1. On the abandonment of the choice of co-ordinates with g = - 1, there remain four functions of space with liberty of choice, corresponding to the four arbitrary functions at our disposal in the choice of co-ordinates.

$$F^{\mu\nu} = g^{\mu\alpha}g^{\nu\beta}F_{\alpha\beta} \tag{62}$$

and also the contravariant vector J^μ of the density of the electric current. Then, taking (40) into consideration, the following equations will be invariant for any substitution whose invariant is unity (in agreement with the chosen co-ordinates):—

$$\frac{\partial}{\partial x_\nu} F^{\mu\nu} = J^\mu \tag{63}$$

Let

$$\left.\begin{array}{ll} F^{23} = H'_x, & F^{14} = -E'_x \\ F^{31} = H'_y, & F^{24} = -E'_y \\ F^{12} = H'_z, & F^{34} = -E'_z \end{array}\right\} \tag{64}$$

which quantities are equal to the quantities $H_x \ldots E^z$ in the special case of the restricted theory of relativity; and in addition

$$J^1 = j_x, \; J^2 = j_y, \; J^3 = j_z, \; J^4 = \rho,$$

we obtain in place of (63)

$$\left.\begin{array}{l} \dfrac{\partial E'}{\partial t} + j = \text{curl } H' \\ \text{div } E' = \rho \end{array}\right\} \tag{63a}$$

The equations (60), (62), and (63) thus form the generalization of Maxwell's field equations for free space, with the convention which we have established with respect to the choice of co-ordinates.

The Energy-components of the Electromagnetic Field.—We form the inner product

$$\kappa_\sigma = F_{\sigma\mu}J^\mu \tag{65}$$

By (61) its components, written in the three-dimensional manner, are

$$\left.\begin{array}{l} \kappa_1 = \rho E_x + \left[j \cdot H\right]^x \\ \vdots \quad \vdots \quad \vdots \quad \vdots \\ \kappa_4 = -\left(jE\right) \end{array}\right\} \tag{65a}$$

κ_σ is a covariant vector the components of which are equal to the negative momentum, or, respectively, the energy, which is transferred from the electric masses to the electromagnetic field per unit of time and volume. If the electric masses are free, that is, under the sole influence of the electromagnetic field, the covariant vector κ_σ will vanish.

To obtain the energy-components T_σ^ν of the electromagnetic field, we need only give to equation $\kappa_\sigma = 0$ the form of equation (57). From (63) and (65) we have in the first place

$$\kappa_\sigma = F_{\sigma\mu}\frac{\partial F^{\mu\nu}}{\partial x_\nu} = \frac{\partial}{\partial x_\nu}\Big(F_{\sigma\mu}F^{\mu\nu}\Big) - F^{\mu\rho}\frac{\partial F_{\sigma\mu}}{\partial x_\nu}.$$

The second term of the right-hand side, by reason of (60), permits the transformation

$$F^{\mu\nu}\frac{\partial F_{\sigma\mu}}{\partial x_\nu} = \frac{1}{2}F^{\mu\nu}\frac{\partial F_{\mu\nu}}{\partial x_\sigma} = -\frac{1}{2}g^{\mu\alpha}g^{\nu\beta}F_{\alpha\beta}\frac{\partial F_{\mu\nu}}{\partial x_\sigma},$$

which latter expression may, for reasons of symmetry, also be written in the form

$$-\frac{1}{4}\Bigg[g^{\mu\alpha}g^{\nu\beta}F_{\alpha\beta}\frac{\partial F_{\mu\nu}}{\partial x_\sigma} + g^{\mu\alpha}g^{\nu\beta}\frac{\partial F_{\mu\beta}}{\partial x_\sigma}F_{\mu\nu}\Bigg].$$

But for this we may set

$$-\frac{1}{4}\frac{\partial}{\partial x_\sigma}\Big(g^{\mu\alpha}g^{\nu\beta}F_{\alpha\beta}F_{\mu\nu}\Big) + -\frac{1}{4}F_{\alpha\beta}F_{\mu\nu}\frac{\partial}{\partial x_\sigma}\Big(g^{\mu\alpha}g^{\nu\beta}\Big).$$

The first of these terms is written more briefly

$$-\frac{1}{4}\frac{\partial}{\partial x_\sigma}\Big(F^{\mu\nu}F_{\mu\nu}\Big);$$

the second, after the differentiation is carried out, and after some reduction, results in

$$-\frac{1}{2}F^{\mu\tau}F_{\mu\nu}g^{\nu\rho}\frac{\partial g_{\sigma\tau}}{\partial x_\sigma}.$$

Taking all three terms together we obtain the relation

$$\kappa_\sigma = \frac{\partial T_\sigma^\nu}{\partial x_\nu} - \frac{1}{2}g^{\tau\mu}\frac{\partial g_{\mu\nu}}{\partial x_\sigma}T_\tau^\nu \qquad (66)$$

where

$$T_\sigma^\nu = -F_{\sigma\alpha}F^{\nu\alpha} + \frac{1}{4}\delta_\sigma^\nu F_{\alpha\beta}F^{\alpha\beta}.$$

Equation (66), if κ_σ vanishes, is, on account of (30), equivalent to (57) or (57a) respectively. Therefore the T_σ^ν are the energy-components of the electromagnetic field. With the help of (61) and (64), it is easy to show that these energy-components of the electromagnetic field in the case of the special theory of relativity give the well-known Maxwell-Poynting expressions.

We have now deduced the general laws which are satisfied by the gravitational field and matter, by consistently using a system of co-ordinates for which $\sqrt{-g}$. We have thereby achieved a considerable simplification of formulæ and calculations, without failing to comply with the requirement of general covariance; for we have drawn our equations from generally covariant equations by specializing the system of co-ordinates.

Still the question is not without a formal interest, whether with a correspondingly generalized definition of the energy-components of gravitational field and matter, even

without specializing the system of co-ordinates, it is possible to formulate laws of conservation in the form of equation (56), and field equations of gravitation of the same nature as (52) or (52a), in such a manner that on the left we have a divergence (in the ordinary sense), and on the right the sum of the energy-components of matter and gravitation. I have found that in both cases this is actually so. But I do not think that the communication of my somewhat extensive reflexions on this subject would be worth while, because after all they do not give us anything that is materially new.

E

§ 21. Newton's Theory as a First Approximation

As has already been mentioned more than once, the special theory of relativity as a special case of the general theory is characterized by the $g_{\mu\nu}$ having the constant values (4). From what has already been said, this means complete neglect of the effects of gravitation. We arrive at a closer approximation to reality by considering the case where the $g_{\mu\nu}$ differ from the values of (4) by quantities which are small compared with 1, and neglecting small quantities of second and higher order. (First point of view of approximation.)

It is further to be assumed that in the space-time territory under consideration the $g_{\mu\nu}$ at spatial infinity, with a suitable choice of co-ordinates, tend toward the values (4); i.e. we are considering gravitational fields which may be regarded as generated exclusively by matter in the finite region.

It might be thought that these approximations must lead us to Newton's theory. But to that end we still need to approximate the fundamental equations from a second point of view. We give our attention to the motion of a material point in accordance with the equations (16). In the case of the special theory of relativity the components

$$\frac{dx_1}{ds}, \frac{dx_2}{ds}, \frac{dx_3}{ds}$$

may take on any values. This signifies that any velocity

$$v = \sqrt{\left(\frac{dx_1}{dx_4}\right)^2 + \left(\frac{dx_2}{dx_4}\right)^2 + \left(\frac{dx_3}{dx_4}\right)^2}$$

may occur, which is less than the velocity of light *in vacuo*. If we restrict ourselves to the case which almost exclusively offers itself to our experience, of v being small as compared with the velocity of light, this denotes that the components

$$\frac{dx_1}{ds}, \frac{dx_2}{ds}, \frac{dx_3}{ds}$$

are to be treated as small quantities, while dx_4/ds, to the second order of small quantities, is equal to one. (Second point of view of approximation.)

Now we remark that from the first point of view of approximation the magnitudes $\Gamma^\tau_{\mu\nu}$ are all small magnitudes of at least the first order. A glance at (46) thus shows that in this equation, from the second point of view of approximation, we have to consider only terms for which $\mu = \nu = 4$. Restricting ourselves to terms of lowest order we first obtain in place of (46) the equations

$$\frac{d^2 x_\tau}{dt^2} = \Gamma^\tau_{44}$$

where we have set $ds = dz_4 = dt$; or with restriction to terms which from the first point of view of approximation are of first order:—

$$\frac{d^2 x_\tau}{dt^2} = \left[44, \tau\right] \quad (\tau = 1, 2, 3)$$

$$\frac{d^2 x_4}{dt^2} = -\left[44, 4\right]$$

If in addition we suppose the gravitational field to be a quasi-static field, by confining ourselves to the case where the motion of the matter generating the gravitational field is but slow (in comparison with the velocity of the propagation of light), we may neglect on the right-hand side differentiations with respect to the time in comparison with those with respect to the space co-ordinates, so that we have

$$\frac{d^2 x_\tau}{dt^2} = -\frac{1}{2}\frac{\partial g_{44}}{\partial x_\tau} \quad (\tau = 1, 2, 3) \qquad (67)$$

This is the equation of motion of the material point according to Newton's theory, in which $\frac{1}{2} g_{44}$ plays the part of the gravitational potential. What is remarkable in this result is that the component g_{44} of the fundamental tensor alone defines, to a first approximation, the motion of the material point.

We now turn to the field equations (53). Here we have to take into consideration that the energy-tensor of "matter" is almost exclusively defined by the density of matter in the narrower sense, i.e. by the second term of the right-hand side of (58) [or, respectively, (58a) or (58b)]. If we form the approximation in question, all the components vanish with the one exception of $T_{44} = \rho = T$. On the left-hand side of (53) the second term is a small quantity of second order; the first yields, to the approximation in question,

$$\frac{\partial}{\partial x_1}\left[\mu\nu, 1\right] + \frac{\partial}{\partial x_2}\left[\mu\nu, 2\right] + \frac{\partial}{\partial x_3}\left[\mu\nu, 3\right] - \frac{\partial}{\partial x_4}\left[\mu\nu, 4\right].$$

For $\mu = \nu = 4$, this gives, with the omission of terms differentiated with respect to time,

$$-\frac{1}{2}\left(\frac{\partial^2 g_{44}}{\partial x_1^2} + \frac{\partial^2 g_{44}}{\partial x_2^2} + \frac{\partial^2 g_{44}}{\partial x_3^2}\right) = -\frac{1}{2}\nabla^2 g_{44}.$$

The last of equations (53) thus yields

$$\nabla^2 g_{44} = \kappa \rho \qquad (68)$$

The equations (67) and (68) together are equivalent to Newton's law of gravitation.

By (67) and (68) the expression for the gravitational potential becomes

$$-\frac{\kappa}{8\pi} \int \frac{\rho d\tau}{r} \qquad (68a)$$

while Newton's theory, with the unit of time which we have chosen, gives

$$-\frac{K}{r^2} \int \frac{\rho d\tau}{r}$$

in which K denotes the constant $6 \cdot 7 \times 10^{-8}$, usually called the constant of gravitation. By comparison we obtain

$$\kappa = \frac{8\pi K}{c^2} = 1 \cdot 87 \times 10^{-27} \qquad (69)$$

§ 22. BEHAVIOUR OF RODS AND CLOCKS IN THE STATIC GRAVITATIONAL FIELD. BENDING OF LIGHT-RAYS. MOTION OF THE PERIHELION OF A PLANETARY ORBIT

To arrive at Newton's theory as a first approximation we had to calculate only one component, g_{44}, of the ten $g_{\mu\nu}$ of the gravitational field, since this component alone enters into the first approximation, (67), of the equation for the motion of the material point in the gravitational field. From this, however, it is already apparent that other components of the $g_{\mu\nu}$ must differ from the values given in (4) by small quantities of the first order. This is required by the condition $g = -1$.

For a field-producing point mass at the origin of co-ordinates, we obtain, to the first approximation, the radially symmetrical solution

$$\left.\begin{array}{l} g_{\rho\sigma} = -\delta_{\rho\sigma} - \alpha \dfrac{x_\rho x_\sigma}{r^3} \quad (\rho, \sigma = 1, 2, 3) \\[2mm] g_{\rho 4} = -\delta_{4\rho} = 0 \qquad (\rho = 1, 2, 3) \\[2mm] g_{44} = 1 - \dfrac{\alpha}{r} \end{array}\right\} \qquad (70)$$

where $\delta_{\rho\sigma}$ is 1 or 0, respectively, accordingly as $\rho = \sigma$ or $\rho \neq \sigma$, and r is the quantity $+\sqrt{x_1^2 + x_2^2 + x_3^2}$ on account of (68a)

$$\alpha = \frac{\kappa M}{4\pi}, \qquad (70a)$$

if M denotes the field-producing mass. It is easy to verify that the field equations (outside the mass) are satisfied to the first order of small quantities.

We now examine the influence exerted by the field of the mass M upon the metrical properties of space. The relation

$$ds^2 = g_{\mu\nu}dx_\mu dx_\nu.$$

always holds between the "locally" (§ 4) measured lengths and times ds on the one hand, and the differences of co-ordinates dx_ν on the other hand.

For a unit-measure of length laid "parallel" to the axis of x, for example, we should have to set $ds^2 = -1; dx_2 = dx_3 = dx_4 = 0$. Therefore $-1 = g_{11}dx_1^2$. If, in addition, the unit-measure lies on the axis of x, the first of equations (70) gives

$$g_{11} = -\left(1 + \frac{\alpha}{r}\right).$$

From these two relations it follows that, correct to a first order of small quantities,

$$dx = 1 - \frac{\alpha}{2r} \tag{71}$$

The unit measuring-rod thus appears a little shortened in relation to the system of co-ordinates by the presence of the gravitational field, if the rod is laid along a radius.

In an analogous manner we obtain the length of co-ordinates in tangential direction if, for example, we set

$$ds^2 = -1; dx_1 = dx_3 = dx_4 = 0; x_1 = r, x_2 = x_3 = 0.$$

The result is

$$-1 = g_{22}dx_2^2 = -dx_2^2 \tag{71a}$$

With the tangential position, therefore, the gravitational field of the point of mass has no influence on the length of rod.

Thus Euclidean geometry does not hold even to a first approximation in the gravitational field, if we wish to take one and the same rod, independently of its place and orientation, as a realization of the same interval; although, to be sure, a glance at (70a) and (69) shows that the deviations to be expected are much too slight to be noticeable in measurements of the earth's surface.

Further, let us examine the rate of a unit clock, which is arranged to be at rest in a static gravitational field. Here we have for a clock period $ds = 1; dx_1 = dx_2 = dx_3 = 0$ Therefore

$$1 = g_{44}dx_4^2;$$

$$dx_4 = \frac{1}{\sqrt{g_{44}}} = \frac{1}{\sqrt{(1 + (g_{44} - 1))}} = 1 - \frac{1}{2}(g_{44} - 1)$$

or

$$dx_4 = 1 + \frac{\kappa}{8\pi}\int\rho\frac{d\tau}{r} \tag{72}$$

Thus the clock goes more slowly if set up in the neighbourhood of ponderable masses.

From this it follows that the spectral lines of light reaching us from the surface of large stars must appear displaced towards the red end of the spectrum.[1]

We now examine the course of light-rays in the static gravitational field. By the special theory of relativity the velocity of light is given by the equation

$$-dx_1^2 - dx_2^2 - dx_3^2 + dx_4^2 = 0$$

and therefore by the general theory of relativity by the equation

$$ds^2 = g_{\mu\nu}dx_\mu dx_\nu = 0 \tag{73}$$

If the direction, i.e. the ratio $dx_1 : dx_2 : dx_3$ is given, equation (73) gives the quantities

$$\frac{dx_1}{dx_4}, \frac{dx_2}{dx_4}, \frac{dx_3}{dx_4}$$

and accordingly the velocity

$$\sqrt{\left(\frac{dx_1}{dx_4}\right)^2 + \left(\frac{dx_2}{dx_4}\right)^2 + \left(\frac{dx_3}{dx_4}\right)^2} = \gamma$$

defined in the sense of Euclidean geometry. We easily recognize that the course of the light-rays must be bent with regard to the system of co-ordinates, if the $g_{\mu\nu}$ are not constant. If n is a direction perpendicular to the propagation of light, the Huyghens principle shows that the light-ray, envisaged in the plane (γ, n), has the curvature $-\partial\gamma/\partial n$.

We examine the curvature undergone by a ray of light passing by a mass M at the distance Δ. If we choose the system of co-ordinates in agreement with the accompanying diagram, the total bending of the ray (calculated positively if concave towards the origin) is given in sufficient approximation by

$$\int_{-\infty}^{+\infty} \frac{\partial\gamma}{\partial x_1} dx_2,$$

while (73) and (70) give

$$\gamma = \sqrt{\left(-\frac{g_{44}}{g_{22}}\right)} = 1 - \frac{a}{2r}\left(1 + \frac{x_2^2}{r^2}\right)$$

Carrying out the calculation, this gives

$$B = \frac{2\alpha}{\Delta} = \frac{\kappa M}{2\pi\Delta} \tag{74}$$

According to this, a ray of light going past the sun undergoes a deflexion of 1.7"; and a ray going past the planet Jupiter a deflexion of about .02".

If we calculate the gravitational field to a higher degree of approximation, and likewise with corresponding accuracy the orbital motion of a material point of relatively

1. According to E. Freundlich, spectroscopical observations on fixed stars of certain types indicate the existence of an effect of this kind, but a crucial test of this consequence his not yet been made.

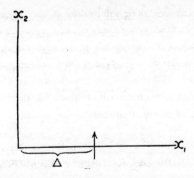

infinitely small mass, we find a deviation of the following kind from the Kepler-Newton laws of planetary motion. The orbital ellipse of a planet undergoes a slow rotation, in the direction of motion, of amount

$$\varepsilon = 24\pi^3 \frac{a^2}{T^2 c^2 \left(1 - e^2\right)}$$ (75)

per revolution. In this formula a denotes the major semi-axis, c the velocity of light in the usual measurement, e the eccentricity, T the time of revolution in seconds.[1]

Calculation gives for the planet Mercury a rotation of the orbit of 43" per century, corresponding exactly to astronomical observation (Leverrier); for the astronomers have discovered in the motion of the perihelion of this planet, after allowing for disturbances by other planets, an inexplicable remainder of this magnitude.

HAMILTON'S PRINCIPLE AND THE GENERAL THEORY OF RELATIVITY

Translated from "Hamiltonsches Princip und allgemeine Relativitätstheorie," Sitzungsberichte der Preussischen Akad. Wissenschaften, 1916.

The general theory of relativity has recently been given in a particularly clear form by H. A. Lorentz and D. Hilbert,[2] who have deduced its equations from one single

1. For the calculation I refer to the original papers: A. Einstein, Sitzungsber. d. Preuss. Akad. d. Wiss., 1915, p. 831; K. Schwarzschild, *ibid.*, 1916, p. 189.
2. Four papers by Lorentz in the Publications of the Koninkl. Akad. van Wetensch. te Amsterdam, 1915 end 1916; D. Hilbert, Göttinger Nachr., 1915, Part 3.

principle of variation. The same thing will be done in the present paper. But my purpose here is to present the fundamental connexions in as perspicuous a manner as possible, and in as general terms as is permissible from the point of view of the general theory of relativity. In particular we shall make as few specializing assumptions as possible, in marked contrast to Hilbert's treatment of the subject. On the other hand, in antithesis to my own most recent treatment of the subject, there is to be complete liberty in the choice of the system of co-ordinates.

§ 1. THE PRINCIPLE OF VARIATION AND THE FIELD-EQUATIONS OF GRAVITATION AND MATTER

Let the gravitational field be described as usual by the tensor[1] of the $g_{\mu\nu}$ (or the $g^{\mu\nu}$); and matter, including the electromagnetic field, by any number of space-time functions $q_{(\rho)}$. How these functions may be characterized in the theory of invariants does not concern us. Further, let $\mathfrak{H}$ be a function of the

$$g^{\mu\nu}, g_\sigma^{\mu\nu}\left(=\frac{\partial g^{\mu\nu}}{\partial x_\sigma}\right) \text{ and } g_{\sigma\tau}^{\mu\tau}\left(=\frac{\partial^2 g^{\mu\nu}}{\partial x_\sigma \partial x_\tau}\right), \text{ the } q_{(\rho)} \text{ and } q_{(\rho)\alpha}\left(=\frac{\partial q_{(\rho)}}{\partial x_\alpha}\right).$$

The principle of variation

$$\delta \int \mathfrak{H} d\tau = 0 \tag{1}$$

then gives us as many differential equations as there are functions $g_{\mu\nu}$, and $q_{(\rho)}$ to be defined, if the $g^{\mu\nu}$ and $q_{(\rho)}$ are varied independently of one another, and in such a way that at the limits of integration the $\delta q_{(\rho)}$, $\delta g^{\mu\nu}$, and $\frac{\partial}{\partial x_\sigma}(\delta g_{\mu\nu})$ all vanish.

We will now assume that $\mathfrak{H}$ is linear in the $g_{\sigma\tau}$, and that the coefficients of the $g_{\sigma\tau}^{\mu\nu}$ depend only on the $g^{\mu\nu}$. We may then replace the principle of variation (1) by one which is more convenient for us. For by appropriate partial integration we obtain

$$\int \mathfrak{H} d\tau = \int \overset{*}{\mathfrak{H}} d\tau + F \tag{2}$$

where F denotes an integral over the boundary of the domain in question, and $\overset{*}{\mathfrak{H}}$ depends only on the $g^{\mu\nu}$, $g_\sigma^{\mu\nu}$, $q_{(\rho)}$, $q_{(\rho)\alpha}$, and no longer on the $g_{\sigma\tau}^{\mu\nu}$. From (2) we obtain, for such variations as are of interest to us,

$$H = \frac{\mathfrak{H}}{\sqrt{-g}} \tag{3}$$

so that we may replace our principle of variation (1) by the more convenient form

$$\delta \int \overset{*}{\mathfrak{H}} d\tau = 0. \tag{1a}$$

---

1. No use is made for the present of the tensor character of the $g_{\mu\nu}$.

By carrying out the variation of the $g^{\mu\nu}$ and the $q_{(\rho)}$, we obtain, as field-equations of gravitation and matter, the equations[1]

$$\frac{\partial}{\partial x_\alpha}\left(\frac{\partial \mathfrak{H}^*}{\partial g_\alpha^{\mu\nu}}\right) - \frac{\partial \mathfrak{H}^*}{\partial g^{\mu\nu}} = 0 \qquad (4)$$

$$\frac{\partial}{\partial x_\alpha}\left(\frac{\partial \mathfrak{H}^*}{\partial q_{(\rho)\alpha}}\right) - \frac{\partial \mathfrak{H}}{\partial q_{(\rho)}} = \qquad (5)$$

§ 2. SEPARATE EXISTENCE OF THE GRAVITATIONAL FIELD

If we make no restrictive assumption as to the manner in which $\mathfrak{H}$ depends on the $g_{\mu\nu}$, $g_\sigma^{\mu\nu}$, $g_\sigma^{\mu\nu}$, $q_{(\rho)}$, $q_{(\rho)\alpha}$, the energy-components cannot be divided into two parts, one belonging to the gravitational field, the other to matter. To ensure this feature of the theory, we make the following assumption

$$\mathfrak{H} = \mathfrak{G} + \mathfrak{M} \qquad (6)$$

where $\mathfrak{G}$ is to depend only on the $g^{\mu\nu}$, $g_\sigma^{\mu\nu}$, $g_\sigma^{\mu\nu}$, and $\mathfrak{M}$ only on $g^{\mu\nu}$, $q_{(\rho)}$, $q_{(\rho)\alpha}$. Equations (4), (4a) then assume the form

$$\frac{\partial}{\partial x_\alpha}\left(\frac{\partial \mathfrak{G}^*}{\partial g_\alpha^{\mu\nu}}\right) - \frac{\partial \mathfrak{G}^*}{\partial g^{\mu\nu}} = \frac{\partial \mathfrak{M}}{\partial g^{\mu\nu}} \qquad (7)$$

$$\frac{\partial}{\partial x_\alpha}\left(\frac{\partial \mathfrak{M}}{\partial q_{(\rho)\alpha}}\right) - \frac{\partial \mathfrak{M}}{\partial q_{(\rho)}} = 0 \qquad (8)$$

Here $\mathfrak{G}^*$ stands in the same relation to $\mathfrak{G}$ as $\mathfrak{H}^*$ to $\mathfrak{H}$.

It is to be noted carefully that equations (8) or (5) would have to give way to others, if we were to assume $\mathfrak{M}$ or $\mathfrak{H}$ to be also dependent on derivatives of the $q_{(\rho)}$ of order higher than the first. Likewise it might be imaginable that the $q_{(\rho)}$ would have to be taken, not as independent of one another, but as connected by conditional equations. All this is of no importance for the following developments, as these are based solely on the equations (7), which have been found by varying our integral with respect to the $g^{\mu\nu}$.

§ 3. PROPERTIES OF THE FIELD EQUATIONS OF GRAVITATION CONDITIONED BY THE THEORY OF INVARIANTS

We now introduce the assumption that

$$ds^2 = g_{\mu\nu}dx_\mu dx_\nu \qquad (9)$$

[1]. For brevity the summation symbols are omitted in the formulæ. Indices occurring twice in a term are always to be taken as summed. Thus in (4), for example, $\dfrac{\partial}{\partial x_\alpha}\left(\dfrac{\partial \mathfrak{i}^*}{\partial g_\alpha^{\mu\nu}}\right)$ denotes the term $\sum_\alpha \dfrac{\partial}{\partial x_\alpha}\left(\dfrac{\partial \mathfrak{H}}{\partial g_\alpha^{\mu\nu}}\right)$

is an invariant. This determines the transformational character of the $g_{\mu\nu}$. As to the transformational character of the $q_{(\rho)}$, which describe matter, we make no supposition. On the other hand, let the functions $H = \dfrac{\mathfrak{H}}{\sqrt{-g}}$, as well as $G = \dfrac{\mathfrak{G}}{\sqrt{-g}}$, and $M = \dfrac{\mathfrak{M}}{\sqrt{-g}}$, be invariants in relation to any substitutions and space-time co-ordinates. From these assumptions follows the general covariance of the equations (7) and (8), deduced from (1). It further follows that G (apart from a constant factor) must be equal to the scalar of Riemann's tensor of curvature; because there is no other invariant with the properties required for G.[1] Thereby $\mathfrak{G}^*$ is also perfectly determined, and consequently the left-hand side of field equation (7) as well.[2]

From the general postulate of relativity there follow certain properties of the function $\mathfrak{G}^*$ which we shall now deduce. For this purpose we carry through an infinitesimal transformation of the co-ordinates, by setting

$$x'_\nu = x_\nu + \Delta x_\nu \tag{10}$$

where the Δx_ν are arbitrary, infinitely small functions of the co-ordinates, and x'_ν are the co-ordinates, in the new system, of the world-point having the co-ordinates x_ν in the original system. As for the co-ordinates, so too for any other magnitude ψ, a law of transformation holds good, of the type

$$\psi' = \psi + \Delta\psi,$$

where $\Delta\psi$ must always be expressible by the Δx_ν. From the covariant property of the $g^{\mu\nu}$ we easily deduce for the $g^{\mu\nu}$ and $g_\sigma^{\mu\nu}$ the laws of transformation

$$\Delta g^{\mu\nu} = g^{\mu\alpha}\frac{\partial(\Delta x_\nu)}{\partial x_\alpha} + g^{\nu\alpha}\frac{\partial(\Delta x_\mu)}{\partial x_\alpha} \tag{11}$$

$$\Delta g_\sigma^{\mu\nu} = \frac{\partial(\Delta g^{\mu\nu})}{\partial x_\sigma} + g_\alpha^{\mu\nu}\frac{\partial(\Delta x_\alpha)}{\partial x_\sigma} \tag{12}$$

Since $\mathfrak{G}^*$ depends only on the $g^{\mu\nu}$ and $g_\sigma^{\mu\nu}$, it is possible, with the help of (11) and (12), to calculate $\Delta\mathfrak{G}^*$. We thus obtain the equation

$$\sqrt{-g}\,\Delta\left(\frac{\mathfrak{G}^*}{\sqrt{-g}}\right) = S_\sigma^\nu\frac{\partial(\Delta x_\sigma)}{\partial x_\nu} + 2\frac{\partial\mathfrak{G}^*}{\partial g_\alpha^{\mu\sigma}}g^{\mu\nu}\frac{\partial^2\Delta x_\sigma}{\partial x_\nu\partial x_\alpha}, \tag{13}$$

where for brevity we have set

$$S_\sigma^\nu = 2\frac{\partial\mathfrak{G}^*}{\partial g^{\mu\sigma}}g^{\mu\nu} + 2\frac{\partial\mathfrak{G}^*}{\partial g_\alpha^{\mu\sigma}}g_\alpha^{\mu\nu} + \mathfrak{G}^*\delta_\sigma^\nu - \frac{\partial\mathfrak{G}^*}{\partial g_\nu^{\mu\alpha}}g_\sigma^{\mu\alpha}. \tag{14}$$

1. Herein is to be found the reason why the general postulate of relativity leads to a very definite theory of gravitation.
2. By performing partial integration we obtain
$$\mathfrak{G}^* = \sqrt{-g}\,g^{\mu\nu}[\{\mu\alpha,\,\beta\}\{\nu\beta,\,\alpha\} - \{\mu\nu,\,\alpha\}\{\alpha\beta,\,\beta\}].$$

From these two equations we draw two inferences which are important for what follows. We know that $\frac{\mathfrak{G}}{\sqrt{-g}}$ is an invariant with respect to any substitution, but we do not know this of $\frac{\mathfrak{G}^*}{\sqrt{-g}}$. It is easy to demonstrate, however, that the latter quantity is an invariant with respect to any *linear* substitutions of the co-ordinates. Hence it follows that the right side of (13) must always vanish if all $\frac{\partial^2 \Delta x_\sigma}{\partial x_\nu \partial x_\alpha}$ vanish. Consequently $\mathfrak{G}^*$ must satisfy the identity

$$S_\sigma^\nu = 0 \tag{15}$$

If, further, we choose the Δx_ν so that they differ from zero only in the interior of a given domain, but in infinitesimal proximity to the boundary they vanish, then, with the transformation in question, the value of the boundary integral occurring in equation (2) does not change. Therefore $\Delta F = 0$, and, in consequence,[1]

$$\Delta \int \mathfrak{G} d\tau = \Delta \int \mathfrak{G}^* d\tau$$

But the left-hand side of the equation must vanish, since both $\dfrac{\mathfrak{G}}{\sqrt{-g}}$ and $\sqrt{-g}\,d\tau$ are invariants. Consequently the right-hand side also vanishes. Thus, taking (14), (15), and (16) into consideration, we obtain, in the first place, the equation

$$\int \frac{\partial \mathfrak{G}^*}{\partial g_\alpha^{\mu\sigma}} g^{\mu\nu} \frac{\partial^2 (\Delta x_\sigma)}{\partial x_\nu \partial x_\alpha} d\tau = 0 \tag{16}$$

Transforming this equation by two partial integrations, and having regard to the liberty of choice of the Δx_σ, we obtain the identity

$$\frac{\partial^2}{\partial x_\nu \partial x_\alpha} \left(g^{\mu\nu} \frac{\partial \mathfrak{G}^*}{\partial g_\alpha^{\mu\sigma}} \right) \equiv 0 \tag{17}$$

From the two identities (16) and (17), which result from the invariance of $\dfrac{\mathfrak{G}}{\sqrt{-g}}$, and therefore from the postulate of general relativity, we now have to draw conclusions.

We first transform the field equations (7) of gravitation by mixed multiplication by $g^{\mu\sigma}$. We then obtain (by interchanging the indices σ and ν), as equivalents of the field equations (7) the equations

$$\frac{\partial}{\partial x_\alpha} \left(g^{\mu\nu} \frac{\partial \mathfrak{G}^*}{\partial g_\alpha^{\mu\sigma}} \right) = -(\mathfrak{T}_\sigma^\nu + t_\sigma^\nu) \tag{18}$$

where we have set

$$\mathfrak{T}_\sigma^\nu = -\frac{\partial \mathfrak{M}}{\partial g^{\mu\sigma}} g^{\mu\nu} \tag{19}$$

---

1. By the introduction of the quantities $\mathfrak{G}$ and $\mathfrak{G}^*$ instead of $\mathfrak{H}$ and $\mathfrak{H}^*$.

$$t^{\nu}_{\sigma} = -\left(\frac{\partial \mathfrak{G}^*}{\partial g^{\mu\sigma}_{\alpha}}g^{\mu\nu}_{\alpha} + \frac{\partial \mathfrak{G}^*}{\partial g^{\mu\nu}}g^{\mu\nu}\right) = \tfrac{1}{2}\left(\mathfrak{G}^*\delta^{\nu}_{\sigma} - \frac{\partial \mathfrak{G}^*}{\partial g^{\mu\alpha}_{\nu}}g^{\mu\alpha}_{\sigma}\right) \quad (20)$$

The last expression for t^{ν}_{μ} is vindicated by (14) and (15). By differentiation of (18) with respect to x_{ν} and summation for ν, there follows, in view of (17),

$$\frac{\partial}{\partial x_{\nu}}(\mathfrak{T}^{\nu}_{\sigma} + t^{\nu}_{\sigma}) = 0 \qquad (21)$$

Equation (21) expresses the conservation of momentum and energy. We call $\mathfrak{T}^{\nu}_{\sigma}$ the components of the energy of matter, t^{ν}_{σ} the components of the energy of the gravitational field.

Having regard to (20), there follows from the field equations (7) of gravitation, by multiplication by $g^{\mu\nu}_{\sigma}$, and summation with respect to μ and ν,

$$\frac{\partial t^{\nu}_{\sigma}}{\partial x_{\nu}} + \tfrac{1}{2}g^{\mu\nu}_{\sigma}\frac{\partial \mathfrak{M}}{\partial g^{\mu\nu}} = 0,$$

or, in view of (19) and (21),

$$\frac{\partial \mathfrak{T}^{\nu}_{\sigma}}{\partial x_{\nu}} + \tfrac{1}{2}g^{\mu\nu}_{\sigma}\mathfrak{T}_{\mu\nu} = 0 \qquad (22)$$

where $\mathfrak{T}_{\mu\nu}$ denotes the quantities $g_{\nu\sigma}\mathfrak{T}_{\mu\nu}$. These are four equations which the energy-components of matter have to satisfy.

It is to be emphasized that the (generally covariant) laws of conservation (21) and (22) are deduced from the field equations (7) of gravitation, in combination with the postulate of general covariance (relativity) *alone*, without using the field equations (8) for material phenomena.

COSMOLOGICAL CONSIDERATIONS ON THE GENERAL THEORY OF RELATIVITY

Translated from "Kosmologische Betrachtungen zur allgemeinen Relativitätstheorie," Sitzungsberichte der Preussischen Akad. d. Wissenschaften, 1917.

It is well known that Poisson's equation

$$\nabla^2 \phi = 4\pi K\rho \qquad (1).$$

in combination with the equations of motion of a material point is not as yet a perfect substitute for Newton's theory of action at a distance. There is still to be taken into account the condition that at spatial infinity the potential φ tends toward a fixed limiting value. There is an analogous state of things in the theory of gravitation in general relativity. Here, too, we must supplement the differential equations by limiting conditions at spatial infinity, if we really have to regard the universe as being of infinite spatial extent.

In my treatment of the planetary problem I chose these limiting conditions in the form of the following assumption: it is possible to select a system of reference so that at spatial infinity all the gravitational potentials $g_{\mu\nu}$ become constant. But it is by no means evident *a priori* that we may lay down the same limiting conditions when we wish to take larger portions of the physical universe into consideration. In the following pages the reflexions will be given which, up to the present, I have made on this fundamentally important question.

§ 1. THE NEWTONIAN THEORY

It is well known that Newton's limiting condition of the constant limit for φ at spatial infinity leads to the view that the density of matter becomes zero at infinity. For we imagine that there may be a place in universal space round about which the gravitational field of matter, viewed on a large scale, possesses spherical symmetry. It then follows from Poisson's equation that, in order that φ may tend to a limit at infinity, the mean density ρ must decrease toward zero more rapidly than $1/r^2$ as the distance r from the centre increases.[1] In this sense, therefore, the universe according to Newton is finite, although it may possess an infinitely great total mass.

From this it follows in the first place that the radiation emitted by the heavenly bodies will, in part, leave the Newtonian system of the universe, passing radially outwards, to become ineffective and lost in the infinite. May not entire heavenly bodies fare likewise? It is hardly possible to give a negative answer to this question. For it follows from the assumption of a finite limit for φ at spatial infinity that a heavenly body with finite kinetic energy is able to reach spatial infinity by overcoming the Newtonian forces of attraction. By statistical mechanics this case must occur from time to time, as long as the total energy of the stellar system—transferred to one single star—is great enough to send that star on its journey to infinity, whence it never can return.

We might try to avoid this peculiar difficulty by assuming a very high value for the limiting potential at infinity. That would be a possible way, if the value of the gravitational potential were not itself necessarily conditioned by the heavenly bodies. The

---

1. ρ is the mean density of matter, calculated for a region which is large as compared with the distance between neighbouring fixed stars, but small in comparison with the dimensions of the whole stellar system.

truth is that we are compelled to regard the occurrence of any great differences of potential of the gravitational field as contradicting the facts. These differences must really be of so low an order of magnitude that the stellar velocities generated by them do not exceed the velocities actually observed.

If we apply Boltzmann's law of distribution for gas molecules to the stars, by comparing the stellar system with a gas in thermal equilibrium, we find that the Newtonian stellar system cannot exist at all. For there is a finite ratio of densities corresponding to the finite difference of potential between the centre and spatial infinity. A vanishing of the density at infinity thus implies a vanishing of the density at the centre.

It seems hardly possible to surmount these difficulties on the basis of the Newtonian theory. We may ask ourselves the question whether they can be removed by a modification of the Newtonian theory. First of all we will indicate a method which does not in itself claim to be taken seriously; it merely serves as a foil for what is to follow. In place of Poisson's equation we write

$$\nabla^2 \phi - \lambda \phi = 4\pi\kappa\rho \tag{2}$$

where λ denotes a universal constant. If ρ_0 be the uniform density of distribution of mass, then

$$\phi = -\frac{4\pi\kappa}{\lambda} \rho_0 \tag{3}$$

is a solution of equation (2). This solution would correspond to the case in which the matter of the fixed stars was distributed uniformly through space, if the density ρ_0 is equal to the actual mean density of the matter in the universe. The solution then corresponds to an infinite extension of the central space, filled uniformly with matter. If, without making any change in the mean density, we imagine matter to be non-uniformly distributed locally, there will be, over and above the ϕ with the constant value of equation (3), an additional ϕ, which in the neighbourhood of denser masses will so much the more resemble the Newtonian field as $\lambda\phi$ is smaller in comparison with $4\pi\kappa\rho$.

A universe so constituted would have, with respect to its gravitational field, no centre. A decrease of density in spatial infinity would not have to be assumed, but both the mean potential and mean density would remain constant to infinity. The conflict with statistical mechanics which we found in the case of the Newtonian theory is not repeated. With a definite but extremely small density, matter is in equilibrium, without any internal material form (pressures) being required to maintain equilibrium.

§ 2. THE BOUNDARY CONDITIONS ACCORDING TO THE GENERAL THEORY OF RELATIVITY

In the present paragraph I shall conduct the reader over the road that I have myself travelled, rather a rough and winding road, because otherwise I cannot hope that he will take much interest in the result at the end of the journey. The conclusion I shall arrive at is that the field equations of gravitation which I have championed hitherto still need a slight modification, so that on the basis of the general theory of relativity those fundamental difficulties may be avoided which have been set forth in § 1 as confronting the Newtonian theory. This modification corresponds perfectly to the transition from Poisson's equation (1) to equation (2) of § 1. We finally infer that boundary conditions in spatial infinity fall away altogether, because the universal continuum in respect of its spatial dimensions is to be viewed as a self-contained continuum of finite spatial (three-dimensional) volume.

The opinion which I entertained until recently, as to the limiting conditions to be laid down in spatial infinity, took its stand on the following considerations. In a consistent theory of relativity there can be no inertia *relatively to "space,"* but only an inertia of masses *relatively to one another*. If, therefore, I have a mass at a sufficient distance from all other masses in the universe, its inertia must fall to zero. We will try to formulate this condition mathematically.

According to the general theory of relativity the negative momentum is given by the first three components, the energy by the last component of the covariant tensor multiplied by $\sqrt{-g}$

$$m\sqrt{-g}\,g_{\mu\alpha}\frac{dx_{\alpha}}{ds} \tag{4}$$

where, as always, we set

$$ds^2 = -g_{\mu\nu}dx_{\mu}dx_{\nu} \tag{5}$$

In the particularly perspicuous case of the possibility of choosing the system of co-ordinates so that the gravitational field at every point is spatially isotropic, we have more simply

$$ds^2 = -A\left(dx_1^2 + dx_2^2 + dx_3^2\right) + Bdx_4^2$$

If, moreover, at the same time

$$\sqrt{-g} = 1 = \sqrt{A^3 B}$$

we obtain from (4), to a first approximation for small velocities,

$$m \frac{A}{\sqrt{B}} \frac{dx_1}{dx_4}, m \frac{A}{\sqrt{B}} \frac{dx_2}{dx_4}, m \frac{A}{\sqrt{B}} \frac{dx_3}{dx_4}$$

for the components of momentum, and for the energy (in the static case)

$$m\sqrt{B}.$$

From the expressions for the momentum, it follows that $m\frac{A}{\sqrt{B}}$ plays the part of the rest mass. As m is a constant peculiar to the point of mass, independently of its position, this expression, if we retain the condition $\sqrt{g} - = 1$ at spatial infinity, can vanish only when A diminishes to zero, while B increases to infinity. It seems, therefore, that such a degeneration of the co-efficients $g_{\mu\nu}$ is required by the postulate of relativity of all inertia. This requirement implies that the potential energy $m\sqrt{B}$ becomes infinitely great at infinity. Thus a point of mass can never leave the system; and a more detailed investigation shows that the same thing applies to light-rays. A system of the universe with such behaviour of the gravitational potentials at infinity would not therefore run the risk of wasting away which was mooted just now in connexion with the Newtonian theory.

I wish to point out that the simplifying assumptions as to the gravitational potentials on which this reasoning is based, have been introduced merely for the sake of lucidity. It is possible to find general formulations for the behaviour of the $g_{\mu\nu}$ at infinity which express the essentials of the question without further restrictive assumptions.

At this stage, with the kind assistance of the mathematician J. Grommer, I investigated centrally symmetrical, static gravitational fields, degenerating at infinity in the way mentioned. The gravitational potentials $g_{\mu\nu}$ were applied, and from them the energy-tensor $T_{\mu\nu}$ of matter was calculated on the basis of the field equations of gravitation. But here it proved that for the system of the fixed stars no boundary conditions of the kind can come into question at all, as was also rightly emphasized by the astronomer de Sitter recently.

For the contravariant energy-tensor $T^{\mu\nu}$ of ponderable matter is given by

$$T^{\mu\nu} = \rho \frac{dx_\mu}{ds} \frac{dx_\nu}{ds},$$

where ρ is the density of matter in natural measure. With an appropriate choice of the system of co-ordinates the stellar velocities are very small in comparison with that of light. We may, therefore, substitute $\sqrt{g_{44}} dx_4$ for ds. This shows us that all components of $T^{\mu\nu}$ must be very small in comparison with the last component T^{44}. But it was quite impossible to reconcile this condition with the chosen boundary conditions. In the retrospect this result does not appear astonishing. The fact of the small velocities of the stars allows the conclusion that wherever there are fixed stars, the gravitational potential (in our case $\sqrt{B}$) can never be much greater than here on earth. This follows from

statistical reasoning, exactly as in the case of the Newtonian theory. At any rate, our calculations have convinced me that such conditions of degeneration for the $g_{\mu\nu}$ in spatial infinity may not be postulated.

After the failure of this attempt, two possibilities next present themselves.

(a) We may require, as in the problem of the planets, that, with a suitable choice of the system of reference, the $g_{\mu\nu}$ in spatial infinity approximate to the values

$$
\begin{array}{cccc}
-1 & 0 & 0 & 0 \\
0 & -1 & 0 & 0 \\
0 & 0 & -1 & 0 \\
0 & 0 & 0 & 1
\end{array}
$$

(b) We may refrain entirely from laying down boundary conditions for spatial infinity claiming general validity; but at the spatial limit of the domain under consideration we have to give the $g_{\mu\nu}$ separately in each individual case, as hitherto we were accustomed to give the initial conditions for time separately.

The possibility (b) holds out no hope of solving the problem, but amounts to giving it up. This is an incontestable position, which is taken up at the present time by de Sitter.[1] But I must confess that such a complete resignation in this fundamental question is for me a difficult thing. I should not make up my mind to it until every effort to make headway toward a satisfactory view had proved to be vain.

Possibility (a) is unsatisfactory in more respects than one. In the first place those boundary conditions pre-suppose a definite choice of the system of reference, which is contrary to the spirit of the relativity principle. Secondly, if we adopt this view, we fail to comply with the requirement of the relativity of inertia. For the inertia of a material point of mass m (in natural measure) depends upon the $g_{\mu\nu}$; but these differ but little from their postulated values, as given above, for spatial infinity. Thus inertia would indeed be *influenced*, but would not be *conditioned* by matter (present in finite space). If only one single point of mass were present, according to this view, it would possess inertia, and in fact an inertia almost as great as when it is surrounded by the other masses of the actual universe. Finally, those statistical objections must be raised against this view which were mentioned in respect of the Newtonian theory.

From what has now been said it will be seen that I have not succeeded in formulating boundary conditions for spatial infinity. Nevertheless, there is still a possible way out, without resigning as suggested under (b). For if it were possible to regard the universe as a continuum which is *finite (closed) with respect to its spatial dimensions*, we should have no need at all of any such boundary conditions. We shall proceed to show that both the general postulate of relativity and the fact of the small stellar velocities

1. de Sitter, Akad. van Wetensch. te Amsterdam, 8 Nov., 1916.

are compatible with the hypothesis of a spatially finite universe; though certainly, in order to carry through this idea, we need a generalizing modification of the field equations of gravitation.

§ 3. THE SPATIALLY FINITE UNIVERSE WITH A UNIFORM DISTRIBUTION OF MATTER

According to the general theory of relativity the metrical character (curvature) of the four-dimensional space-time continuum is defined at every point by the matter at that point and the state of that matter. Therefore, on account of the lack of uniformity in the distribution of matter, the metrical structure of this continuum must necessarily be extremely complicated. But if we are concerned with the structure only on a large scale, we may represent matter to ourselves as being uniformly distributed over enormous spaces, so that its density of distribution is a variable function which varies extremely slowly. Thus our procedure will somewhat resemble that of the geodesists who, by means of an ellipsoid, approximate to the shape of the earth's surface, which on a small scale is extremely complicated.

The most important fact that we draw from experience as to the distribution of matter is that the relative velocities of the stars are very small as compared with the velocity of light. So I think that for the present we may base our reasoning upon the following approximative assumption. There is a system of reference relatively to which matter may be looked upon as being permanently at rest. With respect to this system, therefore, the contravariant energy-tensor $T^{\mu\nu}$ of matter is, by reason of (5), of the simple form

$$\begin{bmatrix} 0 & 0 & 0 & 0 \\ 0 & 0 & 0 & 0 \\ 0 & 0 & 0 & 0 \\ 0 & 0 & 0 & \rho \end{bmatrix}$$

$$(6)$$

The scalar ρ of the (mean) density of distribution may be *a priori* a function of the space co-ordinates. But if we assume the universe to be spatially finite, we are prompted to the hypothesis that ρ is to be independent of locality. On this hypothesis we base the following considerations.

As concerns the gravitational field, it follows from the equation of motion of the material point

$$\frac{d^2 x_\nu}{ds^2} + \{\alpha\beta, \nu\} \frac{dx_\alpha}{ds} \frac{dx_\beta}{ds} = 0$$

that a material point in a static gravitational field can remain at rest only when g_{44} is independent of locality. Since, further, we presuppose independence of the time co-ordinate x_4 for all magnitudes, we may demand for the required solution that, for all x_ν,

$$g_{44} = 1 \tag{7}$$

Further, as always with static problems, we shall have to set

$$g_{14} = g_{24} = g_{34} = 0 \tag{8}$$

It remains now to determine those components of the gravitational potential which define the purely spatial-geometrical relations of our continuum $(g_{11}, g_{12}, \cdots g_{33})$. From our assumption as to the uniformity of distribution of the masses generating the field, it follows that the curvature of the required space must be constant. With this distribution of mass, therefore, the required finite continuum of the x_1, x_2, x_3, with constant x_4, will be a spherical space.

We arrive at such a space, for example, in the following way. We start from a Euclidean space of four dimensions, $\xi_1, \xi_2, \xi_3, \xi_4$, with a linear element $d\sigma$; let, therefore,

$$d\sigma^2 = d\xi_1^2 + d\xi_2^2 + d\xi_3^2 + d\xi_4^2 \tag{9}$$

In this space we consider the hyper-surface

$$R^2 = \xi_1^2 + \xi_2^2 + \xi_3^2 + \xi_4^2, \tag{10}$$

where R denotes a constant. The points of this hyper-surface form a three-dimensional continuum, a spherical space of radius of curvature R.

The four-dimensional Euclidean space with which we started serves only for a convenient definition of our hyper-surface. Only those points of the hyper-surface are of interest to us which have metrical properties in agreement with those of physical space with a uniform distribution of matter. For the description of this three-dimensional continuum we may employ the co-ordinates ξ_1, ξ_2, ξ_3 (the projection upon the hyper-plane $\xi_4 = 0$) since, by reason of (10), ξ_4 can be expressed in terms of ξ_1, ξ_2, ξ_3. Eliminating ξ_4 from (9), we obtain for the linear element of the spherical space the expression

$$\left. \begin{array}{l} d\sigma^2 = \gamma_{\mu\nu} d\xi_\mu d\xi_\nu \\[2mm] \gamma_{\mu\nu} = \delta_{\mu\nu} + \dfrac{\xi_\mu \xi_\nu}{R^2 - \rho^2} \end{array} \right\} \tag{11}$$

where $\delta_{\mu\nu} = 1$, if $\mu = \nu$; $\delta_{\mu\nu} = 0$, if $\mu \neq \nu$, and $\rho^2 = \xi_1^2 + \xi_2^2 + \xi_3^2$. The co-ordinates chosen are convenient when it is a question of examining the environment of one of the two points $\xi_1 = \xi_2 = \xi_3 = 0$.

Now the linear element of the required four-dimensional space-time universe is also given us. For the potential $g_{\mu\nu}$, both indices of which differ from 4, we have to set

$$\rho^2 = \xi_1^2 + \xi_2^2 + \xi_3^2. \tag{12}$$

which equation, in combination with (7) and (8), perfectly defines the behaviour of measuring-rods, clocks, and light-rays.

§ 4. ON AN ADDITIONAL TERM FOR THE FIELD EQUATIONS OF GRAVITATION

My proposed field equations of gravitation for any chosen system of co-ordinates run as follows:—

$$
\left.
\begin{aligned}
G_{\mu\nu} &= -\kappa\left(T_{\mu\nu} - \frac{1}{2}g_{\mu\nu}T\right), \\
G_{\mu\nu} &= -\frac{\partial}{\partial x_\alpha}\{\mu\nu,\alpha\} + \{\mu\nu,\beta\}\{\nu\beta,\alpha\} \\
&\quad + \frac{\partial^2 \log\sqrt{-g}}{\partial x_\mu \partial x_\nu} - \{\mu\nu,\alpha\}\frac{\partial \log\sqrt{-g}}{\partial x_\alpha}
\end{aligned}
\right\}
\tag{13}
$$

The system of equations (13) is by no means satisfied when we insert for the $g_{\mu\nu}$ the values given in (7), (8), and (12), and for the (contravariant) energy-tensor of matter the values indicated in (6). It will be shown in the next paragraph how this calculation may conveniently be made. So that, if it were certain that the field equations (13) which I have hitherto employed were the only ones compatible with the postulate of general relativity, we should probably have to conclude that the theory of relativity does not admit the hypothesis of a spatially finite universe.

However, the system of equations (14) allows a readily suggested extension which is compatible with the relativity postulate, and is perfectly analogous to the extension of Poisson's equation given by equation (2). For on the left-hand side of field equation (13) we may add the fundamental tensor $g_{\mu\nu}$, multiplied by a universal constant, $-\lambda$, at present unknown, without destroying the general covariance. In place of field equation (13) we write

$$
G_{\mu\nu} - \lambda g_{\mu\nu} = -\kappa\left(T_{\mu\nu} - \frac{1}{2}g_{\mu\nu}T\right)
\tag{13a}
$$

This field equation, with λ sufficiently small, is in any case also compatible with the facts of experience derived from the solar system. It also satisfies laws of conservation of momentum and energy, because we arrive at (13a) in place of (13) by introducing into Hamilton's principle, instead of the scalar of Riemann's tensor, this scalar increased by a universal constant; and Hamilton's principle, of course, guarantees the validity of laws of conservation. It will be shown in § 5 that field equation (13a) is compatible with our conjectures on field and matter.

§ 5. CALCULATION AND RESULT

Since all points of our continuum are on an equal footing, it is sufficient to carry through the calculation for *one* point, e.g. for one of the two points with the co-ordinates

$$x_1 = x_2 = x_3 = x_4 = 0.$$

Then for the $g_{\mu\nu}$ in (13a) we have to insert the values

$$\begin{matrix} -1 & 0 & 0 & 0 \\ 0 & -1 & 0 & 0 \\ 0 & 0 & -1 & 0 \\ 0 & 0 & 0 & -1 \end{matrix}$$

wherever they appear differentiated only once or not at all. We thus obtain in the first place

$$G_{\mu\nu} = \frac{\partial}{\partial x_1}[\mu\nu, 1] + \frac{\partial}{\partial x_2}[\mu\nu, 2] + \frac{\partial}{\partial x_3}[\mu\nu, 3] + \frac{\partial^2 \log\sqrt{-g}}{\partial x_\mu \partial x_\nu}.$$

From this we readily discover, taking (7), (8), and (13) into account, that all equations (13a) are satisfied if the two relations

$$-\frac{2}{R^2} + \lambda = -\frac{\kappa\rho}{2}, \quad -\lambda = -\frac{\kappa\rho}{2},$$

or

$$\lambda = \frac{\kappa\rho}{2} = \frac{1}{R^2} \tag{14}$$

are fulfilled.

Thus the newly introduced universal constant λ defines both the mean density of distribution ρ which can remain in equilibrium and also the radius R and the volume $2\pi^2R^3$ of spherical space. The total mass M of the universe, according to our view, is finite, and is in fact

$$M = \rho \cdot 2\pi^2 R^3 = 4\pi^2 \frac{R}{\kappa} = \pi^2 \sqrt{\frac{32}{\kappa^3 \rho}} \tag{15}$$

Thus the theoretical view of the actual universe, if it is in correspondence with our reasoning, is the following. The curvature of space is variable in time and place, according to the distribution of matter, but we may roughly approximate to it by means of a spherical space. At any rate, this view is logically consistent, and from the standpoint of the general theory of relativity lies nearest at hand; whether, from the standpoint of present astronomical knowledge, it is tenable, will not here be discussed. In order to arrive at this consistent view, we admittedly had to introduce an extension of the field equations of gravitation which is not justified by our actual knowledge of gravitation.

It is to be emphasized, however, that a positive curvature of space is given by our results, even if the supplementary term is not introduced. That term is necessary only for the purpose of making possible a quasi-static distribution of matter, as required by the fact of the small velocities of the stars.

DO GRAVITATIONAL FIELDS PLAY AN ESSENTIAL PART IN THE STRUCTURE OF THE ELEMENTARY PARTICLES OF MATTER?

Translated from "Spielen Gravitationsfelder im Aufber der materiellen Elementarteilchen eine wesentliche Rolle?" Sitzungsberichte der Preussischen Akad. d. Wissenschaften,
1919.

Neither the Newtonian nor the relativistic theory of gravitation has so far led to any advance in the theory of the constitution of matter. In view of this fact it will be shown in the following pages that there are reasons for thinking that the elementary formations which go to make up the atom are held together by gravitational forces.

§ 1. DEFECTS OF THE PRESENT VIEW

Great pains have been taken to elaborate a theory which will account for the equilibrium of the electricity constituting the electron. G. Mie, in particular, has devoted deep researches to this question. His theory, which has found considerable support among theoretical physicists, is based mainly on the introduction into the energy-tensor of supplementary terms depending on the components of the electro-dynamic potential, in addition to the energy terms of the Maxwell-Lorentz theory. These new terms, which in outside space are unimportant, are nevertheless effective in the interior of the electrons in maintaining equilibrium against the electric form of repulsion. In spite of the beauty of the formal structure of this theory, as erected by Mie, Hilbert, and Weyl, its physical results have hitherto been unsatisfactory. On the one hand the multiplicity of possibilities is discouraging, and on the other hand those additional terms have not as yet allowed themselves to be framed in such a simple form that the

solution could be satisfactory.

So far the general theory of relativity has made no change in this state of the question. If we for the moment disregard the additional cosmological term, the field equations take the form

$$G_{\mu\nu} = \frac{1}{2} g_{\mu\nu} G = -\kappa T_{\mu\nu} \tag{1}$$

where $G_{\mu\nu}$ denotes the contracted Riemann tensor of curvature, G the scalar of curvature formed by repeated contraction, and $T_{\mu\nu}$ the energy-tensor of "matter." The assumption that the $T_{\mu\nu}$ do *not* depend on the derivatives of the $g_{\mu\nu}$ is in keeping with the historical development of these equations. For these quantities are, of course, the energy-components in the sense of the special theory of relativity, in which variable $g_{\mu\nu}$ do not occur. The second term on the left-hand side of the equation is so chosen that the divergence of the left-hand side of (1) vanishes identically, so that taking the divergence of (1), we obtain the equation

$$\frac{\partial \mathfrak{T}_\mu^\sigma}{\partial x_\sigma} + \tfrac{1}{2} g_{\bar\mu}^{\sigma\tau} \mathfrak{T}_{\sigma\tau} = 0 \tag{2}$$

which in the limiting case of the special theory of relativity gives the complete equations of conservation

$$\frac{\partial T_{\mu\nu}}{\partial x_\nu} = 0.$$

Therein lies the physical foundation for the second term of the left-hand side of (1). It is by no means settled *a priori* that a limiting transition of this kind has any possible meaning. For if gravitational fields do play an essential part in the structure of the particles of matter, the transition to the limiting case of constant $g_{\mu\nu}$ would, for them, lose its justification, for indeed, with constant $g_{\mu\nu}$ there could not be any particles of matter. So if we wish to contemplate the possibility that gravitation may take part in the structure of the fields which constitute the corpuscles, we cannot regard equation (1) as confirmed.

Placing in (1) the Maxwell-Lorentz energy-components of the electromagnetic field $\phi_{\mu\nu}$,

$$T_{\mu\nu} = \frac{1}{4} g_{\mu\nu} \phi_{\sigma\tau} \phi^{\sigma\tau} - \phi_{\mu\sigma} \phi_{\nu\tau} g^{\sigma\tau} \tag{3}$$

we obtain for (2), by taking the divergence, and after some reduction,[1]

$$\phi_{\mu\sigma} \mathfrak{I}^\sigma = 0 \tag{4}$$

where, for brevity, we have set

$$\frac{\partial}{\partial x_\tau}\left(\sqrt{-g}\,\phi_{\mu\nu} g^{\mu\sigma} g^{\nu\tau}\right) = \frac{\partial \mathfrak{f}^{\sigma\tau}}{\partial x_\tau} = \mathfrak{I}^\sigma \tag{5}$$

1. Cf. e.g. A. Einstein, Sitzungsber. d. Preuss. Akad. d. Wiss., 1916, pp. 187, 188.

In the calculation we have employed the second of Maxwell's systems of equations

$$\frac{\partial \phi_{\mu\nu}}{\partial x_\rho} + \frac{\partial \phi_{\nu\rho}}{\partial x_\mu} + \frac{\partial \phi_{\rho\mu}}{\partial x_\nu} = 0 \tag{6}$$

We see from (4) that the current-density $\Im^\sigma$ must everywhere vanish. Therefore, by equation (1), we cannot arrive at a theory of the electron by restricting ourselves to the electromagnetic components of the Maxwell-Lorentz theory, as has long been known. Thus if we hold to (1) we are driven on to the path of Mie's theory.[1]

Not only the problem of matter, but the cosmological problem as well, leads to doubt as to equation (1). As I have shown in the previous paper, the general theory of relativity requires that the universe be spatially finite. But this view of the universe necessitated an extension of equations (1), with the introduction of a new universal constant λ, standing in a fixed relation to the total mass of the universe (or, respectively, to the equilibrium density of matter). This is gravely detrimental to the formal beauty of the theory.

§ 2. THE FIELD EQUATIONS FREED OF SCALARS

The difficulties set forth above are removed by setting in place of field equations (1) the field equations

$$G_{\mu\nu} - \frac{1}{4} g_{\mu\nu} G = -\kappa T_{\mu\nu} \tag{1a}$$

where $T_{\mu\nu}$ denotes the energy-tensor of the electromagnetic field given by (3).

The formal justification for the factor $-1/4$ in the second term of this equation lies in its causing the scalar of the left-hand side,

$$g^{\mu\nu}\left(G_{\mu\nu} - \frac{1}{4} g_{\mu\nu} G \right),$$

to vanish identically, as the scalar $g^{\nu\mu} T_{\mu\nu}$ of the right-hand side does by reason of (3). If we had reasoned on the basis of equations (1) instead of (1a), we should, on the contrary, have obtained the condition $G = 0$, which would have to hold good everywhere for the $g_{\mu\nu}$, independently of the electric field. It is clear that the system of equations [(1a), (3)] is a consequence of the system [(1), (3)], but not conversely.

We might at first sight feel doubtful whether (1a) together with (6) sufficiently define the entire field. In a generally relativistic theory we need $n - 4$ differential equations, independent of one another, for the definition of n independent variables, since in the solution, on account of the liberty of choice of the co-ordinates, four quite arbitrary functions of all co-ordinates must naturally occur. Thus to define the sixteen independent quantities $g_{\mu\nu}$ and $\phi_{\mu\nu}$ we require twelve equations, all independent of one another. But as it happens, nine of the equations (1a), and three of the equations (6) are independent of one another.

1. Cf. D. Hilbert, Göttinger Nachr., 20 Nov., 1915.

Forming the divergence of (1a), and taking into account that the divergence of $G_{\mu\nu} - \frac{1}{2} g_{\mu\nu} G$ vanishes, we obtain

$$\phi_{\mu\alpha} J^\alpha + \frac{1}{4\kappa} \frac{\partial G}{\partial x_\sigma} = 0 \qquad (4a)$$

From this we recognize first of all that the scalar of curvature G in the four-dimensional domains in which the density of electricity vanishes, is constant. If we assume that all these parts of space are connected, and therefore that the density of electricity differs from zero only in separate "world-threads," then the scalar of curvature, everywhere outside these world-threads, possesses a constant value G_0. But equation (4a) also allows an important conclusion as to the behaviour of G within the domains having a density of electricity other than zero. If, as is customary, we regard electricity as a moving density of charge, by setting

$$J^\sigma = \frac{\mathfrak{J}^\sigma}{\sqrt{-g}} = \rho \frac{dx_\sigma}{ds}, \qquad (7)$$

we obtain from (4a) by inner multiplication by J^σ, on account of the antisymmetry of $\phi_{\mu\nu}$, the relation

$$\frac{\partial G}{\partial x_\sigma} \frac{dx_\sigma}{ds} = 0 \qquad (8)$$

Thus the scalar of curvature is constant on every world-line of the motion of electricity. Equation (4a) can be interpreted in a graphic manner by the statement: The scalar of curvature plays the part of a negative pressure which, outside of the electric corpuscles, has a constant value G_0. In the interior of every corpuscle there subsists a negative pressure (positive $G - G_0$) the fall of which maintains the electrodynamic force in equilibrium. The minimum of pressure, or, respectively, the maximum of the scalar of curvature, does not change with time in the interior of the corpuscle.

We now write the field equations (1a) in the form

$$\left(G_{\mu\nu} - \frac{1}{2} g_{\mu\nu} G \right) + \frac{1}{4} g_{\mu\nu} G_0 = -\kappa \left(T_{\mu\nu} + \frac{1}{4\kappa} g_{\mu\nu} (G - G_0) \right) \qquad (9)$$

On the other hand, we transform the equations supplied with the cosmological term as already given

$$G_{\mu\nu} - \lambda g_{\mu\nu} = -\kappa \left(T_{\mu\nu} - \frac{1}{2} g_{\mu\nu} T \right)$$

Subtracting the scalar equation multiplied by $^1/_2$, we next obtain

$$\left(G_{\mu\nu} - \frac{1}{2} g_{\mu\nu} G \right) + g_{\mu\nu} \lambda = -\kappa T.$$

Now in regions where only electrical and gravitational fields are present, the right-hand side of this equation vanishes. For such regions we obtain, by forming the scalar,

$$G + 4\lambda = 0.$$

In such regions, therefore, the scalar of curvature is constant, so that λ may be replaced by $^1/_4G_0$. Thus we may write the earlier field equation (1) in the form

$$G_{\mu\nu} - \frac{1}{2} g_{\mu\nu} G + \frac{1}{4\kappa} g_{\mu\nu} G_0 = -\kappa T_{\mu\nu} \tag{10}$$

Comparing (9) with (10), we see that there is no difference between the new field equations and the earlier ones, except that instead of $T_{\mu\nu}$ as tensor of "gravitating mass" there now occurs $T_{\mu\nu} + \frac{1}{4\kappa} g_{\mu\nu}(G - G_0)$ which is independent of the scalar of curvature. But the new formulation has this great advantage, that the quantity λ appears in the fundamental equations as a constant of integration, and no longer as a universal constant peculiar to the fundamental law.

§ 3. ON THE COSMOLOGICAL QUESTION

The last result already permits the surmise that with our new formulation the universe may be regarded as spatially finite, without any necessity for an additional hypothesis. As in the preceding paper I shall again show that with a uniform distribution of matter, a spherical world is compatible with the equations.

In the first place we set

$$ds^2 = -\gamma_{ik} dx_i dx_k + dx_4^2 \left(i, k = 1, 2, 3\right) \tag{11}$$

Then if P_{ik} and P are, respectively, the curvature tensor of the second rank and the curvature scalar in three-dimensional space, we have

$$G_{ik} = P_{ik} \left(i, k = 1, 2, 3\right)$$
$$G_4^i = G_{4i} = G_{44} = 0$$
$$G = -P$$
$$-g = \gamma.$$

It therefore follows for our case that

$$G_{ik} - \frac{1}{2} g_{ik} G = P_{ik} - \frac{1}{2} \gamma_{ik} P \left(i, k = 1, 2, 3\right)$$
$$G_{44} - \frac{1}{2} g_{44} G = \frac{1}{2} P.$$

We pursue our reflexions, from this point on, in two ways. Firstly, with the support of equation (1a). Here $T_{\mu\nu}$ denotes the energy-tensor of the electro-magnetic field, arising from the electrical particles constituting matter. For this field we have everywhere

$$\mathfrak{T}_1^1 + \mathfrak{T}_2^2 + \mathfrak{T}_3^3 + \mathfrak{T}_4^4 = 0.$$

The individual $\mathfrak{T}_\mu^\nu$ are quantities which vary rapidly with position; but for our purpose we no doubt may replace them by their mean values. We therefore have to choose

$$\left.\begin{array}{l} \mathfrak{T}_1^1 = \mathfrak{T}_2^2 = \mathfrak{T}_3^3 = -\tfrac{1}{3}\mathfrak{T}_4^4 = \text{const.} \\ \mathfrak{T}_\mu^\nu = 0 \ (\text{for } \mu \neq \nu), \end{array}\right\} \tag{12}$$

and therefore

$$T_{ik} = \tfrac{1}{3}\frac{\mathfrak{T}_4^4}{\sqrt{\gamma}}\gamma_{ik}, \ T_{44} = \frac{\mathfrak{T}_4^4}{\sqrt{\gamma}}.$$

In consideration of what has been shown hitherto, we obtain in place of (1a)

$$P_{ik} - \tfrac{1}{2}\gamma_{ik}P = -\tfrac{1}{3}\gamma_{ik}\frac{\kappa\mathfrak{T}_4^4}{\sqrt{\gamma}} \tag{13}$$

$$\tfrac{1}{4}P = -\frac{\kappa\mathfrak{T}_4^4}{\sqrt{\gamma}} \tag{14}$$

The scalar of equation (13) agrees with (14). It is on this account that our fundamental equations permit the idea of a spherical universe. For from (13) and (14) follows

$$P_{ik} + \frac{4}{3}\frac{\kappa\mathfrak{T}_4^4}{\sqrt{\gamma}}\gamma_{ik} = 0 \tag{15}$$

and it is known[1] that this system is satisfied by a (three-dimensional) spherical universe.

But we may also base our reflexions on the equations (9). On the right-hand side of (9) stand those terms which, from the phenomenological point of view, are to be replaced by the energy-tensor of matter; that is, they are to be replaced by

$$\begin{array}{cccc} 0 & 0 & 0 & 0 \\ 0 & 0 & 0 & 0 \\ 0 & 0 & 0 & 0 \\ 0 & 0 & 0 & \rho \end{array}$$

where ρ denotes the mean density of matter assumed to be at rest. We thus obtain the equations

$$P_{ik} + \frac{1}{2}\gamma_{ik}P - \frac{1}{4}\gamma_{ik}G_0 \tag{16}$$

$$\frac{1}{2}P + \frac{1}{4}G_0 = \tag{17}$$

From the scalar of equation (16) and from (17) we obtain

$$G_0 = -\frac{2}{3}P = 2\kappa\rho \tag{18}$$

and consequently from (16)

$$P_{ik} - \kappa\rho\gamma_{ik} = 0 \tag{19}$$

1. Cf. H. Weyl, "Raum, Zeit, Materie," § 33.

which equation, with the exception of the expression for the co-efficient, agrees with (15). By comparison we obtain

$$\mathfrak{T}_4^4 = \tfrac{3}{4}\rho\sqrt{\gamma} \tag{20}$$

This equation signifies that of the energy constituting matter three-quarters is to be ascribed to the electromagnetic field, and one-quarter to the gravitational field.

§ 4. CONCLUDING REMARKS

The above reflexions show the possibility of a theoretical construction of matter out of gravitational field and electromagnetic field alone, without the introduction of hypothetical supplementary terms on the lines of Mie's theory. This possibility appears particularly promising in that it frees us from the necessity of introducing a special constant λ for the solution of the cosmological problem. On the other hand, there is a peculiar difficulty. For, if we specialize (1) for the spherically symmetrical static case we obtain one equation too few for defining the $g_{\mu\nu}$ and $\phi_{\mu\nu}$, with the result that any *spherically symmetrical distribution* of electricity appears capable of remaining in equilibrium. Thus the problem of the constitution of the elementary quanta cannot yet be solved on the immediate basis of the given field equations.

Stephen Hawking

Stephen Hawking is considered the most brilliant theoretical physicist since Einstein. He has also done much to popularize science. His book, *A Brief History of Time*, sold more than 10 million copies in 40 languages, achieving the kind of success almost unheard of in the history of science writing. His subsequent books, *The Universe in A Nutshell*, and *The Future of Spacetime*, with Kip S. Thorne and others, have also been well-received.

He was born in Oxford, England on January 8, 1942 (300 years after the death of Galileo). He studied physics at University College, Oxford, received his Ph.D. in Cosmology at Cambridge and since 1979, has held the post of Lucasian Professor of Mathematics. The chair was founded in 1663 with money left in the will of the Reverend Henry Lucas, who had been the Member of Parliament for the University. It was first held by Isaac Barrow, and then in 1663 by Isaac Newton. It is reserved for those individuals considered the most brilliant thinkers of their time.

Professor Hawking has worked on the basic laws that govern the universe. With Roger Penrose, he showed that Einstein's General Theory of Relativity implied space and time would have a beginning in the Big Bang and an end in black holes. The results indicated it was necessary to unify General Relativity with Quantum Theory, the other great scientific development of the first half of the twentieth century. One consequence of such a unification that he discovered was that black holes should not be completely black but should emit radiation and eventually disappear. Another conjecture is that the universe has no edge or boundary in imaginary time.

Stephen Hawking has twelve honorary degrees, and is the recipient of many awards, medals and prizes. He is a Fellow of the Royal Society and a Member of the US National Academy of Sciences. He continues to combine family life (he has three children and one grandchild) and his research into theoretical physics together with an extensive program of travel and public lectures.

He just wanted a decent book to read ...

Not too much to ask, is it? It was in 1935 when Allen Lane, Managing Director of Bodley Head Publishers, stood on a platform at Exeter railway station looking for something good to read on his journey back to London. His choice was limited to popular magazines and poor-quality paperbacks – the same choice faced every day by the vast majority of readers, few of whom could afford hardbacks. Lane's disappointment and subsequent anger at the range of books generally available led him to found a company – and change the world.

'We believed in the existence in this country of a vast reading public for intelligent books at a low price, and staked everything on it'
Sir Allen Lane, 1902–1970, founder of Penguin Books

The quality paperback had arrived – and not just in bookshops. Lane was adamant that his Penguins should appear in chain stores and tobacconists, and should cost no more than a packet of cigarettes.

Reading habits (and cigarette prices) have changed since 1935, but Penguin still believes in publishing the best books for everybody to enjoy. We still believe that good design costs no more than bad design, and we still believe that quality books published passionately and responsibly make the world a better place.

So wherever you see the little bird – whether it's on a piece of prize-winning literary fiction or a celebrity autobiography, political tour de force or historical masterpiece, a serial-killer thriller, reference book, world classic or a piece of pure escapism – you can bet that it represents the very best that the genre has to offer.

Whatever you like to read – trust Penguin.